Advanced Mathematics for Engineers

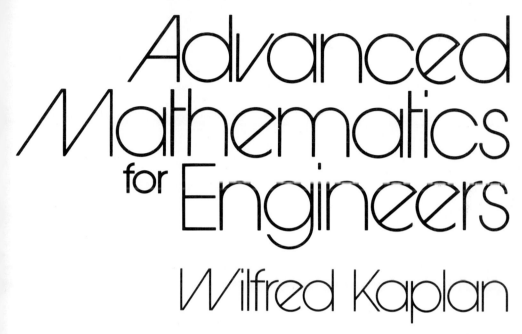

Advanced Mathematics for Engineers

Wilfred Kaplan

University of Michigan

Addison-Wesley Publishing Company

Reading, Massachusetts • Menlo Park, California
London • Amsterdam • Don Mills, Ontario • Sydney

Sponsoring Editor: Steve Quigley
Production Editor: Rima Zolina
Designer: Robert Rose
Illustrator: Oxford Illustrators Ltd.
Cover Design: Ann Scrimgeour Rose

Library of Congress Cataloging in Publication Data

Kaplan, Wilfred, 1915-
  Advanced mathematics for engineers.
  Includes bibliographical references and index.
  1. Engineering mathematics. I. Title.
TA330.K32   515   80-19492

ISBN 0-201-03773-4

DEFGHIJ-HA-8987654

Preface

This book is intended to provide the essential undergraduate mathematics needed by engineering students following a course in calculus. The development is motivated by applications in engineering, and many practical illustrations are given.

The recent phenomenal development of digital computers has greatly changed attitudes towards the role of mathematics in applied fields. Many previously unassailable problems can now be solved routinely by using prepared computer programs. One might easily conclude that less mathematical theory is needed. However, use of the computer without real understanding can easily lead to meaningless solutions to problems. Hence a solid grasp of the relevant mathematics is all the more necessary.

In this book, the computer is taken into account in various ways. Mathematical methods which underlie computer solutions are highlighted. Furthermore, a solid chapter (Chapter 13) is devoted to numerical analysis. Such recently developed tools as Romberg integration and the fast Fourier transform are considered in detail. An introduction to the finite-element method is presented in a separate chapter (Chapter 14).

In view of the variety of courses in the subject area, the author has sought to make the chapters essentially independent. Thus there is considerable freedom in ordering the chapters.

Differential equations are a central topic. Indeed, the first eight chapters form a course in this subject. Chapter 1 covers the elementary aspects of ordinary differential equations. Chapter 2 is devoted to infinite series and ends with solution of differential equations by power series. Chapter 3 has a similar goal based on Fourier series, as does Chapter 4, based on Laplace and Fourier transforms. Chapter 5 develops matrices and basic linear algebra. Chapter 6 uses the linear algebra to handle simultaneous linear differential equations. Chapter 7 stands by itself as a treatment of nonlinear differential equations. Chapter 8 is devoted to partial differential equations, with an emphasis on separation of variables.

These eight chapters are presented in a form that assumes knowledge of only elementary calculus. Partial derivatives do appear, but more advanced topics of calculus, such as chain rules and line integrals, are postponed to Chapters 9 and 10. Thus the full derivation of the heat equation, the wave equation, and similar equations is achieved in Chapter 10. For the purposes of Chapter 8 (partial differential equations), discrete models are used to motivate the equations. These models should give a good physical feeling for what is involved and help to understand the numerical methods commonly used.

Chapter 11 covers the essential theory of functions of a complex variable, with some applications. Chapter 12 uses this theory to give a brief introduction to special functions.

As mentioned above, Chapters 13 and 14 are devoted to numerical analysis. The final Chapter 15 is an introduction to probability and statistics.

The following is a sample five-semester sequence based on this text:

I. Ordinary differential equations, infinite series, Fourier series, Laplace transforms (Chapters 1–4).

II. Matrices, simultaneous linear differential equations, nonlinear differential equations (Chapters 5–7).

III. Partial differential equations and vector calculus (Chapters 8–10, say, in the order 9, 10, 8).

IV. Complex variables, conformal mapping, residues (Chapter 11).

V. Special functions, numerical analysis (Chapters 12–14).

Many exercises are provided throughout. Answers to selected problems are given at the end of the text; those problems for which answers are provided have letters or numbers in boldface type.

References are given at numerous points to provide fuller treatment of topics. Titles cited several times are referred to by an abbreviation or short title as in the list of references following the preface.

In a few chapters (especially Chapters 5 and 11) the author has drawn substantially on his previous books.

The author expresses his appreciation for the fine cooperation provided by the publisher, Addison-Wesley, and its staff, also for the advice of many colleagues, and for the assistance of D. A. Lentz and M. D. Spickenagel, who did a major part of the typing of the manuscript. To his wife he expresses his thanks for help in many details and especially for her patience throughout the long period of preparation of the book.

For their thorough reviewing, the author also wishes to thank James Dowdy (West Virginia University), William J. Firey (Oregon State University), Robert Kadlec (The University of Michigan), Ronald J. Lomax (The University of Michigan), David O. Lomen (The University of Arizona), Francis C. Moon (Cornell University), William L. Perry (Texas A&M University), Stephen M. Pollock (The University of Michigan), David A. Sanchez (The University of New Mexico), Klaus Schmitt (The University of Utah), David F. Ullrich (North Carolina State University at Raleigh), Michael Williams (Virginia Polytechnic Institute and State University).

University of Michigan,
Ann Arbor
December 1980   W. K.

References

AC: Advanced calculus, 2nd. ed., by the author. Reading, Mass.: Addison-Wesley, 1973.

CLA: Calculus and Linear Algebra (2 vols.), by the author and Donald J. Lewis. New York: John Wiley & Sons, Inc., 1970–1971.

IAF: Introduction to Analytic Functions, by the author. Reading, Mass.: Addison-Wesley, 1966.

OMLS: Operational Methods for Linear Systems, by the author. Reading, Mass.: Addison-Wesley, 1962.

ODE: Ordinary Differential Equations, by the author. Reading, Mass.: Addison-Wesley, 1958.

BRUNK: H. D. Brunk, An Introduction to Mathematical Statistics, 2nd ed. Waltham, Mass.: Blaisdell, 1965.

CHURCHILL: Ruel V. Churchill, Operational Mathematics, 3rd. ed. New York: McGraw-Hill, 1972.

CODDINGTON-LEVINSON: Earl A. Coddington and Norman Levinson, Theory of Ordinary Differential Equations. New York: McGraw-Hill, 1955.

COURANT-HILBERT: R. Courant and D. Hilbert, Methods of Mathematical Physics (2 vols.). New York: Interscience, 1953 and 1962.

CRAMÉR: Harald Cramér, Mathematical Methods of Statistics. Princeton: Princeton University Press, 1966.

ERDÉLYI: A. Erdélyi, Editor, and Staff of the Bateman Manuscript Project, Higher Transcendental Functions (3 vols.) and Tables of Integral Transforms (2 vols.). New York: McGraw-Hill, 1953–1955.

GERALD: Curtis F. Gerald, Applied Numerical Analysis, 2nd ed. Reading, Mass: Addison-Wesley, 1978.

HENRICI: Peter Henrici, Elements of Numerical Analysis. New York: John Wiley & Sons, Inc., 1964.

HOCHSTADT: Harry Hochstadt, The Functions of Mathematical Physics. New York: Wiley, Interscience, 1971.

JACKSON: Dunham Jackson, Fourier Series and Orthogonal Polynomials (Carus Mathematical Monographs, No. 6). Menasha, Wisconsin: Mathematical Association of America, 1941.

JAHNKE-EMDE: E. Jahnke and F. Emde, Tables of Functions. Leipzig: B. G. Teubner, 1938.

KELLOGG: Oliver Dimon Kellogg, Foundations of Potential Theory. New York: Springer, Berlin, 1929.

KNOPP: K. Knopp, Theory and Application of Infinite Series, translated by R. C. Young. Glasgow: Blackie & Son, 1928.

LAWDEN: Derek F. Lawden, Mathematics of Engineering Systems. New York: John Wiley & Sons, Inc., 1954.

LIGHTHILL: M. J. Lighthill, An Introduction to Fourier Analysis and Generalized Functions. Cambridge: Cambridge University Press, 1958.

SANSONE-CONTI: G. Sansone and R. Conti, Non-Linear Differential Equations, translated by A. H. Diamond. New York: Macmillan, 1964.

STERN: T. E. Stern, Theory of Nonlinear Networks and Systems. Reading, Mass.: Addison-Wesley, 1965.

WATSON: G. N. Watson, Theory of Bessel Functions, 2nd ed. New York: Macmillan, 1954.

WHITTAKER-WATSON: E. T. Whittaker and G. N. Watson, Modern Analysis, 4th ed. Cambridge: Cambridge University Press, 1940.

Contents

Chapter **4**   **Operational Calculus**_____   179

Chapter **10**   **Vector Integral Calculus**_____   511

Chapter 1 Ordinary Differential Equations

1–1 IMPORTANCE OF DIFFERENTIAL EQUATIONS

Over the centuries the development of science has led to the formulation of many laws governing the behavior of physical objects. When restated in mathematical form the laws often appear as differential equations—that is, as equations involving derivatives. An example is Newton's second law:

$$\text{Force} = \text{Mass} \times \text{Acceleration}.$$

When applied to a particle of mass m moving along a line, the x-axis, subject to a force $f(x)$ depending on position x (Fig. 1–1), the law becomes

$$m\frac{d^2x}{dt^2} = f(x). \qquad (1\text{–}10)$$

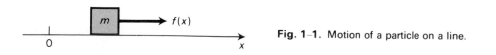

Fig. 1–1. Motion of a particle on a line.

Newton's second law does not say exactly how each particle will move. It rather restricts the possible motions to a certain category. Similarly the differential equation (1–10), for given $f(x)$, does not specify fully the function $x(t)$. The differential equation has many *solutions*, each of which corresponds to an allowed motion of the particle. In each particular instance, the actual solution can be determined by giving appropriate *initial conditions*.

Another example is the application of Kirchhoff's laws for electrical networks. For the simple circuit containing an inductance L, resistance R, capacitance C, and driving electromotive force $E(t)$, the current I satisfies the differential equation

$$L\frac{d^2I}{dt^2} + R\frac{dI}{dt} + \frac{1}{C}I = E'(t). \tag{1–11}$$

Again the equation does not exactly determine the way I depends on time t, but does provide a class of allowed functions $I(t)$, each of which can be specified by initial conditions.

The general theory of electromagnetic phenomena is based on *Maxwell's equations* which involve partial derivatives. In this first chapter we consider only *ordinary* differential equations that do not contain partial derivatives. Later chapters (especially Chapter 8) are devoted to partial differential equations. It is worth observing that in many special cases the solution of these partial differential equations can be reduced to the solution of ordinary differential equations. A similar statement applies to other major physical theories: fluid motion, heat conduction, diffusion, elasticity, plasma physics, quantum mechanics. For each theory the general cases require partial differential equations, but many special cases are treated by ordinary differential equations.

Ordinary differential equations also serve as models for biological and economic systems and for other social sciences. They are able to model a variety of basic phenomena: oscillations, stability and instability, response to inputs, transients, resonance. The fact that particular solutions are determined by initial conditions matches the common experience of mechanisms whose behavior, once they have been put into operation, is determined fully by the conditions under which operations begin.

1–2 BASIC TERMINOLOGY

By an ordinary differential equation we mean an equation such as one of the following:

$$y' + y = 2\sin x, \tag{1–20}$$

$$y'' + 3y' + 2y = 0, \tag{1–21}$$

$$\frac{d^2x}{dt^2} + e^t\frac{dx}{dt} + x = \cos 2t, \tag{1–22}$$

$$EI\frac{d^4w}{dx^4} = p_0\cos ax \quad (E, I, p_0, a \text{ constants}), \tag{1–23}$$

$$y' = x^2 + y^2. \tag{1–24}$$

Thus the equation relates a function and its derivatives up to a certain order.

In general our problem is to find all functions that satisfy the equation over some interval. Each such function is called a *solution* (or *particular solution*) of the differential equation.

For example, $y = \sin x - \cos x$ is a solution of Eq. (1–20), since

$$y' + y = (\cos x + \sin x) + (\sin x - \cos x) = 2 \sin x.$$

But each of the functions $y = \sin x - \cos x + e^{-x}$, $y = \sin x - \cos x + 2e^{-x}$, and, in general,

$$y = \sin x - \cos x + ce^{-x}, \tag{1–25}$$

where c is a constant, is also a solution of Eq. (1–20), as can be verified by substitution.

This example is typical. Each differential equation has infinitely many solutions, in which arbitrary constants appear. By the *general solution* of the equation we mean a formula, containing such constants, for all the solutions. It can be shown that (1–25) is the general solution of Eq. (1–20).

The word *ordinary* is used above to indicate that no partial derivatives appear. We shall omit the word except when needed for emphasis.

By the *order* of a differential equation we mean the order of the highest derivative appearing. Thus Eqs. (1–20) and (1–24) have order *one* or are of *first order*; Eqs. (1–21) and (1–22) have order 2; Eq. (1–23) has order 4.

Equations of the form

$$a(x)y' + b(x)y = f(x),$$

$$a(x)y'' + b(x)y' + c(x)y = f(x), \text{ etc.,}$$

are said to be *linear*; otherwise the equation is said to be *nonlinear*. Of the six examples given above, all are linear except Eq. (1–24).

Very often we are given an equation such as Eq. (1–21) and asked to find a solution satisfying *initial conditions*, that is, prescribed values of y and y' for a chosen x (say, $y = 3$ and $y' = -2$ for $x = 0$). For an equation of order n the natural initial conditions are the values of the function and of all derivatives through order $n - 1$ for the chosen x.

A basic Existence Theorem ensures that, *under appropriate continuity hypotheses, there is exactly one solution satisfying the initial conditions.* For the equation of first order: $y' = F(x, y)$, where $F(x, y)$ is continuous and has a continuous partial derivative $\partial F/\partial y$ for all (x, y), the theorem asserts that for each (x_0, y_0) there exists a solution $y = f(x)$ defined in an interval $a < x < b$ containing x_0 and such that $y_0 = f(x_0)$. Furthermore, the solution is *unique*—that is, if $y = f_1(x)$ is a second solution defined for $a < x < b$ and such that $y_0 = f_1(x_0)$, then $f(x) \equiv f_1(x)$ for $a < x < b$. (See Sections 6–1 and 7–1 for further discussion of the existence theorem).

Instead of initial conditions one may impose *boundary conditions*. These can take a variety of forms. For a second order equation for y, such as Eq. (1–21), one usually gives the values of one linear combination $a_1 y + b_1 y'$ at x_1 and of another such combination $a_2 y + b_2 y'$ at x_2. For example, one might require: $2y + 3y' = 0$ for $x = 0$ and $3y - 2y' = 0$ for $x = 1$. We consider such *boundary-value problems* in Section 3–14 and in Chapter 8.

=============================== **Problems (Section 1-2)** ===============================

1. For each of the following differential equations state the order and whether the equation is linear or nonlinear:

   a) $y' = x + y$ **b)** $y'' - y' + 3y = x^3$ **c)** $y'' = x + \sin y$

   d) $y'' - 2(y')^4 + y^5 = x^6 - 1$

   e) $\dfrac{d^4 x}{dt^4} + t^5 \dfrac{d^2 x}{dt^2} + 3x = \sin 2t$ **f)** $\dfrac{dx}{dt} = xt + t^2$

   g) $\dfrac{dr}{d\theta} + r^2 \cos \theta = 0$

   h) $R\dfrac{dq}{dt} + \dfrac{q}{C} = e_0 \omega \cos \omega t$, where R, C, e_0, ω are constants

2. Verify that the given function is a solution of the given differential equation:

   a) $y = e^{-x}$, Eq. (1–21)   **b)** $y = e^{-2x}$, Eq. (1–21)

   c) $y = c_1 e^{-x} + c_2 e^{-2x}$, where c_1, c_2 are constants, Eq. (1–21)

   d) $x = x_0 \cos \omega t$ for $m\dfrac{d^2 x}{dt^2} + k^2 x = 0$, with x_0, ω, m, and k positive constants and

   $\omega^2 = k^2/m$

   e) $I = I_0 e^{-t/(RC)}$ for $R\dfrac{dI}{dt} + \dfrac{1}{C}I = 0$, where R, C, I_0 are nonzero constants

   f) $x = x_0 + v_0 t$ for $m\dfrac{d^2 x}{dt^2} = 0$, where m, x_0, v_0 are constants $(m \ne 0)$

   g) $x = x_0 + v_0 t + g\left(\dfrac{t^2}{2}\right)$ for $m\dfrac{d^2 x}{dt^2} = mg$, where x_0, v_0, m, g are constants $(m > 0, g > 0)$

3. The *general solution* of the differential equation $y' + y = e^{-x} \cos x$ is known to be $y = e^{-x} \sin x + ce^{-x}$. For each initial condition given find the corresponding solution:

   a) $y = 0$ for $x = 0$ **b)** $y = 1$ for $x = 0$
   c) $y = 0$ for $x = \pi$ **d)** $y = 3$ for $x = \pi/2$

4. The general solution of Eq. (1–21) is given in Problem 2(c) above. For each set of initial conditions given find the corresponding solution:

   a) $y = 1$ and $y' = -3$ for $x = 0$ **b)** $y = 0$ and $y' = 1$ for $x = 0$
   c) $y = 1$ and $y' = 0$ for $x = 1$ **d)** $y = 0$ and $y' = 0$ for $x = 0$

5. *Differential equation with given solutions.* If the solutions of a differential equation are given in the form $f(x, y, c) = 0$ or some equivalent form, then differentiation with respect to x and elimination of c provide the desired equation. Thus from $y = x + ce^{2x}$ we obtain $y' = 1 + 2ce^{2x}$; elimination of c from the two equations gives $y' = 1 + 2(y - x)$. If the solutions are given in the form $f(x, y, c_1, \dots, c_n) = 0$ or some equivalent form, with arbitrary constants $c_1, \dots, c_n$, then differentiation n times and elimination of $c_1, \dots, c_n$ provides the desired differential equation (of order n). Find the differential equation with solutions given:

   a) $y = \sin x + cx^2$ **b)** $y = e^x + c \cos x$ **c)** $x^2 + y^2 = c$
   d) $xy = c$ **e)** $y = c_1 e^x + c_2 e^{-x}$ **f)** $y = c_1 \cos x + c_2 \sin x$

1–3 FIRST-ORDER EQUATIONS

For many first-order equations it is easy to find the general solution explicitly. However, one also encounters such equations for which no formula for the general solution can be obtained. For this reason and also in order to understand the relationship between a differential equation and its solutions, we here consider a graphical interpretation of the problem.

EXAMPLE 1 $y' = 2x$.   The general solution is obtained by integration: $y = x^2 + c$. The solutions are graphed in Fig. 1–2. They are parabolas, one through each point of the plane. The differential equation gives the slope of the tangent to each curve at each point. This slope must always be twice the x-coordinate of the point, as illustrated in the figure. ◄

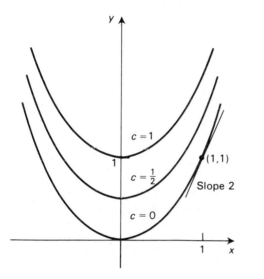

Fig. 1–2. Solutions of $y' = 2x$.

EXAMPLE 2 $y' = x - y$.   By methods to be developed in Section 1–7, the general solution is found to be $y = x - 1 + ce^{-x}$. The solutions are graphed in Fig. 1–3. Again there is one solution through each point. Here the slope of the tangent to a solution at (x, y) is $x - y$, as illustrated in the figure. ◄

In both examples there is one solution through each point (x, y). This fact is in agreement with the theorem about initial conditions: for a first-order equation, a solution is uniquely specified by giving the value of y for one x. Thus here the one initial condition gives one point on the solution curve and that point completely determines the whole curve.

It should be remarked that in some differential equations discontinuities appear, so that certain values of x and y have to be excluded. For example, the differential equation $y' = \ln(x + y)$ has meaning only for $x + y > 0$.

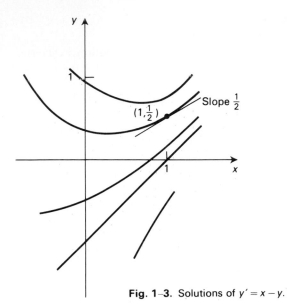

Fig. 1-3. Solutions of $y' = x - y$.

EXAMPLE 3 $y' = x^2 - y^2$. Here there is no simple formula (in terms of elementary functions) for the solutions. However, we can use the graphical ideas of the previous examples to roughly sketch solutions, as in Fig. 1–4. To this end, we compute the slope y' at a number of points and at each point draw a short line segment of that slope—the tangent to the unknown solution through the point. If enough segments have been drawn, the solution curves almost appear by themselves.

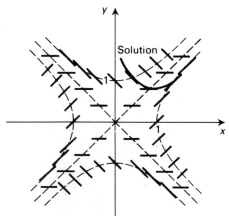

Fig. 1–4. Solutions of $y' = x^2 - y^2$ found graphically.

One can also specify an initial condition, that is, select a starting point and follow the solution through the point by tracing the appropriate very short tangent segments. ◄

The process of sketching the line segments can be simplified by drawing *isoclines*. These are curves (usually *not* solution curves) along which the segments all have the same slope, say *m*. For Example 3, the isoclines are the hyperbolas $x^2 - y^2 = m$, reduced to two straight lines for $m = 0$, shown as dotted lines in Fig. 1–4.

The process of following a particular solution along tangent segments is equivalent to a numerical procedure: one is simply replacing the differential equation by the *difference equation*

$$\Delta y = (x^2 - y^2)\Delta x. \tag{1-30}$$

For the solution starting at (0, 2), that is, $y = 2$ for $x = 0$, we select a Δx, say 0.1, and compute $\Delta y = (0 - 4)(0.1) = -0.4$, so that our next point is (0.1, 1.6). Here, with the same Δx, we find $\Delta y = (0.01 - 2.56)(0.1) = -0.255$, so that our next point is (0.2, 1.345). The results of such a process are shown in Table 1–1 and Fig. 1–5.

Table 1–1
SOLUTION OF $y' = x^2 - y^2$ WITH $\Delta x = 0.1$

x	y	x^2	y^2	$x^2 - y^2$	$\Delta y = (x^2 - y^2)\Delta x$
0	2	0	4	−4	−0.4
0.1	1.6	0.01	2.56	−2.55	−0.255
0.2	1.345	0.04	1.81	−1.77	−0.177
0.3	1.168	0.09	1.36	−1.27	−0.127
0.4	1.041	0.16	1.08	−0.92	−0.092
0.5	0.949	−	−	−	−

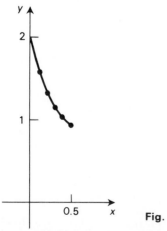

Fig. 1–5. Solutions of $y' = x^2 - y^2$ found numerically, as in Table 1–1.

This simple numerical procedure (or some variant of it) is widely applied. To improve accuracy, one can decrease the size of Δx. By choosing Δx negative, one can follow decreasing x along the solution. There are many refinements that improve the process (see Chapter 13). The procedure can be generalized to higher-order differential equations, as illustrated in Problem 6 below.

1–4 EQUATIONS WITH VARIABLES SEPARABLE

A first order equation of form

$$\frac{dy}{dx} = \frac{f(x)}{g(y)} \tag{1-40}$$

is said to have *variables separable*. To obtain the general solution, one writes the equation as

$$g(y)\,dy = f(x)\,dx \tag{1-41}$$

and integrates:

$$\int g(y)\,dy = \int f(x)\,dx + c. \tag{1-42}$$

One verifies that, under assumptions of continuity, (1–42) provides the general solution in *implicit form*, that is, not solved for y. To obtain the particular solution through the initial point (x_0, y_0) one simply uses definite integrals:

$$\int_{y_0}^{y} g(u)\,du = \int_{x_0}^{x} f(t)\,dt. \tag{1-43}$$

EXAMPLE 1 $y' = x^2 y^3$. We write:

$$\frac{dy}{dx} = x^2 y^3, \qquad \frac{dy}{y^3} = x^2\,dx, \qquad \int \frac{dy}{y^3} = \int x^2\,dx, \qquad -\frac{1}{2y^2} = \frac{x^3}{3} + c.$$

One can solve for y if one desires. It should be noted that because of division by y we lose control of what happens when $y = 0$. In fact, $y = 0$ (the x-axis) is a solution, as we can check by substitution. ◄

EXAMPLE 2 $m\dfrac{dv}{dt} = mg - kv$, where k, m, g are positive constants. This is the equation for a body, such as a parachute, falling against resistance proportional to velocity v (Fig. 1–6). We proceed as above, seeking the solution such that $v = v_0$ for $t = t_0$:

$$\frac{m\,dv}{mg - kv} = dt, \qquad \int_{v_0}^{v} \frac{m\,du}{mg - ku} = \int_{t_0}^{t} d\tau,$$

$$-\frac{m}{k} \ln\,(mg - ku)\Big|_{v_0}^{v} = t - t_0, \qquad \ln \frac{mg - kv}{mg - kv_0} = -\frac{k}{m}(t - t_0),$$

$$v = \frac{mg}{k} - \left(\frac{mg}{k} - v_0\right)\exp\left[-\frac{k}{m}(t - t_0)\right]. \tag{1-44}$$

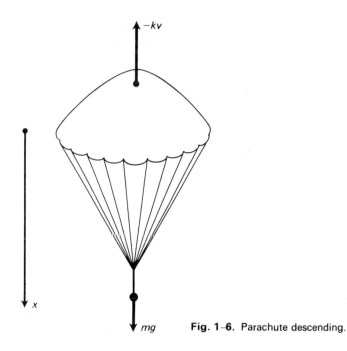

Fig. 1–6. Parachute descending.

Here division by $mg - kv$ could cause trouble. We verify that $v - mg/k$ is in fact a solution and that this is obtained from (1–44) if $v_0 = mg/k$. We observe that in Eq. (1–44) the second term on the right has limit zero as $t \to \infty$. Thus *the falling body has a limiting velocity* $v_{\lim} = mg/k$. ◄

EXAMPLE 3 *Isentropic curves for an ideal gas.* For an isentropic process the pressure p and absolute temperature T are related by the differential equation

$$c_p \frac{dT}{T} = R \frac{dp}{p},$$

where c_p is a constant (specific heat at constant pressure) and R is the gas constant. We assume that $p = p_1$ for $T = T_1$ and integrate:

$$c_p \int_{T_1}^{T} \frac{du}{u} = R \int_{p_1}^{p} \frac{dv}{v}.$$

Hence,

$$c_p \ln \frac{T}{T_1} = R \ln \frac{p}{p_1}.$$

If we take the exponential of both sides, we obtain

$$p = p_1 \left(\frac{T}{T_1} \right)^{c_p/R}.$$

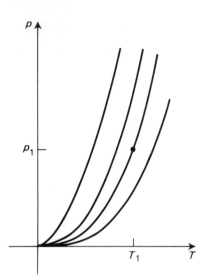

Fig. 1–7. Isentropic curves.

We observe that for the physical problem $p > 0$ and $T > 0$, so that there is no difficulty caused by division by zero. Solution curves for a typical case are graphed in Fig. 1–7. ◄

Problems (Section 1–4)

1. Find the general solution and graph some particular solutions:

   a) $y' = \sin x$ **b)** $y' = e^{-x}$ **c)** $y' = -\dfrac{x}{y}$ **d)** $y' = \dfrac{x}{y}$

   e) $y' = 2xy$ **f)** $y' = 3x^2y$ **g)** $y' = \dfrac{y}{x}$ **h)** $y' = \dfrac{3y}{2x}$

   i) $y' = \dfrac{y^2 - 1}{x}$ **j)** $y' = e^{2x - 3y}$

2. Find the solutions graphically:
   a) $y' = x - y$ (cf. Fig. 1–3)   b) $y' = x + y$

   c) $y' = \dfrac{x - y}{x + y}$ d) $y' = \dfrac{2x - y}{x + 2y}$

   e) $y' = x + y^2$ f) $y' = x^2 + y^2$

3. Find the particular solution satisfying the given initial condition:
   a) $y' = x(y + 1)$, $y = 0$ for $x = 0$
   b) $y' = (x - 1)y$, $y = 2$ for $x = 1$
   c) $y' = e^y \sin 2x$, $y = 0$ for $x = 0$
   d) $y' = (1 + y^2)xe^{x^2}$, $y = 1$ for $x = 2$
   e) $\dfrac{dp}{dy} = -\dfrac{g\rho_0}{p_0}p$, where g, ρ_0, p_0 are positive constants and $p = p_0$ for $y = y_0$ (equation
   for variation of pressure p in an isothermal atmosphere)

f) $\dfrac{dp}{dy} = -\dfrac{g}{R}\dfrac{p}{T_0 + y}$, where g, R, T_0 are positive constants, $p = p_0$ for $y = 0$ (equation for variation of pressure p in an atmosphere with constant temperature gradient)

g) $m\dfrac{dv}{dt} = av - bv^2$, where m, a, b are positive constants, $v = v_0 > 0$ for $t = 0$ (motion of a particle on a horizontal track against nonlinear resistance)*

h) $\dfrac{d\theta}{dt} = -k(\theta - \theta_0)$, $\theta = \theta_1$ for $t = t_1$, where k is a positive constant, θ_0 and θ_1 are constants. (When θ and θ_0 are the temperatures of a solid and of the surrounding medium, respectively, we get Newton's law of cooling.)

i) $\dfrac{dx}{dt} = (1 - x)(2 - x)$, $x = 0$ for $t = 0$. (Equations of form $\dfrac{dx}{dt} = k(a - x)(b - x)$ appear in studying chemical reactions, where k, a, and b are positive constants and x is the amount of a chemical being formed.)*

4. *Motion of a particle on a line subject to force depending only on position.* Here Newton's second law has the form

$$m\frac{d^2 x}{dt^2} = f(x),$$

where $f(x)$ is the force applied at position x. We can write

$$\frac{d^2 x}{dt^2} = \frac{dv}{dt} = \frac{dv}{dx}\frac{dx}{dt} = v\frac{dv}{dx}$$

and hence obtain a first-order differential equation

$$mv\frac{dv}{dx} = f(x),$$

from which we can find the velocity v in terms of position x: say, $v = g(x)$. Then replacement of v by dx/dt gives a first-order differential equation for x in terms of t. Apply the described method to the following cases, writing v in terms of x and x in terms of t:

a) $\dfrac{d^2 x}{dt^2} = 32$, $x = 0$ and $v = 0$ for $t = 0$ (motion under gravity);

b) $\dfrac{d^2 x}{dt^2} = -4x$, $x = 0$ and $v = 1$ for $t = 0$ (spring force, Hooke's law);

c) $\dfrac{d^2 x}{dt^2} = -\dfrac{1}{x^2}$, $x = 2$ and $v = 1$ for $t = 0$ (motion under inverse-square law, for example by a rocket leaving the earth on a straight line passing through the center of the earth; the initial conditions given provide just enough velocity for escape);

d) $\dfrac{d^2 x}{dt^2} = x$, $x = 1$ and $v = 1$ for $t = 0$ (motion under a repulsive force directed away from $x = 0$).

---

* For the integration use the algebraic relation

$$\frac{1}{(u - c)(u - d)} = \frac{1}{c - d}\left(\frac{1}{u - c} - \frac{1}{u - d}\right).$$

This is known as the method of *partial-fraction decomposition* (see Section 4–6).

5. Find the particular solution requested by the numerical procedure of Section 1–3, using the given data:

   a) $y' = x - y$, $y = 0$ for $x = 0$, values of y sought for $x = 0.1, 0.2, 0.3, 0.4, 0.5$ (solution is $y = x - 1 + e^{-x}$).

   b) $y' = 2x - x^2 + y$, $y = 0$ for $x = 0$, values of y sought for $x = 0.1, 0.2, 0.3, 0.4, 0.5$ (solution is $y = x^2$).

   c) $y' = x^2 - y^2$, $y = 0$ for $x = 2$, values of y sought for $x = 1.9, 1.8, 1.7, 1.6$ (see Fig. 1–4).

   d) $y' = x^2 - y^2$, $y = 2$ for $x = 0$, values of y sought for $x = 0.05, 0.10, 0.15, 0.20$ (cf. Table 1–1).

6. To find a particular solution of the second-order differential equation $y'' = f(x, y, y')$ numerically, one writes $y' = v$, $v' = f(x, y, v)$ and then approximates this pair of equations by the pair of difference equations

$$\Delta y = v\,\Delta x, \quad \Delta v = f(x, y, v)\,\Delta x.$$

Given initial values y_0, $y_0' = v_0$ for $x = x_0$ and with Δx specified, one can then compute $\Delta y = v_0 \Delta x$, $\Delta v = f(x_0, y_0, v_0)\Delta x$ to obtain the values of y and v (which equals y') for $x = x_0 + \Delta x$. The process can now be repeated with $x_0 + \Delta x$ as starting value of x. Apply this method to obtain the particular solution requested:

   a) $y'' = -y$, $y = 1$, and $y' = 0$ for $x = 0$, values of y and y' sought for $x = 0.1, 0.2, 0.3, 0.4$ (solution is $y = \cos x$).

   b) $y'' = y' - y^2$, $y = 0$, $y' = 1$ for $x = 0$, values of y, y' sought for $x = 0.1, 0.2, 0.3, 0.4$.

7. *Homogeneous first-order equation.* A first-order equation of form $y' = f(v)$, where $v = y/x$, is said to be *homogeneous*. Examples are

$$y' = \frac{x - y}{2x + y} = \frac{1 - (y/x)}{2 + (y/x)} = \frac{1 - v}{2 + v}, \qquad y' = \frac{x^2 - y^2}{x^2 + y^2} = \frac{1 - v^2}{1 + v^2}.$$

For such an equation the substitution of $v = y/x$ for y as dependent variable leads to an equation in x, v with variables separable. For $y = xv$,

$$y' = xv' + v = f(v), \qquad x\,dv = [f(v) - v]\,dx,$$

and one can now separate variables and integrate; finally one replaces v by y/x.

   Find the general solution of each of the following differential equations:

   a) $y' = \dfrac{y}{x} - e^{y/x}$ b) $y' = \dfrac{y}{x} + \dfrac{x^2 - y^2}{x^2}$

   c) $y' = \dfrac{y - x}{y + x}$ d) $y' = \dfrac{x^2 + y^2}{xy}$

8. *Use of polar coordinates.* For the homogeneous equations of Problem 7, variables can be separated by writing the equation in polar coordinates. One writes $x = r \cos \theta$, $y = r \sin \theta$; then

$$y' = \frac{dy}{dx} = \frac{\sin \theta\, dr + r \cos \theta\, d\theta}{\cos \theta\, dr - r \sin \theta\, d\theta} = f(v) = f(\tan \theta) = g(\theta),$$

so that

$$[-\sin \theta + g(\theta) \cos \theta]\,dr = r[\cos \theta + g(\theta) \sin \theta]\,d\theta,$$

and the variables can be separated. Apply this method to the following parts of Problem 7:

   a) Part (c)   b) Part (d)

1–5 EXACT EQUATIONS

A first-order differential equation $y' = f(x, y)$ can be written in a variety of ways in the form

$$P(x, y)\,dx + Q(x, y)\,dy = 0. \tag{1–50}$$

For example, the equation

$$y' = \frac{x^2 - 2y}{x} \tag{1–51}$$

can be written as

$$(x^2 - 2y)\,dx - x\,dy = 0, \qquad \left(xy - \frac{2y^2}{x}\right)dx - y\,dy = 0,$$

$$(x^3 - 2xy)\,dx - x^2\,dy = 0. \tag{1–52}$$

The second and third equations (1–52) are obtained from the first by multiplying by y/x and by x respectively.

The last equation (1–52) can also be written as

$$x^3\,dx - d(x^2 y) = d\left(\frac{x^4}{4} - x^2 y\right) = 0.$$

This relation can be integrated to yield

$$\frac{x^4}{4} - x^2 y = c,$$

and this gives the general solution of (1–51).

We say that a differential equation in the form (1–50) is *exact* if the left side of the equation is the differential of some function $F(x, y)$:

$$dF(x, y) = \frac{\partial F}{\partial x}\,dx + \frac{\partial F}{\partial y}\,dy = P(x, y)\,dx + Q(x, y)\,dy. \tag{1–53}$$

When this holds, the equation is the same as $dF(x, y) = 0$, and hence the solutions are given in implicit form by $F(x, y) = c$.

EXAMPLES

1. $y\,dx + x\,dy = 0$, $d(xy) = 0$, $xy = c$
2. $\dfrac{-y\,dx + x\,dy}{x^2} = 0$, $d\left(\dfrac{y}{x}\right) = 0$, $\dfrac{y}{x} = c$
3. $e^{y^2}\,dx + (2xye^{y^2} + 1)\,dy = 0$, $d(xe^{y^2} + y) = 0$, $xe^{y^2} + y = c$

Test for exactness

If the equation $P\,dx + Q\,dy = 0$ is exact, then (1–53) holds, so that

$$\frac{\partial F}{\partial x} = P(x, y), \qquad \frac{\partial F}{\partial y} = Q(x, y)$$

and hence

$$\frac{\partial^2 F}{\partial y\,\partial x} = \frac{\partial P}{\partial y}, \qquad \frac{\partial^2 F}{\partial x\,\partial y} = \frac{\partial Q}{\partial x}.$$

But from calculus

$$\frac{\partial^2 F}{\partial x \partial y} = \frac{\partial^2 F}{\partial y \partial x}$$

(appropriate continuity being assumed), so that

$$\frac{\partial P}{\partial y} = \frac{\partial Q}{\partial x}. \tag{1–54}$$

Equation (1–54) is the *test for exactness*. We have shown that, if equation $Pdx + Qdy = 0$ is exact, then Eq. (1–54) must hold. Conversely, if Eq. (1–54) holds, then $Pdx + Qdy = 0$ is exact. This is most easily proved by line integrals (Section 10–9).

EXAMPLE 4 $(2xy^3 - y^2)dx + (3x^2y^2 - 2xy + 2y)dy = 0.$ Here

$$P(x, y) = 2xy^3 - y^2, \qquad Q(x, y) = 3x^2y^2 - 2xy + 2y,$$

$$\frac{\partial P}{\partial y} = 6xy^2 - 2y, \qquad \frac{\partial Q}{\partial x} = 6xy^2 - 2y.$$

Hence the equation is exact. To find the function $F(x, y)$, we can often use inspection. We can also reason as follows:

$$\frac{\partial F}{\partial x} = P = 2xy^3 - y^2, \qquad \frac{\partial F}{\partial y} = Q = 3x^2y^2 - 2xy + 2y. \tag{1–55}$$

From the *first* equation, by integrating with respect to x (and holding y fixed), we find

$$F(x, y) = x^2y^3 - xy^2 + g(y),$$

where $g(y)$ is an *arbitrary function* of y. From this last equation, we get

$$\frac{\partial F}{\partial y} = 3x^2y^2 - 2xy + g'(y).$$

We *compare* this with the second equation (1–55) and conclude that

$$g'(y) = 2y, \qquad g(y) = y^2$$

(a constant is not needed here). Hence,

$$F(x, y) = x^2y^3 - xy^2 + y^2$$

and the general solution is given by

$$x^2y^3 - xy^2 + y^2 = c. \blacktriangleleft$$

For some cases it is easier to first integrate the equation $\partial F/\partial y = Q$ with respect to y and then to differentiate with respect to x.

EXAMPLE 5 $[e^{x^2+y}(1+2x^2) - \sin x]dx + xe^{x^2+y}dy = 0.$ Here

$$\frac{\partial P}{\partial y} = \frac{\partial Q}{\partial x} = e^{x^2+y}(1+2x^2).$$

Thus the equation is exact:

$$\frac{\partial F}{\partial x} = e^{x^2+y}(1+2x^2) - \sin x, \qquad \frac{\partial F}{\partial y} = xe^{x^2+y}.$$

We integrate the second equation with respect to y and then differentiate with respect to x:

$$F(x, y) = xe^{x^2 + y} + g(x), \qquad \frac{\partial F}{\partial x} = e^{x^2 + y}(1 + 2x^2) + g'(x);$$

then, by comparison,

$$g'(x) = -\sin x, \qquad g(x) = \cos x,$$

so that our general solution is

$$xe^{x^2 + y} + \cos x = c. \quad \blacktriangleleft$$

EXAMPLE 6 *Flow past a circular obstacle.* It can be shown that the solution curves of the differential equation

$$\frac{2xy}{(x^2 + y^2)^2} dx + \left[1 + \frac{y^2 - x^2}{(x^2 + y^2)^2} \right] dy = 0$$

are streamlines for a possible two-dimensional flow past a circular obstacle bounded by the circle $x^2 + y^2 = 1$. The equation is exact, since we find

$$\frac{\partial P}{\partial y} = \frac{\partial Q}{\partial x} = (x^2 + y^2)^{-3}(2x^3 - 6xy^2).$$

The equation $\partial F / \partial x = P$ gives

$$F(x, y) = -\frac{y}{x^2 + y^2} + g(y).$$

Substitution of this expression in the equation $\partial F / \partial y = Q$ gives $g'(y) = 1$, so that we can take

$$g(y) = y \qquad \text{and} \qquad F(x, y) = y - \frac{y}{(x^2 + y^2)}.$$

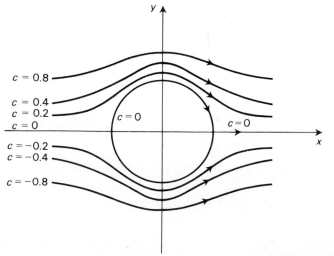

$c = 0.8$

$c = 0.4$
$c = 0.2$
$c = 0$
$c = 0$ $c = 0$

$c = -0.2$
$c = -0.4$

$c = -0.8$

Fig. 1–8. Flow past a circular obstacle.

The solutions are thus given by the equation

$$y - \frac{y}{x^2 + y^2} = c.$$

We observe that for $c = 0$ the equation is satisfied when $y = 0$ and when $x^2 + y^2 = 1$. We can plot other solution curves by solving for x^2:

$$x^2 = 1 - y^2 + \frac{c}{y - c}$$

and selecting various values of c. The results are shown in Fig. 1–8. ◄

The condition (1–54) for exactness has wide physical application. If P and Q are interpreted as the x and y components of a vector (vector field in the xy-plane), then the condition $\partial P/\partial y = \partial Q/\partial x$ means that the vector field is *irrotational*. The conditions $\partial F/\partial x = P$, $\partial F/\partial y = Q$ state that the vector field is the *gradient* of a function. In physics such a function F (or its negative) is interpreted as the *potential* associated with the field. Thus our test for exactness is equivalent to the statement that every irrotational field in the plane is a gradient field or has a potential and, conversely, every gradient field is irrotational. The analogous statement for three-dimensional space is proved in Chapter 10.

1–6 INTEGRATING FACTOR

If the equation $Pdx + Qdy = 0$ is not exact, we can seek a function $\varphi(x, y)$ such that, after multiplication by $\varphi(x, y)$, the equation becomes exact. Such a function $\varphi(x, y)$ is called an *integrating factor*.

EXAMPLE 1 $(3x + 2y)dx + xdy = 0$. Here, after experimentation, we find that x is an integrating factor, for the equation

$$(3x^2 + 2xy)dx + x^2 dy = 0$$

is exact ($\partial P/\partial y = \partial Q/\partial x = 2x$), and the general solution is seen by inspection to be

$$x^3 + x^2 y = c. \quad ◄$$

Finding integrating factors is often difficult, and considerable experience is needed. In fact, there is no guarantee that an integrating factor can be found (except as an infinite series or in some other form that is awkward to use).

Often it is helpful to group the terms and take advantage of known exact differentials such as those of xy, x/y, y/x, and

$$d\left[\tfrac{1}{2}\ln(x^2 + y^2)\right] = \frac{xdx + ydy}{x^2 + y^2}, \qquad d\tan^{-1}\frac{y}{x} = \frac{-ydx + xdy}{x^2 + y^2}.$$

Also, an exact differential $Pdx + Qdy = dF$ remains exact when multiplied by a function of F.

EXAMPLE 2 $[x(x^2 + y^2)^2 - y]dx + [(x^2 + y^2)^2 y + x]dy = 0$. As it stands, this is not an exact equation. We regroup and divide by $x^2 + y^2$:

$$(x^2 + y^2)^2 (x\,dx + y\,dy) - y\,dx + x\,dy = 0,$$

$$(x^2 + y^2)(x\,dx + y\,dy) + \frac{-y\,dx + x\,dy}{x^2 + y^2} = 0,$$

$$d\frac{(x^2 + y^2)^2}{2} + d\tan^{-1}\frac{y}{x} = 0, \qquad \frac{(x^2 + y^2)^2}{2} + \tan^{-1}\frac{y}{x} = c. \blacktriangleleft$$

EXAMPLE 3 $(3xy + 2y^2)\,dx + (4x^2 + 5xy)\,dy = 0$. This is not exact and regrouping does not seem to help. We seek an integrating factor of form $x^m y^n$, with m and n to be found. After multiplication by such a factor, we have

$$P(x, y) = 3x^{m+1}y^{n+1} + 2x^m y^{n+2}, \qquad Q(x, y) = 4x^{m+2}y^n + 5x^{m+1}y^{n+1},$$

$$\frac{\partial P}{\partial y} = 3(n+1)x^{m+1}y^n + 2(n+2)x^m y^{n+1},$$

$$\frac{\partial Q}{\partial x} = 4(m+2)x^{m+1}y^n + 5(m+1)x^m y^{n+1}.$$

Therefore we can make $\partial P/\partial y = \partial Q/\partial x$ if $3(n+1) = 4(m+2)$, $2(n+2) = 5(m+1)$. These are simultaneous equations for n and m: $3n - 4m = 5$, $2n - 5m = 1$, which are satisfied for $n = 3$, $m = 1$. Hence xy^3 is an integrating factor. Using it, we obtain

$$(3x^2 y^4 + 2xy^5)\,dx + (4x^3 y^3 + 5x^2 y^4)\,dy = 0,$$

$$\frac{\partial F}{\partial x} = 3x^2 y^4 + 2xy^5, \qquad F = x^3 y^4 + x^2 y^5 + g(y), \qquad \frac{\partial F}{\partial y} = 4x^3 y^3 + 5x^2 y^4,$$

so that we can take $g(y) = 0$. The solutions are given implicitly by

$$x^3 y^4 + x^2 y^5 = c. \blacktriangleleft$$

===== **Problems (Section 1-6)** =====

1. Verify that the equation is exact and find the general solution:
   a) $(x + 2y)\,dx + (2x + 3y)\,dy = 0$ b) $(5x - 2y)\,dx + (7y - 2x)\,dy = 0$
   c) $(\sin xy + xy\cos xy)\,dx + (x^2 \cos xy + \sin y)\,dy = 0$
   d) $(3x^2 + y^2 e^{xy})\,dx + (e^{xy} + xye^{xy})\,dy = 0$
   e) $\dfrac{x\,dx + y\,dy}{(x^2 + y^2)^2} = 0$ f) $(xy)^3 (y\,dx + x\,dy) = 0$
   g) $(x^4 + 6x^2 y^2 + 2xy^3)\,dx + (4x^3 y + 3x^2 y^2 + y^4)\,dy = 0$
   h) $\dfrac{x^2 - y^2}{x^2}\,dx + \dfrac{2y + 2xy}{x}\,dy = 0$

2. Verify that the equation is exact and find the particular solution requested (if it exists):
   a) $(2x + y)\,dx + (x + 2y)\,dy = 0$, $y = 1$ for $x = 1$
   b) $(4 + 2xy)\,dx + (x^2 - 4)\,dy = 0$, $y = 0$ for $x = 0$
   c) $(x^3 + xy^2)\,dx + (x^2 y + y^3)\,dy = 0$, $y = 0$ for $x = 0$
   d) $\dfrac{-y\,dx + x\,dy}{x^2 + y^2} = 0$, $y = 0$ for $x = 0$

3. For a particle moving in the xy-plane subject to a force with x-component $P(x, y)$ and y-component $Q(x, y)$, the force is said to be derived from a potential $U(x, y)$ if

$P(x, y) = -\partial U/\partial x$, $Q(x, y) = -\partial U/\partial y$. Then $U(x, y)$ is the *potential energy* of the particle. The curves $U(x, y) = c$ are then the *equipotential lines*; these curves are the solutions of $P(x, y)\,dx + Q(x, y)\,dy = 0$. Show that, for the following choices of P and Q, the force is derived from a potential U; find U and sketch the equipotential lines:

a) $P(x, y) = -k_1 x$, $Q(x, y) = -k_2 y$, where k_1, k_2 are positive constants

b) $P(x, y) = -\dfrac{kMmx}{(x^2 + y^2)^{3/2}}$, $Q(x, y) = -\dfrac{kMmy}{(x^2 + y^2)^{3/2}}$, where k, M, m are positive constants (for example, particle of mass m attracted by gravitation to a particle of mass M at the origin).

4. For each of the following differential equations find an integrating factor and hence obtain the general solution:

a) $\dfrac{x(x^2 + y^2) + y}{x^2 + y^2}\,dx + \dfrac{x + (x^2 + y^2)y}{x^2 + y^2}\,dy = 0$ b) $[x^2(x^2 + y^2) - y]\,dx + x\,dy = 0$

c) $(3xy + 2y^2)\,dx + (x^2 + 2xy)\,dy = 0$ d) $3x^2 y\,dx + (3x^3 + 4y)\,dy = 0$

e) $(e^{xy} + x^3 y)\,dx + x^4\,dy = 0$

f) $[\sin(x + 2y) + 3x\cos(x + 2y)]\,dx + 6x\cos(x + 2y)\,dy = 0$.   HINT: Try $\sin''(x + 2y)$.

5. *Orthogonal trajectories.* The orthogonal trajectories of the solutions of a first order equation $y' = f(x, y)$ are the solutions of the equation $y' = -1/f(x, y)$; that is, they are the curves meeting the solutions of $y' = f(x, y)$ at right angles. If the equation $y' = f(x, y)$ is written in the form $P(x, y)\,dx + Q(x, y)\,dy = 0$, then the equation of the orthogonal trajectories can be written in the form $Q(x, y)\,dx - P(x, y)\,dy = 0$. For each of the following families of curves find the family of orthogonal trajectories.

HINT: In each case, find the differential equation of the given family as in Problem 5 of Section 1–2.

a) $x^2 + y^2 = c$ (family of circles)   b) $xy = c$ (family of hyperbolas)
c) $\sin x \cosh y = c$ d) $e^x y = c$

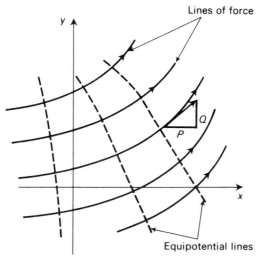

Lines of force

Equipotential lines

Fig. 1–9. Equipotential lines and lines of force.

REMARK: For a force field as in Problem 3, whether it is derived from a potential or not, the solutions of the equation $Q(x, y)dx - P(x, y)dy = 0$ are called the *lines of force*; they have the force vectors as *tangent vectors*. When there is a potential U, the equipotential lines and the lines of force are orthogonal trajectories of each other (see Fig. 1–9).

6. *Two-dimensional flow in a 45° sector* (Fig. 1–10). Here possible streamlines are the solution curves of the differential equation

$$(3x^2y - y^3)\,dx + (x^3 - 3xy^2)\,dy = 0.$$

Find the solutions and show that they can be written in polar coordinates as $r^4\sin 4\theta = c$. Plot a few curves $(c = 0, c = 1, \ldots)$.

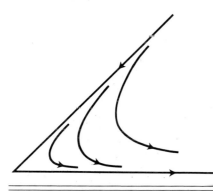

Fig. 1–10. Flow in a 45° sector.

1–7 FIRST-ORDER LINEAR EQUATIONS

Here the equation has the form

$$a(x)y' + b(x)y = f(x) \tag{1–70}$$

or, after division by $a(x)$,

$$y' + p(x)y = q(x). \tag{1–71}$$

If $p(x)$ happens to have the form v'/v for some function $v(x)$, the equation is

$$y' + \frac{v'}{v}y = q(x)$$

or, after multiplying by $v(x)$,

$$vy' + v'y = v(x)q(x) \qquad \text{or} \qquad (vy)' = v(x)q(x).$$

Hence

$$v(x)y = \int v(x)\,q(x)\,dx + c$$

or

$$y = \frac{1}{v(x)}\int v(x)\,q(x)\,dx + \frac{c}{v(x)} \tag{1–72}$$

and Eq. (1–72) is the general solution.

We obtained Eq. (1–72) by assuming that

$$p(x) = \frac{v'(x)}{v(x)}. \tag{1–73}$$

But $p(x)$ can *always* be written this way because Eq. (1–73) is in fact a differential equation for v:

$$p(x) = \frac{1}{v}\frac{dv}{dx}.$$

We separate variables and solve:

$$\int p(x)\,dx = \ln v(x) \qquad \text{or} \qquad v(x) = e^{\int p(x)\,dx}.$$

We indicated no arbitrary constant, since we want just one choice of $v(x)$.

We summarize what we have found: The general solution of the linear differential equation $y' + p(x)y = q(x)$ is

$$y = \frac{1}{v(x)}\int v(x)\,q(x)\,dx + \frac{c}{v(x)}, \tag{1–74}$$

where

$$v(x) = e^{\int p(x)\,dx}. \tag{1–75}$$

We can combine (1–74) and (1–75) and write the general solution as

$$y = e^{-\int p(x)\,dx}\int e^{\int p(x)\,dx}\,q(x)\,dx + ce^{-\int p(x)\,dx}. \tag{1–76}$$

One choice is to be made of each indefinite integral.

EXAMPLE 1 $xy' + y = x^3$. Here $y' + x^{-1}y = x^2$ and $p(x) = x^{-1}$, $q(x) = x^2$. Hence by (1–76), with $e^{\int p\,dx} = e^{\ln x} = x$, we get

$$y = \frac{1}{x}\int x^3\,dx + \frac{c}{x} = \frac{x^3}{4} + \frac{c}{x}. \quad \blacktriangleleft$$

EXAMPLE 2 $\dfrac{dx}{dt} + x = \cos t$. Here $p(t) = 1$, $q(t) = \cos t$, $e^{\int p\,dt} = e^t$, and the solutions are

$$x = e^{-t}\int e^t \cos t\,dt + ce^{-t}.$$

By integral tables we thus obtain

$$x = \frac{1}{2}(\cos t + \sin t) + ce^{-t}. \quad \blacktriangleleft$$

1–8 APPLICATIONS OF THE FIRST-ORDER LINEAR EQUATION

We consider the differential equation in the form

$$\frac{dx}{dt} + ax = f(t) \tag{1–80}$$

and assume that a is a *constant*. In many applications t is *time* and $f(t)$ is an external *input*, yielding the *output*, or *response*, x.

If $f(t)$ is identically 0, we have the equation

$$\frac{dx}{dt} + ax = 0 \qquad \text{or} \qquad \frac{dx}{dt} = -ax. \qquad (1\text{–}81)$$

Here by Eq. (1–76) or by separating variables we obtain the solutions

$$x = ce^{-at}. \qquad (1\text{–}82)$$

These are graphed for the two cases: $a > 0$ and $a < 0$ in Fig. 1–11. When $a > 0$, we have decreasing functions for $x > 0$, increasing functions for $x < 0$, and all solutions approach the *equilibrium solution* $x \equiv 0$ as $t \to \infty$. This is the phenomenon of *exponential decay* represented by many natural processes such as cooling of a hot body (x is temperature), coming to rest of a body moving against high viscosity (x is velocity), discharge of a capacitor (x is charge), decay of radioactive substances such as radium.

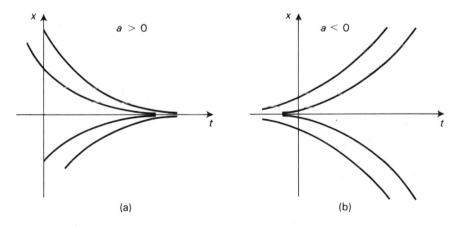

Fig. 1–11. Solutions of $x'(t) + ax = 0$: (a) $a > 0$, exponential decay; (b) $a < 0$, exponential growth.

For $t = 0$, we get $x = c$, so that $c = x_0$, i.e., the *initial value* of x. For $t = a^{-1} = 1/a$, we get $x = ce^{-1} = x_0e^{-1}$; for $t = 2a^{-1}$, we get $x = x_0e^{-2}$; for $t = 3a^{-1}$, we get $x = x_0e^{-3}$, and so on. The constant a^{-1} has the dimension of *time* and is called the *time constant* of the system. It is the time required for a deviation from equilibrium to be reduced to e^{-1} times its starting value.

When $a < 0$, the graphs move away from the t-axis as t increases and we have *exponential growth*; there is an equilibrium solution $x = 0$, but it is *unstable*. This can be illustrated by growth processes, such as that of populations, of bacteria or of chemical compounds being formed.

When $f(t)$ is not identically 0, we can apply Eq. (1–76) to obtain the solutions of Eq. (1–80) in the form

$$x = e^{-at} \int e^{at} f(t)\, dt + c e^{-at}. \tag{1–83}$$

We can choose the indefinite integral here as the integral from 0 to t and hence obtain

$$x = e^{-at} \int_0^t e^{au} f(u)\, du + c e^{-at}$$

or

$$x = \int_0^t e^{-a(t-u)} f(u)\, du + c e^{-at}. \tag{1–84}$$

The integral appearing here is a special case of the *convolution* studied in Chapter 4:

$$\int_0^t g(t-u) f(u)\, du.$$

It is known that the convolution is a smoothing operation: that is, the result is a function smoother than $f(t)$.

We now consider the response to various inputs $f(t)$ and assume that a is a *positive* constant.

Constant input. When $f(t)$ is a constant k, then Eq. (1–83) or Eq. (1–84) gives

$$x = \frac{k}{a}(1 - e^{-at}) + c e^{-at}. \tag{1–85}$$

The solutions are graphed in Fig. 1–12. They all approach the equilibrium solution $x = k/a$ as $t \to \infty$. This can be illustrated by chemical processes approaching completion or by a falling parachute approaching its terminal velocity.

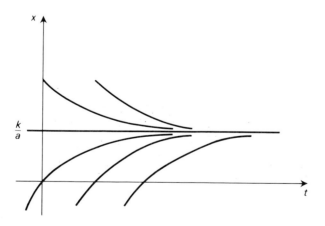

Fig. 1–12. Solutions of $x'(t) + ax = k = $ const.

Ramp input. For $f(t) = kt$, $k = \text{const}$, we find from (1–84) that

$$x = \frac{k}{a^2}(at - 1 + e^{-at}) + ce^{-at}.\tag{1–86}$$

The solutions approach the straight line $x = ka^{-2}(at - 1)$ which can be regarded as a *steady state*. They are graphed in Fig. 1–13.

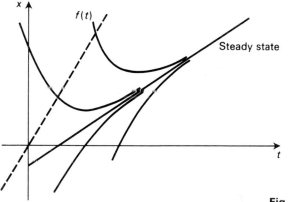

Fig. 1–13. Solutions of $x'(t) + ax = kt$.

Sinusoidal input. For $f(t) = A\sin\omega t$, with A and ω positive constants, we find from Eq. (1–84):

$$x = \frac{A}{a^2 + \omega^2}(a\sin\omega t - \omega\cos\omega t) + \frac{A\omega}{a^2 + \omega^2}e^{-at} + ce^{-at}.\tag{1–87}$$

The next to the last term can be absorbed in the last term. Also by trigonometry:

$$a\sin t - \omega\cos\omega t = \sqrt{a^2 + \omega^2}\left(\frac{a}{\sqrt{a^2 + \omega^2}}\sin\omega t - \frac{\omega}{\sqrt{a^2 + \omega^2}}\cos\omega t\right)$$

$$= \sqrt{a^2 + \omega^2}\sin(\omega t - \alpha),$$

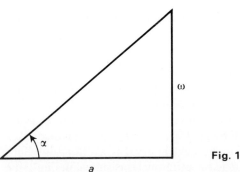

Fig. 1–14. Phase shift α.

where $\cos\alpha = a/\sqrt{a^2 + \omega^2}$ and $\sin\alpha = \omega/\sqrt{a^2 + \omega^2}$ (Fig. 1–14), so that $0 < \alpha < \pi/2$. Therefore the solutions can be written as follows:

$$x = \frac{A}{\sqrt{a^2 + \omega^2}} \sin(\omega t - \alpha) + ce^{-at}. \tag{1–87'}$$

The first term is itself a solution that can be regarded as the *steady state*; as $t \to \infty$, the other solutions approach this one, as shown in Fig. 1–15. The steady state

$$x = \frac{A}{\sqrt{a^2 + \omega^2}} \sin(\omega t - \alpha)$$

is a sinusoidal function with *amplitude* $A/\sqrt{a^2 + \omega^2}$ and *initial phase* $-\alpha$. This solution is obtained from the input $f(t) = A\sin\omega t$ by a phase shift and a multiplication of the amplitude by $1/\sqrt{a^2 + \omega^2}$; if this factor is greater than 1, there is *amplification* of the input; if it is less than 1, there is *attenuation* of the input. Both input and output have the same frequency ω.

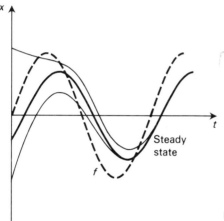

Fig. 1–15. Response to sinusoidal input.

We observe that for large ω attenuation occurs; the larger the ω, the more the attenuation. Thus *high-frequency inputs produce very small outputs*. We can observe this experimentally by repeatedly placing a thermometer in and out of a pan of hot water, thereby creating a rapidly oscillating input; the thermometer will hardly change its reading.

Step-function input. By a step function we mean a function that is constant over successive intervals, as suggested in Fig. 1–16. The response $x(t)$ can be found by the following reasoning. For $0 \leqslant t \leqslant t_1$ we are dealing with an equation with constant input $f(t) \equiv k_1$ and the solutions are given by (1–85). At $t = t_1$ a particular solution has a value that serves as initial value for the next interval $t_1 \leqslant t \leqslant t_2$, in which $f(t) \equiv k_2$. Thus the solutions can be prolonged continuously, as shown in Fig. 1–17.

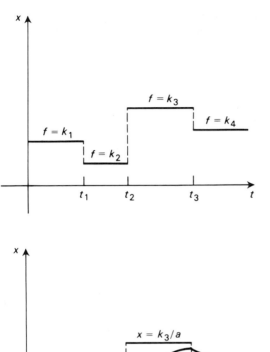

Fig. 1–16. Step function.

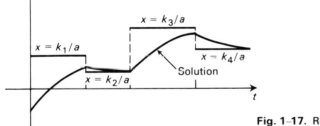

Fig. 1–17. Response to step-function input.

The response appears to be pursuing the step function with successive values k_1/a, k_2/a, ... (that is, the function $f(t)/a$); as the solution gets close to this step function, the function jumps and the pursuit is resumed.

An experiment illustrating such behavior consists of depressing and raising the accelerator pedal in an automobile. For each level there is a corresponding equilibrium speed, but it takes time for this to be reached; if we change the level before the equilibrium speed has been reached, then the motor adjusts to the new level and tries to reach the corresponding speed.

═══════════════ **Problems (Section 1–8)** ═══════════════

1. Imitate the derivation of (1–76) to seek the function $v(x)$ and obtain the general solution:

a) $y' + y = e^x$ b) $y' + 2y = xe^{-2x}$ c) $y' + xy = x$

d) $y' + \dfrac{1}{x}y = x^2 - 1$ e) $(\tan x)y' + y = \cos x$ f) $(x^2 + 1)y' + 2xy = x^2 - 1$

2. Apply the formula (1–76) to obtain the general solution:

 a) $y' + 3y = e^x$ **b)** $y' - y = e^{2x}$ **c)** $xy' + 2y = x - 1$
 d) $(x + 1)y' - y = x$ **e)** $y' + (\sec x)y = \tan x$ **f)** $(1 - x^2)y' - xy = (1 - x^2)^{3/2}$,
 $\qquad\qquad\qquad\qquad\qquad\qquad\qquad\qquad\qquad\qquad -1 < x < 1.$

3. Obtain the solutions given in the text in the equation named (with a a positive constant): (a) Eq. (1–85), (b) Eq. (1–86), (c) Eq. (1–87).

4. Graph the solution requested and discuss it as in Section 1–8:

 a) $\dfrac{dx}{dt} + 3x = 0$, $x = 1$ for $t = 0$ **b)** $\dfrac{dx}{dt} + 5x = 0$, $x = 1$ for $t = 0$

 c) $\dfrac{dx}{dt} - 2x = 0$, $x = 1$ for $t = 0$ **d)** $\dfrac{dx}{dt} - 2x = 0$, $x = 0.001$ for $t = 0$

 e) $\dfrac{dx}{dt} + 2x = 3$, $x = 0$ for $t = 0$ **f)** $\dfrac{dx}{dt} + 2x = 3$, $x = 1$ for $t = 0$

 g) $\dfrac{dx}{dt} + 2x = 2t$, $x = 0$ for $t = 0$ **h)** $\dfrac{dx}{dt} + 3x = t$, $x = -1$ for $t = 0$

 i) $\dfrac{dx}{dt} + x = 3 \sin 2t$, $x = 0$ for $t = 0$ **j)** $\dfrac{dx}{dt} + x = 5 \sin 3t$, $x = 1$ for $t = 0$

 k) $\dfrac{dx}{dt} + x = U(t)$, $U(t)$ is a unit-step function: $U(t) = 0$ for $t < 0$, $U(t) = 1$ for $t > 0$, and
 $x = 0$ for $t = 0$

 l) $\dfrac{dx}{dt} + x = U(t)$ as in (k), $x = 2$ for $t = 0$

 m) $\dfrac{dx}{dt} + x = f(t)$, $f(t) = 0$ for $t < 0$, $f(t) = 1$ for $0 < t < 2$, $f(t) = 3$ for $2 < t < 3$, $f(t) = 0$
 for $t > 3$, and $x = 0$ for $t = 0$

 n) $\dfrac{dx}{dt} + 3x = f(t)$, $f(t) = 0$ for $t < 0$, $f(t) = -1$ for $0 < t < 3$, $f(t) = 1$ for $3 < t < 6$,
 $f(t) = 0$ for $t > 6$ and $x = 0$ for $t = 0$.

5. Discuss the solutions of $\dfrac{dx}{dt} + ax = A \cos \omega t$ for positive constants A, ω, and a.

6. *Transfer of radiation.* For radiation propagated parallel to the x-axis in a medium whose physical properties depend only on x, the intensity y of the radiation at x obeys the differential equation

$$\frac{dy}{dx} = -p(x)y + q(x).$$

Here the term $-p(x)y$ corresponds to absorption, the term $q(x)$, to emission of radiation. Take $p(x) = ax$ and $q(x) = bx$, where a, b are positive constants, and find the solution such that $y = y_0$ for $x = 0$.

7. Radioactive decay is governed by the differential equation

$$\frac{dx}{dt} + ax = 0,$$

where x is the mass, t is time, a is a positive constant. The *half-life* T is the time during which the mass decays to half its initial value. Express T in terms of a and evaluate a for the uranium isotope U^{238}, for which $T = 4.5 \times 10^9$ years.

8. The temperature θ of a body is governed by the differential equation (Newton's law)

$$\frac{d\theta}{dt} = -k[\theta - f(t)],$$

where k is a positive constant and $f(t)$ is the temperature of the surrounding medium at time t. For a certain body, $k = 0.25$ when temperature is measured in degrees Celsius and time t in hours.

a) If $f(t) = 10°$ (constant function) and $\theta = 20°$ for $t = 0$, find θ for $t = 3$

b) If $f(t) = 15 + 0.2t$ and $\theta = 12°$ for $t = 0$, at what time will the temperature θ first come within $1°$ of the temperature of the surrounding medium?

9. *Range of an airplane.* If m is the total mass of the plane, including fuel, then for level flight at constant speed one has the approximating equation

$$\frac{dm}{ds} = -km,$$

where s is distance traveled and k is a positive constant. Find the range if the initial mass, including fuel of mass m_0, is m_1.

1–9 COMPLEX NUMBERS

We represent complex numbers by the points of a plane (the complex plane) as in Fig. 1–18. The complex number $z = a + bi$ is represented by the point with abscissa a and ordinate b (a and b being real, $i = \sqrt{-1}$). We call a and b the *real* and *imaginary parts* of z correspondingly. In symbols:

$$a = \operatorname{Re} z, \qquad b = \operatorname{Im} z. \tag{1–90}$$

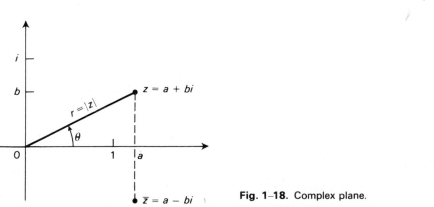

Fig. 1–18. Complex plane.

Two complex numbers are equal precisely when their real parts are equal and their imaginary parts are equal.

The *absolute value* of the complex number $z = a + bi$ is

$$r = |z| = \sqrt{a^2 + b^2}; \tag{1-91}$$

that is, it is the polar distance r from 0 to z. The polar angle of z is called the *argument* of z (see Fig. 1–18). As in trigonometry,

$$z = a + bi = r \cos \theta + ir \sin \theta = r(\cos \theta + i \sin \theta). \tag{1-92}$$

By De Moivre's theorem, for $n = 1, 2, \ldots$, we have

$$z^n = r^n(\cos n\theta + i \sin n\theta). \tag{1-93}$$

We also have the rules:

$$|z_1 z_2| = |z_1||z_2|, \qquad |z_1 + z_2| \leqslant |z_1| + |z_2|. \tag{1-94}$$

The first follows from (1–92) by trigonometry. The second asserts that one side of a triangle is at most equal to the sum of the other two sides (Fig. 1–19). Here we use the fact, illustrated in the figure, that the addition of complex numbers follows the parallelogram rule for vectors.

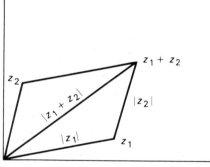

Fig. 1–19. Triangle inequality.

The *conjugate* of $z = a + bi$ is $\bar{z} = a - bi$ (Fig. 1–18). Hence

$$z\bar{z} = |z|^2, \qquad z + \bar{z} = 2a = 2 \operatorname{Re} z, \qquad z - \bar{z} = 2bi = 2i \operatorname{Im} z. \tag{1-95}$$

Also one verifies that

$$\overline{z_1 + z_2} = \bar{z}_1 + \bar{z}_2, \qquad \overline{z_1 z_2} = \bar{z}_1 \bar{z}_2, \qquad |\bar{z}| = |z|. \tag{1-96}$$

1–10 COMPLEX FUNCTIONS

For applications to ordinary differential equations in this chapter, we mainly require certain complex-valued functions of the real variable x. An example of such a function is

$$F(x) = x^2 + ixe^x. \tag{1-100}$$

Here, for example, $F(0) = 0$, $F(1) = 1 + ie$, $F(2) = 4 + 2ie^2$.

In general, we consider

$$F(x) = f(x) + ig(x), \tag{1–101}$$

where f and g are real-valued functions of x, both defined over some interval. We define the derivative of such a function as

$$F'(x) = f'(x) + ig'(x), \tag{1–102}$$

provided that f and g have derivatives at the x considered. Higher derivatives are defined in the same way. For the function of Eq. (1–100), we have

$$F'(x) = 2x + ie^x(x+1), \qquad F''(x) = 2 + ie^x(x+2).$$

From Eq. (1–102) we verify that the usual rules for derivative of sum, product, and quotient are valid.

Similarly we introduce indefinite integrals (antiderivatives) and definite integrals

$$\int_a^b F(x)\,dx = \int_a^b f(x)\,dx + i\int_a^b g(x)\,dx, \tag{1–103}$$

provided that the indicated integrals of f and g exist.

We call F *continuous* at x_0 if f and g are continuous at x_0. Also *limits* can be introduced:

$$\lim_{x \to x_0} F(x) = \lim_{x \to x_0} f(x) + i \lim_{x \to x_0} g(x). \tag{1–104}$$

This is related to continuity as usual.

Exponential function

One writes

$$e^{ix} = \cos x + i\sin x \tag{1–105}$$

(Euler relation). This can be considered as a definition. It can also be justified by power series (Section 2–20). More generally, for a and b real we define:

$$e^{(a+ib)x} = e^{ax}(\cos bx + i\sin bx). \tag{1–106}$$

For $x = 1$, this gives

$$e^{a+ib} = e^a(\cos b + i\sin b). \tag{1–107}$$

This equation defines e^z for each complex number $z = a + ib$. From Eq. (1–107) and trigonometry we verify the familiar rules

$$e^{z_1}e^{z_2} = e^{z_1 + z_2},$$
$$(e^z)^n = e^{nz} \qquad (n = 1, 2, \ldots). \tag{1–108}$$

From Eqs. (1–106) and (1–102) we find

$$\begin{aligned}
[e^{(a+ib)x}]' &= (e^{ax}\cos bx)' + i(e^{ax}\sin bx)' \\
&= e^{ax}(a\cos bx - b\sin bx) + ie^{ax}(a\sin bx + b\cos bx) \\
&= (a + ib)e^{(a+ib)x}.
\end{aligned}$$

Thus the rule

$$(e^{cx})' = ce^{cx} \tag{1–109}$$

is seen to be valid when c is a complex number. Similarly, an antiderivative of e^{cx} is e^{cx}/c and from Eq. (1–103) we verify that

$$\int_{x_1}^{x_2} e^{cx}\,dx = \frac{1}{c}e^{cx}\,\Big|_{x_1}^{x_2} = \frac{e^{cx_2} - e^{cx_1}}{c}$$

for a nonzero constant c.

From this discussion it is clear that functions of the form $e^{ax}p(x)$, where c is a complex constant and $p(x)$ is a polynomial in x (with real or complex coefficients), can be integrated and differentiated as usual.

For example, the derivative of $F(x) = e^{(1-2i)x}(x^2 - ix + 2)$ is

$$F'(x) = e^{(1-2i)x}\big[(1-2i)(x^2 - ix + 2) + 2x - i\big]$$
$$= e^{(1-2i)x}\big[(1-2i)x^2 - ix + 2 - 5i\big].$$

=============== **Problems (Section 1-10)** ===============

1. Evaluate:

   a) $(5 - 3i)(6 + 2i)$ **b)** $\dfrac{1-i}{1+i}$ **c)** $|3 - 4i|$

   d) $\overline{\left(\dfrac{2-3i}{3-4i}\right)}$ **e)** $(\sqrt{2} + \sqrt{2}i)^{15}$ HINT: Use (1–92) and (1–93)

   f) $|(2 - 3i)^7|$ **g)** $e^{i\pi/2}$ **h)** $e^{3+i\pi}$ **i)** $e^{n\pi i}$ $(n = 1, 2, \ldots)$

2. Verify the rules: a) (1–95)   b) (1–96)   c) (1–108)

3. Differentiate:

   a) $F(x) = 3x^2 - 5x + i(x^3 - 7x^2)$
   b) $F(x) = \ln x + i\cos^2 x$ $(x > 0)$
   c) $F(x) = e^{3ix}$ **d)** $F(x) = e^{(2-5i)x}$
   e) $F(x) = e^{2ix}[x^2 + (2+i)x + 1 - i]$
   f) $F(x) = e^{(1-i)x}[ix^3 - (2 + 3i)x]$

4. Evaluate $\int_0^\pi F(x)\,dx$ for $F(x)$

   a) As in Problem 3(a)   b) As in Problem 3(c)   c) As in Problem 3(d)

5. Solve for z:

   a) $z^3 = -\sqrt{2} + \sqrt{2}i$

   HINT: Write $z = r(\cos\theta + i\sin\theta)$, $-\sqrt{2} + \sqrt{2}i = 2(\cos 3\pi/4 + i\sin 3\pi/4)$.

   b) $z^4 = -1$

1–11 SECOND-ORDER LINEAR EQUATIONS WITH CONSTANT COEFFICIENTS (HOMOGENEOUS CASE)

These are equations of the form

$$ay'' + by' + cy = 0, \tag{1–110}$$

where a, b, and c are real constants with $a \neq 0$. Experience has shown that such equations have solutions of the form

$$y = e^{\lambda x} \tag{1–111}$$

for constant λ. Substitution of (1–111) in (1–110) gives the equation

$$a\lambda^2 e^{\lambda x} + b\lambda e^{\lambda x} + c e^{\lambda x} = 0.$$

Here we can divide by $e^{\lambda x}$ (which cannot be 0) and obtain the quadratic equation

$$a\lambda^2 + b\lambda + c = 0 \tag{1–112}$$

called the *auxiliary*, or *characteristic*, *equation*. Thus $y = e^{\lambda x}$ *is a solution of Eq. (1–110) precisely when λ is a solution of the auxiliary equation* (1–112). Such values of λ are called *characteristic roots*.

When Eq. (1–112) has distinct real roots λ_1, λ_2, we obtain two different solutions $y = e^{\lambda_1 x}$ and $y = e^{\lambda_2 x}$. We verify by substitution that each linear combination of two solutions of Eq. (1–110) is also a solution; that is, in this case

$$y = c_1 e^{\lambda_1 x} + c_2 e^{\lambda_2 x} \tag{1–113}$$

is a solution for every choice of the constants c_1, c_2. In fact, Eq. (1–113) is the *general solution*; that is, it provides all solutions of Eq. (1–110). We discuss the reason for this below.

EXAMPLE 1 $y'' - 2y' - 3y = 0$. The characteristic equation is $\lambda^2 - 2\lambda - 3 = 0$; it has roots $\lambda_1 = 3$, $\lambda_2 = -1$. Hence the general solution is

$$y = c_1 e^{3x} + c_2 e^{-x}.$$

We also seek the particular solution such that $y = 7$ and $y' = 5$ for $x = 0$. Since $y' = 3c_1 e^{3x} - c_2 e^{-x}$, we are led to the two equations for c_1, c_2:

$$7 = c_1 + c_2, \qquad 5 = 3c_1 - c_2$$

and solve them to obtain $c_1 = 3$, $c_2 = 4$. Thus

$$y = 3e^{3x} + 4e^{-x} \tag{1–114}$$

is the particular solution sought. ◄

When the auxiliary equation has two equal roots λ_1, λ_1, we obtain only one solution $y = e^{\lambda_1 x}$. However, in this case $y = xe^{\lambda_1 x}$ is also a solution. For here (after dividing by a) the auxiliary equation can be written as

$$(\lambda - \lambda_1)^2 = 0 \qquad \text{or} \qquad \lambda^2 - 2\lambda_1 \lambda + \lambda_1^2 = 0,$$

so that the differential equation must be

$$y'' - 2\lambda_1 y' + \lambda_1^2 y = 0.$$

If we now set $y = xe^{\lambda_1 x}$, the left side becomes

$$e^{\lambda_1 x}(\lambda_1^2 x + 2\lambda_1) - 2\lambda_1 e^{\lambda_1 x}(\lambda_1 x + 1) + \lambda_1^2 x e^{\lambda_1 x}$$
$$= e^{\lambda_1 x}[x(\lambda_1^2 - 2\lambda_1^2 + \lambda_1^2) + 2\lambda_1 - 2\lambda_1] \equiv 0.$$

Thus $y = xe^{\lambda_1 x}$ is a solution, as asserted. The general solution is now given by

$$y = c_1 e^{\lambda_1 x} + c_2 x e^{\lambda_1 x}. \tag{1–115}$$

EXAMPLE 2 $y'' + 4y' + 4y = 0$. The auxiliary equation is $\lambda^2 + 4\lambda + 4 = 0$, the only root is $\lambda_1 = -2$ (repeated), and $y = e^{-2x}$, $y = xe^{-2x}$ are solutions. Thus

$$y = c_1 e^{-2x} + c_2 xe^{-2x}$$

is a solution for each choice of c_1, c_2 and provides the general solution.   ◄

The roots of the auxiliary equation may also be distinct complex numbers $\alpha \pm \beta i$. In that case

$$y = e^{(\alpha + \beta i)x}$$

is a complex-valued solution of the differential equation. For we saw in the previous section that $y = e^{cx}$ for complex c obeys the usual rules of calculus. If we call this function $F(x)$ and write $F(x) = f(x) + ig(x)$ as in Eq. (1–101), then the fact that $F(x)$ is a solution of the differential equation (1–110) gives

$$a[f''(x) + ig''(x)] + b[f'(x) + ig'(x)] + c[f(x) + ig(x)] = 0$$

or

$$af'' + bf' + cf + i(ag'' + bg' + cg) = 0.$$

Thus, upon equating the real and imaginary parts of both sides of this equation, we obtain

$$af'' + bf' + cf = 0 \qquad \text{and} \qquad ag'' + bg' + cg = 0.$$

Accordingly, both $f(x) = \operatorname{Re} F(x)$ and $g(x) = \operatorname{Im} F(x)$ are solutions of the differential equation (1–110). Since

$$F(x) = e^{(\alpha + i\beta)x} = e^{\alpha x} \cos \beta x + i e^{\alpha x} \sin \beta x,$$

the functions

$$f(x) = e^{\alpha x} \cos \beta x \qquad \text{and} \qquad g(x) = e^{\alpha x} \sin \beta x$$

are solutions of the differential equation (1–110). As before, the functions

$$y = c_1 e^{\alpha x} \cos \beta x + c_2 e^{\alpha x} \sin \beta x \tag{1–116}$$

are solutions for each choice of the real constants c_1, c_2, and we can verify that Eq. (1–116) provides the general (real) solution of Eq. (1–110).

In constructing the general solution (1–116), we used only the complex solution $y = e^{(\alpha + \beta i)x}$. If we use the complex solution $y = e^{(\alpha - \beta i)x}$, then we obtain the same general solution (1–116), since

$$\operatorname{Re} e^{(\alpha - \beta i)x} = e^{\alpha x} \cos \beta x, \qquad \operatorname{Im} e^{(\alpha - \beta i)x} = -e^{\alpha x} \sin \beta x$$

(the last minus sign can be absorbed in the arbitrary constant c_2).

Thus we obtain no further real solutions. However, from the two complex solutions $y = e^{(\alpha \pm i\beta)x}$ we can form further complex solutions:

$$y = C_1 e^{(\alpha + i\beta)x} + C_2 e^{(\alpha - i\beta)x}, \tag{1–117}$$

where C_1 and C_2 are arbitrary *complex* constants. We can verify that Eq. (1–117) gives the *general complex solution* of Eq. (1–110) for the case considered.

EXAMPLE 3 $y'' + 2y' + 5y = 0$. The auxiliary equation is $\lambda^2 + 2\lambda + 5 = 0$; the roots are $-1 \pm 2i$. Hence the general solution can be written in complex form as

$$y = c_1 e^{(-1 + 2i)x} + c_2 e^{(-1 - 2i)x}$$

with arbitrary complex constants c_1, c_2, or as

$$y = c_1 e^{-x} \cos 2x + c_2 e^{-x} \sin 2x$$

with arbitrary real constants c_1, c_2. (The former expression gives all *complex* solutions of the differential equation; the latter expression gives all *real* solutions.) ◄

EXAMPLE 4 $\dfrac{d^2x}{dt} + \omega^2 x = 0$, with $\omega > 0$. This differential equation arises in many physical problems. In particular, it describes the motion of a mass–spring system without friction (Fig. 1–20). We assume Hooke's law: the force exerted is $F = -kx$, where k is a positive constant and x is a coordinate measuring displacement of mass m from its equilibrium position. By Newton's second law,

$$m\frac{d^2x}{dt^2} = -kx \qquad \text{or} \qquad \frac{d^2x}{dt^2} + \omega^2 x = 0 \qquad \text{with} \qquad \omega^2 = \frac{k}{m}.$$

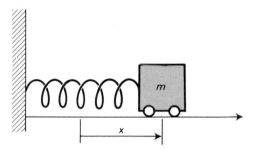

Fig. 1–20. Frictionless mass–spring system.

The solutions of the equation describe *simple harmonic motion*. The auxiliary equation is $\lambda^2 + \omega^2 = 0$, the roots are $\pm \omega i$ (so that $\alpha = 0$, $\beta = \omega$), and the solutions are

$$x = c_1 \cos \omega t + c_2 \sin \omega t.$$

The solutions can also be written thus:

$$x = A \sin(\omega t + \gamma),$$

with new constants $A \geqslant 0$ (the *amplitude*) and γ (initial *phase*). If we compare the two forms, we see that

$$c_1 = A \sin \gamma, \qquad c_2 = A \cos \gamma.$$

These relationships and a typical solution are shown in Fig. 1–21. The graph is like that of a sine function, except for changes in scale and a shift in phase along the t-axis. The vertical scale makes the maximum of the function equal to the amplitude A; the horizontal scale makes the *period* equal to $2\pi/\omega$ and the *frequency* (circular frequency, radians per unit time) equal to ω.

We also seek the particular solution of Example 4 such that $x = x_0$ and $dx/dt = v_0$ for $t = t_0$. We are led to equations for c_1, c_2:

$$x_0 = c_1 \cos \omega t_0 + c_2 \sin \omega t_0, \qquad v_0 = -\omega c_1 \sin \omega t_0 + \omega c_2 \cos \omega t_0,$$

and find by elimination:

$$c_1 = x_0 \cos \omega t_0 - \frac{v_0}{\omega} \sin \omega t_0, \qquad c_2 = x_0 \sin \omega t_0 + \frac{v_0}{\omega} \cos \omega t_0,$$

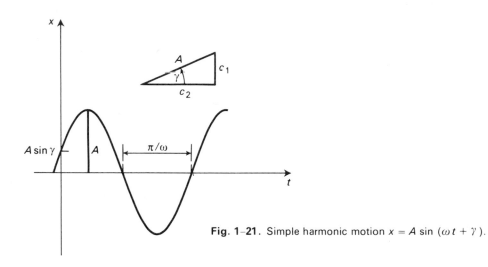

Fig. 1-21. Simple harmonic motion $x = A \sin(\omega t + \gamma)$.

so that the solution sought is

$$x = \left(x_0 \cos \omega t_0 - \frac{v_0}{\omega} \sin \omega t_0 \right) \cos \omega t + \left(x_0 \sin \omega t_0 + \frac{v_0}{\omega} \cos \omega t_0 \right) \sin \omega t$$

$$= x_0 \cos \omega(t - t_0) + \frac{v_0}{\omega} \sin \omega(t - t_0). \quad \blacktriangleleft$$

Remark on general solutions For a second-order differential equation, the Existence Theorem assures us that (under suitable conditions of continuity) there is one and only one solution with given initial value and derivative at a given point. Hence to verify that a formula for solutions containing arbitrary constants provides the general solution, we must show that for each such set of initial conditions we can find corresponding values of the arbitrary constants. That is exactly what we did for special cases in Examples 1 and 4 above. The same method can be applied to the general formulas (1–113), (1–115), (1–116), (1–117).

══════════════════ **Problems (Section 1–11)** ══════════════════

1. Find the general solution:

   a) $y'' - 9y = 0$ **b)** $y'' + 4y' - 5y = 0$ **c)** $2y'' - y' - y = 0$
   d) $y'' + 5y' + 6y = 0$ **e)** $y'' - 8y' + 16y = 0$ **f)** $4y'' + 4y' + y = 0$
   g) $y'' = 0$ **h)** $y'' + 10y' + 25y = 0$ **i)** $y'' + y = 0$
   j) $y'' + 9y = 0$ **k)** $4y'' + y = 0$ **l)** $y'' + 3y = 0$
   m) $y'' + 2y' + 10y = 0$ **n)** $y'' - 2y' + 5y = 0$ **o)** $y'' + y' + y = 0$
   p) $y'' + 7y' + 11y = 0$

2. Find the particular solution requested:

   a) $y'' - 9y = 0$, $y = 4$ and $y' = 0$ for $x = 0$
   b) $4y'' + 4y' + y = 0$, $y = 0$ and $y' = 1$ for $x = 0$

c) $\dfrac{d^2x}{dt^2} + 25x = 0$, $x = 3$ and $\dfrac{dx}{dt} = 0$ for $t = 0$

d) $\dfrac{d^2\theta}{dt^2} + 2\dfrac{d\theta}{dt} + 2\theta = 0$, $\theta = \dfrac{\pi}{4}$ and $\dfrac{d\theta}{dt} = -1$ for $t = 0$

e) $y'' + 2y' - 3y = 0$, $y = 0$ and $y' = 0$ for $x = 0$

f) $y'' + 2y' - 3y = 0$, $y = 1$ and $y' = -2$ for $x = 1$

3. Find the solution requested, determine its amplitude and initial phase and graph:

a) $\dfrac{d^2x}{dt^2} + x = 0$, $x = 3$ and $\dfrac{dx}{dt} = 4$ for $t = 0$

b) $\dfrac{d^2x}{dt^2} + 4x = 0$, $x = 0$ and $\dfrac{dx}{dt} = 5$ for $t = 0$

c) $\dfrac{d^2x}{dt^a} + 9x = 0$, $x = 3$ and $\dfrac{dx}{dt} = 0$ for $t = \dfrac{\pi}{3}$

d) $\dfrac{d^2x}{dt^2} + 9x = 0$, $x = 1$ and $\dfrac{dx}{dt} = 2$ for $t = \pi$

4. The following exercises refer to a frictionless mass–spring system as in Fig. 1–20.
   a) If the mass is 100 grams and the force is 40 dynes when the displacement is 10 centimeters, find the frequency and the period.
   b) If the mass is 10 grams and the frequency is 2 radians per second, find the force when the displacement is 3 centimeters.
   c) If for a particular motion $x = 0$ and $dx/dt = 1$ cm/sec for $t = 0$ and the amplitude is 5 cm, find the frequency.

5. For a vertical frictionless spring (Fig. 1–22), we must also consider the force of gravity. Derive the differential equation

$$m\dfrac{d^2x}{dt^2} = -kx + mg,$$

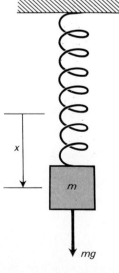

Fig. 1–22. Vertical spring.

where x is measured as usual from the position at which the spring has its natural length. Show that there is a new equilibrium position x_0 (so that $x \equiv x_0$ for all t is a solution) and that the substitution $u = x - x_0$ leads to an equation of form $(d^2u/dt^2) + \omega^2 u = 0$. Conclude that the mass follows simple harmonic motion about the new equilibrium position x_0.

6. Show that bodies floating in water (Fig. 1–23) also describe simple harmonic motion. HINT: Ignore friction and let x measure vertical displacement from the equilibrium position; use Archimedes' law according to which the buoyant force equals the weight of the water displaced.

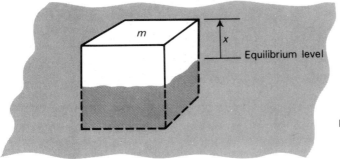

Fig. 1-23. Floating body.

7. Solve the boundary-value problems (if possible):

   a) $y'' - 4y = 0$, $y = 2$ for $x = 0$, $y = 3$ for $x = 1$
   b) $y'' - 4y = 0$, $y = 1$ for $x = 0$, $y = 2$ for $x = 1$
   c) $y'' + 9y = 0$, $y = 0$ for $x = 0$, $y' = 0$ for $x = \pi/6$
   d) $y'' + 9y = 0$, $y' = 0$ for $x = 0$, $y' = 0$ for $x = \pi/2$

1–12 SECOND-ORDER LINEAR EQUATIONS WITH CONSTANT COEFFICIENTS (NONHOMOGENEOUS CASE)

These are equations of the form

$$ay'' + by' + cy = f(x) \tag{1–120}$$

with constant a, b, c, and $a \neq 0$. We assume that $f(x)$ is continuous over a given interval and seek the solutions over that interval. When $f(x) \equiv 0$, the equation is *homogeneous* and we are back to the case of the preceding section. Note that the term *homogeneous* used here is unrelated to that for the first-order equation, as used in Problem 7 of Section 1–4. We call the equation

$$ay'' + by' + cy = 0 \tag{1–121}$$

the *homogeneous equation related to* Eq. (1–120).

Let us suppose that we have somehow found one solution $y^*(x)$ of Eq. (1–120). Then in Eq. (1–120) we let
$$y = v + y^*(x),$$
that is, v is to be a new dependent variable. Accordingly, v must satisfy the equation
$$a[v + y^*(x)]'' + b[v + y^*(x)]' + c[v + y^*(x)] = f(x).$$
If we expand and use the fact that
$$a[y^*(x)]'' + b[y^*(x)]' + cy^*(x) = f(x),$$
since $y^*(x)$ is a solution of (1–120), we conclude that
$$av'' + bv' + cv = 0.$$
Thus v is simply required to be a solution of the related homogeneous equation. If, for example, the auxiliary equation has different roots λ_1, λ_2, then
$$v = c_1 e^{\lambda_1 x} + c_2 e^{\lambda_2 x} \qquad \text{and} \qquad y = c_1 e^{\lambda_1 x} + c_2 e^{\lambda_2 x} + y^*(x).$$
We verify, as in the preceding section, that this formula gives the general solution. We write $y_c(x)$ for the terms in c_1, c_2, giving the general solution of the related homogeneous equation, and call $y_c(x)$ the *complementary function*. Then the general solution of Eq. (1–120) is
$$y = y_c(x) + y^*(x). \tag{1–122}$$

EXAMPLE 1 $y'' - y = 3e^{2x}$. The related homogeneous equation is $y'' - y = 0$, the auxiliary equation is $\lambda^2 - 1 = 0$, the roots are $1, -1$, the complementary function is
$$y_c(x) = c_1 e^x + c_2 e^{-x}.$$
To find the particular solution $y^*(x)$, we use a method of *undetermined coefficients*. In this case the method tells us to *try*
$$y^*(x) = ke^{2x},$$
where k is a constant to be determined. Substitution in the differential equation gives
$$4ke^{2x} - ke^{2x} = 3e^{2x} \qquad \text{or} \qquad 3k = 3,$$
so that $k = 1$ and $y^*(x) = e^{2x}$. Thus as in Eq. (1–122) the general solution is given by
$$y = c_1 e^x + c_2 e^{-x} + e^{2x}.$$
We also seek the particular solution such that $y = 2$ and $y' = 0$ for $x = 0$ and are led to the equations
$$2 = c_1 + c_2 + 1, \qquad 0 = c_1 - c_2 + 2,$$
so that $c_1 = -1/2$, $c_2 = 3/2$, and the solution sought is
$$y = -\frac{1}{2}e^x + \frac{3}{2}e^{-x} + e^{2x}. \quad \blacktriangleleft$$

General rules for method of undetermined coefficients

The method applies when $f(x)$ has the form
$$e^{hx}[p(x)\cos \omega x + q(x)\sin \omega x], \tag{1–123}$$

where h and ω are constants and $p(x)$, $q(x)$ are polynomials of degree at most n. If $\omega = 0$, this reduces to the form

$$e^{hx}p(x). \tag{1–123'}$$

The *trial function* for $f(x)$ of form (1–123), with $\omega \neq 0$ and $h + \omega i$ not a root of the auxiliary equation, is

$$y = e^{hx}[(k_1 + k_2 x + \cdots + k_{n+1}x^n)\cos \omega x \tag{1–124}$$
$$+ (k_{n+2} + k_{n+3}x + \cdots + k_{2n+2}x^n)\sin \omega x],$$

with constants $k_1, \ldots, k_{2n+2}$ to be found. If $h + \omega i$ is a root of the auxiliary equation, the whole trial function (1–124) is to be multiplied by x.

The trial function for $f(x)$ of form (1–123') is

$$y = e^{hx}(k_1 + k_2 x + \cdots + k_{n+1}x^n), \tag{1–125}$$

provided h is not a root of the auxiliary equation. If h is a simple root of the auxiliary equation, the trial function (1–125) is multiplied by x; if h is a double root of the auxiliary equation, the trial function (1–125) is multiplied by x^2. (These rules can all be derived from the Laplace transform method; see Section 4–11.) ◄

EXAMPLE 2 $y'' - y = e^x(x^2 - 1)$. Here (1–125) would give

$$y = e^x(k_1 + k_2 x + k_3 x^2)$$

as trial function. But $h = 1$ and 1 is a simple root of the auxiliary equation $\lambda^2 - 1 = 0$. Hence we multiply by x and use

$$y = e^x(k_1 x + k_2 x^2 + k_3 x^3).$$

Substitution of this function in the differential equation gives

$$e^x[2k_2 + 2k_1 + (6k_3 + 4k_2)x + 6k_3 x^2] = e^x(x^2 - 1).$$

If we divide by e^x and compare terms in like powers of x, we obtain three equations:

$$2k_1 + 2k_2 = -1, \qquad 4k_2 + 6k_3 = 0, \qquad 6k_3 = 1,$$

from which we obtain $k_1 = -1/4$, $k_2 = -1/4$, $k_3 = 1/6$. Thus

$$y^*(x) = e^x\left(-\frac{1}{4}x - \frac{1}{4}x^2 + \frac{1}{6}x^3\right).$$

Since $y_c(x) = c_1 e^x + c_2 e^{-x}$, the general solution is

$$y = c_1 e^x + c_2 e^{-x} + \frac{e^x}{12}(-3x - 3x^2 + 2x^3). \quad ◄$$

EXAMPLE 3 $y'' + 2y' + y = 5\cos 2x$. Here (1–124), with $h = 0$, $\omega = 2$, $n = 0$, gives the trial function

$$y = k_1 \cos 2x + k_2 \sin 2x.$$

Since $h + \omega i = 2i$ is *not* a root of the auxiliary equation, we do not modify this trial function. Substitution in the differential equation gives

$$(-3k_1 + 4k_2)\cos 2x + (-4k_1 - 3k_2)\sin 2x = 5\cos 2x.$$

Hence our equation is satisfied if

$$-3k_1 + 4k_2 = 5, \qquad -4k_1 - 3k_2 = 0,$$

so that $k_1 = -3/5$, $k_2 = 4/5$, and

$$y^*(x) = -\frac{3}{5}\cos 2x + \frac{4}{5}\sin 2x.$$

The complementary function is $y_c(x) = e^{-x}(c_1 + c_2 x)$ and the general solution is

$$y = e^{-x}(c_1 + c_2 x) + \frac{1}{5}(-3\cos 2x + 4\sin 2x). \quad \blacktriangleleft$$

If the function $f(x)$ is a sum of two or more functions $f_1(x)$, $f_2(x)$, ..., and we have found a particular solution for each function $f_1(x)$, $f_2(x)$, ... as right-hand side of the equation, then a particular solution for the function $f(x)$ as right-hand side is obtained by adding the results. Also a particular solution for $cf(x)$ as right-hand side (where $c = $ const) is c times a particular solution for $f(x)$ as right-hand side. These rules, which are together called the *superposition principle*, follow at once from the rules for derivative of a sum and of a constant times a function (see Problem 5 below).

EXAMPLE 4 $y'' + 4y = 3e^{2x} + 7\cos x$. By the previous method we find particular solutions of

$$y'' + 4y = e^{2x} \qquad \text{and} \qquad y'' + 4y = \cos x$$

to be $e^{2x}/8$ and $\frac{1}{3}\cos x$ respectively, so that

$$y^*(x) = \frac{3}{8}e^{2x} + \frac{7}{3}\cos x$$

is a solution of the given equation and the general solution is

$$y = c_1\cos 2x + c_2\sin 2x + \frac{3}{8}e^{2x} + \frac{7}{3}\cos x. \quad \blacktriangleleft$$

The superposition principle is a very important aspect of systems described by a linear differential equation (1–120) (or by a more general linear differential equation (1–170) in Section 1–17). The principle is widely used, often without reflection, in considering *cause* and *effect*: if a certain cause leads to a certain effect, then amplification of the cause by a factor k should lead to amplification of the effect by the same factor k; the joint effect of two different causes is the sum of the effects of the two causes considered separately. Unfortunately, many real-life systems are not truly linear and the application of the superposition principle may not be fully justified.

There are other methods for finding the particular solution. An important one, *variation of parameters*, is discussed in the next section.

Remark on validity of solutions   The general existence theorem assures us that solutions with given initial values exist, but it does not say anything about the *interval* over which the solutions exist. For example, for the *nonlinear* first-order equation $y' = 1 + y^2$, the solution with $y = 0$ for $x = 0$ is $y = \tan x$, and this is valid only for $-\pi/2 < x < \pi/2$. The interval could *not* be predicted from the differential equation. For *linear* differential equations, on the other hand, each solution is valid throughout the whole interval in which the coefficients and right-hand side are continuous and in which the equation is solvable for the highest derivative (see ODE, Chapter 12).

═══════════════════════Problems (Section 1–12)═══════════════

1. Find the general solution:
   a) $y'' - 9y = 2e^{2x}$
   b) $y'' - 9y = 5e^{4x}$
   c) $y'' - 9y = 4e^{3x}$
   d) $y'' - 9y = 7e^{-3x}$
   e) $y'' - 9y = 2e^{2x} + 4e^{3x}$
   f) $y'' - 9y = e^{x} + 6e^{-x}$
   g) $y'' - 9y = e^{x}(8x - 18)$
   h) $y'' - 9y = x^{2}e^{x}$
   i) $y'' - 9y = xe^{3x}$
   j) $y'' - 9y = e^{-3x}(x^{2} + 1)$

2. Find the general solution:*
   a) $\dfrac{d^{2}x}{dt^{2}} + 4x = 3\sin 5t$

   b) $\dfrac{d^{2}x}{dt^{2}} + 4x = 6\cos t,$

   c) $\dfrac{d^{2}x}{dt^{2}} + 4x = 2\cos 3t + 3\sin 3t$

   d) $\dfrac{d^{2}x}{dt^{2}} + 4x = 3\sin t + 5\cos 3t$

   e) $\dfrac{d^{2}x}{dt^{2}} + 4x = 5\sin 2t$

   f) $\dfrac{d^{2}x}{dt^{2}} + 4x = 6\cos 2t + 4\sin 2t$

   g) $\dfrac{d^{2}x}{dt^{2}} + 4x = t\cos 2t$

   h) $\dfrac{d^{2}x}{dt^{2}} + 4x = t\sin 2t + \cos 2t$

   i) $\dfrac{d^{2}x}{dt^{2}} + 4x = e^{-t}\cos 2t$

   j) $\dfrac{d^{2}x}{dt^{2}} + 4x = e^{-t}(t\sin 2t + \cos 2t).$

3. Find the general solution:*
   a) $\dfrac{d^{2}x}{dt^{2}} + 2\dfrac{dx}{dt} + 2x = 5\cos t$

   b) $\dfrac{d^{2}x}{dt^{2}} + 2\dfrac{dx}{dt} + 2x = 5\sin t - \cos t$

   c) $\dfrac{d^{2}x}{dt^{2}} + 2\dfrac{dx}{dt} + 2x = 7e^{-t}$

   d) $\dfrac{d^{2}x}{dt^{2}} + 2\dfrac{dx}{dt} + 2x = e^{-t}(2t + 1)$

   e) $\dfrac{d^{2}x}{dt^{2}} + 2\dfrac{dx}{dt} + 2x = e^{-t}\cos t$

   f) $\dfrac{d^{2}x}{dt^{2}} + 2\dfrac{dx}{dt} + 2x = e^{-t}(2\cos t + \sin t)$

4. Find the particular solution requested:
   a) Eq. of Problem 1 (a), $y = 1$ and $y' = 0$ for $x = 0$

   b) Eq. of Problem 1 (c), $y = 2$ and $y' = 1$ for $x = 0$

   c) Eq. of Problem 2 (a), $x = 0$ and $\dfrac{dx}{dt} = 1$ for $t = 0$

   d) Eq. of Problem 2 (c), $x = 3$ and $\dfrac{dx}{dt} = 2$ for $t = 0$

   e) Eq. of Problem 3 (a), $x = 0$ and $\dfrac{dx}{dt} = 0$ for $t = 0$

   f) Eq. of Problem 3 (e), $x = 0$ and $\dfrac{dx}{dt} = 0$ for $t = 0$

─────────────────

* Problems 2 and 3 can also be solved with the aid of complex-valued functions. For example, for Problem 2(a) we first consider the equation $x''(t) + 4x = 3e^{5it}$ and find a particular solution ke^{5it}. Then we take the *imaginary* part of this solution to obtain the desired particular solution.

5. *Superposition principle.* Prove that if $y_1(x)$ and $y_2(x)$ are solutions over an interval of the linear differential equations

$$a(x)y'' + b(x)y' + c(x)y = f_1(x) \qquad \text{and} \qquad a(x)y'' + b(x)y' + c(x)y = f_2(x),$$

respectively, then $y = k_1 y_1(x) + k_2 y_2(x)$, with constant k_1, k_2, is a solution of the differential equation

$$a(x)y'' + b(x)y' + c(x)y = k_1 f_1(x) + k_2 f_2(x)$$

over the interval.

6. Verify by substitution that the function given is a solution of the differential equation given:
   a) $y = x^2$ for $y'' + 4y = 4x^2 + 2$
   b) $y = xe^{3x}$ for $y'' + 4y = e^{3x}(13x + 6)$
   c) $y = x \sin 2x$ for $y'' + 4y = 4 \cos 2x$

7. Use the results of Problem 6 to find a particular solution of each of the following differential equations:
   a) $y'' + 4y = 8x^2 + 4$ b) $y'' + 4y = 4 \cos 2x + 4x^2 + 2$
   c) $y'' + 4y = 5 \cos 2x + 6 + 12x^2 + e^{3x}(26x + 12)$

1–13 METHOD OF VARIATION OF PARAMETERS

This is another method for finding a particular solution of the nonhomogeneous linear differential equation

$$ay'' + by' + cy = f(x), \qquad a \neq 0. \tag{1–130}$$

It can be applied whenever one knows the complementary function

$$y_c(x) = c_1 y_1(x) + c_2 y_2(x). \tag{1–131}$$

It does not require that $f(x)$ be of the special form (1–123). However, $f(x)$ is assumed to be continuous over the interval considered.

We start by replacing the constants c_1, c_2 in the complementary function (1–131) by unknown functions $v_1(x)$, $v_2(x)$. Thus we seek a particular solution of form

$$y(x) = v_1(x)y_1(x) + v_2(x)y_2(x). \tag{1–132}$$

Two conditions are needed to determine the two functions $v_1(x)$, $v_2(x)$. One condition on these functions is obtained from substitution in the differential equation; the second condition can be imposed arbitrarily. We choose the condition:

$$v_1' y_1(x) + v_2' y_2(x) = 0, \tag{1–133}$$

which simplifies the work to follow. Because of Eq. (1–133), we have

$$y'(x) = v_1(x)y_1'(x) + v_2(x)y_2'(x) + 0,$$
$$y''(x) = v_1 y_1'' + v_2 y_2'' + v_1' y_1' + v_2' y_2'. \tag{1–134}$$

If we now replace y by $v_1(x)y_1(x) + v_2(x)y_2(x)$ in the differential equation (1–130) and use the relations (1–134), we obtain

$$a(v_1 y_1'' + v_2 y_2'' + v_1' y_1' + v_2' y_2') + b(v_1 y_1' + v_2 y_2') + c(v_1 y_1 + v_2 y_2) = f(x)$$

or

$$v_1(ay_1'' + by_1' + cy_1) + v_2(ay_2'' + by_2' + cy_2) + a(v_1' y_1' + v_2' y_2') = f(x).$$

Since $y_1(x)$ and $y_2(x)$ are solutions of the related homogeneous equation, the first two expressions in parentheses reduce to 0 and we finally obtain our second condition:
$$a(v_1' y_1' + v_2' y_2') = f(x). \tag{1-135}$$
Equations (1–133) and (1–135) are two simultaneous linear equations for v_1', v_2'. These can be solved for v_1', v_2', and then $v_1(x)$, $v_2(x)$ can be obtained by integration. Arbitrary constants can be ignored since we seek only one particular solution.

EXAMPLE 1 $y'' - 4y = e^x$. Here the complementary function is
$$y_c(x) = c_1 e^{2x} + c_2 e^{-2x},$$
so that our solution is sought in the form $y = v_1 e^{2x} + v_2 e^{-2x}$. Conditions (1–133) and (1—135) here become
$$v_1' e^{2x} + v_2' e^{-2x} = 0 \quad \text{and} \quad 2v_1' e^{2x} - 2v_2' e^{-2x} = e^x.$$
We solve these by elimination to obtain $4e^{2x} v_1' = e^x$ and $4e^{-x} v_2' = -e^x$, and hence $v_1' = e^{-x}/4$, $v_2' = -e^{3x}/4$. We can take $v_1 = -e^{-x}/4$, $v_2 = -e^{3x}/12$ and obtain finally
$$y = v_1 e^{2x} + v_2 e^{-2x} = -\frac{1}{4} e^{-x} e^{2x} - \frac{1}{12} e^{3x} e^{-2x} = -\frac{1}{3} e^x. \quad \blacktriangleleft$$

REMARK: Since $a \neq 0$, the two simultaneous equations (1–133), (1–135) for v_1', v_2' can be written as
$$y_1 v_1' + y_2 v_2' = 0, \quad y_1' v_1' + y_2' v_2' = \frac{f(x)}{a},$$
where $y_1 = y_1(x)$, $y_2 = y_2(x)$. These equations can be solved uniquely for v_1', v_2', provided the determinant of the coefficients is not 0 (see Section 5–1 for a review of the basic facts on simultaneous linear equations). This determinant is
$$\begin{vmatrix} y_1 & y_2 \\ y_1' & y_2' \end{vmatrix} = y_1 y_2' - y_2 y_1'. \tag{1-136}$$
It is called the *Wronskian determinant* of $y_1(x)$, $y_2(x)$. When the characteristic roots are the distinct real numbers λ_1, λ_2, it becomes
$$\begin{vmatrix} e^{\lambda_1 x} & e^{\lambda_2 x} \\ \lambda_1 e^{\lambda_1 x} & \lambda_2 e^{\lambda_2 x} \end{vmatrix} = (\lambda_2 - \lambda_1) e^{(\lambda_1 + \lambda_2)x}$$
and hence cannot be 0. Similarly one verifies that the determinant is never 0 for the case of equal roots or of complex roots (see Problem 2 below).

═══════════════ **Problems (Section 1–13)** ═══════════════

1. Obtain a solution by variation of parameters:
   a) $y'' - 4y = e^{2x}$ b) $y'' - 4y = e^{-2x}$ c) $y'' + y = \sec x$, $-\pi/2 < x < \pi/2$

   d) $y'' + y = \tan x$, $-\pi/2 < x < \pi/2$ e) $y'' - y = \dfrac{e^x}{1 + e^x}$

   f) $y'' - y = \dfrac{e^{2x} - 1}{e^{2x} + 1}$

2. Show that the Wronskian determinant (1–136) is never 0 for each of the cases:

   a) $y_1 = e^{\lambda x}$, $y_2 = xe^{\lambda x}$ b) $y_1 = e^{\alpha x}\cos\beta x$, $y_2 = e^{\alpha x}\sin\beta x$ $(\beta \neq 0)$

3. Use variation of parameters to obtain the solution

$$y = \frac{e^{ax}}{b}\int_c^x e^{-au}\sin b(x-u)f(u)\,du$$

for the differential equation $y'' - 2ay' + (a^2 + b^2)y = f(x)$, where $b \neq 0$.

HINT: Use the expression $\int_c^x g(u)\,du$ for an antiderivative of g in obtaining v_1, v_2 from their derivatives.

1–14 APPLICATIONS OF SECOND-ORDER LINEAR EQUATIONS WITH CONSTANT COEFFICIENTS

These equations occur in a great variety of physical problems. Here we select two problems that are typical of many others: (1) forced oscillations of a spring; (2) the simple electric circuit containing inductance, resistance, capacitance, and an alternating source of power. These are illustrated in Figs. 1–24 and 1–25.

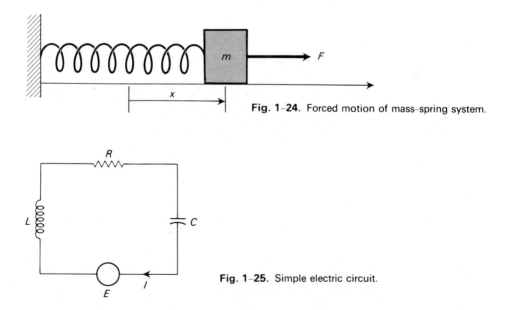

Fig. 1–24. Forced motion of mass–spring system.

Fig. 1–25. Simple electric circuit.

For the mass–spring system we assume three forces to be acting: the spring force $-kx$, where k is a positive constant (Hooke's law) and x is the displacement from equilibrium; a friction force $-h(dx/dt)$, where h is a positive constant; a driving force F varying with time t. Then by Newton's second law, the motion of the particle of mass m is governed by the differential equation

$$m\frac{d^2x}{dt^2} = -kx - h\frac{dx}{dt} + F(t) \tag{1-140}$$

or

$$m\frac{d^2x}{dt^2} + h\frac{dx}{dt} + kx = F(t). \tag{1-140'}$$

For the electric circuit there are an inductance L, a resistance R, and a capacitance C, as well as a driving e.m.f. E depending on time t. The charge q on the capacitance and the current I are related by the equation

$$\frac{dq}{dt} = I. \tag{1-141}$$

By Kirchhoff's laws, we have the voltage relation

$$L\frac{dI}{dt} + RI + \frac{1}{C}q = E(t). \tag{1-142}$$

From Eqs. (1–141) and (1–142) we obtain a second-order linear differential equation for q:

$$L\frac{d^2q}{dt^2} + R\frac{dq}{dt} + \frac{1}{C}q = E(t). \tag{1-143}$$

We can also differentiate Eq. (1–142) with respect to t and use Eq. (1–141) to obtain the differential equation for I:

$$L\frac{d^2I}{dt^2} + R\frac{dI}{dt} + \frac{1}{C}I = E'(t). \tag{1-144}$$

All three differential equations (1–140'), (1–143), (1–144) have the same form. Hence to study the mechanical problem of Fig. 1–24 we can construct a corresponding electric circuit as in Fig. 1–25. This idea, only more elaborated, is the basis of the *analog computer*. In general, to study the solutions of a differential equation, we construct an electric circuit or combination of circuits such that one of its variables (typically, a voltage) obeys the same differential equation. The analog computer makes it possible to achieve the desired circuits quickly for a great variety of differential equations.

It commonly occurs in applications that the driving force or e.m.f. is sinusoidal in character:

$$F = F_0 \sin \omega t \quad \text{or} \quad E(t) = E_0 \sin \omega t. \tag{1-145}$$

We study the typical motion of the mass in response to such a force. Similar arguments then apply to the electric circuit. We are thus considering the equation

$$m\frac{d^2x}{dt^2} + h\frac{dx}{dt} + kx = F_0 \sin \omega t, \tag{1-146}$$

in which m, h, k, F_0, and ω are positive constants. We also consider the limiting cases $h = 0$ (no friction) and $F_0 = 0$ (no driving force).

Study of related homogeneous equation

The related homogeneous equation is

$$m\frac{d^2x}{dt^2} + h\frac{dx}{dt} + kx = 0$$

and the characteristic equation is

$$m\lambda^2 + h\lambda + k = 0.$$

Its roots are

$$\lambda = -\frac{h}{2m} \pm \frac{\sqrt{h^2 - 4\,km}}{2m} \quad \text{or} \quad -\frac{h}{2m} \pm \frac{\sqrt{4\,km - h^2}}{2m}\,i,$$

depending on whether the discriminant $h^2 - 4\,km$ is positive or negative; when it is zero, either form can be used and we have a double root $-h/(2m)$. In the case of distinct real roots we notice that the roots λ_1, λ_2 are both *negative*. This follows from the fact that the characteristic equation has positive coefficients and hence has no positive real roots. Thus the solutions of the related homogeneous equation are given by

$$x = c_1 e^{\lambda_1 t} + c_2 e^{\lambda_2 t}, \qquad \lambda_1 < 0 \text{ and } \lambda_2 < 0.$$

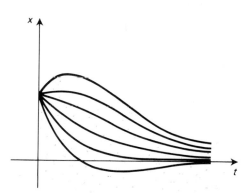

Fig. 1–26. Overdamped motion.

The solutions with fixed initial value of x are shown in Fig. 1–26; all solutions approach the t-axis as $t \to \infty$. In the case of complex roots $\alpha \pm \beta i$, we have $\alpha = -h/(2m) < 0$ and the solutions are

$$x = e^{\alpha t}(c_1 \cos \beta t + c_2 \sin \beta t) = A e^{\alpha t} \sin(\beta t + \gamma), \qquad \alpha < 0.$$

These represent *damped oscillations* as shown in Fig. 1–27.

If we fix m and k and vary h, then we observe the following: for $h = 0$ we have sinusoidal oscillations (simple harmonic motion) as in Example 4 of Section 1–11 (Fig. 1–21); as h increases from 0, the oscillations are damped and die out with increasing rapidity, since α becomes more and more negative. For $h^2 = 4km$ or $h = 2\sqrt{km}$ we have *critical damping*, and the discriminant is 0. Beyond this value, we

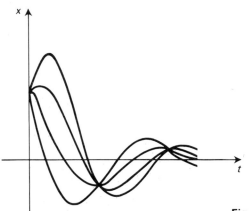

Fig. 1–27. Damped oscillations.

have the overdamped motion depicted in Fig. 1–26. At critical damping, the solutions of the homogeneous equation are given by

$$x = c_1 e^{-ht/(2m)} + c_2 t e^{-ht/(2m)};$$

they resemble those of Fig. 1–26 (see Problem 6 below).

Particular solution of nonhomogeneous equation

We now consider the particular solution of (1–146) with h (and the other constants) positive. By the method of undetermined coefficients we use the trial function

$$x^*(t) = k_1 \cos \omega t + k_2 \sin \omega t$$

and find (see Problem 9 below) that

$$x^*(t) = F_0 \frac{(k - m\omega^2) \sin \omega t - h\omega \cos \omega t}{(k - m\omega^2)^2 + h^2 \omega^2}. \tag{1–147}$$

This function is again sinusoidal, with the same frequency ω as the driving force. The amplitude is F_0/Δ, where $\Delta = [(k - m\omega^2)^2 + h^2 \omega^2]^{1/2}$, and the initial phase $-\gamma$ is given by $\cos \gamma = (k - m\omega^2)/\Delta$, with $0 < \gamma < \pi$.

Since our "input" is $F_0 \sin \omega t$ and our "output" x^* is $(F_0/\Delta) \sin (\omega t - \gamma)$, we can say that a sinusoidal input leads to a sinusoidal output, with a lag in phase of at most π and an amplification (if $\Delta > 1$) or an attenuation (if $\Delta < 1$). The relationships are shown in Fig. 1–28.

For a careful study of the amplification two quantities of the same physical dimension should be compared, such as F and kx (both are forces). Then the ratio of the amplitudes of output kx and input F is $(kF_0/\Delta)/F_0$ or

$$\text{Amplification} = \frac{k}{\Delta} = \frac{1}{[(1 - (m\omega^2/k))^2 + (h^2 \omega^2/k^2)]^{1/2}}.$$

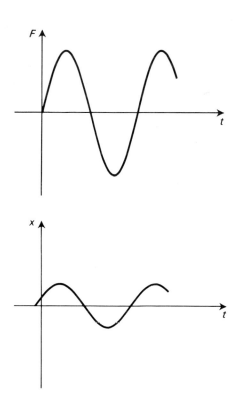

Fig. 1–28. Driving force versus displacement.

Now $k/m = \omega_0^2$, where ω_0 is the natural frequency of the unforced oscillations with no damping. In terms of this frequency and the parameter

$$\eta = \frac{h}{2(km)^{1/2}}$$

we can write

$$\frac{m\omega^2}{k} = \frac{\omega^2}{\omega_0^2} \quad \text{and} \quad \frac{h^2\omega^2}{k} = \frac{h^2}{km}\frac{m\omega^2}{k},$$

so that

$$\text{Amplification} = \frac{1}{\{[1-(\omega/\omega_0)^2]^2 + 4\eta^2(\omega/\omega_0)^2\}^{1/2}}. \qquad (1\text{–}148)$$

The parameter η measures the strength of the damping and is dimensionless. The function of Eq. (1–148) is graphed in Fig. 1–29 for various values of η. We note that $\eta = 1$ corresponds to *critical damping*.

For each fixed positive $\eta < \sqrt{2}/2$, the amplification has a maximum at $\omega = \omega_r = (1-2\eta^2)^{1/2}\omega_0$, which is called the *resonant frequency* under damping. The maximum amplification is found to be $(2\eta)^{-1}(1-\eta^2)^{-1/2}$. For $\eta \geqslant \sqrt{2}/2$, there is no maximum and the amplification decreases steadily as ω increases. In all cases the amplification approaches 0 as $\omega \to \infty$.

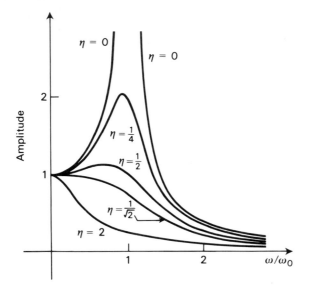

Fig. 1–29. Amplification versus forcing frequency ω
for various values of damping parameter η

If η is close to 0, the maximum amplification is very large. Thus there is exaggerated response to frequencies in a narrow band about the resonant frequency, and much smaller response at other frequencies.

If the driving force is of form $F_1 \sin \omega_1 t + \ldots + F_k \sin \omega_k t$, then by the superposition principle the particular solution $x^*(t)$ is a sum of the particular solutions corresponding to the individual terms. For those terms for which the frequency is close to the resonant frequency, the corresponding terms in $x^*(t)$ will be greatly amplified; for the other terms the amplification is modest or there is attenuation. This *filtering effect* is very important for applications (see Sections 3–8 and 4–20 for further discussion).

General solution

Thus far we have been emphasizing the particular solution of Eq. (1–146). The general solution of the differential equation is as usual given by $x = x_c(t) + x^*(t)$. As shown above, $x_c(t)$ is a damped motion, as in Figs. 1–26 and 1–27, and this approaches 0 as $t \to \infty$. Thus for large positive t the particular solution $x^*(t)$ is predominant. As t increases from 0, there is a *transient period* in which $x_c(t)$ is important. Thereafter the *steady state* $x^*(t)$ is all that matters. Of course, the length of the transient period depends on the choice of the arbitrary constants in $x_c(t)$. The relationships are illustrated in Fig. 1–30.

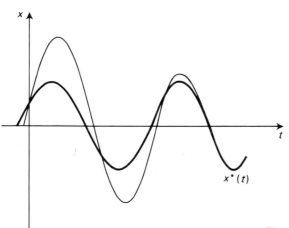

Fig. 1–30. Transients versus steady state.

Stability

The fact that for Eq. (1–146) the solutions of the related homogeneous equation approach 0 as $t \to \infty$ is often described by saying that the system is *stable*. The stability can be traced to the fact that the roots of the auxiliary equation are either negative or have a negative real part. In the limiting case $h = 0$, the roots have zero real part, and the solutions of the related homogeneous equation follow simple harmonic motion. Sometimes this case is referred to as being *neutrally stable*. If h is *negative*, then the solutions of the homogeneous equation become arbitrarily large (positive or negative) for large positive t, and the system is said to be *unstable*.

Case of no damping

In the extreme case when $\eta = 0$, so that $h = 0$, we are dealing with the equation

$$m\frac{d^2x}{dt^2} + kx = F_0 \sin \omega t. \tag{1–146'}$$

The complementary function is itself sinusoidal:

$$x_c(t) = c_1 \cos \omega_0 t + c_2 \sin \omega_0 t = A \sin(\omega_0 t + \delta),$$

and the particular solution is given by

$$x^*(t) = \begin{cases} [F_0/(k - m\omega^2)] \sin \omega t, & \omega \neq \omega_0 \\ -[F_0/(2m\omega_0)] t \cos \omega_0 t, & \omega = \omega_0. \end{cases} \tag{1–147'}$$

Thus for $\omega \neq \omega_0$ the sum $x = x_c(t) + x^*(t)$ is a sum of two sinusoidal functions; it may have a complicated graph if ω and ω_0 are not simply related; in particular, if ω is close to ω_0, but not equal to ω_0, the phenomenon of *beats* occurs, illustrated in Fig. 1–31 (see Problem 5 below).

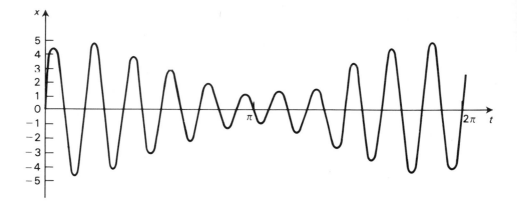

Fig. 1–31. Beats. Response to sinusoidal input at frequency close to natural frequency.

For $\omega = \omega_0$, the particular solution follows increasing oscillations and completely dominates the solution for large t; the amplitude of these oscillations approaches ∞ as $t \to \infty$. This is the phenomenon of *pure resonance* which has toppled bridges and torn machines apart (see Fig. 1–32).

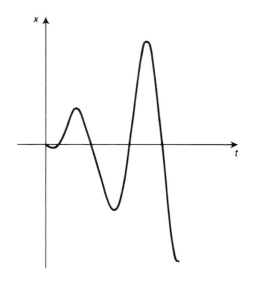

Fig. 1–32. Pure resonance.

Problems (Section 1–14)

1. For the system of Fig. 1–24 let m, h, k, $F(t)$ have the given values. Find the solution with given initial values and graph. (It is assumed throughout that t, x, h, k, m, F_0 are measured in a consistent system of units: for example, t in sec, x in cm, m in g, h in dyne·sec/cm, k in dyne/cm, F_0 in dyne).

a) $m = 2$, $h = 2$, $k = 1$, $F(t) \equiv 0$, $x = 1$ and $dx/dt = 0$ for $t = 0$
b) $m = 4$, $h = 4$, $k = 1$, $F(t) \equiv 0$, $x = 1$ and $dx/dt = 0$ for $t = 0$
c) $m = 1$, $h = 5$, $k = 6$, $F(t) \equiv 0$, $x = 0$ and $dx/dt = 1$ for $t = 0$
d) $m = 2$, $h = 2$, $k = 1$, $F(t) = 3 \sin t$, $x = 0$ and $dx/dt = 0$ for $t = 0$
e) $m = 2$, $h = 2$, $k = 1$, $F(t) = 3 \sin 5t$, $x = 0$ and $dx/dt = 0$ for $t = 0$
f) $m = 1$, $h = 5$, $k = 6$, $F(t) = 3 \sin t$, $x = 0$ and $dx/dt = 0$ for $t = 0$
g) $m = 1$, $h = 5$, $k = 6$, $F(t) = 3 \sin 5t$, $x = 0$ and $dx/dt = 0$ for $t = 0$
h) $m = 1$, $h = 0$, $k = 4$, $F(t) \equiv 0$, $x = 1$ and $dx/dt = 0$ for $t = 0$
i) $m = 1$, $h = 0$, $k = 4$, $F(t) = 3 \sin t$, $x = 0$ and $dx/dt = 0$ for $t = 0$
j) $m = 1$, $h = 0$, $k = 4$, $F(t) = 5 \sin 2t$, $x = 0$ and $dx/dt = 0$ for $t = 0$
k) $m = 1$, $h = 0$, $k = 4$, $F(t) = 2 \sin 5t$, $x = 0$ and $dx/dt = 0$ for $t = 0$

2. *Simple pendulum* (Fig. 1–33).

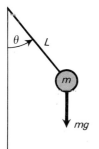

Fig. 1–33. Simple pendulum.

a) Derive the differential equation $mL\dfrac{d^2\theta}{dt^2} = -mg \sin \theta.$

HINT: Equate the tangential component of force to the mass times the tangential component of acceleration, which equals $Ld^2\theta/dt^2$, according to mechanics.

b) For small oscillations, during which θ remains close to 0, one can approximate $\sin \theta$ by θ. Use this approximation to show that the period is $2\pi \sqrt{L/g}$.

3. *Torsional vibrations* A mass is suspended by a wire, as in Fig. 1–34. If one rotates the mass about the wire, the stiffness of the wire provides a restoring torque proportional to θ.

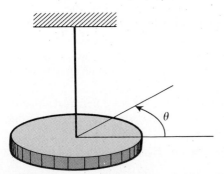

Fig. 1–34. Torsional vibrations.

a) Deduce the differential equation of the motion.

HINT: By mechanics, the torque equals $I\,d^2\theta/dt^2$, where I is the moment of inertia of the body about the axis of rotation.

b) Make a physical experiment to test the equation and to determine the role of friction.

4. In the mass–spring system of Fig. 1–35, let the support be moving sinusoidally, so that its position at time t is $u = c \sin \omega t$, where u is measured from a fixed position on the line of motion with the same positive direction as x. Find the differential equation of the motion.

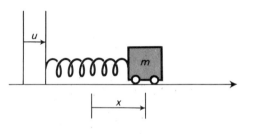

Fig. 1–35. Spring with moving support.

HINT: The effect of moving the support is to change the length of the spring at each instant by adding $-u$ to the length.

5. *Beats.* By Eq. (1–147′), the general solution of Eq. (1–146′) for $\omega \neq \omega_0$ is

$$x = c_1 \cos \omega_0 t + c_2 \sin \omega_0 t + B \sin \omega t,$$

where $B = F_0/(k - m\omega^2)$. Write the last term as

$$B \sin [(\omega - \omega_0)t + \omega_0 t] = B \sin (\omega - \omega_0)t \cos \omega_0 t + B \cos (\omega - \omega_0)t \sin \omega_0 t,$$

and hence write the general solution as

$$x = v \cos \omega_0 t + u \sin \omega_0 t = \sqrt{u^2 + v^2} \sin (\omega_0 t + \theta),$$

where $v = c_1 + B \sin (\omega - \omega_0)t$, $u = c_2 + B \cos (\omega - \omega_0)t$. For ω close to ω_0, so that $\omega - \omega_0$ is small compared to ω_0, we can regard this as a sinusoidal oscillation of frequency ω_0, with a slowly changing phase shift θ and slowly changing amplitude $\sqrt{u^2 + v^2}$. Show that (u, v) traces a circle in the uv-plane and that the maximum value of $\sqrt{u^2 + v^2}$ is $\sqrt{c_1^2 + c_2^2} + B$. Show also that θ is the polar angle of (u, v). Carry out the process described for $x = 2 \sin 10t + 3 \sin 11t$ (say, $\omega_0 = 10$) and graph. (The graph is shown in Fig. 1–31.)

The effect is noticeable, to a sensitive ear, when one note is played on an out-of-tune piano. The sound wavers in a regular fashion.

6. For the case of critical damping ($h^2 = 4km$), the solutions of the differential equation $mx'' + hx' + kx = 0$ have the form $x = e^{\alpha t}(c_1 + c_2 t)$, where $\alpha = -h/(2m) < 0$. Graph typical solutions for which $x = 1$ when $t = 0$.

7. Consider the electric circuit of Fig. 1–25, governed by Eqs. (1–141) to (1–144), where L, R, C are positive constants.

a) Let $E(t) = E_0 \sin \omega t$ and show that a particular solution of Eq. (1–143) is $q = [E_0/(\omega Z)] \sin (\omega t - \gamma)$, where

$$Z = \left[\left(\frac{1}{\omega C} - L\omega \right)^2 + R^2 \right]^{1/2}, \qquad \cos \gamma = Z^{-1} \left(\frac{1}{C\omega} - L\omega \right), \qquad 0 < \gamma < \pi.$$

Here Z is called the *impedance*. Hence show that the steady-state current is

$$I = \frac{E_0}{Z} \cos (\omega t - \gamma).$$

Compare the phases of $E(t)$, q, and I.

b) Reason directly from Eqs. (1–141) and (1–143) that if $E(t)$ in part (a) is replaced by $E_0 \sin (\omega t + \beta)$, then q is replaced by $[E_0/(\omega Z) \sin (\omega t - \gamma + \beta)$ and I is replaced by $(E_0/Z) \cos (\omega t - \gamma + \beta)$.

c) Obtain the solution of part (a) by taking $E(t) = E_0 e^{i\omega t}$ and using undetermined coefficients in Eq. (1–143) to obtain

$$q = \frac{E_0}{\omega} \left[\left(\frac{1}{C\omega} - L\omega \right) + Ri \right]^{-1} e^{i\omega t}$$

and then taking imaginary parts in $E(t)$ and q. The expression in brackets is the *complex impedance*.

8. Show that for fixed η, $0 < \eta < \sqrt{2}/2$, the amplification function has a maximum for $\omega = \omega_r = (1 - 2\eta^2)^{1/2} \omega_0$ and that the maximum amplification is $(2\eta)^{-1}(1 - \eta^2)^{-1/2}$.

9. a) Use undetermined coefficients to obtain the particular solution (1–147) of Eq. (1–146).

b) Obtain the particular solution (1–147') of Eq. (1–146') for the two cases $\omega \neq \omega_0$ and $\omega = \omega_0$.

1–15 LINEAR EQUATIONS OF HIGHER ORDER WITH CONSTANT COEFFICIENTS

We consider here the general linear differential equation

$$a_0 y^{(n)} + a_1 y^{(n-1)} + \cdots + a_{n-1} y' + a_n y = f(x), \qquad (1–150)$$

where $a_0, a_1, \ldots, a_n$ are constants and $a_0 \neq 0$. The equation is said to be *homogeneous* if $f(x) \equiv 0$. For a given equation (1–150) the *related homogeneous equation* is the equation

$$a_0 y^{(n)} + a_1 y^{(n-1)} + \cdots + a_{n-1} y' + a_n y = 0 \qquad (1–151)$$

obtained by replacing $f(x)$ by 0.

The methods of Sections 1–11 and 1–12 extend to the equations considered here. The general solution of Eq. (1–150) has the form

$$y = y_c(x) + y^*(x),$$

where $y_c(x)$ (the complementary function) is the general solution of the related homogeneous equation and $y^*(x)$ is a particular solution of the given equation

(1–150). The complementary function has the form
$$c_1 y_1(x) + \cdots + c_n y_n(x),$$
where $y_1(x), \ldots, y_n(x)$ are solutions of Eq. (1–151) and $c_1, \ldots, c_n$ are arbitrary constants.

To find the complementary function, one again seeks solutions of Eq. (1–151) of form $y = e^{\lambda x}$ and is led to the *auxiliary*, or *characteristic, equation* of degree n:
$$a_0 \lambda^n + \cdots + a_{n-1}\lambda + a_n = 0. \tag{1–152}$$
If this equation has n distinct real roots $\lambda_1, \ldots, \lambda_n$, then we can take
$$y_c(x) = c_1 e^{\lambda_1 x} + \cdots + c_n e^{\lambda_n x}.$$
For each real root λ of multiplicity k, we modify the previous expression by introducing the functions
$$e^{\lambda x}, xe^{\lambda x}, \ldots, x^{k-1}e^{\lambda x}$$
instead of the pure exponentials. If, for example, $n = 6$ and the roots of Eq. (1–152) are 7, 7, 7, 5, 5, -2, then we can take
$$y_c(x) = c_1 e^{7x} + c_2 xe^{7x} + c_3 x^2 e^{7x} + c_4 e^{5x} + c_5 xe^{5x} + c_6 e^{-2x}.$$
If complex roots $\alpha \pm \beta i$ appear, then we can follow the same procedures and allow complex constants, thereby obtaining a general complex solution. Instead, one can use $e^{\alpha x} \cos \beta x$ and $e^{\alpha x} \sin \beta x$ to replace $e^{(\alpha + \beta i)x}$ and $e^{(\alpha - \beta i)x}$. If these complex roots are repeated, then we again use the same functions multiplied by successive powers of x. If, for example, $n = 7$ and the characteristic roots are $2 \pm 3i, 2 \pm 3i, 4, 4, 1$, then we can take
$$y_c(x) = e^{2x}(c_1 \cos 3x + c_2 \sin 3x) + e^{2x}(c_3 x \cos 3x + c_4 x \sin 3x)$$
$$+ c_5 e^{4x} + c_6 xe^{4x} + c_7 e^x.$$
To justify these rules, we must verify that each term in $y_c(x)$ is a solution of Eq. (1–151) and that arbitrary initial conditions can be satisfied. For the Eq. (1–151) of order n, the initial conditions have the form
$$y = y_0, y' = y_0', \ldots, y^{(n-1)} = y_0^{(n-1)} \quad \text{for } x = 0.$$
A justification of the rules can be obtained easily by Laplace transforms (see Chapter 4).

EXAMPLE 1 $y''' - y'' - 4y' + 4y = 0$. The characteristic equation is
$$\lambda^3 - \lambda^2 - 4\lambda + 4 = 0;$$
its roots are 1, 2, -2. Hence
$$y = c_1 e^x + c_2 e^{2x} + c_3 e^{-2x}$$
is the general solution. For the initial conditions $y = 1, y' = 0, y'' = -1$ for $x = 0$ we are led to the equations
$$1 = c_1 + c_2 + c_3, \qquad 0 = c_1 + 2c_2 - 2c_3, \qquad -1 = c_1 + 4c_2 + 4c_3.$$
They have the unique solution $c_1 = 5/3, c_2 = -3/4, c_3 = 1/12$, so that
$$y = \frac{5}{3}e^x - \frac{3}{4}e^{2x} + \frac{1}{12}e^{-2x}$$
is the solution sought. ◄

For the discussion of such differential equations it is convenient to use *operator* notation. We write Dy for y', D^2y for y'', and so on. Also, for example, we write $(2D^3 - 5D^2 + 3D + 2)y$ for

$$2D^3y - 5D^2y + 3Dy + 2y \qquad \text{or} \qquad 2y''' - 5y'' + 3y' + 2y.$$

Thus the differential equation (1–150) becomes

$$(a_0D^n + a_1D^{n-1} + \cdots + a_{n-1}D + a_n)y = f(x).$$

The expression in parentheses is called a *differential operator*. We are here considering only differential operators with constant coefficients $a_0, \ldots, a_n$.

Two operators can be multiplied as polynomials and the resulting relation is correct when both sides are applied to a function. For example,

$$(D+1)(D+2) = D^2 + 3D + 2$$

and

$$
\begin{aligned}
(D+1)(D+2)y - (D+1)[(D+2)y] &= (D+1)(y'+2y) \\
&= (y'+2y)' + y' + 2y = y'' + 3y' + 2y \\
&= (D^2 + 3D + 2)y
\end{aligned}
$$

as asserted.

The differential equation of Example 1 can now be written as

$$(D^3 - D^2 - 4D + 4)y = 0 \qquad \text{or} \qquad (D-1)(D+2)(D-2)y = 0.$$

The factored form reveals the characteristic roots.

EXAMPLE 2 $(D^2+9)^2y = 0$. The characteristic equation is $(\lambda^2+9)^2 = 0$, the characteristic roots are $\pm 3i, \pm 3i$, the general solution is

$$y = c_1 \cos 3x + c_2 \sin 3x + c_3 x \cos 3x + c_4 x \sin 3x. \quad \blacktriangleleft$$

To obtain the particular solution $y^*(x)$ of the nonhomogeneous equation (1–150), one can employ the method of undetermined coefficients when $f(x)$ is of the allowed form, or the method of variation of parameters (Problem 3 following Section 1–18). For the method of undetermined coefficients, the rules are the same as those of Section 1–12 except that $h + \omega i$, for form (1–123), or h, for form (1–123′), may be a multiple root of multiplicity k of the auxiliary equation. In that case, the trial function is modified by multiplication by x^k.

EXAMPLE 3 $(D^4 + 4D^3 + 6D^2 + 4D + 1)y = x^2e^{-x}$ or $(D+1)^4y = x^2e^{-x}$. The characteristic equation is $(\lambda+1)^4 = 0$, so that the roots are $-1, -1, -1, -1$. Since the right-hand side is of form $e^{hx}p(x)$ with $h = -1$, we must multiply the basic trial function by x^4. Thus we use

$$x^4e^{-x}(k_1 + k_2x + k_3x^2) \qquad \text{or} \qquad e^{-x}(k_1x^4 + k_2x^5 + k_3x^6).$$

Substitution in the differential equation gives (after a lengthy calculation)

$$e^{-x}(24k_1 + 120k_2x + 360k_3x^2) = e^{-x}x^2.$$

Hence we can take $k_1 = 0, k_2 = 0, k_3 = 1/360$. The general solution is

$$y = e^{-x}\left(c_1 + c_2x + c_3x^2 + c_4x^3 + \frac{x^6}{360}\right). \quad \blacktriangleleft$$

The *Laplace transform method* (Chapter 4) provides one further way of finding the solutions of the linear differential equation with constant coefficients. It is exceptionally valuable in solving initial-value problems.

Remark on applications The applications of the nth order linear equation with constant coefficients can be studied as in Sections 1–7 and 1–14. If x is the dependent variable and t is the independent variable, then a response x to a sinusoidal input $A \sin \omega t$ is again sinusoidal, except in the case of resonance, when ωi is a characteristic root. The physical system is *stable*, so that the solutions of the related homogeneous equation are transients, precisely when all characteristic roots have negative real part.

$=============$ **Problems (Section 1–15)** $==========$

1. Find the general solution:

   a) $(D-1)(D+2)(D+5)y = 0$ b) $(D^2-9)(D^2+4)y = 0$

   c) $(D-1)^2(D+1)y = 0$ d) $(D-2)^2(D^2+1)y = 0$

   e) $(2D^2+3D+1)(D^2+2D+2)y = 0$ f) $(9D^2+4)^2y = 0$

   g) $(D^2+16)^3y = 0$ h) $(D-5)^4(D^2+1)^2y = 0$

2. Find the particular solution satisfying the stated initial conditions:

   a) $(D^3+3D^2+3D+1)y = 0$, $y = 1$, $y' = 2$, $y'' = -1$ for $x = 0$

   b) $(D-1)(D+2)^2y = 0$, $y = 3$, $y' = 6$, $y'' = 0$ for $x = 0$

   c) $y^{(5)} = 0$, $y = 3$, $y' = 7$, $y'' = 1$, $y''' = 4$, $y^{(iv)} = -2$ for $x = 1$

   d) $(D-1)(D^2+4)y = 0$, $y = 0$, $y' = 0$, $y'' = 1$ for $x = 0$

3. Find the general solution:

   a) $(D-1)(D+2)(D+4)y = 5e^{2x}$ b) $(D-1)(D+2)(D+4)y = 200\cos 2x$

   c) $(D-1)(D+2)(D+4)y = xe^{2x}$ d) $(D-1)(D+2)(D+4)y = 5e^x$

   e) $(D-1)(D+2)(D+4)y = xe^x$ f) $(D-1)^2(D+3)y = 2e^{-3x}$

   g) $(D-1)^2(D+3)y = x^2e^x$ h) $(D-1)(D^2+4)y = 3\cos x$

   i) $(D-1)(D^2+4)y = 5\sin 3x$ j) $(D-1)(D^2+4)y = 7\cos 2x$

4. *Beam on elastic foundation.* It is shown in mechanics that the vertical deflection w of such a beam is governed by the differential equation

$$EI\frac{d^4 w}{dx^4} + kw = p(x),$$

where x is a horizontal coordinate along the beam, $p(x)$ is the load per unit length at x, E (Young's modulus), I (moment of inertia of a cross-section with respect to an axis perpendicular to the x-axis), and k (modulus of the foundation) are positive constants (Fig. 1–36).

    Assume that, in a consistent system of units, $EI = 1$, $k = 16$, and $p(x) = 16x^2$ for $-1 \leqslant x \leqslant 1$, and find the deflection under the boundary conditions: $w = 0$ and $dw/dx = 0$ for $x = -1$ and for $x = 1$. Here the beam is clamped at both ends: $x = \pm 1$.

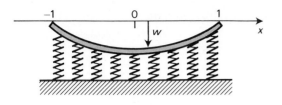

Fig. 1–36. Deflection of beam on elastic foundation.

HINT: By symmetry, one expects the solution $w(x)$ to be an even function of x, that is, $w(x) = w(-x)$. Verify that the general solution can be written in the form

$$w = c_1 y_1(x) + \cdots + c_4 y_4(x) + x^2,$$

where

$$y_1(x) = \sin bx \sinh bx, \qquad y_2(x) = \cos bx \cosh bx,$$
$$y_3(x) = \sin bx \cosh bx, \qquad y_4(x) = \cos bx \sinh bx,$$

but that only $y_1(x)$ and $y_2(x)$ are even. It can be shown that the only even solutions are those with $c_3 = 0$ and $c_4 = 0$. Now take $c_3 = 0$ and $c_4 = 0$ and choose c_1, c_2 to satisfy the boundary conditions at $x = 1$. Show that those at $x = -1$ are then also satisfied.

5. *Coupled springs.* Let two masses m_1, m_2 move along a horizontal line as in Fig. 1–37. Here m_1 is attached to a rigid support by a spring with a spring constant k_1, and m_2 is attached to m_1 by a spring with a constant k_2. Coordinates x_1 and x_2 measure the displacements of m_1, m_2, respectively, from their equilibrium positions. By Newton's second law, the differential equations are:

$$m_1 \frac{d^2 x_1}{dt^2} = -k_1 x_1 + k_2(x_2 - x_1), \tag{1–153}$$

$$m_2 \frac{d^2 x_2}{dt^2} = -k_2(x_2 - x_1). \tag{1–154}$$

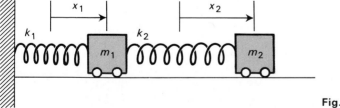

Fig. 1–37. Coupled springs.

If Eq. (1–153) is differentiated twice with respect to t and then combined with (1–153) and (1–154), we can eliminate x_2 and $d^2 x_2/dt^2$ to obtain a fourth-order differential equation for x_1. If x_1 has been found, x_2 can then be obtained from Eq. (1–153). In appropriate units, let $k_1 = k_2 = 2$, $m_1 = 8$, $m_2 = 3$; find the general solutions x_1, x_2.

1–16 LINEAR SPACES AND LINEAR OPERATORS

Our results for linear differential equations can be described more simply by using the language of linear-space theory.

A *linear space* (or *vector space*) is a collection of objects that can be added and multiplied by numbers, in accordance with the usual rules for such operations.

EXAMPLE 1   All vectors in the xy-plane form a linear space, for any two vectors can be added to yield a third, and any vector can be multiplied by a number c (see Fig. 1–38). Also

$$\mathbf{u} + \mathbf{v} = \mathbf{v} + \mathbf{u}, \qquad \mathbf{u} + (\mathbf{v} + \mathbf{w}) = (\mathbf{u} + \mathbf{v}) + \mathbf{w}, \qquad c(\mathbf{u} + \mathbf{v}) = c\mathbf{u} + c\mathbf{v}. \quad \blacktriangleleft$$

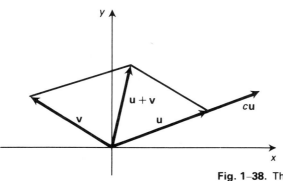

Fig. 1–38. The linear space of vectors in the plane.

EXAMPLE 2   All polynomials in x of degree at most two form a linear space. Addition of two such polynomials yields a third one, as does multiplication of such a polynomial by a number. For example,

$$(2 + x + 3x^2) + (3 - x^2) = 5 + x + 2x^2, \qquad 6(1 + 2x + 4x^2) = 6 + 12x + 24x^2.$$

Again the familiar rules hold. ◀

EXAMPLE 3   All functions of x defined on a given interval and having continuous first and second derivatives form a linear space. For $0 \leqslant x \leqslant 1$ we can use the functions e^x, $\sin x$, xe^x, e^{-x}, $\cosh x$, $\ln(1 + x)$, etc.

The objects in a linear space are commonly vectors (as in Example 1) or functions (as in Examples 2 and 3). Here we are mainly concerned with linear spaces of functions, hence refer to the objects as functions (though occasionally we do refer to functions as vectors); we denote them by symbols such as $f, g, \ldots, F, G, \ldots$. The function equal to zero on a given interval will be denoted simply by 0.

Let $f_1, \ldots, f_n$ be functions in a given linear space. An expression $c_1 f_1 + \cdots + c_n f_n$, where $c_1, \ldots, c_n$ are numbers, is called a *linear combination* of $f_1, \ldots, f_n$. If all of $c_1, \ldots, c_n$ are 0, we call it the *trivial* linear combination; if not all are 0, we call it a *nontrivial* linear combination.

We say that $f_1, \ldots, f_n$ are *linearly dependent* if some nontrivial linear combination of $f_1, \ldots, f_n$ equals 0. Otherwise, we say that $f_1, \ldots, f_n$ are *linearly independent*.

In the linear space of Example 2, the functions $x - 1$, $x + 1$, and x are linearly dependent since
$$1(x-1) + 1(x+1) + (-2)x = 0.$$
In the same linear space, the functions 1, x, x^2 are linearly independent since
$$c_1 1 + c_2 x + c_3 x^2 = 0 \quad \text{(for all } x\text{)}$$
only if $c_1 = 0, c_2 = 0, c_3 = 0$; no *nontrivial* linear combination of $1, x, x^2$ reduces to 0.

In a given linear space, a *linear operator* (or *linear mapping*) L is an assignment to each function f of the linear space of a function $g = L[f]$ in a second linear space in such a way that in general

$$L[c_1 f_1 + c_2 f_2] = c_1 L[f_1] + c_2 L[f_2]. \quad \blacktriangleleft \qquad (1\text{–}160)$$

EXAMPLE 4 In the linear space of Example 3 let $L[f] = f'$, the derivative of f. Thus $L[x^2] = 2x$, $L[e^x] = e^x$, $L[\sin x] = \cos x$. In general, the rule (1–160) is obeyed, since

$$L[c_1 f_1 + c_2 f_2] = (c_1 f_1 + c_2 f_2)' = c_1 f_1' + c_2 f_2' = c_1 L[f_1] + c_2 L[f_2].$$

In this example, the second linear space consists of all functions with continuous first derivatives on the chosen interval. $\blacktriangleleft$

EXAMPLE 5 In the linear space of Example 3, let $L[f] = (aD^2 + bD + c)f$, where a, b, c are constants. Here L is a *differential operator*, as in Section 1–15.

We see at once that L is a linear operator. We can even allow a, b, c to be functions of x, say continuous on the chosen interval. Thus $xD^2 + e^x D + 1$ is a linear operator:

$$L[f] = (xD^2 + e^x D + 1)f = xf'' + e^x f' + f,$$
$$L[c_1 f_1 + c_2 f_2] = x(c_1 f_1'' + c_2 f_2'') + e^x(c_1 f_1' + c_2 f_2') + c_1 f_1 + c_2 f_2$$
$$= c_1(xf_1'' + e^x f_1' + f_1) + c_2(xf_2'' + e^x f_2' + f_2)$$
$$= c_1 L[f_1] + c_2 L[f_2]. \quad \blacktriangleleft$$

EXAMPLE 6 In the linear space of all continuous functions $f(t)$ for all t, let $L[f]$ be the solution $x(t)$ of the differential equation

$$\frac{dx}{dt} + 2x = f(t)$$

such that $x = 0$ for $t = 0$. Then L is a linear operator. For if $L[f_1] = x_1(t)$, $L[f_2] = x_2(t)$, then $x_1(0) = 0$, $x_2(0) = 0$. Hence $c_1 x_1 + c_2 x_2 = 0$ for $t = 0$ and

$$\frac{dx_1}{dt} + 2x_1 = f_1(t), \qquad \frac{dx_2}{dt} + 2x_2 = f_2(t),$$

so that

$$\frac{d}{dt}(c_1 x_1 + c_2 x_2) + 2(c_1 x_1 + c_2 x_2) = c_1 f_1(t) + c_2 f_2(t).$$

It follows that $c_1 x_1(t) + c_2 x_2(t) = c_1 L[f_1] + c_2 L[f_2] = L[c_1 f_1 + c_2 f_2]$.

REMARK: Here the operator L can be given explicitly by Eq. (1–84) with $a = 2$ and $c = 0$:

$$x = \int_0^t e^{-2(t-u)} f(u)\, du. \blacktriangleleft \qquad (1\text{–}161)$$

1–17 THE GENERAL LINEAR EQUATION

The general linear equation has the form

$$a_0(x)y^{(n)} + a_1(x)y^{(n-1)} + \cdots + a_{n-1}(x)y' + a_n(x)y = f(x). \qquad (1\text{–}170)$$

Here the coefficients $a_0(x), \ldots, a_n(x)$ are permitted to be functions of x. Usually these functions and f are continuous in an interval over which solutions are sought, and $a_0(x) \neq 0$ on this interval. The related homogeneous equation is the equation

$$a_0(x)y^{(n)} + \cdots + a_n(x)y = 0. \qquad (1\text{–}171)$$

With Eq. (1–170) we associate the differential operator

$$L = a_0(x)D^n + \cdots + a_{n-1}(x)D + a_n(x). \qquad (1\text{–}172)$$

Thus

$$L[y] = a_0(x)y^{(n)} + a_1(x)y^{(n-1)} + \cdots + a_{n-1}(x)y' + a_n(x)y \qquad (1\text{–}173)$$

and we can write Eq. (1–170) and the related homogeneous equation (1–171) concisely as

$$L[y] = f, \qquad (1\text{–}170')$$
$$L[y] = 0. \qquad (1\text{–}171')$$

The operator L is a linear operator in the linear space of all functions with continuous derivatives through order n on the given interval.

If $y_1(x), \ldots, y_n(x)$ are solutions of the homogeneous equation (1–170), then

$$L[y_1] = 0, \ldots, L[y_n] = 0,$$

so that, for every choice of constants $c_1, \ldots, c_n$, we have

$$L[c_1 y_1 + \cdots + c_n y_n] = c_1 L[y_1] + \cdots + c_n L[y_n] = 0.$$

Thus, for all choices of $c_1, \ldots, c_n$, the function

$$y = c_1 y_1(x) + \cdots + c_n y_n(x) \qquad (1\text{–}174)$$

is a solution of the homogeneous equation. If $y_1(x), \ldots, y_n(x)$ are *linearly independent* on the given interval, then (1–174) is the general solution; that is, $c_1, \ldots, c_n$ can be chosen to satisfy arbitrary initial conditions

$$y(x_0) = y_0, \quad y'(x_0) = y_0', \ldots, y^{(n-1)}(x_0) = y_0^{(n-1)} \qquad (1\text{–}175)$$

at a point x_0 of the given interval.

We call the function (1–174), with $y_1(x), \ldots, y_n(x)$ linearly independent solutions of the homogeneous equation, the *complementary function* and denote it as usual by $y_c(x)$. If $y^*(x)$ is a particular solution of the nonhomogeneous equation, then

$$L[y^*] = f.$$

Hence by linearity
$$L[y_c + y^*] = L[y_c] + L[y^*] = 0 + f = f.$$
Thus
$$y = y_c + y^* = c_1 y_1(x) + \cdots + c_n y_n(x) + y^*(x) \qquad (1\text{–}176)$$
is a solution of the nonhomogeneous equation, no matter how the constants $c_1, \ldots, c_n$ are chosen. Every solution of the nonhomogeneous equation has this form. For if $L[y] = f$, then
$$L[y - y^*] = L[y] - L[y^*] = f - f = 0.$$
Thus $y - y^*$ is a solution of the related homogeneous equation, so that for some set of $c_1, \ldots, c_n$:
$$y - y^* = c_1 y_1(x) + \cdots + c_n y_n(x).$$
Accordingly, Eq. (1–176) *gives the general solution of the nonhomogeneous equation.*

Superposition principle

If $L[u_1] = f_1, \ldots, L[u_k] = f_k$, then by linearity
$$L[c_1 u_1 + \cdots + c_k u_k] = c_1 f_1 + \cdots + c_k f_k.$$
Thus a particular solution of $L[y] = f$, where f is a linear combination of functions $f_1, \ldots, f_k$, is given by the corresponding linear combination of particular solutions $u_1, \ldots, u_k$ for $f = f_1, \ldots, f_k$ respectively.

We can also phrase the superposition principle in another way, closer to the physical point of view. For given continuous *inputs* $f_1, \ldots, f_k$, we choose the *outputs* $u_1, \ldots, u_k$ as the particular solutions of $L[y] = f_1, \ldots, L[y] = f_k$, respectively, such that $y(x_0) = 0, y'(x_0) = 0, \ldots, y^{(n-1)}(x_0) = 0$ (zero initial conditions). Then to the linear combination $c_1 f_1 + \cdots + c_k f_k$ as input there corresponds the output $y = c_1 u_1 + \cdots + c_k u_k$ satisfying the same zero initial conditions at x_0.

We can say: the operator that assigns to each input f the output y with zero initial conditions at x_0 is a linear operator T. Thus $y = T[f]$.

An illustration is provided by Example 6 in Section 1–16.

EXAMPLE 1 $(x + 1)y'' - (x + 2)y' + y = f$. We verify that here $y = x + 2$ and $y = e^x$ are solutions of the related homogeneous equation and are linearly independent for $x \geqslant 0$. Also $L[x] = -2$, $L[x^2] = -x^2 - 2x + 2$. Hence for $f = 2x^2 + 4x + 6 = -2(-x^2 - 2x + 2) - 5(-2)$, a particular solution is $-2x^2 - 5x$:
$$L[-2x^2 - 5x] = -2L[x^2] - 5L[x] = 2x^2 + 4x + 6.$$
The general solution is
$$y = c_1(x + 2) + c_2 e^x + y^*(x).$$
For $f = -2$ as above, we can take $y^* = x$ and then obtain the particular solution with zero initial values at $x = 0$:
$$y = c_1(x + 2) + c_2 e^x + x, \qquad y' = c_1 + c_2 e^x + 1,$$
$$0 = 2c_1 + c_2, \qquad 0 = c_1 + c_2 + 1, \qquad c_1 = 1, c_2 = -2,$$
$$y = 2x + 2 - 2e^x.$$

Similarly, for $f = -x^2 - 2x + 2$, the particular solution with zero initial values at $x = 0$ is found to be $y = x^2$. Accordingly, for the input $f = c_1(-2) + c_2(-x^2 - 2x + 2)$, the output with zero initial values at $x = 0$ is

$$y = c_1(2x + 2 - 2e^x) + c_2 x^2. \quad \blacktriangleleft$$

In this example, the chosen equation had a known complementary function, and the particular solutions were constructed artificially. In practice, a linear equation with variable coefficients (except for order 1, as in Section 1–7) is difficult to solve. Power series are useful for this purpose (see Sections 2–22 to 2–23 of Chapter 2). Once $y_c(x)$ has been found, $y^*(x)$ can be found by *variation of parameters* (Problem 3 below).

For further theoretical background on linear differential equations the reader is referred to Chapter 6 of this volume and to Chapters 4 and 12 of ODE.

1–18 NONLINEAR DIFFERENTIAL EQUATIONS

For nonlinear equations, the general solution can no longer be expressed in the form $y_c(x) + y^*(x)$ and only in special cases (such as those of Sections 1–5 and 1–6) can a simple formula be given for the general solution. Power series and numerical methods can be used to obtain particular solutions. Nonlinear equations are considered in detail in Chapter 7.

═══════════════════════ **Problems (Section 1–18)** ═══════════════════════

1. Find the general solution for $x \geq 0$:
   a) $(x + 1)^2 y'' - 3(x + 1)y' + 3y = 0$ **b)** $4(x + 1)^2 y'' - 4(x + 1)y' + 3y = 0$

   HINT: Seek solutions of form $y = (x + 1)^m$.

2. Let $L[y] = y'' + xy$
   a) Verify that $L[1] = x$, $L[x] = x^2$, $L[x^2] = x^3 + 2$.
   b) Find a particular solution of $y'' + xy = 3x - 2x^2$.
   c) Find a particular solution of $y'' + xy = 2x^3 - 3x^2 + 4x + 4$.

3. *Variation of parameters* (cf. Section 1–13). To find a particular solution $y^*(x)$ of Eq. (1–170), find the complementary function and then replace the constants $c_1, \ldots, c_n$ by functions $v_1(x), \ldots, v_n(x)$ to be determined, thus obtaining the form

$$y^*(x) = v_1(x) y_1(x) + \cdots + v_n(x) y_n(x)$$

   for the solution sought. Now n conditions must be imposed on $v_1, \ldots, v_n$; one of these conditions is that the differential equation (1–170) must be satisfied; the others are chosen to simplify the calculation of $y^{*\prime}, \ldots, y^{*(n)}$. The conditions are

$$y_1(x)v_1' + \cdots + y_n(x)v_n' = 0,$$
$$y_1'(x)v_1' + \cdots + y_n'(x)v_n' = 0,$$
$$\vdots \qquad \vdots \qquad \qquad (1\text{--}180)$$
$$y_1^{(n-2)}(x)v_1' + \cdots + y_n^{(n-2)}(x)v_n' = 0.$$

Because of these conditions, we now find:

$$y*' = y_1'v_1 + \cdots + y_n'v_1,$$
$$y*'' = y_1''v_1 + \cdots + y_n''v_n,$$

$$\vdots \qquad \vdots$$

$$y*^{(n-1)} = y_1^{(n-1)}v_1 + \cdots + y_n^{(n-1)}v_n$$
$$y*^{(n)} = y_1^{(n)}v_1 + \cdots + y_n^{(n)}v_n + y_1^{(n-1)}v_1' + \cdots + y_n^{(n-1)}v_n'.$$

If we now substitute the expression for $y*$ in Eq. (1–170) and use the fact that $y_1(x), \ldots, y_n(x)$ are solutions of the related homogeneous equation, we obtain the final condition:

$$a_0[y_1^{(n-1)}(x)v_1' + \cdots + y_n^{(n-1)}(x)v_n'] = f(x). \qquad (1\text{–}181)$$

Equations (1–180) and (1–181) provide n equations for $v_1', \ldots, v_n'$, which can be solved for these functions. Integration then yields $v_1(x), \ldots, v_n(x)$. (For complete justification of the method, see ODE, pp. 145–149.)

Apply the method described to find a particular solution:

a) $(D-1)(D-2)(D-3)y = 2e^{5x}$ **b)** $(D-1)(D-2)(D-3)y = \cos x$
c) $(D-1)(D-2)(D-3)y = xe^x$ **d)** $(D-1)(D^2+4)y = e^x$
e) $(D-3)(D^2+1)y = f(x)$ **f)** $(D^2-1)(D^2+1)y = f(x)$
g) $(x+1)y'' - (x+2)y' + y = (x+1)^2$
h) $(x+1)y'' - (x+2)y' + y = e^x(x+1)^2$

HINT: For (g) and (h) see Example 1 in Section 1–17.

4. Linear operators with variable coefficients do not usually commute, that is, in general $L_1 L_2 \neq L_2 L_1$ for such operators. Determine whether the following pairs commute:
a) $L_1 = xD + 1$, $L_2 = xD - 1$ **b)** $L_1 = e^x D^2$, $L_2 = D + 1$

5. a) Show that the solutions of a homogeneous linear differential equation $L[y] = 0$ on a given interval form a linear space.
   b) Do the solutions of a nonhomogeneous linear differential equation $L[y] = f$ on a given interval form a linear space?
   c) Do the solutions of the nonlinear differential equation $y' - y^2 = 0$ for $x \geq 0$ form a linear space?

Chapter 2 Infinite Series

2-1 INTRODUCTION

Infinite series arise naturally in a variety of contexts. For example, let a rubber ball be dropped from a height of 32 ft. onto a pavement. If the ball rebounds and continues to bounce and, say, loses half its kinetic energy at each impact, then (from the rule $v^2 = 2gs$, where v is velocity, g is acceleration due to gravity, s is distance covered) we find easily that the successive heights reached are 16 ft., 8 ft., 4 ft., . . . Thus the total distance travelled after hitting the pavement is, in feet,

$$16 + 8 + 4 + 2 + 1 + \frac{1}{2} + \frac{1}{4} + \frac{1}{8} + \cdots$$

This is an infinite series whose sum we shall find to be 32. We can also ask for the time elapsed after first hitting the pavement. We find (with the help of the rule $s = \frac{1}{2}gt^2$, where t is time), that the time between the first and second impacts is 2 sec, between the second and third is $\sqrt{2}$ sec, between the third and fourth is 1 sec, etc. The total time, in seconds, is

$$2 + \sqrt{2} + 1 + \frac{1}{\sqrt{2}} + \frac{1}{2} + \frac{1}{2\sqrt{2}} + \cdots$$

This is again an infinite series, whose sum we shall show to be $4 + 2\sqrt{2}$.

The solution of differential equations often leads us to new functions, not obtainable by simple algebraic operations or substitution from the elementary functions of calculus. To have an explicit representation of such functions, we may find infinite series of value. Even the simple differential equation $y' = y$ leads us to the elementary function $y = e^x$ which can be obtained by a power series:

$$e^x = 1 + x + \frac{x^2}{2!} + \cdots + \frac{x^n}{n!} + \cdots$$

In the study of *periodic* phenomena it is of great value to represent functions in the form of *Fourier series:*

$$\frac{a_0}{2} + a_1 \cos \omega t + b_1 \sin \omega t + \cdots + a_n \cos n\omega t + b_n \sin n\omega t + \cdots$$

Such a representation is particularly valuable in the study of response of linear systems (Sections 1–8 and 1–14). The frequencies ω, 2ω, 3ω, . . . appearing here form the *spectrum* for this case.

In the present chapter we first consider infinite series of constant terms and conditions for their convergence or divergence. We then consider series of functions and their convergence. Special results are given for power series. Later chapters provide further discussions of series and their role in solving problems that arise in applications.

2–2   SEQUENCES: CONVERGENCE AND DIVERGENCE

By an infinite sequence we mean the assignment of a number s_n to each integer $n = 1, 2, 3, \ldots$ Thus the formulas

$$s_n = \frac{1}{n^2} \qquad \text{and} \qquad s_n = \frac{\ln n}{n!}$$

define the sequences

$$1, \frac{1}{4}, \frac{1}{9}, \cdots, \frac{1}{n^2}, \cdots \qquad \text{and} \qquad 0, \frac{\ln 2}{2}, \frac{\ln 3}{6}, \cdots, \frac{\ln n}{n!}, \cdots$$

The sequence as a whole is denoted by $\{s_n\}$ or often simply by s_n. We may start with $n = 0$ or with $n = 2$ or some other value.

The sequence $\{s_n\}$ is said to *converge* if for some number S

$$\lim_{n \to \infty} s_n = S.$$

This means s_n remains within a prescribed distance ε from S for sufficiently large n, that is $S - \varepsilon < s_n < S + \varepsilon$ for $n > N$ for appropriate N. We call S the *limit* of the sequence and say that the sequence *converges to* S. For example, the sequence $\{1/n^2\}$ converges and has limit 0; the sequence $\{(n+1)/n\}$ converges to 1, since, by familiar rules for limits,

$$\lim_{n \to \infty} \frac{n+1}{n} = \lim_{n \to \infty} \left(\frac{1}{n} + 1 \right) = 0 + 1 = 1.$$

Here clearly the error in approximation of the limit by s_n is precisely $1/n$, and this is less than 0.001 for $n > 1000$, less than 0.0001 for $n > 10\,000$, and so on.

Many practical problems are solved by processes which yield sequences known to converge to the desired value. Typically, a computer produces successive numbers of a sequence:

$$3.7015670, \ 3.7015721, \ 3.7015733, \ 3.7015737, \ldots$$

Often one relies on inspection to decide when to stop and how much accuracy has been achieved. However, appearance can be deceptive, and whenever possible some mathematical estimate should be obtained for the error made in stopping at a particular n.

A sequence that does not converge is said to *diverge*. For example, the sequence $n!$ diverges. In this case we can also write:

$$\lim_{n \to \infty} n! = \infty$$

and state that the sequence *diverges to* ∞. Similarly, the sequence $\{\ln (1/n)\}$ diverges to $-\infty$. The sequence $\{s_n\}$ with

$$s_n = 1 + \frac{1}{2} + \cdots + \frac{1}{n}$$

can be shown to diverge to ∞ (see Example 1 in Section 2-8). For $n = 1, 2, 3, \ldots, 10$ we obtain the values

$$1, \ 1.5, \ 1.83, \ 2.083, \ 2.283, \ldots, 2.929;$$

we might be misled by the appearance of the sequence and call it convergent, even though this is not true. The sequence $(-1)^n$ diverges, but not to ∞ or $-\infty$: the successive numbers $-1, \ 1, \ -1, \ 1, \ldots$ are clearly approaching no limit.

In determining convergence and finding the limit it is often useful to write $s_n = f(n)$ and consider the corresponding $f(x)$. If $f(x)$ is defined for all sufficiently large real numbers x and has a limit S as $x \to \infty$, then $\{s_n\}$ converges to S. For example, for $s_n = n^2 \ln (n + 1)/(n^3 + 1)$ we write

$$f(x) = \frac{x^2 \ln (x + 1)}{x^3 + 1} = \frac{\ln (x + 1)}{x + (1/x^2)}.$$

Then by L'Hospital's rule (for the form ∞/∞), we get

$$\lim_{x \to \infty} f(x) = \lim_{x \to \infty} \frac{1/(x + 1)}{1 - (2/x^3)} = \frac{0}{1 - 0} = 0.$$

Thus the sequence has limit 0.

EXAMPLE   The value of π can be determined by seeking the area of a circle of radius 1 as the limit of the area of inscribed regular polygons with 4, 8, 16, $\ldots$, 2^n, $\ldots$ sides, where $n = 2, 3, 4, \ldots$ (Fig. 2-1). For the square the area is 2, for the octagon it is $2\sqrt{2} = 2.828$, for the 16-gon it is

$$8\sqrt{\frac{\sqrt{2} - 1}{2\sqrt{2}}} = 3.06.$$

The process yields an infinite sequence converging to π. ◄

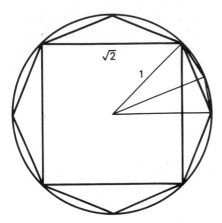

Fig. 2−1. π as a limit of areas of regular polygons.

2–3 BOUNDED SEQUENCES, MONOTONE SEQUENCES

A sequence $\{s_n\}$ is said to be *bounded above* if $s_n \leqslant B =$ const for all n, *bounded below* if $s_n \geqslant A =$ const for all n, *bounded* if $A \leqslant s_n \leqslant B$ for all n, with A and B constant.

Thus a sequence is bounded precisely when it is both bounded above and bounded below. The three conditions are illustrated in Fig. 2–2.

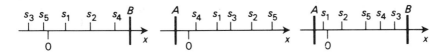

Fig. 2–2. Boundedness for sequences.

EXAMPLES
1. $\{1/n\}$ is bounded, with $A = 0$, $B = 1$.
2. $\{\sin (n\pi/4)\}$ is bounded, with $A = -1$, $B = 1$.
3. $\{n^2\}$ is bounded below with $A = 0$, but is not bounded above.
4. $\{(-1)^{n+1} - n\}$ is bounded above with $B = 0$, but is not bounded below.
5. $\{n \sin (n\pi/3)\}$ is not bounded above or below. ◄

Every convergent sequence $\{s_n\}$ is bounded, since $S - \varepsilon < s_n < S + \varepsilon$ when $n > N$ for some N and a particular $\varepsilon > 0$. Then A and B can be chosen to satisfy the finite set of conditions

$$A < s_1 < B, \ A < s_2 < B, \ldots, \ A < s_N < B, \ A < S - \varepsilon < S + \varepsilon < B,$$

and such a choice of A and B meets the requirements.

However, not every bounded sequence is convergent. This is illustrated by Example 2 above, with the successive values

$$\sqrt{2}/2,\ 1,\ \sqrt{2}/2,\ 0,\ -\sqrt{2}/2,\ -1,\ -\sqrt{2}/2,\ 0,\ \sqrt{2}/2,\ \ldots,$$

which clearly approach no limit. A simpler example is the bounded sequence $(-1)^n$ with successive values $-1,\ 1,\ -1,\ 1,\ \ldots$

A sequence $\{s_n\}$ is said to be *monotone increasing* if for all n

$$s_1 \leqslant s_2 \leqslant \cdots \leqslant s_n \leqslant s_{n+1} \leqslant \cdots$$

A sequence is said to be *monotone decreasing* if for all n

$$s_1 \geqslant s_2 \geqslant \cdots \geqslant s_n \geqslant s_{n+1} \geqslant \cdots$$

The sequence in Example 1 above is monotone decreasing, since

$$1 > \frac{1}{2} > \frac{1}{3} > \cdots > \frac{1}{n} > \frac{1}{n+1} \cdots,$$

while that in Example 3 is clearly monotone increasing. ◀

THEOREM ON MONOTONE SEQUENCES　*If a sequence $\{s_n\}$ is bounded and monotone (increasing or decreasing), then the sequence $\{s_n\}$ converges.*

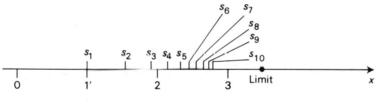

Fig. 2–3. Example for theorem on monotone sequences.

The theorem is illustrated in Fig. 2–3 for the sequence $\{7n/(2n+5)\}$, which can be shown to be bounded and monotone increasing, with limit $7/2$. The example is typical; the successive members of the sequence are forced to "pile up" against some number, and hence have that number as limit. (For a proof of the Theorem see AC, p. 374.)

A sequence $\{s_n\}$ that is monotone increasing but *unbounded* cannot converge (since, as remarked above, every convergent sequence is bounded). Here necessarily $\lim\limits_{n \to \infty} s_n = \infty$, so that $\{s_n\}$ diverges to ∞. Thus a monotone increasing sequence is either bounded above and convergent or unbounded above and divergent to ∞. Similarly a monotone decreasing sequence is either convergent or divergent to $-\infty$.

REMARK:· The convergence or divergence of an infinite sequence (and the limit in the case of convergence) clearly depend only on the behavior of s_n for large n. Hence we may disregard or alter the first few members of the sequence without affecting convergence, divergence, or the limit.

========================== **Problems (Section 2–3)** ==========================

1. Test for convergence and find the limit if it exists or is infinite:

   a) $\dfrac{1}{n^2}$ **b)** $\sqrt{n}$ **c)** $\dfrac{\sin n}{n}$ **d)** ne^{-n} **e)** $\dfrac{(-1)^n}{n^2}$

   f) $(-1)^n \dfrac{\ln n}{n}$ **g)** $\dfrac{n^2 - 3n - 2}{n^2 + 5n + 1}$ **h)** $\dfrac{n^3 - 1}{n^3 + 1}$

2. *Newton's method* for solution of the equation $f(x) = 0$ produces the infinite sequence $x_1, x_2, \ldots, x_n, \ldots$, which under appropriate conditions converges to a solution. (See CLA, pp. 446–447.) Here x_1 is a first estimate, and in general

   $$x_n = x_{n-1} - \frac{f(x_{n-1})}{f'(x_{n-1})}.$$

   a) Take $f(x) = x^2 - 3$ and $x_1 = 2$ to obtain a sequence converging to $\sqrt{3}$. Find x_n for $n = 1, 2, 3, 4, 5$.

   b) Take $f(x) = x^2 - 5$ and $x_1 = 2$ to obtain a sequence converging to $\sqrt{5}$. Find x_n for $n = 1, \ldots, 5$.

   c) Take $f(x) = e^x + x - 2$ and $x_1 = 0$ to obtain a sequence converging to a solution of the equation $e^x + x - 2 = 0$. Show by graphing $y = e^x$ and $y = 2 - x$ that the solution is unique. Find x_n for $n = 1, 2, 3, 4$.

   d) Take $f(x) = e^x$ and $x_1 = 0$. Discuss the resulting sequence.

3. The number e can be obtained as a limit of a sequence $\{s_n\}$ in several ways:

   a) Take $s_n = \left(1 + \dfrac{1}{n}\right)^n$ and estimate by trial how large n must be to approximate e with error less than 0.1.

   b) Take $s_n = 1 + \dfrac{1}{1!} + \cdots + \dfrac{1}{n!}$ and estimate by trial how large n must be to approximate e with an error less than 0.0001.

4. Determine whether the sequence is bounded above or below or both:

   a) $\dfrac{1}{n^2 + 1}$ **b)** $\dfrac{n}{3n + 1}$ **c)** $\cos \dfrac{n\pi}{4}$ **d)** $\dfrac{e^n}{n^2}$ **e)** $\dfrac{\ln (n + 1)}{1 + \ln n}$

   f) Arc $\tan n$ **g)** $\dfrac{2n + 1}{(3n - 5)(2n - 9)}$ **h)** $e^n \sin n + \dfrac{1}{n}$

5. Determine whether the sequence is monotone increasing or decreasing.

   HINT: For more difficult sequences one may find it helpful (a) to calculate $s_{n+1} - s_n$ and show that this is positive or 0 for an increasing sequence, and negative or 0 for a decreasing sequence, (b) if $s_n > 0$ for all n, to calculate s_{n+1}/s_n and show that this is greater than or equal to 1 for increasing sequences, less than or equal to 1 for decreasing sequences (if $s_n < 0$ for all n, proceed in the same way with reversed inequalities), (c) to write $s_n = f(n)$ for differentiable f if possible, and show that $f'(x) \geq 0$ for $x \geq 1$ for increasing sequences, $f'(x) \leq 0$ for $x \geq 1$ for decreasing sequences.

   a) $\dfrac{1}{n}$ **b)** $\ln n$ **c)** e^{-n} **d)** $\dfrac{1}{n!}$ **e)** $\dfrac{n}{n + 1}$ **f)** $\dfrac{n + 2}{n + 3}$

g) $\dfrac{2n-1}{3n+1}$ h) $\dfrac{1-n}{5+n}$ i) $\dfrac{n!}{2^n}$ j) $\dfrac{\ln n}{n}$, $n \geqslant 2$ k) $1+\dfrac{1}{2}+\cdots+\dfrac{1}{2^n}$

l) $\dfrac{3}{2} \cdot \dfrac{3}{4} \cdot \dfrac{5}{4} \cdot \dfrac{5}{6} \cdots \dfrac{2n+1}{2n} \cdot \dfrac{2n+1}{2n+2}$

6. Show that the sequence is monotone and bounded and therefore convergent:

a) $\dfrac{n-1}{n}$ b) e^{-n^2} c) $\operatorname{Arc\,tan} n$ d) $\dfrac{1}{2}+\dfrac{1}{4}+\cdots+\dfrac{1}{2^n}$

e) $n \sin \dfrac{1}{n}$ f) $\dfrac{1}{1+\dfrac{1}{2}+\cdots+\dfrac{1}{n}}$

2–4 INFINITE SERIES

An infinite series is an indicated sum of infinitely many terms, such as

$$a_1 + a_2 + \cdots + a_n + \cdots \qquad \text{or} \qquad b_2 + b_3 + \cdots + b_n + \cdots$$

These are commonly abbreviated as

$$\sum_{n=1}^{\infty} a_n \qquad \text{or} \qquad \sum_{n=2}^{\infty} b_n.$$

The terms of the series form an infinite sequence, such as $\{a_n\}$. However, the convergence or divergence of this sequence is not the same as convergence or divergence of the infinite series.

EXAMPLE 1 $\sum_{n=1}^{\infty} \dfrac{n+1}{n}$. Here the terms form the sequence $\left\{\dfrac{n+1}{n}\right\}$, which converges to 1. However, the infinite series diverges, as will be seen. ◀

The infinite series $\sum_{n=1}^{\infty} a_n$ determines another sequence: the sequence $\{S_n\}$ of *partial sums*. Here

$$S_n = a_1 + \cdots + a_n, \quad \text{where } n = 1, 2, \ldots$$

Thus S_n is the sum of the first n terms of the series. The series is said to *converge* if the sequence $\{S_n\}$ converges and to *diverge* if the sequence $\{S_n\}$ diverges. If $\{S_n\}$ converges to S, then we write

$$S = \sum_{n=1}^{\infty} a_n$$

and call S the *sum* of the series. Thus

$$S = \lim_{n \to \infty} S_n = \lim_{n \to \infty} (a_1 + \cdots + a_n).$$

EXAMPLE 2 $\sum_{n=1}^{\infty} \dfrac{1}{10^n}$. Here $S_1 = 0.1, S_2 = 0.1 + 0.01 = 0.11, S_3 = 0.111, \ldots,$ $S_n = 0.11 \ldots 1$ (to n places). The sequence $\{S_n\}$ of partial sums is monotone increasing and bounded (since $S_n < 0.2$ for all n, for example). Therefore the sequence converges. We observe that S_n is simply the decimal expansion of $1/9$ to n places, and therefore

$$S = \sum_{n=1}^{\infty} \frac{1}{10^n} = 0.111 \ldots 1 \ldots = \frac{1}{9}.$$

As this last example illustrates, we are using an infinite series every time we use an infinite decimal expansion of a number, such as

$$\pi = 3.1415927 \ldots, \quad e = 2.7182818 \ldots, \quad \sqrt{2} = 1.414214 \ldots \quad \blacktriangleleft$$

EXAMPLE 3 $\dfrac{1}{3} + \left(\dfrac{2}{5} - \dfrac{1}{3}\right) + \left(\dfrac{3}{7} - \dfrac{2}{5}\right) + \cdots + \left(\dfrac{n}{2n+1} - \dfrac{n-1}{2n-1}\right) + \cdots$

Due to cancellations, we have here

$$S_1 = \frac{1}{3}, S_2 = \frac{2}{5}, \cdots, \quad S_n = \frac{n}{2n+1}, \cdots,$$

so that the sequence $\{S_n\}$ converges to $1/2$ and

$$\frac{1}{2} = \sum_{n=1}^{\infty} \left(\frac{n}{2n+1} - \frac{n-1}{2n-1}\right).$$

Because of the cancellations, such a series can be called a *telescoping* series. $\blacktriangleleft$

REMARKS: As Example 3 suggests, *every infinite sequence $\{S_n\}$ can be represented as the sequence of partial sums of an infinite series:* namely, the series

$$S_1 + (S_2 - S_1) + (S_3 - S_2) + \cdots + (S_n - S_{n-1}) + \cdots$$

For a series $\sum_{n=0}^{\infty} a_n$, one commonly writes:

$$S_n = a_0 + \cdots + a_n.$$

2–5 THE GEOMETRIC SERIES

This is the series:

$$1 + r + \cdots + r^n + \cdots = \sum_{n=0}^{\infty} r^n. \tag{2–50}$$

By algebra, $1 - r^{n+1} = (1-r)(1 + r + \cdots + r^n)$. Hence (for $r \neq 1$)

$$S_n = 1 + r + \cdots + r^n = \frac{1 - r^{n+1}}{1-r} = \frac{1}{1-r} - \frac{r^{n+1}}{1-r}. \tag{2–51}$$

When $r = 1$, we get $S_n = n + 1$, which diverges to ∞. When $r > 1$, the value of r^{n+1} tends to ∞ and by (2–51) the series diverges to ∞. When $r = -1$, we have $S_1 = 0, S_2 = 1, S_3 = 0, \ldots$, and $S_n = 0$ or 1 depending on whether n is odd or even, so that there is no limit. When $r < -1$, the value of r^{n+1} oscillates between larger and larger positive and negative values approaching $+\infty$ or $-\infty$ as $n \to \infty$. Hence by (2–51) there is no limit for $r < -1$. Thus *the series diverges when* $r \leqslant -1$ *and when* $r \geqslant 1$. When $-1 < r < 1$, the series converges, since $r^{n+1} \to 0$ as $n \to \infty$ for such r. For example,

$$\left(\frac{1}{2}\right)^{n+1} = \frac{1}{2^{n+1}} \to 0 \qquad \text{and} \qquad \left(-\frac{1}{2}\right)^{n+1} = (-1)^{n+1}\frac{1}{2^{n+1}} \to 0 \text{ as } n \to \infty.$$

Therefore, from Eq. (2–51),

$$S = \sum_{n=0}^{\infty} r^n = \lim_{n \to \infty} (1 + r + \cdots + r^n) = \lim_{n \to \infty} \left(\frac{1}{1-r} - \frac{r^{n+1}}{1-r}\right)$$

$$= \frac{1}{1-r} \text{ when } -1 < r < 1. \tag{2–52}$$

This is the *basic rule for the geometric series.*

2–6 THE nTH-TERM TEST

In this and the following sections, we shall formulate several tests for convergence or divergence of infinite series. For a particular series, a test may be applicable and allow us to conclude convergence or divergence. But often the test is inapplicable, and we say: *The test fails.* When this happens, we have no conclusion: we simply don't know, from trying the selected test, whether the series converges or diverges. We can try other tests and may be able to find a decisive one. Finding a successful test is not always easy!

For a series $\sum_{n=1}^{\infty} a_n$ let us consider the sequence $\{a_n\}$ formed by the terms of the series $a_1, a_2, \ldots, a_n, \ldots$ We can write:

$$S_n = a_1 + \cdots + a_n, \qquad S_{n-1} = a_1 + \cdots + a_{n-1}.$$

Therefore,

$$a_n = S_n - S_{n-1}, \text{ where } n = 2, 3, \ldots.$$

If the series converges, with sum S, then

$$\lim_{n \to \infty} S_n = S, \qquad \lim_{n \to \infty} S_{n-1} = S.$$

Therefore, if the series converges,

$$\lim_{n \to \infty} a_n = \lim_{n \to \infty} S_n - \lim_{n \to \infty} S_{n-1} = 0.$$

For a convergent series, the nth term must have limit 0 as $n \to \infty$.

From this rule we obtain the *nth-term test: if the sequence* $\{a_n\}$ *does not converge to 0, then the series* $\sum_{n=1}^{\infty} a_n$ *diverges.*

Note that the test is a test for *divergence only*. If $\{a_n\}$ does have limit 0, then the test fails and no conclusion can be drawn.

EXAMPLE 1 $\sum_{n=1}^{\infty} (n+1)/n$. Here $a_n = (n+1)/n \to 1$ as $n \to \infty$, and the series diverges. ◄

EXAMPLE 2 $\sum_{n=1}^{\infty} 2^n$. Here $a_n = 2^n \to \infty$ as $n \to \infty$, and the series diverges. The series is a geometric series (minus the first term), as in (2–50), with $r = 2 > 1$, so that we expect divergence. ◄

EXAMPLE 3 $\sum_{n=0}^{\infty} e^{-n}$. Here $a_n = e^{-n} \to 0$ as $n \to \infty$ and the test fails. The series could converge or diverge. However, the series is again a geometric series, with $r = 1/e$, so that $-1 < r < 1$, and as in (2–52) the series converges, with sum $1/(1-r) = e/(e-1)$. ◄

EXAMPLE 4 $\sum_{n=1}^{\infty} 1/n$. Here $a_n = 1/n \to 0$ as $n \to \infty$ and the test fails. The series is called the *harmonic series* and the series *diverges*, as in Ex. 1 of Section 2–8 below. ◄

EXAMPLE 5 $\sum_{n=1}^{\infty} 1/n^2$. Here $a_n = 1/n^2 \to 0$ as $n \to \infty$. The test fails. The series is a *p-series*, with $p = 2$, and is *convergent*, as in Ex. 2 of Section 2–8. ◄

EXAMPLE 6 $\sum_{n=1}^{\infty} (-1)^n$. Here $a_n = (-1)^n$ and a_n has no limit as $n \to \infty$. Therefore, the series diverges. (The series is geometric, with $r = -1$.) ◄

=========================== **Problems (Section 2–6)** ===========================

1. Find the partial sums $S_1, \ldots, S_5$ for each of the following infinite series:

a) $\displaystyle\sum_{n=0}^{\infty} \frac{1}{n!}$ b) $\displaystyle\sum_{n=1}^{\infty} \frac{n+1}{n}$ c) $\displaystyle\sum_{n=1}^{\infty} (-1)^n n$

d) $\displaystyle\sum_{n=1}^{\infty} 2^n$ e) $\displaystyle\sum_{n=1}^{\infty} (-1)^n n^3$

2. Show that the series converges and find the sum:

a) $\displaystyle\sum_{n=0}^{\infty} \frac{1}{3^n}$ b) $\displaystyle\sum_{n=0}^{\infty} (-1)^n e^{-n}$ c) $\displaystyle\sum_{n=0}^{\infty} \frac{3^n}{5^n}$ d) $\displaystyle\sum_{n=0}^{\infty} \frac{e^n}{\pi^n}$

e) $\displaystyle\sum_{n=1}^{\infty} \frac{1}{2^n}$ f) $\displaystyle\sum_{n=2}^{\infty} \frac{1}{3^n}$ g) $\displaystyle\sum_{n=0}^{\infty} \frac{5}{7^n}$ h) $\displaystyle\sum_{n=0}^{\infty} \frac{\pi}{2^n}$

i) $\displaystyle\sum_{n=1}^{\infty} \left(\frac{n+1}{n+2} - \frac{n}{n+1} \right)$ j) $\displaystyle\sum_{n=1}^{\infty} \left(\frac{\ln(n+1)}{n+1} - \frac{\ln n}{n} \right)$

k) $\displaystyle\sum_{n=1}^{\infty} \frac{-2}{n(n+1)}$

l) $\displaystyle\sum_{n=1}^{\infty} \ln\left[\frac{(n+1)^{1/(n+1)^2}}{n^{1/n^2}}\right]$

3. Apply the nth-term test if possible:

a) $\displaystyle\sum_{n=1}^{\infty} \frac{3n+5}{2n+1}$

b) $\displaystyle\sum_{n=1}^{\infty} \frac{3^n}{2^n+1}$

c) $\displaystyle\sum_{n=1}^{\infty} \frac{2^n}{n!}$

d) $\displaystyle\sum_{n=1}^{\infty} \frac{(-1)^n n}{n+1}$

e) $\displaystyle\sum_{n=1}^{\infty} \frac{\ln n}{\ln (n+1)}$

f) $\displaystyle\sum_{n=1}^{\infty} \frac{1}{n^3}$

g) $\displaystyle\sum_{n=1}^{\infty} \frac{1}{\sqrt{n}}$

h) $\displaystyle\sum_{n=1}^{\infty} \frac{\ln n}{n}$

2–7 GENERAL REMARKS ON SERIES

As for sequences, one may disregard or modify a finite number of terms of a series in testing for convergence. Of course these terms will affect the sum, if the series converges. For example,

$$\sum_{n=0}^{\infty} r^{n-3} = \frac{1}{r^3}+\frac{1}{r^2}+\frac{1}{r}+1+r+r^2+\cdots$$

Apart from the first three terms, this is the geometric series. Therefore the series converges and has sum

$$\frac{1}{r^3}+\frac{1}{r^2}+\frac{1}{r}+\frac{1}{1-r},$$

provided $-1 < r < 1$ and $r \neq 0$ (because of the first three terms).

We may freely insert or remove zero terms in a series without affecting convergence or the sum. For example,

$$\sum_{n=1}^{\infty} \frac{\sin (n\pi/2)}{n} = 1+\frac{0}{2}-\frac{1}{3}+\frac{0}{4}+\frac{1}{5}+\frac{0}{6}-\cdots$$

can be replaced by

$$1-\frac{1}{3}+\frac{1}{5}-\frac{1}{7}+\cdots = \sum_{n=1}^{\infty} \frac{(-1)^{n+1}}{2n-1}.$$

A series may be multiplied by a nonzero constant:

$$k\sum_{n=1}^{\infty} a_n = \sum_{n=1}^{\infty} ka_n;$$

that is, if either side converges, then the other side also converges and the equality holds.

Convergent series may be added or subtracted:

$$\sum_{n=1}^{\infty} a_n + \sum_{n=1}^{\infty} b_n = \sum_{n=1}^{\infty} (a_n + b_n), \qquad \sum_{n=1}^{\infty} a_n - \sum_{n=1}^{\infty} b_n = \sum_{n=1}^{\infty} (a_n - b_n).$$

Thus the series

$$\sum_{n=0}^{\infty} \left(\frac{1}{2^n} + \frac{1}{3^n} \right)$$

converges since it is obtained by adding two convergent geometric series.

It will be seen that series with *positive* terms are of special importance. The following few sections are devoted to such series. In Section 2–10 it will be shown how to decide on the convergence of series with mixed signs by referring to the appropriate series of positive terms.

In a positive-term series, $S_n = a_1 + \cdots + a_n$ is necessarily monotone increasing. Hence by the theorem on monotone sequences (Section 2–3), either $S_n \to \infty$ as $n \to \infty$ or else $\{S_n\}$ is bounded and $S_n \to S$, so that the series converges and has sum S. Thus *to show convergence it is sufficient to show that the sequence of partial sums is bounded.* When $S_n \to \infty$, the series is called *properly divergent.*

2–8 THE INTEGRAL TEST

This test reduces the question of convergence or divergence of a positive-term series to that of the improper integral $\int_1^{\infty} f(x)\,dx$.

We recall that such an integral converges and has a finite value L if

$$\lim_{b \to \infty} \int_1^b f(x)\,dx = L.$$

Thus

$$\int_1^{\infty} \frac{dx}{x} = \lim_{b \to \infty} \ln b = \infty, \qquad \int_1^{\infty} \frac{dx}{\sqrt{x}} = \lim_{b \to \infty} (2\sqrt{b} - 2) = \infty,$$

$$\int_1^{\infty} \frac{dx}{x^{3/2}} = \lim_{b \to \infty} \left(2 - \frac{2}{b^{1/2}} \right) = 2, \qquad \int_1^{\infty} e^{-x}\,dx = \lim_{b \to \infty} (e^{-1} - e^{-b}) = e^{-1}.$$

Accordingly, the first two integrals diverge, the last two converge. In general, we find that

$$\int_1^{\infty} \frac{dx}{x^p} \begin{cases} \text{converges to } \dfrac{1}{p-1} \text{ for } p > 1 \\[2mm] \text{diverges for } p \leqslant 1 \end{cases}$$

$$(2\text{–}80)$$

INTEGRAL TEST   *In the series $\sum_{n=1}^{\infty} a_n$ let $a_n > 0$ for all n and let $f(x)$ be a function defined and continuous for $1 \leqslant x < \infty$ and such that $f(n) = a_n$; let $f(x)$ decrease as x increases. Then the series converges or diverges depending on whether the improper integral $\int_1^{\infty} f(x)\, dx$ converges or diverges.*

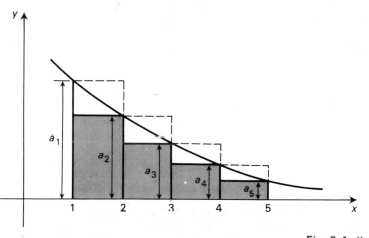

Fig. 2–4. Proof of integral test.

To prove the test, we assume first that the improper integral converges. Then (as suggested in Fig. 2–4), a_2 equals the area of a rectangle beneath the graph of $f(x)$ from $x = 1$ to $x = 2$; a_3 equals the area of a rectangle beneath the graph of $f(x)$ from $x = 2$ to $x = 3$, and so on, so that

$$a_2 < \int_1^2 f(x)\, dx, \qquad a_3 < \int_2^3 f(x)\, dx, \qquad \cdots$$

and hence

$$a_2 + \cdots + a_n < \int_1^n f(x)\, dx.$$

Since the integral converges, the right side has a limit as $n \to \infty$. Hence the sequence $\{a_2 + \cdots + a_n\}$ is bounded, as is $\{S_n\}$ with $S_n = a_1 + \cdots + a_n$. Therefore, as remarked at the end of Section 2–7, the series converges.

If the integral diverges, we reason similarly using the circumscribed rectangles (upper edges shown as broken lines in Fig. 2–4) to conclude that

$$a_1 + a_2 + \cdots + a_n > \int_1^{n+1} f(x)\, dx.$$

Since $f(x) > 0$ and the improper integral diverges, it must diverge to $+\infty$, so that the right side approaches ∞ as $n \to \infty$. It follows that $S_n \to \infty$ also, so that the series is properly divergent.

EXAMPLE 1 The *harmonic series*

$$\sum_{n=1}^{\infty} \frac{1}{n} = 1 + \frac{1}{2} + \cdots + \frac{1}{n} + \cdots \tag{2-81}$$

diverges, since $\int_1^{\infty} (1/x)\, dx = \ln x \big|_1^{\infty} = \infty$, as in the example above; this is the case of $p = 1$ in the rule (2–80). ◄

EXAMPLE 2 We consider the *p-series*:

$$\sum_{n=1}^{\infty} \frac{1}{n^p} = 1 + \frac{1}{2^p} + \cdots + \frac{1}{n^p} + \cdots . \tag{2-82}$$

For $p > 0$, the function $f(x) = 1/x^p$ satisfies the conditions for the integral test (observe that $f(x)$ is required to be monotone *decreasing*), and hence by the rule (2–80) the series converges for $p > 1$ and diverges for $0 < p \leqslant 1$ (for $p = 1$ we have the harmonic series again). For $p \leqslant 0$, the nth term n^{-p} does not have limit 0 when $n \to \infty$. Hence by the nth-term test the series diverges. We summarize: *the p-series (2–82) converges for $p > 1$ and diverges for $p \leqslant 1$.* ◄

EXAMPLE 3 The series $\sum_{n=1}^{\infty} n/e^n$ converges because $f(x) = xe^{-x}$ is continuous, positive, and decreasing for $x \geqslant 1$ (its derivative $e^{-x}(1-x)$ is negative for $x > 1$), and

$$\int_1^{\infty} xe^{-x}\, dx = -e^{-x}(x+1) \bigg|_1^{\infty} = 5e^{-1}. ◄$$

EXAMPLE 4 The series $\sum_{n=1}^{\infty} n^2/e^n$ converges. Here we take $f(x) = x^2 e^{-x}$ and find $f'(x) = x(2-x)e^{-x}$, so that $f'(x) < 0$ when $x > 2$. The fact that the conditions for the integral test hold only when $x > 2$ clearly has no effect, just as the first few terms of a series have no effect on convergence. We apply the test and deduce convergence:

$$\int_1^{\infty} x^2 e^{-x}\, dx = -e^{-x}(x^2 + 2x + 2) \bigg|_1^{\infty} = 5e^{-1}.$$

By the same reasoning, f might even have several discontinuities. We require only that $f(x)$ be continuous, positive, and decreasing when $x \geqslant a$ for some a, and that $f(n) = a_n$ when n is sufficiently large. To apply the test, we consider $\int_a^{\infty} f(x)\, dx$. ◄

EXAMPLE 5 $\sum_{n=2}^{\infty} 1/(n \ln n)$. Here we can take $f(x) = 1/(x \ln x)$ for $x \geqslant 2$, and f is continuous, positive, and decreasing. The series diverges, since

$$\int_2^{\infty} \frac{dx}{x \ln x} = \ln \ln x \bigg|_2^{\infty} = \infty. ◄$$

2–9 THE COMPARISON TEST

The proof of the integral test is based on the comparison of one monotone increasing sequence, A_n, with another, B_n. If $A_n \leqslant B_n$ for all n and $\{B_n\}$ converges, then so does $\{A_n\}$ since $\{B_n\}$ is bounded and hence so is $\{A_n\}$. If $\{A_n\}$ diverges, then so does $\{B_n\}$ since $\{A_n\}$ must diverge to ∞ and hence $\{B_n\}$ also does. We now apply this idea to positive-term series.

COMPARISON TEST   *Let $0 < a_n \leqslant b_n$ for all n. If $\sum_{n=1}^{\infty} b_n$ converges, then so does $\sum_{n=1}^{\infty} a_n$. If $\sum_{n=1}^{\infty} a_n$ diverges, then so does $\sum_{n=1}^{\infty} b_n$.*

This can be proved by applying the reasoning above to the monotone increasing sequences

$$A_n = a_1 + \cdots + a_n, \quad B_n = b_1 + \cdots + b_n.$$

EXAMPLE 1   The series $\sum_{n=1}^{\infty} 1/(n2^n)$ converges since $a_n = 1/(n2^n) \leqslant b_n = 1/2^n$ for all n and $\sum b_n$ is a convergent geometric series.   ◄

EXAMPLE 2 $\sum_{n=1}^{\infty} \dfrac{n+1}{n^3 + 2n^2 + 1}$ converges, since

$$a_n = \frac{n+1}{n^3 + 2n^2 + 1} < \frac{n+1}{n^3} \leqslant \frac{2n}{n^3} = \frac{2}{n^2} = b_n.$$

Since $\sum (1/n^2)$ converges, so does $\sum (2/n^2)$, and the comparison test applies.   ◄

Sometimes a *second form of the comparison test* is convenient: if $0 < a_n$ and $0 < b_n$ for all n and

$$\lim_{n \to \infty} \frac{a_n}{b_n} = k, \quad \text{with } 0 < k < \infty, \tag{2–90}$$

then convergence of $\sum b_n$ implies convergence of $\sum a_n$; divergence of $\sum b_n$ implies divergence of $\sum a_n$. By (2–90), $a_n/b_n < k+1$ or $a_n < (k+1)b_n$ for n sufficiently large and $b_n/a_n \to k' = 1/k$, so that $b_n < (k'+1)a_n$ for n sufficiently large. Thus convergence (divergence) of $\sum b_n$ implies convergence (divergence) of $\sum a_n$.   ◄

EXAMPLE 3 $\sum_{n=1}^{\infty} \dfrac{n^2 + n - 2}{n^3 - n^2 + n + 1}$. We take a_n to be the nth term of the given series and $b_n = 1/n$. (The value of b_n is chosen by retaining only the highest power in numerator and denominator.) Now,

$$\lim_{n \to \infty} \frac{n^2 + n - 2}{n^3 - n^2 + n + 1} \div \frac{1}{n} = \lim_{n \to \infty} \frac{n^3 + n^2 - 2n}{n^3 - n^2 + n + 1} = \lim_{n \to \infty} \frac{1 + \dfrac{1}{n} - \dfrac{2}{n^2}}{1 - \dfrac{1}{n} + \dfrac{1}{n^2} + \dfrac{1}{n^3}} = 1.$$

Since $\sum b_n$ diverges, the given series diverges. ◄

If $k = \infty$ in (2–90), then b_n/a_n has limit 0 and we can conclude that if $\sum b_n$ diverges, then so does $\sum a_n$; if $\sum a_n$ converges, then so does $\sum b_n$.

=== **Problems (Section 2–9)** ===

1. Test for convergence by the integral test (see Section 4–6 for (c) and (d)):

a) $\displaystyle\sum_{n=1}^{\infty} \frac{1}{n^2+1}$

b) $\displaystyle\sum_{n=1}^{\infty} \frac{n}{n^2+1}$

c) $\displaystyle\sum_{n=1}^{\infty} \frac{1}{(n+1)(n+2)}$

d) $\displaystyle\sum_{n=1}^{\infty} \frac{1}{(n+1)(n^2+1)}$

e) $\displaystyle\sum_{n=2}^{\infty} \frac{1}{n \ln^2 n}$

f) $\displaystyle\sum_{n=2}^{\infty} \frac{1}{n \ln n \ln \ln n}$

g) $\displaystyle\sum_{n=1}^{\infty} \frac{1}{\cosh^2 n}$

h) $\displaystyle\sum_{n=1}^{\infty} \frac{n^3}{2^n}$

2. Test for convergence by the comparison test:

a) $\displaystyle\sum_{n=1}^{\infty} \frac{n}{n^3+1}$

b) $\displaystyle\sum_{n=1}^{\infty} \frac{n+2}{n^2}$

c) $\displaystyle\sum_{n=1}^{\infty} \frac{n(n+1)}{(n^3+1)^2}$

d) $\displaystyle\sum_{n=1}^{\infty} \frac{n+1}{n^{3/2}}$

e) $\displaystyle\sum_{n=1}^{\infty} \frac{3+\sin n}{n^2}$

f) $\displaystyle\sum_{n=1}^{\infty} \frac{2+\cos n}{n}$

g) $\displaystyle\sum_{n=1}^{\infty} \frac{n^2-n\sin n}{n^3+1}$

h) $\displaystyle\sum_{n=1}^{\infty} \frac{n \ln n}{n^4+1}$

i) $\displaystyle\sum_{n=1}^{\infty} \frac{1+\sin n}{2^n}$

j) $\displaystyle\sum_{n=1}^{\infty} \frac{1}{2^n n!}$

k) $\displaystyle\sum_{n=1}^{\infty} \frac{n^3}{2^n+1}$

3. Test for convergence:

a) $\displaystyle\sum_{n=1}^{\infty} \frac{\pi^n}{2^n e^n}$

b) $\displaystyle\sum_{n=1}^{\infty} \frac{(-1)^n n}{n+1}$

c) $\displaystyle\sum_{n=1}^{\infty} \frac{1}{n^{1.1}}$

d) $\displaystyle\sum_{n=1}^{\infty} \frac{1}{n^{0.9}}$

e) $\displaystyle\sum_{n=1}^{\infty} \frac{n^3+1}{2^n}$

f) $\displaystyle\sum_{n=2}^{\infty} \frac{1}{n^2 \ln^2 n}$

g) $\displaystyle\sum_{n=1}^{\infty} \frac{\ln n}{n 2^n}$

h) $\displaystyle\sum_{n=1}^{\infty} \frac{\sin^2 n}{n^{3/2}}$

2–10 ABSOLUTE CONVERGENCE AND CONDITIONAL CONVERGENCE; ALTERNATING SERIES

A series $\sum_{n=1}^{\infty} a_n$ is said to be *absolutely convergent* if $\sum_{n=1}^{\infty} |a_n|$ converges. For example, the series

$$\sum_{n=1}^{\infty} \frac{(-1)^{n+1}}{n^2}$$

is absolutely convergent since $\sum_{n=1}^{\infty} 1/n^2$ converges. It will be shown below that the series $\sum (-1)^{n+1}/n$ converges. But this series is not absolutely convergent since $\sum 1/n$ is the divergent harmonic series. We have the important rule:

Absolute convergence implies convergence.

The example of $\sum (-1)^{n+1}/n$ above shows that the converse rule is not true.

To prove the rule, we assume that $\sum |a_n|$ converges. We can assume that no $a_n = 0$ (since zero terms can be dropped). Some of the first n terms of the series $\sum a_n$ are positive, and we let their sum be P_n; some are negative, and we let their sum be $-N_n$. Thus for the series $\sum a_n$

$$S_n = P_n - N_n.$$

Both P_n and N_n are monotone increasing sequences. For example, in the series

$$1 - \frac{1}{2} + \frac{1}{3} - \frac{1}{5} + \frac{1}{9} + \cdots$$

we have $P_1 = 1, P_2 = 1, P_3 = 4/3, P_4 = 4/3, P_5 = 13/9, N_1 = 0, N_2 = 1/2, N_3 = 1/2, N_4 = 7/10, N_5 = 7/10, \ldots$ If we can show that P_n and N_n are bounded sequences, then P_n and N_n have limits, so that S_n has a limit. Now the n th partial sum of $\sum |a_n|$ is $P_n + N_n$ and, since this series converges, $P_n + N_n$ is a bounded sequence. But $P_n \leqslant P_n + N_n$ and $N_n \leqslant P_n + N_n$ since P_n and N_n are positive or 0. Hence P_n and N_n are bounded also, and the conclusion follows.

EXAMPLE 1 $\sum_{n=1}^{\infty} (-1)^{n+1}/n^2$. Since $\sum_{n=1}^{\infty} 1/n^2$ converges, the given series is absolutely convergent, and therefore convergent. ◄

EXAMPLE 2 $\sum_{n=1}^{\infty} (\sin n)/2^n$. Here $\sin n$ changes sign in irregular fashion. However,

$$\left| \frac{\sin n}{2^n} \right| \leqslant \frac{1}{2^n}$$

for all n, and $\sum 1/2^n$ is a convergent geometric series. Therefore, by the comparison test, $\sum |(\sin n)/2^n|$ converges; the given series is absolutely convergent and therefore convergent. ◄

Since absolute convergence implies convergence, we usually state only that a series is absolutely convergent, the ordinary convergence being then understood.

EXAMPLE 3 $\sum_{n=1}^{\infty} (-1)^{n+1}/n$. As remarked above, this series is not absolutely convergent, since $\sum 1/n$ diverges. However, the given series converges. In order to see this, we consider the successive partial sums:

$$S_1 = 1, \quad S_2 = 1 - \frac{1}{2} = \frac{1}{2} = 0.5, \quad S_3 = 1 - \frac{1}{2} + \frac{1}{3} = 0.833,$$

$$S_4 = 1 - \frac{1}{2} + \frac{1}{3} - \frac{1}{4} = 0.5833, \quad S_5 = 1 - \cdots + \frac{1}{5} = 0.7833, \quad S_6 = 0.6167.$$

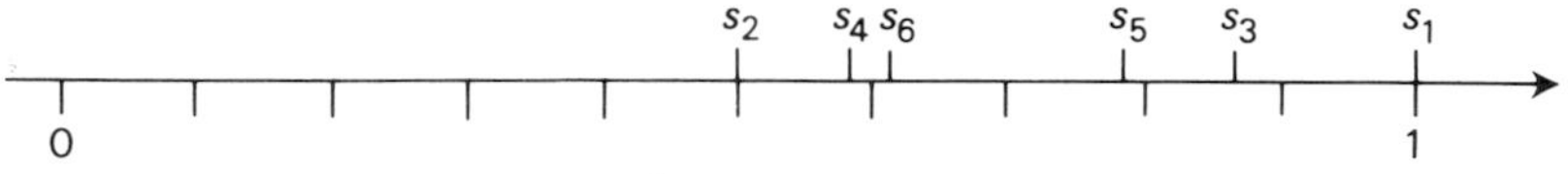

Fig. 2–5. Convergence of alternating series.

We observe that $S_1 > S_3 > S_5$ and $S_2 < S_4 < S_6$, as in Fig. 2–5. In general, the odd partial sums form a monotone decreasing sequence, since to go from one to the next we first subtract a number and then add a *smaller* number than the one subtracted. For a similar reason, the even partial sums form a monotone increasing sequence. Also it can be shown that each odd partial sum is greater than all even partial sums (by similar reasoning on how much is added and subtracted). Therefore, the odd partial sums form a bounded monotone decreasing sequence and have a limit. By analogy, the even partial sums form a bounded monotone increasing sequence and have a limit. The two limits must be equal because each odd partial sum differs from the next even partial sum by $1/n$ for appropriate n, and this difference has limit 0 as $n \to \infty$. For this example, it can be shown that the limit is

$$\ln 2 = 0.6931471806 \ldots \quad \blacktriangleleft$$

The above arguments apply to an arbitrary series with alternating signs subject to similar conditions.

ALTERNATING-SERIES TEST *In the series* $\sum_{n=1}^{\infty} a_n$, *let* $a_n = (-1)^n b_n$, *where* $\{b_n\}$ *is a monotone sequence converging to* 0. *Then* $\sum_{n=1}^{\infty} a_n$ *converges.*

The following series all converge according to the alternating-series test:

$$\sum_{n=1}^{\infty} \frac{(-1)^{n+1}}{\sqrt{n}}, \quad \sum_{n=1}^{\infty} \frac{(-1)^n \ln n}{n}, \quad \sum_{n=3}^{\infty} \frac{(-1)^n}{\ln \ln n}, \quad \sum_{n=1}^{\infty} \frac{(-1)^n}{2^n},$$

and only the last one converges absolutely.

EXAMPLE 4 *A physical example.* We consider charged particles with charges $\pm e_0$ at positions $x = 0, \pm 1, \pm 2, \ldots$ along the x-axis, as in Fig. 2–6. This is an example of a *one-dimensional crystal*. We seek the potential energy for the total electrostatic force exerted on the particle at $x = 0$. By physics, the potential energy of two particles with charges e_1 and e_2 spaced at a distance r is $e_1 e_2/r$ (in appropriate units). Hence by summing the potentials due to the particles at $x = \pm 1$, at $x = \pm 2$, and so on, we obtain the total potential energy:

$$e_0^2 \left(-\frac{1}{1} - \frac{1}{1} + \frac{1}{2} + \frac{1}{2} - \frac{1}{3} - \frac{1}{3} + \cdots \right) = -2e_0^2 \sum_{n=1}^{\infty} \frac{(-1)^{n+1}}{n}.$$

Thus, as in Example 3, the total potential energy is $-2e_0^2 \ln 2$. ◄

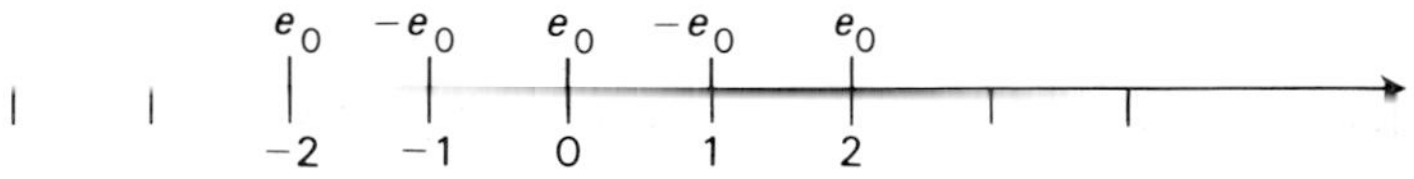

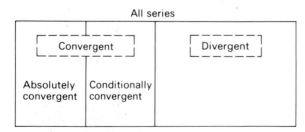

Fig. 2–6. One-dimensional crystal.

A series that converges but is not absolutely convergent is called *conditionally convergent*. The alternating-series test covers the most common conditionally convergent series. However, there are other conditionally convergent series, such as the series $\sum_{n=1}^{\infty} (\sin n)/n$ (see Example 2 of Section 3–4). The sum is $(\pi - 1)/2 = 1.0708 \ldots$

In Fig. 2–7 we give a diagram indicating the classification of series.

All series

Convergent	Divergent
Absolutely convergent Conditionally convergent	

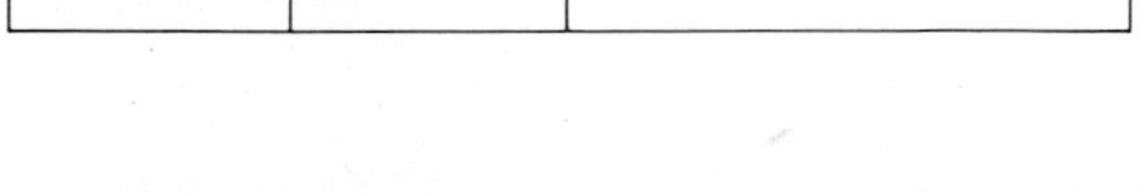

Fig. 2–7. Classification of series.

REMARK ON THE ERROR FOR ALTERNATING SERIES   It follows from the reasoning of Example 3, applied to $\sum (-1)^n b_n$, that the sum S lies between S_n and S_{n+1} for every n. Thus, in computing S, the error made by stopping at S_n is at most $(-1)^{n+1} b_{n+1}$, that is, *at most the first term omitted*. This rule gives the sign of the error and an upper estimate of its absolute value.

REMARK ON IMPORTANCE OF ABSOLUTE CONVERGENCE   Certain operations are permitted for absolutely convergent series that are not generally permitted for conditionally convergent series. In particular, the terms of an absolutely convergent series may be *rearranged* without affecting absolute convergence or the sum. For example,

$$1 + \frac{1}{2} + \frac{1}{4} + \frac{1}{8} + \frac{1}{16} + \frac{1}{32} + \frac{1}{64} + \cdots = 1 + \frac{1}{4} + \frac{1}{2} + \frac{1}{32} + \frac{1}{8} + \frac{1}{64} + \frac{1}{16} + \cdots$$

If $\sum a_n$ and $\sum b_n$ are absolutely convergent, then they may be added as usual, and the sum is absolutely convergent. Furthermore, the series may be *multiplied*:

$$\sum_{n=1}^{\infty} a_n \sum_{n=1}^{\infty} b_n = a_1 b_1 + a_1 b_2 + \cdots + a_i b_j + \cdots$$

Here a definite order must be chosen for the terms on the right. However, each order yields the same sum and provides an absolutely convergent series. One operation permitted for conditionally convergent as well as absolutely convergent series is insertion of parentheses to group terms. Thus

$$1 - \frac{1}{2} + \frac{1}{3} - \frac{1}{4} + \cdots = \left(1 - \frac{1}{2}\right) + \left(\frac{1}{3} - \frac{1}{4}\right) + \cdots = \frac{1}{2} + \frac{1}{12} + \cdots$$

However, the fact that a series converges *after* insertion of parentheses does not generally imply that it converges to start with; it does imply it if all terms of the series are of the same sign. It is common practice in multiplying two absolutely convergent series to insert parentheses as follows:

$$(a_0 + a_1 + \cdots + a_n + \cdots)(b_0 + b_1 + \cdots + b_n + \cdots)$$
$$= a_0 b_0 + (a_0 b_1 + a_1 b_0) + \cdots + (a_0 b_n + a_1 b_{n-1} + \cdots + a_n b_0) + \cdots$$

The last series is called the *Cauchy product* of the given series. (For further discussion of the procedures considered here see AC, pp. 401–405.)

=========== **Problems (Section 2-10)** ===========

1. Test for absolute convergence:

a) $\displaystyle\sum_{n=1}^{\infty} \frac{(-1)^{n+1}}{10^n}$

b) $\displaystyle\sum_{n=1}^{\infty} \frac{(-1)^{n+1}}{n^5}$

c) $\displaystyle\sum_{n=2}^{\infty} \frac{(-1)^n}{n \ln n}$

d) $\displaystyle\sum_{n=1}^{\infty} \frac{(-1)^n n}{n+1}$

e) $\displaystyle\sum_{n=1}^{\infty} \frac{\sin(n^2 \pi / 2)}{n^2}$

f) $\displaystyle\sum_{n=1}^{\infty} \frac{n \cos(n^2 \pi / 2)}{n^2 + 1}$

g) $\displaystyle\sum_{n=1}^{\infty} \frac{(-1)^n (n+1)}{e^{n^2}}$

h) $\displaystyle\sum_{n=1}^{\infty} \frac{(-1)^n}{n+5}$

2. Test for convergence:

a) $\displaystyle\sum_{n=1}^{\infty} \frac{(-1)^{n+1}\,n}{n^2+1}$ b) $\displaystyle\sum_{n=2}^{\infty} \frac{(-1)^n}{n\ln n}$ c) $\displaystyle\sum_{n=1}^{\infty} \frac{(-1)^n n}{3n+1}$

d) $\displaystyle\sum_{n=1}^{\infty} \frac{(-1)^n\,n^3}{e^n}$ e) $\displaystyle\sum_{n=1}^{\infty} \frac{(-1)^n\,7^n}{5^n}$ f) $\displaystyle\sum_{n=1}^{\infty} (-1)^{n+1} e^{-n^2}$

g) $\displaystyle\sum_{n=1}^{\infty} (-1)^n \left(\frac{\pi}{2} - \operatorname{Arctan} n\right)$ h) $\displaystyle\sum_{n=1}^{\infty} \frac{\cos n\pi}{n+2}$ i) $\displaystyle\sum_{n=1}^{\infty} \frac{\sin n}{n^2}$

3. Compute the sum with an error less than 0.01:

a) $\displaystyle\sum_{n=1}^{\infty} \frac{(-1)^{n+1}}{n\,5^n}$ b) $\displaystyle\sum_{n=1}^{\infty} \frac{(-1)^{n+1}}{n^3}$

c) $\displaystyle\sum_{n=1}^{\infty} \frac{(-1)^{n+1}}{10n}$ d) $\displaystyle\sum_{n=1}^{\infty} \frac{(-1)^n n}{10n^2+1}$

2–11 THE RATIO TEST

This is a test for absolute convergence:

RATIO TEST *For the series $\sum_{n=1}^{\infty} a_n$, let no term be 0 and let the sequence of* test ratios

$$\left|\frac{a_{n+1}}{a_n}\right|, \quad \text{where } n = 1, 2, \dots,$$

have a limit L. If $L < 1$, the series converges absolutely; if $L > 1$, the series diverges; if $L = 1$, the test fails.

We illustrate the test first and then prove the rule.

EXAMPLE 1 $\sum_{n=1}^{\infty} (-1)^n/n!$. The general test ratio is

$$\frac{1}{(n+1)!} \div \frac{1}{n!} = \frac{n!}{(n+1)!} = \frac{1}{n+1}.$$

(We dropped the $(-1)^n$ since we consider only absolute values.) This sequence has limit $L = 0$. Therefore the series converges absolutely. ◄

EXAMPLE 2 $\sum_{n=1}^{\infty} n!/2^n$. The general test ratio is

$$\frac{(n+1)!}{2^{n+1}} \div \frac{n!}{2^n} = \frac{2^n(n+1)!}{2^{n+1}n!} = \frac{n+1}{2}.$$

This has limit ∞. Therefore the series diverges. ◄

EXAMPLE 3 $\sum_{n=1}^{\infty} (-1)^n \dfrac{1\cdot3\cdot5 \cdots (2n-1)}{4\cdot7\cdot10 \cdots (3n+1)}.$ The general test ratio is

$$\frac{1\cdot3\cdot5 \cdots (2n-1)\,(2n+1)}{4\cdot7\cdot10 \cdots (3n+1)\,(3n+4)} \cdot \frac{4\cdot7\cdot10 \cdots (3n+1)}{1\cdot3\cdot5 \cdots (2n-1)} = \frac{2n+1}{3n+4}.$$

This has limit 2/3. Therefore the series converges absolutely.

This last example illustrates a series whose terms, in absolute values, are expanding products:

$$\frac{1}{4}, \quad \frac{1}{4}\cdot\frac{3}{7}, \quad \frac{1}{4}\cdot\frac{3}{7}\cdot\frac{5}{10}, \quad \cdots \quad \frac{1}{4}\cdot\frac{3}{7}\cdot\frac{5}{10}\cdots\frac{2n-1}{3n+1}, \quad \frac{1}{4}\cdot\frac{3}{7}\cdot\frac{5}{10}\cdots\frac{(2n+1)}{(3n+4)}, \quad \cdots$$

For such a series, the test ratio is simply the new factor inserted in going from the nth term to the next term; that is, in the example, $(2n+1)/(3n+4)$. ◀

Proof of ratio test

Let $L < 1$. Then we choose r so that $L < r < 1$. Since the test ratios have limit L, we can be sure that they are less than r for n sufficiently large, so that

$$\left|\frac{a_{n+1}}{a_n}\right| < r, \quad \left|\frac{a_{n+2}}{a_{n+1}}\right| < r, \quad \left|\frac{a_{n+3}}{a_{n+2}}\right| < r, \quad \cdots$$

and hence

$$|a_{n+1}| < r|a_n|, \quad |a_{n+2}| < r|a_{n+1}| < r^2|a_n|, \quad |a_{n+3}| < r^3|a_n|, \quad \cdots$$

and therefore the series $|a_n| + |a_{n+1}| + |a_{n+2}| + |a_{n+3}| + \cdots$ converges. This follows from comparison with the series

$$|a_n| + r|a_n| + r^2|a_n| + r^3|a_n| + \cdots,$$

which is a constant $|a_n|$ times a convergent geometric series, since $r < 1$. The terms $|a_1|, \ldots, |a_{n-1}|$ cannot affect convergence. Hence $\Sigma|a_n|$ converges.

Next let $L > 1$. Then for n sufficiently large, we have

$$\left|\frac{a_{n+1}}{a_n}\right| > 1, \quad \left|\frac{a_{n+2}}{a_{n+1}}\right| > 1, \quad \left|\frac{a_{n+3}}{a_{n+2}}\right| > 1, \quad \cdots$$

so that

$$|a_{n+1}| > |a_n|, \quad |a_{n+2}| > |a_{n+1}|, \quad |a_{n+3}| > |a_{n+2}|, \quad \cdots,$$

that is, the sequence $|a_n|, |a_{n+1}|, |a_{n+2}|, \ldots$ is monotone increasing and, since $|a_n| > 0$, the sequence cannot have limit 0. Therefore a_n cannot have limit 0 and by the nth-term test (Section 2–6) the series diverges.

To show that the test fails for $L = 1$, we need only consider the examples

$$\sum_{n=1}^{\infty} \frac{1}{n^2} \quad \text{and} \quad \sum_{n=1}^{\infty} \frac{1}{n}, \qquad (2\text{--}110)$$

both of which have $L = 1$ (Problems 1(i), (j) below, following Section 2–12). The first converges, the second diverges.

2–12 THE ROOT TEST

This is similar to the ratio test.

ROOT TEST: *For the series $\sum_{n=1}^{\infty} a_n$, let the sequence $|a_n|^{1/n}$ have limit L. If $L < 1$, the series converges; if $L > 1$, the series diverges; if $L = 1$, the test fails.*

The proof is similar to that for the ratio test. If $L < 1$, we choose r so that $L < r < 1$ and have $|a_n|^{1/n} < r$ for large r, so that $|a_n| < r^n$, and we get convergence by comparison with a converging geometric series. If $L > 1$, then $|a_n|^{1/n} > 1$ for large n, so that $|a_n| > 1$, and we get divergence by the nth-term test. To show that the test fails for $L = 1$, it suffices to consider the same two examples (2–110) (see Problems 2(g), (h) below).

=== **Problems (Section 2–12)** ===

1. Test for convergence by the ratio test:

a) $\displaystyle\sum_{n=1}^{\infty} \frac{(-1)^n(n^2+1)}{n!}$ b) $\displaystyle\sum_{n=1}^{\infty} \frac{(-1)^n(n+3)}{n!\,n}$

c) $\displaystyle\sum_{n=1}^{\infty} \frac{n+1}{1\cdot3\cdot5\,\cdots\,(2n-1)}$ d) $\displaystyle\sum_{n=1}^{\infty} \frac{n!}{2\cdot4\cdot6\,\cdots\,(2n)}$

e) $\displaystyle\sum_{n=1}^{\infty} \frac{1\cdot6\,\cdots\,(5n-4)}{3\cdot7\,\cdots\,(4n-1)}$ f) $\displaystyle\sum_{n=1}^{\infty} \frac{2\cdot4\,\cdots\,(2n)}{n^2(1\cdot3\cdot5\,\cdots\,(2n-1)}$

g) $\displaystyle\sum_{n=1}^{\infty} \frac{(2n)!}{(n!)^2}$ h) $\displaystyle\sum_{n=1}^{\infty} \frac{n!}{n!+n}$

i) $\displaystyle\sum_{n=1}^{\infty} \frac{1}{n^2}$ j) $\displaystyle\sum_{n=1}^{\infty} \frac{1}{n}$

2. Test for convergence by the root test:

a) $\displaystyle\sum_{n=1}^{\infty} (-1)^n \left(\frac{n}{2n+1}\right)^n$ b) $\displaystyle\sum_{n=1}^{\infty} (-1)^n \left(\frac{2n}{3n-1}\right)^n$

c) $\displaystyle\sum_{n=1}^{\infty} \left(\frac{n}{n+1}\right)^{n^2}$ d) $\displaystyle\sum_{n=1}^{\infty} \frac{n^2+1}{n^n}$

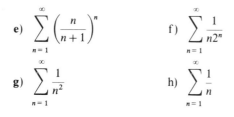

e) $\displaystyle\sum_{n=1}^{\infty}\left(\frac{n}{n+1}\right)^{n}$ **f)** $\displaystyle\sum_{n=1}^{\infty}\frac{1}{n2^{n}}$

g) $\displaystyle\sum_{n=1}^{\infty}\frac{1}{n^{2}}$ **h)** $\displaystyle\sum_{n=1}^{\infty}\frac{1}{n}$

2–13 ESTIMATION OF ERROR IN CALCULATING THE SUM

In Section 2–10 we pointed out how the error in stopping at the nth term can be estimated, in computing the sum of a convergent alternating series. There are similar estimates for series whose convergence is proved by one of the other tests. When the series $\sum a_n$ converges, we can write

$$S = \sum_{n=1}^{\infty} a_n = (a_1 + \cdots + a_n) + (a_{n+1} + a_{n+2} + \cdots) = S_n + R_n,$$

where R_n is the *remainder* after n terms. Since S_n has limit S, the remainder R_n has limit 0 as $n \to \infty$. We wish to estimate R_n and give several rules to that end (for simplicity we assume that no $a_n = 0$):

I. If the series converges by the integral test, with $f(n) = a_n$, then

$$0 < R_n < \int_n^{\infty} f(x)\, dx.$$

II. If the series converges by the comparison test, with $0 < a_n \leqslant b_n$ for all n and Σb_n convergent, then

$$0 < R_n \leqslant b_{n+1} + b_{n+2} + \cdots = \sum_{k=n+1}^{\infty} b_k.$$

III. If the series is absolutely convergent, then

$$|R_n| \leqslant |a_{n+1}| + |a_{n+2}| + \cdots = \sum_{k=n+1}^{\infty} |a_k|.$$

IV. If the series converges by the ratio test, then

$$|R_n| \leqslant \frac{a_{n+1}}{1-r},$$

provided that $|a_{n+2}/a_{n+1}| \leqslant r,\ |a_{n+3}/a_{n+2}| \leqslant r, \cdots, 0 < r < 1.$

V. If the series converges by the root test, then

$$|R_n| \leqslant \frac{r^{n+1}}{1-r},$$

provided that $|a_{n+1}| \leqslant r^{n+1}, |a_{n+2}| \leqslant r^{n+2}, \ldots, 0 < r < 1.$

These rules all follow at once from the proofs of the corresponding tests. We illustrate their application:

EXAMPLE 1 $\sum_{n=1}^{\infty} (1/n^2)$. We seek the error after 20 terms. By Rule I,

$$0 < R_{20} < \int_{20}^{\infty} \frac{1}{x^2}\, dx = -\frac{1}{x}\Big|_{20}^{\infty} = \frac{1}{20} = 0.05.$$

Hence $S = S_{20} = 1 + (1/4) + (1/9) + \cdots + (1/400)$, with an error less than 0.05. Upon calculation, we find $S_{20} = 1.5964$. The sum of the series is $\pi^2/6 = 1.64493 \ldots$, as shown in Problem 1(d) following Section 3–8. ◄

EXAMPLE 2 $\sum_{n=1}^{\infty} 1/(2^n + n)$. This converges by comparison with the geometric series $\sum 2^{-n}$. Hence by Rule II,

$$0 < R_n < \frac{1}{2^{n+1}} + \frac{1}{2^{n+2}} + \cdots = \frac{1}{2^{n+1}}\left(1 + \frac{1}{2} + \cdots\right).$$

The sum of the last series in parentheses is 2. Thus $0 < R_n < 1/2^n$. To calculate S with an error less than 0.001, it suffices to add the first ten terms. ◄

EXAMPLE 3 $\sum_{n=1}^{\infty} (-1)^{n+1}/n^2$. By Rule III, $|R_{20}|$ is less than the remainder after 20 terms of the series $\sum 1/n^2$ and, by Example 1, this remainder is less than 0.05. However, here it would be better to use the rule for alternating series, which would give $0 < R_{20} < 1/21^2 = 0.0023.$ ◄

EXAMPLE 4 $\sum_{n=0}^{\infty} 1/n!$. Here $|a_{n+1}/a_n| = 1/(n+1)$. We estimate R_6, the remainder after the term of index 6. Now $|a_8/a_7| = 1/8, |a_9/a_8| = 1/9, \ldots$, so that we can use $r = 1/8$ in Rule IV. Hence

$$|R_6| < \frac{1}{7!}\frac{1}{1-\frac{1}{8}} = 0.000\,2267.$$

Upon calculating, we find $S_6 = 2.718055556\ldots$. The sum of the series is $e = 2.7182818285\ldots$, as shown in Section 2–18. Thus the actual error is $0.00022627.$ ◄

REMARK: For a positive term series, the error in stopping at the nth term is clearly negative (S_n is less than S), and R_n is at least equal to a_{n+1}, the first term omitted. Thus in Example 4:

$$\frac{1}{11!} < R_{10} < \frac{1}{11!}\frac{12}{11}.$$

═══════════════════════**Problems (Section 2-13)**═══════════════

1. Compute the sum of the series by stopping at the term of index n for the given n and estimate the error made.

a) $\displaystyle\sum_{n=1}^{\infty} \frac{1}{n^3}, n = 4$
b) $\displaystyle\sum_{n=1}^{\infty} \frac{1}{n^3 + 1}, n = 4$

c) $\displaystyle\sum_{n=1}^{\infty} \frac{(-1)^{n+1}}{3^n + 1}, n = 5,$
d) $\displaystyle\sum_{n=1}^{\infty} \frac{\sin n}{n^7}, n = 5$

e) $\displaystyle\sum_{n=1}^{\infty} \frac{1}{3^n n!}, n = 4$
f) $\displaystyle\sum_{n=1}^{\infty} \frac{1 \cdot 3 \cdot 5 \cdots (2n-1)}{4 \cdot 7 \cdot 10 \cdots (3n+1)}, n = 5$

g) $\displaystyle\sum_{n=2}^{\infty} \frac{1}{n \ln^2 n + 1}, n = 4$
h) $\displaystyle\sum_{n=2}^{\infty} \left(\frac{n}{2n+1}\right)^n, n = 4$

2. Determine N so that, for $n \geqslant N$, the nth partial sum S_n approximates S with an error less than ε in absolute value, for the given ε.

a) $\displaystyle\sum_{n=1}^{\infty} \frac{1}{n^4}, \varepsilon = 0.01$
b) $\displaystyle\sum_{n=1}^{\infty} \frac{1}{n2^n}, \varepsilon = 0.001$

c) $\displaystyle\sum_{n=1}^{\infty} \frac{n}{n^3 + 1}, \varepsilon = 0.01$
d) $\displaystyle\sum_{n=1}^{\infty} \frac{(-1)^n}{n! + 1}, \varepsilon = 0.0001$

───

2-14 SEQUENCES AND SERIES OF FUNCTIONS, POWER SERIES

Instead of sequences of numbers, one may consider sequences of functions, such as

$$1, x, x^2, \ldots x^n, \ldots$$

$$\sin x, \cos x, \sin 2x, \cos 2x, \ldots, \sin kx, \cos kx, \ldots$$

$$1, 1 + \frac{x}{1!}, 1 + \frac{x}{1!} + \frac{x^2}{2!}, \cdots, 1 + \frac{x}{1!} + \cdots + \frac{x^n}{n!}, \cdots$$

In general, sequences $\{f_n(x)\}$ may be considered, where all functions $f_n(x)$ are defined on some interval of the x-axis. For each fixed x, the sequence $\{f_n(x)\}$ may converge or diverge. The first problem is always to determine, if possible, the set of values of x for which the sequence converges. Most commonly, this set is an interval, part or all of the interval on which all the functions $f_n(x)$ are defined. For each x of such a convergence interval, the limit of the sequence is a number that can be denoted by $f(x)$. Thus

$$\lim_{n \to \infty} f_n(x) = f(x),$$

for each x of the convergence interval.

EXAMPLE 1 $f_n(x) = x^n$. For $-1 < x < 1$, $x^n \to 0$ when $n \to \infty$ (as, for example, $(1/2)^n \to 0$ or $(-1/2)^n \to 0$). When $x = 1$, we get $x^n = 1$ for all n, so that $x^n \to 1$; when $x = -1$, x^n is $(-1)^n$, which has no limit. When $x > 1$, $x^n \to \infty$ as $n \to \infty$, and when $x < -1$, $|x^n| \to \infty$ as $n \to \infty$; thus in both cases there is no limit. Therefore,

$$\lim_{n \to \infty} x^n = \begin{cases} 0, & -1 < x < 1 \\ 1, & x = 1 \\ \text{nonexistent otherwise} \end{cases} \tag{2–140}$$
◀

EXAMPLE 2 $f_n(x) = 1 + x + \cdots + x^n$. Here $f_n(x)$ is simply the nth partial sum of the geometric series $1 + x + \cdots + x^n + \cdots$. We know (Section 2–5) that this series converges when $-1 < x < 1$ and we even know the sum:

$$f(x) = \lim_{n \to \infty} f_n(x) = \lim_{n \to \infty} (1 + x + \cdots + x^n) = \sum_{n=0}^{\infty} x^n = \frac{1}{1-x}. \tag{2–141}$$

Outside of the interval $-1 < x < 1$ the sequence diverges. ◀

Series of functions

The sequences we are mainly interested in can be derived from infinite series of functions by forming the nth partial sum, as in Example 2. Such a series has the form

$$\sum_{n=1}^{\infty} u_n(x),$$

where the $u_n(x)$ are all defined on some interval. The corresponding nth partial sum is

$$S_n(x) = u_1(x) + \cdots + u_n(x),$$

and the series converges precisely when the sequence $\{S_n(x)\}$ converges. If

$$\lim_{n \to \infty} S_n(x) = f(x)$$

for an interval (or set) of values of x, then

$$\sum_{n=1}^{\infty} u_n(x) = f(x)$$

for x on the interval (or set). This is illustrated by Eq. (2–141), with $u_n(x) = x^n$, where $n = 0, 1, 2, \ldots$; here the sum starts at 0 instead of at 1.

Power series

Generalizing Example 2, we can consider a series

$$\sum_{n=0}^{\infty} c_n x^n \qquad (2\text{–}142)$$

with constant coefficients $c_0, c_1, \ldots, c_n, \ldots$ This is the simplest form of a *power series*. More generally, we also term

$$\sum_{n=0}^{\infty} c_n (x - a)^n \qquad (2\text{–}142')$$

a power series (in powers of $x - a$). A simple substitution $x' = x - a$ reduces $(2\text{–}142')$ to $(2\text{–}142)$, and hence we emphasize $(2\text{–}142)$.

Convergence set of a power series

The power series $(2\text{–}142)$ converges for $x = 0$ in any case, and this may be the only value for which it converges. It may also converge for all x. If it converges for some x other than 0 but not for all x, then the series has a *radius of convergence* r^* such that the series converges for $-r^* < x < r^*$ and diverges for $x > r^*$ and for $x < -r^*$. The three cases are illustrated in Fig. 2–8. As suggested in the figure, we assign the radius of convergence 0 to the first case and the radius of convergence ∞ to the second.

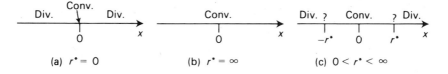

Fig. 2–8. The three cases for the power series.

We illustrate the cases here and refer to AC, pp. 419, 420, for a proof. We notice that in the case of Fig. 2–8 (c) nothing is said about what happens when $x = r^*$ or $x = -r^*$. It may converge or diverge at either point, depending on the particular series.

The geometric series of Example 2 above illustrates a series with $r^* = 1$, with divergence at both endpoints.

EXAMPLE 3 $\sum_{n=1}^{\infty} x^n / n$.   We apply the ratio test:

$$\lim_{n \to \infty} \left| \frac{x^{n+1}}{n+1} \div \frac{x^n}{n} \right| = \lim_{n \to \infty} \frac{n|x|}{n+1} = |x|.$$

Hence the series converges for $|x| < 1$, that is, for $-1 < x < 1$, and diverges for

$x > 1$ and for $x < -1$. Here $r^* = 1$. For $x = 1$ we obtain the harmonic series, which diverges. For $x = -1$, we obtain the convergent alternating series $\Sigma(-1)^n/n$. Thus the series converges for $-1 \leqslant x < 1$ and diverges otherwise.

If we replace x by $-x$ in this example, we obtain the series

$$\sum_{n=1}^{\infty}(-1)^n\frac{x^n}{n}$$

that converges for $-1 < x \leqslant 1$ and diverges otherwise. ◄

EXAMPLE 4 $\sum_{n=1}^{\infty} x^n/(2^n n^2)$. We apply the ratio test:

$$\lim_{n \to \infty}\left|\frac{x^{n+1}}{2^{n+1}(n+1)^2} \div \frac{x^n}{2^n n^2}\right| = \lim_{n \to \infty}\frac{n^2}{(n+1)^2}\frac{|x|}{2} = \frac{|x|}{2}.$$

Hence the series converges for $|x|/2 < 1$ and diverges for $|x|/2 > 1$; that is, $r^* = 2$. For $x = 2$, we obtain the convergent series $\Sigma 1/n^2$; for $x = -2$ we obtain the convergent alternating series $\Sigma(-1)^n/n^2$. Hence the series converges for $-2 \leqslant x \leqslant 2$ and diverges otherwise. ◄

EXAMPLE 5 $\sum_{n=0}^{\infty} x^n/n!$. We apply the ratio test:

$$\lim_{n \to \infty}\left|\frac{x^{n+1}}{(n+1)!} \div \frac{x^n}{n!}\right| = \lim_{n \to \infty}\frac{|x|}{n+1} = 0.$$

Hence the series converges for all x. ◄

In all these examples we showed convergence for $-r^* < x < r^*$ by the ratio test. Therefore for each such x the series is *absolutely convergent*. This property can be shown to hold generally (see AC, pp. 419–420). At each of the endpoints $\pm r^*$ the series may converge, when it does, only conditionally (as in Example 3; in Example 4 the convergence is absolute at both endpoints).

For the more general form (2–142′) the substitution $x' = x - a$ shows that we have the three cases: convergence only for $x = a$ (that is, $r^* = 0$), convergence for all x (that is, $r^* = \infty$), convergence for $a - r^* < x < a + r^*$ and divergence for $x > a + r^*$ and for $x < a - r^*$ (that is, $0 < r^* < \infty$).

EXAMPLE 6 $\sum_{n=0}^{\infty} \dfrac{n(x-3)^n}{5n^2 + 1}$. We use the ratio test:

$$\lim_{n \to \infty}\left|\frac{(n+1)(x-3)^{n+1}}{5n^2 + 10n + 6} \div \frac{n(x-3)^n}{5n^2 + 1}\right| = \lim_{n \to \infty}\frac{n+1}{n}\frac{5n^2 + 1}{5n^2 + 10n + 6}|x-3| = |x-3|.$$

Hence $r^* = 1$ and the series converges for $|x - 3| < 1$ and diverges for $|x - 3| > 1$; that is, it converges for $2 < x < 4$ and diverges for $x > 4$ and for $x < 2$. For $x = 4$ we find that the series diverges (by the integral test or by comparison with a harmonic series), and for $x = 2$ we find that the series converges by the alternating-series test. ◄

===Problems (Section 2–14)===

1. Determine the set of values of x for which the sequence converges and find the limit for these values:

a) $\{e^{-nx}\}$ b) $\{\ln nx\}$ c) $\left\{\dfrac{1}{x^n}\right\}$ d) $\left\{\dfrac{x^n+1}{x^n+2}\right\}$

e) $\left\{1+\dfrac{x}{2}+\cdots+\left(\dfrac{x}{2}\right)^n\right\}$ f) $\{1+e^{-x}+e^{-2x}+\cdots+e^{-nx}\}$

2. Find the set of x for which the series converges and state the radius of convergence:

a) $\displaystyle\sum_{n=0}^{\infty}\left(\frac{2}{3}\right)^n x^n$ b) $\displaystyle\sum_{n=0}^{\infty}\left(\frac{x}{e}\right)^n$ c) $\displaystyle\sum_{n=1}^{\infty}\frac{x^n}{n^3}$

d) $\displaystyle\sum_{n=2}^{\infty}\frac{x^n}{n\ln n}$ e) $\displaystyle\sum_{n=0}^{\infty}\frac{(n+1)x^n}{n!}$ f) $\displaystyle\sum_{n=0}^{\infty}\frac{1+(-1)^n}{n!}x^n$

g) $\displaystyle\sum_{n=0}^{\infty}\frac{2n-1}{n^2+1}\frac{x^n}{2^n}$ h) $\displaystyle\sum_{n=2}^{\infty}\frac{n+1}{n\ln n}x^n$ i) $\displaystyle\sum_{n=0}^{\infty}n!\,x^n$

j) $\displaystyle\sum_{n=1}^{\infty}\frac{x^n}{n^n}$ k) $\displaystyle\sum_{n=0}^{\infty}\frac{(x-3)^n}{n^2+1}$ l) $\displaystyle\sum_{n=0}^{\infty}\frac{(x+1)^n}{5^n}$

m) $\displaystyle\sum_{n=1}^{\infty}\frac{2\cdot4\cdots(2n)}{4\cdot7\cdots(3n+1)}(x-2)^n$ n) $\displaystyle\sum_{n=0}^{\infty}\frac{n!+1}{(n!+2)(n!+3)}(x-1)^n$

3. Find the set of x for which the series converges:

a) $\displaystyle\sum_{n=0}^{\infty}\frac{1}{x^n}$ b) $\displaystyle\sum_{n=0}^{\infty}\frac{3^n}{x^n}$ c) $\displaystyle\sum_{n=0}^{\infty}e^{-nx}$

d) $\displaystyle\sum_{n=1}^{\infty}\frac{\sin nx}{n^2}$ e) $\displaystyle\sum_{n=1}^{\infty}\left(\frac{x+1}{x+2}\right)^n$ f) $\displaystyle\sum_{n=0}^{\infty}n\left(\frac{x}{x-1}\right)^n$

5. Let the power series $\sum_{n=0}^{\infty}c_n x^n$ be given. Show that if the limit exists, then

a) $r^* = \lim_{n\to\infty}\left|\dfrac{c_n}{c_{n+1}}\right|$ b) $r^* = \lim_{n\to\infty}\dfrac{1}{|c_n|^{1/n}}$

2–15 UNIFORM CONVERGENCE

If a sequence or series of functions converges over an interval, it may converge much more rapidly at some points than at others. We here explore this question for a series $\sum_{n=1}^{\infty}u_n(x)$ with sum $f(x)$ and corresponding partial sums $S_n(x)$. The discussion for an arbitrary sequence of functions $\{f_n(x)\}$ is the same as that for the sequence $\{S_n(x)\}$.

We introduce the idea of uniform convergence by considering two figures that illustrate two ways in which $S_n(x)$ can approach $f(x)$ (see Fig. 2–9). In each case we are especially interested in the maximum error E_n in the approximation of $f(x)$ by $S_n(x)$; this is shown for $n = 1, 2, 3$ for the two cases. In case (a), $S_2(x)$ is much closer to $f(x)$ than $S_1(x)$ for *all x in the interval*, and $S_3(x)$ is even closer throughout. For case (b), $S_2(x)$ is closer to $f(x)$ than $S_1(x)$ for some (but not all) x in the interval, and $S_3(x)$ is closer still for some (but not all) x. In case (a), the maximum errors E_1, E_2, E_3 are decreasing, and with the process continuing as suggested, approach 0 as $n \to \infty$. In case (b), the maximum errors E_1, E_2, E_3 are all about the same size, however we can see that as n increases the sharp dips in the graphs move over to the left, becoming narrower and narrower, so that $S_n(x)$ does have $f(x)$ as limit for each fixed x. Uniform convergence is illustrated by case (a), nonuniform convergence by case (b).

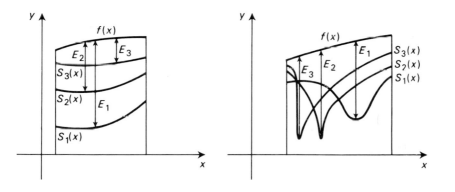

Fig. 2–9. (a) Uniform convergence; (b) nonuniform convergence.

We can define *uniform convergence over an interval* $a \leqslant x \leqslant b$ as convergence such that the *maximum error over the interval approaches* 0 *as* $n \to \infty$; that is,

$$\max_{a \leqslant x \leqslant b} |f(x) - S_n(x)| \to 0 \quad \text{as} \quad n \to \infty. \tag{2–150}$$

This definition is adequate for most applications, namely those in which $f(x)$ and $S_n(x)$ are continuous for all n on the closed interval $a \leqslant x \leqslant b$. For more general situations, the definition becomes as follows: for every $\varepsilon > 0$, there is an N, independent of x, such that $|f(x) - S_n(x)| < \varepsilon$ for $n \geqslant N$ for all x on the given interval. It can be shown that the sum of a uniformly convergent series of continuous functions is continuous. Thus for a convergent series $\sum u_n(x)$ of continuous functions on a closed interval $a \leqslant x \leqslant b$ the two definitions of uniform convergence agree.

Uniform convergence is very important for computer calculations. When a series converges uniformly over a given interval, the sum can be computed to desired accuracy by choosing an appropriate n and then computing $S_n(x)$ for all x of interest. When the convergence is not uniform, one is forced to vary n with x, perhaps in a complicated way.

EXAMPLE 1 $\sum_{n=0}^{\infty} x^n$ on the interval $0 \leqslant x \leqslant k$ for $0 < k < 1$. This is the geometric series, for which we know both $f(x)$ and $S_n(x)$:

$$f(x) = \frac{1}{1-x}, \qquad S_n(x) = \frac{1-x^{n+1}}{1-x}$$

(see Fig. 2–10). Thus,

$$f(x) - S_n(x) = \frac{x^{n+1}}{1-x} = x^{n+1} \frac{1}{1-x}.$$

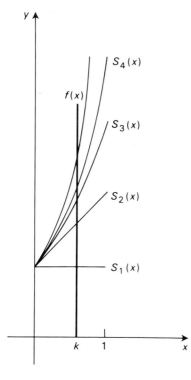

Fig. 2-10. Sequence of partial sums of the geometric series.

On the interval $0 \leqslant x \leqslant k$, the last expression is positive or 0, so that we do not need absolute-value signs. For fixed n, both factors x^{n+1} and $1/(1-x)$ increase as x increases. Therefore the maximum error occurs for $x = k$ and has the value

$$E_n = \frac{k^{n+1}}{1-k}.$$

Since $0 < k < 1$, $E_n \to 0$ as $n \to \infty$ and we have uniform convergence. If, for example, $k = 1/2$ and $n = 10$, then $E_n = 1/2^{10} = 0.000976$, so that $S_{10}(x)$ approximates $f(x)$ with an error less than 0.001 *for all x of the interval $0 \leqslant x \leqslant 1/2$.* ◄

We have a useful test for uniform convergence:

WEIERSTRASS M-TEST   Let $|u_n(x)| \leqslant M_n$ on a given interval, and let the series of constants $\sum_{n=1}^{\infty} M_n$ converge. Then $\sum_{n=1}^{\infty} u_n(x)$ is uniformly convergent on the given interval.

PROOF   By the comparison test, the hypotheses made imply that $\sum u_n(x)$ converges absolutely for each x on the given interval. Let the sum be $f(x)$. Then

$$|f(x) - S_n(x)| = |u_{n+1}(x) + u_{n+2}(x) + \cdots| \leqslant |u_{n+1}(x)| + |u_{n+2}(x)| + \cdots$$
$$\leqslant M_{n+1} + M_{n+2} + \cdots$$

The last series is the remainder R_n for the convergent series $\sum M_n$. Hence we can choose N so that $R_n < \varepsilon$ for $n \geqslant N$ for a given positive ε, and then

$$|f(x) - S_n(x)| < \varepsilon \quad \text{for} \quad n \geqslant N,$$

as required, with N independent of x.

EXAMPLE 2 $\sum_{n=0}^{\infty} x^n/n!$ (power series for e^x) on the interval $0 \leqslant x \leqslant 2$. Here

$$|u_n(x)| = \left|\frac{x^n}{n!}\right| \leqslant \frac{2^n}{n!} = M_n \quad \text{for} \quad 0 \leqslant x \leqslant 2$$

and $\sum M_n = \sum 2^n/n!$ converges by the ratio test. Thus, by the Weierstrass M-test, the given series converges uniformly for $0 \leqslant x \leqslant 2$. ◄

EXAMPLE 3 $\sum_{n=1}^{\infty} (\cos nx)/n^2$ on the interval $0 \leqslant x \leqslant 2\pi$. Here

$$|u_n(x)| = \left|\frac{\cos nx}{n^2}\right| \leqslant \frac{1}{n^2} = M_n$$

and $\sum M_n$ converges. Therefore, by the Weierstrass M-test, the series converges uniformly for $0 \leqslant x \leqslant 2$. ◄

EXAMPLE 4 $\sum_{n=0}^{\infty} x^n$ for $-1 < x \leqslant 0$. This is the geometric series again. We have to omit $x = -1$ because the series diverges at this point. However, the nth partial sum $S_n(x)$ and the sum $f(x) = 1/(1-x)$ are continuous for $-1 \leqslant x \leqslant 0$. According to Example 1,

$$|f(x) - S_n(x)| = \left|\frac{x^{n+1}}{1-x}\right| = |x^{n+1}| \frac{1}{1-x}.$$

When $x \to -1$, this expression has limit 1/2. Thus, in the interval $-1 < x \leqslant 0$, we can find x such that $|f(x) - S_n(x)|$ is as close to 1/2 as we desire. Therefore we cannot make $|f(x) - S_n(x)| < \varepsilon$, say for $\varepsilon = 0.1$, for *all* x on the given interval, no matter how large we choose n. It follows that the given series is *not* uniformly convergent for $-1 < x \leqslant 0$.

However, if we confine attention to a slightly smaller interval, say $-0.9 \leqslant x \leqslant 0$, then on this interval $|f(x) - S_n(x)|$ has its maximum E_n at $x = -0.9$, so that

$$E_n = \max |f(x) - S_n(x)| = \frac{0.9^{n+1}}{1.9} \to 0 \quad \text{as} \quad n \to \infty.$$

Accordingly, on the new interval we have uniform convergence. ◀

We list three important *properties of uniformly convergent series*. We state these for the case of a series $\sum u_n(x)$ that converges uniformly when $a \leqslant x \leqslant b$, and let $f(x)$ be the sum.

I. If all $u_n(x)$ are continuous for $a \leqslant x \leqslant b$, then so also is $f(x)$.
II. Under the hypotheses of Rule I, the series may be integrated term by term; that is, if $a \leqslant x_1 < x_2 \leqslant b$ or $a \leqslant x_2 < x_1 \leqslant b$, then

$$\int_{x_1}^{x_2} f(x)\, dx = \sum_{n=1}^{\infty} \int_{x_1}^{x_2} u_n(x)\, dx.$$

III. If $h(x)$ is continuous for $a \leqslant x \leqslant b$, then the series $\Sigma h(x) u_n(x)$ is also uniformly convergent for $a \leqslant x \leqslant b$.

The first property was pointed out above (for proofs see AC, pp. 414–418).

══════════════════════ **Problems (Section 2-15)** ══════════════════════

1. Show that the series is uniformly convergent on the given interval:

a) $\displaystyle\sum_{n=1}^{\infty} \frac{x^n}{2^n n^3},\ 0 \leqslant x \leqslant 1$

b) $\displaystyle\sum_{n=1}^{\infty} \frac{x^n}{n},\ 0 \leqslant x \leqslant \frac{1}{2}$

c) $\displaystyle\sum_{n=0}^{\infty} \frac{2^n x^n}{n!},\ -1 \leqslant x \leqslant 1$

d) $\displaystyle\sum_{n=2}^{\infty} \frac{x^n}{n \ln n},\ -\frac{1}{2} \leqslant x \leqslant \frac{1}{2}$

e) $\displaystyle\sum_{n=0}^{\infty} \frac{(x-1)^n}{(n!)^2},\ 0 \leqslant x \leqslant 2$

f) $\displaystyle\sum_{n=0}^{\infty} \frac{(x+3)^n}{5^n(2n+1)},\ -6 \leqslant x \leqslant 0$

g) $\displaystyle\sum_{n=1}^{\infty} \frac{\sin nx}{n^2},\ 0 \leqslant x \leqslant \pi$

h) $\displaystyle\sum_{n=0}^{\infty} \frac{\cos^3 nx}{n^3+1},\ 0 \leqslant x \leqslant 2\pi$

2. Choose n so that $S_n(x)$ approximates the sum of the series with error less than 0.001 over the given interval:

a) $\displaystyle\sum_{n=0}^{\infty} \frac{x^n}{2^n},\ 0 \leqslant x \leqslant 1$

b) $\displaystyle\sum_{n=0}^{\infty} \frac{x^n}{n!},\ 0 \leqslant x \leqslant 1$

c) $\displaystyle\sum_{n=1}^{\infty} \frac{\sin nx}{n^2}, 0 \leqslant x \leqslant \pi$ d) $\displaystyle\sum_{n=1}^{\infty} \frac{1}{nx^n}, 2 \leqslant x \leqslant 3$

3. For $n = 1, 2, \ldots$ let the function $f_n(x)$ be defined on the interval $0 \leqslant x \leqslant 1$ as follows: $f(0) = 0, f(1/n) = 1, f(x) = 0$ for $2/n \leqslant x \leqslant 1, f(x)$ is linear between $x = 0$ and $x = 1/n$ and between $x = 1/n$ and $x = 2/n$. Show that the *sequence* $f_n(x)$ converges to 0 for $0 \leqslant x \leqslant 1$ but not uniformly.

2–16 PROPERTIES OF POWER SERIES

We consider a power series $\sum_{n=0}^{\infty} c_n x^n$ with nonzero radius of convergence r^*, so that the series converges absolutely for $-r^* < x < r^*$. Let $f(x)$ be the sum.
   The convergence is uniform for $|x| \leqslant k$, provided $0 < k < r^*$, since

$$|c_n x^n| \leqslant |c_n k^n| = M_n \quad \text{for} \quad |x| \leqslant k,$$

and $\sum M_n$ converges, so that the M-test applies.
   Therefore, by Property I of Section 2–15, the sum $f(x)$ of the series is continuous when $|x| \leqslant k$ for every positive $k < r^*$; that is, $f(x)$ is *continuous for $-r^* < x < r^*$.*
   By Property II of Section 2–15, *we can integrate the power series term by term,* say from 0 to x:

$$\int_0^x f(t)\, dt = \sum_{n=0}^{\infty} c_n \int_0^x t^n\, dt = \sum_{n=0}^{\infty} c_n \frac{x^{n+1}}{n+1}, \quad -r^* < x < r^*. \quad (2\text{–}160)$$

   From this result, we can prove that *we are also permitted to differentiate the power series term by term:*

$$f'(x) = \sum_{n=1}^{\infty} nc_n x^{n-1}, \quad -r^* < x < r^*. \quad (2\text{–}161)$$

(For proof see CLA, pp. 612–614.)
   By repeated differentiation, we now conclude that

$$f''(x) = \sum_{n=2}^{\infty} n(n-1)c_n x^{n-2},$$

$$f'''(x) = \sum_{n=3}^{\infty} n(n-1)(n-2)c_n x^{n-3}, \quad (2\text{–}162)$$

$$\vdots$$

$$f^{(k)}(x) = \sum_{n=k}^{\infty} n(n-1)\cdots(n-k+1)c_n x^{n-k},$$

$$\vdots$$

If we put $x = 0$ in these series and in the series $\sum c_n x^n$ for f and the series $\sum c_n n x^{n-1}$ for f', we obtain

$$f(0) = c_0, \quad f'(0) = 1! c_1, \quad f''(0) = 2! c_2, \quad \cdots, \quad f^{(k)}(0) = k! c_k,$$

and hence

$$c_k = \frac{f^{(k)}(0)}{k!}. \tag{2–163}$$

Thus

$$f(x) = \sum_{n=0}^{\infty} c_n x^n = \sum_{n=0}^{\infty} \frac{f^{(n)}(0)}{n!} x^n, \quad |x| < r^*. \tag{2–164}$$

The series

$$\sum_{n=0}^{\infty} \frac{f^{(n)}(0)}{n!} x^n$$

is called the *Maclaurin series of* $f(x)$ or the *Taylor series of* $f(x)$ *at* $x = 0$.

By reasoning in the same manner for the case of a more general series $\sum c_n (x - a)^n$ we conclude that if it has nonzero radius of convergence r^* and sum $f(x)$, then $f(x)$ is continuous for $|x - a| < r^*$ and

$$f(x) = \sum_{n=0}^{\infty} c_n (x - a)^n = \sum_{n=0}^{\infty} \frac{f^{(n)}(a)}{n!} (x - a)^n. \tag{2–165}$$

The last series is the *Taylor series of* $f(x)$ *at* $x = a$. Thus *every convergent power series is the Taylor series of its sum.*

It follows that *if two power series have the same sum, then their coefficients are the same*; that is, if

$$\sum_{n=0}^{\infty} c_n (x - a)^n = \sum_{n=0}^{\infty} b_n (x - a)^n = f(x)$$

over an interval including a, then $c_n = b_n = f^{(n)}(a)/n!$ for all n. This is the basis for the much used *method of comparing coefficients*.

REMARK: For a given function $f(x)$, we may be able to compute all derivatives of f at $x = a$ and hence to obtain all the coefficients in the Taylor series of f at $x = a$. This does not ensure that the Taylor series converges in an interval about $x = a$ or, if it converges in such an interval, that it converges to $f(x)$. However, for many familiar functions this is the case, as will be shown in Section 2–18 and Chapter 11. Also, starting with known Taylor series representations, other such representations can be obtained by the operations of this and the following sections.

A function $f(x)$ which, for each point a where it is defined, is equal to the sum of its Taylor series at a in some interval about a, is said to be *analytic*. It can be shown that e^x, $\sin x$, $\cos x$, $\ln x$, and rational functions of x are analytic.

2–17 FURTHER OPERATIONS ON POWER SERIES

As absolutely convergent series, convergent power series can be *added, subtracted,* and *multiplied*: that is, if

$$f(x) = \sum_{n=0}^{\infty} c_n x^n \quad \text{and} \quad g(x) = \sum_{n=0}^{\infty} d_n x^n \tag{2–170}$$

for $|x| < r^*$, then

$$f(x) \pm g(x) = \sum_{n=0}^{\infty} (c_n \pm d_n) x^n,$$

$$f(x)g(x) = \sum_{n=0}^{\infty} (c_0 d_n + c_1 d_{n-1} + \cdots + c_{n-1} d_1 + c_n d_0) x^n.$$

The multiplication rule follows from the results of Section 2–10.

Thus, by squaring the geometric series formula

$$\frac{1}{1-x} = 1 + x + \cdots + x^n + \cdots, \quad |x| < 1, \tag{2–171}$$

we obtain

$$\frac{1}{(1-x)^2} = (1 + x + \cdots + x^n + \cdots)(1 + x + \cdots + x^n + \cdots)$$

$$= 1 + 2x + 3x^2 + \cdots + (n+1)x^n + \cdots, \quad |x| < 1.$$

The same result could be achieved by differentiating both sides of Eq. (2–171).

Also we observe that from (2–171) it follows for $c \neq 0$:

$$\frac{1}{x-c} = \frac{-1}{c(1-x/c)} = -\frac{1}{c}\left(1 + \frac{x}{c} + \cdots + \frac{x^n}{c^n} + \cdots\right), \quad |x| < |c|,$$

and hence by partial fractions (see Section 4–6)

$$\frac{1}{(x-3)(x-5)} = \frac{1}{2}\frac{1}{x-5} - \frac{1}{2}\frac{1}{x-3} = -\frac{1}{10}\frac{1}{1-\frac{x}{5}} + \frac{1}{6}\frac{1}{1-\frac{x}{3}}$$

$$= -\frac{1}{10}\left(1 + \frac{x}{5} + \cdots + \frac{x^n}{5^n} + \cdots\right) + \frac{1}{6}\left(1 + \frac{x}{3} + \cdots + \frac{x^n}{3^n} + \cdots\right)$$

$$= \frac{1}{6} - \frac{1}{10} + x\left(\frac{1}{18} - \frac{1}{50}\right) + \cdots + x^n\left(\frac{1}{6 \cdot 3^n} - \frac{1}{10 \cdot 5^n}\right) + \cdots$$

The result is valid if $|x| < 5$ and $|x| < 3$, that is, for $|x| < 3$.

The operations of division, substitution, and forming the inverse function for power series are illustrated in Problems 5–7 below (for more information, see KNOPP, pp. 184–188).

===================**Problems (Section 2–17)**===============

1. Find the Maclaurin series representation for $\ln(1-x)$ from that for $1/(1-x)$.

2. Find the Maclaurin series representation for each function:

   a) $\dfrac{1}{1+x}$ b) $\dfrac{1}{1+x^2}$ c) Arctan x

   d) $\dfrac{1}{1-x^2}$ e) $\ln\dfrac{1+x}{1-x}$ f) $\dfrac{1}{x^4-1}$

   g) $\sin^2 x$ h) $\cosh x$ i) $\cos 3x$

3. Obtain the terms through x^3 for the Maclaurin series representation of each function:

   a) $e^x \sin x$ b) $\dfrac{e^x}{1-x}$ c) $\dfrac{\sin x}{1+e^x}$ d) $\sec x$

   e) $\tanh x$ f) $\dfrac{1}{(1+x^2)\cos x}$ g) e^{-x^2} h) e^{2x-x^2}

   i) $e^{x/(1-x)}$ j) $\sin(e^x-1)$ k) Arcsin x l) Arctan x

4. Compute the Taylor series of the given function through the term in $(x-a)^3$ for the given a:

   a) $\sqrt{x},\ a=1$ b) $\sqrt[3]{x},\ a=2$

   c) $\sqrt[3]{\dfrac{1-x}{1+x}},\ a=0$ d) $e^{\sqrt{x}},\ a=1$

   For each of these examples it can be shown that the Taylor series does converge over an interval and has the given function as sum.

5. *Division of power series.* From the series expressions

   $$\sin x = x - \frac{x^3}{3!} + \cdots,\quad \cos x = 1 - \frac{x^2}{2!} + \cdots,$$

   we find the Maclaurin series for $\tan x$ by writing

   $$\frac{\sin x}{\cos x} = \tan x = c_0 + c_1 x + c_2 x^2 + \cdots,$$

   and hence

   $$x - \frac{x^3}{3!} + \cdots = (c_0 + c_1 x + c_2 x^2 + \cdots)\left(1 - \frac{x^2}{2!} + \cdots\right).$$

   Multiplying out on the right and comparing terms of same degree on both sides, equations can be obtained for the unknown coefficients $c_0, c_1, \ldots$: $0 = c_0$, $1 = c_1$, $0 = c_2 - (c_0/2!)$, ... From these we can determine the coefficients.
   Follow the procedure described to get

   $$\tan x = x + \frac{x^3}{3} + \frac{2x^5}{15} + \frac{17x^7}{315} + \cdots$$

   This can be shown to be valid for $|x| < \pi/2$.

6. *Substitution in power series.* We can write

$$e^{\sin x} = 1 + \frac{\sin x}{1!} + \frac{\sin^2 x}{2!} + \cdots$$

$$= 1 + \frac{1}{1!}\left(x - \frac{x^3}{3!} + \cdots\right) + \frac{1}{2!}\left(x - \frac{x^3}{3!} + \cdots\right)^2 + \cdots$$

Follow the procedure suggested to obtain

$$e^{\sin x} = 1 + x + \frac{x^2}{2} + 0x^3 - \frac{x^4}{8} + \cdots$$

7. *Series for inverse function.* Let $y = xe^x = f(x)$. We seek a series for the inverse function $x = g(y)$. For $x = 0$, $y = 0$. Hence we try $x = c_1 y + c_2 y^2 + \cdots$ Now

$$y = xe^x = (c_1 y + c_2 y^2 + \cdots)e^{c_1 y + c_2 y^2 + \cdots}$$

$$= (c_1 y + c_2 y^2 + \cdots)\left[1 + (c_1 y + c_2 y^2 + \cdots) + \frac{1}{2!}(c_1 y + c_2 y^2 + \cdots)^2 + \cdots\right].$$

Compare coefficients of like powers of y on both sides to find $c_1 = 1$, $c_2 = -1, \ldots$ and hence $x = y - y^2 + (3/2)y^3 + \cdots$

2–18 TAYLOR'S FORMULA WITH REMAINDER

This is the formula

$$f(x) = f(a) + f'(a)(x - a) + \frac{f''(a)}{2!}(x - a)^2 + \cdots + \frac{f^{(n)}(a)}{n!}(x - a)^n + R_n, \qquad (2\text{--}180)$$

where the *remainder* R_n can be given in several forms. The following two are commonly used: the *Lagrange form*

$$R_n = f^{(n+1)}(x_1)\frac{(x - a)^{n+1}}{(n+1)!}, \quad x_1 \text{ between } a \text{ and } x, \qquad (2\text{--}181)$$

and the *integral form*

$$R_n = \int_a^x \frac{(x - t)^n}{n!} f^{(n+1)}(t)\,dt \qquad (2\text{--}182)$$

(for proofs see CLA, vol. I, pp. 616–618).

Formula (2–180) is valid whenever $f, f', \ldots, f^{(n+1)}$ are continuous on the interval from a to x (x may also be less than a). Hence it is available even when f may not be representable by its Taylor series and serves as an important substitute for such a series. For effective use of the formula, some estimate for the size of the remainder R_n is needed; this in turn depends on having an estimate for the size of $|f^{(n+1)}|$ on the interval.

EXAMPLE 1 To show how the formula can be used to justify the Taylor series representations, we prove that e^x can be represented by its Maclaurin series. (A similar proof shows that e^x can be represented by its Taylor series at $x = a$ for all x; this shows that e^x is analytic.) We must prove that

$$e^x = 1 + \frac{x}{1!} + \cdots + \frac{x^n}{n!} + \cdots, \quad |x| < \infty. \tag{2-183}$$

We take a fixed x (not 0, where no proof is needed) and apply (2–180) and (2–181) for a general n to the interval from 0 to x:

$$e^x = 1 + \frac{x}{1!} + \cdots + \frac{x^n}{n!} + R_n, \quad R_n = \frac{e^{x_1} x^{n+1}}{(n+1)!}. \tag{2-184}$$

If x is positive, $0 < R_n < e^x x^{n+1}/(n+1)!$ and the last expression approaches 0 as $n \to \infty$ (since $x^n/n!$ is the general term of the absolutely convergent series $\sum x^n/n!$), so that $R_n \to 0$. If x is negative, then $|R_n| < |x|^{n+1}/(n+1)!$, and again $R_n \to 0$ as $n \to \infty$. Therefore (2–184) implies (2–183). ◄

Similarly, by showing that the remainder in Taylor's formula has limit 0 as $n \to \infty$, we can derive other representations:

$$\sin x = x - \frac{x^3}{3!} + \cdots + (-1)^{n+1} \frac{x^{2n-1}}{(2n-1)!} + \cdots, \quad |x| < \infty,$$

$$\cos x = 1 - \frac{x^2}{2!} + \cdots + (-1)^n \frac{x^{2n}}{(2n)!} + \cdots, \quad |x| < \infty$$

$$\ln x = (x-1) - \frac{(x-1)^2}{2} + \cdots + (-1)^{n+1} \frac{(x-1)^n}{n} + \cdots, |x-1| < 1 \tag{2-185}$$

$$(1+x)^r = 1 + \frac{rx}{1!} + \frac{r(r-1)}{2!} x^2 + \cdots + \frac{r(r-1) \cdots (r-n+1)}{n!} x^n$$

$$+ \cdots, \quad |x| < 1.$$

The last equation is the *general binomial formula*. From these, by the methods of Sections 2–16 and 2–17 we can derive other important representations:

$$\text{Arcsin } x = x + \frac{x^3}{6} + \cdots + \frac{1 \cdot 3 \cdots (2n-3)}{2 \cdot 4 \cdots (2n-2)} \frac{x^{2n-1}}{2n-1} + \cdots, \quad |x| < 1,$$

$$\text{Arctan } x = x - \frac{x^3}{3} + \cdots + (-1)^{n+1} \frac{x^{2n-1}}{2n-1} + \cdots, \quad |x| < 1,$$

$$\tan x = x + \frac{x^3}{3} + \frac{2x^5}{15} + \frac{17x^7}{315} + \frac{62x^9}{2835} + \cdots, \quad |x| < \pi/2, \tag{2-186}$$

$$\cosh x = 1 + \frac{x^2}{2!} + \cdots + \frac{x^{2n}}{(2n)!} + \cdots, \quad |x| < \infty,$$

$$\sinh x = x + \frac{x^3}{3!} + \cdots + \frac{x^{2n-1}}{(2n-1)!} + \cdots, \quad |x| < \infty.$$

2–19 POWER SERIES AND TAYLOR'S FORMULA FOR FUNCTIONS OF SEVERAL VARIABLES

A power series in x and y has the form:

$$a_{00} + a_{10}x + a_{01}y + a_{20}x^2 + a_{11}xy + a_{02}y^2 + \cdots + a_{mn}x^m y^n + \cdots \qquad (2\text{–}190)$$

or, for expansion about x_0, y_0:

$$a_{00} + a_{10}(x - x_0) + a_{01}(y - y_0) + a_{20}(x - x_0)^2 + \cdots \qquad (2\text{–}191)$$

We emphasize the case of (2–190), since the general case (2–191) can be reduced to the special case (2–190) by the substitutions $x' = x - x_0$, $y' = y - y_0$.

We consider only absolute convergence. The set on which a series (2–190) converges absolutely may not be easy to describe. However, it is usually feasible to find a set $|x| < A$, $|y| < B$ (where A and B are constants, possibly ∞) in which the series converges absolutely (Fig. 2–11). For example, the series

$$2 + x + y + x^2 + y^2 + x^3 + y^3 + \cdots$$

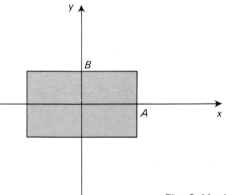

Fig. 2–11. A set of absolute convergence for series (2–190).

converges absolutely when $|x| < 1$ and $|y| < 1$. It can be rewritten as

$$(1 + x + x^2 + \cdots) + (1 + y + y^2 + \cdots)$$

and hence its sum is

$$\frac{1}{1 - x} + \frac{1}{1 - y}.$$

When the series (2–190) is absolutely convergent for $|x| < A$, $|y| < B$, we may rearrange the series and sum in any convenient order, for example, as $\sum_{m=0}^{\infty} \sum_{n=0}^{\infty} a_{mn} x^m y^n$ (see the Remark at the close of Section 2–10). As for power series in one variable, we can show that the sum $f(x, y)$ is continuous for $|x| < A$,

$|y| < B$ and that the series can be differentiated repeatedly term by term; that is,

$$\frac{\partial f}{\partial x} = a_{10} + 2a_{20}x + \cdots + ma_{mn}x^{m-1}y^n + \cdots,$$

$$\frac{\partial f}{\partial y} = a_{01} + a_{11}x + \cdots + na_{mn}x^m y^{n-1} + \cdots,$$

$$\frac{\partial^2 f}{\partial x^2} = 2a_{20} + \cdots + m(m-1)a_{mn}x^{m-2}y^n + \cdots$$

for $|x| < A$, $|y| < B$, where the series converge absolutely. It follows that

$$a_{00} = f(0, 0), \quad a_{10} = \frac{\partial f}{\partial x}(0, 0), \quad a_{01} = \frac{\partial f}{\partial y}(0, 0), \quad \cdots$$

and in general the coefficients are given by the formula

$$a_{mn} = \frac{1}{m! \, n!} \frac{\partial^{m+n} f}{\partial x^m \, \partial y^n}(0, 0). \tag{2-192}$$

With coefficients evaluated in this way, Eq. (2–190) is called the *Taylor series* of $f(x, y)$ at $(0, 0)$.

The Taylor series expansion can be written as follows:

$$f(x, y) = f(0, 0) + x\frac{\partial f}{\partial x}(0, 0) + y\frac{\partial f}{\partial y}(0, 0)$$

$$+ \frac{1}{2!}\left[x^2 \frac{\partial^2 f}{\partial x^2}(0, 0) + 2xy \frac{\partial^2 f}{\partial x \, \partial y}(0, 0) + y^2 \frac{\partial^2 f}{\partial y^2}(0, 0) \right] \tag{2-193}$$

$$+ \frac{1}{3!}\left[x^3 \frac{\partial^3 f}{\partial x^3}(0, 0) + 3x^2 y \frac{\partial^3 f}{\partial x^2 \, \partial y}(0, 0) + 3xy^2 \frac{\partial^3 f}{\partial x \, \partial y^2}(0, 0) \right.$$

$$\left. + y^3 \frac{\partial^3 f}{\partial y^3}(0, 0) \right] + \cdots$$

The numerical coefficients in each bracket are *binomial coefficients*. As for functions of one variable, this expansion is valid for many common functions. The following are examples:

$$e^{x+y} = 1 + x + y + \frac{1}{2!}(x^2 + 2xy + y^2) + \cdots, \quad \text{all } (x, y), \tag{2-194}$$

$$\sin x \cos y = x - \frac{x^3}{6} - \frac{xy^2}{2} + \frac{x^5}{120} + \frac{x^3 y^2}{12} + \frac{xy^4}{24} + \cdots, \quad \text{all } (x, y), \tag{2-195}$$

$$\frac{1}{1 - x - y} = 1 + x + y + x^2 + 2xy + y^2 + \cdots, \quad |x| + |y| < 1. \tag{2-196}$$

As for functions of one variable, there is also a *Taylor's formula with remainder* for functions of two variables. The simplest case is the *mean-value theorem*:

$$f(x, y) = f(0, 0) + x\frac{\partial f}{\partial x}(x^*, y^*) + y\frac{\partial f}{\partial y}(x^*, y^*). \tag{2-197}$$

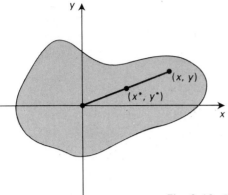

Fig. 2–12. Remainder formula for functions of two variables.

Here f is assumed to be continuous, with continuous first partial derivatives, in an open region containing the line segment from $(0, 0)$ to (x, y), as in Fig. 2–12, and (x^*, y^*) is a point of the segment between the endpoints.

The general case (with the Lagrange form of the remainder) is

$$f(x, y) = f(0, 0) + x\frac{\partial f}{\partial x}(0, 0) + y\frac{\partial f}{\partial y}(0, 0) + \cdots + \frac{1}{n!}\left[x^n\frac{\partial^n f}{\partial x^n}(0, 0) + \cdots\right]$$

$$+ \frac{1}{(n+1)!}\left[x^{n+1}\frac{\partial^{n+1} f}{\partial x^{n+1}}(x^*, y^*) + (n+1)x^n y\frac{\partial^{n+1} f}{\partial x^n \partial y} + \cdots\right], \quad (2\text{–}198)$$

which is valid under analogous hypotheses. The derivation of the formula requires the chain rule of Section 9–4 (see AC, pp. 438–441).

The discussion of this section can be generalized in a natural way to power series in three or more variables.

2–20 COMPLEX SERIES

The power series expansions of the preceding sections can also be very easily justified by the theory of functions of a complex variable, as developed in Chapter 11. Here we comment briefly on some of the basic notions. As suggested by the appearance of the function $e^{(\alpha + i\beta)x}$ in Section 1–10, complex numbers, functions and series are widely used in applications.

As in Section 1–9, complex numbers $a + bi$ can be represented by the points (a, b) of a plane, the *complex plane*, as in Fig. 2–13. We write $z = x + yi$ for a general point of this plane. As in Section 1–9, the *absolute value* of z is defined as

$$|z| = \sqrt{x^2 + y^2}. \quad (2\text{–}200)$$

This is the distance r from z to the origin 0 (or $0 + 0i$), while $|z_1 - z_2|$ equals the *distance between z_1 and z_2.*

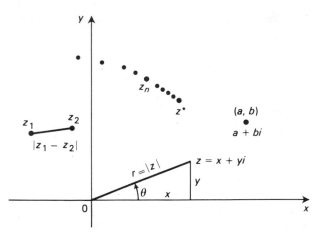

Fig. 2-13. The complex plane.

A complex sequence z_n is said to converge and have limit z^* if

$$\lim_{n \to \infty} |z_n - z^*| = 0. \tag{2-201}$$

Thus the distance from z_n to z^* must approach 0 as $n \to \infty$.

The basic ideas of convergence and divergence for complex sequences and series are developed exactly as for real series. In fact, convergence questions for the complex case can be reduced to those for the real case. One way of doing this is by taking *real and imaginary parts*.

For a sequence z_n we write $z_n = x_n + iy_n$, $z^* = x^* + iy^*$. Then z_n *converges to* z^* *precisely when* x_n *converges to* x^* *and* y_n *to* y^*. This is easily seen from the geometric meaning of convergence, as suggested in Fig. 2-13. Similarly, a series $\sum_{n=1}^{\infty} (a_n + ib_n)$ converges and has sum $a^* + ib^*$ if and only if $\sum_{n=1}^{\infty} a_n = a^*$ and $\sum_{n=1}^{\infty} b_n = b^*$.

EXAMPLE 1 Let x be real. Then

$$\sum_{n=0}^{\infty} \frac{(ix)^n}{n!} = 1 + \frac{ix}{1!} - \frac{x^2}{2!} - i\frac{x^3}{3!} + \frac{x^4}{4!} + \cdots$$

By taking real and imaginary parts (and dropping zero terms) we obtain the real series

$$1 - \frac{x^2}{2!} + \frac{x^4}{4!} + \cdots \qquad \text{and} \qquad x - \frac{x^3}{3!} + \cdots$$

for $\cos x$ and $\sin x$ respectively. The function e^{ix} is commonly defined as the sum of the series $\sum (ix)^n/n!$. From this definition it follows that

$$e^{ix} = \cos x + i \sin x, \qquad -\infty < x < \infty. \tag{2-202}$$

This is the *Euler identity* (see Section 1-10). ◀

Another method of reducing a question for complex series to one for real series is by taking *absolute values*, for the basic rule that *absolute convergence implies convergence* continues to hold in the complex case. To justify this, we let $\sum (a_n + i b_n)$ be absolutely convergent, so that

$$\sum_{n=1}^{\infty} |a_n + i b_n| = \sum_{n=1}^{\infty} (a_n^2 + b_n^2)^{1/2}$$

converges. Now $|a_n| \leqslant (a_n^2 + b_n^2)^{1/2}$ and $|b_n| \leqslant (a_n^2 + b_n^2)^{1/2}$, so that, by the comparison test, $\sum |a_n|$ and $\sum |b_n|$ converge. Therefore $\sum a_n$ and $\sum b_n$ converge, and so does $\sum (a_n + i b_n)$.

EXAMPLE 2 We consider the complex series $\sum z^n/n$. By De Moivre's theorem,

$$z^n = [r(\cos \theta + i \sin \theta)]^n = r^n (\cos n\theta + i \sin n\theta).$$

Hence $|z^n| = r^n = |z|^n$ and therefore $|z^n/n| = |z|^n/n$. But the real series $\sum |z|^n/n$ is convergent for $|z| < 1$ (by the ratio test). Hence the series $\sum z^n/n$ is *absolutely convergent for* $|z| < 1$. Also, for $|z| > 1$, $|z|^n/n \to \infty$ as $n \to \infty$. Hence (by the nth-term test) the series $\sum z^n/n$ diverges for $|z| > 1$. Thus there is a circular region of convergence of radius $r^* = 1$ (Fig. 2–14). Inside the circle we have absolute convergence, outside we have divergence. On the circle itself, we write $z = \cos \theta + i \sin \theta$, so that

$$\sum_{n=1}^{\infty} \frac{z^n}{n} = \sum_{n=1}^{\infty} \frac{\cos n\theta + i \sin n\theta}{n}.$$

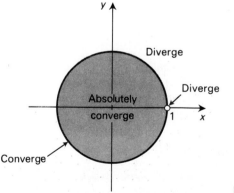

Fig. 2–14. Convergence set for $\sum z^n/n$.

By the theory of Fourier series (Chapter 3), it can be shown that the real series

$$\sum_{n=1}^{\infty} \frac{\cos n\theta}{n} \quad \text{and} \quad \sum_{n=1}^{\infty} \frac{\sin n\theta}{n}$$

converge for all θ except, in the case of the first series, for $\theta = 0$. Thus the given series converges for $|z| \leqslant 1$ except for $z = 1$.

This example illustrates how the theory of power series carries over to the complex case. In general, a series $\sum_{n=0}^{\infty} c_n (z - z_0)^n$ has a radius of convergence r^*, which may be 0, finite and positive, or infinite. When $r^* = 0$, the series converges only for $z = z_0$. When r^* is positive and finite, the series converges absolutely in the circular region of convergence $|z - z_0| < r^*$, with center at z_0 and radius r^*; outside the circle (that is, for $|z - z_0| > r^*$) the series diverges. On the circular boundary (often called the *circle of convergence*) the series may converge at some points and diverge at others. When $r^* = \infty$, the series converges absolutely for all z. The proof of these rules is similar to that for real power series. ◀

The functions e^z, $\sin z$, and others can all be defined for complex z by the Maclaurin series of Eqs. (2–183), (2–185), (2–186), with x replaced by z. The functions $\ln z$, $(1 + z)^r$, $\text{Arcsin } z$, and $\text{Arctan } z$ are multivalued, and the series representations mentioned provide a "branch" of each function (for more details see Section 11–23). For $\tan z$, the equation $\tan z = \sin z / \cos z$ defines the function for $z \neq (\pi/2) + n\pi$, where $n = 0, \pm 1, \pm 2, \ldots$

=======================Problems (Section 2–20)=======================

1. Use the information given about $f(x)$, assuming all derivatives continuous as needed, to estimate the value requested:

   a) $f(0) = 1$, $f'(0) = 2$, $0.2 \leqslant f''(x) \leqslant 0.5$ for $0 \leqslant x \leqslant 1$; value $f(1)$ sought.
   b) $f(0) = 3$, $f'(0) = 7$, $-1 \leqslant f''(x) \leqslant 1$ for $0 \leqslant x \leqslant 2$; value $f(2)$ sought.
   c) $f(0) = 1$, $f'(0) = 2$, $f''(0) = 5$, $f'''(x) = 1/(1 + x^3)$; value $f(3)$ sought.
   d) $f(0) = 5$, $f'(0) = -1$, $f''(0) = -2$, $f'''(x) = e^{-x^2}$; value $f(1)$ sought.

   e) $f(1) = 2$, $f'(1) = 2$, $f''(x) = \displaystyle\int_0^1 e^{-xt^2} \, dt$; value $f(2)$ sought.

   f) $f(1) = 7$, $f'(1) = -1$, $f''(x) = \dfrac{1}{10} \displaystyle\int_0^x \dfrac{t-1}{\sqrt{1+t^4}} \, dt$; value $f(3)$ sought.

2. Write out Taylor's formula with remainder in the Lagrange form using $a = 0$ for each of the following functions and prove that $R_n \to 0$ as $n \to \infty$, thereby justifying the series expansion given in Section 2–18.

   a) $f(x) = \sin x$ **b)** $f(x) = \cos x$

3. **a)** Derive the expansion (2–194) as a Taylor series.
   b) Derive the expansion (2–194) by replacing u by $x + y$ in the series $e^u = 1 + u + \cdots$
   c) Derive the series (2–195) as a Taylor series, verifying the terms shown.
   d) Derive the series (2–195) by multiplying the expansions

   $$\sin x = x - (x^3/3!) + \cdots \qquad \text{and} \qquad \cos y = 1 - (y^2/2!) + \cdots$$

   e) Derive the expansion (2–196) by any convenient method.

4. Determine whether these complex sequences converge:

   a) $\left\{ \left(\dfrac{1+i}{2} \right)^n \right\}$ **b)** $\left\{ \left(\dfrac{3+4i}{5} \right)^n \right\}$ **c)** $\{ e^{-n-(i/n)} \}$ **d)** $\{ e^{(1/n) + [(n\pi i)/2]} \}$

5. Test for convergence:

a) $\displaystyle\sum_{n=1}^{\infty} \frac{(3+4i)^n}{5^n n^2}$

b) $\displaystyle\sum_{n=0}^{\infty} \frac{(1+i)^n}{n!}$

c) $\displaystyle\sum_{n=2}^{\infty} \left(\frac{i^n}{n} + \frac{i^{n+1}}{n \ln n}\right)$

d) $\displaystyle\sum_{n=0}^{\infty} \frac{n e^{in\pi/3}}{n^3 + 1}$

e) $\displaystyle\sum_{n=1}^{\infty} \frac{2 \cdot 4 \cdots (2n)}{4 \cdot 7 \cdots (3n+1)} \frac{(1+i)^n}{2^n}$

f) $\displaystyle\sum_{n=1}^{\infty} \left(\frac{1+2ni}{n^2}\right)^n$

6. Show that the rule for the geometric series:

$$\frac{1}{1-z} = 1 + z + \cdots + z^n + \cdots, \qquad |z| < 1,$$

is valid for complex z.

7. Find the radius of convergence:

a) $\displaystyle\sum_{n=0}^{\infty} \frac{3n}{2n+5} z^n$

b) $\displaystyle\sum_{n=0}^{\infty} \frac{z^n}{3^n (n^2 + 1)}$

c) $\displaystyle\sum_{n=0}^{\infty} \frac{2^n z^n}{n! + n + 1}$

d) $\displaystyle\sum_{n=0}^{\infty} n! z^n$

e) $\displaystyle\sum_{n=3}^{\infty} \frac{(z-3)^n}{5^n \ln n}$

f) $\displaystyle\sum_{n=0}^{\infty} \frac{n!(z+i)^n}{n! + 3n^2}$

8. Prove each of the following identities from the power series representations of the functions:

a) $\sin(-z) = -\sin z$

b) $\cos(-z) = \cos z$

c) $e^{iz} = \cos z + i \sin z$

d) $\cos iz = \cosh z$

e) $\sin iz = i \sinh z$

f) $\cos^2 z + \sin^2 z = 1$

2-21 APPLICATION OF POWER SERIES TO INTEGRATION AND OTHER PROBLEMS

If an integration is difficult, numerical methods (Chapter 13) or power series may be used.

EXAMPLE 1 Find the area under the normal error curve $y = e^{-x^2}$ from $x = 0$ to $x = 1$.

Solution. We must evaluate $\int_0^1 e^{-x^2}\, dx$. Since $e^{-x^2} = 1 - x^2 + (x^4/2) + \cdots$ (by replacement of x by $-x^2$ in the Maclaurin series for e^x), we get

$$\int_0^1 e^{-x^2}\,dx = \int_0^1 \left(1 - x^2 + \frac{x^4}{2} + \cdots\right)dx = 1 - \frac{1}{3} + \frac{1}{10} - \frac{1}{7\cdot3!} + \frac{1}{9\cdot4!} - \frac{1}{11\cdot5!} + \cdots$$

If we sum the terms shown, we obtain the value 0.7467. ◀

EXAMPLE 2   *Find the circumference of an ellipse of small eccentricity*

Solution. The standard ellipse (Fig. 2–15) can be represented parametrically by the equations $x = a\cos t$, $y = b\sin t$, with $a \geqslant b$. The eccentricity is $e = \sqrt{a^2 - b^2}/a$, so that $e^2 = (a^2 - b^2)/a^2 = 1 - (b/a)^2$. By calculus and the symmetry of the ellipse, the arc length is

$$L = 4\int_0^{\pi/2} \sqrt{\left(\frac{dx}{dt}\right)^2 + \left(\frac{dy}{dt}\right)^2}\,dt = 4\int_0^{\pi/2}\sqrt{a^2\sin^2 t + b^2\cos^2 t}\,dt$$

$$= 4\int_0^{\pi/2}\sqrt{b^2 + (a^2 - b^2)\sin^2 t}\,dt = 4b\int_0^{\pi/2}\sqrt{1 + \frac{e^2}{1 - e^2}\sin^2 t}\,dt.$$

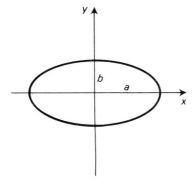

Fig. 2-15. Standard ellipse.

For e small, we use the binomial series (2–185):

$$(1 + u)^{1/2} = 1 + \frac{1}{2}u - \frac{1}{8}u^2 + \cdots$$

and approximate the arc length by

$$4b\int_0^{\pi/2}\left[1 + \frac{1}{2}\frac{e^2}{1 - e^2}\sin^2 t - \frac{1}{8}\left(\frac{e^2}{1 - e^2}\right)^2\sin^4 t\right]dt.$$

We evaluate the integrals with the aid of integral tables to obtain

$$L = \pi b\left[2 + \frac{1}{2}\frac{e^2}{1 - e^2} - \frac{3}{32}\left(\frac{e^2}{1 - e^2}\right)^2\right]. \tag{2–210}$$

If $a = 10$ and $b = 9$, then $e^2 = 19/100$ and $L = 59.14$. To check, we verify that the value found lies about halfway between the circumferences of two circles of radii 9 and 10. ($18\pi = 56.55$ and $20\pi = 62.83$.)

We observe that Eq. (2–210) itself is the beginning of a power series for L in powers of $e^2/(1 - e^2)$. Power series can be used to advantage for many other problems. ◄

EXAMPLE 3 *Kepler's equation.* In describing the motion of a planet P in an ellipse about the sun S (or of a satellite about a planet) we are led to the equations

$$M = n(t - T) = E - e \sin E.$$

Here n is the mean angular velocity, E the eccentric anomaly (Fig. 2–16), e the eccentricity of the ellipse. The equation $M = E - e \sin E$ is called Kepler's equation. We wish to solve it for E in terms of e, for fixed M. The convergence of the Maclaurin series of E in terms of e can be justified for e close to 0. When $e = 0$, we have $E = M$.

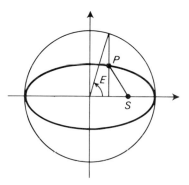

Fig. 2–16. Planetary motion.

Next, by differentiation, we obtain

$$0 = \frac{dE}{de} - \sin E - e \cos E \frac{dE}{de},$$

$$0 = \frac{d^2 E}{de^2} - 2 \cos E \frac{dE}{de} + e \sin E \frac{d^2 E}{de^2}.$$

Thus for $e = 0$, we get $dE/de = \sin M$ and $d^2 E/de^2 = 2 \sin M \cos M = \sin 2M$. Therefore,

$$E = M + e \sin M + \frac{e^2}{2} \sin 2M + \cdots$$

For most applications, e is so small (orbits nearly circular), that the three terms shown suffice. ◄

EXAMPLE 4 Find the solution of the differential equation

$$y' = y(1 + e^{-x^2}) \text{ such that } y = 1 \text{ when } x = 0.$$

Solution. We separate variables and obtain

$$\int \frac{dy}{y} = \int (1 + e^{-x^2})\,dx, \qquad \ln y = x + \int e^{-x^2}\,dx + c.$$

As in Example 1, we replace e^{-x^2} by its power series $1 - x^2 + \cdots$ and obtain

$$\ln y = x + x - \frac{x^3}{3} + \frac{x^5}{5 \cdot 2!} + \cdots + (-1)^n \frac{x^{2n+1}}{(2n+1)\,n!} + \cdots + c.$$

From the initial condition, $c = 0$ and hence

$$y = \exp\left[2x - \frac{x^3}{3} + \cdots + (-1)^n \frac{x^{2n+1}}{(2n+1)\,n!} + \cdots \right]. \quad \blacktriangleleft$$

2–22 POWER SERIES SOLUTIONS OF DIFFERENTIAL EQUATIONS

We illustrate the method by examples.

EXAMPLE 1 Find the solution of $y' = y$ such that $y = 1$ for $x = 0$.

Solution. We assume that the solution can be represented by its Maclaurin series:

$$y = c_0 + c_1 x + \cdots + c_n x^n + \cdots$$

By the initial condition, $c_0 = 1$. Now,

$$y' = c_1 + 2c_2 x + \cdots + nc_n x^{n-1} + \cdots$$

If we substitute in the differential equation, we obtain:

$$c_1 + 2c_2 x + \cdots + nc_n x^{n-1} + \cdots = c_0 + c_1 x + \cdots + c_n x^n + \cdots$$

We compare coefficients of like powers of x:

$$c_1 = c_0, \quad c_2 = \frac{c_1}{2} = \frac{c_0}{2}, \quad c_3 = \frac{c_2}{3} = \frac{c_0}{2 \cdot 3}, \quad \cdots, \quad c_{n+1} = \frac{c_n}{n+1} = \frac{c_0}{2 \cdot 3 \cdots (n+1)}$$

and hence, with $c_0 = 1$, we get $c_{n+1} = 1/(n+1)!$, so that

$$y = 1 + x + \cdots + \frac{x^n}{n!} + \cdots = e^x,$$

as we should have expected. ◀

EXAMPLE 2 Find the general solution of the equation $y'' - xy = 0$.

Solution. We again assume that $y = c_0 + c_1 x + \cdots + c_n x^n + \cdots$ and substitute in the differential equation, obtaining

$$2c_2 + 6c_3 x + \cdots + n(n-1)c_n x^{n-2} + \cdots - x(c_0 + c_1 x + \cdots + c_n x^n + \cdots) = 0.$$

After multiplying in the factor x, we compare coefficients on both sides; the coefficient of each power of x on the left must be 0. Thus

$$2c_2 = 0, \quad 6c_3 - c_0 = 0, \quad 12c_4 - c_1 = 0, \cdots, \quad (n+2)(n+1)c_{n+2} - c_{n-1} = 0, \quad \text{etc.}$$

and hence

$$c_2 = 0, \quad c_3 = \frac{c_0}{6}, \quad c_4 = \frac{c_1}{12}, \quad c_5 = \frac{c_2}{20} = 0, \quad c_6 = \frac{c_3}{6 \cdot 5} = \frac{c_0}{6 \cdot 5 \cdot 3 \cdot 2},$$

$$c_7 = \frac{c_4}{7 \cdot 6} = \frac{c_1}{7 \cdot 6 \cdot 4 \cdot 3}, \quad c_8 = \frac{c_5}{8 \cdot 7} = 0, \text{ etc.}$$

Clearly $c_2, c_5, c_8, c_{11}, \ldots$ are all 0 and for $n = 1, 2, \ldots$

$$c_{3n} = \frac{c_0}{(3n)(3n-1)(3n-3)(3n-4) \cdots (6 \cdot 5)(3 \cdot 2)},$$

$$c_{3n+1} = \frac{c_1}{(3n+1)(3n)(3n-2)(3n-3) \cdots (7 \cdot 6)(4 \cdot 3)},$$

so that

$$y = c_0 + c_1 x + \frac{c_0 x^3}{6} + \frac{c_1 x^4}{12} + \cdots$$

$$+ \frac{c_0 x^{3n}}{(3n)(3n-1) \cdots 3 \cdot 2} + \frac{c_1 x^{3n+1}}{(3n+1)(3n) \cdots 4 \cdot 3} + \cdots.$$

We can write this in the standard form:

$$y = c_0 y_1(x) + c_1 y_2(x), \tag{2–220}$$

with arbitrary constants c_0, c_1 and

$$y_1(x) = 1 + \frac{x^3}{6} + \cdots + \frac{x^{3n}}{3n(3n-3) \cdots 3 \cdot (3n-1)(3n-4) \cdots 2} + \cdots,$$

$$y_2(x) = x + \frac{x^4}{12} + \cdots + \frac{x^{3n+1}}{(3n+1)(3n-2) \cdots 4 \cdot 3n(3n-3) \cdots 3} + \cdots.$$

By the ratio test, both series converge for all x. Hence all the above steps are justified, so that, for every choice of c_0 and c_1, Eq. (2–220) defines a solution of the differential equation for all x. Furthermore, the functions $y_1(x)$, $y_2(x)$ are linearly independent for all x (Section 1–16), so that Eq. (2–220) provides the general solution of the differential equation (Problem 5 below). To satisfy initial conditions for $x = 0$, we are led to equations

$$y_0 = c_0 y_1(0) + c_1 y_2(0) = c_0, \qquad y_0' = c_0 y_1'(0) + c_1 y_2'(0) = c_1.$$

Thus $c_0 = y_0$ and $c_1 = y_0'$; with these choices, the initial conditions are satisfied. ◀

The method of Examples 1 and 2 is applicable quite widely to linear equations

$$a_0(x)y'' + a_1(x)y' + a_2(x)y = f(x)$$

(and analogous equations of higher order), provided the coefficients $a_0(x)$, $a_1(x)$, $a_2(x)$ and right-hand member $f(x)$ are representable by power series in an interval about x_0 and $a_0(x_0) \neq 0$. The general solution is obtainable in the form

$$y = c_1 y_1(x) + c_2 y_2(x) + y*(x), \tag{2-221}$$

where $y_1(x)$, $y_2(x)$, $y*(x)$ are power series in $x - x_0$ and Eq. (2–221) provides all solutions in some interval about $x = x_0$.

Among the important equations of this type are the following homogeneous equations:

the Legendre equation of order N:

$$(1 - x^2)y'' - 2xy' + N(N+1)y = 0; \tag{2-222}$$

the hypergeometric equation:

$$x(1 - x)y'' + [c - (a + b + 1)x]y' - aby = 0, \quad \text{with} \quad a, b, c \text{ constants}; \tag{2-223}$$

the Bessel equation of order N:

$$x^2 y'' + xy' + (x^2 - N^2)y = 0. \tag{2-224}$$

These give rise to the Legendre polynomials and functions, the hypergeometric functions and the Bessel functions (see Chapter 12).

Above we required that $a_0(x_0) \neq 0$. When $a_0(x_0) = 0$, the value x_0 is called a *singular point* of the differential equation. Thus $x = 1$ and $x = -1$ are singular points of the Legendre equation. One can avoid the singular points by simply choosing a value x_0 such that $a_0(x_0) \neq 0$ in forming the power series in powers of $x - x_0$. Thus we can use $x_0 = 0$ for the Legendre equation. However, for many applications it is desirable to have the solution somehow representable in powers of $x - x_0$ even though x_0 is a singular point. In the next section we describe cases in which modified power series can be used for the solutions near a singular point.

It should be remarked that the power series method gives a *recursion formula* for the nth coefficient c_n in terms of preceding coefficients. For Examples 1 and 2, the recursion formula is quite simple, involving only two coefficients. In other examples, one might be forced to work with a recursion formula expressing c_n in terms of two or more (even all) of the preceding coefficients. This causes no difficulty in computing successive coefficients (in particular, on a digital computer), but makes it hard to obtain a general formula for c_n and to study convergence of the series.

EXAMPLE 3 Find the solution of $y' = 1 + y^2$ such that $y = 0$ for $x = 0$.

Solution. This differential equation is nonlinear. We could use the same method of comparing coefficients as in Examples 1 and 2; however, we shall illustrate another method. We *differentiate the differential equation* successively to obtain

$$y'' = 2yy', \quad y''' = 2y'^2 + 2yy'', \quad y^{(4)} = 6y'y'' + 2yy''',$$
$$y^{(5)} = 8y'y''' + 6y''^2 + 2yy^{(4)}, \text{ etc.}$$

By hypothesis, $y = 0$ for $x = 0$. By the differential equation, $y' = 1 + y^2 = 1$ for $x = 0$. Hence by the above equations we obtain successively (for $x = 0$):

$$y'' = 0, \quad y''' = 2, \quad y^{(4)} = 0, \ y^{(5)} = 16, \quad \text{etc.,}$$

so that, by the formula for Taylor series (Section 2–16),

$$y = y(0) + y'(0)x + y''(0)\frac{x^2}{2!} + \cdots = x + \frac{x^3}{3} + \frac{2}{15}x^5 + \cdots$$

Here we can find the solution by separating the variables: we obtain $y = \tan x$. The series found agrees (up to the term in x^5) with the known Maclaurin series for $\tan x$, which converges for $|x| < \pi/2$.

This example illustrates the greater complexity of nonlinear differential equations. The power series method produces more complicated formulas and it is rarely possible to find an explicit formula for the general term. However, the method is of great practical importance. It is closely related to the numerical methods of Chapter 13. Furthermore, there are methods for estimating the radius of convergence for a power series solution and for estimating the error in stopping at the nth term. For more information on the theoretical basis of the method and on the estimates referred to, see ODE, Chapter 12. ◀

EXAMPLE 4   *Pendulum.*   From mechanics, the motion of a pendulum (Fig. 2–17) is governed by the differential equation

$$mL\frac{d^2\theta}{dt^2} = -mg\sin\theta.$$

For small oscillations, $\sin\theta$ can be replaced by θ and a linear differential equation is obtained whose solutions represent simple harmonic motion (see Example 4 of Section 1–11). If we do not make this approximation, we have a nonlinear differential equation. We seek the solution $\theta(t)$ such that $\theta = \pi/6$ and $d\theta/dt = 0$ for $t = 0$ in the form of a series: $\theta = c_0 + c_1 t + \cdots$ through terms of degree 4.

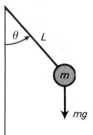

Fig. 2–17. Pendulum.

Solution. Here $c_0 = \pi/6$ and $c_1 = 0$ are given. Now,

$$\theta'' = -\frac{g}{L}\sin\theta, \quad \theta''' = -\frac{g}{L}\theta'\cos\theta, \quad \theta^{iv} = \frac{g}{L}(\theta'^2\sin\theta - \theta''\cos\theta).$$

From the given initial values we thus successively find for $t = 0$:

$$\theta'' = -\frac{g}{2L}, \quad \theta''' = 0, \quad \theta^{iv} = \frac{\sqrt{3}g^2}{4L^2}.$$

Now $c_2 = \theta''(0)/2!$, $c_3 = \theta'''(0)/3!$, $c_4 = \theta^{iv}(0)/4!$, ... Thus the solution sought is

$$\theta = \frac{\pi}{6} - \frac{g}{4L}t^2 + \frac{\sqrt{3}g^2}{96L^2}t^4 + \cdots$$

REMARK: As shown in Section 7–2, the exact solution of the nonlinear differential equation with the given initial values describes an oscillation. ◄

================ **Problems (Section 2–22)** ================

1. Use power series to evaluate the following integrals:

   a) $\displaystyle\int_0^\pi \frac{\sin x}{x}\,dx$ b) $\displaystyle\int_0^1 x^{1/3}e^x\,dx$

2. Use infinite series to find the arc length of the following curves:

   a) $y = \cos x$, $0 \le x \le \pi/4$ b) $y = \frac{1}{2}e^{-x}$, $0 \le x \le 1$

3. Obtain the general solution of the given differential equation in terms of power series in powers of x; verify convergence of the power series:

   a) $y'' + y = 0$ b) $y'' + 2y' + y = 0$ c) $y' - y = e^x = 1 + x + \dfrac{x^2}{2!} + \cdots$

   d) $y'' - y = e^x$ e) $y'' + xy' + y = 0$ f) $y'' + 2x^2 y' + xy = 0$

4. Obtain the solution such that $y = 3$ and $y' = 7$ for $x = 0$:

   a) $(x^2 + 1)y'' - 2y = 0$ b) $y'' - x^2 y = 0$

5. Show that the functions $y_1(x)$, $y_2(x)$ found in the solution of Example 2 in Section 2–21 are linearly independent for all x, that is, if $k_1 f_1(x) + k_2 f_2(x) \equiv 0$ for all x, with constant k_1, k_2, then $k_1 = 0$ and $k_2 = 0$.

6. In the Legendre equation (2–212) let N be an integer.
   a) Show that if N is a positive even integer or 0, then the equation has a solution of the form $1 + b_2 x^2 + b_4 x^4 + \cdots + b_N x^N$.
   b) Show that if N is a positive odd integer, the equation has a solution of form $x + b_3 x^3 + b_5 x^5 + \cdots + b_N x^N$.

   REMARK: The solutions of parts (a) and (b) are constants times the Legendre polynomials $P_N(x)$. For parts (a) and (b) the constants are

   $$(-1)^{N/2}\frac{2\cdot 4 \cdots N}{1\cdot 3 \cdots (N-1)} \quad \text{and} \quad (-1)^{(N-1)/2}\frac{2\cdot 4 \cdots (N-1)}{1\cdot 3 \cdots N}$$

   respectively (see Chapter 12). Exceptionally, for $N = 0$ or 1 the constant is 1.

7. The differential equation

   $$\frac{dx}{dt} = kx(at + b - x)$$

   describes the growth of population when the environment is steadily improving. The population can grow as long as $x < at + b$. Take $k = 1$, $a = 1$, $b = 2$ and use power series to find the solution such that $x = 1$ for $t = 0$.

8. Find the terms through the one in x^5 for a power series solution of the given differential equation satisfying the given initial conditions:

   a) $y' = 1 - y^2$, $y = 0$ for $x = 0$
   b) $y' = e^y$, $y = 1$ for $x = 0$
   c) $y'' + y^3 = 0$, $y = 1$ and $y' = 0$ for $x = 0$
   d) $yy'' + 1 = 0$, $y = 1$ and $y' = 1$ for $x = 0$

2–23 SOLUTIONS OF LINEAR DIFFERENTIAL EQUATIONS AT A SINGULAR POINT

We confine attention here to the homogeneous second-order linear equation

$$a_0(x)y'' + a_1(x)y' + a_2(x)y = 0 \tag{2–230}$$

and assume that the singular point is at $x = 0$. (A substitution $x' = x - x_0$ reduces the case of a singular point at x_0 to the special case considered.) We assume that $a_0(x)$, $a_1(x)$, $a_2(x)$ are analytic in an interval about $x = 0$, hence representable by their Maclaurin series:

$$a_0(x) = \sum_{n=0}^{\infty} \alpha_n x^n, \quad a_1(x) = \sum_{n=0}^{\infty} \beta_n x^n, \quad a_2(x) = \sum_{n=0}^{\infty} \gamma_n x^n, \tag{2–231}$$

say for $|x| < b$. Finally, we must assume that $a_0(0) = 0$; that is, $\alpha_0 = 0$, since 0 is a singular point.

    Our discussion also covers equations written in a form such as the following:

$$y'' + \frac{1}{x}y' + y = 0.$$

Here multiplication of the equation by x gives an equation of form (2–230) with $a_0(x) = x$, so that $a_0(0) = 0$ and 0 is a singular point:

$$xy'' + y' + xy = 0.$$

An example will illustrate why singular points give trouble.

EXAMPLE 1 $4x^2 y'' + 4xy' + (4x^2 - 1)y = 0$. We seek the solution such that $y = 1$ and $y' = 0$ for $x = 0$. From the differential equation we obtain the expression for y'':

$$y'' = \frac{-4xy' - (4x^2 - 1)y}{4x^2}.$$

For $x = 0$, $y = 1$, $y' = 0$, the right-hand side becomes $1/0$; thus y'' is not defined at the initial point and the existence theorem does not apply.

    We could nevertheless try to obtain a series solution:

$$y = c_0 + c_1 x + \cdots + c_n x^n + \cdots$$

The method of the preceding section leads to the relations

$$c_0 = 0, \quad c_1 = 0, \quad 4c_0 + 15c_2 = 0, \quad \ldots, \quad (4n^2 - 1)c_n + 4c_{n-2} = 0, \quad \ldots$$

From these we find $c_0 = 0, c_1 = 0, c_2 = 0, \cdots$, so that only the one solution $y = 0$ is obtained.

So far the situation looks hopeless. However, it was discovered over a century ago that a series solution may be salvaged by a simple modification. We try

$$y = x^m (c_0 + c_1 x + \cdots + c_n x^n + \cdots) = c_0 x^m + c_1 x^{m+1} + \cdots + c_n x^{m+n} + \cdots,$$

where m is a number to be found (not necessarily an integer). Now

$$y' = mc_0 x^{m-1} + (m+1)c_1 x^m + \cdots + (m+n)c_n x^{m+n-1} + \cdots,$$

$$y'' = m(m-1)c_0 x^{m-2} + (m+1)mc_1 x^{m-1} + \cdots$$
$$+ (m+n)(m+n-1)c_n x^{m+n-2} + \cdots,$$

so that substitution in the differential equation, division by x^m, and then comparing coefficients gives the relations

$$(4m^2 - 1)c_0 = 0, \quad (4m^2 + 8m + 3)c_1 = 0, \quad (4m^2 + 16m + 15)c_2 + 4c_0 = 0, \quad \ldots,$$

$$(4m^2 + 8mn + 4n^2 - 1)c_n + 4c_{n-2} = 0, \quad \ldots$$

We want c_0 to be different from 0 (if $c_0 = 0$ and $y \not\equiv 0$, we could factor out a higher power than x^m and again achieve a solution of the same form with $c_0 \neq 0$). Hence we must have

$$4m^2 - 1 = 0. \tag{2–232}$$

This equation, called the *indicial equation*, tells us which exponents m may be used. Here m may be $\pm 1/2$. With $m = 1/2$, our recursion formulas are

$$0c_0 = 0, \quad 8c_1 = 0, \quad 24c_2 + 4c_0 = 0, \quad \ldots, \quad (4n + 4n^2)c_n + 4c_{n-2} = 0, \quad \ldots$$

Thus c_0 is *arbitrary*. We may as well make it equal to 1 and then multiply by an arbitrary constant at the end. With this choice, we find

$$c_1 = 0, \quad c_2 = -\frac{1}{6}, \quad c_3 = 0, \quad c_4 = -\frac{c_2}{20} = \frac{1}{20 \cdot 6}, \quad c_5 = 0, \cdots$$

Thus $c_n = 0$ for n odd, and for even n

$$c_n = -\frac{c_{n-2}}{(n+1)n} = \frac{c_{n-4}}{(n+1)n(n-1)(n-2)} = \cdots = \frac{(-1)^{n/2}}{(n+1)n \cdots 3 \cdot 2}.$$

Hence our solution is

$$y_1(x) = x^{1/2} \left[1 - \frac{x^2}{6} + \frac{x^4}{120} + \cdots + \frac{(-1)^n x^{2n}}{(2n+1)!} + \cdots \right].$$

The series converges for all x, so that we appear to have a solution for all x. However, because of the factor $x^{1/2}$ the derivative y' does not exist for $x = 0$. In fact, $x^{1/2}$ is imaginary for x negative. This last complication can be taken care of by using

$$y_1(x) = (-x)^{1/2} \left(1 - \frac{x^2}{6} + \frac{x^4}{120} + \cdots \right)$$

for negative x. (This is a constant times the previous solution; the constant is the *imaginary* number $i = \sqrt{-1}$.)

We proceed in exactly the same way with the root $-1/2$ of the indicial equation. Our recursion formulas are

$$0c_1 = 0, \quad 0c_1 = 0, \quad 8c_2 + 4c_0 = 0, \quad \ldots, \quad (4n^2 - 4n)c_n + 4c_{n-2} = 0, \quad \ldots$$

Here c_0 and c_1 are both arbitrary. We take $c_0 = 1$ and $c_1 = 0$ and comment below on the significance of these choices. We find

$$c_0 = 1, \quad c_1 = 0, \quad c_2 = -\frac{c_0}{2} = -\frac{1}{2}, \quad c_4 = -\frac{c_2}{12} = \frac{1}{4 \cdot 3 \cdot 2}, \quad c_5 = 0, \ldots$$

and hence obtain the solution

$$y_2(x) = |x|^{-1/2} \left[1 - \frac{x^2}{2} + \frac{x^4}{24} + \cdots + (-1)^n \frac{x^{2n}}{(2n)!} + \cdots \right],$$

the absolute-value sign being used to take care of positive and negative x. The general solution is

$$y = c_1 y_1(x) + c_2 y_2(x). \tag{2-233}$$

Because of the discontinuity at $x = 0$, Eq. (2–233) gives all solutions for $x > 0$ and all solutions for $x < 0$. There are no solutions, other than $y = 0$, bridging the gap between positive and negative x, so that initial conditions at $x = 0$ (other than $y = 0$, $y' = 0$ for $x = 0$) cannot be satisfied.

If for $m = -1/2$ we leave both c_0, c_1 arbitrary, then we obtain

$$y = c_0 y_2(x) + c_1 y_1(x),$$

which is the same general solution.

REMARK: The differential equation of Example 1 is a special case of the Bessel equation (2–224); here $N = 1/2$. The solutions we have found are constants times Bessel functions:

$$y_1(x) = \sqrt{\pi/2}\, J_{1/2}(x), \quad y_2(x) = \sqrt{\pi/2}\, J_{-1/2}(x),$$

as one can verify (see Section 12–11). Furthermore, $y_1(x)$, $y_2(x)$ are related to $\sin x$ and $\cos x$, as we see from the series found:

$$y_1(x) = |x|^{-1/2} \sin x, \quad y_2(x) = |x|^{-1/2} \cos x.$$

(For $N = \pm 1/2, \pm 3/2, \ldots$, the Bessel functions $J_N(x)$ are simply related to $\sin x$ and $\cos x$; see Problem 7 following Section 12–12.) ◄

The method illustrated above can be verified to work for equations of the forms

$$(\alpha_1 x + \alpha_2 x^2 + \cdots)y'' + (\beta_0 + \beta_1 x + \cdots)y' + (\gamma_0 + \gamma_1 x + \cdots)y = 0, \quad \alpha_1 \neq 0, \tag{2-234a}$$

$$(\alpha_2 x^2 + \alpha_3 x^3 + \cdots)y'' + (\beta_1 x + \beta_2 x^2 + \cdots)y' + (\gamma_0 + \gamma_1 x + \cdots)y = 0, \quad \alpha_2 \neq 0, \tag{2-234b}$$

where the coefficients are power series converging for $|x| < b$ (they may reduce to polynomials), and we assume that we have cancelled common factors of form x^q (q a positive integer). Such equations are said to have a *regular singular point* at $x = 0$. In Example 1 we have form (2–234b) with $\alpha_2 = 4$, $\beta_1 = 4$, $\gamma_0 = -1$. When Eq. (2–230) has a singular point at $x = 0$ but cannot be written in one of the two forms (2–234a) or (2–234b), the equation is said to have an *irregular singular point* at $x = 0$.

For equations with regular singular points at $x = 0$, we can always obtain at least one solution $y_1(x)$ of form $y = x^m(1 + c_1 x + \cdots)$, and the series converges for $|x| < r^*$, for some $r^* > 0$. However, there may be difficulty in obtaining a second solution $y_2(x)$ (linearly independent of the first). We can show that $y_2(x)$ is always obtainable in the form

$$y_2(x) = cy_1(x)\ln x + x^m(b_0 + b_1 x + \cdots),\qquad(2\text{–}235)$$

where c is a constant, which may be 0.

In general, substitution of $y = x^m(c_0 + c_1 x + \cdots)$ in the differential equation leads to an *indicial equation* $f(m) = 0$ of second degree for m. When the roots m_1, m_2 of this equation are distinct and do not differ by an integer, two linearly independent solutions of form $x^m(1 + c_1 x + \cdots)$ are obtained: one has $m = m_1$, the other has $m = m_2$. Thus for this case the constant c in (2–235) is 0. When $m_1 = m_2$, we get $y_1(x)$ of form $x^m(1 + c_1 x + \cdots)$ as before and $y_2(x)$ as in (2–235) with $c \neq 0$. When m_1 and m_2 are distinct but $m_1 - m_2$ is an integer, we obtain $y_1(x) = x^{m_1}(1 + c_1 x + \cdots)$, where m_1 is the larger of the two roots, and $y_2(x)$ as in (2–235) with $m = m_2$ and c possibly 0. This last case is illustrated by Example 1, with $m_1 = 1/2$ and $m_2 = -1/2$.

Most commonly, the differential equation has polynomial coefficients and we get a *two-term* recursion formula for the coefficients. Then the procedures can be simplified as follows:

We write our differential equation in operator notation as

$$[a_0(x)D^2 + a_1(x)D + a_2(x)]\,y = 0.$$

Now consider the effect of replacing y by x^m on the left side. We assume that

$$[a_0(x)D^2 + a_1(x)D + a_2(x)]x^m = f(m)x^{m+h} + g(m)x^{m+h+k}\qquad(2\text{–}236)$$

for certain integers h, k, with $k > 0$. It can be verified that this assumption is equivalent to assuming that we have a two-term recursion formula. Then the equation

$$f(m) = 0\qquad(2\text{–}237)$$

is a quadratic equation in m, the *indicial equation*.

For Example 1 we have

$$(4x^2 D^2 + 4xD + 4x^2 - 1)x^m = (4m^2 - 1)x^m + 4x^{m+2},\qquad(2\text{–}238)$$

so that $f(m) = 4m^2 - 1$ and (2–236) becomes the indicial equation $4m^2 - 1 = 0$.

We now assume that the indicial equation has distinct roots, whose difference is not divisible by k. We can then obtain linearly independent solutions $y_1(x)$, $y_2(x)$ of form $x^m(1 + c_1 x + \cdots)$ as follows. We let

$$\varphi(x, m) = x^m \left[1 - \frac{g(m)}{f(m+k)}x^k + \frac{g(m)g(m+k)}{f(m+k)f(m+2k)}x^{2k} + \cdots \right.$$
$$\left. + (-1)^n x^{nk} \frac{g(m)\,g(m+k)\cdots g(m+(n-1)k)}{f(m+k)f(m+2k)\cdots f(m+nk)} + \cdots \right]$$

$$(2\text{–}239)$$

Then
$$y_1(x) = \varphi(x, m_1), \quad y_2(x) = \varphi(x, m_2).$$

For Example 1, we saw in Eq. (2–238) that $f(m) = 4m^2 - 1 = (2m+1)(2m-1)$, $g(m) = 4, k = 2$. Thus $m_1 = 1/2, m_2 = -1/2$, and $m_1 - m_2 = 1$, which is not divisible by k. We form

$$\varphi(x, m) = x^m \left\{ 1 - \frac{4}{[2(m+2)+1][2(m+2)-1]} x^2 \right.$$
$$+ \frac{16}{[2(m+2)+1][2(m+2)-1][2(m+4)+1][2(m+4)-1]} x^4 + \cdots$$
$$+ \frac{(-1)^n 4^n x^{2n}}{[2(m+2)+1][2(m+2)-1][2(m+4)+1][2(m+4)-1]}$$
$$+ \frac{1}{\cdots [2(m+2n)+1][2(m+2n)-1]} + \cdots \left. \right\}.$$

We take $m = m_1 = 1/2$ and obtain

$$y_1(x) = \varphi(x, m_1)$$
$$= x^{1/2} \left[1 - \frac{4}{6 \cdot 4} x^2 + \frac{16}{6 \cdot 4 \cdot 10 \cdot 8} x^4 + \cdots + \frac{(-1)^n 4^n x^{2n}}{6 \cdot 4 \cdot 10 \cdot 8 \cdots (4n+2)(4n)} + \cdots \right].$$

This is the same as the solution $y_1(x)$ found above. Similarly, we take $m = m_2 = -1/2$ and obtain $y_2(x)$ as above.

In general, the series for $y_1(x), y_2(x)$ can be shown to converge for $|x| < r^*$ by the ratio test, where $r^* = \infty$ if $g(m)$ is of degree 0 or 1 in m. When $g(m)$ is of degree 2,

$$r^* = \left| \frac{f_0}{g_0} \right|^{1/k}, \quad \text{where} \quad f(m) = f_0 m^2 + \cdots, \ g(m) = g_0 m^2 + \cdots$$

Because of the factor x^m, the solutions may not be valid for $x = 0$, as in Example 1. However, the roots m_1, m_2 may even be positive integers, so that the expected trouble for $x = 0$ may disappear.

REMARK: The methods described apply also when $x = 0$ is not a singular point, provided only that (2–236) is valid. We find that the roots of the indicial equation are 0 and 1. A point that is not a singular point is called an *ordinary point*.

EXAMPLE 2 $2x^2 y'' + (x^2 - x)y' + y = 0$. There is a regular singular point at $x = 0$. We find

$$[2x^2 D^2 + (x^2 - x)D + 1] x^m = x^m (2m^2 - 3m + 1) + m x^{m+1}.$$

Hence
$$f(m) = 2m^2 - 3m + 1 = (2m-1)(m-1), \quad g(m) = m, \quad k = 1, \quad h = 0,$$

$$\varphi(x, m) = x^m \left[1 + \sum_{s=1}^{\infty} (-1)^s x^s \frac{m(m+1) \cdots (m+s-1)}{(2m+1)(m) \cdots (2m+2s-1)(m+s-1)} \right].$$

The indicial equation has roots $1/2$ and 1. Their difference is not an integral multiple of k, hence we obtain two solutions:

$$y_1 = \varphi\left(x, \frac{1}{2}\right) = x^{1/2}\left[1 + \sum_{s=1}^{\infty}(-1)^s x^s \frac{\frac{1\cdot3}{2\cdot2}\cdots\left(\frac{2s-1}{2}\right)}{(2\cdot\frac{1}{2})(4\cdot\frac{3}{2})\cdots(2s)\left(\frac{2s-1}{2}\right)}\right]$$

$$= x^{1/2}\left[1 + \sum_{s=1}^{\infty}(-1)^s x^s \frac{1}{2^s s!}\right];$$

$$y_2 = \varphi(x, 1) = x\left[1 + \sum_{s=1}^{\infty}(-1)^s x^s \frac{1\cdot2\cdots s}{(3\cdot1)(5\cdot2)\cdots(2s+1)s}\right]$$

$$= x\left[1 + \sum_{s=1}^{\infty}(-1)^s x^s \frac{1}{3\cdot5\cdots(2s+1)}\right].$$

Since $g(m)$ is of degree 1, the series converge for all x and the general solution is

$$y = c_1 y_1(x) + c_2 y_2(x)$$

for all x except 0. We note that $y_2(x)$ is also valid as a solution at $x = 0$. (For further discussion see ODE, pp. 366–386, and WHITTAKER–WATSON, Chapter X.) ◄

==================================**Problems (Section 2-23)**==================================

1. Determine whether there is a singular point at $x = 0$ and, if so, whether it is regular or not:

   a) $x^2 y'' + (1 + x)y' + 2y = 0$ **b)** $(x + x^3)y'' + (2 - x)y' + xy = 0$
   c) $(1 + x^2)y'' + (x + 2x^2)y' + y = 0$
   d) $(2x^2 - x^3)y' + (x + 3x^2)y' + (1 + x)y = 0$
   e) $e^x y'' + xy' + y = 0$ **f)** $\sin xy'' - y' + 2y = 0$

2. To determine whether an equation $a_0(x)y'' + a_1(x)y' + a_2(x)y = 0$ has a regular singular point at $x = x_0$, where $a_0(x_0) = 0$, we can set $t = x - x_0$ and rewrite the equation in terms of y as a function of t; the resulting equation has a regular singular point at $t = 0$ if and only if the original equation has a regular singular point at x_0. Equivalently, we can expand the coefficients in the given equation in powers of $x - x_0$:

$$a_0(x) = \sum_{n=0}^{\infty}\alpha_n(x - x_0)^n, \quad a_1(x) = \sum_{n=0}^{\infty}\beta_n(x - x_0)^n, \quad a_2(x) = \sum_{n=0}^{\infty}\gamma_n(x - x_0)^n.$$

   Then x_0 is a regular singular point if $\alpha_0 = 0$, $\alpha_1 \neq 0$ or if $\alpha_0 = 0$, $\alpha_1 = 0$, $\alpha_2 \neq 0$, and $\beta_0 = 0$ (common factor $(x - x_0)^q$ having been cancelled).

   Determine whether there is a regular singular point at the points named:

   a) $(x - 1)y'' + xy' + y = 0$, $x = 1$
   b) $(x - 1)^2 y'' + (1 + x - 2x^3)y' + (1 + x)y = 0$ at $x = 1$
   c) the Legendre equation (2–222) at $x = 1$ and $x = -1$
   d) the hypergeometric equation (2–223) at $x = 0$ and $x = 1$

  e) the Bessel equation (2–224) at $x = 0$

  f) $\cos^2 xy'' + \sin xy' + y = 0$ at $x = \pi/2$

3. Verify that the equation has a regular singular point at $x = 0$ and that replacement of y by x^m on the left leads to $f(m)x^{m+h} + g(m)x^{m+h+k}$ for the given $f(m)$, $g(m)$, h, and k; obtain linearly independent solutions $y_1(x)$, $y_2(x)$ of the form x^m times a power series in x:

  a) $x^2 y'' + (x + 2x^4)y' + (x^3 - 1)y = 0$, $f(m) = m^2 - 1$, $g(m) = 2m + 1$, $h = 0$, $k = 3$

  b) $(x + x^3)y'' + (4 + x^2)y' + xy = 0$, $f(m) = m^2 + 3m$, $g(m) = m^2 + 1$, $h = -1$, $k = 2$

  c) $xy'' + (2 + 2x^3)y' + 3x^2 y = 0$, $f(m) = m^2 + m$, $g(m) = 2m + 3$, $h = -1$, $k = 3$

  d) $x^2 y'' + (x^3 - 4x)y' + 4y = 0$, $f(m) = (m - 4)(m - 1)$, $g(m) = m$, $h = 0$, $k = 2$

4. Show that the equation $x^3 y'' + y' - 2xy = 0$ has an irregular singular point at $x = 0$. Try to obtain a solution of the form $y = x^m$ times a power series and show that the power series diverges for all x.

5. *Cauchy's linear equation of second order.* This is the equation $a_0 x^2 y'' + a_1 xy' + a_2 y = 0$, where a_0, a_1, a_2 are constants and $a_0 \neq 0$.

  a) Show that there are solutions of the form $y = x^m$, provided m satisfies a certain quadratic equation.

  b) When the roots of this quadratic equation are equal, show that there are solutions of the form $y = x^m$ and $y = x^m \ln x$.

  c) Show that the substitution $x = e^t$ converts the equation into one with constant coefficients for y as a function of t.

6. Find the general solution (see Problem 5):

  a) $x^2 y'' + 4xy' + 2y = 0$ b) $x^2 y'' + xy' - y = 0$

  c) $x^2 y'' - 3xy' + 4y = 0$ d) $x^2 y'' + xy' + y = 0$

7. *Solutions valid for large x.* For many problems of the type considered it is important to have series solutions in powers of $1/x$. They are in general easier to evaluate for large positive or negative x than are the series in powers of x. To obtain such solutions we can set $t = 1/x$ in the differential equation and obtain a new equation for y in terms of t. If this has an ordinary point or a regular singular point at $t = 0$, then we can obtain two linearly independent solutions in terms of power series in t. Finally we can replace t by $1/x$ to obtain the desired solutions in terms of x. Apply the method described to obtain such solutions:

  a) $x^5 y'' + 2x^4 y' + (1 - x^3)y = 0$

  b) $x^4 y'' + (2x^3 + x)y' + y = 0$

Chapter 3 Fourier Series

3–1 INTRODUCTION

Fourier series are involved in a common phenomenon—musical sound. The basic element is simple harmonic motion described by a function

$$x = A \sin (\omega t + \alpha).$$

Here x can be the displacement of one end of a tuning fork from equilibrium; as usual, t is time, ω is frequency (radians per unit time), A is amplitude, α is the initial phase. The sound is a pure tone of frequency ω. For $\omega = 440 \times 2\pi$ radians per second there are 440 cycles per second and the sound is the standard "A" of the violin A string.

We could make a visual image of the vibration by attaching a pen to the tuning fork and moving a recording paper along, so that the pen traces the sine curve on the paper. A phonograph record, when examined under a microscope, also displays a visual record of musical notes.

The sound of a musical instrument is normally much more complex than that of a simple sine wave. A sensitive ear detects a fundamental tone and "overtones" of varying strength. The overtones are notes related to the fundamental tone in a simple way: octave, octave plus a fifth, two octaves, etc. The corresponding frequencies are $2\omega, 3\omega, 4\omega$, etc., where ω is the fundamental frequency. Accordingly, a visual record of the sound is the graph of a function

$$x = A_1 \sin (\omega t + \alpha_1) + A_2 \sin (2\omega t + \alpha_2) + \cdots + A_n \sin (n\omega t + \alpha_n) + \cdots \tag{3-10}$$

This is illustrated in Fig. 3–1 for the function

$$x = 10 \sin t + 5 \sin 2t + \sin 3t. \tag{3-11}$$

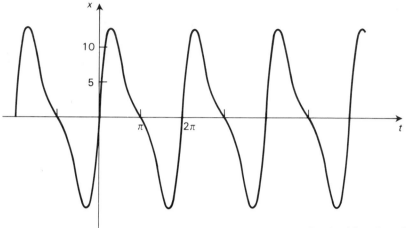

Fig. 3–1. Graph of function of Eq. (3–11).

As the graph of Fig. 3–1 illustrates, the function $x(t)$ is *periodic*; values are repeated after a certain time interval, the *period*. For Eq. (3–11), the period is 2π; for the general vibration of Eq. (3–10), the period is $2\pi/\omega$, since each term repeats after this interval.

Periodic functions occur in many other physical situations. For example, the motion of the earth about the sun is periodic (to very high accuracy), with a period of one year; accordingly, the coordinates of the earth in an appropriate coordinate system are periodic functions of time. In biology, the heart is a mechanism which, under normal conditions, performs periodically.

There are also functions with spatial periodicity, that is, periodic with respect to a space coordinate, say x. This is illustrated by the electrostatic potential in a crystal, which is periodic along each line. Functions defined on a circle will automatically be periodic functions of an angular coordinate θ.

The experience with musical sound suggests that every periodic function can be represented as a series similar to (3–10). It is convenient to standardize the period as 2π and to write the function, say $f(x)$, in the form

$$\frac{a_0}{2} + a_1 \cos x + b_1 \sin x + \cdots + a_n \cos nx + b_n \sin nx + \cdots, \qquad (3\text{–}12)$$

where the a_n and b_n are constants. If x is time, (3–12) represents a vibration with period 2π units about the equilibrium state $y = a_0/2$.

The series (3–12) is called the *Fourier series* of $f(x)$. We shall see that, under very general conditions, an arbitrary function $f(x)$ of period 2π can be represented by a convergent Fourier series.

However, the value of Fourier series goes far beyond representation of periodic functions. An essentially arbitrary function $f(x)$ defined on the interval $0 \leqslant x \leqslant 2\pi$ can be represented by a Fourier series (3–12). This is not surprising, since such a function can always be extended to all x by repeating it periodically outside the given interval. For a variety of physical problems, such as vibrations of strings, heat conduction along a rod, wave motion, the Fourier series representation of a function $f(x)$ arises naturally and is most helpful in understanding the physics.

3–2 TRIGONOMETRIC SERIES AND FOURIER SERIES

A series of form (3–12) is called a *trigonometric series*. If for all x the series converges to $f(x)$, then

$$f(x) = \frac{a_0}{2} + a_1 \cos x + b_1 \sin x + \cdots + a_n \cos nx + b_n \sin nx + \cdots \quad (3\text{–}20)$$

If we multiply both sides by $\cos mx$ (with m a positive integer or 0) and integrate from $-\pi$ to π, then we obtain

$$\int_{-\pi}^{\pi} f(x) \cos mx \, dx = \frac{a_0}{2} \int_{-\pi}^{\pi} \cos mx \, dx + a_1 \int_{-\pi}^{\pi} \cos x \cos mx \, dx + \cdots$$

The integrals on the right are easily evaluated with the aid of the identities:

$$\sin A \cos B = \frac{1}{2} \Big[\sin (A + B) + \sin (A - B) \Big],$$

$$\cos A \cos B = \frac{1}{2} \Big[\cos (A + B) + \cos (A - B) \Big], \quad (3\text{–}21)$$

$$\sin A \sin B = \frac{1}{2} \Big[\cos (A - B) - \cos (A + B) \Big],$$

and we find (see Problems 2 and 3 below) that

$$\int_{-\pi}^{\pi} \cos nx \cos mx \, dx = \begin{cases} 0 & \text{for } m \neq n, \\ \pi & \text{for } m = n > 0, \end{cases}$$

$$\int_{-\pi}^{\pi} \sin nx \cos mx \, dx = 0 \quad \text{for all integers } m, n.$$

Accordingly, for $m = 1, 2, \ldots,$

$$\int_{-\pi}^{\pi} f(x) \cos mx \, dx = \pi a_m,$$

or

$$a_n = \frac{1}{\pi} \int_{-\pi}^{\pi} f(x) \cos nx \, dx, \quad n = 1, 2, \ldots \quad (3\text{–}22)$$

If we multiply (3–20) by $\cos 0x = 1$ and proceed in the same way, we find that (3–22) is valid also for $n = 0$. (The notation $a_0/2$ for the first term in (3–20) is chosen to make (3–22) valid for all n.) If we multiply (3–20) by $\sin mx$ and use the rules above and the rule (Problem 2 below)

$$\int_{-\pi}^{\pi} \sin nx \sin mx \, dx = \begin{cases} 0 & \text{for } n \neq m, \\ \pi & \text{for } n = m, \end{cases}$$

we find that

$$b_n = \frac{1}{\pi} \int_{-\pi}^{\pi} f(x) \sin nx \, dx, \quad n = 1, 2, \ldots . \tag{3–23}$$

DEFINITION: Let $f(x)$ be defined for $-\pi \leqslant x \leqslant \pi$. Then by the *Fourier series* of f we mean the series

$$\frac{a_0}{2} + a_1 \cos x + b_1 \sin x + \cdots + a_n \cos nx + b_n \sin nx + \cdots,$$

where

$$a_n = \frac{1}{\pi} \int_{-\pi}^{\pi} f(x) \cos nx \, dx, \quad b_n = \frac{1}{\pi} \int_{-\pi}^{\pi} f(x) \sin nx \, dx \tag{3–24}$$

for $n = 0, 1, 2, \ldots$ (except that b_0 is undefined).

Thus the Fourier series of a function f is a trigonometric series obtained from f by special formulas, just as a Taylor series of a function f is a power series obtained from f by special formulas. We saw that every power series with a nonzero convergence radius is a Taylor series. Similarly, *every trigonometric series satisfying reasonable convergence conditions is a Fourier series*. In particular, every trigonometric series converging uniformly to $f(x)$ for $-\pi \leqslant x \leqslant \pi$ is the Fourier series of $f(x)$. This follows at once from the theorems on uniformly convergent series in Section 2–15, since the operations leading from (3–20) to (3–22) and (3–23) are then valid. Just as for power series, in Section 2–16, we now conclude that two trigonometric series (say, uniformly convergent) with the same sum $f(x)$ must be the same series, that is, we have a *rule for comparing coefficients*:

$$\frac{a_0}{2} + a_1 \cos x + b_1 \sin x + \cdots + a_n \cos nx + b_n \sin x + \cdots$$

$$\equiv \frac{a_0'}{2} + a_1' \cos x + b_1' \sin x + \cdots + a_n' \cos nx + b_n' \sin nx + \cdots$$

implies that

$$a_0 = a_0', \quad a_n = a_n', \quad b_n = b_n' \text{ for } n = 1, 2, \ldots$$

The Fourier series of a function $f(x)$ is defined whenever all the integrals in (3–24) have meaning. This is certainly the case if $f(x)$ is continuous on the interval $-\pi \leqslant x \leqslant \pi$. However, the integrals also have meaning when f has *jump discontinuities*, as suggested in Fig. 3–2. Here f is continuous except at $x_1, \ldots x_N$, at which f has limits from the left and from the right. We call such a function *piecewise continuous*. At the jump points it is not obvious what value to assign to f. We shall always use the *average of left and right limits*, so that

$$f(x_k) = \frac{1}{2}\left[\lim_{x \to x_k^+} f(x) + \lim_{x \to x_k^-} f(x) \right], \quad k = 1, \ldots, N. \qquad (3\text{–}25)$$

Not only does the Fourier series of such a function have meaning. Under quite general conditions the series even converges to the function, as we shall see.

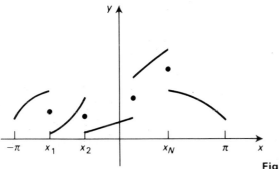

Fig. 3–2. Piecewise continuous function.

EXAMPLE 1   *Square wave*: Let $f(x) = 0$ for $-\pi \leqslant x < -\pi/2$, $f(x) = 1$ for $-\pi/2 < x < \pi/2$, $f(x) = 0$ for $\pi/2 < x \leqslant \pi$. We let $f(-\pi/2) = 1/2$, $f(\pi/2) = 1/2$ in accordance with (3–25). We can define f outside the given interval to obtain a function of period 2π, as suggested in Fig. 3–3.

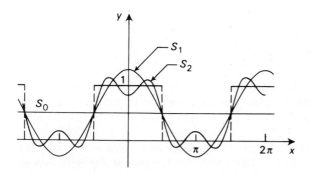

Fig. 3–3. Function of Example 1.

Here we find

$$a_0 = \frac{1}{\pi} \int_{-\pi}^{\pi} f(x)\, dx = \frac{1}{\pi} \int_{-\pi/2}^{\pi/2} dx = 1,$$

$$a_n = \frac{1}{\pi} \int_{-\pi/2}^{\pi/2} \cos nx\, dx = \frac{1}{\pi} \frac{\sin nx}{n} \Big|_{-\pi/2}^{\pi/2} = \frac{1}{\pi} \frac{\sin(n\pi/2) - \sin(-n\pi/2)}{n}$$

Now $\sin n\pi/2 = 1$ for $n = 1,\ 5,\ 9,\ \ldots$, $\sin n\pi/2 = -1$ for $n = 3,\ 7,\ 11,\ \ldots$, $\sin n\pi/2 = 0$ otherwise. Also $\sin(-n\pi/2) = -\sin(n\pi/2)$. Therefore,

$$a_n = \begin{cases} -\dfrac{2}{n\pi}, & \text{for } n = 3,\ 7,\ 11,\ \ldots \\[2mm] \dfrac{2}{n\pi}, & \text{for } n = 1,\ 5,\ 9,\ \ldots \\[2mm] 0, & \text{for } n = 2,\ 4,\ 6,\ \ldots \end{cases}$$

Next, for $n = 1,\ 2,\ \ldots$, we have

$$b_n = \frac{1}{\pi} \int_{-\pi/2}^{\pi/2} \sin nx\, dx = \frac{1}{\pi} \frac{-\cos nx}{n} \Big|_{-\pi/2}^{\pi/2} = \frac{-\cos(n\pi/2) + \cos(-n\pi/2)}{n\pi};$$

since $\cos\alpha = \cos(-\alpha)$ for all α, we get $b_n = 0$ for all n. Thus the Fourier series of f is the series

$$\frac{1}{2} + \frac{2}{\pi}\cos x - \frac{2}{3\pi}\cos 3x + \frac{2}{5\pi}\cos 5x - \frac{2}{7\pi}\cos 7x + \cdots$$

or

$$\frac{1}{2} - \frac{2}{\pi} \sum_{n=1}^{\infty} (-1)^n \frac{\cos(2n-1)x}{2n-1}.$$

We denote by S_n the partial sum of the Fourier series, through the terms in $\cos nx$, $\sin nx$, so that

$$S_0 = \frac{1}{2}, \quad S_1 = \frac{1}{2} + \frac{2}{\pi}\cos x, \quad S_3 = \frac{1}{2} + \frac{2}{\pi}\cos x - \frac{2}{3\pi}\cos 3x, \quad \text{etc.}$$

In Fig. 3–3, several partial sums are graphed, and we can see how they are approximating f. ◀

We remark that by (3–24) the constant term $a_0/2$ is the *average value* of $f(x)$ over the interval $-\pi \leqslant x \leqslant \pi$. In many cases this value can be found at once, as in Example 1, where it is clearly $1/2$.

EXAMPLE 2 Let $f(x) = \pi x - x^2$ for $0 \leqslant x \leqslant \pi$ and $f(x) = x^2 + \pi x$ for $-\pi \leqslant x \leqslant 0$. The function and its periodic extension (of period 2π) is shown in Fig. 3–4. The function resembles a sine wave, and hence we expect a good

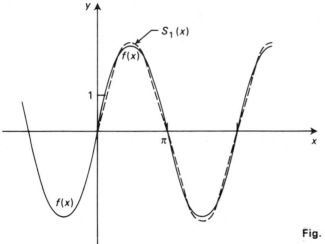

Fig. 3–4. Function of Example 2.

approximation by partial sums of the Fourier series. We find

$$a_0 = \frac{1}{\pi} \int_{-\pi}^{0} (x^2 + \pi x)\, dx + \frac{1}{\pi} \int_{0}^{\pi} (\pi x - x^2)\, dx = 0,$$

$$a_n = \frac{1}{\pi} \int_{-\pi}^{0} (x^2 + \pi x) \cos nx\, dx + \frac{1}{\pi} \int_{0}^{\pi} (\pi x - x^2) \cos nx\, dx = 0, \quad n = 1, 2, \ldots,$$

$$b_n = \frac{1}{\pi} \int_{-\pi}^{0} (x^2 + \pi x) \sin nx\, dx + \frac{1}{\pi} \int_{0}^{\pi} (\pi x - x^2) \sin nx\, dx$$

$$= \begin{cases} 0 & \text{for } n \text{ even}, \\ \dfrac{8}{\pi n^3} & \text{for } n \text{ odd}. \end{cases}$$

(We use integration by parts twice to evaluate the integrals.) Accordingly, the Fourier series of f is the series

$$\frac{8}{\pi} \sin x + \frac{8}{\pi} \frac{\sin 3x}{27} + \frac{8}{\pi} \frac{\sin 5x}{125} + \cdots + \frac{8}{\pi} \frac{\sin (2n+1)x}{(2n+1)^3} + \cdots$$

The first partial sum is shown in Fig. 3–4. ◀

Odd and even functions

A function f defined for $-a \leqslant x \leqslant a$ is said to be *even* if $f(-x) = f(x)$ on the interval, and to be *odd* if $f(-x) = -f(x)$. By symmetry we find that

$$\int_{-a}^{a} f(x)\, dx = \begin{cases} 0 & \text{for } f \text{ odd}, \\ 2 \int_{0}^{a} f(x)\, dx & \text{for } f \text{ even}. \end{cases}$$

Now if f is defined for $-\pi \leqslant x \leqslant \pi$ and is *even*, then so is $f(x)\cos nx$, whereas $f(x)\sin nx$ is odd. We conclude that, for all n,

$$b_n = 0 \qquad \text{and} \qquad a_n = \frac{2}{\pi} \int_0^\pi f(x)\cos nx\, dx \quad \text{(for } f \text{ even).} \qquad (3\text{-}26)$$

Similarly, we find that, for all n,

$$a_n = 0 \qquad \text{and} \qquad b_n = \frac{2}{\pi} \int_0^\pi f(x)\sin nx\, dx \quad \text{(for } f \text{ odd).} \qquad (3\text{-}27)$$

This shortens the calculations. Examples 1 and 2 illustrate the cases of an even and an odd function, respectively. If

$$f(x) = \text{const} + \text{Odd function,}$$

then we see that (3–27) still holds except that a_0 need not be 0 (in fact, $a_0/2$ equals the constant).

Rate of convergence

It will be seen that the smoother the graph of f, the more rapidly do the partial sums S_n approach f. This is illustrated by Examples 1 and 2. In Example 1 the jumps lead to slow convergence, especially near the jump points; at the jump point itself, the series must converge to the average value (3–25), and that is clearly satisfied for Example 1, where $S_n(x) = 1/2$ (the average value) at each jump point. In Example 2, functions f and f' are continuous and periodic (f'' has a jump at 0, $\pm\pi$, . . .), and this leads to rapid convergence. It should be said that if f is continuous for $-\pi < x \leqslant \pi$, but $f(-\pi) \neq f(\pi)$, then the periodic extension of f must have jumps at π, $-\pi$, 3π, -3π, . . . (see Fig. 3–5), and these will slow down the convergence, especially near the jump points. The value of the function at each point must be replaced by the average of the left and right limits, as before.

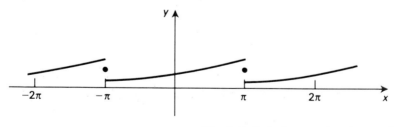

Fig. 3–5. Periodic extension showing how jumps arise.

=================== **Problems (Section 3–2)** ===================

1. Graph each of the following functions and the periodic extension of period 2π; find the Fourier series of the function and graph the partial sums named:
   a) $f(x) = \pi - x$, $-\pi \leqslant x \leqslant \pi$; S_1, S_2, S_3

b) $f(x) = 0$ for $-\pi \leqslant x < 0$, $f(x) = 1$ for $0 \leqslant x \leqslant \pi$; S_1, S_2, S_3

c) $f(x) = x^2$ for $-\pi \leqslant x \leqslant \pi$; S_1, S_2

d) $f(x) = 0$ for $-\pi \leqslant x \leqslant 0$, $f(x) = x$ for $0 \leqslant x \leqslant \pi$; S_1, S_2

e) $f(x) = e^x$, $-\pi \leqslant x \leqslant \pi$; S_1, S_2

f) $f(x) = \cos^3 x$, $-\pi \leqslant x \leqslant \pi$; S_1, S_3

2. Prove that for integers m, n:

a) $\displaystyle\int_{-\pi}^{\pi} \cos nx \cos mx \, dx = 0,\ m \neq n$

b) $\displaystyle\int_{-\pi}^{\pi} \sin nx \sin mx \, dx = 0,\ m \neq n$

c) $\displaystyle\int_{-\pi}^{\pi} \sin nx \cos mx \, dx = 0$

3. Prove that for a positive integer n:

$$\int_{-\pi}^{\pi} \sin^2 nx \, dx = \int_{-\pi}^{\pi} \cos^2 nx \, dx = \pi$$

4. Show that the Fourier series of each function is uniformly convergent for all x:
a) Function of Example 2 above   b) Function of Problem 1(c)

5. Convergent power series $\Sigma a_n z^n$ give rise to convergent Fourier series by substituting

$$z = r(\cos\theta + i\sin\theta),\quad z^2 = r^2(\cos 2\theta + i\sin 2\theta),\ \ldots$$

with r any fixed number less than r^*, the radius of convergence, and by taking real and imaginary parts. Obtain the following Fourier series representations

a) $e^{r\cos\theta}\cos(r\sin\theta) = 1 + \dfrac{r\cos\theta}{1!} + \dfrac{r^2\cos 2\theta}{2!} + \cdots + \dfrac{r^n\cos n\theta}{n!} + \cdots$

and

$$e^{r\cos\theta}\sin(r\sin\theta) = \frac{r\sin\theta}{1!} + \frac{r^2\sin 2\theta}{2!} + \cdots + \frac{r^n\sin n\theta}{n!} + \cdots$$

for $0 \leqslant r < \infty$.

HINT: Start with the power series for e^z.

b) $\dfrac{1 - r\cos\theta}{1 + r^2 - 2r\cos\theta} = 1 + r\cos\theta + r^2\cos 2\theta + \cdots + r^n\cos n\theta + \cdots,$

$\dfrac{r\sin\theta}{1 + r^2 - 2r\cos\theta} = r\sin\theta + r^2\sin 2\theta + \cdots + r^n\sin n\theta + \cdots,$

for $0 \leqslant r < 1$.

HINT: Start with the series for $1/(1 - z)$.

c) $\dfrac{1 - \cos(n+1)\theta - \cos\theta + \cos n\theta}{2 - 2\cos\theta} = 1 + \cos\theta + \cdots + \cos n\theta,$

$\dfrac{\sin n\theta + \sin\theta - \sin(n+1)\theta}{2 - 2\cos\theta} = \sin\theta + \cdots + \sin n\theta$

for $\theta \neq 0, \pm 2\pi, \ldots$.

HINT: Start with the formula for $1 + z + \ldots + z^n$.

d) From the first result of part (c) deduce that for $\theta \neq 0, \pm 2\pi, \ldots$

$$\frac{\sin(n+\frac{1}{2})\theta}{2\sin\frac{1}{2}\theta} = \frac{1}{2} + \cos\theta + \cdots + \cos n\theta.$$

6. *Musical experiments*

   a) Press, without sounding, the piano keys for C above middle C, G above the C just pressed, and C above this G. With all three keys pressed down, strike middle C. The other notes should then also be heard. They correspond to the frequencies $2\omega, 3\omega, 4\omega$, where ω is the frequency of middle C (262 cycles per sec). The response of the other notes is a resonance effect (see Section 1–14) due to the terms in the Fourier series for middle C of the corresponding frequencies.

   b) A flute has a sound with very weak overtones; an oboe, by contrast, has very strong overtones. What does this imply about the Fourier series of a flute sound and that of an oboe sound? If possible, listen to these instruments and try to hear the difference in overtones.

   c) The human ear responds to musical sound by decomposing it as a Fourier series and reacting in a different way to each frequency. What does this imply about the way different persons respond to the same sound? What does the *auditory limit* mean?

$3\text{--}3$ CONVERGENCE OF FOURIER SERIES

Let $f(x)$ be defined for all x and have period 2π. Then, under very general conditions, the Fourier series of f converges to $f(x)$ for all x. Here we describe a set of conditions that assures such convergence. We also give some indications as to *why* the convergence is to be expected and references for detailed proofs.

The conditions are illustrated in Fig. 3–6. Function f is continuous in each interval of length 2π except for a finite number of jump discontinuities (as at x_1 and x_3 in the figure), at which the value of f is the average of its limits from the left and from the right. Furthermore, in each interval of length 2π, function f has a continuous derivative, except at the jump points and at a finite number of corners (as at x_2 and x_4 in the figure). At the jump points and at the corners there is a limiting

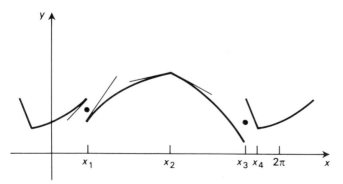

Fig. 3–6. Piecewise-smooth periodic function.

value for the derivative from the right and from the left (suggested by the tangent lines drawn for $x = x_1$ and $x = x_2$). Function f satisfying these conditions is called a *piecewise smooth function*. Thus *the Fourier series of a piecewise smooth function $f(x)$ of period 2π converges to $f(x)$ for all x.*

Uniform convergence

For a function f as in Fig. 3–6 it is too much to expect that the Fourier series of f converges uniformly when $0 \leqslant x \leqslant 2\pi$, since the sum of a uniformly convergent series of continuous functions must itself be continuous (Section 2–15), while our function f in Fig. 3–6 is allowed to have jump discontinuities. However, if f happens to have no jump discontinuities (even though f has corners), then the convergence must be uniform. Furthermore, even when f has jumps, the convergence is uniform in each closed interval $a \leqslant x \leqslant b$ containing no jump points.

These results are illustrated by Examples 1 and 2 of Section 3–2. In Example 1 (Fig. 3–3), the partial sums all agree with f at the jump points (average value); in an interval such as $-\pi/4 \leqslant x \leqslant \pi/4$, containing no jump points, the graphs suggest that the maximum error is approaching 0 as the number of terms becomes infinite, that is, they suggest uniform convergence. In Example 2 (Fig. 3–4) there are no jumps, and the convergence appears to be uniform. In fact, the general term of the Fourier series is

$$\frac{8}{\pi} \frac{\sin (2n+1)x}{(2n+1)^3}$$

and hence we can apply the Weierstrass M-test (Section 2–15), with

$$M_n = \frac{8}{\pi (2n+1)^3}$$

to conclude that the convergence is uniform.

Idea of convergence proof

The idea is based on the physical concept of the *Dirac delta function* $\delta(t)$. This function arises naturally in discussing density. For mass distributed along the x-axis, there would be a density $\rho(x)$ such that

$$\int_a^b \rho(x)\,dx = \text{Mass between } a \text{ and } b.$$

We now assume that the total mass is 1 (in appropriate units) and follow a limit process, concentrating the mass more and more near $x = 0$. The corresponding density is then as in Fig. 3–7. The delta function $\delta(x)$ is defined as the limiting density when the mass approaches concentration at the point $x = 0$. Since the total mass is 1, for $a < 0 < b$ we should have

$$\int_a^b \delta(x)\,dx = 1. \qquad\qquad (3\text{–}30)$$

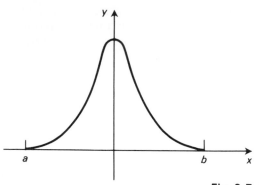

Fig. 3-7. Approximation $\rho(x)$ to delta function $\delta(x)$.

However, $\delta(x) = 0$ for all x except 0 and $\delta(0)$ must be $+\infty$. These somewhat remarkable properties can be given a sound basis. We can think of $\delta(x)$ as the density for a particle of unit mass at $x = 0$, or as the limiting case of the density $\rho(x)$ when the width of the pulse goes to 0.

For a particle of unit mass at x_0, the corresponding density is $\delta(x - x_0)$. For several particles of masses $m_1, \ldots, m_n$ at $x_1, \ldots, x_n$, respectively, the density is $m_1 \delta(x - x_1) + \cdots + m_n \delta(x - x_n)$.

In mechanics we deal with moments of such mass distributions. The integral

$$\int_a^b x^k \rho(x)\, dx$$

is the kth moment about the origin. In general, there is a need for integrals of form

$$\int_a^b f(x)\rho(x)\, dx$$

with continuous f. If we now have only the single particle of unit mass at x_0, on the interval $x_0 - c < x < x_0 + c$, then $\rho(x) = \delta(x - x_0)$ and the kth moment about 0 is simply x_0^k (times the unit of mass); this is the kth moment of a particle of unit mass. Thus

$$\int_{x_0 - c}^{x_0 + c} x^k \delta(x - x_0)\, dx = x_0^k.$$

In the same way, for a general continuous $f(x)$ and every $c > 0$ we have

$$\int_{x_0 - c}^{x_0 + c} f(x)\delta(x - x_0)\, dx = f(x_0)$$

or, if we set $t = x - x_0$,

$$\int_{-c}^{c} f(x_0 + t)\delta(t)\, dt = f(x_0). \tag{3-31}$$

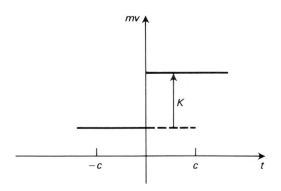

Fig. 3–8. Effect of impulse force.

Instead of integrating from $-c$ to c, we can integrate from a to b, where $a < 0 < b$, as above.

A further example of the delta function is an impulse caused by a force that acts instantaneously. If the force acts at $t = 0$ and is applied along the x-axis to a particle of mass m, we get an equation

$$m\frac{dv}{dt} = K\delta(t),$$

where K is a constant, the magnitude of the impulse. Hence

$$\int_{-c}^{c} m\frac{dv}{dt}\,dt = \int_{-c}^{c} K\,\delta(t)\,dt$$

or

$$mv(c) - mv(-c) = K \quad \text{for every } c > 0.$$

Thus the impulse of magnitude K leads to a jump of K in momentum (Fig. 3–8). Because of this application, $\delta(t)$ is often called the *unit impulse function*.

We now turn to Fourier series making use of the fact that the function

$$P_n(t) = \frac{1}{2\pi} + \frac{1}{\pi}\cos t + \cdots + \frac{1}{\pi}\cos nt \tag{3–32}$$

can be considered to approach $\delta(t)$, for $-\pi \leqslant t \leqslant \pi$, as a limit when $n \to \infty$. More precisely, it can be shown that

$$\lim_{n \to \infty} \int_{-\pi}^{\pi} f(x+t)P_n(t)\,dt = \int_{-\pi}^{\pi} f(x+t)\delta(t)\,dt = f(x), \tag{3–33}$$

provided f is continuous at x and $f'(x)$ exists. A graph of $P_n(t)$ for $n = 5$ is given in Fig. 3–9.

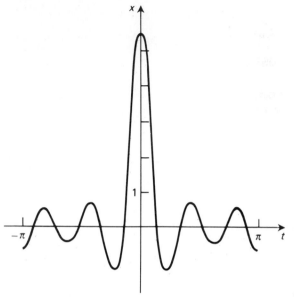

Fig. 3–9. Function $P_n(t)$ for $n = 5$.

Now, the nth partial sum of the Fourier series of f is

$$S_n(x) = \frac{a_0}{2} + a_1 \cos x + b_1 \sin x + \cdots + a_n \cos nx + b_n \sin nx$$

$$= \frac{1}{2\pi} \int_{-\pi}^{\pi} f(u) du + \frac{\cos x}{\pi} \int_{-\pi}^{\pi} f(u) \cos u \, du + \cdots$$

$$= \frac{1}{\pi} \int_{-\pi}^{\pi} f(u) \left(\frac{1}{2} + \cos x \cos u + \sin x \sin u + \cdots \right) du$$

$$= \frac{1}{\pi} \int_{-\pi}^{\pi} f(u) \left[\frac{1}{2} + \cos (u - x) + \cos 2(u - x) + \cdots + \cos n(u - x) \right] du.$$

Accordingly,

$$S_n(x) = \int_{-\pi}^{\pi} f(u) P_n(u - x) du. \tag{3–34}$$

If we set $u = x + t$ and take advantage of periodicity, we conclude (Problem 3 below) that

$$S_n(x) = \int_{-\pi}^{\pi} f(x + t) P_n(t) dt. \tag{3–35}$$

Thus, as in Eq. (3–33),

$$\lim_{n \to \infty} S_n(x) = \int_{-\pi}^{\pi} f(x+t)\delta(t)dt = f(x). \tag{3-36}$$

This gives the asserted convergence, at least wherever f is continuous and $f'(x)$ exists.

A suggestion that $P_n(t)$ does approach $\delta(t)$ as limit, for $-\pi \leqslant t \leqslant \pi$, is provided by the following formal calculation. We seek the Fourier coefficients of the function equal to $\delta(t)$ for $-\pi \leqslant t \leqslant \pi$. By the rule (3–31), with $x_0 = 0$,

$$a_0 = \frac{1}{\pi}\int_{-\pi}^{\pi} \delta(t)dt = \frac{1}{\pi},$$

$$a_n = \frac{1}{\pi}\int_{-\pi}^{\pi} \delta(t)\cos nt\,dt = \frac{1}{\pi}, \quad n = 1, 2, \ldots ,$$

$$b_n = \frac{1}{\pi}\int_{-\pi}^{\pi} \delta(t)\sin nt\,dt = 0, \quad n = 1, 2, \ldots .$$

Thus we expect:

$$\delta(t) = \frac{1}{2\pi} + \frac{1}{\pi}\cos t + \frac{1}{\pi}\cos 2t + \cdots$$

$$= \lim_{n \to \infty}\left(\frac{1}{2\pi} + \frac{1}{\pi}\cos t + \cdots + \frac{1}{\pi}\cos nt\right) = \lim_{n \to \infty} P_n(t).$$

It should be remarked that since $P_n(t)$ has period 2π, its limit must be a *periodic delta function* agreeing with $\delta(t)$ for $-\pi \leqslant t \leqslant \pi$. For proofs of convergence of Fourier series, see AC, pp. 461–483, and JACKSON, Chapter I. For more information on the delta function, see Section 4–9.

═══════════════ **Problems (Section 3–3)** ═══════════════

1. Let $f(x)$ have period 2π and be as given below for $-\pi \leqslant x \leqslant \pi$. Verify that f is piecewise smooth and point out all jump discontinuities and corners in the interval $-\pi \leqslant x \leqslant \pi$.
   a) $f(x) = x$ for $-\pi < x < \pi$, $f(\pi) = f(-\pi) = 0$
   b) $f(x) = 0$ for $-\pi < x \leqslant 0$, $f(x) = x$ for $0 \leqslant x < \pi$, $f(\pi) = f(-\pi) = \pi/2$
   c) $f(x) = |\sin x|$ for $-\pi \leqslant x \leqslant \pi$
   d) $f(x) = x^2$ for $-\pi \leqslant x \leqslant \pi$

2. Let $f(x)$ be piecewise continuous for all x and let f have period $\tau > 0$. Show that

$$\int_a^{a+\tau} f(x)dx = \int_0^\tau f(x)dx,$$

so that the integral on the left has a value independent of a.

HINT: Interpret the integrals as areas under a curve.

3. Derive Eq. (3–35) from Eq. (3–34).

HINT: Set $u = x+t$ as suggested, for fixed x, and apply the result of Problem 2, noting that f and P_n have period 2π.

4. Evaluate the following integrals:

a) $\displaystyle\int_{-1}^{1} e^x \, \delta(x) \, dx$ b) $\displaystyle\int_{-2}^{3} (x^2 - 5x + 7)\delta(x)\, dx$

c) $\displaystyle\int_{0}^{2} e^x \, \delta(x - 1)\, dx$ d) $\displaystyle\int_{-3}^{5} x^2 \, \delta(x - 2)\, dx$

5. a) Prove the trigonometric identity

$$\frac{1}{2} + \cos t + \cdots + \cos nt = \frac{\sin (n + \frac{1}{2})t}{2 \sin (t/2)}.$$

HINT: Multiply the left side by $\sin (t/2)$ and use the identities (3–21) to replace each term $\sin (t/2)\cos t, \ldots, \sin (t/2)\cos nt$ by a sum of two terms. Another proof is given in Problem 5(d) following Section 3–2.

b) With the aid of the result of part (a), graph the function (3–32) for $-\pi \leqslant t \leqslant \pi$ for $n = 5$ and then for $n = 10$ (see Fig. 3–9).

6. Discuss in a qualitative way how the concept of the delta function appears in the following contexts:
   a) a point load in mechanics
   b) a charged particle
   c) a point source of energy (say on a line)
   d) an instantaneous signal

$3-4$ FOURIER COSINE SERIES, FOURIER SINE SERIES

In Section 3–2 we pointed out the special features of the Fourier series of even and odd functions. We now take advantage of these to obtain new types of Fourier series representations.

Let $f(x)$ be given only for $0 \leqslant x \leqslant \pi$ and let f be piecewise smooth on this interval. By reflection of the graph of f in the y-axis, we obtain an *even* function on the interval $-\pi \leqslant x \leqslant \pi$, agreeing with f for $0 \leqslant x \leqslant \pi$. We extend the new function by periodicity to all x to obtain a function f_1 that is *even*, has *period* 2π, and agrees with f for $0 \leqslant x \leqslant \pi$. We call f_1 the *even periodic extension* of f (Fig. 3–10).

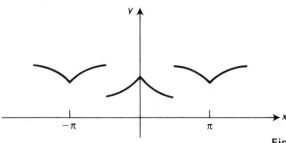

Fig. 3–10. Even periodic extension of f.

Function f_1 satisfies the hypotheses of Section 3–3. Hence the Fourier series of f_1 converges to f_1 for all x. Since f_1 is even, $b_n = 0$ for all n and

$$a_n = \frac{2}{\pi} \int_0^\pi f_1(x) \cos nx \, dx, \quad n = 0, 1, 2, \ldots$$

$$f_1(x) = \frac{a_0}{2} + \sum_{n=1}^\infty a_n \cos nx, \quad \text{all } x.$$

In the formula for a_n, function f_1 can be replaced by f, since $f_1 = f$ for $0 \leqslant x \leqslant \pi$. Also, by the same reason,

$$f(x) = f_1(x) = \frac{a_0}{2} + \sum_{n=1}^\infty a_n \cos nx, \quad \text{for} \quad 0 \leqslant x \leqslant \pi.$$

We therefore conclude that each function $f(x)$, defined and piecewise smooth for $0 \leqslant x \leqslant \pi$, can be expanded in a series involving only cosines:

$$f(x) = \frac{a_0}{2} + \sum_{n=1}^\infty a_n \cos nx, \quad 0 \leqslant x \leqslant \pi,$$

$$\tag{3-40}$$

$$a_n - \frac{2}{\pi} \int_0^\pi f(x) \cos nx \, dx, \quad n - 0, 1, 2, \ldots$$

The series appearing here is called the *Fourier cosine series of f*. The series converges to f for $0 \leqslant x \leqslant \pi$ and converges to the *even* periodic extension of f outside the interval.

In exactly the same way, we can expand the same function f in a series involving only sines:

$$f(x) = \sum_{n=1}^\infty b_n \sin nx,$$

$$\tag{3-41}$$

$$b_n = \frac{2}{\pi} \int_0^\pi f(x) \sin nx \, dx, \quad n = 1, 2, \ldots$$

The series appearing here is called the *Fourier sine series of f*. The series converges to $f(x)$ for $0 \leqslant x \leqslant \pi$ and converges to the *odd periodic extension* of f outside this interval (see Fig. 3–11). It should be noted that jumps may appear at 0 and $\pm\pi$, forcing us to use a new value at these points. Because of oddness and periodicity, the new value is 0. This is also forced on us by the form of the series $\Sigma b_n \sin nx$ that converges to 0 for $x = 0, \pi,$ or $-\pi$.

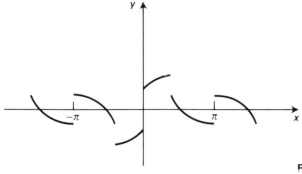

Fig. 3-11. Odd periodic extension.

EXAMPLE 1 Let $f(x) = \pi - x$ for $0 \leqslant x \leqslant \pi$. Then the even periodic extension of f is as shown in Fig. 3–12. We find

$$a_0 = \frac{2}{\pi} \int_0^\pi (\pi - x)\,dx = \pi,$$

$$a_n = \frac{2}{\pi} \int_0^\pi (\pi - x)\cos nx\,dx = \frac{2}{\pi}\frac{1-(-1)^n}{n^2},$$

so that $a_1 = 4/\pi$, $a_2 = 0$, $a_3 = 4/(\pi \cdot 3^2)$, ..., and

$$f(x) = \frac{\pi}{2} + \frac{4}{\pi}\left(\cos x + \frac{\cos 3x}{3^2} + \cdots + \frac{\cos(2n-1)x}{(2n-1)^2} + \cdots\right).$$

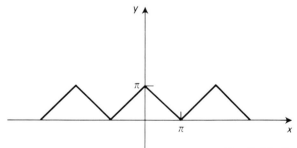

Fig. 3-12. Even periodic extension for Example 1.

The series converges to f for $0 \leqslant x \leqslant \pi$ and to the even periodic extension otherwise. Since there are no jumps, the convergence is uniform for all x. ◄

EXAMPLE 2 We can expand the same function in its Fourier sine series:

$$b_n = \frac{2}{\pi} \int_0^\pi (\pi - x)\sin nx\,dx = \frac{2}{n}, \quad n = 1, 2, \ldots$$

Thus

$$f(x) = 2 \sum_{n=1}^{\infty} \frac{\sin nx}{n}, \quad 0 \leqslant x \leqslant \pi,$$

with $f(0) = 0$ and $f(\pi) = 0$. The series converges to the odd periodic extension of f for all x (Fig. 3–13). The convergence is uniform in each interval $a \leqslant x \leqslant b$ containing no jump point, for example in the interval $\pi/10 \leqslant x \leqslant 2\pi - \pi/10$. ◀

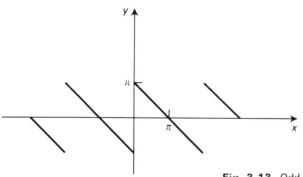

Fig. 3–13. Odd periodic extension for Example 2.

3–5 CHANGE OF PERIOD

If $f(x)$ has period τ, not necessarily 2π, we can reduce f to a function of period 2π by making a change of scale:

$$x_1 = \frac{2\pi}{\tau} x = \omega x, \quad \omega = \frac{2\pi}{\tau},$$

and expressing f in terms of x_1: $f(x) = f(x_1/\omega)$. We obtain a Fourier series:

$$\frac{a_0}{2} + a_1 \cos x_1 + b_1 \sin x_1 + \cdots = \frac{a_0}{2} + a_1 \cos \omega x + b_1 \sin \omega x + \cdots$$

Here

$$a_n = \frac{1}{\pi} \int_{-\pi}^{\pi} f\left(\frac{x_1}{\omega}\right) \cos nx_1 \, dx_1, \quad b_n = \frac{1}{\pi} \int_{-\pi}^{\pi} f\left(\frac{x_1}{\omega}\right) \sin nx_1 \, dx_1.$$

We can also use $x = x_1/\omega$ as integration variable:

$$a_n = \frac{\omega}{\pi} \int_{-\pi/\omega}^{\pi/\omega} f(x) \cos n\omega x \, dx = \frac{2}{\tau} \int_{-\tau/2}^{\tau/2} f(x) \cos n\omega x \, dx,$$

and there is a similar formula for b_n. Thus we define *the Fourier series of f, for period τ*, as the series

$$\frac{a_0}{2} + a_1 \cos \omega x + b_1 \sin \omega x + \cdots + a_n \cos n\omega x + b_n \sin n\omega x + \cdots, \quad (3\text{--}50)$$

where $\omega = 2\pi/\tau$, and

$$a_n = \frac{2}{\tau} \int_{-\tau/2}^{\tau/2} f(x) \cos n\omega x \, dx, \quad b_n = \frac{2}{\tau} \int_{-\tau/2}^{\tau/2} f(x) \sin n\omega x \, dx. \quad (3\text{--}51)$$

Since the integral of a function of period τ over an interval of length τ has the same value for all such intervals (see Problem 2 following Section 3–3), in (3–51) we can integrate from 0 to τ or over any other interval of length τ. Thus Eqs. (3–51) can be replaced by

$$a_n = \frac{2}{\tau} \int_0^{\tau} f(x) \cos \frac{2n\pi x}{\tau} \, dx, \quad b_n = \frac{2}{\tau} \int_0^{\tau} f(x) \sin \frac{2n\pi x}{\tau} \, dx. \quad (3\text{--}51')$$

The series (3–50) converges to f under conditions analogous to those of Section 3–3.

EXAMPLE 1 Let $f(x) = \pi - x$ for $0 < x < \pi$ and let f have period π, as in Fig. 3–14. Then we apply (3–50), (3–51′) with $\tau = \pi$, obtaining

$$a_0 = \frac{2}{\pi} \int_0^{\pi} (\pi - x) \, dx = \pi,$$

$$a_n = \frac{2}{\pi} \int_0^{\pi} (\pi - x) \cos 2nx \, dx = 0, \quad n = 1, 2, \ldots,$$

$$b_n = \frac{2}{\pi} \int_0^{\pi} (\pi - x) \sin 2nx \, dx = \frac{1}{n}, \quad n = 1, 2, \ldots,$$

so that

$$f(x) = \frac{\pi}{2} + \sum_{n=1}^{\infty} \frac{\sin 2nx}{n}.$$

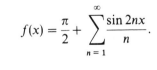

Fig. 3–14. Function of Example 1 with period π.

We observe that this series and those of Examples 1 and 2 of the preceding section all have the same sum for $0 < x < \pi$. Outside this interval the sums differ, as shown in Figs. 3–12, 3–13, and 3–14. ◄

REMARK: A function $f(x)$ defined for $0 \leqslant x \leqslant c$ can be expanded in a Fourier cosine series for period $\tau = 2c$. By change of scale, as above, we obtain the series

$$\frac{a_0}{2} + \sum_{n=1}^{\infty} a_n \cos n\omega x, \quad \omega = \frac{2\pi}{\tau}, \quad \tau = 2c,$$

$$a_n = \frac{4}{\tau} \int_0^{\tau/2} f(x) \cos n\omega x \, dx = \frac{2}{c} \int_0^c f(x) \cos \frac{n\pi x}{c} dx.$$

There is a similar sine series:

$$\sum_{n=1}^{\infty} b_n \sin n\omega x, \quad \omega = \frac{2\pi}{\tau}, \quad \tau = 2c,$$

$$b_n = \frac{4}{\tau} \int_0^{\tau/2} f(x) \sin n\omega x \, dx = \frac{2}{c} \int_0^c f(x) \sin \frac{n\pi x}{c} dx.$$

EXAMPLE 2   Let $f(x) = x^2$ for $0 \leqslant x \leqslant 1$. Then we can expand f in a sine series for $\tau = 2$, $\omega = \pi$. We find

$$b_n = 2 \int_0^1 x^2 \sin n\pi x \, dx = 2\frac{(-1)^n (2 - n^2\pi^2) - 2}{n^3\pi^3},$$

so that

$$f(x) = 2 \sum_{n=1}^{\infty} \frac{(-1)^n (2 - n^2\pi^2) - 2}{n^3\pi^3} \sin n\pi x.$$

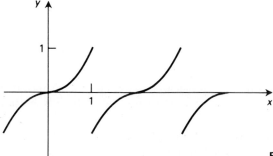

Fig. 3-15. Sum of series for Example 2.

The series converges for all x to the odd periodic extension of $f(x)$ with period 2 (Fig. 3–15). ◄

═══════════════════ **Problems (Section 3-5)** ═══════════════════

1. Represent the function by a Fourier cosine series and graph* the sum of the series:
   a) $f(x) = 0$ for $0 \leqslant x \leqslant \pi/2$, $f(x) = 1$ for $\pi/2 < x \leqslant \pi$
   b) $f(x) = x/\pi$ for $0 \leqslant x \leqslant \pi$
   c) $f(x) - \sin x$ for $0 \leqslant x \leqslant \pi$
   d) $f(x) = e^x$ for $0 \leqslant x \leqslant \pi$

2. Represent the function by a Fourier sine series and graph* the sum of the series:
   a) $f(x) = 1$, $0 \leqslant x \leqslant \pi/2$, $f(x) = 0$, $\pi/2 < x \leqslant \pi$
   b) $f(x) = x$, $0 \leqslant x \leqslant \pi/2$, $f(x) = \pi - x$, $\pi/2 \leqslant x \leqslant \pi$
   c) $f(x) = \cos x$, $0 \leqslant x \leqslant \pi$
   d) $f(x) = \sin^2 x$

3. For each of the following periodic functions of period τ given, expand in a Fourier series
   for period τ and graph the sum.
   a) $f(x) = x$ for $-1 \leqslant x \leqslant 1$, $\tau = 2$
   b) $f(x) = x^2$ for $0 \leqslant x \leqslant 1$, $\tau = 1$
   c) $f(x) = e^{-x}$ for $-1 \leqslant x \leqslant 1$, $\tau = 2$
   d) $f(x) = \sin x$ for $0 \leqslant x \leqslant 1$, $\tau = 1$

4. Let $f(x) = 1 + x$ for $0 \leqslant x \leqslant 1$.
   a) Expand f in a Fourier cosine series for period 2 and graph the sum.
   b) Expand f in a Fourier sine series for period 2 and graph the sum.

5. *Heat conduction.* Let a thin rod lie along the x-axis from 0 to π and let heat conduction
   from or into the rod occur only at the ends of the rod, which are maintained at
   temperature 0 (for an appropriate scale). It is shown in physics that if, at time $t = 0$, the
   temperature u along the rod equals $b_n \sin nx$ (where $b_n = $ const and n is a positive integer),
   then at time $t > 0$ it equals $b_n(\sin nx)e^{-cn^2t}$, where c is a positive constant (see Section 8–6
   for further discussion). There is also a superposition principle, which allows us to add
   effects of different initial temperature distributions. Hence, if the initial temperature is

$$f(x) = \sum_{n=1}^{\infty} b_n \sin nx,$$

then at time $t > 0$

$$u = \sum_{n=1}^{\infty} b_n(\sin nx)e^{-cn^2t}.$$

Accordingly, to find the temperature for $t > 0$ for an arbitrary given initial temperature
$f(x)$, we should expand $f(x)$ in a Fourier sine series and then proceed as above. Apply this
method for the following choices of $f(x)$:
   a) $3 \sin x + 5 \sin 2x$ **b)** $5 \sin x - 10 \sin 3x + 15 \sin 5x$
   c) $e^x \sin x$ **d)** As in Problem 2(b) above

---

*   The sum can be graphed easily from the information given about the function (without knowledge of
     the series), as in Fig. 3–12 above.

3-6 CONVERGENCE IN THE MEAN, BESSEL'S INEQUALITY, PARSEVAL RELATION

The convergence of Fourier series can be approached in another way, using an idea important in statistics.

If $g(x)$ is an approximation to $f(x)$ over the interval $a \leqslant x \leqslant b$, then the *mean square error* of this approximation is

$$\frac{1}{b-a} \int_a^b [g(x) - f(x)]^2 dx.$$

In statistics we also consider the square root of this quantity, the *root mean square error*, or RMS error. When this error is zero, $g(x) \equiv f(x)$ (except perhaps at discontinuity points). When the error is small $g(x)$ must remain close to f for most values of x; large values of $|g(x) - f(x)|$ may occur, but only for intervals of x of very small total length.

We now proceed to show that, *for each n*, the coefficients of the Fourier series of f are such that $a_0, a_1, \ldots, a_n, b_1, \ldots, b_n$ *are the choices of $p_0, p_1, \ldots, p_n, q_1, \ldots, q_n$ which give the smallest mean square error* in the approximation of f over the interval $-\pi \leqslant x \leqslant \pi$ by a function

$$g(x) = \frac{p_0}{2} + p_1 \cos x + q_1 \sin x + \cdots + p_n \cos nx + q_n \sin nx.$$

Such a function is called a *trigonometric polynomial*. To show this, we consider the total square error

$$\int_{-\pi}^{\pi} [g(x) - f(x)]^2 dx.$$

(To obtain the mean square error, this is divided by 2π. If we minimize the total square error, we also minimize the mean square error.) We find

$$\int_{-\pi}^{\pi} [g(x) - f(x)]^2 dx = \int_{-\pi}^{\pi} \left[\frac{p_0}{2} + p_1 \cos x + \cdots - f(x) \right]^2 dx$$

$$= \int_{-\pi}^{\pi} \left\{ \frac{p_0^2}{4} + p_1^2 \cos^2 x + \cdots - p_0 f(x) - 2p_1 f(x) \cos x - \cdots + [f(x)]^2 \right\} dx.$$

Here we omitted the terms from products of two of the terms of $g(x)$ (say, $p_m \cos mx$ and $q_s \sin sx$), since each such term has integral 0. Integrating the terms in p_0^2, p_1^2, ... and grouping the result, we obtain

$$\int_{-\pi}^{\pi} [g(x) - f(x)]^2 dx = \int_{-\pi}^{\pi} [f(x)]^2 dx + \left[\frac{\pi p_0^2}{2} - p_0 \int_{-\pi}^{\pi} f(x) dx \right]$$

$$+ \left[\pi p_1^2 - 2p_1 \int_{-\pi}^{\pi} f(x) \cos x \, dx \right] + \ldots + \left[\pi q_n^2 - 2q_n \int_{-\pi}^{\pi} f(x) \sin nx \, dx \right]. \quad (3-60)$$

Hence we can achieve the smallest total square error if we choose p_0 to minimize the term containing p_0 on the right, p_1 to minimize the term containing p_1, and so on.

The term in p_0 is a quadratic function of p_0, whose graph would be a parabola opening upward. Hence the minimum occurs where the derivative with respect to p_0 is 0:

$$\pi p_0 - \int_{-\pi}^{\pi} f(x)\,dx = 0 \qquad \text{or} \qquad p_0 = \frac{1}{\pi}\int_{-\pi}^{\pi} f(x)\,dx = a_0.$$

Similarly, for the term in p_1 we obtain

$$2\pi p_1 - 2\int_{-\pi}^{\pi} f(x)\cos x\,dx = 0 \qquad \text{or} \qquad p_1 = \frac{1}{\pi}\int_{-\pi}^{\pi} f(x)\cos x\,dx = a_1,$$

and in general $p_2 = a_2, \ldots, p_n = a_n, q_1 = b_1, \ldots, q_n = b_n$.

Thus the Fourier coefficients give the smallest total square error.

We denote the minimum total square error of this approximation by E_n. Thus

$$E_n = \int_{-\pi}^{\pi} [S_n(x) - f(x)]^2\,dx,$$

where S_n is the nth partial sum of the Fourier series of f. By taking $p_0 = a_0, p_1 = a_1$, $q_1 = b_1, \ldots$ in (3–60) we obtain

$$E_n = \int_{-\pi}^{\pi} [f(x)]^2\,dx - \pi\left(\frac{a_0^2}{2} + a_1^2 + b_1^2 + \cdots + a_n^2 + b_n^2\right). \qquad (3\text{-}61)$$

Since the total square error is positive or zero, we deduce that $E_n \geqslant 0$, and hence from (3–61) we get

$$\frac{a_0^2}{2} + a_1^2 + b_1^2 + \cdots + a_n^2 + b_n^2 \leqslant \frac{1}{\pi}\int_{-\pi}^{\pi} [f(x)]^2\,dx. \qquad (3\text{–}62)$$

This is *Bessel's inequality*; it implies that the infinite series

$$\frac{a_0^2}{2} + \sum_{n=1}^{\infty} (a_n^2 + b_n^2)$$

converges, since it is a positive term series whose partial sums form a bounded sequence by (3–62) (Section 2–7).

We can go further. As n increases, E_n decreases, as we can see from (3–61), and in fact

$$\lim_{n \to \infty} E_n = 0. \qquad (3\text{–}63)$$

By (3–61), this is equivalent to the statement that

$$\int_{-\pi}^{\pi} [f(x)]^2\,dx = \frac{\pi a_0^2}{2} + \pi \sum_{n=1}^{\infty} (a_n^2 + b_n^2), \qquad (3\text{–}64)$$

which is known as *Parseval's relation*. We prove this under the assumption that the Fourier series of f converges uniformly to f for all x. Then

$$f(x) = \frac{a_0}{2} + \sum_{n=1}^{\infty} (a_n \cos nx + b_n \sin nx),$$

and after multiplication of both sides by $f(x)$ we get

$$[f(x)]^2 = \frac{a_0}{2} f(x) + \sum_{n=1}^{\infty} [a_n f(x) \cos nx + b_n f(x) \sin nx].$$

The series remains uniformly convergent by the results of Section 2–15. Therefore we may integrate term by term:

$$\int_{-\pi}^{\pi} [f(x)]^2 dx$$

$$= \frac{a_0}{2} \int_{-\pi}^{\pi} f(x) dx + \sum_{n=1}^{\infty} \left[a_n \int_{-\pi}^{\pi} f(x) \cos nx \, dx + b_n \int_{-\pi}^{\pi} f(x) \sin nx \, dx \right]$$

$$= \frac{a_0}{2} \pi a_0 + \pi \sum_{n=1}^{\infty} (a_n a_n + b_n b_n),$$

and this is the same as (3–64). (For a proof of Parseval's relation under more general conditions, see AC, Sections 7–11 and 7–12).

The results found can be phrased in terms of *convergence in the mean*. In general, we say that the sequence $\{g_n(x)\}$ *converges in the mean to* $f(x)$ on the interval $a \leqslant x \leqslant b$ if

$$\lim_{n \to \infty} \int_a^b [g_n(x) - f(x)]^2 dx = 0.$$

Then we write

$$\text{l.i.m.} \; g_n(x) = f(x),$$
$$\scriptstyle n \to \infty$$

where l.i.m. stands for *limit in the mean*. Thus by (3–63), if $S_n(x)$ is the nth partial sum of the Fourier series of f, then

$$\text{l.i.m.} \; S_n(x) = f(x).$$
$$\scriptstyle n \to \infty$$

Convergence in the mean is important because for many physical problems it is the natural kind of convergence and we are far less concerned with the errors at particular values of x than with mean errors, or mean square errors over an interval. Also, for Fourier series, convergence in the mean is valid under considerably more general conditions than is uniform convergence or even nonuniform convergence for all x. The fact that $S_n(x)$ is a best approximation to f in the sense of total square error can also be used to estimate the coefficients $a_1, b_1, \ldots$ successively, $a_0/2$ being the average value of f (see AC, Section 7–4).

3–7 COMPLEX FORM OF FOURIER SERIES

We now use complex functions as in Section 1–10. The Euler identity

$$e^{i\beta} = \cos \beta + i \sin \beta \qquad (3\text{–}70)$$

(Sections 1–10 and 2–20) allows us to write Fourier series in a very useful form. By replacing β with $-\beta$ in (3–70) we get

$$e^{-i\beta} = \cos \beta - i \sin \beta, \qquad (3\text{–}70')$$

and hence from these two equations

$$\cos \beta = \frac{e^{i\beta} + e^{-i\beta}}{2} \qquad \text{and} \qquad \sin \beta = \frac{e^{i\beta} - e^{-i\beta}}{2i}. \qquad (3\text{–}71)$$

Thus the Fourier series

$$\frac{a_0}{2} + \sum_{n=1}^{\infty} (a_n \cos nx + b_n \sin nx)$$

can be written as

$$\frac{a_0}{2} + \sum_{n=1}^{\infty} \left[\frac{a_n(e^{inx} + e^{-inx})}{2} + \frac{b_n(e^{inx} - e^{-inx})}{2i} \right]$$

$$= \frac{a_0}{2} + \sum_{n=1}^{\infty} \frac{a_n - ib_n}{2} e^{inx} + \sum_{n=1}^{\infty} \frac{a_n + ib_n}{2} e^{-inx}.$$

We can combine these expressions into one series

$$\sum_{n=-\infty}^{\infty} c_n e^{inx}, \qquad (3\text{–}72)$$

where

$$c_0 = \frac{a_0}{2}, \quad c_n = \frac{a_n - ib_n}{2} \text{ for } n \geqslant 1, \quad c_n = \frac{a_{-n} + ib_{-n}}{2} \text{ for } n \leqslant -1. \qquad (3\text{–}73)$$

The series (3–72) is known as the *complex form* of the Fourier series. The coefficients c_n can be calculated directly from f:

$$c_n = \frac{1}{2\pi} \int_{-\pi}^{\pi} f(x) e^{-inx} \, dx, \quad n = 0, \pm 1, \pm 2, \cdots, \qquad (3\text{–}74)$$

since we can verify that (3–74) is in agreement with (3–73) (see Problem 6(a) below). As usual, we can integrate from 0 to 2π or over any other interval of length 2π.

Our original partial sum $S_n(x)$ corresponds to the terms of index $0, \pm 1, \cdots, \pm n$ in the series (3–72) and hence this series should be regarded as

$$\lim_{n \to \infty} \sum_{k=-n}^{n} c_k e^{ikx}.$$

EXAMPLE 1   Let $f(x) = x^2$ for $0 \leqslant x \leqslant 2\pi$ and let f have period 2π. Then through integration by parts

$$c_n = \frac{1}{2\pi} \int_0^{2\pi} x^2 e^{-inx} \, dx = \frac{1}{2\pi} \left[\frac{x^2 e^{-inx}}{-in} \bigg|_0^{2\pi} + \frac{2}{in} \int_0^{2\pi} x e^{-inx} \, dx \right]$$

$$= \frac{1}{2\pi} \left\{ 4 \frac{i\pi^2 e^{-in2\pi}}{n} + \frac{2}{in} \left[\frac{x e^{-inx}}{-in} \bigg|_0^{2\pi} + \frac{1}{in} \int_0^{2\pi} e^{-inx} \, dx \right] \right\}$$

$$= \frac{1}{2\pi} \left\{ 4 \frac{i\pi^2 e^{-in2\pi}}{n} + \frac{2}{in} \left[\frac{2\pi e^{-in2\pi}}{-in} + \frac{e^{-inx}}{n^2} \bigg|_0^{2\pi} \right] \right\}.$$

Now $e^{-in2\pi} = \cos 2n\pi - i \sin 2n\pi = 1$. Hence the above simplifies to

$$c_n = \frac{2 + 2\pi n i}{n^2}.$$

Because of the n in the denominator, the evaluation is not correct for $n = 0$. We find

$$c_0 = \frac{1}{2\pi} \int_0^{2\pi} x^2 \, dx = \frac{4\pi^2}{3}.$$

Hence

$$f(x) = \frac{4\pi^2}{3} + \sum_{n=-\infty}^{\infty} {}' \frac{2 + 2\pi n i}{n^2} e^{inx},$$

where $\sum'$ denotes a sum excluding the value for $n = 0$. We observe that f, as a periodic function, has jump discontinuities at $0, \pm 2\pi, \ldots$; at these points the series converges to $2\pi^2$. ◄

In working the example we took advantage of the fact that the complex function of x, e^{aix}, obeys the usual rules of differentiation and integration (Section 1–10).

As in Section 3–5, we can extend the complex form of Fourier series to functions of general period τ. With $\omega = 2\pi/\tau$, the series becomes

$$\sum_{n=-\infty}^{\infty} c_n e^{in\omega x}, \tag{3–75}$$

where

$$c_n = \frac{1}{\tau} \int_0^{\tau} f(x) e^{-in\omega x} \, dx, \quad n = 0, \pm 1, \pm 2, \cdots \tag{3–76}$$

We can integrate from $-\tau/2$ to $\tau/2$ or over any interval of length τ.

REMARK: Throughout this section $f(x)$ may be complex-valued. If $f(x)$ is real-valued, then $c_{-n} = \bar{c}_n$ in Eqs. (3–74) and (3–76) (see Problem 6(b) below).

3–8   APPLICATION OF FOURIER SERIES TO FREQUENCY RESPONSE

In Section 1–14 we pointed out the application of second-order differential equations to the study of forced oscillations in mechanical or electrical systems. We here extend this discussion to mth order differential equations and take advantage of Fourier series. We consider the differential equation

$$a_0 \frac{d^m x}{dt^m} + a_1 \frac{d^{m-1} x}{dt^{m-1}} + \cdots + a_{m-1} \frac{dx}{dt} + a_m x = f(t) \qquad (3\text{–}80)$$

and assume that coefficients $a_0, a_1, \ldots, a_m$ are real constants, with $a_0 \neq 0$. The general solution of (3–80) then has the form

$$x = x_c(t) + x^*(t).$$

Here $x_c(t)$ is the complementary function, the general solution of the related homogeneous equation; this is found as in Section 1–15 with the aid of the auxiliary equation

$$a_0 \lambda^m + a_1 \lambda^{m-1} + \cdots + a_{m-1}\lambda + a_m = 0. \qquad (3\text{–}81)$$

The function $x^*(t)$ is a particular solution of the given differential equation (3–80).

As in Sections 1–14 and 1–15 (see the Remark at the end of Section 1–15), Eq. (3–80) describes a *stable* system when all characteristic roots have negative real part, so that all solutions of the related homogeneous equation are *transients* (approach 0 as $t \to \infty$). We here consider only this stable case, so that $\operatorname{Re} \lambda < 0$ for each solution λ of Eq. (3–81).

Now let $f(t)$ be periodic, of period τ, and let f be piecewise smooth, so that f is represented by its Fourier series. It is simpler to write this in complex form:

$$f(t) = \sum_{n=-\infty}^{\infty} c_n e^{in\omega t}, \qquad \omega = \frac{2\pi}{\tau},$$

as in Section 3–7 above.

To obtain a particular solution $x^*(t)$ of Eq. (3–80), we then use the idea of superposition (Section 1–12). We proceed formally and then comment on the results. For each term $c_n e^{in\omega t}$ of $f(t)$ we find a solution of the equation

$$a_0 \frac{d^m x}{dt^m} + \cdots + a_n x = c_m e^{in\omega t}. \qquad (3\text{–}82)$$

By undetermined coefficients (Section 1–12) this is found to be

$$x = \frac{c_n e^{in\omega t}}{a_0 (in\omega)^m + \cdots + a_m}. \qquad (3\text{–}83)$$

Here the denominator is not zero, since by hypothesis our equation is *stable*, so that $in\omega$ is not a characteristic root. We write

$$Y(s) = \frac{1}{a_0 s^m + \cdots + a_m}. \tag{3–84}$$

This function is the *transfer function* for the system considered. For ω real, the function $Y(i\omega)$ is called the *frequency response function* of the system. Our particular solution (3–83) can be written as

$$x = c_n Y(in\omega)e^{in\omega t}. \tag{3–83'}$$

Accordingly, by superposition, a solution of the equation

$$a_0 \frac{d^m x}{dt^m} + \cdots + a_m x = f(t) = \sum_{n=-\infty}^{\infty} c_n e^{in\omega t}$$

is given by

$$x^*(t) = \sum_{n=-\infty}^{\infty} c_n Y(in\omega)e^{in\omega t}. \tag{3–85}$$

Thus corresponding to the *periodic input* $f(t)$ there is a *periodic output* $x^*(t)$. For large t, only $x^*(t)$ will be observed, since the terms of $x_c(t)$ are transients, and hence $x^*(t)$ is called the *steady state*. We observe that $x^*(t)$ is the *unique periodic* solution of the differential equation.

EXAMPLE 1 Let $f(t) = 0$ for $-\pi/2 \leqslant t < -\pi/2$, $f(t) = 1$ for $-\pi/2 \leqslant t < \pi/2$, $f(t) = 0$ for $\pi/2 \leqslant t \leqslant \pi$, and let f have period 2π. This is the function of Example 1 in Section 3–2, i.e., a square wave. Thus $\tau = 2\pi$, $\omega = 1$ and

$$c_n = \frac{1}{2\pi} \int_{-\pi/2}^{\pi/2} e^{-int}\, dt = \frac{e^{-int}}{-2\pi in} \Big|_{-\pi/2}^{\pi/2}$$

$$= \frac{e^{-in\pi/2} - e^{in\pi/2}}{-2\pi in} = \frac{\sin n\pi/2}{n\pi}, \quad n \neq 0.$$

For c_0 we find the value $1/2$. Thus

$$f(t) = \frac{1}{2} + \frac{1}{\pi} \sum_{n=-\infty}^{\infty} {}' \frac{\sin(n\pi/2)}{n} e^{int},$$

where $\sum'$ denotes a sum without the term for $n = 0$. We now use this function as input in the equation

$$\frac{d^3 x}{dt^3} + 3\frac{d^2 x}{dt^2} + 4\frac{dx}{dt} + 2x = f(t).$$

The auxiliary equation is found to have roots $-1, -1 \pm i$, so that the system is stable. We have, with $\omega = 1$,

$$Y(s) = \frac{1}{s^3 + 3s^2 + 4s + 2}; \quad Y(in\omega) = \frac{1}{2 - 3n^2 + i(4n - 4n^3)},$$

and hence

$$x^*(t) = \frac{1}{4} + \sum_{n=-\infty}^{\infty}{}' \frac{\sin(n\pi/2)}{n\pi} \frac{e^{int}}{(2 - 3n^2) + i(4n - n^3)}.$$

We examine the terms for $n = 1$ and $n = -1$, observing that

$$\frac{e^{it}}{-1 + 3i} + \frac{e^{-it}}{-1 - 3i} = \frac{(-1 - 3i)e^{it} + (-1 + 3i)e^{-it}}{10} = \frac{3\sin t - \cos t}{5}.$$

The terms for $n = 2$ and -2 are 0. Hence

$$x^*(t) \sim \frac{1}{4} + \frac{3\sin t - \cos t}{5\pi}.$$

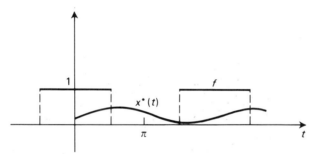

Fig. 3–16. Input and steady-state response for Example 1.

This approximation is quite good, since the series converges rather rapidly. Input f and steady-state output x^* are graphed in Fig. 3–16. ◀

A physical illustration

The behavior of the following mechanical system is governed by the equations of this example. It has coupled springs, as shown in Fig. 3–17, but the effect of friction has

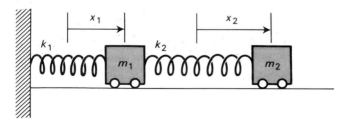

Fig. 3–17. Coupled springs.

to be considered. Further it is assumed that the second mass is extremely small, so that the corresponding inertia term (mass times acceleration) can be neglected. With appropriate spring constants and friction coefficients we get the equations

$$\frac{d^2x_1}{dt^2} + \frac{dx_1}{dt} + 2x_1 = x_2 \quad \text{and} \quad \frac{1}{2}\frac{dx_2}{dt} + x_2 = x_1 + \frac{1}{2}F.$$

Here x_1 and x_2 are measured from equilibrium positions, at which both spring forces are 0, and $(1/2)F$ is an external force. The equations can be written as

$$(D^2 + D + 2)x_1 = x_2 \quad \text{and} \quad (\tfrac{1}{2}D + 1)x_2 = x_1 + \tfrac{1}{2}F,$$

where $D = d/dt$, and hence lead to the equation of motion for the first mass:

$$(D^3 + 3D^2 + 4D + 2)x_1 = F.$$

We see at once that this is the same as the differential equation for Example 1, with $x = x_1$. Thus if the applied force F is the square wave $f(t)$, as in Fig. 3–16, then the steady state $x^*(t)$ gives the motion of the first mass, after transients have died out.

In the general formula (3–85) we observe that the constant term $c_0 Y(0)$ is real and that the terms for $n\omega$ and $-n\omega$ are complex conjugates of each other (see Problem 6 below), so that their sum is real and is twice the real part of either term. Thus these terms together give

$$2\,\mathrm{Re}\, c_n\, Y(in\omega)e^{in\omega t}.$$

If we write (see Section 2–20)

$$c_n = |c_n|\,(\cos\gamma_n + i\sin\gamma_n) = |c_n|\,e^{i\gamma_n},$$
$$Y(in\omega) = |Y(in\omega)|\,e^{i\beta_n},$$

then the sum of the two terms is

$$2\,\mathrm{Re}\,|c_n|\,e^{i\gamma_n}\,|Y(in\omega)|\,e^{i\beta_n}e^{in\omega t}$$
$$= 2\,\mathrm{Re}\,|c_n|\,|Y(in\omega)|\,e^{i(\beta_n + \gamma_n + n\omega t)}$$
$$= 2|c_n|\,|Y(in\omega)|\cos(\beta_n + \gamma_n + n\omega t),$$

that is, a sinusoidal oscillation of amplitude $2|c_n|\,|Y(in\omega)|$. These terms are the output corresponding to the sum of the terms $c_n e^{in\omega t}$, $\bar{c}_n e^{-in\omega t}$ in the series for $f(t)$. We find as above that this sum is

$$2|c_n|\cos(\gamma_n + n\omega t),$$

that is, a sinusoidal oscillation of amplitude $2|c_n|$. Thus *the term of frequency $n\omega$ in the input $f(t)$ produces a term of the same frequency $n\omega$ in the output $x^*(t)$, with an amplification of $|Y(in\omega)|$ and a phase shift of β_n, the polar angle of $Y(in\omega)$.*

In Example 1 we have $Y(in\omega) = 1/(-1 + 3i)$ for $n = 1$, so that $|Y(in\omega)| = 1/\sqrt{10}$ for $n = 1$. For $n = 2$, $c_n = 0$ and hence we disregard this value. For $n = 3$, $Y(in\omega) = 1/(-25 - 15i)$, so that $|Y(in\omega)| = 1/\sqrt{850} \approx 0.03$, so that this frequency receives very little weight in the output. The higher frequencies receive even smaller weights, and $|Y(in\omega)| \to 0$ when $n \to \infty$, as one sees from the expression

$$|Y(in\omega)| = \frac{1}{\sqrt{n^6 + n^4 + 4n^2 + 4}}.$$

Validity of method

It can be shown that the series for $x^*(t)$ converges uniformly for all t, so that $x^*(t)$ is continuous, and $x^*(t)$ is a solution of the differential equation (except at points where $f(t)$ has jumps). In general, multiplication of the coefficients c_n by $Y(in\omega)$ *improves convergence*, since, as noted above, $|Y(in\omega)| \to 0$ as $n \to \infty$. (For more details, see OMLS, Chapter 4.)

═══════════════════ **Problems (Section 3–8)** ═══════════════════

1. Let $f(x)$ have period 2π and let $f(x) = \pi - x$ for $0 < x < 2\pi$ as in Example 2 of Section 3–4, so that

$$f(x) = 2 \sin x + \cdots + [(2 \sin nx)/n] + \cdots$$

a) Find the total square error in approximating f by $S_n(x) = 2 \sin x + \cdots + (2 \sin nx)/n$ and evaluate for $n = 1, 2, 3$.

b) Show by graphing that $y = \sin x$ and $y = 3 \sin x$ give larger total square errors than $y = 2 \sin x$ in approximating f for $0 \leqslant x \leqslant 2\pi$.

c) Show by graphing that

$$y = \sin x + 2 \sin 2x \qquad \text{and} \qquad y = 2 \sin x + \frac{1}{2} \sin 2x$$

give larger total square errors than $y = 2 \sin x + \sin 2x$ in approximating f for $0 \leqslant x \leqslant 2\pi$.

d) Show with the aid of Parseval's relation that

$$\frac{1}{1^2} + \frac{1}{2^2} + \cdots + \frac{1}{n^2} + \cdots = \frac{\pi^2}{6}.$$

2. Let $f(x) = \pi - x$ for $0 \leqslant x \leqslant \pi$ and let f be even and have period 2π, as in Example 1 of Section 3–4, so that

$$f(x) = \frac{\pi}{2} + \frac{4}{\pi} \sum_{n=1}^{\infty} \frac{\cos(2n-1)}{(2n-1)^2} x.$$

a) Verify that the constant term $\pi/2$ is the average value of f over a period.

b) Show by graphing that

$$y = \frac{\pi}{2} + 2 \cos x \qquad \text{and} \qquad y = \frac{\pi}{2} + \frac{1}{2} \cos x$$

have larger total square errors than $y = \pi/2 + (4/\pi) \cos x$ in approximating f over the interval $-\pi \leqslant x \leqslant \pi$.

c) Find the total square error in approximating f by

$$S_{2n-1}(x) = \frac{\pi}{2} + \frac{4}{\pi} \left[\cos x + \cdots + \frac{\cos(2n-1)x}{(2n-1)^2} \right]$$

and evaluate for $n = 1, 2, 3$.

d) Show by Parseval's relation that

$$1 + \frac{1}{3^4} + \frac{1}{5^4} + \cdots + \frac{1}{(2n-1)^4} + \cdots = \frac{\pi^4}{96}.$$

3. Expand f in a Fourier series in complex form:

   a) $f(x) = 2\pi x - x^2$, $0 \leqslant x \leqslant 2\pi$, f of period 2π.

   b) $f(x) = e^x$, $0 \leqslant x \leqslant 1$, f of period 1.

   c) $f(x) = \sin \dfrac{x}{2} = \dfrac{e^{ix/2} - e^{-ix/2}}{2i}$, $0 \leqslant x \leqslant 2\pi$, f of period 2π.

   d) $f(x) = 1 - x^2$, $-1 \leqslant x \leqslant 1$, $f(x) = 0$ for $-2 \leqslant x \leqslant -1$ and for $1 \leqslant x \leqslant 2$, f of period 4.

4. Let the various constants for the coupled springs of Fig. 3–17 be such that the differential equations are

$$(D^2 + 5D + 2)x_1 = x_2 \qquad \text{and} \qquad (D + 1)x_2 = x_1 + F,$$

so that $x = x_1$ is governed by the equation

$$(D^3 + 6D^2 + 7D + 1)x = F. \tag{3–86}$$

Let $F = f(t)$, where $f(t) = 2\pi t - t^2$ for $0 \leqslant t \leqslant 2\pi$ and $f(t)$ has period 2π, as in Problem 3(a).

   a) Show that the differential equation (3–86) is stable.

   b) Find the transfer function and the frequency response function.

   c) Find the steady state response $x^*(t)$ and graph the partial sum $S_1(t)$. Comment on how well $S_1(t)$ approximates $x^*(t)$.

5. Let the differential equation

$$\left(\frac{1}{10}D^2 + \frac{1}{5}D + 1 \right)^2 x = f(t)$$

be given, with $D = d/dt$, $f(t) = \sin(t/2)$ for $0 \leqslant t \leqslant 2\pi$, and f having period 2π, as in Problem 3(c).

   a) Show that the equation is stable.

   b) Find the transfer function and the frequency response function.

   c) Find the steady state response $x^*(t)$ and graph the partial sum $S_1(t)$. Comment on how well $S_1(t)$ approximates $x^*(t)$.

6. **a)** Show that the expression (3–74) for c_n is in agreement with (3–73).

   b) Show from Eq. (3–73) that, if $f(x)$ is real-valued, then $c_{-n} = \bar{c}_n$ (the bar denoting complex conjugation) for $n = 1, 2, \ldots$

   c) Prove that in general $\overline{zw} = \bar{z}\,\bar{w}$ for two complex numbers z and w.

   d) Show from Eq. (3–84) that $Y(-in\omega) = \overline{Y(in\omega)}$ for $n = 1, 2, \ldots$

   e) Show that for $n = 1, 2, 3, \ldots$ in Eq. (3–85), $c_{-n}Y(-in\omega)e^{-in\omega t}$ is the complex conjugate of $c_n Y(in\omega)e^{in\omega t}$.

7. Let

$$f(x) = \frac{a_0}{2} + \sum_{n=1}^{\infty} (a_n \cos nx + b_n \sin nx),$$

$$g(x) = \frac{a_0'}{2} + \sum_{n=1}^{\infty} (a_n' \cos nx + b_n' \sin nx).$$

Prove under appropriate hypotheses that

$$\int_{-\pi}^{\pi} f(x)g(x)\,dx = \pi\left[\frac{a_0 a_0'}{2} + \sum_{n=1}^{\infty} (a_n a_n' + b_n b_n')\right].$$

This is another *Parseval relation*.

3–9 ORTHOGONAL FUNCTIONS

If we review the development of Fourier series, we see that a crucial role is played by the formulas:

$$\int_{-\pi}^{\pi} \sin nx \cos mx\,dx = 0,$$

$$\int_{-\pi}^{\pi} \cos nx \cos mx\,dx = 0, \qquad m \neq n,$$

$$\int_{-\pi}^{\pi} \sin nx \sin mx\,dx = 0, \qquad m \neq n,$$

where m and n are positive integers. This suggests a generalization: to expand $f(x)$ in a series, we write

$$f(x) = c_1 \varphi_1(x) + \cdots + c_n \varphi_n(x) + \cdots, \tag{3–90}$$

where the functions $\varphi_n(x)$ are continuous and are such that

$$\int_a^b \varphi_n(x)\varphi_m(x)\,dx = 0, \quad n \neq m. \tag{3–91}$$

We say that φ_n, φ_m are *orthogonal to each other* over the interval $a \leqslant x \leqslant b$. As will be seen, we also want

$$\int_a^b [\varphi_n(x)]^2\,dx = k_n > 0, \quad n = 1, 2, \ldots \tag{3–92}$$

[If k_n were 0 for some n, then $\varphi_n(x)$ would have to be identically 0, which is clearly of no use here.] When (3–91) and (3–92) hold, we say that the sequence $\{\varphi_n(x)\}$ forms an *orthogonal system* over the interval $a \leqslant x \leqslant b$. It is often convenient to have $k_n = 1$ for all n. In that case we say that $\{\varphi_n(x)\}$ forms an *orthonormal system*.

Orthogonal systems turn out to arise naturally in many physical problems. For Fourier series, we are dealing with the orthogonal system

$$1, \quad \cos x, \quad \sin x, \quad \ldots, \quad \cos nx, \quad \sin nx, \quad \ldots$$

on the interval $-\pi \leqslant x \leqslant \pi$ (the constant function 1 being needed because of the constant term in the series). But the Fourier cosine series (Section 3–4) provides us with the orthogonal system

$$1, \quad \cos x, \quad \cos 2x, \quad \ldots, \quad \cos nx, \quad \ldots$$

on the interval $0 \leqslant x \leqslant \pi$. Indeed, for $n \neq m$,

$$\int_0^\pi 1 \cdot \cos nx \, dx = 0 \qquad \text{and} \qquad \int_0^\pi \cos nx \cos mx \, dx = 0,$$

as one verifies easily. Similarly, the Fourier sine series provides the orthogonal system

$$\sin x, \quad \sin 2x, \quad \ldots, \quad \sin nx, \quad \ldots.$$

on the interval $0 \leqslant x \leqslant \pi$.

Physical problems concerning *spheres* often require the system of *Legendre polynomials*

$$1, \quad x, \quad \frac{3}{2}x^2 - \frac{1}{2}, \quad \cdots$$

(Section 3–12) which form an orthogonal system on the interval $-1 \leqslant x \leqslant 1$. Problems for circles and cylinders lead similarly to orthogonal systems related to *Bessel functions* (Section 3–13).

As illustrations of the use of these functions, we mention the following: the gravitational field of the earth can be represented by an infinite series, whose terms are expressible in terms of Legendre polynomials. The analysis of frequency modulation leads us to integrals expressible in terms of Bessel functions. The conduction of heat in a spherical solid can be described by an infinite series whose terms involve both Bessel functions and Legendre polynomials (See Problem 6 at the end of Chapter 8).

Now let $\{\varphi_n(x)\}$ be an orthogonal system on the interval $a \leqslant x \leqslant b$ and let Eq. (3–90) be valid where the series, converges uniformly. Then we multiply both sides by $\varphi_m(x)$ and integrate from a to b. By orthogonality, all terms on the right drop out, except the one for $n = m$, and we obtain

$$\int_a^b f(x)\varphi_m(x)\,dx = c_m \int_a^b [\varphi_m(x)]^2 \, dx = c_m k_m.$$

Thus we obtain the rule for the coefficients:

$$c_n = \frac{1}{k_n} \int_a^b f(x)\varphi_n(x)\,dx, \quad n = 1, 2, \ldots \qquad (3\text{–}93)$$

Now for an arbitrary $f(x)$, say piecewise continuous on $a \leqslant x \leqslant b$, we can calculate the coefficients c_n by Eq. (3–93) and form the corresponding series

$$c_1 \varphi_1(x) + \cdots + c_n \varphi_n(x) + \cdots$$

This is called the *Fourier series of f with respect to the orthogonal system* $\{\varphi_n(x)\}$.

If the system happens to be orthonormal, the rule (3–93) simplifies to

$$c_n = \int_a^b f(x)\varphi_n(x)\,dx, \quad n = 1, 2, \ldots \qquad (3\text{–}93')$$

We observe that a general system $\{\varphi_n(x)\}$ can be converted into an orthonormal system by multiplying each $\varphi_n(x)$ by $1/\sqrt{k_n}$.

For an arbitrary orthogonal system we cannot expect a theorem (like the one of Section 3–3) which asserts that the series $\sum c_n \varphi_n(x)$ converges to $f(x)$ for all f satisfying certain continuity conditions. For example, the system

$$\sin x, \quad \sin 2x, \quad \ldots, \quad \sin nx, \quad \ldots$$

is an orthogonal system for the interval $-\pi \leqslant x \leqslant \pi$. However, the corresponding Fourier series $\sum_{n=1}^{\infty} b_n \sin nx$ can converge to $f(x)$ only if f is odd. In fact, *omission of one function* from our trigonometric system causes trouble: the system

$$1, \quad \cos x, \quad \sin x, \quad \sin 2x, \quad \cos 3x, \quad \sin 3x, \quad \ldots$$

is orthogonal on the interval $-\pi \leqslant x \leqslant \pi$. But the function $f(x) = \cos 2x$ cannot be represented in a corresponding Fourier series, because

$$\int_{-\pi}^{\pi} \cos 2x \, dx = 0, \qquad \int_{-\pi}^{\pi} \cos 2x \cos x \, dx = 0, \quad \text{etc.}$$

by orthogonality, and the series reduces to $0 + \cdots + 0 + \cdots$ which cannot converge to $\cos 2x$.

This discussion suggests that, to be useful, an orthogonal system must contain *enough functions*. The technical term is that the system must be *complete*. For a complete system there are theorems on convergence for large classes of functions f (we return to this point in Section 3–10). The orthogonal systems that arise in many important physical problems are complete. In particular, this holds for the system of Legendre polynomials.

Weight function

We say that functions $\varphi(x)$ and $\psi(x)$ are *orthogonal with respect to the weight function* $w(x)$ on the interval $a \leqslant x \leqslant b$ if

$$\int_a^b \varphi(x) \psi(x) w(x) \, dx = 0.$$

Here $w(x)$ is usually required to be positive (except perhaps at a and b). The concept of orthogonal system $\{\varphi_n\}$ extends to this case also. The corresponding Fourier series for f has the form $\sum c_n \varphi_n(x)$, where

$$c_n = \frac{1}{k_n} \int_a^b f(x) \varphi_n(x) w(x) \, dx,$$

$$k_n = \int_a^b \varphi_n^2(x) w(x) \, dx. \tag{3–94}$$

This case can in fact be reduced to the previous one by replacing $\varphi_n(x)$ with $\varphi_n(x) \sqrt{w(x)}$ and expanding $f(x) \sqrt{w(x)}$ in the corresponding series.

Infinite intervals

For some applications it is necessary to extend the theory to the case of infinite intervals. For example, the Laguerre polynomials (Section 3–13) are orthogonal over the interval $0 \leqslant x < \infty$ with respect to the weight function $x^{\alpha}e^{-x}$. The infinite interval leads to improper integrals for the corresponding coefficients c_n (and for the k_n). We study only those cases for which these improper integrals have meaning.

Even for the finite interval, improper integrals may arise, for example, if the weight function $w(x)$ becomes infinite at a or b. Again we deal only with cases for which this causes no trouble.

3–10 INNER PRODUCT AND NORM

We consider a fixed interval $a \leqslant x \leqslant b$, and all functions considered will be continuous or piecewise continuous on this interval. These functions form a linear space, as in Section 1–16. Hence the familiar concept of linear independence can be applied to sets of such functions. If $f(x)$ and $g(x)$ are two such functions, then we write

$$(f, g) = \int_a^b f(x)g(x)\,dx \qquad (3\text{–}100)$$

and call (f, g) the *inner product* of f and g. The term is based on the familiar product (also called dot product, or scalar product) for vectors, and there are great similarities between the inner product for functions and for vectors. In particular, the following rules apply:

$$\begin{aligned}
(f, g) &= (g, f), \\
(f, g+h) &= (f, g) + (g, h), \\
(f, cg) &= (cf, g) = c(f, g), \quad c = \text{const.}
\end{aligned} \qquad (3\text{–}101)$$

We also observe that

$$(f, f) = \int_a^b [f(x)]^2\,dx \geqslant 0 \qquad (3\text{–}102)$$

and hence $(f, f) = 0$ only when $f(x) \equiv 0$ (except perhaps at a finite number of points). It is convenient here to consider two functions to be the same if they agree except perhaps on a finite set; also to denote by 0 the function $f(x) \equiv 0$. Then we simply write:

$$(f, f) = 0 \text{ if and only if } f = 0. \qquad (3\text{–}103)$$

We now introduce the *norm* $\|f\|$ by the equation

$$\|f\| = \sqrt{(f, f)}. \qquad (3\text{–}104)$$

Thus by (3–102) and (3–103),

$$\|f\| \geqslant 0 \text{ and } \|f\| = 0 \text{ if and only if } f = 0. \qquad (3\text{–}103')$$

Also from (3–104) and (3–101),

$$\|cf\| = |c|\,\|f\|, \quad \text{where } c = \text{const,} \qquad (3\text{–}105)$$

since $\|cf\| = (cf, cf)^{1/2} = [c^2(f, f)]^{1/2} = |c| (f, f)^{1/2} = |c| \|f\|$. The norm corresponds to the *length* of a vector and has the familiar properties of length.

Orthogonality can now be defined in terms of the inner product: f and g are *orthogonal* if $(f, g) = 0$. When this holds, we have a Pythagorean theorem:

$$\|f + g\|^2 = \|f\|^2 + \|g\|^2,$$

since by the rules (3–101)

$$\|f + g\|^2 = (f + g, f + g) = (f, f) + (g, f) + (f, g) + (g, g)$$
$$= \|f\|^2 + \|g\|^2.$$

Convergence in the mean of a sequence $\{f_n\}$ to f is then defined by the condition

$$\lim_{n \to \infty} \|f - f_n\| = 0.$$

This is the same as requiring that the total square error has limit 0, since

$$\int_a^b (f_n - f)^2 \, dx = (f_n - f, f_n - f) = \|f_n - f\|^2.$$

The Fourier series of a function f with respect to the orthogonal system $\{\varphi_n\}$ can now be defined as the series

$$\sum_{n=1}^{\infty} c_n \varphi_n(x), \quad \text{where} \quad c_n = \frac{1}{\|\varphi_n\|^2} (f, \varphi_n) \qquad (3\text{–}106)$$

(see Eq. (3–93)).

The discussion of approximation in terms of square error (Section 3–6) carries over without essential change, and we find that, for each n, the total square error

$$\|f - (p_1 \varphi_1 + \cdots + p_n \varphi_n)\|^2, \quad \text{where} \quad p_1, \ldots, p_n \text{ are constants,}$$

has its smallest value when $p_1 = c_1, \ldots, p_n = c_n$ and that this smallest value is

$$E_n = \|f - (c_1 \varphi_1 + \cdots + c_n \varphi_n)\|^2 = \|f\|^2 - (c_1^2 \|\varphi_1\|^2 + \cdots + c_n^2 \|\varphi_n\|^2) \geq 0. \qquad (3\text{–}107)$$

Again we have a *Bessel's inequality*:

$$c_1^2 \|\varphi_1\|^2 + \cdots + c_n^2 \|\varphi_n\|^2 \leq \|f\|^2, \qquad (3\text{–}108)$$

and the infinite series $\sum c_n^2 \|\varphi_n\|^2$ must converge.

We call the orthogonal system $\{\varphi_n\}$ *complete* if $E_n \to 0$ as $n \to \infty$ for every piecewise continuous f, that is, if the Fourier series $\sum c_n \varphi_n(x)$ converges in the mean to f. This condition is also equivalent to *Parseval's relation*

$$\|f\|^2 = \sum_{n=1}^{\infty} c_n^2 \|\varphi_n\|^2. \qquad (3\text{–}109)$$

Generalizations

All the results of this section continue to hold if a weight function $w(x)$ is used, so that the inner product is defined by the equation

$$(f, g) = \int_a^b f(x)g(x)w(x)\,dx. \tag{3-100'}$$

The weight function must satisfy the conditions described in the previous section. It can also be extended to infinite intervals, provided our attention is restricted to functions f, g, etc., for which the corresponding improper integrals exist.

3–11 GRAM-SCHMIDT PROCESS

The three functions $f_1 = 1, f_2 = x, f_3 = e^x$ do not form an orthogonal system on the interval $0 \leqslant x \leqslant 1$, since

$$(f_1, f_2) = \int_0^1 x\,dx = \frac{1}{2}, \quad (f_1, f_3) = \int_0^1 e^x\,dx = e - 1, \quad (f_2, f_3) = \int_0^1 xe^x\,dx = 1.$$

However, we can use the three functions to *build* an orthogonal system. We let

$$g_1 = f_1, \quad g_2 = f_2 + af_1, \quad g_3 = f_3 + bf_1 + cf_2,$$

where a, b, c are numbers to be found. Then we want $(g_1, g_2) = 0$; that is,

$$(f_1, f_2 + af_1) = 0 \quad \text{or} \quad (f_1, f_2) + a(f_1, f_1) = 0.$$

Hence we take

$$a = -\frac{(f_1, f_2)}{(f_1, f_1)}.$$

We found (f_1, f_2) to be $1/2$. Since $(f_1, f_1) = \int_0^1 dx = 1$, we have $a = -1/2$.
Similarly, we want (g_1, g_3) and (g_2, g_3) to be 0. This gives

$$(f_1, f_3 + bf_1 + cf_2) = 0, \quad (f_2 - \tfrac{1}{2}f_1, f_3 + bf_1 + cf_2) = 0$$

or

$$(f_1, f_3) + b(f_1, f_1) + c(f_1, f_2) = 0,$$
$$(f_2, f_3) - \tfrac{1}{2}(f_1, f_3) + b[(f_2, f_1) - \tfrac{1}{2}(f_1, f_1)] + c[(f_2, f_2) - \tfrac{1}{2}(f_1, f_2)] = 0.$$

Here we know all the inner products except $(f_2, f_2) = \int_0^1 x^2\,dx = 1/3$. Hence we obtain the equations

$$e - 1 + b + \frac{1}{2}c = 0, \quad \frac{3-e}{2} + \frac{1}{12}c = 0.$$

They have a unique solution: $b = 10 - 4e$, $c = 6e - 18$. Accordingly,

$$g_1(x) = 1, \quad g_2(x) = x - \frac{1}{2}, \quad g_3(x) = e^x + 10 - 4e + (6e - 18)x$$

form an orthogonal system on the interval $0 \leqslant x \leqslant 1$.

The process illustrated can be generalized as follows. We let $f_0, \ldots, f_n$ be linearly independent functions on the interval $a \leqslant x \leqslant b$, that is, let

$$c_0 f_0 + c_1 f_1 + \cdots + c_n f_n = 0 \qquad (3\text{–}110)$$

(in the sense of equaling the zero function, as above) only for $c_0 = 0, c_1 = 0, \ldots,$ $c_n = 0$. Then there are unique constants k_{ij} such that the functions

$$\begin{aligned} g_0 &= f_0, \\ g_1 &= k_{10} f_0 + f_1, \cdots, \\ &\;\vdots \\ g_n &= k_{n0} f_0 + \cdots + k_{n,n-1} f_{n-1} + f_n \end{aligned} \qquad (3\text{–}111)$$

form an orthogonal system on the interval. Solving for the k_{ij}, as in the example above, we find

$$\begin{aligned} g_0 &= f_0, \\ g_1 &= f_1 - \frac{(f_1, g_0)}{(g_0, g_0)} g_0, \\ g_2 &= f_2 - \frac{(f_2, g_0)}{(g_0, g_0)} g_0 - \frac{(f_2, g_1)}{(g_1, g_1)} g_1, \cdots, \\ g_n &= f_n - \frac{(f_n, g_0)}{(g_0, g_0)} g_0 - \cdots - \frac{(f_n, g_{n-1})}{(g_{n-1}, g_{n-1})} g_{n-1}. \end{aligned} \qquad (3\text{–}112)$$

The process just described is the *Gram–Schmidt orthogonalization process*. It has a simple geometric interpretation in terms of vectors, with dot products for inner products, as in Fig. 3–18.

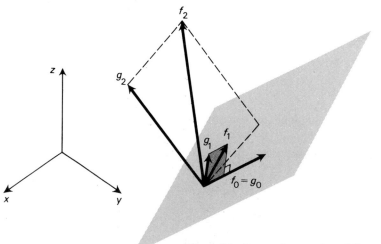

Fig. 3–18. Geometric meaning of Gram–Schmidt process.

Uniqueness of the orthogonal system

As stated above, the requirement that $g_0, g_1, \ldots, g_n$ form an orthogonal system uniquely determines the constants k_{ij} in (3–111).

Once we have the orthogonal system $g_0, g_1, \ldots, g_n$, we can of course multiply each g_j by a nonzero constant and still have an orthogonal system. The constants can be chosen uniquely (up to $\pm$ sign) to achieve an orthonormal system. In general we can say: given the linearly independent functions $f_0, f_1, \ldots, f_n$, there are functions $g_0, g_1, \ldots, g_n$, unique up to constant factors, that form an orthogonal system and are such that each g_j is a linear combination of $f_0, f_1, \ldots, f_j$.

The process can be extended to an infinite sequence of linearly independent functions $\{f_n\}$, that is, a sequence such that for each n, functions $f_0, \ldots, f_n$ are linearly independent. We obtain an orthogonal system $\{g_n\}$ related to f_i by equations of form (3–111). Hence if $\{f_n\}$ are polynomials, with f_n of degree n, each g_n is also a polynomial of degree n.

3–12 THE LEGENDRE POLYNOMIALS

We now illustrate the process described to obtain a very important orthogonal system $\{P_n\}$ whose functions are polynomials, the nth one being of degree n.

We take the interval to be $-1 \leqslant x \leqslant 1$, so that

$$(f, g) = \int_{-1}^{1} f(x) g(x) \, dx, \quad \|f\|^2 = \int_{-1}^{1} [f(x)]^2 \, dx,$$

and the weight function $w(x) \equiv 1$. We take f_0 to be 1, f_1 to be x, $\ldots, f_n(x)$ to be x^n for all n. Then $f_0, f_1, \ldots, f_n$ are linearly independent for $-1 \leqslant x \leqslant 1$, since by algebra an equation $c_0 + c_1 x + \cdots + c_n x^n = 0$ can be valid for at most n values of x, except for $c_0 = 0, c_1 = 0, \ldots, c_n = 0$. Then, by (3–112), $g_0 = f_0 = 1$, so that

$$(g_0, g_0) = \int_{-1}^{1} dx = 2, \quad (f_1, g_0) = \int_{-1}^{1} x \, dx = 0,$$

$$g_1 = f_1 - \frac{(f_1, g_0)}{(g_0, g_0)} g_0 = x - 0 = x,$$

and hence $(g_1, g_1) = \dfrac{2}{3}$, $(f_2, g_0) = \displaystyle\int_{-1}^{1} x^2 \, dx = \dfrac{2}{3}$, $(f_2, g_1) = 0$, so that

$$g_2 = f_2 - \frac{(f_2, g_0)}{(g_0, g_0)} g_0 - \frac{(f_2, g_1)}{(g_1, g_1)} g_1 = x^2 - \frac{2}{3},$$

and so on. We find

$$g_3(x) = x^3 - \frac{3}{5} x,$$

$$g_4(x) = x^4 - \frac{6}{7} x^2 + \frac{3}{35},$$

$$g_5(x) = x^5 - \frac{10}{9} x^3 + \frac{5}{21} x, \quad \text{etc.}$$

Here for odd index $g_n(x)$ contains only odd powers and is an odd function; while for even index $g_n(x)$ contains only even powers and is even.

Except for constant factors, these are the *Legendre polynomials* $P_n(x)$. We define

$$P_0(x) = g_0(x), \quad P_n(x) = \frac{1 \cdot 3 \cdots (2n-1)}{n!} g_n(x), \quad n = 1, 2, \ldots, \quad (3\text{-}120)$$

so that

$$P_0(x) = 1, \quad P_1(x) = x, \quad P_2(x) = \frac{3}{2}x^2 - \frac{1}{2},$$

$$P_3(x) = \frac{5}{2}x^3 - \frac{3}{2}x, \quad P_4(x) = \frac{35}{8}x^4 - \frac{15}{4}x^2 + \frac{3}{8}, \quad (3\text{-}121)$$

$$P_5(x) = \frac{63}{8}x^5 - \frac{70}{8}x^3 + \frac{15}{8}x, \quad \text{etc.}$$

The first few polynomials are graphed in Fig. 3–19.

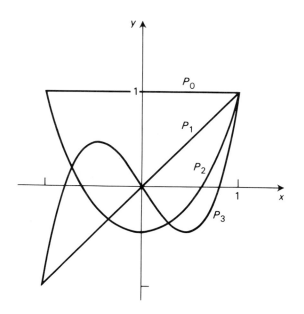

Fig. 3-19. Legendre polynomials.

We remark that such an orthogonal system of polynomials with given weight function, the nth one being of degree n, is determined uniquely up to scale factors. This follows from the uniqueness rule of the previous section.

The Legendre polynomials have many special properties. We list a few of these here (for more details and proofs, see Chapter 12).

Rodrigues' formula:

$$P_0(x) = 1, \quad P_n(x) = \frac{1}{2^n n!} \frac{d^n}{dx^n}(x^2 - 1)^n, \quad n = 1, 2, \ldots \quad (3\text{-}122)$$

Differential equation:
$$(1 - x^2) P_n''(x) - 2x P_n'(x) + n(n + 1) P_n(x) = 0. \tag{3–123}$$

Norm:
$$\|P_n\|^2 = \int_{-1}^{1} [P_n(x)]^2 \, dx = \frac{2}{2n + 1}. \tag{3–124}$$

Bound:
$$|P_n(x)| \leq 1 \quad \text{for} \quad -1 \leq x \leq 1. \tag{3–125}$$

Recursion formula:
$$(n + 1) P_{n+1}(x) = (2n + 1) x P_n(x) - n P_{n-1}(x), \quad n = 1, 2, \ldots \tag{3–126}$$

An explicit formula for $P_n(x)$ is given in Problem 6 following Section 2–22 and in the answer to this problem.

In accordance with Section 3–9, we can now form the *Fourier–Legendre series* of an arbitrary function $f(x)$. This is the series:

$$\sum_{n=0}^{\infty} c_n P_n(x), \quad \text{with} \quad c_n = \frac{2n + 1}{2} \int_{-1}^{1} f(x) P_n(x) \, dx. \tag{3–127}$$

It can be shown that the Legendre polynomials form a *complete* orthogonal system, so that the *Fourier–Legendre* series converges in the mean to f for every f piecewise continuous on $-1 \leq x \leq 1$. That is,

$$\lim_{n \to \infty} \|f(x) - \{c_0 P_0(x) + \cdots + c_n P_x(x)\}\| = 0.$$

There are also theorems on uniform convergence similar to those for Fourier series. For example, the Fourier–Legendre series of f converges uniformly to f for $-1 \leq x \leq 1$ if f has continuous first and second derivatives for $-1 \leq x \leq 1$. (For details see AC, pp. 499–503, and JACKSON, Chapter II.)

Since $P_n(x)$ is even for n even and odd for n odd, we verify, as for Fourier series, that when f is even $c_n = 0$ for n odd, and when f is odd, $c_n = 0$ for n even.

EXAMPLE 1 Let $f(x) = |x|$, so that $f(x) = x$ for $0 \leq x \leq 1$ and $f(x) = -x$ for $-1 \leq x \leq 0$. Since f is even, $c_n = 0$ for n odd, and for n even

$$c_n = 2 \cdot \frac{2n + 1}{2} \int_0^1 f(x) P_n(x) \, dx = (2n + 1) \int_0^1 x P_n(x) \, dx.$$

We find:

$$c_0 = \int_0^1 x \, dx = \frac{1}{2}, \quad c_2 = 5 \int_0^1 \left(\frac{3x^3}{2} - \frac{x}{2} \right) dx = \frac{5}{8},$$

$$c_4 = 9 \int_0^1 \left(\frac{35x^5}{8} - \frac{15}{4} x^3 + \frac{3}{8} x \right) dx = -\frac{3}{16}.$$

Hence

$$f(x) \approx \frac{1}{2} P_0(x) + \frac{5}{8} P_2(x) - \frac{3}{16} P_4(x) = \frac{15 + 210x^2 - 105x^4}{128}.$$

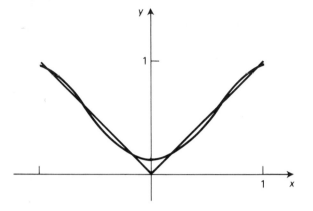

Fig. 3–20. Approximation of $|x|$ by partial sum of its Fourier–Legendre series.

The function and approximating partial sum are shown in Fig. 3–20. By (3–107), the total square error is

$$E_4 = ||f||^2 - \left(\frac{1}{4}||P_0||^2 + \frac{25}{64}||P_2||^2 + \frac{9}{256}||P_4^2||\right) = \frac{1}{384}.$$

(The mean square error is one half of this and the RMS error is 0.036.) Thus we have obtained a very good approximation to $|x|$ for $-1 \leqslant x \leqslant 1$ by a polynomial. ◄

EXAMPLE 2 In studying gravitational forces, we encounter the function $1/r$, where r is one side of a triangle expressed in terms of the other two sides ρ, ρ' and the angle γ between them (see Fig. 3–21).

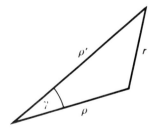

Fig. 3–21. Relationships in a triangle.

By the law of cosines,

$$r^2 = \rho^2 + \rho'^2 - 2\rho\rho'\cos\gamma = \rho'^2(1 + v^2 - 2uv),$$

where we have written v for ρ/ρ' and u for $\cos\gamma$. Hence,

$$\frac{1}{r} = \frac{1}{\rho'}(1 + v^2 - 2uv)^{-1/2}.$$

We now use the binomial series (Section 2–18) and can write, for $|x| < 1$,

$$(1-x)^{-1/2} = 1 + \frac{1}{2}x + \frac{1}{2}\cdot\frac{3}{4}x^2 + \cdots + \frac{1}{2}\cdot\frac{3}{4}\cdots\frac{2n-1}{2n}x^n + \cdots$$

Therefore, for $|2uv - v^2| < 1$,

$$\frac{1}{r} = \frac{1}{\rho'}\left[1 + \frac{1}{2}(2uv - v^2) + \frac{3}{8}(2uv - v^2)^2 + \cdots\right].$$

We collect terms of the same degree in v on the right and obtain:

$$\frac{1}{r} = \frac{1}{\rho'}\left[P_0(u) + P_1(u)v + P_2(u)v^2 + \cdots + P_n(u)v^n + \cdots\right],$$

where $P_0(u) = 1$, $P_1(u) = u$, $P_2(u) = (3/2)[u^2 - (1/3)]$, ...It can be verified that, as the notation suggests, $P_n(u)$ is the nth Legendre polynomial (see Section 12–7 for a proof).

The result found implies that for fixed v such that $|2uv - v^2| < 1$ for $-1 \leqslant u \leqslant 1$, the function $(1 + v^2 - 2uv)^{-1/2}$ has the Fourier–Legendre series $\sum_{n=0}^{\infty} P_n(u)v^n$. Therefore, by Eq. (3–127), for $n = 0, 1, 2, \ldots$

$$\frac{2n+1}{2}\int_{-1}^{1}(1 + v^2 - 2uv)^{-1/2}P_n(u)\,du = v^n. \quad \blacktriangleleft$$

========= **Problems (Section 3–12)** =========

1. Verify that each of the following finite or infinite sets of functions forms an orthogonal system over the interval given:
   a) $\{\sin 2nx\}$, $n = 1, 2, \ldots, 0 \leqslant x \leqslant \pi/2$
   b) $\{\cos 3nx\}$, $n = 0, 1, 2, \ldots, 0 \leqslant x \leqslant \pi/3$
   c) $x + 1$, $9x - 5$, $6x^2 - 6x + 1$, $0 \leqslant x \leqslant 1$
   d) x, $\cos x$, $\cos 2x$, $-\pi \leqslant x \leqslant \pi$

2. Verify that each of the following sets of function forms an orthogonal system for the given interval and weight function:
   a) 1, x, $5x^2 - 1$, $w(x) = 1 - x^2$, $-1 \leqslant x \leqslant 1$
   b) 1, x, $2x^2 - 1$, $w(x) = (1 - x^2)^{-1/2}$, $-1 \leqslant x \leqslant 1$
   c) 1, $2 - x$, $x^2 - 6x + 6$, $w(x) = xe^{-x}$, $0 \leqslant x < \infty$

   HINT: First show that $\displaystyle\int_0^{\infty} x^n e^{-x}\,dx = n!$

   d) 1, x, $x^2 - 1$, $w(x) = e^{-x^2/2}$, $-\infty \leqslant x < \infty$.

3. Let $(f, g) = \displaystyle\int_0^1 f(x)g(x)\,dx$, $f_1(x) = x$, $f_2(x) = x^2$, $f_3(x) = x^3$. Evaluate:
   a) (f_1, f_2) b) (f_1, f_3)
   c) $\|f_1\|^2$ d) $\|f_2\|$
   e) $(f_1 + f_2, f_1 + f_3)$ f) $(2f_1 - f_2, f_2 - f_3)$

4. In the notations of Problem 3, find the following:
   a) g, a nonzero linear combination of f_1 and f_2 such that $(f_1, g) = 0$
   b) h, a nonzero linear combination of f_1, f_2, f_3 such that $(h, f_1) = 0$ and $(h, f_2) = 0$

5. Carry out the Gram–Schmidt process for the interval $0 \leqslant x \leqslant 1$ using as f_n the function x^n for $n = 0, 1, 2, \ldots$ and obtain g_0, g_1, g_2, g_3.

6. Carry out the Gram–Schmidt process for the interval $0 \leqslant x \leqslant 1$, using as f_n the function e^{nx} for $n = 0, 1, 2, \ldots$ and obtain g_0, g_1, g_2, g_3.

7. Expand $f(x)$ in a Fourier–Legendre series, obtaining the terms up to the one in $P_5(x)$, and graph f and the corresponding partial sum:

   a) $f(x) = 1$ for $0 \leqslant x \leqslant 1, f(x) = -1$ for $-1 \leqslant x < 0$
   b) $f(x) = x^2$ for $0 \leqslant x \leqslant 1, f(x) = -x^2$ for $-1 \leqslant x \leqslant 0$
   c) $f(x) = e^x$ for $-1 \leqslant x \leqslant 1$.

   HINT: By integration by parts,

   $$\int x^n e^x \, dx = e^x [x^n - nx^{n-1} + n(n-1)x^{n-2} + \cdots + (-1)^n n!] + C.$$

   d) $f(x) = \sin x$ for $-\pi \leqslant x \leqslant \pi$

8. **a)** Let $\{Q_n(x)\}$, where $n = 0, 1, 2, \ldots$, be a sequence of polynomials and Q_n have degree n. Show that x^n can be expressed as a linear combination of $Q_0(x), Q_1(x), \ldots, Q_n(x)$ for each n.

   b) Show that if the sequence $\{Q_n(x)\}$ of part (a) is an orthogonal system on the interval $a \leqslant x \leqslant b$, then $Q_n(x)$ is orthogonal to each polynomial of degree less than n.

9. On the basis of Problem 8(a),
   a) Express x^2 as a linear combination of $P_0(x), P_1(x), P_2(x)$.
   b) Express x^5 as a linear combination of $P_0(x), P_1(x), \ldots, P_5(x)$.

   REMARK: These give the Fourier–Legendre series for x^2 and x^5 respectively.

10. Verify for the Legendre polynomials $P_2(x)$ and $P_3(x)$:
    a) Rodrigues' formula (3–122)
    b) the differential equation (3–123)
    c) the norm (3–124)
    d) the bound (3–125)

11. Use the recursion formula (3–126) to calculate $P_2(x), \ldots, P_5(x)$ from $P_0(x)$ and $P_1(x)$.

3–13 OTHER ORTHOGONAL SYSTEMS

For convenience we list here several other important orthogonal systems. In each case we give the interval, the weighting function, a general formula, a differential equation of second order satisfied by the functions, a recursion formula when available.

   Chebyshev polynomials of the first kind: $T_n(x), \ -1 \leqslant x \leqslant 1,$

   $$w(x) = (1 - x^2)^{-1/2},$$

   $$T_n(x) = \cos[n \cos^{-1} x],$$

   $$\left.\begin{array}{l} (1 - x^2) \dfrac{d^2}{dx^2} T_n(x) - x \dfrac{d}{dx} T_n(x) + n^2 T_n(x) = 0, \\[2mm] T_{n+1}(x) = 2x \, T_n(x) - T_{n-1}(x). \end{array}\right\} \qquad (3\text{–}130)$$

Laguerre polynomials: $L_n^{\alpha}(x),\ 0 \leqslant x < \infty,\ w(x) = x^{\alpha} e^{-x},\ \alpha > -1,$

$$L_n^{\alpha}(x) = \frac{1}{n!} e^x x^{-\alpha} \frac{d^n}{dx^n} (e^{-x} x^{n+\alpha}),$$

$$x \frac{d^2}{dx^2} L_n^{\alpha}(x) + (\alpha + 1 - x) \frac{d}{dx} L_n^{\alpha}(x) + n L_n^{\alpha}(x) = 0, \qquad (3\text{–}131)$$

$$L_{n+1}^{\alpha}(x) = \left(\frac{2n + \alpha + 1}{n+1} - \frac{x}{n+1} \right) L_n^{\alpha}(x) - \frac{n+\alpha}{n+1} L_{n-1}^{\alpha}(x).$$

Hermite polynomials: $H_n(x),\ -\infty < x < \infty,\ w(x) = e^{-x^2/2},$

$$H_n(x) = (-1)^n e^{x^2/2} \frac{d^n}{dx^n} \left[e^{-x^2/2} \right],$$

$$\frac{d^2}{dx^2} H_n(x) - x \frac{d}{dx} H_n(x) + n H_n(x) = 0 \qquad (3\text{–}132)$$

$$H_{n+1}(x) = x H_n(x) - n H_{n-1}(x).$$

Bessel functions of first kind of order m: $J_m(\lambda_{mn} x),\ m \geqslant 0,\ 0 \leqslant x \leqslant 1,\ w(x) = x,$
where $\lambda_{m0},\ \lambda_{m1},\ \lambda_{m2},\ \ldots$ are the successive positive zeros of the Bessel function

$$J_m(x) = \sum_{k=0}^{\infty} \frac{(-1)^k x^{m+2k}}{2^{m+2k} k!\, \Gamma(m+k+1)}, \qquad (3\text{–}133)$$

where $\Gamma(z)$ is the gamma function (Section 12–3),

$$x^2 \frac{d^2}{dx^2} J_m(x) + x \frac{dJ_m}{dx} + (x^2 - m^2) J_m(x) = 0. \qquad (3\text{–}134)$$

For further information on these functions see Chapter 12.

3–14 STURM–LIOUVILLE PROBLEMS

For the vibrations of a single particle (simple harmonic motion) there is a single frequency. For a system of two particles on a line, coupled by springs (as in Problem 5 following Section 1–15), two frequencies occur. For n such particles, there are n frequencies. For the vibrations of a continuous medium, such as a string or a membrane, there is an infinite sequence of frequencies $\lambda_1, \lambda_2, \ldots, \lambda_m, \ldots$ A detailed analysis of some typical cases is given in Chapter 8. The methods of Chapter 8 lead to boundary-value problems, whose solutions give the desired frequencies. As a side-product, these boundary value problems produce a number of orthogonal systems. For one-dimensional problems, such as the vibrating string, the boundary-value problems have a special form; they are called Sturm–Liouville problems. We introduce these by a specific example.

EXAMPLE 1   Find all λ, $y(x)$, where $\lambda = $ const and $y(x) \not\equiv 0$, satisfying the set of conditions:

$$y'' + \lambda y = 0 \quad \text{for} \quad 0 \leqslant x \leqslant 1,$$
$$y'(0) = 0, \quad y(1) + y'(1) = 0. \tag{3-140}$$

Solution.   We must find λ such that the differential equation has a nonzero solution satisfying the boundary conditions. If $\lambda = 0$, then $y = c_1 + c_2 x$, and the boundary conditions give $c_2 = 0$, $c_1 + c_2 + c_2 = 0$, so that $y \equiv 0$. Hence $\lambda = 0$ is not allowed.
  If $\lambda > 0$, we can write $\lambda = \omega^2$ for a positive ω and

$$y = c_1 \cos \omega x + c_2 \sin \omega x.$$

The boundary conditions give $c_2 = 0$ and

$$c_1 \cos \omega + c_2 \sin \omega - \omega c_1 \sin \omega + \omega c_2 \cos \omega = 0.$$

Since $c_2 = 0$, $c_1 \neq 0$, and therefore we must have

$$\cos \omega - \omega \sin \omega = 0 \qquad \text{or} \qquad \tan \omega = \frac{1}{\omega}.$$

  From a graph in Fig. 3–22 we see that $\tan \omega = 1/\omega$ has solutions $\omega_0, \omega_1, \ldots,$ with $0 < \omega_0 < \omega_1 < \ldots$ and with $\omega_n/n \to \pi$ as $n \to \infty$. Thus for $\lambda > 0$

$$y = \varphi_n(x) = \cos \omega_n(x),$$
$$\lambda = \omega_n^2, \quad n = 0, 1, 2, \ldots \tag{3-141}$$

gives all solutions of our problem up to multiplication of $\varphi_n(x)$ by nonzero constants. We can show that for $\lambda < 0$ there are no solutions (see Problem 5(a) below). Hence (3–141) gives all solutions up to the nonzero constant factors.

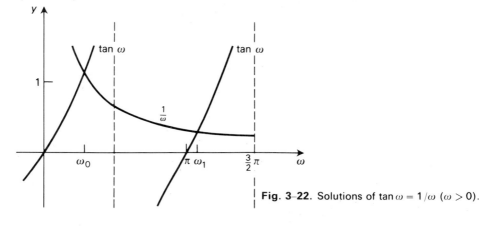

Fig. 3-22. Solutions of $\tan \omega = 1/\omega$ ($\omega > 0$).

  The system $\{\varphi_n(x)\}$ is an *orthogonal system* on the interval $0 \leqslant x \leqslant 1$, since

$$(\varphi_n', \varphi_m) = \int_0^1 \cos \omega_n x \cos \omega_m x \, dx \begin{cases} = 0, & n \neq m, \\ > 0, & n = m, \end{cases} \tag{3-142}$$

(see Problem 3(b) below). ◄

Our Example 1 is an illustration of a *Sturm–Liouville boundary-value problem*; the solutions $\{\varphi_n(x)\}$ are called *eigenfunctions* of the problem, the numbers $\lambda_n = \omega_n^2$ are called the *eigenvalues* of the problem. A common form of a Sturm–Liouville boundary-value problem is the following:

$$p(x)y'' + p'(x)y' + [\lambda\rho(x) - q(x)]y = 0, \quad a \leqslant x \leqslant b, \tag{3-143}$$

$$\alpha_1 y(a) + \alpha_2 y'(a) = 0, \quad \beta_1 y(b) + \beta_2 y'(b) = 0, \tag{3-144}$$

where $p(x)$, $p'(x)$, $q(x)$, $\rho(x)$ are continuous on the interval $a \leqslant x \leqslant b$, $p(x) > 0$ and $\rho(x) > 0$ on the interval, $\alpha_1, \alpha_2, \beta_1, \beta_2$ are constants, and $\alpha_1^2 + \alpha_2^2 \neq 0$, $\beta_1^2 + \beta_2^2 \neq 0$. The boundary conditions (3–144) also appear in other forms, for example, as periodicity conditions:

$$y(a) - y(b) = 0, \quad y'(a) - y'(b) = 0 \tag{3-144'}$$

or, most generally, in the form

$$\alpha_1 y(a) + \alpha_2 y'(a) + \alpha_3 y(b) + \alpha_4 y'(b) = 0,$$
$$\beta_1 y(a) + \beta_2 y'(a) + \beta_3 y(b) + \beta_4 y'(b) = 0, \tag{3-144''}$$

where the two conditions are independent in the sense that neither one can be obtained from the other by multiplying by a constant.

With the problem (3–143)–(3–144) we associate the inner product

$$(f, g) = \int_a^b f(x)g(x)\rho(x)\,dx. \tag{3-145}$$

It can be shown that the eigenvalues form a sequence $\{\lambda_n\}$, where $\lambda_0 < \lambda_1 < \lambda_2 < \ldots < \lambda_n < \ldots$, with $\lambda_n \to \infty$ as $n \to \infty$. For each λ_n there is a nonzero eigenfunction φ_n (unique up to multiplication by a nonzero constant) and all solutions of the problem for $\lambda = \lambda_n$ are given by $y = \text{const} \cdot \varphi_n(x)$. The functions $\{\varphi_n(x)\}$ form a complete orthogonal system on the interval $a \leqslant x \leqslant b$ with inner product (3–145).

For the general boundary conditions (3–144'') we have similar results, provided that the following extra condition is satisfied:

$$p(a)(\alpha_3\beta_4 - \alpha_4\beta_3) = p(b)(\alpha_1\beta_2 - \alpha_2\beta_1). \tag{3-146}$$

However, some eigenvalues may now be double, so that we repeat them, say, $\lambda_n = \lambda_{n+1}$. For such a double eigenvalue we can choose linearly independent eigenfunctions φ_n, φ_{n+1} such that all solutions of the problem for $\lambda = \lambda_n$ are given by $y = c_1\varphi_n(x) + c_2\varphi_{n+1}(x)$. All the φ_n together again form an orthogonal system on the interval $a \leqslant x \leqslant b$ with the inner product (3–145).

EXAMPLE 2 $y'' + \lambda y = 0$, $0 \leqslant x \leqslant 2\pi$, $y(0) = y(2\pi)$, $y'(0) = y'(2\pi)$. It can be verified, as in Example 1, that the eigenvalues are $0, 1, 1, 4, 4, \ldots, n^2, n^2, \ldots$, so that $\lambda_0 = 0$, $\lambda_1 = \lambda_2 = 1$, $\lambda_3 = \lambda_4 = 4, \ldots$, and the corresponding eigenfunctions are $1, \cos x, \sin x, \cos 2x, \sin 2x, \ldots, \cos nx, \sin nx, \ldots$ Thus we simply recover our *Fourier series*. ◀

EXAMPLE 3 $y'' + \lambda y = 0$, $0 \leqslant x \leqslant \pi$; $y(0) = 0$, $y(\pi) = 0$. It can be verified (see Problem 5(b) below) that all eigenvalues are positive, say $\lambda = \omega^2$, with $\omega > 0$. Then

$$y = c_1 \cos \omega x + c_2 \sin \omega x,$$

and by the boundary conditions

$$c_1 = 0, \ c_1 \cos \omega \pi + c_2 \sin \omega \pi = 0.$$

Hence $\omega = 1, 2, \ldots, n, \ldots$ is allowed, and our eigenvalues are $1^2, 2^2, \ldots, n^2, \ldots,$ with corresponding eigenfunctions $\sin x$, $\sin 2x$, ..., $\sin nx$, ... We recover the *Fourier sine series.* ◀

EXAMPLE 4 $y'' + \lambda y = 0$, $0 \leqslant x \leqslant \pi$, $y'(0) = 0$, $y'(\pi) = 0$. It is left as an exercise to show that this problem leads to *Fourier cosine series* (see Problem 6(a) below). ◀

If in our Sturm–Liouville problem the coefficients $p(x)$, $p'(x)$, $\rho(x)$, or $q(x)$ have discontinuities (usually at a or b), or $p(a) = 0$ or $p(b) = 0$, or if the interval is infinite, then the problem becomes a *singular Sturm–Liouville problem*. The boundary conditions are then usually modified, especially for the infinite interval. The theory of eigenvalues and eigenfunctions can be extended to such singular problems under suitable restrictions. In certain cases the results obtained are like those stated above for (3–143) and (3–144). These cases include problems leading to the Legendre polynomials, Hermite polynomials, and other orthogonal systems described in Section 3–13. For example, in the case of Legendre polynomials we have a singular problem because $p(x) = 1 - x^2$ in the differential equation (3–123), so that $p(x) = 0$ at *both* endpoints ± 1.

For other singular Sturm–Liouville problems there may be no sequence of eigenvalues and we may have a *continuous spectrum*. The expansions in series are replaced by expansions in *integrals*, such as the *Fourier* integral (Chapter 4).

For more background on the Sturm–Liouville problems see the following references: CODDINGTON–LEVINSON; COURANT–HILBERT; R. H. COLE, *Theory of Ordinary Differential Equations*, Appleton–Century–Crofts, New York, 1968; G. HELLWIG, *Differential Operators of Mathematical Physics*, Addison–Wesley, Reading, Mass., 1964.

Sturm–Liouville problems and all the associated systems of orthogonal functions arise in physical problems of great variety: mechanical vibrations, wave motion, heat conduction, diffusion, electromagnetism, fluid motion, elasticity, quantum mechanics. The eigenvalues appear at times as crucial frequencies of vibrations, also as determining wave lengths. The physical problems are usually first formulated as *partial* differential equations. Then simplifying assumptions or symmetry properties lead to ordinary differential equations, such as those in Sections 3–13 and 3–14. In Chapter 8 we treat partial differential equations and illustrate the procedures referred to.

3–15 ORTHOGONAL SYSTEMS IN HIGHER DIMENSIONS

The concept of orthogonal system extends immediately to functions of two or more variables. For example, given a region R in the xy-plane, such as a square or a rectangle, we introduce an inner product

$$(f, g) = \iint_R f(x, y)g(x, y)w(x, y)\,dx\,dy,$$

where $w(x, y)$ is a given function on R, usually continuous and positive except perhaps on the boundary of R.

With this inner product, the discussion of Section 3–10 can be repeated, and we are led as before to the concept of orthogonal system: a system $\{\varphi_n(x, y)\}$ such that $(\varphi_n, \varphi_m) = 0$ for $n \neq m$, but $(\varphi_n, \varphi_n) > 0$. Completeness can also be defined as before.

Such orthogonal systems arise naturally in physical problems. For example, the vibrations of a membrane (Section 8–13) filling the square $0 \leqslant x \leqslant \pi, 0 \leqslant y \leqslant \pi$ lead to the complete orthogonal system with weight $w(x, y) \equiv 1$:

$$\sin x \sin y, \quad \sin 2x \sin y, \quad \sin x \sin 2y, \quad \ldots, \quad \sin mx \sin ny, \quad \ldots$$

(These can be numbered as a single sequence.)

In general, if $\{f_m(x)\}$ is an orthogonal system for weight function $w_1(x)$ on the interval $a \leqslant x \leqslant b$ and $g_n(y)$ is an orthogonal system for weight function $w_2(y)$ on the interval $c \leqslant y \leqslant d$, then the functions $\varphi_{mn}(x, y) = f_m(x)g_n(y)$, when numbered to form a single sequence, form an orthogonal system with weight function $w(x, y) = w_1(x)w_2(y)$ on the rectangle $a \leqslant x \leqslant b, c \leqslant y \leqslant d$. For

$$(\varphi_{mn}, \varphi_{m'n'}) = \int_a^b \int_c^d f_m(x)g_n(y)f_{m'}(x)g_{n'}(y)w_1(x)w_2(y)\,dy\,dx$$

$$= \int_a^b f_m(x)f_{m'}(x)w_1(x)\,dx \int_c^d g_n(y)g_{n'}(y)w_2(y)\,dy,$$

and this is 0 except for $m = m', n = n'$. It can also be verified that if $\{f_m\}$ and $\{g_n\}$ are complete, then so is $\{\varphi_{mn}(x, y)\}$.

We also observe that, if $\{f_m(r)\}$ is an orthogonal system with weight function r for the interval $0 \leqslant r \leqslant 1$ and $\{g_n(\theta)\}$ is an orthogonal system with weight function 1 for the interval $0 \leqslant \theta \leqslant 2\pi$, then $\{\varphi_{mn}(r, \theta)\} = \{f_m(r)g_n(\theta)\}$ is an orthogonal system over the set $0 \leqslant r \leqslant 1, 0 \leqslant \theta \leqslant 2\pi$ with weight function r. The reasoning is as above. However, if the $g_n(\theta)$ are periodic in θ, with period 2π, we can interpret the φ_{mn} as an orthogonal system of functions over the circular region $x^2 + y^2 \leqslant 1$ with weight function 1. (See Section 10–4 for double integrals in polar coordinates.)

$=$ **Problems** $=$

1. With reference to the Chebyshev polynomials of Eqs. (3–130),
   a) show that $T_0(x) = 1$, $T_1(x) = x$,
   b) use the recursion formula to find $T_2(x)$, $T_3(x)$,
   c) verify that $T_0(x)$, $T_1(x)$, $T_2(x)$, $T_3(x)$ satisfy the differential equation for $n = 0, 1, 2, 3$ respectively.

2. With the reference to the Laguerre polynomials of Eqs. (3–131),
   a) show that $L_0^a(x) = 1$, $L_1^a(x) = 1 + \alpha - x$,
   b) use the recursion formula to find $L_2^a(x)$ and $L_3^a(x)$,
   c) verify that $L_1^a(x)$, $L_2^a(x)$, $L_3^a(x)$ satisfy the differential equation for $n = 1, 2, 3$ respectively.

3. **a)** With the aid of tables find the solutions ω_1, ω_2, ω_3 of the equation $\tan \omega = 1/\omega$ to two decimal places and hence find the eigenvalues λ_0, λ_1, λ_2 for the boundary-value problem (3–140).
   b) Show that (3–142) is valid for positive solutions ω_n and ω_m of $\tan \omega = 1/\omega$.

4. For the Sturm–Liouville problem of Example 2 in Section 3–14, verify that (3–146) is satisfied.

5. Show that if $\lambda = 0$ or $\lambda < 0$, then λ cannot be an eigenvalue for
   a) Example 1 in Section 3–14
   b) Example 3 in Section 3–14.

6. Find the eigenvalues and eigenfunctions for the following Sturm–Liouville boundary-value problems:

   a) Example 4 in Section 3–14
   b) $y'' + \lambda y = 0$, $0 \leqslant x \leqslant \pi$, $y(0) = 0$, $y'(\pi) = 0$
   c) $y'' + \lambda y = 0$, $0 \leqslant x \leqslant 1$, $y(0) = 0$, $y(1) + y'(1) = 0$
   d) $y'' + \lambda y = 0$, $0 \leqslant x \leqslant 1$, $y(0) + y'(0) = 0$, $y'(1) = 0$.

7. Verify that each of the following is an orthogonal system over the set indicated with the weight function w given:

   a) $\varphi_{mn}(x, y) = \cos mx \cos ny$, $0 \leqslant x \leqslant \pi$, $0 \leqslant y \leqslant \pi$, $m = 0, 1, 2, \ldots$, $n = 0, 1, 2, \ldots$,
   $w(x, y) = 1$
   b) $\varphi_{mn}(x, y) = \sin mx \cos ny$, $0 \leqslant x \leqslant \pi$, $0 \leqslant y \leqslant \pi$, $m = 1, 2, \ldots$, $n = 0, 1, 2, \ldots$,
   $w(x, y) = 1$

8. Expand each function in a series of orthogonal functions as requested:
   a) $f(x, y) = x + y$, $0 \leqslant x \leqslant \pi$, $0 \leqslant y \leqslant \pi$, in the system of Problem 7(a)
   b) $f(x, y) = xe^y$, $0 \leqslant x \leqslant \pi$, $0 \leqslant y \leqslant \pi$, in the system of Problem 7(b)

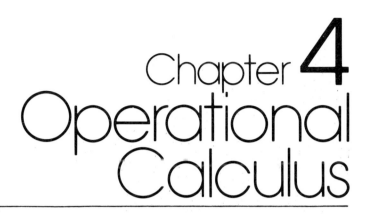

Chapter 4
Operational Calculus

4–1 INTRODUCTION

In this chapter we develop the theory of Laplace transforms, Fourier transforms, and several related concepts. These turn out to be powerful tools for solving both ordinary and partial differential equations.

One source of the ideas studied here is the concept of differential operator, such as the operators D, $D+5$, $2D^2+3D+1, \ldots$, studied in Section 1–15. The transforms to be defined lead to procedures paralleling those for differential operators, and hence the transform theory is called *operational calculus*. It will be seen that the principal idea is that of *operator*: a procedure for assigning functions to functions.

In engineering the transforms play a decisive role in a great variety of problems. In particular, they are heavily used in the study of *feedback control systems*. A simple example of such a system is the control of temperature in a building by a thermostat (Fig. 4–1). The desired temperature is chosen; at each instant this is compared to the actual temperature, and the thermostat responds by turning on the furnace when the actual temperature is below the desired temperature. The furnace provides heat and this, along with heat gain or loss from the outside, determines the actual temperature. The feedback consists of comparing the actual to the desired temperature.

In this age of automatic devices such systems as speed governors, automatic pilots, chemical-process control, control of water level in boilers and others are widely used. Economic theory prescribes ways of responding to inflation, unemployment and other crucial indicators by adjusting taxes and interest rates; this is also a feedback control system. The human body involves many control systems: for body

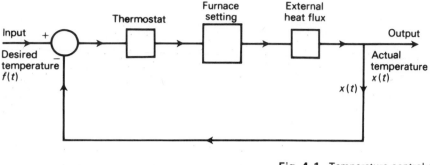

Fig. 4–1. Temperature control system.

temperature, heart beat, eye motion, muscle response, breathing rate, etc.

A mathematical model appropriate to many of these systems is a linear differential equation with constant coefficients connecting an input $f(t)$ and an output $x(t)$:

$$a_0 D^n x + \cdots + a_n x = f(t).$$

The Laplace transform provides a simple language for studying these equations and especially for highlighting important physical concepts, such as *stability*. The closely related Fourier transform brings out the frequency aspects of the process. In essence it allows us to predict behavior under general conditions from a knowledge of how the system responds to a sinusoidal input of arbitrary frequency.

4–2 **THE LAPLACE TRANSFORM**

We consider functions $f(t)$, $g(t)$, ... of the real variable t defined for $0 \leqslant t < \infty$. To each such function $f(t)$ we assign a function $F(s)$ by the rule

$$F(s) = \int_0^\infty f(t)e^{-st}dt, \tag{4-20}$$

provided the integral has meaning for some values of s. We call the function $F(s)$ the *Laplace transform of $f(t)$* and write

$$F = \mathscr{L}[f]. \tag{4-21}$$

Thus $\mathscr{L}$ is an operator that assigns functions of s to functions of t. The operator $\mathscr{L}$ is also called the *Laplace transform*.

EXAMPLE 1   Let $f(t) = 1$ for $0 \leqslant t < \infty$. Then

$$F(s) = \int_0^\infty e^{-st}dt = \frac{e^{-st}}{-s}\bigg|_0^\infty = \frac{1}{s}.$$

The steps are valid for $s > 0$, since $e^{-st} \to 0$ as $t \to \infty$ if $s > 0$. If $s \leqslant 0$, the integral has no meaning. ◄

EXAMPLE 2   Let $f(t) = e^{at}$, $0 \leqslant t < \infty$. Then

$$F(s) = \int_0^\infty e^{-st} e^{at}\, dt = \int_0^\infty e^{(a-s)t}\, dt = \frac{e^{(a-s)t}}{a-s}\bigg|_0^\infty = \frac{1}{s-a},$$

provided $s > a$. ◄

EXAMPLE 3   Let $f(t) = \sin kt$, $0 \leqslant t < \infty$. Then

$$F(s) = \int_0^\infty e^{-st} \sin kt\, dt = \frac{e^{-st}}{s^2 + k^2}(-s \sin kt - k \cos kt)\bigg|_0^\infty = \frac{k}{s^2 + k^2},$$

provided $s > 0$. ◄

We observe that the appearing functions of s are very simple rational functions. Other types of functions can also be produced.

EXAMPLE 4   Let $f(t) = 1$ for $0 \leqslant t \leqslant 1$, $f(t) = 0$ for $t > 1$. Then

$$F(s) = \int_0^1 e^{-st}\, dt = \frac{1-e^{-s}}{s}.$$

Here the result is valid for all s except 0. For $s = 0$, the integral equals 1. Since $(1-e^{-s})/s$ has limit 1 as $s \to 0$, we do not usually give the value at 0 separately. ◄

Extension to complex s

We observe that the functions $F(s)$ of our examples all have meaning for complex s. It is advantageous to let s be complex in the definition (4–20). The integration of the complex-valued function of t is carried out as in Section 1–10. For simplicity, we shall for the most part treat s as real.

Allowed values of s

For real positive s, function e^{-st} approaches 0 as $t \to \infty$; the larger the value of s, the more rapidly it approaches 0. Hence, typically, the integral $\int_0^\infty e^{-st} f(t)\, dt$ converges, if at all, for sufficiently large s. For most applications we need to know only this much and do not need to know precisely what the restrictions on s are. We shall usually give the restrictions in the following discussion, since occasionally it is necessary to know them. However, this detail can often be ignored. We give more information on this question in Section 4–5.

The functions $f(t)$ will generally be assumed to be *piecewise continuous* in each finite interval $0 \leqslant t \leqslant a$; that is, continuous except for a finite number of jump discontinuities (see Section 3–2). For certain results it may be required that f be *piecewise smooth* in each finite interval (See Section 3–3).

4–3 PROPERTIES OF LAPLACE TRANSFORM

The Laplace transform $\mathscr{L}$ is a *linear operator*: if c_1 and c_2 are constants, then

$$\mathscr{L}[c_1 f_1 + c_2 f_2] = c_1 \mathscr{L}[f_1] + c_2 \mathscr{L}[f_2]. \tag{4-30}$$

More precisely, if $\mathscr{L}[f_1]$ exists for $s > a_1$ and $\mathscr{L}[f_2]$ exists for $s > a_2$, then $\mathscr{L}[c_1 f_1 + c_2 f_2]$ exists for $s > a$, where a is the larger of a_1, a_2, and Eq. (4–30) is valid. The rule (4–30) follows at once from the definition (4–20).

Transform of derivative

The rule is

$$\mathscr{L}[f'] = s\mathscr{L}[f] - f(0). \tag{4-31}$$

This is valid if f and f' are continuous for $t \geq 0$ and $\mathscr{L}[f]$, $\mathscr{L}[f']$ both exist for s sufficiently large, say for $s > a$. To prove (4–31) we take $s > a$, integrate from 0 to t, and then use integration by parts:

$$\int_0^t f'(u)e^{-su}\,du = f(u)e^{-su}\Big|_0^t + s\int_0^t f(u)e^{-su}\,du$$

$$= f(t)e^{-st} - f(0) + s\int_0^t f(u)e^{-su}\,du. \tag{4-32}$$

If we now let $t \to \infty$, then both integrals have limits, so that $f(t)e^{-st}$ also has a limit. This limit must be 0, since otherwise $\int_0^\infty f(t)e^{-st}\,dt$ could not exist (see Problem 5 following Section 4–4). Therefore $f(t)e^{-st} \to 0$ as $t \to \infty$ and (4–31) follows from (4–32) by letting $t \to \infty$.

By repeated application of rule (4–31), we obtain rules for transforms of higher derivatives:

$$\mathscr{L}[f''] = s\mathscr{L}[f'] - f'(0) = s\{s\mathscr{L}[f] - f(0)\} - f'(0)$$
$$= s^2 \mathscr{L}[f] - sf(0) - f'(0),$$
$$\mathscr{L}[f'''] = s\mathscr{L}[f''] - f''(0) = s^3 \mathscr{L}[f] - s^2 f(0) - sf'(0) - f''(0),$$

and in general

$$\mathscr{L}[f^{(n)}] = s^n \mathscr{L}[f] - [s^{n-1} f(0) + s^{n-2} f'(0) + \cdots + f^{(n-1)}(0)]. \tag{4-33}$$

These rules hold under the corresponding assumptions of continuity and existence of the transforms. Rule (4–33) is basic for the application of Laplace transforms to differential equations.

Transform of integral

Let $f(t)$ be continuous for $0 \leq t < \infty$ and let

$$g(t) = \int_0^t f(u)\,du.$$

Then $g(0) = 0$ and by calculus $g'(t) = f(t)$, so that by (4-31)

$$\mathscr{L}[f] = \mathscr{L}[g'] = s\mathscr{L}[g].$$

Therefore we have the rule

$$\mathscr{L}[g] = \mathscr{L}\left[\int_0^t f(u)\,du\right] = \frac{1}{s}\mathscr{L}[f]. \qquad (4\text{-}34)$$

This is valid if $\mathscr{L}[f]$, $\mathscr{L}[g]$ exist for sufficiently large s. It can be shown that existence of $\mathscr{L}[f]$ for $s > a \geqslant 0$ implies existence of $\mathscr{L}[g]$ for $s > a$ (see OMLS, pp. 313–314).

Multiplication by special functions

We have the rules

$$\mathscr{L}[e^{bt}f] = F(s - b), \qquad (4\text{-}35)$$
$$\mathscr{L}[t^n f] = (-1)^n F^{(n)}(s), \quad n = 1, 2, \ldots, \qquad (4\text{-}36)$$

where $\mathscr{L}[f] = F$. For the first of these we have

$$\mathscr{L}[e^{bt}f] = \int_0^\infty e^{bt}e^{-st}f(t)\,dt = \int_0^\infty e^{-(s-b)}f(t)\,dt = F(s-b).$$

If $\mathscr{L}[f]$ exists for $s > a$, then $\mathscr{L}[e^{bt}f]$ exists for $s - b > a$. For the second rule, we have, for $s > a$,

$$\mathscr{L}[tf] = \int_0^\infty te^{-st}f(t)\,dt = \int_0^\infty -\frac{\partial}{\partial s}e^{-st}f(t)\,dt$$
$$= -\frac{d}{ds}\int_0^\infty e^{-st}f(t)\,dt = -F'(s). \qquad (4\text{-}37)$$

The justification of these formal steps requires a deeper analysis (see OMLS, pp. 315–317). By repeated application of (4-37), we obtain (4-36).

From (4-37) we can deduce a rule for $\mathscr{L}[f/t]$, provided that $f(t) = tg(t)$, where $\mathscr{L}[g]$ exists for $s > a$:

$$\mathscr{L}\left[\frac{1}{t}f\right] = \int_s^\infty F(\sigma)\,d\sigma, \qquad (4\text{-}37')$$

where $F(s) = \mathscr{L}[f]$ (see CHURCHILL, p. 66).

Translation

Given $f(t)$ for $t \geqslant 0$, we first define $f(t)$ to be 0 for negative t. Then for $c > 0$, function $f(t - c)$ is obtained from f by translating the graph of f a distance c to the *right*, as shown in Fig. 4-2. We now have

$$\mathscr{L}[f(t-c)] = e^{-cs}\mathscr{L}[f], \quad c > 0, \qquad (4\text{-}38)$$

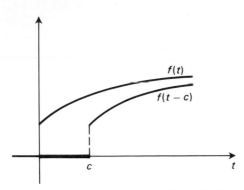

Fig. 4-2. Translation.

since, by the substitution $u = t - c$,

$$\mathscr{L}[f(t-c)] = \int_c^\infty f(t-c)e^{-st}\,dt = \int_0^\infty f(u)e^{-s(u+c)}\,du$$

$$= e^{-cs}\int_0^\infty f(u)e^{-su}\,du = e^{-cs}\,\mathscr{L}[f].$$

The result is valid for $s > a$, as above.

Periodic functions

Let $f(t + \tau) = f(t)$ for $t \geqslant 0$, τ being a positive constant, so that f has period τ for $t \geqslant 0$ (Fig. 4–3). Then

$$\mathscr{L}[f] = \frac{1}{1 - e^{-s\tau}}\int_0^\tau f(t)e^{-st}\,dt, \quad s > 0. \tag{4-39}$$

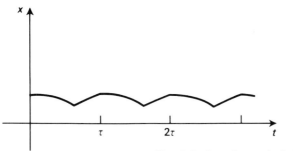

Fig. 4-3. Function periodic for $t \geqslant 0$.

To prove this, we use the rule for translation (4–38):

$$\mathscr{L}[f(t-\tau)] = e^{-s\tau}\,\mathscr{L}[f].$$

However, $f(t - \tau) = f(t)$ for $t \geqslant \tau$ by periodicity and $f(t - \tau) = 0$ for $0 \leqslant t \leqslant \tau$, as in Fig. 4–2. Hence

$$e^{-s\tau}\mathscr{L}[f] = \mathscr{L}[f(t-\tau)] = \int_{\tau}^{\infty} f(t)e^{-st}\,dt = \mathscr{L}[f] - \int_{0}^{\tau} f(t)e^{-st}\,dt,$$

so that (4–39) follows. If f is piecewise continuous here, then because of the periodicity $\mathscr{L}[f]$ exists for $s > 0$ and all steps are valid for $s > 0$.

4–4 EXAMPLES

The following examples illustrate application of the rules.

EXAMPLE 1 $\mathscr{L}[\cosh at] = \mathscr{L}[\frac{1}{2}e^{at} + \frac{1}{2}e^{-at}] = \frac{1}{2}\mathscr{L}[e^{at}] + \frac{1}{2}\mathscr{L}[e^{-at}]$ by linearity. Therefore, by Example 2 of Section 4–2,

$$\mathscr{L}[\cosh at] = \frac{1}{2}\frac{1}{s-a} + \frac{1}{2}\frac{1}{s+a} = \frac{s}{s^2 - a^2}, \qquad (4\text{–}40)$$

provided $s > a$ and $s > -a$ (that is, $s > |a|$. ◀

EXAMPLE 2

$$\mathscr{L}[\cos kt] = \mathscr{L}\left[\frac{d}{dt}\frac{\sin kt}{k}\right] = \frac{1}{k}\mathscr{L}\left[\frac{d}{dt}\sin kt\right] = \frac{1}{k}\{s\mathscr{L}[\sin kt] - \sin 0\}.$$

Here we used linearity to factor out $1/k$ and the rule (4–31) for $\mathscr{L}[f']$. Hence,

$$\mathscr{L}[\cos kt] = \frac{1}{k}\frac{sk}{s^2 + k^2} = \frac{s}{s^2 + k^2}. \qquad (4\text{–}41)$$

The result is valid for $s > 0$. ◀

EXAMPLE 3 $\mathscr{L}[t] = \mathscr{L}[\int_0^t 1\,du] = (1/s)\mathscr{L}[1]$ by the rule (4–34). Hence, by Example 1 of Section 4–2,

$$\mathscr{L}[t] = \frac{1}{s} \cdot \frac{1}{s} = \frac{1}{s^2}.$$

Similarly, $\mathscr{L}[t^2/2] = (1/s)\mathscr{L}[t] = 1/s^3$ and by induction $\mathscr{L}[t^n/n!] = 1/s^{n+1}$, so that by linearity

$$\mathscr{L}[t^n] = \frac{n!}{s^{n+1}}, \qquad n = 0, 1, 2, \ldots \qquad (4\text{–}42)$$

We can also deduce this result from rule (4–36):

$$\mathscr{L}[t^n] = \mathscr{L}[t^n \cdot 1] = (-1)^n \frac{d^n}{ds^n}\mathscr{L}[1] = (-1)^n \frac{d^n}{ds^n}\frac{1}{s} = \frac{n!}{s^{n+1}}.$$

We can verify that (4–42) is valid for $s > 0$. ◀

EXAMPLE 4 $\mathscr{L}[t^n e^{bt}] = F(s-b)$, where $F(s) = \mathscr{L}[t^n]$, by rule (4–35). Hence, by Eq. (4–42),

$$\mathscr{L}[t^n e^{bt}] = \frac{n!}{(s-b)^{n+1}}.\tag{4–43}$$

(This is valid for $s > b$.) ◄

EXAMPLE 5 Let $g(t) = 1$ for $a \leqslant t < b$, where $a \geqslant 0$; let $g(t) = 0$ otherwise. If we let $f(t) = 1$ for $t \geqslant 0$, $f(t) = 0$ for $t < 0$, then

$$g(t) = f(t-a) - f(t-b),$$

as suggested in Fig. 4-4. Hence, by linearity and the translation rule (4–38),

$$\mathscr{L}[g] = e^{-as}\mathscr{L}[f] - e^{-bs}\mathscr{L}[f] = \frac{e^{-as} - e^{-bs}}{s}.\tag{4–44}$$

(This is valid for all s, if we assign the limit value $b-a$ for $s = 0$.) ◄

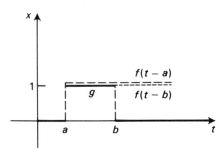

Fig. 4-4. Rectangular pulse.

EXAMPLE 6 Let $f(t) = t$ for $0 \leqslant t \leqslant 1$, $f(t) = 2 - t$ for $1 \leqslant t \leqslant 2$, and let f have period 2 for $t \geqslant 0$, as in Fig. 4–5, so that f is a triangular wave. Now,

$$\int_0^2 f(t)e^{-st}\, dt = \int_0^1 te^{-st}\, dt + \int_1^2 (2-t)e^{-st}dt = \frac{e^{-2s} - 2e^{-s} + 1}{s^2}.$$

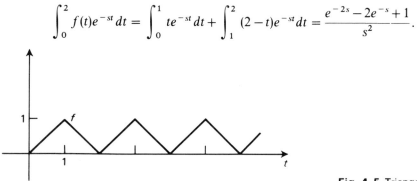

Fig. 4-5 Triangular wave.

Therefore, by rule (4–39),

$$\mathscr{L}[f] = \frac{e^{-2s} - 2e^{-s} + 1}{s^2(1 - e^{-2s})} = \frac{1 - e^{-s}}{s^2(1 + e^{-s})}.\quad ◄$$

A table of transforms

Table 4–1 lists commonly used Laplace transforms, including those of the examples studied thus far. (For an extensive table of Laplace transforms see ERDELYI, Vol. I of Tables.)

Table 4–1
LAPLACE TRANSFORMS
In the table, a, b, and c are real constants

No.	$f(t)$ for $t \geqslant 0$	$f = F(s) = \int_0^\infty f(t)e^{-st}dt$	Restrictions on s (s real)
1	1	$\dfrac{1}{s}$	$s > 0$
2	e^{at}	$\dfrac{1}{s-a}$	$s > a$
3*	t^n, $n > -1$	$\begin{cases} \dfrac{n!}{s^{n+1}}, & n = 0, 1, \ldots \\[2ex] \dfrac{\Gamma(n+1)}{s^{n+1}}, & \text{any } n > -1 \end{cases}$	$s > 0$
4*	$t^n e^{at}$, $n > -1$	$\begin{cases} \dfrac{n!}{(s-a)^{n+1}}, & n = 0, 1, \ldots \\[2ex] \dfrac{\Gamma(n+1)}{(s-a)^{n+1}}, & \text{any } n > -1 \end{cases}$	$s > a$
5	$\cosh at$	$\dfrac{s}{s^2 - a^2}$	$s > \lvert a \rvert$
6	$\sinh at$	$\dfrac{a}{s^2 - a^2}$	$s > \lvert a \rvert$
7	$\cos at$	$\dfrac{s}{s^2 + a^2}$	$s > 0$
8	$\sin at$	$\dfrac{a}{s^2 + a^2}$	$s > 0$
9	$t \cos at$	$\dfrac{s^2 - a^2}{(s^2 + a^2)^2}$	$s > 0$
10*	$t^n \cos at$, $n > -1$	$\begin{cases} \dfrac{n![(s+ai)^{n+1} + (s-ai)^{n+1}]}{2(s^2 + a^2)^{n+1}}, & n = 0, 1, \ldots \\[2ex] \dfrac{\Gamma(n+1)[(s+ai)^{n+1} + (s-ai)^{n+1}]}{2(s^2 + a^2)^{n+1}}, & \text{any } n > -1 \end{cases}$	$s > 0$
11	$t \sin at$	$\dfrac{2as}{(s^2 + a^2)^2}$	$s > 0$
12*	$t^n \sin at$, $n > -1$	$\begin{cases} \dfrac{n![(s+ai)^{n+1} - (s-ai)^{n+1}]}{2i(s^2 + a^2)^{n+1}}, & n = 0, 1, \ldots \\[2ex] \dfrac{\Gamma(n+1)[(s+ai)^{n+1} - (s-ai)^{n+1}]}{2i(s^2 + a^2)^{n+1}}, & \text{any } n > -1 \end{cases}$	$s > 0$

Table 4-1 (*contd*)

No.	$f(t)$ for $t \geqslant 0$	$f = F(s) = \displaystyle\int_0^\infty f(t)e^{-st}dt$	Restrictions on s (s real)
13	$\sin at \sin bt$	$\dfrac{2abs}{[s^2+(a+b)^2][s^2+(a-b)^2]}$	$s > 0$
14	$e^{at}\sin(bt+c)$	$\dfrac{(s-a)\sin c + b\cos c}{(s-a)^2+b^2}$	$s > a$
15	$e^{at}\sin bt$	$\dfrac{b}{(s-a)^2+b^2}$	$s > a$
16	$e^{at}\cos bt$	$\dfrac{s-a}{(s-a)^2+b^2}$	$s > a$
17	1 for $0 \leqslant t < c$, 0 for $c \leqslant t < 2c$ period $2c$ (square wave)	$\dfrac{1}{s(1+e^{-cs})}$	$s > 0$
18	1 for $0 \leqslant t < c$, 0 for $t \geqslant c$,	$\dfrac{1-e^{-cs}}{s}$	none
19	0 for $0 \leqslant t < c$, 1 for $t \geqslant c$,	$\dfrac{e^{-cs}}{s}$	$s > 0$
20	1 for $c_1 \leqslant t \leqslant c_2$, 0 otherwise, $0 \leqslant c_1 < c_2$	$\dfrac{e^{-c_1 s} - e^{-c_2 s}}{s}$	none
21	e^{at} for $c_1 \leqslant t \leqslant c_2$, 0 otherwise, $0 \leqslant c_1 < c_2$	$\dfrac{e^{-c_1(s-a)} - e^{-c_2(s-a)}}{s-a}$	none
22	0 for $0 \leqslant t < c$, e^{at} for $c \leqslant t < \infty$	$\dfrac{e^{-c(s-a)}}{s-a}$	$s > a$
23	0 for $0 \leqslant t < c$, t^n for $c \leqslant t < \infty$, $n = 1, 2, \ldots$	$\dfrac{e^{-cs}[(cs)^n + n(cs)^{n-1} + \cdots + n!]}{s^{n+1}}$	$s > 0$
24	t^n for $0 \leqslant t < c$, 0 for $t \geqslant c$, $n = 1, 2, \ldots$	$\dfrac{n! - e^{-cs}[(cs)^n + n(cs)^{n-1} + \cdots + n!]}{s^{n+1}}$	none
25	t for $0 \leqslant t \leqslant c$, $2c - t$ for $c \leqslant t \leqslant 2c$, 0 for $t \geqslant 2c$	$\dfrac{1 - 2e^{-cs} + e^{-2cs}}{s^2}$	none
26	t for $0 \leqslant t \leqslant c$, $2c - t$ for $c \leqslant t \leqslant 2c$, period $2c$ (triangular wave)	$\dfrac{1-e^{-cs}}{s^2(1+e^{-cs})}$	$s > 0$
27	at for $0 \leqslant t < c$, period c (sawtooth wave)	$\dfrac{a(1 + cs - e^{-cs})}{s^2(1 - e^{cs})}$	$s > 0$

* The gamma function $\Gamma(s)$ is discussed in Chapter 12. We have $\Gamma(n+1) = n!$ for $n = 0, 1, 2, \ldots$

=============================== **Problems (Section 4–4)** ===============================

1. Find the Laplace transform of each function:
   a) $f(t) = t$ for $0 \leqslant t \leqslant 1$, $f(t) = 1$ for $1 \leqslant t < \infty$
   b) $f(t) = 1 - t$ for $0 \leqslant t \leqslant 1$, $f(t) = 0$ for $1 \leqslant t < \infty$
   c) $f(t) = t^2$ for $0 \leqslant t \leqslant 1$, $f(t) = e^{1-t}$ for $1 \leqslant t < \infty$
   d) $f(t) = 2t - t^2$ for $0 \leqslant t \leqslant 2$, $f(t) = 0$ for $t \geqslant 2$
   è) $f(t) = e^{-t} \sin 2t$ for $0 \leqslant t \leqslant 2\pi$, $f(t) = 0$ for $t \geqslant 2\pi$
   f) $f(t) = 2t/\pi$ for $0 \leqslant t \leqslant \pi/2$, $f(t) = \sin t$ for $t \geqslant \pi/2$

2. With the aid of the rules of Section 4–3, obtain the Laplace transforms of the functions given from transforms previously found or from those of Table 4–1:
   a) $f(t) = 3e^{-t} + 2e^{-2t}$ b) $f(t) = 3 \cos 2t + 5 \cos 7t$
   c) $f(t) = g''(t)$, where $g(t) = e^{2t} \sin 3t$
   d) $f(t) = g'''(t)$, where $g(t) = t^2 e^{-t}$

   e) $f(t) = \displaystyle\int_0^t u e^{2u} \cos 3u \, du$

   f) $f(t) = \displaystyle\int_0^t (u^5 - u^3 + 1) e^{2u} \, du$

   g) $f(t) = g(t-1)$, where $g(t) = 0$ for $t < 0$, $g(t) = t^2 e^{-t}$ for $t \geqslant 0$
   h) $f(t) = g(t-\pi)$, where $g(t) = 0$ for $t < 0$, $g(t) = \sin(t/5)$ for $t \geqslant 0$
   i) $f(t) = \sin t$ for $0 \leqslant t \leqslant \pi$, f of period π for $t \geqslant 0$
   j) $f(t) = t$ for $0 \leqslant t \leqslant 1$, $f(t) = 1$ for $1 \leqslant t \leqslant 2$, f of period 2 for $t \geqslant 0$

3. Derive the specified transform of Table 4–1, using previous entries in the table and rules of Section 4–3:
a) No. 6	b) No. 9	c) No. 11	d) No. 13	e) No. 14	f) No. 15
g) No. 17	h) No. 19	i) No. 20	j) No. 21	k) No. 26	l) No. 27

4. Prove the rule: if $\mathscr{L}[f] = F(s)$ and $g(t) = f(kt)$, where k is a positive constant, then $\mathscr{L}[g] = (1/k)F(s/k)$.

5. Prove that if $g(t)$ is continuous for $0 \leqslant t < \infty$ and $\lim\limits_{t \to \infty} g(t) = b \neq 0$, then $\int_0^\infty g(t)\,dt$ does not exist.

   HINT: Assume that b is positive. Then for t sufficiently large, say $t > T$, we have $g(t) > b/2$. Accordingly, for $t > T$, we have

   $$\int_0^t g(u)\,du = \int_0^T g(u)\,du + \int_T^t g(u)\,du > \int_0^T g(u)\,du + \frac{b}{2}(t - T).$$

   Now let $t \to \infty$. If b is negative, we obtain the same result by considering $-g(t)$.

4–5 MORE ON THE DEFINITION OF THE LAPLACE TRANSFORM; COMPLEX ASPECTS

Let $g(t)$ be continuous for $0 \leqslant t < \infty$. We recall that the improper integral $\int_0^\infty g(t)\,dt$ is said to be *convergent* and have value k if

$$\lim_{b \to \infty} \int_0^b g(t)\,dt = k.$$

The integral is said to be *absolutely convergent* if $\int_0^\infty |g(t)|\,dt$ converges. As for infinite series, *absolute convergence implies convergence* (see Section 2–10), and the proof parallels that for series.

Now let $f(t)$ be real valued and continuous for $0 \leqslant t < \infty$ and let

$$|f(t)| \leqslant Ae^{at}, \quad 0 \leqslant t < \infty, \tag{4–50}$$

where A and a are constants, with $A > 0$. The commonly used functions f satisfy an inequality of form (4–50); for example, polynomials do, as do rational functions, trigonometric polynomials, exponential functions, logarithmic functions. For example, for k fixed,

$$0 \leqslant t^k \leqslant Ae^t, \quad \text{when} \quad 0 \leqslant t < \infty \quad \text{and} \quad k = 1, 2, 3, \ldots,$$

since $t^k e^{-t} \to 0$ as $t \to \infty$; hence $t^k e^{-t}$ must have a finite positive maximum A. We show in the same way that for every $\varepsilon > 0$ there is a constant A such that

$$0 \leqslant t^k \leqslant Ae^{\varepsilon t} \quad \text{for} \quad 0 \leqslant t < \infty.$$

There are of course functions that satisfy no condition of form (4–50); for example, the function e^{t^2}, since

$$e^{t^2} e^{-at} \to \infty \quad \text{as} \quad t \to \infty,$$

no matter how a is chosen.

For functions f satisfying (4–50) *the Laplace transform*

$$\int_0^\infty f(t)e^{-st}\,dt \tag{4–51}$$

is absolutely convergent for s real and $s > a$, since

$$|f(t)e^{-st}| = |f(t)e^{-at}| |e^{-(s-a)t}| \leqslant Ae^{-(s-a)t}.$$

If $s > a$, then $\int_0^\infty Ae^{-(s-a)t}\,dt$ converges and hence so does $\int_0^\infty |f(t)e^{-st}|\,dt$. We are here using a comparison test for integrals; it also parallels the test for infinite series (see Section 2–9).

To get the best possible information in this way for a given f, we choose a as small as possible in (4–50). It can be shown that for an arbitrary $f(t)$ there is an *abscissa of absolute convergence* σ^* such that the Laplace transform of f is absolutely convergent for $s > \sigma^*$ (s real) and fails to be absolutely convergent for $s < \sigma^*$. If $\sigma^* = \infty$, there is no absolute convergence; if $\sigma^* = -\infty$, there is absolute convergence for all s (for details see OMLS, pp. 308–310). In Examples 1 to 4 of

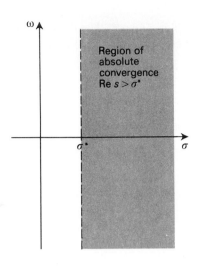

Fig. 4–6. Complex s-plane and region of absolute convergence.

Section 4–2 the abscissa of absolute convergence is given in each case (as $0, a, 0, -\infty$, respectively).

We now allow s to be complex and write $s = \sigma + i\omega$, as suggested in Fig. 4–6. Then

$$\int_0^\infty f(t)e^{-st}\,dt = \int_0^\infty f(t)e^{-(\sigma + i\omega)t}\,dt = \int_0^\infty f(t)e^{-\sigma t}e^{-i\omega t}\,dt,$$

where, as usual,

$$e^{-i\omega t} = \cos \omega t - i\sin \omega t.$$

Thus we can decompose the Laplace transform into real and imaginary parts:

$$\int_0^\infty f(t)e^{-st}\,dt = \int_0^\infty f(t)e^{-\sigma t}\cos \omega t\,dt - i\int_0^\infty f(t)e^{-\sigma t}\sin \omega t\,dt. \qquad (4\text{–}52)$$

If both integrals on the right exist, then the Laplace transform of f is defined for this $s = \sigma + i\omega$. The value of the transform for each s is in general complex, and we are thus dealing with a *complex-valued function of the complex variable s*.

It can be verified that for s complex the Laplace transform is absolutely convergent for $\sigma = \mathrm{Re}\,s > \sigma^*$, with the same σ^* as before. Thus we obtain a region of absolute convergence in the complex s-plane, as in Fig. 4–6. In Table 4–1, the last column gives the abscissa of absolute convergence.

We can also allow $f(t)$ to be complex valued, say $f(t) = p(t) + iq(t)$, where $p(t)$ and $q(t)$ are real valued. This does not affect the above discussion and we again obtain the abscissa of absolute convergence. We remark that here, as in Section 2–20,

$$|f(t)| = |p(t) + iq(t)| = [p(t)^2 + q(t)^2]^{1/2}.$$

For example, let $f(t) = e^{at}$, where a is the complex number $\alpha + i\beta$. Then

$$\mathscr{L}[f] = \int_0^\infty e^{at}e^{-st}\,dt = \int_0^\infty e^{(a-s)t}\,dt = \frac{e^{(a-s)t}}{a-s}\Big|_0^\infty = \frac{1}{s-a}.$$

As in Section 1–10, all steps are valid for s complex and $e^{(a-s)t} \to 0$ as $t \to \infty$, provided that $\mathrm{Re}(a-s) < 0$, that is, $\alpha < \sigma$. Thus here $\alpha = \mathrm{Re}\ a$ is the abscissa of absolute convergence.

Similarly, the rule

$$\mathscr{L}\left[t^n e^{at}\right] = \frac{n!}{(s-a)^{n+1}}, \quad n = 0, 1, 2, \ldots \tag{4–53}$$

is valid for complex $a = \alpha + i\beta$, provided that $\sigma = \mathrm{Re}\ s > \alpha$.

4–6 PARTIAL FRACTIONS

For working with Laplace transforms and for other purposes it is important to know the procedure for decomposing a rational function into partial fractions. For example, we have the identity

$$\frac{s+1}{s^2 - 3s + 2} = \frac{3}{s-2} - \frac{2}{s-1}. \tag{4–60}$$

Its correctness can be verified by putting the right side over a common denominator.

The procedure applies only to *proper* rational functions: that is, to those for which the degree of the denominator exceeds that of the numerator.

If the denominator has only real roots $s_1, \ldots, s_k$, then the rational function has the form

$$\frac{P(s)}{(s-s_1)^{m_1} \cdots (s-s_k)^{m_k}}.$$

If each root is simple, then $m_1 = 1, \ldots, m_k = 1$, and the corresponding partial-fraction expression has the form

$$\frac{A_1}{s-s_1} + \cdots + \frac{A_k}{s-s_k},$$

where $A_1, \ldots, A_k$ are constants. We indicate below how the constants can be found. This case of simple real roots is illustrated by Eq. (4–60).

If any of the roots $s_1, \ldots, s_k$ is repeated several times, then there is a corresponding number of terms in the partial-fraction expression. For example, if $m_1 = 2$, then for s_1 we have two terms

$$\frac{A_1}{s-s_1} + \frac{A_2}{(s-s_1)^2}.$$

EXAMPLE 1

$$\frac{s^2 + 3}{(s-1)^2(s+1)} = \frac{A_1}{s-1} + \frac{A_2}{(s-1)^2} + \frac{A_3}{s+1}.$$

We now indicate a way of finding the constants. We cross multiply by the given denominator, thereby getting rid of the denominators. We obtain the equation:

$$s^2 + 3 = A_1(s-1)(s+1) + A_2(s+1) + A_3(s-1)^2$$

or

$$s^2 + 3 = A_1(s^2 - 1) + A_2(s+1) + A_3(s^2 - 2s + 1).$$

We then compare the coefficients of terms of the same degree on both sides. Thus,

$$0 \text{ degree:} \quad 3 = -A_1 + A_2 + A_3,$$
$$1\text{st degree:} \quad 0 = A_2 - 2A_3,$$
$$2\text{nd degree:} \quad 1 = A_1 + A_3.$$

We thus have three simultaneous linear equations for A_1, A_2, A_3. They are easily solved to give $A_1 = 0$, $A_2 = 2$, $A_3 = 1$. Thus,

$$\frac{s^2 + 3}{(s-1)^2(s+1)} = \frac{2}{(s-1)^2} + \frac{1}{s+1},$$

and again we can verify the result by cross multiplying. ◄

If some of the roots are complex, the *same* procedure can be followed. However, it does force us to work with complex numbers.

EXAMPLE 2

$$\frac{3s+1}{(s^2+2s+2)(s-1)} = \frac{3s+1}{(s+1+i)(s+1-i)(s-1)}$$
$$= \frac{A_1}{s+1+i} + \frac{A_2}{s+1-i} + \frac{A_3}{s-1}.$$

The factorization can be achieved by solving the quadratic equation $s^2 + 2s + 2 = 0$ to obtain $s = -1 \pm i$ and then writing

$$s^2 + 2s + 2 = (s - s_1)(s - s_2),$$

where s_1, s_2 are the two roots $-1 \pm i$.

Proceeding as in Example 1, we get:

$$3s + 1 = A_1(s+1-i)(s-1) + A_2(s+1+i)(s-1) + A_3(s^2+2s+2)$$
$$= A_1(s^2 - is - 1 + i) + A_2(s^2 + is - 1 - i) + A_3(s^2 + 2s + 2),$$
$$1 = (-1+i)A_1 + (-1-i)A_2 + 2A_3,$$
$$3 = -iA_1 + iA_2 + 2A_3,$$
$$0 = A_1 + A_2 + A_3,$$
$$A_1 = \frac{-4+7i}{10}, \quad A_2 = \frac{-4-7i}{10}, \quad A_3 = \frac{4}{5},$$
$$\frac{3s+1}{(s^2+2s+2)(s-1)} = \frac{-4+7i}{10}\frac{1}{s+1+i} + \frac{-4-7i}{10}\frac{1}{s+1-i} + \frac{4}{5}\frac{1}{s-1}.$$

If we put the first two terms on the right over a common denominator, the i disappears. We find

$$\frac{-4+7i}{10}\frac{1}{s+1+i} + \frac{-4-7i}{10}\frac{1}{s+1-i} = -\frac{1}{5}\frac{4s-3}{s^2+2s+2}. \quad ◄$$

In general, as this example suggests, we can avoid working with complex numbers (if only real numbers appear in the given rational function) by replacing two terms

$$\frac{A_1}{s - s_1} + \frac{A_2}{s - s_2}$$

for a pair of conjugate complex roots by an expression

$$\frac{B_1 s + B_2}{s^2 + ps + q},$$

where $s^2 + ps + q$ is the corresponding quadratic factor. Similarly, when there is a repeated pair of conjugate complex roots, we can also combine terms; for example,

$$\frac{A_1}{s - s_1} + \frac{A_2}{s - s_2} + \frac{A_3}{(s - s_1)^2} + \frac{A_4}{(s - s_2)^2} = \frac{B_1 s + B_2}{s^2 + ps + q} + \frac{B_3 s + B_4}{(s^2 + ps + q)^2}.$$

EXAMPLE 3

$$\frac{2s^2 + 5s + 7}{(2s^2 + 2s + 1)^2 (s - 3)} = \frac{(1/4)(2s^2 + 5s + 7)}{(s^2 + s + 1/2)^2 (s - 3)}$$

$$= \frac{A_1 s + A_2}{s^2 + s + 1/2} + \frac{A_3 s + A_4}{(s^2 + s + 1/2)^2} + \frac{A_5}{s - 3}.$$

The five constants can be found by getting rid of the denominators and proceeding as above. ◄

When only simple roots appear, there is a very quick procedure, since we have the rule

$$\frac{P(s)}{Q(s)} = \frac{P(s)}{(s - s_1)(s - s_2) \cdots (s - s_k)}$$

$$\tag{4-61}$$

$$= \frac{P(s_1)}{Q_1(s_1)} \frac{1}{s - s_1} + \cdots + \frac{P(s_k)}{Q_k(s_k)} \frac{1}{s - s_k},$$

where

$$Q_1(s) = (s - s_2) \cdots (s - s_k),$$
$$Q_2(s) = (s - s_1)(s - s_3) \cdots (s - s_k),$$
$$\vdots$$
$$Q_k(s) = (s - s_1) \cdots (s - s_{k-1}).$$

For the example (4–60), $Q(s) = (s - 2)(s - 1)$, $Q_1(s) = s - 1$, $Q_2(s) = s - 2$, $P(s) = s + 1$. Thus we find

$$\frac{P(2)}{Q_1(2)} = \frac{3}{1}, \quad \frac{P(1)}{Q_2(1)} = \frac{2}{-1}$$

and obtain the numerators 3 and -2 of Eq. (4–60).

A justification of all these rules is most easily obtained by complex variables (Section 11–18).

4–7 INVERSE LAPLACE TRANSFORM

In practice it is of much importance to be able to recover $f(t)$ from its Laplace transform $F(s)$. We shall see that $f(t)$ is *uniquely determined by its Laplace transform*: more precisely, if $f_1(t)$ and $f_2(t)$ are continuous and have absolutely convergent Laplace transforms $F_1(s)$, $F_2(s)$, with $F_1(s) \equiv F_2(s)$ for sufficiently large Re s, then $f_1(t) \equiv f_2(t)$. (The result remains true for piecewise continuous functions if we assign the average of left and right limits at jump points as the value of the function at the points.)

Because of this result, we can speak of *the inverse Laplace transform* of a function $F(s)$: this is the unique $f(t)$ such that $\mathscr{L}[f] = F(s)$. We write

$$f = \mathscr{L}^{-1}[F]. \tag{4–70}$$

EXAMPLES

1. $\mathscr{L}^{-1}[1/s] = 1$
2. $\mathscr{L}^{-1}[1/(s+2)^2] = te^{-2t}$
3. $\mathscr{L}^{-1}[1/(s^2+1)] = \sin t$

These all follow from Table 4–1. ◄

It is important to know which functions $F(s)$ have inverse transforms, that is, which functions $F(s)$ are Laplace transforms. We give some information on this here and in later sections of this chapter, also in Chapter 11. We first observe that if $F_1(s)$ and $F_2(s)$ are Laplace transforms, then so also is $c_1 F_1(s) + c_2 F_2(s)$ for constant c_1 and c_2 and

$$\mathscr{L}^{-1}[c_1 F_1(s) + c_2 F_2(s)] = c_1 \mathscr{L}^{-1}[F_1] + c_2 \mathscr{L}^{-1}[F_2].$$

This follows from the linearity of the operator [Section 4–3]. Thus $\mathscr{L}^{-1}$ is also a *linear operator*.

Next we observe that, by entry No. 4 of Table 4–1, for $k = 1, 2, \ldots$,

$$\mathscr{L}^{-1}\left[\frac{1}{(s-s_0)^k}\right] = \frac{t^{k-1}}{(k-1)!} e^{s_0 t}. \tag{4–71}$$

If now $P(s)$ and $Q(s)$ are polynomials such that

$$F(s) = \frac{P(s)}{Q(s)}$$

is a proper rational function of s (degree of P less than degree of Q), then $F(s)$ can be expanded in partial fractions as in Section 4–6:

$$F(s) = \frac{P(s)}{Q(s)} = \frac{A_1}{(s-s_1)} + \frac{A_2}{(s-s_1)^2} + \cdots + \frac{B_1}{(s-s_2)} + \frac{B_2}{(s-s_2)^2} + \cdots,$$

where $A_1, A_2, \ldots, B_1, B_2, \ldots$ are constants and $s_1, s_2, \ldots$ are the roots of $Q(s)$. It follows that

$$\mathscr{L}^{-1}[F(s)] = A_1 e^{s_1 t} + A_2 t e^{s_1 t} + \cdots$$

Thus *every proper rational function of s has an inverse Laplace tansform.*

In calculating the inverse Laplace transform of a proper rational function, the following rule is useful:

If $P(s) = c_0 s^m + c_1 s^{m-1} + \cdots + c_m$ and $\mathcal{L}^{-1}[1/Q(s)] = f(t)$, where $Q(s)$ has degree greater than m, then

$$\mathcal{L}^{-1}\left[\frac{P(s)}{Q(s)}\right] = c_0 f^{(m)}(t) + c_1 f^{(m-1)}(t) + \cdots + c_m f(t). \qquad (4\text{-}72)$$

To prove the rule, we first let $\mathcal{L}[g] = s/Q(s)$. Then, by the integration rule (4–34),

$$\mathcal{L}\left[\int_0^t g(u)du\right] = \frac{1}{Q(s)} = \mathcal{L}[f].$$

Hence, on taking inverse transforms, $f(t) = \int_0^t g(u)\,du$, so that $g(t) = f'(t)$ and $f(0) = 0$. Thus

$$\mathcal{L}[f'(t)] = \frac{s}{Q(s)}.$$

Similarly, if $\mathcal{L}[h] = s^2/Q(s)$, then $\mathcal{L}[\int_0^t h(u)\,du] = s/Q(s) = \mathcal{L}[f'(t)]$, so that $\int_0^t h(u)\,du = f'(t)$ and $h(t) = f''(t)$ (and $f'(0) = 0$), so that

$$\mathcal{L}[f''(t)] = \frac{s^2}{Q(s)}.$$

In general, as long as k is less than the degree of $Q(s)$,

$$\mathcal{L}[f^{(k)}(t)] = \frac{s^k}{Q(s)}.$$

The rule (4–72) now follows by linearity.

We have incidentally shown that, if $\mathcal{L}^{-1}[1/Q(s)] = f(t)$ and $Q(s)$ is a polynomial of degree n, then

$$f(0) = 0, \quad f'(0) = 0, \quad \ldots, \quad f^{(n-2)}(0) = 0. \qquad (4\text{-}73) \ \blacktriangleleft$$

EXAMPLE 4   To   find $\mathcal{L}^{-1}\left[\dfrac{3s^2 + 5s - 2}{(s-1)\,(s-2)\,(s-3)}\right]$ we   first   find

$\mathcal{L}^{-1}\left[\dfrac{1}{(s-1)\,(s-2)\,(s-3)}\right] = f(t)$. Since

$$\frac{1}{(s-1)(s-2)(s-3)} = \frac{1}{2}\frac{1}{s-1} - \frac{1}{s-2} + \frac{1}{2}\frac{1}{s-3},$$

we have $f(t) = e^t/2 - e^{2t} + e^{3t}/2$. Therefore, the desired inverse is

$$3f'' + 5f' - 2f = 3e^t - 20e^{2t} + 20e^{3t}.$$

We also verify that $f(0) = 0$ and $f'(0) = 0$, in accordance with rule (4–73). $\blacktriangleleft$

========= **Problems (Section 4-7)** =========

1. Verify that $|f(t)| \leqslant Ae^{at}$ for $t \geqslant 0$ and A and a as given and that the Laplace transform of f is absolutely convergent for $s > a$:

a) $f(t) = 1$, $A = 1$, $a = 0$
b) $f(t) = t$, $A = e^{-1}$, $a = 1$
c) $f(t) = t^2$, $A = 4e^{-2}/\varepsilon^2$, $a = \varepsilon > 0$
d) $f(t) = \cos t$, $A = 1$, $a = \varepsilon > 0$
e) $f(t) = te^t$, $A = e^{-1}$, $a = 2$
f) $f(t) = \ln(t+1)$, $A = e^{-1}$, $a = 1$. HINT: $\ln(t+1) \leqslant t$ for $t \geqslant 0$
g) $f(t) = e^{-t}$, $A = 1$, $a = 0$

2. Find the inverse transform with the aid of Table 4–1 and linearity:

a) $\dfrac{5}{s} + \dfrac{7}{s^2}$
b) $\dfrac{8}{s^3} + \dfrac{12}{s^4}$
c) $\dfrac{3}{s-1} + \dfrac{2}{s+1}$

d) $\dfrac{5}{s+2} + \dfrac{9}{s+4}$
e) $\dfrac{1}{(s+1)^2} + \dfrac{3}{s+1}$
f) $\dfrac{2}{s-5} + \dfrac{3}{(s-5)^2} + \dfrac{7}{(s-5)^3}$

g) $\dfrac{2+3s}{s^2+1}$
h) $\dfrac{s^2+3s-1}{(s^2+1)^2}$
i) $\dfrac{2e^{-s}}{s} + \dfrac{3e^{-2s}}{s-1}$

j) $\dfrac{3(e^{-(s-1)} - e^{-2(s-1)})}{s-1} + \dfrac{5(e^{-3(s-2)} - e^{-4(s-2)})}{s-2}$

3. Decompose into partial fractions and then find the inverse transform:

a) $\dfrac{1}{(s-1)(s-2)}$
b) $\dfrac{s}{s^2-4}$
c) $\dfrac{9s-6}{(s-1)(s^2-4)}$
d) $\dfrac{s^2+1}{(s-1)(s+3)(s+5)}$

e) $\dfrac{s+1}{s^2(s-1)}$
f) $\dfrac{s^2-s+1}{(s-1)^2(s+3)}$
g) $\dfrac{s-2}{s(s+2)^3}$
h) $\dfrac{s^3+1}{(s^2-1)^2}$

i) $\dfrac{1}{s(s^2+1)}$. HINT: Use entries 7 and 8 in Table 4–1
j) $\dfrac{s+1}{(s-2)(s^2+9)}$

k) $\dfrac{3s^2+2}{s(s^2+2s+5)}$. HINT: Use entries 15 and 16 in Table 4–1.

l) $\dfrac{s}{(s^2+1)^2(s^2+2s+5)}$. HINT: Use complex form.

4. Given that $\mathscr{L}^{-1}\left[\dfrac{1}{s^3-s^2-9s+9}\right] = \dfrac{1}{24}(2e^{3t} + e^{-3t} - 3e^t)$, find

a) $\mathscr{L}^{-1}\left[\dfrac{3s+2}{s^3-s^2-9s+9}\right]$
b) $\mathscr{L}^{-1}\left[\dfrac{s^2-1}{s^3-s^2-9s+9}\right]$

5. Given that $\mathscr{L}^{-1}\left[\dfrac{1}{s^4+5s^2+4}\right] = \dfrac{1}{6}(2\sin t - \sin 2t)$, find

a) $\mathscr{L}^{-1}\left[\dfrac{2s-1}{s^4+5s^2+4}\right]$
b) $\mathscr{L}^{-1}\left[\dfrac{s^3-s^2}{s^4+5s^2+4}\right]$

c) $\mathscr{L}^{-1}\left[\dfrac{2s-1}{s(s^4+5s^2+4)}\right]$
d) $\mathscr{L}^{-1}\left[\dfrac{e^{-\pi s}}{s(s^4+5s^2+4)}\right]$

4–8 CONVOLUTION

Let $f(t)$ and $g(t)$ be defined for $t \geqslant 0$ and be piecewise continuous. Then the *convolution* $f * g$ of f and g is defined by the equation

$$f * g = \int_0^t f(u)g(t-u)\, du, \quad 0 \leqslant t < \infty. \tag{4-80}$$

Thus the convolution of f and g is another function of t defined for $0 \leqslant t < \infty$.

EXAMPLE 1 Let $f(t) = t$, $g(t) = e^t$. Then

$$f * g = \int_0^t u e^{t-u}\, du = e^t \int_0^t u e^{-u}\, du = e^t - t - 1. \ \blacktriangleleft$$

EXAMPLE 2 Let $f(t) = 0$ for $0 \leqslant t < 1$, $f(t) = 1$ for $t \geqslant 1$. Let $g(t) = t$. Then $f * g = 0$ for $0 \leqslant t \leqslant 1$ and

$$f * g = \int_1^t (t-u)\, du \quad \text{for} \quad 1 \leqslant t < \infty.$$

Hence $f * g = 0$ for $0 \leqslant t \leqslant 1$ and $f * g = (t^2 - 2t + 1)/2$ for $1 \leqslant t < \infty$. ◀

In Example 2, function f is discontinuous, but $f * g$ is *continuous* (with value 0 at $t = 1$). In general, *the convolution of two continuous or piecewise continuous functions is continuous* (see OMLS, Section 2–8).

The convolution has a number of special properties and, as will be seen, is of value in solving differential equations. The most striking property is the following:

$$\mathscr{L}[f * g] = \mathscr{L}[f]\mathscr{L}[g]. \tag{4-81}$$

In words: *the Laplace transform of the convolution of two functions equals the product of the Laplace transforms of the functions.* This is valid (for Re s sufficiently large) if f and g have absolutely convergent Laplace transforms (for $\sigma =$ Re s sufficiently large).

The following formal steps are the heart of the proof of rule (4–81). We have

$$\mathscr{L}[f * g] = \int_0^\infty (f * g) e^{-st}\, dt = \int_0^\infty \int_0^t f(u)g(t-u) e^{-st}\, du\, dt.$$

The double integral is over the region in the ut-plane shown in Fig. 4–7. If the order of integration is interchanged, we obtain

$$\mathscr{L}[f * g] = \int_0^\infty \int_u^\infty f(u)g(t-u) e^{-st}\, dt\, du$$

$$= \int_0^\infty f(u) e^{-su} \left(\int_u^\infty g(t-u) e^{-s(t-u)}\, dt \right) du.$$

If we replace $t - u$ by v in the inner integral, then it no longer depends on u:

$$\mathscr{L}[f * g] = \int_0^\infty f(u) e^{-su} \left(\int_0^\infty g(v) e^{-sv}\, dv \right) du = \mathscr{L}[f]\mathscr{L}[g].$$

For justification of the formal steps, see OMLS, Section 5–11.

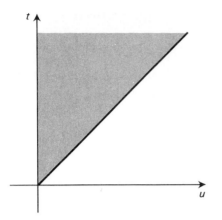

Fig. 4–7. Proof of convolution theorem.

EXAMPLE 3 Let f and g be as in Example 1. Then $\mathscr{L}[f] = 1/s^2$, $\mathscr{L}[g] = 1/(s-1)$, so that $\mathscr{L}[f]\mathscr{L}[g] = 1/[s^2(s-1)]$. Also

$$\mathscr{L}[f * g] = \mathscr{L}[e^t - t - 1] = \frac{1}{s-1} - \frac{1}{s^2} - \frac{1}{s} = \frac{1}{s^2(s-1)}.$$

Hence (4–81) is verified. ◀

We list other properties of the convolution:

$$f * g - g * f, \tag{4-82}$$
$$f * cg = (cf) * g = c(f * g), \quad c = \text{const}, \tag{4-83}$$
$$f * (g_1 + g_2) = f * g_1 + f * g_2, \tag{4-84}$$
$$f_1 * (f_2 * f_3) = (f_1 * f_2) * f_3, \tag{4-85}$$
$$e^{at}(f * g) = e^{at}f * e^{at}g. \tag{4-86}$$

These follow easily from the definition or from rule (4–81). Thus for (4–82),

$$\mathscr{L}[f * g] = \mathscr{L}[f]\mathscr{L}[g] = \mathscr{L}[g]\mathscr{L}[f] = \mathscr{L}[g * f].$$

By the uniqueness of the inverse transform we have $f * g = g * f$. The proofs of the other rules are left as exercises (see Problems 3–5 below).

The convolution is helpful in finding inverse transforms. If a given function of s can be factored as a product $F(s)G(s)$, where F and G are known transforms of f and g respectively, then, by Eq. (4–81),

$$\mathscr{L}^{-1}[F(s)G(s)] = f * g. \tag{4-87}$$

EXAMPLE 4 Given the function e^{-s}/s^3, we examine Table 4–1 and by entries 3 and 19 are led to write

$$\frac{e^{-s}}{s^3} = \frac{e^{-s}}{s} \cdot \frac{1}{s^2} = \mathscr{L}[f]\mathscr{L}[g],$$

where $f(t) = 0$ for $0 \leqslant t < 1$ and $f(t) = 1$ for $t \geqslant 1$ and $g(t) = t$. Hence

$$\frac{e^{-s}}{s^3} = \mathscr{L}[f * g] \quad \text{or} \quad \mathscr{L}^{-1}\left[\frac{e^{-s}}{s^3}\right] = f * g.$$

Now $f * g$ was found in Example 2, so that we have the desired inverse transform. ◀

EXAMPLE 5   To find the inverse transform of $(s^2 + 1)^{-2}$, we write

$$\frac{1}{(s^2 + 1)^2} = \frac{1}{s^2 + 1} \frac{1}{s^2 + 1} = \mathscr{L}[\sin t]\mathscr{L}[\sin t]$$

$$= \mathscr{L}[\sin t * \sin t] = \mathscr{L}[f(t)].$$

Hence,

$$f(t) = \int_0^t \sin u \sin (t - u)\, du = \frac{1}{2}\int_0^t [\cos (2u - t) - \cos t]\, du$$

$$= \frac{1}{2}(\sin t - t \cos t). \quad ◀$$

4–9 GENERALIZED FUNCTIONS

The Dirac delta function $\delta(t)$ is defined as a function that is equal to 0 for $t \neq 0$ and becomes infinite at $t = 0$ in such a manner that

$$\int_{-\infty}^{\infty} \delta(t)\, dt = 1.$$

No ordinary function can have these properties, and for a long time the delta function and related functions were considered to be pure fiction, despite their admitted value in physics and engineering. However, from about 1950 on, mathematicians have developed several rigorous theories of these objects called *generalized functions* (or *distributions*).

The delta function is introduced in Section 3–3 where it is shown to have the physical interpretation of the density associated with a particle of unit mass at the point $t = 0$ on the t-axis.

Here we give a further intuitive discussion of the delta function and some related functions, emphasizing their importance for Laplace and Fourier transforms and their applications. (For a fuller discussion the reader is referred to OMLS, Chapter 1, and to LIGHTHILL.)

We consider functions such as

$$g(t) + 3\delta(t) - 2\delta(t - 1) + 5\delta'(t + 3) + \delta''(t),$$

that is, linear combinations of an ordinary function $g(t)$ with the delta function, its derivatives, and the translations of these functions.

The delta function itself can be thought of as a very narrow and very large pulse, as in Fig. 4–8(a). It can also be thought of as the derivative of the *unit-step function* $U(t)$ which is equal to 0 for $t < 0$ and equal to 1 for $t \geqslant 0$ (Fig. 4–9). A smooth approximation to $U(t)$, as shown in Fig. 4–9, has as first derivative a pulse similar to

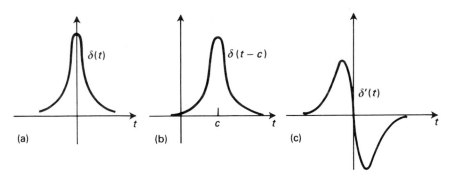

Fig. 4–8. Generalized functions: (a) $\delta(t)$; (b) $\delta(t-c)$; (c) $\delta'(t)$.

the one in Fig. 4–8(a). The function $\delta(t-c)$ is obtained by translation of $\delta(t)$, as suggested in Fig. 4–8(b); it can be treated as the derivative of $U(t-c)$. (Here c is positive for translation to the right, as in the figure; for translation to the left, c is negative.) The function $\delta'(t)$ can be thought of as a large positive pulse just to the left of zero followed by a large negative pulse just to the right of zero (Fig. 4–8(c)); it can also be interpreted as the *second* derivative of $U(t)$, and a smooth approximation to $U(t)$ has a second derivative as in Fig. 4–8(c).

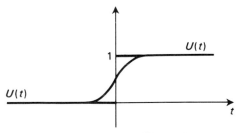

Fig. 4–9. Unit-step function $U(t)$ and smooth approximation to it.

A piecewise smooth ordinary function $g(t)$ can be thought of as a continuous function $g_0(t)$ plus terms of form $kU(t-c)$ corresponding to the jumps, k being the size of a jump (positive for an increase, negative for a decrease), and c being the time t at which it occurs. Thus in Fig. 4–10

$$g(t) = g_0(t) + k_1 U(t-c_1) + k_2 U(t-c_2),$$

where k_1 is positive and k_2 is negative. Accordingly,

$$g'(t) = g_0'(t) + k_1 \delta(t-c_1) + k_2 \delta(t-c_2).$$

The process can be repeated and usually permits us to obtain derivatives of arbitrarily high order. Except at the jumps, the derivatives are taken as usual; at each

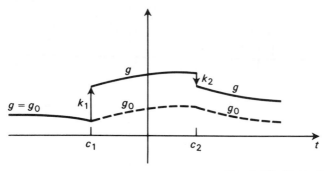

Fig. 4-10. Piecewise-smooth function.

jump there is a corresponding delta-function contribution to the derivative. Also at the jump, the derivative of the continuous function $g_0(t)$ may have a jump.

EXAMPLE 1 Let $g(t) = e^{-t}$ for $-\infty < t < 0$, $g(t) = \sin t$ for $0 < t < \pi/2$, and $g(t) = 3e^{-[t-(\pi/2)]}$ for $t > \pi/2$, as in Fig. 4-11. Then

$$g'(t) = g_0'(t) - \delta(t) + 2\delta(t - \pi/2),$$

where $g_0'(t) = -e^{-t}$ for $-\infty < t < 0$, $g_0'(t) = \cos t$ for $0 < t < \pi/2$, and $g_0'(t) = -3e^{-[t-(\pi/2)]}$ for $t > \pi/2$. Similarly,

$$g''(t) = g_1''(t) + 2\delta(t) - 3\delta\left(t - \frac{\pi}{2}\right) - \delta'(t) + 2\delta'\left(t - \frac{\pi}{2}\right),$$

where $g_1''(t) = e^{-t}$ for $-\infty < t < 0$, $g_1''(t) = -\sin t$ for $0 < t < \pi/2$, $g_1''(t) = 3e^{-[t-(\pi/2)]}$ for $t > \pi/2$. ◄

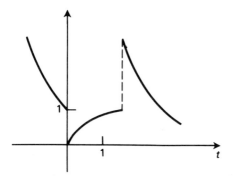

Fig. 4-11. Function of Example 1.

Generalized functions may be added, subtracted, and multiplied by constants. Under appropriate conditions, an ordinary function can also be multiplied by a generalized function. Thus

$$\delta(t)f(t) = f(t)\delta(t) = f(0)\delta(t),$$
$$\delta(t-c)f(t) = f(t)\delta(t-c) = f(c)\delta(t-c), \tag{4-90}$$

provided f is continuous at $t = 0$ and at $t = c$ respectively. These rules are suggested by the interpretation of $\delta(t)$ and $\delta(t-c)$ as pulses (see Fig. 4–8). They are also suggested by the product rule for derivatives, since $f(t)U(t)$ should have derivative $f'(t)U(t) + f(t)\delta(t)$. But $f(t)U(t) = 0$ for $t < 0$ and $f(t)U(t) = f(t)$ for $t > 0$, so that there is a jump of $f(0)$ at $t = 0$. Hence, as above, the derivative of $f(t)U(t)$ should be $f'_0(t) + f(0)\delta(t)$, where $f'_0(t) = f'(t)U(t)$. If we equate the two expressions, we obtain

$$f'(t)U(t) + f(t)\delta(t) = f'(t)U(t) + f(0)\delta(t),$$

so that $f(t)\delta(t) = f(0)\delta(t)$.

Similarly, an expression for $f(t)\delta'(t)$ can be deduced from the product rule, since we are led to the two expressions for $[f(t)\delta(t)]'$:

$$[f(t)\delta(t)]' = [f(0)\delta(t)]' = f(0)\delta'(t),$$
$$[f(t)\delta(t)]' = f(t)\delta'(t) + f'(t)\delta(t) = f(t)\delta'(t) + f'(0)\delta(t).$$

If we equate the two expressions, we conclude that

$$f(t)\delta'(t) = f(0)\delta'(t) - f'(0)\delta(t), \tag{4-91}$$

provided that f and f' are continuous at $t = 0$.

In the same way we can deduce the rules:

$$f(t)\delta'(t-c) = f(c)\delta'(t-c) - f'(c)\delta(t-c),$$
$$f(t)\delta''(t-c) = f(c)\delta''(t-c) - 2f'(c)\delta'(t-c) + f''(c)\delta(t-c). \tag{4-92}$$

Generalized functions can also be integrated. For example, if $a < b$ and $a \neq 0$, $b \neq 0$, then

$$\int_a^b \delta(t)\,dt = \int_a^b U'(t)\,dt = U(b) - U(a) = \begin{cases} 1, & a < 0 < b \\ 0, & \text{otherwise.} \end{cases} \tag{4-93}$$

$$\int_a^b \delta'(t)\,dt = \delta(b) - \delta(a) = 0, \quad \int_a^b \delta''(t)\,dt = 0, \dots, \tag{4-94}$$

Also, if $a < 0 < b$,

$$\int_a^b g(t)\delta(t)\,dt = \int_a^b g(0)\delta(t)\,dt = g(0)\int_a^b \delta(t)\,dt = g(0), \tag{4-95}$$

$$\int_a^b g(t)\delta'(t)\,dt = \int_a^b [g(0)\delta'(t) - g'(0)\delta(t)]\,dt = -g'(0), \tag{4-96}$$

and in general for $a < c < b$

$$\int_a^b g(t)\delta^{(k)}(t-c)\,dt = (-1)^k g^{(k)}(c), \tag{4-97}$$

provided $g^{(k)}(t)$ is continuous at $t = c$.

Finally we remark that $\delta(t)$ is an even function, $\delta'(t)$ is an odd function, $\delta''(t)$ is even, and so on, that is, $\delta(-t) = \delta(t)$, $\delta'(-t) = -\delta'(t)$, etc. These rules are suggested by the nature of the smooth approximations to $\delta(t)$, $\delta'(t)$, ... as in Fig. 4–8.

4–10 LAPLACE TRANSFORMS OF GENERALIZED FUNCTIONS

We find from Eq. (4–97) for $c > 0$:

$$\mathscr{L}[\delta(t-c)] = \int_0^\infty \delta(t-c)e^{-st}\,dt = \lim_{b\to\infty} \int_0^b \delta(t-c)e^{-st}\,dt = e^{-cs}. \quad (4\text{–}100)$$

For $c = 0$ it is better to interpret the transform in a special way:

$$\mathscr{L}[\delta(t)] = \lim_{\varepsilon\to 0^-} \int_\varepsilon^\infty \delta(t)e^{-st}\,dt = e^{-s0} = 1. \quad (4\text{–}101)$$

Similarly, we find for $c > 0$, by (4–97), that

$$\mathscr{L}[\delta'(t-c)] = \int_0^\infty \delta'(t-c)e^{-st}\,dt = -\frac{d}{dt}(e^{-st})\bigg|_{t=c} = se^{-cs}, \quad (4\text{–}102)$$

and this is valid for $c = 0$ also. In general,

$$\mathscr{L}[\delta^{(k)}(t-c)] = s^k e^{-cs}, \quad c \geq 0. \quad (4\text{–}103)$$

If we use generalized functions in our work with Laplace transforms, then it is best to consider each function $f(t)$ as defined for *all* t, $-\infty < t < \infty$, and required to equal 0 for $t < 0$. Then for an ordinary function $f(t)$, $f'(t)$ contains a delta function if $f(0) \neq 0$, since there is a jump at 0. For example, if $f(t) = e^{3t}$, then we consider this to mean $e^{3t}U(t)$ and hence

$$f'(t) = 3e^{3t}U(t) + \delta(t)$$
$$f''(t) = 9e^{3t}U(t) + \delta'(t) + 3\delta(t), \quad \text{etc.}$$

If we use generalized functions in this way, then we achieve an important simplification in our rules, namely:

$$\mathscr{L}[f'] = s\mathscr{L}[f], \quad \mathscr{L}[f''] = s^2\mathscr{L}[f], \quad \ldots, \quad \mathscr{L}[f^{(k)}] = s^k\mathscr{L}[f]. \quad (4\text{–}104)$$

For $f(t) = f(t)U(t)$, we have $f'(t) = f'(t)U(t) + f(0)\delta(t)$, and hence

$$\mathscr{L}[f'] = \int_0^\infty f'(t)e^{-st}\,dt + f(0) = s\mathscr{L}[f] - f(0) + f(0) = s\mathscr{L}[f].$$

The other rules are proved in the same way.

EXAMPLE 1 Let $f(t) = e^{3t} = e^{3t}U(t)$ as above. Then $f'(t) = 3e^{3t}U(t) + \delta(t)$, so that

$$\mathscr{L}[f] = \frac{1}{s-3}, \quad \mathscr{L}[f'] = \frac{3}{s-3} + 1 = \frac{s}{s-3} = s\mathscr{L}[f]. \quad \blacktriangleleft$$

We can also form *convolutions* using generalized functions. One way of doing this is by requiring the rule $\mathscr{L}[f*g] = \mathscr{L}[f]\mathscr{L}[g]$ to hold. Thus, since $\mathscr{L}[\delta] = 1$, we define for each ordinary function f:

$$\delta(t) * f = f * \delta(t) = f; \qquad (4\text{--}105)$$

all three expressions then have Laplace transform $\mathscr{L}[f]$. Similarly, we define:

$$\delta^{(k)}(t-c) * f = f * \delta^{(k)}(t-c) = f^{(k)}(t-c), \quad c \geqslant 0; \qquad (4\text{--}106)$$

all three expressions then have transform $s^k e^{-cs}\mathscr{L}[f]$. The last expression in (4–106) will normally be a generalized function. For example,

$$\delta'(t) * U(t) = U'(t) = \delta(t),$$

and all three have transform 1. We can also form convolutions of derivatives of delta functions. We define generally, for $k, l = 0, 1, 2, \ldots$:

$$\delta^{(k)}(t-c_1) * \delta^{(l)}(t-c_2) = \delta^{(k+l)}(t-c_1-c_2), \qquad (4\text{--}107)$$

where $c_1 \geqslant 0$, $c_2 \geqslant 0$. Here by the rule (4–103) both sides have transforms $s^{k+l}e^{-(c_1+c_2)s}$. The rules (4–105) and (4–106) can also be obtained from the expression of the convolution as an integral (see Section 4–8). From these rules and linearity, convolutions of arbitrary generalized functions can be formed.

EXAMPLE 2 $[e^{2t} + 2\delta(t) + 3\delta'(t)] * [\delta(t-1) + \delta''(t)]$ is the sum of six convolutions:

$$e^{2t} * \delta(t-1) + 2\delta(t) * \delta(t-1) + 3\delta'(t) * \delta(t-1) + e^{2t} * \delta''(t) + 2\delta(t) * \delta''(t) + 3\delta'(t) * \delta''(t).$$

Here we interpret e^{2t} throughout as $e^{2t}U(t)$. Then by the rules given above all six terms can be evaluated to yield

$$e^{2(t-1)}U(t-1) + 2\delta(t-1) + 3\delta'(t-1) + [4e^{2t}U(t) + 2\delta(t) + \delta'(t)] + 2\delta''(t) + 3\delta'''(t).$$

Inverse Laplace transforms involving generalized functions now appear. The new transforms defined permit us to obtain unique inverse transforms of polynomials in s or polynomials times e^{-cs} for $c \geqslant 0$. Thus

$$\mathscr{L}^{-1}[3 + 5s + 2s^2] = 3\delta(t) + 5\delta'(t) + 2\delta''(t),$$

$$\mathscr{L}^{-1}[e^{-s}(7 + 2s) + e^{-2s}(3s - s^3)] = 7\delta(t-1) + 2\delta'(t-1) + 3\delta''(t-2) - \delta'''(t-2).$$

════════════════ **Problems (Section 4-10)** ════════════════

1. Evaluate the convolution $f * g$ and verify that the rule $\mathscr{L}[f * g] = \mathscr{L}[f]\mathscr{L}[g]$ is obeyed:

   a) $f(t) = 1$ for $t \geqslant 0$, $g(t) = t$ for $t \geqslant 0$
   b) $f(t) = t$ for $t \geqslant 0$, $g(t) = t^2$ for $t \geqslant 0$
   c) $f(t) = e^{2t}$ for $t \geqslant 0$, $g(t) = e^{5t}$ for $t \geqslant 0$
   d) $f(t) = \sin 3t$ for $t \geqslant 0$, $g(t) = \sin 5t$ for $t \geqslant 0$
   e) $f(t) = 1$ for $0 \leqslant t < 1$, $f(t) = 0$ for $t \geqslant 1$, $g(t) = e^t$ for $t \geqslant 0$
   f) $f(t) = t$ for $0 \leqslant t < 1$, $f(t) = 0$ for $t \geqslant 1$, $g(t) = t^2$ for $t \geqslant 0$

2. Write the inverse Laplace transform of the given function $F(s)$ as a convolution, but do not evaluate:

a) $\dfrac{1}{(s-1)(s-2)}$ b) $\dfrac{1}{s^2(s-2)^2}$ c) $\dfrac{e^{-s}}{(s-3)(s-5)}$ d) $\dfrac{e^{-s}}{s(s-2)^2}$

e) $\dfrac{1-e^{-2s}}{s(s-1)}$ f) $\dfrac{1}{s^2(1+e^{-3s})}$ g) $\dfrac{1-e^{-3s}}{s^2(s-2)(1+e^{-3s})}$

3. Prove the rule (4–82) directly from the definition (4–80). HINT: set $v=t-u$ in the integral in (4–80).

4. From the definition (4–80) prove:
   a) (4–83)   b) (4–84)   c) (4–86)

5. Prove (4–85) with the aid of (4–81).

6. Graph each function and find the first and second derivatives as generalized functions:
   a) $e^{2t}U(t)$ b) $(3-t)U(t)$ c) $U(t)-U(t-1)$ d) $t[U(t-1)-U(t-2)]$
   e) $\sin t\left[U(t)-U\left(t-\dfrac{\pi}{2}\right)\right]$ f) $(2t-t^2)[U(t)-U(t-2)]$

7. Evaluate the products:
   a) $e^{2t}\cdot\delta(t)$ b) $\delta(t)\cdot[\sin^3 t\,U(t)]$
   c) $[e^{3t}U(t)]\cdot\delta(t-2)$ d) $\delta(t-1)\cdot[t^2U(t)]$
   e) $e^{3t}\cdot\delta'(t)$ f) $[t^2U(t)]\cdot\delta'(t-2)$
   g) $[e^{2t}U(t-1)]\cdot\delta''(t-2)$ h) $[t^2U(t-2)]\cdot\delta''(t-4)$

8. Evaluate the integrals:
   a) $\displaystyle\int_{-1}^{1}\delta(t)\,dt$ b) $\displaystyle\int_{-5}^{3}[2\delta(t)+3\delta'(t)]\,dt$ c) $\displaystyle\int_{-1}^{1}e^{5t}\delta(t)\,dt$

   d) $\displaystyle\int_{-1}^{1}(3+t^2)\delta(t)\,dt$ e) $\displaystyle\int_{-1}^{1}e^{2t}\,\delta'(t)\,dt$ f) $\displaystyle\int_{0}^{3}t^2\delta'(t-2)\,dt$

   g) $\displaystyle\int_{-1}^{1}\cos^2 t\,\delta''(t)\,dt$ h) $\displaystyle\int_{-1}^{1}(t^2-3t+5)\delta''(t)\,dt$

9. For $f(t)$ as given, find $f'(t)$ as a generalized function and verify that $\mathscr{L}[f']=s\mathscr{L}[f]$:
   a) $(2t+3)U(t)$ b) $\cos t\,U(t)$ c) $U(t)-U(t-1)$ d) $t[U(t)-U(t-1)]$
   e) $2U(t)-3\delta(t)$ f) $3U(t-1)-\delta(t)+\delta'(t-1)$

10. Evaluate the convolution:
    a) $\delta(t)*e^{3t}U(t)$ b) $\delta(t)*\sin t\,U(t)$ c) $\delta'(t)*e^{2t}U(t)$ d) $\delta'(t)*\cos t\,U(t)$

    e) $\delta''(t-1)*t^2U(t)$ f) $\delta''(t-2)*\dfrac{1}{1+t^2}U(t)$

    g) $[(1+t^2)U(t)+\delta(t)]*[t^3U(t)+2\delta''(t)]$
    h) $[\sin t\,U(t)+2\delta(t)-\delta'(t-1)]*\delta''(t)$

11. Find the inverse transform:

    a) $1-s$ b) $2+s+3s^2$ c) $\dfrac{s^2-1}{s^2+1}$ d) $\dfrac{s^2+3s+5}{(s-1)(s-2)}$

    e) $e^{-2s}(2+5s)$ f) $e^{-s}\dfrac{s+2}{s+1}$

4–11 APPLICATION OF LAPLACE TRANSFORMS TO LINEAR DIFFERENTIAL EQUATIONS

We first proceed without generalized functions and then, in the next section, point out the benefit of using them.

We recall the rule for the transform of the kth derivative (Section 4–3):

$$\mathcal{L}[f^{(k)}] = s^k \mathcal{L}[f] - f^{(k-1)}(0) - sf^{(k-2)}(0) - \cdots - s^{k-1}f(0). \quad (4\text{–}110)$$

This rule provides a standard method for solving the *initial-value problem* for the linear differential equation with constant coefficients (Section 1–15):

$$a_0 \frac{d^n x}{dt^n} + \cdots + a_n x = g(t). \quad (4\text{–}111)$$

The method consists in taking the Laplace transform of both sides, applying rule (4–110), solving for $X(s)$, the transform of the desired solution $x(t)$, and finally taking the inverse transform to obtain $x(t)$ for $t \geq 0$. We here emphasize the formal procedures (for a complete justification, see OMLS, pp. 358–359).

EXAMPLE 1 $x'' + 3x' + 2x = e^{5t}$, $x = 1$ and $x' = 3$ for $t = 0$. Since

$$\mathcal{L}[x'] = sX(s) - x(0) = sX(s) - 1,$$
$$\mathcal{L}[x''] = s^2 X(s) - x'(0) - sx(0) = s^2 X(s) - 3 - s,$$

applying the Laplace transform to the given equation gives successively

$$s^2 X(s) - 3 - s + 3[sX(s) - 1] + 2X(s) = \frac{1}{s-5},$$

$$X(s) = \frac{s^2 + s - 29}{(s-5)(s^2 + 3s + 2)} = \frac{s^2 + s - 29}{(s-5)(s+1)(s+2)}$$

$$= \frac{1}{42} \frac{1}{s-5} + \frac{29}{6(s+1)} - \frac{27}{7(s+2)},$$

$$x(t) = \frac{1}{42} e^{5t} + \frac{29}{6} e^{-t} - \frac{27}{7} e^{-2t}, \quad t \geq 0.$$

This method is clearly simpler than the method of undetermined coefficients (Section 1–12); it can also be applied more generally. ◀

EXAMPLE 2 $x'' + 2x' + 2x = g(t)$, $x(0) = 0$, $x'(0) = 0$, where $g(t) = 1$ for $0 \leq t \leq 1$, $g(t) = 0$ for $t > 1$. By entry 18 of Table 4–1 and rule (4–110),

$$s^2 X(s) + 2sX(s) + 2X(s) = \frac{1 - e^{-s}}{s},$$

$$X(s) = \frac{1 - e^{-s}}{s(s^2 + 2s + 2)}.$$

Therefore by linearity and the translation rule (4–38)

$$x(t) = f(t) - f(t-1), \quad t \geqslant 0,$$

where

$$f(t) = \mathscr{L}^{-1} \left[\frac{1}{s(s+2s+2)} \right] - \mathscr{L}^{-1} \left[\frac{1}{2s} + \frac{-1+i}{4} \frac{1}{s-(-1+i)} \right.$$
$$\left. + \frac{-1-i}{4} \frac{1}{s-(-1-i)} \right]$$

$$= \frac{1}{2} + \frac{-1+i}{4} e^{(-1+i)t} + \frac{-1-i}{4} e^{(-1-i)t} = \frac{1}{2} - \frac{1}{2} (\cos t + \sin t) e^{-t} \quad \text{for} \quad t \geqslant 0$$

and $f(t) = 0$ for $t < 0$. Thus $f(t-1) = 0$ for $t \leqslant 1$ and

$$f(t-1) = \frac{1}{2} - \frac{1}{2} \left[\cos (t-1) + \sin (t-1)\right] e^{-(t-1)} \quad \text{for} \quad t \geqslant 1. \quad \blacktriangleleft$$

EXAMPLE 3 $x'' + 3x' + 2x = 1/(1+t^2)$, $x = 0$, $x' = 1$ for $t = 0$. If we let $G(s) = \mathscr{L}\left[1/(1+t^2)\right]$, then

$$s^2 X(s) - 1 + 3s X(s) + 2X(s) = G(s),$$

$$X(s) = \frac{1}{s^2+3s+2} + \frac{G(s)}{s^2+3s+2} = \frac{1}{s+1} - \frac{1}{s+2} + G(s) \left(\frac{1}{s+1} . \frac{1}{s+2} \right)$$

$$= \mathscr{L}\left[e^{-t} - e^{-2t}\right] + \mathscr{L}\left[1/(1+t^2)\right] \mathscr{L}\left[e^{-t} - e^{-2t}\right].$$

Thus, by the rule (4–81) for the transform of a convolution,

$$x(t) = e^{-t} - e^{-2t} + \frac{1}{1+t^2} * (e^{-t} - e^{-2t})$$

$$= e^{-t} - e^{-2t} + \int_0^t \frac{1}{1+u^2} \left(e^{-(t-u)} - e^{-2(t-u)} \right) du$$

$$= e^{-t} - e^{-2t} + e^{-t} \int_0^t \frac{e^u}{1+u^2} du - e^{-2t} \int_0^t \frac{e^{2u}}{1+u^2} du, \quad t \geqslant 0.$$

For each fixed t, the integrals can be evaluated numerically and hence $x(t)$ can be found. $\blacktriangleleft$

The method of Example 3 can be generalized. For the Eq. (4–111) we let

$$Y(s) = \frac{1}{a_0 s^n + \cdots + a_n}. \tag{4–112}$$

This is the *transfer function* (Section 3–8). Since $Y(s)$ is a proper rational function, it can be decomposed into partial fractions and hence has an inverse Laplace transform. We denote this inverse transform by $W(t)$:

$$W(t) = \mathscr{L}^{-1}\left[Y(s) \right], \tag{4–113}$$

and call $W(t)$ the *weight function*.

If now *all the initial values* $x(0)$, $x'(0)$, . . . , $x^{(n-1)}(0)$ *are* 0, then application of the Laplace transform to Eq. (4–111) gives

$$a_0 s^n X(s) + a_1 s^{n-1} X(s) + \cdots + a_n X(s) = G(s),$$

$$X(s) = \frac{1}{a_0 s^n + \cdots + a_n} G(s) = Y(s) G(s).$$

Therefore, our solution is

$$x(t) = \mathscr{L}^{-1}[\, Y(s) G(s)], \quad t \geqslant 0. \qquad (4\text{–}114)$$

We can also write: $X(s) = \mathscr{L}[W]\mathscr{L}[g]$ and hence

$$x = W(t) * g(t) = \int_0^t W(u) g(t-u) \, du, \quad t \geqslant 0. \qquad (4\text{–}114')$$

It is this form of the solution that gives rise to the term weight function. We can think of the *response*, or *output*, $x(t)$ being obtained from the *input* $q(t)$ by "summing" the past values of the input weighted, for time u in the past, by the factor $W(u)$. This could be calculated graphically, as suggested in Fig. 4–12. For a fixed t we place the graph of W (going backwards and with origin above t) above that of g. Then for each u between 0 and t we multiply $g(t-u)$ by $W(u)$ (which appears on the W-graph above $t - u$) and integrate the resulting function from 0 to t. The shape of the W-graph is very important. For the case depicted in Fig. 4–12, values of u near 1 receive major weight. Therefore the output x "remembers" the input especially well for times about 1 unit in the past.

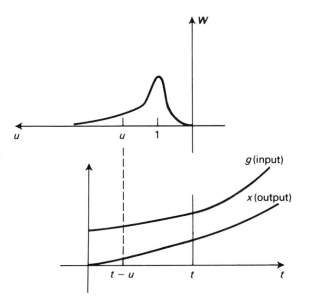

Fig. 4-12. $W(t)$ as weight function.

Now let us take $g(t) \equiv 0$ but allow arbitrary initial values $x_0, x'_0, \ldots, x_0^{(n-1)}$. Then application of the Laplace transform to (4–111) gives, by rule (4–110),

$$a_0\left[s^n X(s) - x_0^{(n-1)} - sx_0^{(n-2)} - \cdots - s^{n-1}x_0\right] + a_1\left[s^{(n-1)}X(s) - x_0^{(n-2)} - \cdots\right]$$
$$+ \cdots + a_{n-1}\left[sX(s) - x_0\right] + a_n X(s) = 0,$$

$$\left[a_0 s^n + a_1 s^{n-1} + \cdots + a_n\right]X(s) = a_0 x_0^{(n-1)} + \cdots + a_{n-1}x_0 + (a_0 x_0^{(n-2)} + \cdots$$
$$+ a_{n-2}x_0)s + \cdots + a_0 x_0 s^{n-1} = P(s).$$

Here $P(s)$ is a polynomial of degree $n-1$. Therefore we can apply the rule (4–72) for inverse transforms, using $W(t) = \mathcal{L}^{-1}[Y(s)]$. If we write

$$P(s) = c_0 + c_1 s + \cdots + c_{n-1}s^{n-1},$$

then

$$x(t) = c_0 W(t) + c_1 W'(t) + \cdots + c_{n-1}W^{(n-1)}(t). \tag{4–115}$$

Here

$$c_0 = a_0 x_0^{(n-1)} + \cdots + a_{n-1}x_0, \quad c_1 = a_0 x_0^{(n-2)} + \cdots + a_{n-2}x_0, \ldots, c_{n-1} = a_0 x_0. \tag{4–116}$$

Equation (4–115) expresses the general solution of the *homogeneous equation* for $t \geqslant 0$ in terms of the weight function and its derivatives, while Eq. (4–116) expresses this solution in terms of the initial values. We can also write

$$P(s) = x_0(a_{n-1} + a_{n-2}s + \cdots + a_0 s^{n-1}) + x'_0(a_{n-2} + \cdots + a_0 s^{n-2}) + \cdots + x_0^{(n-1)}a_0$$
$$= x_0 p_0(s) + x'_0 p_1(s) + \cdots + x_0^{(n-1)}p_{n-1}(s),$$

and hence

$$X(s) = x_0\frac{p_0(s)}{a_0 s^n + \cdots + a_n} + x'_0\frac{p_1(s)}{a_0 s^n + \cdots + a_n} + \cdots + x_0^{(n-1)}\frac{p_{n-1}(s)}{a_0 s^n + \cdots + a_n},$$
$$x(t) = x_0 f_0(t) + x'_0 f_1(t) + \cdots + x_0^{(n-1)}f_{n-1}(t). \tag{4–117}$$

This shows that the general solution of the homogeneous equation depends *linearly* on the initial values for $x(t)$.

Finally we allow arbitrary initial values and allow $g(t)$ to be different from 0. We obtain in the same way:

$$X(s) = Y(s)P(s) + Y(s)G(s),$$
$$x(t) = c_0 W(t) + c_1 W'(t) + \cdots + c_{n-1}W^{(n-1)}(t) + W(t)*g(t) \tag{4–118}$$

or, using (4–117),

$$x(t) = x_0 f_0(t) + x'_0 f_1(t) + \cdots + x_0^{(n-1)}f_{n-1}(t) + W(t)*g(t). \tag{4–118'}$$

We remark that if the considered system is *stable* (Section 1–15), then the solutions of the homogeneous equation are *transients*, and for large t only the particular solution (steady state) $x = W(t)*g(t)$ is important.

EXAMPLE 4   *The direct-current motor.* As suggested in Fig. 4–13, a shaft is driven at angular velocity ω by forces due to an input field voltage V_f. The corresponding differential equation is

$$\frac{JL_f}{K_m}\left(D + \frac{f}{J}\right)\left(D + \frac{R_f}{L_f}\right)\omega = V_f(t),$$

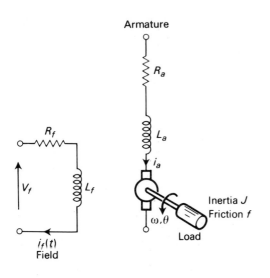

Armature

R_a

R_f

V_f L_f

L_a

i_a

ω, θ

Inertia J
Friction f

Load

$i_f(t)$
Field

Fig. 4-13. The dc-motor.

where J, f, R_f, L_f are positive constants, as shown in Fig. 4–13, and K_m is the motor constant. We take Laplace transforms and obtain the transfer function

$$Y(s) = \frac{K_m/(JL_f)}{(s+f/J)\,(s+R_f/L_f)}.$$

The system is clearly stable, since the characteristic roots are the negative numbers $-f/J, \, -R_f/L_f$. We can also interpret ω as $d\theta/dt$, as in Fig. 4–13, and then obtain the differential equation

$$\frac{JL_f}{K_m}\left(D+\frac{f}{J}\right)\left(D+\frac{R_f}{L_f}\right)D\theta = V_f\,(t),$$

with corresponding transfer function

$$\frac{K_m/(JL_f)}{s(s+f/J)\,(s+R_f/L_f)}.$$

Now there is a 0 characteristic root, so that we have lost stability. However, there is no benefit in controlling $\theta(t)$; it is the angular velocity $\omega(t)$ that is significant in engineering. (For more discussion of this example and further examples see: R. C. Dorf, *Modern Control Systems*, 2nd ed., Chapter 2, Addison-Wesley, Reading, Mass., 1974.) ◄

4–12 APPLICATION OF LAPLACE TRANSFORMS OF GENERALIZED FUNCTIONS TO LINEAR DIFFERENTIAL EQUATIONS

In working with Laplace transforms of generalized functions, as in Section 4–10, we focus our attention on functions that are identically 0 for $t < 0$. This implies that the solution of a differential equation has zero initial values at $t = 0$. However, as will be seen, the function and its derivatives may have *jumps* at $t = 0$, and thus in effect really have initial values other than 0.

As in Section 4–10, our basic rule is now

$$\mathcal{L}[f^{(k)}] = s^k \mathcal{L}[f] \quad (k = 0, 1, 2, \ldots). \tag{4–120}$$

EXAMPLE 1 $x'' + 5x' + 4x = 3 + 2\delta(t)$. We take Laplace transforms, applying rule (4–120):

$$(s^2 + 5s + 4) X(s) = \frac{3}{s} + 2,$$

$$X(s) = \frac{3 + 2s}{s(s+1)(s+4)} = \frac{3}{4}\frac{1}{s} - \frac{1}{3}\frac{1}{s+1} - \frac{5}{12}\frac{1}{s+4},$$

$$x(t) = \left(\frac{3}{4} - \frac{1}{3}e^{-t} - \frac{5}{12}e^{-4t}\right) U(t).$$

We multiplied by $U(t)$ to emphasize that $x = 0$ for $t < 0$. The solution is graphed in Fig. 4–14. We see that, $x_0 = 0$, but x' jumps from 0 to 2 in passing through $t = 0$. This can easily be traced to the $\delta(t)$ term in the differential equation. ◄

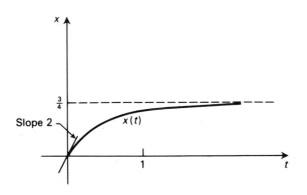

Fig. 4-14. Solution to Example 1.

EXAMPLE 2 $x'' + 5x' + 4x = 3 + a\delta(t) + b\delta'(t)$. This is like Example 1, except for general terms in $\delta(t)$ and $\delta'(t)$. We proceed as before:

$$(s^2 + 5s + 4) X(s) = \frac{3}{s} + a + bs,$$

$$X(s) = \frac{3 + as + bs^2}{s(s^2 + 5s + 4)} = \frac{3}{4}\frac{1}{s} + \frac{a - b - 3}{3}\frac{1}{s+1} + \frac{3 - 4a + 16b}{12}\frac{1}{s+4},$$

$$x(t) = \left(\frac{3}{4} + \frac{a - b - 3}{3}e^{-t} + \frac{3 - 4a + 16b}{12}e^{-4t}\right) U(t).$$

Therefore, for $t \to 0^+$, $x(t)$ tends to limit

$$\frac{3}{4} + \frac{a-b-3}{3} + \frac{3-4a+16b}{12} = b,$$

while for $t > 0$

$$x'(t) = \frac{-a+b+3}{3}e^{-t} + \frac{-3+4a-16b}{3}e^{-4t},$$

and hence, for $t \to 0^+$, $x'(t)$ has limit $a - 5b$. Thus addition of the terms in $\delta(t)$ and $\delta'(t)$ allows the initial values of x and x' to jump from 0 to b and $a - 5b$ respectively. By adjusting a and b we can make these effective initial values (limits from the right) equal to given numbers x_0 and x_0'.

Our problem can be solved on an analog computer. As a last step before letting the computer produce the solution we turn knobs for x and x' to adjust to the given initial values. In this act we are in effect applying an input $a\delta(t) + b\delta'(t)$ that causes the initial values to jump from 0 to assigned values.

Similar arguments hold for an equation of order n with terms in $\delta(t), \delta'(t), \ldots, \delta^{(n-1)}(t)$. ◄

If we allow terms in $\delta^{(n)}, \delta^{(n+1)}, \ldots$, then the solution itself becomes a generalized function.

EXAMPLE 3 $x'' + 4x = \delta''(t)$. We obtain

$$(s^2 + 4)X(s) = s^2,$$

$$X(s) = \frac{s^2}{s^2+4} = 1 - \frac{4}{s^2+4},$$

$$x(t) = \delta(t) - 2\sin 2t. ◄$$

Relation to weight function and step response

For the general equation of order n we consider the response to a delta-function input:

$$a_0 x^{(n)} + \cdots + a_n x = \delta(t),$$

$$(a_0 s^n + \cdots + a_n)X(s) = 1,$$

$$X(s) = Y(s), \quad x(t) = \mathcal{L}^{-1}[Y(s)] = W(t).$$

Hence the *weight function* $W(t)$ is the response to $\delta(t)$ or, as it is often called, the *impulse response*.

Similarly, for a unit step input, we have

$$a_0 x^{(n)} + \cdots + a_n x = U(t),$$

$$(a_0 s^n + \cdots + a_n)X(s) = \frac{1}{s},$$

$$X(s) = \frac{1}{s}Y(s).$$

Hence by the rule (4–34) for integration:

$$x(t) = \int_0^t W(u)\, du.$$

Thus the step response is the integral of the impulse response, or *the impulse response is the derivative of the step response*.

We observe that here, since there is no delta term on the right, $x(t)$ is simply the solution with 0 initial values. There is no jump in $x, x', \ldots, x^{(n-1)}$ at $t = 0$. However, if we let t approach 0^+ in the differential equation, we obtain:

$$\lim_{t \to 0^+} a_0 x^{(n)} + 0 + \cdots + 0 = 1.$$

Hence $x^{(n)}$ has limit $1/a_0$ from the right, so that there is a jump in the nth derivative. Since $x' = W(t)$, $x'' = W'(t)$, $\ldots$, $x^{(n)} = W^{(n-1)}(t)$, we see that $W(t), W'(t), \ldots, W^{(n-2)}(t)$ have value 0 at $t = 0$, whereas $W^{(n-1)}(t)$ has value $1/a_0$ (taken as limit from the right). Thus $W(t)$ can be described as the solution of the homogeneous equation for $t \geqslant 0$ with these initial values at $t = 0$.

The results obtained here suggest an experimental way of determining the weight function $W(t)$. Given a physical system governed by a linear differential equation with constant coefficients and such that arbitrary inputs can be applied, we can use a unit step input $U(t)$ with zero initial values and record the solution $x(t)$. Then $W(t)$ is simply dx/dt.

An alternative procedure is to approximate a delta-function input by a very large impulse of unit area, as in Fig. 4–8(a), and record the response for zero initial values.

══════════════════════ **Problems (Section 4-12)** ══════════════════════

1. Find the solution for $t \geqslant 0$ with given initial values:

   a) $x' + 2x = 3 + 6t$, $x = 1$ for $t = 0$
   b) $x' + x = 2e^{-t}$, $x = 3$ for $t = 0$
   c) $x'' - x = 2 \sin t$, $x = 1$ and $x' = -1$ for $t = 0$
   d) $x'' - x = 1 + 2t$, $x = 2$ and $x' = 0$ for $t = 0$
   e) $x'' + 2x' + 2x = 1$, $x = 0$ and $x' = 0$ for $t = 0$
   f) $x'' + 4x = 3t$, $x = 0$ and $x' = 1$ for $t = 0$
   g) $x''' + 6x'' + 11x' + 6x = 0$, $x = 1$, $x' = 2$ and $x'' = 0$ for $t = 0$
   h) $x''' + 2x'' - x' - 2x = 0$, $x = 0$, $x' = 1$, $x'' = -1$ for $t = 0$
   i) $x'' - 4x = g(t)$, $g(t) = t$ for $0 \leqslant t \leqslant 2$, $g(t) = 0$ for $t > 2$, $x = 0$, $x' = 0$ for $t = 0$
   j) $x'' - 4x = g(t)$, $g(t) = 0$ for $0 \leqslant t \leqslant 3$, $g(t) = e^{-t}$ for $t \geqslant 3$, $x = x' = 0$ for $t = 0$
   k) $x'' + 2x' + 2x = e^{-t^2}$, $x = x' = 0$ for $t = 0$
   l) $x'' + 3x' + 2x = \ln(1 + t)$, $x = x' = 0$ for $t = 0$

2. Find the transfer and function for each equation of Problem 1 named:

   a) Part (a)   **b)** Part (c)   **c)** Part (e)   **d)** Part (f)   **e)** Part (g)   **f)** Part (h)

3. A body losing heat by cooling but also having heat provided by an external source has temperature u governed by the equation

$$\frac{du}{dt} + ku = g(t),$$

   where k is a positive constant.

a) Find a general formula for $u(t)$, if $u_0 = u(0)$ is given.

b) Take $k = 2$, $u_0 = 20$, $g(t) = 3t^2$ in consistent units and find $u(t)$.

4. The torsion pendulum (Fig. 4–15) is formed of a mass suspended by a wire. The mass rotates about an axis along the wire. In addition, an external torque $g(t)$ may be applied. If friction is ignored, the angular motion is governed by the equation

$$I\frac{d^2\theta}{dt^2} + k\theta = g(t),$$

where I is the moment of inertia of the solid about the axis of rotation and k is a positive constant.

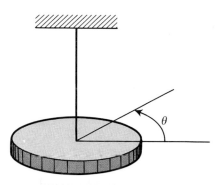

Fig. 4–15. Torsion pendulum.

a) Obtain a general formula for $\theta(t)$, if $\theta(0) = \theta_0$ and $\theta'(0) = \theta'_0$ are given.

b) Take $\theta_0 = 0$, $\theta'_0 = 0$, $I = 3$, $k = 12$, $g(t) = 2te^{-t}$ in consistent units and find $\theta(t)$.

5. A hydraulic actuator (Fig. 4–16) makes it possible to apply great force to position an object (see page 37 of the book by DORF cited at the end of Section 4–12). An input displacement $x(t)$ leads to a displacement $y(t)$ of the mass M. We find that

$$M\frac{d^2y}{dt^2} + \left(f + \frac{A^2}{k_p}\right)\frac{dy}{dt} = \frac{Ak_x}{k_p}x(t),$$

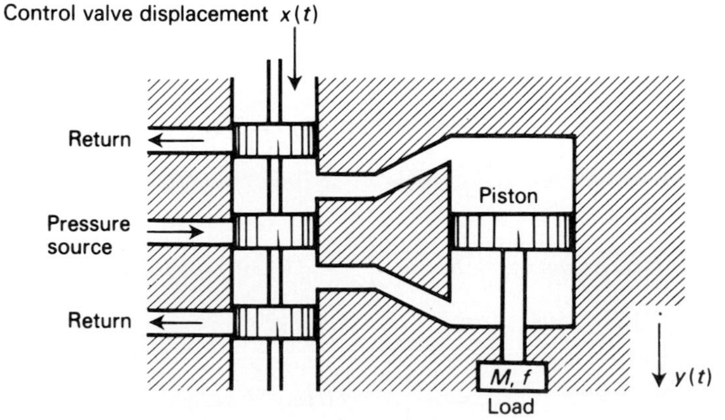

Control valve displacement $x(t)$

Return

Pressure source

Piston

Return

M, f
Load

$y(t)$

Fig. 4–16. Hydraulic actuator.

where f is a friction coefficient, A is the area of the piston, and k_x, k_p are positive constants:
a) Find the transfer function, if $x(t)$ is regarded as input.
b) Assume that, in appropriate units, the equation has the particular form

$$\frac{d^2y}{dt^2} + 2\frac{dy}{dt} = 5\sin t.$$

Find $y(t)$ if $y(0) = 0$, $y'(0) = 0$.

6. Find the solution as in Section 4–12 and in each case discuss initial values at $t = 0$ as limits from the right:

a) $x' + 3x = 5\delta(t)$ b) $x' + 2x = 1 - 2\delta(t)$ c) $x'' - 4x = 2\delta'(t)$
d) $x'' - 4x = 3\delta(t)$ e) $x'' + x' - 2x = 3e^{-t} + 2\delta(t) - 5\delta'(t)$
f) $x'' - x' + 2x = 1 - 2t + 3\delta(t) + \delta'(t)$ g) $x'' + 9x = 3 + 2\delta'(t) + 5\delta''(t)$
h) $x''' - x = \delta'''(t)$

7. Find the impulse response and step response:

a) $x' + 3x = g(t)$ b) $x' + 2x = g(t)$ c) $x'' - 4x = g(t)$
d) $x'' + x' - 2x = g(t)$ e) $x'' + 9x = g(t)$ f) $x''' - x = g(t)$

4–13 FOURIER TRANSFORMS

We now turn to functions $f(t)$, $g(t)$, . . . defined for $-\infty < t < \infty$ and continuous (or piecewise continuous). The Fourier transform of such a function, denoted by $\Phi[f]$, is the new function $\varphi(\omega)$, $-\infty < \omega < \infty$, defined by the equation

$$\Phi[f] = \int_{-\infty}^{\infty} f(t)e^{-i\omega t}\, dt = \varphi(\omega), \qquad -\infty < \omega < \infty, \tag{4–130}$$

provided the integral has meaning for all ω.

We here proceed formally and return in Section 4–15 to a discussion of the validity of the results.

We first recall, for the purpose of motivation, basic formulas for Fourier series for the interval $-\tau/2 \leqslant t \leqslant \tau/2$ (Sections 3–4, 3–5):

$$f(t) = \frac{a_0}{2} + \sum_{n=1}^{\infty} (a_n \cos n\omega t + b_n \sin n\omega t), \qquad \omega = 2\pi/\tau,$$

$$\tag{4–131}$$

$$a_n = \frac{2}{\tau}\int_{-\tau/2}^{\tau/2} f(t)\cos n\omega t\, dt, \qquad b_n = \frac{2}{\tau}\int_{-\tau/2}^{\tau/2} f(t)\sin n\omega t\, dt,$$

$$f(t) = \sum_{n=-\infty}^{\infty} c_n e^{in\omega t}, \qquad c_n = \frac{1}{\tau}\int_{-\tau/2}^{\tau/2} f(t)e^{-in\omega t}\, dt, \tag{4–132}$$

$$f(t) = \frac{a_0}{2} + \sum_{n=1}^{\infty} a_n \cos n\omega t, \quad f \text{ even}, \qquad a_n = \frac{4}{\tau} \int_0^{\tau/2} f(t) \cos n\omega t \, dt, \quad (4\text{--}133)$$

$$f(t) = \sum_{n=1}^{\infty} b_n \sin n\omega t, \quad f \text{ odd}, \qquad b_n = \frac{4}{\tau} \int_0^{\tau/2} f(t) \sin n\omega t \, dt. \quad (4\text{--}134)$$

We now let $\tau \to \infty$, so that $\omega \to 0$. The frequencies $n\omega$ then become ever more narrowly spaced. It is plausible that in the limit the series in (4–131) becomes an *integral* over all ω from 0 to ∞ and that the coefficients a_n, b_n become functions of ω defined by integrals of $f(t) \cos \omega t$, $f(t) \sin \omega t$ over the interval $-\infty < t < \infty$. We in fact obtain

$$f(t) = \int_0^{\infty} \left[a(\omega) \cos \omega t + b(\omega) \sin \omega t \right] d\omega,$$

$$(4\text{--}131')$$

$$a(\omega) = \frac{1}{\pi} \int_{-\infty}^{\infty} f(t) \cos \omega t \, dt, \qquad b(\omega) = \frac{1}{\pi} \int_{-\infty}^{\infty} f(t) \sin \omega t \, dt.$$

This is a representation of f as a *Fourier integral in real form*. From (4–132) we similarly obtain

$$f(t) = \int_{-\infty}^{\infty} A(\omega) e^{i\omega t} d\omega, \qquad A(\omega) = \frac{1}{2\pi} \int_{-\infty}^{\infty} f(t) e^{-i\omega t} dt. \qquad (4\text{--}132')$$

This is a representation of $f(t)$ as a *Fourier integral in complex form*. Because of the greater convenience, the *complex* form is preferred and the term *Fourier integral* usually refers to this form.

In a similar manner we obtain from (4–133) and (4–134) for f even and odd, respectively:

$$f(t) = \int_0^{\infty} a(\omega) \cos \omega t \, d\omega, \qquad a(\omega) = \frac{2}{\pi} \int_0^{\infty} f(t) \cos \omega t \, dt, \qquad (4\text{--}133')$$

$$f(t) = \int_0^{\infty} b(\omega) \sin \omega t \, d\omega, \qquad b(\omega) = \frac{2}{\pi} \int_0^{\infty} f(t) \sin \omega t \, dt. \qquad (4\text{--}134')$$

These are representations of f as a *Fourier cosine integral* and *Fourier sine integral*, respectively.

We now examine the Fourier integral representation (4–132′). This is closely related to the Fourier transform of Eq. (4–130). In fact, $\varphi(\omega) = 2\pi A(\omega) = \Phi[f]$. Hence we can write Eq. (4–132′) as follows:

$$f(t) = \frac{1}{2\pi} \int_{-\infty}^{\infty} \varphi(\omega) e^{i\omega t} d\omega, \qquad \varphi(\omega) = \int_{-\infty}^{\infty} f(t) e^{-i\omega t} dt = \Phi[f]. \quad (4\text{--}135)$$

We shall use this form in preference to that of Eq. (4–132′).

REMARK: In general, the Fourier integral in (4–135) should be interpreted as a *principal value*, that is, as

$$\lim_{b \to \infty} \int_{-b}^{b} \varphi(\omega)\, e^{i\omega t}\, d\omega.$$

To indicate this, we write the integral as

$$(P) \int_{-\infty}^{\infty} \varphi(\omega)\, e^{i\omega t}\, d\omega.$$

We often integrate an *odd* function of ω in this way. For such a function the integral from $-b$ to b is zero and hence the principal value of the integral is zero.

The principal value arises naturally in passing from the real form (4–131′) to the complex form. The representation in terms of positive and negative frequencies ω is an artificiality of the complex form. By taking the principal value, we in effect bring the positive and negative frequencies together again.

EXAMPLE 1   Let $f(t) = 1$ for $0 < t < 1$ and $f(t) = 0$ otherwise. Then

$$\varphi(\omega) = \int_{0}^{1} e^{-i\omega t}\, dt = \frac{e^{-i\omega t}}{-i\omega}\bigg|_{t=0}^{t=1} = \frac{i}{\omega}(e^{-i\omega} - 1).$$

Hence,

$$f(t) = \frac{i}{2\pi} \int_{-\infty}^{\infty} \frac{e^{-i\omega} - 1}{\omega}\, e^{i\omega t}\, d\omega.$$

The ω in the denominator causes no trouble, since

$$e^{-i\omega} - 1 = -i\omega + (-i\omega)^2/2! + \cdots$$

is divisible by ω. The Fourier integral converges to the average of left and right limits at jump points, like Fourier series. Thus for $t = 0$ we obtain

$$\frac{1}{2} = \frac{i}{2\pi} \int_{-\infty}^{\infty} \frac{e^{-i\omega} - 1}{\omega}\, d\omega = \frac{1}{2\pi} \int_{-\infty}^{\infty} \left[\frac{\sin \omega}{\omega} + i\, \frac{\cos \omega - 1}{\omega} \right] d\omega.$$

The function $(\cos \omega - 1)/\omega$ is odd, so that the imaginary term (*principal value*) is zero (as it should be) and

$$\frac{1}{2} = \frac{1}{2\pi} \int_{-\infty}^{\infty} \frac{\sin \omega}{\omega}\, d\omega.$$

This equality gives the value of an important integral that cannot be found by elementary means. (No principal value is needed for it, since the integrand is *even*.) ◄

EXAMPLE 2 Let $f(t) = e^{-t}$ for $0 < t < \infty$ and $f(t) = 0$ for $t < 0$. Then

$$\varphi(\omega) = \int_0^\infty e^{-t} e^{-i\omega t}\, dt = \int_0^\infty e^{-(1+i\omega)t}\, dt = \frac{e^{-(1+i\omega)t}}{-(1+i\omega)}\Big|_{t=0}^{t=\infty} = \frac{1}{1+i\omega}.$$

Hence

$$f(t) = \frac{1}{2\pi} \int_{-\infty}^\infty \frac{e^{i\omega t}}{1+i\omega}\, d\omega.$$

Here if we put $t = 0$, we can check the result:

$$\frac{1}{2} = \frac{1}{2\pi} \int_{-\infty}^\infty \frac{1}{1+i\omega}\, d\omega = \frac{1}{2\pi} \int_{-\infty}^\infty \frac{1-i\omega}{1+\omega^2}\, d\omega.$$

Again the imaginary part on the right is zero, since we have an odd function and take the principal value. Thus

$$\frac{1}{2} = \frac{1}{2\pi} \int_{-\infty}^\infty \frac{d\omega}{1+\omega^2} = \frac{1}{2\pi} \text{Arc tan } \omega \Big|_{-\infty}^\infty,$$

and this is correct. For $t \neq 0$, we are dealing with new integrals. ◄

Validity of the formulas

We return to this question in detail in Section 4–15. For the present we remark that the Fourier transform of f is well-defined if f is piecewise continuous on each finite interval and f is absolutely integrable from $-\infty$ to ∞, that is, if

$$\int_{-\infty}^\infty |f(t)|\, dt$$

is finite. For the representation of f by its Fourier integral (principal value), it is then sufficient that f be piecewise smooth on each finite interval and equal to the average value at jumps.

Transform definitions

The Fourier transform is sometimes defined in other ways, differing from (4–130) by scale factors. For example, we may define the transform as

$$\psi(\omega) = \frac{1}{\sqrt{2\pi}} \int_{-\infty}^\infty f(t) e^{-i\omega t}\, dt.$$

Then the Fourier integral representation becomes

$$f(t) = \frac{1}{\sqrt{2\pi}} (P) \int_{-\infty}^\infty \psi(\omega) e^{i\omega t}\, d\omega.$$

The Fourier cosine and sine integrals (see (4–133′) and (4–134′) above) also give rise to transforms. The *Fourier cosine transform* is commonly defined as

$$\Phi_c[f] = \sqrt{\frac{2}{\pi}} \int_0^\infty f(t) \cos \omega t\, dt = \psi_c(\omega). \tag{4–136}$$

Then we have complete symmetry:

$$f(t) = \sqrt{\frac{2}{\pi}} \int_0^\infty \psi_c(\omega) \cos \omega t \, d\omega. \tag{4-137}$$

Similarly, the *Fourier sine transform* of f is defined as

$$\Phi_s[f] = \sqrt{\frac{2}{\pi}} \int_0^\infty f(t) \sin \omega t \, dt = \psi_s(\omega) \tag{4-138}$$

and we obtain

$$f(t) = \sqrt{\frac{2}{\pi}} \int_0^\infty \psi_s(\omega) \sin \omega t \, d\omega. \tag{4-139}$$

In this chapter we shall emphasize the Fourier transform (4–130). (For more information on the Fourier cosine and sine transforms see ERDELYI, Vol. I of Tables.)

4–14 PROPERTIES OF THE FOURIER TRANSFORM

To each suitable function $f(t)$, $-\infty < t < \infty$, the Fourier transform assigns a function $\varphi(\omega)$, $-\infty < \omega < \infty$. Usually $f(t)$ is real valued, but $\varphi(\omega)$ is normally complex valued (as in Examples 1 and 2 of the preceding section). We can also allow $f(t)$ to be complex valued; the theory covers this case.

Linearity If f_1 and f_2 have Fourier transforms, then so does $c_1 f_1 + c_2 f_2$ for each choice of the constants c_1, c_2 (real or complex) and

$$\Phi[c_1 f_1 + c_2 f_2] = c_1 \Phi[f_1] + c_2 \Phi[f_2]. \tag{4-140}$$

This follows from the definition (4–130).

Transform of derivative Here the rule is simply

$$\Phi[f'] = i\omega \, \Phi[f], \tag{4-141}$$

provided f and f' are continuous and have absolutely convergent integrals from $-\infty$ to ∞. Under these assumptions, f must have limit 0 as $t \to \pm \infty$ (cf. the proof of Eq. (4–31) and Problem 5 following Section 4-4). Now, through integration by parts we get

$$\Phi[f'] = \int_{-\infty}^\infty f'(t) e^{-i\omega t} \, dt = f(t) e^{-i\omega t} \Big|_{-\infty}^\infty + i\omega \int_{-\infty}^\infty f(t) e^{-i\omega t} \, dt$$

$$= 0 + i\omega \, \Phi[f].$$

Hence the rule (4–141) is proved. By repeatedly applying the rule, we deduce that

$$\Phi[f^{(k)}] = (i\omega)^k \Phi[f], \quad k = 1, 2, \dots, \tag{4-141'}$$

under similar assumptions.

Transform of integral It may happen that $f(t)$ has a Fourier transform, whereas $\int_0^t f(u) \, du$ does not. However, $g(t) = \int_0^t f(u) \, du + c$, for an appropriate constant c, may have a transform. When this happens,

$$\Phi[g] = \frac{1}{i\omega}\Phi[f], \quad \omega \neq 0. \tag{4-142}$$

Indeed, $g' = f$, so that (4–142) follows from (4–141).

Translation If f has a Fourier transform, then so does $f(t-c)$, and

$$\Phi[f(t-c)] = e^{-i\omega c}\Phi[f]. \tag{4-143}$$

The proof is left as an exercise (see Problem 4 below).

Multiplication by $e^{i\alpha t}$ If α is a real constant and $\Phi[f] = \varphi(\omega)$, then

$$\Phi[e^{i\alpha t}f(t)] = \varphi(\omega - \alpha). \tag{4-144}$$

The proof is left as an exercise (see Problem 4 below).

Iterated transform Since $\varphi(\omega) = \Phi[f]$ is a function of the real variable ω, $-\infty < \omega < \infty$, we can write its Fourier transform (assumed to exist) as:

$$\Phi[\varphi] = \int_{-\infty}^{\infty} \varphi(\omega)e^{-it\omega}\,d\omega = \psi(t), \quad -\infty < t < \infty.$$

We here take advantage of the fact that we can label the independent variable as t or ω, as we choose. We now have the rule:

$$\Phi[\varphi] = \Phi[\Phi[f]] = 2\pi f(-t). \tag{4-145}$$

This rule is simply a restatement of the basic formulas (4 135), for by those formulas

$$f(-t) = \frac{1}{2\pi}\int_{-\infty}^{\infty} \varphi(\omega)e^{-i\omega t}\,d\omega = \frac{1}{2\pi}\Phi[\varphi].$$

This rule can be rewritten with the aid of a *reflection operator* Θ that takes $f(t)$ to $f(-t)$. Thus $\Theta[1 + t - t^2] = 1 - t - t^2$. The rule becomes

$$\Phi[\Phi[f]] = 2\pi\Theta[f]. \tag{4 145'}$$

It is also useful to introduce a conjugation operator $\mathscr{C}$ that takes a complex-valued function $f(t)$ to its conjugate $\overline{f(t)}$; thus $\mathscr{C}[e^{it}] = e^{-it}$. Then we have the rules:

$$\Theta[\Theta[f]] = f, \tag{4-146}$$

$$\Phi[\Theta[f]] = \Theta[\Phi[f]], \tag{4-147}$$

$$\Theta[f]\cdot\Theta[g] = \Theta[f\cdot g], \tag{4-148}$$

$$\Phi[\mathscr{C}[f]] = \mathscr{C}[\Theta[\Phi[f]]]. \tag{4-149}$$

Their proofs are left as exercises (see Problem 4 below). Several of the preceding rules can be written very concisely as operator equalities:

$$\Phi^2 = 2\pi\Theta, \quad \Theta^2 = I \text{ (identity)}, \quad \Phi\Theta = \Theta\Phi, \quad \Phi\mathscr{C} = \mathscr{C}\Theta\Phi.$$

Thus Φ and Θ are *commuting* operators, but Φ and $\mathscr{C}$ are not.

Table of Fourier transforms

In the accompanying Table 4–2 we list some useful Fourier transforms. (For a more extensive table, see ERDELYI, vol. I of Tables.)

Table 4-2
FOURIER TRANSFORMS

No.	$f(t)$ $[f(t) = 0$ outside interval given$]$	$\Phi[f] = \displaystyle\int_{-\infty}^{\infty} f(t)e^{-i\omega t}dt = \varphi(\omega)$
1	1 for $c_1 < t < c_2$	$\dfrac{e^{-i\omega c_1} - e^{-i\omega c_2}}{i\omega}$
2	1 for $-c < t < c$	$\dfrac{2\sin c\omega}{\omega}$
3	e^{at} for $c_1 < t < c_2$	$\dfrac{e^{(a-i\omega)c_2} - e^{(a-i\omega)c_1}}{a - i\omega}$
4	e^{at} for $t > 0$, Re $a < 0$	$\dfrac{1}{i\omega - a}$
5	e^{at} for $t > c$, Re $a < 0$	$\dfrac{e^{(a-i\omega)c}}{i\omega - a}$
6	e^{-at} for $t < 0$, Re $a < 0$	$\dfrac{-1}{i\omega + a}$
7	$e^{a|t|}$, Re $a < 0$	$\dfrac{-2a}{\omega^2 + a^2}$
8	$-e^{-at}$ for $t < 0$, e^{at} for $t > 0$, Re $a < 0$	$\dfrac{-2i\omega}{\omega^2 + a^2}$
9	$t^k e^{at}$ for $t > 0$, $k = 1, 2, \ldots$, Re $a < 0$	$\dfrac{k!}{(i\omega - a)^{k+1}}$
10	e^{ibt} for $-c < t < c$	$\dfrac{2\sin c(\omega - b)}{\omega - b}$
11	$\dfrac{1}{a^2 + t^2}$, Re $a < 0$	$-\dfrac{\pi}{a}e^{a|\omega|}$
12	$\dfrac{t}{(a^2 + t^2)^2}$, Re $a < 0$	$\dfrac{i\omega\pi}{2a}e^{a|\omega|}$
13	$\dfrac{e^{ibt}}{a^2 + t^2}$, Re $a < 0$, b real	$-\dfrac{\pi}{a}e^{a|\omega - b|}$
14	$\dfrac{\cos bt}{a^2 + t^2}$, Re $a < 0$, b real	$-\dfrac{\pi}{2a}\left(e^{a|\omega - b|} + e^{a|\omega + b|}\right)$
15	$\dfrac{\sin bt}{a^2 + t^2}$, Re $a < 0$, b real	$-\dfrac{\pi}{2ai}\left(e^{a|\omega - b|} - e^{a|\omega + b|}\right)$
16	t for $0 < t < c$	$\dfrac{1 - e^{-i\omega c}(1 + i\omega c)}{-\omega^2}$
17	t^k for $0 < t < c$, $k = 1, 2, \ldots$	$\dfrac{k! - e^{-i\omega c}g_k(i\omega c)}{(i\omega)^{k+1}}$, $g_k(x) = x^k + kx^{k-1} + \cdots + k!$

Table 4–2 (*contd*)

No.	$f(t)$ [$f(t) = 0$ outside interval given]	$\Phi[f] = \int_{-\infty}^{\infty} f(t)e^{-i\omega t}dt = \varphi(\omega)$
18	t for $0 < t < c$, $2t - c$ for $c < t < 2c$	$\dfrac{1 - 2e^{-i\omega c} + e^{-2i\omega c}}{-\omega^2}$
19	e^{-at^2}, $a > 0$	$\sqrt{\dfrac{\pi}{a}}\, e^{-\omega^2/(4a)}$

═══════════════ **Problems (Section 4–15)** ═══════════════

1. Verify the Fourier transform given for each of the following entries in Table 4–2:
   a) No. 1   b) No. 2   c) No. 4   d) No. 7   e) No. 9   f) No. 16

2. Apply the rule (4–145) to the function f of entry 7 in Table 4–2 and hence deduce entry 11.

   HINT: The rule gives us the transform of $-2a/(\omega^2 + a^2)$; interchange ω and t to obtain the transform of $-2a/(t^2 + a^2)$; use linearity to obtain entry 11.

3. Deduce each entry named in Table 4–2 as suggested:
   a) Entry 2 from entry 1   b) Entry 7 from entries 4 and 6
   c) Entry 13 from entry 11 by rule (4–144)   d) Entries 14 and 15 from entry 13
   e) Entry 12 from entry 11 by rule (4–141)

4. Prove the following rules:
   a) (4–143)   b) (4–144)   c) (4–146)   d) (4–147)   e) (4–148)   f) (4–149)

4–15 THEORY OF THE FOURIER TRANSFORM

We here outline the main ideas involved in deriving the key formulas of the preceding section. (For a detailed discussion the reader is referred to OMLS, pp. 245–250.)

The properties of integrals of form

$$\int_0^\infty f(t)\,dt \tag{4–150}$$

parallel those of infinite series (Chapter 2). Here $f(t)$ may be a complex-valued function $f(t) = f_1(t) + if_2(t)$, and we assume that $f(t)$ is piecewise continuous (that is, $f(t)$ has at most a finite number of jump discontinuities) on each finite interval. In particular, as in Section 4–5, the integral converges if it is absolutely convergent: that is, if

$$\int_0^\infty |f(t)|\,dt \tag{4–151}$$

is finite. If $f(t)$ is complex valued, then

$$|f(t)| = \{[f_1(t)]^2 + [f_2(t)]^2\}^{1/2}.$$

We now let $f(t)$ be as in the preceding paragraph, so that the integral (4–151) is convergent. We let

$$\psi(\omega) = \int_0^\infty f(t)e^{-i\omega t}\,dt, \quad -\infty < \omega < \infty. \tag{4–152}$$

Then $\psi(\omega)$ exists for all ω because $\left|f(t)e^{-i\omega t}\right| = |f(t)|\left|e^{-i\omega t}\right| = |f(t)|$, since $\left|e^{-i\omega t}\right| = |\cos \omega t - i \sin \omega t| = 1$, and therefore the integral in (4–152) is absolutely convergent. We can write $\psi(\omega)$ as the sum of an infinite series:

$$\psi(\omega) = \sum_{n=1}^\infty g_n(\omega), \quad g_n(\omega) = \int_{n-1}^n f(t)e^{-i\omega t}\,dt, \tag{4–153}$$

because the nth partial sum of the series is

$$\int_0^n f(t)\,e^{-i\omega t}\,dt,$$

and this has $\psi(\omega)$ as limit when $n \to \infty$, according to Eq. (4–152). Each function $g_n(\omega)$ in (4–153) is continuous in ω for all ω. This follows from basic results of the calculus if $f(t)$ is continuous. If $f(t)$ is only piecewise continuous, we get the same result by writing the integral from $n-1$ to n as the sum of a finite number of integrals of continuous functions.

The infinite series (4–153) converges *uniformly* for all ω, because

$$|g_n(\omega)| \leqslant \int_{n-1}^n \left|f(t)e^{-i\omega t}\right|dt = \int_{n-1}^n |f(t)|\,dt = M_n,$$

and $\sum M_n$ converges, since

$$\sum_{n=1}^\infty M_n = \int_0^\infty |f(t)|\,dt.$$

Therefore, the Weierstrass M-test (see Section 2–15) is applicable and the uniform convergence follows. The function $\psi(\omega)$, as defined by Eq. (4–153), is thus the sum of a uniformly convergent series of continuous functions and is *continuous* for all ω (see Section 2–15).

A similar discussion applies to $\int_{-\infty}^0 f(t)\,e^{-i\omega t}\,dt$ and hence, by addition, to

$$\int_{-\infty}^\infty f(t)e^{-i\omega t}\,dt = \int_{-\infty}^0 f(t)\,e^{-i\omega t}\,dt + \int_0^\infty f(t)\,e^{-i\omega t}\,dt.$$

Therefore, if $f(t)$ is piecewise continuous and

$$\int_{-\infty}^\infty |f(t)|\,dt = \int_{-\infty}^0 |f(t)|\,dt + \int_0^\infty |f(t)|\,dt$$

converges, then

$$\Phi[f] = \varphi(\omega) = \int_{-\infty}^{\infty} f(t)\, e^{-i\omega t}\, dt$$

is continuous for all ω.

We now wish to show that $f(t)$ is represented by its Fourier integral (as a principal value); more precisely, that

$$f(t) = \frac{1}{2\pi} \lim_{b \to \infty} \int_{-b}^{b} \varphi(\omega)\, e^{i\omega t}\, d\omega. \tag{4–154}$$

As in Section 3–3, a proof of this equation can be based on the idea of the Dirac delta function, and we again suggest the main ideas. Here we use

$$P_b(v) = \frac{1}{2\pi} \int_{-b}^{b} e^{-i\omega v}\, d\omega = \frac{1}{2\pi} \int_{-b}^{b} \cos \omega v\, d\omega \tag{4–155}$$

(since $\sin \omega v$ is an *odd* function of ω). Integrating out, we obtain

$$P_b(v) = \frac{\sin bv}{\pi v}. \tag{4–155'}$$

As for the analogous function $P_n(t)$ of Section 3–3, $P_b(v)$ turns out to have the delta function $\delta(v)$ as limit when $b \to \infty$, in the sense that

$$\lim_{b \to \infty} \int_{-\infty}^{\infty} f(t + v) P_b(v)\, dv = \int_{-\infty}^{\infty} f(t + v)\, \delta(v)\, dv = f(t), \tag{4–156}$$

provided that $f'(t)$ exists at the t considered (see Fig. 4–17). Now,

$$\int_{-b}^{b} \varphi(\omega)\, e^{i\omega t}\, d\omega = \int_{-b}^{b} \left(\int_{-\infty}^{\infty} f(u)\, e^{-i\omega u}\, du \right) e^{i\omega t}\, d\omega$$

$$= \int_{-b}^{b} \int_{-\infty}^{\infty} f(u)\, e^{-i\omega (u-t)}\, du\, d\omega$$

$$= \int_{-b}^{b} \int_{-\infty}^{\infty} f(t + v)\, e^{-i\omega v}\, dv\, d\omega$$

$$= \int_{-\infty}^{\infty} \int_{-b}^{b} f(t + v)\, e^{-i\omega v}\, d\omega\, dv$$

$$= \int_{-\infty}^{\infty} f(t + v) \left(\int_{-b}^{b} e^{-i\omega v}\, d\omega \right) dv.$$

Here on the third line we made the substitution $v = u - t$ and on the fourth line we interchanged the order of integration; both steps can be justified from the convergence of the integral (4–151). We have thus

$$\frac{1}{2\pi} \int_{-b}^{b} \varphi(\omega)\, e^{i\omega t}\, d\omega = \int_{-\infty}^{\infty} f(t + v) P_b(v)\, dv,$$

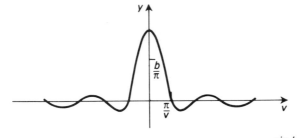

Fig. 4-17. The function $\dfrac{\sin bv}{\pi v}$ as approximation to $\delta(v)$.

and thus by Eq. (4-156)

$$\lim_{b \to \infty} \frac{1}{2\pi} \int_{-b}^{b} \varphi(\omega) e^{i\omega t} \, d\omega = \int_{-\infty}^{\infty} f(t+v) \delta(v) \, dv = f(t),$$

provided that $f'(t)$ exists at the t considered. This establishes the basic rule (4-154), at least at points where f has a derivative. As for Fourier series, it can be shown that (4-154) is valid for all t if $f(t)$ is *piecewise smooth* in each finite interval (Section 3-3), $f(t)$ is equal to the average of left and right limits at each jump point, and the integral (4-151) converges. (For a complete proof, see OMLS, pp. 245-250.)

REMARK 1: An indication that $P_b(v)$ has $\delta(v)$ as limit is given by the fact that the Fourier transform of $\delta(t)$ is (see Section 4-19 below):

$$\varphi(\omega) = \int_{-\infty}^{\infty} \delta(t) e^{-i\omega t} \, dt = 1, \quad -\infty < \omega < \infty.$$

Thus the Fourier integral representation of $\delta(t)$ should be

$$\delta(t) = \frac{1}{2\pi} \int_{-\infty}^{\infty} e^{i\omega t} \, d\omega = \lim_{b \to \infty} \frac{1}{2\pi} \int_{-b}^{b} e^{i\omega t} \, d\omega = \lim_{b \to \infty} P_b(t).$$

REMARK 2: From Eqs. (4-155′) and (4-156) we deduce a useful equation:

$$f(t) = \frac{1}{\pi} \lim_{b \to \infty} \int_{-\infty}^{\infty} f(t+v) \frac{\sin bv}{v} \, dv. \tag{4-157}$$

This is known as the *Fourier single integral representation of $f(t)$.*

REMARK 3: The reasoning given at the beginning of this section can be applied to Laplace transforms. Let $\mathscr{L}[f] = F(s)$, where the transform is absolutely convergent for s real and $s > \sigma^*$. Then for s real

$$\mathscr{L}[f] = F(s) = \sum_{n=1}^{\infty} g_n(s), \quad g_n(s) = \int_{n-1}^{n} f(t) e^{-st} \, dt,$$

so that each $g_n(s)$ is continuous for all s. For $s \geqslant \sigma_0 > \sigma^*$ we have

$$|g_n(s)| \leqslant \int_{n-1}^{n} |f(t)|e^{-\sigma_0 t}\, dt = M_n,$$

and $\sum M_n$ converges, since its sum equals the convergent integral

$$\int_0^\infty |f(t)|\, e^{-\sigma_0 t}\, dt.$$

Thus the Weierstrass M-test applies and we conclude that the series $\sum g_n(s)$ is uniformly convergent for $s \geqslant \sigma_0$. Hence $F(s)$ is continuous for $s \geqslant \sigma_0$ for each $\sigma_0 > \sigma^*$; accordingly, $F(s)$ is continuous for $s > \sigma^*$.

A similar reasoning applies to the complex case, and we conclude that $F(s)$ is continuous for $\operatorname{Re} s > \sigma^*$ (see Section 11–21).

4 16 INVERSE FOURIER TRANSFORM

Given $\varphi(\omega)$, the function f such that $\Phi[f] = \varphi(\omega)$ is uniquely determined by Eq. (4–154). Thus there is a unique inverse Fourier transform $f(t) = \Phi^{-1}[\varphi]$. If φ is absolutely integrable from $-\infty$ to ∞, as often occurs, then we can write

$$f = \Phi^{-1}[\varphi] = \frac{1}{2\pi}\int_{-\infty}^{\infty}\varphi(\omega)e^{i\omega t}\, d\omega, \qquad (4\text{–}160)$$

and this can be interpreted in terms of Fourier transforms, since

$$2\pi f(-t) = \int_{-\infty}^{\infty}\varphi(\omega)e^{-i\omega t}\, d\omega = \Phi[\varphi].$$

This is simply another way of stating the relation (4–145'). We can then write

$$f(t) = \frac{1}{2\pi}\Theta\Phi[\varphi],$$

that is,

$$\Phi^{-1} = \frac{1}{2\pi}\Theta\Phi. \qquad (4\text{–}160')$$

Since Φ is linear, we deduce that Φ^{-1} is also linear. Particular inverse transforms can be read off Table 4–2 and others can be found by linearity. If $\varphi(\omega)$ is absolutely integrable from $-\infty$ to ∞, then it has a Fourier transform and we can then choose $f(t)$ so that $f(t) = \Theta\Phi[\varphi]/2\pi$, as in (4–160'). If f is also absolutely integrable from $-\infty$ to ∞, then f has a Fourier transform that must be φ.

EXAMPLE 1   Find $\Phi^{-1}[2e^{-|\omega|} + 3e^{-2|\omega|}]$. By linearity and entry 11 of Table 4–2, the inverse transform is

$$2\cdot\frac{1/\pi}{t^2+1} + 3\cdot\frac{2/\pi}{t^2+4}. \quad \blacktriangleleft$$

EXAMPLE 2   Find $\Phi^{-1}[\omega/(1+\omega^2)^2]$. By entry 12 of Table 4–2, with t and ω interchanged,

$$\Phi\left[\frac{\omega}{(1+\omega^2)^2}\right] = \frac{it\pi}{-2}\, e^{-|t|}.$$

Hence by the rule (4–160′),

$$\Phi^{-1}\left[\frac{\omega}{(1+\omega^2)^2}\right] = \frac{1}{2\pi}\left(\frac{-it\pi}{-2}\, e^{-|t|}\right) = \frac{it}{4}\, e^{-|t|}. \quad \blacktriangleleft$$

In applications we often need the inverse Fourier transform of a proper rational function of ω. This can always be written as a rational function of $i\omega$ and then decomposed into partial fractions (see Section 4–6) and hence into a sum of terms of the form const$/(i\omega - a)^k$, where k is an integer. To obtain the inverse transform, we have the rules:

$$\Phi^{-1}\left[\frac{1}{(i\omega - a)^k}\right] = \begin{cases} \dfrac{t^{k-1}\, e^{at}}{(k-1)!}, & t > 0 \\ 0, & t < 0 \end{cases} \quad \text{for Re } a < 0, k = 1, 2, \ldots$$

$$(4\text{–}161)$$

$$\Phi^{-1}\left[\frac{1}{(i\omega - a)^k}\right] = \begin{cases} 0, & t > 0 \\ -\dfrac{t^{k-1}\, e^{at}}{(k-1)!}, & t < 0 \end{cases} \quad \text{for Re } a > 0, k = 1, 2, \ldots$$

$$(4\text{–}162)$$

The first rule is given by entry 9 in Table 4–2. The second rule follows from the first by replacing ω with $-\omega$, t with $-t$, a with $-a$, and applying rule (4–147).

We remark that for Re $a = 0$, there is no rule. In fact, for Re $a = 0$, $\varphi(\omega) = 1/(i\omega - a)^k$ is discontinuous for $\omega = a/i$ (a real number), and $\varphi(\omega)$ is not absolutely integrable from $-\infty$ to ∞. Thus there appears to be no inverse transform. If we work with generalized functions, then there is an inverse transform (see Section 4–19); this is given by entry 23 in Table 4–3 below.)

EXAMPLE 3   Find $\Phi^{-1}[1/(\omega^2 + i\omega + 2)]$. We write

$$\frac{1}{\omega^2 + i\omega + 2} = \frac{1}{-(i\omega)^2 + i\omega + 2} = \frac{-1}{(i\omega - 2)(i\omega + 1)} = \frac{1}{3}\frac{1}{i\omega + 1} - \frac{1}{3}\frac{1}{i\omega - 2}.$$

Hence by (4–161) with $k = 1$ and $a = -1$ and by (4–162) with $k = 1$ and $a = 2$,

$$\Phi^{-1}\left[\frac{1}{\omega^2 + i\omega + 2}\right] = \begin{cases} \frac{1}{3}e^{-t} & \text{for } t > 0 \\ \frac{1}{3}e^{2t} & \text{for } t < 0. \end{cases} \quad \blacktriangleleft$$

4–17 RELATIONSHIPS BETWEEN FOURIER TRANSFORMS AND LAPLACE TRANSFORMS

With $s = \sigma + i\omega$, we can write

$$\mathscr{L}[f] = \int_0^\infty f(t)e^{-(\sigma+i\omega)t}\,dt = \int_0^\infty [f(t)e^{-\sigma t}]e^{-i\omega t}\,dt.$$

Hence

$$\mathscr{L}[f] = \Phi[f(t)e^{-\sigma t}], \qquad (4\text{–}170)$$

where $f(t) = 0$ for $t < 0$.

If $\mathscr{L}[f]$ happens to be absolutely convergent for $\sigma = 0$, then (4–170) can be applied with $\sigma = 0$ to give

$$\Phi[f] = \mathscr{L}[f]\Big|_{\sigma=0}. \qquad (4\text{–}171)$$

EXAMPLE 1 Let $f(t) = t^k, k = 1, 2, \ldots$ for $t > 0$; let $f(t) = 0$ for $t < 0$. Then by entry 9 of Table 4–2:

$$\Phi[f(t)e^{-\sigma t}] = \frac{k!}{(i\omega + \sigma)^{k+1}} = \frac{k!}{s^{k+1}} = \mathscr{L}[f],$$

as in entry 3 of Table 4–1. ◀

EXAMPLE 2 Let $f(t) = 1$ for $0 \leq t < 1$, $f(t) = 0$ for $t \geq 1$. Then by entry 18 of Table 4–1, $\mathscr{L}[f] = (1 - e^{-s})/s$ for all σ. Hence

$$\Phi[f] = \mathscr{L}[f]\Big|_{\sigma=0} = \frac{1 - e^{-i\omega}}{i\omega} = i\,\frac{e^{-i\omega} - 1}{\omega}$$

as in Example 1 of Section 4–13. ◀

By (4–170), every Laplace transform, with σ fixed, can be regarded as a Fourier transform. Hence we can take advantage of our Fourier integral representation (4–154), applied to $f(t)e^{-\sigma t}$. If $\mathscr{L}[f] = F(s)$, then

$$\Phi[f(t)e^{-\sigma t}] = F(\sigma + i\omega), \quad \text{for } \sigma \text{ fixed,}$$

and hence

$$f(t)e^{-\sigma t} = \frac{1}{2\pi} \int_{-\infty}^\infty F(\sigma + i\omega)e^{i\omega t}\,d\omega, \qquad (4\text{–}172)$$

where the integral is taken as a principal value, as usual, and σ is chosen greater than the abscissa of absolute convergence of the Laplace transform. From Eq. (4–172),

$$f(t) = \frac{e^{\sigma t}}{2\pi} \int_{-\infty}^\infty F(\sigma + i\omega)e^{i\omega t}\,d\omega = \frac{1}{2\pi} \int_{-\infty}^\infty F(\sigma + i\omega)e^{st}\,d\omega. \qquad (4\text{–}173)$$

This is the *inversion formula for the Laplace transform.*

The complex function $F(\sigma + i\omega)$ can be decomposed into real and imaginary parts:

$$F(\sigma + i\omega) = P(\sigma, \omega) + iQ(\sigma, \omega).$$

Then Eq. (4–173) becomes

$$f(t) = \frac{e^{\sigma t}}{2\pi} \int_{-\infty}^{\infty} (P + iQ)(\cos \omega t + i \sin \omega t)\, d\omega$$

or

$$f(t) = \frac{e^{\sigma t}}{2\pi} \int_{-\infty}^{\infty} (P \cos \omega t - Q \sin \omega t)\, d\omega + i \frac{e^{\sigma t}}{2\pi} \int_{-\infty}^{\infty} (P \sin \omega t + Q \cos \omega t)\, d\omega.$$

(4–174)

If $f(t)$ is real valued, then the second term drops out. Further we verify that in this case $P \cos \omega t$ and $Q \sin \omega t$ are even functions of ω, so that

$$f(t) = \frac{e^{\sigma t}}{\pi} \int_{0}^{\infty} [P(\sigma, \omega) \cos \omega t - Q(\sigma, \omega) \sin \omega t]\, d\omega.$$

(4–175)

This is the inversion formula for the Laplace transform for real-valued functions $f(t)$. In it, σ can be chosen as any number greater than the abscissa of absolute convergence of $\mathscr{L}[f]$. ◀

EXAMPLE 3 Let $f(t) = e^{t}$, $t \geq 0$, where $f(t) = 0$ for $t < 0$. Then

$$\mathscr{L}[f] = \frac{1}{s-1} = \frac{1}{\sigma + i\omega - 1} = \frac{\sigma - 1 - i\omega}{(\sigma - 1)^2 + \omega^2}, \qquad \sigma > 1,$$

so that

$$P(\sigma, \omega) = \frac{\sigma - 1}{(\sigma - 1)^2 + \omega^2}, \qquad Q(\sigma, \omega) = \frac{-\omega}{(\sigma - 1)^2 + \omega^2}.$$

Therefore, for each $\sigma > 1$ and $t > 0$ we have

$$e^{t} = \frac{e^{\sigma t}}{\pi} \int_{0}^{\infty} \frac{(\sigma - 1) \cos \omega t + \omega \sin \omega t}{(\sigma - 1)^2 + \omega^2}\, d\omega.$$

For $t < 0$, the right side equals 0; for $t = 0$, it equals the average of the limits from left and right, which is 1/2. ◀

════════════════ **Problems (Section 4-17)** ════════════════

1. Find the inverse Fourier transform:

a) $\dfrac{\sin 3\omega}{\omega}$ b) $\dfrac{1}{i\omega + 2}$ c) $\dfrac{1}{i\omega - 3}$ d) $\dfrac{5\omega}{\omega^2 + 1}$

e) $\dfrac{3}{\omega + 1 + i} - \dfrac{7}{\omega + 1 - i}$ f) $\dfrac{5 \sin(\omega - \pi/3)}{3\omega - \pi} + \dfrac{\sin(\omega - \pi/6)}{6\omega - \pi}$

g) $\dfrac{2}{(i\omega - 3)^3} - \dfrac{5}{(i\omega + 3)^3}$ h) $\dfrac{4}{(\omega - 2i)^5} + \dfrac{7}{(\omega - 3i)^2}$

i) $\dfrac{i\omega}{(i\omega + 4)(i\omega + 1)}$ j) $\dfrac{2i\omega + 1}{(i\omega + 3)(i\omega + 2)}$ k) $\dfrac{1}{6\omega^2 - 13i\omega - 6}$

l) $\dfrac{1}{\omega^2 - 2i\omega - 1}$ m) $\dfrac{2}{(i\omega + 1)(\omega^2 - 2i\omega - 2)}$ n) $\dfrac{i\omega + 2}{i\omega^3 + 5\omega^2 - 7i\omega - 3}$

2. Represent the inverse Fourier transform by an integral:

a) $\dfrac{e^{-\omega^2}}{\omega^2 + 1}$ **b)** $e^{-\omega^2} \sin \omega$ **c)** $\dfrac{\cos \omega}{\omega^2 + 3\omega + 11}$ **d)** $\dfrac{1}{e^\omega + \omega^2}$

3. Evaluate the limits:

a) $\displaystyle\lim_{b \to \infty} \frac{1}{2\pi} \int_{-b}^{b} \frac{\sin \omega}{\omega} e^{i\omega t} \, d\omega$ **b)** $\displaystyle\lim_{b \to \infty} \int_{-b}^{b} \frac{e^{i\omega t}}{(i\omega + 1)^3} \, d\omega$

c) $\displaystyle\lim_{b \to \infty} \frac{1}{\pi} \int_{-\infty}^{\infty} \frac{1}{(t+v)^2 + 1} \frac{\sin bv}{v} \, dv$ **d)** $\displaystyle\lim_{b \to \infty} \int_{-\infty}^{\infty} e^{-|t+v|} \frac{\sin bv}{v} \, dv$

4–18 CONVOLUTION

For $f(t)$ and $g(t)$ defined on $-\infty < t < \infty$, the convolution $f * g = h$ is defined by the equation

$$h(t) = \int_{-\infty}^{\infty} f(u)g(t - u) \, du. \tag{4–180}$$

This has meaning if f and g are piecewise continuous, one of the two functions, say $f(t)$, is absolutely integrable on $-\infty < t < \infty$, and the other one of the two functions is bounded (for example, $|g(t)| < \text{const}$ for all t). If $f(t) = 0$ and $g(t) = 0$ for $t < 0$, then Eq. (4–180) reduces to

$$h(t) = \int_{0}^{t} f(u)g(t - u) \, du,$$

as in Section 4–8.

The convolution has the crucial property:

$$\Phi[f * g] = \Phi[f]\Phi[g], \tag{4–181}$$

provided f and g are piecewise continuous and absolutely integrable on $-\infty < t < \infty$ and one of the two functions is bounded. The proof is like the proof of the analogous rule (4–81) for Laplace transforms in Section 4–8.

From (4–181) we deduce a rule for the Fourier transform of a product:

$$\Phi[f \cdot g] = \frac{1}{2\pi} \Phi[f] * \Phi[g]. \tag{4–182}$$

To derive this, we let $\Phi[f] = \varphi$ and $\Phi[g] = \psi$. Then

$$\Phi[\varphi * \psi] = \Phi[\varphi]\Phi[\psi].$$

But $\Phi[\varphi] = \Phi[\Phi[f]] = 2\pi\Theta[f]$ and similarly $\Phi[\psi] = 2\pi\Theta[g]$ by (4–145′), so that by rule (4–148)

$$\Phi[\varphi * \psi] = 2\pi\Theta[f] \cdot 2\pi\Theta[g] = 4\pi^2\Theta[f \cdot g].$$

We apply Φ to these expressions and deduce by rule (4–147) that

$$\Phi[\Phi[\varphi * \psi]] = 4\pi^2\Phi\Theta[f \cdot g] = 4\pi^2\Theta\Phi[f \cdot g]$$

and hence by rule (4–146) again

$$2\pi\Theta[\varphi * \psi] = 4\pi^2\Theta\Phi(f\cdot g),$$

from which Eq. (4–182) follows. (To justify all steps we need to assume that $f,\ g,\ \varphi,\ \psi$ are all absolutely integrable from $-\infty$ to ∞.)

The convolution obeys other formal rules, as in Section 4–8:

$$f * (g_1 + g_2) = f * g_1 + f * g_2, \tag{4–183}$$

$$f * (cg) = (cf) * g = c(f * g), \quad c = \text{const}, \tag{4–184}$$

$$(f * g) * h = f * (g * h), \tag{4–185}$$

$$e^{at}(f*g) = (e^{at}f)*(e^{at}g), \quad a = \text{const}. \tag{4–186}$$

These are valid if, for example, all the functions $f,\ g,\ g_1,\ g_2$ are bounded and absolutely integrable from $-\infty$ to ∞. The proofs are left as exercises (see Problem 3 below).

4–19 FOURIER TRANSFORMS OF GENERALIZED FUNCTIONS

We proceed as in Sections 4–9 and 4–10. However, now all functions are defined for $-\infty < t < \infty$, and some of the restrictions disappear. Thus, by (4–97),

$$\Phi[\delta(t)] = \int_{-\infty}^{\infty}\delta(t)e^{-i\omega t}\,dt = e^{-i\omega t}\Big|_{t=0} = 1, \tag{4–190}$$

$$\Phi[\delta(t-c)] = \int_{-\infty}^{\infty}\delta(t-c)e^{-i\omega t}\,dt = e^{-i\omega t}\Big|_{t=c} = e^{-ic\omega}, \tag{4–191}$$

$$\Phi[\delta^{(k)}(t-c)] = \int_{-\infty}^{\infty}\delta^{(k)}(t-c)e^{-i\omega t}\,dt = (-1)^k\frac{d^k}{dt^k}e^{-i\omega t}\Big|_{t=c} = e^{-ic\omega}(i\omega)^k. \tag{4–192}$$

We can extend the Fourier transform to new functions by requiring that standard rules continue to hold. Thus if we require that $\Phi^2 = 2\pi\Theta$, then

$$\Phi[\Phi[\delta(t)]] = 2\pi\delta(-t) = 2\pi\delta(t),$$

since $\delta(t)$ is even. But $\Phi[\delta(t)] = 1$ (as a function of ω) by Eq. (4–190). Hence, if we interchange t and ω,

$$\Phi[1] = 2\pi\delta(\omega). \tag{4–193}$$

Here 1 is the constant function, equal to 1 for all t. Since this function is not absolutely integrable for $-\infty < t < \infty$, Eq. (4–193) defines a new transform. By similar requirements that rules continue to hold, we deduce a variety of other new transforms, as in Table 4–3. In particular, by virtue of entries 5, 6, 7 and linearity, each *polynomial* in t now has a transform.

Entry 10 gives a transform for the unit step function $U(t)$. We note that

$$\Phi[U(t)] = (i\omega)^{-1} + \pi\delta(\omega)$$

is a sum of a function discontinuous at $\omega = 0$ and of a delta term.

Table 4-3

FOURIER TRANSFORMS OF GENERALIZED FUNCTIONS*

No.	$f(t)$	$\varphi(\omega)=\Phi[f]$	No.	$f(t)$	$\varphi(\omega)=\Phi[f]$
1	$\delta(t)$	1	15	$e^{i\alpha t}t^n U(t)$	$\dfrac{n!}{[i(\omega-\alpha)]^{n+1}}+\pi i^n\delta^{(n)}(\omega-\alpha)$
2	$\delta(t-c)$	$e^{-ic\omega}$	16	$e^{i\alpha t}(t-c)^n U(t-c)$	$\dfrac{n!e^{-ic(\omega-\alpha)}}{[i(\omega-\alpha)]^{n+1}}+\pi i^n\displaystyle\sum_{r=0}^{n}\binom{n}{r}$ $\times (ic)^{n-r}\delta^{(r)}(\omega-\alpha)$
3	$\delta^{(n)}(t)$	$(i\omega)^n$	17	$\dfrac{1}{t}$	$\pi i-2\pi i U(\omega)$
4	$\delta^{(n)}(t-c)$	$e^{-ic\omega}(i\omega)^n$	18	$\dfrac{1}{t^n}$	$\dfrac{(-i\omega)^{n-1}}{(n-1)!}[\pi i-2\pi i U(\omega)]$
5	1	$2\pi\delta(\omega)$	19	$\dfrac{1}{t-c}$	$e^{-ic\omega}[\pi i-2\pi i U(\omega)]$
6	t	$2\pi i\delta'(\omega)$	20	$\dfrac{1}{(t-c)^n}$	$\dfrac{e^{-ic\omega}(-i\omega)^{n-1}}{(n-1)!}[\pi i-2\pi i U(\omega)]$
7	t^n	$2\pi i^n\delta^{(n)}(\omega)$	21	$\dfrac{e^{i\alpha t}}{t-c}$	$e^{-ic(\omega-\alpha)}[\pi i-2\pi i U(\omega-\alpha)]$
8	$e^{i\alpha t}$	$2\pi\delta(\omega-\alpha)$	22	$\dfrac{e^{i\alpha t}}{(t-c)^n}$	$\dfrac{e^{-ic(\omega-\alpha)}[-i(\omega-\alpha)]^{n-1}}{(n-1)!}[\pi i-2\pi i U(\omega-\alpha)]$
9	$t^n e^{i\alpha t}$	$2\pi i^n\delta^{(n)}(\omega-\alpha)$	23	$\dfrac{e^{i\alpha t}t^{n-1}}{(n-1)!}[U(t)-\tfrac{1}{2}]$	$\dfrac{1}{(i\omega-i\alpha)^n}$
10	$U(t)$	$\dfrac{1}{i\omega}+\pi\delta(\omega)$			
11	$t^n U(t)$	$\dfrac{n!}{(i\omega)^{n+1}}+\pi i^n\delta^{(n)}(\omega)$			
12	$U(t-c)$	$\dfrac{e^{-i\omega c}}{i\omega}+\pi\delta(\omega)$			
13	$(t-c)^n U(t-c)$	$\dfrac{n!e^{-ic\omega}}{(i\omega)^{n+1}}+\pi i^n\displaystyle\sum_{r=0}^{n}\binom{n}{r}(ic)^{n-r}\delta^{(r)}(\omega)$			
14	$e^{i\alpha t}U(t-c)$	$\dfrac{e^{-ic(\omega-\alpha)}}{i(\omega-\alpha)}+\pi\delta(\omega-\alpha)$			

* *Note:* Throughout, $n = 1, 2, 3, \ldots$, and c and α are real constants.

The rule $\Phi[g'] = i\omega\Phi[g]$ continues to hold for all entries in the table. Thus,

$$\Phi[U'(t)] = \Phi[\delta(t)] = 1 = i\omega\Phi[U(t)] = i\omega\left[\frac{1}{i\omega} + \pi\delta(\omega)\right]$$
$$= 1 + \pi i\omega\delta(\omega) = 1,$$

since $\omega\delta(\omega) = 0$ by the rule (4–90). The rule for the transform of a derivative can also be applied now to a variety of ordinary functions with jump discontinuities. For example, if $g(t) = e^{-2t}U(t)$, then $g'(t) = -2e^{-2t}U(t) + \delta(t)$ and

$$\Phi[g'] = \frac{-2}{i\omega + 2} + 1 = i\omega\Phi[g] = i\omega\frac{1}{i\omega + 2}.$$

We can also introduce convolutions involving generalized functions, as in Section 4–9. The rule $\Phi[f*g] = \Phi[f]\Phi[g]$ tells us what value to assign in various cases in which the definition (4–180) breaks down. Thus we define:

$$\delta^{(k)}(t-a) * \delta^{(l)}(t-b) = \delta^{(k+l)}(t-a-b), \tag{4–194}$$

since the product of the transforms of $\delta^{(k)}(t-a)$ and $\delta^{(l)}(t-b)$ is

$$e^{-ia\omega}(i\omega)^k \cdot e^{-ib\omega}(i\omega)^l = e^{-i(a+b)\omega}(i\omega)^{k+l}.$$

Also, as in Section 4–10, for an ordinary function $f(t)$ with continuous kth derivative we have

$$\delta^{(k)}(t-c) * f(t) = f(t) * \delta^{(k)}(t-c) = f^{(k)}(t-c). \tag{4–195}$$

However, certain convolutions remain undefined: for example, $1 * 1$, since $\delta(\omega) \cdot \delta(\omega)$ is not defined. (For a thorough discussion of this topic, see LIGHTHILL.)

========================= **Problems (Section 4-19)** =========================

1. Evaluate the convolution $f * g$ and verify that Eq. (4–181) is satisfied:
   a) $f(t) = e^{-t}$ for $t > 0$, $f(t) = 0$ for $t < 0$; $g(t) = e^{-2t}$ for $t > 0$, $g(t) = 0$ for $t < 0$;
   b) $f(t) = 1$ for $-1 < t < 1$, $f(t) = 0$ otherwise; $g(t) = e^{-t}$ for $t > 0$, $g(t) = 0$ otherwise;
   c) $f(t) = e^{-t}$ for $t > 0$, $f(t) = 0$ for $t < 0$; $g(t) = e^{t}$ for $t < 0$, $g(t) = 0$ for $t > 0$;
   d) $f(t) = 1$ for $-1 < t < 1$, $f(t) = 0$ otherwise; $g(t) = e^{-|t|}$.

2. Represent $\Phi[f \cdot g]$ as a convolution, as in Eq. (4–182):
   a) $f = \dfrac{1}{1+t^2}$, $g = e^{-|t|}$ b) $f = e^{-|t|}$, $g = e^{-t^2}$

3. Prove each of the rules:
   a) (4–183)   b) (4–184)
   c) (4–185)   d) (4–186)

4. Verify with the aid of Table 4–3 and the rules (4–90) to (4–92) for multiplication that the rule $\Phi[g'] = i\omega\Phi[g]$ is satisfied for the following entries in the table:

a) No. 5   b) No. 6
c) No. 7   d) No. 8
e) No. 9   f) No. 17

5. Verify that the rule $\Phi[\Phi[f]] = 2\pi\Theta(f)$ is satisfied for the following entries in Table 4–3:

a) No. 8   b) No. 12
c) No. 17   d) No. 21

6. Verify that the rule $\Phi[e^{i\alpha t}f] = \varphi(\omega - \alpha)$, for $\varphi = \Phi[f]$, is satisfied for the following entries in Table 4–3:

a) No. 1   b) No. 5   c) No. 10   d) No. 17

4–20 APPLICATION OF FOURIER TRANSFORMS TO LINEAR DIFFERENTIAL EQUATIONS

As in Section 4–11, we proceed formally (for justification of the procedures, see OMLS, pp. 284–294). We consider the differential equation of order n with constant coefficients:

$$a_0 \frac{d^n x}{dt^n} + \cdots + a_n x = g(t). \qquad (4\text{–}200)$$

We apply the Fourier transform to both sides, using the rule $\Phi[f^{(k)}] = (i\omega)^k \Phi[f]$, and obtain

$$[a_0(i\omega)^n + \cdots + a_{n-1}(i\omega) + a_n]\Phi[x] = \Phi[g],$$

$$\Phi[x] = \frac{1}{a_0(i\omega)^n + \cdots + a_n}\Phi[g].$$

We again introduce the transfer function

$$Y(s) = \frac{1}{a_0 s^n + \cdots + a_n}.$$

Thus we have

$$\Phi[x] = Y(i\omega)\Phi[g]. \qquad (4\text{–}201)$$

For applications the main interest is in the *stable case*, in which all characteristic roots have negative real parts (Section 1–15). Thus the factors of the denominator of $Y(s)$ are of the form $(s - a)^k$ with $\operatorname{Re} a < 0$. If $Y(s)$ is decomposed into partial fractions and then s is replaced by $i\omega$, then a linear combination is obtained of terms of form

$$\frac{1}{(i\omega - a)^k}, \qquad \operatorname{Re} a < 0.$$

Each such term has an inverse Fourier transform, given by Eq. (4–161). Hence we can write

$$Y(i\omega) = \Phi[W(t)].$$

In fact, from Eq. (4–161) we see that $Y(s) = \mathscr{L}[W(t)]$. Thus $W(t)$ is the same weight function as the one introduced in Section 4–11, and $W(t) = 0$ for $t < 0$.

Assuming stability, we can now write Eq. (4–201) in the form

$$\Phi[x] = \Phi[W]\Phi[g],$$

so that $\Phi[x] = \Phi[W]\Phi[g]$ and

$$x = W * g. \tag{4–202}$$

If $g(t) \equiv 0$ for $t < 0$, then this is the same as the solution obtained in Section 4–11, with all initial values 0 for $t = 0$. However, Eq. (4–202) is valid more generally, in particular, whenever $g(t)$ is bounded for $t < 0$. It provides the unique solution of the differential equation which has *small initial values as $t \to -\infty$*. For example, if $g(t)$ is bounded when $t < 0$, it is the unique solution that is also bounded when $t < 0$ (see OMLS, pp. 264–285 for a detailed discussion).

EXAMPLE 1 $x'' + 3x' + 2x = e^t$. We obtain

$$Y(s) = \frac{1}{s^2 + 3s + 2} = \frac{1}{(s+1)(s+2)} = \frac{1}{s+1} - \frac{1}{s+2},$$

$$Y(i\omega) = \frac{1}{i\omega + 1} - \frac{1}{i\omega + 2}.$$

The equation is stable, since the characteristic roots are -1 and -2. We now find:

$$W(t) = e^{-t} - e^{-2t} \quad \text{for} \quad t > 0 \qquad \text{and} \qquad W(t) = 0 \quad \text{for } t < 0.$$

Thus

$$x = W(t) * e^t = \int_0^\infty (e^{-u} - e^{-2u}) e^{t-u} \, du = \frac{e^t}{6}.$$

Here $g(t) \to 0$ as $t \to -\infty$ and $x \to 0$ as $t \to -\infty$. We observe that $\Phi[g]$ is not defined, since $g(t) \to +\infty$ as $t \to +\infty$. There are other solutions of the differential equation:

$$x = \frac{e^t}{6} + c_1 e^{-t} + c_2 e^{-2t},$$

but, except for $c_1 = 0$ and $c_2 = 0$, none is bounded for $t < 0$. Thus the method leads to the unique solution which is bounded for $t < 0$. ◄

EXAMPLE 2 $x'' + 3x' + 2x = 1/(1 + t^2)$. As in Example 1, we obtain

$$x = W(t) * \frac{1}{1 + t^2} = \int_0^\infty \frac{e^{-u} - e^{-2u}}{1 + (t-u)^2} \, du.$$

This cannot be evaluated explicitly, but can be evaluated numerically for each t (see Chapter 13). We can see that $x(t)$ is bounded for $t < 0$. In fact, for all t:

$$0 < x(t) < \int_0^\infty (e^{-u} - e^{-2u}) \, du = \frac{1}{2},$$

and we can show that $x(t) \to 0$ as $t \to \pm \infty$ (see Problem 6 below, following Section 4–21.)

In this example, $g(t) = (1 + t^2)^{-1}$ has a Fourier transform, and we can therefore write, as in Eq. (4–201),

$$\Phi[x] = \frac{1}{-\omega^2 + 3i\omega + 2} \pi e^{-|\omega|}.$$

Therefore,

$$x = \Phi^{-1} \left[\frac{\pi e^{-|\omega|}}{-\omega^2 + 3i\omega + 2} \right] = \frac{1}{2\pi} \int_{-\infty}^{\infty} \frac{\pi e^{-|\omega|}}{-\omega^2 + 3i\omega + 2} e^{i\omega t} \, d\omega.$$

This gives another representation of the desired solution. Again numerical evaluation would be required. ◄

The described method, although useful for finding solutions, is more important for the insight it gives. First of all, $x = W * g$ can be interpreted as in Fig. 4–12, except that $g(t)$ is not required to be 0 for $t < 0$. Thus again the shape of the graph of $W(t)$ describes the kind of *memory* the system has. Secondly, the equation $\Phi[x] = Y(i\omega) \Phi[g]$ shows the relative importance of the various frequencies ω in the solution $x(t)$. The Fourier integral

$$x = \frac{1}{2\pi} \int_{-\infty}^{\infty} Y(i\omega) \, \varphi(\omega) \, e^{i\omega t} \, d\omega$$

with $\varphi(\omega) = \Phi[g]$ can be regarded as a representation of x as a *linear combination* of the various sinusoidal oscillations $e^{i\omega t}$ and $Y(i\omega) \, \varphi(\omega)$ is the *complex amplitude*. We can write

$$Y(i\omega) \, \varphi(\omega) = A(\omega) e^{i\beta(\omega)},$$

where A and β are the polar coordinates of the complex number $Y(i\omega)\varphi(\omega)$, as in Fig. 4–18. Then

$$Y(i\omega) \, \varphi(\omega) \, e^{i\omega t} = A(\omega) \, e^{i\beta(\omega)} \, e^{i\omega t} = A(\omega) \, e^{i[\beta(\omega) + \omega t]}$$
$$= A(\omega) \{ \cos[\beta(\omega) + \omega t] + i \sin[\beta(\omega) + \omega t] \}.$$

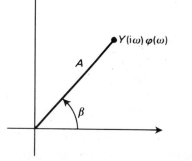

Fig. 4-18. Meaning of A and β.

Thus $A(\omega)$ is the (real) amplitude of the oscillation and $\beta(\omega)$ is the initial phase. The variation of $A(\omega) = |Y(i\omega)\,\varphi(\omega)| = |Y(i\omega)|\,|\varphi(\omega)|$ gives the relative weight assigned to the frequencies ω.

For Example 2, $|\varphi(\omega)| = \pi e^{-|\omega|}$ and

$$|Y(i\omega)| = \left|\frac{1}{-\omega^2 + 3i\omega + 2}\right| = \frac{1}{(\omega^4 + 5\omega^2 + 4)^{1/2}},$$

so that

$$A(\omega) = \frac{\pi e^{-|\omega|}}{(\omega^4 + 5\omega^2 + 4)^{1/2}},$$

as graphed in Fig. 4–19. The graph shows that low frequencies have by far the greatest weight.

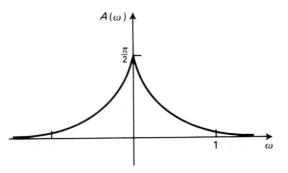

Fig. 4-19. Graph of $A(\omega)$ for Example 2.

In general $|Y(i\omega)|$ measures the extent to which the oscillation $\varphi(\omega)e^{i\omega t}$ in $g(t)$ is strengthened or diminished in $x(t)$. For this reason, $Y(i\omega)$ is called the *frequency response function* (see Section 3–8). A possible shape for the graph of $|Y(i\omega)|$ is depicted in Fig. 4–20. Here frequencies near ω_0 are greatly emphasized, while other

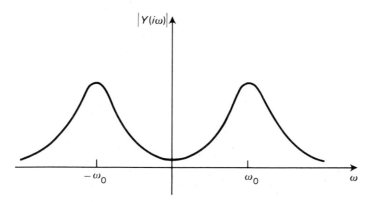

Fig. 4-20. Absolute value of typical frequency response function $Y(i\omega)$.

frequencies are de-emphasized. This reasoning is the basis for various filtering
devices, which effectively cut out selected frequency intervals in the input, so that the
output x responds mainly to the remaining frequencies.

EXAMPLE 3   We consider the equation $(D+2)(D+0.1)x = f(t)$. This is of the
form of the differential equation for a dc-motor (Example 4 in Section 4-11 above).
For a constant field voltage, $f(t)$ is constant and the steady-state output x (shaft
angular velocity) is constant. It is natural to ask what happens when $f(t)$ fluctuates.
The frequency response function $Y(i\omega)$ tells how different sinusoidal terms in $f(t)$ are
amplified. We can see this graphically by plotting the curve $z = Y(i\omega), 0 \leqslant \omega < \infty$,
with ω as a parameter, in a complex z-plane. In this example,

$$Y(s) = 1/(s^2 + 2.1s + 0.2),$$

so that we find

$$Y(i\omega) = \frac{0.2 - \omega^2 - 2.1i\omega}{\omega^4 + 4.01\omega^2 + 0.04}.$$

The graph (Fig. 4-21) shows that as ω increases from 0, the amplification $|Y(i\omega)|$
decreases from 5 and approaches 0 as $\omega \to +\infty$. Hence, low frequency terms in $f(t)$
are significant, high frequency terms have little effect. From the graph we can also
read off arg $Y(i\omega)$, which gives the phase shift. This is 0 for $\omega = 0$ and decreases to
$-\pi/2$ at $\omega = 0.447$, approaches $-\pi$ as $\omega \to \infty$. Thus the response to a sinusoidal
input, say $A \sin \omega t$, lags behind in phase, but by at most π for high frequencies.
    We can also graph $\ln |Y(i\omega)|$ and arg $Y(i\omega)$ as functions of ω. The *Bode diagram*
uses $20 \log_{10}|Y(i\omega)|$ instead of $\ln|Y(i\omega)|$. Since

$$20 \log_{10} x = 0.86858 \ln x,$$

this is simply a change of scale. We interpret $20 \log_{10}|Y(i\omega)|$ as *logarithmic gain*
measured in units of decibels. When $|Y(i\omega)|$ grows by a factor of 10, the logarithm to

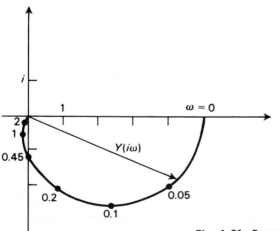

Fig. 4-21. Frequency response function for Example 2.

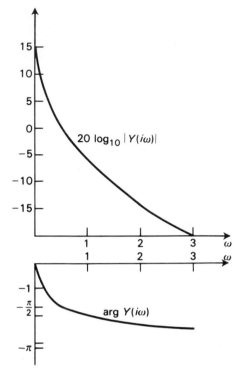

Fig. 4–22. Bode diagram for Example 2.

base 10 grows by 1 and the logarithmic gain increases by 20. The Bode diagram for Example 3 is shown in Fig. 4–22. ◄

We recall (Sections 1–9, 1–10) that for $z_1 = r_1 e^{i\theta_1}$, $z_2 = r_2 e^{i\theta_2}$ we have

$$z_1 z_2 = r_1 r_2 e^{i(\theta_1 + \theta_2)},$$

so that

$$|z_1 z_2| = r_1 r_2 = |z_1| |z_2|, \qquad \arg(z_1 z_2) = \theta_1 + \theta_2 = \arg z_1 + \arg z_2,$$

where the angles are determined only up to multiples of 2π. Therefore, for

$$Y(s) = \frac{1}{(s+2)(s+0.1)}$$

we have

$$20 \log_{10}|Y(i\omega)| = 20 \log_{10}\frac{1}{|i\omega + 2|} + 20 \log_{10}\frac{1}{|i\omega + 0.1|},$$

$$\arg Y(i\omega) = \arg \frac{1}{i\omega + 2} + \arg \frac{1}{i\omega + 0.1}.$$

Accordingly, the Bode diagram for a product, such as $(s+2)^{-1}(s+0.1)^{-1}$, can be obtained from the Bode diagrams of the factors by addition. This clearly applies to any number of factors.

Extension of concept of transfer function

In many applications, an input function $g(t)$ and output function $x(t)$ are related by
an equation more general than (4–200) in that $g(t)$ is replaced by $(b_0 D^m + \cdots + b_m)g$
for an appropriate differential operator $b_0 D^m + \cdots + b_m$. The equation is commonly the result of an elimination process applied to simultaneous differential
equations (see Chapter 6, especially Section 6–11). For the equation

$$(a_0 D^n + \cdots + a_n)x = (b_0 D^m + \cdots + b_m)g(t)$$

we define the transfer function as

$$Y(s) = \frac{b_0 s^m + \cdots + b_m}{a_0 s^n + \cdots + a_n}.$$

Most commonly, m is less than n. We verify that this transfer function has the same
significance as the previous one. In particular, $Y(i\omega)$ is again a frequency response
function: $x = A\,Y(i\omega)e^{i\omega t}$ is the response to a sinusoidal input $g(t) = Ae^{i\omega t}$.

EXAMPLE 4 In a standard LRC circuit let $L = 0.05$, $R = 0.2$, $C = 4$ in
consistent units (Fig. 4–23). Let the input be the applied electromotive force
$E = g(t)$. Then the circuit equation for I as output is

$$L\frac{d^2 I}{dt^2} + R\frac{dI}{dt} + \frac{1}{C}I = g'(t).$$

With the numbers given, the transfer function is

$$Y(s) = \frac{s}{0.05 s^2 + 0.2 s + 0.25}.$$

Hence, for $\omega > 0$,

$$Y(i\omega) = \left| \frac{i\omega}{(0.25 - 0.05\,\omega^2) + 0.2i\omega} \right| = \frac{\omega}{[(0.25 - 0.05\omega^2)^2 + 0.04\omega^2]^{1/2}}.$$

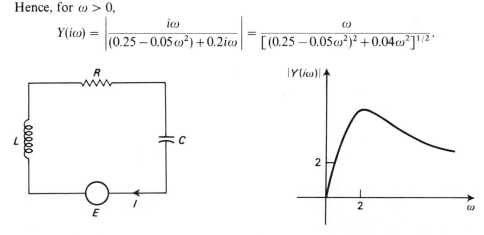

Fig. 4–23. LRC circuit.   **Fig. 4–24.** Frequency response for Example 4.

This function is graphed in Fig. 4–24. Here frequencies close to $\sqrt{5}$ are favored,
those far from $\sqrt{5}$ are filtered out. Thus this circuit acts somewhat like a *band pass
filter.* ◄

4–21 APPLICATION OF FOURIER TRANSFORMS OF GENERALIZED FUNCTIONS TO LINEAR DIFFERENTIAL EQUATIONS

By allowing generalized functions, we can extend the formalism of the preceding section even further.

EXAMPLE 1 $x'' + 3x' + 2x = \cos 5t = (1/2)(e^{5it} + e^{-5it})$. Here by entry 8 of Table 4–3 we have

$$\Phi[g] = \Phi[\cos 5t] = \pi[\delta(\omega - 5) + \delta(\omega + 5)].$$

This shows that only the frequencies 5 and -5 are present in $g(t)$. Thus

$$\Phi[x] = Y(i\omega)\Phi[g] = Y(i\omega)\pi[\delta(\omega - 5) + \delta(\omega + 5)]$$
$$= \pi[Y(5i)\delta(\omega - 5) + Y(-5i)\delta(\omega + 5)]$$

and again, by entry 8 of Table 4–3 we get

$$x = \frac{1}{2}\left[Y(5i)e^{5it} + Y(-5i)e^{-5it}\right] = \frac{1}{2}\left(\frac{e^{5it}}{-23 + 15i} + \frac{e^{-5it}}{-23 - 15i}\right)$$

$$= \frac{15\sin 5t - 23\cos 5t}{754}. \quad \blacktriangleleft$$

EXAMPLE 2 $x'' + 3x' + 2x = t$. By entries 5, 6 of Table 4–3 and (4–91),

$$\Phi[x] = Y(i\omega)2\pi i\delta'(\omega) = \frac{2\pi i}{-\omega^2 + 3i\omega + 2}\delta'(\omega) = \pi i\delta'(\omega) - \frac{3\pi}{2}\delta(\omega),$$

$$x = \frac{t}{2} - \frac{3}{4}. \quad \blacktriangleleft$$

EXAMPLE 3 $x'' + 3x' + 2x = \delta(t + 1)$. We find

$$\Phi[x] = Y(i\omega)e^{i\omega} = \frac{e^{i\omega}}{i\omega + 1} - \frac{e^{i\omega}}{i\omega + 2} = \Phi[e^{-1-t}U(t+1) - e^{-2-2t}U(t+1)].$$

Here we used entry 5 of Table 4–2. Finally,

$$x = \left(e^{-1-t} - e^{-2-2t}\right)U(t+1)$$

We verify that our answer is simply $x = W(t + 1)$. In general, the response to $\delta(t)$ is $\Phi^{-1}[Y(i\omega)] = W(t)$, the response to $\delta(t - c)$ is $\Phi^{-1}[Y(i\omega)e^{-i\omega c}] = W(t - c)$, by rule (4–143); the response to $\delta^{(k)}(t - c)$ is $W^{(k)}(t - c)$, which is a generalized function for $k \geqslant n$, the order of the differential equation. These results also follow from the representation $x = W(t) * \delta^{(k)}(t - c)$ and the rule (4–195).

=========== **Problems (Section 4–21)** ===========

1. Find a solution with the aid of Eq. (4–202):

   a) $x'' + 5x' + 6x = e^{3t}$ **b)** $x'' + 5x' + 6x = \cos t$ **c)** $x'' + 5x' + 6x = t$
   d) $x'' + 5x' + 6x = g(t)$, $g(t) = e^t$ for $t < 0$, $g(t) = 0$ for $t > 0$
   e) $x'' + 4x' + 4x = g(t)$, $g(t) = e^{2t}$ for $t < 0$, $g(t) = 1$ for $t > 0$
   f) $x'' + 4x' + 4x = g(t)$, $g(t) = \sin t$ for $t < 0$, $g(t) = 0$ for $t > 0$

2. Represent a solution as a Fourier integral:

 a) $x'' + 2x' + 3x = g(t)$, $g(t) = 1$ for $-1 < t < 1$, $g(t) = 0$ otherwise
 b) $x'' + 3x' + 4x = g(t)$, $g(t) = e^{(1 + 2i)t}$ for $t < 0$, $g(t) = 0$ for $t > 0$

 c) $x'' + x' + x = \dfrac{t}{(4 + t^2)^2}$ d) $x'' + 2x' + x = \dfrac{e^{3it}}{4 + t^2}$

 e) $x''' + 3x'' + 4x' + 2x = g(t)$, $g(t) = e^{5t}$ for $-2 < t < 2$, $g(t) = 0$ otherwise
 f) $x''' + 3x'' + 4x' + 2x = e^{-t^2/4}$

 g) $x'' + 2x' + x = g(t)$, $g(t) = \displaystyle\int_{-\infty}^{\infty} \dfrac{e^{i\omega t}}{\omega^4 + 1} \, d\omega$

 h) $x'' + 2x' + x = g(t)$, $g(t) = \displaystyle\int_{-\infty}^{\infty} \dfrac{e^{i\omega t} e^{-|\omega|}}{\omega^2 + 1} \, d\omega$

3. Find a solution by Fourier transforms, allowing generalized functions:

 a) $x'' + 4x' + 3x = 2 \sin 5t$ b) $x'' + 4x' + 3x = 3e^{2it}$

 c) $x'' + 4x' + 3x - 1$ d) $x'' + 6x' + 9x - t^2$

 e) $x'' + 4x' + 3x = 3\delta(t)$ f) $x'' + 4x' + 3x = 7\delta'(t)$

4. For the electric network of Fig. 4–25, Kirchhoff's laws give the equations

$$R_1 I_1 + \frac{q}{C} = E, \quad \frac{dq}{dt} = I, \quad L\frac{dI_2}{dt} + R_2 I_2 + R_1 I_1 = E, \quad I_1 = I + I_2.$$

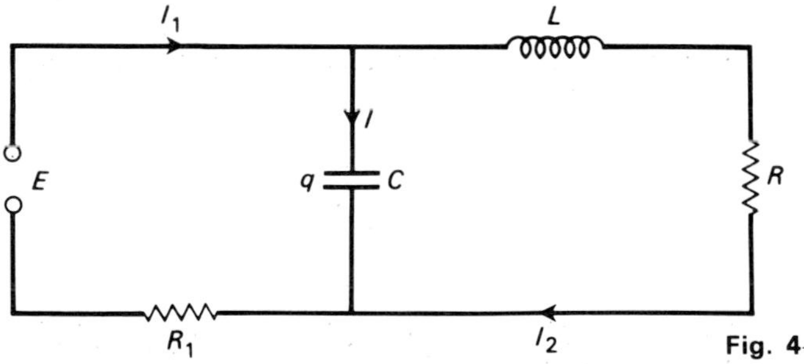

Fig. **4-25** Network of Problem 4.

a) Through differentiation and elimination obtain the differential equation relating $E = E(t)$ and I_1:

$$LCR_1 \frac{d^2 I_1}{dt^2} + (L + R_1 R_2 C)\frac{dI_1}{dt} + (R_1 + R_2)I_1 = LC\frac{d^2 E}{dt^2} + R_2 C\frac{dE}{dt} + E.$$

b) Obtain the transfer function $Y(s)$ for $E(t)$ as input and $I_1(t)$ as output.
c) Assume that $L = 100$, $C = 10$, $R_1 = 10$, $R_2 = 1$ in consistent units and graph $|Y(i\omega)|$. Discuss the filtering effect of this network.

5. a) For Example 3 in Section 4–20 verify the graph of $Y(i\omega)$ in Fig. 4–21.
 b) Find the Bode diagrams for $Y(s) = 1/(s + 2)$ and for $Y(s) = 1/(s + 0.1)$ and add to obtain the Bode diagrams of Fig. 4–22.

6. Show that the solution $x(t)$ obtained for Example 2 in Section 4–20 has limit 0 as $t \to \pm\infty$. HINT: For $t < 0$ show that the integral is less than $\int_0^\infty [1 + (t - u)^2]^{-1} \, du$; evaluate this integral and show that it has limit 0 as $t \to -\infty$. For $t > 0$ divide the integral

into one from 0 to $t/2$ and one from $t/2$ to ∞. Show that the first integral is less than $4(1+t^2)^{-1} \int_0^{t/2} (e^{-u} - e^{-2u})du$ and hence less than $4(1+t^2)^{-1} \int_0^{t/2} (e^{-u} - e^{-2u})du$. Show that the second integral is less than $\int_{t/2}^{\infty} e^{-u} - e^{-2u})du$.

4–22 PARSEVAL'S THEOREM AND ENERGY SPECTRUM

Let $f(t)$ have Fourier transform $\varphi(\omega)$. Parseval's theorem states that

$$\int_{-\infty}^{\infty} |\varphi(\omega)|^2 \, d\omega = 2\pi \int_{-\infty}^{\infty} |f(t)|^2 \, dt. \qquad (4\text{–}220)$$

To prove this, we observe that $|f(t)|^2 = f(t)\overline{f(t)}$ (Section 1–9). We let $g(t) = \overline{f(-t)}$. Then $\Phi[f*g] = \Phi[f]\Phi[g]$. But

$$\Phi[g] = \Phi[\mathscr{C}\Theta[f]] = \mathscr{C}\Theta\Phi[\Theta[f]] = \mathscr{C}\Theta^2\Phi[f] = \mathscr{C}\Phi[f]$$

by the rules (4–146), (4–147), (4–149). Thus $\Phi[g] = \overline{\varphi(\omega)}$ and

$$\Phi[f*g] = \varphi(\omega)\overline{\varphi(\omega)} = |\varphi(\omega)|^2.$$

Therefore,

$$f*g = \frac{1}{2\pi} \int_{-\infty}^{\infty} |\varphi(\omega)|^2 \, e^{i\omega t} \, d\omega$$

or

$$\int_{-\infty}^{\infty} f(u)g(t-u) \, du = \frac{1}{2\pi} \int_{-\infty}^{\infty} |\varphi(\omega)|^2 \, e^{i\omega t} \, d\omega.$$

If we now set $t = 0$ on both sides and use the fact that $g(-u) = \overline{f(u)}$, we obtain Eq (4–220). The steps are valid if, say, $f(t)$ is continuous, bounded, and absolutely integrable from $-\infty$ to ∞.

For a particle of mass m moving on the x-axis subject to a spring force, we have the differential equation (Section 1–14)

$$m\frac{d^2x}{dt} + kx = 0.$$

If we multiply both sides by dx/dt, then we can integrate to obtain

$$m\frac{1}{2}\left(\frac{dx}{dt}\right)^2 + k\frac{1}{2}x^2 = \text{const.}$$

The first term on the left is $mv^2/2$, that is, the *kinetic energy* of the particle, the second term is interpreted as *potential energy*, and the constant as *total energy*. We know from Section 1–14 that $x = A \sin(\omega t + \alpha)$ for appropriate constants A, α, where $\omega^2 = k/m$. Hence the total energy is

$$\tfrac{1}{2}mA^2\omega^2 \cos^2(\omega t + \alpha) + \tfrac{1}{2}kA^2 \sin^2(\omega t + \alpha)$$
$$= \tfrac{1}{2}kA^2 (\cos^2(\omega t + \alpha) + \sin^2(\omega t + \alpha)) = \tfrac{1}{2}kA^2.$$

Hence *the total energy is proportional to the square of the amplitude.*

We can write our oscillations in term of complex functions as

$$x = \frac{A}{2i}\left[e^{i(\omega t + \alpha)} - e^{-i(\omega t + \alpha)}\right] = \frac{Ae^{i\alpha}}{2i}e^{i\omega t} - \frac{Ae^{-i\alpha}}{2i}e^{-i\omega t}.$$

Since

$$\left|\frac{Ae^{i\alpha}}{2i}\right| = \frac{A}{2} \quad \text{and} \quad \left|\frac{Ae^{-i\alpha}}{2i}\right| = \frac{A}{2}.$$

the real amplitude A is twice the amplitude assigned to each of the frequencies $\omega, -\omega$ in the complex representation.

For a more general physical system (say a nonlinear spring, not governed by Hooke's law), the motion may be given by $x = f(t)$, $-\infty < t < \infty$. If we can represent $f(t)$ by a Fourier integral

$$x = f(t) = \frac{1}{2\pi}\int_{-\infty}^{\infty}\varphi(\omega)e^{i\omega t}\,d\omega,$$

then we can think of x as a superposition of sinusoidal oscillations of all possible frequencies. For a real function f, we get $\varphi(-\omega) = \overline{\varphi(\omega)}$, as follows from the rule (4-140) (see Problem 4 below). Hence the same amplitude $(2\pi)^{-1}|\varphi(\omega)|$ is assigned to ω and $-\omega$. We now interpret the *square of the amplitude as a measure of the energy for that frequency*. This is motivated by the case of simple harmonic motion considered above. Then we can define the *energy spectrum* of $f(t)$ as the function $|\varphi(\omega)|^2$. By Parseval's theorem, the *total energy* is then $2\pi\int_{-\infty}^{\infty}|f(t)|^2\,dt$.

For a system described by a linear differential equation with constant coefficients, as in Section 4-20, with $\Phi[g] = \varphi(\omega)$ and $\Phi[x] = \psi(\omega)$, we have $\psi(\omega) = Y(i\omega)\varphi(\omega)$, and hence

$$|\psi(\omega)|^2 = |Y(i\omega)|^2\,|\varphi(\omega)|^2.$$

Thus $|Y(i\omega)|^2$ is the ratio of output energy spectrum to input energy spectrum. For this reason $|Y(i\omega)|^2$ is called the *energy transfer function*.

========================== **Problems (Section 4-22)** ==========================

1. Verify the correctness of Parseval's equation for each of the following choices of $f(t)$:

   a) $1/(1 + t^2)$ b) $e^{-t}U(t)$

2. Use Parseval's equation to evaluate the integrals:

   a) $\displaystyle\int_{-\infty}^{\infty}\frac{\sin^2\omega}{\omega^2}\,d\omega$ b) $\displaystyle\int_{-\infty}^{\infty}\frac{\cos^2 t}{(1 + t^2)^2}\,dt$

3. Prove the rule (also called Parseval's theorem): if $\Phi[f] = \varphi(\omega)$ and $\Phi[h] = \psi(\omega)$, then

   $$\int_{-\infty}^{\infty}\varphi(\omega)\overline{\psi(\omega)}\,d\omega = 2\pi\int_{-\infty}^{\infty}f(t)\overline{h(t)}\,dt.$$

   HINT: As in the proof of Eq. (4-220), let $g(t) = \overline{h(-t)}$.

4. Prove from rule (4-149): if f is real valued and $\Phi[f] = \varphi(\omega)$, then $\varphi(-\omega) = \overline{\varphi(\omega)}$ for all ω.

Chapter 5 Matrices and Linear Algebra

5–1 INTRODUCTION

In this chapter we study linear equations and related concepts. Such equations arise in many physical problems—in particular, in finding equilibrium conditions.

EXAMPLE 1 Forces $\mathbf{F}_1, \mathbf{F}_2, \mathbf{F}_3$ act on a particle in equilibrium at the origin, as in Fig. 5–1. It is known that $\mathbf{F}_1$ has magnitude 10 lbs. and that $\mathbf{F}_1, \mathbf{F}_2, \mathbf{F}_3$ have the directions shown. Find $\mathbf{F}_2, \mathbf{F}_3$.

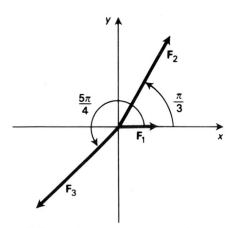

Fig. 5–1. Particle in equilibrium.

Solution. Let $\mathbf{F}_2$ have x-component u_2, so that its y-component is $\sqrt{3}u_2$. Similarly, $\mathbf{F}_3$ has x-component u_3 and y-component u_3 (negative). At equilibrium, the sum of all x-components is zero and the sum of all y-components is zero. Therefore,

$$10 + u_2 + u_3 = 0 \qquad \text{and} \qquad 0 + \sqrt{3}u_2 + u_3 = 0.$$

Solving these linear equations, we find:

$$u_2 = 5\sqrt{3} + 5 \qquad \text{and} \qquad u_3 = -15 - 5\sqrt{3}. \quad \blacktriangleleft$$

In general, we shall be concerned with linear equations in n unknowns. When there are n equations in n unknowns, they have the form

$$
\begin{aligned}
a_{11}x_1 + \cdots + a_{1n}x_n &= k_1, \\
&\vdots \\
a_{n1}x_1 + \cdots + a_{nn}x_n &= k_n.
\end{aligned}
\tag{5-10}
$$

When $k_1, \ldots, k_n$ are all zero, they become a *homogeneous system*

$$
\begin{aligned}
a_{11}x_1 + \cdots + a_{1n}x_n &= 0, \\
&\vdots \\
a_{n1}x_1 + \cdots + a_{nn}x_n &= 0.
\end{aligned}
\tag{5-11}
$$

We state here some basic rules about such sets of equations. The rules make use of determinants, which are reviewed in the next section. We denote by D the *determinant of the coefficients* for each of these systems:

$$
D = \begin{vmatrix} a_{11} & \cdots & a_{1n} \\ \vdots & & \vdots \\ a_{n1} & \cdots & a_{nn} \end{vmatrix}.
\tag{5-12}
$$

We also denote by $D_1, \ldots, D_n$ the determinants obtained from D by replacing the first, second, etc., nth column by the numbers $k_1, \ldots, k_n$ on the right of Eqs. (5–10):

$$
D_1 = \begin{vmatrix} k_1 & a_{12} & \cdots & a_{1n} \\ \vdots & & & \vdots \\ k_n & a_{n2} & \cdots & a_{nn} \end{vmatrix}, \qquad
D_2 = \begin{vmatrix} a_{11} & k_1 & a_{13} & \cdots & a_{1n} \\ \vdots & & & & \vdots \\ a_{n1} & k_n & a_{n3} & \cdots & a_{nn} \end{vmatrix}, \ldots
\tag{5-13}
$$

Then, *if $D \neq 0$, Eqs. (5–10) have the unique solution*

$$
x_1 = \frac{D_1}{D}, \quad \ldots, \quad x_n = \frac{D_n}{D}.
\tag{5-14}
$$

This is *Cramer's rule.*

For the homogeneous system (5–11) we have two cases:

a) $D \neq 0$. Then, as a special case of (5–14), we obtain as the only solution the *trivial solution*

$$x_1 = 0, \quad x_2 = 0, \quad \ldots, \quad x_n = 0.$$

b) $D = 0$. Then there are *infinitely many solutions*. To obtain these, r of the unknowns ($r < n$) can be expressed in terms of the remaining unknowns, which can be assigned arbitrary values (when every coefficient a_{ij} is 0, each unknown can be given an arbitrary value).

These rules are illustrated in Problem 10 below. (For proofs, the reader is referred to CLA, Chapter 10.)

EXAMPLE 2 Weights w_1, w_2, w_3 are supported as shown in Fig. 5–2, and it is known that $w_1 = w_2 + w_3$. Can the system be in static equilibrium and, if so, what are the weights?

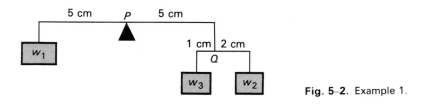

Fig. 5–2. Example 1.

Solution. Taking moments of all weights about P and of w_2 and w_3 about Q, we conclude that for equilibrium we must have $5w_1 = 7w_2 + 4w_3$ and $2w_2 = w_3$; we know also that $w_1 = w_2 + w_3$. These are homogeneous equations with determinant

$$D = \begin{vmatrix} 5 & -7 & -4 \\ 0 & 2 & -1 \\ 1 & -1 & -1 \end{vmatrix} = 0.$$

Accordingly, infinitely many solutions exist. The value of w_2 can be chosen arbitrarily and then $w_1 = 3w_2$, $w_3 = 2w_2$. ◄

Much of the theory has a geometric aspect concerning vectors in n-dimensional space, linear independence of sets of such vectors, linear mappings from n-dimensional space to m-dimensional space. We explore some of the geometrical concepts in later sections.

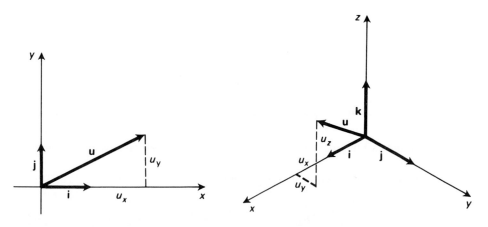

Fig. 5–3. Vectors in the plane.   Fig. 5–4. Vectors in space.

In many cases we illustrate our geometrical results by examples in the plane (2-dimensional case) or in space (3-dimensional case). For the plane we use the standard xy-coordinates and the corresponding unit vectors $\mathbf{i}$, $\mathbf{j}$ as in Fig. 5–3; each vector $\mathbf{u}$ in the plane can be represented as

$$\mathbf{u} = u_x\mathbf{i} + u_y\mathbf{j} \tag{5-15}$$

in terms of the components u_x, u_y. Similarly, vectors in space can be represented as

$$\mathbf{u} = u_x\mathbf{i} + u_y\mathbf{j} + u_z\mathbf{k} \tag{5-16}$$

in terms of the unit vectors $\mathbf{i}$, $\mathbf{j}$, $\mathbf{k}$, as in Fig. 5–4.

5–2 DETERMINANTS

We list important properties of determinants. Each determinant has a numerical value (and hence determinants must always be distinguished from matrices, which are arrays of numbers). For example,

$$\begin{vmatrix} 3 & 6 \\ 1 & 5 \end{vmatrix} = 9.$$

The properties are as follows:

$$\begin{vmatrix} a & b \\ c & d \end{vmatrix} = ad - bc; \tag{5-20}$$

$$\begin{vmatrix} a_1 & b_1 & c_1 \\ a_2 & b_2 & c_2 \\ a_3 & b_3 & c_3 \end{vmatrix} = a_1 \begin{vmatrix} b_2 & c_2 \\ b_3 & c_3 \end{vmatrix} - b_1 \begin{vmatrix} a_2 & c_2 \\ a_3 & c_3 \end{vmatrix} + c_1 \begin{vmatrix} a_2 & b_2 \\ a_3 & b_3 \end{vmatrix}; \tag{5-21}$$

$$\begin{vmatrix} a_1 & b_1 & c_1 & d_1 \\ a_2 & b_2 & c_2 & d_2 \\ a_3 & b_3 & c_3 & d_3 \\ a_4 & b_4 & c_4 & d_4 \end{vmatrix} = a_1 \begin{vmatrix} b_2 & c_2 & d_2 \\ b_3 & c_3 & d_3 \\ b_4 & c_4 & d_4 \end{vmatrix} - b_1 \begin{vmatrix} a_2 & c_2 & d_2 \\ a_3 & c_3 & d_3 \\ a_4 & c_4 & d_4 \end{vmatrix} + \cdots \tag{5-22}$$

Thus, in general, a determinant of nth order is defined in terms of determinants of $(n-1)$st order; the coefficients of $a_1, b_1, \ldots$ on the right side of (5–21) and (5–22) are the "cofactors" of these elements. In general, determinants obtained from a given determinant by deleting certain rows and columns are called *minors* of the given determinant. The rules for expansion by minors of the first row are indicated in (5–21) and (5–22).

In general, rows and columns can be interchanged:

$$\begin{vmatrix} a_1 & b_1 & c_1 \\ a_2 & b_2 & c_2 \\ a_3 & b_3 & c_3 \end{vmatrix} = \begin{vmatrix} a_1 & a_2 & a_3 \\ b_1 & b_2 & b_3 \\ c_1 & c_2 & c_3 \end{vmatrix}. \tag{5-23}$$

In general, interchanging two rows (or columns) multiplies the determinant by -1:

$$\begin{vmatrix} a_1 & b_1 & c_1 \\ a_2 & b_2 & c_2 \\ a_3 & b_3 & c_3 \end{vmatrix} = - \begin{vmatrix} a_2 & b_2 & c_2 \\ a_1 & b_1 & c_1 \\ a_3 & b_3 & c_3 \end{vmatrix}. \qquad (5\text{–}24)$$

Because of these rules, we can use any row or column in expanding by minors, provided we use the checkerboard pattern of $+$ and $-$ signs:

$$\begin{matrix} + & - & + & - & \cdots \\ - & + & - & + & \cdots \\ + & - & + & - & \cdots \\ \vdots & & \vdots & & \end{matrix}$$

A factor of any row (or column) can be placed before the determinant:

$$\begin{vmatrix} ka_1 & kb_1 & kc_1 \\ a_2 & b_2 & c_2 \\ a_3 & b_3 & c_3 \end{vmatrix} = k \begin{vmatrix} a_1 & b_1 & c_1 \\ a_2 & b_2 & c_2 \\ a_3 & b_3 & c_3 \end{vmatrix}. \qquad (5\text{–}25)$$

If two rows (or columns) are proportional, the determinant equals 0:

$$\begin{vmatrix} ka_1 & kb_1 & kc_1 \\ a_1 & b_1 & c_1 \\ a_2 & b_2 & c_2 \end{vmatrix} = 0. \qquad (5\text{–}26)$$

Two determinants differing only in one row or column can be added in the following way:

$$\begin{vmatrix} a_1 & b_1 & c_1 \\ a_2 & b_2 & c_2 \\ a_3 & b_3 & c_3 \end{vmatrix} + \begin{vmatrix} A_1 & b_1 & c_1 \\ A_2 & b_2 & c_2 \\ A_3 & b_3 & c_3 \end{vmatrix} = \begin{vmatrix} a_1+A_1 & b_1 & c_1 \\ a_2+A_2 & b_2 & c_2 \\ a_3+A_3 & b_3 & c_3 \end{vmatrix}. \qquad (5\text{–}27)$$

The value of the determinant is unchanged if the elements of one row are multiplied by the same quantity k and added to the corresponding elements of another row.

$$\begin{vmatrix} a_1 & b_1 & c_1 \\ a_2 & b_2 & c_2 \\ a_3 & b_3 & c_3 \end{vmatrix} = \begin{vmatrix} a_1+ka_2 & b_1+kb_2 & c_1+kc_2 \\ a_2 & b_2 & c_2 \\ a_3 & b_3 & c_3 \end{vmatrix}. \qquad (5\text{–}28)$$

By suitable choice of k, this rule can be used to introduce zeros; by repetition of the process, all elements but one in a chosen row can be reduced to zero. This procedure is basic for numerical evaluation of determinants. (For proofs of these rules, see CLA, Chapter 10.)

5–3 MATRICES

By a *matrix* we mean a rectangular array of m rows and n columns:

$$\begin{bmatrix} a_{11} & \cdots & a_{1n} \\ \vdots & & \vdots \\ a_{m1} & \cdots & a_{mn} \end{bmatrix}. \tag{5–30}$$

For this chapter (with a very few exceptions), the objects $a_{11}, a_{12}, \ldots, a_{mn}$ will be real numbers. In some applications they are complex numbers, and in some they are functions of one or more variables. We call each a_{ij} an *entry* of the matrix; more specifically, a_{ij} is the *ij-entry*.

We can denote a matrix by a single letter such as $A, B, C, X, Y, \ldots$ If A denotes the matrix (5–30), then we write also, concisely, $A = (a_{ij})$.

Let A be the matrix (a_{ij}) of (5–30). We say that A is an $m \times n$ matrix. When $m = n$, we say that A is a *square matrix of order n*. The following are examples of matrices:

$$A = \begin{bmatrix} 2 & 3 & 5 \\ 1 & 2 & 3 \end{bmatrix}, \quad B = \begin{bmatrix} 1 & 2 \\ 4 & 3 \end{bmatrix}, \quad C = \begin{bmatrix} 1 & 4 \\ 2 & -3 \end{bmatrix}. \tag{5–31}$$

Here A is 2×3, B and C are 2×2; B and C are *square matrices of order* 2.

An important square matrix is the *identity matrix* of order n, denoted by I:

$$I = \begin{bmatrix} 1 & 0 & \cdots & 0 \\ 0 & 1 & \cdots & 0 \\ \vdots & & & \vdots \\ 0 & 0 & \cdots & 1 \end{bmatrix} = (\delta_{ij}), \quad \delta_{ij} = \begin{cases} 1 & \text{for } i = j, \\ 0 & \text{for } i \neq j. \end{cases} \tag{5–32}$$

We call δ_{ij} the *Kronecker delta symbol*. We sometimes write I_n to indicate the order of I, but normally the context makes this unnecessary.

For each m and n we define the $m \times n$ *zero matrix* as

$$O = \begin{bmatrix} 0 & \cdots & 0 \\ \vdots & & \vdots \\ 0 & \cdots & 0 \end{bmatrix}. \tag{5–33}$$

Sometimes this matrix is denoted by O_{mn} to indicate the *size*, that is, the number of rows and columns.

In general, two matrices $A = (a_{ij})$ and $B = (b_{ij})$ are said to be *equal*, $A = B$, when A and B have the same size and $a_{ij} = b_{ij}$ for all i and j.

A $1 \times n$ matrix A is formed of one row: $A = (a_{11}, \ldots, a_{1n})$. We call such a matrix a *row vector*. In a general $m \times n$ matrix (5–30), each of the successive rows forms a row vector. We often denote a row vector by a boldface symbol: $\mathbf{u}, \mathbf{v}, \ldots$ (or in handwriting, by an arrow). Thus the matrix A in (5–31) has the row vectors $\mathbf{u}_1 = (2, 3, 5)$ and $\mathbf{u}_2 = (1, 2, 3)$.

Similarly, an $m \times 1$ matrix A is formed of one column:

$$A = \begin{bmatrix} a_{11} \\ \vdots \\ a_{m1} \end{bmatrix}. \tag{5–34}$$

We call such a matrix a *column vector*. For typographical reasons, we sometimes denote this matrix by col $(a_{11}, \ldots, a_{m1})$ or even by $(a_{11}, \ldots, a_{m1})$, if the context makes clear that a column vector is intended. We also denote column vectors by boldface letters: $\mathbf{u}, \mathbf{v}, \ldots$ The matrix B in (5–31) has the column vectors $\mathbf{v}_1 = \mathrm{col}\,(1, 4)$ and $\mathbf{v}_2 = \mathrm{col}\,(2, 3)$.

We denote by $\mathbf{0}$ the row vector or column vector $(0, \ldots, 0)$. The context will make clear whether $\mathbf{0}$ is a row vector or a column vector and the number of entries.

The vectors occurring here can be interpreted geometrically as vectors in k-dimensional space, for appropriate k. For example, the row vectors or column vectors with three entries are simply ordered triples of numbers and, as in Fig. 5–4, they can be represented as vectors $a\mathbf{i} + b\mathbf{j} + c\mathbf{k}$ in 3-dimensional space.

Matrices often arise in connection with simultaneous linear equations. Let such a set of equations be given:

$$a_{11}x_1 + \ldots a_{1n}x_n = y_1,$$
$$\vdots \qquad\qquad \vdots \tag{5–35}$$
$$a_{m1}x_1 + \ldots a_{mn}x_n = y_m.$$

Here we may think of $y_1, \ldots, y_m$ as given numbers and $x_1, \ldots, x_n$ as unknown numbers to be found; however, we may also think of $x_1, \ldots, x_n$ as variable numbers and $y_1, \ldots, y_m$ as *dependent variables* whose values are determined by the values chosen for the *independent variables* $x_1, \ldots, x_n$. Both points of view will be important in this chapter. In either case, we call $A = (a_{ij})$ the *coefficient matrix* of the set of equations. The numbers $y_1, \ldots, y_m$ can be considered the entries in a column vector $\mathbf{y} = \mathrm{col}\,(y_1, \ldots, y_m)$. The numbers $x_1, \ldots, x_n$ can be thought of as the entries in a row vector or column vector $\mathbf{x}$; in this chapter, usually we write $\mathbf{x} = \mathrm{col}\,(x_1, \ldots, x_n)$.

5–4 ADDITION OF MATRICES. SCALAR TIMES MATRIX

Let $A = (a_{ij})$ and $B = (b_{ij})$ be matrices of the *same size*, both $m \times n$. Then we define the sum $A + B$ to be the $m \times n$ matrix $C = (c_{ij})$ such that $c_{ij} = a_{ij} + b_{ij}$ for all i and j; that is, *two matrices can be added by adding corresponding entries*. For example,

$$\begin{bmatrix} 3 & 5 & 1 \\ 1 & 0 & -2 \end{bmatrix} + \begin{bmatrix} 0 & 1 & 0 \\ 2 & 3 & 5 \end{bmatrix} = \begin{bmatrix} 3 & 6 & 1 \\ 3 & 3 & 3 \end{bmatrix}.$$

Let c be a number (scalar); let $A = (a_{ij})$ be an $m \times n$ matrix. Then we define cA to be the $m \times n$ matrix $B = (b_{ij})$ such that $b_{ij} = ca_{ij}$ for all i and j; that is, *cA is obtained from A by multiplying each entry of A by c*. For example,

$$5\begin{bmatrix} 2 & 0 \\ 1 & -2 \end{bmatrix} = \begin{bmatrix} 10 & 0 \\ 5 & -10 \end{bmatrix}.$$

We denote $(-1)A$ by $-A$ and $B+(-A)$ by $B-A$.

From these definitions we can deduce the following rules governing the two operations:

1. $A + B = B + A.$ 2. $A + (B+C) = (A+B) + C.$
3. $c(A+B) = cA + cB.$ 4. $(a+b)C = aC + bC.$
5. $a(bC) = (ab)C.$ 6. $1A = A.$ (5–40)
7. $0A = O.$ 8. $A + O = A.$
9. $A + C = B$ if and only if $C = B - A.$

Throughout we assume that the sizes of the matrices are such that the operations have meaning. The proofs are obtained by simply applying the definitions. For example, for Rule 1 we write

$$A + B = (a_{ij}) + (b_{ij}) = (a_{ij} + b_{ij}) = (b_{ij} + a_{ij}) = (b_{ij}) + (a_{ij}) = B + A.$$

═══════════════════ **Problems (Section 5-4)** ═══════════════════

In these problems, the following matrices are given:

$$A = \begin{bmatrix} 1 \\ 3 \end{bmatrix}, \qquad B = \begin{bmatrix} 2 \\ 0 \end{bmatrix}, \qquad C = \begin{bmatrix} 2 & 3 \\ 4 & 1 \end{bmatrix}, \qquad D = \begin{bmatrix} 1 & -1 \\ 2 & 0 \end{bmatrix},$$

$$E = \begin{bmatrix} 1 & 2 \\ 2 & 4 \end{bmatrix}, \qquad F = \begin{bmatrix} 1 & 4 & 5 \\ 2 & 0 & 7 \end{bmatrix}, \qquad G = \begin{bmatrix} 3 & 1 & 4 \\ -1 & 0 & -1 \end{bmatrix},$$

$$H = (1, 0, 1), \qquad J = (3, 5, 2), \qquad K = (3, 5),$$

$$L = \begin{bmatrix} 3 & 1 & 0 \\ 2 & 5 & 6 \\ 1 & 4 & 3 \end{bmatrix}, \quad M = \begin{bmatrix} 2 & -1 & 0 \\ 1 & 2 & 1 \\ 3 & 2 & -1 \end{bmatrix}, \quad N = \begin{bmatrix} 1 & 4 \\ 0 & 3 \\ 7 & 1 \end{bmatrix}, \quad P = \begin{bmatrix} 2 & 2 \\ -1 & -1 \\ 3 & 3 \end{bmatrix}.$$

1. **a)** Give the number of rows and columns for each of the matrices A, F, H, L, and P.
   b) Writing $A = (a_{ij})$, $B = (b_{ij})$ and so on, give the values of the following entries: $a_{11}, a_{21},$
   $c_{21}, c_{22}, d_{12}, e_{21}, f_{11}, g_{23}, g_{21}, h_{12}, m_{23}.$
   c) Give the row vectors of C, G, L, and P.
   d) Give the column vectors of D, F, L, and N.

2. Evaluate each expression which is meaningful:
   a) $A + B$ **b)** $C + D$ **c)** $E + F$ **d)** $L + M$
   e) $N - P$ **f)** $G - F$ **g)** $5C$ **h)** $2E$
   i) $3E + 4D$ **j)** $2C + D - E$ **k)** $3L - N$

3. Solve for X:
   a) $C + X = D$ **b)** $F - 5X = G$

4. Solve for X and Y:
   a) $X + Y = N$, $X - Y = P$
   b) $2X - 3Y = L$, $X - 2Y = M$

5. Prove each of the following rules of (5–40):
   a) Rule 2   b) Rule 3   c) Rule 4   d) Rule 5
   e) Rule 6   f) Rule 7   g) Rule 8   h) Rule 9

6. A particle on a frictionless inclined plane, as in Fig. 5–5, is in equilibrium under a force **F** of 10 lbs. Find its weight **W** and the normal reaction **N** of the plane.

7. In Problem 6 we allow for friction, with coefficient of friction equal to 0.1; find the weight and normal reaction.

8. Find the currents I_1, I_2, I_3 flowing in the network of Fig. 5–6.

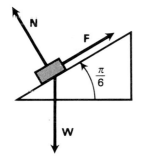

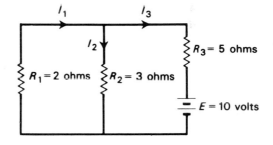

Fig. 5–5. Problem 6.   **Fig. 5–6.** Problem 8.

9. Evaluate the determinants:

a) $\begin{vmatrix} 3 & 1 \\ 2 & 2 \end{vmatrix}$ b) $\begin{vmatrix} 6 & 4 \\ 3 & 2 \end{vmatrix}$ c) $\begin{vmatrix} 1 & 7 & 1 \\ 3 & 5 & 1 \\ 2 & 0 & 5 \end{vmatrix}$

d) $\begin{vmatrix} 2 & 1 & -1 \\ 1 & 2 & 2 \\ 5 & 4 & 0 \end{vmatrix}$ e) $\begin{vmatrix} 1 & 3 & 2 & 2 \\ 0 & 1 & 5 & 4 \\ 0 & 3 & 0 & 7 \\ 0 & 1 & 5 & 2 \end{vmatrix}$ f) $\begin{vmatrix} 1 & 0 & 1 & 0 \\ 0 & 1 & 0 & 1 \\ 1 & 0 & 0 & 1 \\ 0 & 1 & 1 & 0 \end{vmatrix}$

10. Find all solutions:

a) $2x - y = 3$, $3x + y = 7$
b) $x + 3y = 8$, $x - y = 0$
c) $2x - y + z = 0$, $x + y + z = 1$, $3x + y - 2z = 6$
d) $x + y - z = 4$, $x - 2y - 2z = -1$, $x + y - 3z = 0$
e) $x - y = 0$, $x + 2y = 0$
f) $2x - y = 0$, $4x - 2y = 0$
g) $x + y + z = 0$, $2x - y - 3z = 0$, $x + 4y + 6z = 0$
h) $2x - y - z = 0$, $4x - 2y - 2z = 0$, $6x - 3y - 3z = 0$
i) $x + y + z = 0$, $x + y - z = 0$, $x - y + z = 0$
j) $x - y - z = 0$, $2x + y - w = 0$, $3x - z - w = 0$, $x + 2y + z - w = 0$

5–5 MULTIPLICATION OF MATRICES

In order to motivate the definition of the product AB of two matrices A and B, we consider two systems of simultaneous equations:

$$\begin{cases} a_{11}u_1 + \cdots + a_{1p}u_p = y_1, \\ \phantom{a_{11}u_1} \vdots \phantom{+ \cdots + a_{1p}u_p} \vdots \\ a_{m1}u_1 + \cdots + a_{mp}u_p = y_m. \end{cases} \tag{5–50}$$

$$\begin{cases} b_{11}x_1 + \cdots + b_{1n}x_n = u_1, \\ \phantom{b_{11}x_1} \vdots \phantom{+ \cdots + b_{1n}x_n} \vdots \\ b_{p1}x_1 + \cdots + b_{pn}x_n = u_p. \end{cases} \tag{5–51}$$

Such pairs of systems arise in many practical problems. A typical situation is that in which $x_1, \ldots, x_n$ are known numbers and $y_1, \ldots, y_m$ are sought, all coefficients b_{ij} and a_{ij} being known. The second set of equations allows us to compute $u_1, \ldots, u_p$; if we substitute the values found in the first set of equations, we can then find $y_1, \ldots, y_m$. Thus

$$\begin{aligned} y_1 &= a_{11}u_1 + \cdots + a_{1p}u_p \\ &= a_{11}(b_{11}x_1 + \cdots + b_{1n}x_n) + \cdots + a_{1p}(b_{p1}x_1 + \cdots + b_{pn}x_n) \\ &= (a_{11}b_{11} + \cdots + a_{1p}b_{p1})x_1 + \cdots + (a_{11}b_{1n} + \cdots + a_{1p}b_{pn})x_n; \end{aligned}$$

and, in general, for $i = 1, \ldots, m$,

$$y_i = (a_{i1}b_{11} + \cdots + a_{ip}b_{p1})x_1 + \cdots + (a_{i1}b_{1n} + \cdots + a_{ip}b_{pn})x_n$$

or

$$y_i = c_{i1}x_1 + \cdots + c_{in}x_n, \quad i = 1, \ldots, m$$

where

$$c_{ij} = a_{i1}b_{1j} + \cdots + a_{ip}b_{pj} \quad \text{for } i = 1, \ldots, m, \quad j = 1, \ldots, n. \tag{5–52}$$

Thus, from the coefficient matrix $A = (a_{ij})$ and the coefficient matrix $B = (b_{ij})$ we obtain the coefficient matrix $C = (c_{ij})$ by the rule (5–52). We write $C = AB$ and have thereby defined the *product of the matrices A and B*.

We observe that $(a_{i1}, \ldots, a_{ip})$ is the ith *row* vector of A and that col $(b_{1j}, \ldots, b_{pj})$ is the jth *column* vector of B. Hence, to form the product $AB = C = (c_{ij})$, we obtain each c_{ij} by multiplying corresponding entries of the ith row of A and the jth column of B and adding. The process is illustrated in Fig. 5–7.

We remark that the product AB is defined only when the number of *columns* of A equals the number of *rows* of B; that is, when A is $m \times p$ and B is $p \times n$, AB is defined and is $m \times n$, as in Fig. 5–7. Also, when AB is defined, BA need not be defined and even when it is, AB is generally *not equal* to BA; that is, there is *no commutative law* for multiplication.

EXAMPLE 1 $\begin{bmatrix} a & b \\ c & d \end{bmatrix} \begin{bmatrix} e & f & g \\ h & i & j \end{bmatrix} = \begin{bmatrix} ae+bh & af+bi & ag+bj \\ ce+dh & cf+di & cg+dj \end{bmatrix}.$

Here $m = 2$, $p = 2$, $n = 3$. ◄

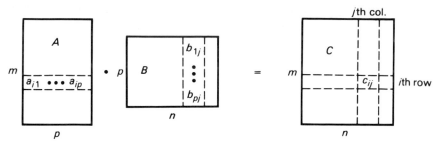

Fig. 5–7. Product of two matrices.

EXAMPLE 2 $\begin{bmatrix} 1 & 2 & 1 \\ 3 & 0 & 5 \\ 2 & 1 & -7 \end{bmatrix} \begin{bmatrix} 1 \\ 1 \\ 2 \end{bmatrix} = \begin{bmatrix} 1+8+2 \\ 3+0+10 \\ 2+4-14 \end{bmatrix} = \begin{bmatrix} 11 \\ 13 \\ -8 \end{bmatrix}$

Here $m = 3$, $p = 3$, $n = 1$. ◄

The second example illustrates the important case of the product $A\mathbf{v}$, where A is an $m \times n$ matrix and $\mathbf{v}$ is an $n \times 1$ column vector. The product $A\mathbf{v}$ is now an $m \times 1$ column vector $\mathbf{u}$.

In the general product $AB = C$, as defined above, we note that the jth column vector of C is formed from A and the jth column vector of B, for the jth column vector of C is

$$\begin{bmatrix} a_{11}b_{1j} + \cdots + a_{1p}b_{pj} \\ a_{21}b_{1j} + \cdots + a_{2p}b_{pj} \\ \vdots \\ a_{m1}b_{1j} + \cdots + a_{mp}b_{pj} \end{bmatrix} = \begin{bmatrix} a_{11} & \cdots & a_{1p} \\ a_{21} & \cdots & a_{2p} \\ \vdots & & \vdots \\ a_{m1} & \cdots & a_{mp} \end{bmatrix} \cdot \begin{bmatrix} b_{1j} \\ \vdots \\ b_{pj} \end{bmatrix}.$$

Hence, if we denote the successive column vectors of B by $\mathbf{u}_1, \ldots, \mathbf{u}_n$, then the column vectors of $C = AB$ are $A\mathbf{u}_1, \ldots, A\mathbf{u}_n$. Symbolically,

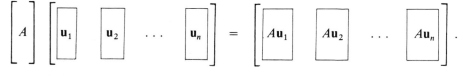

EXAMPLE 3 To calculate AB, where

$$A = \begin{bmatrix} 3 & 1 & 2 \\ 1 & 0 & 5 \end{bmatrix} \quad \text{and} \quad B = \begin{bmatrix} 1 & 0 & 3 & 1 & 3 \\ 5 & 2 & 1 & 5 & 1 \\ -1 & 2 & 4 & -1 & 4 \end{bmatrix},$$
$$\qquad\qquad\qquad\qquad\qquad\; \mathbf{u}_1 \;\; \mathbf{u}_2 \;\; \mathbf{u}_3 \;\; \mathbf{u}_4 \;\; \mathbf{u}_5$$

we calculate

$$A\mathbf{u}_1 = \begin{bmatrix} 6 \\ -4 \end{bmatrix}, \quad A\mathbf{u}_2 = \begin{bmatrix} 6 \\ 10 \end{bmatrix}, \quad A\mathbf{u}_3 = \begin{bmatrix} 18 \\ 23 \end{bmatrix},$$

and then note that $\mathbf{u}_4 = \mathbf{u}_1$ and $\mathbf{u}_5 = \mathbf{u}_3$, so that $A\mathbf{u}_4 = A\mathbf{u}_1$ and $A\mathbf{u}_5 = A\mathbf{u}_3$. Therefore,

$$AB = \begin{bmatrix} 6 & 6 & 18 & 6 & 18 \\ -4 & 10 & 23 & -4 & 23 \end{bmatrix}. \quad \blacktriangleleft$$

EXAMPLE 4 The simultaneous equations
$$\begin{aligned} 3x + 2y + 5z &= u, \\ 4x - 5y - 8z &= v, \\ 7x + 2y + 9z &= w \end{aligned}$$
are equivalent to the matrix equation
$$\begin{bmatrix} 3 & 2 & 5 \\ 4 & -5 & -8 \\ 7 & 2 & 9 \end{bmatrix} \begin{bmatrix} x \\ y \\ z \end{bmatrix} = \begin{bmatrix} u \\ v \\ w \end{bmatrix},$$
for the product on the left equals the column vector
$$\begin{bmatrix} 3x + 2y + 5z \\ 4x - 5y - 8z \\ 7x + 2y + 9z \end{bmatrix},$$
and this equals col (u, v, w) precisely when the given simultaneous equations hold. $\blacktriangleleft$

In the same way, the two sets of simultaneous equations (5–50) and (5–51) can be replaced by the equations
$$A\mathbf{u} = \mathbf{y} \quad \text{and} \quad B\mathbf{x} = \mathbf{u}.$$
The elimination process at the beginning of this section is equivalent to replacing $\mathbf{u}$ by $B\mathbf{x}$ in the first equation to obtain $\mathbf{y} = A(B\mathbf{x})$. *Our definition of the product AB is then such that* $\mathbf{y} = A(B\mathbf{x}) = (AB)\mathbf{x}$. Therefore, *for every column vector* $\mathbf{x}$,
$$A(B\mathbf{x}) = (AB)\mathbf{x}.$$

Powers of a square matrix

If A is a square matrix of order n, then the product AA has meaning and is again $n \times n$; we write A^2 for this product. Similarly, $A^3 = A^2A$, $A^4 = A^3A, \ldots,$ $A^{s+1} = A^sA, \ldots$ We also define A^0 to be the $n \times n$ identity matrix I. Negative powers can also be defined for certain square matrices A (see Section 5–6 below).

Rules for multiplication

Multiplication of matrices obeys a set of rules, which we adjoin to those of the preceding section:

10. $A(BC) = (AB)C$.   11. $AI = A$.   12. $IA = A$.
13. $A(B+C) = AB + AC$.   14. $c(AB) = A(cB)$.   (5–53)
15. $AO = O$.   16. $OA = O$.   17. $A^0 = I$.   18. $A^kA^l = A^{k+l}$.
19. $(A^k)^l = A^{kl}$.   20. $A\mathbf{x} = B\mathbf{x}$ for all $\mathbf{x}$ if and only if $A = B$.

Here, the sizes of the matrices must again be such that the operations are defined. For example, in Rule 13, if A is $m \times p$, then B and C must be $p \times n$.

To prove Rule 10 (associative law), we let C have the column vectors $\mathbf{u}_1, \ldots, \mathbf{u}_k$. Then BC has the column vectors $B\mathbf{u}_1, \ldots, B\mathbf{u}_k$, and hence $A(BC)$ has the column vectors $A(B\mathbf{u}_1), \ldots, A(B\mathbf{u}_k)$. But, as remarked above, $A(B\mathbf{x}) = (AB)\mathbf{x}$ for every $\mathbf{x}$. Therefore $A(BC)$ has the column vectors $(AB)\mathbf{u}_1, \ldots, (AB)\mathbf{u}_k$. But these are the column vectors of $(AB)C$. Hence $A(BC) = (AB)C$.

For Rule 11, A is, say, $m \times p$ and I is $p \times p$, so that

$$
AI = \begin{bmatrix} a_{11} & \cdots & a_{1p} \\ \vdots & & \vdots \\ a_{m1} & \cdots & a_{mp} \end{bmatrix} \begin{bmatrix} 1 & 0 & \cdots & 0 \\ 0 & 1 & \cdots & 0 \\ \vdots & \vdots & & \vdots \\ 0 & 0 & \cdots & 1 \end{bmatrix} = \begin{bmatrix} a_{11} & \cdots & a_{1p} \\ \vdots & & \vdots \\ a_{m1} & \cdots & a_{mp} \end{bmatrix} = A.
$$

We can also write $AI = C = (c_{ij})$, where

$$
c_{ij} = a_{i1}\delta_{1j} + \cdots + a_{ip}\delta_{pj} = a_{ij},
$$

since $\delta_{jj} = 1$ but $\delta_{ij} = 0$ for $i \neq j$. Rule 12 is proved similarly. For Rule 13 we have $A(B+C) = D$, where

$$
d_{ij} = a_{i1}(b_{1j}+c_{1j}) + \cdots + a_{ip}(b_{pj}+c_{pj})
$$
$$
= (a_{i1}b_{1j} + \cdots + a_{ip}b_{pj}) + (a_{i1}c_{1j} + \cdots + a_{ip}c_{pj}),
$$

and hence $D = AB + AC$. Rule 14 is proved similarly. Rules 15 and 16 follow from the fact that all entries of O are 0; here again the size of O must be such that the products have meaning.

In Rules 17, 18 and 19, A is a square matrix and k and l are nonnegative integers. Rule 17 is true by definition of A^0; and Rules 18 and 19 are true for $l = 0$ and $l = 1$ by definition. They can be proved for general l by induction (Problem 5 below).

For Rule 20, let A and B be $m \times n$ and let $\mathbf{e}_1, \ldots, \mathbf{e}_n$ be the column vectors of the identity matrix I of order n. Then AI is a matrix whose columns are $A\mathbf{e}_1, \ldots, A\mathbf{e}_n$, and BI is a matrix whose column vectors are $B\mathbf{e}_1, \ldots, B\mathbf{e}_n$. If $A\mathbf{x} = B\mathbf{x}$ for all $\mathbf{x}$, then we have

$$
A\mathbf{e}_1 = B\mathbf{e}_1, \ldots, A\mathbf{e}_n = B\mathbf{e}_n.
$$

Therefore $AI = BI$ or, by Rule 11, $A = B$. Conversely, if $A = B$, then $A\mathbf{x} = B\mathbf{x}$ for all $\mathbf{x}$, by the definition of equality of matrices.

REMARK: Because of the associative law (Rule 10), we can generally drop parentheses in multiple products of matrices. For example, we replace $[A(BC)]D$ by $ABCD$. No matter how we group the factors, the same result is obtained.

EXAMPLE 5 *A primitive model for heat conduction.* We consider a rod made of four similar pieces (Fig. 5–8) that are perfect conductors of heat, so that the temperature is the same throughout each piece. The two end pieces are maintained at

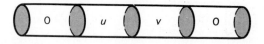

Fig. 5–8. Example 5.

temperature 0. It is assumed that heat can flow across the faces of adjacent pieces according to Newton's law of cooling, so that the rate of heat flow is proportional to the temperature difference; it is also assumed that there is no other gain or loss of heat for the inner pieces. We are led to differential equations

$$\frac{du}{dt} = -k^2(2u - v) \quad \text{and} \quad \frac{dv}{dt} = -k^2(2v - u)$$

for the temperatures u and v of the inner pieces; k^2 is a positive constant. We solve these equations numerically, letting $u_n = u(n\Delta t)$ and $v_n = v(n\Delta t)$, $n = 0, 1, 2, \ldots$, where $u_0 = u(0)$ and $v_0 = v(0)$ are assumed known. We replace du/dt by $\Delta u/\Delta t = (u_{n+1} - u_n)/\Delta t$ and dv/dt by $\Delta v/\Delta t = (v_{n+1} - v_n)/\Delta t$; then we get equations

$$u_{n+1} = (1 - 2k^2\Delta t)u_n + k^2\Delta t v_n \quad \text{and} \quad v_{n+1} = k^2\Delta t u_n + (1 - 2k^2\Delta t)v_n.$$

We assume that, in appropriate units, $k^2\Delta t = 1/9$ and then obtain the matrix equation

$$\begin{bmatrix} u_{n+1} \\ v_{n+1} \end{bmatrix} = A \begin{bmatrix} u_n \\ v_n \end{bmatrix} \quad \text{with} \quad A = \begin{bmatrix} \frac{7}{9} & \frac{1}{9} \\ \frac{1}{9} & \frac{7}{9} \end{bmatrix}.$$

Hence

$$\begin{bmatrix} u_1 \\ v_1 \end{bmatrix} = A \begin{bmatrix} u_0 \\ v_0 \end{bmatrix}, \quad \begin{bmatrix} u_2 \\ v_2 \end{bmatrix} = A \begin{bmatrix} u_1 \\ v_1 \end{bmatrix} = A^2 \begin{bmatrix} u_0 \\ v_0 \end{bmatrix}, \quad \ldots, \quad \begin{bmatrix} u_n \\ v_n \end{bmatrix} = A^n \begin{bmatrix} u_0 \\ v_0 \end{bmatrix}.$$

We find

$$A^2 = \frac{1}{81} \begin{bmatrix} 50 & 14 \\ 14 & 50 \end{bmatrix} \quad \text{and} \quad A^3 = \frac{1}{729} \begin{bmatrix} 364 & 148 \\ 148 & 364 \end{bmatrix}, \ldots$$

Thus if $u_0 = 2$ and $v_0 = 1$, then $u_1 = 1.67$, $v_1 = 1$, $u_2 = 1.4$, $v_2 = 0.96$, $u_3 = 1.2$, $v_3 = 0.91$. These calculations suggest that A^n differs very little from O for large n and hence that, no matter what the initial temperature distribution is, u_n and v_n approach 0 as $n \to \infty$. For the original physical problem it is clear that the temperatures should approach 0 as $t \to \infty$ (see Section 8–6). ◄

====================== **Problems (Section 5-5)** ======================

Let the matrices $A, \ldots, P$ be given as at the beginning of the set of problems following Section 5–4.

1. Evaluate each expression which is meaningful:

   a) AB b) CA c) AC d) CD and DC e) CE and EC
   f) AI g) IL h) $2I + 3GL$ i) $C^2 - 3C - 10C^0$ j) $E(E - 5I)$
   k) LNI l) MPG m) $HL + J$ n) KA o) $OC + N$
   p) E^2 q) E^3 r) E^4 s) $HELP$

2. Calculate RS for each of the following choices of R and S:

   a) $R = \begin{bmatrix} 3 & 1 \\ 5 & 2 \end{bmatrix}$, $S = \begin{bmatrix} 1 & 0 & 2 & 0 & 2 \\ 0 & 1 & 3 & 1 & 3 \end{bmatrix}$

   b) $R = \begin{bmatrix} 1 & 4 \\ 2 & 1 \\ 5 & 0 \end{bmatrix}$, $S = \begin{bmatrix} 2 & 1 & 1 & 2 & 1 \\ 3 & 2 & 2 & 3 & 2 \end{bmatrix}$

3. Consider each of the following pairs of simultaneous equations as cases of (5–50) and (5–51) and express $y_1, \ldots$ in terms of $x_1, \ldots$ (i) by eliminating $u_1, \ldots$ and (ii) by multiplying the coefficient matrices:

a) $\begin{cases} 3u_1 + 2u_2 = y_1 \\ 5u_1 + 6u_2 = y_2 \end{cases}$ and $\begin{cases} 5x_1 - x_2 = u_1 \\ x_1 + 2x_2 = u_2 \end{cases}$

b) $\begin{cases} 2u_1 - u_2 = y_1 \\ 5u_1 + u_2 = y_2 \end{cases}$ and $\begin{cases} x_1 + 2x_2 - x_3 = u_1 \\ 2x_1 + 3x_2 + x_3 = u_2 \end{cases}$

4. Many physical systems are described by differential equations of the form

$$\frac{d^2 x}{dt^2} + a(t)x = 0.$$

When $a(t)$ is periodic, this is *Hill's equation*, which arose in studying the motion of the moon. A numerical approximation to a solution with given initial values $x(0)$ and $x'(0)$ is obtained by solving the equation for $x_n = x(n\Delta t)$:

$$x_{n+1} - 2x_n + x_{n-1} + k_n x_n = 0,$$

where $k_n = \overline{\Delta t^2}\, a(n\Delta t)$, and $x_0 = x(0)$, $x_1 = x(\Delta t)$ are assumed known (we take $x_1 = x_0 + x'(0)\Delta t$ to get an approximation to x_1).

a) Show that the values x_n satisfy the matrix equation

$$\begin{bmatrix} x_n \\ x_{n+1} \end{bmatrix} = A_n \begin{bmatrix} x_{n-1} \\ x_n \end{bmatrix}, \quad n = 1, 2, 3, \ldots,$$

where

$$A_n = \begin{bmatrix} 0 & 1 \\ -1 & 2 - k_n \end{bmatrix},$$

and that accordingly

$$\begin{bmatrix} x_n \\ x_{n+1} \end{bmatrix} = A_n A_{n-1} \cdots A_1 \begin{bmatrix} x_0 \\ x_1 \end{bmatrix}.$$

b) Apply the method with $\Delta t = 0.25$ to obtain $x(1/2)$, $x(3/4)$, $x(1)$ for the solution of the Hill equation (here a *Mathieu equation*)

$$\frac{d^2 x}{dt^2} + (2 - \cos \pi t)x = 0$$

with $x(0) = 0$, $x'(0) = 4$.

c) To check the validity of the method, use it on the equation with constant $a(t)$

$$\frac{d^2 x}{dt^2} + 4x = 0.$$

Show first that the solution is given by

$$\begin{bmatrix} x_n \\ x_{n-1} \end{bmatrix} = A^n \begin{bmatrix} x_0 \\ x_1 \end{bmatrix}, \quad A = \begin{bmatrix} 0 & 1 \\ -1 & 2 - 4\Delta t^2 \end{bmatrix}.$$

Take $x_0 = 0$, $x'(0) = 2$, $\Delta t = 0.1$ and obtain $x(0.2)$, $x(0.3)$, $x(0.4)$, $x(0.5)$. Compare with the exact solution $x = \sin 2t$.

5. Prove each of the following rules of Section 5–5:
   a) Rule 12   b) Rule 14
   c) Rule 18, by induction with respect to l
   d) Rule 19, by induction with respect to l and Rule 18

6. Let A be a square matrix. Prove that
   a) $A^2 - I = (A + I)(A - I)$
   b) $A^3 - I = (A - I)(A^2 + A + I)$
   c) $A^2 - 2A - 3I = (A - 3I)(A + I)$
   d) $6A^2 - A - 2I = (2A + I)(3A - 2I)$

7. If A and B are $n \times n$ matrices, is $A^2 - B^2$ necessarily equal to $(A - B) \times (A + B)$? When must this be true?

8. Find nonzero 2×2 matrices A and B such that $A^2 + B^2 = O$.

5–6 INVERSE OF A SQUARE MATRIX

Let A be an $n \times n$ matrix. If an $n \times n$ matrix B exists such that $AB = I$, then we call B an *inverse of A*. We shall see below that A can have at most one inverse. Hence, if $AB = I$, we call B *the* inverse of A and write $B = A^{-1}$.

For a general $n \times n$ matrix A we denote by det A the determinant formed from A; that is,

$$\det A = \begin{vmatrix} a_{11} & \cdots & a_{1n} \\ \vdots & & \vdots \\ a_{n1} & \cdots & a_{nn} \end{vmatrix}. \tag{5–60}$$

We stress that det A is a number, whereas A itself is a square array, that is, a matrix. If A and B are $n \times n$ matrices, then we have the rule

$$\det A \, \det B = \det (AB). \tag{5–61}$$

For example,

$$\begin{vmatrix} a & b \\ c & d \end{vmatrix} \begin{vmatrix} p & q \\ r & s \end{vmatrix} = (ad - bc)(ps - qr)$$

$$= adps + bcqr - adqr - bcps$$

$$= \begin{vmatrix} ap + br & aq + bs \\ cp + dr & cq + ds \end{vmatrix},$$

as can be verified by multiplying out (four terms cancel). (For a proof of rule (5–61) for any n, see CLA, Sections 10–13 and 10–14.)

From this rule it follows that if A has an inverse, then det $A \neq 0$, since $AB = I$ implies that

$$\det A \, \det B = \det I = 1,$$

so that det $A \neq 0$; also det $B = \det A^{-1} \neq 0$ and, in fact,

$$\det B = \det A^{-1} = \frac{1}{\det A}. \tag{5–62}$$

Conversely, *if* det $A \neq 0$, *then A has an inverse*. For if det $A \neq 0$, then the simultaneous linear equations

$$a_{11}x_1 + \cdots + a_{1n}x_n = y_1,$$
$$\vdots \qquad\qquad \vdots \tag{5–63}$$
$$a_{n1}x_1 + \cdots + a_{nn}x_n = y_n$$

can be solved for $x_1, \ldots, x_n$ by Cramer's rule (Section 5–1). For example,

$$
x_1 = \begin{vmatrix} y_1 & a_{12} & \cdots & a_{1n} \\ \vdots & \vdots & & \vdots \\ y_n & a_{n2} & \cdots & a_{nn} \end{vmatrix} \div D,
$$

where $D = \det A$. Upon expanding the first determinant on the right by minors of the first column (see Section 5–2), we obtain an expression of the form

$$x_1 = b_{11}y_1 + \cdots + b_{1n}y_n,$$

with appropriate constants $b_{11}, \ldots, b_{1n}$. In general,

$$x_i = b_{i1}y_1 + \cdots + b_{in}y_n, \quad i = 1, \ldots, n. \tag{5–64}$$

Now our given equations (5–63) are equivalent to the matrix equation

$$A\mathbf{x} = \mathbf{y},$$

and the solution (5–64) is given by

$$\mathbf{x} = B\mathbf{y}.$$

The fact that this is a solution is expressed by the relation

$$AB\mathbf{y} = \mathbf{y} \quad \text{or} \quad AB\mathbf{y} = I\mathbf{y}.$$

This relation holds for all $\mathbf{y}$. Hence by Rule 20 of Section 5–5 we must have $AB = I$, so that B is an inverse of A.

The reasoning just given also provides a constructive way of finding A^{-1}. We simply form the equations (5–63) and solve for $x_1, \ldots, x_n$. The solution can be written as $\mathbf{x} = B\mathbf{y}$, where $B = A^{-1}$.

EXAMPLE 1 $A = \begin{bmatrix} 2 & 5 \\ 1 & 3 \end{bmatrix}$. The simultaneous equations are

$$2x_1 + 5x_2 = y_1, \qquad x_1 + 3x_2 = y_2.$$

We solve by elimination and find

$$x_1 = 3y_1 - 5y_2, \qquad x_2 = -y_1 + 2y_2.$$

Therefore, $A^{-1} = \begin{bmatrix} 3 & -5 \\ -1 & 2 \end{bmatrix}$. We check by verifying that $AA^{-1} = I$. ◄

Nonsingular matrices

A square matrix A having an inverse is said to be *nonsingular*. Hence we have shown that A is nonsingular precisely when $\det A \neq 0$. A square matrix having *no* inverse is said to be *singular*.

Now let A have an inverse B, so that $AB = I$. Then as remarked above, also $\det B \neq 0$, so that B also has an inverse B^{-1}, and $BB^{-1} = I$. We can now write

$$BA = BAI = BABB^{-1} = B(AB)B^{-1} = BIB^{-1} = BB^{-1} = I.$$

Therefore, also $BA = I$. Furthermore, if $AC = I$, then

$$C = IC = BAC = BI = B.$$

This shows that the inverse of A is unique. Furthermore, if $CA = I$, then
$$C = CI = CAB = IB = B.$$
These results can be summarized as follows: *the inverse $B = A^{-1}$ of A is unique and B satisfies the two equations*
$$AB = I \quad \text{and} \quad BA = I;$$
furthermore, if a matrix B satisfies either one of these two equations, then B must satisfy the other equation and $B = A^{-1}$.

The inverse satisfies several additional rules:

21. $(AD)^{-1} = D^{-1}A^{-1}.$ 22. $(cA)^{-1} = c^{-1}A^{-1} \ (c \neq 0).$
23. $(A^{-1})^{-1} = A.$

Here A and D are assumed to be nonsingular $n \times n$ matrices. To prove Rule 21, we write
$$(AD)(D^{-1}A^{-1}) = A(DD^{-1})A^{-1} = AIA^{-1} = AA^{-1} = I.$$
Therefore $D^{-1}A^{-1}$ must be the inverse of AD. The proof of Rule 22 is left as an exercise (Problem 5 below). For Rule 23, we reason that A^{-1} is nonsingular and hence A^{-1} has an inverse. But $A^{-1}A = I$, so that A is the inverse of A^{-1}; that is, $A = (A^{-1})^{-1}.$

Rule 21 extends to more than two factors: for example,
$$(ABCD)^{-1} = D^{-1}C^{-1}B^{-1}A^{-1}.$$
The proof is as above. In this way we see that the product of two or more nonsingular matrices is nonsingular.

Negative powers of a square matrix

Let A be nonsingular, so that A^{-1} exists. For each positive integer p, we now define A^{-p} to mean $(A^{-1})^p$. Since A is nonsingular, A^p is also nonsingular; in fact, A^p has the inverse
$$(AA \cdots A)^{-1} = A^{-1}A^{-1} \cdots A^{-1} = (A^{-1})^p.$$
Therefore,
$$A^{-p} = (A^p)^{-1} \quad \text{or} \quad A^p A^{-p} = I. \tag{5–65}$$
Rules 18 and 19,
$$A^p A^q = A^{p+q}, \tag{5–66}$$
$$(A^p)^q = A^{pq}, \tag{5–67}$$
are now satisfied, for A nonsingular, for arbitrary integers p and q, positive, negative, or zero. These rules are proved for p and q nonnegative in Section 5–5. To prove Rule (5–66) in the general case, we observe first that if p or q is 0, one factor on the left equals I and hence the rule is correct. If for example $p < 0, q > 0$, and $p + q > 0$, then
$$A^{-p}A^{p+q} = A^q$$
by the rule for the case of positive exponents. By (5–65) it follows that
$$A^p A^{-p} A^{p+q} = A^p A^q \quad \text{or} \quad IA^{p+q} = A^p A^q.$$
Thus Eq. (5–66) is verified in this case. The other cases are handled in the same way.

We note that, by (5–66), $A^p A^q = A^q A^p$. Hence A^p and A^q commute (that is, obey the commutative law) under multiplication. The proof of (5–67) is left as an exercise (Problem 6 below).

The procedure of Example 1 for finding the inverse of a matrix can be much improved. The calculation of inverses of matrices is very important for applications and various procedures have been devised for this purpose, especially adapted to digital computers (see Section 5–14).

Inverses can be used to solve matrix equations, for example, equations of the form $AX = B$ or $XA = B$, where A and B are known and X is sought; if A is $n \times n$ and nonsingular, then we find, respectively,

$$X = A^{-1}B \qquad \text{and} \qquad X = BA^{-1},$$

and verify in each case that the equation is satisfied. As a special case, we solve the equation $Ax = y$ (A being $n \times n$) to obtain $x = A^{-1}y$.

EXAMPLE 2 In seeking the equilibrium temperature $u(x, y)$ in a plane region (see Section 8–11), we have to solve a boundary-value problem for the partial-differential equation (Laplace equation)

$$\frac{\partial^2 u}{\partial x^2} + \frac{\partial^2 u}{\partial y^2} = 0.$$

A numerical method for such a problem replaces the partial-differential equation by the equation

$$\frac{u(x+h, y) + u(x, y+h) + u(x-h, y) + u(x, y-h) - 4u(x, y)}{h^2} = 0.$$

If the region concerned is a square, as in Fig. 5–9, then the unknown values u_1, u_2, u_3, u_4 are related to the boundary values $v_1, \ldots, v_8$ by the equations

$$4u_1 - u_2 - u_4 = v_1 + v_8 = w_1,$$
$$-u_1 + 4u_2 - u_3 = v_2 + v_3 = w_2,$$
$$-u_2 + 4u_3 - u_4 = v_4 + v_5 = w_3,$$
$$-u_1 - u_3 - 4u_4 = v_6 + v_7 = w_4.$$

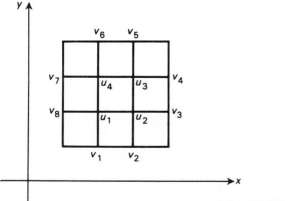

Fig. 5–9. Numerical solution of Laplace equation.

Hence we have an equation of the form $A\mathbf{u} = \mathbf{w}$, where $\mathbf{w}$ is known and, if A is nonsingular, the solution is $\mathbf{u} = A^{-1}\mathbf{w}$. We find in fact

$$A^{-1} = \frac{1}{24}\begin{bmatrix} 7 & 2 & 1 & 2 \\ 2 & 7 & 2 & 1 \\ 1 & 2 & 7 & 2 \\ 2 & 1 & 2 & 7 \end{bmatrix}.$$

Thus the approximating boundary-value problem has a unique solution. ◄

=========== **Problems (Section 5–6)** ===========

1. Verify that each of the following matrices is nonsingular and find the inverse of each:

a) $A = \begin{bmatrix} 3 & 5 \\ 2 & 4 \end{bmatrix}$ b) $B = \begin{bmatrix} 4 & 7 \\ 1 & 6 \end{bmatrix}$ c) $C = \begin{bmatrix} 1 & 0 & 1 \\ 2 & 2 & 1 \\ 0 & 1 & -1 \end{bmatrix}$ d) $D = \begin{bmatrix} 2 & 0 & 1 \\ 3 & 1 & 2 \\ 4 & 0 & 3 \end{bmatrix}$

2. Let $A, \ldots, D$ be as in Problem 1. With the aid of the answers to Problem 1, solve for X or $\mathbf{x}$:

a) $AX = \begin{bmatrix} 0 & 1 \\ 1 & 0 \end{bmatrix}$ b) $AX = \begin{bmatrix} 1 & 6 \\ 5 & 2 \end{bmatrix}$ c) $BXA = \begin{bmatrix} 7 & 2 \\ 0 & 5 \end{bmatrix}$

d) $B^2XA^2 = \begin{bmatrix} 4 & 3 \\ 2 & 1 \end{bmatrix}$ e) $A\mathbf{x} = \mathrm{col}\,(4, 3)$ f) $B\mathbf{x} = \mathrm{col}\,(3, 6)$

g) $BX = \begin{bmatrix} 1 & 6 & 2 \\ 5 & 0 & -3 \end{bmatrix}$ h) $XC = \begin{bmatrix} 0 & 4 & 5 \\ 4 & 0 & 1 \end{bmatrix}$ i) $\mathbf{x}C = (1, 2, 3)$

j) $\mathbf{x}D = (1, 2, 0)$

3. Simplify:

a) $[(AB)^{-1}A^{-1}]^{-1}$ b) $[C(ABC)^{-1}ABC^{-1}]^{-1}$
c) $\{[(A^{-1})^2B]^{-1}A^{-2}B^{-1}\}^{-2}$

4. Prove:
   a) If A and B are nonsingular and $AB = BA$, then $A^{-1}B^{-1} = B^{-1}A^{-1}$.
   b) If $ABC = I$, then $BCA = I$ and $CAB = I$.

5. Prove the Rule 22.

6. Prove the rule (5–67) for arbitrary integers p, q.   HINT: For $s \geq 0, t \geq 0$, show by (5–65) that $(A^s)^{-t} = A^{-st}$ and that $(A^{-s})^{-t} = A^{st}$.

7. Let $A = (a_{ij})$ be a nonsingular square matrix of order n. Prove:

a) If $n = 2$, then $A^{-1} = \dfrac{1}{\det A}\begin{bmatrix} a_{22} & -a_{12} \\ -a_{21} & a_{11} \end{bmatrix}.$

b) If $n = 3$, then A^{-1} is the matrix

$$\frac{1}{\det A}
\begin{bmatrix}
\begin{vmatrix} a_{22} & a_{23} \\ a_{32} & a_{33} \end{vmatrix} & \begin{vmatrix} a_{13} & a_{12} \\ a_{33} & a_{32} \end{vmatrix} & \begin{vmatrix} a_{12} & a_{13} \\ a_{22} & a_{23} \end{vmatrix} \\[12pt]
\begin{vmatrix} a_{23} & a_{21} \\ a_{33} & a_{31} \end{vmatrix} & \begin{vmatrix} a_{11} & a_{13} \\ a_{31} & a_{33} \end{vmatrix} & \begin{vmatrix} a_{13} & a_{11} \\ a_{23} & a_{21} \end{vmatrix} \\[12pt]
\begin{vmatrix} a_{21} & a_{22} \\ a_{31} & a_{32} \end{vmatrix} & \begin{vmatrix} a_{12} & a_{11} \\ a_{32} & a_{31} \end{vmatrix} & \begin{vmatrix} a_{11} & a_{12} \\ a_{21} & a_{22} \end{vmatrix}
\end{bmatrix}.$$

c) Let A_{ij} denote the minor determinant of A obtained by deleting the ith row and jth column from A. Let $b_{ij} = (-1)^{i+j} A_{ji}$. The matrix $B = (b_{ij})$ is called the *adjoint* of A and is denoted by adj A. Show that

$$A^{-1} = \frac{1}{\det A} \, \text{adj } A.$$

8. Let all matrices occurring in the following equations be *square* of order n. Solve for X and Y, stating which matrices are assumed to be nonsingular:

a) $X + Y = A, \quad X - Y = B$ b) $X - Y = A, \quad X + B Y = C$
c) $X + A Y = B, \quad X + C Y = D$ d) $2AX - Y = B, \quad X + 3A Y = C$
e) $AX + B Y = C, \quad DX + E Y = F$ f) $X B + YA = C, \quad X E + YD = F$

9. With reference to Example 2 in Section 5–6,

a) Assume $v_1 = 2, v_2 = 0, v_3 = -1, v_4 = 5, v_5 = 3, v_6 = 2, v_7 = 0, v_8 = -2$ and find u_1, u_2, u_3, u_4.

b) Show that we can write $\mathbf{u} = A^{-1} B\mathbf{v}$ in terms of an appropriate matrix B.

c) Show that the average $u_1, \ldots, u_4$ equals the average of $v_1, \ldots, v_8$. Also show that none of $u_1, \ldots, u_4$ can exceed the maximum of $v_1, \ldots, v_8$. When can one of $u_1, \ldots, u_4$ equal the maximum of $v_1, \ldots, v_8$?

d) Verify that A^{-1} has the value given.

10. Various physical problems lead to boundary-value problems of the form

$$y''(x) + p(x)y'(x) + q(x)y(x) = r(x), \quad a \leqslant x \leqslant b, \tag{5–68}$$
$$y(a) = y_a, \qquad y(b) = y_b,$$

where y_a, y_b are given. For example, an ideal transmission line with no inductance or capacitance satisfies the equation

$$V''(x) - RGV(x) = 0,$$

where V is the voltage drop at x, R the resistance, G the leakage conductance. Here R, G could vary with x.

A numerical solution of the problem (5–68) can be sought by replacing the derivatives by difference quotients (see Chapter 13 for more discussion). In this way we are led to the equation

$$\frac{y_{n+1} - 2y_n + y_{n-1}}{\Delta x^2} + p(x_n)\frac{y_{n+1} - y_{n-1}}{2\Delta x} + q(x_n)y_n = r(x_n), \quad n = 1, \ldots, N,$$

where $y_n = y(a + n\Delta x) = y(x_n), 0 \leqslant n \leqslant N+1, \Delta x = (b-a)/(N+1)$. These are N simultaneous linear equations for N unknowns $y_1, \ldots, y_N$; they can be written in the form

$A\mathbf{y} = \mathbf{c}$. Solution for arbitrary $\mathbf{c}$ requires finding A^{-1}.

Apply the method to the problem

$$y'' - x^2 y = 0, \qquad y(0) = \alpha, \qquad y(4) = \beta, \qquad 0 \leqslant x \leqslant 4,$$

taking $\Delta x = 1$. Show that

$$A = \begin{bmatrix} -3 & 1 & 0 \\ 1 & -6 & 1 \\ 0 & 1 & -11 \end{bmatrix},$$

find A^{-1} and find the approximate solution for $\alpha = 0$, $\beta = 1$.

5–7 EIGENVALUES OF A SQUARE MATRIX

Let A be an $n \times n$ matrix. For some *nonzero* column vector $\mathbf{v} = \text{col}\,(v_1, \ldots, v_n)$ it may happen that, for some scalar λ,

$$A\mathbf{v} = \lambda\mathbf{v}. \tag{5–70}$$

If this occurs, we say that λ is an *eigenvalue* of A and that $\mathbf{v}$ is an *eigenvector* of A, associated with the eigenvalue λ. The concept of eigenvalue has important applications in engineering. In vibrations problems, eigenvalues (or their imaginary parts) arise as the allowed frequencies of motion, as, for example, in a function

$$x = A_1 \sin(\lambda_1 t + \alpha_1) + \cdots + A_k \sin(\lambda_k t + \alpha_k).$$

The boundary problems of Chapter 8 typically lead to problems of form (5–70) but in the more general setting of differential operators applied to functions; however, numerical methods for finding approximate solutions of such problems usually reduce them to problems of form (5–70) for matrices and vectors (see Chapter 13).

We can write Eq. (5–70) in the form $A\mathbf{v} = \lambda I\mathbf{v}$ or in the form

$$(A - \lambda I)\mathbf{v} = \mathbf{0}.$$

Thus $\mathbf{v} = \text{col}\,(v_1, \ldots, v_n)$ is an eigenvector of A associated with the eigenvalue λ precisely when $v_1, \ldots, v_n$ form a nontrivial solution of the set of homogeneous linear equations

$$
\begin{aligned}
(a_{11} - \lambda)v_1 + a_{12}v_2 + \cdots + \quad & a_{1n}v_n = 0, \\
a_{21}v_1 + (a_{22} - \lambda)v_2 + \cdots + \quad & a_{2n}v_n = 0, \\
\vdots \qquad\qquad\qquad & \quad \vdots \\
a_{n1}v_1 + a_{n2}v_2 \qquad + \cdots + (a_{nn} - \lambda)v_n &= 0.
\end{aligned}
\tag{5–71}
$$

Now we know (Section 5–1) that Eqs. (5–71) have a nontrivial solution precisely when the determinant of the coefficients is 0; that is, when $\det(A - \lambda I) = 0$ or

$$
\begin{vmatrix}
a_{11} - \lambda & a_{12} & \cdots & a_{1n} \\
a_{21} & a_{22} - \lambda & \cdots & a_{2n} \\
\vdots & \vdots & & \vdots \\
a_{n1} & a_{n2} & \cdots & a_{nn} - \lambda
\end{vmatrix} = 0.
\tag{5–72}
$$

When expanded, (5–72) becomes an algebraic equation of degree n for λ, called the *characteristic equation* of the matrix A. The eigenvalues of A are simply the real roots of the characteristic equation. (The complex roots of the characteristic equation can also be interpreted as eigenvalues of A; see below.)

EXAMPLE 1   Let $A = \begin{bmatrix} 1 & 2 \\ 3 & 2 \end{bmatrix}$. Then the characteristic equation is

$$\begin{vmatrix} 1-\lambda & 2 \\ 3 & 2-\lambda \end{vmatrix} = 0 \quad \text{or} \quad \lambda^2 - 3\lambda - 4 = 0 \quad \text{or} \quad (\lambda - 4)(\lambda + 1) = 0.$$

The roots are $\lambda_1 = 4$ and $\lambda_2 = -1$. To find $\mathbf{v}$ for $\lambda = \lambda_1 = 4$, we form the equations

$$(1-4)v_1 + 2v_2 = 0, \qquad 3v_1 + (2-4)v_2 = 0,$$

and find that the solutions are all vectors $k(2, 3)$. Hence the eigenvectors associated with the eigenvalue $\lambda_1 = 4$ are all vectors $k(2, 3)$ for nonzero k. In the same way we find that all eigenvectors associated with the eigenvalue $\lambda_2 = -1$ are all vectors $k(1, -1)$ for nonzero k. ◄

EXAMPLE 2   Let

$$B = (\lambda_i \delta_{ij}) = \begin{bmatrix} \lambda_1 & 0 & \cdots & 0 \\ \vdots & \vdots & & \vdots \\ 0 & 0 & \cdots & \lambda_n \end{bmatrix}.$$

We call B a *diagonal* matrix and write $B = \mathrm{diag}\,(\lambda_1, \ldots, \lambda_n)$. Then B has the characteristic equation

$$\begin{vmatrix} \lambda_1 - \lambda & \cdots & 0 \\ \vdots & & \vdots \\ 0 & \cdots & \lambda_n - \lambda \end{vmatrix} = 0 \quad \text{or} \quad (\lambda_1 - \lambda)(\lambda_2 - \lambda) \ldots (\lambda_n - \lambda) = 0.$$

Hence, B has the eigenvalues $\lambda_1, \ldots, \lambda_n$. We leave to Problem 5 below the discussion of the eigenvectors of B. We remark here that, when $\lambda_1, \ldots, \lambda_n$ are n different numbers, the eigenvectors associated with the eigenvalue λ_k are the vectors $c(0, \ldots, 0, 1, 0, \ldots, 0)$, with 1 as kth entry and c nonzero; that is, the vectors $c\mathbf{e}_k$ for $c \neq 0$, where $\mathbf{e}_k$ is the kth column vector of the identity matrix I. ◄

EXAMPLE 3   We consider coupled springs depicted in Fig. 5–10. As in Problem 5 of Section 1–15, we obtain the differential equations

$$m_1 \frac{d^2 x_1}{dt^2} = -k_1 x_1 + k_2 (x_2 - x_1),$$

$$m_2 \frac{d^2 x_2}{dt^2} = -k_2 (x_2 - x_1),$$

where m_1, m_2, k_1, k_2 are positive constants. It is observed that the two masses can oscillate synchronously in simple harmonic motion, so that

$$x_1 = p \sin (\lambda t + \varepsilon) \qquad \text{and} \qquad x_2 = q \sin (\lambda t + \varepsilon)$$

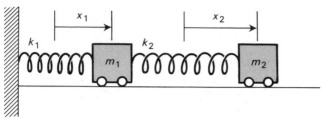

Fig. 5–10. Coupled springs.

for appropriate constants p, q, λ and arbitrary phase shift ε. By substituting these expressions in the differential equations, we find that

$$(k_1 + k_2 - m_1 \lambda^2)p - k_2 q = 0,$$
$$-k_2 p + (k_2 - m_2 \lambda^2)q = 0.$$

These are equivalent to the matrix equation

$$\begin{bmatrix} \dfrac{k_1 + k_2}{m_1} & -\dfrac{k_2}{m_1} \\[2mm] -\dfrac{k_2}{m_2} & \dfrac{k_2}{m_2} \end{bmatrix} \begin{bmatrix} p \\ q \end{bmatrix} = \lambda^2 \begin{bmatrix} p \\ q \end{bmatrix},$$

that is, col (p, q) is an eigenvector of the matrix, corresponding to the eigenvalue λ^2. The characteristic equation (with λ^2 instead of λ) is

$$m_1 m_2 \lambda^4 - (k_2 m_1 + k_1 m_2 + k_2 m_2)\lambda^2 + k_1 k_2 = 0,$$

so that

$$\lambda^2 = \frac{k_2 m_1 + k_1 m_2 + k_2 m_2 \pm \sqrt{(k_2 m_1 + k_1 m_2 + k_2 m_2)^2 - 4k_1 k_2 m_1 m_2}}{2m_1 m_2}.$$

The discriminant (under the square-root sign) can be written as

$$(k_2 m_1 - k_1 m_2)^2 + 2(k_2 m_1 + k_1 m_2)k_2 m_2 + k_2^2 m_2^2$$

and hence is positive, so that the roots are real and distinct. The square root of the discriminant is also less than $k_2 m_1 + k_1 m_2 + k_2 m_2$, so that both roots are positive. Hence we have two positive choices of λ: λ_1, λ_2 (and their negatives, $-\lambda_1$, $-\lambda_2$, which can be ignored, since they are taken care of by phase shifts). We have two corresponding eigenvectors col (p_1, q_1), col (p_2, q_2). This gives two different oscillatory motions of the system:

$$x_1 = p_1 \sin(\lambda_1 t + \varepsilon_1), \qquad x_2 = q_1 \sin(\lambda_1 t + \varepsilon_1);$$
$$x_1 = p_2 \sin(\lambda_2 t + \varepsilon_2), \qquad x_2 = q_2 \sin(\lambda_2 t + \varepsilon_2).$$

The phase constants ε_1, ε_2 are arbitrary. The two synchronous oscillations are called *normal modes* of oscillation. It can be verified that an arbitrary motion of the system is a (vector) linear combination of the two normal modes, that is, it can be written as

$$x_1 = c_1 p_1 \sin(\lambda_1 t + \varepsilon_1) + c_2 p_2 \sin(\lambda_2 t + \varepsilon_2),$$
$$x_2 = c_1 q_1 \sin(\lambda_1 t + \varepsilon_1) + c_2 q_2 \sin(\lambda_2 t + \varepsilon_2).$$

The normal modes are discussed further in Section 6–8. ◀

Similar matrices

Let A and B be $n \times n$ matrices. We say that B is *similar* to A if

$$B = C^{-1}AC$$

for some nonsingular $n \times n$ matrix C. If this holds, then

$$A = CBC^{-1} = (C^{-1})^{-1}BC^{-1},$$

so that A is also similar to B. Hence we speak of similar matrices A, B.

If A and B are similar, then A and B have the same characteristic equation. For

$$\det (B - \lambda I) = \det (C^{-1}AC - \lambda I) = \det (C^{-1}AC - \lambda C^{-1}IC)$$
$$= \det C^{-1}(A - \lambda I)C$$
$$= \det C^{-1} \det (A - \lambda I) \det C = \det (A - \lambda I),$$

since $\det C^{-1} \det C = 1$. It follows also that A and B have the same eigenvalues.

Matrices with n distinct real eigenvalues

Now let the $n \times n$ matrix A have n distinct (real) eigenvalues $\lambda_1, \ldots, \lambda_n$ and let $\mathbf{v}_1, \ldots, \mathbf{v}_n$ be eigenvectors associated with $\lambda_1, \ldots, \lambda_n$, respectively: $A\mathbf{v}_i = \lambda_i \mathbf{v}_i$, $i = 1, \ldots, n$. Now let C be the matrix whose column vectors are $\mathbf{v}_1, \ldots, \mathbf{v}_n$, respectively. Write $\mathbf{v}_j = \text{col } (v_{1j}, \ldots, v_{nj})$ for $j = 1, \ldots, n$. Then

$$AC = A \begin{bmatrix} v_{11} & \cdots & v_{1n} \\ \vdots & & \vdots \\ v_{n1} & \cdots & v_{nn} \end{bmatrix} = \begin{bmatrix} \lambda_1 v_{11} & \cdots & \lambda_n v_{1n} \\ \vdots & & \vdots \\ \lambda_1 v_{n1} & \cdots & \lambda_n v_{nn} \end{bmatrix}$$

$$= \begin{bmatrix} v_{11} & \cdots & v_{1n} \\ \vdots & & \vdots \\ v_{n1} & \cdots & v_{nn} \end{bmatrix} \begin{bmatrix} \lambda_1 & \cdots & 0 \\ \vdots & & \vdots \\ 0 & \cdots & \lambda_n \end{bmatrix} = CB,$$

where B is a diagonal matrix, $B = \text{diag } (\lambda_1, \ldots, \lambda_n)$. It can be shown that C must be nonsingular (see Problem 12 following Section 5–9). Hence $B = C^{-1}AC$ and A is similar to the diagonal matrix $\text{diag } (\lambda_1, \ldots, \lambda_n)$. ◄

EXAMPLE 4 $A = \begin{bmatrix} 1 & 2 & 2 \\ 2 & 3 & -2 \\ -5 & 3 & 8 \end{bmatrix}$. Here A has the characteristic equation

$$\begin{vmatrix} 1-\lambda & 2 & 2 \\ 2 & 3-\lambda & -2 \\ -5 & 3 & 8-\lambda \end{vmatrix} = 0 \quad \text{or} \quad (\lambda - 3)(\lambda - 4)(\lambda - 5) = 0.$$

For $\lambda = 3$, the eigenvector $\mathbf{v} = (v_1, v_2, v_3)$ must satisfy the equations

$$-2v_1 + 2v_2 + 2v_3 = 0, \quad 2v_1 + 0v_2 - 2v_3 = 0, \quad -5v_1 + 3v_2 + 5v_3 = 0.$$

Hence $v_1 = (1, 0, 1)$ is an associated eigenvector. Similarly, for $\lambda_2 = 4$ an associated eigenvector is $(2, 2, 1)$ and for $\lambda_3 = 5$ an associated eigenvector is $(0, 1, -1)$. Hence $C^{-1}AC = B$, with

$$C = \begin{bmatrix} 1 & 2 & 0 \\ 0 & 2 & 1 \\ 1 & 1 & -1 \end{bmatrix}, \qquad B = \begin{bmatrix} 3 & 0 & 0 \\ 0 & 4 & 0 \\ 0 & 0 & 5 \end{bmatrix} = \operatorname{diag}(3, 4, 5).$$

We check by verifying that $AC = CB$. Thus A is similar to the diagonal matrix B. ◄

In general, some of the eigenvalues may coincide (multiple roots of the characteristic equation), and there may be less than n real roots of the characteristic equation (5–72). To handle the general case, it is best to extend the theory of matrices to the case of *complex matrices*, whose elements are complex numbers. All concepts can be easily generalized to this case and, in particular, complex eigenvalues and complex eigenvectors can be defined as above. For example, the matrix

$$\begin{bmatrix} 1 & -2 \\ 1 & -1 \end{bmatrix}$$

has the characteristic equation $\lambda^2 + 1 = 0$. The eigenvalues are $\pm i$, and associated eigenvectors are $(2, 1 \mp i)$. We can prove, exactly as above, that when the matrix A has n distinct complex eigenvalues, A is similar to a diagonal matrix.

EXAMPLE 5   In Example 5 of Section 5–5, a primitive model for heat conduction led to the equation

$$\begin{bmatrix} u_n \\ v_n \end{bmatrix} = A^n \begin{bmatrix} u_0 \\ v_0 \end{bmatrix}, \qquad n = 1, 2, \ldots, \qquad A = \begin{bmatrix} \frac{7}{9} & \frac{1}{9} \\ \frac{1}{9} & \frac{7}{9} \end{bmatrix},$$

for temperatures $u_n = u(n\Delta t)$ and $v_n = v(n\Delta t)$. In order to show that $u_n \to 0$ and $v_n \to 0$ as $n \to \infty$, we can reason as follows. We verify that A has eigenvalues $8/9, 2/3$. Hence A is similar to $B = \operatorname{diag}(8/9, 2/3)$, that is, $A = CBC^{-1}$. Therefore,

$$A^2 = (CBC^{-1})(CBC^{-1}) = CB^2C^{-1}, \qquad A^3 = (CBC^{-1})(CB^2C^{-1}) = CB^3C^{-1},$$

and, in general, $A^n = CB^nC^{-1}$. Since B is a diagonal matrix, $B^n = \operatorname{diag}[(8/9)^n, (2/3)^n]$ and, since $(8/9)^n \to 0$ and $(2/3)^n \to 0$ when $n \to \infty$, the elements of B^n approach 0 as limit when $n \to \infty$. Therefore, since C and C^{-1} are fixed, so do the elements of $A^n = CB^nC^{-1}$. Therefore, $u_n \to 0$ and $v_n \to 0$ when $n \to \infty$. ◄

═══════════════════════ **Problems (Section 5–7)** ═══════════════════════

1. Find all eigenvalues and associated eigenvectors:

a) $\begin{bmatrix} 3 & 1 \\ 4 & 3 \end{bmatrix}$ b) $\begin{bmatrix} 4 & 1 \\ 8 & 2 \end{bmatrix}$ c) $\begin{bmatrix} 0 & 1 & -2 \\ 2 & 1 & 0 \\ 4 & -2 & 5 \end{bmatrix}$ d) $\begin{bmatrix} 11 & -4 & 8 \\ 10 & -5 & -12 \\ -4 & 2 & -3 \end{bmatrix}$

2. (a), (b), (c), (d). For each of the matrices of Problem 1, call the matrix A and find a nonsingular matrix C and a diagonal matrix B such that $A = C^{-1}BC$.

3. *Complex case.* For each of the following choices of matrix A, find all eigenvalues and associated eigenvectors, and find a diagonal matrix B such that $A = C^{-1}BC$:

a) $\begin{bmatrix} 1 & -1 \\ 4 & 1 \end{bmatrix}$ b) $\begin{bmatrix} 4 & -2 \\ 1 & 6 \end{bmatrix}$ c) $\begin{bmatrix} 0 & 0 & 0 \\ 0 & 1 & -2 \\ 0 & 1 & -1 \end{bmatrix}$

4. *Repeated roots.* Find all eigenvalues and associated eigenvectors:

a) I_3 b) O_{44} c) $\begin{bmatrix} 3 & -4 \\ 4 & -5 \end{bmatrix}$ d) $\begin{bmatrix} 0 & 1 & -2 \\ -6 & 5 & -4 \\ 0 & 0 & 3 \end{bmatrix}$

5. *Diagonal matrices.*
   a) Let $B = \text{diag}\,(\lambda, \mu)$. Show that, if $\lambda \neq \mu$, then the eigenvectors associated with λ are all nonzero vectors $c(1, 0)$ and those associated with μ are all nonzero vectors $c(0, 1)$; show that, if $\lambda = \mu$, then the eigenvectors associated with λ are all nonzero vectors (v_1, v_2).
   b) Let $B = \text{diag}\,(\lambda_1, \lambda_2, \lambda_3)$ and let $\mathbf{e}_1 = (1, 0, 0)$, $\mathbf{e}_2 = (0, 1, 0)$, $\mathbf{e}_3 = (0, 0, 1)$ (column vectors). Show that if $\lambda_1, \lambda_2, \lambda_3$ are distinct, then for each λ_k the associated eigenvectors are the vectors $c\mathbf{e}_k$ for $c \neq 0$; show that if $\lambda_1 = \lambda_2 \neq \lambda_3$, then the eigenvectors associated with λ_1 are all nonzero vectors $c_1\mathbf{e}_1 + c_2\mathbf{e}_2$ and those associated with λ_3 are all nonzero vectors $c\mathbf{e}_3$; show that if $\lambda_1 = \lambda_2 = \lambda_3$, then the eigenvectors associated with λ_1 are all nonzero vectors $\mathbf{v} = (v_1, v_2, v_3)$.
   c) Let $B = \text{diag}\,(\lambda_1, \ldots, \lambda_n)$. Show that the eigenvectors associated with the eigenvalue λ_k are all nonzero vectors $\mathbf{v} = (v_1, \ldots, v_n)$ such that $v_i = 0$ for all i such that $\lambda_i \neq \lambda_k$.

6. Let A and B be similar $n \times n$ matrices with $A = C^{-1}BC$.
   a) Prove that $\det A = \det B$.
   b) The *trace* of a square matrix is defined as the sum of the diagonal terms; the trace of $A = (a_{ij})$ is $a_{11} + a_{22} + \cdots + a_{nn}$. Show that A and B have equal traces.

   HINT: Consider the coefficient of λ^{n-1} in the characteristic equation.

   c) Prove: If $\mathbf{v}$ is an eigenvector of A associated with the eigenvalue λ, then $C\mathbf{v}$ is an eigenvector of B associated with the eigenvalue λ.

7. Prove that the matrix $A = \begin{bmatrix} 1 & 1 \\ 0 & 1 \end{bmatrix}$ is not similar to a diagonal matrix.

   HINT: Show by the results of Problem 6 that the diagonal matrix would have to be I.

8. Prove:
   a) Every square matrix is similar to itself.
   b) If A is similar to B and B is similar to C, then A is similar to C.

9. Find the normal modes and graph as paths $x_1 = x_1(t)$, $x_2 = x_2(t)$ in an $x_1 x_2$-plane for Example 3 of Section 5–7, using the following choices of the constants:
   a) $k_1 = 4$, $k_2 = 9$, $m_1 = 4$, $m_2 = 9$
   b) $k_1 = 9$, $k_2 = 62$, $m_1 = 9$, $m_2 = 62$

10. *Vibrations of a square membrane.* These are governed by a partial differential equation. The occurring frequencies λ are given by the solutions of the equation

$$c^2 \left(\frac{\partial^2 u}{\partial x^2} + \frac{\partial^2 u}{\partial y^2} \right) + \lambda^2 u = 0,$$

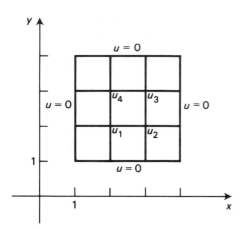

Fig. 5–11. Membrane problem.

satisfying suitable boundary conditions. For the square membrane $1 \leqslant x \leqslant 4, 1 \leqslant y \leqslant 4$ clamped along the edges (Fig. 5–11), we require that $u = 0$ along the edges. As in Example 2 of Section 5–6, we can approximate the partial-differential equation by a difference equation. If $c = 1$ and $h = 1$, we are led to the equations

$$u_2 + u_4 + (\lambda^2 - 4)u_1 = 0,$$
$$u_1 + u_3 + (\lambda^2 - 4)u_2 = 0,$$
$$u_2 + u_4 + (\lambda_2 - 4)u_3 = 0,$$
$$u_1 + u_3 + (\lambda^2 - 4)u_4 = 0.$$

a) Interpret λ^2 as eigenvalue of a matrix A.

b) Find the eigenvalues λ^2 of A and corresponding eigenvectors.

The exact solution of the partial differential equation with $u = 0$ on the boundary leads to a sequence of frequencies λ such that $\lambda^2 = 2\pi^2/9$, $\lambda^2 = 5\pi^2/9$, $\lambda^2 = 5\pi^2/9$, $\lambda^2 = 8\pi^2/9, \ldots, \lambda^2 = (m^2 + n^2)\pi^2/9, \ldots$, where m, n are positive integers. The crude numerical approximation gives surprisingly good approximations to the lowest frequencies.

5–8 VECTORS IN n-DIMENSIONAL SPACE R^n

The concepts of geometry can be extended to n-dimensional space denoted by R^n. The *points* of R^n are the n-tuples $(x_1, \ldots, x_n)$ of real numbers. The point P has *coordinates* $x_1, \ldots, x_n$. The *origin* O is the point $(0, \ldots, 0)$. The *distance* between $P(x_1, \ldots, x_n)$ and $Q(y_1, \ldots, y_n)$ is

$$\sqrt{(x_1 - y_1)^2 + \cdots + (x_n - y_n)^2}.$$

A *straight line* in R^n is a set given by all points $(x_1, \ldots, x_n)$ such that

$$x_1 = x_1^0 + a_1 t, \ldots, x_n = x_n^0 + a_n t, \quad -\infty < t < \infty,$$

where $x_1^0, \ldots, x_n^0$ are given coordinates of P_0 and $a_1, \ldots, a_n$ are given and are not all 0.

A vector **v** in R^n is also given by an n-tuple $(v_1, \ldots, v_n)$. This can be thought of as the vector from the origin O to the *point* $(v_1, \ldots, v_n)$. In general, for each pair of points $P(x_1, \ldots, x_n)$, $Q(y_1, \ldots, y_n)$, there is a vector

$$\overrightarrow{PQ} = (y_1 - x_1, \ldots, y_n - x_n)$$

that can be thought of as the vector represented by the directed line segment from P to Q.

The fact that vectors and points are both represented by n-tuples causes no difficulty. The context makes clear what is meant in each case. Further, it is often convenient to change from one to the other, and the notation is then of help.

The vector operations are defined as in 3-dimensional space R^3. Thus for $\mathbf{u} = (u_1, \ldots, u_n)$ and $\mathbf{v} = (v_1, \ldots, v_n)$ we have

$$\mathbf{u} \pm \mathbf{v} = (u_1 \pm v_1, \ldots, u_n \pm v_n),$$
$$c\mathbf{u} = (cu_1, \ldots, cu_n),$$
$$\mathbf{u} \cdot \mathbf{v} = u_1 v_1 + \cdots + u_n v_n,$$
$$\|\mathbf{u}\| = \sqrt{u_1^2 + \cdots + u_n^2} = \sqrt{\mathbf{u} \cdot \mathbf{u}}.$$

Thus the distance between P and Q, as defined above, is the same as $\|\overrightarrow{PQ}\|$, the length, or *norm*, of PQ.

These definitions satisfy the usual rules, such as the following:

$$\mathbf{u} + \mathbf{v} = \mathbf{v} + \mathbf{u}, \qquad \mathbf{u} + (\mathbf{v} + \mathbf{w}) = (\mathbf{u} + \mathbf{v}) + \mathbf{w},$$
$$c(\mathbf{u} + \mathbf{v}) = c\mathbf{u} + c\mathbf{v}, \qquad \mathbf{u} \cdot \mathbf{v} = \mathbf{v} \cdot \mathbf{u},$$
$$(\mathbf{u} + \mathbf{v}) \cdot \mathbf{w} = \mathbf{u} \cdot \mathbf{w} + \mathbf{v} \cdot \mathbf{w}. \tag{5-80}$$

For the dot product we have the important *Schwarz inequality*:

$$|\mathbf{u} \cdot \mathbf{v}| \leqslant \|\mathbf{u}\| \, \|\mathbf{v}\| \tag{5-81}$$

and the closely related *triangle inequality*:

$$\|\mathbf{u} + \mathbf{v}\| \leqslant \|\mathbf{u}\| + \|\mathbf{v}\|. \tag{5-82}$$

In (5-81) equality holds if and only if $\mathbf{u} = h\mathbf{v}$ or $\mathbf{v} = h\mathbf{u}$; in (5-82) equality holds if and only if $\mathbf{u} = h\mathbf{v}$ or $\mathbf{v} = h\mathbf{u}$ with $h \geqslant 0$. (The proofs of these inequalities are left as Problems 10 and 11 below.)

From the triangle inequality applied to $\mathbf{u} = \mathbf{x}$ and $\mathbf{v} = -\mathbf{y}$, we obtain the rule

$$\|\mathbf{x} - \mathbf{y}\| \leqslant \|\mathbf{x}\| + \|\mathbf{y}\|. \tag{5-83}$$

The left side can be interpreted as the distance between the points $(x_1, \ldots, x_n)$ and $(y_1, \ldots, y_n)$. The right side is the sum of the distances from $(x_1, \ldots, x_n)$ and $(y_1, \ldots, y_n)$ to the origin $(0, \ldots, 0)$. Inequality (5-83) thus asserts that the length of one side of a triangle is at most equal to the sum of the lengths of the other two sides, as in Fig. 5-12.

Because of the Schwarz inequality (5-81), the angle θ between two nonzero vectors **u** and **v** can be defined by the conditions

$$\cos \theta = \frac{\mathbf{u} \cdot \mathbf{v}}{\|\mathbf{u}\| \, \|\mathbf{v}\|}, \qquad 0 \leqslant \theta \leqslant \pi, \tag{5-84}$$

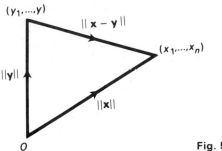

Fig. 5-12. Triangle inequality.

for by (5–81), the right-hand side lies between -1 and 1 inclusive. With the aid of distance and angle, we can now develop analytic geometry in R^n along familiar lines.

We observe from (5–84) that $\mathbf{u}$ is perpendicular to $\mathbf{v}$, that is, $\theta = \pi/2$, precisely when $\mathbf{u} \cdot \mathbf{v} = 0$.

5–9 LINEAR INDEPENDENCE

Vectors $\mathbf{v}_1, \ldots, \mathbf{v}_k$ in R^n are said to be *linearly dependent* if

$$c_1 \mathbf{v}_1 + \cdots + c_k \mathbf{v}_k = \mathbf{0} \tag{5–90}$$

for appropriate scalars $c_1, \ldots, c_k$, not all 0. If, for example, $c_1 \neq 0$, then Eq. (5–90) can be written:

$$\mathbf{v}_1 = -\frac{c_2}{c_1}\mathbf{v}_2 - \cdots - \frac{c_k}{c_1}\mathbf{v}_k.$$

Thus, when the vectors are linearly dependent, one of the vectors can be expressed as a linear combination of the others. Conversely, if one of the vectors, say $\mathbf{v}_1$, is expressible as a linear combination of the others, then the vectors $\mathbf{v}_1, \ldots, \mathbf{v}_k$ are linearly dependent since

$$\mathbf{v}_1 = k_2 \mathbf{v}_2 + \cdots + k_n \mathbf{v}_n$$

for appropriate scalars $k_2, \ldots, k_n$, and hence

$$\mathbf{v}_1 - k_2 \mathbf{v}_2 - \cdots - k_n \mathbf{v}_n = \mathbf{0},$$

which is a relation of form (5–90) with $c_1 = 1, c_2 = -k_2, \ldots, c_n = -k_n$, so that not all c_j are 0. Thus *vectors* $\mathbf{v}_1, \ldots, \mathbf{v}_k$ *are linearly dependent precisely when one of these vectors is expressible as a linear combination of the others.*

EXAMPLE 1 In R^2 (Fig. 5–13) the three vectors $\mathbf{v}_1 = 3\mathbf{i} + \mathbf{j}$, $\mathbf{v}_2 = \mathbf{i} + 3\mathbf{j}$, $\mathbf{v}_3 = 7\mathbf{i} + 5\mathbf{j}$ are linearly dependent since $\mathbf{v}_3 = 2\mathbf{v}_1 + \mathbf{v}_2$ or $2\mathbf{v}_1 + \mathbf{v}_2 - \mathbf{v}_3 = 0$.

Vectors $\mathbf{v}_1, \ldots, \mathbf{v}_k$ are said to be *linearly independent* when they are not linearly dependent. In this case relation (5–90) can hold only for $c_1 = 0, \ldots, c_k = 0$ and not one of the vectors is expressible as a linear combination of the others. Vectors $\mathbf{v}_1, \mathbf{v}_2$ of Fig. 5–13 are linearly independent since $\mathbf{v}_1 = k\mathbf{v}_2$ or $\mathbf{v}_2 = k\mathbf{v}_1$ would mean that $\mathbf{v}_1, \mathbf{v}_2$ have the same or opposite directions, and that is not the case.

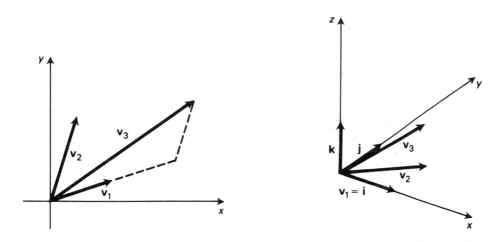

Fig. 5-13. Linearly dependent vectors.   **Fig. 5-14.** Linearly independent vectors in space.

Vectors $v_1 = i$, $v_2 = i + j$, $v_3 = i + j + k$ in space (Fig. 5–14) are linearly independent since $c_1 v_1 + c_2 v_2 + c_3 v_3 = 0$ means that

$$c_1 i + c_2(i + j) + c_3(i + j + k) = 0$$

or

$$(c_1 + c_2 + c_3)i + (c_2 + c_3)j + c_3 k = 0$$

and hence

$$c_1 + c_2 + c_3 = 0, c_2 + c_3 = 0, c_3 = 0,$$

so that $c_1 = c_2 = c_3 = 0$. We also see that v_3 cannot be a linear combination of v_1, v_2 since all such linear combinations are representable by vectors in the xy-plane; for a similar reason, v_1 is not representable as a linear combination of v_2, v_3.

The definition of linear dependence can also be applied when $k = 1$: a single vector v_1 is linearly dependent when $c_1 v_1 = 0$ with $c_1 \neq 0$, that is, *when v_1 is the zero vector*.

If any *one* of the vectors $v_1, \ldots, v_k$ is the zero vector, then $v_1, \ldots, v_k$ are linearly dependent. For if, say $v_k = 0$, then

$$0 v_1 + \cdots + 0 v_{k-1} + 1 v_k = 0.$$

In general, if some of the vectors $v_1, \ldots, v_k$ form a linearly dependent set, then $v_1, \ldots, v_k$ are linearly dependent. For example, if v_1, v_2, v_3 are linearly dependent and $k > 3$, then $c_1 v_1 + c_2 v_2 + c_3 v_3 = 0$ with not all of c_1, c_2, c_3 equal to 0, so that

$$c_1 v_1 + c_2 v_2 + c_3 v_3 + 0 v_4 + \cdots + 0 v_k = 0.$$

A very common application of linear independence is the process of *comparing coefficients.* If $\mathbf{v}_1, \ldots, \mathbf{v}_k$ are linearly independent, then

$$a_1 \mathbf{v}_1 + \cdots + a_k \mathbf{v}_k = b_1 \mathbf{v}_1 + \cdots + b_k \mathbf{v}_k \tag{5-91}$$

implies that

$$a_1 = b_1, \quad a_2 = b_2, \quad \ldots, \quad a_k = b_k,$$

for Eq. (5–91) is the same as

$$(a_1 - b_1)\mathbf{v}_1 + \cdots + (a_k - b_k)\mathbf{v}_k = \mathbf{0},$$

and by linear independence this can hold only when all coefficients are zero.

We can always find a set of n linearly independent vectors in R^n. For example, the vectors

$$\mathbf{e}_1 = (1, 0, \ldots, 0), \quad \mathbf{e}_2 = (0, 1, 0, \ldots, 0) \quad, \ldots, \quad \mathbf{e}_n = (0, \ldots, 0, 1) \tag{5-92}$$

are linearly independent, since

$$c_1 \mathbf{e}_1 + c_2 \mathbf{e}_2 + \cdots + c_n \mathbf{e}_n = (c_1, c_2, \ldots, c_n),$$

and this equals $\mathbf{0} = (0, 0, \ldots, 0)$ precisely when $c_1 = 0, \ldots, c_n = 0$. In particular, $\mathbf{i}, \mathbf{j}$ in R^2 are linearly independent and $\mathbf{i}, \mathbf{j}, \mathbf{k}$ in R^3 are linearly independent.

For a set of n vectors $\mathbf{v}_1, \ldots, \mathbf{v}_n$ in R^n we form the square matrix A whose column vectors are $\mathbf{v}_1, \ldots, \mathbf{v}_n$. Then the equation

$$c_1 \mathbf{v}_1 + \cdots + c_n \mathbf{v}_n = \mathbf{0}$$

is equivalent to a set of n homogeneous linear equations for $c_1, \ldots, c_n$, whose coefficient matrix is A; that is, it is equivalent to the matrix equation

$$A \begin{bmatrix} c_1 \\ \vdots \\ c_n \end{bmatrix} = \begin{bmatrix} 0 \\ \vdots \\ 0 \end{bmatrix} \qquad \text{or} \qquad A\mathbf{c} = \mathbf{0},$$

with $\mathbf{c} = \mathrm{col}\,(c_1, \ldots, c_n)$. But such a homogeneous system has only the trivial solution $c_1 = 0, \ldots, c_n = 0$ precisely when $\det A \neq 0$ (see Section 5–1). Thus, $\mathbf{v}_1, \ldots, \mathbf{v}_n$ *are linearly independent if and only if A is nonsingular.*

Now let $\mathbf{v}_1, \ldots, \mathbf{v}_n$ be linearly independent. Then A is nonsingular and hence $\mathbf{v}_1, \ldots, \mathbf{v}_n$ form a *basis* for vectors in R^n; that is, every vector $\mathbf{v}$ in R^n is uniquely expressible as a linear combination of $\mathbf{v}_1, \ldots, \mathbf{v}_n$. Indeed, the equation

$$\mathbf{v} = k_1 \mathbf{v}_1 + \cdots + k_n \mathbf{v}_n$$

is equivalent to

$$\mathbf{v} = A\mathbf{k}, \quad \mathbf{k} = \mathrm{col}(k_1, \ldots, k_n),$$

so that, since A is nonsingular,

$$\mathbf{k} = A^{-1}\mathbf{v}.$$

As seen above, vectors $\mathbf{e}_1 = (1, 0, \ldots, 0), \ldots, \mathbf{e}_n = (0, 0, \ldots, 0, 1)$ are linearly independent. Hence they form a basis called the *standard basis* for vectors in R^n. To express $\mathbf{v} = (v_1, \ldots, v_n)$ in terms of this basis we write simply

$$\mathbf{v} = (v_1, 0, \ldots, 0) + (0, v_2, 0, \ldots, 0) + \cdots + (0, \ldots, 0, v_n)$$

$$= v_1 \mathbf{e}_1 + v_2 \mathbf{e}_2 + \cdots + v_n \mathbf{e}_n.$$

For $n = 3$, this reduces to the familiar representation of $\mathbf{v}$ as $v_1 \mathbf{i} + v_2 \mathbf{j} + v_3 \mathbf{k}$.

We can now show that *every set of more than n vectors in R^n is linearly dependent*, for if $v_1, \ldots, v_k$ is such a set (with $k > n$), then the set is surely linearly dependent if the first n vectors $v_1, \ldots, v_n$ form a dependent set, by the reasoning above. If $v_1, \ldots, v_n$ are linearly independent, then, as just seen, they form a basis for vectors in R^n. Thus, in particular, v_{n+1} is expressible as a linear combination of $v_1, \ldots, v_n$. Accordingly, $v_1, \ldots, v_{n+1}$ form a linearly dependent set and therefore the whole set $v_1, \ldots, v_k$ is linearly dependent.

EXAMPLE 2 Test for linear independence of $v_1 = (3, 0, 2)$, $v_2 = (2, -1, 1)$, $v_3 = (5, 2, 4)$ in R^3.

Solution. Since these are three vectors in R^3, we form the matrix A whose column vectors are v_1, v_2, v_3:

$$A = \begin{bmatrix} 3 & 2 & 5 \\ 0 & -1 & 2 \\ 2 & 1 & 4 \end{bmatrix}.$$

We find that $\det A = 0$. Therefore, the vectors are linearly dependent. The equation $Ac = 0$ becomes

$$3c_1 + 2c_2 + 5c_3 = 0,$$
$$- c_2 + 2c_3 = 0,$$
$$2c_1 + c_2 + 4c_3 = 0.$$

These are satisfied if $c_1 = -3c_3$, $c_2 = 2c_3$, hence, in particular, if $c_1 = 3$, $c_2 = -2$, $c_3 = -1$. Therefore

$$3v_1 - 2v_2 - v_3 = 0. \quad \blacktriangleleft$$

EXAMPLE 3 Test for linear independence of $v_1 = (1, 1, 0, 1)$, $v_2 = (0, 1, 1, 1)$, $v_3 = (1, 0, 1, 0)$.

Solution. Equation $c_1 v_1 + c_2 v_2 + c_3 v_3 = 0$ is equivalent to the set of equations

$$c_1 + c_3 = 0, \qquad c_1 + c_2 = 0, \qquad c_2 + c_3 = 0, \qquad c_1 + c_2 = 0.$$

(These are again equivalent to $Ac = 0$, where A is the 4×3 matrix whose column vectors are v_1, v_2, v_3.) The first and second equations imply $c_2 = c_3$; the third equation then gives $c_2 = 0$, $c_3 = 0$, and hence by the first equation $c_1 = 0$. Therefore the vectors are linearly independent. ◄

We saw above that each set of n linearly independent vectors in R^n forms a basis for vectors in R^n. We can show further that these are the only bases in R^n: that is, *every basis must consist of n linearly independent vectors* (see AC, pp. 72–73).

Now let $v_1, \ldots, v_k$ be linearly independent, with $k < n$. Since $v_1, \ldots, v_k$ do not form a basis, we can find a v_{k+1} not expressible as a linear combination of $v_1, \ldots, v_k$; thus $v_1, \ldots, v_{k+1}$ are linearly independent. Continuing the process, we obtain $v_1, \ldots, v_k, v_{k+1}, \ldots, v_n$, which are linearly independent and hence do form a basis. Thus every set of linearly independent vectors in R^n can be enlarged to form a basis.

We summarize our results (including the definitions) on linear independence of vectors in R^n:

a) $\mathbf{v}_1, \ldots, \mathbf{v}_k$ are linearly dependent if
$$c_1 \mathbf{v}_1 + \cdots + c_k \mathbf{v}_k = \mathbf{0}$$
with not all of $c_1, \ldots, c_k$ equal to 0.

b) $\mathbf{v}_1, \ldots, \mathbf{v}_k$ are linearly independent (not linearly dependent) if
$$c_1 \mathbf{v}_1 + \cdots + c_k \mathbf{v}_k = \mathbf{0}$$
implies $c_1 = 0, \ldots, c_k = 0$.

c) $\mathbf{v}_1, \ldots, \mathbf{v}_k$ are linearly dependent precisely when one of these vectors is expressible as a linear combination of the others.

d) If any one of $\mathbf{v}_1, \ldots, \mathbf{v}_k$ is $\mathbf{0}$, then $\mathbf{v}_1, \ldots, \mathbf{v}_k$ are linearly dependent.

e) If some of $\mathbf{v}_1, \ldots, \mathbf{v}_k$ are linearly dependent, then the whole set $\mathbf{v}_1, \ldots, \mathbf{v}_k$ is linearly dependent.

f) If $\mathbf{v}_1, \ldots, \mathbf{v}_k$ are linearly independent, then $a_1 \mathbf{v}_1 + \cdots + a_k \mathbf{v}_k = b_1 \mathbf{v}_1 + \cdots + b_k \mathbf{v}_k$ implies $a_1 = b_1, \ldots, a_k = b_k$.

g) $\mathbf{v}_1, \ldots, \mathbf{v}_n$ are linearly independent if and only if the matrix A whose column vectors are $\mathbf{v}_1, \ldots, \mathbf{v}_n$ is nonsingular.

h) A basis for vectors consists of a set of vectors such that every vector of R^n is uniquely expressible as a linear combination of the set.

i) Every basis consists of n linearly independent vectors and every set of n linearly independent vectors forms a basis.

j) No set of linearly independent vectors in R^n has more than n vectors.

k) If $\mathbf{v}_1, \ldots, \mathbf{v}_k$ are linearly independent and $k < n$, then $\mathbf{v}_{k+1}, \ldots, \mathbf{v}_n$ can be found, so that $\mathbf{v}_1, \ldots, \mathbf{v}_n$ form a basis.

l) The vectors $\mathbf{e}_1 = (1, 0, \ldots, 0)$, $\mathbf{e}_2 = (0, 1, 0, \ldots, 0), \ldots$, $\mathbf{e}_n = (0, \ldots, 0, 1)$ form a basis called the *standard basis* for vectors in R^n.

=============== **Problems (Section 5–9)** ===============

1. Let $\mathbf{u} = (3, 2, 1, 0)$, $\mathbf{v} = (1, 0, 1, 2)$, $\mathbf{w} = (5, 4, 1, -2)$ be vectors in R^4.

   a) Find $\mathbf{u} + \mathbf{v}$, $\mathbf{u} + \mathbf{w}$, $2\mathbf{u}$, $-3\mathbf{v}$, $0\mathbf{w}$.

   b) Find $3\mathbf{u} - 2\mathbf{v}$, $2\mathbf{v} + 3\mathbf{w}$, $\mathbf{u} - \mathbf{w}$, $\mathbf{u} + \mathbf{v} - 2\mathbf{w}$.

   c) Find $\mathbf{u} \cdot \mathbf{v}$, $\mathbf{u} \cdot \mathbf{w}$, $\|\mathbf{u}\|$, $\|\mathbf{v}\|$.

   d) Show that $\mathbf{w}$ can be expressed as a linear combination of $\mathbf{u}$ and $\mathbf{v}$ and hence that $\mathbf{u}, \mathbf{v}, \mathbf{w}$ are linearly dependent.

2. In R^n the line segment $P_1 P_2$ joining points P_1, P_2 is formed of all points P such that $\overrightarrow{P_1 P} = t\overrightarrow{P_1 P_2}$, where $0 \leqslant t \leqslant 1$.

   a) In R^4, show that $(5, 10, -1, 8)$ is on the line segment from $(1, 2, 3, 4)$ to $(7, 14, -3, 10)$.

   b) In R^4, is $(2, 8, 1, 6)$ on the line segment joining $(1, 7, 2, 5)$ to $(9, 2, 0, 7)$?

   c) In R^5, find the *midpoint* of the line segment from $P_1(2, 0, 1, 3, 7)$ to $P_2(10, 4, -3, 3, 5)$, that is, find the point P on the segment such that $\|\overrightarrow{P_1 P}\| = \|\overrightarrow{P_2 P}\|$.

3. Prove the Pythagorean theorem in R^n: if $\overrightarrow{P_1 P_2}$ is orthogonal to $\overrightarrow{P_2 P_3}$, then
$$\|\overrightarrow{P_1 P_3}\|^2 = \|\overrightarrow{P_1 P_2}\|^2 + \|\overrightarrow{P_2 P_3}\|^2.$$

   HINT: Write the left side as $(\overrightarrow{P_1 P_2} + \overrightarrow{P_2 P_3}) \cdot (\overrightarrow{P_1 P_2} + \overrightarrow{P_2 P_3})$.

4. Prove the Law of cosines in R^n: if $\overrightarrow{P_1P_2}$ and $\overrightarrow{P_1P_3}$ are nonzero vectors, then

$$\|\overrightarrow{P_2P_3}\|^2 = \|\overrightarrow{P_1P_2}\|^2 + \|\overrightarrow{P_1P_3}\|^2 - 2\|\overrightarrow{P_1P_2}\|\,\|\overrightarrow{P_1P_3}\|\cos\theta,$$

where θ is the angle between $\overrightarrow{P_1P_2}$ and $\overrightarrow{P_1P_3}$.

5. Let $\mathbf{u}$, $\mathbf{v}$, $\mathbf{w}$ be linearly independent vectors in R^n. Test for linear independence:

a) $\mathbf{u} - \mathbf{v}$, $\mathbf{v} - \mathbf{w}$, $\mathbf{w} - \mathbf{u}$

b) $\mathbf{u} + \mathbf{v} + \mathbf{w}$, $\mathbf{u} + 2\mathbf{v} + 3\mathbf{w}$, $\mathbf{u} + 3\mathbf{v} + 4\mathbf{w}$

c) $\mathbf{u} + \mathbf{v}$, $\mathbf{u} - \mathbf{w}$, $\mathbf{0}$

d) $\mathbf{u} + \mathbf{v}$, $\mathbf{u} + 2\mathbf{v} + \mathbf{w}$, $\mathbf{u} - \mathbf{v} + \mathbf{w}$, $3\mathbf{u} - \mathbf{w}$

6. Let $\mathbf{u}$, $\mathbf{v}$, $\mathbf{w}$ be linearly independent vectors in R^n. Find a, b, c if

a) $3\mathbf{u} + 2\mathbf{v} + \mathbf{w} = (a - b)\mathbf{u} + (b - c)\mathbf{v} + (a + c)\mathbf{w}$

b) $(a + b)\mathbf{u} + (a + b + c)\mathbf{v} + (3a - c)\mathbf{w} = \mathbf{u} + b\mathbf{v} + c\mathbf{w}$

7. In R^n, vectors $\mathbf{v}_1, \ldots, \mathbf{v}_k$ are said to form an *orthogonal system* if none is zero and $\mathbf{v}_j \cdot \mathbf{v}_k = 0$ for $j \neq k$. Let $\mathbf{v}_1, \ldots, \mathbf{v}_k$ be such a system.

a) Show that $\mathbf{v}_1, \ldots, \mathbf{v}_k$ are linearly independent.

b) Show that $k \leqslant n$.

c) Show that if $k = n$, then $\mathbf{v}_1, \ldots, \mathbf{v}_n$ form a basis for vectors in R^n (called an *orthogonal basis*).

d) Show that $\mathbf{e}_1, \ldots, \mathbf{e}_n$ is an orthogonal basis for R^n.

8. Determine whether each set of vectors is a basis for vectors in R^n as specified:

a) In R^3: $(2, 1, 0)$, $(1, 1, 3)$, $(0, 1, 0)$

b) In R^3: $(3, 5, 2)$, $(1, 0, 4)$, $(0, 1, 1)$

c) In R^4: $(1, 0, 1, 1)$, $(1, 1, 0, 2)$, $(1, 3, 1, 2)$

d) In R^4: $(1, 3, 2, 4)$, $(1, -2, 0, 1,)$, $(0, 0, 1, 2)$, $(1, 3, 5, 6)$, $(0, 2, 4, 0)$

e) In R^4: $(1, 0, 1, 0)$, $(0, 1, 0, 1)$, $(1, 1, 1, 1)$, $(3, 1, 4, 2)$

f) In R^4: $(3, 0, 1, 2)$, $(1, 1, 0, -1)$, $(1, 4, 3, 7)$, $(0, 1, 0, 1)$

9. Prove the identities in R^n:

a) $\|\mathbf{u} - \mathbf{v}\|^2 + \|\mathbf{u} + \mathbf{v}\|^2 = 2\|\mathbf{u}\|^2 + 2\|\mathbf{v}\|^2$

b) $\|\,\|\mathbf{v}\|^2\mathbf{u} - (\mathbf{u} \cdot \mathbf{v})\mathbf{v}\|^2 = [\|\mathbf{u}\|^2\,\|\mathbf{v}\|^2 - (\mathbf{u} \cdot \mathbf{v})^2]\,\|\mathbf{v}\|^2$

10. Prove the Schwarz inequality (5–81).

HINT: If $\mathbf{v} = \mathbf{0}$, show that both sides are 0. If $\mathbf{v} \neq \mathbf{0}$, use the result of Problem 9(b) by noting that the left side in Problem 9(b) is greater than or equal to zero.

11. Prove the triangle inequality (5–82).

HINT: Apply the Schwarz inequality to $\|\mathbf{u} + \mathbf{v}\|^2 = (\mathbf{u} + \mathbf{v}) \cdot (\mathbf{u} + \mathbf{v})$.

12. Let A be an $n \times n$ square matrix with n distinct real eigenvalues $\lambda_1, \ldots, \lambda_n$ and corresponding eigenvectors $\mathbf{v}_1, \ldots, \mathbf{v}_n$ (Section 5–7). Prove that $\mathbf{v}_1, \ldots, \mathbf{v}_n$ are linearly independent.

HINT: By hypothesis $A\mathbf{v}_i = \lambda_i\mathbf{v}_i$ for $i = 1, \ldots, n$ and none of $\mathbf{v}_1, \ldots, \mathbf{v}_n$ is $\mathbf{0}$. Suppose that $c_1\mathbf{v}_1 + \cdots + c_n\mathbf{v}_n = \mathbf{0}$, apply A to this equation and deduce that $c_1(\lambda_1 - \lambda_n)\mathbf{v}_1 + \cdots + c_{n-1}(\lambda_{n-1} - \lambda_n)\mathbf{v}_{n-1} = \mathbf{0}$. Repeat the process to conclude finally that $c_1 = 0$ and similarly that $c_2 = 0, \ldots, c_n = 0$.

13. Let A be an $n \times n$ matrix with real eigenvalues $\lambda_1, \ldots, \lambda_n$, some of which may be repeated. Let $\mathbf{v}_1, \ldots, \mathbf{v}_n$ be corresponding eigenvectors and let $\mathbf{v}_1, \ldots, \mathbf{v}_n$ be linearly independent. Show that A is similar to the diagonal matrix $\text{diag}(\lambda_1, \ldots, \lambda_n)$. HINT: Follow the procedure for the case of distinct eigenvalues in Section 5–7.

5–10 LINEAR MAPPINGS FROM R^n TO R^m

At many points in the preceeding chapters and the present one, linear equations and linear operators, or mappings, have appeared. In Chapter 1, differential operators and related ones are considered (Section 1–16). In Chapter 3, the functions are expressed as linear combinations of the special functions $1, \cos x, \sin x, \ldots$ Chapter 4 is devoted to two linear operators: the Laplace transform and the Fourier transform.

Here we concentrate on linear mappings from n-dimensional space to m-dimensional space. A simple example is provided by the motion of a rigid body about a fixed point, say the origin O in 3-dimensional space R^3. For each point P of the body there is an initial position (x_1, x_2, x_3) and a final position (x_1', x_2', x_3') after the motion is completed (see Fig. 5–15). The following are typical equations relating the two sets of coordinates:

$$x_1' = \frac{2}{3}x_1 + \frac{2}{3}x_2 + \frac{1}{3}x_3,$$

$$x_2' = \frac{2}{3}x_1 - \frac{1}{3}x_2 - \frac{2}{3}x_3, \tag{5–100}$$

$$x_3' = \frac{1}{3}x_1 - \frac{2}{3}x_2 + \frac{2}{3}x_3.$$

Thus the point P originally at $(1, 1, 1)$ moves to $(5/3, -1/3, 1/3)$; point P at $(2, 5, 7)$ moves to $(7, -5, 2)$; the point O at $(0, 0, 0)$ moves to $(0, 0, 0)$, that is, the origin is

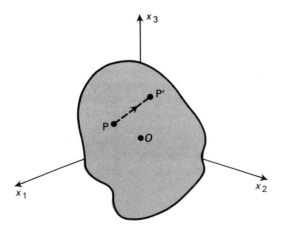

Fig. 5–15. Rigid-body motion.

fixed, as assumed. The matrix of coefficients is an example of an orthogonal matrix, (see Section 5–16); such a motion of a rigid body is always described by an orthogonal matrix.

The general case to be considered here is the system

$$a_{11}x_1 + \cdots + a_{1n}x_n = y_1,$$
$$\vdots \qquad\qquad \vdots \qquad\qquad (5\text{–}101)$$
$$a_{m1}x_1 + \cdots + a_{mn}x_n = y_m$$

or, in matrix form,

$$A\mathbf{x} = \mathbf{y}, \qquad\qquad (5\text{–}101')$$

where $\mathbf{x} = \text{col}\,(x_1, \ldots, x_n)$, $\mathbf{y} = \text{col}\,(y_1, \ldots, y_m)$. (In this section all vectors will be written as column vectors.) Through (5–101) or (5–101') a vector $\mathbf{y}$ of R^m is assigned to each vector $\mathbf{x}$ of R^n. We speak of such a correspondence as a *mapping* of R^n into R^m. If $\mathbf{x}_1$ and $\mathbf{x}_2$ are in R^n and c_1, c_2 are scalars, then

$$A(c_1\mathbf{x}_1 + c_2\mathbf{x}_2) = c_1(A\mathbf{x}_1) + c_2(A\mathbf{x}_2); \qquad\qquad (5\text{–}102)$$

that is, *the mapping (5–101') assigns to each linear combination $c_1\mathbf{x}_1 + c_2\mathbf{x}_2$ the corresponding linear combination of the values assigned to $\mathbf{x}_1$ and $\mathbf{x}_2$.* We call such a mapping linear:

DEFINITION Let T be a mapping of R^n into R^m. Then T is *linear* if, for every choice of $\mathbf{x}_1$ and $\mathbf{x}_2$ in R^n and every pair of scalars c_1, c_2,

$$T(c_1\mathbf{x}_1 + c_2\mathbf{x}_2) = c_1 T(\mathbf{x}_1) + c_2 T(\mathbf{x}_2). \qquad\qquad (5\text{–}103)$$

Thus for every $m \times n$ matrix A, the equation $\mathbf{y} = A\mathbf{x}$ defines a linear mapping T of R^n into R^m with $T(\mathbf{x}) = A\mathbf{x}$.

Conversely, *if T is a linear mapping of R^n into R^m, then there is an $m \times n$ matrix A such that $T(\mathbf{x}) = A\mathbf{x}$ for all $\mathbf{x}$ in R^n.* To prove this assertion, we first note that (5–103) implies the general rule

$$T(c_1\mathbf{x}_1 + \cdots + c_k\mathbf{x}_k) = c_1 T(\mathbf{x}_1) + \cdots + c_k T(\mathbf{x}_k). \qquad\qquad (5\text{–}104)$$

Now, let $\mathbf{e}_1, \ldots, \mathbf{e}_n$ be the standard basis in R^n and let

$$T(\mathbf{e}_j) = \mathbf{u}_j, \quad j = 1, \ldots, n. \qquad\qquad (5\text{–}105)$$

For each vector $\mathbf{x} = (x_1, \ldots, x_n) = x_1\mathbf{e}_1 + \cdots + x_n\mathbf{e}_n$ in R^n, it then follows from (5–104) that

$$T(\mathbf{x}) = x_1 T(\mathbf{e}_1) + \cdots + x_n T(\mathbf{e}_n) = x_1\mathbf{u}_1 + \cdots + x_n\mathbf{u}_n$$
$$= A\,\text{col}\,(x_1, \ldots, x_n) = A\mathbf{x},$$

where A is the $m \times n$ matrix whose column vectors are $\mathbf{u}_1, \ldots, \mathbf{u}_n$.

It follows that the study of linear mappings of R^n into R^m is equivalent to the study of mappings of the form $\mathbf{y} = A\mathbf{x}$, where A is an $m \times n$ matrix. For each linear mapping T, we call A *the matrix of T,* or *the matrix representing T.* For each matrix A, we call the corresponding linear mapping T the *linear mapping T determined by A;* we also write more concisely; the linear mapping $\mathbf{y} = A\mathbf{x}$, or even, the linear mapping A.

From the proof just given we see that for each linear mapping T, the column vectors $\mathbf{u}_1, \ldots, \mathbf{u}_n$ of the matrix A representing T are the vectors $T(\mathbf{e}_1), \ldots, T(\mathbf{e}_n)$ of R^m. Thus if $A = (a_{ij})$, then

$$A\mathbf{e}_j = \operatorname{col}(a_{1j}, \ldots, a_{mj}) = \mathbf{u}_j. \tag{5-106}$$

Furthermore, as in the proof, we can write

$$\mathbf{y} = A\mathbf{x} = A\operatorname{col}(x_1, \ldots, x_n) = x_1\mathbf{u}_1 + \cdots + x_n\mathbf{u}_n.$$

Thus, the linear mapping A assigns to each vector $\mathbf{x} = \operatorname{col}(x_1, \ldots, x_n)$ the linear combination $x_1\mathbf{u}_1 + \cdots + x_n\mathbf{u}_n$ of the column vectors

$$\mathbf{u}_1, \ldots, \mathbf{u}_n \quad \text{(vectors of } R^m\text{)}.$$

It is very helpful to think of a linear mapping as suggested in Fig. 5–16, in which the vectors are represented by directed segments from the origin; here $n = 2, m = 3$.

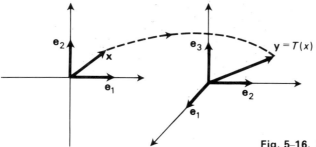

Fig. 5–16. Linear mapping of R^2 into R^3.

EXAMPLE 1 Let the linear mapping T have the matrix $A = \begin{bmatrix} 2 & 3 \\ 1 & 2 \end{bmatrix}$. Evaluate $T(\mathbf{x})$ for $\mathbf{x} = (1, 0) = \mathbf{x}_1$, $\mathbf{x} = (0, 1) = \mathbf{x}_2$, $\mathbf{x} = (2, -1) = \mathbf{x}_3$, $\mathbf{x} = (1, -1) = \mathbf{x}_4$ and graph.

Solution (see Fig. 5–17):

$$T(\mathbf{x}_1) = A\operatorname{col}(1, 0) = \operatorname{col}(2, 1), \qquad T(\mathbf{x}_2) = A\operatorname{col}(0, 1) = \operatorname{col}(3, 2),$$
$$T(\mathbf{x}_3) = A\operatorname{col}(2, -1) = \operatorname{col}(1, 0), \qquad T(\mathbf{x}_4) = A\operatorname{col}(1, -1) = \operatorname{col}(-1, -1).$$

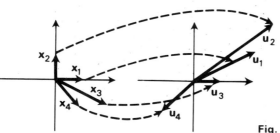

Fig. 5–17. Linear mapping of R^2 into R^2.

We note that $x_1 = e_1$ and $x_2 = e_2$, so that
$$T(x_1) = T(e_1) = u_1 = \text{col}\,(2, 1),$$
the first column vector of A. Similarly, $T(x_2) = u_2 = \text{col}\,(3, 2)$, the second column vector of A. Also $x_3 = 2e_1 - e_2$, so that
$$T(x_3) = 2u_1 - u_2 = 2(2, 1) - (3, 2) = (1, 0).$$
Similarly, $T(x_4) = T(e_1 - e_2) = u_1 - u_2 = (-1, -1)$. ◀

EXAMPLE 2   Find a linear mapping T of R^2 into R^3 such that
$$T(1, 0) = (2, 1, 2) \quad \text{and} \quad T(0, 1) = (5, 3, 7).$$

Solution. T has the matrix $A = \begin{bmatrix} 2 & 5 \\ 1 & 3 \\ 2 & 7 \end{bmatrix}$. ◀

For each particular linear mapping T of R^n into R^m, we are concerned with such properties as the following:
a) the *range* of T: the set of all y for which $T(x) = y$ for at least one x;
b) whether T maps R^n *onto* R^m: that is, whether the range of T is all of R^m;
c) whether T is one-to-one: that is, whether, for every y, $T(x) = y$ has at most one solution x;
d) the set of all x for which $T(x) = 0$; this set is called the *kernel* of T;
e) the set of all x for which $T(x) = y_0$, where y_0 is a given element of R^m.

We remark that $T(0)$ must be 0. For $T(0) = T(00) = 0T(0) = 0$. Hence *the kernel of T always contains the zero vector of R^n.* With regard to (c) and (e) we have a useful rule:

THEOREM   *Let T be a linear mapping of R^n into R^m and let $T(x_0) = y_0$ for a particular x_0 and y_0. Then all solutions x of the equation $T(x) = y_0$ are given by*
$$x = x_0 + z,$$
where z is an arbitrary vector in the kernel of T. Hence T is one-to-one precisely when the kernel of T consists of 0 alone.

PROOF: We are given that $T(x_0) = y_0$. If also $T(x) = y_0$, then
$$T(x - x_0) = T(x) - T(x_0) = y_0 - y_0 = 0.$$
Hence $x - x_0$ is an element z of the kernel of T:
$$x - x_0 = z, \quad \text{or} \quad x = x_0 + z.$$
Conversely, if $x = x_0 + z$, where z is in the kernel of T, then
$$T(x) = T(x_0) + T(z) = T(x_0) + 0 = y_0.$$
The last sentence of the theorem follows from the previous result, since T is one-to-one precisely when the equation $T(x) = y_0$ has exactly one solution for every y_0 for which there is a solution.

EXAMPLE 3 Let T map R^2 into R^3 and have the matrix

$$A = \begin{bmatrix} 2 & 4 \\ 3 & 6 \\ 1 & 2 \end{bmatrix}.$$

Then $A\mathbf{x} = \mathbf{0}$ is equivalent to

$$2x_1 + 4x_2 = 0, \qquad 3x_1 + 6x_2 = 0, \qquad x_1 + 2x_2 = 0.$$

This is satisfied for $x_1 = -2x_2$, that is, by all vectors $t(2, -1)$ of R^2. We note that A col $(2, 3) = (16, 24, 8)$; that is, $T(\mathbf{x}) = (16, 24, 8)$ is satisfied for $\mathbf{x} = \mathbf{x}_0 = (2, 3)$. Hence, all vectors $\mathbf{x}$ such that $T(\mathbf{x}) = (16, 24, 8)$ are given by

$$\mathbf{x} = (2, 3) + t(2, -1),$$

where t is arbitrary. ◄

EXAMPLE 4 T maps R^3 into R^2 and

$$T(x_1, x_2, x_3) = (y_1, y_2) = (x_1 - x_2, x_2 - x_3).$$

To find the kernel, we must solve the equations $x_1 - x_2 = 0$, $x_2 - x_3 = 0$. The solutions are given by all (x_1, x_2, x_3) such that $x_1 = x_2 = x_3$, hence, by all vectors $t(1, 1, 1)$, where t is an arbitrary scalar. Here T is not one-to-one. However, the range of T is all of R^2, since the equations

$$x_1 - x_2 = y_1, \qquad x_2 - x_3 = y_2$$

can always be solved for x_1, x_2, x_3, say, by taking $x_3 = 0$, $x_2 = y_2$, and $x_1 = y_1 + y_2$. Therefore, T maps R^3 *onto* R^2 but is not one-to-one. ◄

EXAMPLE 5 T maps R^n into R^m and $T(\mathbf{x}) = \mathbf{0}$ for every $\mathbf{x}$. Thus T has the matrix $O = O_{mn}$. We call T the *zero mapping*, and sometimes also denote this mapping by O. Here T is clearly linear and T is neither one-to-one nor onto. ◄

EXAMPLE 6 T maps R^n into R^n and $T(\mathbf{x}) = \mathbf{x}$ for every $\mathbf{x}$. Thus T has the matrix $I = I_n$. We call T the *identity mapping* and sometimes also denote this mapping by I. Here T is clearly linear, is one-to-one and onto. ◄

EXAMPLE 7 Let T map R^3 into R^3 and have the matrix

$$A = \begin{bmatrix} 2 & 3 & 1 \\ 1 & 0 & 2 \\ 1 & 2 & 0 \end{bmatrix}.$$

Find the range of T.

Solution. The range of T consists of all $\mathbf{y} = x_1\mathbf{u}_1 + x_2\mathbf{u}_2 + x_3\mathbf{u}_3$, where $\mathbf{u}_1, \mathbf{u}_2, \mathbf{u}_3$ are the column vectors of A:

$$\mathbf{u}_1 = \text{col}(2, 1, 1), \qquad \mathbf{u}_2 = \text{col}(3, 0, 2), \qquad \mathbf{u}_3 = \text{col}(1, 2, 0).$$

Thus the range consists of all linear combinations of $\mathbf{u}_1, \mathbf{u}_2, \mathbf{u}_3$. If $\mathbf{u}_1, \mathbf{u}_2, \mathbf{u}_3$ are linearly independent, then they form a basis for R^3 and the range is R^3. However, we verify that $2\mathbf{u}_1 - \mathbf{u}_2 - \mathbf{u}_3 = \mathbf{0}$, so that these vectors are linearly dependent. Furthermore, we see that $\mathbf{u}_1, \mathbf{u}_2$ are linearly independent and that $\mathbf{u}_3$ is expressible as a linear combination of $\mathbf{u}_1$ and $\mathbf{u}_2$. Hence the range is given by all linear combinations of $\mathbf{u}_1$ and $\mathbf{u}_2$, and this set is not all of R^3. Thus T does not map R^3 onto R^3 and the equation $T(\mathbf{x}) = \mathbf{y}_0$ has no solution for $\mathbf{x}$ for some choices of $\mathbf{y}_0$; for example, $T(\mathbf{x}) = (1, 0, 0)$ has no solution, as can be verified (the vector $(1, 0, 0)$ is not a linear combination of $\mathbf{u}_1$ and $\mathbf{u}_2$). ◄

Subspaces

A collection of vectors in R^n is called a *subspace* (of the *space* of all vectors in R^n), if it contains $\mathbf{0}$ and if, for each choice of scalar c and vectors $\mathbf{u}$, $\mathbf{v}$ in the collection, vectors $c\mathbf{u}$ and $\mathbf{u} + \mathbf{v}$ are also in the collection. The kernel of a linear mapping T from R^n to R^m is such a subspace, for $\mathbf{0}$ is in the kernel and, if $\mathbf{u}$ and $\mathbf{v}$ are in the kernel, then $T\mathbf{u} = \mathbf{0}$ and $T\mathbf{v} = \mathbf{0}$, so that also

$$T(\mathbf{u} + \mathbf{v}) = T\mathbf{u} + T\mathbf{v} = \mathbf{0}, \qquad T(c\mathbf{u}) = cT\mathbf{u} = \mathbf{0}.$$

Similarly, we see that the range of T is a subspace of the vectors of R^m.

The *dimension* d of a subspace is the maximum number d of linearly independent vectors in the subspace. Thus $0 \leqslant d \leqslant n$. If $\mathbf{v}_1, \ldots, \mathbf{v}_d$ are such linearly independent vectors in the subspace, then they serve as a *basis* for the subspace: every vector of the form $c_1\mathbf{v}_1 + \cdots + c_d\mathbf{v}_d$ is in the subspace and every vector of the subspace can be so represented for a unique choice of $c_1, \ldots, c_d$. Every basis of the subspace consists of d linearly independent vectors, as in property (i) of Section 5–9. If $d = 0$, the subspace consists of $\mathbf{0}$ alone. If $d = n$, the subspace consists of *all* vectors in R^n, since then we have $\mathbf{v}_1, \ldots, \mathbf{v}_n$, a basis for R^n.

═══════════════════════ **Problems (Section 5–10)** ═══════════════════════

In these problems all vectors are written as column vectors. Also, the following matrices are referred to:

$$A = \begin{bmatrix} 2 & 1 \\ 3 & 5 \end{bmatrix}, \qquad B = \begin{bmatrix} 2 & 3 \\ 4 & 6 \end{bmatrix}, \qquad C = \begin{bmatrix} 1/\sqrt{2} & -1/\sqrt{2} \\ 1/\sqrt{2} & 1/\sqrt{2} \end{bmatrix}, \qquad D = \begin{bmatrix} 1 & 0 \\ 0 & -1 \end{bmatrix},$$

$$E = \begin{bmatrix} -1 & 0 \\ 0 & -1 \end{bmatrix}, \qquad F = \begin{bmatrix} 3 & 1 & 2 \\ 0 & 2 & 4 \end{bmatrix}, \qquad G = \begin{bmatrix} 1 & 4 & 3 \\ 2 & 8 & 6 \end{bmatrix},$$

$$H = \begin{bmatrix} 2 & 1 \\ 4 & 2 \\ -6 & -3 \end{bmatrix}, \qquad J = \begin{bmatrix} 2 & 1 \\ 1 & 2 \\ 1 & 2 \end{bmatrix}, \qquad K = \begin{bmatrix} 3 & 1 & 2 \\ 1 & 0 & 1 \\ 5 & 2 & 3 \end{bmatrix}, \qquad L = \begin{bmatrix} 1 & 0 & 1 \\ 0 & 1 & 0 \\ 0 & 1 & 1 \end{bmatrix},$$

$$M = \begin{bmatrix} 2 & 0 & 0 \\ 0 & 2 & 0 \\ 0 & 0 & 2 \end{bmatrix} = 2I, \qquad N = \begin{bmatrix} 1 & 0 & 0 \\ 0 & 2 & 0 \\ 0 & 0 & 3 \end{bmatrix}.$$

1. For the rigid-body motion described by Eqs. (5–100), find $\mathbf{x}' = (x_1', x_2', x_3')$ for the following choices of $\mathbf{x} = (x_1, x_2, x_3)$: $\mathbf{x} = (7, 1, 2)$, $\mathbf{x} = (1, -1, 0)$, $\mathbf{x} = (1, 0, 0)$, $\mathbf{x} = (0, 1, 0)$, $\mathbf{x} = (0, 0, 1)$. From these results deduce that the final position of the lines of the body initially along the coordinate axes are three mutually perpendicular lines through $(0, 0, 0)$.

2. Let the linear mapping T have the matrix A.
   a) Evaluate $T(1, 0)$, $T(0, 1)$, $T(2, -1)$, $T(-1, 1)$ and graph.
   b) Find the kernel of T, determine whether T is one-to-one, and find all $\mathbf{x}$ such that $T(\mathbf{x}) = (2, 3)$.
   c) Find the range of T and determine whether T maps R^2 onto R^2.

3. Let the linear mapping T have the matrix B.
   a) Evaluate $T(1, 0)$, $T(0, 1)$, $T(1, -1)$, $T(-1, -1)$ and graph.
   b) Find the kernel of T, determine whether T is one-to-one, and find all $\mathbf{x}$ such that $T(\mathbf{x}) = (2, 4)$
   c) Find the range of T and determine whether T maps R^2 onto R^2.

4. Let T map R^n into R^m and have the matrix F.
   a) Find n and m.
   b) Find the kernel of T and determine whether T is one-to-one.
   c) Find the range of T and determine whether T maps R^n onto R^m.

5. (a), (b), (c) Proceed as in Problem 4 with matrix G.

6. (a), (b), (c) Proceed as in Problem 4 with matrix H.

7. (a), (b), (c) Proceed as in Problem 4 with matrix J.

8. (a), (b), (c) Proceed as in Problem 4 with matrix K.

9. (a), (b), (c) Proceed as in Problem 4 with matrix L.

10. Let the linear mapping T have the equation $\mathbf{y} = C\mathbf{x}$. For general $\mathbf{x}$, find the angle between $\mathbf{x}$ and $T(\mathbf{x}) = \mathbf{y}$, as vectors in R^2, and also compare $\|x\|$ and $\|T(\mathbf{x})\|$. From these results, interpret T geometrically.   HINT: Consider $\mathbf{x}$ as $\overrightarrow{OP}$ and $\mathbf{y}$ as $\overrightarrow{OQ}$, where O is the origin of R^2 and P and Q are points of R^2.

11. Let the linear mapping T have the equation $\mathbf{y} = D\mathbf{x}$. Regard $\mathbf{x}$ as $\overrightarrow{OP}$, $\mathbf{y}$ as $\overrightarrow{OQ}$, as in Problem 10, and describe geometrically the relation between $\mathbf{x}$ and $\mathbf{y} = T(\mathbf{x})$.

12. Interpret each of the following linear mappings geometrically, as in Problems 10 and 11:
    a) $\mathbf{y} = E\mathbf{x}$ b) $\mathbf{y} = M\mathbf{x}$

13. Let T be a linear mapping of R^n into R^m. Prove: If $\mathbf{v}_1, \ldots, \mathbf{v}_k$ are linearly dependent vectors of R^n, then $T(\mathbf{v}_1), \ldots, T(\mathbf{v}_k)$ are linearly dependent vectors of R^m. Is the converse true? Explain.

5–11 RANK OF A MATRIX OR LINEAR MAPPING

In this section we assign to each matrix an integer r, its rank. For an $m \times n$ matrix, the rank is most commonly equal to the smaller of m, n (or equal to both, when $m = n$). However, r can be smaller than m and n. When this occurs, we often have to modify procedures in a special way. For example, in solving n linear equations in n unknowns, the matrix of coefficients A is $n \times n$, but A may have rank r less than n. If r is less than n, the equations may have no solution and, if they have a solution, then they have infinitely many solutions. On the other hand, if $r = n$, then there is exactly one solution.

We now define the *rank of matrix A* as *the largest number r such that A has r linearly independent column vectors.* For example, the matrix

$$B = \begin{bmatrix} 2 & 3 \\ 1 & 5 \end{bmatrix}$$

has rank 2, since the two column vectors, col (2, 1) and col (3, 5), are linearly independent. The matrix

$$C = \begin{bmatrix} 5 & 2 & 7 \\ 1 & 2 & 3 \\ 3 & 7 & 10 \end{bmatrix}$$

has rank 2, since the column vectors $\mathbf{u}_1 = $ col (5, 1, 3) and $\mathbf{u}_2 = $ col (2, 2, 7) are linearly independent, but $\mathbf{u}_3 = $ col (7, 3, 10) $= \mathbf{u}_1 + \mathbf{u}_2$. If $A = O$, then r must be chosen as 0.

The rank can be calculated in other ways. It can be shown that the rank of a matrix equals *the largest number r such that the matrix has a nonzero minor of order r* (see CLA, p. 768). Here a minor of a matrix is the determinant of a square array obtained from the matrix by deleting some rows and columns. If the matrix is square, its determinant is also considered to be a minor of the matrix. For example, the matrix B above has minors

$$\begin{vmatrix} 2 & 3 \\ 1 & 5 \end{vmatrix} = 7, \qquad |2| = 2, \qquad |3| = 3, \qquad |1| = 1, \qquad |5| = 5.$$

Hence its rank should be 2, as above. The matrix

$$D = \begin{bmatrix} 2 & 4 & 6 \\ 1 & 2 & 3 \end{bmatrix}$$

has the minors

$$\begin{vmatrix} 2 & 4 \\ 1 & 2 \end{vmatrix} = 0, \quad \begin{vmatrix} 4 & 6 \\ 2 & 3 \end{vmatrix} = 0, \quad \begin{vmatrix} 2 & 6 \\ 1 & 3 \end{vmatrix} = 0, \quad |2| = 2, \quad \dots, \quad |3| = 3.$$

Here all the minors of order 2 are 0, but some of order 1 are not 0. Therefore D has rank 1. We see also that the only set of linearly independent column vectors of D consists of one of these vectors.

Since the determinant method clearly treats row vectors and column vectors in the same way, we can also state that *the rank of a matrix A is the largest number r such that the matrix has r linearly independent row vectors.* We verify that $r = 2$ for $B, r = 2$ for C, and $r = 1$ for D by this rule, in agreement with the results of the other two methods.

We can also interpret the rank geometrically. Consider matrix A as the matrix of a linear mapping $\mathbf{y} = A\mathbf{x}$ from R^n to R^m, as in the previous section. Then, as in that section, the range of this linear mapping is a subspace of the vectors of R^m consisting of all vectors $\mathbf{y}$ (column vectors) expressible as linear combinations of the column vectors $\mathbf{u}_1, \dots, \mathbf{u}_n$ of A. If, say, $\mathbf{u}_1, \dots, \mathbf{u}_r$ are linearly independent, but

$\mathbf{u}_{r+1}, \ldots, \mathbf{u}_n$ are expressible as linear combinations of $\mathbf{u}_1, \ldots, \mathbf{u}_r$, then all vectors in the range are expressible as linear combinations of $\mathbf{u}_1, \ldots, \mathbf{u}_r$. Thus $\mathbf{u}_1, \ldots, \mathbf{u}_r$ serve as a basis for the range of A and *the rank r is the dimension of the range* (Section 5–10).

If $r = 1$ here, the vectors $\mathbf{y}$ are all vectors $t\mathbf{u}_1$, where t is arbitrary, and the corresponding points $(y_1, \ldots, y_m)$ fill out a *line* in R^m. If $r = 2$, the vectors $\mathbf{y}$ are all vectors $t_1\mathbf{u}_1 + t_2\mathbf{u}_2$ and the corresponding points fill a *plane* in R^m.

It is useful here to consider the *augmented matrix* denoted by $(A\ \mathbf{y})$, obtained from A by adjoining the column vector $\mathbf{y}$. If $\mathbf{y}$ is in the range of the linear mapping, then $\mathbf{y}$ is expressible as a linear combination of the other column vectors, and $(A\ \mathbf{y})$ has the same rank r as A. If $\mathbf{y}$ is not in the range, $\mathbf{y}$ is not so expressible, and the rank of $(A\ \mathbf{y})$ is $r + 1$. Therefore, we deduce the useful rule:

The equation $\mathbf{y} = A\mathbf{x}$ *has a solution for* $\mathbf{x}$ *precisely when the matrices A and $(A\ \mathbf{y})$ have the same rank.*

For example, with D as above and $\mathbf{y} = \text{col}\,(3, 5)$, we have

$$(D\ \mathbf{y}) = \begin{bmatrix} 2 & 4 & 6 & 3 \\ 1 & 2 & 3 & 5 \end{bmatrix},$$

and this matrix has rank 2. But D has rank 1. Therefore the equation $D\mathbf{x} = \mathbf{y}$ has *no* solution. Hence the equations

$$2x_1 + 4x_2 + 6x_3 = 3$$
$$x_1 + 2x_2 + 3x_3 = 5$$

cannot be solved for x_1, x_2, x_3.

Rank and nullity

We recall (Section 5–10) that the kernel of matrix A consists of all $\mathbf{x}$ such that $A\mathbf{x} = \mathbf{0}$. Thus to find the kernel of A we must find all solutions of the set of homogeneous linear equations

$$a_{11}x_1 + \cdots + a_{1n}x_n = 0,$$
$$\vdots \qquad\qquad \vdots \qquad\qquad\qquad (5\text{–}110)$$
$$a_{m1}x_1 + \cdots + a_{mn}x_n = 0.$$

Now if the row vectors $\mathbf{v}_1, \ldots, \mathbf{v}_m$ of A are linearly dependent, some of these equations can be obtained from the others and hence can be omitted; for example, if $\mathbf{v}_3 = \mathbf{v}_1 + 2\mathbf{v}_2$, then the third equation is a consequence of the first two and can be omitted. Thus to find the kernel, we need choose only r of the equations such that the corresponding row vectors are linearly independent, with r chosen as large as possible, that is, with r equal to the rank of A. For example, if the first r row vectors are linearly independent, we solve the equations

$$a_{11}x_1 + \cdots + a_{1n}x_n = 0,$$
$$\vdots \qquad\qquad \vdots \qquad\qquad\qquad (5\text{–}111)$$
$$a_{r1}x_1 + \cdots + a_{rn}x_n = 0.$$

Since r is the rank of A, we can find a nonzero minor of A of order r, for example, the minor

$$\Delta = \begin{vmatrix} a_{11} & \cdots & a_{1r} \\ \vdots & & \vdots \\ a_{r1} & \cdots & a_{rr} \end{vmatrix}.$$

We can then solve Eqs. (5–111) for $x_1, \ldots, x_r$ by Cramer's rule (Section 5–1). For example,

$$x_1 = \begin{vmatrix} -a_{1,r+1}x_{r+1} - \cdots - a_{1n}x_n & a_{12} & \cdots & a_{1r} \\ \vdots & & & \vdots \\ -a_{r,r+1}x_{r+1} - \cdots - a_{rn}x_n & a_{r2} & \cdots & a_{rr} \end{vmatrix} \div \Delta.$$

Similar expressions are found for $x_2, \ldots, x_r$. Thus we obtain

$$x_k = b_{k1}x_{r+1} + \cdots + b_{k,n-r}x_n, \quad k = 1, \ldots, r,$$

for appropriate numbers b_{kj}. The numbers $x_{r+1}, \ldots, x_n$ can be chosen arbitrarily. Thus we can describe the kernel of A as all vectors $\mathbf{x}$ such that

$$\left\{ \begin{aligned} x_1 &= b_{11}t_1 + \cdots + b_{1,n-r}t_{n-r}, \\ &\vdots \\ x_r &= b_{r1}t_1 + \cdots + b_{r,n-r}t_{n-r}, \\ x_{r+1} &= t_1, \\ &\vdots \\ x_n &= t_{n-r}, \end{aligned} \right. \tag{5–112}$$

where $t_1, \ldots, t_{n-r}$ are arbitrary. If we write

$$\mathbf{w}_1 = \text{col}\,(b_{11}, \ldots, b_{r1}, 1, 0, \ldots, 0),$$
$$\mathbf{w}_2 = \text{col}\,(b_{12}, \ldots, b_{r2}, 0, 1, \ldots, 0).$$
$$\mathbf{w}_{n-r} = \text{col}\,(b_{1,n-r}, \ldots, b_{r,n-r}, 0, \ldots, 0, 1),$$

then the kernel is given as all vectors

$$\mathbf{x} = t_1\mathbf{w}_1 + \cdots + t_{n-r}\mathbf{w}_{n-r}. \tag{5–113}$$

The vectors $\mathbf{w}_1, \ldots, \mathbf{w}_{n-r}$ are easily seen to be linearly independent (see Problem 5 below). Thus we can say: *the kernel of A is a k-dimensional subspace of the vectors of R^n*, where

$$k = n - r. \tag{5–114}$$

The number k is called the *nullity* of A.

It can happen that $r = n$. Then Eqs. (5–111) have only the solution $\mathbf{x} = \mathbf{0}$ (Section 5–1), the nullity k is 0, and we say that the kernel is 0-dimensional. It can also happen that $r = 0$. This would mean that every minor of A is 0 and hence $A = O$. The kernel of O consists of *all vectors* $\mathbf{x}$ of R^n, since $O\mathbf{x} = \mathbf{0}$ for all $\mathbf{x}$. The range of O is only the $\mathbf{0}$ vector of R^m.

EXAMPLE 1   The matrix C above has rank 2, hence its nullity k must be 1. To find the kernel, we observe that the first two row vectors of C are linearly independent, and hence we use only the two equations

$$5x_1 + 2x_2 + 7x_3 = 0,$$
$$x_1 + 2x_2 + 3x_3 = 0.$$

Here the minor $\Delta = \begin{vmatrix} 5 & 2 \\ 1 & 2 \end{vmatrix} = 8 \neq 0$, so that we can solve for x_1 and x_2 in terms of x_3:

$$x_1 = \begin{vmatrix} -7x_3 & 2 \\ -3x_3 & 2 \end{vmatrix} \div 8, \qquad x_2 = \begin{vmatrix} 5 & -7x_3 \\ 1 & -3x_3 \end{vmatrix} \div 8,$$

so that $x_1 = -x_3, x_2 = -x_3$, and the kernel is given by $x_1 = -t, x_2 = -t, x_3 = t$ or by

$$\mathbf{x} = t(-1, -1, 1).$$

This corresponds to a line in space. The kernel is 1-dimensional. ◄

REMARK: The rank r of the $m \times n$ matrix A is at most equal to the number m of rows of A and at most equal to the number n of columns of A; hence r is at most equal to the smaller of the numbers m, n (or $r \leqslant m$ if $m = n$). When r has the largest possible value allowed by these conditions, we say that A has *maximum rank*.

EXAMPLE 2 The application of Kirchhoff's first law to the network of Fig. 5–18 gives the three equations

$$L\frac{dI}{dt} + R_1 I_1 - E = 0,$$

$$R_2 I_2 + \frac{q_2}{C} - R_1 I_1 = 0,$$

$$L\frac{dI}{dt} + R_2 I_2 + \frac{q_2}{C} - E = 0.$$

We can consider these as three homogeneous linear equations for $q_2, I_1, I_2, dI/dt, E$, with coefficient matrix

$$\begin{bmatrix} 0 & R_1 & 0 & L & -1 \\ \dfrac{1}{C} & -R_1 & R_2 & 0 & 0 \\ \dfrac{1}{C} & 0 & R_2 & L & -1 \end{bmatrix}.$$

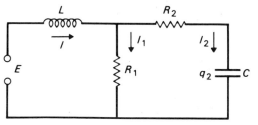

Fig. 5–18. Example 2.

We verify that the matrix has rank 2. In fact, the third equation is just the sum of the first two equations and can therefore be dropped.

This network is discussed more fully in Section 6–5, with the aid of Kirchhoff's second law and the relation $dq_2/dt = I_2$. ◀

EXAMPLE 3 *Statical indeterminacy.* Many problems in statics cannot be fully solved by using standard equilibrium conditions; the problems are said to be indeterminate. We illustrate this in Fig. 5–19, in which a 10 lb. weight is supported by three wires in a plane. The tensions in the wires are the forces shown. Taking components along the x and y directions, we find

$$\frac{3}{5}u - \frac{4}{5}w = 0, \qquad \frac{4}{5}u + v + \frac{3}{5}w = 10.$$

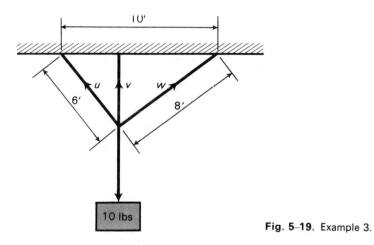

Fig. 5–19. Example 3.

These are two equations in three unknowns u, v, w (the magnitudes of the forces). The coefficient matrix has rank 2, so that we cannot disregard either equation; the augumented matrix also has rank 2, so that solutions exist. From the equations, we can express u and v in terms of w. For example, $u = (4/3)w$, $v = 30 - 5w$, but w can be chosen arbitrarily. ◀

═══════════════════════════ **Problems (Section 5–11)** ═══════════════════════════

1. Find the rank and nullity of each matrix:

a) $A = \begin{bmatrix} 5 & 3 \\ -10 & 6 \end{bmatrix}$ b) $B = \begin{bmatrix} 1 & 0 \\ 2 & 0 \end{bmatrix}$ c) $C = \begin{bmatrix} 2 & 1 & 1 \\ 1 & 2 & 1 \\ 1 & 1 & 2 \end{bmatrix}$

d) $D = \begin{bmatrix} 3 & 1 & 2 \\ 1 & 3 & 2 \\ 7 & -3 & 2 \end{bmatrix}$ e) $E = \begin{bmatrix} 2 & 10 \\ 3 & 15 \\ 6 & 30 \end{bmatrix}$ f) $F = \begin{bmatrix} 3 & 0 & 0 \\ 0 & 0 & 2 \end{bmatrix}$

g) $G = \begin{bmatrix} 2 & 1 & 0 & 1 \\ 0 & 2 & 1 & 0 \\ 3 & 1 & 0 & 2 \end{bmatrix}$ h) $H = \begin{bmatrix} 1 & 3 & 4 & 2 \\ 2 & 1 & 3 & -1 \\ 2 & 3 & 5 & 1 \end{bmatrix}$

2. Using the matrices of Problem 1, verify that each of the following mappings has a range of dimension equal to the rank of the matrix:

   a) $y = Ax$ b) $y = Cx$ c) $y = Ex$ d) $y = Gx$

3. (a), (b), (c) For each of the mappings of Problem 2, find the kernel and verify that its dimension equals the nullity of the matrix.

4. Using the matrices of Problem 1, determine whether the equations have a solution:

   a) $Ax = \text{col}(1, 1)$ b) $Cx = \text{col}(2, 0, 1)$

   c) $Ex = \text{col}(4, 6, 12)$ d) $Gx = \text{col}(1, 2, 1)$

REMARK: When the equation has a solution, all solutions are found as in the theorem and Example 3 of Section 5–10.

5. Show that, for $0 < k < n$ and $r = n - k$, the k vectors

$$\mathbf{w}_1 = \text{col}(b_{11}, \ldots, b_{r1}, 1, 0, \ldots, 0),$$
$$\vdots$$
$$\mathbf{w}_k = \text{col}(b_{1k}, \ldots, b_{rk}, 0, \ldots, 0, 1),$$

of R^n are linearly independent.

6. Explain why the counterpart of Example 3 in Section 5–11, in which a weight is suspended by three wires *not in a plane*, is statically determinate, and find the forces when the three wires are suspended at $A(0,0,1)$, $B(0,2,1)$, $C(3,2,1)$, and the weight of 10 lbs, is suspended at $D(1,1,0)$. Here the coordinates (x, y, z) are chosen with the z-axis pointing up.

5–12 SYSTEMS OF LINEAR EQUATIONS: THEORY

The results of the preceding two sections now give complete information on the solutions of simultaneous linear equations

$$a_{11} x_1 + \cdots + a_{1n} x_n = b_1,$$
$$\vdots \qquad\qquad \vdots \qquad\qquad (5\text{–}120)$$
$$a_{m1} x_1 + \cdots + a_{mn} x_n = b_m,$$

or, in matrix form, $Ax = b$, where $x = \text{col}(x_1, \ldots, x_n)$, $b = \text{col}(b_1, \ldots, b_m)$.

Case I: $m = n$. Here A is a square matrix. If $\det A \neq 0$, the equations have a unique solution given by Cramer's rule, as in Section 5–1. If in particular $b = 0$, then that unique solution is $x = 0$. If $\det A = 0$, there may be no solution. If there is a solution $\mathbf{x}_0$, then all solutions are given by

$$\mathbf{x} = \mathbf{x}_0 + \mathbf{z},$$

where $\mathbf{z}$ is an arbitrary vector of the kernel of A; hence the solutions can be represented in the form

$$\mathbf{x} = \mathbf{x}_0 + t_1 \mathbf{w}_1 + \cdots + t_k \mathbf{w}_k, \qquad -\infty < t_k < \infty, \qquad (5\text{–}121)$$

where $\mathbf{w}_1, \ldots, \mathbf{w}_k$ are linearly independent vectors in the kernel of A, and A has nullity k. If in particular $\mathbf{b}_0 = \mathbf{0}$, then $\mathbf{x}_0$ can be chosen as $\mathbf{0}$ and the solutions are given by

$$\mathbf{x} = t_1 \mathbf{w}_1 + \cdots + t_k \mathbf{w}_k, \qquad (5\text{–}122)$$

filling out the kernel of A. Whether or not there is a solution can be determined by examining the ranks of A and of $(A\ \mathbf{b})$. When these ranks are equal, there is a solution; when they are unequal, there is none.

Case II: $m \neq n$, A has rank r, nullity k. If $r = m$, then the range of A is all of R^m, so that there is a solution for every choice of $\mathbf{b}$. The solutions are again given by (5–121) or, if $\mathbf{b} = \mathbf{0}$, by (5–122). If $r < m$, then the range of A is of lower dimension than m, so that for some choices of $\mathbf{b}$ there is no solution. Whether or not there is a solution can be decided as in Case I.

5–13 SYSTEMS OF LINEAR EQUATIONS: TECHNIQUE

Computation of determinants of high order is lengthy. Accordingly, Cramer's rule of Section 5–1 is to be avoided in computing solutions, except when only determinants of small size (say, of order less than four) are involved. The familiar method of elimination is much faster. Furthermore, it can be improved and systematized by matrix techniques. Such procedures are explained in this section.

We recall the essential ideas in the elimination process. Let us consider an example:

$$\begin{aligned} 2x_1 - x_2 + x_3 &= 2, \\ x_1 + x_2 + 2x_3 &= 7, \\ 3x_1 - x_2 - 4x_3 &= -1. \end{aligned} \qquad (5\text{–}130)$$

As a first step, we add the first two equations to eliminate x_2 and add the last two equations for the same reason. We obtain two new equations:

$$\begin{aligned} 3x_1 + 3x_3 &= 9, \\ 4x_1 - 2x_3 &= 6. \end{aligned}$$

We can solve these two equations for x_1 and x_3. Then we use one of the given equations, say the second, to find x_2. Thus we are really using the three equations

$$\begin{aligned} x_1 + x_2 + 2x_3 &= 7, \\ 3x_1 + 3x_3 &= 9, \\ 4x_1 - 2x_3 &= 6 \end{aligned} \qquad (5\text{–}131)$$

to replace the given equations. We could further divide the second equation here by 3 (to obtain $x_1 + x_3 = 3$) and the third equation by 2 (to obtain $2x_1 - x_3 = 3$). The essential idea in the elimination method is to replace the given set of equations by an *equivalent* set of equations, that is, by a set of equations having the *same* solutions. It is easy to see that in going from (5–130) to (5–131) we have done just that since equations (5–131) were obtained from (5–130) by adding certain pairs, and the equations (5–130) can be recovered from (5–131) by reversing the process. Thus subtracting the first of (5–131) from the second gives the first of (5–130); subtracting the first of (5–131) from the third gives the third of (5–130); and the second of (5–130) is the same as the first of (5–131). In general, the following three steps are used:

a) Interchanging two equations;
b) Multiplying or dividing an equation by a number, not 0;
c) Replacing one equation by another one obtained from it by adding to it a nonzero multiple of some other equation of the system.

By the reasoning of the example, each of these steps replaces the given system by an equivalent one. The steps may be used several times, and we end up with an equivalent system.

These steps can be carried out very systematically by matrices. We replace the equations (5–130), for example, by the array

$$\begin{bmatrix} 2 & -1 & 1 & | & 2 \\ 1 & 1 & 2 & | & 7 \\ 3 & -1 & -4 & | & -1 \end{bmatrix}$$

The vertical bar sets off the coefficients from the right side. The steps to follow replace this array successively by different ones, each giving the coefficients and right-hand members for an equivalent system.

We want a 1 in the $(1, 1)$ position (first row, first column). To achieve this, interchange the first and second rows of the array (step (a)) to obtain

$$\begin{bmatrix} 1 & 1 & 2 & | & 7 \\ 2 & -1 & 1 & | & 2 \\ 3 & -1 & -4 & | & -1 \end{bmatrix}.$$

Then add -2 times the first row to the second, and -3 times the first row to the third (step (c) used twice) to obtain

$$\begin{bmatrix} 1 & 1 & 2 & | & 7 \\ 0 & -3 & -3 & | & -12 \\ 0 & -4 & -10 & | & -22 \end{bmatrix}.$$

Then divide the second row by -3, the third by -2 (step (b) used twice):

$$\begin{bmatrix} 1 & 1 & 2 & | & 7 \\ 0 & 1 & 1 & | & 4 \\ 0 & 2 & 5 & | & 11 \end{bmatrix}.$$

Then add -1 times the second row to the first, add -2 times the second row to the third:

$$\begin{bmatrix} 1 & 0 & 1 & 3 \\ 0 & 1 & 1 & 4 \\ 0 & 0 & 3 & 3 \end{bmatrix}.$$

Divide the last row by 3 and then add -1 times this row to the first and second rows:

$$\begin{bmatrix} 1 & 0 & 1 & 3 \\ 0 & 1 & 1 & 4 \\ 0 & 0 & 1 & -1 \end{bmatrix}, \qquad \begin{bmatrix} 1 & 0 & 0 & 2 \\ 0 & 1 & 0 & 3 \\ 0 & 0 & 1 & 1 \end{bmatrix}.$$

Thus we finally have the system $x_1 = 2$, $x_2 = 3$, $x_3 = 1$, that gives the desired solution.

This method can be applied quite generally to m equations in n unknowns It may happen that at some stage so many zeros appear in a column that the operations cannot be used to produce the next desired 1 along the diagonal. In that case, we simply move on to the next column or a later one to achieve a 1. Thus we may end up with an array such as

$$\begin{bmatrix} 1 & 3 & 0 & 0 & 7 \\ 0 & 0 & 1 & 0 & 2 \\ 0 & 0 & 0 & 1 & 5 \end{bmatrix}.$$

Here we read off: $x_1 + 3x_2 = 7$, $x_3 = 2$, $x_4 = 5$. Thus x_2 can be chosen arbitrarily. If we denote its value by t, then the solutions are given by $x_1 = 7 - 3t$, $x_2 = t$, $x_3 = 2$, $x_4 = 5$, or by

$$\mathbf{x} = \mathrm{col}(7, 0, 2, 5) + t \, \mathrm{col}(-3, 1, 0, 0).$$

The process may also lead to a row of form $(0, 0, \ldots, 0, b)$, corresponding to the equation $0x_1 + \cdots + 0x_n = b$. If $b \neq 0$, this is a contradiction and the system has no solution. If $b = 0$, we can disregard the row. This situation arises when the original matrix of coefficients A has linearly dependent row vectors.

In fact, it can be seen that our process replaces the original matrix of coefficients by one of the same rank. The rank of the final matrix is simply the number of rows with not all entries 0.

EXAMPLE 1 $\quad 2x_1 + 2x_2 + 3x_3 - x_4 + x_5 = 7,$
$$x_1 - x_2 + 2x_3 + x_4 - x_5 = 2,$$
$$2x_1 + 3x_2 - x_3 + 3x_4 - 2x_5 = 5.$$

The successive arrays are as follows:

1. $\begin{bmatrix} 2 & 2 & 3 & -1 & 1 & 7 \\ 1 & -1 & 2 & 1 & -1 & 2 \\ 2 & 3 & -1 & 3 & -2 & 5 \end{bmatrix}$ 2. $\begin{bmatrix} 1 & -1 & 2 & 1 & -1 & 2 \\ 2 & 2 & 3 & -1 & 1 & 7 \\ 2 & 3 & -1 & 3 & -2 & 5 \end{bmatrix}$

3. $\begin{bmatrix} 1 & -1 & 2 & 1 & -1 & | & 2 \\ 0 & 4 & -1 & -3 & 3 & | & 3 \\ 0 & 5 & -5 & 1 & 0 & | & 1 \end{bmatrix}$ 4. $\begin{bmatrix} 1 & -1 & 2 & 1 & -1 & | & 2 \\ 0 & 1 & -\frac{1}{4} & -\frac{3}{4} & \frac{3}{4} & | & \frac{3}{4} \\ 0 & 5 & -5 & 1 & 0 & | & 1 \end{bmatrix}$

5. $\begin{bmatrix} 1 & 0 & \frac{7}{4} & \frac{1}{4} & -\frac{1}{4} & | & \frac{11}{4} \\ 0 & 1 & -\frac{1}{4} & -\frac{3}{4} & \frac{3}{4} & | & \frac{3}{4} \\ 0 & 0 & -\frac{15}{4} & \frac{19}{4} & -\frac{15}{4} & | & -\frac{11}{4} \end{bmatrix}$ 6. $\begin{bmatrix} 1 & 0 & \frac{7}{4} & \frac{1}{4} & \frac{1}{4} & | & \frac{11}{4} \\ 0 & 1 & -\frac{1}{4} & -\frac{3}{4} & \frac{3}{4} & | & \frac{3}{4} \\ 0 & 0 & 1 & -\frac{19}{15} & 1 & | & \frac{11}{15} \end{bmatrix}$

7. $\begin{bmatrix} 1 & 0 & 0 & \frac{37}{15} & -2 & | & \frac{22}{15} \\ 0 & 1 & 0 & -\frac{16}{15} & 1 & | & \frac{14}{15} \\ 0 & 0 & 1 & -\frac{19}{15} & 1 & | & \frac{11}{15} \end{bmatrix}$

The seventh array gives the equations $x_1 + \dfrac{37}{15}x_4 - 2x_5 = \dfrac{22}{15}$, etc., from which we obtain x_1 x_2, x_3 in terms of x_4 and x_5:

$$x_1 = -\tfrac{37}{15}x_4 + 2x_5 + \tfrac{22}{15},$$
$$x_2 = \tfrac{16}{15}x_4 - x_5 + \tfrac{14}{15},$$
$$x_3 = \tfrac{19}{15}x_4 - x_5 + \tfrac{11}{15}.$$

Thus x_4, x_5 can be chosen arbitrarily; each choice determines the values of x_1, x_2, x_3, that satisfy the given equations. ◄

EXAMPLE 2 $x_1 + 2x_2 + x_3 + 3x_4 = 5$
$\qquad\qquad\quad 2x_1 + 4x_2 + 3x_3 + 12x_4 = 12.$

We obtain successively:

$$\begin{bmatrix} 1 & 2 & 1 & 3 & | & 5 \\ 2 & 4 & 3 & 12 & | & 12 \end{bmatrix}, \quad \begin{bmatrix} 1 & 2 & 1 & 3 & | & 5 \\ 0 & 0 & 1 & 6 & | & 2 \end{bmatrix}, \quad \begin{bmatrix} 1 & 2 & 0 & -3 & | & 3 \\ 0 & 0 & 1 & 6 & | & 2 \end{bmatrix}.$$

Our final display gives $x_1 + 2x_2 - 3x_4 = 3$, $x_3 + 6x_4 = 2$, from which we obtain x_1 and x_3 in terms of x_2 and x_4. ◄

EXAMPLE 3 $x_1 + 2x_2 - x_3 = 1,$
$\qquad\qquad\quad 2x_1 + x_2 + x_3 = 4,$
$\qquad\qquad\quad 3x_1 - x_2 - x_3 = 1,$
$\qquad\qquad\quad x_1 + x_2 + 2x_3 = 7.$

The work follows:

1. $\begin{bmatrix} 1 & 2 & -1 & | & 1 \\ 2 & 1 & 1 & | & 4 \\ 3 & -1 & -1 & | & 1 \\ 1 & 1 & 2 & | & 7 \end{bmatrix}$ 2. $\begin{bmatrix} 1 & 2 & -1 & | & 1 \\ 0 & -3 & 3 & | & 2 \\ 0 & -7 & 2 & | & -2 \\ 0 & -1 & 3 & | & 6 \end{bmatrix}$

3. $\begin{bmatrix} 1 & 2 & -1 & | & 1 \\ 0 & 1 & -3 & | & -6 \\ 0 & -3 & 3 & | & 2 \\ 0 & -7 & 2 & | & -2 \end{bmatrix}$ 4. $\begin{bmatrix} 1 & 0 & 5 & | & 13 \\ 0 & 1 & -3 & | & -6 \\ 0 & 0 & -6 & | & -16 \\ 0 & 0 & -19 & | & -44 \end{bmatrix}$

5. $\begin{bmatrix} 1 & 0 & 5 & | & 13 \\ 0 & 1 & -3 & | & -6 \\ 0 & 0 & 1 & | & \frac{8}{3} \\ 0 & 0 & -19 & | & -44 \end{bmatrix}$ 6. $\begin{bmatrix} 1 & 0 & 0 & | & -\frac{1}{3} \\ 0 & 1 & 0 & | & 2 \\ 0 & 0 & 1 & | & \frac{8}{3} \\ 0 & 0 & 0 & | & \frac{20}{3} \end{bmatrix}$

The last line of the last display gives $0 = 20/3$. This is a contradiction. Therefore the equations have no solution. The absence of a solution is typically revealed in this way. If in the last display the 20/3 were replaced by a 0, the last line would give simply $0 = 0$ and could be ignored, while the other lines would give the solution $x_1 = -1/3$, $x_2 = 2$, $x_3 = 8/3$. ◄

The methods illustrated can all be programmed for computer solution and are in common use (with certain refinements). In computer work the fractions are replaced by decimals, often converted to base 2 (see Section 13–2).

5–14   FINDING THE INVERSE OF A SQUARE MATRIX

If B is the inverse of A, then $AB - I$. Hence, as in Section 5–5, if B has column vectors $\mathbf{u}_1, \mathbf{u}_2, \ldots$, then $A\mathbf{u}_1$ equals the first column vector of I, that is, $\text{col}(1, 0, \ldots, 0)$, and similarly $A\mathbf{u}_2 = \text{col}(0, 1, 0, \ldots, 0)$, and so on. Thus to solve for the column vectors $\mathbf{u}_1, \mathbf{u}_2, \ldots$ of B, we have to solve several systems of simultaneous equations. We can use the elimination method to solve all systems at once by simply listing all the right-hand members successively.

EXAMPLE 1 To find the inverse of $A = \begin{bmatrix} 3 & 1 \\ 5 & 2 \end{bmatrix}$, we follow the arrays given:

1. $\begin{bmatrix} 3 & 1 & | & 1 & 0 \\ 5 & 2 & | & 0 & 1 \end{bmatrix}$ 2. $\begin{bmatrix} 1 & \frac{1}{3} & | & \frac{1}{3} & 0 \\ 5 & 2 & | & 0 & 1 \end{bmatrix}$

3. $\begin{bmatrix} 1 & \frac{1}{3} & | & \frac{1}{3} & 0 \\ 0 & \frac{1}{3} & | & -\frac{5}{3} & 1 \end{bmatrix}$ 4. $\begin{bmatrix} 1 & \frac{1}{3} & | & \frac{1}{3} & 0 \\ 0 & 1 & | & -5 & 3 \end{bmatrix}$

5. $\begin{bmatrix} 1 & 0 & | & 2 & -1 \\ 0 & 1 & | & -5 & 3 \end{bmatrix}$

Here we start with A on the left and I on the right, and end with I on the left and $B = A^{-1}$ on the right. The result is easily checked. ◄

EXAMPLE 2

1. $\begin{bmatrix} 1 & 2 & 3 \\ 2 & 3 & 1 \\ 3 & 1 & 2 \end{bmatrix} \begin{array}{|ccc} 1 & 0 & 0 \\ 0 & 1 & 0 \\ 0 & 0 & 1 \end{array}$
2. $\begin{bmatrix} 1 & 2 & 3 \\ 0 & -1 & -5 \\ 0 & -5 & -7 \end{bmatrix} \begin{array}{|ccc} 1 & 0 & 0 \\ -2 & 1 & 0 \\ -3 & 0 & 1 \end{array}$

3. $\begin{bmatrix} 1 & 2 & 3 \\ 0 & 1 & 5 \\ 0 & 5 & 7 \end{bmatrix} \begin{array}{|ccc} 1 & 0 & 0 \\ 2 & -1 & 0 \\ 3 & 0 & 1 \end{array}$
4. $\begin{bmatrix} 1 & 0 & -7 \\ 0 & 1 & 5 \\ 0 & 0 & -18 \end{bmatrix} \begin{array}{|cccc} -3 & 2 & 0 \\ 2 & -1 & 0 \\ -7 & 5 & 1 \end{array}$

5. $\begin{bmatrix} 1 & 0 & -7 \\ 0 & 1 & 5 \\ 0 & 0 & 1 \end{bmatrix} \begin{array}{|ccc} -3 & 2 & 0 \\ 2 & -1 & 0 \\ \frac{7}{18} & -\frac{5}{18} & \frac{1}{18} \end{array}$

6. $\begin{bmatrix} 1 & 0 & 0 \\ 0 & 1 & 0 \\ 0 & 0 & 1 \end{bmatrix} \begin{array}{|ccc} -\frac{5}{18} & \frac{1}{18} & \frac{7}{18} \\ \frac{1}{18} & \frac{7}{18} & -\frac{5}{18} \\ \frac{7}{18} & -\frac{5}{18} & \frac{1}{18} \end{array}$

Again the matrix A appears on the left in the first display and A^{-1} on the right in the last display. ◄

The same method can be applied if we have to solve several systems of form $A\mathbf{x} = \mathbf{v}_1, A\mathbf{x} = \mathbf{v}_2, \ldots, A\mathbf{x} = \mathbf{v}_k$, with the same square coefficient matrix A for all. We simply replace I in the first display by a matrix whose successive columns are $\mathbf{v}_1, \ldots, \mathbf{v}_k$. The process then yields the solution $\mathbf{x}$ of $A\mathbf{x} = \mathbf{v}_1$ in the position occupied initially by $\mathbf{v}_1$, the solution of $A\mathbf{x} = \mathbf{v}_2$ in the next column, and so on. Of course, A^{-1} can be found as above; then $A\mathbf{x} = \mathbf{v}_1, A\mathbf{x} = \mathbf{v}_2, \ldots$ are solved by $\mathbf{x} = A^{-1}\mathbf{v}_1, \mathbf{x} = A^{-1}\mathbf{v}_2, \ldots$

=========== **Problems (Section 5-14)** ===========

1. Find all solutions:
   a) $2x_1 - x_2 = 4, x_1 + 3x_2 = 9$
   b) $x_1 + x_2 = 5, 3x_1 + 4x_2 = 17$
   c) $2x_1 - x_2 + 3x_3 = 0, x_1 + x_2 - 3x_3 = 3, 4x_1 - 2x_2 + 3x_3 = -3$
   d) $x_1 + 2x_2 - x_3 = 0, 2x_1 - x_2 + x_3 = 0, 7x_1 + 4x_2 - x_3 = 0$
   e) $3x_1 - x_2 + x_3 = 1, x_1 + x_2 + x_3 = 2, 5x_1 - 3x_2 + x_3 = 1$
   f) $x_1 - x_2 + 2x_3 = 2, 3x_1 - x_2 + 5x_3 = 11, x_1 + x_2 - 4x_3 = 2$
   g) $x_1 - x_2 + x_3 = 2, 2x_1 - x_2 + 3x_3 + 5x_4 = 7, 6x_1 - x_2 + x_3 - 5x_4 = 17$
   h) $x_1 + 3x_2 + 5x_3 + 13x_4 = 7, 2x_1 + 6x_2 + 11x_3 + 3x_4 = 15,$
      $4x_1 + 12x_2 + 20x_3 + 13x_4 = 28$
   i) $2x_1 + 2x_2 + 3x_3 + x_4 = 1, 2x_1 - x_2 + x_3 + 3x_4 = 2, 2x_1 + 5x_2 + 5x_3 - x_4 = 1$
   j) $x_1 - 3x_2 + x_3 - x_4 = 1, 3x_1 + x_2 - x_3 + x_4 = 2, x_1 - 13x_2 + 5x_3 - 5x_4 = 2$
   k) $x_1 + x_2 + x_3 = 3, x_1 - x_2 + x_3 = 1, 2x_1 + x_2 + 3x_3 = 4, x_1 + 3x_2 + 4x_3 = 8$
   l) $2x_1 + 3x_2 + x_3 = 4, x_1 + 3x_2 + 2x_3 = 3, 3x_1 - x_2 + 2x_3 = 7, x_1 + 2x_2 + 3x_3 = 1$
   m) $x_1 + x_2 + x_3 - 6x_4 - 2x_5 = 0, 2x_1 - x_2 - x_3 + 5x_5 = 0,$
      $3x_1 + 2x_2 - x_3 - 5x_4 + 3x_5 = 0$
   n) $x_1 + 2x_2 + x_3 - 3x_4 + x_5 = 4, 2x_1 + 4x_2 - x_3 + 3x_4 + 2x_5 = 2,$
      $3x_1 + 6x_2 + 3x_3 - 9x_4 + x_5 = 6$

o) $2x_1 + x_2 - x_3 + x_4 + x_5 = 1$, $3x_1 - x_2 - x_3 + 2x_4 - x_5 = 2$,
$3x_1 + 4x_2 - 2x_3 + x_4 + 4x_5 = 2$

p) $2x_1 + x_2 - x_3 + x_4 + x_5 = 1$, $3x_1 - x_2 - x_3 + 2x_4 - x_5 = 2$,
$3x_1 + 4x_2 - 2x_3 + x_4 + 4x_5 = 1$.

2. For each matrix find the inverse, if possible, and state the rank of the matrix:

a) $\begin{bmatrix} 3 & 3 \\ 5 & 2 \end{bmatrix}$ b) $\begin{bmatrix} 2 & 3 \\ 1 & 1 \end{bmatrix}$ c) $\begin{bmatrix} 1 & 2 & 2 \\ 2 & 2 & 1 \\ 1 & 4 & 5 \end{bmatrix}$ d) $\begin{bmatrix} 1 & 1 & 1 \\ 2 & 1 & 1 \\ 1 & 1 & 2 \end{bmatrix}$

e) $\begin{bmatrix} 3 & 2 & 1 \\ 2 & 1 & 0 \\ 1 & 0 & 1 \end{bmatrix}$ f) $\begin{bmatrix} 1 & 2 & 2 \\ 5 & 1 & 1 \\ 3 & -3 & -3 \end{bmatrix}$ g) $\begin{bmatrix} 1 & 2 & 1 & 2 \\ 1 & 2 & 2 & 1 \\ 2 & 1 & 1 & 0 \\ 1 & 0 & 1 & 1 \end{bmatrix}$

h) $\begin{bmatrix} 1 & 1 & 0 & 1 \\ 3 & 1 & 2 & -1 \\ 1 & 2 & 1 & 0 \\ 0 & 3 & -1 & 3 \end{bmatrix}$ i) $\begin{bmatrix} 1 & 3 & 4 & 2 \\ 2 & 1 & 1 & 3 \\ 1 & 2 & 1 & 4 \\ 0 & 6 & 10 & -1 \end{bmatrix}$ j) $\begin{bmatrix} 1 & 0 & 1 & 0 \\ 0 & 1 & 1 & 1 \\ 1 & 0 & 0 & 1 \\ 1 & 0 & 1 & 2 \end{bmatrix}$

k) $\begin{bmatrix} 1 & 1 & 1 & 2 & 0 \\ 0 & 1 & 2 & 3 & 0 \\ 0 & 0 & 1 & 1 & 2 \\ 0 & 0 & 0 & 2 & 3 \\ 0 & 0 & 0 & 0 & 1 \end{bmatrix}$ l) $\begin{bmatrix} 1 & 0 & 0 & 0 & 0 \\ 1 & 2 & 0 & 0 & 0 \\ 0 & 1 & 3 & 0 & 0 \\ 0 & 0 & 1 & 4 & 0 \\ 0 & 0 & 0 & 1 & 5 \end{bmatrix}$

3. Let $A = \begin{bmatrix} 1 & 1 & 1 \\ 3 & 2 & 3 \\ 2 & 1 & 4 \end{bmatrix}$, $B = \begin{bmatrix} 1 & 0 & 1 \\ 2 & 2 & 1 \\ 3 & 1 & 3 \end{bmatrix}$, $\mathbf{v}_1 = \begin{bmatrix} 1 \\ 1 \\ 1 \end{bmatrix}$, $\mathbf{v}_2 = \begin{bmatrix} 2 \\ 1 \\ 2 \end{bmatrix}$, $\mathbf{v}_3 = \begin{bmatrix} 1 \\ 3 \\ 3 \end{bmatrix}$.

a) Solve $A\mathbf{x} = \mathbf{v}_1$, $A\mathbf{x} = \mathbf{v}_2$, $A\mathbf{x} = \mathbf{v}_3$.

b) Solve $B\mathbf{x} = \mathbf{v}_1$, $B\mathbf{x} = \mathbf{v}_2$, $B\mathbf{x} = \mathbf{v}_3$.

4. The solutions of a system may be given in different forms. Let the equation $A\mathbf{x} = \mathbf{b}$ have as solutions all

$$\mathbf{x} = \mathbf{x}_0 + t_1 \mathbf{v}_1 + t_2 \mathbf{v}_2$$

($\mathbf{v}_1$, $\mathbf{v}_2$ being linearly independent) and also all

$$\mathbf{x} = \mathbf{x}_1 + t_1 \mathbf{u}_1 + t_2 \mathbf{u}_2$$

($\mathbf{u}_1$, $\mathbf{u}_2$ being linearly independent). Show that these two sets of solutions agree if and only if the matrix whose columns are $\mathbf{x}_1 - \mathbf{x}_0$, $\mathbf{v}_1$, $\mathbf{v}_2$, $\mathbf{u}_1$, $\mathbf{u}_2$ has rank 2.

5–15 THE TRANSPOSE

Let $\mathbf{A} = (a_{ij})$ be an $m \times n$ matrix. We denote by A' the $n \times m$ matrix $B = (b_{ij})$ such that $b_{ij} = a_{ji}$ for $i = 1, \ldots, n$, $j = 1, \ldots, m$. Thus $B = A'$ is obtained from A by interchanging rows and columns. The following pair is an illustration:

$$A = \begin{bmatrix} 3 & 1 & 2 \\ 5 & 0 & 7 \end{bmatrix}, \qquad A' = \begin{bmatrix} 3 & 5 \\ 1 & 0 \\ 2 & 7 \end{bmatrix}.$$

The first row of A becomes the first column of A'; the second row of A becomes the second column of A'. In general, we call A' the *transpose* of A. We observe that $I' = I$. The transpose of a matrix obeys several rules, which we adjoin to our list:

24. $(A + B)' = A' + B'$.   25. $(cA)' = cA'$.   26. $(A')' = A$.

27. $(AB)' = B'A'$.   28. If A is nonsingular, then $(A^{-1})' = (A')^{-1}$.   (5–150)

To prove Rule 24, we write $D = A + B = (d_{ij})$, so that

$$d_{ij} = a_{ij} + b_{ij}$$

for all i and j. Then $D' = E = (e_{ij})$, where $e_{ij} = d_{ji}$ for all i and j, or

$$e_{ij} = a_{ji} + b_{ji}.$$

Thus $E = A' + B'$ or $D' = A' + B'$. The proofs of Rules 25 and 26 are left as exercises (Problem 4 below).

To prove Rule 27, we let A be $m \times p$, B be $p \times n$. We then write $C = AB = (c_{ij})$, $D = A' = (d_{ij})$, $E = B' = (e_{ij})$. Then $B'A' = ED = F = (f_{ij})$, where

$$f_{ij} = e_{i1} d_{1j} + \cdots + e_{ip} d_{pj} = b_{1i} a_{j1} + \cdots + b_{pi} a_{jp}$$
$$= a_{j1} b_{1i} + \cdots + a_{jp} b_{pi} = c_{ji}, \quad i = 1, \ldots, n, \quad j = 1, \ldots, m.$$

Hence, $F = C'$ or $B'A' = (AB)'$.

To prove Rule 28, we write: $AA^{-1} = I$. Then by Rule 27,

$$(A^{-1})'A' = I' = I.$$

From this equation it follows, as in Section 5–6, that $(A')^{-1} = (A^{-1})'$.

A matrix A such that $A = A'$ is called a *symmetric* matrix. Here A must be a square matrix. The matrix I is symmetric, as are the following matrices:

$$\begin{bmatrix} 1 & 2 \\ 2 & 3 \end{bmatrix}, \qquad \begin{bmatrix} 3 & -1 & 0 \\ -1 & 7 & 2 \\ 0 & 2 & 4 \end{bmatrix}$$

Also, every diagonal matrix is symmetric.

Symmetric matrices are useful in discussing quadratic forms; that is, algebraic expressions of the form

$$\sum_{i=1}^{n} \sum_{j=1}^{n} a_{ij} x_i x_j. \tag{5–151}$$

For $n = 2$, the expression is

$$a_{11} x_1^2 + a_{12} x_1 x_2 + a_{21} x_2 x_1 + a_{22} x_2^2.$$

Here $x_1 x_2$ is the same as $x_2 x_1$, so that we could combine the second and third terms. However, it is preferable to split the combined term into two equal terms, each having as coefficient the average of a_{12} and a_{21}. For example, $3x_1^2 + 5x_1 x_2 + 7x_2 x_1 + 4x_2^2$ is replaced by

$$3x_1^2 + 6x_1 x_2 + 6x_2 x_1 + 4x_2^2. \tag{5-152}$$

By proceeding similarly for the general quadratic form (5–151), we can always assume that the *coefficient matrix* (a_{ij}) is symmetric, and it is standard practice to write quadratic forms in this way. For the above example this matrix is $\begin{bmatrix} 3 & 6 \\ 6 & 4 \end{bmatrix}$.

Now let the $n \times n$ symmetric matrix $A = (a_{ij})$ be given, and consider the quadratic form (5–151). Here we can take $x_1, \ldots, x_n$ as variables. For each assignment of numerical values to $x_1, \ldots, x_n$, the form (5–151) has a numerical value Q. Hence Q is a function of the n variables $x_1, \ldots, x_n$. However, it is simpler to think of Q as a function of the vector $\mathbf{x} = \text{col}(x_1, \ldots, x_n)$:

$$Q(\mathbf{x}) = \sum_{i=1}^{n} \sum_{j=1}^{n} a_{ij} x_i x_j. \tag{5-153}$$

Furthermore, for each $\mathbf{x}$ we can compute the number $Q(\mathbf{x})$ by matrix multiplications:

$$Q(\mathbf{x}) = \mathbf{x}' A \mathbf{x}. \tag{5-154}$$

Here $\mathbf{x}'$ is the transpose of the column vector $\mathbf{x}$ and is therefore the row vector $(x_1, \ldots, x_n)$. Accordingly, (5–154) is the same as

$$Q(\mathbf{x}) = (x_1, \ldots, x_n) \begin{bmatrix} a_{11} & \cdots & a_{1n} \\ \vdots & & \vdots \\ a_{n1} & \cdots & a_{nn} \end{bmatrix} \begin{bmatrix} x_1 \\ \vdots \\ x_n \end{bmatrix}.$$

The product of the last two factors is an $n \times 1$ column vector whose ith entry is $a_{i1} x_1 + \cdots + a_{in} x_n$. The product of the $1 \times n$ row vector $(x_1, \ldots, x_n)$ and this $n \times 1$ column vector is a 1×1 matrix, that is, a number; in fact, it is precisely the number on the right of (5–153). Therefore, (5–154) is indeed another way of writing (5–153).

As an illustration, we write the quadratic form $Q(\mathbf{x})$ of (5–151) as follows:

$$Q(\mathbf{x}) = \mathbf{x}' \begin{bmatrix} 3 & 6 \\ 6 & 4 \end{bmatrix} \mathbf{x}, \qquad \mathbf{x} = \text{col}(x_1, x_2).$$

When expanded, this becomes (as expected)

$$Q(\mathbf{x}) = (x_1, x_2) \begin{bmatrix} 3 & 6 \\ 6 & 4 \end{bmatrix} \begin{bmatrix} x_1 \\ x_2 \end{bmatrix} = (x_1, x_2) \begin{bmatrix} 3x_1 + 6x_2 \\ 6x_1 + 4x_2 \end{bmatrix}$$

$$= x_1(3x_1 + 6x_2) + x_2(6x_1 + 4x_2) = 3x_1^2 + 6x_1 x_2 + 6x_2 x_1 + 4x_2^2.$$

5–16 ORTHOGONAL MATRICES

Let A be a real $n \times n$ matrix. Then A is said to be *orthogonal* if

$$AA' = I. \tag{5–160}$$

Hence, A is orthogonal if and only if $A^{-1} = A'$, that is, if and only if the inverse of A equals the transpose of A. Thus every orthogonal matrix is nonsingular. The following are examples of orthogonal matrices:

$$A = \begin{bmatrix} \frac{3}{5} & \frac{4}{5} \\ -\frac{4}{5} & \frac{3}{5} \end{bmatrix}, \qquad B = \begin{bmatrix} \frac{2}{3} & \frac{2}{3} & \frac{1}{3} \\ \frac{2}{3} & -\frac{1}{3} & -\frac{2}{3} \\ \frac{1}{3} & -\frac{2}{3} & \frac{2}{3} \end{bmatrix}.$$

Let us consider the row vectors $\mathbf{u}_1$, $\mathbf{u}_2$ of A as vectors in the xy-plane:

$$\mathbf{u}_1 = \tfrac{3}{5}\mathbf{i} + \tfrac{4}{5}\mathbf{j}, \qquad \mathbf{u}_2 = -\tfrac{4}{5}\mathbf{i} + \tfrac{3}{5}\mathbf{j}.$$

Then we observe that $\mathbf{u}_1$ and $\mathbf{u}_2$ are both unit vectors and that $\mathbf{u}_1 \cdot \mathbf{u}_2 = 0$, so that $\mathbf{u}_1$, $\mathbf{u}_2$ are perpendicular. A similar statement applies to the column vectors of A:

$$\mathbf{v}_1 = \tfrac{3}{5}\mathbf{i} - \tfrac{4}{5}\mathbf{j}, \qquad \mathbf{v}_2 = \tfrac{4}{5}\mathbf{i} + \tfrac{3}{5}\mathbf{j}.$$

We can proceed similarly with the row vectors (or column vectors) of B, regarding them as vectors in space:

$$\mathbf{u}_1 = \tfrac{2}{3}\mathbf{i} + \tfrac{2}{3}\mathbf{j} + \tfrac{1}{3}\mathbf{k}, \ldots .$$

Again we verify that the row vectors (or column vectors) are mutually perpendicular unit vectors.

The geometrical concepts used here can be generalized to n dimensions, as in Section 5–8 above. Here we phrase the conditions algebraically. For an $n \times n$ matrix $A = (a_{ij})$, the crucial conditions are as follows:

$$a_{i1}^2 + \cdots + a_{in}^2 = 1, \quad i = 1, \ldots, n, \tag{5–161}$$

$$a_{i1}a_{j1} + \cdots + a_{in}a_{jn} = 0, \quad i \neq j, \ i, j = 1, \ldots, n, \tag{5–162}$$

$$a_{1j}^2 + \cdots + a_{nj}^2 = 1, \quad j = 1, \ldots, n, \tag{5–163}$$

$$a_{1j}a_{1k} + \cdots + a_{nj}a_{nk} = 0, \quad j \neq k, \ j, k = 1, \ldots, n. \tag{5–164}$$

Here (5–161) says that the row vectors are unit vectors; (5–162) states that different row vectors are orthogonal; (5–163) and (5–164) express analogous conditions on the column vectors.

Every $n \times n$ orthogonal matrix $A = (a_{ij})$ satisfies all four conditions (5–161) through (5–164), because if A is orthogonal, then $AA' = I$. By the definition of matrix multiplication, this means that

$$a_{i1}a_{j1} + \cdots + a_{in}a_{jn} = \delta_{ij} = \begin{cases} 1, & i = j \\ 0, & i \neq j \end{cases}. \tag{5–165}$$

Thus (5–161) and (5–162) follow.

By the properties of inverses (Section 5–6), $AA' = I$ also implies that $A'A = I$, and hence

$$a_{1i}a_{1j} + \cdots + a_{ni}a_{nj} = \delta_{ij}, \tag{5–166}$$

and this implies (5–163) and (5–164).

If A satisfies either pair of conditions (5–161), (5–162) or (5–163), (5–164), then it satisfies the other, so that A is orthogonal. The reasoning just given shows that the first pair is equivalent to $AA' = I$ and the second to $A'A = I$. Since $AA' = I$ implies $A'A = I$, and conversely, the conclusion follows.

Orthogonal matrices arise naturally in changing coordinates in R^n. Up till now we have used the standard basis $e_1 = (1, 0, \ldots, 0), \ldots, e_n = (0, \ldots, 1)$ for vectors, and the coordinates of the point P are the components of $\overrightarrow{OP}$ with respect to this basis. This basis is *orthonormal*, that is, formed of perpendicular unit vectors.

Now let $e_1^*, \ldots, e_n^*$ be a second orthonormal basis for R^n, so that these vectors are mutually perpendicular unit vectors. Then an arbitrary vector v has components $(v_1, \ldots, v_n)$ with respect to the old basis $e_1, \ldots, e_n$ and components $(v_1^*, \ldots, v_n^*)$ with respect to the new basis. We ask how the two sets of components are related.

To answer our question, we introduce the matrix $A = (a_{ij})$, where

$$a_{ij} = e_i^* \cdot e_j, \quad i, j = 1, \ldots, n. \tag{5–167}$$

Thus the ith row of A (with i fixed) gives the components of e_i^* with respect to the old basis. Since the e_i^* are mutually perpendicular vectors, A is an orthogonal matrix. From (5–167) we also see that the jth column of A gives the components of e_j with respect to the new basis.

We can now write (using orthogonality)

$$v = v_1 e_1 + \cdots + v_n e_n, \qquad v_i = v \cdot e_i \quad \text{for } i = 1, \ldots, n;$$
$$v = v_1^* e_1^* + \cdots + v_n^* e_n^*, \qquad v_j^* = v \cdot e_j^* \quad \text{for } j = 1, \ldots, n.$$

Therefore

$$v_i^* = (v_1 e_1 + \cdots + v_n e_n) \cdot e_i^* = v_1 e_i^* \cdot e_1 + \cdots + v_n e_i^* \cdot e_n$$
$$= a_{i1} v_1 + \cdots + a_{in} v_n, \quad i = 1, \ldots, n.$$

Thus

$$\begin{bmatrix} v_1^* \\ \vdots \\ v_n^* \end{bmatrix} = \begin{bmatrix} a_{11} & \cdots & a_{1n} \\ \vdots & & \vdots \\ a_{n1} & \cdots & a_{nn} \end{bmatrix} \begin{bmatrix} v_1 \\ \vdots \\ v_n \end{bmatrix} = A \begin{bmatrix} v_1 \\ \vdots \\ v_n \end{bmatrix}. \tag{5–168}$$

Accordingly, the orthogonal matrix A relates the two sets of components. Since $A^{-1} = A'$, from (5–168) we have also

$$\begin{bmatrix} v_1 \\ \vdots \\ v_n \end{bmatrix} = A' \begin{bmatrix} v_1^* \\ \vdots \\ v_n^* \end{bmatrix}. \tag{5–169}$$

If we use the new basis to assign new coordinates $(x_1^*, \ldots, x_n^*)$ to each point P, with the same origin O, then we conclude that new and old coordinates are related by a similar equation

$$\begin{bmatrix} x_1^* \\ \vdots \\ x_n^* \end{bmatrix} = A \begin{bmatrix} x_1 \\ \vdots \\ x_n \end{bmatrix}.$$

If we also choose a new origin O^*, then the new coordinates are the components of $\overrightarrow{O^*P}$ with respect to the new basis. Since

$$\overrightarrow{OP} = \overrightarrow{OO^*} + \overrightarrow{O^*P},$$

we conclude that

$$\begin{bmatrix} x_1^* \\ \vdots \\ x_n^* \end{bmatrix} = A \begin{bmatrix} x_1 - h_1 \\ \vdots \\ x_n - h_n \end{bmatrix},$$

where $(h_1, \ldots, h_n)$ are the coordinates of O^* with respect to the old basis and origin O. If we change the origin but not the basis, then $A = I$, so that

$$x_1^* = x_1 - h_1, \ldots, x_n^* = x_n - h_n.$$

Here we deal with a *translation of axes*.

EXAMPLE We introduce the new basis

$$\mathbf{e}_1^* = \tfrac{3}{5}\mathbf{i} + \tfrac{4}{5}\mathbf{j}, \qquad \mathbf{e}_2^* = -\tfrac{4}{5}\mathbf{i} + \tfrac{3}{5}\mathbf{j}$$

for vectors in the plane, as in Fig. 5–20. Thus matrix A is

$$\begin{bmatrix} \tfrac{3}{5} & \tfrac{4}{5} \\ -\tfrac{4}{5} & \tfrac{3}{5} \end{bmatrix}.$$

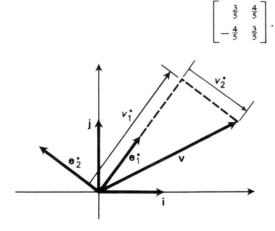

Fig. 5–20. Change of basis in the plane.

If $\mathbf{v} = 2\mathbf{i} + \mathbf{j}$, then (as in the figure)

$$\begin{bmatrix} v_1^* \\ v_2^* \end{bmatrix} = A \begin{bmatrix} 2 \\ 1 \end{bmatrix} = \begin{bmatrix} \tfrac{3}{5} & \tfrac{4}{5} \\ -\tfrac{4}{5} & \tfrac{3}{5} \end{bmatrix} \begin{bmatrix} 2 \\ 1 \end{bmatrix} = \begin{bmatrix} 2 \\ -1 \end{bmatrix}. \blacktriangleleft$$

Application to symmetric matrices

Let A be an $n \times n$ symmetric matrix. Then it can be shown that the eigenvalues $\lambda_1, \ldots, \lambda_n$ of A are *real* and that corresponding eigenvectors $\mathbf{v}_1, \ldots, \mathbf{v}_n$ of A can be chosen to be pairwise orthogonal. We can further require that $\mathbf{v}_1, \ldots, \mathbf{v}_n$ all have length 1. Thus the matrix C, whose column vectors are $\mathbf{v}_1, \ldots, \mathbf{v}_n$, is *orthogonal*. We saw in Section 5–7 that

$$AC = CB, \qquad B = \text{diag}(\lambda_1, \ldots, \lambda_n)$$

(see also Problem 13 following Section 5–9). Hence

$$B = C^{-1}AC = C'AC,$$

where $C' = C^{-1}$ by orthogonality. Thus A is similar to a diagonal matrix: $C'AC = B = \text{diag}(\lambda_1, \ldots, \lambda_n)$, where C is orthogonal.

This result has implications for quadratic forms. Let

$$Q(\mathbf{x}) = \sum_{i=1}^{n} \sum_{j=1}^{n} a_{ij} x_i x_j = \mathbf{x}' A \mathbf{x},$$

as in Section 5–15. Now we express Q in terms of new variables $y_1, \ldots, y_n$ by setting $\mathbf{x} = C\mathbf{y}$, with C chosen as above. Then $\mathbf{x}' = \mathbf{y}'C'$ by Rule 27 and hence

$$Q(\mathbf{x}) = \mathbf{x}' A \mathbf{x} = \mathbf{y}'C'AC\mathbf{y} = \mathbf{y}'B\mathbf{y}$$

$$= \sum_{i-1}^{n} \sum_{j-1}^{n} b_{ij} y_i y_j.$$

Since $B = \text{diag}(\lambda_1, \ldots, \lambda_n)$, the expression for $Q(\mathbf{x})$ in terms of $y_1, \ldots, y_n$ is simply

$$\lambda_1 y_1^2 + \lambda_2 y_2^2 + \cdots + \lambda_n y_n^2.$$

Thus every quadratic form can be written, in terms of new variables, in a form containing only terms in the squares of the variables. Furthermore, since $\mathbf{x} = C\mathbf{y}$ and C is orthogonal, $y_1, \ldots, y_n$ can be regarded as new coordinates in R^n obtained from $x_1, \ldots, x_n$ by a change of basis as above (the new basis vectors are eigenvectors of A). (For proofs of these results see Section 5–3 of *Theory of Matrices* by S. Perlis, Addison-Wesley Pub. Co., Reading, Mass. 1952.)

Potential energy in physics

In many physical problems we introduce potential energy U which is a quadratic form $\sum a_{ij} x_i x_j$ (as above) and has its minimum value (zero) only for $x_1 = 0, \ldots, x_n = 0$. By changing coordinates as above, we can write

$$U = \lambda_1 y_1^2 + \cdots + \lambda_n y_n^2 = F(y_1, \ldots, y_n)$$

in terms of the new coordinates $y_1, \ldots, y_n$. Now the unique minimum occurs for $y_1 = 0, \ldots, y_n = 0$. If any of $\lambda_1, \ldots, \lambda_n$ is negative or 0, this condition is violated. For example, if $\lambda_1 \leqslant 0$, then $F(y_1, 0, \ldots, 0) = \lambda_1 y_1^2 \leqslant 0$ for all y_1. Therefore, we must have $\lambda_1 > 0, \ldots, \lambda_n > 0$. Thus *for a potential energy U as described all eigenvalues λ_j of the matrix (a_{ij}) must be positive.* Such a potential energy U is said to be *positive definite.*

We remark that under an *arbitrary* change of variables $\mathbf{x} = C\mathbf{y}$ with an orthogonal $n \times n$ matrix C (as above) the positive definite quadratic form

$U = \sum a_{ij}x_i x_j$ becomes $\sum b_{ij}y_i y_j = F(y_i, \ldots, y_n)$, with $B = C'AC$. Now B is not necessarily diagonal, but the new form must still be positive definite, for $F(y_1, \ldots, y_n)$ is simply the function U expressed in terms of new coordinates and hence $F(y_1, \ldots, y_n) \geqslant 0$ for all $\mathbf{y}$. Also U can equal 0 only when $\mathbf{x} = \mathbf{0}$, that is, in the new coordinates, only when $\mathbf{y} = C^{-1}\mathbf{x} = C^{-1}\mathbf{0} = \mathbf{0}$. Therefore, the new quadratic form is positive definite and $B = C'AC$ has all eigenvalues positive.

========================= **Problems (Section 5–16)** =========================

1. Find the transpose of each of the matrices:

   a) $\begin{bmatrix} 1 & 2 & 3 \\ 3 & 0 & 5 \end{bmatrix}$
   b) $\begin{bmatrix} 3 & 1 \\ 0 & 2 \\ 1 & 0 \end{bmatrix}$
   c) $(1, 5, 0, 4)$
   d) $\begin{bmatrix} 1 \\ 0 \\ 7 \end{bmatrix}$

2. Choose a and b so that each of the following matrices becomes symmetric:

   a) $\begin{bmatrix} 1 & 3a-1 \\ 2a & 3 \end{bmatrix}$
   b) $\begin{bmatrix} 2 & a & 3 \\ b-a & 0 & 4+a \\ 3 & b & 5 \end{bmatrix}$

3. For each of the following quadratic forms obtain the coefficient matrix when the form is written in such a way that the coefficient matrix is symmetric:

   a) $5x_1^2 + 4x_1 x_2 + 3x_2^2$ b) $7x_1^2 + 2x_1 x_2 - x_2^2$
   c) $x_1^2 + 3x_2^2 - x_3^2 + 4x_1 x_2 + 6x_1 x_3 + 2x_2 x_3$
   d) $2x_1^2 + x_2^2 + x_3^2 + 2x_1 x_3 - 4x_2 x_3$

4. Prove each of the following parts of (5–150):
   a) Rule 25   b) Rule 26

5. Show that each of the following matrices is orthogonal:

   a) $\dfrac{1}{13}\begin{bmatrix} 5 & 12 \\ -12 & 5 \end{bmatrix}$
   b) $\begin{bmatrix} \cos\omega & \sin\omega \\ -\sin\omega & \cos\omega \end{bmatrix}$

   c) $\dfrac{1}{7}\begin{bmatrix} 2 & 3 & 6 \\ 6 & 2 & -3 \\ 3 & -6 & 2 \end{bmatrix}$
   d) $\dfrac{1}{2}\begin{bmatrix} 1 & 1 & 1 & 1 \\ 1 & -1 & -1 & 1 \\ 1 & 1 & -1 & -1 \\ 1 & -1 & 1 & -1 \end{bmatrix}$

6. (a), (b), (c). Represent the row vectors of each of the matrices in Problems 5 (a), (b), (c) as vectors in the plane or in space, graph them, and verify that they are mutually perpendicular unit vectors.

7. (a), (b), (c). Proceed as in Problem 6 with column vectors instead of row vectors.

8. Let A and B be $n \times n$ orthogonal matrices. Prove:

   a) det $A = \pm 1$ b) AB is also orthogonal   c) A' and A^{-1} are orthogonal

9. Let $A = \begin{bmatrix} 7 & 24 \\ 24 & -7 \end{bmatrix}$.
   a) Verify that A is symmetric.

b) Find the eigenvalues of A and verify that they are real. Show that $\mathbf{v}_1 = \text{col}(4/5, 3/5)$ and $\mathbf{v}_2 = \text{col}(3/5, -4/5)$ are eigenvectors of A, which are unit vectors and are perpendicular.

c) Use the eigenvectors of Part (b) as successive columns of matrix C. Verify that C is orthogonal and that $B = C'AC$ is diagonal.

d) Verify that the quadratic form

$$Q(x) = 7x_1^2 + 24x_1x_2 + 24x_2x_1 - 7x_2^2$$

corresponding to A becomes $25\,y_1^2 - 25\,y_2^2$ if we set $\mathbf{x} = C\mathbf{y}$, with C as in Part (c). Show the y_1, y_2 axes in the x_1x_2-plane.

5–17 GENERAL VECTOR SPACES

Although we have emphasized the geometric aspects of vectors, the same concepts arise in other contexts, and we can interpret a surprising variety of objects as "vectors" in a "vector space" or "linear space". This was pointed out in Section 1–16. Here we expand the discussion of that section, emphasizing the dimension concepts.

By a *vector space* (or *linear space*) V we mean a collection of objects, called *vectors*, that can be added and multiplied by scalars (real numbers) subject to the usual rules:

$$\mathbf{u} + \mathbf{v} = \mathbf{v} + \mathbf{u}, \qquad \mathbf{u} \dotplus (\mathbf{v} + \mathbf{w}) = (\mathbf{u} + \mathbf{v}) + \mathbf{w},$$

$$c(\mathbf{u} + \mathbf{v}) = c\mathbf{u} + c\mathbf{v}, \qquad a(b\mathbf{u}) = (ab)\mathbf{u}, \qquad (a + b)\mathbf{u} = a\mathbf{u} + b\mathbf{u}, \qquad (5\text{–}170)$$

$$1\mathbf{u} = \mathbf{u}, \quad 0\mathbf{u} = \mathbf{0}.$$

From these rules we derive a variety of others. For example, with $-\mathbf{u}$ standing for $(-1)\mathbf{u}$, we have $\mathbf{u} + (-\mathbf{u}) = \mathbf{0}$, and the equation $\mathbf{u} + \mathbf{w} = \mathbf{v}$ has a unique solution for $\mathbf{w}$: namely $\mathbf{w} = \mathbf{v} + (-\mathbf{u})$, which we usually write as $\mathbf{w} = \mathbf{v} - \mathbf{u}$. Note that the rules (5–170) require that V has a zero vector $\mathbf{0}$; in each particular case of a vector space, we must specify which is the $\mathbf{0}$ vector.

In vector space V, vectors $\mathbf{u}_1, \ldots, \mathbf{u}_k$ are said to be *linearly independent* if the equation

$$c_1\mathbf{u}_1 + \cdots + c_k\mathbf{u}_k = \mathbf{0} \qquad (5\text{–}171)$$

can hold only for $c_1 = 0, \ldots, c_k = 0$. Otherwise, $\mathbf{u}_1, \ldots, \mathbf{u}_k$ are said to be *linearly dependent*; in this case, a relation (5–171) will hold for certain scalars $c_1, \ldots, c_k$ (not all 0) or, what is equivalent, one of the vectors $\mathbf{u}_1, \ldots, \mathbf{u}_k$ will be expressible as a linear combination of the others. For example, if $\mathbf{v}$, $\mathbf{w}$ are vectors of V, then $\mathbf{u}_1 = \mathbf{v} + \mathbf{w}$, $\mathbf{u}_2 = \mathbf{v} - \mathbf{w}$ and $\mathbf{u}_3 = 2\mathbf{v} + 3\mathbf{w}$ are linearly dependent, since

$$5\mathbf{u}_1 - \mathbf{u}_2 - 2\mathbf{u}_3 = 5(\mathbf{v} + \mathbf{w}) - (\mathbf{v} - \mathbf{w}) - 2(2\mathbf{v} + 3\mathbf{w}) = \mathbf{0},$$

or, equivalently, $\mathbf{u}_2 = 5\mathbf{u}_1 - 2\mathbf{u}_3$.

The *dimension* of vector space V is the largest number n such that V contains n linearly independent vectors. However, it may happen that, for *every* positive integer n, space V contains n linearly independent vectors, so that there is no largest n; in this case, V is said to have *infinite dimension*.

We shall now illustrate these definitions.

EXAMPLE 1 Let V represent all vectors in 3-dimensional space (Figure 5–4). This is a 3-dimensional vector space. ◄

EXAMPLE 2 Let V represent all vectors in n-dimensional space R^n. As shown in Section 5–9, V is an n-dimensional vector space. ◄

EXAMPLE 3 Let V represent all polynomials in x of degree at most 2. It is clear that the sum of two such polynomials is a polynomial of V and that a number times a polynomial of V equals a polynomial of V:

$$(1 + 2x - 3x^2) + (2 + x + x^2) = 3 + 3x - 2x^2,$$
$$3(2 + x - x^2) = 6 + 3x - 3x^2.$$

Thus the basic operations are defined and the rules (5–170) can easily be verified. The **0** vector is the polynomial 0 (equal to 0 for all x). Space V contains the polynomials 1, x, x^2, and these three polynomials are linearly independent, for in V

$$c_1 \cdot 1 + c_2 x + c_3 x^2 = 0$$

means that the polynomial on the left is to be *identically* 0. By algebra, this is possible only when $c_1 = 0, c_2 = 0, c_3 = 0$. However, there can be no more than three linearly independent polynomials in V. For example, each set of four polynomials:

$$\begin{aligned}
p_1(x) &= a_1 + a_2 x + a_3 x^2, \\
p_2(x) &= b_1 + b_2 x + b_3 x^2, \\
p_3(x) &= c_1 + c_2 x + c_3 x^2, \\
p_4(x) &= d_1 + d_2 x + d_3 x^2
\end{aligned} \qquad (5\text{–}172)$$

must be linearly dependent. To prove this, let

$$\Delta = \begin{vmatrix} a_1 & a_2 & a_3 \\ b_1 & b_2 & b_3 \\ c_1 & c_2 & \cdot\, c_3 \end{vmatrix}. \qquad (5\text{–}173)$$

Then, if $\Delta \neq 0$, we can solve the first three equations of (5–172) for 1, x, x^2 in terms of $p_1(x), p_2(x), p_3(x)$ by the usual method for simultaneous equations (Section 5–1). Thus

$$1 = \begin{vmatrix} p_1(x) & a_2 & a_3 \\ p_2(x) & b_2 & b_3 \\ p_3(x) & c_2 & c_3 \end{vmatrix} \div \Delta, \qquad x = \begin{vmatrix} a_1 & p_1(x) & a_3 \\ a_2 & p_2(x) & b_3 \\ a_3 & p_3(x) & c_3 \end{vmatrix} \div \Delta, \qquad x^2 = \dots$$

If we substitute the results in the expression for $p_4(x)$, we have $p_4(x)$ as a linear combination of $p_1(x), p_2(x), p_3(x)$. Thus the polynomials are linearly dependent. If $\Delta = 0$, then the equations

$$\begin{aligned}
k_1 a_1 + k_2 b_1 + k_3 c_1 &= 0, \\
k_1 a_2 + k_2 b_2 + k_3 c_2 &= 0, \\
k_1 a_3 + k_2 b_3 + k_3 c_3 &= 0
\end{aligned} \qquad (5\text{–}174)$$

have a solution for k_1, k_2, k_3, with not all of these numbers equal to 0 (Section 5–1). Then

$$k_1 p_1(x) + k_2 p_2(x) + k_3 p_3(x) + 0\,p_4(x) = 0,$$

as we see from (5–172) and (5–174). Hence again we have linear dependence. In the same way we can show that any five polynomials of V, or any six polynomials of V, etc., are linearly dependent. Thus the dimension of V is 3. ◄

EXAMPLE 4   Let V represent all polynomials in x of degree at most 4. Here V contains the linearly independent set $1, x, x^2, x^3, x^4$, and V has dimension 5, since by reasoning as in Example 3 we can show that any set of more than five polynomials in V is linearly dependent. ◄

In general, a set of vectors $\mathbf{u}_1, \ldots, \mathbf{u}_n$ is said to form a *basis* for vector space V if each vector $\mathbf{u}$ of V can be expressed in *unique* fashion as a linear combination of $\mathbf{u}_1, \ldots, \mathbf{u}_n$, that is,

$$\mathbf{u} = c_1 \mathbf{u}_1 + \cdots + c_n \mathbf{u}_n$$

for unique scalars $c_1, \ldots, c_n$. *The vectors* $\mathbf{u}_1, \ldots, \mathbf{u}_n$ *of a basis must be linearly independent*, for if not, we would have

$$\mathbf{0} = 0\mathbf{u}_1 + \cdots + 0\mathbf{u}_n \qquad \text{and} \qquad \mathbf{0} = c_1 \mathbf{u}_1 + \cdots + c_n \mathbf{u}_n,$$

with $c_1, \ldots, c_n$ not all 0. Thus $\mathbf{0}$ would have two *different* expressions as a linear combination of $\mathbf{u}_1, \ldots, \mathbf{u}_n$, contrary to assumption.

In Example 1 the vectors $\mathbf{i}, \mathbf{j}, \mathbf{k}$ form a basis; in Example 2, it is the vectors $\mathbf{e}_1 = (0, 0, \ldots, 0) \ldots, \mathbf{e}_n = (0, \ldots, 0, 1)$; in Example 3, it is the polynomials $1, x, x^2$; in Example 4, the polynomials $1, x, x^2, x^3, x^4$.

As these examples illustrate and the definition of basis shows, the *number of vectors in a basis equals the dimension of V.*

EXAMPLE 5   Let V represent all polynomials in x. We see at once that V is a vector space. However, V has infinite dimension since polynomials $1, x, \ldots, x^{n-1}$ are in V and are linearly independent for each n. ◄

EXAMPLE 6   Let V represent all continuous functions of x on the interval $a \leqslant x \leqslant b$. The sum of two continuous functions is continuous, a scalar times a continuous function is continuous. Rules (5–170) are all satisfied, where the $\mathbf{0}$ vector is the identically 0 function 0. Again, V has infinite dimension, since $1, x, \ldots, x^{n-1}, \ldots$ are in V. Also $e^x, e^{2x}, \ldots, e^{nx}, \ldots$ are in V and, for each n, the first n of these are linearly independent, since if

$$c_1 e^x + c_2 e^{2x} + \cdots + c_n e^{nx} = 0$$

(identically 0), then by differentiating repeatedly we obtain $(n - 1)$ further identities:

$$c_1 e^x + 2c_2 e^{2x} + \cdots + nc_n e^{nx} = 0,$$

$$\vdots \qquad\qquad\qquad \vdots$$

$$c_1 e^x + 2^{n-1} c_2 e^{2x} + \cdots + n^{n-1} c_n e^{nx} = 0.$$

Thus we have n homogeneous linear equations in $c_1, \ldots, c_n$. If these can be satisfied with not all of $c_1, \ldots, c_n$ equal to 0, then the determinant of the coefficients must be 0 (Section 5–1); that is,

$$
\begin{vmatrix}
e^x & e^{2x} & \cdots & e^{nx} \\
e^x & 2e^{2x} & \cdots & ne^{nx} \\
\vdots & & & \vdots \\
e^x & 2^{n-1}e^{2x} & \cdots & n^{n-1}e^{nx}
\end{vmatrix} = 0
\quad\text{or}\quad
\begin{vmatrix}
1 & 1 & \cdots & 1 \\
1 & 2 & \cdots & n \\
\vdots & & & \vdots \\
1 & 2^{n-1} & \cdots & n^{n-1}
\end{vmatrix} = 0.
$$

The last determinant is a Vandermonde determinant. It is shown in algebra (CLA, page 779), that it has the value

$$
1^{n-1} \quad 2^{n-2} \quad \cdots \quad (n-1)^1,
$$

which is not 0. Therefore $c_1, \ldots, c_n$ must all be 0 and the functions $e^x, \ldots, e^{nx}$ are linearly independent.

In the same way, we can show that $e^{\lambda_1 x}, \ldots, e^{\lambda_n x}$ are linearly independent if $\lambda_1, \ldots, \lambda_n$ are distinct; also that

$$
\cos\beta_1 x, \quad \sin\beta_1 x, \quad \cos\beta_2 x, \quad \sin\beta_2 x, \ldots, \cos\beta_n x, \sin\beta_n x
$$

are $2n$ linearly independent members of V, provided that $\beta_1, \ldots, \beta_n$ are all different and positive. ◄

Such sets occur in the solution of homogeneous linear differential equations with constant coefficients (Sections 1–11 and 1–15). By methods similar to those used here, it can be shown that the procedures of Chapter 1 always produce linearly independent sets. (For a full discussion, see ODE, Sections 4–2, 4–3, 4–8.)

Subspaces

As done at the end of Section 5–10, we define a subspace of V as a collection of vectors in V such that $\mathbf{0}$ is in the collection and if $\mathbf{u}$ and $\mathbf{v}$ are in the collection, then so are $\mathbf{u} + \mathbf{v}$ and $c\mathbf{u}$ for every scalar c. If V consists of all polynomials of degree at most 3, then the polynomials of degree at most 2 form a subspace. In general, a subspace of a vector space itself forms a vector space.

If $\mathbf{v}_1, \ldots, \mathbf{v}_n$ are linearly independent vectors of V, then the set W of all linear combinations of $\mathbf{v}_1, \ldots, \mathbf{v}_n$ forms a subspace of V, as can be seen from the definition of subspace. Furthermore, $\mathbf{v}_1, \ldots, \mathbf{v}_n$ clearly form a basis for W, so that W has dimension n.

5–18 LINEAR MAPPINGS

Let U and V be vector spaces. Then by a linear mapping T of U into V we mean an assignment of a vector $\mathbf{v} = T(\mathbf{u})$ in V to each vector $\mathbf{u}$ of U, whereby in general

$$
T(c_1\mathbf{u}_1 + c_2\mathbf{u}_2) = c_1 T(\mathbf{u}_1) + c_2 T(\mathbf{u}_2) \tag{5–180}
$$

A major part of mathematics and its applications is concerned with such mappings.

For a linear mapping T, we define the *range* of T as the set of all $\mathbf{v}$ in V such that $\mathbf{v} = T(\mathbf{u})$ for some $\mathbf{u}$ in U. The *kernel* of T is the set of all $\mathbf{u}$ in U such that $T(\mathbf{u}) = \mathbf{0}$.

From the rule (5–180) we conclude that $T(\mathbf{0}) = \mathbf{0}$, since
$$T(\mathbf{0}) = T(0\mathbf{0}) = 0T(\mathbf{0}) = \mathbf{0}.$$
Hence the range and the kernel of T include $\mathbf{0}$ (as vectors in V and U respectively). Both the range and kernel are subspaces of V and U, respectively, since if $\mathbf{v}_1$ and $\mathbf{v}_2$ are in the range, then $T(\mathbf{u}_1) = \mathbf{v}_1$ and $T(\mathbf{u}_2) = \mathbf{v}_2$ for appropriate $\mathbf{u}_1$ and $\mathbf{u}_2$, respectively. Therefore,
$$T(c_1\mathbf{u}_1 + c_2\mathbf{u}_2) = c_1 T(\mathbf{u}_1) + c_2 T(\mathbf{u}_2) = c_1\mathbf{v}_1 + c_2\mathbf{v}_2,$$
so that $c_1\mathbf{v}_1 + c_2\mathbf{v}_2$ is also in the range. Similar reasoning shows that if $T(\mathbf{u}_1) = \mathbf{0}$ and $T(\mathbf{u}_2) = \mathbf{0}$, then $T(c_1\mathbf{u}_1 + c_2\mathbf{u}_2) = \mathbf{0}$.

EXAMPLE 1 Let U be the vector space of all functions $f(x)$ with a continuous derivative for all x. Let V be the vector space of all functions continuous for all x. (That U and V are vector spaces follows at once from the definition.) Let $T(f) = f'$ (the derivative of f) for each f in U. Then T is linear, since
$$T(c_1 f_1 + c_2 f_2) = (c_1 f_1 + c_2 f_2)' = c_1 f_1' + c_2 f_2'$$
$$= c_1 T(f_1) + c_2 T(f_2),$$
by basic properties of the derivative. The range of T is all of V, since every continuous function g is the derivative of some function f (antiderivative of g), as shown in the calculus. The kernel of T consists of all f such that $f' = 0$, that is, of all constant functions. ◄

EXAMPLE 2 Let U be the vector space of all functions $f(x)$ with continuous first and second derivatives for all x. Let V be the vector space of all continuous functions, as in Example 1. Let $Tf = f'' - 2f' - 3f$ for each f in U. Then the kernel of T consists of all f such that
$$f'' - 2f' - 3f = 0.$$

Thus the kernel consists of all solutions of a homogeneous linear differential equation of second order with constant coefficients. As in Section 1–12, we find
$$f = c_1 e^{3x} + c_2 e^{-x}.$$
This shows that the kernel of T is 2-dimensional with basis e^{3x}, e^{-x}.

To find the range of T, we ask for the continuous functions g such that
$$f'' - 2f' - 3f = g$$
for some f. The method of variation of parameters (Section 1–13) shows that for *every* continuous g a corresponding f can be found. Hence the range of T is all of V.

One-to-one mappings; inverse mapping

Let T be a linear mapping of U into V. Then T is said to be *one-to-one* if $T(\mathbf{u}_1) = T(\mathbf{u}_2)$ implies $\mathbf{u}_1 = \mathbf{u}_2$. Thus each vector $\mathbf{v}$ in the range of T equals $T(\mathbf{u})$ for one and only one $\mathbf{u}$. In particular, $T(\mathbf{u}) = \mathbf{0}$ only for $\mathbf{u} = \mathbf{0}$, so that the kernel of T consists of $\mathbf{0}$ alone. Conversely, if the kernel of T contains only $\mathbf{0}$, then T must be one-to-one, for $T(\mathbf{u}_1) = T(\mathbf{u}_2)$ implies $T(\mathbf{u}_1 - \mathbf{u}_2) = \mathbf{0}$ and hence $\mathbf{u}_1 - \mathbf{u}_2 = \mathbf{0}$ or $\mathbf{u}_1 = \mathbf{u}_2$.

When T is one-to-one, T has an inverse T^{-1}, mapping the range of T onto U. Thus $T^{-1}(T\mathbf{u}) = \mathbf{u}$ for each $\mathbf{u}$ in U.

If T is not one-to-one, the equation $T(\mathbf{u}) = \mathbf{v}$ for given $\mathbf{v}$ in the range of T has many solutions $\mathbf{u}$. *If $\mathbf{u}_0$ is one solution, then all solutions are given by*

$$\mathbf{u} = \mathbf{u}_0 + \mathbf{w},$$

where $\mathbf{w}$ is an arbitrary vector of the kernel of T. For

$$T(\mathbf{u}) = T(\mathbf{u}_0 + \mathbf{w}) = T(\mathbf{u}_0) + T(\mathbf{w}) = \mathbf{v} + \mathbf{0} = \mathbf{v},$$

since $\mathbf{u}_0$ is one solution of $T(\mathbf{u}) = \mathbf{v}$ and $\mathbf{w}$ is in the kernel. Thus every vector $\mathbf{u} = \mathbf{u}_0 + \mathbf{w}$ is a solution. Further, if $\mathbf{u}$ is a solution, then

$$T(\mathbf{u} - \mathbf{u}_0) = T(\mathbf{u}) - T(\mathbf{u}_0) = \mathbf{v} - \mathbf{v} = \mathbf{0},$$

so that $\mathbf{u} - \mathbf{u}_0$ is in the kernel, that is, $\mathbf{u} - \mathbf{u}_0 = \mathbf{w}$ or $\mathbf{u} = \mathbf{u}_0 + \mathbf{w}$, as asserted.

EXAMPLE 3 Let U and V be all continuous functions of x for $0 \leqslant x \leqslant 1$. For $\mathbf{u}$, i.e., the function $u(x)$, in U let $T(\mathbf{u}) = \mathbf{v}$, i.e., the function $v(x)$, where

$$v(x) = \int_0^x u(t)\,dt. \tag{5–181}$$

Then T is linear, since

$$T(c_1\mathbf{u}_1 + c_2\mathbf{u}_2) = \int_0^x (c_1 u_1(t) + c_2 u_2(t))\,dt$$

$$= c_1 \int_0^x u_1(t)\,dt + c_2 \int_0^x u_2(t)\,dt = c_1 T(\mathbf{u}_1) + c_2 T(\mathbf{u}_2).$$

Here T is one-to-one, for $T(u(x)) = \mathbf{0}$ means

$$\int_0^x u(t)\,dt = 0, \quad 0 \leqslant x \leqslant 1.$$

If we differentiate this equation and use the rule

$$\frac{d}{dx} \int_0^x u(t)\,dt = u(x), \tag{5–182}$$

we conclude that $u(x) \equiv 0$. Hence the kernel of T consists of the zero vector only. By the rule (5–182), we see that every function $v(x)$ in the range of T has a continuous derivative for $0 \leqslant x \leqslant 1$. Further, by (5–181), we get $v(0) = 0$. In fact, the range of T consists precisely of the functions $v(x)$ such that $v'(x)$ is continuous for $0 \leqslant x \leqslant 1$ and $v(0) = 0$, since every such function can be represented as in (5–181) with $u(t) = v'(t)$.

Equation (5–181) shows that the inverse mapping T^{-1} assigns $v'(x)$ to $v(x)$ for each $v(x)$ in the range of T. For example,

$$T\{x^2\} = \frac{x^3}{3}, \qquad T(\cos 2x) = \frac{1}{2}\sin 2x, \quad \text{etc.,}$$

$$T^{-1}\{x^3\} = 3x^2, \qquad T^{-1}\{\sin 2x\} = 2\cos 2x, \quad \text{etc.} \blacktriangleleft$$

EXAMPLE 4 Let U denote the vector space of all functions $u(x)$ with continuous first and second derivatives for all x and let $T(\mathbf{u}) = u'' + 4u$, for

$\mathbf{u} = \{u(x)\}$ in U. Then T is linear, from the basic properties of the derivative. Further, as in Section 1–11, the kernel of T consists of all $u(x)$ such that

$$u'' + 4u = 0;$$

that is, of all linear combinations $c_1 \cos 2x + c_2 \sin 2x$. This is a *two-dimensional* subspace of U. A basis for this space consists of the two functions, $\cos 2x$ and $\sin 2x$. Since the kernel of T does not consist of $\mathbf{0}$ alone, T is not one-to-one. For example, to solve $T(\mathbf{u}) = e^{3x}$, that is,

$$u'' + 4u = e^{3x}, \tag{5–183}$$

we find one solution: $u = u_1 = e^{3x}/13$ (see Section 1–12) and then, as above, add an arbitrary vector of the kernel of T. Thus, as above,

$$u = \frac{e^{3x}}{13} + c_1 \cos 2x + c_2 \sin 2x.$$

This is the general solution of the differential equation. ◀

EXAMPLE 5   Let T be the *Laplace transform*

$$T(u) = \int_0^\infty e^{-st}u(t)\,dt = U(s). \tag{5–184}$$

Here T can be defined, for example, in the space V of all continuous functions $u(t)$ for $0 \leqslant t < \infty$ such that the integral in (5–184) converges absolutely when $s \geqslant k$ for some constant k. We verify that V is a vector space and that T is a linear mapping into the vector space W of all functions $U(s)$ continuous when $s \geqslant k$ (for further details see Sections 4–1 to 4–12). ◀

EXAMPLE 6   Let T be the *Fourier transform*:

$$T(u) = \int_{-\infty}^\infty u(t)e^{-i\omega t}\,dt = \varphi(\omega). \tag{5–185}$$

Again this can be interpreted as a linear mapping from an appropriate vector space U into vector space V. For example, U can be chosen as the vector space of all $u(t)$ that are continuous and absolutely integrable from $-\infty$ to ∞:

$$\int_{-\infty}^\infty |u(t)|\,dt = \text{Finite value.} \tag{5–186}$$

Then V can be chosen as the vector space of complex-valued continuous functions $\varphi(\omega)$, $-\infty < \omega < \infty$ (for further details, see Sections 4–13 to 4–21). ◀

   This discussion and the examples show that the concepts of vector space and linear mapping are important in a great variety of mathematical theories and their applications.

Matrix representation of linear mappings

When U and V are finite dimensional, with U of dimension n and V of dimension m, we can represent the linear mapping T by a matrix (in many ways).

Let $\mathbf{u}_1, \ldots, \mathbf{u}_n$ be a basis for U and $\mathbf{v}_1, \ldots, \mathbf{v}_m$ be a basis for V. Then for each i we can write

$$T(\mathbf{u}_i) = a_{1i}\mathbf{v}_1 + \cdots + a_{mi}\mathbf{v}_m \tag{5–187}$$

for appropriate scalars $a_{1i}, \ldots, a_{mi}$. If $\mathbf{u}$ is an arbitrary vector of U, we can write

$$\mathbf{u} = x_1\mathbf{u}_1 + \cdots + x_n\mathbf{u}_n$$

and hence associate with $\mathbf{u}$ a column vector $\text{col}\,(x_1, \ldots, x_n)$. Similarly,

$$T(\mathbf{u}) = y_1\mathbf{v}_1 + \cdots + y_m\mathbf{v}_m,$$

and we associate with $T(\mathbf{u})$ the column vector $\text{col}\,(y_1, \ldots, y_m)$. But by linearity,

$$T(\mathbf{u}) = T(x_1\mathbf{u}_1 + \cdots + x_n\mathbf{u}_n) = x_1 T(\mathbf{u}_1) + \cdots + x_n T(\mathbf{u}_n)$$
$$= x_1(a_{11}\mathbf{v}_1 + \cdots + a_{m1}\mathbf{v}_m) + \cdots + x_n(a_{1n}\mathbf{v}_1 + \cdots + a_{mn}\mathbf{v}_m)$$
$$= (a_{11}x_1 + \cdots + a_{1n}x_n)\mathbf{v}_1 + \cdots + (a_{m1}x_1 + \cdots + a_{mn}x_n)\mathbf{v}_m.$$

Therefore,

$$y_1 = a_{11}x_1 + \cdots + a_{1n}x_n,$$
$$\vdots \qquad\qquad\qquad \vdots \tag{5–188}$$
$$y_m = a_{m1}x_1 + \cdots + a_{mn}x_n.$$

We can interpret $x_1, \ldots, x_n$ as coordinates for vectors in U referred to the basis $\mathbf{u}_1, \ldots, \mathbf{u}_n$, and interpret $y_1, \ldots, y_m$ as coordinates for vectors in V referred to the basis $\mathbf{v}_1, \ldots, \mathbf{v}_m$. By Eqs. (5–188), the linear mapping T is given explicitly by linear equations with an $m \times n$ matrix $A = (a_{ij})$.

Thus the full power of matrices can be brought to bear whenever U and V are both finite dimensional. In particular, in this case both the kernel and the range of T must have finite dimensions, say k and r as usual, and the familiar equation (5–114)

$$k = n - r \tag{5–189}$$

must hold.

═══════════════ **Problems (Section 5-18)** ═══════════════

1. Let V be the vector space of all polynomials in x. Test each of the following sets of polynomials for linear independence:

   a) $3x - 2,\ 2x + 5,\ 7x - 1$ \qquad b) $x + 1,\ x + 2,\ x + 3,\ x + 4$

   c) $x^2 - 1,\ 2x^2 + x,\ x + 3$ \qquad d) $3x^2 + 1,\ x^2 - x,\ 2x^2 + x + 2$

   e) $x^2,\ x^4,\ x^6,\ x^8$ \qquad\qquad f) $1 - x^3,\ x,\ x^2,\ 1 + x^3$

2. Show that each of the following sets of objects, with the usual operations of addition and multiplication by scalars, forms a vector space. Give the dimension in each case and, if the dimension is finite, give a basis.

   a) All polynomials containing no term of odd degree: $3 + 5x^2 + x^4,\ x^2 - x^{10}, \ldots$
   b) All functions of the form $ae^x + be^{-x}$: for example, $2e^x + 3e^{-x},\ 5e^x - 2e^{-x}, \ldots$
   c) All functions $y = f(x)$, $-\infty < x < \infty$, such that $y'' + 9y = 0$.
   d) All trigonometric polynomials:

   $$a_0 + a_1 \cos x + b_1 \sin x + \cdots + a_n \cos nx + b_n \sin nx.$$

   e) All functions $f(x)$ defined and continuous on $0 \leqslant x \leqslant 1$.

f) All functions $f(x)$ that are defined and have a continuous derivative on $0 \leqslant x \leqslant 1$.
g) All infinite sequences $\{a_n\}, \{b_n\}, \ldots$, where $n = 1, 2, \ldots$
h) All convergent sequences.

3. a) Show that all 2×2 matrices form a vector space of dimension 4 with basis $\begin{bmatrix} 1 & 0 \\ 0 & 0 \end{bmatrix}$,

$$\begin{bmatrix} 0 & 1 \\ 0 & 0 \end{bmatrix}, \begin{bmatrix} 0 & 0 \\ 1 & 0 \end{bmatrix}, \begin{bmatrix} 0 & 0 \\ 0 & 1 \end{bmatrix}.$$

b) Show that another basis for this vector space consists of the matrices: $\begin{bmatrix} 1 & 2 \\ 1 & -3 \end{bmatrix}$,

$$\begin{bmatrix} 0 & 1 \\ 2 & -1 \end{bmatrix}, \begin{bmatrix} 5 & 2 \\ 7 & 1 \end{bmatrix}, \begin{bmatrix} 0 & 1 \\ 1 & 0 \end{bmatrix}.$$

c) Show that all 3×3 diagonal matrices form a vector space and give a basis.
d) Show that all 4×4 symmetric matrices form a vector space and give a basis.

e) Show that the matrices of form $\begin{bmatrix} a & b \\ -b & a \end{bmatrix}$ form a 2–dimensional subspace of the

vector space of part (a).

4. Show that the polynomials of type $ax^3 + bx$ form a subspace of the vector space of polynomials of degree at most 3.

5. Show that each of the following is a linear mapping T; describe the range and kernel and state whether T is one-to-one:

a) $U = V =$ All linear functions $u(x) = ax + b$; $T(u) = v$, where $v(x) = (a + b)x + 2a$.
b) $U = V =$ All functions $u(x) = ax^2 + bx + c$ of degree at most 2; $T(u) = v$, where $v(x) = 2ax^2 + bx$.
c) $U = V =$ All polynomials $u(x) = a_0x^3 + a_1x^2 + a_2x + a_3$ of degree at most 3; $T(u) = v$, where

$$v(x) = (a_0 + a_1)x^3 + (a_1 + a_2)x^2 + (a_2 + a_3)x + a_3 + a_0.$$

d) $U = V =$ All functions of form $ae^{2x} + be^{3x}$; $T(u) = v$, where $v(x) = u'(x) + u(x)$.
e) $U = V =$ All continuous functions $u(x)$, $-\infty < x < \infty$; $T(u) = v$, where

$$v(x) = \int_0^x e^t u(t)\, dt.$$

f) $U = V =$ All continuous functions $u(x)$, $-\infty < x < \infty$; $T(u) = v$, where $v(x)$ is the solution of the differential equation $v' + xv = u(x)$, with $v(0) = 0$.

Chapter 6 Systems of Ordinary Differential Equations

6–1 GENERAL CONCEPTS

In this chapter we consider systems of simultaneous ordinary differential equations such as the following:

$$\begin{cases} \dfrac{dx}{dt} = 2x + y, \\[2mm] \dfrac{dy}{dt} = x + 2y; \end{cases} \tag{6-10}$$

$$\begin{cases} \dfrac{dx}{dt} = x + 2y - z + \cos 3t, \\[2mm] \dfrac{dy}{dt} = 2x - y + z - \sin 3t, \\[2mm] \dfrac{dz}{dt} = x + y + 2z. \end{cases} \tag{6-11}$$

In general, we consider n differential equations in n unknowns (to be found as functions of t). For the most part, we restrict attention to *linear* differential equations, as illustrated by (6–10) and (6–11) (see Chapter 7 for nonlinear differential equations).

Such systems arise in a great variety of physical problems. Most commonly, t is *time*. The differential equations express physical laws governing the rates of change of these quantities. Often a solution of the system for given *initial data* is sought; this

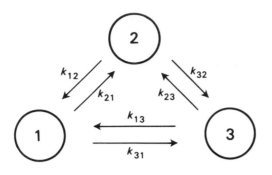

Fig. 6–1. Three-component chemical reaction.

solution may be obtained by formulas or by a computer. We may also be concerned about certain qualitative features of the solutions: do they describe approach to an equilibrium state as $t \to +\infty$ (stable case) or departure from equilibrium (unstable case)? Do they describe oscillations? With what frequencies?

EXAMPLE 1 A typical 3-component monomolecular chemical reaction (Fig. 6–1) is governed by differential equations of the form

$$\frac{dx_1}{dt} = -(k_{21} + k_{31})x_1 + k_{12}x_2 + k_{13}x_3,$$

$$\frac{dx_2}{dt} = k_{21}x_1 - (k_{12} + k_{32})x_2 + k_{23}x_3, \qquad (6\text{-}12)$$

$$\frac{dx_3}{dt} = k_{31}x_1 + k_{32}x_2 - (k_{13} + k_{23})x_3,$$

where (k_{ij}) is a 3×3 matrix of *rate constants* k_{ij}; x_1, x_2, x_3 are the concentrations of the three reacting substances. We think of k_{ij} as the rate by which substance i grows at the expense of substance j. ◄

The general system of n first-order equations in n unknowns has the form

$$\frac{dx_i}{dt} = F_i(x_1, \ldots, x_n, t), \quad i = 1, \ldots, n. \qquad (6\text{-}13)$$

As explained below, higher-order equations and systems of higher-order equations can be reduced to this form, so that (6–13) can be considered a standard form. In (6–13) the functions F_i are usually defined and continuous for all values of the variables, and we shall assume this except when pointed out. A *solution* of Eq. (6–13) is a set of functions

$$x_1 = g_1(t), \ldots, x_n = g_n(t), \quad \alpha < t < \beta, \qquad (6\text{-}14)$$

that satisfy the differential equations identically. For example,

$$x = e^{3t}, \qquad y = e^{3t}$$

is a solution of Eqs. (6–10) for all t. The *general solution* of (6–13) is a set of formulas giving all solutions. For example, the general solution of (6–10) is

$$x = c_1 e^{3t} + c_2 e^t, \qquad y = c_1 e^{3t} - c_2 e^t,$$

where c_1, c_2 are arbitrary constants. The general solution of system (6–13) typically involves n arbitrary constants.

EXISTENCE THEOREM *Let the functions F_i in (6–13) have continuous first partial derivatives for all values of the variables. Then, for each choice of t_0, $x_1^0, \ldots, x_n^0$, there is a unique solution (6–14) of (6–13) satisfying the initial conditions*

$$x_1 = x_1^0, \ldots, x_n = x_n^0 \quad for \quad t = t_0. \tag{6–15}$$

(For a proof of this theorem, see ODE, Chapter 12.)

System (6–13) is said to be *linear* if it has the form

$$\frac{dx_i}{dt} = a_{i1} x_1 + \cdots + a_{in} x_n + f_i(t), \quad i = 1, \ldots, n. \tag{6–16}$$

Here the a_{ij} are constants or functions of t; when they are all constants, we say that the system has *constant coefficients*. Both (6–10) and (6–11) are linear systems with constant coefficients. When the $f_i(t)$ are identically 0, the system is said to be *homogeneous*. System (6–12) is linear and homogeneous.

For a linear system the existence theorem can be extended to state that each solution (6–14) is defined for $-\infty < t < \infty$. (We continue to assume that all functions in the differential equations are defined for all values of the variables; if this is not the case, then the solutions are correspondingly restricted.)

A single differential equation of order n can be replaced by a system of form (6–13) or, if the equation is linear, of form (6–16). For example, the second-order equation (nonlinear)

$$\frac{d^2 x}{dt^2} = x^2 \frac{dx}{dt} + t^2 x$$

can be replaced by

$$\frac{dx_1}{dt} = x_2, \qquad \frac{dx_2}{dt} = x_1^2 x_2 + t^2 x_1.$$

Here x_1 stands for x and x_2 for dx/dt.

EXAMPLE 2 The simple mass–spring system (Section 1–14) is governed by the equation

$$m \frac{d^2 x}{dt^2} + h \frac{dx}{dt} + kx = F(t).$$

This can be replaced by the system

$$\frac{dx_1}{dt} = x_2, \qquad \frac{dx_2}{dt} = -\frac{k}{m} x_1 - \frac{h}{m} x_2 + \frac{1}{m} F(t).$$

We see that the system is linear and nonhomogeneous. ◀

Similarly, the fourth-order equation (linear)

$$\frac{d^4 x}{dt^4} + 3\frac{d^3 x}{dt^3} - \frac{d^2 x}{dt^2} + 2\frac{dx}{dt} - x = \sin 3t$$

can be replaced by the system (linear)

$$\frac{dx_1}{dt} = x_2, \qquad \frac{dx_2}{dt} = x_3, \qquad \frac{dx_3}{dt} = x_4, \qquad \frac{dx_4}{dt} = -3x_4 + x_3 - 2x_2 + x_1 + \sin 3t.$$

Here x_1 stands for x, x_2 for dx/dt, x_3 for $d^2 x/dt^2$, x_4 for $d^3 x/dt^3$. In both these examples, finding all solutions of the given single equation is equivalent to finding all solutions of the corresponding system. Assigning initial values to x, dx/dt, ... is equivalent to giving initial values of x_1, x_2, ... for the corresponding system, as in Eq. (6–15).

The illustrated procedure applies also to simultaneous higher-order equations. For example, the system

$$\frac{d^2 x}{dt^2} + 2\frac{dx}{dt} - 3\frac{dy}{dt} + x = t^2,$$

$$\frac{d^2 y}{dt^2} - 5\frac{dx}{dt} + 2\frac{dy}{dt} - 2x - y = 0$$

can be replaced by the system of four first-order equations

$$\frac{dx_1}{dt} = x_2, \frac{dx_3}{dt} = x_4,$$

$$\frac{dx_2}{dt} = -2x_2 + 3x_4 - x_1 + t^2,$$

$$\frac{dx_4}{dt} = 5x_2 - 2x_4 + 2x_1 + x_3.$$

Here x_1 stands for x, x_2 for dx/dt, x_3 for y, x_4 for dy/dt.

The procedure extends in principle to general systems of higher-order equations. To carry it out, we must be able to solve for the highest-order derivatives of all unknowns.

We refer to the first-order system (6–13) as a system of *order n*. The process just illustrated shows that a single equation of order n can be replaced by a system of order n. Simultaneous equations solvable for highest-order derivatives $d^m x/dt^m$, $d^p y/dt^p$, ... can be replaced by a system (6–13) of order $n = m + p + \cdots$, that is, of order n which is the sum of the orders of the highest-order derivatives appearing.

EXAMPLE 3   The motion of a satellite of mass m (Fig. 6–2) in a fixed xy-plane is governed by the equations

$$m\frac{d^2 x}{dt^2} = -\frac{kmM}{r^3}x, \qquad m\frac{d^2 y}{dt^2} = -\frac{kmM}{r^3}y,$$

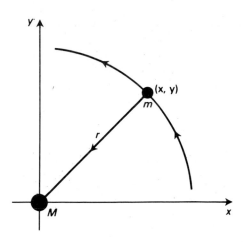

Fig. 6-2. Motion of satellite.

where k is the gravitation constant, M is the mass of the attracting large body (Jupiter, for example) assumed to be fixed at $(0, 0)$, and $r = (x^2 + y^2)^{1/2}$. The differential equations can be replaced by the fourth-order system

$$\frac{dx_1}{dt} = x_2, \qquad \frac{dx_2}{dt} = -\frac{kM}{r^3}x_1, \qquad \frac{dx_3}{dt} = x_4, \qquad \frac{dx_4}{dt} = -\frac{kM}{r^3}x_3,$$

where now $r = (x_1^2 + x_3^2)^{1/2}$. The system is *nonlinear*. ◄

6–2 LINEAR SYSTEMS: GENERAL THEORY

Here matrices prove to be of exceptional value. In particular, we consider *matrix functions* of t: $A = (a_{ij})$, where the a_{ij} are functions of t. For example,

$$A = \begin{bmatrix} t^2 & e^{3t} \\ \sin t & \cos t \end{bmatrix} \tag{6–20}$$

is such a function.

We define the derivative dA/dt of such a matrix function as the matrix whose entries are da_{ij}/dt. For (6–20),

$$\frac{dA}{dt} = \begin{bmatrix} 2t & 3e^{3t} \\ \cos t & -\sin t \end{bmatrix}.$$

We verify that the usual rules for derivative of a sum and of a product hold:

$$\frac{d}{dt}(A + B) = \frac{dA}{dt} + \frac{dB}{dt}, \tag{6–21}$$

$$\frac{d}{dt}(AB) = A\frac{dB}{dt} + \frac{dA}{dt}B. \tag{6–22}$$

For (6–22) the *order* of the factors is important, since matrices do not obey the commutative law of multiplication.

As a particular case of this definition, we shall often use the derivative of a vector function $\mathbf{x}(t)$, usually a column vector $\mathrm{col}(x_1(t), \ldots, x_n(t))$:

$$\frac{d\mathbf{x}}{dt} = \mathrm{col}\left(\frac{dx_1}{dt}, \ldots, \frac{dx_n}{dt}\right).$$

The general linear system (6–16) can now be written concisely as follows:

$$\frac{d\mathbf{x}}{dt} = A\mathbf{x} + \mathbf{f}(t), \tag{6–23}$$

where $\mathbf{f}(t) = \mathrm{col}(f_1(t), \ldots, f_n(t))$ and $A = (a_{ij})$. Each solution of the system is then a vector function $\mathbf{x}(t)$ that satisfies Eq. (6–23). Thus the system (6–10) becomes

$$\frac{d\mathbf{x}}{dt} = \begin{bmatrix} 2 & 1 \\ 1 & 2 \end{bmatrix}\mathbf{x}$$

and the particular solutions are

$$\mathbf{x} = \begin{bmatrix} e^{3t} \\ e^{3t} \end{bmatrix} = \mathbf{x}_1(t) \quad \text{and} \quad \mathbf{x} = \begin{bmatrix} e^t \\ -e^t \end{bmatrix} = \mathbf{x}_2(t),$$

while the general solution is

$$\mathbf{x} = c_1\mathbf{x}_1(t) + c_2\mathbf{x}_2(t),$$

where c_1, c_2 are arbitrary constants.

A set of vector functions $\mathbf{x}_1(t), \ldots, \mathbf{x}_k(t)$, defined over the interval $\alpha < t < \beta$, is said to be *linearly independent* over this interval if

$$c_1\mathbf{x}_1(t) + \cdots + c_k\mathbf{x}_k(t) \equiv \mathbf{0}, \quad \alpha < t < \beta,$$

is possible (with constant scalars $c_1, \ldots, c_k$) only for $c_1 = 0, \ldots, c_k = 0$. Otherwise the functions are said to be *linearly dependent* over the interval.

For the homogeneous system

$$\frac{d\mathbf{x}}{dt} = A\mathbf{x}, \quad A = (a_{ij}(t)), \quad -\infty < t < \infty, \tag{6–24}$$

the general solution has the form

$$\mathbf{x} = c_1\mathbf{x}_1(t) + \cdots + c_n\mathbf{x}_n(t). \tag{6–25}$$

Here $\mathbf{x}_1(t), \ldots, \mathbf{x}_n(t)$ are solutions of the differential equation $d\mathbf{x}/dt = A\mathbf{x}$ and are linearly independent for $-\infty < t < \infty$, while $c_1, \ldots, c_n$ are arbitrary constants. The functions $\mathbf{x}_1(t), \ldots, \mathbf{x}_n(t)$ can be chosen as any such linearly independent set of solutions.

To justify this rule, we first remark that, if $\mathbf{x}_1(t), \ldots, \mathbf{x}_n(t)$ are solutions of $d\mathbf{x}/dt = A\mathbf{x}$, $-\infty < t < \infty$, then Eq. (6–25) defines a solution for each choice of the constants $c_1, \ldots, c_n$, since

$$\frac{d\mathbf{x}}{dt} = c_1\frac{d\mathbf{x}_1}{dt} + \cdots + c_n\frac{d\mathbf{x}_n}{dt} = c_1 A\mathbf{x}_1 + \cdots + c_n A\mathbf{x}_n$$

$$= A(c_1\mathbf{x}_1 + \cdots + c_n\mathbf{x}_n) = A\mathbf{x}.$$

Furthermore, if these solutions are such that the vectors $x_1(t_0), \ldots, x_n(t_0)$ are linearly independent for one value of t_0, then the vectors $x_1(t), \ldots, x_n(t)$ are linearly independent for every t and $x_1(t), \ldots, x_n(t)$ are linearly independent vector functions for all t. For suppose

$$c_1 x_1(t_1) + \cdots + c_n x_n(t_1) = 0 \tag{6-26}$$

for some t_1 and some choices of $c_1, \ldots, c_n$, not all 0. Then

$$x(t) = c_1 x_1(t) + \cdots + c_n x_n(t)$$

would be a solution of the differential equation such that $x(t_1) = 0$. By the existence theorem of Section 6–1, there is one and only one solution $x(t)$ such that $x(t_1) = 0$, namely $x \equiv 0$. Thus we must have

$$c_1 x_1(t) + \cdots + c_n x_n(t) \equiv 0, \quad -\infty < t < \infty.$$

This implies that $x_1(t_0), \ldots, x_n(t_0)$ are linearly dependent, contrary to assumption. Hence (6–26) cannot hold and $x_1(t), \ldots, x_n(t)$ are linearly independent vectors for each t. The same argument shows that $x_1(t), \ldots, x_n(t)$ are linearly independent functions for $-\infty < t < \infty$.

With such a linearly independent set, the formula (6–25) defines solutions. To show that *all* solutions are included in this formula, we must show that the initial condition $x(t_0) = x_0$ (the same as (6–14)) can be satisfied by appropriate choice of $c_1, \ldots, c_n$, that is,

$$c_1 x_1(t_0) + \cdots + c_n x_n(t_0) = x_0. \tag{6-27}$$

Since $x_1(t_0), \ldots, x_n(t_0)$ are linearly independent, they form a basis for vectors in R^n (Section 5–9). Therefore Eq. (6–27) can be solved uniquely for $c_1, \ldots, c_n$, and the initial condition is satisfied.

It remains to be shown that there exist n linearly independent solutions $x_1(t), \ldots, x_n(t)$. But the existence theorem of Section 6–1 assures us that we can find a solution $x_k(t)$ such that $x_k(t_0) = (0, 0, \ldots, 0, 1, 0, \ldots, 0) = e_k$ (the kth basis vector of R^n with 1 as kth component). If we do this for $k = 1, \ldots, n$, we obtain n solutions $x_1(t), \ldots, x_n(t)$. They are linearly independent vectors for $t = t_0$, since for $t = t_0$ they reduce to the basis vectors $e_1, \ldots, e_n$. Hence by the reasoning above they are linearly independent for $-\infty < t < \infty$.

Nonhomogeneous system

This is the general system (6–23). By the existence theorem, it has solutions. Let $x = x^*(t)$ be one solution. Then the general solution is given by

$$x = x^*(t) + c_1 x_1(t) + \cdots + c_n x_n(t), \quad -\infty < t < \infty, \tag{6-28}$$

where the *complementary function*

$$c_1 x_1(t) + \cdots + c_n x_n(t) = x_c(t)$$

is the general solution of the corresponding homogeneous equation (6–24), since from (6–28) it follows that

$$\frac{dx}{dt} = \frac{dx^*}{dt} + c_1 \frac{dx_1}{dt} + \cdots + c_n \frac{dx_n}{dt}$$

$$= Ax^* + f(t) + c_1 Ax_1 + \cdots + c_n Ax_n$$

$$= A(x^* + c_1 x_1 + \cdots + c_n x_n) + f(t)$$

$$= Ax + f(t).$$

Thus the formula (6–28) defines solutions of Eq. (6–23). It gives all solutions. For we again need only show that $c_1, \ldots, c_n$ can be chosen to satisfy the initial condition $x(t_0) = x_0$. This leads to the equation

$$x^*(t_0) + c_1 x_1(t_0) + \cdots + c_n x_n(t_0) = x_0$$

or

$$c_1 x_1(t_0) + \cdots + c_n x_n(t_0) = x_0 - x^*(t_0).$$

As above, $x_1(t_0), \ldots, x_n(t_0)$ are linearly independent and form a basis for vectors in R^n. Thus the last equation can be solved uniquely for $c_1, \ldots, c_n$.

EXAMPLE $\dfrac{dx}{dt} = \begin{bmatrix} 2 & 1 \\ 1 & 2 \end{bmatrix} x + \begin{bmatrix} 1 \\ t \end{bmatrix}.$

The related homogeneous system is

$$\frac{dx}{dt} = \begin{bmatrix} 2 & 1 \\ 1 & 2 \end{bmatrix} x.$$

As above, this has the solutions

$$x_1(t) = \begin{bmatrix} e^{3t} \\ e^{3t} \end{bmatrix}, \qquad x_2(t) = \begin{bmatrix} e^{t} \\ -e^{t} \end{bmatrix}.$$

These are linearly independent, since $c_1 x_1(t) + c_2 x_2(t) \equiv 0$ would mean that

$$c_1 e^{3t} + c_2 e^{t} \equiv 0, \qquad c_1 e^{3t} - c_2 e^{t} \equiv 0,$$

from which we find at once $c_1 = 0$, $c_2 = 0$. Thus the complementary function is $c_1 x_1(t) + c_2 x_2(t)$. To find a particular solution $x^*(t)$, we try

$$x(t) = \begin{bmatrix} a_1 t + b_1 \\ a_2 t + b_2 \end{bmatrix}$$

(method of undetermined coefficients). We substitute in the given nonhomogeneous differential equation:

$$\frac{dx}{dt} = \begin{bmatrix} a_1 \\ a_2 \end{bmatrix} = \begin{bmatrix} 2 & 1 \\ 1 & 2 \end{bmatrix} \begin{bmatrix} a_1 t + b_1 \\ a_2 t + b_2 \end{bmatrix} + \begin{bmatrix} 1 \\ t \end{bmatrix},$$

$$a_1 = 2(a_1 t + b_1) + a_2 t + b_2 + 1,$$

$$a_2 = a_1 t + b_1 + 2(a_2 t + b_2) + t.$$

We compare terms of same degree in t:

$$a_1 = 2b_1 + b_2 + 1, \qquad 0 = 2a_1 + a_2,$$

$$a_2 = b_1 + 2b_2, \qquad 0 = a_1 + 2a_2 + 1.$$

These equations are easily solved to give $a_1 = 1/3$, $a_2 = -2/9$, $b_1 = -2/3$, $b_2 = -2/9$. Therefore,

$$\mathbf{x}^*(t) = \begin{bmatrix} \frac{1}{9}(t-2) \\ \frac{1}{9}(-6t-2) \end{bmatrix},$$

and the general solution is

$$\mathbf{x} = c_1 \begin{bmatrix} e^{3t} \\ e^{3t} \end{bmatrix} + c_2 \begin{bmatrix} e^t \\ -e^t \end{bmatrix} + \begin{bmatrix} \frac{1}{9}(t-2) \\ \frac{1}{9}(-6t-2) \end{bmatrix}. \blacktriangleleft$$

In the following sections we consider in detail methods for finding the complementary function and particular solution.

========= **Problems (Section 6-2)** =========

1. Verify that the given functions define a solution of the given system:

a) $x = \sin 3t$, $y = \cos 3t$ for $\dfrac{dx}{dt} = 3y$, $\dfrac{dy}{dt} = -3x$

b) $x = 1 + 3t$, $y = t^2$ for $\dfrac{dx}{dt} = x^2 - 9y - 6t + 2$, $\dfrac{dy}{dt} = tx - 3y + t$

c) $x = 3e^{2t}$, $y = e^{2t}$, $z = e^{2t}$ for $\dfrac{dx}{dt} = x + 2y + z$, $\dfrac{dy}{dt} = 2x - 3y - z$, $\dfrac{dz}{dt} = y + z$

d) $x = 1 + te^t$, $y = t - e^t$, $z = 2t + e^t$ for

$\dfrac{dx}{dt} = x + y + z - 1 - 3t + e^t$, $\dfrac{dy}{dt} = 2x - y + z - 1 + t - e^t(3 + 2t)$,

$\dfrac{dz}{dt} = x + 2y \quad z - 1 - e^t(t - 4)$

2. Write the following equations as a system of first-order differential equations:

a) $\dfrac{d^2 x}{dt^2} - 3\dfrac{dx}{dt} + 5x = 5\sin 2t$

b) $\dfrac{d^3 x}{dt^3} + t\dfrac{d^2 x}{dt^2} - x = 0$

c) $\dfrac{dx}{dt} - 5x + 2y = 0$, $\dfrac{d^2 y}{dt^2} + 3\dfrac{dy}{dt} + 2x + 4y = 0$

d) $2\dfrac{dx}{dt} + 3\dfrac{dy}{dt} + x - y = 0$, $\dfrac{dx}{dt} + 2\dfrac{dy}{dt} - x + 2y = 0$

e) $m\dfrac{dv}{dt} = -D + mg\sin\theta$, $mv\dfrac{d\theta}{dt} = -L + mg\cos\theta$, $I\dfrac{d^2\varphi}{dt^2} = -M$.

These are the equations of motion of an airplane. Here v is the speed, m the mass, D is the drag minus the component of engine thrust in the direction of flight, L is the lift, M is the moment of lift and drag forces about an axis perpendicular to both, θ is the angle of inclination of the path with respect to the horizontal (measured positive for a descending path), φ is the angle of attack, I the moment of inertia about the axis previously mentioned.

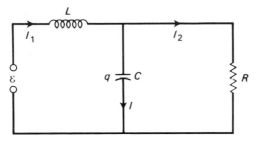

Fig. 6-3. Network for Problem 2 (f).

f) The differential equations of the network shown, with $x_1 = q$, $x_2 = I_1$.

3. Test for linear independence over the interval $-\infty < t < \infty$:

a) $\mathbf{x}_1 = \begin{bmatrix} t^2 \\ 1-t \end{bmatrix}$, $\qquad \mathbf{x}_2 = \begin{bmatrix} 1+t \\ t^2 \end{bmatrix}$

b) $\mathbf{x}_1 = \begin{bmatrix} 2e^t \\ 3e^t \end{bmatrix}$, $\qquad \mathbf{x}_2 = \begin{bmatrix} 5e^t \\ 2e^t \end{bmatrix}$

c) $\mathbf{x}_1 = \begin{bmatrix} \sin 2t \\ \cos 2t \\ 2\cos 2t \end{bmatrix}$, $\qquad \mathbf{x}_2 = \begin{bmatrix} \cos 2t - \sin 2t \\ -\sin 2t \\ \cos 2t + \sin 2t \end{bmatrix}$,

$\mathbf{x}_3 = \begin{bmatrix} \cos 2t + \sin 2t \\ 2\cos 2t - \sin 2t \\ 5\cos 2t + \sin 2t \end{bmatrix}$

4. Show that the vector functions

$$\mathbf{x}_1 = \begin{bmatrix} 1 \\ t \end{bmatrix}, \qquad \mathbf{x}_2 = \begin{bmatrix} e^t \\ te^t \end{bmatrix}$$

are linearly independent for $-\infty < t < \infty$, but $\mathbf{x}_1(t)$ and $\mathbf{x}_2(t)$ are linearly dependent vectors for each fixed t. (The discussion in Section 6-2 shows that this cannot happen for solutions of a linear differential equation $d\mathbf{x}/dt = A\mathbf{x}$.)

5. Show that the given vector functions are linearly independent solutions of the given differential equation; obtain the general solution and the particular solution satisfying the given initial condition:

a) $\dfrac{d\mathbf{x}}{dt} = \begin{bmatrix} -2 & 2 \\ -15 & 9 \end{bmatrix} \mathbf{x}$, $\quad \mathbf{x}_1 = \begin{bmatrix} 2e^{3t} \\ 5e^{3t} \end{bmatrix}$, $\quad \mathbf{x}_2 = \begin{bmatrix} e^{4t} \\ 3e^{4t} \end{bmatrix}$, $\quad \mathbf{x}(0) = \begin{bmatrix} 1 \\ 1 \end{bmatrix}$.

b) $\dfrac{d\mathbf{x}}{dt} = \begin{bmatrix} 7 & 3 \\ -10 & -4 \end{bmatrix} \mathbf{x}$, $\quad \mathbf{x}_1 = \begin{bmatrix} 3e^{2t} \\ -5e^{2t} \end{bmatrix}$, $\quad \mathbf{x}_2 = \begin{bmatrix} -e^t \\ 2e^t \end{bmatrix}$, $\quad \mathbf{x}(0) = \begin{bmatrix} -7 \\ -8 \end{bmatrix}$.

c) $\dfrac{d\mathbf{x}}{dt} = \begin{bmatrix} 9 & -11 & -5 \\ 4 & -4 & -3 \\ 2 & -4 & 1 \end{bmatrix} \mathbf{x}$, $\quad \mathbf{x}_1 = \begin{bmatrix} 3e^{2t} \\ e^{2t} \\ 2e^{2t} \end{bmatrix}$, $\quad \mathbf{x}_2 = \begin{bmatrix} e^{3t} \\ e^{3t} \\ -e^{3t} \end{bmatrix}$, $\quad \mathbf{x}_3 = \begin{bmatrix} 2e^t \\ e^t \\ e^t \end{bmatrix}$,

$\mathbf{x}(0) = \begin{bmatrix} 3 \\ 0 \\ 5 \end{bmatrix}$.

d) $\dfrac{d\mathbf{x}}{dt} = \begin{bmatrix} -5 & 5 & 7 \\ -2 & 2 & 3 \\ -4 & 5 & 4 \end{bmatrix} \mathbf{x}, \qquad \mathbf{x}_1 = \begin{bmatrix} 2e^t \\ e^t \\ e^t \end{bmatrix}, \qquad \mathbf{x}_2 = \begin{bmatrix} \cos t - 7\sin t \\ \cos t - 3\sin t \\ -\cos t - 3\sin t \end{bmatrix},$

$\mathbf{x}_3 = \begin{bmatrix} 7\cos t + \sin t \\ 3\cos t + \sin t \\ 3\cos t - \sin t \end{bmatrix}, \qquad \mathbf{x}(0) = \begin{bmatrix} 0 \\ 0 \\ 0 \end{bmatrix}.$

6. Verify that the given vector function is a solution of the given differential equation and, with the aid of the results of Problem 5, find the general solution:

a) $\dfrac{d\mathbf{x}}{dt} = \begin{bmatrix} -2 & 2 \\ -15 & 9 \end{bmatrix} \mathbf{x} + \begin{bmatrix} 11e^t \\ 59e^t \end{bmatrix}, \qquad \mathbf{x} = \begin{bmatrix} 5e^t \\ 2e^t \end{bmatrix}$

b) $\dfrac{d\mathbf{x}}{dt} = \begin{bmatrix} 9 & -11 & -5 \\ 4 & -4 & -3 \\ 2 & -4 & 1 \end{bmatrix} \mathbf{x} + \begin{bmatrix} 17 + 30t \\ 8 + 14t \\ 7 + 4t \end{bmatrix}, \qquad \mathbf{x} = \begin{bmatrix} 1 - t \\ 2 + t \\ 1 + 2t \end{bmatrix}$

6–3 HOMOGENEOUS LINEAR SYSTEMS WITH CONSTANT COEFFICIENTS

Here our system has the form

$$\frac{d\mathbf{x}}{dt} = A\mathbf{x}, \quad A = (a_{ij}), \qquad (6\text{--}30)$$

where A is a constant $n \times n$ matrix. The examples of the preceding sections suggest that we can find solutions of the form $\mathbf{x} = e^{\lambda t}\mathbf{u}$, where $\mathbf{u}$ is a nonzero constant vector, λ an appropriate constant. Substitution in Eq. (6–30) gives successively

$$\lambda e^{\lambda t}\mathbf{u} = Ae^{\lambda t}\mathbf{u} = e^{\lambda t}A\mathbf{u},$$

$$\lambda \mathbf{u} = A\mathbf{u},$$

$$(A - \lambda I)\mathbf{u} = \mathbf{0}.$$

Thus $\mathbf{u}$ must be an *eigenvector* of A, associated with the eigenvalue λ, as discussed in Section 5–7. The eigenvalues λ are the roots of the *characteristic equation* $\det(A - \lambda I) = 0$, that is, the equation

$$\begin{vmatrix} a_{11} - \lambda & a_{12} & \cdots & a_{1n} \\ a_{21} & a_{22} - \lambda & \cdots & a_{2n} \\ \vdots & & & \vdots \\ a_{n1} & a_{n2} & \cdots & a_{nn} - \lambda \end{vmatrix} = 0. \qquad (6\text{--}31)$$

This is an nth degree equation with roots $\lambda_1, \ldots, \lambda_n$; some of these may be repeated. Also, complex roots may occur.

Let us assume that $\lambda_1, \ldots, \lambda_n$ are distinct real numbers. Then we can find associated eigenvectors $\mathbf{u}_1, \ldots, \mathbf{u}_n$. (The computation of these vectors is discussed below.) *These n vectors are necessarily linearly independent* (see Problem 12 following Section 5–9).

We now have n solutions of the differential equation (6–30):

$$\mathbf{x}_1 = e^{\lambda_1 t}\mathbf{u}_1, \quad \ldots, \quad \mathbf{x}_n = e^{\lambda_n t}\mathbf{u}_n. \tag{6-32}$$

These n solutions are linearly independent, for when $t = 0$, the solutions reduce to $\mathbf{u}_1, \ldots, \mathbf{u}_n$. Since these vectors are linearly independent, the solutions (6–32) are linearly independent for all t, as shown in Section 6–2. Therefore the general solution of Eq. (6–30) is

$$\mathbf{x} = c_1 e^{\lambda_1 t}\mathbf{u}_1 + \cdots + c_n e^{\lambda_n t}\mathbf{u}_n,$$

where $c_1, \ldots, c_n$ are arbitrary constants.

Computation of eigenvalues and eigenvectors

The eigenvalues are the solution of the nth degree algebraic equation (6–31) and can be solved by standard methods for such equations. Once the eigenvalues have been determined, the eigenvector $\mathbf{u}_j$ is found by setting $\lambda = \lambda_j$ in the equation $(A - \lambda I)\mathbf{u} = \mathbf{0}$ and solving for $\mathbf{u}$ (not $\mathbf{0}$). As illustrated in Section 5–7, the equation $(A - \lambda I)\mathbf{u} = \mathbf{0}$ is equivalent to n linear equations in n unknowns. They can be solved by the elimination procedures of Section 5–13 (for more information on this topic, see Section 13–5).

EXAMPLE 1 $\dfrac{d\mathbf{x}}{dt} = \begin{bmatrix} 2 & 1 \\ 1 & 2 \end{bmatrix}\mathbf{x}$ (see the Example in Section 6–2). The characteristic equation is

$$\begin{vmatrix} 2 - \lambda & 1 \\ 1 & 2 - \lambda \end{vmatrix} = 0 \quad \text{or} \quad \lambda^2 - 4\lambda + 3 = 0.$$

The eigenvalues are $\lambda_1 = 3$, $\lambda_2 = 1$. For $\lambda_1 = 3$ we have the equation

$$\begin{bmatrix} -1 & 1 \\ 1 & -1 \end{bmatrix}\mathbf{u} = \mathbf{0}$$

for the eigenvector $\mathbf{u} = \text{col}(u_1, u_2)$. Hence

$$-u_1 + u_2 = 0, \quad u_1 - u_2 = 0,$$

and a nonzero solution is $\mathbf{u} = \mathbf{u}_1 = \text{col}(1, 1)$. In the same way, for $\lambda = \lambda_2$ we obtain the equations

$$u_1 + u_2 = 0, \quad u_1 + u_2 = 0$$

with the nonzero solution $\mathbf{u} = \mathbf{u}_2 = \text{col}(1, -1)$. Hence the general solution is

$$\mathbf{x} = c_1 e^{3t}\begin{bmatrix} 1 \\ 1 \end{bmatrix} + c_2 e^{t}\begin{bmatrix} 1 \\ -1 \end{bmatrix},$$

as in Section 6–2. ◄

When the roots $\lambda_1, \ldots, \lambda_n$ are distinct, but some are complex, we can proceed in exactly the same way and obtain a general complex solution as in Section 1–11. From this we can obtain the general real solution by simply taking the real part of the general complex solution (allowing complex arbitrary constants). Another method makes use of the fact that the complex roots come in conjugate pairs $\alpha \pm \beta i$. Corresponding to $\lambda = \alpha + \beta i$, we obtain an eigenvector $\mathbf{u}$ with complex components and the complex solution

$$\mathbf{x} = e^{\lambda t}\mathbf{u} = e^{(\alpha + \beta i)t}\mathbf{u}.$$

We can verify that the real and imaginary parts of this complex solution provide two solutions corresponding to the roots $\alpha + \beta i, \alpha - \beta i$, as desired. (A justification of this procedure is similar to that in Section 1–11.) ◄

EXAMPLE 2 $\dfrac{d\mathbf{x}}{dt} = \begin{bmatrix} -5 & 5 & 7 \\ -2 & 2 & 3 \\ -4 & 5 & 4 \end{bmatrix} \mathbf{x}$ (see Problem 5(d) following Section 6–2).

The characteristic equation is

$$\begin{vmatrix} -5-\lambda & 5 & 7 \\ -2 & 2-\lambda & 3 \\ -4 & 5 & 4-\lambda \end{vmatrix} = 0 \quad \text{or} \quad -\lambda^3 + \lambda^2 - \lambda + 1 = 0.$$

The eigenvalues are 1 and $\pm i$. Corresponding to $\lambda_1 = 1$, we obtain the eigenvector $\mathbf{u}_1 = \text{col}(2, 1, 1)$, as for Example 1 above. Corresponding to $\lambda_2 = i$, we have the equations

$$\begin{aligned}
(-5-i)u_1 + 5u_2 \quad &+ \quad 7u_3 = 0, \\
-2u_1 + (2-i)u_2 + \quad &3u_3 = 0, \\
-4u_1 + 5u_2 \quad &+ (4-i)u_3 = 0.
\end{aligned}$$

By elimination, we obtain the eigenvector

$$\mathbf{u}_2 = \text{col}(1 + 7i, 1 + 3i, -1 + 3i).$$

In the same way, from $\lambda_3 = -i$, we obtain

$$\mathbf{u}_3 = \text{col}(1 - 7i, 1 - 3i, -1 - 3i).$$

Hence the general *complex* solution is

$$\mathbf{x} = c_1 e^t \begin{bmatrix} 2 \\ 1 \\ 1 \end{bmatrix} + c_2 e^{it} \begin{bmatrix} 1+7i \\ 1+3i \\ -1+3i \end{bmatrix} + c_3 e^{-it} \begin{bmatrix} 1-7i \\ 1-3i \\ -1-3i \end{bmatrix},$$

where c_1, c_2, c_3 are arbitrary complex constants. Alternatively, from the solution

$$e^{it} \begin{bmatrix} 1+7i \\ 1+3i \\ -1+3i \end{bmatrix} = (\cos t + i \sin t) \begin{bmatrix} 1+7i \\ 1+3i \\ -1+3i \end{bmatrix}$$

$$= \begin{bmatrix} \cos t - 7\sin t + i\,(\sin t + 7\cos t) \\ \cos t - 3\sin t + i\,(\sin t + 3\cos t) \\ -\cos t - 3\sin t + i\,(-\sin t + 3\cos t) \end{bmatrix}$$

we obtain, by taking real and imaginary parts, the two solutions

$$\begin{bmatrix} \cos t - 7 \sin t \\ \cos t - 3 \sin t \\ -\cos t - 3 \sin t \end{bmatrix} \quad \text{and} \quad \begin{bmatrix} \sin t + 7 \cos t \\ \sin t + 3 \cos t \\ -\sin t + 3 \cos t \end{bmatrix}$$

and the general real solution

$$\mathbf{x} = c_1 e^t \begin{bmatrix} 2 \\ 1 \\ 1 \end{bmatrix} + c_2 \begin{bmatrix} \cos t - 7 \sin t \\ \cos t - 3 \sin t \\ -\cos t - 3 \sin t \end{bmatrix} + c_3 \begin{bmatrix} \sin t + 7 \cos t \\ \sin t + 3 \cos t \\ -\sin t + 3 \cos t \end{bmatrix}. \quad \blacktriangleleft$$

If multiple roots occur, then the described procedures may provide the desired n linearly independent solutions. For example, a double root λ_1, λ_1 may have two linearly independent eigenvectors $\mathbf{u}_1$, $\mathbf{u}_2$, and these can be used as before in constructing the general solution. However, the procedure may also fail to provide enough solutions. In that case, it can be shown that the missing solutions can be found in the form

$$e^{\lambda t}(\mathbf{v}_1 + t \mathbf{v}_2 + \cdots + t^{k-1}\mathbf{v}_k), \tag{6-33}$$

where λ is an eigenvalue of multiplicity k and $\mathbf{v}_1, \ldots, \mathbf{v}_k$ are appropriate constant vectors. The vectors $\mathbf{v}_1, \ldots, \mathbf{v}_k$ are found by substituting the expression (6-33) in the differential equation, and we are led to simultaneous linear equations (for details see ODE, Chapter 6). An alternative is to use Laplace transforms, as in Section 6-6 below. $\blacktriangleleft$

In engineering problems the matrix A normally depends on measured numbers known only with limited accuracy. Occurrence of repeated eigenvalues is possible only for special choices of the a_{ij}, and the slightest change in the a_{ij} will normally eliminate the repetition of eigenvalues, hence in practice they do not really arise. In the case when they "almost" arise, the computation of eigenvalues and eigenvectors is more difficult, since we deal with two or more eigenvalues that are very close.

6-4 NONHOMOGENEOUS LINEAR SYSTEMS WITH CONSTANT COEFFICIENTS

We consider here the system

$$\frac{d\mathbf{x}}{dt} = A\mathbf{x} + \mathbf{f}(t), \tag{6-40}$$

where $A = (a_{ij})$ is a constant $n \times n$ matrix and $\mathbf{f}(t) = \text{col}(f_1(t), \ldots, f_n(t))$ is a given vector function $(-\infty < t < \infty)$. By the results of Section 6-2, the general solution of Eq. (6-40) is

$$\mathbf{x} = \mathbf{x}_c(t) + \mathbf{x}^*(t), \tag{6-41}$$

where the complementary function $\mathbf{x}_c(t)$ is the general solution of the related homogeneous system

$$\frac{d\mathbf{x}}{dt} = A\mathbf{x} \tag{6-42}$$

and $\mathbf{x}^*(t)$ is one solution of the given nonhomogeneous equation (6-40). The function $\mathbf{x}_c(t)$ is found as in the preceding section:

$$\mathbf{x}_c(t) = c_1\,\mathbf{x}_1(t) + \cdots + c_n\mathbf{x}_n(t), \tag{6-43}$$

where typically $\mathbf{x}_1(t) = e^{\lambda t}\mathbf{u}_1, \ldots$ To find the particular solution $\mathbf{x}^*(t)$, we use the method of *variation of parameters*: we replace the constants $c_1, \ldots, c_n$ by unknown functions $v_1(t), \ldots, v_n(t)$ and seek a solution $\mathbf{x}(t)$ of (6-40) in the form

$$\mathbf{x} = v_1(t)\,\mathbf{x}_1(t) + \cdots + v_n(t)\,\mathbf{x}_n(t). \tag{6-44}$$

We can also write this as

$$\mathbf{x} = X(t)\,\mathbf{v}(t), \tag{6-44'}$$

where $X(t)$ is the matrix whose column vectors are $\mathbf{x}_1(t), \ldots, \mathbf{x}_n(t)$. Since these vectors are linearly independent for each t, matrix $X(t)$ is *nonsingular*. We substitute Eq. (6-44') in Eq. (6-40):

$$X(t)\frac{d\mathbf{v}}{dt} + \frac{dX}{dt}\mathbf{v} = AX(t)\,\mathbf{v}(t) + \mathbf{f}(t). \tag{6-45}$$

Now dX/dt is the matrix whose column vectors are $d\mathbf{x}_1/dt, \ldots, d\mathbf{x}_n/dt$. Since $\mathbf{x}_1(t), \ldots, \mathbf{x}_n(t)$ are solutions of the homogeneous system (6-42), these column vectors equal $A\mathbf{x}_1(t), \ldots, A\mathbf{x}_n(t)$ respectively. Therefore

$$\frac{dX}{dt} = AX(t) \tag{6-46}$$

(see the discussion of multiplication of matrices in Section 5-5). Accordingly, Eq. (6-45) becomes

$$X(t)\frac{d\mathbf{v}}{dt} + AX(t)\,\mathbf{v}(t) = AX(t)\mathbf{v}(t) + \mathbf{f}(t)$$

or

$$X(t)\frac{d\mathbf{v}}{dt} = \mathbf{f}(t) \qquad \text{or} \qquad \frac{d\mathbf{v}}{dt} = X^{-1}(t)\,\mathbf{f}(t).$$

We can thus write

$$\mathbf{v} = \int X^{-1}(t)\,\mathbf{f}(t)\,dt.$$

Here the integration is simply applied to each component of the vector $X^{-1}(t)\,\mathbf{f}(t)$ to obtain the corresponding component of $\mathbf{v}$. From (6-44') we now obtain our desired particular solution

$$\mathbf{x} = \mathbf{x}^*(t) = X(t)\int X^{-1}(t)\,\mathbf{f}(t)\,dt. \tag{6-47}$$

This is the *variation-of-parameters formula*.

EXAMPLE 1 $\dfrac{dx}{dt} = \begin{bmatrix} 2 & 1 \\ 1 & 2 \end{bmatrix} x + \begin{bmatrix} 2e^{5t} \\ 3e^{2t} \end{bmatrix}.$

The related homogeneous system is that of Example 1 in Sec. 6–3. Hence

$$x_1(t) = e^{3t} \begin{bmatrix} 1 \\ 1 \end{bmatrix} = \begin{bmatrix} e^{3t} \\ e^{3t} \end{bmatrix}, \quad x_2(t) = e^t \begin{bmatrix} 1 \\ -1 \end{bmatrix} = \begin{bmatrix} e^t \\ -e^t \end{bmatrix}$$

and

$$X(t) = \begin{bmatrix} e^{3t} & e^t \\ e^{3t} & -e^t \end{bmatrix}.$$

As in Section 5–6, we find the inverse of this matrix to be

$$X^{-1}(t) = \frac{1}{2} \begin{bmatrix} e^{-3t} & e^{-3t} \\ e^{-t} & -e^{-t} \end{bmatrix}.$$

Thus

$$X^{-1}(t) f(t) = \frac{1}{2} \begin{bmatrix} e^{-3t} & e^{-3t} \\ e^{-t} & -e^{-t} \end{bmatrix} \begin{bmatrix} 2e^{5t} \\ 3e^{2t} \end{bmatrix} = \frac{1}{2} \begin{bmatrix} 2e^{2t} + 3e^{-t} \\ 2e^{4t} - 3e^{t} \end{bmatrix}$$

and

$$\int X^{-1}(t) f(t)\, dt = \frac{1}{2} \begin{bmatrix} e^{2t} - 3e^{-t} \\ \frac{1}{2}e^{4t} - 3e^{t} \end{bmatrix},$$

(so that by (6–47)

$$x^*(t) = \begin{bmatrix} e^{3t} & e^t \\ e^{3t} & -e^t \end{bmatrix} \cdot \frac{1}{2} \begin{bmatrix} e^{2t} - 3e^{-t} \\ \frac{1}{2}e^{4t} - 3e^{t} \end{bmatrix} = \begin{bmatrix} \frac{3}{4}e^{5t} - 3e^{2t} \\ \frac{1}{4}e^{5t} \end{bmatrix}. \blacktriangleleft$$

REMARK 1: In the integration of $X^{-1}(t) f(t)$ we ignored the arbitrary constants. If we carried these along as c_1, c_2, we simply add $c_1 x_1(t) + c_2 x_2(t) = x_c(t)$ to $x^*(t)$. At times it is convenient to choose these constants in a special way. By treating the integral as

$$\int_{t_0}^{t} X^{-1}(\tau) f(\tau)\, d\tau,$$

we in effect make such a special choice. In fact it is this choice that makes $x^*(t_0) = 0$.

REMARK 2: The method of variation of parameters is applicable as above to equations with variable coefficients: $A = (a_{ij}(t))$. Finding $x_c(t)$ is in general far more difficult; often power series can be used, as in Section 2–22. However, once $x_c(t)$ has been found, $x^*(t)$ can then be determined as above.

===== **Problems (Section 6-4)** =====

1. Find the general solution:

a) $\dfrac{dx}{dt} = \begin{bmatrix} 1 & 2 \\ 12 & -1 \end{bmatrix} x$

b) $\dfrac{dx}{dt} = \begin{bmatrix} 1 & -2 \\ -2 & 1 \end{bmatrix} x$

c) $\dfrac{dx}{dt} = \begin{bmatrix} 2 & -1 & 3 \\ -1 & 1 & -1 \\ 0 & 1 & -1 \end{bmatrix} x$
d) $\dfrac{dx}{dt} = \begin{bmatrix} 9 & 4 & 2 \\ -11 & -4 & -4 \\ -5 & -3 & 1 \end{bmatrix} x$

e) $\dfrac{dx}{dt} = \begin{bmatrix} 1 & -2 \\ 5 & -1 \end{bmatrix} x$
f) $\dfrac{dx}{dt} = \begin{bmatrix} 0 & 1 & 0 \\ 0 & 0 & 1 \\ 1 & -1 & 1 \end{bmatrix} x$

2. With the aid of the answers to Problem 1, find the general solution:

a) $\dfrac{dx}{dt} = \begin{bmatrix} 1 & 2 \\ 12 & -1 \end{bmatrix} x + \begin{bmatrix} e^t \\ 2e^t \end{bmatrix}$

b) $\dfrac{dx}{dt} = \begin{bmatrix} 1 & 2 \\ 12 & -1 \end{bmatrix} x + \begin{bmatrix} te^{5t} \\ e^{-5t} \end{bmatrix}$

c) $\dfrac{dx}{dt} = \begin{bmatrix} 2 & -1 & 3 \\ 1 & 1 & -1 \\ 0 & 1 & -1 \end{bmatrix} x + \begin{bmatrix} 1-t \\ 1+t \\ t \end{bmatrix}$

d) $\dfrac{dx}{dt} = \begin{bmatrix} 2 & -1 & 3 \\ -1 & 0 & -1 \\ 0 & 1 & -1 \end{bmatrix} x + \begin{bmatrix} \sin t \\ 3\sin t \\ 2\sin t \end{bmatrix}$

e) $\dfrac{dx}{dt} = \begin{bmatrix} 1 & -2 \\ 5 & -1 \end{bmatrix} x + \begin{bmatrix} \sin t \\ \cos t \end{bmatrix}$

f) $\dfrac{dx}{dt} = \begin{bmatrix} 1 & -2 \\ 5 & -1 \end{bmatrix} x + \begin{bmatrix} \sin 3t \\ \cos 3t \end{bmatrix}$

3. With the aid of the answers to Problem 2, find a solution satisfying the given initial condition:

a) Equation of Problem 2(a), $x(0) = 0$
b) Equation of Problem 2(a), $x(0) = \mathrm{col}(5, 2)$
c) Equation of Problem 2(c), $x(0) = 0$
d) Equation of Problem 2(c), $x(0) = \mathrm{col}(1, 1, 1)$

4. A particle of mass m moves in a plane subject to linear restoring forces towards the axes, so that its equations of motion are

$$m\frac{d^2 x}{dt^2} = -ax - by, \quad m\frac{d^2 y}{dt^2} = -bx - cy,$$

where a, b, c are constants. Such equations arise in the study of the motion of a particle subject to nonlinear forces near an equilibrium point at $(0, 0)$.

a) Take $m = 1$, $a = 1$, $b = 1$, $c = 5$ (in appropriate units) and find the general solution.
b) With constants as in (a), find the particular solution such that $x = 1$, $dx/dt = 0$, $y = 0$, and $dy/dt = 0$ for $t = 0$.

5. A famous physics experiment by Thomson to determine the ratio of charge e to mass m for an electron uses the differential equations

$$m\frac{d^2 x}{dt^2} + He\frac{dy}{dt} = Ee, \quad m\frac{d^2 y}{dt^2} - He\frac{dx}{dt} = 0.$$

Here H is the intensity of the applied magnetic field, E the intensity of the applied electric field. Find the general solution.

6. The matrix function $X(t)$ used in Section 6–4 (whose column vectors are n linearly independent solutions of the equation $dx/dt = Ax$) is called a *fundamental matrix* for the differential equation (homogeneous or nonhomogeneous) considered.
   a) Show that $x_c(t) = X(t)c$, where $c = \mathrm{col}(c_1, \ldots, c_n)$.
   b) Show that if $X(t_0) = I$ (identity matrix), then $x = X(t)x_0$ is the solution of $dx/dt = Ax$ such that $x(t_0) = x_0$.
   c) Show that if $X(t_0) = I$, then

$$x = X(t)x_0 + X(t) \int_0^t X^{-1}(\tau)\, f(\tau)\, d\tau$$

   is the solution of $dx/dt = Ax + f(t)$ such that $x(t_0) = x_0$.

6–5 APPLICATION TO ELECTRICAL NETWORKS

The behavior of electrical networks is governed by Kirchhoff's laws:

I. The total potential drop around each closed circuit is 0.

II. The total current entering each junction (node) is zero.

In addition we have the following rules: the potential drop across an inductance through which current I is flowing is $L\, dI/dt$; across a resistance R it is RI; across a capacitor of charge q and capacitance C it is q/C.

We illustrate these laws and rules for the network of Fig. 6–4. This network contains an inductance L (measured in henries), two resistances R_1 and R_2 (in ohms), a capacitance C (in farads), an applied electromotive force $\mathscr{E}$ (in volts). In each branch a positive direction for currents and voltage drops is chosen; the corresponding currents are I, I_1, I_2 (in amperes). The charge q_2 on the capacitor is measured in coulombs, time t in seconds and

$$\frac{dq_2}{dt} = I_2. \tag{6–50}$$

Kirchhoff's second law gives

$$I = I_1 + I_2. \tag{6–51}$$

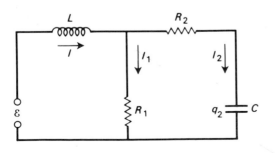

Fig. 6-4. Electrical network.

The first law gives

$$L\frac{dI}{dt} + R_1I_1 = \mathcal{E}, \tag{6-52}$$

$$R_2I_2 + \frac{q_2}{C} - R_1I_1 = 0, \tag{6-53}$$

$$L\frac{dI}{dt} + R_2I_2 + \frac{q_2}{C} = \mathcal{E}; \tag{6-54}$$

the first equation corresponds to the left-hand loop, the second to the right-hand loop, and the third to the exterior circuit. However, the third equation is simply the result of adding the first two, so that only two independent equations are obtained.

We can use Eqs. (6–51) and (6–53) to express I_1 and I_2 in terms of I and q_2:

$$I_1 = \frac{R_2I + C^{-1}q_2}{R_1 + R_2}, \qquad I_2 = \frac{R_1I - C^{-1}q_2}{R_1 + R_2} \tag{6 55}$$

If we use these expressions in Eqs. (6–50) and (6–52), we obtain the matrix equation

$$\frac{d}{dt}\begin{bmatrix} I \\ q_2 \end{bmatrix} = A\begin{bmatrix} I \\ q_2 \end{bmatrix} + \begin{bmatrix} \frac{\mathcal{E}}{L} \\ 0 \end{bmatrix}, \quad A = \frac{1}{CL(R_1 + R_2)}\begin{bmatrix} -CR_1R_2 & -R_1 \\ CLR_1 & -L \end{bmatrix}. \tag{6-56}$$

We now introduce numerical values: $R_1 = 100$ ohms, $R_2 = 200$ ohms, $C = 500^{-1}$ farad, $L = 10$ henries, $\mathcal{E} = 20$ volts. Then

$$A = \frac{1}{3}\begin{bmatrix} -20 & -50 \\ 1 & -5 \end{bmatrix}, \tag{6-57}$$

and we find easily that A has the eigenvalue -5 with corresponding eigenvector col$(10, -1)$ and the eigenvalue $-10/3$ with corresponding eigenvector $(5, -1)$. Also there is clearly a steady-state particular solution $I =$ const, $q_2 =$ const given by

$$\begin{bmatrix} I \\ q_2 \end{bmatrix} = A^{-1}\begin{bmatrix} -\frac{\mathcal{E}}{L} \\ 0 \end{bmatrix} = \begin{bmatrix} \frac{1}{5} \\ \frac{1}{25} \end{bmatrix}.$$

Therefore

$$\begin{bmatrix} I \\ q_2 \end{bmatrix} = c_1e^{-5t}\begin{bmatrix} 10 \\ -1 \end{bmatrix} + c_2e^{-(10/3)t}\begin{bmatrix} 5 \\ -1 \end{bmatrix} + \begin{bmatrix} \frac{1}{5} \\ \frac{1}{25} \end{bmatrix}. \tag{6-58}$$

Accordingly, there is an exponential decay to the steady state. The other two currents are given by Eqs. (6–51) and (6–55). In the steady state, $I_2 = 0$ since q_2 is constant.

================ **Problems (Section 6-5)** ================

1. For the network of Fig. 6-4 show that the matrix A in Eq. (6-56) has characteristic roots with negative real part, so that the system is stable. (See Section 6-11.)

2. For the network of Fig. 6-4, carry out the details leading to the general solution (6-58) for I and q_2.

3. For the network of Fig. 6-4, use the numerical values given, so that A is as in Eq. (6–57). Find a particular solution by variation of parameters when $\mathscr{E} = \mathscr{E}_0 e^{i\omega t}$.

4. For the network of Fig. 6-5, obtain simultaneous differential equations for I, q_1, and q_2.

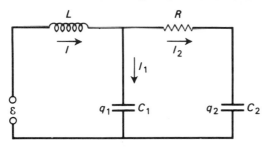

Fig. 6–5. Network for Problem 4.

5. *Mutual inductance.* When two circuits or networks have adjacent inductances, potential drops $M \, dI/dt$ are caused, where M is the mutual inductance (positive or negative, depending on how the coils are wound) and I is the current in the adjacent branch. Thus for the two RLC circuits of Fig. 6-6, Kirchhoff's first law gives the equations:

$$R_1 I_1 + L_1 \frac{dI_1}{dt} + \frac{q_1}{C_1} + M \frac{dI_2}{dt} = \mathscr{E},$$

$$R_2 I_2 + L_2 \frac{dI_2}{dt} + \frac{q_2}{C_2} \cdot M \frac{dI_1}{dt} = 0.$$

These equations can be solved for dI_1/dt and dI_2/dt, and the resulting equations, together with the relation $dq_1/dt = I_1$, $dq_2/dt = I_2$, yield a fourth-order system

$$\frac{d\mathbf{x}}{dt} = A\mathbf{x} + \mathbf{b},$$

where $\mathbf{x} = \text{col}(I_1, I_2, q_1, q_2)$. Find the matrix A and the vector $\mathbf{b}$.

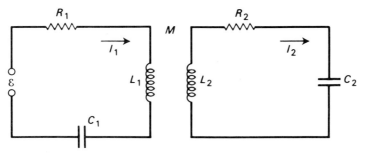

Fig. 6-6. Mutual inductance.

6-6 APPLICATION OF THE LAPLACE TRANSFORM

As in Section 4–11, we apply the rule for the transform of the derivative:

$$\mathscr{L}[f'(t)] = s\mathscr{L}[f] - f(0). \tag{6-60}$$

EXAMPLE 1 $\dfrac{d\mathbf{x}}{dt} = \begin{bmatrix} 2 & 1 \\ 1 & 2 \end{bmatrix} \mathbf{x}$. The equations are

$$\frac{dx_1}{dt} = 2x_1 + x_2, \qquad \frac{dx_2}{dt} = x_1 + 2x_2.$$

We apply the Laplace transform to each, writing

$$X_1(s) = \mathscr{L}[x_1(t)], \qquad X_2(s) = \mathscr{L}[x_2(t)], \qquad c_1 = x_1(0), \qquad c_2 = x_2(0).$$

We obtain

$$sX_1(s) - c_1 = 2X_1(s) + X_2(s)$$
$$sX_2(s) - c_2 = X_1(s) + 2X_2(s).$$

These are two simultaneous equations for $X_1(s)$ and $X_2(s)$ that can be solved by elimination to give

$$X_1(s) = \frac{c_1(s-2) + c_2}{s^2 - 4s + 3}, \qquad X_2(s) = \frac{c_1 + c_2(s-2)}{s^2 - 4s + 3}.$$

We decompose the coefficients of c_1 and c_2 into partial fractions:

$$X_1(s) = c_1 \left(\frac{1}{2}\frac{1}{s-1} + \frac{1}{2}\frac{1}{s-3} \right) + c_2 \left(\frac{1}{2}\frac{1}{s-3} - \frac{1}{2}\frac{1}{s-1} \right)$$

$$X_2(s) = c_1 \left(\frac{1}{2}\frac{1}{s-3} - \frac{1}{2}\frac{1}{s-1} \right) + c_2 \left(\frac{1}{2}\frac{1}{s-1} + \frac{1}{2}\frac{1}{s-3} \right).$$

We then take inverse transforms using the rule $\mathscr{L}[e^{at}] = 1/(s-a)$ (entry 2 in Table 4–1):

$$x_1(t) = c_1 \left(\frac{1}{2}e^t + \frac{1}{2}e^{3t} \right) + c_2 \left(\frac{1}{2}e^{3t} - \frac{1}{2}e^t \right),$$

$$x_2(t) = c_1 \left(\frac{1}{2}e^{3t} - \frac{1}{2}e^t \right) + c_2 \left(\frac{1}{2}e^t + \frac{1}{2}e^{3t} \right). \tag{6-61}$$

This gives the general solution $\mathbf{x}(t) = c_1\mathbf{x}_1(t) + c_2\mathbf{x}_2(t)$ for $t \geqslant 0$ in terms of the arbitrary constants $c_1 = x_1(0)$, $c_2 = x_2(0)$.

The same equation is considered in Example 1 of Section 6–3. To find the particular solutions obtained there:

$$\begin{bmatrix} e^{3t} \\ e^{3t} \end{bmatrix} \quad \text{and} \quad \begin{bmatrix} e^t \\ -e^t \end{bmatrix},$$

we take $c_1 = 1$, $c_2 = 1$ and then $c_1 = 1$, $c_2 = -1$ in Eqs. (6–61). ◀

EXAMPLE 2 $\dfrac{d\mathbf{x}}{dt} = \begin{bmatrix} 3 & 1 \\ -1 & 1 \end{bmatrix} \mathbf{x}$. The characteristic equation is $\lambda^2 - 4\lambda + 4 = 0$,
so that we have the more difficult case of repeated roots: 2, 2. The method of Laplace transforms proceeds without difficulty:

$$\frac{dx_1}{dt} = 3x_1 + x_2, \qquad \frac{dx_2}{dt} = -x_1 + x_2,$$

$$sX_1(s) - c_1 = 3X_1(s) + X_2(s), \qquad sX_2(s) - c_2 = -X_1(s) + X_2(s),$$

$$X_1(s) = \frac{c_1(s-1) - c_2}{s^2 - 4s + 4}, \qquad X_2(s) = \frac{-c_1 + c_2(s-3)}{s^2 - 4s + 4}.$$

To find the inverse transforms we use the rule (4–72) of Section 4–7:

$$\mathscr{L}^{-1}\left[\frac{c_0 s^m + c_1 s^{m-1} + \cdots + c_m}{Q(s)} \right] = c_0 f^{(m)}(t) + \cdots + c_m f(t), \qquad (6\text{-}62)$$

where $Q(s)$ is a polynomial of degree greater than m and $\mathscr{L}[f] = 1/Q(s)$. Here we take $Q(s) = (s-2)^2$, so that $\mathscr{L}^{-1}[1/Q(s)] = te^{2t}$ (by entry 4 of Table 4–1). Thus by (6-62), $\mathscr{L}^{-1}[s/Q(s)] = (te^{2t})' = e^{2t}(2t+1)$. Accordingly,

$$x_1(t) = c_1[e^{2t}(2t+1) - te^{2t}] + c_2 te^{2t} = c_1 e^{2t}(t+1) - c_2 te^{2t},$$

$$x_2(t) = -c_1 te^{2t} + c_2[e^{2t}(2t+1) - 3te^{2t}] = -c_1 te^{2t} + c_2 e^{2t}(1-t),$$

for $x_1(0) = c_1$, $x_2(0) = c_2$. ◄

The Laplace transform can also be applied to nonhomogeneous equations, provided we can obtain the transforms of the nonhomogeneous terms:

EXAMPLE 3 $\dfrac{d\mathbf{x}}{dt} = \begin{bmatrix} 2 & 1 \\ 1 & 2 \end{bmatrix} \mathbf{x} + \begin{bmatrix} 2e^{5t} \\ 3e^{2t} \end{bmatrix}.$

The equations are

$$\frac{dx_1}{dt} = 2x_1 + x_2 + 2e^{5t}, \qquad \frac{dx_2}{dt} = x_1 + 2x_2 + 3e^{2t}.$$

We seek the solution with $x_1(0) = 0$, $x_2(0) = 0$:

$$sX_1(s) = 2X_1(s) + X_2(s) + \frac{2}{s-5},$$

$$sX_2(s) = X_1(s) + 2X_2(s) + \frac{3}{s-2},$$

$$X_1(s) = \frac{2s^2 - 5s - 7}{(s-1)(s-2)(s-3)(s-5)},$$

$$X_2(s) = \frac{3s - 13}{(s-1)(s-3)(s-5)},$$

$$X_1(s) = \frac{5}{4}\frac{1}{s-1} - \frac{3}{s-2} + \frac{1}{s-3} + \frac{3}{4}\frac{1}{s-5}.$$

$$X_2(s) = -\frac{5}{4}\frac{1}{s-1}+\frac{1}{s-3}+\frac{1}{4}\frac{1}{s-5},$$

$$x_1(t) = \frac{5}{4}e^t - 3e^{2t} + e^{3t} + \frac{3}{4}e^{5t},$$

$$x_2(t) = -\frac{5}{4}e^t + e^{3t} + \frac{1}{4}e^{5t}.$$

This problem of finding a particular solution from a known complementary function is treated as Example 1 in Section 6–4. ◀

EXAMPLE 4   For the network of Fig. 6–5, the chosen numerical values lead to the matrix equation

$$\frac{d}{dt}\begin{bmatrix} I \\ q_2 \end{bmatrix} = \frac{1}{3}\begin{bmatrix} -20 & -50 \\ 1 & -5 \end{bmatrix}\begin{bmatrix} I \\ q_2 \end{bmatrix} + \begin{bmatrix} 2 \\ 0 \end{bmatrix}.$$

We apply Laplace transforms to seek the solution with $I = 0$ and $q_2 = 0$ for $t = 0$:

$$s\mathscr{L}[I] = \frac{1}{3}\left(-20\mathscr{L}[I] - 50\mathscr{L}[q_2]\right) + \frac{2}{s},$$

$$s\mathscr{L}[q_2] = \frac{1}{3}\left(\mathscr{L}[I] - 5\mathscr{L}[q_2]\right).$$

We solve for $\mathscr{L}[I]$ and $\mathscr{L}[q_2]$:

$$\mathscr{L}[I] = \frac{6s+10}{s(s+5)(3s+10)}, \quad \mathscr{L}[q_2] = \frac{2}{s(s+5)(3s+10)}.$$

Accordingly,

$$I = \frac{1}{5}[1 - 4e^{-5t} + 3e^{-(10/3)t}], \quad q_2 = \frac{1}{25}[1 + 2e^{-5t} - 3e^{-(10/3)t}]. \quad ◀$$

The Laplace transform can also be applied to simultaneous equations without reducing the equations to standard form as in Section 6–1.

EXAMPLE 5 $\dfrac{d^2x_1}{dt^2} + 2x_1 - x_2 = 0, \quad \dfrac{d^2x_2}{dt^2} + 2x_2 - x_1 = 0.$

We allow for initial conditions: $x_1(0) = c_1, x_1'(0) = c_2, x_2(0) = c_3, x_2'(0) = c_4$. From (6-60),

$$\mathscr{L}[f''(t)] = s\mathscr{L}[f'] - f'(0) = s^2\mathscr{L}[f] - sf(0) - f'(0). \tag{6-63}$$

Hence we obtain

$$s^2X_1(s) - sc_1 - c_2 + 2X_1(s) - X_2(s) = 0,$$

$$s^2X_2(s) - sc_3 - c_4 + 2X_2(s) - X_1(s) = 0,$$

$$X_1(s) = \frac{c_1(s^3 + 2s) + c_2(s^2 + 2) + c_3s + c_4}{s^4 + 4s^2 + 3},$$

$$X_2(s) = \frac{c_1s + c_2 + c_3(s^3 + 2s) + c_4(s^2 + 2)}{s^4 + 4s^2 + 3}.$$

Now

$$\frac{1}{s^4 + 4s^2 + 3} = \frac{1}{(s^2 + 1)(s^2 + 3)} = \frac{1}{2}\left(\frac{1}{s^2 + 1} - \frac{1}{s^2 + 3}\right)$$

$$= \frac{1}{2}\mathscr{L}\left[\sin t - \frac{1}{\sqrt{3}}\sin\sqrt{3}t\right].$$

Thus we can apply rule (6–62) to find the inverse transforms:

$$x_t(t) = c_1\left(-\frac{1}{2}\cos t + \frac{3}{2}\cos\sqrt{3}t + \cos t - \cos\sqrt{3}t\right) + \cdots$$

$$= \frac{1}{6}\left[c_1\left(3\cos t + 3\cos\sqrt{3}t\right) + c_2\left(3\sin t + \sqrt{3}\sin\sqrt{3}t\right)\right.$$

$$\left. + c_3\left(3\cos t - 3\cos\sqrt{3}t\right) + c_4\left(3\sin t - \sqrt{3}\sin\sqrt{3}t\right)\right], \qquad (6\text{--}64)$$

$$x_2(t) = \frac{1}{6}\left[c_1\left(3\cos t - 3\cos\sqrt{3}t\right) + c_2\left(3\sin t - \sqrt{3}\sin\sqrt{3}t\right)\right.$$

$$\left. + c_3\left(3\cos t + 3\cos\sqrt{3}t\right) + c_4\left(3\sin t + \sqrt{3}\sin\sqrt{3}t\right)\right]$$

DISCUSSION: The general solution depends on four arbitrary constants, since this is a fourth-order system. An equivalent system of four first-order equations can be obtained as in Section 6–1 in terms of x_1, x_2, $x_3 = dx_1/dt$, $x_4 = dx_2/dt$. The solutions for x_3 and x_4 can be obtained by differentiating $x_1(t)$ and $x_2(t)$, as given in Eqs. (6–64) above.

In this example we also obtain the following particular solutions by making special choices of $c_1, \ldots, c_4$:

$$x_1(t) = \sin t, \qquad x_2(t) = \sin t, \qquad c_1 = 0, \quad c_2 = 1, \quad c_3 = 0, \quad c_4 = 1;$$
$$x_1(t) = \cos t, \qquad x_2(t) = \cos t, \qquad c_1 = 1, \quad c_2 = 0, \quad c_3 = 1, \quad c_4 = 0;$$
$$x_1(t) = \sin\sqrt{3}t, \quad x_2(t) = -\sin\sqrt{3}t, \quad c_1 = 0, \quad c_2 = \sqrt{3}, \quad c_3 = 0, \quad c_4 = -\sqrt{3};$$
$$x_1(t) = \cos\sqrt{3}t, \quad x_2(t) = -\cos\sqrt{3}t, \quad c_1 = 1, \quad c_2 = 0, \quad c_3 = -1, \quad c_4 = 0.$$

The expressions for $x_3(t) = dx_1/dt$ and $x_4(t) = dx_2/dt$ can be derived from these, and we obtain four vector solutions $\mathbf{x}(t)$. They are linearly independent, since the corresponding matrix $X(t)$ has determinant,

$$\det X(t) = \begin{vmatrix} \sin t & \cos t & \sin\sqrt{3}t & \cos\sqrt{3}t \\ \sin t & \cos t & -\sin\sqrt{3}t & -\cos\sqrt{3}t \\ \cos t & -\sin t & \sqrt{3}\cos\sqrt{3}t & -\sqrt{3}\sin\sqrt{3}t \\ \cos t & -\sin t & -\sqrt{3}\cos\sqrt{3}t & \sqrt{3}\sin\sqrt{3}t \end{vmatrix} = -4\sqrt{3}.$$

It follows that these four solutions can also be used to form the general solution. Thus for x_1, x_2 we obtain (with new arbitrary constants) as in alternative to (6–64):

$$x_1 = c_1 \sin t + c_2 \cos t + c_3 \sin \sqrt{3}t + c_4 \cos \sqrt{3}t, \qquad (6\text{–}65)$$
$$x_2 = c_1 \sin t + c_2 \cos t - c_3 \sin \sqrt{3}t - c_4 \cos \sqrt{3}t. \quad \blacktriangleleft$$

(See also Section 6–8.)

═══════════════════ **Problems (Section 6–6)** ═══════════════════

1. Find the solution for $t \geq 0$ with given initial values:

   a) $\dfrac{d\mathbf{x}}{dt} = \begin{bmatrix} -7 & 20 \\ -6 & 15 \end{bmatrix} \mathbf{x}, \quad \mathbf{x}(0) = \mathrm{col}\,(2, 1)$

   b) $\dfrac{d\mathbf{x}}{dt} = \begin{bmatrix} 7 & 10 \\ -3 & -4 \end{bmatrix} \mathbf{x} + \begin{bmatrix} e^t \\ 3e^t \end{bmatrix}, \quad \mathbf{x}(0) = \mathbf{0}$

   c) $\dfrac{d\mathbf{x}}{dt} = \begin{bmatrix} -3 & -4 \\ 1 & 1 \end{bmatrix} \mathbf{x}, \quad \mathbf{x}(0) = \mathrm{col}\,(1, 1)$

   d) $\dfrac{d\mathbf{x}}{dt} = \begin{bmatrix} -2 & 4 \\ -1 & 2 \end{bmatrix} \mathbf{x} + \begin{bmatrix} 2 \\ 1 \end{bmatrix}, \quad \mathbf{x}(0) = \mathrm{col}\,(0, 0)$

   e) $\dfrac{d\mathbf{x}}{dt} = \begin{bmatrix} 2 & 1 & 2 \\ -1 & 0 & -2 \\ 0 & 0 & 1 \end{bmatrix} \mathbf{x} + \begin{bmatrix} e^t \\ 0 \\ 2e^t \end{bmatrix}, \quad \mathbf{x}(0) = \mathbf{0}$

   f) $\dfrac{d\mathbf{x}}{dt} = \begin{bmatrix} -8 & 47 & -8 \\ -4 & 18 & -2 \\ -8 & 39 & -5 \end{bmatrix} \mathbf{x}, \quad \mathbf{x}(0) = \mathrm{col}\,(1, 1, 0)$

   g) $\dfrac{d^2 x_1}{dt^2} = x_1 - x_2 + e^t, \quad \dfrac{d^2 x_2}{dt^2} = -x_1 + x_2,$

   $x_1 = 0,\ x_2 = 0,\ dx_1/dt = 1,\ dx_2/dt = 1$ for $t = 0$

   h) $(2D - 3)x_1 + (3D - 6)x_2 + (D^2 + D + 5)x_3 = 0,$

   $(7D - 12)x_1 + (11D - 24)x_2 + (3D^2 + 4D + 12)x_3 = 0,$

   $(D - 3)x_1 + (2D - 6)x_2 + (D^2 + 3D - 1)x_3 = e^{3t},$

   where $D = d/dt$ and $x_1 = 0,\ x_2 = 0,\ x_3 = 0,\ Dx_3 = 0$ for $t = 0$.

2. Find the general solution by Laplace transforms:

   a) $\dfrac{d\mathbf{x}}{dt} = \begin{bmatrix} -1 & 1 \\ -5 & 3 \end{bmatrix} \mathbf{x}$ b) $\dfrac{d\mathbf{x}}{dt} = \begin{bmatrix} 1 & 1 \\ -1 & 3 \end{bmatrix} \mathbf{x}$

   c) Equation of Problem 1(a)   d) Equation of Problem 1(c)

═══

6–7 METHOD OF CHANGE OF VARIABLE

Another way of attacking the problem

$$\frac{d\mathbf{x}}{dt} = A\mathbf{x} + \mathbf{f}(t) \qquad (6\text{–}70)$$

is to make a change of variable

$$\mathbf{x} = C\mathbf{y}, \tag{6-71}$$

where C is a nonsingular constant $n \times n$ matrix. The differential equation (6-70) is transformed into a differential equation for $\mathbf{y}$ as follows: $\mathbf{y} = C^{-1}\mathbf{x}$, so that

$$\frac{d\mathbf{y}}{dt} = C^{-1}\frac{d\mathbf{x}}{dt} = C^{-1}[A\mathbf{x} + \mathbf{f}(t)]$$

$$= C^{-1}AC\mathbf{y} + C^{-1}\mathbf{f}(t).$$

Therefore, the new equation is

$$\frac{d\mathbf{y}}{dt} = B\mathbf{y} + \mathbf{g}(t), \tag{6-72}$$

where $B = C^{-1}AC$ and $\mathbf{g}(t) = C^{-1}\mathbf{f}(t)$, so that B is similar to A (Section 5–7).

It may happen that B is simpler than A, in which case we have made progress. In particular, if A has distinct eigenvalues $\lambda_1, \ldots, \lambda_n$ with corresponding eigenvectors $\mathbf{v}_1, \ldots, \mathbf{v}_n$, then we can choose C to be the matrix whose column vectors are $\mathbf{v}_1, \ldots, \mathbf{v}_n$. Then, as in Section 5–7, B is the diagonal matrix diag $(\lambda_1, \ldots, \lambda_n)$. Thus (6–72) is equivalent to the system

$$\frac{dy_i}{dt} = \lambda_i y_i + g_i(t), \quad (i = 1, \ldots, n), \tag{6-73}$$

and can be solved for each y_i separately to obtain $\mathbf{y}(t)$ and hence $\mathbf{x}(t) = C\mathbf{y}(t)$. Since we have to solve for the λ_j and $\mathbf{v}_j$, this method is essentially equivalent to that of Sections 6-3 and 6-4.

It is shown in algebra that if the eigenvalues are not distinct, C can be chosen so that B has the *Jordan normal form*. This form is illustrated by the following 6×6 matrix:

$$\begin{bmatrix} \lambda_1 & 1 & 0 & 0 & 0 & 0 \\ 0 & \lambda_1 & 1 & 0 & 0 & 0 \\ 0 & 0 & \lambda_1 & 0 & 0 & 0 \\ 0 & 0 & 0 & \lambda_2 & 0 & 0 \\ 0 & 0 & 0 & 0 & \lambda_2 & 0 \\ 0 & 0 & 0 & 0 & 0 & \lambda_3 \end{bmatrix}.$$

In general, the eigenvalues, with repetition according to multiplicity, appear along the diagonal. Whenever two successive diagonal elements are the same, say b_{ii} and $b_{i+1,i+1}$, the element $b_{i,i+1}$ is 1 or 0. Otherwise all off-diagonal elements are 0. With such a matrix B, the differential equations (6-72) can be solved as illustrated below.

EXAMPLE 1 $\dfrac{d\mathbf{y}}{dt} = \begin{bmatrix} 3 & 1 & 0 \\ 0 & 3 & 0 \\ 0 & 0 & 2 \end{bmatrix} \mathbf{y}$. Here we have the system

$$\frac{dy_1}{dt} = 3y_1 + y_2, \qquad \frac{dy_2}{dt} = 3y_2, \qquad \frac{dy_3}{dt} = 2y_3.$$

Hence $y_2 = c_2 e^{3t}$, $y_3 = c_3 e^{2t}$, with arbitrary constants c_2, c_3. Then

$$\frac{dy_1}{dt} - 3y_1 = c_2 e^{3t},$$

so that (as in Section 1–7),

$$y_1 = e^{3t} \int c_2 e^{-3t} e^{3t} \, dt = e^{3t} (c_2 t + c_1).$$

These equations give the general solution, which can also be written as follows:

$$\mathbf{y} = c_1 \operatorname{col} (e^{3t}, 0, 0) + c_2 \operatorname{col} (te^{3t}, e^{3t}, 0) + c_3 \operatorname{col} (0, 0, e^{2t}). \blacktriangleleft$$

(The technique of reducing A to Jordan normal form is described in ODE, pp. 287–290.)

If matrix A happens to be symmetric, that is, $a_{ij} = a_{ji}$ for all i and j (see Section 5–15), then it can be shown that A has real eigenvalues $\lambda_1, \ldots, \lambda_n$ (with some repetitions allowed) and that, even when repetitions occur, n linearly independent eigenvectors $\mathbf{v}_1, \ldots, \mathbf{v}_n$ can be found. With these vectors, C can be formed as above, and again $C^{-1} AC = B = \operatorname{diag} (\lambda_1, \ldots, \lambda_n)$; thus A is similar to a diagonal matrix (see Problem 13 following Section 5–9). The substitution (6–71) is thus again successful.

EXAMPLE 2 $\dfrac{d\mathbf{x}}{dt} = \begin{bmatrix} 7 & 4 & -4 \\ 4 & 1 & 8 \\ -4 & 8 & 1 \end{bmatrix} \mathbf{x} = A\mathbf{x}.$

The characteristic equation is

$$\begin{bmatrix} 7-\lambda & 4 & -4 \\ 4 & 1-\lambda & 8 \\ -4 & 8 & 1-\lambda \end{bmatrix} = -(\lambda - 9)^2 (\lambda + 9).$$

For $\lambda = 9$, the eigenvectors $\mathbf{v}$ must satisfy

$$\begin{bmatrix} -2 & 4 & -4 \\ 4 & -8 & 8 \\ -4 & 8 & -8 \end{bmatrix} \begin{bmatrix} v_1 \\ v_2 \\ v_3 \end{bmatrix} = 0$$

or

$$-2v_1 + 4v_2 - 4v_3 = 0,$$
$$4v_1 - 8v_2 + 8v_3 = 0,$$
$$-4v_1 + 8v_2 - 8v_3 = 0.$$

These equations are equivalent. They are satisfied, for example, for $v_1 = 2$, $v_2 = 1$, $v_3 = 0$ and for $v_1 = 2$, $v_2 = 0$, $v_3 = -1$. Thus we have the linearly independent vectors

$$\mathbf{v}_1 = \operatorname{col} (2, 1, 0) \quad \text{and} \quad \mathbf{v}_2 = \operatorname{col} (2, 0, -1).$$

For $\lambda = -9$, we find $\mathbf{v}_3 = (1, -2, 2)$. Thus with

$$C = \begin{bmatrix} 2 & 2 & 1 \\ 1 & 0 & -2 \\ 0 & -1 & 2 \end{bmatrix}$$

we must have $C^{-1}AC = \text{diag}\,(9, 9, -9)$, as can be verified. The substitution $\mathbf{x} = C\mathbf{y}$ thus gives

$$\frac{d\mathbf{y}}{dt} = B\mathbf{y}, \quad B = \text{diag}\,(9, 9, -9).$$

Hence

$$\frac{dy_1}{dt} = 9y_1, \quad \frac{dy_2}{dt} = 9y_2, \quad \frac{dy_3}{dt} = -9y_3$$

and $y_1 = c_1 e^{9t}$, $y_2 = c_2 e^{9t}$, $y_3 = c_3 e^{-9t}$. Finally,

$$\mathbf{x} = C\mathbf{y} = \begin{bmatrix} 2 & 2 & 1 \\ 1 & 0 & -2 \\ 0 & -1 & 2 \end{bmatrix} \begin{bmatrix} c_1 e^{9t} \\ c_2 e^{9t} \\ c_3 e^{-9t} \end{bmatrix}$$

$$= c_1 e^{9t} \,\text{col}\,(2, 1, 0) + c_2 e^{9t} \,\text{col}\,(2, 0, -1) + c_3 e^{-9t}(1, -2, 2)$$

$$= c_1 e^{9t}\mathbf{v}_1 + c_2 e^{9t}\mathbf{v}_2 + c_3 e^{-9t}\mathbf{v}_3. \quad \blacktriangleleft$$

6–8 NORMAL MODES OF VIBRATION

We know that a simple frictionless mass–spring system leads to simple harmonic motion with a certain frequency of oscillation. It can be shown that a system of two masses coupled by springs (as in Fig. 6–7) is capable of oscillating as a whole in two different ways, called *normal modes*, with two different frequencies.

    In one normal mode (the *push–push mode*), both masses execute simple harmonic motions that are in phase, so that both move to the right at the same time and both move to the left at the same time. In the other normal mode (the *push–pull mode*), the masses are 180° out of phase, so that one moves to the left when the other moves to the right. In general, the two masses have different amplitudes of oscillation. Thus the two normal modes might lead to functions $x_1(t)$, $x_2(t)$ as in Fig. 6–8 (a) or (b).

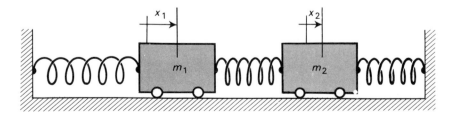

Fig. 6–7. Vibrating systems of two masses.

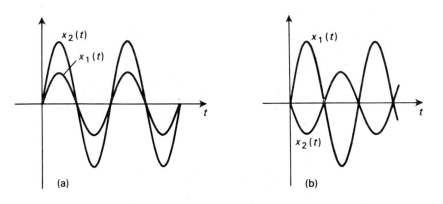

Fig. 6–8. Normal modes for coupled masses as in Fig. 6–7: (a) push–push mode; (b) push–pull mode.

The normal modes can be easily demonstrated by a physical experiment. A similar effect is provided by two persons nodding their heads to the left and right in time to music.

In the case of three masses there are three normal modes with three different frequencies. In general, n masses on a line have n normal modes with n different frequencies.

There are also other vibrating physical systems which can be thought of as formed of particles and rigid bodies restrained by interconnecting springs (like the springs in a cot). The phenomenon of normal modes turns out to be general; it is the basic concept of molecular spectra (the spectrum is simply the record of frequencies which occur).

We shall explain why the normal modes occur. In this section we consider the differential equations of motion of the form

$$m_i \frac{d^2 x_i}{dt^2} + \sum_{j=1}^{n} a_{ij} x_j = 0, \quad i = 1, \ldots, n, \tag{6–80}$$

where the m_i are all positive and $A = (a_{ij})$ is an $n \times n$ symmetric matrix (Section 5–15) whose eigenvalues are real and positive. It will be seen that this includes the case of the coupled masses of Fig. 6–7; it also includes the case of n masses on a line. In the next section we consider a more general form that covers all the cases of physical importance.

We can write Eqs. (6–80) in vector form by writing $N = \mathrm{diag}\,(m_1^{1/2}, \ldots, m_n^{1/2})$; the equations become

$$N^2 \frac{d^2 \mathbf{x}}{dt^2} + A\mathbf{x} = 0. \tag{6–81}$$

We now set $\mathbf{y} = N\mathbf{x}$, as a first change of variable:

$$\frac{d^2\mathbf{y}}{dt^2} = N\frac{d^2\mathbf{x}}{dt^2} = -N^{-1}A\mathbf{x} = -N^{-1}AN^{-1}\mathbf{y} = -B\mathbf{y}, \qquad (6\text{--}82)$$

where $B = N^{-1}AN^{-1}$. Since A is symmetric and N is diagonal, B is also symmetric. Furthermore, the eigenvalues of B are also positive (see Problem 7 below; see also AC, pp. 567–569).

As in the preceding section, we can now find a nonsingular matrix C such that

$$C^{-1}BC = D = \text{diag}(\lambda_1, \ldots, \lambda_n),$$

where $\lambda_1, \ldots, \lambda_n$ are the eigenvalues of B. We set $\mathbf{y} = C\mathbf{z}$ in (6–82); then $\mathbf{z} = C^{-1}\mathbf{y}$, so that

$$\frac{d^2\mathbf{z}}{dt^2} = C^{-1}\frac{d^2\mathbf{y}}{dt^2} = -C^{-1}B\mathbf{y} = -C^{-1}BC\mathbf{z} = -D\mathbf{z}, \qquad (6\text{--}83)$$

that is,

$$\frac{d^2z_i}{dt^2} + \lambda_i z_i = 0, \quad i = 1, \ldots, n. \qquad (6\text{--}84)$$

Since each λ_i is positive, we can write $\lambda_i = \omega_i^2$, and the solutions of Eq. (6–84) are given by

$$z_i = A_i \sin(\omega_i t + \alpha_i), \quad i = 1, \ldots, n, \qquad (6\text{--}85)$$

where $A_1, \ldots, A_n, \alpha_1, \ldots, \alpha_n$ are arbitrary real constants. Now,

$$\mathbf{x} = N^{-1}\mathbf{y} = N^{-1}C\mathbf{z} = E\mathbf{z},$$

where $E = N^{-1}C = (e_{ij})$. Hence,

$$x_i = \sum_{j=1}^{n} e_{ij}z_j = \sum_{j=1}^{n} e_{ij}A_j \sin(\omega_j t + \alpha_j), \quad i = 1, \ldots, n. \qquad (6\text{--}86)$$

This is the general solution of the system (6–80). If, for example, $A_2 = \ldots = A_n = 0$, then

$$x_i = e_{i1}A_1 \sin(\omega_1 t + \alpha_1), \quad i = 1, \ldots, n, \qquad (6\text{--}87)$$

and all coordinates oscillate synchronously with frequency ω_1. This motion is called a *normal mode of vibration*. The general solution can be regarded as a superposition of normal modes.

Since we know the form of the general solution (6–86) and of the particular solutions (6–87), we can find these solutions more directly by substituting

$$x_i = e_i \sin\omega t, \quad i = 1, \ldots, n, \qquad (6\text{--}88)$$

in the given equations (6–80). In effect we are thus seeking the normal modes. Once we have found them, the general solution can be reconstructed in form (6–86).

EXAMPLE 1 $\quad 5\dfrac{d^2x_1}{dt^2} + 5x_1 - 2x_2 = 0, \quad 6\dfrac{d^2x_2}{dt^2} - 2x_1 + 6x_2 - 2x_3 = 0,$

$$7\frac{d^2x_3}{dt^2} - 2x_2 + 7x_3 = 0.$$

It can be verified that the eigenvalues of A are positive. That this is so is also revealed by the fact that the equation for ω^2 to follow has only positive roots. We make the substitution (6–88) with $n = 3$ and, after cancelling the common factor $\sin \omega t$, are led to the equations

$$(5 - 5\omega^2)e_1 - 2e_2 = 0, \quad -2e_1 + (6 - 6\omega^2)e_2 - 2e_3 = 0,$$
$$-2e_2 + (7 - 7\omega^2)e_3 = 0.$$

These are three homogeneous equations for e_1, e_2, e_3, whose determinant is

$$\begin{vmatrix} 5 - 5\omega^2 & -2 & 0 \\ -2 & 6 - 6\omega^2 & -2 \\ 0 & -2 & 7 - 7\omega^2 \end{vmatrix} = -210\,\omega^6 + 630\omega^4 - 582\omega^2 + 162.$$

If we equate this to 0, we obtain a cubic equation for ω^2, whose roots are 1 and $1 \pm (8/35)^{1/2}$. We can take ω to be positive, since a negative ω can be accounted for by a phase shift. Thus we obtain $\omega_1 = 1$, $\omega_2 = 0.72$, $\omega_3 = 1.22$. Corresponding to $\omega_1 = 1$, we obtain $e_1 = 1$, $e_2 = 0$, $e_3 = -1$; for $\omega_2 = 0.72$ we obtain $e_1 = 1$, $e_2 = 1.2$, $e_3 = 0.728$; for $\omega_3 = 1.22$ we obtain $e_1 = 1$, $e_2 = -1.2$, $e_3 = 0.728$. Accordingly, the general solution is

$$x_1 = A_1 \sin(t + \alpha_1) + A_2 \sin(0.72t + \alpha_2) + A_3 \sin(1.22t + \alpha_3),$$
$$x_2 = 1.2\,A_2 \sin(0.72t + \alpha_2) - 1.2\,A_3 \sin(1.22t + \alpha_3),$$
$$x_3 = -A_1 \sin(t + \alpha_1) + 0.73\,A_2 \sin(0.72t + \alpha_2) + 0.73\,A_3 \sin(1.22t + \alpha_3),$$

where A_1, A_2, A_3, α_1, α_2, α_3 are arbitrary constants. ◀

We now turn to the system of two masses coupled by springs, as in Fig. 6–7. We assume that all three springs have natural length L and spring constant k. We let x_1, x_2 measure the displacements of m_1, m_2 from their equilibrium positions, at which the springs all have length L_1. The mass m_1 is subject to spring forces

$$-k(L_1 + x_1 - L) \quad \text{and} \quad k(L_1 + x_2 - x_1 - L)$$

to the left and to the right, respectively. A similar statement applies to m_2, and we obtain the differential equation

$$m_1 \frac{d^2 x_1}{dt^2} = -k(L_1 + x_1 - L) + k(L_1 + x_2 - x_1 - L) = -2kx_1 + kx_2$$

$$(6-89)$$

$$m_2 \frac{d^2 x_2}{dt^2} = -k(L_1 + x_2 - x_1 - L) + k(L_1 - x_2 - L) = kx_1 - 2kx_2.$$

These are clearly of form (6–80), with $A = k \begin{bmatrix} 2 & -1 \\ -1 & 2 \end{bmatrix}$, so that A has the positive eigenvalues k, $3k$. A special case of this system is solved by Laplace transforms as Example 5 in Section 6–6. Four particular normal modes are used in forming the general solution (6–65).

6–9 A GENERALIZATION

The equations leading to normal modes sometimes appear in a more general form:

$$\sum_{j=1}^{n} p_{ij}\frac{d^2x_j}{dt^2} + \sum_{j=1}^{n} g_{ij}x_j = 0 \qquad \text{or} \qquad P\frac{d^2\mathbf{x}}{dt^2} + G\mathbf{x} = \mathbf{0}, \qquad (6\text{-}90)$$

where $P = (p_{ij})$ and $G = (g_{ij})$ are both symmetric matrices whose eigenvalues are real and positive. In this case, $P = CMC^{-1}$, where $M = \operatorname{diag}(m_1, \ldots, m_n)$, $m_1, \ldots, m_n$ are the eigenvalues of P, and C is orthogonal (Section 5–16), so that $C' = C^{-1}$. If we set $\mathbf{x} = C\mathbf{u}$, then the differential equation becomes

$$PC\frac{d^2\mathbf{u}}{dt^2} + GC\mathbf{u} = \mathbf{0} \qquad \text{or} \qquad CM\frac{d^2\mathbf{u}}{dt^2} + GC\mathbf{u} = \mathbf{0}$$

or

$$M\frac{d^2\mathbf{u}}{dt^2} + A\mathbf{u} = \mathbf{0}, \qquad (6\text{-}91)$$

with $M = \operatorname{diag}(m_1, \ldots, m_n)$ and $A = C^{-1}GC = C'GC$. Hence A is symmetric (see Section 5–15) and has the same eigenvalues as G (Section 5–7). Therefore, we have reduced the differential equations (6–90) to the form (6–80) and again obtain normal modes.

As in Section 6–8, we can obtain the normal modes directly by setting

$$x_i = e_i \sin \omega t, \quad i = 1, \ldots, n$$

in the differential equations (6–90).

=============== **Problems (Section 6–9)** ===============

1. Verify that C is nonsingular and that $C^{-1}AC = B$ (or equivalently $AC = CB$) for the following choices of A, B, C:

a) $A = \begin{bmatrix} 4 & -3 \\ 5 & -4 \end{bmatrix}$, $B = \begin{bmatrix} 1 & 0 \\ 0 & -1 \end{bmatrix}$, $C = \begin{bmatrix} 1 & 3 \\ 1 & 5 \end{bmatrix}$

b) $A = \begin{bmatrix} 7 & 6 \\ 2 & 6 \end{bmatrix}$, $B = \begin{bmatrix} 10 & 0 \\ 0 & 3 \end{bmatrix}$, $C = \begin{bmatrix} 2 & 3 \\ 1 & -2 \end{bmatrix}$

c) $A = \begin{bmatrix} 4 & -9 & 5 \\ 1 & -10 & 7 \\ 1 & -17 & 12 \end{bmatrix}$, $B = \begin{bmatrix} 1 & 0 & 0 \\ 0 & 2 & 0 \\ 0 & 0 & 3 \end{bmatrix}$, $C = \begin{bmatrix} 1 & 1 & 1 \\ 2 & 3 & -1 \\ 3 & 5 & -2 \end{bmatrix}$

d) $A = \begin{bmatrix} -9 & 19 & 4 \\ -3 & 7 & 1 \\ -7 & 17 & 2 \end{bmatrix}$, $B = \begin{bmatrix} 0 & 0 & 0 \\ 0 & i & 0 \\ 0 & 0 & -i \end{bmatrix}$, $C = \begin{bmatrix} 3 & 9-4i & 9+4i \\ 1 & 3-i & 3+i \\ 2 & 7-2i & 7+2i \end{bmatrix}$

e) $A = \begin{bmatrix} 4 & 1 \\ -1 & 2 \end{bmatrix}$, $B = \begin{bmatrix} 3 & 1 \\ 0 & 3 \end{bmatrix}$, $C = \begin{bmatrix} 1 & 1 \\ -1 & 0 \end{bmatrix}$

f) $A = \begin{bmatrix} 2 & 1 & 2 \\ -1 & 0 & -2 \\ 0 & 0 & 1 \end{bmatrix}$, $B = \begin{bmatrix} 1 & 1 & 0 \\ 0 & 1 & 0 \\ 0 & 0 & 1 \end{bmatrix}$, $C = \begin{bmatrix} 1 & 2 & 2 \\ -1 & -1 & 0 \\ 0 & 0 & -1 \end{bmatrix}$

2. (a), (b), (c), (d), (e), (f). With the aid of the results of Problem 1, find the general solution of the equation $dx/dt = Ax$ for A as in the corresponding parts of Problem 1.

3. Verify that A is symmetric, find C so that $C^{-1}AC$ is a diagonal matrix, and find the general solution of the differential equation $dx/dt = Ax$:

a) $A = \begin{bmatrix} 4 & 2 \\ 2 & 1 \end{bmatrix}$ b) $A = \begin{bmatrix} 3 & 2 \\ 2 & 3 \end{bmatrix}$

c) $A = \begin{bmatrix} 8 & 2 & -2 \\ 2 & 5 & 4 \\ -2 & 4 & 5 \end{bmatrix}$ d) $A = \begin{bmatrix} 85 & 12 & -18 \\ 12 & 53 & -6 \\ -18 & -6 & 64 \end{bmatrix}$

4. Find the normal modes of oscillation for the system described by Eqs. (6–89) for the following cases:
   a) $m_1 = 4$, $m_2 = 1$, $k = 1$ b) $m_1 = 1$, $m_2 = 1$, $k = 2$.

5. Generalize the system of Fig. 6–7 to one with three masses, as in Fig. 6–9. Again assume that all springs have the same natural length L and spring constant k and that at equilibrium all have length L_1.
   a) Obtain the differential equations of motion and show that they have the form (6–80).
   b) Take all masses equal to 1 and $k = 1$ (in appropriate units) and find the normal modes.

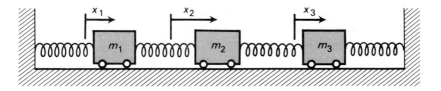

Fig. 6–9. Coupled system of three masses.

6. For the electrical network of Fig. 6–10, in which we assume that there was an applied e.m.f. that has been suddenly switched off, show that I_1, I_2 satisfy the differential equations

$$L_1 \frac{d^2 I_1}{dt^2} + \frac{I_1 - I_2}{C} = 0, \quad L_2 \frac{d^2 I_2}{dt^2} - \frac{I_1 - I_2}{C} = 0,$$

and that there is only one normal mode of oscillation.

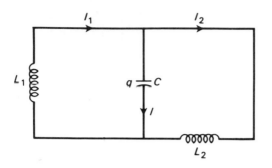

Fig. 6-10. Network for Problem 6.

7. Let A be a symmetric $n \times n$ matrix, let N be a nonsingular diagonal matrix diag$(c_1, \ldots, c_n)$, and let $B = N^{-1} A N^{-1}$. Show that B is symmetric.

HINT: Show that $b_{ij} = a_{ij} d_i d_j$, where $d_i = c_i^{-1}$.

6–10 FURTHER APPLICATIONS TO ELECTRIC NETWORKS

As in Section 6-5, the analysis of a general electric network is based on Kirchhoff's laws:

I. The total voltage drop about each closed circuit (mesh) is zero.
II. The total current entering each junction (node) is zero.

A general network has a scheme as shown in Fig. 6–11. The nodes are represented by large dots, the branches by lines (possibly curved), each of which connects two nodes. Each branch has a positive direction selected, with respect to which currents and voltage drops are measured.

If there are N branches, then the N currents $I_1, \ldots, I_N$ can be taken as unknowns. By virtue of the second law, some of these can be eliminated, and n currents remain, in terms of which all others can be expressed.

An alternative procedure is to introduce n *mesh currents* $J_1, \ldots, J_n$. These are certain linear combinations of $I_1, \ldots, I_N$, in terms of which $I_1, \ldots, I_N$ can all be expressed. The term *mesh* refers to a closed circuit in the network. For simplicity we

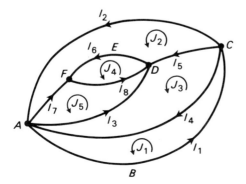

Fig. 6–11. Mesh currents in a planar network.

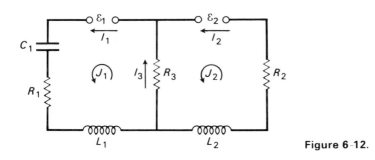

Figure 6-12.

restrict attention to planar networks (i.e., those that can be placed in a plane without forcing two branches to cross). For a network in a plane the boundaries of the regions into which the interior of the network is subdivided form n meshes with which $J_1, \ldots, J_n$ can be associated; each is assumed to flow in the counterclockwise direction. If the kth mesh contains a branch on the outermost boundary, we can choose the current J_k as the current I through that branch. (We can assume the positive direction on the outermost boundary to be counterclockwise, as in Fig. 6–11). The other mesh currents are then defined by systematically applying the rule that *if the lth branch is common to two meshes, say the pth and the qth, then* $I = \pm (J_p - J_q)$, *with a plus sign when the chosen direction on the branch is the same as the counterclockwise direction in the pth mesh, and a minus sign otherwise.*

For the network of Fig. 6–11 we choose $J_1 = I_1, J_2 = I_2$; then

$$
\begin{array}{ll}
I_4 = J_1 - J_3, & I_5 = J_3 - J_2, \\
I_6 = J_4 - J_2, & I_8 = J_4 - J_5, \\
I_7 = J_2 - J_5, & I_3 = J_5 - J_3.
\end{array}
\tag{6-100}
$$

Hence

$$
\begin{array}{ll}
J_3 = J_1 - I_4 = I_1 - I_4, & J_4 = J_2 + I_6 = I_2 + I_6, \\
J_5 = J_4 - I_8 = I_2 + I_6 - I_8.
\end{array}
\tag{6-101}
$$

We can verify that by virtue of Kirchhoff's second law, all the equations (6–100) are satisfied. Thus $I_1, \ldots, I_8$ can be expressed in terms of $J_1, \ldots, J_5$, and conversely. For the network of Fig. 6–12 we have

$$
J_1 = I_1, \qquad J_2 = I_2, \qquad I_3 = J_1 - J_2,
\tag{6-102}
$$

in agreement with Kirchhoff's second law.

To obtain the differential equations for the network, we now apply Kirchhoff's first law to each mesh separately and express the results in terms of the mesh currents. We illustrate this for the network of Fig. 6–12. For the first mesh,

$$
L_1 \frac{dI_1}{dt} + R_1 I_1 + \frac{q_1}{C_1} + R_3 I_3 = \mathcal{E}_1,
$$

where $q_1 = \int I_1 \, dt$ is the charge on the capacitor. By (6–102) this can be rewritten:

$$
L_1 \frac{dJ_1}{dt} + R_1 J_1 + \frac{1}{C_1} \int J_1 \, dt + R_3 (J_1 - J_2) = \mathcal{E}_1.
\tag{6-103}
$$

Similarly, for the second mesh,

$$L_2 \frac{dJ_2}{dt} + R_2 J_2 + \frac{1}{C_2} \int J_2 dt + R_3 (J_2 - J_1) = \mathscr{E}_2. \tag{6-104}$$

By differentiating (6–103), (6 104), we eliminate the integral signs:

$$L_1 \frac{d^2 J_1}{dt^2} + R_1 \frac{dJ_1}{dt} + \frac{J_1}{C_1} + R_3 \left(\frac{dJ_1}{dt} - \frac{dJ_2}{dt} \right) = \frac{d\mathscr{E}_1}{dt},$$

$$L_2 \frac{d^2 J_2}{dt^2} + R_2 \frac{dJ_2}{dt} + \frac{J_2}{C_2} + R_3 \left(\frac{dJ_2}{dt} - \frac{dJ_1}{dt} \right) = \frac{d\mathscr{E}_2}{dt}. \tag{6-104'}$$

We can also introduce the *mesh charges* Q_1, Q_2:

$$Q_1 = \int J_1 \, dt, \qquad Q_2 = \int J_2 \, dt, \tag{6-105}$$

for appropriate choices of the indefinite integrals. The equations (6–104′) then become

$$L_1 \frac{d^2 Q_1}{dt^2} + R_1 \frac{dQ_1}{dt} + \frac{Q_1}{C_1} + R_3 \left(\frac{dQ_1}{dt} - \frac{dQ_2}{dt} \right) = \mathscr{E}_1,$$

$$L_2 \frac{d^2 Q_2}{dt^2} + R_2 \frac{dQ_2}{dt} + \frac{Q_2}{C_2} + R_3 \left(\frac{dQ_2}{dt} - \frac{dQ_1}{dt} \right) = \mathscr{E}_2. \tag{6-104''}$$

For a general planar network containing inductance, resistance, capacitance, and driving electromotive forces, we obtain a system of equations of second order

$$\sum_{\beta = 1}^{n} (L_{\alpha\beta} D^2 + R_{\alpha\beta} D + \gamma_{\alpha\beta}) J_\beta = \frac{d\mathscr{E}_\alpha}{dt}, \tag{6-106}$$

for $\alpha = 1, \ldots, n$. In terms of mesh charges $Q_\alpha = \int J_\alpha \, dt$, the system becomes

$$\sum_{\beta = 1}^{n} (L_{\alpha\beta} D^2 + R_{\alpha\beta} D + \gamma_{\alpha\beta}) Q_\beta = \mathscr{E}_\alpha. \tag{6-106'}$$

The $\mathscr{E}_\alpha$ on the right-hand side is the total driving e.m.f. in the αth mesh. From the way in which the mesh currents are defined, we can verify the *reciprocity law*:

$$L_{\alpha\beta} = L_{\beta\alpha}, \qquad R_{\alpha\beta} = R_{\beta\alpha}, \qquad \gamma_{\alpha\beta} = \gamma_{\beta\alpha}. \tag{6-107}$$

These equations follow from the fact that $L_{\alpha\beta}$ corresponds to the inductance contributed to the αth mesh along the branch shared by the αth mesh and βth mesh; by the above definitions, interchanging α and β describes exactly the same inductance. A similar argument holds for the other terms.

We now introduce the two energy functions

$$T = \frac{1}{2} \sum_{\alpha,\beta=1}^{n} L_{\alpha\beta} J_\alpha J_\beta = \frac{1}{2} \sum_{\alpha,\beta=1}^{n} L_{\alpha\beta} \frac{dQ_\alpha}{dt} \frac{dQ_\beta}{dt},$$

$$V = \frac{1}{2} \sum_{\alpha,\beta=1}^{n} \gamma_{\alpha\beta} Q_\alpha Q_\beta.$$

$$(6\text{--}108)$$

The first is the *total electromagnetic energy*; it is analogous to the total kinetic energy in a mechanical system. The function V is the total *electrostatic energy*; it is analogous to potential energy. When all $R_{\alpha\beta}$ and $\mathscr{E}_\alpha$ are 0, Eqs. (6–106′) have the form of Eqs. (6–90), with symmetric coefficient matrices. Accordingly, we obtain normal modes of oscillation, as in Sections 6–8 and 6–9.

The terms in the $R_{\alpha\beta}$ correspond to *dissipation of energy*. The function

$$F = \tfrac{1}{2} \sum R_{\alpha\beta} Q'_\alpha Q'_\beta,$$

$$(6\text{--}108')$$

where $(') = d/dt$, is termed the *dissipation function*; it is one-half the rate of loss of energy (converted to heat) per unit time. It can be shown that $T \geqslant 0, V \geqslant 0, F \geqslant 0$ for all values of the variables.

From these inequalities we can conclude that the characteristic roots have real parts which are 0 or negative, so that there is some form of stability; the case of 0 real part is exceptional (see Problem 4 below).

For further information on network theory see *Linear Circuits* by R. E. Scott, Addison-Wesley Publishing Co., Reading, Mass., 1960.

Finally, let us consider the network of Fig. 6–13. The mesh currents $J_1, \ldots, J_4$ are chosen as $I_1, \ldots, I_4$, respectively. Then

$$I_5 = J_1 - J_2, \qquad I_6 = J_1 - J_3,$$
$$I_7 = J_3 - J_4, \qquad I_8 = J_4 - J_2.$$

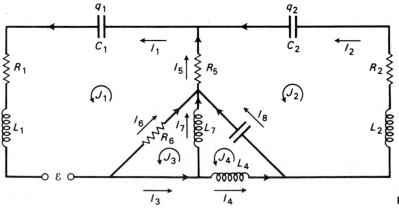

Figure 6-13.

Application of Kirchhoff's first law to the four meshes gives the equations

$$L_1 \frac{dI_1}{dt} + R_1 I_1 + \frac{q_1}{C_1} + R_6 I_6 + R_5 I_5 = \mathcal{E},$$

$$L_2 \frac{dI_2}{dt} + R_2 I_2 + \frac{q_2}{C_2} - R_5 I_5 - \frac{q_8}{C_8} = 0,$$

$$L_7 \frac{dI_7}{dt} - R_6 I_6 = 0, \tag{6-109}$$

$$L_4 \frac{dI_4}{dt} + \frac{q_8}{C_8} - I_7 \frac{dI_7}{dt} = 0.$$

We write $q_\alpha = \int I_\alpha \, dt$, $Q_\alpha = \int J_\alpha \, dt$, as above, so that $q_\alpha = Q_\alpha$ for $\alpha = 1, \ldots, 4$, and $q_5 = Q_1 - Q_2$, $q_6 = Q_1 - Q_3$, $q_7 = Q_3 - Q_4$, $q_8 = Q_4 - Q_2$. The differential equations can be written as

$$\left[L_1 D^2 + (R_1 + R_5 + R_6)D + \frac{1}{C_1} \right] Q_1 - R_5 DQ_2 - R_6 DQ_3 = \mathcal{E},$$

$$- R_5 DQ_1 + \left[L_2 D^2 + (R_2 + R_5)D + \left(\frac{1}{C_2} + \frac{1}{C_8} \right) \right] Q_2 - \frac{Q_4}{C_8} = 0,$$

$$- R_6 DQ_1 + (L_7 D^2 + R_6 D)Q_3 - L_7 D^2 Q_4 = 0, \tag{6-109'}$$

$$- \frac{Q_2}{C_8} - L_7 D^2 Q_3 + \left[(L_4 + L_7)D^2 + \frac{1}{C_8} \right] Q_4 = 0.$$

═══════════════════ **Problems (Section 6-10)** ═══════════════════

1. **a)** For the network of Fig. 6-14 define mesh currents in terms of branch currents, and express all branch currents in terms of mesh currents. Note that $I_2 = I_6$ by Kirchhoff's second law.

   b) Obtain the differential equations satisfied by the mesh charges associated with the mesh currents of part (a).

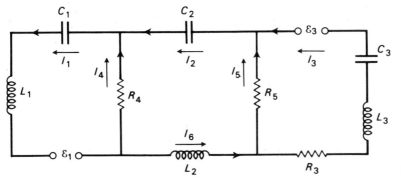

Figure 6-14.

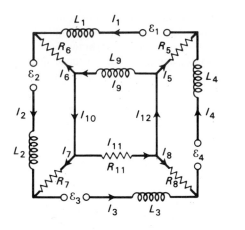

Figure 6-15.

2. a) Proceed as in Problem 1 (a) for the network of Fig. 6-15.
   b) Obtain the differential equations for the mesh currents of part (a).

3. a) Verify the reciprocity law (6-107) for the system (6-109').
   b) Find the functions T, F, V for the system (6-109').

4. *Stability of passive networks.* Let the functions T, F, V of Eqs. (6-108), (6-108') be *positive definite*, that is, let all three be ≥ 0 and let T and F reduce to 0 only when all Q'_α are 0; let V reduce to 0 only when all Q_α are 0. Show that the system (6-106') is stable.
   HINT: Set $Q_\beta = c_\beta e^{\lambda t}$ in the related homogeneous system to obtain the system

$$\lambda^2 \sum_{\beta=1}^{n} L_{\alpha\beta} c_\beta + \lambda \sum_{\beta=1}^{n} R_{\alpha\beta} c_\beta + \sum_{\beta=1}^{n} \gamma_{\alpha\beta} c_\beta = 0,$$

where $\alpha = 1, \ldots, n$. Let $c_\beta = a_\beta + ib_\beta$, multiply by $\bar{c}_\alpha = a_\alpha - ib_\alpha$, and sum from $\alpha = 1$ through n to obtain

$$\lambda^2 \sum_{\alpha=1}^{n} \sum_{\beta=1}^{n} L_{\alpha\beta} c_\beta \bar{c}_\alpha + \lambda \sum_{\alpha=1}^{n} \sum_{\beta=1}^{n} R_{\alpha\beta} c_\beta \bar{c}_\alpha + \sum_{\alpha=1}^{n} \sum_{\beta=1}^{n} \gamma_{\alpha\beta} c_\beta \bar{c}_\alpha = 0.$$

Use the reciprocity law (6-107) to show that this equation reduces to

$$\lambda^2 \sum_{\alpha=1}^{n} \sum_{\beta=1}^{n} L_{\alpha\beta}(a_\alpha a_\beta + b_\alpha b_\beta) + \lambda \sum_{\alpha=1}^{n} \sum_{\beta=1}^{n} R_{\alpha\beta}(a_\alpha a_\beta + b_\alpha b_\beta)$$

$$+ \sum_{\alpha=1}^{n} \sum_{\beta=1}^{n} \gamma_{\alpha\beta}(a_\alpha a_\beta + b_\alpha b_\beta) = 0.$$

Apply the positive definiteness of T, F, V to conclude that unless $c_1 = \cdots = c_n = 0$, all coefficients in the last equation are positive. Hence show that Re $(\lambda) < 0$, so that the system is stable. (See Section 6-11.)

REMARK: The condition of positive definiteness is not always satisfied. For example, if many of the branches contain no inductance, T may reduce to 0 even though not all I_α are zero. We can of course reason that in a physical network each branch does contain at least a small inductance. Similar reasoning applies to the resistance and capacitance terms and justifies the conclusion that in a physical network positive definiteness is satisfied.

6–11 STABILITY. TRANSFER MATRIX AND FREQUENCY RESPONSE MATRIX

The Laplace transform method helps to provide insight into the way a system responds to different forcing functions.

We again consider our basic system

$$\frac{d\mathbf{x}}{dt} = A\mathbf{x} + \mathbf{f}(t), \tag{6-110}$$

where A is a constant $n \times n$ matrix. We know that the general solution has the form

$$\mathbf{x} = \mathbf{x}_c(t) + \mathbf{x}^*(t), \tag{6-111}$$

where the *complementary function*

$$\mathbf{x}_c(t) = c_1 \mathbf{x}_1(t) + \cdots + c_n \mathbf{x}_n(t) \tag{6-112}$$

is the general solution of the related homogeneous equation

$$\frac{d\mathbf{x}}{dt} = A\mathbf{x}. \tag{6-110'}$$

The system (6–110) is called stable if each solution of (6–110′) has limit $\mathbf{0}$ as $t \to +\infty$ (see Sections 1–14 and 1–15). Thus for a stable system the solution (6–111) is formed of a *transient* $\mathbf{x}_c(t)$ and a *steady state* $\mathbf{x}^*(t)$. For each choice of $c_1, \ldots, c_n$ only $\mathbf{x}^*(t)$ is significant for large t.

In Section 1–15 we saw that a simple equation of order n is stable precisely when all characteristic roots have negative real part. *This rule remains valid for the system* (6-110). As for the equation of order n, the rule follows from the fact that $\mathbf{x}_c(t)$ is a sum of terms of the form

$$\text{Const} \times t^k \times e^{\lambda t},$$

where λ is a characteristic root and k is 0 or a positive integer. A complete justification of this follows from the Laplace transform method of Section 6-6 (see Examples 1 and 2 in that section). Thus to test for stability we must write the *characteristic equation*

$$\det(A - \lambda I) = 0$$

and determine whether all roots λ of this equation have negative real part. It is possible to settle this question without solving for all roots λ. (For more information see OMLS, Chapter 7, and F. R. GANTMACHER, *The Theory of Matrices*, vol. 2, Chap. 15, Chelsea Publishing Co., New York 1959.)

Let us assume now that Eq. (6–110) describes a stable system and let us seek a particular solution $\mathbf{x}^*(t)$ for the case when $\mathbf{f}(t)$ has the form $e^{st}\mathbf{f}_0$, where $\mathbf{f}_0$ is a constant vector, s a constant scalar (perhaps, complex). It is reasonable to guess that $\mathbf{x}(t)$ has a similar form: $\mathbf{x}(t) = e^{st}\mathbf{k}$, where $\mathbf{k}$ is a constant vector (method of undetermined coefficients). If so, then from (6–110) it follows that

$$se^{st}\mathbf{k} = e^{st}A\mathbf{k} + e^{st}\mathbf{f}_0,$$

so that

$$(sI - A)\mathbf{k} = \mathbf{f}_0$$

and hence

$$\mathbf{k} = (sI - A)^{-1}\mathbf{f}_0.$$

This choice of $\mathbf{k}$ provides our desired solution $\mathbf{x} = e^{st}\mathbf{k}$. However, it is meaningful only if $sI - A$ is not a singular matrix: that is, only if s is not a characteristic root. This leads us to the definition:

DEFINITION: *The transfer matrix* of system (6–110) is the matrix

$$Y(s) = (sI - A)^{-1},$$

where s is not a characteristic root of A (see Section 4–11).

Thus, when s is not a characteristic root, a particular solution of Eq. (6–110) for $\mathbf{f}(t) = e^{st}\mathbf{f}_0$ is $e^{st}Y(s)\mathbf{f}_0$ or, more simply, $Y(s)\mathbf{f}(t)$. Since we assumed stability, this result is valid for $s \geqslant 0$ (when s is real) and for Re $s \geqslant 0$ (when s is complex).

In particular, we can take $s = i\omega$, with ω real. Thus, corresponding to the "sinusoidal input" $e^{i\omega t}\mathbf{f}_0$, we obtain a sinusoidal output $Y(i\omega)e^{i\omega t}\mathbf{f}_0$. The matrix A is assumed to have real coefficients. From this we can easily see that the input Re$(e^{i\omega t}\mathbf{f}_0) = (\cos\omega t)\mathbf{f}_0$ corresponds to the output Re$[Y(i\omega)e^{i\omega t}\mathbf{f}_0]$ while, Im$(e^{i\omega t}\mathbf{f}_0) = (\sin\omega t)\mathbf{f}_0$ corresponds to the output Im$[Y(i\omega)e^{i\omega t}\mathbf{f}_0]$. Thus the matrix $Y(i\omega)\cdot$ (Function of ω) gives us the response to sinusoidal inputs of different frequencies. For this reason, it is called the *frequency response matrix* (see Section 4–20).

If $\mathbf{f}(t)$ is a periodic function of period $2\pi/\omega$, each of its components can be expanded in a Fourier series. Thus we obtain a Fourier series for $\mathbf{f}(t)$. For simplicity we write this in complex form (Section 3–7):

$$\mathbf{f}(t) = \sum_{n=-\infty}^{\infty} e^{in\omega t}\mathbf{f}_n.$$

By the superposition principle (see Section 1–17), a corresponding output should be

$$\mathbf{x}(t) = \sum_{n=-\infty}^{\infty} e^{in\omega t}Y(in\omega)\mathbf{f}_n.$$

Under other conditions $\mathbf{f}(t)$ can be represented as a Fourier integral (see Section 4-13):

$$\mathbf{f}(t) = \frac{1}{2\pi} \int_{-\infty}^{\infty} \varphi(\omega) e^{i\omega t} \, d\omega. \tag{6-113}$$

Here we are expanding each component of the vector $\mathbf{f}(t)$ in such an integral. Thus Eq. (6-113) is a shorthand for

$$\mathbf{f}(t) = \text{col}\left[\frac{1}{2\pi} \int_{-\infty}^{\infty} \varphi_1(\omega) e^{i\omega t} \, d\omega, \ldots, \frac{1}{2\pi} \int_{-\infty}^{\infty} \varphi_n(\omega) e^{i\omega t} \, d\omega \right].$$

By analogy with the preceding results, a corresponding solution $\mathbf{x}(t)$ should be

$$\mathbf{x}(t) = \frac{1}{2\pi} \int_{-\infty}^{\infty} Y(i\omega) \varphi(\omega) e^{i\omega t} \, d\omega.$$

These results can all be justified under reasonable conditions that assure the convergence of the corresponding Fourier series or integral for $\mathbf{f}(t)$.

Finally, we consider an arbitrary continuous $\mathbf{f}(t)$ for $t \geqslant 0$. Then we form $\mathscr{L}[\mathbf{f}(t)]$, the Laplace transform of $\mathbf{f}(t)$, that is, the vector

$$\text{col}\{\mathscr{L}[f_1(t)], \ldots, \mathscr{L}[f_n(t)]\},$$

which we denote by $\mathbf{F}(s) = \text{col}[F_1(s), \ldots, F_n(s)]$. If $\mathbf{x}(t)$ is the solution of (6-110) with 0 initial values for $t = 0$, we form its Laplace transform $\mathscr{L}[\mathbf{x}(t)] = \mathbf{X}(s) = \text{col}[X_1(s), \ldots, X_n(s)]$. Then, by (6-110),

$$s\mathbf{X}(s) = A\mathbf{X}(s) + \mathbf{F}(s),$$

so that

$$(sI - A)\mathbf{X}(s) = \mathbf{F}(s),$$
$$\mathbf{X}(s) = (sI - A)^{-1} \mathbf{F}(s) = Y(s)\mathbf{F}(s).$$

The entries $Y_{ij}(s)$ of the matrix $Y(s)$ are rational functions of s. Hence we can write $Y(s) = \mathscr{L}[W(t)]$, that is, $Y_{ij}(s) = \mathscr{L}[W_{ij}(t)]$, thereby defining the *weight function matrix* $W_{ij}(t)$ (see Section 4-11). Hence,

$$\mathbf{X}(s) = Y(s)\mathbf{F}(s) = \mathscr{L}[W(t)]\mathscr{L}[\mathbf{f}(t)].$$

This suggests using the convolution (see Sections 4-8, 4-11) and writing:

$$\mathbf{X}(s) = \mathscr{L}[W(t) * \mathbf{f}(t)],$$

so that

$$\mathbf{x}(t) = W(t) * \mathbf{f}(t). \tag{6-114}$$

In detail, we have

$$\mathbf{X}(s) = Y(s)\mathbf{F}(s) = [Y_{ij}(s)][F_j(S)],$$

so that

$$X_i(s) = \sum_{j=1}^{n} Y_{ij}(s) F_j(s) = \sum_{j=1}^{n} \mathscr{L}[W_{ij}(t)] \mathscr{L}[f_j(t)]$$

$$= \sum_{j=1}^{n} \mathscr{L}[W_{ij}(t) * f_j(t)], \quad i = 1, \ldots, n,$$

and

$$x_i(t) = \sum_{j=1}^{n} W_{ij}(t) * f_j(t),$$

as in (6-114). From the definition of the convolution this means

$$x_i(t) = \sum_{j=1}^{n} \int_0^t W_{ij}(u) f_j(t-u)\, du, \quad i = 1, \ldots, n. \tag{6-115}$$

This formula can also be deduced from the variation of parameters formula (6-47). It is valid if $\mathbf{f}(t)$ is continuous for $t \geq 0$, even though the derivation above seemed to use stronger hypotheses on $\mathbf{f}(t)$.

The weight function matrix can be interpreted in terms of memory, as in Section 4-11. The following simple example illustrates the results we have obtained.

EXAMPLE 1 $\dfrac{d\mathbf{x}}{dt} = \begin{bmatrix} -2 & -1 \\ -1 & -2 \end{bmatrix} \mathbf{x} + \mathbf{f}(t).$

The characteristic equation is

$$\begin{bmatrix} -2-\lambda & -1 \\ -1 & -2-\lambda \end{bmatrix} = 0 \quad \text{or} \quad \lambda^2 + 4\lambda + 3 = 0,$$

so that the characteristic roots are -1 and -3. These are negative, therefore the system is *stable*. The solutions of the related homogeneous equation

$$\frac{d\mathbf{x}}{dt} = \begin{bmatrix} -2 & -1 \\ -1 & -2 \end{bmatrix} \mathbf{x}$$

approach $\mathbf{0}$ as $t \to \infty$, and are *transients*.

The transfer matrix is

$$Y(s) = (sI - A)^{-1} = \begin{bmatrix} s+2 & 1 \\ 2 & s+1 \end{bmatrix}^{-1} = \begin{bmatrix} \dfrac{s+2}{s^2+4s+3} & \dfrac{-1}{s^2+4s+3} \\[2mm] \dfrac{-1}{s^2+4s+3} & \dfrac{s+2}{s^2+4s+3} \end{bmatrix}.$$

Thus if $\mathbf{f}(t) = e^{2t}\mathrm{col}\,(3, 5)$, then a particular solution is

$$\mathbf{x} = Y(2)\mathbf{f}(t) = e^{2t} \begin{bmatrix} \frac{4}{15} & \frac{-1}{15} \\[1mm] \frac{-1}{15} & \frac{4}{15} \end{bmatrix} \begin{bmatrix} 3 \\ 5 \end{bmatrix} = e^{2t}\,\mathrm{col}\left(\tfrac{7}{15}, \tfrac{17}{15}\right).$$

If $\mathbf{f}(t) = (\sin 3t)\,\text{col}\,(1,\,-2) = \text{Im}[e^{3it}\,\text{col}\,(1,\,-2)]$, then a particular solution is

$$\mathbf{x}(t) = \text{Im}[Y(3i)e^{3it}\text{col}(1,\,-2)] = \text{Im}\left\{e^{3it}\left[\begin{array}{cc} \dfrac{2+3i}{-6+12i} & \dfrac{-1}{-6+12i} \\[2mm] \dfrac{-1}{-6+12i} & \dfrac{2+3i}{-6+12i} \end{array}\right]\left[\begin{array}{c} 1 \\ -2 \end{array}\right]\right\}$$

$$= \text{Im}\left[e^{3it}\,\text{col}\left(\frac{4+3i}{-6+12i},\,\frac{-5-6i}{-6+12i}\right)\right]$$

$$= \tfrac{1}{30}\,\text{col}\,(2\sin 3t - 11\cos 3t,\ 16\cos 3t - 17\sin 3t).$$

By taking inverse Laplace transforms of the entries of $Y(s)$, we obtain the weight function matrix

$$W(t) = \left[\begin{array}{cc} \dfrac{1}{2}e^{-t} + \dfrac{1}{2}e^{-3t} & -\dfrac{1}{2}e^{-t} + \dfrac{1}{2}e^{-3t} \\[3mm] -\dfrac{1}{2}e^{-t} + \dfrac{1}{2}e^{-3t} & \dfrac{1}{2}e^{-t} + \dfrac{1}{2}e^{-3t} \end{array}\right]$$

$$= \frac{1}{2}e^{-t}\left[\begin{array}{cc} 1 & -1 \\ -1 & 1 \end{array}\right] + \frac{1}{2}e^{-3t}\left[\begin{array}{cc} 1 & 1 \\ 1 & 1 \end{array}\right].$$

Therefore, for general $\mathbf{f}(t) = \text{col}[f_1(t), f_2(t)]$, the solution with zero initial values for $t = 0$ is

$$\mathbf{x} = W(t) * \mathbf{f}(t),$$

that is,

$$x_1 = \frac{1}{2}\int_0^t \{e^{-u}[f_1(t-u) - f_2(t-u)] + e^{-3u}[f_1(t-u) + f_2(t-u)]\}\,du,$$

$$x_2 = \frac{1}{2}\int_0^t \{e^{-u}[-f_1(t-u) + f_2(t-u)] + e^{-3u}[f_1(t-u) + f_2(t-u)]\}\,du.$$

6–12 RESONANCE

We saw in Section 1–14 that for an undamped mass–spring system a sinusoidal input $\sin \omega t$ at the natural frequency leads to resonance: the response contains a term in $t\cos\omega t$ and this term grows indefinitely in amplitude. Several coupled springs yield a similar result, except that now there are several natural frequencies (corresponding to different normal modes, as in Section 6–8). We illustrate this for a particular case and point out how resonance can *fail* to occur, even when one forces at the natural frequency.

We take the system of Example 5 in Section 6–6, but introduce a sinusoidal input. Thus our equations are

$$\frac{d^2 x_1}{dt^2} + 2x_1 - x_2 = a \sin t, \qquad \frac{d^2 x_2}{dt^2} + 2x_2 - x_1 = b \sin t, \qquad (6\text{--}120)$$

where a and b are constants. We know that 1 and $\sqrt{3}$ are the natural frequencies, so that resonance is expected. Taking Laplace transforms (for zero initial values) and solving for $X_1(s)$, $X_2(s)$, we obtain

$$X_1(s) = \frac{as^2 + 2a + b}{(s^2 + 1)^2 (s^2 + 3)}, \qquad X_2(s) = \frac{bs^2 + 2b + a}{(s^2 + 1)^2 (s^2 + 3)}.$$

We expect the factor $(s^2 + 1)^2$ in the denominator to lead (via partial fractions, as in Section 4–6) to a term $(ps + q)/(s^2 + 1)^2$ in $X_1(s)$ and a similar term in $X_2(s)$, hence to terms in $t \sin t$ or $t \cos t$ in $x_1(t)$, $x_2(t)$. These terms indicate the presence of resonance. However, if $b = -a$, then $X_1(s)$ and $X_2(s)$ can be simplified:

$$X_1(s) = \frac{a}{(s^2 + 1)(s^2 + 3)}, \qquad X_2(s) = \frac{-a}{(s^2 + 1)(s^2 + 3)}.$$

A factor $s^2 + 1$ has been canceled in numerator and denominator. Thus for special inputs of the form $\mathrm{col}\,(a \sin t,\ -a \sin t)$ resonance does not occur.

In practice, making b exactly equal to $-a$ is usually not achievable. However, if b is very close to $-a$, the terms in $t \cos t$ and $t \sin t$ have very small coefficients and hence grow slowly. Furthermore, some friction will always be present, preventing true resonance.

Our result can be explained if the system is thought of as formed of two parts, each corresponding to the two normal modes. When the choices of a and b are correct, the input reaches only one part, and if the natural frequency of that part is different from the input frequency, there is no resonance.

An exaggerated effect is provided by a system consisting of two *uncoupled* mass–spring systems with different natural frequencies ω_1, ω_2 (Fig. 6–16). If one part is forced at the frequency of the other, as suggested in the figure, then no resonance occurs.

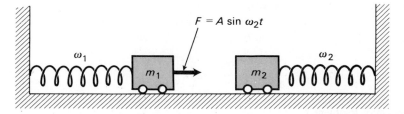

Fig. 6-16. Uncoupled systems.

In a very complicated system, such as a large machine with many parts, there are many normal modes with corresponding frequencies. If we apply a sinusoidal input to one portion of the system, it may be so loosely coupled to the rest that, just as for the separated mass–spring systems of Fig. 6–16, no resonance will occur.

Vibration absorbers

In another approach to the resonance problem we observe that if inputs at the natural frequencies (or close to them) are expected, then resonance can be avoided by modifying the given system to shift the natural frequencies. The modification can be achieved simply by introducing an extra mass and a spring. Such a system is illustrated in Fig. 6–17. Here M_2 has been added to change the frequency of M_1.

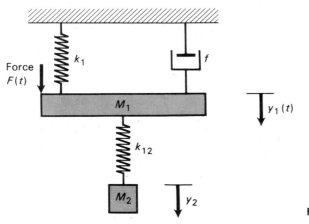

Fig. 6-17. Vibration absorber.

====== **Problems (Section 6-12)** ======

1. Given the system

$$\frac{d\mathbf{x}}{dt} = \begin{bmatrix} -5 & 4 \\ -3 & 2 \end{bmatrix} \mathbf{x} + \mathbf{f}(t),$$

   a) Test for stability   **b)** Find the transfer matrix
   c) Find the frequency response matrix

2. For the system of Problem 1, find a particular solution for the following choices of $\mathbf{f}(t)$:
   a) $e^{3t}\,\text{col}\,(2, 1)$ **b)** $e^{t}\,\text{col}\,(1, 2) + e^{-3t}\,\text{col}\,(1, 3)$
   c) $(\sin 2t)\,\text{col}\,(5, 3)$ **d)** $\text{col}\,(\sin 2t + \cos 2t, 3\sin 2t - \cos 2t)$

   e) $\displaystyle\sum_{n=-\infty}^{\infty} \frac{e^{int}}{n^2 + 1}\,\text{col}\,(1, 3)$ **f)** $\displaystyle\frac{1}{2\pi}\int_{-\infty}^{\infty} \frac{e^{i\omega t}}{1 + \omega^2}\,\text{col}\left(1, \frac{1}{1 + 2\omega^2}\right)d\omega$

3. For the system of Problem 1,
   a) Find the weight-function matrix

b) Write an expression in terms of convolutions for the solution $x(t)$ with zero initial values for $t = 0$.

4. Given the system

$$\frac{d\mathbf{x}}{dt} = \begin{bmatrix} 6 & -4 & 5 \\ 0 & -1 & 0 \\ -10 & 6 & -8 \end{bmatrix} \mathbf{x} + \mathbf{f}(t),$$

  a) Test for stability   **b)** Find the transfer matrix
  c) Find the frequency response matrix

5. For the system of Problem 4, find a particular solution for the following choices of $\mathbf{f}(t)$:
  a) $e^t \operatorname{col}(0, 1, 1)$ **b)** $e^{3t} \operatorname{col}(1, 5, 3) + e^{-3t} \operatorname{col}(5, 1, 2)$
  c) $(\cos 5t)\operatorname{col}(1, 7, 3)$ **d)** $\operatorname{col}(\sin 4t - \cos 4t, 2\sin 4t + \cos 4t, 3\sin 4t)$

  e) $\displaystyle\sum_{n=-\infty}^{\infty} \frac{e^{int}}{n^4 + 1} \operatorname{col}(3, 2, 5)$ **f)** $\displaystyle\frac{1}{2\pi} \int_{-\infty}^{\infty} e^{-|\omega|} e^{i\omega t} \operatorname{col}(5, 3, 2)\, d\omega$

6. For the system of Problem 4,
  a) Find the weight-function matrix;
  b) Write an expression in terms of convolutions for the solution $x(t)$ with zero initial values for $t = 0$.

7. In the equations (6–120) of forced motion for a system of two masses, replace the right sides by $a \sin \omega t$ and $b \sin \omega t$, where $\omega > 0$.
  a) Find the response for zero initial values when $\omega \neq 1$ and $\omega \neq \sqrt{3}$, where 1 and $\sqrt{3}$ are the natural frequencies.
  b) Take $\omega = \sqrt{3}$ and find the response; also show how a and b can be chosen to eliminate resonance.

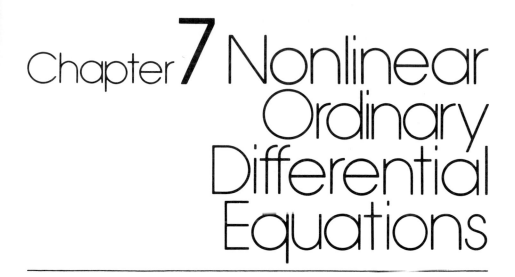

Chapter 7 Nonlinear Ordinary Differential Equations

7–1 THE PHASE PLANE

If a differential equation is not linear, then with a few exceptions, explicit solution is not possible. Many particular solutions can be found by numerical methods (see Chapter 13). These solutions may or may not shed light on the behavior of the system being studied. In the present chapter we consider several simple types of nonlinear differential equations and point out ways of finding the crucial properties of their solutions. For the most part we reduce the equations to a system of two first-order equations:

$$\frac{dx}{dt} = F(x, y), \qquad \frac{dy}{dt} = G(x, y). \tag{7–10}$$

Such a system is called *autonomous*. We shall see that physical mechanisms without external inputs lead to such systems. On the other hand, problems of forced motion or of electric circuits with a driving e.m.f. lead to *nonautonomous* systems; for these the functions F and G in (7–10) depend on t as well as on x and y. We shall also consider nonlinear nonautonomous systems briefly.

The form (7–10) arises commonly from studying second-order differential equations with the aid of the *phase plane*. We introduce this idea with an example.

EXAMPLE 1 $\dfrac{d^2x}{dt^2} + \dfrac{dx}{dt} + x + x^3 = 0.$ This can be interpreted as the equation of motion of a damped mass–spring system with a *nonlinear* force (Fig. 7–1). In order to study the solutions, we reduce the equation to a pair of first-order equations, as in

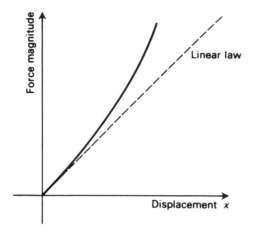

Fig. 7–1. Nonlinear spring force.

Section 6–1. We can write

$$\frac{dx}{dt} = v, \qquad \frac{dv}{dt} = -x - x^3 - v. \qquad (7\text{–}11)$$

(These are of form (7–10) with y replaced by v.) We can also eliminate t to obtain a first-order equation in x and v:

$$\frac{dv}{dx} = \frac{-x - x^3 - v}{v}. \qquad (7\text{–}12)$$

This shows that our solutions can be represented as curves in an xv-plane, the *phase plane*. The variable t can be interpreted as a *parameter* along the curves. We call each solution curve a *trajectory*.

A rough picture of the trajectories can be obtained by the method of *isoclines* discussed in Section 1–3. These are the curves along which dv/dx has a constant value m. Thus here we plot the curves

$$\frac{-x - x^3 - v}{v} = m$$

for various choices of m. Once we have plotted a number of isoclines, we can draw many short line segments of slope m through the points of the isocline for that m-value. Together the line segments suggest the trajectories. The results are shown in Fig. 7–2.

We observe that, since $dx/dt = v$, x increases with t along the trajectory in the upper half-plane, where v is positive; similarly, x decreases with t in the lower half-plane. This leads to the assignment of arrows showing the direction of increasing t, as in Fig. 7–2.

From the phase–plane diagram (or *phase portrait*, as it is often called) we see that x alternately increases and decreases on each solution, swinging back and forth

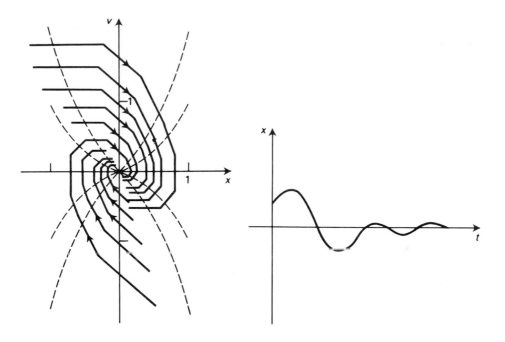

Fig. 7–2. Trajectories for Example 1 (nonlinear spring).

Fig. 7–3. Damped oscillation of nonlinear spring.

betwccn ncgative and positive values that approach 0. Thus $x(t)$ is a *damped oscillation*, as in Fig. 7–3.

It should be remarked that the actual t-values are not given by the solution curves of Fig. 7–2. As Eq. (7–11) shows, we know only derivatives with respect to t. However, we can arbitrarily assign $t = 0$, say, to a particular point (x_0, v_0) on a trajectory. The curve in effect gives v as a function of x. Then we can write

$$\frac{dx}{v(x)} = dt, \qquad \int_{x_0}^{x_1} \frac{dx}{v(x)} = \int_{t_0}^{t_1} dt,$$

and, by a numerical integration, obtain the t-value t_1 at the point (x_1, v_1). For precise results, we can simply solve the differential equation numerically, as in Chapter 13.

The curves in the phase plane are spiraling into the origin. The origin itself is a point at which $dx/dt = 0$ and $dv/dt = 0$ by Eqs. (7–11). We call such a point an *equilibrium point* (or *critical point*). An equilibrium point always represents an equilibrium solution of the differential equation, that is, a solution $x \equiv \text{const}$ for all t. Here $x \equiv 0$ is an equilibrium solution.

We can seek other equilibrium points. By (7–11), they are all (x, v) such that

$$v = 0, \qquad -x - x^3 - v = 0.$$

The only solution of these equations is $x = 0$, $v = 0$. Thus there is only one equilibrium point.

In Fig. 7–2 we observe that all solutions approach the equilibrium point as $t \to \infty$. We say that the equilibrium point is *stable* or that the solution $x \equiv 0$ of the given differential equation is a *stable solution*.

We also name the configuration of the trajectories at the equilibrium point. We call it a *focus* and, because it is stable, a *stable focus*. The term is used whenever the trajectories spiral into the equilibrium point, as in this case. ◄

EXAMPLE 2 $\dfrac{d^2x}{dt^2} + x = 0.$ This is a linear differential equation that describes the familiar simple harmonic motion of a linear mass–spring system. In order to gain confidence in the methods, we try our new methods on this example, even though we can solve explicitly.

We have the corresponding system

$$\frac{dx}{dt} = v, \qquad \frac{dv}{dt} = -x$$

and the corresponding first-order equation

$$\frac{dv}{dx} = -\frac{x}{v},$$

which can be of course integrated to obtain the trajectories:

$$x^2 + v^2 = c.$$

They are *circles* in the xv-plane. The isoclines are the straight lines

$$\frac{-x}{v} = m = \text{const}$$

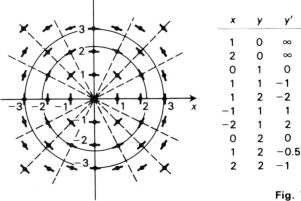

x	y	y'
1	0	∞
2	0	∞
0	1	0
1	1	−1
1	2	−2
−1	1	1
−2	1	2
0	2	0
1	2	−0.5
2	2	−1

Fig. 7–4. Trajectories for Example 2 (linear mass–spring system).

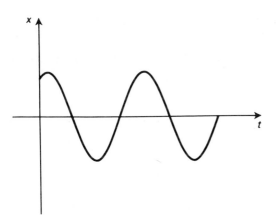

Fig. 7–5. Solution $x(t)$ for Example 2 (simple harmonic motion).

in the xv-plane. They are simply the lines of slope $-1/m$ passing through the origin. Hence the trajectory crosses each at right angles, i.e., the trajectories are *orthogonal trajectories* of the isoclines (see Problem 5 following Section 1–6). Thus even a rough sketching of trajectories built of short segments gives us a good approximation to the circular trajectories (Fig. 7–4). We can also obtain $x(t)$ by integration as above; of course we know that, as in Fig. 7–5, the solutions are sinusoidal functions $x = A \sin(t + \alpha)$.

In this example, each trajectory is a *closed curve* and each solution $x(t)$ must be a *periodic* function of t, since $x(t)$ must retrace the same pattern over and over again as we follow increasing t around the trajectory.

There is again an equilibrium point at the origin and $x = 0$ is again an equilibrium solution, the only one, as we see at once. This time the trajectories do not approach the equilibrium point, nor do they recede from it. The equilibrium point is said to be *neutrally stable*. This is interpreted physically to mean that if a small initial deviation from the equilibrium state is not corrected, then it leads to a motion that remains close to the equilibrium solution as t increases indefinitely.

The configuration formed about the equilibrium point here is called a *center*. We use this term whenever the trajectories near the equilibrium point are closed curves about the point, as in Fig. 7–6. ◄

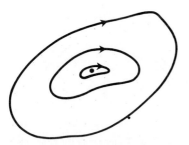

Fig. 7–6. Center.

EXAMPLE 3 $\dfrac{d^2x}{dt^2} - x = 0$. An equation of this form arises if we consider a pendulum near the vertical position (Fig. 7–7). For if θ is measured from the vertical line (as in Fig. 7–7), the differential equation is the nonlinear equation

$$mL\frac{d^2\theta}{dt^2} = mg \sin \theta.$$

If θ is small, $\sin \theta$ can be replaced by θ; also $g/L = 1$ in appropriate units. Thus

$$\frac{d^2\theta}{dt^2} - \theta = 0.$$

If we replace θ by x, we obtain the given equation. The application of this equation to the pendulum is valid only when x is close to 0. The pendulum with arbitrary angle θ is studied in Section 7–2. ◀

Here our system is $\dfrac{dx}{dt} = v$, $\dfrac{dv}{dt} = x$ and the first-order equation is $\dfrac{dv}{dx} = \dfrac{x}{v}$.
Again we can integrate to obtain the trajectories:

$$x^2 - v^2 = c.$$

They are hyperbolas. The isoclines are the lines

$$\frac{x}{v} = m = \text{const.}$$

They again are lines, of slope, $1/m$ passing through the origin. Thus the trajectories cross them at varying angles. For $m = 1$ and $m = -1$ the angle is 0; in fact, the lines $v = \pm x$ are trajectories, as we see at once (corresponding to $x^2 - v^2 = 0$).

We graph the actual trajectories and isoclines in Fig. 7–8. We see that there is an equilibrium point at the origin and that two solutions approach it as $t \to \infty$.

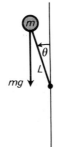

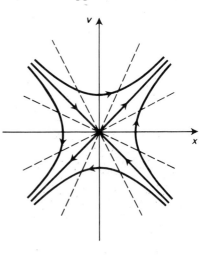

Fig. 7–7. Unstable equilibrium of pendulum.   **Fig. 7–8.** Trajectories for Example 3.

However, all other solutions depart from it. Hence the equilibrium point is said to be *unstable.*

The solutions of the differential equation are found as in Chapter 1 to be

$$x = c_1 e^t + c_2 e^{-t}.$$

For $c_1 = 0$ we have $x = c_2 e^{-t}$ and $v = -c_2 e^{-t}$, that is, a special trajectory $x = -v$ leading to the origin. However, for $c_1 \neq 0$, x becomes infinite (positive or negative) as $t \to \infty$ (in Fig. 7–9).

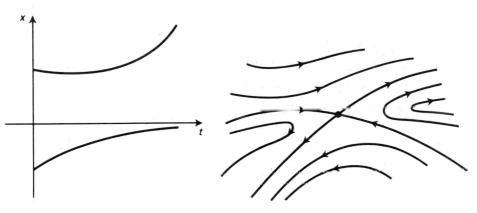

Fig. 7–9. Solutions for Example 3.   **Fig. 7–10.** Saddle point.

The configuration at the equilibrium point in this case is called a *saddle point.* We use this term whenever the qualitative features are the same as in Fig. 7–10.

========== **Problems (Section 7–1)** ==========

1. With the aid of isoclines, graph the trajectories in the phase plane for each of the following equations. In each case comment on equilibrium points and their stability and also on the character of a typical solution $x = x(t)$.

   a) $\dfrac{d^2 x}{dt^2} + 4x = 0$ b) $\dfrac{d^2 x}{dt^2} + 9x = 0$ c) $\dfrac{d^2 x}{dt^2} + \dfrac{dx}{dt} + x = 0$

   d) $\dfrac{d^2 x}{dt^2} + 2\dfrac{dx}{dt} + 2x = 0$ e) $\dfrac{d^2 x}{dt^2} + 2\dfrac{dx}{dt} + x = 0$ f) $\dfrac{d^2 x}{dt^2} + 6\dfrac{dx}{dt} + 9x = 0$

   g) $\dfrac{d^2 x}{dt^2} + 3\dfrac{dx}{dt} + 4x = 0$ h) $\dfrac{d^2 x}{dt^2} + \dfrac{dx}{dt} - 2x = 0$ i) $\dfrac{d^2 x}{dt^2} + x + x^3 = 0$

   j) $\dfrac{d^2 x}{dt^2} + 4x + x^3 = 0$

   REMARK Parts (a) to (g) describe motions of a (linear) mass–spring system; parts (i) and (j) describe the motion of an undamped nonlinear mass–spring system.

7–2 CONSERVATIVE SYSTEMS

Let a particle of unit mass move on the x-axis subject to a force depending only on position. We can represent the force as $-U'(x)$, where $U(x)$ is a suitable function: the negative of an indefinite integral of the function $F(x)$ representing the force. Thus, by Newton's second law, the motion is governed by the differential equation

$$\frac{d^2x}{dt^2} = -U'(x). \tag{7–20}$$

As usual, we interpret this in the phase plane and write y for the velocity dx/dt:

$$\frac{dx}{dt} = y, \qquad \frac{dy}{dt} = -U'(x). \tag{7–21}$$

Thus we have an autonomous system. Along each solution we have

$$\frac{y^2}{2} + U(x) = c = \text{const}, \tag{7–22}$$

since along a solution $x = x(t)$, $y = y(t)$,

$$\frac{d}{dt}\left[\frac{y^2}{2} + U(x)\right] = y\frac{dy}{dt} + U'(x)\frac{dx}{dt} = -yU'(x) + U'(x)y = 0.$$

Equation (7–22) expresses the *conservation of energy* for these motions; $y^2/2$ is the *kinetic energy* $mv^2/2$, since $m = 1$ and $y = v$, and $U(x)$ is interpreted as *potential energy*. We thus call (7–21) a *conservative system*.

The trajectories of such a system are the curves (7–22), which are the level curves of the function $(y^2/2) + U(x)$. For given $U(x)$, these can usually be plotted without much difficulty.

EXAMPLE 1 $U(x) = kx^2/2$, $k = \text{const} > 0$. Here $F(x) = -kx$ and we have the case of a spring force with no friction. The trajectories are ellipses, as shown in Fig. 7–11, except for the equilibrium point $(0, 0)$. Hence all solutions are periodic (corresponding to the simple harmonic motions of Example 4 in Section 1–11). The equilibrium point is a center and is neutrally stable. ◄

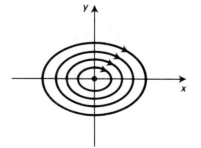

Fig. 7–11. Example 1.

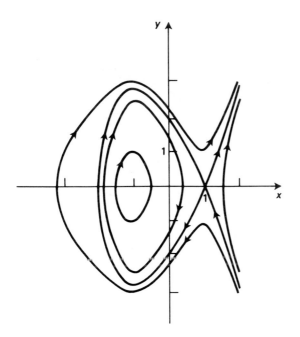

Fig. 7–12. Example 2: center at $(-1, 0)$, saddle point at $(1, 0)$.

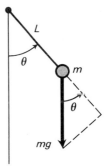

Fig. 7–13. Simple pendulum.

EXAMPLE 2 $U(x) = 3x - x^3$. The trajectories are graphed in Fig. 7–12. The two equilibrium points are at $(\pm 1, 0)$. Near $(-1, 0)$ the trajectories are closed, near $(1, 0)$ they resemble a family of hyperbolas with common asymptotes; in particular, there are two trajectories approaching $(1, 0)$ as $t \to \infty$ and two as $t \to -\infty$. ◄

From Eqs. (7–21) we see that in general the equilibrium points are the points $(x_0, 0)$ for which $U'(x_0) = 0$, so that x_0 is a critical point for $U(x)$. If U has a strict relative minimum at the point, so that $U(x) > U(x_0)$ for all x near but not at x_0, then the nearby trajectories are closed curves, as at $(0, 0)$ for Example 1 and at $(-1, 0)$ for Example 2. Thus the trajectories have a *center* at the equilibrium point. If U has a strict relative maximum at the point, then the nearby solutions are like hyperbolas, as in Example 2 near $(1, 0)$, so that the trajectories have a *saddle point* at the equilibrium point (see Problem 4 below).

EXAMPLE 3 $U(x) = -k \cos x$, $k > 0$.   The corresponding Eq. (7–20) arises in studying the motion of a simple pendulum, with x replaced by the angle θ of departure from the vertical, as in Fig. 7–13 (friction is ignored). As in the figure, the tangential component of force is $-mg \sin \theta$ and the tangential component of acceleration is $L d^2\theta/dt^2$ (see Problem 2 below). Hence, by Newton's second law,

$$mL \frac{d^2\theta}{dt^2} = -mg \sin \theta. \tag{7–23}$$

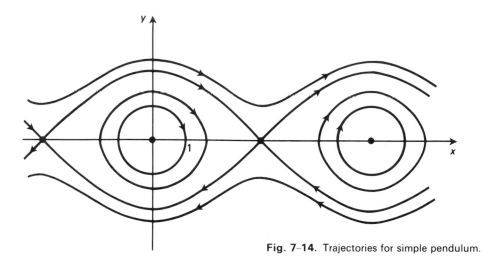

Fig. 7-14. Trajectories for simple pendulum.

Thus, with $k = g/L$ and $\theta = x$, the equation is

$$\frac{d^2x}{dt^2} = -U'(x) = -k\sin x.$$

The trajectories in the phase plane are shown in Fig. 7–14. The function $U(x) = -k\cos x$ has minima at $0, \pm 2\pi, \ldots$, and maxima at $\pm \pi, \pm 3\pi, \ldots$, and these are all its critical points. We see that the equilibrium points corresponding to these values are centers and saddle points, as predicted. We observe that there are solutions on which $y(t)$ remains positive for all t (and symmetric ones on which y remains negative for all t); these correspond to motions of the pendulum such that the bob rotates very quickly in the same direction. The equilibrium points that are centers correspond to stationary positions of the bob at its lowest position; they are clearly stable. The saddle-points are theoretical stationary positions with the bob at its highest point; they are unstable and hence cannot be effectively achieved physically. ◄

REMARK 1: Conservative systems are usually an idealization of real physical problems. When friction or similar effects are taken into account, energy is no longer conserved. However, a conservative system is a good initial model, a first approximation. As in the above examples, this approximation can be used to plot trajectories. Then we can study the effect on the trajectories of correcting the differential equations for a more accurate presentation of the physical situation.

REMARK 2: We have emphasized the trajectories rather than the solutions $x = x(t)$, $y = y(t)$. As in Section 7–1, we can use the equation $dx/dt = y$

and the energy relation (7–22) to obtain $x = x(t)$ on each trajectory:

$$\frac{dx}{dt} = y = \pm \sqrt{2[c - U(x)]}, \qquad \frac{dx}{\sqrt{2[c - U(x)]}} = \pm \, dt.$$

Integration gives t in terms of x, and hence x in terms of t, along the portions of the trajectory in the upper half-plane ($+$ sign) or lower half-plane ($-$ sign). Explicit integration is often awkward, but numerical procedures can always be used (see Chapter 13).

REMARK 3: An autonomous system (7–10), not necessarily of form (7–21), may also have a *first integral*, i.e., a function $V(x, y)$ that is constant along solutions, as is the function $(y^2/2) + U(x)$ for the system (7–21). Then the trajectories are given by $V(x, y) = \text{const} = c$, the level curves of $V(x, y)$. The methods of Sections 1–5 and 1–6, relating to exact equations and integrating factors, provide ways of finding such first integrals. For example, the system

$$\frac{dx}{dt} = -2xy, \qquad \frac{dy}{dt} = 1 + y^2$$

leads to

$$\frac{dy}{dx} = \frac{1 + y^2}{-2xy} \qquad \text{or} \qquad (1 + y^2)dx + 2xy \, dy = 0.$$

The trajectories are given by $x + xy^2 = c$. Thus $V = x + xy^2$ is a first integral.

=========================== **Problems (Section 7-2)** ===========================

1. Verify the graph of the trajectories for Example 2 in Section 7–2 (Fig. 7–12).

2. For the pendulum problem of Example 3 in Section 7–2
   a) Give a complete justification of the differential equation (7–23),
   b) Verify the graph of the trajectories as in Fig. 7–14.

3. Graph the trajectories (7–22) of the system (7–21) for the following choices of $U(x)$:
   a) gx, $g = \text{const} > 0$ (falling body, x is altitude above earth's surface)
   b) $\dfrac{x^2}{2} + \dfrac{x^4}{4}$ (nonlinear spring)
   c) $x^5 - 5x$
   d) $-\dfrac{k}{x}$, $k = \text{const} > 0$, $x > 0$ (body falling under gravitational inverse-square law).

4. Let $U(x)$ have a continuous derivative for all x and let $U'(x) = 0$ only at x_1, x_2, at which $U(x)$ has a relative minimum and a relative maximum respectively. Graph the trajectories (7–22) of the corresponding system (7–21) and explain why there is a center at the equilibrium point $(x_1, 0)$ and a saddle point at the equilibrium point $(x_2, 0)$.
   HINT: Relate the trajectories to the graphs of $y = \pm \sqrt{2[c - U(x)]}$.

5. Possible behavior of the trajectories (7–22) of Eqs. (7–21) at an equilibrium point is illustrated by the cases $U(x) = \pm x^n$, where $n = 2, 3, 4, \ldots$. For n even we have a maximum or minimum at $x = 0$, as in Problem 4. For n odd the trajectories leading to the

equilibrium point form a *cusp* and the configuration is called a *zero-order saddle point* (it is clearly unstable). *Graph the trajectories* $(y^2/2) + U(x) = \text{const} = c$ *for each of the following choices of* $U(x)$:

a) x^3 b) $-x^3$ c) x^4 d) $-x^4$ e) x^5 f) $-x^5$

6. Show that the system has the given first integral $V(x, y)$ and plot the trajectories, showing directions:

a) $\dfrac{dx}{dt} = 3y^2$, $\dfrac{dy}{dt} = 2x$, $V(x, y) = y^3 - x^2$

b) $\dfrac{dx}{dt} = x^2$, $\dfrac{dy}{dt} = -2xy$, $V(x, y) = x^2 y$

c) $\dfrac{dx}{dt} = xe^y + 1$, $\dfrac{dy}{dt} = -e^y$, $V(x, y) = xe^y + y$

7–3 STRUCTURE OF TRAJECTORIES NEAR AN EQUILIBRIUM POINT

Let (x_0, y_0) be an equilibrium point of the system

$$\frac{dx}{dt} = F(x, y), \qquad \frac{dy}{dt} = G(x, y), \tag{7–30}$$

so that

$$F(x_0, y_0) = 0, \qquad G(x_0, y_0) = 0. \tag{7–31}$$

As in Section 2–19, we can expand F and G in Taylor series about (x_0, y_0):

$$F(x, y) = F(x_0, y_0) + F_x(x_0, y_0)(x - x_0) + F_y(x_0, y_0)(y - y_0) + \cdots,$$
$$G(x, y) = G(x_0, y_0) + G_x(x_0, y_0)(x - x_0) + G_y(x_0, y_0)(y - y_0) + \cdots.$$

Because of (7–31) the first term in each series is 0. If we replace $F(x, y)$, $G(x, y)$ by these series in (7–30) and then drop all terms except those shown, we are left with the *approximating linear system*

$$\frac{dx}{dt} = a_{11}(x - x_0) + a_{12}(y - y_0),$$
$$\frac{dy}{dt} = a_{21}(x - x_0) + a_{22}(y - y_0), \tag{7–32}$$

where

$$a_{11} = F_x(x_0, y_0), \qquad a_{12} = F_y(x_0, y_0),$$
$$a_{21} = G_x(x_0, y_0), \qquad a_{22} = G_y(x_0, y_0). \tag{7–33}$$

The appearance of the trajectories of the given nonlinear system (7–30) *near the equilibrium point is generally the same as that for the approximating linear system.*

If, for example, the linear system (7–32) has a focus at (x_0, y_0), then so does the nonlinear system (7–30).

There are some exceptions to this assertion, which we point out below.

EXAMPLE 1 $\dfrac{dx}{dt} = y$, $\dfrac{dy}{dt} = -k \sin x$, $k = \text{const} > 0$. As in Example 3 of Section 7–2, these are the equations for the motion of a pendulum. There are equilibrium points where $y = 0$ and $\sin x = 0$, that is, the points $(n\pi, 0)$, where $n = 0$, ± 1, ± 2, ...

At $(0, 0)$, the approximating linear system is

$$\frac{dx}{dt} = y, \qquad \frac{dy}{dt} = -kx.$$

The trajectories are the ellipses $kx^2 + y^2 = c$ and there is a center at $(0, 0)$. As in Fig. 7–14, the nonlinear system also has a center at $(0, 0)$.

At $(\pi, 0)$, the approximating linear system is

$$\frac{dx}{dt} = y, \qquad \frac{dy}{dt} = k(x - \pi).$$

Its trajectories are the hyperbolas $k(x - \pi)^2 - y^2 = c$. Thus there is a saddle point at $(\pi, 0)$, and the nonlinear system also has a saddle point at $(\pi, 0)$, as in Fig. 7–14.

Similar results are found at all equilibrium points $(n\pi, 0)$. Thus our assertion is verified for this example. ◄

In order to take full advantage of the assertion made above, we have to know the possible configurations of the trajectories of a linear system (7 32) near the equilibrium point (x_0, y_0) and be able to recognize them.

We have already encountered some of these configurations: center (Fig. 7–15), saddle point (Fig. 7–16), focus. There are some others. We take stability properties

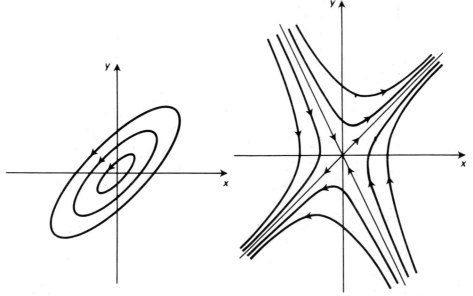

Fig. 7–15. Center.   **Fig. 7–16.** Saddle point.

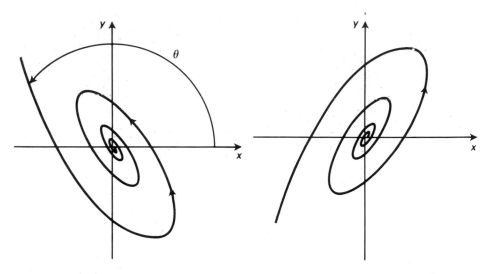

Fig. **7–17**. Stable focus.   Fig. **7–18**. Unstable focus.

into account. We met the stable focus (Fig. 7–17); if we reverse all directions on trajectories, we obtain the unstable focus (Fig. 7–18). In addition, the linear system can have a node (stable or unstable), as in Figs. 7–19 and 7–20. There are other possibilities that should be considered as exceptional cases. Those illustrated are the principal ones for applications.

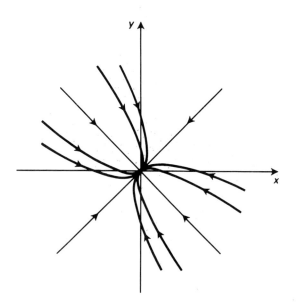

Fig. **7–19**. Stable node.

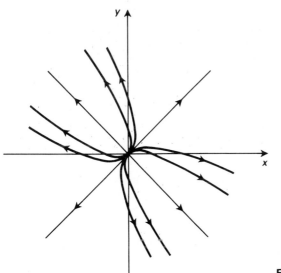

Fig. 7–20. Unstable node.

We have seen the center, saddle-point, and stable focus in examples. The unstable focus, stable node, and unstable node are illustrated by the following systems, respectively:

$$\frac{dx}{dt} = -x - 5y, \qquad \frac{dy}{dt} = x + 3y; \tag{7–34}$$

$$\frac{dx}{dt} = -4x - y, \qquad \frac{dy}{dt} = 2x - y; \tag{7–35}$$

$$\frac{dx}{dt} = 5x + y, \qquad \frac{dy}{dt} = -3x + y. \tag{7–36}$$

The verification of these configurations is left to the reader as Problem 1 below. It should be said that at a node the trajectories (with two exceptions) are all tangent to one line at (x_0, y_0).

In classifying the configurations, we do not distinguish between clockwise and anticlockwise directions around the equilibrium point. Thus a stable focus may have trajectories spiraling-in clockwise or anticlockwise.

Criteria for configurations of the linear system (7–32)

These can be stated very simply in terms of the eigenvalues λ_1, λ_2 of the matrix $A = (a_{ij})$. As in Section 5–7, the eigenvalues are the roots of the equation

$$\begin{vmatrix} a_{11} - \lambda & a_{12} \\ a_{21} & a_{22} - \lambda \end{vmatrix} = 0 \qquad \text{or} \qquad \lambda^2 - (a_{11} + a_{22})\lambda + \det A = 0.$$

The criteria are as follows:

Center: pure imaginary eigenvalues $\pm bi$, $b \neq 0$;

Saddle point: eigenvalues of opposite sign, say $\lambda_1 < 0 < \lambda_2$;

Stable focus: complex eigenvalues $a \pm bi$ with $a < 0$;

Unstable focus: complex eigenvalues $a \pm bi$ with $a > 0$;

Stable node: eigenvalues real, unequal, negative;

Unstable node: eigenvalues real, unequal, positive.

Determination of configuration for nonlinear system by its linear approximation

Our assertion above applies to all the above cases except the center. The nonlinear equations may fail to have a center at a point at which the linear approximation has one. Our Example 1 (pendulum equation) does not suggest this possibility, but other examples can be given (Problem 8 below) to demonstrate that the center cannot be established by the linear approximation. However, when the nonlinear system happens to have a first integral (and that is the case for the pendulum which is a conservative system) the linear approximation does again reveal the presence of a center.

Stability of nonlinear system versus that of the linear approximation

The results stated above about the saddle point, focus, node, and (with qualification) the center, indicate that there is a strong connection between the stability of the nonlinear system and that of its linear approximation. We can state the conclusions most simply in terms of the eigenvalues of the linear approximation:

If both eigenvalues are negative or have negative real part, then the equilibrium point is stable for both the linear approximation and the given nonlinear system.

If at least one eigenvalue is positive or if both have positive real parts, then the equilibrium point is unstable for both the linear approximation and the given nonlinear system.

Neutral stability is assured for the nonlinear system if there is a center at the equilibrium point. As said above, this may or may not happen when the approximating linear system has a center, that is, when both eigenvalues are pure imaginary. (For a more detailed discussion of these results see ODE, pp. 416-437.)

EXAMPLE 2 $\dfrac{d^2x}{dt^2} + 4(x^2 - 1)\dfrac{dx}{dt} + x = 0.$ This is a case of the van der Pol equation studied in Section 7–5. The phase-plane equations are

$$\frac{dx}{dt} = y, \qquad \frac{dy}{dt} = -x - 10(x^2 - 1)y.$$

The only equilibrium point is the origin $(0, 0)$. The linear approximation at $(0, 0)$ is

$$\frac{dx}{dt} = y, \qquad \frac{dy}{dt} = -x + 10y.$$

The eigenvalues are the solutions of the equation

$$\begin{vmatrix} -\lambda & 1 \\ -1 & 10-\lambda \end{vmatrix} = \lambda^2 - 10\lambda + 1 = 0.$$

We find $\lambda = 5 \pm \sqrt{24}$. Both roots are positive, hence there is an *unstable node* at $(0, 0)$ for the linear approximation and for the nonlinear system. The equilibrium point is *unstable* for both systems. ◄

EXAMPLE 3 $\dfrac{d^2 x}{dt^2} + \dfrac{dx}{dt} + \sin x = 0.$ This equation can be considered as one describing the motion of a pendulum with damping due to friction. The phase-plane equations are

$$\frac{dx}{dt} = y, \qquad \frac{dy}{dt} = -\sin x - y.$$

The equilibrium points are $(n\pi, 0)$ with $n = 0, \pm 1, \pm 2, \ldots$ We consider only the one at $(0, 0)$. The linear approximation is

$$\frac{dx}{dt} = y, \qquad \frac{dy}{dt} = -x - y.$$

For the eigenvalues we have the equation

$$\begin{vmatrix} -\lambda & 1 \\ -1 & -1-\lambda \end{vmatrix} = 0 \qquad \text{or} \qquad \lambda^2 + \lambda + 1 = 0.$$

The roots are $(-1 \pm \sqrt{3}i)/2$. Both have negative real parts. Therefore we have a *stable focus* for both the linear approximation and for the given nonlinear system; the equilibrium is *stable* for both. The result agrees with experience, for a real pendulum will execute damped oscillations gradually coming to rest at $x = \theta = 0$. Such pendulum problems are discussed further in Section 7–6. ◄

EXAMPLE 4 *Alternating gradient synchrotron.* Here we have a system such as the following:

$$\frac{dx}{dt} = -5x + 1600 \sin y - 790,$$

$$\frac{dy}{dt} = -2000 x - 4000 y + 4000.$$

The variable x is a distance (radial deviation of an orbit from a reference value), y is the phase of an accelerating voltage $E \cos y$.

To find the equilibrium points, we must solve the equations

$$-5x + 1600 \sin y - 790 = 0, \qquad -2000 x - 4000 y + 4000 = 0.$$

We eliminate x to obtain the equation $80 - y = 160 \sin y$. For small y the left side is approximately 80, and we get simply $\sin y = 1/2$ or $y = \pi/6 = 0.52$, $y = 5\pi/6 = 2.62$, and so on. We consider only the first two y values and use the approximate values given (they turn out to be correct to two significant figures). We thus have the equilibrium points $(0.96, 0.52)$, $(-3.24, 2.62)$.

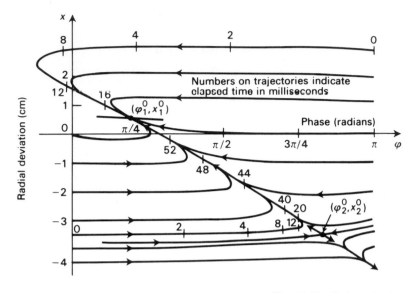

Fig. 7–21. Trajectories for alternating gradient synchrotron ($\varphi = y$).

To study the linear approximations we have to solve the characteristic equation

$$\begin{vmatrix} -5-\lambda & 1600\cos y \\ -2000 & -4000-\lambda \end{vmatrix} = 0$$

for $y = \pi/6$ and then for $y = 5\pi/6$. We find that for the first value there are two unequal negative roots λ_1, λ_2, while for the second value there are two real roots of opposite sign. Hence we have a stable node and a saddle point. (This example, with slightly different numerical values, is considered in detail on pages 297 ff. of STERN, and the trajectories are found with the aid of numerical methods as in Fig. 7–21.) ◄

═══════════════════ **Problems (Section 7–3)** ═══════════════════

1. For each of the following linear systems graph the trajectories with the aid of isoclines:

a) $\dfrac{dx}{dt} = -x - 5y$, $\quad \dfrac{dy}{dt} = x + 3y$ (cf. Fig. 7–18)

b) $\dfrac{dx}{dt} = -4x - y$, $\quad \dfrac{dy}{dt} = 2x - y$ (cf. Fig. 7–19)

c) $\dfrac{dx}{dt} = 5x + y$, $\quad \dfrac{dy}{dt} = -3x + y$ (cf. Fig. 7–20)

REMARK: It is easy to see from the differential equations or from the isoclines that for such a linear system the trajectories are *similar curves*. Hence here we can obtain all of them by first graphing one trajectory (not a line through the origin) and then replacing each point (x, y) on it by (kx, ky) for a fixed k (positive or negative).

Equations for the trajectories can be found as in Section 6–3. For example, in part (a) one trajectory is

$$x = 5e^t \cos t, \qquad y = e^t(-2\cos t + \sin t).$$

2. Determine the nature of the equilibrium point for each of the following linear systems:
   a) $dx/dt = 3x - 2y$, $dy/dt = 2x + 3y$
   b) $dx/dt = 2x + y$, $dy/dt = x + 2y$
   c) $dx/dt = 2x + 5y$, $dy/dt = x + 2y$
   d) $dx/dt = -3x + 6y$, $dy/dt = -2x$
   e) $dx/dt = x - y$, $dy/dt = 2x - y$

3. Consider the equation for the damped *vibration of a spring* (Section 1–14):

$$m\frac{d^2x}{dt^2} + h\frac{dx}{dt} + kx = 0, \quad m > 0,\ h \geqslant 0,\ k > 0.$$

Replace the second-order equation by the two first-order equations:

$$\frac{dx}{dt} = y, \qquad \frac{dy}{dt} = -\frac{k}{m}x - \frac{h}{m}y,$$

which have an equilibrium point at $(0, 0)$. Determine the nature of the equilibrium point for each of the following cases:

a) No friction $(h = 0)$
b) Undercritical damping
c) Overcritical damping

4. Let the linear system

$$\frac{dx}{dt} = a_{11}x + a_{12}y, \qquad \frac{dy}{dt} = a_{21}x + a_{22}y$$

have eigenvalues $\pm bi$ with $b > 0$. Show that

$$a_{11} + a_{22} = 0 \qquad \text{and} \qquad a_{11}a_{22} - a_{21}a_{12} = b^2.$$

Conclude that the equations have a first integral $a_{21}x^2 - 2a_{11}xy - a_{12}y^2 = \text{const}$ and that the trajectories are ellipses.

HINT: From analytic geometry, $Ax^2 + Bxy + Cy^2 = k$ is an ellipse if $B^2 - 4AC < 0$.

5. Describe the nature of the equilibrium point at $(0, 0)$:
   a) $dx/dt = x + 2y + x^3 - y^3$, $\quad dy/dt = 5x + y - x^3 + y^3$
   b) $dx/dt = 3x - y + x^2 - y^4$, $\quad dy/dt = x + xy - xy^3$
   c) $dx/dt = \sin x - 3\sin y$, $\quad dy/dt = 2x - 2y$
   d) $dx/dt = e^{-x-2y} - 1$, $dy/dt = -x(1-y)^2$

6. Find all equilibrium points and describe the stability of each:
   a) $dx/dt = x - 2y$, $\quad dy/dt = x^2 + 2y^2 - 6$
   b) $dx/dt = x^2 - y^2 - 7$, $\quad dy/dt = x^2 + y^2 - 25$

7. For each of the following second-order equations, consider the corresponding system in the phase plane, determine the nature of the equilibrium point at $(0, 0)$, and graph trajectories near $(0, 0)$:

   a) $\dfrac{d^2x}{dt^2} + \dfrac{dx}{dt} + x - x^3 = 0$ (nonlinear spring)

b) $\dfrac{d^2x}{dt^2} + 4\dfrac{dx}{dt} + x - x^3 = 0$ (nonlinear spring)

c) $\dfrac{d^2x}{dt^2} + \dfrac{dx}{dt} + \left(\dfrac{dx}{dt}\right)^3 + x = 0$ (spring with nonlinear damping)

d) $\dfrac{d^2x}{dt^2} + 4\dfrac{dx}{dt} + \left(\dfrac{dx}{dt}\right)^3 + x = 0$ (spring with nonlinear damping)

8. Show that for the system

$$\frac{dx}{dt} = y + x(x^2 + y^2), \qquad \frac{dy}{dt} = -x + y(x^2 + y^2)$$

there is an equilibrium point at $(0, 0)$, at which the linear approximation has a center, but that the nonlinear system does not have a center at the point and that the equilibrium point is unstable for the nonlinear system.

HINT: Show that along the trajectories of the nonlinear system the polar coordinates (r, θ) of the point (x, y) satisfy $dr/dt = r^3$, $d\theta/dt = -1$.

$7-4$ PERIODIC SOLUTIONS; LIMIT CYCLES; STABILITY

We consider the autonomous system

$$\frac{dx}{dt} = F(x, y), \qquad \frac{dy}{dt} = G(x, y). \tag{7-40}$$

A particular trajectory of (7–40) may be a closed curve near a center. As in Section 7–1, the corresponding solution functions $x = f(t)$, $y = g(t)$ must be *periodic*, that is, for all t, $f(t + \tau) = f(t)$, $g(t + \tau) = g(t)$ for some $\tau > 0$. (We always choose τ as the shortest period, i.e., the time needed to trace the closed curve once.)

Such a closed trajectory is called a *limit cycle*. We observe that there may also be closed curves leading to equilibrium points as for C_1 in Fig. 7–22; also a trajectory may start and end at an equilibrium point, as for C_3 in Fig. 7–22. Only C_2 in Fig. 7–22 is a limit cycle.

A given system (7–40) may have no limit cycles, as shown by the simplest examples. However, under certain conditions, we can be sure that a limit cycle exists. A very important theorem of Poincaré and Bendixson is helpful for this purpose. We

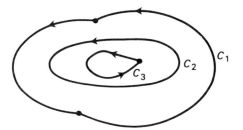

Fig. 7–22. Limit cycle versus closed curves leading to equilibrium points.

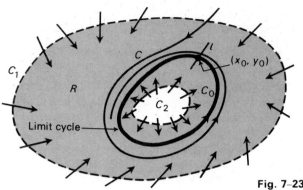

Fig. 7–23. Poincare–Bendixson theorems.

state the conditions of the theorem for the configuration most commonly used. We have a region R in the xy-plane bounded by two closed curves C_1, C_2 that are *not* trajectories (see Fig. 7–23). It is assumed that there is no equilibrium point in R, further that along C_1 and C_2, the trajectories all *enter* R as t increases. Then there is a limit cycle C_0 in R. In particular, if we follow one trajectory C as it enters R, then C must wind around a limit cycle C_0, as suggested in Fig. 7–23.

A typical application of the Poincaré–Bendixson theorem is to the van der Pol equation, which we consider in the next section.

Configurations near a limit cycle

There are a variety of possibilities, but we mention only the three most common ones:
a) a family of concentric limit cycles, as at a center;
b) a limit cycle approached by spirals as $t \to \infty$;
c) a limit cycle approached by spirals as $t \to -\infty$.

The three cases are illustrated in Fig. 7–24. In case (b) we call the limit cycle *stable* (more precisely, *orbitally stable*), since a slight displacement from the limit cycle puts us on a trajectory that brings us back to the limit cycle as t increases. In case (c) we call the limit cycle *unstable*; in case (a) it is called *neutrally stable*.

In the case of Fig. 7–23, if there is only one limit cycle in R, then it must be stable, for, as noted above, each trajectory C entering R must wind around the limit cycle.

Another way of testing for stability is to use *characteristic exponents*. Without going into detail on this topic, we give one simple criterion. If the limit cycle $x = f(t)$, $y = g(t)$ is known (perhaps, numerically), then one characteristic exponent is

$$\lambda = \frac{1}{\tau} \int_0^\tau \left[A_1(t) + B_2(t) \right] dt, \tag{7–41}$$

where τ is the period and

$$A_1(t) = F_x[f(t), g(t)], \qquad B_2(t) = G_y[f(t), g(t)].$$

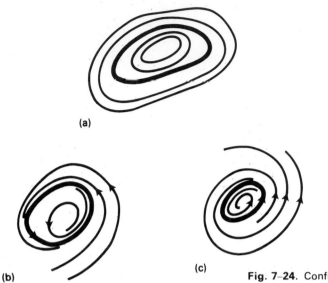

(a)

(b)   (c)

Fig. 7–24. Configurations near a limit cycle.

If $\lambda < 0$, then the limit cycle is stable; if $\lambda > 0$, then it is unstable; if $\lambda = 0$, no conclusion can be drawn.

The following example illustrates both the Poincaré–Bendixson theorem and the calculation of the characteristic exponent.

EXAMPLE 1 $\dfrac{dx}{dt} = x - y - x^3 - xy^2,$ $\dfrac{dy}{dt} = x + y - x^2y - y^3.$

On a circle $x^2 + y^2 = a^2$ we have

$$\frac{dx}{dt} = x(1 - a^2) - y, \qquad \frac{dy}{dt} = x + y(1 - a^2).$$

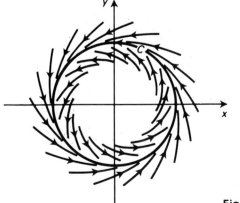

Fig. 7–25. Example 1.

From this we easily see that for $a = 2$ and $a = 1/2$ the trajectories enter the region R bounded by the circles $x^2 + y^2 = 4$, $x^2 + y^2 = 1/4$, as in Fig. 7–25. Also, R contains no equilibrium point. Hence, by the Poincaré–Bendixson theorem, R contains a limit cycle. In fact, the circle C of radius 1 is a limit cycle corresponding to the solution $x = \cos t$, $y = \sin t$, as we verify. We find $F_x = 1 - 3x^2 - y^2$, $G_y = 1 - x^2 - 3y^2$. Thus

$$A_1'(t) = -2\cos^2 t, \qquad B_2(t) = -2\sin^2 t,$$

and by Eq. (7–41) we get

$$\lambda = \frac{1}{2\pi}\int_0^{2\pi}(-2\cos^2 t - 2\sin^2 t)dt = -2.$$

Therefore, the limit cycle is stable. (For more information on limit cycles, see ODE, pp. 451–458.) ◄

7–5 THE VAN DER POL EQUATION

This is the differential equation

$$\frac{d^2x}{dt^2} + \mu(x^2 - 1)\frac{dx}{dt} + x = 0, \tag{7–50}$$

where μ is a positive constant. Certain electric networks containing vacuum tubes are governed by such an equation (see Fig. 7–26; also see LAWDEN, pp. 305–313).

A graphical analysis in the phase plane reveals a single stable limit cycle. A similar result is found for arbitrary μ, though the shape of the limit cycle varies considerably (Fig. 7–27). With $y = dx/dt$, the differential equations in the phase plane are

$$\frac{dx}{dt} = y, \qquad \frac{dy}{dt} = \mu(1 - x^2)y - x. \tag{7–51}$$

From these we see that there is one equilibrium point, at $x = 0$, $y = 0$. The approximating linear equations are

$$\frac{dx}{dt} = y, \qquad \frac{dy}{dt} = \mu y - x. \tag{7–52}$$

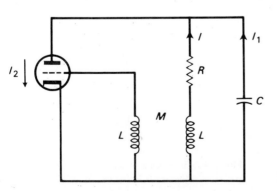

Fig. 7–26. Network with vacuum tube.

The eigenvalues are given by the equation

$$\lambda^2 - \mu\lambda + 1 = 0.$$

For $\mu > 2$, the roots are unequal, real and positive; for $\mu = 2$, they are both 1; for $0 < \mu < 2$, they are complex with positive real part. We conclude that the origin is unstable (an unstable node for $\mu > 2$, an unstable focus for $\mu < 2$). From a study of the isoclines we can find two simple closed curves C_1, C_2 such that every solution meeting C_1 or C_2 enters the region R bounded by C_1, C_2 and never leaves, while R contains no equilibrium point. Such curves are sketched in Fig. 7–27 (a) and (c). From the Poincaré–Bendixson theorem we conclude that there must be at least one limit cycle. It is not so easy to prove that there is only one.

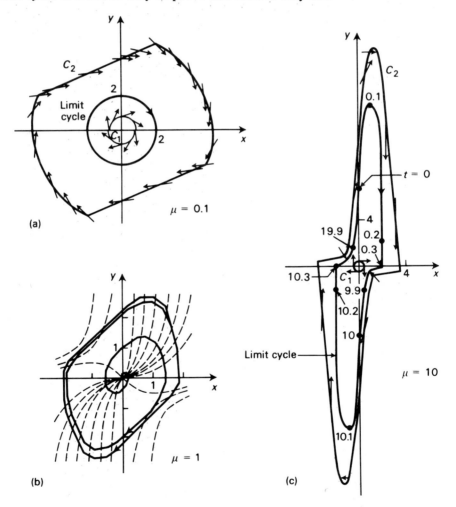

Fig. 7-27. Solutions of van der Pol equation for (a) $\mu = 0.1$; (b) $\overline{\mu} = 1$; (c) $\mu = 10$.

THEOREM   (*Liénard*) *For every positive value of* μ *the van der Pol equation*
(7–50) *has exactly one stable limit cycle in the phase plane.*

(For proof, see ODE, pp. 460–464.) The conclusion is also valid for the more
general equation

$$x''(t) + f(x)x'(t) + p(x) = 0,$$

where $f(x)$, $p(x)$, and $p'(x)$ are continuous for all x, function $f(x)$ is even and $p(x)$ is
odd, $g(x) = \int_0^x f(u)du$ and $q(x) = \int_0^x p(u)du$ approach $+\infty$ as $x \to +\infty$, $p(x) > 0$
for $x > 0$ and $g(x) < 0$ for $0 < x < \bar{x}$, $f(x) > 0$ and $g(x) > 0$ for $x > \bar{x}$. (For proof,
see CODDINGTON–LEVINSON, pp. 402–403.)

7–6 PENDULUM-TYPE EQUATIONS

The simple pendulum is studied in Section 7–2 (Example 3, Fig. 7–14).
  In various problems we encounter generalizations of the pendulum equation
included in the form

$$\frac{d^2\theta}{dt^2} + \alpha\frac{d\theta}{dt} + \gamma\sin\theta - \beta = 0, \tag{7–60}$$

where α, β, γ are constants, with $\gamma \neq 0$. By proper choice of the time scale we can
ensure that $\gamma = 1$. Hence we are led to the equation

$$\frac{d^2\theta}{dt^2} + \alpha\frac{d\theta}{dt} + \sin\theta - \beta = 0. \tag{7–61}$$

We observe that in Eq. (7–61), for $\alpha > 0$, the term $\alpha\, d\theta/dt$ is a typical friction term,
which will tend to damp the oscillations. The β-term corresponds to a constant
driving force. Thus Eq. (7–61) describes a pendulum with friction and constant
external force.
  In Section 7–2 we have already considered the case $\alpha = 0$, $\beta = 0$. The case $\alpha = 0$,
$\beta \neq 0$ can be similarly analyzed as a conservative system. With θ replaced by x, the
corresponding 2-dimensional autonomous system becomes

$$\frac{dx}{dt} = y, \qquad \frac{dy}{dt} = \beta - \sin x, \tag{7–62}$$

and a first integral is

$$\frac{1}{2}y^2 - \beta x - \cos x = c. \tag{7–63}$$

In Fig. 7–28 the corresponding trajectories are shown for the three cases: $0 < \beta < 1$,
$\beta = 1$, $\beta > 1$. For negative β, the angle θ in (7–61) can be replaced by $-\theta$ to obtain a
similar equation with positive β.
  As the figures suggest, for $0 < \beta < 1$ we have a structure much like that for
$\beta = 0$, whereas for $\beta = 1$ the equilibrium points at $(\pi/2 + 2n\pi, 0)$ are all 0-order saddle
points (see Problem 5 following Section 7–2), and for $\beta > 1$ there are no equilibrium points.

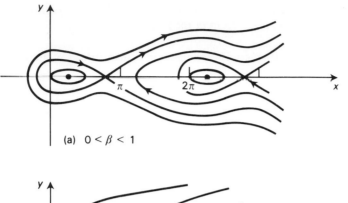

(a) $0 < \beta < 1$

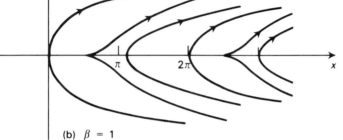

(b) $\beta = 1$

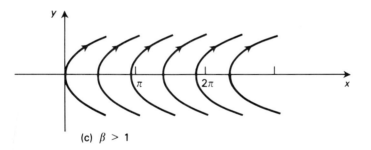

(c) $\beta > 1$

Fig. 7-28. Trajectories for $dx/dt = y$, $dy/dt = \beta - \sin x$ for the three cases: (a) $0 < \beta < 1$; (b) $\beta = 1$; (c) $\beta > 1$.

We next turn to the general equation (7–61) with $\alpha \neq 0$, $\beta \neq 0$. Here we can assume $\alpha > 0$, $\beta > 0$, for if $\alpha > 0$, a replacement of t by $-t$ changes our equation to a similar one with term $-\alpha d\theta/dt$ in $d\theta/dt$. And if $\beta < 0$, we can as above replace θ by $-\theta$ to reverse the sign of the last term. Thus we are led to the 2-dimensional autonomous system

$$\frac{dx}{dt} = y, \qquad \frac{dy}{dt} = \beta - \alpha y - \sin x, \qquad \alpha > 0, \quad \beta > 0. \qquad (7\text{–}64)$$

Let $0 < \beta < 1$; then we can write $\beta = \sin\theta_0$, $0 < \theta_0 < \pi/2$. The equilibrium points are at $(\theta_0 + 2n\pi, 0)$, which are stable foci, and at $((2n+1)\pi - \theta_0, 0)$, which are saddle points ($n = 0, \pm 1, +2, \ldots$). It is convenient to consider, along with (7–64), the differential equation

$$\frac{dy}{dx} = \frac{\beta - \alpha y - \sin x}{y} \tag{7–65}$$

whose solutions, for $y > 0$ or for $y < 0$, give particular whole trajectories of (7–64) as

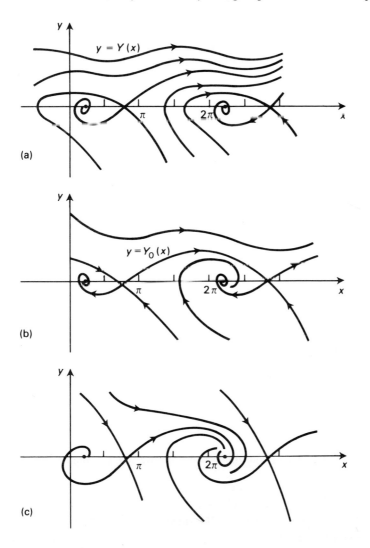

(a)

(b)

(c)

Fig. 7-29. Trajectories for $dx/dt = y$; $dy/dt = \beta - \alpha y - \sin x$, $\beta = \sin\theta_0$, $0 < \theta_0 < \pi/2$: (a) case $0 < \alpha < \alpha(\theta_0)$; (b) case $\alpha = \alpha(\theta_0)$; (c) case $\alpha = \alpha(\theta_0)$.

curves. It can then be shown that there exists a critical value $\alpha(\theta_0) > 0$ such that, for $0 < \alpha < \alpha(\theta_0)$, Eq. (7–65) has a *unique* periodic solution $y = Y(x)$ of period 2π with $0 < Y(x) < (1 + \beta)/\alpha$ for $-\infty < x < \infty$. Each nonequilibrium trajectory of (7–64) either approaches an equilibrium point as $t \to \infty$ or else, for t sufficiently large, remains in the region $y > 0$ and hence provides a solution of (7–65) approaching the periodic solution $y = Y(x)$ as x (and t) $\to \infty$. The relationships are suggested in Fig. 7–29 (a). For $\alpha = \alpha(\theta_0)$ there is no periodic solution of (7–65), but there is a *pseudoperiodic solution* $y = Y_0(x)$ formed of solutions whose graphs go from one saddle point to the next. Solutions of (7–65) starting above this curve approach it asymptotically as x (and t) $\to \infty$, as in Fig. 7–29 (b). For $\alpha > \alpha(\theta_0)$ every trajectory of (7–64) approaches an equilibrium point, as suggested in Fig. 7–29 (c).

Finally, let $\beta > 1$. Then (7–64) has no equilibrium points. It can be shown however that (7–65) again has a unique periodic solution $y = Y(x)$ of period 2π with $Y(x) > 0$ for all x, and every trajectory of (7–64) remains in the region $y > 0$ for t sufficiently large, and approaches the curve $y = Y(x)$ as above.

(For proofs and further details on all these cases the reader is referred to SANSONE–CONTI, pp. 278–303.)

========================= Problems (Section 7–6) =========================

1. Find all limit cycles for each of the following systems:
   a) $dx/dt = 2x - 3y$, $dy/dt = 3x - 2y$
   b) $dx/dt = -y^3$, $dy/dt = x^3$.   HINT: Eliminate t.
   c) $dx/dt = 2y$, $dy/dt = -\sin x$

2. Apply the Poincaré–Bendixson theorem to show that there is a limit cycle between the circles $x^2 + y^2 = 1$ and $x^2 + y^2 = 16$:

   a) $\dfrac{dx}{dt} = 4x - 4y - x(x^2 + y^2)$, $\dfrac{dy}{dt} = 4x + 4y - y(x^2 + y^2)$

   b) $\dfrac{dx}{dt} = x \sin \dfrac{x^2 + y^2}{4} - y$, $\dfrac{dy}{dt} = x + y \sin \dfrac{x^2 + y^2}{4}$

3. For each of the following systems verify that the path given is a limit cycle, and apply (7–41) to determine λ and stability of the limit cycle.
   a) Equations of Problem 2(a), $x = 2 \cos 4t$, $y = 2 \sin 4t$
   b) Equations of Problem 2(b), $x = 2\sqrt{\pi} \cos t$, $y = 2\sqrt{\pi} \sin t$

4. Find the trajectories of the van der Pol equation in the phase plane for $\mu = 1$ with the aid of isoclines (see Fig. 7–27(b)).

5. Apply numerical integration to obtain the t-values shown on the limit cycle of Fig. 7–27(c).

   HINT: Use the equations $dt = \dfrac{dx}{v} = \dfrac{dv}{10(1 - x^2)v - x}$.

   Show that the graphs of x and v versus t have the appearance of Fig. 7–30.

6. Show that the differential equation

   $$a\frac{d^2x}{dt^2} + b(x^2 - 1)\frac{dx}{dt} + cx = 0, \quad a > 0, \quad b > 0, \quad c > 0,$$

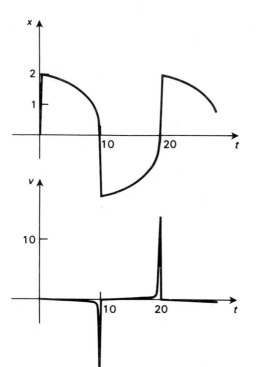

Fig. 7–30. Periodic solution of van der Pol equation ($v = dx/dt$).

where a, b, c are constants, can be reduced to the van der Pol equation (7–50) by an appropriate change of time scale.

7. Consider Eqs. (7–64) for $\alpha = 0$:

$$\frac{dx}{dt} = y, \qquad \frac{dy}{dt} = \beta - \sin x;$$

these are the phase-plane equations equivalent to the second-order equation

$$\frac{d^2x}{dt^2} + \sin x - \beta = 0,$$

which is Eq. (7–61) with $\alpha = 0$ and θ replaced by x.

a) Show that $U(x) = -\beta x - \cos x$ can be chosen as the potential energy of the system, so that, as in Section 7–2, the trajectories are the curves $(y^2/2) + U(x) = \text{const.}$

b) Show that for $0 < \beta < 1$ the critical points of $U(x)$ are $x_0 + 2n\pi$, $\pi - x_0 + 2n\pi$ for a unique x_0 with $0 < x_0 < \pi/2$, $n = 0, \pm 1, \pm 2, \ldots$, and that the former correspond to minima of $U(x)$, the latter to maxima, so that the trajectories are as in Fig. 7–28(a).

c) Show that for $\beta = 1$ the critical points of $U(x)$ are at $(\pi/2) + 2n\pi$, at which the graph of $U(x)$ has horizontal inflection points, so that there are 0-order saddle points at the corresponding equilibrium points (see Problem 5 following Section 7–2), and the trajectories are as in Fig. 7–28(b).

d) Show that for $\beta > 1$ $U(x)$ has no critical points and the trajectories are as in Fig. 7–28(c).

7-7 COMPETING POPULATIONS. VOLTERRA MODEL

For two competing animal populations, measured by x and y respectively, a useful model is that of Volterra:

$$\frac{dx}{dt} = a_1 x - k_1 xy, \qquad \frac{dy}{dt} = -a_2 y + k_2 xy. \tag{7-70}$$

Here a_1, a_2, k_1, k_2 are positive constants and we are interested only in the region $x \geqslant 0$, $y \geqslant 0$. Here y is often called the *predator* population, x the *prey* population.

We gain some insight into the behavior of the solutions by studying the case in which all the constants are equal to 1:

$$\frac{dx}{dt} = x - xy, \qquad \frac{dy}{dt} = -y + xy. \tag{7-71}$$

We see that there are equilibrium points at $(0, 0)$ and $(1, 1)$. At the first point the linear approximation reveals a saddle; in fact, two solutions leading to this point as $t \to +\infty$ and as $t \to -\infty$ are, respectively,

$$x = 0, \; y = e^{-t} \qquad \text{and} \qquad x = e^t, \; y = 0.$$

At $(1, 1)$, the linear approximation is found to be

$$\frac{dx}{dt} = -(y - 1), \qquad \frac{dy}{dt} = x - 1,$$

so that there is a center. This does not usually ensure a center for the given nonlinear equations, however, Eqs. (7-71) have a first integral

$$\varphi(x, y) = xye^{-x-y} = \text{const}, \tag{7-72}$$

as can be verified. Accordingly, as pointed out in Section 7-3, the nonlinear equations (7-71) do have a center at $(1, 1)$. Thus near $(1, 1)$ the trajectories are simple closed curves enclosing the point. In fact, all the trajectories in the first quadrant are such closed curves. We can see this by verifying that along each straight line $y - 1 = m(x - 1)$ passing through the point $(1, 1)$ the function $\varphi(x, y)$ has the following behavior as x increases: it rises from the value 0, where the line meets the coordinate axis on the boundary of the first quadrant, to the value e^{-2} at $(1, 1)$ and then decreases, approaching 0 as the line approaches the boundary of the quadrant or $|x| + |y| \to \infty$. For example, when $m = 1$, then $y = x$, and along this line φ becomes $g(x) = x^2 e^{-2x}$ for $0 \leqslant x < \infty$. We can verify that $g(x)$ increases from 0 at $x = 0$ to e^{-2} at $x = 1$ and then decreases, approaching 0 as $x \to +\infty$. From this behavior along these lines in the first quadrant it follows that φ takes the value e^{-2} only at $(1, 1)$ and takes each value between 0 and e^{-2} exactly twice on each line (once below $x = 1$, once above $x = 1$). Thus the level curve $\varphi(x, y) = c$, $0 < c < e^{-2}$, is necessarily a closed curve enclosing $(1, 1)$. These level curves fill the quadrant, since $0 < \varphi(x, y) < e^{-2}$ in the quadrant. For $c = 0$, one obtains the rays $x = 0$, $y = 0$ bounding the quadrant.

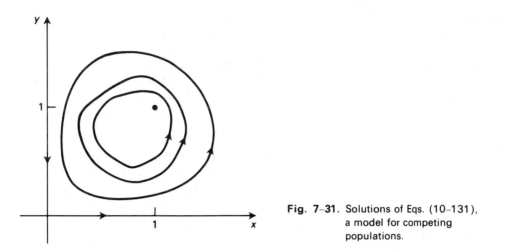

Fig. 7–31. Solutions of Eqs. (10-131), a model for competing populations.

The trajectories are graphed in Fig. 7–31. They show that typically the two populations rise and drop in periodic manner. If at any time the prey x becomes 0 (extinction), then the predator y must gradually also become extinct for lack of food. However, if y becomes extinct first, then the prey x would grow indefinitely, according to the model.

The verification of some of the assertions made above is left to the problems. The results of the special case (7–71) continue to hold for the general case (7–70). The first integral becomes

$$(ue^{-u})^{1/a_1}(ve^{-v})^{1/a_2} = c, \tag{7–73}$$

where $u = k_2x/a_2$, $v = k_1y/a_1$.

The model has been generalized in various ways; for example, in the form

$$\frac{dx}{dt} = k_1x(a_1 - x - b_1y), \qquad \frac{dy}{dt} = -k_2y(a_2 - y - b_2x), \tag{7–74}$$

where a_1, a_2, b_1, b_2 are positive constants and k_1, k_2 are nonzero constants.

=======================**Problems (Section 7–7)**=======================

1. Let the system (7–71) be given.
   a) Verify that at $(0, 0)$ the linear approximation has a saddle point and that at $(1, 1)$ it has a center.
   b) Verify that Eq. (7–72) does define a first integral.
   c) Show that along the line $y = x/2 + 1/2$, $0 \leqslant x < \infty$, function φ becomes a function $g(x)$ that rises from 0 at $x = 0$ to e^{-2} at $x = 1$ and then decreases with limit 0 as $x \to +\infty$.
   d) Extend the result of part (b) to the line $y = 2x - 1$, $1/2 \leqslant x < \infty$.
   e) Extend the result of part (c) to an arbitrary line $y - 1 = m(x - 1)$ restricted to the portion in the first quadrant.

2. Let the system (7–70) be given as above.
   a) Show that the equilibrium points are $(0, 0)$ and $(a_2/k_2, a_1/k_1)$.
   b) Determine the nature of the linear approximations to (7–70) as the equilibrium points.
   c) Verify that Eq. (7–73) defines a first integral of the system.

3. For the system (7–74) let $a_1 = 2$, $a_2 = 3$, $b_1 = 1$, $b_2 = 2$, $k_1 = 3$, $k_2 = 2$. Find all equilibrium points and classify on the basis of the linear approximations.

7–8 FORCED OSCILLATIONS OF NONLINEAR SYSTEMS

We know that a linear system with constant coefficients and a periodic forcing function usually has a periodic solution; for example, equation

$$\frac{d^2x}{dt^2} + 2\frac{dx}{dt} + 4x = A \cos \omega t$$

has a solution of form $x = k_1 \cos \omega t + k_2 \sin \omega t$ (see Section 3–8). It is natural to expect a similar result for a nonlinear system, such as the one described by the equation

$$\frac{d^2x}{dt^2} + \frac{dx}{dt} + x + x^3 = A \cos \omega t. \tag{7–80}$$

This is a case of the *Duffing equation*

$$\frac{d^2x}{dt^2} + c\frac{dx}{dt} + x + \beta x^3 = p(t), \tag{7–81}$$

where c and β are constants and $p(t)$ is periodic. These and more general nonlinear systems have been much studied and there are a variety of theorems on existence of periodic solutions. Also a number of methods have been developed for finding the periodic solutions. Among the most effective methods are *perturbation* methods. These apply to a general system

$$\frac{dx}{dt} = F_0(x, y, t) + \mu F_1(x, y, t, \mu),$$

$$\frac{dy}{dt} = G_0(x, y, t) + \mu G_1(x, y, t, \mu). \tag{7–82}$$

Here μ is a parameter. We assume that the right side has period τ in t and that for $\mu = 0$ the system has one or more periodic solutions of period τ. We then seek a periodic solution $x = x(t, \mu)$, $y = y(t, \mu)$ of (7–82) for μ close to 0. Under appropriate conditions, this solution can be found in the form of a power series in μ with coefficients depending on t. To determine the coefficients, we substitute the series, with unknown coefficients, in (7–82) and express both sides as power series in μ. A comparison of coefficients of like powers of μ leads to successive differential equations for the unknown coefficients sought. We then solve these differential equations to obtain the coefficients as functions of t of period τ. In the process, we are forced to choose the initial conditions for the differential equations in a special way to maintain the desired periodicity.

For the general theory on which this process is based, the reader is referred to CODDINGTON–LEVINSON, Chapter 14. Here we consider only the example (7–80) for $\omega = 1$. Our first step is to write the equation as

$$\frac{d^2x}{dt^2} + x = \mu\left(A\cos t - \frac{dx}{dt} - x^3\right) \tag{7–83}$$

and to seek a solution of period 2π when $\mu = 1$. We write

$$x = \sum_{n=0}^{\infty} g_n(t)\mu^n \tag{7–84}$$

and substitute in Eq. (7–83) to obtain

$$\sum_{n=0}^{\infty}(g_n'' + g_n)\mu^n = \mu\left[A\cos t - \sum_{n=0}^{\infty}g_n'(t)\mu^n - (g_0 + g_1\mu + \cdots)^3\right]$$

We compare terms of same degree in μ on both sides:

$$g_0'' + g_0 = 0,$$
$$g_1'' + g_1 = A\cos t - g_0'(t) - (g_0(t))^3,$$
$$g_2'' + g_2 = -g_1'(t) - 3(g_0(t))^2 g_1(t), \text{ etc.}$$

These are successive equations for $g_0(t)$, $g_1(t)$, ... The first is satisfied by

$$g_0(t) = a_0\cos t + b_0\sin t,$$

where a_0 and b_0 are to be found. The second then becomes (after some trigonometry)

$$g_1'' + g_1 = A\cos t + a_0\sin t - b_0\cos t$$

$$-\frac{1}{4}[a_0^3(\cos 3t + 3\cos t) + 3a_0^2 b_0(\sin t + \sin 3t) \tag{7–85}$$

$$+ 3a_0 b_0^2(\cos t - \cos 3t) + b_0^3(3\sin t - \sin 3t)].$$

We know that terms in $\cos t$ and $\sin t$ on the right would give rise to terms in $t\cos t$ and $t\sin t$ in $g_1(t)$ due to the resonance effect and hence vitiate the periodicity of $g_1(t)$. Therefore we require that these terms vanish. This leads to the equations

$$A - b_0 - \frac{3}{4}(a_0^3 + a_0 b_0^2) = 0, \qquad a_0 - \frac{3}{4}(a_0^2 b_0 + b_0^3) = 0; \tag{7–86}$$

these are obtained by equating to zero the coefficients of $\cos t$ and $\sin t$ in the Fourier series (in a finite number of terms) on the right of Eq. (7–85). Equations (7–86) are simultaneous equations for a_0, b_0. When they are solved (numerically, for given A), the results can be used to obtain the new equation for $g_1(t)$:

$$g_1'' + g_1 = -\frac{1}{4}(a_0^3\cos 3t + 3a_0^2 b_0\sin 3t - 3a_0 b_0^2\cos 3t - b_0^3\sin 3t). \tag{7–85'}$$

Here the terms in $\cos t$ and $\sin t$ have dropped out because of Eqs. (7–86), and the solution is

$$g_1(t) = \frac{1}{32}[(a_0^3 - 3a_0 b_0^2)\cos 3t + (3a_0^2 b_0 - b_0^3)\sin 3t] + a_1 \cos t + b_1 \sin t.$$

(7–87)

Here a_1, b_1 will be determined to ensure that the equation for $g_2(t)$ has a periodic solution. Continuing in this manner, we can obtain all coefficients $g_n(t)$ and hence obtain the series for $x(t)$. As a first approximation,

$$x(t) = g_0(t) = a_0 \cos t + b_0 \sin t,$$

(7–88)

where a_0, b_0 are determined by Eqs. (7–86). This first approximation is called the *generating solution*. For small μ, this should give a good indication of the form of the solution. Its amplitude is $A_0 = (a_0^2 + b_0^2)^{1/2}$. From Eqs. (7–86), by transposing A, squaring, and adding, we find:

$$A^2 = A_0^2 + \frac{9}{16}A_0^6.$$

(7–89)

This equation shows that the input $A \cos t$ is amplified (or, better, attenuated, since clearly $A_0 < A$) for small μ (see Fig. 7–32).

The general theory ensures the validity of the procedure we have used, but does not give an interval of μ for which the series converges.

In Eq. (7–80) we have been seeking a periodic solution only for $\omega = 1$. A modification of the procedure makes it possible to find a periodic solution for variable ω. Again we obtain a first approximation like (7–88):

$$x = a_0 \cos \omega t + b_0 \sin \omega t,$$

where a_0, b_0 depend on A and ω. Hence the output amplitude A_0 depends on A and ω. By plotting A_0 as a function of ω for various values of A we get a picture of the frequency response of the nonlinear system.

If this is done for the Duffing equation (7–81) with $p(t) = A \cos \omega t$, we obtain response curves as in Fig. 7–33. The cases $\beta > 0$ and $\beta < 0$ differ. We can interpret the term $x + \beta x^3$ as representing a spring force at displacement x. For $\beta > 0$ the force increases more rapidly than in the linear case; for $\beta < 0$, it increases less rapidly than

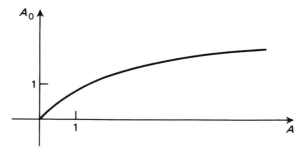

Fig. 7–32. Amplification curve for Eq. (7–83) based on first approximation.

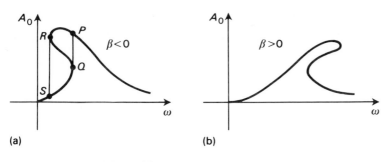

Fig. 7–33. Frequency response for Duffing equation: (a) $\beta < 0$, soft spring; (b) $\beta > 0$, hard spring.

in the linear case (Fig. 7–34). We refer to the former as a *hard spring*, to the latter as a *soft spring*.

In Fig. 7–33 (a) we see another phenomenon. If the frequency ω is being gradually increased from 0, then we move on the response curve from $(0,0)$ to Q, but we cannot go any further than Q as ω increases except by jumping to P. This suggests a sudden jump in amplitude at this frequency. If we continue further, the amplitude slowly decreases and there is no problem. If we now reverse the process and work back to P, then we can continue on the upper curve to R but are there forced to jump down to S and then continue on to $(0,0)$. Thus there is a *hysteresis* effect in addition to the jumps. There are similar effects for the hard spring, as shown in Fig. 7–33 (b).

It should be stressed that these results hold only for small values of the perturbation parameter (in particular for β close to 0). Furthermore, the solutions found do not exhaust the known periodic solutions of the Duffing equation. For example, it can be shown that in certain cases there is *subharmonic resonance*, that is,

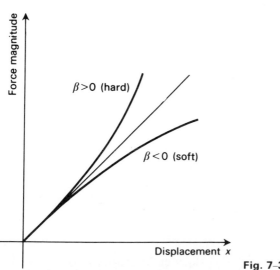

Fig. 7–34. Force laws for nonlinear springs.

there is a periodic solution whose period is an integral multiple of the input period $\tau = 2\pi/\omega$, say one of period 3τ.

For further discussion see LAWDEN, Chapter 5; STERN, Chapter 11; and SANSONE–CONTI, Chapter VII.

7–9 SYSTEMS WITH DISCONTINUITIES

We illustrate the effect of discontinuities by two examples.

EXAMPLE 1 Consider a mass-spring system with *Coulomb friction*, typical of that between unlubricated surfaces. The friction force F is constant in magnitude for the range of velocities occurring. It reverses direction when the direction of motion reverses. Hence the force can be taken to be $-f_0$ for $v > 0$ and f_0 for $v < 0$, as in Fig. 7–35. For $v = 0$, the force is taken to be 0 as long as the spring force is less than f_0; however, for $v = 0$ and $kx > f_0$, where kx is the spring force, the friction will act to reduce the spring force, so that the combined force is taken to be $-(kx - f_0)$; similarly, for $v = 0$ and $kx < -f_0$, the combined force is $-(kx + f_0)$. The equation of motion is

$$m\frac{d^2x}{dt^2} = -kx + F.$$

The phase–plane equations (with v replaced by y) are

$$\frac{dx}{dt} = y, \qquad \frac{dy}{dt} = \frac{-kx + F}{m} = G(x, y),$$

where

$$G(x, y) = \begin{cases} \dfrac{f_0 - kx}{m} & \text{for } y < 0 \text{ and for } y = 0 \text{ when } kx > f_0, \\[2ex] \dfrac{-f_0 - kx}{m} & \text{for } y \geq 0 \text{ and for } y = 0 \text{ when } kx < -f_0, \\[2ex] 0 & \text{for } y = 0 \text{ and } -f_0 \leqslant kx \leqslant f_0. \end{cases}$$

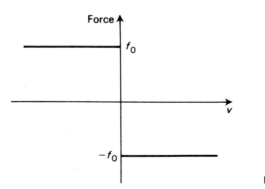

Fig. **7–35.** Coulomb friction.

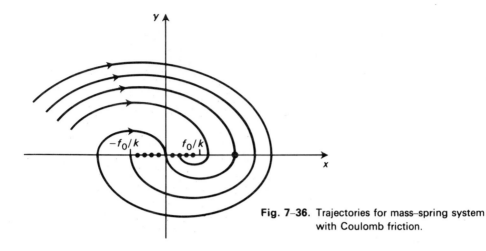

Fig. 7–36. Trajectories for mass–spring system with Coulomb friction.

Here all the points $(x, 0)$ with $-f_0 \leqslant kx \leqslant f_0$ are equilibrium points. In the half-plane $y > 0$ we have a first integral

$$kx^2 + 2f_0 x + my^2 = c.$$

Thus the trajectories are semiellipses. Similarly in the half-plane $y < 0$ the trajectories are semiellipses

$$kx^2 - 2f_0 x + my^2 = c.$$

When such a trajectory leads, with increasing t, to an equilibrium point, the trajectory reaches the equilibrium point in finite time and then remains at the equilibrium point. When it leads to a point $(x, 0)$, which is not an equilibrium point, the trajectory crosses from one half-plane to the other (hence from one semiellipse to another). Each individual trajectory reaches an equilibrium point in finite time. The trajectories are shown in Fig. 7–36. ◄

EXAMPLE 2 Under some simplifying assumptions about the performance characteristics of the vacuum tube, the network of Fig. 7–26 leads to the equation

$$\frac{d^2 I}{dt^2} + \frac{R}{L}\frac{dI}{dt} + \frac{1}{LC}I = \begin{cases} 0 & \text{for } dI/dt < 0, \\ \dfrac{I_0}{LC} & \text{for } dI/dt \geqslant 0. \end{cases}$$

If we set $x = I/I_0$ and $y = dx/dt$, we obtain the phase–plane equations

$$\frac{dx}{dt} = y, \qquad \frac{dy}{dt} = G(x, y),$$

where

$$G(x, y) = \begin{cases} -\omega^2 x - 2hy & \text{for } y < 0, \\ \omega^2 - \omega^2 x - 2hy & \text{for } y \geqslant 0, \end{cases}$$

with $\omega^2 = 1/(LC)$ and $2h = R/L$. We assume that the resistance R is small, so that

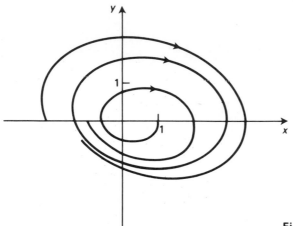

Fig. 7–37. Trajectories for Example 2.

$0 < h < \omega$. Here in the lower half-plane we have portions of the trajectories for a system with a focus at $(0, 0)$ (corresponding to damped vibrations), and in the upper half-plane similar trajectories for a system with a focus at $(1, 0)$. The only equilibrium point is $(1, 0)$. We find that trajectories starting close to $(1, 0)$ spiral outwards as t increase, while those starting far from $(1, 0)$ spiral inwards, much as for the van der Pol equation. There is a unique limit cycle, meeting the x-axis at $((1 - e^{-d/2})^{-1}, 0)$, where $d = 2\pi h(\omega^2 - h^2)^{-1/2}$, shown in Fig. 7–37. ◄

===================== **Problems (Section 7–9)** =====================

1. Follow the procedures described in the text to find the equations that determine a_1, b_1 for Eq. (7–87).

2. Follow the procedure described in the text to find a first approximation to a periodic solution:

   a) $\dfrac{d^2x}{dt^2} + x = \mu\left[(x^2 - 1)\dfrac{dx}{dt} + A\cos t\right]$ (van der Pol equation with a force term);

   b) $\dfrac{d^2x}{dt^2} + x = \mu\left(A\sin t - 2\dfrac{dx}{dt} - x^3\right)$.

3. In Example 1 of Section 7–9, take $f_0 = 4$, $k = 4$, $m = 1$ and draw the trajectories.

4. In Example 2 of Section 7–9, take $h = 1$, $\omega^2 = 2$ and draw the trajectories. In particular, locate the limit cycle.

Chapter 8
Introduction to Partial Differential Equations

8–1 ORIGIN OF PARTIAL DIFFERENTIAL EQUATIONS IN PHYSICS

Just as ordinary differential equations provide mathematical formulations of physical laws governing the mechanics of particles and electrical networks, so do partial differential equations describe the phenomena of continuous media: for example, fluids, solids in states of deformation or vibration, heat conduction, electromagnetic fields.

A typical example is the *one-dimensional wave equation*

$$\frac{\partial^2 z}{\partial t^2} - c^2 \frac{\partial^2 z}{\partial x^2} = 0, \quad c = \text{const} > 0; \tag{8–10}$$

it describes the vibrations of a string, as suggested in Fig. 8–1.

Here the string is stretched between $(0, 0)$ and $(L, 0)$ in an xz-plane, and function $z(x, t)$ gives the shape of the string at time t. The same equation describes the longitudinal vibrations of a thin rod, as in Fig. 8–2, where the cross section at equilibrium position x is displaced by $z(x, t)$ at time t.

There are several alternatives to justify such an equation. One of them is to show that the equation is an inevitable consequence of some basic principle of physics, such as conservation of energy. Another is to show that the equation is the well-determined limit of accepted equations that describe simpler physical problems, such as the motion of particles. Or we may simply postulate the equation itself as a law of physics and verify it by experiment. In any case, this last approach is normally required before an equation can be accepted with confidence as a proper mathematical model of the relevant physical phenomenon.

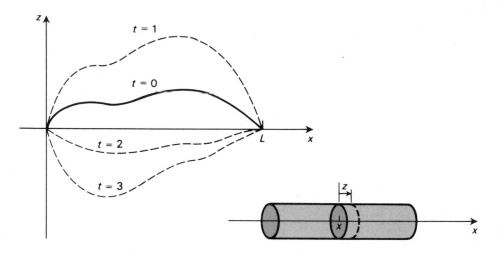

Fig. 8–1. Vibrating string.

Fig. 8–2. Longitudinal vibrations of a thin rod.

Here we briefly describe a limit process for the Eq. (8–10). In Chapter 10 it will be shown that other partial differential equations are consequences of physical laws and properties of line and surface integrals.

The limit process is based on the following *discrete model* of a vibrating string. Particles (all of the same mass m) are constrained to move on equally spaced lines $x = x_1, x = x_2, \ldots, x = x_n$ as in Fig. 8–3. The particles are connected, as shown, by identical springs, all stretched beyond their natural length l. They are also attached to the fixed points $(0, 0)$ and $(L, 0)$ at the ends.

The displacement of the jth particle is z_j. The spring to the right of this particle has been stretched to length $\Delta x \sec \alpha_j$, where the angle α_j is as shown in Fig. 8–3.

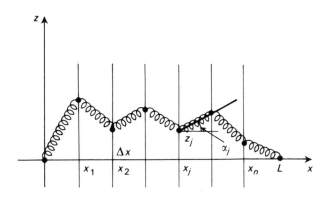

Fig. 8–3. Discrete approximation to vibrating string.

Therefore the spring force is $k(\Delta x \sec \alpha_j - l)$, where k is the spring constant. The component of force in the z-direction is $k(\Delta x \sec \alpha_j - l) \sin \alpha_j$. Similarly, the force on the jth particle due to the spring to the left has a z-component $-k(\Delta x \sec \alpha_{j-1} - l) \sin \alpha_{j-1}$. Therefore, the total force on the jth particle has a z-component

$$k(\Delta x \tan \alpha_j - \Delta x \tan \alpha_{j-1} - l \sin \alpha_j + l \sin \alpha_{j-1})$$

and the motion of the particles is governed by the differential equations

$$m \frac{d^2 z_j}{dt^2} = k \Delta x (\tan \alpha_j - \tan \alpha_{j-1}) - kl(\sin \alpha_j - \sin \alpha_{j-1})$$

for $j = 1, \ldots, n$. Now $\Delta x \tan \alpha_j = z_{j+1} - z_j$. If we assume that all angles are small, we can replace $\sin \alpha_j$ by $\tan \alpha_j$ in the differential equations. Thus the equations become

$$m \frac{d^2 z_j}{dt^2} = k_0 (z_{j+1} - 2z_j + z_{j-1}), \quad j = 1, \ldots, n, \tag{8-11}$$

where $k_0 = k[1 - (l/\Delta x)]$. In (8-11), $z_0 = 0$ and $z_{n+1} = 0$, since the ends are fixed, i.e., their displacement is 0.

We now pass to the limit, letting $n \to \infty$, $m \to 0$, $\Delta x \to 0$. For that purpose we divide by Δx and write the differential equations as follows:

$$\frac{m}{\Delta x} \frac{d^2 z_j}{dt^2} = k_0 \Delta x \frac{z_{j+1} - 2z_j + z_{j-1}}{\Delta x^2}.$$

As $n \to \infty$, the value of $m/\Delta x$ (which represents the average density) can be assumed to approach a limit ρ (the density of the string). Next, $k_0 \Delta x = k (\Delta x - l)$ is the spring force on each side when all $z_j = 0$. Hence this can be assumed to approach a positive limit K (which is the tension in the string when $z = 0$ throughout). Next, $(z_{j+1} - z_j)/\Delta x$ is a slope and

$$\frac{z_{j+1} - 2z_j + z_{j-1}}{\Delta x^2} = \left(\frac{z_{j+1} - z_j}{\Delta x} - \frac{z_j - z_{j-1}}{\Delta x} \right) \frac{1}{\Delta x}$$

represents a rate of change of the slope. Therefore in the limit it should approach the second derivative $\partial^2 z / \partial x^2$. Thus we are led from (8-11) to the limiting equation

$$\rho \frac{\partial^2 z}{\partial t^2} = K \frac{\partial^2 z}{\partial x^2}.$$

If we replace K/ρ by c^2, we obtain the wave equation (8-10).

The derivation we have sketched can be completely justified (see *Partial Differential Equations* by J. D. Tamarkin and W. Feller, pp. 160–169 mimeographed notes of lectures at Brown University, 1941; see also AC, pp. 645–660). It provides a very useful model (8-11) for the differential equation (8-10); much insight into the solutions of the partial differential equation can be gained from the model, by experiment or calculation. It is sufficient to study the motion of only two or three particles to realize what is happening. Experiments reveal that the masses can oscillate synchronously at certain frequencies; a system with n masses has n different frequencies. The corresponding motions are called *normal modes* (see Section 6–8). For the limiting case of the wave equation we expect an infinite sequence of frequencies, and that is found to be correct.

8–2 GENERAL CONCEPTS AND TYPICAL PROBLEMS

We now begin a systematic study of partial differential equations. A *partial differential equation* is an equation such as one of the following:

$$x^2 \frac{\partial z}{\partial x} + y^2 \frac{\partial z}{\partial y} = 0,$$

$$\frac{\partial^2 z}{\partial x^2} - \frac{\partial^2 z}{\partial y^2} = 2xy, \tag{8–20}$$

$$\frac{\partial^2 w}{\partial x^2} + \frac{\partial^2 w}{\partial y^2} + \frac{\partial^2 w}{\partial z^2} - \frac{\partial^2 w}{\partial t^2} = 0,$$

relating a function of several variables and its partial derivatives up to a certain order. The *order* of the differential equation is that of the highest derivative present.

A *solution* of a partial differential equation is a function that satisfies the equation in some region of the space of the independent variables. Thus $z = x^{-1} - y^{-1}$ is a solution of the first equation in (8–20) in the region $x > 0$, $y > 0$, since

$$x^2 \frac{\partial z}{\partial x} + y^2 \frac{\partial z}{\partial y} = x^2 \frac{-1}{x^2} + y^2 \frac{1}{y^2} \equiv 0.$$

The general solution of an ordinary differential equation of order n typically depends on n arbitrary constants, where n is the order of the equation. For a partial differential equation the situation is more complicated. The general solution of certain equations can be given in terms of one or more *arbitrary functions*. For example, the equation

$$\frac{\partial z}{\partial y} = 0,$$

where $z = f(x, y)$, is satisfied by $z = g(x)$, where g is an arbitrary function of x, and every solution is of this form. For the equation

$$\frac{\partial^2 z}{\partial x \, \partial y} = 0, \tag{8–21}$$

where $z = f(x, y)$, we can write the equation thus:

$$\frac{\partial}{\partial x} \left(\frac{\partial z}{\partial y} \right) = 0$$

and reason that $\partial z/\partial y$ can be an arbitrary function of y:

$$\frac{\partial z}{\partial y} = g(y).$$

Hence, upon integrating,

$$z = \int g(y)\,dy + q(x),$$

where $q(x)$ is an arbitrary function of x. Equivalently, we can write

$$z = p(y) + q(x) \tag{8-22}$$

for the general solution, where $p(y)$ and $q(x)$ are arbitrary (differentiable) functions of y and x, respectively.

In applications, we are less concerned with finding general solutions than with finding solutions satisfying *initial conditions* or *boundary conditions*. The initial conditions concern values or derivatives of the solution, say at $t = 0$, where t is one of the independent variables (usually time), and the solution is sought for $t = 0$. The boundary conditions concern the values or derivatives of the unknown function on the boundary of the region (or interval) of the independent space variables. We often consider initial conditions as a form of boundary conditions and refer to a partial differential equation with all the conditions as a *boundary-value problem*.

We shall illustrate these conditions for several commonly occurring equations:

I. The *wave equation* in one dimension:

$$c^2 \frac{\partial^2 z}{\partial x^2} - \frac{\partial^2 z}{\partial t^2} = 0, \tag{8-23}$$

where c is a positive constant, in the region $0 < x < L, 0 < t < \infty$ of the xt-plane, with *boundary conditions*

$$z = 0 \quad \text{for} \quad x = 0, \quad x = L, \quad 0 \leqslant t < \infty, \tag{8-23'}$$

and *initial conditions*

$$z = f(x), \quad \frac{\partial z}{\partial t} = g(x), \quad \text{for} \quad t = 0, \quad 0 \leqslant x \leqslant L, \tag{8-23''}$$

where $f(x)$ and $g(x)$ are given.

As in Section 8–1, the differential equation governs the vibrations of a string. The boundary conditions (8–23') state that the ends of the string are fixed. The initial conditions (8–23'') give the initial displacement and velocity along the string.

II. The *heat equation* in one dimension:

$$c \frac{\partial^2 z}{\partial x^2} - \frac{\partial z}{\partial t} = 0, \tag{8-24}$$

where c is a positive constant, in the region $0 < x < L, 0 < t < \infty$ of the xt-plane, with boundary conditions

$$z = z_0 \quad \text{for} \quad x = 0 \quad \text{and} \quad z = z_1 \quad \text{for } x = L, \quad 0 \leqslant t < \infty, \tag{8-24'}$$

and initial conditions

$$z = f(x) \quad \text{for} \quad t = 0, \quad 0 \leqslant x \leqslant L, \tag{8-24''}$$

where z_0 and z_1 are given numbers, $f(x)$ is a given function.

The differential equation governs the variation of temperature z in a thin rod (see Section 8–6 for a derivation). The boundary conditions state that the temperatures at the ends of the rod have fixed values. The initial condition (8–24″) gives the initial temperature along the rod.

III. The *Laplace equation* in two dimensions

$$\frac{\partial^2 z}{\partial x^2} + \frac{\partial^2 z}{\partial y^2} = 0, \tag{8–25}$$

in the region $0 < x < L$, $0 < y < M$, with boundary conditions

$$
\begin{array}{llll}
z = f(x) & \text{for} \quad y = 0, & z = g(x) & \text{for} \quad y = M, \quad 0 \leqslant x \leqslant L, \\
z = p(y) & \text{for} \quad x = 0, & z = q(y) & \text{for} \quad x = L, \quad 0 \leqslant y \leqslant M,
\end{array} \tag{8–25'}
$$

where $f(x)$, $g(x)$, $p(y)$, and $q(y)$ are given functions.

Here the differential equation describes the equilibrium temperature z in a thin rectangular plate. The boundary conditions give the temperatures along the edges of the plate.

REMARK: The solutions of Eq. (8–25) are called *harmonic functions*. The left side is called the *Laplacian* of z and is denoted by $\nabla^2 z$.

The partial differential equations occurring here have generalizations for up to three (or more) space variables. For the *wave equation* with three space variables the form is

$$c^2 \left(\frac{\partial^2 w}{\partial x^2} + \frac{\partial^2 w}{\partial y^2} + \frac{\partial^2 w}{\partial z^2} \right) - \frac{\partial^2 w}{\partial t^2} = 0. \tag{8–23'''}$$

For the *heat equation* it is

$$c \left(\frac{\partial^2 w}{\partial x^2} + \frac{\partial^2 w}{\partial y^2} + \frac{\partial^2 w}{\partial z^2} \right) - \frac{\partial w}{\partial t} = 0. \tag{8–24'''}$$

For the *Laplace equation* it is

$$\frac{\partial^2 w}{\partial x^2} + \frac{\partial^2 w}{\partial y^2} + \frac{\partial^2 w}{\partial z^2} = 0. \tag{8–25''}$$

There are analogous extensions of the initial and boundary conditions to higher dimensions.

The solutions $w(x, y, z)$ of the Laplace equation (8–25″) are again called *harmonic functions* and Eq. (8–25) is written $\nabla^2 w = 0$.

The initial conditions and boundary conditions are understood in terms of limiting values of the solution as the boundary is approached. Equivalently, we can require that the solution be defined in the region plus boundary (closed region), that it be continuous in the closed region, and that it have values or derivatives (taken from one side) along the boundary that satisfy the prescribed conditions. With proper continuity requirements, each of the problems I, II, III has a *unique solution* satisfying the boundary and initial conditions. We shall not prove this, but shall explain methods for obtaining the explicit solution, often in the form of an infinite series. (For proofs of uniqueness and validity of the solutions obtained, see AC, Chapter 10, and COURANT–HILBERT.)

The three basic types of equations illustrated are of *second order*. They are also *linear*, if we extend the concept of linearity in the natural way from ordinary to partial differential equations. Thus the most general second-order linear partial differential equation in two independent variables has the form

$$A(x, y)z_{xx} + B(x, y)z_{xy} + C(x, y)z_{yy} + D(x, y)z_x + E(x, y)z_y + F(x, y)z = G(x, y). \tag{8–26}$$

Here we write z_x for $\partial z/\partial x$, z_{xy} for $\partial^2 z/\partial x \partial y$, and so on; this notation will be much used in this chapter. Equation (8–26) is said to be

$$\begin{aligned} elliptic &\quad \text{if } B^2 - 4AC < 0, \\ parabolic &\quad \text{if } B^2 - 4AC = 0, \\ hyperbolic &\quad \text{if } B^2 - 4AC > 0 \end{aligned}$$

in the region considered. Thus the wave equation (8–23) is hyperbolic (with t replacing y), since $A = c^2$, $B = 0$, $C = -1$, so that

$$B^2 - 4AC = 4c^2 > 0.$$

The heat equation (8–24) is parabolic, since $A = c$, $B = 0$, $C = 0$, so that $B^2 - 4AC = 0$. The Laplace equation (8–25) is elliptic, since $B^2 - 4AC = -4$.

This classification of second-order linear partial differential equations can be extended to more independent variables. The generalizations (8–23‴), (8–24‴), and (8–25″) are then respectively hyperbolic, parabolic, and elliptic.

The classification is important in that the type of boundary condition and nature of the solution are quite different in the three different cases. They correspond to three different classes of physical phenomena: the wave equation to vibrations and wave propagation; the heat equation to diffusion; the Laplace equation to equilibrium states. The wave equation involves a second derivative with respect to time t, that is, *acceleration*. Hence typical initial conditions involve both the value and derivative (velocity) of the unknown function. Similarly, the heat equation involves only a first derivative with respect to t, and the typical initial condition gives only the initial value of the unknown function. The Laplace equation does not involve t, and no initial values are prescribed.

In Eq. (8–26), if A, B, and C are all identically 0, the equation reduces to a linear *first-order* partial differential equation

$$D(x, y)z_x + E(x, y)z_y + F(x, y)z = G(x, y). \tag{8–27}$$

Such equations are also important in applications; for example, in chemical process control (see *Mathematical Methods in Chemical Engineering*, vol 2, by R. ARIS and N. R. AMUNDSON, Prentice-Hall, Inc., Englewood Cliffs, New Jersey, 1973).

Equations of higher order are also important: for example, the *biharmonic equation*

$$\nabla^4 w = \nabla^2(\nabla^2 w) = \frac{\partial^4 w}{\partial x^4} + 2\frac{\partial^4 w}{\partial x^2 \partial y^2} + \frac{\partial^4 w}{\partial y^4} = 0 \tag{8–28}$$

in elasticity theory. Also, *systems* of partial differential equations occur. A common example is the set of Cauchy–Riemann equations

$$\frac{\partial u}{\partial x} - \frac{\partial v}{\partial y} = 0, \qquad \frac{\partial u}{\partial y} + \frac{\partial v}{\partial x} = 0, \tag{8-29}$$

which are important for incompressible-fluid, heat conduction (equilibrium state), electrostatic fields. They are studied in Chapter 11.

8–3 THE METHOD OF SEPARATION OF VARIABLES: THE ONE-DIMENSIONAL WAVE EQUATION

We here describe a method that provides the desired solution, in series form, for a number of second-order linear partial differential equations. For the general equation (8–26) we first seek solutions of form

$$z = X(x)\, Y(y).$$

We often find that such solutions exist, where $X(x)$ and $Y(y)$ are solutions of *ordinary* second-order linear differential equations in x and y respectively. Further, we find that imposing some of the boundary conditions on z restricts one of the functions, say $X(x)$, to a sequence of possible choices: $X_1(x), X_2(x), \ldots, X_n(x), \ldots$ For each $X_n(x)$ there is then a $Y_n(y)$ such that $X_n(x)\,Y_n(y)$ satisfies the partial differential equation; here $Y_n(y)$ generally contains one or more arbitrary constants. Then the infinite series

$$\sum_{n=1}^{\infty} X_n(x)\, Y_n(y)$$

provides (under appropriate convergence conditions) a solution of the partial differential equation that satisfies some of the boundary conditions. By adjusting the arbitrary constants in the functions $Y_n(y)$, we may be able to satisfy *all* the boundary conditions and thus obtain our desired solution. This, in outline, is the method of *separation of variables*.

We now apply the method to a typical boundary-value problem for the one-dimensional wave equation. The dependent variable z is replaced by y, the independent variables are taken to be x and t.

EXAMPLE 1 $c^2 y_{xx} - y_{tt} = 0, \quad 0 < x < \pi, \quad 0 < t < \infty;$

$y(x, t) = 0 \quad$ for $x = 0$ and for $x = \pi, 0 \leqslant t < \infty;$

$y = f(x)$ and $y_t = g(x)$ for $t = 0, 0 \leqslant x \leqslant \pi.$

This problem describes the *vibrations of a string* whose ends are fixed (Fig. 8–4). This is the case of a violin string vibrating close to an equilibrium position. In general, y is the vertical displacement of the string from equilibrium at point x at time t, so that $y = y(x, t); y \equiv 0$ is the equilibrium state. As in Section 8–1, we show that, for small displacements, y obeys the one-dimensional wave equation.

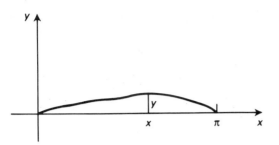

Fig. 8-4. Vibrating string.

The first set of boundary conditions ($y = 0$ for $x = 0$, $x = \pi$) states that the string is fixed at both ends ($x = 0$ and $x = \pi$). The second set of boundary conditions provides initial conditions: the initial displacement and velocity at each x. The length of the string is chosen as π for convenience, this is equivalent to a special choice of unit of length. The general case of length L can be reduced to that of length π by the substitution $x' = \pi x/L$.

Following our program above, we now seek solutions (not identically 0) of form $y = X(x)T(t)$. Substitution in the differential equation gives

$$c^2 X''(x)T(t) - X(x)T''(t) = 0 \qquad \text{or} \qquad \frac{X''}{X} - \frac{1}{c^2}\frac{T''}{T} = 0.$$

Here the first term depends only on x, the second only on t. This is possible only if each term is a constant (otherwise, by varying x, for example, we could change the value of the first term, while the second term is unchanged, and this would destroy the validity of the equation). We are thus led to two ordinary differential equations in x and t, respectively:

$$\frac{X''}{X} = -\lambda, \qquad -\frac{1}{c^2}\frac{T''}{T} = \lambda,$$

where λ is a constant to be determined. The first equation can be written as

$$X'' + \lambda X = 0. \tag{8-30}$$

Now $y = X(x)T(t)$ will satisfy the first set of boundary conditions if

$$X(0) = 0 \qquad \text{and} \qquad X(\pi) = 0. \tag{8-31}$$

We now try to find all solutions $X(x)$, not identically 0, of Eq. (8-30) satisfying Eqs. (8-31). Here λ is at our disposal and we may succeed only for certain values of λ. For $\lambda = 0$, Eq. (8-30) becomes $X'' = 0$: the general solution of this equation is $X = c_1 x + c_2$.

The condition (8-31) gives

$$c_2 = 0, \qquad c_1 \pi + c_2 = 0,$$

so that $c_1 = 0$ and $c_2 = 0$, and hence $X(x) \equiv 0$, which we exclude. Thus $\lambda = 0$ is ruled out. We find out similarly that for $\lambda < 0$ there is only the solution $X(x) \equiv 0$

(Problem 5(a) below). For $\lambda > 0$, we can write $\lambda = \omega^2$ ($\omega > 0$) and Eqs. (8–30), (8–31) become

$$X'' + \omega^2 X = 0, \qquad X(0) = 0, \qquad X(\pi) = 0.$$

The general solution of the differential equation (see Section 1–11) is

$$X = c_1 \cos \omega x + c_2 \sin \omega x.$$

The boundary conditions give

$$c_1 = 0, \qquad c_1 \cos \omega \pi + c_2 \sin \omega \pi = 0.$$

Thus we avoid an identically zero solution by choosing ω so that $\sin \omega \pi = 0$, that is,

$$\omega = 1, \quad \omega = 2, \quad \ldots, \quad \omega = n, \quad \ldots$$

Accordingly, for $\lambda = \lambda_n = n^2$, we have the solution

$$X = X_n(x) = \sin nx, \quad n = 1, 2, \ldots$$

For this choice of λ, function $T(t)$ must satisfy the equation

$$T'' + c^2 \lambda_n T = 0 \qquad \text{or} \qquad T'' + c^2 n^2 T = 0.$$

Hence

$$T = T_n(t) = p_n \cos nct + q_n \sin nct, \tag{8–32}$$

and for each choice of the constants p_n, q_n (not both zero),

$$y = X_n(x) T_n(t) = \sin nx (p_n \cos nct + q_n \sin nct), \quad n = 1, 2, \ldots,$$

is a solution of the partial differential equation, not identically 0, satisfying the first set of boundary conditions. The sum of solutions of the differential equation is also a solution (as follows from linearity). Accordingly, the series

$$y(x, t) = \sum_{n=1}^{\infty} X_n(x) T_n(t) = \sum_{n=1}^{\infty} (p_n \sin nx \cos nct + q_n \sin nx \sin nct) \tag{8–33}$$

(under appropriate convergence conditions) should also be a solution; furthermore, it also satisfies the first set of boundary conditions.

We finally adjust the constants p_n and q_n to satisfy the second set of boundary conditions (initial conditions): $y(x, 0) = f(x)$ and $y_t(x, 0) = g(x)$ for $0 \leqslant x \leqslant \pi$. By Eq. (8–33) we must have

$$\sum_{n=1}^{\infty} p_n \sin nx = f(x), \qquad \sum_{n=1}^{\infty} ncq_n \sin nx = g(x), \quad 0 \leqslant x \leqslant \pi. \tag{8–34}$$

Here the left-hand side of the first equation is obtained by setting $t = 0$ in Eq. (8–33) and that of the second equation is obtained by differentiating the series term by term with respect to t and then setting $t = 0$. In Eqs. (8–34), the first equation requires that the p_n are the coefficients in the expansion of $f(x)$ in a *Fourier sine series* (Section 3–4). Thus

$$p_n = \frac{2}{\pi} \int_0^{\pi} f(x) \sin nx \, dx, \quad n = 1, 2, \ldots \tag{8–35}$$

and similarly

$$ncq_n = \frac{2}{\pi} \int_0^\pi g(x) \sin nx\, dx, \quad n = 1, 2, \ldots \tag{8–36}$$

Accordingly, *the solution of the boundary-value problem of* Example 1 *is given by* Eq.(8–33), *where the* p_n *and* q_n *are given by* Eqs. (8–35), (8–36), provided the series converges and appropriate term-by-term differentiation is allowed. The conditions for validity of the solution are met if $f(x)$ and $g(x)$ are sufficiently smooth, say $f^{(iv)}(x)$ and $g'''(x)$ are continuous and $f(0) = f(\pi) = 0$, $g(0) = g(\pi) = 0$ (see AC, pp. 664–667). ◄

For the method of separation of variables the form of the boundary conditions is crucial. We observe that in Example 1 the boundary conditions relating to $x = 0$ and $x = \pi$ are *homogeneous*, that is, they could be satisfied by the identically zero solution of the partial differential equation. On the other hand, the boundary conditions relating to $t = 0$ are not homogeneous. The partial differential equation of Example 1 would also be called homogeneous, as would every equation of form (8–26) with $G(x, y) = 0$.

EXAMPLE 2 $c^2 y_{xx} - y_{tt} = 0, \quad 0 < x < \pi, \quad 0 < t < \infty;$

$$y_x(0, t) = 0 \quad \text{and} \quad y_x(\pi, t) = 0, \quad 0 \le t < \infty;$$

$$y(x, 0) = f(x) \quad \text{and} \quad y_t(x, 0) = g(x), \quad 0 \le x \le \pi.$$

This corresponds to a vibrating string whose ends at $x = 0$ and $x = \pi$ are not fixed and can move, but only so as to have zero slope at the ends, as in Fig. 8–5. We set $y = X(x)T(t)$ and again are led to the equations

$$X'' + \lambda X = 0, \qquad T'' + \lambda c^2 T = 0.$$

In order to satisfy the boundary conditions for $x = 0$ and $x = \pi$ we must require that

$$X'(0) = 0, \qquad X'(\pi) = 0.$$

We verify that for λ negative these conditions can be satisfied only by $X(x) \equiv 0$ (see Problem 5(b) below). For $\lambda = 0$, the conditions on $X(x)$ are satisfied by

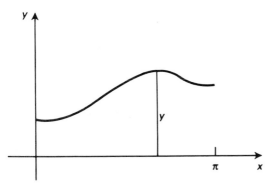

Fig. 8–5. Vibrating-string problem of Example 2.

$X(x) = $ const, say by $X_0(x) \equiv 1$. For λ positive, we can write $\lambda = \omega^2$ with $\omega > 0$ and

$$X(x) = c_1 \cos \omega x + c_2 \sin \omega x.$$

The conditions at $x = 0$ and $x = \pi$ lead us to the equations

$$c_2 = 0, \qquad -c_1 \sin \omega \pi + c_2 \cos \omega \pi = 0.$$

Thus the possibilities are $\omega = 1, 2, \ldots, n, \ldots$ Corresponding to $\omega = n$, we obtain

$$X = X_n(x) = \cos nx, \qquad \lambda = \lambda_n = n^2.$$

This is correct for $n = 0$ also, by our choice above for $\lambda = 0$. To determine $T_n(t)$ for $\lambda = \lambda_n$ we have the equation

$$T'' + n^2 c^2 T = 0, \quad n = 0, 1, 2, \ldots .$$

Accordingly,

$$T_0(t) = \frac{p_0}{2} + \frac{q_0}{2} t, \qquad T_n(t) = p_n \cos nct + q_n \sin nct \quad \text{for} \quad n \geq 1.$$

(The notation for T_0 is chosen to simplify later formulas.) We have thus obtained our solutions $X_n(t) T_n(t)$ and add them to get

$$y = \sum_{n=0}^{\infty} X_n(t) T_n(t) = \frac{p_0}{2} + \frac{q_0}{2} t + \sum_{n=1}^{\infty} (p_n \cos nx \cos nct + q_n \cos nx \sin nct).$$

$$(8\text{--}37)$$

This satisfies the partial differential equation and the boundary conditions at $x = 0$, $x = \pi$ (if we assume convergence as in Example 1). To satisfy the boundary conditions for $t = 0$ we must have

$$\frac{p_0}{2} + \sum_{n=1}^{\infty} p_n \cos nx = f(x), \qquad \frac{q_0}{2} + \sum_{n=1}^{\infty} ncq_n \cos nx = g(x).$$

Here we are dealing with Fourier cosine series and thus, as in Section 3–4,

$$p_n = \frac{2}{\pi} \int_0^\pi f(x) \cos nx \, dx, \quad n = 0, 1, 2, \ldots ,$$

$$(8\text{--}38)$$

$$q_0 = \frac{2}{\pi} \int_0^\pi g(x) \, dx, \qquad ncq_n = \frac{2}{\pi} \int_0^\pi g(x) \cos nx \, dx \quad \text{for} \quad n \geq 1.$$

With these choices *the series (8–37) gives the solution of the boundary-value problem of Example 2* (with the same conditions for convergence as in Example 1).

===== **Problems (Section 8–3)** =====

1. Verify that the given function is a solution of the given partial differential equation. The function $f(u)$ is assumed differentiable in some interval.

a) $z = 3x - 2y$ for $2z_x + 3z_y = 0$ b) $z = e^{x-y}$ for $z_x + z_y = 0$

c) $z = x^2 - y^2$ for $z_{xx} + z_{yy} = 0$ d) $z = e^x \cos y$ for $z_{xx} + z_{yy} = 0$

e) $z = f(x + y)$ for $z_x - z_y = 0$ f) $z = f(x - t)$ for $z_{xx} - z_{tt} = 0$

2. State whether the given second-order equation is hyperbolic, parabolic, or elliptic:

a) $5z_{xx} + 7z_{yy} = 0$ b) $3z_{xx} - 2z_{yy} = 0$

c) $4z_{xx} + z_y = 0$ d) $z_{xx} + 3z_{xy} - z_{yy} = 0$

e) $z_{xx} + z_{xy} + z_{yy} = 0$

f) $y^2 z_{xx} + xyz_{xy} + x^2 z_{yy} = 0$ for $x > 0$, $y > 0$

g) $e^{2y} z_{xx} + e^{x+y} z_{xy} + e^{2x} z_{yy} + z_x - z_y = 0$

3. Carry out the solution by separation of variables in the boundary-value problem of Example 1 for the wave equation in the following special cases:

a) $c = 1$, $f(x) = \sin^3 x$, $g(x) = 0$

b) $c = 1$, $f(x) = \pi x - x^2$, $g(x) = 0$

c) $c = 1$, $f(x) = 0$, $g(x) = x \sin x$

d) $c = 1$, $f(x) = \sin^3 x$, $g(x) = x \sin x$

4. Carry out the solution by separation of variables in the boundary-value problem of Example 2 for the wave equation in the following special cases:

a) $c = 1$, $f(x) = \cos^2 x$, $g(x) = 0$ b) $c = 1$, $f(x) = 0$, $g(x) = x^2 (\pi - x)^2$

5. Show that $X(x) \equiv 0$ is the only solution of the boundary-value problem for $\lambda < 0$:

a) $X'' + \lambda X = 0$, $X(0) = X(\pi) = 0$ b) $X'' + \lambda X = 0$, $X'(0) - X'(\pi) = 0$

HINT: Write $\lambda = -k^2$, with $k > 0$, so that the solutions of the differential equation have the form $X = c_1 e^{kt} + c_2 e^{-kt}$.

6. *Longitudinal vibrations* As remarked in Section 8–1, the wave equation also describes longitudinal vibrations of a thin rod (also of an air column). To obtain this as a limit of a problem for particles, consider n equal particles on a line, coupled by identical springs, as in Fig. 8–6. At equilibrium all particles are equally spaced at a distance $\Delta x = L/(n+1)$. Let z_j be the displacement of the jth particle from equilibrium along the x-axis (measured positively to the right). Show that the motions are governed by the differential equations

$$m \frac{d^2 z_j}{dt^2} = k(z_{j+1} - 2z_j + z_{j-1}), \quad j = 1, \ldots, n,$$

where $z_0 = 0$, $z_{n+1} = 0$. Let $n \to \infty$, $m \to 0$, $\Delta x \to 0$; obtain the wave equation (8–1) as a limit.

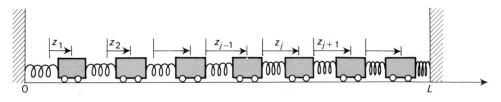

Fig. 8–6. Model for longitudinal vibrations.

8–4 SOLUTIONS OF THE WAVE EQUATION; TRAVELING AND STANDING WAVES

We reexamine our solution (8–33) to the boundary-value problem of Example 1 in Section 8–3 and first assume that $g(x) \equiv 0$, so that $q_n = 0$ for all n. Hence

$$y(x, t) = \sum_{n=1}^{\infty} p_n \sin nx \cos nct, \qquad (8\text{–}40)$$

where

$$f(x) = \sum_{n=1}^{\infty} p_n \sin nx. \qquad (8\text{–}41)$$

Now we can write Eq. (8–40) in the form

$$y = \sum_{n=1}^{\infty} \frac{1}{2} p_n [\sin(nx + nct) + \sin(nx - nct)]$$

$$= \frac{1}{2} \sum_{n=1}^{\infty} p_n [\sin n(x + ct) + \sin n(x - ct)],$$

as follows by trigonometry. By Eq. (8–41), we can now also write

$$y = \frac{1}{2}[f(x + ct) + f(x - ct)]. \qquad (8\text{–}42)$$

Here we are using Eq. (8–41) to represent $f(x)$ for all x as the odd periodic extension of the given function (Section 3–4).

Equation (8–42) represents our solution as a sum of two *traveling waves*: one going to the left, the other to the right, as t increases (Fig. 8–7). The wave velocity is c.

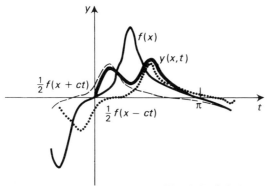

Fig. 8–7. Solution to wave equation as sum of traveling waves.

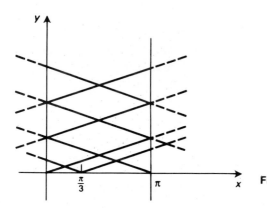

Fig. 8–8. Characteristics of one-dimensional wave equation.

For each of the two traveling waves, we can also follow the wave motion by observing that in the xt-plane $(1/2)f(x+ct)$ is constant along each line $x+ct =$ const and, similarly, $(1/2)f(x-ct)$ is constant along each line $x-ct =$ const. Thus there are two families of parallel lines, called *characteristics*, along which the waves are propagated (Fig. 8–8). We can assume that $f(x)$ is chosen to have its maximum at $x = \pi/3$, and hence the peaks of the traveling waves move along the characteristics $x \pm ct = \pi/3$. By noting the significance of the odd periodic extension of f, we can see that, when each of these characteristics meets $x = 0$ or $x = \pi$, the peak gradually disappears and then seems to reappear with a reversal of sign and going in the opposite direction. Thus there is a reflection, with reversal of sign, at each boundary. This effect can be observed experimentally when a violin string is released from a stretched position.

If we now consider Example 1 of Section 8–3 with $f(x) \equiv 0$ and $g(x) \neq 0$, then similar reasoning shows that the solution can be written as

$$y = \frac{1}{2c}[G(x+ct) - G(x-ct)], \qquad (8\text{--}43)$$

where $G'(x) = g(x)$, so that $G(x)$ is an indefinite integral of $g(x)$:

$$g(x) = \sum_{n=1}^{\infty} ncq_n \sin nx, \qquad G(x) = -\sum_{n=1}^{\infty} cq_n \cos nx,$$

both functions being periodic. By Eq. (8–43), the solution is the *difference* of two traveling waves.

For general f and g, we verify that the solution is the sum of the solutions (8–42) and (8–43):

$$y = \frac{1}{2}[f(x+ct) + f(x-ct)] + \frac{1}{2c}[G(x+ct) - G(x-ct)]. \qquad (8\text{--}44)$$

This shows that the *general solution* of the wave equation can be written in the form

$$y = P(x+ct) + Q(x-ct), \qquad (8\text{--}45)$$

and we can verify that every function (8–45), where P and Q have continuous second derivatives, is a solution of the wave equation (see Problem 4 below).

Representation (8–45) shows that each solution of the wave equation is a sum of two traveling waves: one moving to the left, one moving to the right.

We observe that there are also *standing-wave solutions*. For example, each term of the solution (8–33) can be written as

$$y_n(x, t) = A_n \sin (nct + \alpha_n) \sin nx. \tag{8–46}$$

This term is itself a solution of the wave equation: it is precisely the solution $X_n(x)T_n(t)$ we obtained by separating the variables. For each t, function $y_n(x, t)$ is a constant times $\sin nx$; as t varies, the "constant" varies sinusoidally between A_n and $-A_n$ with a frequency of nc. Typical shapes are shown in Fig. 8–9.

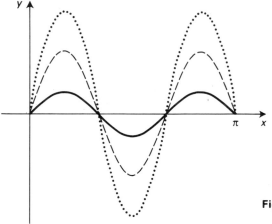

Fig. 8–9. Standing-wave solution (8–36) graphed for $n = 3$.

These solutions are familiar to violinists. The wave motion (8–46) for $n = 1$, with time frequency c, corresponds to the *fundamental tone* of the violin string; that for $n = 2$, with frequency $2c$, to the *first overtone* (octave); those for $n = 3, 4, \ldots$ to the higher overtones. The general solution (8–33) is a superposition of fundamental tone and overtones.

REMARK: The standing-wave solutions can be regarded as *normal modes* of motion of the string. If the string is approximated by a system of n particles, as in Section 8–1, then we obtain a mechanical system with n normal modes that approximate those of the string, in fact, the first n frequencies for the string are well approximated by the n frequencies of the normal modes (see AC, Chapter 10).

8–5 OTHER BOUNDARY CONDITIONS FOR THE WAVE EQUATION

The method of separation of variables can be applied to the wave equation with a variety of boundary conditions. For example, the boundary conditions may be of the form

$$\alpha_1 y(0, t) + \alpha_2 y_x(0, t) = 0, \qquad \beta_1 y(\pi, t) + \beta_2 y_x(\pi, t) = 0,$$
$$y(x, 0) = f(x), \qquad y_t(x, 0) = g(x), \tag{8–50}$$

where $\alpha_1, \alpha_2, \beta_1, \beta_2$ are constants and $\alpha_1^2 + \alpha_2^2 \neq 0, \beta_1^2 + \beta_2^2 \neq 0$. These correspond to the physical requirement that each end of the string is free to move when subjected to a spring-like restoring force (see Problem 9 below).

The method of separation of variables leads to the problem

$$X'' + \lambda X = 0, \; \alpha_1 X(0) + \alpha_2 X'(0) = 0, \; \beta_1 X(\pi) + \beta_2 X'(\pi) = 0. \tag{8–51}$$

This is a Sturm–Liouville problem and, as pointed out in Section 3 14, the solutions (eigenfunctions) are determined up to constant factors and form a complete orthogonal system $\{\psi_n(x)\}$ on the interval $0 \leqslant x \leqslant \pi$. Therefore the procedures followed in Examples 1 and 2 of Section 8–3 can be applied and a unique solution to the problem (8–50) can be obtained, based on expansions in Fourier series with respect to the orthogonal system $\{\psi_n(x)\}$. For Examples 1 and 2 the systems are, respectively,

$$\{\sin nx\}, n = 1, 2, \ldots, \qquad \text{and} \qquad \{\cos nx\}, n = 0, 1, 2, \ldots$$

Similar remarks apply to more general boundary conditions

$$\alpha_1 y(0, t) + \alpha_2 y_x(0, t) + \alpha_3 y(\pi, t) + \alpha_4 y_x(\pi, t) = 0,$$
$$\beta_1 y(0, t) + \beta_2 y_x(0, t) + \beta_3 y(\pi, t) + \beta_4 y_x(\pi, t) = 0, \tag{8–52}$$
$$y(x, 0) = f(x), \qquad y_t(x, 0) = g(x),$$

corresponding to Eqs. (3–144″) in Section 3–14 (see the references cited in that section).

All these problems have homogeneous boundary conditions for $x = 0$ and $x = \pi$. Nonhomogeneous boundary conditions may appear, as in the following problem:

$$c^2 y_{xx} - y_{tt} = 0,$$
$$y(0, t) = p(t), \qquad y(\pi, t) = q(t), \tag{8–53}$$
$$y(x, 0) = f(x), \qquad y_t(x, 0) = g(x).$$

This corresponds to a vibrating string (Figs. 8–1 and 8–4) whose ends are being forced to move in a definite manner, with position and velocity given for $t = 0$. The method of separation of variables is not successful for these boundary conditions (see Problem 5 below). However, as we shall see, the solution can be obtained in a related form.

We first consider a problem involving a *nonhomogeneous wave equation*:

$$y_{tt} - c^2 y_{xx} = F(x, t), \quad 0 < x < \pi, \quad 0 < t < \infty,$$
$$y(0, t) = y(\pi, t) = 0, \qquad y(x, 0) = f_1(x), \qquad y_t(x, 0) = g_1(x). \tag{8–54}$$

Here our string is being forced to move by an external force F varying with time and with position along the string. Motivated by the form of the solution when $F(x, t) = 0$ (see Eq. (8–33)), we seek a solution of form

$$y = \sum_{n=1}^{\infty} h_n(t) \sin nx. \tag{8–55}$$

This will satisfy the homogeneous boundary conditions for $x = 0$, $x = \pi$. It will also satisfy the boundary conditions for $t = 0$ if, for each n,

$$h_n(0) = b_n = \frac{2}{\pi} \int_0^{\pi} f_1(x) \sin nx \, dx, \qquad h'_n(0) = \beta_n = \frac{2}{\pi} \int_0^{\pi} g_1(x) \sin nx \, dx.$$
$$\tag{8–56}$$

If we substitute (8–55) in the partial differential equation in (8–54), we obtain the equation

$$\sum_{n=1}^{\infty} \left[h''_n(t) + c^2 n^2 h_n(t) \right] \sin nx = F(x, t).$$

Thus for each fixed t the left side is to be the Fourier sine series expansion of $F(x, t)$. Hence, for each n,

$$h''_n(t) + c^2 n^2 h_n(t) = F_n(t) = \frac{2}{\pi} \int_0^{\pi} F(x, t) \sin nx \, dx.$$

By (8–56) we now have an initial-value problem for $h_n(t)$:

$$h''_n + n^2 c^2 h_n(t) = F_n(t), \quad 0 < t < \infty, \quad h_n(0) = b_n, \quad h'_n(0) = \beta_n.$$

This can be solved by variation of parameters or by Laplace transforms (see Section 4–11 and Problem 3 following Section 1–13). We find

$$h_n(t) = b_n \cos cnt + \frac{\beta_n}{cn} \sin cnt + \frac{1}{cn} \int_0^t F_n(t - u) \sin cnu \, du$$

and thus

$$y = \sum_{n=1}^{\infty} \left[b_n \cos cnt + \frac{\beta_n}{cn} \sin cnt + \frac{1}{cn} \int_0^t F_n(t - u) \sin cnu \, du \right] \sin nx$$

is the desired solution of the problem (8–54).

Now return to the problem of forced motion, Eqs. (8–53). We reduce this to a problem of form (8–54) by the following procedure. Let

$$\psi(x, t) = p(t)\left(1 - \frac{x}{\pi}\right) + \frac{x}{\pi} q(t).$$

The function ψ satisfies the boundary conditions for $x = 0$, $x = \pi$:

$$\psi(0, t) = p(t), \qquad \psi(\pi, t) = q(t).$$

We then make a *change of dependent variable* in the problem, letting

$$v = y - \psi(x, t).$$

From Eqs. (8–53), the new problem for v is then

$$v_{tt} - c^2 v_{xx} = -\psi_{tt} + c^2 \psi_{xx} = -p''(t)\left(1 - \frac{x}{\pi}\right) - \frac{x}{\pi} q''(t),$$

$$v(0, t) = 0, \quad v(\pi, t) = 0,$$

$$v(x, 0) = f(x) - \psi(x, 0) = f(x) - p(0)\left(1 - \frac{x}{\pi}\right) - \frac{x}{\pi} q(0), \qquad (8\text{–}57)$$

$$v_t(x, 0) = g(x) - \psi_t(x, 0) = g(x) - p'(0)\left(1 - \frac{x}{\pi}\right) - \frac{x}{\pi} q'(0).$$

This problem is of the form (8–54), with

$$F(x, t) = -p''(t)\left(1 - \frac{x}{\pi}\right) - \frac{x}{\pi} q''(t),$$

$$f_1(x) = f(x) - p(0)\left(1 - \frac{x}{\pi}\right) - \frac{x}{\pi} q(0),$$

$$g_1(x) = g(x) - p'(0)\left(1 - \frac{x}{\pi}\right) - \frac{x}{\pi} q'(0).$$

Hence $v(x, t)$ can be found as above as $\sum h_n(t) \sin nx$ and

$$y = v + \psi(x, t) = \sum_{n=1}^{\infty} h_n(t) \sin nx + \psi(x, t)$$

is the desired solution of problem (8–53).

The same procedure is applicable to the nonhomogeneous equation $y_{tt} - c^2 y_{xx} = F(x, t)$ with nonhomogeneous boundary conditions as in (8–52) or as follows:

$$\alpha_1 y(0, t) + \alpha_2 y_x(0, t) = p(t), \qquad \beta_1 y(\pi, t) + \beta_2 y_x(\pi, t) = q(t),$$
$$y(x, 0) = f(x), \qquad y_t(x, 0) = g(x), \qquad (8\text{–}58)$$

obtained from Eqs. (8–50) by introducing $p(t), q(t)$ on the right sides, or for boundary conditions obtained from Eqs. (8–52) by similarly introducing $p(t), q(t)$. In each case we first solve the related problem analogous to (8–54) by a series $y = \sum h_n(t) \varphi_n(x)$, where the $\varphi_n(x)$ are the eigenfunctions obtained from the corresponding Sturm–Liouville boundary-value problem (for homogeneous boundary conditions at $x = 0$ and $x = \pi$). The substitution $v = y - \psi(x, t)$, for appropriate ψ, reduces the given problem to a related problem. Here $\psi(x, t)$ is chosen to satisfy the given boundary conditions at $x = 0$ and $x = \pi$; often ψ can be chosen to be linear in x: $\psi = a(t)x + b(t)$.

Superposition of solutions

If we consider the problems with nonhomogeneous wave equation $y_{tt} - c^2 y_{xx} = F(x, t)$ and nonhomogeneous boundary conditions as in (8–58) (with fixed $\alpha_1, \alpha_2, \beta_1, \beta_2$), then a superposition principle is valid. If $y_1(x, t), y_2(x, t)$ are the solutions corresponding to F_1, p_1, q_1, f_1, g_1 and F_2, p_2, q_2, f_2, g_2 as right-hand members of the respective equations, then $c_1 y_1(x, t) + c_2 y_2(x, t)$, for constant c_1, c_2, is the solution corresponding to right-hand members

$$c_1 F_1(x, t) + c_2 F_2(x, t), \quad c_1 p_1(t) + c_2 p_2(t), \quad \ldots$$

This follows at once from the linearity of the operations concerned. Hence, for a given problem with given $F(x, t), p(t), q(t), f(x), g(x)$, the solution can be obtained as the sum of the *five* solutions of *five* separate problems: one with the given $F(x, t)$ and with $p(t), \ldots, g(x)$ all replaced by 0; one with the given $p(t)$ and with $F(x, t), q(t), \ldots, g(x)$ all replaced by 0, and so on.

=========================**Problems (Section 8–5)**=========================

1. Let $f(x)$ be odd and have period 2π; let $f(x) = x$ for $0 \leqslant x \leqslant \pi/2$ and $f(x) = \pi - x$ for $\pi/2 \leqslant x \leqslant \pi$. In Eq. (8–42) take $c = \pi/2$ and graph the terms $(1/2)f(x + ct), (1/2)f(x - ct)$ and their sum for the following values of t: $0, \pi/4, \pi/2, 3\pi/4, \pi$. Also show the lines $(1/2)f(x \pm ct) = \text{const} = k$ (characteristics) in the xt-plane.

2. Proceed as in Problem 1, using $f(x, t) = x(\pi - x)$ for $0 \leqslant x \leqslant \pi$ and $c = 1$.

3. Graph $y = \sin 2t \sin 2x$ (standing wave) for $t = 0, \pi/8, \pi/4, \pi/2, 3\pi/4, \pi$.

4. Verify that every suitably differentiable function of form (8–45) is a solution of the wave equation $c^2 y_{xx} - y_{tt} = 0$.

5. Try to solve the boundary-value problem (8–53) by separation of variables and explain why the process does not work.

6. Use separation of variables to find the solution of each boundary-value problem (c is a positive constant):
   a) $c^2 y_{xx} - y_{tt} = 0$, $0 < x < 2\pi$, $0 < t < \infty$; $y(0, t) = y(2\pi, t)$ and $y_x(0, t) = y_x(2\pi, t)$; $y(x, 0) = f(x)$ and $y_t(x, 0) = g(x)$.
   b) $c^2 y_{xx} - y_{tt} = 0$, $0 < x < \pi$, $0 < t < \infty$; $y(0, t) = 0$ and $y_x(\pi, t) = 0$; $y(x, 0) = f(x)$ and $y_t(x, 0) = g(x)$.
   c) $c^2 y_{xx} - y_{tt} = 0$, $0 < x < 1$, $0 < t < \infty$; $y(0, t) = 0$ and $y(1, t) + y_x(1, t) = 0$; $y(x, 0) = f(x)$, $y_t(x, 0) = g(x)$.
   HINT: See Problem 6(c) following Section 3–15.

7. Find the solution of the boundary-value problem: $y_{tt} - c^2 y_{xx} = \sum_{n=1}^{\infty} e^{-n-n^2 t} \sin nx$, $0 < x < \pi$, $0 < t < \infty$; $y(0, t) = y(\pi, t) = 0$; $y(x, 0) = \sum_{n=1}^{\infty} n^{-3} \sin nx$, $y_t(x, 0) = 0$. Here c is a positive constant.

8. Find the solution of the boundary-value problem: $c^2 y_{xx} - y_{tt} = 0, 0 < x < \pi, 0 < t < \infty$; $y(0, t) = \sin t$, $y(\pi, t) = 2 \sin t$; $y(x, 0) = 0$, $y_t(x, 0) = 1 + (x/\pi)$.

HINT: Assume that c is a positive constant, with c^{-1} not an integer.

9. In the vibrating-string model of Fig. 8–3 let there be a particle of mass m at $x = 0$ with coordinate z_0, free to move on the z-axis. Let this particle be attached to the one at x_1 by a spring as previously, but let it also be attached to the origin by a spring of spring constant k_1. Show that the approximations used in Section 8–1 lead to the differential equation for z_0:

$$m \frac{d^2 z_0}{dt^2} = k_0(z_1 - z_0) - k_1 z_0.$$

Write this as

$$\Delta x \frac{m}{\Delta x}\frac{d^2 z_0}{dt^2} = k_0 \Delta x \frac{z_1 - z_0}{\Delta x} - k_1 z_0$$

and pass to the limit as in Section 8–1, assuming that $m/\Delta x$ and $k_0 \Delta x$ have limits ρ and K, correspondingly. Conclude that for the wave equation we have the boundary condition $K z_x - k_1 z = 0$ at $x = 0$.

8–6 THE HEAT EQUATION IN ONE DIMENSION

We consider a typical problem for the heat equation in one dimension:

$$c u_{xx} - u_t = 0, \quad 0 < x < \pi, \quad 0 < t < \infty,$$
$$u(0, t) = 0, \; u(\pi, t) = 0, \qquad u(x, 0) = f(x), \tag{8–60}$$

where c is a positive constant. It can be shown that the solution $u(x, t)$ describes the temperature variation u along a thin rod of length π (Fig. 8–10), whose ends are kept at constant temperature 0 (this is an arbitrary reference temperature), and such that heat is exchanged with the environment only at the ends of the rod. The initial temperature along the rod is given by $f(x)$. (As noted near the end of Section 8–2, the initial u but not the initial u_t are given, since the partial-differential equation has only a first derivative with respect to t.)

$u = 0$ $u = 0$

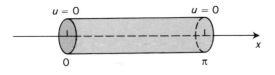

0   π **Fig. 8–10.** Heat-conduction problem.

One way of obtaining the heat equation is to consider the rod as formed of n small pieces, each of which has a definite temperature at any time t, and then let $n \to \infty$, as for the vibrating-string model in Section 8–1 (Fig. 8–11). Let u_j be the temperature in the jth piece ($j = 1, \ldots, n$). We assume that heat is exchanged with the adjacent pieces only, in accordance with the law that the heat loss per unit time from one piece to another is proportional to the temperature difference. The heat loss then causes a corresponding drop in temperature. Thus one is led to the equations

$$\frac{du_j}{dt} = -k(u_j - u_{j+1}) - k(u_j - u_{j-1}), \quad j = 1, \ldots, n,$$

where k is a positive constant; the first term on the right corresponds to the heat loss to the piece to the right; the second term corresponds to the heat loss to the piece to the left. Thus

$$\frac{du_j}{dt} = k(u_{j+1} - 2u_j + u_{j-1}), \quad j = 1, \ldots, n,$$

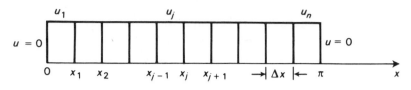

Fig. 8-11. Model for heat equation.

with $u_0 = 0$, $u_{n+1} = 0$. If we write this as

$$\frac{du_j}{dt} = k\overline{\Delta x}^2 \frac{u_{j+1} - 2u_j + u_{j-1}}{\Delta x^2}$$

and let $\Delta x \to 0$, $n \to \infty$, assuming that $k\overline{\Delta x}^2$ has a constant limit c, we obtain the heat equation (8–60).

We seek the solution of the heat conduction problem (8–60). The method of separation of variables is again successful. We seek nonzero solutions of the partial differential equation of the form $u = X(x)T(t)$ and find successively:

$$cX''T - XT' = 0, \qquad \frac{X''}{X} - \frac{1}{c}\frac{T'}{T} = 0,$$

$$X'' + \lambda X = 0, \qquad T' + c\lambda T = 0, \qquad \lambda = \text{const.}$$

We adjust the constant λ so that $X(x)$ can be chosen to satisfy the boundary conditions: $X(0) = 0$, $X(\pi) = 0$. This is the same problem as that encountered in Example 1 of Section 8–3 for the wave equation. Hence we obtain the same solutions $X_n(x) = \sin nx$ and $\lambda_n = n^2$ $(n = 1, 2, \ldots)$. For each n, $T_n(t)$ must then satisfy the equation

$$T' + cn^2 T = 0,$$

so that

$$T_n(t) = p_n e^{-cn^2 t}, \qquad n = 1, 2, \ldots,$$

where p_n is an arbitrary constant. Hence, if we assume convergence as required,

$$u = \sum_{n=1}^{\infty} p_n e^{-cn^2 t} \sin nx \qquad (8\text{–}61)$$

is a solution of the heat equation and $u(0, t) = 0$, $u(\pi, t) = 0$ as required. It remains to satisfy the initial condition $u(x, 0) = f(x)$. This leads us to the condition

$$\sum_{n=1}^{\infty} p_n \sin nx = f(x), \qquad 0 \leqslant x \leqslant \pi,$$

so that the p_n are the Fourier sine coefficients of f:

$$p_n = \frac{2}{\pi} \int_0^\pi f(x) \sin nx \, dx, \qquad n = 1, 2, \ldots \qquad (8\text{–}62)$$

With the p_n so chosen, Eq. (8–61) gives the required solution. The convergence conditions are satisfied if, for example, $f''(x)$ is continuous and $f(0) = 0$, $f(\pi) = 0$; furthermore, under these conditions, the solution is unique (see AC, pp. 671–673).

Because of the factors e^{-cn^2t}, the series (8–61) converges very rapidly for $t > 0$ and can in fact be differentiated arbitrarily many times with respect to x and t. Thus, even though $f(x) = u(x, 0)$ may not be very smooth, the solution $u(x, t)$ is exceptionally smooth for $t > 0$. This corresponds to common experience with temperatures: irregularities in temperature distribution are quickly smoothed out. This description contrasts sharply with that for the wave equation for which the initial disturbances are propagated (to the left and right) with no change, i.e., there is no smoothing (Section 8–4).

Another way of looking at the smoothing property is to observe that even if $f(x) = u(x, 0)$ is 0 except near one point x_0, function $u(x, t)$ will generally fail to be 0 for $t > 0$, as suggested in Fig. 8–12. Thus the disturbance is propagated instantaneously to all points of the rod. This property follows from the fact that, for fixed positive t, $u(x, t)$ is *analytic* in x (Section 2–16). (For a discussion of this question see AC, pp. 673–674.)

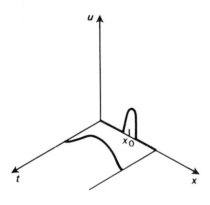

Fig. **8–12.** Smoothness of solutions
of heat equation.

It is also useful to estimate the magnitude of the temperature. Let $|f(x)| \leqslant M$ for $0 \leqslant x \leqslant \pi$. Then from Eq. (8–62):

$$|p_n| \leqslant \frac{2}{\pi} \int_0^\pi |f(x)| \, |\sin nx| \, dx \leqslant \frac{2}{\pi} \int_0^\pi M \, dx = 2M.$$

From (8–61) we can now obtain an upper estimate for $|u(x, t)|$ as the sum of the absolute values of terms in the series, Since

$$|e^{-cn^2t}| = e^{-cn^2t} \leqslant e^{-cnt} \quad \text{for} \quad n = 1, 2, \dots$$

and $|\sin nx| \leqslant 1$, we find for $t > 0$

$$|u(x, t)| \leqslant 2M \sum_{n=1}^{\infty} e^{-cnt} = 2M \frac{e^{-ct}}{1 - e^{-ct}}. \tag{8–63}$$

In the last step we took advantage of the fact that the series is a geometric series with the first term missing and the ratio $r = e^{-ct}$ and

$$\sum_{n=1}^{\infty} r^n = r \sum_{n=0}^{\infty} r^n = \frac{r}{1-r}, \quad -1 < r < 1,$$

as in Section 2–5. From (8–63) we conclude that the *temperature decays exponentially to 0 as* $t \to +\infty$. This conclusion is not surprising, since the ends of the rod are kept at temperature 0 and the only contact with the environment is at the ends.

8–7 OTHER PROBLEMS FOR THE HEAT EQUATION

As for the wave equation, we can vary the boundary conditions. Other homogeneous boundary conditions for $x = 0$ and $x = \pi$ lead to Sturm–Liouville boundary-value problems and results similar to those of the preceding sections.

EXAMPLE 1 $cu_{xx} - u_t = 0$, $0 < x < \pi$, $0 < t < \infty$; $u_x(0, t) = 0$, $u_x(\pi, t) = 0$, $u(x, 0) = f(x)$.

Here the boundary conditions for $x = 0$ and $x = \pi$ correspond to insulating the ends of the rod. This can be seen from the model of Fig. 8–11. The condition $\partial u/\partial x = 0$ at $x = 0$ corresponds to the condition $u_1 - u_0 = 0$. But $u_1 = u_0$ means no temperature difference between the first piece and the adjacent medium, so that no heat flows; effectively, there is complete insulation at $x = 0$. The same reasoning applies to $x = \pi$.

Separation of variables as before leads to the boundary-value problem for $X(x)$:

$$X'' + \lambda X = 0, \qquad X'(0) = 0, \qquad X'(\pi) = 0.$$

This is the same problem as that arising from the problem of Example 2 in Section 8–3 for the wave equation. Hence we again obtain the solutions $\{\cos nx\}$, $n = 0, 1, 2, \ldots$, corresponding to the eigenvalues $\lambda_n = n^2$. As in Section 8–3, we then have the equation for $T_n(t)$:

$$T' + cn^2 T = 0, \quad n = 0, 1, 2, \ldots.$$

Thus, with arbitrary constants p_n, we have $T_0(t) = p_0/2$, $T_n(t) = p_n e^{-cn^2 t}$ for $n = 1, 2, \ldots$ Accordingly, we are led to the expansion

$$u(x, t) = \frac{p_0}{2} + \sum_{n=1}^{\infty} p_n e^{-cn^2 t} \cos nx. \tag{8–70}$$

For $t = 0$ we must have $u(x, 0) = f(x)$, so that

$$\frac{p_0}{2} + \sum_{n=1}^{\infty} p_n \cos nx = f(x).$$

We are again dealing with the Fourier cosine series representation of f:

$$p_n = \frac{2}{\pi} \int_0^\pi f(x) \cos nx \, dx, \quad n = 0, 1, 2, \ldots$$

With these values for p_n, Eq. (8–70) gives our solution (convergence conditions being assumed).

The solution can be discussed as in Section 8–6. We observe that

$$u(x, t) \to \frac{p_0}{2} = \frac{1}{\pi} \int_0^\pi f(x) \, dx \quad \text{as} \quad t \to \infty.$$

Thus *the temperature approaches the mean value of the initial temperature as t becomes infinite.* ◀

The problem with nonhomogeneous boundary conditions at $x = 0$ or $x = \pi$ or both can be reduced to a problem for the nonhomogeneous heat equation (as was found for the wave equation in Section 8–5). We handle the latter problem first.

EXAMPLE 2 $u_t - c^2 u_{xx} = F(x, t), u(x, 0) = 0, u(\pi, t) = 0, u(x, 0) = f_1(x)$. Here we have an extra internal source of heat, which alone would make the temperature at x rise at the rate $F(x, t)$.

Motivated by the form of the solution when $F(x, t) \equiv 0$ (Eq. (8–61)), we seek a solution of form

$$u = \sum_{n=1}^\infty h_n(t) \sin nx. \tag{8–71}$$

Here the boundary conditions at $x = 0$ and $x = \pi$ are satisfied. The one for $t = 0$ is satisfied if

$$\sum_{n=1}^\infty h_n(0) \sin nx = f_1(x),$$

that is, if

$$h_n(0) = b_n = \frac{2}{\pi} \int_0^\pi f_1(x) \sin nx \, dx.$$

If we substitute the function (8–71) in the differential equation, we obtain

$$\sum_{n=1}^\infty [h_n'(t) + cn^2 h_n(t)] \sin nx = F(x, t).$$

We conclude that for each fixed t the left side is the Fourier sine series expansion of $F(x, t)$. Therefore

$$h_n'(t) + cn^2 h_n(t) = F_n(t) = \frac{2}{\pi} \int_0^\pi F(x, t) \sin nx \, dx.$$

We have therefore an initial-value problem for $h_n(t)$:

$$h_n' + cn^2 h_n = F_n(t), \qquad h_n(0) = b_n.$$

This is easily solved (as in Section 1–7) to give

$$h_n(t) = \int_0^t e^{-cn^2(t-u)} F_n(u)\,du + b_n e^{-cn^2 t}.$$

Thus by Eq. (8–71) the desired solution is

$$u(x,\,t) = \sum_{n=1}^{\infty} \sin nx \left[\int_0^t e^{-cn^2(t-u)} F_n(u)\,du + b_n e^{-cn^2 t} \right]. \quad \blacktriangleleft$$

Now we turn to the heat equation with nonhomogeneous boundary conditions for $x = 0$ and $x = \pi$.

EXAMPLE 3 $u_t - cu_{xx} = 0,\ 0 < x < \pi,\ 0 < t < \infty;\ u(0,\,t) = p(t),\ u(\pi,\,t) = q(t),$ $u(x,\,0) = f(x).$

Here the temperatures at the end of the rod (Fig. 8–10) are not fixed at 0 but are changing in a controlled fashion.

We imitate the treatment of the corresponding problem for the wave equation (Eqs. (8–53)). First let

$$\psi(t) = p(t)\left(1 - \frac{x}{\pi}\right) + \frac{x}{\pi}\, q(t),$$

so that $\psi(t)$ satisfies the boundary conditions at $x = 0$ and $x = \pi$. Then let

$$v = u - \psi(x,\,t)$$

and obtain a new problem for v (see the wave-equation problem (8–57)):

$$v_t - cv_{xx} = -\psi_t + c\psi_{xx} = -p'(t)\left(1 - \frac{x}{\pi}\right) - \frac{x}{\pi}\, q'(t),$$

$$v(0,\,t) = 0, \qquad v(\pi,\,t) = 0, \tag{8–72}$$

$$v(x,\,0) = f(x) - \psi(x,\,0) = f(x) - p(0)\left(1 - \frac{x}{\pi}\right) - \frac{x}{\pi}\, q(0).$$

This is of the form of Example 2, with

$$F(x,\,t) = -p'(t)\left(1 - \frac{x}{\pi}\right) - \frac{x}{\pi}\, q'(t),$$

$$f_1(x) = f(x) - p(0)\left(1 - \frac{x}{\pi}\right) - \frac{x}{\pi}\, q(0).$$

Therefore, as above for Example 2, we obtain the solution

$$v = \sum_{n=1}^{\infty} h_n(t) \sin nx$$

and finally

$$u = v + \psi(x,\,t) = \psi(x,\,t) + \sum_{n=1}^{\infty} h_n(t) \sin nx. \quad \blacktriangleleft \tag{8–73}$$

As remarked at the end of Section 8–5, similar methods can be applied to the homogeneous or nonhomogeneous heat equation with more general nonhomogeneous boundary conditions. Further, the discussion of superposition of solutions at the end of Section 8–5 extends to the heat equation. For example, the solution of the problem

$$cu_{xx} - u_t = F(x, t), \qquad u(0, t) = p(t), \qquad u(\pi, t) = q(t), \qquad u(x, 0) = f(x),$$

is the sum of the solutions of the *four* problems obtained from the given one by replacing all but $F(x, t)$ by 0 on the right-hand sides, then all but $p(t)$ by 0, then all but $q(t)$ by 0, finally all but $f(x)$ by 0.

================== **Problems (Section 8–7)** ==================

1. Find the solution $u(x, t)$ of the boundary-value problem: $u_{xx} - u_t = 0$, $0 < x < \pi$, $0 < t < \infty$, $u(0, t) = 0$, $u(\pi, t) = 0$, $u(x, 0) = f(x)$ for $f(x)$ as given and graph $u(x, 0), u(x, 1)$, $u(x, 2)$ as functions of x:

   a) $f(x) = \sin 3x$ b) $f(x) = \sin x + 3 \sin 2x$
   c) $f(x) = x$ for $0 \leqslant x \leqslant \pi/2$, $f(x) = \pi - x$ for $\pi/2 \leqslant x \leqslant \pi$
   d) $f(x) = 1$ for $\pi/3 < x < 2\pi/3$, $f(x) = 0$ otherwise

2. Find the solution $u(x, t)$ of the boundary-value problem for the region $0 < x < \pi$, $0 < t < \infty$:

   a) $u_{xx} - u_t = 0$, $u_x(0, t) = 0$, $u(\pi, t) = 0$, $u(x, 0) = \sin x$
   b) $u_{xx} - u_t = 0$, $u(0, t) = 0$, $u(\pi, t) - u_x(\pi, t) = 0$, $u(x, 0) = x(\pi - x)$

   REMARK: The boundary condition at $x = \pi$ can be interpreted as stating that the adjacent medium is maintained at temperature 0 but that at $x = \pi$ the temperature of the rod obeys Newton's law of cooling.

   c) $u_t - 2u_{xx} = \sum_{n=1}^{\infty} n^{-2} e^{-nt} \sin nx$, $u(0, t) = 0$, $u(\pi, t) = 0$, $u(x, 0) = \sin 2x$
   d) $u_t - u_{xx} = \sin x/(1 + t^2)$, $u(0, t) = 0$, $u(\pi, t) = 0$, $u(x, 0) = \sin 3x$
   e) $u_t - u_{xx} = 0$, $u(0, t) = 0$, $u(\pi, t) = \sin t$, $u(x, 0) = x(\pi - x)$
   f) $u_t - u_{xx} = 0$, $u(0, t) = e^{-t}$, $u(\pi, t) = 0$, $u(x, 0) = \sin x$
   g) $u_t - u_{xx} = -e^{-t}\sin x$, $u(0, t) = e^{-t}$, $u(\pi, t) = 1$, $u(x, 0) = \sin x$

3. Consider the boundary-value problem: $u_t - u_{xx} = a_1 F(x, t)$, $0 < x < \pi$, $0 < t < \infty$; $u(0, t) = a_2 p(t)$, $u(\pi, t) = a_3 q(t)$, $u(x, 0) = a_4 f(x)$. Let $u_j(x, t), j = 1, 2, 3, 4$, be the solution for $a_j = 1$ and $a_i = 0$ for $i \neq j$. Express in terms of $u_1(x, t), \ldots, u_4(x, t)$ the solution of the problem for the following choices of a_1, a_2, a_3, a_4:

   a) $a_1 = 1$, $a_2 = 0$, $a_3 = 1$, $a_4 = 0$ b) $a_1 = 0$, $a_2 = 5$, $a_3 = -2$, $a_4 = 0$
   c) $a_1 = 1$, $a_2 = 1$, $a_3 = 1$, $a_4 = 1$,   d) $a_1, \ldots, a_4$ arbitrary constants

8–8 THE LAPLACE EQUATION FOR A RECTANGLE

We here consider the Laplace equation

$$\frac{\partial^2 u}{\partial x^2} + \frac{\partial^2 u}{\partial y^2} = 0 \tag{8–80}$$

in a region R in the xy-plane (Fig. 8–13). This equation arises in a great variety of

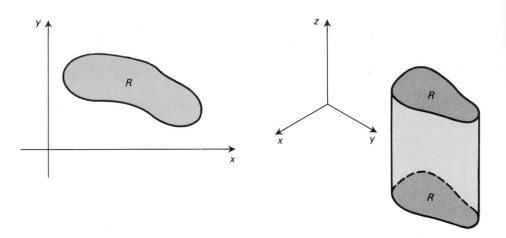

Fig. 8–13. Region for Laplace equation.   **Fig. 8–14.** Problem for solid cylinder.

problems in physics, such as heat conduction, electrostatic fields, fluid motion, etc. It commonly describes an *equilibrium state*. For example, if R represents a flat plate so insulated that heat is exchanged with the environment only along the boundary of R, then the temperature in R is governed by the *two-dimensional heat equation*:

$$c(u_{xx} + u_{yy}) - u_t = 0. \qquad (8\text{–}81)$$

If the boundary temperatures are kept fixed, then it can be shown that the temperature $u(x, y, t)$ approaches an equilibrium distribution $u_0(x, y)$ as $t \to \infty$. This equilibrium distribution satisfies the Laplace equation (8–80); it is a solution of (8–81) independent of t.

The Laplace equation (8–80) also arises from *three-dimensional* problems. For example, consider a cylindrical solid for which each cross section $z = $ const is the region R (Fig. 8–14). The equilibrium temperature distribution $u(x, y, z)$ for such a solid satisfies the *three-dimensional Laplace equation*:

$$\frac{\partial^2 u}{\partial x^2} + \frac{\partial^2 u}{\partial y^2} + \frac{\partial^2 u}{\partial z^2} = 0. \qquad (8\text{–}82)$$

If the temperatures along the boundary of the solid are kept fixed at values depending only on x and y (hence are the same for each cross section $z = $ const), then the equilibrium temperature should be independent of z, so that (8–82) reduces to (8–80).

The heat-conduction problems we have mentioned suggest that a sensible boundary-value problem associated with Eq. (8–80) consists of Eq. (8–80) and the assigned values of u along the boundary of R. This is known as the *Dirichlet problem*. Other boundary-value problems associated with Eq. (8–80) are important, especially the *Neumann problem* and the *mixed-boundary-value problem*. These involve normal derivatives of u along the boundary. Here we consider only the Dirichlet problem.

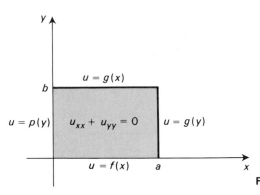

Fig. 8–15. Dirichlet problem for a rectangle.

In this section we study only the case of a rectangular region R. In the next sections we deal with other regions.

We thus consider the problem (see Fig. 8–15):

$$u_{xx} + u_{yy} = 0, \qquad 0 < x < a, \qquad 0 < y < b;$$
$$u(x, 0) = f(x), \qquad u(x, b) = g(x); \qquad\qquad (8\text{–}83)$$
$$u(0, y) = p(y), \qquad u(a, y) = q(y).$$

We can again use separation of variables. Since all boundary conditions are nonhomogeneous, we use superposition to reduce the problem to two problems each having one set of homogeneous boundary conditions. We first consider the special case $p(y) \equiv 0$, $q(y) \equiv 0$:

$$u_{xx} + u_{yy} = 0, \qquad 0 < x < a, \qquad 0 < y < b;$$
$$u(x, 0) = f(x), \qquad u(x, b) = g(x); \qquad\qquad (8\text{–}84)$$
$$u(0, y) = 0, \qquad u(a, y) = 0.$$

We now set $u = X(x)\,Y(y)$ and try to choose $X(x)$, $Y(y)$ so that $u = XY$ is not identically 0 and u satisfies the Laplace equation and the boundary conditions for $x = 0$, $x = a$. We obtain:

$$X''Y + XY'' = 0, \qquad \frac{X''}{X} + \frac{Y''}{Y} = 0, \qquad X'' + \lambda X = 0, \qquad Y'' - \lambda Y = 0.$$

Because of the boundary conditions, we want $X(0) = 0$, $X(a) = 0$. As usual we find that $\lambda \leqslant 0$ is excluded. For $\lambda > 0$ we set $\lambda = \omega^2$ and have

$$X = c_1 \cos \omega x + c_2 \sin \omega x.$$

The boundary conditions give $c_1 = 0$, $\sin \omega a = 0$. Hence $\omega = n\pi/a$, $n = 1, 2, \ldots$, so that we have the eigenfunctions and eigenvalues

$$X_n(x) = \sin \frac{n\pi x}{a}, \qquad \lambda_n = \frac{n^2\pi^2}{a^2}.$$

For $Y_n(y)$ we then have

$$Y_n'' - \frac{n^2\pi^2}{a^2} Y_n = 0.$$

Thus

$$Y_n(y) = p_n e^{n\pi y/a} + q_n e^{-n\pi y/a}$$

and accordingly

$$u = \sum_{n=1}^{\infty} X_n(x)Y_n(y) = \sum_{n=1}^{\infty} (p_n e^{n\pi y/a} + q_n e^{-n\pi y/a}) \sin \frac{n\pi x}{a}$$

is a solution of the Laplace equation satisfying the boundary conditions $u = 0$ for $x = 0$ and for $x = a$. To satisfy the conditions for $y = 0$ and for $y = b$ we must have

$$\sum_{n=1}^{\infty} (p_n + q_n) \sin \frac{n\pi x}{a} = f(x), \qquad \sum_{n=1}^{\infty} (p_n e^{n\pi b/a} + q_n e^{-n\pi b/a}) \sin \frac{n\pi x}{a} = g(x).$$

Again we are dealing with Fourier sine series (with a change of scale, as in Section 3–5). Hence

$$p_n + q_n = b_n = \frac{2}{a} \int_0^a f(x) \sin \frac{n\pi x}{a} dx,$$

$$p_n e^{n\pi b/a} + q_n e^{-n\pi b/a} = \beta_n = \frac{2}{a} \int_0^a g(x) \sin \frac{n\pi x}{a} dx.$$

These are simultaneous equations for p_n, q_n. Solving and substituting in the expression for $Y_n(y)$, we obtain

$$Y_n(y) = \frac{\beta_n \sinh (n\pi y/a) + b_n \sinh [(b-y)n\pi/a]}{\sinh (n\pi b/a)},$$

so that

$$u = u_1(x, y) = \sum_{n=1}^{\infty} \frac{\beta_n \sinh (n\pi y/a) + b_n \sinh [(b-y)n\pi/a]}{\sinh (n\pi b/a)} \sin \frac{n\pi x}{a} \qquad (8\text{--}85)$$

is the desired solution of the problem (8–84) (convergence conditions being assumed).

Next we turn to the special case of (8–83) in which $f(x) = 0$ and $g(x) = 0$:

$$\begin{cases} u_{xx} + u_{yy} = 0, & 0 < x < a, & 0 < y < b; \\ u(x, 0) = 0, & u(x, b) = 0; & u(0, y) = p(y), \quad u(a, y) = q(y). \end{cases} \qquad (8\text{--}86)$$

This problem is analogous to (8–84), with x and y interchanged, a and b interchanged, $f(x)$ replaced by $p(y)$, and $g(x)$ by $q(y)$. Therefore, by analogy with Eq. (8–85), the solution of problem (8–86) is

$$u = u_2(x, y) = \sum_{n=1}^{\infty} \frac{\gamma_n \sinh (n\pi x/b) + c_n \sinh [(a-x)n\pi/b]}{\sinh (n\pi a/b)} \sin \frac{n\pi y}{b}, \qquad (8\text{--}87)$$

where

$$c_n = \frac{2}{b} \int_0^b p(y) \sin \frac{n\pi y}{b} \, dy,$$

$$\gamma_n = \frac{2}{b} \int_0^b q(y) \sin \frac{n\pi y}{b} \, dy.$$

Finally, by superposition,

$$u(x, y) = u_1(x, y) + u_2(x, y)$$

is the solution of the given problem (8–83). This expression is valid, providing the unique solution, if f, g, p, q have continuous fourth derivatives and assign consistent values of u on the boundary: $f(0) = p(0)$, $f(a) = q(0)$, $g(0) = p(b)$, $g(a) = q(b)$ (see AC, chapter 10).

EXAMPLE 1 $u_{xx} + u_{yy} = 0$, $0 < x < \pi$, $0 < y < \pi$, $u(x, 0) = \pi x - x^2$, $u(x, \pi) = 0$, $u(0, y) = 0$, $u(\pi, y) = 0$.

This is a special case of problem (8–84) with $a = \pi$, $b = \pi$, $f(x) = \pi x - x^2$, $g(x) = 0$. Hence the solution is given by Eq. (8–85) with the corresponding values of the constants: $\beta_n = 0$ and

$$b_n = \frac{2}{\pi} \int_0^\pi (\pi x - x^2) \sin nx \, dx = \begin{cases} 8/(\pi n^3), & n \text{ odd,} \\ 0 & n \text{ even.} \end{cases}$$

Thus

$$u(x, y) = \frac{8}{\pi} \sum_{n=1}^{\infty} \frac{\sinh\left[(2n-1)(\pi - y)\right]}{(2n-1)^3 \sinh\left[(2n-1)\pi\right]} \sin\left[(2n-1)x\right].$$

Because of the $(2n-1)^3$ in the denominator, the first two terms of the series give a very good approximation to the solution:

$$u \sim \frac{8}{\pi} \left[\frac{\sinh(\pi - y)}{\sinh \pi} \sin x + \frac{\sinh 3(\pi - y)}{27 \sinh 3\pi} \sin 3x \right].$$

Even the first term is a good approximation to the solution; this is graphed in Fig. 8–16. The graph shows how the solution provides a smooth interpolation of the

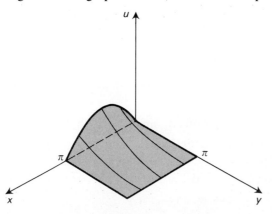

Fig. 8–16. Solution of Example 1.

boundary values in the interior of the rectangle. This is precisely the way we expect an equilibrium temperature to be distributed with respect to fixed boundary values. ◄

8–9 THE DIRICHLET PROBLEM FOR A CIRCULAR REGION

The method of separation of variables is again successful in the circular case, after introduction of polar coordinates. It is shown in Section 9–17 that in polar coordinates r, θ the Laplacian $u_{xx} + u_{yy}$ becomes

$$u_{rr} + \frac{1}{r^2} u_{\theta\theta} + \frac{1}{r} u_r.$$

Thus for the circular region $x^2 + y^2 = 1$ the problem is

$$u_{rr} + \frac{1}{r^2} u_{\theta\theta} + \frac{1}{r} u_r = 0, \quad \theta \text{ unrestricted}, \quad 0 \leqslant r \leqslant 1,$$

$$u(1, \theta) = f(\theta). \tag{8–90}$$

Here $u(r, \theta)$ is to be found. Since we want u to be continuous inside the circle, we also need periodicity in θ:

$$u(r, \theta + 2\pi) = u(r, \theta), \quad 0 \leqslant r \leqslant 1. \tag{8–91}$$

The r and r^2 in the denominator cause trouble for $r = 0$; we avoid this by multiplying the differential equation by r^2:

$$r^2 u_{rr} + u_{\theta\theta} + r u_r = 0. \tag{8–90'}$$

We try separation of variables, seeking solutions of form $u = R(r)\Theta(\theta)$. Substitution in Eq. (8–90') gives

$$r^2 R''\Theta + R\Theta'' + rR'\Theta = 0.$$

We divide by $R\Theta$:

$$r^2 \frac{R''}{R} + \frac{\Theta''}{\Theta} + r\frac{R'}{R} = 0.$$

The second term depends on θ alone, the other two terms depend only on r. Hence

$$\frac{\Theta''}{\Theta} = -\lambda = \text{const}, \qquad r^2\frac{R''}{R} + r\frac{R'}{R} = \lambda. \tag{8–92}$$

From the first equation,

$$\Theta'' + \lambda\Theta = 0. \tag{8–93}$$

Here the only condition imposed on Θ is periodicity implied by Eq. (8–91):

$$\Theta(\theta + 2\pi) = \Theta(\theta).$$

We verify that Eq. (8–93) has solutions of period 2π only for $\lambda = 0, 1^2, \ldots,$ $n^2, \ldots$ The corresponding orthogonal system is

$$1, \cos\theta, \sin\theta, \ldots, \cos n\theta, \sin n\theta, \ldots,$$

as for Fourier series. For $\lambda = n^2$, the second equation in (8–92) becomes

$$r^2 R'' + rR' - n^2 R = 0.$$

This is a Cauchy linear equation (see Problem 5 following Section 2–23). As shown in this problem, the general solution has the form

$$R = c_1 r^{m_1} + c_2 r^{m_2}$$

for appropriate constants m_1, m_2 (except in certain special cases). To find the constants m_1, m_2, we try $R = r^m$ in the differential equation and obtain

$$m(m-1)r^m + mr^m - n^2 r^m = 0 \qquad \text{or} \qquad r^m(m^2 - n^2) = 0.$$

Thus $m = \pm n$ and $R = c_1 r^n + c_2 r^{-n}$. For $n = 0$, this cannot be the general solution. In this case the differential equation can be written as

$$r^2 \frac{dR'}{dr} + rR' = 0 \qquad \text{or} \qquad \frac{dR'}{R'} + \frac{dr}{r} = 0,$$

so that

$$R' = \frac{c_1}{r}, \qquad R = c_1 \ln r + c_2.$$

Thus we have

$$R = c_1 \ln r + c_2 \qquad \text{for} \quad n = 0,$$
$$R - c_1 r^n + c_2 r^{-n} \qquad \text{for} \quad n = 1, 2, \ldots$$

However, we want $R(r)$ to be continuous for $r = 0$. This excludes the terms in $\ln r$ and in r^{-n} (that is, $c_1 = 0$ for $n = 0$, $c_2 = 0$ for $n = 1, 2, \ldots$). Thus finally our separation of variables yields the solutions

$$\frac{a_0}{2}, \; a_1 r \cos\theta, \; b_1 r \sin\theta, \; \ldots, \; a_n r^n \cos n\theta, \; b_n r^n \sin n\theta, \; \ldots,$$

for arbitrary constants $a_0, a_1, \ldots, b_1, \ldots$. Accordingly we are led to the expression

$$u = \frac{a_0}{2} + \sum_{n=1}^{\infty} (a_n r^n \cos n\theta + b_n r^n \sin n\theta) \tag{8–94}$$

for the solution of our Dirichlet problem (8–90). The boundary condition for $r = 1$ gives

$$\frac{a_0}{2} + \sum_{n=1}^{\infty} (a_n \cos n\theta + b_n \sin n\theta) = f(\theta).$$

Thus the left-hand side is the Fourier series of $f(\theta)$ and

$$a_n = \frac{1}{\pi} \int_0^{2\pi} f(\theta) \cos n\theta \, d\theta, \qquad b_n = \frac{1}{\pi} \int_0^{2\pi} f(\theta) \sin n\theta \, d\theta. \tag{8–95}$$

With these choices, Eq. (8–94) is the desired solution. It is valid, providing the unique solution, if f has a continuous second derivative and period 2π. In fact, the series (8–94) converges for $r < 1$ if f is merely continuous and periodic, and the series defines the unique solution of the problem (8–90) even in this case. This is proved in advanced books on Fourier series (see for example, Chapter III of *Trigonometric Series* by A. ZYGMUND, Cambridge University Press, 1959).

8–10 POISSON INTEGRAL FORMULA

It is of interest to know that the solution (8–94)–(8–95) of the Dirichlet problem (8–90) for the circle can be written as follows:

$$u(r, \theta) = \frac{1}{2\pi} \int_0^{2\pi} f(\varphi) \frac{1 - r^2}{1 + r^2 - 2r \cos(\varphi - \theta)} \, d\varphi, \quad r < 1. \tag{8–100}$$

This is known as the *Poisson integral formula* for the solution.

To derive this, we use the result of Problem 5 following Section 3–2:

$$\frac{1 - r \cos \theta}{1 + r^2 - 2r \cos \theta} = 1 + r \cos \theta + \cdots + r^n \cos n\theta + \cdots, \quad 0 \leqslant r < 1.$$

We subtract $1/2$ from both sides to obtain

$$\frac{1 - r^2}{2(1 + r^2 - 2r \cos \theta)} = \frac{1}{2} + r \cos \theta + \cdots + r^n \cos n\theta + \cdots, \quad 0 \leqslant r < 1. \tag{8–101}$$

By Eq. (8–95), our solution u of Eq. (8–94) can be written as

$$u = \frac{1}{2\pi} \int_0^{2\pi} f(\varphi) d\varphi + \frac{1}{\pi} \sum_{n=1}^{\infty} r^n \left[\cos n\theta \int_0^{2\pi} f(\varphi) \cos n\varphi \, d\varphi + \sin n\theta \int_0^{2\pi} f(\varphi) \sin n\varphi \, d\varphi \right]$$

$$= \frac{1}{2\pi} \int_0^{2\pi} f(\varphi) \, d\varphi + \frac{1}{\pi} \sum_{n=1}^{\infty} \int_0^{2\pi} f(\varphi) r^n \cos n(\varphi - \theta) \, d\varphi$$

$$= \frac{1}{\pi} \int_0^{2\pi} f(\varphi) \left[\frac{1}{2} + \sum_{n=1}^{\infty} r^n \cos n(\varphi - \theta) \right] d\varphi.$$

The term-by-term integration of the series is justified by uniform convergence (Section 2–15). If we now apply Eq. (8–101) to the expression in brackets, Eq. (8–100) follows.

8–11 DIRICHLET PROBLEM FOR OTHER REGIONS

With the aid of conformal mapping, the Dirichlet problem for a great variety of regions can be reduced to the Dirichlet problem for the circular region $r \leqslant 1$ and hence solved by a series (8–94) or the Poisson integral formula (8–100) (see Section 11–27 for details). Many numerical techniques have also been developed to solve the problem by computation (Chapters 13, 14).

==================================Problems (Section 8-11)================================

1. Solve the following Dirichlet problems:
   a) $u_{xx} + u_{yy} = 0$, $0 < x < \pi$, $0 < y < 2\pi$; $u(x, 0) = 0$, $u(x, 2\pi) = 0$,
      $u(0, y) = \sin y + 2 \sin 3y$, $u(\pi, y) = 0$
   b) $u_{xx} + u_{yy} = 0$, $0 < x < \pi$, $0 < y < \pi$; $u(x, 0) = \sin^2 x$, $u(x, \pi) = 0$, $u(0, y) = 0$,
      $u(\pi, y) = 0$
   c) $u_{xx} + u_{yy} = 0$, $0 < x < 1$, $0 < y < 1$; $u(x, 0) = x$, $u(x, 1) = 1$, $u(0, y) = y$, $u(1, y) = 1$
   d) $u_{xx} + u_{yy} = 0$, $0 < x < 2$, $0 < y < 1$; $u(x, 0) = e^x$, $u(x, 1) = e^2$, $u(0, y) = e^{2y}$, $u(2, y) = e^2$
   e) $r^2 u_{rr} + u_{\theta\theta} + r u_r = 0$, $0 \leqslant r < 1$, $u(1, \theta) = 5 \sin \theta + \cos 2\theta$
   f) $r^2 u_{rr} + u_{\theta\theta} + r u_r = 0$, $0 \leqslant r < 1$, $u(1, \theta) = 2\pi\theta - \theta^2$, $0 \leqslant \theta \leqslant 2\pi$
   g) $r^2 u_{rr} + u_{\theta\theta} + r u_r = 0$, $0 \leqslant r < 1$, $u(1, \theta) = 1$ for $0 < \theta < \pi$, $u(1, \theta) = 0$ for $\pi < \theta < 2\pi$.

   REMARK: Here the boundary function $f(\theta)$ is discontinuous. The methods given above still provide a solution, agreeing with these boundary values except where $f(\theta)$ is discontinuous. The solution is not unique, but it is the only *bounded* solution.

   h) $r^2 u_{rr} + u_{\theta\theta} + r u_r = 0$, $0 \leqslant r < 1$, $u(1, \theta) = 0$ for $0 \leqslant \theta \leqslant \pi$,
      $u(1, \theta) = \pi - \theta$ for $\pi \leqslant \theta \leqslant 2\pi$

2. Represent the solution by the Poisson integral formula:
   a) Part (g) of Problem 1   b) Part (h) of Problem 1

3. Find a solution by separation of variables: $u_{xx} + u_{yy} = 0$, $0 < x < \pi$, $0 < y < \infty$;
   $u(x, 0) = f(x)$, $u(0, y) = 0$, $u(\pi, y) = 0$, $u(x, y) \to 0$ as $y \to \infty$.

   REMARK: The solution is unique provided the last limit is a uniform one that is, for each $\varepsilon > 0$, there is a c such that $|u(x, y)| < \varepsilon$ for $y > c$ and $0 < x < \pi$, and provided that $f(x)$ is continuous for $0 \leqslant x \leqslant \pi$, with $f(0) = 0$, $f(\pi) = 0$.

4. a) Show by the aid of the Poisson integral formula that the solution $u(r, \theta)$ of the Dirichlet problem (8–90) has the property: the value of u at the center of the circle equals the mean value of f on the circumference.
   b) Show more generally: if $u(x, y)$ is harmonic in a region containing a circle C and its interior, then the value of u at the center of the circle equals the mean value of u on the circumference.
   c) Show that if a function u is harmonic in a region G and takes on its absolute maximum at a unique point (x_0, y_0), then (x_0, y_0) cannot be interior to G, that is, (x_0, y_0) must be on the boundary of G.

8–12 APPLICATION OF LAPLACE AND FOURIER TRANSFORMS

The Laplace and Fourier transforms are powerful tools for obtaining the solution of a variety of boundary-value problems for infinite regions. We give several illustrations here. The procedures will be formal, without concern about the validity of intermediate steps. The main goal is to obtain a formula for a solution. Having obtained it, we can then make a careful study of its properties and determine whether it really solves the problem.

Among the formal steps are certain interchanges of operations. For example, for a function $u(x, t)$ we form the Laplace transform with respect to t:

$$U(x, s) = \mathscr{L}[u] = \int_0^\infty e^{-st} u(x, t)\, dt. \qquad (8\text{–}120)$$

We then also need the Laplace transform of the derivative u_x or u_{xx}. We assume that these are the corresponding derivatives of the transforms:

$$\mathscr{L}[u_x] = \frac{\partial}{\partial x} U(x, s), \qquad \mathscr{L}[u_{xx}] = \frac{\partial^2}{\partial x^2} U(x, s). \qquad (8\text{–}121).$$

Also a common boundary condition is the requirement that $u(x, t) \to 0$ as $x \to \infty$. We shall assume that this implies that $U(x, s) \to 0$ as $x \to \infty$.

EXAMPLE 1 *Wave equation for a semi-infinite string.* The problem is as follows:

$$u_{xx} - u_{tt} = 0, \qquad 0 < x < \infty, \qquad 0 < t < \infty,$$
$$u(0, t) = 0, \qquad u(x, t) \to 0 \quad \text{as} \quad x \to \infty,$$
$$u(x, 0) = xe^{-x}, \qquad u_t(x, 0) = 0.$$

We form the Laplace transform as in Eq. (8–120); then by (8–121), $\mathscr{L}[u_{xx}] = \partial^2 U/\partial x^2$. For $\mathscr{L}[u_{tt}]$ we use the rule for transform of a derivative (see Eq. (4–33):

$$\mathscr{L}[u_{tt}] = s^2 \mathscr{L}[u] - su(x, 0) - u_t(x, 0).$$

Here $u(x, 0) = xe^{-x}$ and $u_t(x, 0) = 0$ by the given initial conditions. Therefore the transformed differential equation is

$$U'' - s^2 U + sxe^{-x} = 0. \qquad (8\text{–}122)$$

Here we write U'' for $\partial^2 U/\partial x^2$, since s can be treated as a parameter in the equation; it is a second-order ordinary differential equation for U as a function of x. We write it in the form:

$$U'' - s^2 U = -sxe^{-x}$$

and obtain the general solution by the methods of Section 1–12:

$$U = c_1 e^{sx} + c_2 e^{-sx} + se^{-x}\left[\frac{x}{s^2 - 1} - \frac{2}{(s^2 - 1)^2}\right].$$

Here c_1, c_2 are arbitrary "constants" but, as we shall see, they may depend on s.

Since $u(x, t) \to 0$ as $x \to \infty$, we want also $U \to 0$ as $x \to \infty$, as remarked above. The variable s is complex but, as usual for Laplace transforms, we are concerned with the behavior for Re s large positive. For such s, all terms in the expression for U have limit 0 as $x \to \infty$ except the first one. Hence we must have $c_1 = 0$.

We have one more boundary condition: $u(0, t) = 0$. This clearly gives $U(0, s) = 0$ and therefore

$$c_1 + c_2 - \frac{2s}{(s^2 - 1)^2} = 0.$$

The preceding equations give c_1 and c_2, so that

$$U(x, s) = 2e^{-sx}\frac{s}{(s^2 - 1)^2} + se^{-x}\left[\frac{x}{s^2 - 1} - \frac{2}{(s^2 - 1)^2}\right]. \qquad (8\text{--}123)$$

With the aid of Table 4–1 in Section 4–4, we find the inverse Laplace transforms of the various terms appearing. We observe that

$$\mathscr{L}[\cosh t] = \frac{s}{s^2 - 1}, \qquad \mathscr{L}[t \sinh t] = \frac{2s}{(s^2 - 1)^2}.$$

The second equation follows from entry 11 of the table if we take $a = i$ and note that

$$\sinh t = \frac{e^t - e^{-t}}{2} = -\frac{e^{i(it)} - e^{-i(it)}}{2} = -i \sin it.$$

For the inverse transform of the term in e^{-sx} we use the translation rule (4–38). We let

$$p(t) = 0 \text{ for } t < 0, \qquad p(t) = t \sinh t \text{ for } t \geq 0.$$

Then for each $x > 0$

$$\mathscr{L}[p(t - x)] = 2e^{-sx}\frac{s}{(s^2 - 1)^2}.$$

Thus if we take inverse transforms in (8–123) we obtain our solution:

$$u(x, t) = p(t - x) + xe^{-x} \cosh t - e^{-x} t \sinh t.$$

It is interesting to observe that the solution can be written in the form

$$u(x, t) = \frac{1}{2}[f(x + ct) + f(x - ct)],$$

as in Section 8–4. Here $c = 1$, $f(x) = xe^{-x}$ for $x \geq 0$ (the initial value of u), and f is extended to all x to be an odd function, so that

$$f(x) = xe^x \quad \text{for} \quad x \leq 0.$$

The details are left as an exercise (see Problem 10 below). ◄

EXAMPLE 2 *Heat conduction on a semi-infinite rod.* The problem is the following:

$$cu_{xx} - u_t = 0, \quad 0 < x < \infty, \quad 0 < t < \infty;$$

$$u(x, 0) = 0, \qquad u(x, t) \to 0 \text{ as } x \to \infty, \qquad u(0, t) = f(t).$$

We take Laplace transforms with respect to t, noting the 0 initial value. We obtain the equation

$$cU'' - sU = 0, \qquad (8\text{--}124)$$

where U is required to satisfy the conditions

$$U(x, s) \to 0 \quad \text{as} \quad x \to \infty, \qquad U(0, s) = F(s) = \mathscr{L}[f]. \qquad (8\text{--}125)$$

Hence

$$U = c_1 e^{x\sqrt{s/c}} + c_2 e^{-x\sqrt{s/c}},$$

where the arbitrary "constants" c_1, c_2 may again depend on s. By the first condition (8–125), c_1 must be 0, and the second then gives $c_2 = F(s)$. Thus

$$U(x, s) = F(s)e^{-x\sqrt{s/c}}.$$

The solution $u(x, t)$ can now be obtained as the inverse Laplace transform of $U(x, s)$, hence by the inversion formula of Section 4–17:

$$u(x, t) = \frac{1}{2\pi} \int_{-\infty}^{\infty} F(s)e^{-x\sqrt{s/c}}e^{st}\,d\omega,$$

where $s = \sigma + i\omega$ and σ is sufficiently large.

The solution can also be obtained in another form. It can be shown that

$$e^{-k\sqrt{s}} = \mathscr{L}\left[\frac{k}{2\sqrt{\pi}\,t^{3/2}}e^{-k^2/(4t)}\right], \qquad \mathrm{Re}\,s > 0,$$

where k is a positive constant (see Churchill, pp. 78–80). Hence

$$U(x, s) = \mathscr{L}[f]\,\mathscr{L}\left[\frac{x}{2\sqrt{\pi c}\,t^{3/2}}e^{-x^2/(4ct)}\right].$$

Therefore by the theorem on convolutions (Section 4–8, Eq. (4–81)),

$$u(x, t) = f * \frac{x}{2\sqrt{\pi c}\,t^{3/2}}e^{-x^2/(4ct)},$$

where the convolution is with respect to t. Thus

$$u(x, t) = \frac{x}{2\sqrt{\pi c}} \int_0^t \frac{f(t-\tau)}{\tau^{3/2}}e^{-x^2/(4c\tau)}\,d\tau. \quad \blacktriangleleft$$

EXAMPLE 3   *Heat equation for an infinite rod.* The problem is:

$$cu_{xx} - u_t = 0, \quad -\infty < x < \infty, \quad 0 < t < \infty; \qquad u(x, 0) = f(x).$$

We have given no boundary condition for $x \to \pm\infty$. However, the method to be used implies some smallness condition for large $|x|$, and there is no uniqueness without some such condition.

This time we use Fourier transforms *with respect to x* and write

$$U(\omega, t) = \Phi[u] = \int_{-\infty}^{\infty} u(x, t)e^{-i\omega x}\,dx.$$

By the rule (4–141'), u_{xx} has transform $-\omega^2 U(\omega, t)$ and, as above for Laplace transforms, u_t is assumed to have transform $\partial U/\partial t$, which we write as U'. Thus our transformed problem is

$$-c\omega^2 U - U' = 0, \qquad U(\omega, 0) = \varphi(\omega);$$

the second equation is the transform of the initial condition $u(x, 0) = f(x)$. From the differential equation for U,

$$U = c_1 e^{-c\omega^2 t}$$

and, by the initial condition, $c_1 = \varphi(\omega)$. Thus

$$U(\omega, t) = \varphi(\omega)e^{-c\omega^2 t}. \tag{8–126}$$

If we take inverse transforms, we obtain

$$u(x, t) = \frac{1}{2\pi} \int_{-\infty}^{\infty} \varphi(\omega) e^{-c\omega^2 t} e^{i\omega t} \, d\omega,$$

as in Eq. (4–135) of Section 4–13.

As for Laplace transforms, we can also write the solution as a convolution (Section 4–18, especially Eqs. (4–180), (4–181)). We first observe that e^{-ax^2} has transform $\sqrt{\pi/a} \, e^{-\omega^2/(4a)}$ by entry 19 of Table 4–2. Hence

$$\Phi^{-1}[e^{-c\omega^2 t}] = \frac{1}{2\sqrt{\pi c t}} e^{-x^2/(4ct)} = g(x, t)$$

and by Eq. (8–126)

$$U(\omega, t) = \Phi[f]\Phi[g] = \Phi[f * g].$$

Accordingly,

$$u(x, t) = f * g = \frac{1}{2\sqrt{\pi c t}} \int_{-\infty}^{\infty} f(x - v) e^{-v^2/(4ct)} \, dv. \quad \blacktriangleleft$$

These methods and related ones, using other transforms, have a great range of applications. For further discussion, see CHURCHILL and COURANT–HILBERT.

════════════════════════**Problems (Section 8–12)**════════════════════════

Solve the boundary-value problems 1 through 9. In Problems 1 through 7 the Laplace transform can be used; in Problems 8 and 9 the Fourier transform with respect to x can be used. In all problems, c is a positive constant.

1. $u_{xx} - u_{tt} = 0$, $0 < x < \pi$, $0 < t < \infty$; $u(0, t) = 0$, $u(\pi, t) = 0$, $u(x, 0) = 0$,
   $u_t(x, 0) = \sin x + 2 \sin 3x$

2. $u_{xx} - u_{tt} = 0$, $0 < x < \infty$, $0 < t < \infty$; $u(0, t) = 0$, $u(x, t) \to 0$ as $x \to \infty$, $u(x, 0) = 0$,
   $u_t(x, 0) = xe^{-x}$

3. $c^2 u_{xx} - u_{tt} = 0$, $0 < x < \infty$, $0 < t < \infty$; $u(0, t) = f(t)$, $u(x, t) \to 0$ as $x \to \infty$, $u(x, 0) = 0$,
   $u_t(x, 0) = 0$

4. $c^2 u_{xx} - u_{tt} = 0$, $0 < x < \infty$, $0 < t < \infty$; $u_x(0, t) = g(t)$, $u(x, t) \to 0$ as $x \to \infty$; $u(x, 0) = 0$,
   $u_t(x, 0) = 0$

5. $u_{xx} - u_t = 0$, $0 < x < \pi$, $0 < t < \infty$; $u(0, t) = 0$, $u(\pi, t) = 0$, $u(x, 0) = \sin 2x - 7 \sin 5x$.

6. $u_{xx} - u_t = e^{-x}$, $0 < x < \infty$, $0 < t < \infty$; $u(0, t) = 0$, $u(x, t) \to 0$ as $x \to \infty$, $u(x, 0) = 0$

7. $u_{xx} - u_t = 0$, $0 < x < \infty$, $0 < t < \infty$, $u(0, t) = 0$, $u(x, t) \to 0$ as $x \to \infty$, $u(x, 0) = e^{-x}$

8. $c^2 u_{xx} - u_{tt} = 0$, $-\infty < x < \infty$, $0 < t < \infty$; $u(x, 0) = f(x)$, $u_t(x, 0) = g(x)$

9. $u_{xx} + u_{yy} = 0$, $-\infty < x < \infty$, $0 < y < \infty$; $u(x, 0) = f(x)$, $u(x, y) \to 0$ as $y \to \infty$

10. Verify that the solution of Example 1 in Section 8–12 can be written in the form
    $u = (1/2)[f(x + ct) + f(x - ct)]$, as remarked just before Example 2.

8–13 PROBLEMS IN HIGHER DIMENSIONS

All of the preceding methods extend to higher dimensions. We give several examples.

EXAMPLE 1 *The wave equation for a rectangular membrane.* The problem for $u(x, y, t)$ is as follows:

$$u_{tt} - c^2(u_{xx} + u_{yy}) = 0, \qquad 0 < x < a, \qquad 0 < y < b, \qquad 0 < t < \infty,$$
$$u(x, 0, t) = 0, \qquad u(x, b, t) = 0, \qquad u(0, y, t) = 0, \qquad u(a, y, t) = 0,$$
$$u(x, y, 0) = f(x, y), \qquad u_t(x, y, 0) = g(x, y).$$

This describes the vibrations of a rectangular membrane whose edges are kept fixed.

The differential equation can be obtained as a limit of equations for a particle problem like that of Fig. 8–3. Now the particles move on rods perpendicular to the xy-plane and are connected by springs in two directions, as in Fig. 8–17, which shows the case of four rods. The analysis is the same as for Fig. 8–3, except that there are terms from both the x and y directions. In the limit these give the term $c^2(u_{xx} + u_{yy})$.

We can solve the problem by separation of variables. We seek solutions $u = X(x) Y(y) T(t)$ of the differential equation satisfying the (homogeneous) boundary conditions for $x = 0$, $x = a$, $y = 0$, $y = b$. We substitute in the differential equation and then divide by $c^2 X Y T$:

$$X Y T'' - c^2(X'' Y T + X Y'' T) = 0, \qquad \frac{1}{c^2} \frac{T''}{T} - \frac{X''}{X} - \frac{Y''}{Y} = 0.$$

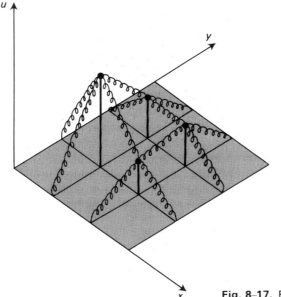

Fig. 8-17. Particle model for vibrating membrane.

Hence

$$\frac{X''}{X} = \text{const} = -\mu, \qquad \frac{Y''}{Y} = \text{const} = -\nu, \qquad \frac{T''}{c^2 T} = \text{const} = -\lambda = -\mu - \nu.$$

To satisfy the boundary conditions mentioned, we must have

$$X'' + \mu X = 0, \qquad X(0) = X(a) = 0,$$
$$Y'' + \nu Y = 0, \qquad Y(0) = Y(b) = 0.$$

Hence, as before, we obtain the eigenvalues $\mu_m = m^2 \pi^2 / a^2$, $\nu_n = n^2 \pi^2 / b^2$, $m = 1, 2, \ldots$, $n = 1, 2, \ldots$, and the corresponding eigenfunctions

$$X_m(x) = \sin(m\pi x / a), \qquad Y_n(y) = \sin(n\pi y / b).$$

For each choice of m and n, $T(t)$ must satisfy the problem

$$T'' + c^2 \lambda_{mn} T = 0, \quad \lambda_{mn} = \mu_m + \nu_n = \pi^2 \left(\frac{m^2}{a^2} + \frac{n^2}{b^2} \right).$$

Therefore, with arbitrary constants p_{mn} and q_{mn},

$$T = T_{mn} = p_{mn} \cos(\sqrt{\lambda_{mn}} \, ct) + q_{mn} \sin(\sqrt{\lambda_{mn}} \, ct)$$

and our solutions are

$$u = X_m Y_n T_{mn} = \sin \frac{m\pi x}{a} \sin \frac{n\pi y}{b} [p_{mn} \cos(\sqrt{\lambda_{mn}} \, ct) + q_{mn} \sin(\sqrt{\lambda_{mn}} \, ct)].$$

We sum these to obtain

$$u = \sum_{m=1}^{\infty} \sum_{n=1}^{\infty} \sin \frac{m\pi x}{a} \sin \frac{n\pi y}{b} [p_{mn} \cos(\sqrt{\lambda_{mn}} \, ct) + q_{mn} \sin(\sqrt{\lambda_{mn}} \, ct)]. \quad (8\text{–}130)$$

Finally, we apply the initial conditions:

$$f(x, y) = \sum_{m=1}^{\infty} \sum_{n=1}^{\infty} p_{mn} \sin \frac{m\pi x}{a} \sin \frac{n\pi y}{b},$$

$$g(x, y) = c \sum_{m=1}^{\infty} \sum_{n=1}^{\infty} \sqrt{\lambda_{mn}} \, q_{mn} \sin \frac{m\pi x}{a} \sin \frac{n\pi y}{b}.$$

As in Section 3–15, the functions $\sin(m\pi x / a) \sin(n\pi y / b)$ form a complete orthogonal system for the rectangle $0 \leqslant x \leqslant a$, $0 \leqslant y \leqslant b$. Hence the p_{mn} and q_{mn} can be chosen so that $f(x, y)$ and $g(x, y)$ have the above series representations, provided f and g satisfy the usual smoothness conditions. According to Eq. (3–106), the coefficients p_{mn}, q_{mn} are given by the equations

$$p_{mn} = \frac{4}{ab} \int_0^b \int_0^a f(x, y) \sin \frac{m\pi x}{a} \sin \frac{n\pi y}{b} \, dx \, dy,$$

$$c \sqrt{\lambda_{mn}} \, q_{mn} = \frac{4}{ab} \int_0^b \int_0^a g(x, y) \sin \frac{m\pi x}{a} \sin \frac{n\pi y}{b} \, dx \, dy. \qquad (8\text{–}131)$$

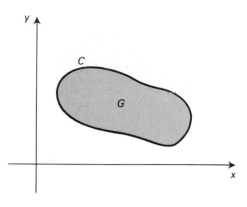

Fig. 8-18. Vibrating membrane.

The solution $u(x, y, t)$, given by Eqs. (8–130) and (8–131), is more complicated than that for the vibrating string in Section 8–3. It is also difficult to interpret the solution in terms of traveling waves, as in Section 8–4. For each m, n the corresponding term in Eq. (8–130) represents a standing wave of frequency $c \sqrt{\lambda_{mn}}$. The frequencies occurring (the spectrum of the vibrating system) form a complicated pattern, with some frequencies repeated (for example, $c \sqrt{\lambda_{mn}} = c \sqrt{\lambda_{nm}}$). For $a = b = \pi, c = 1$, one has the frequencies $c \sqrt{\lambda_{mn}} = \sqrt{m^2 + n^2}$; that is, $\sqrt{\lambda_{11}} = \sqrt{2}$, $\sqrt{\lambda_{12}} = \sqrt{5}, \sqrt{\lambda_{21}} = \sqrt{5}, \sqrt{\lambda_{22}} = \sqrt{8}, \ldots$ ◄

These results can be generalized to a membrane of arbitrary shape: circular, elliptical, etc. In each case the membrane can be considered to fill a region G in the xy-plane, bounded by a smooth curve C, as in Fig. 8–18. The problem of vibrations of such a membrane with boundary fixed is then as follows:

$$\begin{cases} u_{tt} - c^2(u_{xx} + u_{yy}) = 0 \text{ for } (x, y) \text{ in } G, 0 < t < \infty, \\ u(x, y, t) = 0 \text{ for } (x, y) \text{ on } C, \\ u(x, y, 0) = f(x, y), \qquad u_t(x, y, 0) = g(x, y). \end{cases} \qquad (8\text{–}132)$$

We can make a partial separation of variables by setting $u = T(t)V(x, y)$ and seeking solutions such that V and hence u are 0 on C. Substitution in the differential equation and division by $c^2 TV$ lead to the equations

$$\frac{1}{c^2} \frac{T''}{T} - \frac{V_{xx} + V_{yy}}{V} = 0,$$

$$\frac{V_{xx} + V_{yy}}{V} = \text{const} = -\lambda, \qquad \frac{1}{c^2} \frac{T''}{T} = -\lambda,$$

$$V_{xx} + V_{yy} + \lambda V = 0, \quad T'' + \lambda c^2 T = 0.$$

Thus we have the eigenvalue problem for $V(x, y)$:

$$\begin{cases} V_{xx} + V_{yy} + \lambda V = 0 \text{ for } (x, y) \text{ in } G, \\ V(x, y) = 0 \text{ on } C. \end{cases} \qquad (8\text{–}133)$$

This is an example of a *two-dimensional Sturm–Liouville problem*. As for the one-dimensional Sturm–Liouville problem (Section 3–14), it can be shown that there is a sequence of eigenvalues λ_n and a corresponding sequence of eigenfunctions $V_n(x, y)$ that form a complete orthogonal system for the region G.

It can be shown that the eigenvalues λ_n in our membrane problem are all positive. For each λ_n and V_n, there is a T_n determined by the equation

$$T_n'' + \lambda_n c^2 T_n = 0,$$

so that $T_n = p_n \cos(\sqrt{\lambda_n} ct) + q_n \sin(\sqrt{\lambda_n} ct)$. Then as usual we sum to obtain

$$u(x, y, t) = \sum_{n=1}^{\infty} V_n(x, y) \left[p_n \cos(\sqrt{\lambda_n} ct) + q_n(\sqrt{\lambda_n} ct) \right]. \tag{8–134}$$

The initial conditions give

$$f(x, y) = \sum_{n=1}^{\infty} p_n V_n(x, y), \qquad g(x, y) = \sum_{n=1}^{\infty} c \sqrt{\lambda_n} q_n V_n(x, y).$$

Since $\{V_n(x, y)\}$ is a complete orthogonal system for G, the coefficients p_n, q_n are found (see Eq. (3–106) in Section 3–10) by

$$p_n = \frac{1}{k_n} \iint_G f(x, y) V_n(x, y) \, dx \, dy,$$

$$c \sqrt{\lambda_n} q_n = \frac{1}{k_n} \iint_G g(x, y) V_n(x, y) \, dx \, dy, \tag{8–135}$$

$$k_n = \left\{ \iint_G [V_n(x, y)]^2 \, dx \, dy \right\}^{1/2}.$$

(For a justification of these results, see COURANT–HILBERT and G. HELLWIG, *Methods of Mathematical Physics*, Addison–Wesley Publishing Co., Reading, Mass., 1964.)

It should be remarked that for special regions G the problem (8–132) may be solvable by separation of variables. This is the case when G is a rectangle, as in Example 1. It is also the case when G is a circle, as illustrated in Example 2 below.

The previous discussion extends without essential change to the wave equation in three-dimensional space:

$$u_{tt} - c^2(u_{xx} + u_{yy} + u_{zz}) = 0 \text{ in } G,$$

with u required to be 0 on the boundary of G. It also extends to the heat equation in two or three dimensions. To illustrate the process, we consider the case of heat conduction in a circular disk.

EXAMPLE 2 $u_t - c(u_{xx} + u_{yy}) = 0$, $x^2 + y^2 < 1$, $u(x, y, t) = 0$ for $x^2 + y^2 = 1$,
$u(x, y, 0) = f(x, y)$ for $x^2 + y^2 \leqslant 1$.

To illustrate a general approach to such problems, we first separate t from the
space variables by seeking solutions $u = T(t) V(x, y)$. Then, just as for the wave
equation above, we obtain successively

$$T'V - cT(V_{xx} + V_{yy}) = 0, \qquad \frac{1}{c}\frac{T'}{T} - \frac{V_{xx} + V_{yy}}{V} = 0,$$

$$\frac{V_{xx} + V_{yy}}{V} = -\lambda, \qquad \frac{1}{c}\frac{T'}{T} = -\lambda,$$

$$V_{xx} + V_{yy} + \lambda V = 0, \qquad T' + \lambda c T = 0.$$

Here $V(x, y)$ is required to be 0 for $x^2 + y^2 = 1$. We now rewrite the equation for V in
polar coordinates r, θ. As in Section 8–9, we obtain

$$V_{rr} + \frac{1}{r^2} V_{\theta\theta} + \frac{1}{r} V_r + \lambda V = 0, \tag{8–136}$$

where we again write V for our function $V(x, y)$ expressed in polar coordinates r, θ.
Here V is required to be 0 for $r = 1$ and to have period 2π in θ.

We now separate variables in Eq. (8–136) as in Section 8–9:

$$V = R(r) \Theta(\theta),$$

$$R''\Theta + \frac{1}{r^2} R\Theta'' + \frac{1}{r} R'\Theta + \lambda R\Theta = 0,$$

$$r^2 \frac{R''}{R} + \frac{\Theta''}{\Theta} + r\frac{R'}{R} + \lambda r^2 = 0,$$

$$r^2 \frac{R''}{R} + r\frac{R'}{R} + \lambda r^2 = \mu, \qquad \frac{\Theta''}{\Theta} = -\mu,$$

$$\Theta'' + \mu\Theta = 0, \qquad r^2 R'' + rR' + (\lambda r^2 - \mu)R = 0.$$

Here $R(r)$ is sought for $0 \leqslant r \leqslant 1$ and $R(1) = 0$. The function $\Theta(\theta)$ must have period
2π in θ. Therefore $\mu = 0, 1, \ldots, m^2, \ldots$ and, corresponding to the eigenvalues
$0, 1, 1, 4, 4, \ldots$, we have the eigenfunctions

$$1, \cos\theta, \sin\theta, \ldots, \cos m\theta, \sin m\theta, \ldots$$

for $\Theta(\theta)$; they form a complete orthogonal system for the interval $0 \leqslant \theta \leqslant 2\pi$.

When $\mu = m^2$, our problem for $R(r)$ becomes

$$r^2 R'' + rR' + (\lambda r^2 - m^2)R = 0, \qquad R(1) = 0.$$

The differential equation for R is closely related to the Bessel differential equation

$$x^2 y'' + xy' + (x^2 - m^2)y = 0. \tag{8–137}$$

It is pointed out in Section 3–13 that the function $J_m(x)$, that satisfies this differential
equation, gives rise to a complete orthogonal system $\{J_m(s_{mn}x)\}$, with weight

function $w(x) = x$, for $0 \leqslant x \leqslant 1$; here $s_{m1}, s_{m2}, \ldots$ are the successive positive zeros of $J_m(x)$. If we set $x = s_{mn}r$ in Eq. (8–137) and $W(r) = J_m(s_{mn}r)$, then we obtain

$$r^2 W'' + rW' + (s_{mn}^2 r^2 - m^2)W = 0.$$

This is the same as the equation for $R(r)$ if we take $\lambda = s_{mn}^2$; also $W(1) = J_m(s_{mn}) = 0$. Hence the functions $R(r) = J_m(s_{mn}r)$ solve the boundary-value problem for $R(r)$. It can be shown that, up to constant factors, they provide all the solutions. [The Bessel equation (8–137) has a regular singular point at $x = 0$. The methods of Sections 2–23 and 12–2 show that the only solutions that are continuous at $x = 0$ are the functions $\text{Const} \cdot J_m(x)$.]

The functions

$$V = J_m(s_{mn}r)\cos m\theta, \quad m = 0, 1, 2, \ldots, \quad n = 1, 2, \ldots,$$
$$V = J_m(s_{mn}r)\sin m\theta, \quad m = 1, 2, 3, \ldots, \quad n = 1, 2, \ldots$$

then provide solutions of Eq. (8–136) with $\lambda = s_{mn}^2$ and, by the theory described above, they form a complete orthogonal system with a weight function $w(r) = r$ for the origin $0 \leqslant \theta \leqslant 2\pi, 0 \leqslant r \leqslant 1$. Finally,

$$T(t) = \text{const} \cdot e^{-s_{mn}^2 ct}$$

gives the corresponding solutions of the equation for $T(t)$ and

$$u = \frac{1}{2} \sum_{n=1}^{\infty} p_{0n} e^{-s_{0n}^2 ct} J_0(s_{0n}r)$$

$$+ \sum_{m=1}^{\infty} \sum_{n=1}^{\infty} \{p_{mn} e^{-s_{mn}^2 ct} J_m(s_{mn}r)\cos m\theta + q_{mn} e^{-s_{mn}^2 ct} J_m(s_{mn}r)\sin m\theta\}.$$

Coefficients p_{mn}, q_{mn} are found from the initial condition, which can be written as

$$u = f(r, \theta) \quad \text{for} \quad t = 0.$$

Thus

$$f(r, \theta) = \frac{1}{2} \sum_{n=1}^{\infty} p_{0n} J_0(s_{0n}r) + \sum_{m=1}^{\infty} \sum_{n=1}^{\infty} \{p_{mn} J_m(s_{mn}r)\cos m\theta + q_{mn} J_m(s_{mn}r)\sin m\theta\}.$$

By orthogonality (not forgetting the weight function r), the coefficients p_{mn} are found as follows:

$$\int_0^{2\pi} \int_0^1 f(r, \theta)\cos m\theta \, J_m(s_{mn}r)r \, dr \, d\theta$$

$$= p_{mn} \int_0^{2\pi} \int_0^1 \cos^2 m\theta \, J_m^2(s_{mn}r) r \, dr \, d\theta$$

$$= \frac{\pi}{2} p_{mn}[J_{m+1}(s_{mn})]^2.$$

Here we took advantage of the rule

$$\int_0^1 r J_m^2 (s_{mn} r) dr = \frac{1}{2} \left[J_{m+1} (s_{mn}) \right]^2$$

(see Section 12–11). A similar equation is found for q_{mn}, with $\cos m\theta$ replaced by $\sin m\theta$. ◄

Transform methods

Both the Laplace and Fourier transforms can be applied to appropriate boundary-value problems in higher dimensions. The Laplace transform is employed most commonly for initial-value problems, with a transform with respect to t. The Fourier transform is used both with respect to t (on the interval $-\infty < t < \infty$) and with respect to one or more space variables x, y, ... on infinite intervals. The Fourier transform is often used with respect to one space variable and then another. This is equivalent to using a higher-dimension Fourier transform, such as

$$\Phi_2[f(x, y)] = \int_{-\infty}^{\infty} \int_{-\infty}^{\infty} f(x, y) e^{-i\omega_1 x - i\omega_2 y} dx \, dy.$$

The result is a function $\varphi(\omega_1, \omega_2)$. (For a discussion of this topic see COURANT–HILBERT.)

═══════════════ **Problems (Section 8-13)** ═══════════════

Solve the boundary-value problems 1 through 5.

1. $u_t - (u_{xx} + u_{yy}) = 0$ in the square region $0 < x < \pi, 0 < y < \pi$, for $0 < t < \infty$; $u = 0$ on the edges of the square; $u(x, y, 0) = 5 \sin x \sin 3y$.

2. $u_{tt} - (u_{xx} + u_{yy}) = 0$ in the square region $0 < x < \pi, 0 < y < \pi$, for $0 < t < \infty$; $u = 0$ on the edges of the square, $u(x, y, 0) = x(\pi - x) \sin y$, $u_t(x, y, 0) = 0$.

3. $u_{tt} - (u_{xx} + u_{yy}) = 0$ in the square region $0 < x < \pi$, $0 < y < \pi$, for $0 < t < \infty$; $u_x(x, 0, t) = 0$, $u_x(x, \pi, t) = 0$, $u_y(0, y, t) = 0$, $u_y(\pi, y, t) = 0$; $u(x, y, 0) = (3\pi x^2 - 2x^3) \cos y$, $u_t(x, y, 0) = 0$.

4. $u_t - (u_{xx} + u_{yy} + u_{zz}) = 0$ in the cubical region $0 < x < \pi$, $0 < y < \pi$, $0 < z < \pi$, for $0 < t < \infty$; $u = 0$ on the faces of the cube, $u(x, y, 0) = \sin x \sin 2y \sin 3z$.

5. $u_{tt} - (u_{rr} + r^{-2} u_{\theta\theta} + r^{-1} u_r) = 0$, θ unrestricted, $0 \leqslant r < 1$; $u(r, \theta, t)$ has period 2π in θ; $u(1, \theta, t) = 0$; $u(r, \theta, 0) = (1 - r) \sin \theta$, $u_t(r, \theta, 0) = 0$.

6. The Laplace equation in spherical coordinates ρ, φ, θ (where $x = \rho \sin \varphi \cos \theta$, $y = \rho \sin \varphi \sin \theta$, $z = \rho \cos \varphi$) has the form

$$u_{\rho\rho} + \rho^{-2} u_{\varphi\varphi} + (\rho^2 \sin^2 \varphi)^{-1} u_{\theta\theta} + 2\rho^{-1} u_\rho + \rho^{-2} \cot \varphi u_\varphi = 0.$$

(See Section 9–17). Find a solution $u(\rho, \varphi, \theta)$ of this equation in the spherical region $0 \leqslant \rho < 1$ such that $u(1, \varphi, \theta) = \cos \varphi + 2 \cos^2 \varphi$.

HINT: Since the boundary values are independent of θ, seek a solution independent of θ, so that $u_{\theta\theta} = 0$. Separate variables in ρ and φ for $0 \leqslant \rho < 1, 0 \leqslant \varphi \leqslant \pi$; note that the equation for $R(\rho)$ has solutions $R = \rho^n$ for appropriate choices of an eigenvalue λ. Only the choices $n = 0, 1, 2, \ldots$ give rise to solutions of the partial differential equation for $\rho \leqslant 1$, as can be verified. Hence deduce that $\lambda = n(n+1)$. Show that the corresponding equation for Φ, after the substitution $x = \cos\varphi$, becomes the Legendre equation

$$(1 - x^2)\frac{d^2\Phi}{dx^2} - 2x\frac{d\Phi}{dx} + n(n+1)\Phi = 0, \quad -1 \leqslant x \leqslant 1.$$

The only solutions that give rise to solutions u of the partial differential equation are the Legendre polynomials $P_n(x)$ (times constants). Conclude that $u = \sum_{n=0}^{\infty} c_n \rho^n P_n(\cos\varphi)$. Now adjust the constants c_n to satisfy the boundary conditions (again use $x = \cos\varphi$, and use the orthogonality of the Legendre polynomials).

Chapter 9 Vector Differential Calculus

9 – 1 INTRODUCTION

Calculus has its simplest applications to problems in one dimension. For example, for the motion of a particle on a straight line the first derivative can be interpreted as velocity, the second derivative as acceleration. If a particle is subject to a force varying with position x (as in the case of gravitation), then the force itself is described by a single function of x and the work done by the force when the particle moves from x_1 to x_2 is obtained by integrating the function from x_1 to x_2.

In the present chapter the calculus is extended to two- and three-dimensional space and to the general case of n-dimensional space R^n. Here the vector concept plays a central role. From a single function of several variables we obtain the first partial derivatives and hence a vector field, the *gradient field*, that assigns a vector to each point of a region in the space considered (Fig. 9–1). In many physical problems

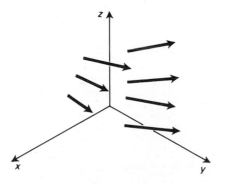

Fig. 9–1. Vector field in space.

the variation of force with position in space is described by such a gradient field.

The simplest cases for vector calculus arise when all functions are linear. A major contribution of the differential calculus is to provide a systematic way of approximating general functions by linear ones. We can regard this as a process of *linearization* of nonlinear problems. The process is crucial in formulating physical laws and will often be stressed in this chapter. Once a problem is linearized, we have only questions of linear algebra to deal with. Thus the methods of Chapter 5 are applicable.

$9-2$ VECTORS IN 3-DIMENSIONAL SPACE

We here review briefly the dot product, cross product, and associated geometry for 3-dimensional space. In the following section we review vector functions of one variable. Most of the ideas and results can be generalized to n-dimensional space (for a detailed discussion see CLA, Chapter 11).

We assume the usual Cartesian coordinate system, with origin O, coordinates (x, y, z), and corresponding basis vectors $\mathbf{i}, \mathbf{j}, \mathbf{k}$, as in Fig. 9–2. For simplicity, we shall assume that we have a right-handed coordinate system, that is, $\mathbf{i}, \mathbf{j}, \mathbf{k}$ are possible directions of the thumb, index finger, and middle finger of the right hand, as in the figure.

Point $P(x, y, z)$ has position vector $\mathbf{r} = \overrightarrow{OP} = x\mathbf{i} + y\mathbf{j} + z\mathbf{k}$. Vector $\mathbf{r}$ has magnitude $\|\mathbf{r}\|$ or $|\mathbf{r}|$:

$$|\mathbf{r}| = r = \sqrt{x^2 + y^2 + z^2}, \tag{9–20}$$

where r is the distance from O to P. Here x, y, z are components of $\mathbf{r}$ (with respect to the chosen basis). In general, a vector $\mathbf{v}$ has components v_x, v_y, v_z with respect to this basis and $\mathbf{v} = v_x\mathbf{i} + v_y\mathbf{j} + v_z\mathbf{k}$. In particular, the vector $\overrightarrow{P_1P_2}$ from point $P_1(x_1, y_1, z_1)$ to point $P_2(x_2, y_2, z_2)$ equals $\overrightarrow{OP_2} - \overrightarrow{OP_1}$ and hence has components $x_2 - x_1, y_2 - y_1, z_2 - z_1$. As in Chapter 5, we can write $\mathbf{v} = (v_x, v_y, v_z)$ and treat this as a row vector or column vector, as convenient.

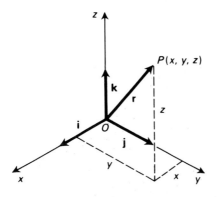

Fig. 9–2. Vectors in space.

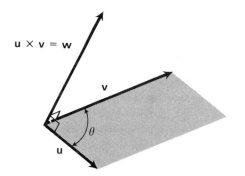

u × v = w

v

θ

u

Fig. 9–3. Dot product and cross product.

The dot product (inner product, scalar product) of two vectors **u**, **v** (Fig. 9–3) is

$$\mathbf{u} \cdot \mathbf{v} = |\mathbf{u}|\,|\mathbf{v}|\cos\theta, \tag{9–21}$$

where θ is the angle between **u** and **v** ($0 \leqslant \theta \leqslant \pi$). If **u** or **v** is $\mathbf{0} = 0\mathbf{i} + 0\mathbf{j} + 0\mathbf{k}$, the angle θ is not defined and, by definition, $\mathbf{u} \cdot \mathbf{v} = 0$ in this case. The general rule is:

$\mathbf{u} \cdot \mathbf{v} = 0$ *if and only if* **u**, **v** *are perpendicular.*

Here the **0** vector is considered to be perpendicular to (and parallel to) all vectors.

The factor $|\mathbf{v}|\cos\theta$ in Eq. (9–21) can be interpreted as the component of **v** in the direction of **u**, written as $\text{comp}_u\,\mathbf{v}$. Thus

$$\mathbf{u} \cdot \mathbf{v} = |\mathbf{u}|\,\text{comp}_u\,\mathbf{v}. \tag{9–21'}$$

The cross product (vector product) of **u** and **v**, denoted by $\mathbf{u} \times \mathbf{v}$, is a vector $\mathbf{w} = \mathbf{u} \times \mathbf{v}$ such that

$$\begin{cases} |\mathbf{w}| = |\mathbf{u}|\,|\mathbf{v}|\sin\theta, \\ \mathbf{w} \cdot \mathbf{u} = 0,\ \mathbf{w} \cdot \mathbf{v} = 0, \\ \mathbf{u},\ \mathbf{v},\ \mathbf{w}\ \text{form a positive triple} \end{cases} \tag{9–22}$$

(see Fig. 9–3). Here **u**, **v**, **w** (in that order) form a positive triple if in the order given they roughly determine a right-handed coordinate system; more precisely if **u**, **v**, **w** in the order given can be obtained from **i**, **j**, **k** by continuously changing the lengths and directions of these vectors without ever reducing one vector to **0** or making the vectors coplanar (linearly dependent). If **u**, **v**, **w** form a positive triple, then so do **w**, **u**, **v** and **v**, **w**, **u**; the triple **u**, **w**, **v**, the triple **w**, **v**, **u**, and the triple **v**, **u**, **w** are all termed negative. Every ordered triple of linearly independent vectors in space is either positive or negative, but not both.

Definition (9–22) breaks down if **u** or **v** is **0** or if $\theta = 0$ or π. In all these cases we set $\mathbf{u} \times \mathbf{v} = \mathbf{0}$. Then the rule is:

$\mathbf{u} \times \mathbf{v} = \mathbf{0}$ *if and only if* **u**, **v** *are parallel.*

Also, as illustrated in Fig. 9–3,

$$|\mathbf{u} \times \mathbf{v}| = \text{Area of parallelogram with sides } \mathbf{u},\ \mathbf{v}.$$

The dot and cross products obey a number of algebraic rules:

$$\mathbf{u}\cdot\mathbf{v} = \mathbf{v}\cdot\mathbf{u}, \qquad \mathbf{u}\cdot(c\mathbf{v}) = (c\mathbf{u})\cdot\mathbf{v} = c(\mathbf{u}\cdot\mathbf{v}),$$

$$\mathbf{u}\cdot(\mathbf{v}+\mathbf{w}) = \mathbf{u}\cdot\mathbf{v}+\mathbf{u}\cdot\mathbf{w}, \qquad \mathbf{u}\cdot\mathbf{u} = |\mathbf{u}|^2,$$

$$\mathbf{u}\times\mathbf{v} = -\mathbf{v}\times\mathbf{u}, \qquad \mathbf{u}\times(c\mathbf{v}) = (c\mathbf{u})\times\mathbf{v} = c(\mathbf{u}\times\mathbf{v}) \tag{9-23}$$

$$\mathbf{u}\times(\mathbf{v}+\mathbf{w}) = \mathbf{u}\times\mathbf{v}+\mathbf{u}\times\mathbf{w}, \qquad \mathbf{u}\times\mathbf{u} = \mathbf{0}.$$

Here c is a number (scalar).

The dot and cross products are related to the components of $\mathbf{u}$ and $\mathbf{v}$ as follows:

$$\left\{ \begin{aligned} &\mathbf{u}\cdot\mathbf{v} = u_x v_x + u_y v_y + u_z v_z, \\ &u_x = \mathbf{u}\cdot\mathbf{i}, \qquad u_y = \mathbf{u}\cdot\mathbf{j}, \qquad u_z = \mathbf{u}\cdot\mathbf{k}, \\ &\mathbf{u}\times\mathbf{v} = (u_y v_z - u_z v_y)\mathbf{i} + (u_z v_x - u_x v_z)\mathbf{j} + (u_x v_y - u_y v_x)\mathbf{k} \\ &\qquad = \begin{vmatrix} \mathbf{i} & \mathbf{j} & \mathbf{k} \\ u_x & u_y & u_z \\ v_x & v_y & v_z \end{vmatrix}. \end{aligned} \right. \tag{9-24}$$

Here the determinant is to be expanded by minors of the first row (see Section 5–2).

A vector $\mathbf{u}$ is called a *unit vector* if $|\mathbf{u}| = 1$. For such a vector the components are *direction cosines:* $u_x = \cos\alpha, u_y = \cos\beta, u_z = \cos\gamma$, where $\alpha = \measuredangle(\mathbf{u}, \mathbf{i})$, and so on (see Problem 3 below).

The *scalar triple product* of $\mathbf{u}, \mathbf{v}, \mathbf{w}$ (in that order) is $\mathbf{u}\times\mathbf{v}\cdot\mathbf{w}$ (no parentheses are needed here, since $\mathbf{u}\times(\mathbf{v}\cdot\mathbf{w})$ makes no sense). This obeys the rules

$$\left\{ \begin{aligned} \mathbf{u}\times\mathbf{v}\cdot\mathbf{w} &= \mathbf{u}\cdot\mathbf{v}\times\mathbf{w} = \mathbf{w}\times\mathbf{u}\cdot\mathbf{v} = \mathbf{w}\cdot\mathbf{u}\times\mathbf{v} \\ &= \mathbf{v}\times\mathbf{w}\cdot\mathbf{u} = \mathbf{v}\cdot\mathbf{w}\times\mathbf{u} \\ &= -\mathbf{u}\times\mathbf{w}\cdot\mathbf{v} = -\mathbf{u}\cdot\mathbf{w}\times\mathbf{v}, \text{ etc.,} \\ \mathbf{u}\times\mathbf{v}\cdot\mathbf{w} &= \begin{vmatrix} u_x & u_y & u_z \\ v_x & v_y & v_z \\ w_x & w_y & w_z \end{vmatrix}, \\ \mathbf{u}\times\mathbf{v}\cdot\mathbf{w} &= \pm\text{Volume of parallelepiped with edges } \mathbf{u}, \mathbf{v}, \mathbf{w}. \end{aligned} \right. \tag{9-25}$$

For the last rule, see Fig. 9-4. The use of the $+$ or $-$ sign depends on whether $\mathbf{u}, \mathbf{v}, \mathbf{w}$ (in that order) form a positive or negative triple; when they are linearly dependent, the determinant equals zero (Section 5–9).

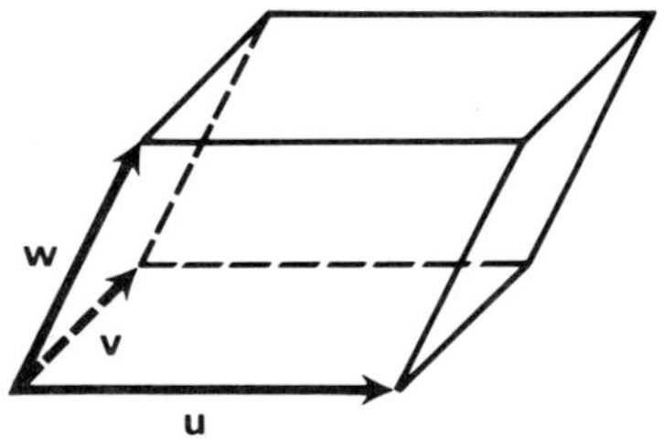

Fig. 9-4. Scalar triple product and volume.

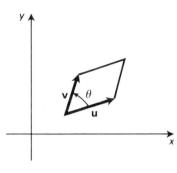

Fig. 9-5. Determinant and area.

The rules (9–25) show that *a third-order determinant can be interpreted as volume* (up to $\pm$ sign). In the same way we can verify that the second-order determinant

$$\begin{vmatrix} u_x & u_y \\ v_x & v_y \end{vmatrix}$$

equals $\pm$ the *area* of the parallelogram in the xy-plane having edges $\mathbf{u} = u_x\mathbf{i} + u_y\mathbf{j}$, $\mathbf{v} = v_x\mathbf{i} + v_x\mathbf{j}$, as in Fig. 9–5. The $+$ sign is used if the directed angle θ from $\mathbf{u}$ to $\mathbf{v}$ is between 0 and π, the $-$ sign is used if $\pi < \theta < 2\pi$. In the first case $\mathbf{u}$, $\mathbf{v}$ form a *positive pair* (as do $\mathbf{i}$, $\mathbf{j}$); in the second case they form a negative pair (as do $\mathbf{j}$, $\mathbf{i}$). When $\mathbf{u}$, $\mathbf{v}$ are linearly dependent, the determinant is 0.

There is a similar rule for determinants of arbitrary order n. If $A = (a_{ij})$ is an $n \times n$ matrix, then det A equals $+$ or $-$ the volume of the *n-dimensional parallelotope* in R^n, whose edges are the row-vectors of A. When these vectors form an orthogonal system (Sections 3–11 and 5–16), we have a *rectangular parallelotope* whose volume equals the product of the lengths of the n row vectors (see CLA, Chapter 10).

The vector triple products of $\mathbf{u}$, $\mathbf{v}$, $\mathbf{w}$ are the expressions $(\mathbf{u} \times \mathbf{v}) \times \mathbf{w}$ and $\mathbf{u} \times (\mathbf{v} \times \mathbf{w})$. Here the parentheses are needed. We have the rules

$$(\mathbf{u} \times \mathbf{v}) \times \mathbf{w} = (\mathbf{u} \cdot \mathbf{w})\mathbf{v} - (\mathbf{v} \cdot \mathbf{w})\mathbf{u},$$

$$\mathbf{u} \times (\mathbf{v} \times \mathbf{w}) = (\mathbf{u} \cdot \mathbf{w})\mathbf{v} - (\mathbf{u} \cdot \mathbf{v})\mathbf{w}.$$

These can be summarized as follows:

Triple product $=$ (Outer dot Remote) Adjacent $-$ (Outer dot Adjacent) Remote.

9 – 3   CALCULUS OF VECTOR FUNCTIONS OF ONE VARIABLE

We consider vector functions $\mathbf{r} = \mathbf{F}(t)$, where $\mathbf{r}$ is a vector of R^3 and t varies over an interval. If we represent $\mathbf{r}$ as $\overrightarrow{OP}$, as in Fig. 9–2, so that P is the point (x, y, z) and $\mathbf{r}$ has components x, y, z, then the vector function $\mathbf{r} = \mathbf{F}(t)$ is equivalent to three scalar functions

$$x = f(t), \qquad y = g(t), \qquad z = h(t). \tag{9–30}$$

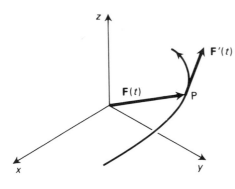

Fig. 9-6. Vector function and its derivative.

For example,

$$\mathbf{r} = (3 + t)\mathbf{i} + e^t\mathbf{j} + (\cos 2t)\mathbf{k} \qquad (9\text{--}31)$$

is such a vector function, with

$$x = 3 + t, \qquad y = e^t, \qquad z = \cos 2t.$$

In general, Eqs. (9–30) describe the motion of the point $P(x, y, z)$ in space in terms of the parameter t interpreted as time (Fig. 9–6).

The vector function $\mathbf{F}(t)$ is said to have limit $\mathbf{u}$ at a particular t_0 if t_0 is a point of an interval on which $\mathbf{F}(t)$ is defined, except perhaps at t_0 itself, and for each $\varepsilon > 0$ there is a $\delta > 0$ such that $|\mathbf{F}(t) - \mathbf{u}| < \varepsilon$ for $0 < |t - t_0| < \delta$. We say that $\mathbf{F}(t)$ is *continuous* at t_0 if $\mathbf{F}(t)$ has limit $\mathbf{F}(t_0)$ at t_0. We verify that $\mathbf{F}(t) = f(t)\mathbf{i} + g(t)\mathbf{j} + h(t)\mathbf{k}$ is continuous at t_0 if and only if all three functions $f(t)$, $g(t)$, $h(t)$ are continuous at t_0. Thus the function (9–31) is continuous for all t.

The *derivative* of a function $\mathbf{r} = \mathbf{F}(t)$ is defined in the familiar way:

$$\frac{d\mathbf{r}}{dt} = \mathbf{F}'(t) = \lim_{h \to 0} \frac{1}{h}[\mathbf{F}(t + h) - \mathbf{F}(t)].$$

We then verify that, for $\mathbf{F}(t) = f(t)\mathbf{i} + g(t)\mathbf{j} + h(t)\mathbf{k}$,

$$\mathbf{F}'(t) = f'(t)\mathbf{i} + g'(t)\mathbf{j} + h'(t)\mathbf{k}. \qquad (9\text{--}32)$$

Thus the function (9–31) has derivative $\mathbf{i} + e^t\mathbf{j} - 2\sin 2t\,\mathbf{k}$.

For the case of the motion of a point P, as in Fig. 9–6, the derivative $\mathbf{F}'(t)$ can be interpreted as the *velocity vector* $\mathbf{v}$ of the motion. It is tangent to the path and has magnitude

$$\mathbf{v} = \left|\frac{d\mathbf{r}}{dt}\right| = \sqrt{\left(\frac{dx}{dt}\right)^2 + \left(\frac{dy}{dt}\right)^2 + \left(\frac{dz}{dt}\right)^2} = \frac{ds}{dt}, \qquad (9\text{--}33)$$

where s is distance along the path, increasing in the direction of motion. Thus $|\mathbf{v}|$ is speed and

$$\mathbf{T} = \frac{\mathbf{v}}{|\mathbf{v}|} = \frac{dx}{ds}\mathbf{i} + \frac{dy}{ds}\mathbf{j} + \frac{dz}{ds}\mathbf{k}$$

is a unit tangent vector to the path.

Second and higher derivatives can be formed in the same way. For the case of motion of point P, the *acceleration vector* of the motion is $\mathbf{F}''(t) = d\mathbf{v}/dt$. Hence, by Newton's second law, if P is a particle of mass m subject to force $\mathbf{G}$, then

$$m\frac{d^2\mathbf{r}}{dt^2} = \mathbf{G}. \tag{9-34}$$

The derivative obeys the usual rules for derivative of a sum and of a constant times a function. Also the following rules can be verified:

$$\begin{cases} [h(t)\,\mathbf{F}(t)]' = h(t)\,\mathbf{F}'(t) + h'(t)\,\mathbf{F}(t), \\ (\mathbf{F}(t)\cdot\mathbf{G}(t))' = \mathbf{F}(t)\cdot\mathbf{G}'(t) + \mathbf{F}'(t)\cdot\mathbf{G}(t), \\ (\mathbf{F}(t)\times\mathbf{G}(t))' = \mathbf{F}(t)\times\mathbf{G}'(t) + \mathbf{F}'(t)\times\mathbf{G}(t). \end{cases} \tag{9-35}$$

Here the indicated derivatives are assumed to exist.

For a *linear function*,

$$\mathbf{r} = t\mathbf{a} + \mathbf{b}, \quad \mathbf{a} \neq \mathbf{0},$$

the corresponding path is a straight line, with velocity vector $\mathbf{a}$ along the line. When $\mathbf{a}$ is a unit vector, t can be interpreted as distance s along the line, so that $\mathbf{r} = s\mathbf{a} + \mathbf{b}$.

We remark that the derivative $\mathbf{F}'(t)$ can exist only at a t value at which $\mathbf{F}$ is continuous. However, continuity alone does not ensure the existence of the derivative.

The *definite integral* of $\mathbf{F}(t)$ from a to b can be defined in the usual way as limit of a sum. The integral exists if $\mathbf{F}(t)$ is continuous for $a \leqslant t \leqslant b$, and

$$\int_a^b \mathbf{F}(t)\,dt = \int_a^b f(t)\,dt\,\mathbf{i} + \int_a^b g(t)\,dt\,\mathbf{j} + \int_a^b h(t)\,dt\,\mathbf{k}. \tag{9-36}$$

We have the familiar rules on integral of a sum, integral of a constant times a function, sum of integrals from a to b and from b to c. Also,

$$\int_a^b \mathbf{u}\cdot\mathbf{F}(t)\,dt = \mathbf{u}\cdot\int_a^b \mathbf{F}(t)\,dt,$$

$$\int_a^b \mathbf{u}\times\mathbf{F}(t)\,dt = \mathbf{u}\times\int_a^b \mathbf{F}(t)\,dt, \tag{9-37}$$

if $\mathbf{u}$ is a *constant* vector. The integral is related to the derivative in the familiar ways:

$$\frac{d}{d\tau}\int_0^\tau \mathbf{F}(t)\,dt = \mathbf{F}(\tau), \tag{9-38}$$

$$\int_a^b \mathbf{F}'(t)\,dt = \mathbf{F}(b) - \mathbf{F}(a), \tag{9-39}$$

if $\mathbf{F}'(t)$ is continuous for $a \leqslant t \leqslant b$.

For the moving point P with velocity vector $\mathbf{v} = d\mathbf{r}/dt$, Eq. (9-39) becomes

$$\int_a^b \mathbf{v}\,dt = \int_a^b \frac{d\mathbf{r}}{dt}\,dt = \overrightarrow{OP_b} - \overrightarrow{OP_a} = \overrightarrow{P_a P_b},$$

where P_a is the position of P for $t = a$, P_b the position for $t = b$. Thus the *integral of the velocity vector of P is the total net displacement of P.*

Similarly, if P has acceleration vector $\mathbf{a} = d\mathbf{v}/dt$ and P is a particle of mass m,

$$\int_a^b m\mathbf{a}\, dt = \int_a^b m\frac{d\mathbf{v}}{dt}\, dt = m\mathbf{v}_b - m\mathbf{v}_a.$$

Since $m\mathbf{v}$ is the *linear momentum* of P, we can say: *the integral of* $m\mathbf{a}$ *is the net gain in momentum.*

==================== **Problems (Section 9–3)** ====================

1. Given the points $A\,(1, 1, 1)$, $B\,(2, 1, 2)$, $C\,(5, 0, 3)$ in space, find the following:
   a) the sides of the triangle ABC,
   b) the angles of triangle ABC,
   c) a vector normal to the plane of the triangle,
   d) the area of the triangle.

2. Proceed as in Problem 1 with $A\,(1, 2, 4)$, $B\,(3, 1, 0)$, $C\,(2, -1, 1)$.

3. a) Show that, if $\mathbf{u}$ is a unit vector, then the direction cosines of $\mathbf{u}$ are the components of $\mathbf{u}$.
   b) Find the direction cosines of the vectors $3\mathbf{i} + 2\mathbf{j} + 6\mathbf{k}$ and $(2/3)\mathbf{i} + (2/3)\mathbf{j} + (1/3)\mathbf{k}$.

4. Let points $A(3, 2, 2)$, $B(5, 7, 4)$, $C(8, 5, 6)$, $D(6, 0, 4)$ be given.
   a) Show that $ABCD$ is a parallelogram.
   b) Show that $ABCD$ is in fact a rectangle.

5. A *plane* in space is determined by a point $P_1\,(x_1, y_1, z_1)$ in the plane and a vector $\mathbf{n} = A\mathbf{i} + B\mathbf{j} + C\mathbf{k}$ normal to the plane.

   a) Show that $P\,(x, y, z)$ is in the plane if and only if $\overrightarrow{P_1 P} \cdot \mathbf{n} = 0$ and hence deduce the equation
   $$A(x - x_1) + B(y - y_1) + C(z - z_1) = 0 \text{ for the plane.}$$

   b) Find the equation of the plane passing through point $(5, 2, -3)$ with normal vector $\mathbf{i} - 2\mathbf{j} + \mathbf{k}$.
   c) Find the equation of the plane passing through points $(1, 0, 1)$, $(1, 2, 2)$, $(3, -1, 5)$.

6. The *work* done by a constant force $\mathbf{F}$ applied to a particle moving from A to B along the segment AB is defined as the component of $\mathbf{F}$ in the direction of motion times the distance moved.
   a) Conclude that the work done equals $\mathbf{F} \cdot \overrightarrow{AB}$.
   b) Find the work done by gravity on a 100 gram mass sliding 3 cm down an inclined plane at an angle of $30°$ with the horizontal.

7. Let a rigid body rotate about an axis L through O at constant angular velocity ω. Justify the expression $\mathbf{v} = \omega \times \overrightarrow{OP}$ for the velocity vector of the point P of the body, where ω is a properly chosen vector along L with magnitude ω (Fig. 9–7).

8. Find the shortest distance from point $(1, 1, 2)$ to the plane determined by points $(0, 1, 0)$, $(1, 1, -3)$, $(5, 0, 1)$.

9. Let $\mathbf{F}(t) = t^2\mathbf{i} + (3t - t^3)\mathbf{j} + (2 + t + t^2)\mathbf{k}$.
   a) Find $\mathbf{F}'(t)$ b) Find $\mathbf{F}''(t)$ c) Evaluate $\int_0^1 \mathbf{F}(t)\, dt$.

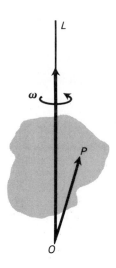

L

ω

P

O

Fig. 9–7. Angular-velocity vector of rotating rigid body.

10. Let $\mathbf{F}(t)$ be a differentiable vector function such that $|\mathbf{F}(t)| \equiv 1$. Show that $\mathbf{F}(t) \cdot \mathbf{F}'(t) \equiv 0$. Interpret the result geometrically.

    HINT: For $\mathbf{r} = \mathbf{F}(t) = \overrightarrow{OP}$, P moves on a sphere.

11. Prove that if $\mathbf{F}'(t) \equiv \mathbf{0}$, then $\mathbf{F}(t) \equiv \text{const}$.

12. Let the point P move in space in such a way that, with $\mathbf{r} = \overrightarrow{OP}$ and $\mathbf{v} = d\mathbf{r}/dt$,

$$\frac{d\mathbf{v}}{dt} = -k\mathbf{r}, \quad k = \text{const} > 0.$$

    Show that $|\mathbf{v}|^2 + k|\mathbf{r}|^2 = \text{const}$.

    HINT: Write the left side of the last equation as $\mathbf{v} \cdot \mathbf{v} + k\mathbf{r} \cdot \mathbf{r}$ and differentiate this expression.

    REMARK: The equation $d\mathbf{v}/dt = -k\mathbf{r}$ is the differential equation for the motion of a particle of unit mass in space subject to a springlike attractive force directed toward the origin. The equation $|\mathbf{v}|^2 + k|\mathbf{r}|^2 = \text{const}$ expresses the *conservation of energy* for each motion: the kinetic energy is $|\mathbf{v}|^2/2$, the potential energy is $k|\mathbf{r}|^2/2$.

13. The motion of a satellite of mass m is governed by the equation

$$m\frac{d\mathbf{v}}{dt} = -\frac{k}{r^3}\mathbf{r},$$

    where k is a positive constant, $\mathbf{r} = \overrightarrow{OP}$, and O is the position of the attracting large body (treated as fixed here). The angular-momentum vector of the satellite about O is the vector $\mathbf{r} \times m\mathbf{v}$.
    a) Show that the angular-momentum vector has a constant value $\mathbf{h}$ for any particular motion of the satellite (see Problem 11).
    b) Conclude from the result of part (a) that if $\mathbf{h} \neq \mathbf{0}$, then the satellite moves in a plane that passes through O and has $\mathbf{h}$ as normal vector.

$9-4$ FUNCTIONS OF TWO VARIABLES, PARTIAL DERIVATIVES, DIFFERENTIALS

We here begin our study of functions of two or more variables, such as

$$z = f(x, y), \qquad w = f(x, y, z), \qquad y = f(x_1, \ldots, x_n).$$

Such functions occur in many applications of mathematics. For example, the variation of temperature u in space is described by a function $u = f(x, y, z, t)$, where x, y, z are coordinates in space and t is time. A similar function describes the variation of barometric pressure or density of the atmosphere. The velocity components of a fluid motion are similar functions of space and time.

We shall emphasize the typical case $z = f(x, y)$ and (as in the next section) point out the generalization to functions of three or more variables. Typically the graph of a function $z = f(x, y)$ is a surface in space, as in·Fig. 9–8.

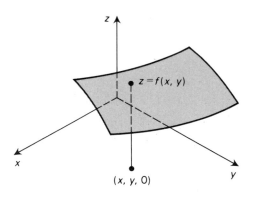

$(x, y, 0)$ **Fig. 9-8.** Graph of $z = f(x,y)$.

If function f is linear, then it can be written as $a(x - x_1) + b(y - y_1) + z_1$ (in various ways). The corresponding equation

$$z = a(x - x_1) + b(y - y_1) + z_1$$

has as graph (Fig. 9–9) a plane through (x_1, y_1, z_1) with normal vector $-a\mathbf{i} - b\mathbf{j} + \mathbf{k}$ (see Problem 5 following Section 9–3).

We need the concept of *open region* (or *domain*). In general, a set in the xy-plane is said to be *open* if, whenever (x_0, y_0) is in the set, for some $\delta > 0$, all points (x, y) at distance less than δ from (x_0, y_0) are in the set. For each δ, the set of these points (x, y) forms a *neighborhood* of (x_0, y_0) of radius δ, or δ-neighborhood of (x_0, y_0). An open region is an open set D that is *connected*, i.e., whenever (x_1, y_1) and (x_2, y_2) are in the set, there is a broken line in the set joining (x_1, y_1) to (x_2, y_2). These concepts are illustrated in Fig. 9–10. Here the open region is bounded by a closed curve C_2. Such open regions, bounded by one or more closed curves, are common. Also the points of a half-plane, say all (x, y) with $y > 0$, form an open region, as do the

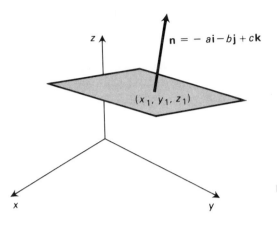

Fig. 9-9. Graph of linear function
$$z = a(x - x_1) + b(y - y_1) + z_1.$$

points between two parallel lines (an infinite strip). The points (x, y) such that $|x| > 1$ form an open set but not an open region (Problem 7 below).

In Fig. 9–10 the points on the bounding curve are not considered to be points of the open region D. In fact each such point fails to have a δ-neighborhood contained in D. We call such points *boundary points* of D; together they form the *boundary of D*. In general a point (x^*, y^*) is said to be a boundary point of a set if every δ-neighborhood of (x^*, y^*) contains at least one point in the set and at least one point not in the set.

An open region plus its boundary points is called a *closed region*. For example, a circle plus its interior forms a closed region R. A set is termed *bounded* if it can be enclosed in a circle. Thus the open region D of Fig. 9–10 is bounded.

Let a function $z = f(x, y)$ be defined in an open region D. The function f is said to be *continuous* at the point (x_0, y_0) of D if, given $\varepsilon > 0$, we can choose $\delta > 0$ so that $|f(x, y) - f(x_0, y_0)| < \varepsilon$ for every point (x, y) of D in the δ-neighborhood of (x_0, y_0).

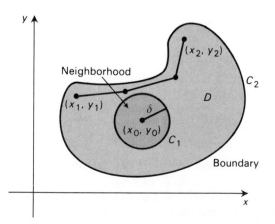

Fig. 9-10. Open region (or domain) D.

We say that f is *continuous* in D if f is continuous at every point of D. We can verify the usual rules on the continuity of the sum, product, and quotient of continuous functions (except for division by 0). Also the composition of continuous functions is continuous. For example, e^z is continuous for all z and hence $e^{f(x,y)}$ is continuous in D if f is continuous in D. By these rules we conclude that each *polynomial* in x, y:

$$z = a_{00} + a_{10} + a_{01}y + a_{20}x^2 + a_{11}xy + \cdots + a_{n0}x^n + \cdots + a_{0n}y^n \quad (9\text{-}40)$$

is continuous for all (x, y), as is each *rational function* (ratio of two polynomials) in each open region in which the denominator is never 0.

> REMARK: We also refer to a function continuous in a closed region, function continuous on a curve, function continuous on the boundary of a region. The definition is the same as that given above for an open region; in general, f is continuous at (x_0, y_0) if $|f(x, y) - f(x_0, y_0)| < \varepsilon$ for every point (x, y) at which f is defined, in some δ-neighborhood of (x_0, y_0).

If $z = f(x, y)$ is defined in the open region D, then we can seek *partial derivatives* of f: at (x_0, y_0), $\partial z/\partial x$ is obtained by differentiating $f(x, y_0)$ with respect to x (y being held constant, equal to y_0), and $\partial z/\partial y$ is obtained similarly. If these derivatives exist in D, then they are also functions defined in D, commonly denoted by f_x and f_y, or z_x and z_y. To emphasize which variable is held constant, we sometimes write, for example, $(\partial z/\partial x)_y$ for $\partial z/\partial x$.

Similarly, we can seek higher-order partial derivatives:

$$\frac{\partial^2 z}{\partial x^2} = \frac{\partial}{\partial x}\left(\frac{\partial z}{\partial x}\right), \qquad \frac{\partial^2 z}{\partial x \partial y} = \frac{\partial}{\partial x}\left(\frac{\partial z}{\partial y}\right), \text{ etc.}$$

It can be shown (see CLA, pp. 957–958) that if all derivatives concerned are continuous, then the order of differentiation does not matter. Thus

$$\frac{\partial^2 z}{\partial x \partial y} = \frac{\partial^2 z}{\partial y \partial x}, \qquad \frac{\partial^3 z}{\partial x^2 \partial y} = \frac{\partial^3 z}{\partial x \partial y \partial z} = \frac{\partial^3 z}{\partial y \partial x^2}, \text{ etc.} \quad (9\text{-}41)$$

These derivatives are denoted by f_{xy}, $f_{xxy}, \ldots$ or z_{xy}, $z_{xxy}, \ldots$

The differential

Let $z = f(x, y)$ be defined in the open region D and have continuous first partial derivatives f_x, f_y in D. Then we can show that for a fixed (x, y) in D and arbitrary increments h, k,

$$\Delta z = f(x + h, y + k) - f(x, y)$$
$$= f_x h + f_y k + h\varepsilon_1(h, k) + k\varepsilon_2(h, k), \quad (9\text{-}42)$$

where f_x, f_y are evaluated at (x, y) and $\varepsilon_1(h, k)$, $\varepsilon_2(h, k)$ are continuous at $(h, k) = (0, 0)$, with $\varepsilon_1(0, 0) = \varepsilon_2(0, 0) = 0$. The terms in ε_1, ε_2 can hence be regarded as small with respect to the first two terms (for small h and k) and Δz can be well approximated, for (h, k) near $(0, 0)$, by the *linear function* $f_x h + f_y k$. This uniquely

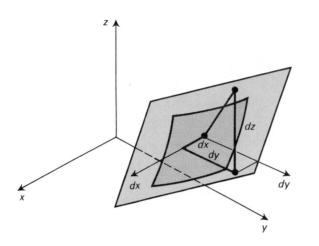

Fig. 9-11. Tangent plane and differentials.

determined* linear function is called the *differential* of the function $z = f(x, y)$ at the point (x, y), and we write:

$$dz = f_x h + f_y k. \tag{9–43}$$

We commonly write also $h = dx$, $k = dy$, so that

$$dz = f_x dx + f_y dy. \tag{9–43'}$$

This notation is consistent with the rules for operating with differentials.

For a function of one variable $y = f(x)$ it is well known that the differential relation $dy = f'(x)dx$ relates the increments dx, dy on the *tangent line* to the graph of f at the point (x, y), with $y = f(x)$. Similarly it is seen that Eq. (9–43') relates the increments dx, dy, dz for the *tangent plane* to the graph of f at (x, y, z), with $z = f(x, y)$ (see Fig. 9–11). Equivalently we can state that

$$z - z_1 = f_x(x_1, y_1)(x - x_1) + f_y(x_1, y_1)(y - y_1) \tag{9–44}$$

is the equation of the tangent plane to the graph of $z = f(x, y)$ at (x_1, y_1) (see Fig. 9–9).

EXAMPLE 1   Let $z = 3xy^2 - y^3$, so that $\partial z/\partial x = 3y^2$, $\partial z/\partial y = 6xy - 3y^2$. At the point $(2, 1)$ we have $z = 5$. To evaluate z at $(2.1, 1.3)$, we can apply Eq. (9–42) with $h = dx = 0.1$, $k = dy = 0.3$. At the point $(2, 1)$, $\partial z/\partial x = 3$ and $\partial z/\partial y = 9$. Hence Eq. (9–43) gives

$$dz = 3 \cdot 0.1 + 9 \cdot 0.3 = 3.0.$$

Thus, approximately, $z = 5 + 3.0 = 8.0$ for $x = 2.1$, $y = 1.3$. The exact value is 8.45. Accordingly, $\Delta z = 3.45$, whereas $dz = 3.0$. ◀

REMARK: From (9–42) we deduce that f is necessarily *continuous* at (x, y), since the continuity of $\varepsilon_1(h, k)$, $\varepsilon_2(h, k)$ at $(0, 0)$ implies the continuity of $f(x + h, y + k)$ at $(h, k) = (0, 0)$, which is the same as the continuity of f at (x, y).

---

* The uniqueness means the following: at a fixed (x, y) we have $\Delta z = ah + bk + h\varepsilon_1(h, k) + k\varepsilon_2(h, k)$ for ε_1, ε_2 as above and for appropriate constants a, b if and only if $a = f_x(x, y)$, $b = f_y(x, y)$.

The chain rule

Let $z = f(x, y)$, as above, and let differentiable functions $x = p(t)$, $y = q(t)$, say for $a < t < b$, be given, such that the composite function

$$z = f(p(t), q(t))$$

is defined. The chain rule asserts that this composite function has derivative

$$\frac{dz}{dt} = f_x \frac{dx}{dt} + f_y \frac{dy}{dt} = \frac{\partial z}{\partial x}\frac{dx}{dt} + \frac{\partial z}{\partial y}\frac{dy}{dt}, \qquad (9\text{-}45)$$

where f_x, f_y are evaluated at $x = p(t)$, $y = q(t)$. The rule follows from the increment formula (9–42), for if we start at a fixed t, with corresponding $x = p(t)$, $y = q(t)$, and assign an increment Δt to t, then x, y have increments $\Delta x, \Delta y$. We can now apply Eq. (9–42) with $h = \Delta x$, $k = \Delta y$ to find Δz. If we do so and then divide by Δt, we obtain

$$\frac{\Delta z}{\Delta t} = f_x \frac{\Delta x}{\Delta t} + f_y \frac{\Delta y}{\Delta t} + \frac{\Delta x}{\Delta t}\varepsilon_1(\Delta x, \Delta y) + \frac{\Delta y}{\Delta t}\varepsilon_2(\Delta x, \Delta y).$$

If we let $\Delta t \to 0$, then $\Delta x \to 0$ and $\Delta y \to 0$, so that the last two terms approach

$$\frac{dx}{dt}\cdot 0 + \frac{dy}{dt}\cdot 0 = 0.$$

By taking the limit of the first two terms we obtain the rule (9–45).

The chain rule extends to more complicated composite functions. For example, if $z = f(x, y)$ and $x = p(u, v)$, $y = q(u, v)$, all functions having continuous first partial derivatives, then we can form the composite function

$$z = f(p(u, v), q(u, v)).$$

If this function is well defined, then its partial derivatives are given by

$$\frac{\partial z}{\partial u} = f_x \frac{\partial x}{\partial u} + f_y \frac{\partial y}{\partial u} = \frac{\partial z}{\partial x}\frac{\partial x}{\partial u} + \frac{\partial z}{\partial y}\frac{\partial y}{\partial u},$$

$$\frac{\partial z}{\partial v} = f_x \frac{\partial x}{\partial v} + f_y \frac{\partial y}{\partial v} = \frac{\partial z}{\partial x}\frac{\partial x}{\partial v} + \frac{\partial z}{\partial y}\frac{\partial y}{\partial v}, \qquad (9\text{-}46)$$

where f_x, f_y are evaluated at $x = p(u, v)$, $y = q(u, v)$. This rule follows from the rule (9–45), since when, say, $\partial z/\partial u$, is formed, v is treated as a constant, so that $x = p(u, v)$, $y = q(u, v)$, are regarded as functions of *one variable* u. Then the first line of (9–46) is obtained from (9–45) by replacing t by u, the ordinary derivatives becoming partial derivatives.

EXAMPLE 2   Let $z = e^x \sin y$, where x and y are functions of t. Then by Eq. (9–45)

$$\frac{dz}{dt} = e^x \sin y \frac{dx}{dt} + e^x \cos y \frac{dy}{dt}. \quad \blacktriangleleft$$

EXAMPLE 3   Let $z = f(x, y)$, where $x = u^2 - v^2$, $y = 2uv$. Then by the rule (9–46)

$$\frac{\partial z}{\partial u} = 2u\frac{\partial z}{\partial x} + 2v\frac{\partial z}{\partial y}, \qquad \frac{\partial z}{\partial v} = -2v\frac{\partial z}{\partial x} + 2u\frac{\partial z}{\partial y}. \quad \blacktriangleleft$$

It should be remarked that these results can be obtained by simply taking differentials and applying Eq. (9–43′) and the rules of one-variable calculus. For Example 2,

$$dz = e^x \sin y \, dx + e^x \cos y \, dy = e^x \sin y \frac{dx}{dt} dt + e^x \cos y \frac{dy}{dt} dt,$$

and division by dt gives the result. For Example 3,

$$dz = \frac{\partial z}{\partial x} dx + \frac{\partial z}{\partial y} dy = \frac{\partial z}{\partial x}(2u \, du - 2v \, dv) + \frac{\partial z}{\partial y}(2v \, du + 2u \, dv)$$

$$= \left(2u \frac{\partial z}{\partial x} + 2v \frac{\partial z}{\partial y}\right) du + \left(-2v \frac{\partial z}{\partial x} + 2u \frac{\partial z}{\partial y}\right) dv$$

and comparison with the formula (for z as function of u and v)

$$dz = \frac{\partial z}{\partial u} du + \frac{\partial z}{\partial v} dv$$

gives the result.

Relation to tangent plane

The basic chain rule (9–45) has the following interpretation. Let there be a surface $z = f(x, y)$ and on this surface a curve C with parametric equations $x = p(t)$, $y = q(t)$, $z = f(p(t), q(t))$; the last equation forces C to lie on the surface $z = f(x, y)$ (see Fig. 9–12). The corresponding velocity vector, tangent to the curve, has components $dx/dt, dy/dt, dz/dt$ (Eq. (9–32)). The chain rule (9–45) can be written as

$$-\frac{\partial z}{\partial x}\frac{dx}{dt} - \frac{\partial z}{\partial y}\frac{dy}{dt} + 1\frac{dz}{dt} = 0$$

or

$$\left(-\frac{\partial z}{\partial x}\mathbf{i} - \frac{\partial z}{\partial y}\mathbf{j} + \mathbf{k}\right) \cdot \left(\frac{dx}{dt}\mathbf{i} + \frac{dy}{dt}\mathbf{j} + \frac{dz}{dt}\mathbf{k}\right) = 0.$$

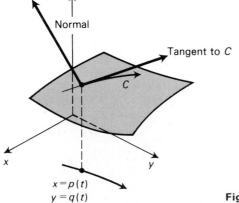

$$x = p(t)$$
$$y = q(t)$$

Fig. 9–12. Chain rule and tangent plane.

Thus at each point P on the path the tangent vector is perpendicular to the vector $-z_x\mathbf{i} - z_y\mathbf{j} + \mathbf{k}$. But by Eq. (9–44), this last vector is normal to the tangent plane at P. Hence the tangent vector to the curve lies in the tangent plane to the surface. We can thus describe the tangent plane at P as the plane containing all tangent lines at P to curves on the surface passing through P. The vector $-z_x\mathbf{i} - z_y\mathbf{j} + \mathbf{k}$ is perpendicular to the tangent plane and is a *normal* vector for the surface at P.

$9-5$ GENERALIZATION TO FUNCTIONS OF THREE OR MORE VARIABLES

All the preceding discussion extends in a natural manner to functions of more than two variables. For example, an open region in xyz-space is defined as before; a δ-neighborhood of (x_0, y_0, z_0) is now given by

$$(x - x_0)^2 + (y - y_0)^2 + (z - z_0)^2 < \delta^2.$$

Continuity and partial derivatives of $f(x, y, z)$ are defined as for $f(x, y)$. A function $w = f(x, y, z)$ has three first partial derivatives

$$\frac{\partial w}{\partial x}, \quad \frac{\partial w}{\partial y}, \quad \frac{\partial w}{\partial z}$$

and six different second partial derivatives

$$\frac{\partial^2 w}{\partial x^2}, \quad \frac{\partial^2 w}{\partial y^2}, \quad \frac{\partial^2 w}{\partial z^2}, \quad \frac{\partial^2 w}{\partial x \, \partial y} = \frac{\partial^2 w}{\partial y \, \partial x},$$

$$\frac{\partial^2 w}{\partial y \, \partial z} = \frac{\partial^2 w}{\partial z \, \partial y}, \quad \frac{\partial^2 w}{\partial x \, \partial z} = \frac{\partial^2 w}{\partial z \, \partial x}$$

(all derivatives being assumed continuous in an open region).

The differential of $w = f(x, y, z)$ is

$$dw = \frac{\partial w}{\partial x} dx + \frac{\partial w}{\partial y} dy + \frac{\partial w}{\partial z} dz,$$

and this gives an approximation to the increment

$$\Delta w = f(x + dx, \, y + dy, \, z + dz) - f(x, y, z)$$

in the sense of the preceding section.

There are also further chain rules. For example, if $w = f(x, y, z)$ and x, y, z are functions of t, then

$$\frac{dw}{dt} = \frac{\partial w}{\partial x}\frac{dx}{dt} + \frac{\partial w}{\partial y}\frac{dy}{dt} + \frac{\partial w}{\partial z}\frac{dz}{dt}. \tag{9–50}$$

This rule can be simplified with the aid of the *gradient vector* of $w = f(x, y, z)$, denoted by ∇f or grad f (or ∇w or grad w):

$$\text{grad } f = \nabla f = \frac{\partial w}{\partial x}\mathbf{i} + \frac{\partial w}{\partial y}\mathbf{j} + \frac{\partial w}{\partial z}\mathbf{k}. \tag{9–51}$$

The chain rule now reads:

$$\frac{dw}{dt} = \text{grad } f \cdot \frac{d\mathbf{r}}{dt}.$$

We can also use a gradient vector for a function of two variables $z = f(x, y)$. This is the vector

$$\nabla z = \frac{\partial z}{\partial x}\mathbf{i} + \frac{\partial z}{\partial y}\mathbf{j}. \tag{9–52}$$

When x and y depend on t, the chain rule becomes

$$\frac{dz}{dt} = \operatorname{grad} z \cdot \frac{d\mathbf{r}}{dt}.$$

The gradient vector is studied further in the next section.

Occasionally we have to deal with several functions of several variables. The general case is

$$y_1 = f_1(x_1, \ldots, x_n),$$

$$y_2 = f_2(x_1, \ldots, x_n),$$

$$\vdots \qquad \qquad \vdots \tag{9–53}$$

$$y_m = f_m(x_1, \ldots, x_n).$$

Here additional vector notation is helpful. We can write $\mathbf{y}$ for $(y_1, \ldots, y_m)$ and $\mathbf{f}$ for $(f_1, \ldots, f_m)$, so that (9–53) becomes simply

$$\mathbf{y} = \mathbf{f}(\mathbf{x}). \tag{9–53'}$$

In general we can think of (9–53') as a *function* or *mapping* from R^n to R^m. The equations assign to each point $\mathbf{x} = (x_1, \ldots, x_n)$ of some set in R^n a point $\mathbf{y} = (y_1, \ldots, y_m)$ in R^m. Such mappings are considered further in Section 9–9 below.

The cases when $m = n$ are especially important. If $m = n = 2$, we are dealing with a pair of functions of two variables, say

$$u = f(x, y), \qquad v = g(x, y).$$

They can be interpreted as a *vector field* in the xy-plane; to each point (x, y) we assign a vector $u\mathbf{i} + v\mathbf{j}$, as in Fig. 9–13(a). Similarly, when $m = n = 3$, we are dealing with functions

$$u = f(x, y, z), \qquad v = g(x, y, z), \qquad w = h(x, y, z).$$

They describe a vector field in xyz-space; to each point (x, y, z) we assign a vector $u\mathbf{i} + v\mathbf{j} + w\mathbf{k}$, as in Fig. 9–13(b). (In the figure, the coordinates are numbered.)

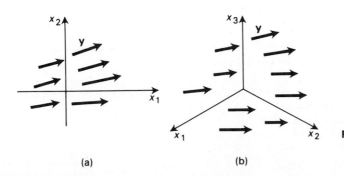

(a)   (b)

Fig. 9–13. Vector fields (a) in the plane; (b) in space.

We remark that the gradient of a function defines a vector field. For $z = f(x, y)$, ∇f is a vector field in the xy-plane. For $w = f(x, y, z)$, ∇f is a vector field in xyz-space. In each case ∇f is assigned to each point.

=========================Problems (Section 9–5)=========================

1. Find the partial derivatives requested:

   a) z_x and z_y for $z = \dfrac{x^2 - y^2}{x^2 + y^2}$

   b) x_u and x_v for $x = u\sqrt{u^2 + v^2}$

   c) z_{xx} and z_{yy} for $z = x^4 - 4x^3 y + 4xy^3 - y^4$

   d) z_{xx} and z_{tt} for $z = e^{-t}(x^2 + t^2)$

2. Use the differential to obtain an approximate value:
   a) $z = xy$ for $x = 2.01$, $y = 18.03$
   b) $z = e^{-x^2 - y^2} \sin(x + 2y)$ for $x = 0.07$, $y = 0.02$

3. Find the tangent plane to the surface:

   a) $z = x^4 + 3x^2 y^2 - y^4$ at $(1, 1, 3)$

   b) $z = \dfrac{1}{x - y}$ at $(2, 1, 1)$

4. Verify that

   a) $z_{xy} = z_{yx}$ for $z = x^m y^n$ (and hence for z a polynomial in x, y)
   b) $z_{xxy} = z_{yxx}$ for $z = e^{x^2 - y^2} \ln(3x + y)$

5. a) Find an expression for dz/dt if

$$z = \frac{x^2 - y^2}{x^2 + y^2}, \qquad x = t^2 + 3t + 2, \qquad y = 4t^2 - 5t + 7.$$

   HINT: See Problem 1(a).

   b) Find an expression for dz/du if

$$z = \ln(x^2 + y^2 + 1), \qquad x = ue^{3u}, \qquad y = ue^{-3u}.$$

6. a) Let $z = f(3x - 2y)$. Show that $2\dfrac{\partial z}{\partial x} + 3\dfrac{\partial z}{\partial y} = 0$.

   HINT: Let $u = 3x - 2y$.

   b) If $z = f(x, y)$ satisfies the equation $z_x + 2z_y = 0$ and $x = 3u + v$, $y = u + 2v$, show that $z_v = 0$ when z is expressed in terms of u and v.

7. For each of the following sets in the xy-plane state whether the set is an open region and describe the boundary of the set:
   a) All (x, y) such that $x > 0$ and $y > 0$
   b) All (x, y) such that $x > 0$ and $x < y < 2x$
   c) All (x, y) such that $0 < x^2 + y^2 < 1$
   d) All (x, y) such that $xy > 0$

e) All (x, y) such that $|x| > 1$

f) All (x, y) such that $x > 0$, $y > 0$ and $x + y > 1$

8. Find the partial derivatives requested:

   a) w_x and w_y for $w = 4x^3 y^2 z - 5xy^3 z^2$

   b) w_u and w_v for $w = s^2 t^2 u^3 v - st^3 u^2 v^2$

9. *Euler's theorem on homogeneous functions.* The function $z = f(x, y)$ is said to be *homogeneous of degree α* if

$$f(tx, ty) \equiv t^\alpha f(x, y). \tag{9–54}$$

Show that for such a function

$$x \frac{\partial f}{\partial x} + y \frac{\partial f}{\partial y} = \alpha f.$$

HINT: Differentiate the relation (9–54) with respect to t with the aid of the chain rule and then set $t = 1$.

10. *The Stokes total time derivative in hydrodynamics.* Let $w = F(x, y, z, t)$, where $x = f(t)$, $y = g(t)$, $z = h(t)$, so that w can be expressed in terms of t alone. Show that

$$\frac{dw}{dt} = \frac{\partial w}{\partial x}\frac{dx}{dt} + \frac{\partial w}{\partial y}\frac{dy}{dt} + \frac{\partial w}{\partial z}\frac{dz}{dt} + \frac{\partial w}{\partial t}.$$

Here both dw/dt and $\partial w/\partial t = F_t(x, y, z, t)$ have meaning and are in general unequal. In hydrodynamics dx/dt, dy/dt, dz/dt are the velocity components of a moving fluid particle and dw/dt describes the variation of w "following the motion of the fluid." It is customary, following Stokes, to write Dw/Dt for dw/dt. (See H. LAMB, *Hydrodynamics*, 6th Edition Cambridge University Press, 1932, p. 3.)

11. Graph as vector fields in the xy-plane:

    a) $u = 2x - y$, $v = x + 2y$

    b) $u = \dfrac{x}{x^2 + y^2}$, $v = \dfrac{y}{x^2 + y^2}$

12. Graph the gradient vector of the function as a vector field in the xy-plane:

    a) $z = x^2 + y^2$

    b) $z = xy$

$9-6$ THE DIRECTIONAL DERIVATIVE AND THE GRADIENT VECTOR

For a function $z = f(x, y)$ the partial derivatives $\partial z/\partial x$ and $\partial z/\partial y$ give the rates of change of z along lines parallel to the x-axis and y-axis respectively. It is natural to generalize by considering the rate of change of z along a line having an arbitrary direction. This leads us to the *directional derivative* of the function.

The idea involved here is a common one. For example, a person moving about in a heated room is sensitive to the rate of change of temperature when moving in a particular direction. Similarly, a person wading into a lake of unknown depths is very conscious of the rate of change of depth as a step is taken in one direction or another. In both cases, the rate is a directional derivative.

In order to obtain an expression for the directional derivative of $z = f(x, y)$, we assume that f has continuous first partial derivatives in the open region D and choose a point $P_0(x_0, y_0)$ of D. We specify a direction at P_0 by choosing a unit vector

$$\mathbf{u} = \cos \alpha \mathbf{i} + \sin \alpha \mathbf{j},$$

so that $x - x_0 = s \cos \alpha$, $y - y_0 = s \sin \alpha$ along the line L through (x_0, y_0) in direction $\mathbf{u}$; that is, L has the vector equation

$$\mathbf{r} = \mathbf{r}_0 + s\mathbf{u},$$

where s is directed distance along L, as in Fig. 9–14 (see Section 9–3). Along L, $z = f(x, y) = f(x_0 + s \cos \alpha, y_0 + s \sin \alpha)$, so that z becomes a function of s. The rate of change of z with respect to s at $s = 0$ (that is, at P_0) is our directional derivative of $z = f(x, y)$ at P_0. We denote the directional derivative by $\nabla_\alpha z$ or $\nabla_u z$ (or $\nabla_\alpha f$ or $\nabla_u f$). By the chain rule,

$$\frac{dz}{ds} = \frac{\partial z}{\partial x}\frac{dx}{ds} + \frac{\partial z}{\partial y}\frac{dy}{ds} = \frac{\partial z}{\partial x}\cos \alpha + \frac{\partial z}{\partial y}\sin \alpha.$$

Therefore, the directional derivative of $z = f(x, y)$ at P_0 is given by

$$\nabla_\alpha z = \frac{\partial z}{\partial x}\cos \alpha + \frac{\partial z}{\partial y}\sin \alpha, \tag{9–60}$$

where $\partial z/\partial x$ and $\partial z/\partial y$ are evaluated at P_0.

By (9–60), for $\alpha = 0$ the directional derivative reduces to $\partial z/\partial x$; for $\alpha = \pi/2$, it reduces to $\partial z/\partial y$. These results are to be expected from the way we defined the directional derivative.

EXAMPLE 1   Let $z = x^3 - x^2 y$ (Fig. 9–15). Then in general

$$\nabla_\alpha z = (3x^2 - 2xy)\cos \alpha - x^2 \sin \alpha.$$

At the point $(2, 1)$ in the direction $\alpha = \pi/4$ this gives

$$\nabla_\alpha z = 8\frac{\sqrt{2}}{2} - 4\frac{\sqrt{2}}{2} = 2\sqrt{2} = 2.818.$$

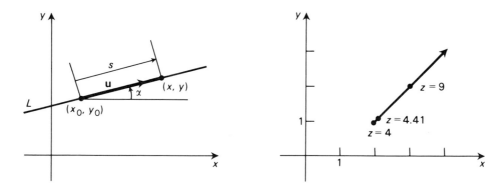

Fig. 9-14. Directional derivative.   **Fig. 9-15.** Example 1.

We check the result by evaluating $z = f(x, y)$ at $(2, 1)$ and $(3, 2)$ (displacement in the direction of $\mathbf{u}$): $f(2, 1) = 4$, $f(3, 2) = 9$. Thus $\Delta z = 5$ and $\Delta s = \sqrt{2}$. Hence $\Delta z / \Delta s = 5/\sqrt{2} = 3.59$. The error is over 25%. To get a better result, we choose a point closer to $(2, 1)$, say $(2.1, 1.1)$. Now $f(2.1, 1.1) = 4.41$, so that $\Delta z = 0.41$ and $\Delta s = 0.1414$, so that $\Delta z / \Delta s = 2.90$, and the agreement is much better. ◄

Relation to gradient vector

The gradient vector of $z = f(x, y)$ (Section 9–5) is the vector

$$\nabla z = \operatorname{grad} z = \frac{\partial z}{\partial x}\mathbf{i} + \frac{\partial z}{\partial y}\mathbf{j}.$$

Therefore, by Eq. (9–60),

$$\nabla_\alpha z = \nabla z \cdot (\cos \alpha \mathbf{i} + \sin \alpha \mathbf{j}) = \nabla z \cdot \mathbf{u}.$$

Since $\mathbf{u}$ is a unit vector, $\nabla z \cdot \mathbf{u} = |\nabla z| \cos \theta$, where θ is the angle between $\mathbf{u}$ and ∇z. Thus (Section 9–2, Eq. (9–21'))

$$\nabla_\alpha z = \operatorname{comp}_\mathbf{u} \nabla z.$$

The directional derivative in a given direction equals the component of the gradient vector in that direction. This explains the notation $\nabla_\alpha z$ or $\nabla_\mathbf{u} z$ for the directional derivative.

The rule just derived allows us to find the directional derivatives of $z = f(x, y)$ at a point by simply finding the component of ∇z in each direction (Fig. 9–16). For a direction orthogonal to ∇z, this component is 0 and hence the directional derivative in such a direction is 0. For the direction of ∇z itself, the component equals the length of the vector ∇z and this is clearly the largest directional derivative at the chosen point. Thus the *magnitude of the gradient vector is the largest directional derivative at the chosen point; the direction of the gradient vector is the direction giving the maximal directional derivative.*

This last result allows us to find the gradient vector experimentally by seeking the direction of *maximum increase* of the function. A person climbing a hill can

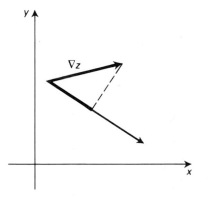

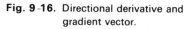

Fig. 9-16. Directional derivative and gradient vector.

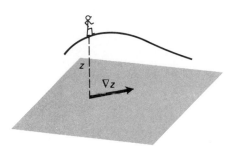

Fig. 9–17. Gradient vector for altitude function.

observe the change of altitude z with position. Climbing in the direction of steepest ascent, the person is following the direction of the gradient vector; more precisely, the person's projection on a horizontal plane follows the direction of the gradient vector ∇z (Fig. 9–17). The rate of climb lim $(\Delta z/\Delta s)$, i.e., vertical distance gained per horizonal distance moved, equals the magnitude of ∇z.

Generalization to higher dimensions

The previous discussion carries over to functions of three or more variables. For a function $w = f(x, y, z)$, the directional derivative in the direction of the unit vector $\mathbf{u} = \cos \alpha \mathbf{i} + \cos \beta \mathbf{j} + \cos \gamma \mathbf{k}$ equals

$$\nabla_u w = \frac{\partial w}{\partial x} \cos \alpha + \frac{\partial w}{\partial y} \cos \beta + \frac{\partial w}{\partial z} \cos \gamma = \nabla w \cdot \mathbf{u},$$

where

$$\nabla w = \operatorname{grad} w = \frac{\partial w}{\partial x} \mathbf{i} + \frac{\partial w}{\partial y} \mathbf{j} + \frac{\partial w}{\partial z} \mathbf{k}.$$

Again $\nabla_u w$ is the component of grad w in the direction of $\mathbf{u}$.

$9-7$ DIRECTIONAL DERIVATIVE ALONG A PATH; NORMAL DERIVATIVE

In our definition of the directional derivative of $z = f(x, y)$, we considered the way in which z varies when (x, y) varies along a straight line. Instead of a straight line, we can use an arbitrary smooth path through our initial point (x_0, y_0). The path can be given by parametric equations $x = x(t)$, $y = y(t)$, where these functions have continuous derivatives in an interval. We can always change to arc length s as parameter, so that the path becomes $x = x(s)$, $y = y(s)$. We recall (Section 9–3) that for a path $x = x(t)$, $y = y(t)$:

$$|\mathbf{v}| = \frac{ds}{dt} = \sqrt{\left(\frac{dx}{dt}\right)^2 + \left(\frac{dy}{dt}\right)^2},$$

so that

$$\mathbf{T} = \frac{\mathbf{v}}{|\mathbf{v}|} = \frac{(dx/dt)\mathbf{i} + (dy/dt)\mathbf{j}}{ds/dt} = \frac{dx}{ds}\mathbf{i} + \frac{dy}{ds}\mathbf{j}$$

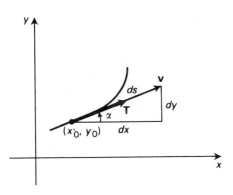

Fig. 9–18. Arc length s as parameter along a path.

is a unit tangent vector that can be written as

$$\mathbf{T} = \cos \alpha \, \mathbf{i} + \sin \alpha \, \mathbf{j}$$

(see Fig. 9–18). We have assumed $\mathbf{v} \neq \mathbf{0}$ at the points considered. The angle α is the angle between the positive x-axis and the direction of the tangent vector $\mathbf{T}$ in the direction of increasing s

Along the path, $z = f(x, y)$ can be expressed in terms of s through the parametric equations $x = x(s)$, $y = y(s)$. Then by the chain rule

$$\frac{dz}{ds} = \frac{\partial z}{\partial x}\frac{dx}{ds} + \frac{\partial z}{\partial y}\frac{dy}{ds} = \frac{\partial z}{\partial x}\cos\alpha + \frac{\partial z}{\partial y}\sin\alpha. \tag{9–70}$$

Thus *the rate of change of z with respect to s along the path, at the point $P_0(x_0, y_0)$, equals the directional derivative of z at P_0 in the direction of the tangent vector to the path in the direction of increasing s.* Thus we can say: the directional derivative along an arbitrary path is the same as the directional derivative along a straight line tangent to the path.

This result applies without change to functions of three or more variables.

Normal derivative

In various applications the direction along which a directional derivative is formed is that of a specified unit normal vector $\mathbf{n}$ to a curve in the plane or to a surface in space. The directional derivative is then denoted by $\partial z/\partial n$ or $\partial w/\partial n$ for $z = f(x, y)$ or $w = f(x, y, z)$ respectively. From our formulas above we have

$$\frac{\partial z}{\partial n} = \nabla z \cdot \mathbf{n}, \qquad \frac{\partial w}{\partial n} = \nabla w \cdot \mathbf{n}. \tag{9–71}$$

For a curve in the plane, $\mathbf{n}$ is simply a unit vector perpendicular to the tangent vector, as in Fig. 9–18. If, as in that figure, $\mathbf{T} = \cos \alpha \, \mathbf{i} + \sin \alpha \, \mathbf{j}$ is the unit tangent vector and $\mathbf{n}$ is obtained by rotating $\mathbf{T}$ by $\pi/2$ in the clockwise direction, then

$$\mathbf{n} = \cos\left(\alpha - \frac{\pi}{2}\right)\mathbf{i} + \sin\left(\alpha - \frac{\pi}{2}\right)\mathbf{j} = \sin\alpha\,\mathbf{i} - \cos\alpha\,\mathbf{j}.$$

Therefore, by (9–71), for $z = f(x, y)$ we have

$$\frac{\partial z}{\partial n} = \frac{\partial z}{\partial x} \sin \alpha - \frac{\partial z}{\partial y} \cos \alpha. \tag{9–72}$$

If the direction of the normal vector is reversed, the right side is multiplied by -1.

For a surface in space, given by equation $z = f(x, y)$, a normal vector is given by $-z_x \mathbf{i} - z_y \mathbf{j} + \mathbf{k}$, as pointed out at the end of Section 9–4. Hence a unit normal vector is

$$\mathbf{n} = \frac{-z_x \mathbf{i} - z_y \mathbf{j} + \mathbf{k}}{\sqrt{z_x^2 + z_y^2 + 1}}$$

and, for $w = f(x, y, z)$,

$$\frac{\partial w}{\partial n} = \frac{1}{\sqrt{z_x^2 + z_y^2 + 1}} \left(-\frac{\partial w}{\partial x} \frac{\partial z}{\partial x} - \frac{\partial w}{\partial y} \frac{\partial z}{\partial y} + \frac{\partial w}{\partial z} \right). \tag{9–73}$$

We observe that $\mathbf{n}$ is an *upper* normal if the positive z-axis determines the upward direction. Again, if we reverse the direction of $\mathbf{n}$, the right side is multiplied by -1.

EXAMPLE 1 The temperature distribution in a homogeneous spherical solid filling the closed region $x^2 + y^2 + z^2 \leqslant 1$ at time t is given by $u = (z^2 - z)e^{-2t}$ (in appropriate units). Let $\mathbf{n}$ be the unit outer normal on the boundary of the sphere. Find the points at which $\partial u / \partial n$ is smallest and interpret the result.

Solution. Here t is held constant. We can represent the spherical surface as $z = \pm \sqrt{1 - x^2 - y^2}$ and apply Eq. (9–73). However, it is simpler to observe that at each point (x, y, z) of the surface the radius vector $x\mathbf{i} + y\mathbf{j} + z\mathbf{k}$ is a unit vector normal to the tangent plane (by geometry) and therefore can be chosen as the normal $\mathbf{n}$; the vector points to the exterior and is hence the outer normal. By Eq. (9–71),

$$\frac{\partial u}{\partial n} = \left(\frac{\partial u}{\partial x} \mathbf{i} + \frac{\partial u}{\partial y} \mathbf{j} + \frac{\partial u}{\partial z} \mathbf{k} \right) \cdot (x\mathbf{i} + y\mathbf{j} + z\mathbf{k}) = (2z^2 - z)e^{-2t}.$$

Function $2z^2 - z$ has a minimum of $-1/8$ at $z = 1/4$. Since $e^{-2t} > 0$ for all t, $\partial u / \partial n$ has a minimum of $-(1/8)e^{-2t}$ for each t. A negative value of $\partial u / \partial n$ indicates that the temperature is decreasing as the boundary is crossed and hence that there is a heat loss at such points; the fact that this negative value is the minimum of $\partial u / \partial n$ indicates that the maximum heat loss occurs at these points of the spherical boundary, that is, at the points where $z = 1/4$. ◀

9 – 8 CONSERVATION OF ENERGY FOR MOTION OF A PARTICLE IN A FORCE FIELD

In many physical problems we deal with a particle of mass m moving in space subject to a force that depends only on the position of the particle. Thus there is a vector field, a force field, giving for each (x, y, z) the force vector $\mathbf{F}$ for that position; thus, $\mathbf{F} = \mathbf{F}(x, y, z)$, or, more concisely, $\mathbf{F} = \mathbf{F}(\mathbf{r})$, is the force field. By Newton's second law, the motion of the particle is then governed by the vector equation

$$m\frac{d^2\mathbf{r}}{dt^2} = \mathbf{F}(\mathbf{r}).$$ (9–80)

If $\mathbf{F} = X\mathbf{i}+Y\mathbf{j}+Z\mathbf{k}$, then in terms of components Eq. (9–80) becomes

$$m\frac{d^2x}{dt^2} = X(x, y, z),\qquad m\frac{d^2y}{dt^2} = Y(x, y, z),\qquad m\frac{d^2z}{dt^2} = Z(x, y, z).$$ (9–81)

Thus we have a system of three second-order differential equations (see Section 6–1).

In many cases the force field $\mathbf{F}$ is a gradient field, that is, $\mathbf{F} = \operatorname{grad} f$ for some function $f(x, y, z)$. This is the case, in particular, for gravitational forces and electrostatic forces. It is convenient here to replace f by $-U$, so that $\mathbf{F} = -\operatorname{grad} U$. Then Eqs. (9–81) become

$$m\frac{d^2x}{dt^2} = -\frac{\partial U}{\partial x},\qquad m\frac{d^2y}{dt^2} = -\frac{\partial U}{\partial y},\qquad m\frac{d^2z}{dt^2} = -\frac{\partial U}{\partial z}.$$ (9–82)

From these equations it follows that *for each particular motion of the particle*

$$\frac{1}{2}m\left[\left(\frac{dx}{dt}\right)^2 + \left(\frac{dy}{dt}\right)^2 + \left(\frac{dz}{dt}\right)^2\right] + U(x, y, z) \equiv \text{const.}$$ (9–83)

Here the first term is $m|\mathbf{v}|^2/2$, the *kinetic energy* of the particle; the second term, $U(x, y, z)$, is interpreted as the *potential energy* of the particle. The sum is then the *total energy* of the particle and we have a law of *conservation of energy*.

In order to prove the assertion, we differentiate the left side of Eq. (9–83) with respect to t:

$$\frac{d}{dt}\left\{\frac{1}{2}m\left[\left(\frac{dx}{dt}\right)^2 + \left(\frac{dy}{dt}\right)^2 + \left(\frac{dz}{dt}\right)^2\right] + U(x, y, z)\right\}$$

$$= m\left(\frac{dx}{dt}\frac{d^2x}{dt^2} + \frac{dy}{dt}\frac{d^2y}{dt^2} + \frac{dz}{dt}\frac{d^2z}{dt^2}\right) + \frac{\partial U}{\partial x}\frac{dx}{dt} + \frac{\partial U}{\partial y}\frac{dy}{dt} + \frac{\partial U}{\partial z}\frac{dz}{dt}.$$

Here we used the chain rule for $(d/dt)\,U(x, y, z)$. If we now use Eqs. (9–82), then the terms in dx/dt cancel, as do those in dy/dt and dz/dt. Thus the time derivative of the total energy along the path is 0 and the total energy remains constant during the motion.

REMARK: The potential energy U is not uniquely determined, since the replacement of $U(x, y, z)$ by $U(x, y, z)+\text{const}$ has no effect on the differential equations (9–82). In general, it can be shown that (in a given open region) the functions $U(x, y, z)+\text{const}$ are all the functions with given gradient $-\mathbf{F}$; however, not every vector field is a gradient field (Section 10–17).

===============================Problems (Section 9-8)===============================

1. Find the directional derivative at the given point in the given directions:

a) $z = x^2+y^2$ at $(3, 4)$ in the directions $\alpha = 0$, $\alpha = \operatorname{Arctan} 4/3$, $\alpha = \operatorname{Arctan}(-3/4)$
b) $z = xy$ at $(2, 1)$ in the directions $\alpha = \pi/4$, $\alpha = \pi/2$, $\alpha = \pi$

   c) $z = x/y$ at $(1, 1)$ in the direction of the unit vector $\mathbf{u} = (1/|\mathbf{v}|)\mathbf{v}$, where $\mathbf{v} = \mathbf{i} + \mathbf{j}$, $\mathbf{v} = \mathbf{i} - \mathbf{j}$, $\mathbf{v} = 3\mathbf{i} + 4\mathbf{j}$

   d) $w = x^2 + y^2 - z^2$ at $(3, 5, 1)$ in the direction of the unit vector $(1/|\mathbf{v}|)\mathbf{v}$, where $\mathbf{v} = \mathbf{i} + 2\mathbf{j} + 2\mathbf{k}$

2. Use the given values to find an estimate for the directional derivative requested:

   a) $z = f(x, y)$ at $(1, 1)$ in the direction $\alpha = \pi/4$, given $f(1, 1) = 3$, $f(2, 1) = 5$, $f(1, 2) = 6$, $f(2, 2) = 8$

   b) $z = f(x, y)$ at $(2, 3)$ in the direction $\alpha = \text{Arctan } 2$, given that $f(2, 3) = 7.02$, $f(2.1, 3.1) = 7.57$, $f(1.9, 2.8) = 6.83$

3. Show that for $z = f(x, y)$ knowledge of $\nabla_\alpha z$ at P_0 for a pair of values of α generally makes it possible to find $\nabla_\alpha z$ at P_0 for all α. What pairs of values of α must be ruled out here?

4. The velocity pattern of a steady planar fluid motion is described by a vector field $u(x, y)\mathbf{i} + v(x, y)\mathbf{j}$ in the xy-plane. If the flow is incompressible and irrotational (see Sections 10–6, 10–7), then at each point (x, y)

$$\frac{\partial u}{\partial x} + \frac{\partial v}{\partial y} = 0, \qquad \frac{\partial u}{\partial y} - \frac{\partial v}{\partial x} = 0,$$

   Show that this implies that $\nabla_\alpha u + \nabla_{\alpha + \pi/2} v = 0$ for all α.

5. Find the directional derivative specified:

   a) of $z = x^4 - 3x^3 y + x^2 y^2$ at $(2, 1)$ along the curve $x = t^2 + 1$, $y = t^3$ in the direction of increasing t

   b) of $w = xyz$ at $(1, 1, 0)$ along the curve $x = e^t$, $y = e^{2t}$, $z = t$ in the direction of increasing t

6. a) Show that the directional derivative of $z = f(x, y)$ along the curve $y = g(x)$ in the direction of increasing x is $[f_x + f_y g'(x)][g(x)^2 + 1]^{-1/2}$.

   b) Show that the normal derivative of $z = f(x, y)$ for the upper normal of the curve $y = g(x)$ is $[f_y - f_x g'(x)] [g'(x)^2 + 1]^{-1/2}$.

7. Find the points of the circle $x = \cos t$, $y = \sin t$ at which the directional derivative of $z = 3x^2 - 4y^2 = f(x, y)$ along the circle, in the direction of increasing t, is 0. Use this information to find the points at which f has its maximum and minimum values on the circle.

8. Find the normal derivative specified:

   a) of $z = 3x - 5y$ at $(1, \sqrt{3})$ for the exterior normal of the circle $x = 2\cos t$, $y = 2\sin t$

   b) of $z = x^2 + y^2$ at $(1, 0)$ for the upper normal of the curve $y = 2x^2 + x - 3$ (see Problem 6(b))

   c) of $w = 2x^2 - y^2 + z^2$ at $(2, 2, 1)$ for the exterior normal of the spherical surface $x^2 + y^2 + z^2 = 9$

   d) of $w = xyz$ at $(1, 0, 1)$ for the upper normal of the surface $z = x \cos y$

9. A particle of mass m moves in space subject to the given force field $\mathbf{F}$. Verify that $\mathbf{F} = -\text{grad } U$ for $U(x, y, z)$ as given and write out the law of conservation of energy for the motions:

   a) $\mathbf{F} = 2ax\,\mathbf{i} + 3ay\,\mathbf{j} + az\,\mathbf{k}$, $U = -(a/2)(2x^2 + 3y^2 + z^2)$, $a = \text{const} > 0$. This is a force field typical of spring forces in three mutually perpendicular directions.

   b) $\mathbf{F} = -\dfrac{kMm}{r^3}\mathbf{r}$, $U = -\dfrac{kMm}{r}$, where k, M, m are positive constants, $r = |\mathbf{r}|$. This is gravitational force due to mass M at the origin.

$9-9$ MAXIMA AND MINIMA

We know that relative maxima or minima of a function of one variable occur inside the interval of definition only where $f'(x) = 0$. A point x_0 such that $f'(x_0) = 0$ is called a *critical point* of f. At a critical point, f may or may not have a maximum or minimum. If $f''(x_0) < 0$, we can conclude that there is a maximum (the graph of f is concave downward at x_0); if $f''(x_0) > 0$, there is a minimum.

These ideas carry over to functions of several variables. A function $f(x, y)$ defined in the open region D is said to have a relative maximum at (x_0, y_0) if $f(x, y) \leqslant f(x_0, y_0)$ in some δ-neighborhood of (x_0, y_0); f has a relative minimum at (x_0, y_0) if $f(x, y) \geqslant f(x_0, y_0)$ in some δ-neighborhood of (x_0, y_0).

Let f have a relative maximum at (x_0, y_0). Then for fixed y, say $y = y_0$, function $f(x, y_0)$ also has a relative maximum at x_0. Therefore $f_x(x_0, y_0) = 0$. Similarly, $f_y(x_0, y_0) = 0$. Therefore, grad $f = \mathbf{0}$ at (x_0, y_0). We call (x_0, y_0) a *critical point* of f if grad $f = \mathbf{0}$ at (x_0, y_0). Therefore, *a relative maximum can occur only at a critical point*. A similar reasoning and conclusion apply to relative minima.

Let f have a critical point at (x_0, y_0). Then f may or may not have a relative maximum or minimum at (x_0, y_0), as we proceed to illustrate.

EXAMPLE 1 $z = x^2 + y^2$. Here $\nabla z = 2x\mathbf{i} + 2y\mathbf{j}$ and therefore $(0, 0)$ is the only critical point. Since $z = 0$ at this point and $z = x^2 + y^2 > 0$ otherwise, z has a relative minimum at the point. ◄

EXAMPLE 2 $z = 1 - x^2 - y^2$. Again we find $(0, 0)$ to be the only critical point; $z = 1$ at the point and $z < 1$ otherwise, so that there is a relative maximum at the point. ◄

EXAMPLE 3 $z = x^2 - y^2$. Again $(0, 0)$ is the only critical point. However, for $y = 0, z = x^2$, so that for this function of x there is a relative *minimum* at $x = 0$; on the other hand, for $x = 0, z = -y^2$, with a relative *maximum* at $y = 0$. This shows that there is no relative minimum or maximum of the given function at $(0, 0)$. In particular, $z = 0$ at $(0, 0)$, z is positive at points (x, y) close to $(0, 0)$ with $x^2 > y^2$; z is negative at points (x, y) close to $(0, 0)$ with $x^2 < y^2$. ◄

The functions of Examples 1, 2, 3 are graphed in Fig. 9–19. The function of Example 3 is said to have a *saddle point* at $(0, 0)$. This is typical of the formation at the top of a mountain pass.

In order to obtain tests for maxima and minima, we expand $z = f(x, y)$ in a Taylor series about the critical point (x_0, y_0). Since $z_x = z_y = 0$ at the point, the series is

$$f(x, y) = f(x_0, y_0) + \frac{1}{2}z_{xx}(x_0, y_0)(x - x_0)^2$$

$$+ z_{xy}(x_0, y_0)(x - x_0)(y - y_0) + \frac{1}{2}z_{yy}(x_0, y_0)(y - y_0)^2 + \cdots$$

(see Section 2–19). Near (x_0, y_0), the terms shown should determine the behavior of

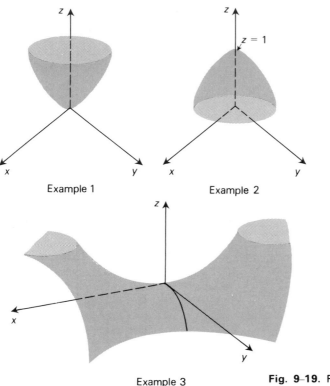

Example 1

Example 2

Example 3

Fig. 9-19. Functions of Examples 1, 2, and 3.

the function. We write

$$x - x_0 = r \cos \theta, \qquad y - y_0 = r \sin \theta,$$

so that r, θ are polar coordinates relative to (x_0, y_0) as origin. Then

$$f(x, y) = f(x_0, y_0) + \frac{r^2}{2}(A \cos^2 \theta + 2B \cos \theta \sin \theta + C \sin^2 \theta) + \cdots, \quad (9\text{--}90)$$

where

$$A = z_{xx}(x_0, y_0), \qquad B = z_{xy}(x_0, y_0), \qquad C = z_{yy}(x_0, y_0). \quad (9\text{--}91)$$

If now

$$A \cos^2 \theta + 2B \cos \theta \sin \theta + C \sin^2 \theta < 0, \qquad 0 \leqslant \theta \leqslant 2\pi,$$

then in (9–90) the second term is *negative*, so that $f(x, y)$ should have a *maximum* at (x_0, y_0). Similarly, if

$$A \cos^2 \theta + 2B \cos \theta \sin \theta + C \sin^2 \theta > 0, \qquad 0 \leqslant \theta \leqslant 2\pi,$$

then in (9–90) the second term is *positive*, so that $f(x, y)$ should have a *minimum* at (x_0, y_0).

By studying the function $A \cos^2 \theta + 2B \cos \theta \sin \theta + C \sin^2 \theta$ further (see Problem 8 below), we deduce the following criteria (see AC, pp. 181–182). The function $z = f(x, y)$ has

$$a \text{ relative minimum at } (x_0, y_0) \text{ if } B^2 - AC < 0 \text{ and } A > 0; \qquad (9\text{–}92)$$

$$a \text{ relative maximum at } (x_0, y_0) \text{ if } B^2 - AC < 0 \text{ and } A < 0; \qquad (9\text{–}93)$$

$$\text{neither minimum nor maximum at } (x_0, y_0) \text{ if } B^2 - AC > 0. \qquad (9\text{–}94)$$

Examples 1, 2, 3 above illustrate the three cases of (9–92), (9–93), (9–94) respectively. For Example 1, $A = 1$, $B = 0$, $C = 1$, so that (9–92) holds; for Example 2, $A = -1$, $B = 0$, $C = -1$, and (9–93) holds; for Example 3, $A = 1$, $B = 0$, $C = -1$, and (9–94) holds.

Generalization to functions of n variables

For $y = f(x_1, \ldots, x_n)$ with continuous first partial derivatives in D, a critical point is a point $x^0 = (x_1^0, \ldots, x_n^0)$ at which $\nabla y = 0$. A relative maximum or minimum of y can occur only at a critical point, but a critical point may fail to yield a maximum or a minimum. A relative minimum is ensured if

$$\sum_{i,j=1}^{n} a_{ij} u_i u_j > 0 \quad \text{for all } (u_1, \ldots, u_n) \text{ other than } (0, 0, \ldots, 0), \qquad (9\text{–}95)$$

where

$$a_{ij} = \frac{\partial^2 y}{\partial x_i \partial x_j} (x_1^0, \ldots, x_n^0), \quad i = 1, \ldots, n, \quad j = 1, \ldots, n. \qquad (9\text{–}96)$$

A relative maximum is ensured if

$$\sum_{i,j=1}^{n} a_{ij} u_i u_j < 0 \quad \text{for all } (u_1, \ldots, u_n) \text{ other than } (0, \ldots, 0). \qquad (9\text{–}97)$$

In the case of (9–95), the *quadratic form* $\sum a_{ij} u_i u_j$ is said to be *positive definite*; for (9–97) the same form is *negative definite*.

Again the conditions can be expressed in terms of the second derivatives a_{ij}. We let A be the $n \times n$ matrix (a_{ij}), unrelated to the A of Eq. (9–91), and observe that by (9–96) A is *symmetric*. Hence all eigenvalues of A are *real* (Section 5–16). *If these eigenvalues are all greater than 0, a relative minimum is ensured; if all are less than 0, a relative maximum is ensured; if some are less than 0 and some are greater than 0, then there is neither maximum nor minimum at* x^0 (see Section 5–16).

Absolute maximum and minimum

For a differentiable function $f(x)$, $a \leqslant x \leqslant b$, a basic theorem asserts that f has its absolute maximum at some point x_0 of the interval, so that $f(x) \leqslant f(x_0)$ for $a \leqslant x \leqslant b$ (CLA, Section 2–14). If x_0 is inside the interval, $a < x_0 < b$, then x_0 must be a critical point of f. However, x_0 may be an endpoint a or b. Similarly, for $f(x, y)$ defined and continuous on a bounded closed region R function f must have an absolute maximum at some point (x_0, y_0) of R. If (x_0, y_0) is interior to R, that is, if

(x_0, y_0) has a δ-neighborhood in R, then (x_0, y_0) must be a critical point. However, (x_0, y_0) may be on the boundary of R. Hence, in seeking the absolute maximum of f, we are generally forced to study separately the values of f on the boundary of R. The largest such value must then be compared with the value of f at every critical point inside R in order to find the absolute maximum. There is a similar discussion for the absolute minimum of f. The ideas also carry over at once to functions of n variables.

====================**Problems (Section 9-9)**====================

1. Find the critical points of the following functions and test for relative maxima and minima:

   a) $z = \sqrt{1 - x^2 - y^2}$ d) $z = x^2 - 5xy - y^2$
   b) $z = 1 + x^2 + y^2$ e) $z = x^2 - 2xy + y^2$
   c) $z = 2x^2 - xy - 3y^2 - 3x + 7y$ f) $z = x^3 - 3xy^2 + y^3$

2. Find the absolute minimum and maximum, if they exist, of the following functions:

   a) $z = \dfrac{1}{1 + x^2 + y^2}$, all (x, y) b) $z = xy$, $x^2 + y^2 \leqslant 1$
   c) $w = x + y + z$, $x^2 + y^2 + z^2 \leqslant 1$.

3. Determine whether the given quadratic form is positive definite:

   a) $3x^2 + 2xy + y^2$ b) $x^2 - xy - 2y^2$
   c) $5x_1^2 + 4x_1 x_2 + 6x_2^2 + 4x_2 x_3 + 7x_3^2$

4. Find the minimum surface area of a rectangular parallelepiped of volume 1 m³.

5. A light source moves in the triangle of vertices $(1, 0, 0)$, $(0, 1, 0)$, $(0, 0, 1)$ in xyz-space. At what position does the light source provide maximum illumination to small region R in the xy-plane around the origin O?

   HINT: When the light is at P (Fig. 9-20), the light reaching R fills a cone of base R and apex P. The illumination is proportional to the ratio U of the volume of this cone to the volume of the sphere of center P and radius OP, for the ratio U gives the fraction of available light reaching R. The volume of the cone is one third of the product of the area of R and its altitude z.

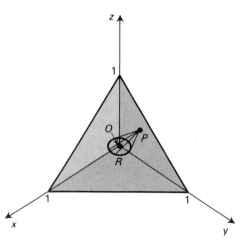

Fig. 9-20. Problem 5.

6. In mechanics a particle P is said to move in a conservative force field if the force $\mathbf{F}$ acting on P depends only on position and

$$\mathbf{F} = -\nabla U,$$

where U is the *potential energy of* P. At each critical point of U, the force is $\mathbf{0}$, and the particle is in *equilibrium position*. It can be shown that the equilibrium is stable if U has a relative minimum at the point; otherwise the equilibrium is unstable.

a) Let P move in the xy-plane with potential energy $k(x^2+y^2)$. Determine the equilibrium position and whether it is stable. Assume $k > 0$.

b) Let P move in the xy-plane with potential energy $U = k(r_1^2 + r_2^2 + r_3^2)$, where k is a positive constant, $r_i = \|\overrightarrow{PP_i}\|$ and P_1 is $(0, 2)$, P_2 is $(\sqrt{3}, -1)$, P_3 is $(-\sqrt{3}, -1)$. Show that the only equilibrium position is $(0, 0)$ and determine whether it is stable. (Here P can be considered to be attached to P_1, P_2 and P_3 by three springs, in accordance with Hooke's law.)

c) Let P move in the xy-plane with potential energy $U = -k[(1/r_1)+(1/r_2)]$, where r_1 is the distance from P to $(1, 0)$, r_2 is the distance from P to $(-1, 0)$, and k is a positive constant. Show that $(0, 0)$ is an equilibrium position and determine whether it is stable. (This problem corresponds to the motion of a particle in a plane subject to the gravitational attraction of two fixed equal masses.)

d) Consider the effect on part (c) of replacing k by a negative constant h. [The forces become repulsive, as would be the case if there were electric charges of same sign at $(1, 0)$, $(-1, 0)$, and P.]

7. *Linear programming.* Find the absolute maximum of $z = 3x + 5y$ when $x \geqslant 0$, $y \geqslant 0$, $x + y \leqslant 1$, $x + 3y \leqslant 2$. Here the inequalities describe a polygonal region R (Fig. 9–21). Verify that $z = f(x, y)$ has no critical point, so that the (absolute) maximum and minimum of z must be taken on the boundary of R. Since z varies linearly on the boundary, conclude that the maximum and minimum must be taken at the vertices of R.

This is one of a great number of problems occurring in industrial engineering. The aim is to maximize or minimize a linear function subject to linear inequalities. A typical application is to the transportation problem: to minimize the cost of shipping goods from several origins to several destinations, subject to restraints on goods available at the origins and on goods needed at the destinations (see *Linear Programming* by G. HADLEY, Addison-Wesley Publishing Co., Reading, Mass., 1962).

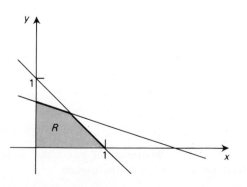

Fig. 9–21. Problem 7.

8. Let $g(\theta) = A\cos^2\theta + 2B\cos\theta\sin\theta + C\sin^2\theta,\ 0 \leqslant \theta \leqslant 2\pi$.

   a) Show that $g(\theta) = \dfrac{A+C}{2} + \dfrac{A-C}{2}\cos 2\theta + B\sin 2\theta$.

   b) Show that, for appropriate α, $g(\theta) = \dfrac{A+C}{2} + \dfrac{1}{2}\sqrt{(A-C)^2 + 4B^2}\,\sin(2\theta+\alpha)$, and

   hence show that

   $$\frac{A+C}{2} \pm \frac{1}{2}\sqrt{(A+C)^2 + 4(B^2 - AC)}$$

   are the absolute maximum and minimum of $g(\theta)$.

   c) From the result of (b) show that when $B^2 - AC < 0$, A and C have the same sign and $g(\theta)$ has the same sign as A and C. Thus $g(\theta) < 0$ for $0 \leqslant \theta \leqslant 2\pi$ if $A < 0$ and $g(\theta) > 0$ for $0 \leqslant \theta \leqslant 2\pi$ if $A > 0$.

   d) Show that if $B^2 - AC > 0$, then $g(\theta)$ changes sign on the interval $0 \leqslant \theta \leqslant 2\pi$.

9–10 FUNCTIONS AND MAPPINGS, LINEARIZATION, JACOBIAN MATRIX, JACOBIAN DETERMINANT

In many processes there is a certain set of adjustable parameters and, for each assignment of values to these parameters, there is a corresponding set of performance indices. For example, in a chemical process the temperature and rate of introduction of various chemical ingredients can be adjusted and the rates of formation of various products can then be measured as performance indices. In economics interest rates and levels of taxation are adjusted and then indices of inflation and of gross national product are measured.

A corresponding mathematical formulation is as follows. We have functions

$$y_1 = F_1(x_1, \ldots, x_n), \quad \ldots, \quad y_m = F_m(x_1, \ldots, x_n) \tag{9–100}$$

defined in an open region D of n-dimensional space R^n. The x_k correspond to the adjustable parameters, $y_1, \ldots, y_m$ to the performance indices. Thus we are dealing with a *mapping* from R^n to R^m, as discussed in Section 9–5. We shall here assume that all F_k are continuous and have continuous first partial derivatives in D.

The simplest case of such a mapping is that in which the F_k are all linear functions. For proper choices of origin in R^n and R^m, the equations (9–100) then take the form

$$y_i = \sum_{k=1}^{n} a_{ik}x_k, \quad i = 1, \ldots, m, \tag{9–101}$$

or, in matrix notation,

$$\mathbf{y} = A\mathbf{x}. \tag{9–101'}$$

The methods of linear algebra (Chapter 5) can then be applied, in particular, to find the range of the mapping and to determine whether it is one-to-one.

When the mapping is not linear, the differential calculus provides a way of approximating the mapping by a linear one, namely by taking differentials in Eqs. (9–100):

$$dy_i = \sum_{k=1}^{n} \frac{\partial F_i}{\partial x_k} dx_k, \quad i = 1, \ldots, m. \tag{9-102}$$

Here there is an initial point $(x_1^0, \ldots, x_n^0)$ of D at which all $\partial F_i/\partial x_k$ are to be evaluated; to the initial point corresponds a point $(y_1^0, \ldots, y_m^0)$ in R^m. We can consider $dx_1^0, \ldots, dx_n^0$ as coordinates in R^n relative to an origin at $(x_1^0, \ldots, x_n^0)$ and $dy_1, \ldots, dy_m$ as coordinates in R^m relative to an origin at $(y_1^0, \ldots, y_m^0)$. In terms of these coordinates, Eqs. (9–102) then describe a linear mapping that approximates the given nonlinear mapping (9–100). Here, as in Section 9-4, $dy_1, \ldots, dy_m$ are the given approximations to the increments $\Delta y_1, \ldots, \Delta y_m$ corresponding to increments $dx_1, \ldots, dx_n$; the relative accuracy of the approximations improves as the $dx_1, \ldots, dx_n$ all approach 0.

The coefficient matrix of the linear mapping (9–102) is the $m \times n$ matrix $(\partial F_i/\partial x_k)$, where i is the row index, k is the column index. This matrix is called the *Jacobian matrix* of the mapping. Thus the Jacobian matrix is the matrix

$$\begin{bmatrix} \dfrac{\partial F_1}{\partial x_1} & \cdots & \dfrac{\partial F_1}{\partial x_n} \\ \vdots & & \vdots \\ \dfrac{\partial F_m}{\partial x_1} & \cdots & \dfrac{\partial F_m}{\partial x_n} \end{bmatrix}.$$

Here, as noted above, all partial derivatives are evaluated at a chosen reference point. By taking differentials we have *linearized* the mapping, that is, approximated it by a linear mapping. The Jacobian matrix is the matrix of the approximating linear mapping.

This approximation procedure underlies many of the linear formulas that appear in all the sciences. Nature rarely works in linear ways. However, if we consider only a small range of variation of the parameters, linear formulas are sufficiently accurate. A simple example is Hooke's law for a spring: $F = kx$ (force equals a constant times displacement); this formula is valid only for sufficiently small displacements.

EXAMPLE 1 $y_1 = x_1^2 - x_1 x_2$, $y_2 = x_1 x_2 + x_2^2$. We study this mapping near the point $\mathbf{x} = (2, 1)$ at which $\mathbf{y} = (2, 3)$. We have at this point

$$dy_1 = (2x_1 - x_2)dx_1 - x_1 dx_2 = 3dx_1 - 2dx_2,$$
$$dy_2 = x_2 dx_1 + (x_1 + 2x_2)dx_2 = dx_1 + 4dx_2,$$

so that at the point the Jacobian matrix is

$$\begin{bmatrix} 3 & -2 \\ 1 & 4 \end{bmatrix}.$$

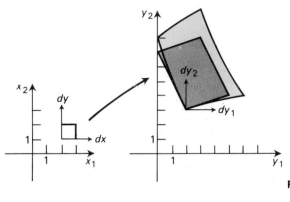

Fig. 9–22. Mapping of Example 1 and linear approximation.

We study the linear mapping in more detail (Fig. 9–22). For $dx_2 = 0, dy_1 = 3dx_1$ and $dy_2 = dx_1$, so that $dy_2/dy_1 = 1/3$ and the points (dy_1, dy_2) follow a line of slope $1/3$. Similarly, for $dx_1 = 0$, $dy_1 = -2dx_2$ and $dy_2 = 4dx_2$, so that the points (dy_1, dy_2) follow a line of slope -2. By similar reasoning we find that the points of the square $0 \leqslant dx_1 \leqslant 1, 0 \leqslant dx_2 \leqslant 1$ correspond (in one-to-one fashion) to the points of the parallelogram shaded in Fig. 9–22. The figure also shows the image of the same square under the original nonlinear mapping. We observe that the image of the line $x_2 = 1$ (that is, $dx_2 = 0$) is a curve passing through the point $\mathbf{y} = (2, 3)$ whose *tangent line* at the point is the edge of the parallelogram obtained from the linear mapping (see Problem 4 below). A similar statement applies to the image of the line $x_1 = 2$, that is, $dx_1 = 0$. In general, point-by-point comparisons show that the linear mapping closely approximates the nonlinear one near the initial point $\mathbf{x} = (2, 1)$. For example, $\mathbf{x} = (2.1, 1.3)$ corresponds to $\mathbf{y} = (1.68, 4.42)$ under the nonlinear mapping, whereas for $d\mathbf{x} = (0.1, 0.3)$ the linear mapping gives $d\mathbf{y} = (-0.3, 1.3)$, so that the corresponding point $\mathbf{y}$ is $(2, 3) + d\mathbf{y} = (1.7, 4.3)$. The approximation process is of course the same as that discussed in Section 9–4. ◄

When $m = n$ in the mapping (9–100), the Jacobian matrix is *square*; then the special results of Chapter 5 relating to square matrices can be applied. For example, we may consider invertibility and eigenvalues; in addition, we can form the determinant of the Jacobian matrix. This is called the *Jacobian determinant* (or simply the *Jacobian*) of the mapping. We write

$$\frac{\partial(y_1, \ldots, y_n)}{\partial(x_1, \ldots, x_n)} = \frac{\partial(F_1, \ldots, F_n)}{\partial(x_1, \ldots, x_n)} = \det\left(\frac{\partial F_i}{\partial x_k}\right) \qquad (9\text{–}103)$$

for the Jacobian determinant.

In Example 1 we have $m = n = 2$, so that the Jacobian determinant can be formed. At the point considered, this is

$$\frac{\partial(y_1, y_2)}{\partial(x_1, x_2)} = \begin{vmatrix} 3 & -2 \\ 1 & 4 \end{vmatrix} = 14.$$

We observe that in Example 1

$$dy = \begin{bmatrix} 3 & -2 \\ 1 & 4 \end{bmatrix} dx,$$

so that for $dx = \text{col}\,(1, 0)$, $dy = \text{col}\,(3, 1)$, the first column of the Jacobian matrix. Similarly, for $dx = \text{col}\,(0, 1)$, we find $dy = \text{col}\,(-2, 4)$, the second column of the Jacobian matrix. These column vectors are edge vectors for the parallelogram of Fig. 9–22. As in Section 9–2, the absolute value of the corresponding determinant gives the *area* of the parallelogram. In Section 9–2, the vectors appear as row vectors, rather than column vectors, but this has no effect on the determinant (see Section 5–2). Thus the Jacobian determinant gives (up to $\pm$ sign) the area of the parallelogram corresponding to the *unit square* of edge vectors $(1, 0)$, $(0, 1)$. The area of the square is 1, the area of the parallelogram is 14; thus the area is multiplied by 14 in going from the $x_1 x_2$-plane to the $y_1 y_2$-plane. We see at once that the same coefficient of 14 applies to arbitrary squares in the $x_1 x_2$-plane with sides parallel to the axes and hence, by a limiting process, to *all* figures. *Under the approximating linear mapping, all areas are multiplied by the same factor in going from the $x_1 x_2$-plane to the $y_1 y_2$-plane. This factor is the absolute value of the Jacobian determinant of the given mapping at the point.* If we write $dx_1 dx_2$ and $dy_1 dy_2$ for corresponding *area elements* as in double integrals, we obtain the formula

$$dy_1 dy_2 = \left| \frac{\partial(y_1, y_2)}{\partial(x_1, x_2)} \right| dx_1 dx_2. \tag{9–104}$$

This formula is considered further in Section 10–4 below. It is basic for double integrals.

We have deduced the rule concerning multiplication of areas only in connection with Example 1. However, the same reasoning applies whenever $m = n = 2$. Furthermore, similar reasoning applies when $m = n = 3$, with areas replaced by volumes (see Eq. (9–25)); and it applies to higher dimensions when we are dealing with n-dimensional volumes.

The multiplier rule uses only the absolute value of the Jacobian determinant. The *sign* of this determinant is related to preservation ($+$ sign) or reversal ($-$ sign) of orientation in going from the $x_1 x_2$-plane to the $y_1 y_2$-plane. When the sign is plus, each positive pair (Section 9–2) of vectors in the $x_1 x_2$-plane corresponds, under the linear mapping, to a positive pair of vectors in the $y_1 y_2$-plane, and we say that *orientation is preserved* by the linear mapping. When the sign is minus, a positive pair in the $x_1 x_2$-plane corresponds to a negative pair in the $y_1 y_2$-plane, and we say that *orientation is reversed* by the linear mapping. Similar arguments hold for n-dimensional space, with positive or negative pairs replaced by positive or negative n-tuples of vectors. As in Section 9–2, an n-tuple $\mathbf{u}_1, \ldots, \mathbf{u}_n$ in R^n is positive or negative according as $\det B > 0$ or $\det B < 0$, where B is the matrix whose successive column vectors are $\mathbf{u}_1, \ldots, \mathbf{u}_n$.

When the Jacobian determinant is 0 at the point, the Jacobian matrix is singular and complications arise (see Section 9–15).

The Jacobian determinant has been defined only for the case $m = n$. However, for arbitrary m, n, various Jacobian determinants can be formed by forming minors of the Jacobian matrix. For example, when $m = 2, n = 3$, the Jacobian matrix is

$$\begin{bmatrix} \dfrac{\partial F_1}{\partial x_1} & \dfrac{\partial F_1}{\partial x_2} & \dfrac{\partial F_1}{\partial x_3} \\[2mm] \dfrac{\partial F_2}{\partial x_1} & \dfrac{\partial F_2}{\partial x_2} & \dfrac{\partial F_2}{\partial x_3} \end{bmatrix},$$

and we can write, for example,

$$\frac{\partial(F_1, F_2)}{\partial(x_1, x_2)} = \begin{vmatrix} \dfrac{\partial F_1}{\partial x_1} & \dfrac{\partial F_1}{\partial x_2} \\[2mm] \dfrac{\partial F_2}{\partial x_1} & \dfrac{\partial F_2}{\partial x_2} \end{vmatrix}, \qquad \frac{\partial(F_1, F_2)}{\partial(x_2, x_3)} = \begin{vmatrix} \dfrac{\partial F_1}{\partial x_2} & \dfrac{\partial F_1}{\partial x_3} \\[2mm] \dfrac{\partial F_2}{\partial x_2} & \dfrac{\partial F_2}{\partial x_3} \end{vmatrix}.$$

Such Jacobian determinants are applied in the next section to the study of surfaces.

One-to-one mappings

When $m = n$ in the mapping (9–100), it is natural to ask whether the equations define a one-to-one correspondence between points $(x_1, \ldots, x_n)$ and $(y_1, \ldots, y_n)$ of appropriate regions in the x-space and the y-space. This is equivalent to the requirement that the equations be solvable uniquely for $x_1, \ldots, x_n$ as functions of $y_1, \ldots, y_n$, to yield the *inverse functions* of the given functions or, more simply, the *inverse mapping* of the given mapping. In the linear case of (9–101) or (9–101'), the inverse mapping exists when A is nonsingular or, equivalently, det $A \neq 0$. For the general nonlinear case it is natural to expect that the given mapping will have an inverse mapping *near* corresponding points $(x_1^0, \ldots, x_n^0), (y_1^0, \ldots, y_n^0)$ if the linearized mapping (9–102) is invertible, that is if the Jacobian determinant at the point $(x_1^0, \ldots, x_n^0)$ is not 0. This can be shown to be correct (see Section 9–15).

Curvilinear coordinates

An important application of the mapping concept is to polar coordinates in the xy-plane. Here the familiar equations

$$x = r \cos \theta, \qquad y = r \sin \theta \tag{9–105}$$

can be regarded as the equations of a mapping from a region in the $r\theta$-plane. For appropriate choice of this region, the mapping is one-to-one and assigns unique polar coordinates r, θ to each point (x, y) in a corresponding region in the xy-plane. For example, r, θ can be restricted thus: $0 < \theta < \pi/2, 1 < r < 2$. To this rectangle in the $r\theta$-plane corresponds a quarter-annulus in the xy-plane, as in Fig. 9–23. As usual, the lines $r = \text{const}, \theta = \text{const}$ in the xy-plane help to show how the polar coordinates are assigned.

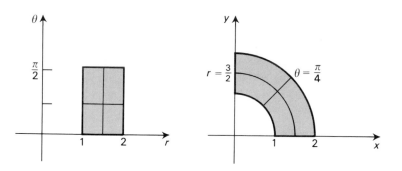

Fig. 9-23. Polar coordinates and mapping (9-105).

For the equations (9-105) we have the Jacobian determinant

$$\frac{\partial(x, y)}{\partial(r, \theta)} = \begin{vmatrix} \cos\theta & -r\sin\theta \\ \sin\theta & r\cos\theta \end{vmatrix} = r.$$

Accordingly, we have the area formula, as in (9-104):

$$dx\,dy = r\,dr\,d\theta. \tag{9-106}$$

This formula is basic for converting double integrals to polar coordinates.

Other mappings of form

$$x = f(u, v), \qquad y = g(u, v) \tag{9-107}$$

can serve in similar fashion to assign curvilinear coordinates u, v to the points (x, y) of a region, as suggested in Fig. 9-24.

There are similar procedures for introducing curvilinear coordinates in higher-dimensional space. For three-dimensional space, the most common cases are the following:

Cylindrical coordinates r, θ, z:

$$x = r\cos\theta, \qquad y = r\sin\theta, \qquad z = z; \tag{9-108}$$

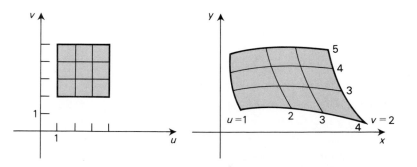

Fig. 9-24. Curvilinear coordinates u, v and mapping (9-107).

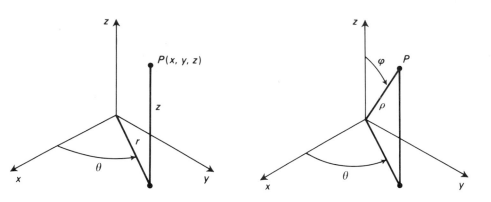

Fig. 9-25. Cylindrical coordinates.   **Fig. 9-26.** Spherical coordinates.

Spherical coordinates ρ, φ, θ:
$$x = \rho \sin\varphi \cos\theta, \qquad y = \rho \sin\varphi \sin\theta, \qquad z = \rho \cos\varphi. \qquad (9\text{--}109)$$
These are suggested in Figs. 9–25 and 9–26.

9 – 11 PARAMETRIC REPRESENTATION OF SURFACES

We consider the special case $m = 3$, $n = 2$ of the mappings of the preceding section and write the mapping as follows:
$$x = f(u, v), \qquad y = g(u, v), \qquad z = h(u, v). \qquad (9\text{--}110)$$
Thus u, v replace x_1, x_2, while x, y, z replace y_1, y_2, y_3. The notation (9–110) is commonly used for the *parametric representation of a surface in space*, with u, v as parameters. As in Section 9–10, (u, v) varies over an open region D in the uv-plane (Fig. 9–27).

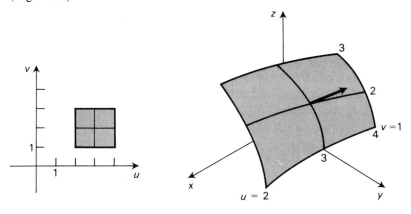

Fig. 9-27. Parametric representation of a surface.

For $v = \text{const}$, Eqs. (9–110) define x, y, z as functions of the parameter u. As in Section 9–3, when u varies, (x, y, z) follows a curve in xyz-space. If we change the constant value of v, we get a different curve in xyz-space. It is plausible that these curves together form a surface S in xyz-space. The surface S is formed of the various curves $v = \text{const}$. It can also be formed, in similar fashion, from curves $u = \text{const}$. In Fig. 9–27 we suggest a typical case and show the curves $u = \text{const}$, $v = \text{const}$ on S. Giving the values of u and v serves to locate a point on S and hence u and v are interpreted as *curvilinear coordinates* on S.

Another way of seeing that Eqs. (9–110) should represent a surface is as follows. Under reasonable hypotheses, the equations $x = f(u, v)$, $y = g(u, v)$ can be solved for u and v in terms of x and y. If we substitute the resulting functions in $z = h(u, v)$, then z becomes a function $F(x, y)$; we thus obtain a surface S in the familiar form $z = F(x, y)$. The question of solvability of the equations $x = f(u, v)$, $y = g(u, v)$ for u and v is considered in Section 9–13 below.

We now linearize Eqs. (9–110) at a particular point (u_0, v_0) by taking differentials:

$$\begin{bmatrix} dx \\ dy \\ dz \end{bmatrix} = \begin{bmatrix} f_u & f_v \\ g_u & g_v \\ h_u & h_v \end{bmatrix} \begin{bmatrix} du \\ dv \end{bmatrix}. \tag{9–111}$$

The Jacobian matrix of the mapping (9–110) is the 3×2 matrix on the right. We observe that the first column of this matrix is the vector $\text{col}(f_u, g_u, h_u)$ which is tangent to the path $v = \text{const} = v_0$; this is the same as the velocity vector $d\mathbf{r}/dt$ of Section 9–3, with t replaced by u. This vector is shown in Fig. 9–27 for $v = 2$. Similarly, the second column of the Jacobian matrix, $\text{col}(f_v, g_v, h_v)$, provides a tangent vector to the path $u = u_0$. Together these vectors determine the *tangent plane* to S, and Eq. (9–111) can be written:

$$\begin{bmatrix} dx \\ dy \\ dz \end{bmatrix} = du \begin{bmatrix} f_u \\ g_u \\ h_u \end{bmatrix} + dv \begin{bmatrix} f_v \\ g_v \\ h_v \end{bmatrix}. \tag{9–111'}$$

This shows that, as du, dv vary, the corresponding linearized displacement (dx, dy, dz) is a linear combination of the two tangent vectors and hence determines a point of the tangent plane (Fig. 9–28). Thus the linearized mapping is a mapping from the uv-plane to the tangent plane.

The cross product of the two tangent vectors provides a normal vector $\mathbf{n}$ to S:

$$\mathbf{n} = (f_u \mathbf{i} + g_u \mathbf{j} + h_u \mathbf{k}) \times (f_v \mathbf{i} + g_v \mathbf{j} + h_v \mathbf{k}). \tag{9–112}$$

The magnitude of $\mathbf{n}$ equals the area of the parallelogram with the two tangent vectors as edge vectors. From Eq. (9–111') we see that this parallelogram corresponds to the square $0 \leqslant du \leqslant 1, 0 \leqslant dv \leqslant 1$ in the uv-plane (Fig. 9–28). Thus $|\mathbf{n}|$ plays the role of the absolute value of the Jacobian determinant in Section 9–10: in the linearized mapping all areas are multiplied by $|\mathbf{n}|$ in going from the uv-plane to the tangent plane of the surface S at the chosen point.

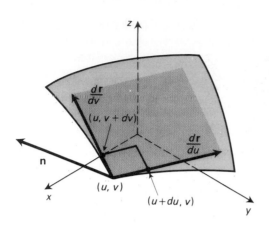

Fig. 9–28. Displacements in the tangent plane.

In fact, $|\mathbf{n}|$ can be expressed in terms of Jacobian determinants, for from (9–112)

$$\mathbf{n} = \begin{vmatrix} g_u & g_v \\ h_u & h_v \end{vmatrix}\mathbf{i} - \begin{vmatrix} f_u & f_v \\ h_u & h_v \end{vmatrix}\mathbf{j} + \begin{vmatrix} f_u & f_v \\ g_u & g_v \end{vmatrix}\mathbf{k}$$

$$= \frac{\partial(g,\,h)}{\partial(u,\,v)}\mathbf{i} - \frac{\partial(f,\,h)}{\partial(u,\,v)}\mathbf{j} + \frac{\partial(f,\,g)}{\partial(u,\,v)}\mathbf{k};$$

it follows that

$$|\mathbf{n}| = \sqrt{\left|\frac{\partial(g,\,h)}{\partial(u,\,v)}\right|^2 + \left|\frac{\partial(f,\,h)}{\partial(u,\,v)}\right|^2 + \left|\frac{\partial(f,\,g)}{\partial(u,\,v)}\right|^2}. \tag{9–113}$$

For small du, dv the rectangle of sides du, dv corresponds to a small piece of the surface S, whose area we can take as *surface area element* $d\sigma$. By using the linearized mapping, we obtain an expression for $d\sigma$:

$$d\sigma = |\mathbf{n}|\,du\,dv, \tag{9–114}$$

where $|\mathbf{n}|$ is given by (9–113). This formula, which has been obtained intuitively, can be fully justified (see *Treatise on Advanced Calculus* by P. FRANKLIN, Wiley, New York, 1940, pp. 371–378).

EXAMPLE 1 $x = \rho \sin \varphi \cos \theta$, $y = \rho \sin \varphi \sin \theta$, $z = \rho \cos \varphi$, where $\rho = \text{const} > 0$. The equations are the same as Eqs. (9–109) for spherical coordinates on a sphere of radius ρ. The parameters u and v are replaced by φ and θ. The typical curves $\varphi = \text{const}$ and $\theta = \text{const}$ (circles) are shown in Fig. 9–29. Here the Jacobian matrix is

$$\begin{bmatrix} \rho \cos\varphi \cos \theta & -\rho \sin\varphi \sin \theta \\ \rho \cos\varphi \sin \theta & \rho \sin\varphi \cos \theta \\ -\rho \sin\varphi & 0 \end{bmatrix}.$$

The two column vectors provide tangent vectors to the curves $\varphi = \text{const}$ and

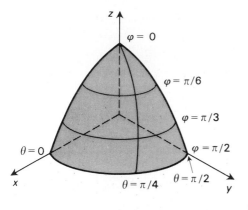

Fig. 9–29. φ and θ as curvilinear coordinates on a sphere.

$\theta = $ const in Fig. 9–29. Their cross product is the normal vector **n**. After calculation we find

$$\mathbf{n} = \rho^2 \sin\varphi\, (\sin\varphi \cos\theta\, \mathbf{i} + \sin\varphi \sin\theta\, \mathbf{j} + \cos\varphi\, \mathbf{k})$$
$$= \rho \sin\varphi\, (x\,\mathbf{i} + y\,\mathbf{j} + z\,\mathbf{k}).$$

Thus **n** has the direction of the radius vector, as expected. The magnitude of **n** is

$$|\mathbf{n}| = \rho \sin\varphi\, \sqrt{x^2 + y^2 + z^2} = \rho^2 \sin\varphi.$$

Hence the surface area element is $d\sigma = \rho^2 \sin\varphi\, d\varphi\, d\theta$. ◄

> REMARK: In our discussion we have tacitly assumed that **n** is not **0** or, what is the same, that the two tangent vectors (the column vectors of the Jacobian matrix) are linearly independent. Points at which **n** = **0** are points at which either the surface has some peculiar property (as at the vertex of a cone) or the curvilinear coordinates do not behave properly. In Example 1, **n** = **0** when $\sin\varphi = 0$, that is, when $\varphi = 0$ or π. The curvilinear coordinates do have a peculiarity at these points (the "north and south poles" of the sphere); the locus $\varphi = 0$ or $\varphi = \pi$ is a point, not a curve, and θ becomes ambiguous at the points. In general, where **n** ≠ **0**, there is a one-to-one correspondence between points of the surface and points of a small region in the uv-plane.

9–12 CHAIN RULE FOR MAPPINGS

Occasionally we may have compositions involving several functions of several variables: for example,

$$z = F(x, y), \qquad w = G(x, y) \tag{9–120}$$

and

$$x = f(u, v), \qquad y = g(u, v), \tag{9–121}$$

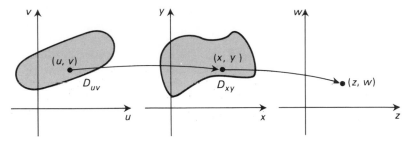

Fig. 9-30. Composition of two mappings.

so that the composite functions can be formed:

$$z = F(f(u, v), g(u, v)), \qquad w = G(f(u, v), g(u, v)). \qquad (9\text{-}122)$$

We can interpret the process as forming the composition of two mappings, as suggested in Fig. 9–30. Here the mapping (9–120) goes from an open region D_{xy} in the xy-plane to the zw-plane. The mapping (9–121) goes from the open region D_{uv} in the uv-plane to the xy-plane, and it is assumed that the range of this mapping is contained in D_{xy}. The two mappings can then be composed: from (u, v) we obtain (x, y) and hence (z, w). The final result is a mapping from D_{uv} to the zw-plane, given by Eqs. (9–122).

We assume all the functions in Eqs. (9–120), (9–121) have continuous first partial derivatives in the open regions concerned. Then we seek the first partial derivatives of the composite functions (9–122). These can be obtained from the chain rule of Section 9–4:

$$\frac{\partial z}{\partial u} = F_x f_u + F_y g_u, \qquad \frac{\partial z}{\partial v} = F_x f_v + F_y g_v,$$

$$\frac{\partial w}{\partial u} = G_x f_u + G_y g_u, \qquad \frac{\partial w}{\partial v} = G_x f_v + G_y g_v, \qquad (9\text{-}123)$$

where F_x, F_y, G_x, G_y are evaluated at $x = f(u, v)$, $y = g(u, v)$. These relations can be expressed in matrix form:

$$\begin{bmatrix} z_u & z_v \\ w_u & w_v \end{bmatrix} = \begin{bmatrix} F_x & F_y \\ G_x & G_y \end{bmatrix} \begin{bmatrix} f_u & f_v \\ g_u & g_v \end{bmatrix}, \qquad (9\text{-}123')$$

as can be verified by multiplying out on the right. Thus *the Jacobian matrix of the composite mapping is the product of the Jacobian matrices of the mappings* (9–120), (9–121) (in that order). We regard (9–123') as a chain rule for the mappings (9–120), (9–121). The rule generalizes to higher dimensions.

═══════════════════ **Problems (Section 9-12)** ═══════════════════

1. Obtain the Jacobian matrix for each of the following mappings:
   a) $y_1 = 2x_1 - 3x_2, \quad y_2 = x_1 + 2x_2$
   b) $y_1 = x_1^2 - x_2^2, \quad y_2 = 2x_1 x_2$

c) $y_1 = x_1 x_2 x_3$, $y_2 = x_1^2 x_3$

d) $u = x \cos y$, $v = x \sin y$, $w = x^2$

2. Obtain the Jacobian determinant requested:

a) $\dfrac{\partial(u, v)}{\partial(x, y)}$ for $u = x^3 - 3xy^2$, $v = 3x^2 y - y^3$

b) $\dfrac{\partial(u, v, w)}{\partial(x, y, z)}$ for $u = xe^y \cos z$, $v = xe^y \sin z$, $w = xe^y$

c) $\dfrac{\partial(f, g)}{\partial(u, v)}$ for $f(u, v, w) = u^2 vw$, $g(u, v, w) = u^2 v^2 - w^4$

d) $\dfrac{\partial(f, g, h)}{\partial(x, y, z)}$ for $f(x, y, z, t) = x^2 + 2y + z^2 - t^2$,

$g(x, y, z, t) = x y z + t^2$, $h(x, y, z, t) = z^2 - t^2$.

3. For the mapping $u = e^x \cos y$, $v = e^x \sin y$ from the xy-plane to the uv-plane carry out the following steps:

a) Evaluate the Jacobian determinant at $(1, 0)$.

b) Show that the square $R_{xy}(0.9 \leqslant x \leqslant 1.1, -0.1 \leqslant y \leqslant 0.1)$ corresponds to the region R_{uv} bounded by arcs of the circles $u^2 + v^2 = e^{1.8}$, $u^2 + v^2 = e^{2.2}$ and the rays $v = \pm \tan 0.1\, u$, $u \geqslant 0$, and find the ratio of the area of R_{uv} to that of R_{xy}. Compare with the result of (a).

c) Obtain the approximating linear mapping at $(1, 0)$ and find the region R'_{uv} corresponding to the square R_{xy} of part (b) under this linear mapping. Find the ratio of the area of R'_{uv} to that of R_{xy} and compare with the results of parts (a) and (b).

4. In Example 1 of Section 9–10, the line $x_2 = 1$ corresponds to the curve $y_1 = x_1^2 - x_1$, $y_2 = x_1 + 1$ in the $y_1 y_2$-plane, with x_1 as parameter. Show that this curve has slope $1/3$ at the point $(2, 3)$ and hence has as tangent line the image of the line $dx_2 = 0$ under the linear approximation found in Section 9–10.

5. Find the Jacobian determinant and hence obtain the corresponding volume element for triple integrals:

a) For cylindrical coordinates (9–108).

b) For spherical coordinates (9–109).

6. Verify the calculation leading to the expression $\mathbf{n} = \rho \sin\varphi\,(x\mathbf{i} + y\mathbf{j} + z\mathbf{k})$ in Example 1 of Section 9–11.

7. For each of the following surfaces, in parametric representation, graph and find a normal vector $\mathbf{n}$ and the corresponding surface area element $d\sigma$:

a) Cylinder: $x = a \cos u$, $y = a \sin u$, $z = v$ ($a = \text{const} > 0$).

b) Cone: $x = av \cos u$, $y = av \sin u$, $z = v$ ($a = \text{const} > 0$).

c) Torus: $x = \cos u\ (b + a \cos v)$, $y = \sin u\ (b + a \cos v)$, $z = a \sin v$ (where a and b are constant, $0 < a < b$).

8. Find the Jacobian matrix $(\partial y_i / \partial u_j)$ in the form of a product of two matrices and evaluate the matrix for the given values of $u_1, u_2, \ldots$ (see HINT on next page).

a) $y_1 = x_1 x_2 - 3x_1$, $y_2 = x_2^2 + 2x_1 x_2 + 2x_1 - x_2$, $x_1 = u_1 \cos 3u_2$, $x_2 = u_1 \sin 3u_2$; $u_1 = 0$, $u_2 = 0$

b) $y_1 = x_1^2 + x_2^2 - 3x_1 + x_3$, $y_2 = x_1^2 - x_2^2 + 2x_1 - 3x_3$; $x_1 = u_1 u_2 u_3^2$, $x_2 = u_1 u_2^2 u_3$, $x_3 = u_1^2 u_2 u_3$; $u_1 = 1$, $u_2 = 1$, $u_3 = 1$

c) $y_1 = x_1 e^{x_2}, y_2 = x_1 e^{-x_2}, y_3 = x_1^2; x_1 = u_1^2 + u_2, x_2 = 2u_1^2 - u_2; u_1 = 1, u_2 = 0$

HINT: Substitute the values *before* multiplying the matrices.

9. Show that if $z = F(x, y)$, $w = G(x, y)$, and $x = f(u, v)$, $y = g(u, v)$, then

$$\frac{\partial(z, w)}{\partial(u, v)} = \frac{\partial(z, w)}{\partial(x, y)} \frac{\partial(x, y)}{\partial(u, v)}.$$

HINT: Use the rule (9–123′) and the rule det $AB = $ det A det B of Section 5–6.

9 – 13   IMPLICIT FUNCTIONS

If $F(x, y, z)$ is a given function of x, y, and z, then the equation

$$F(x, y, z) = 0 \qquad (9\text{–}130)$$

is a relation that may describe one or several functions z of x and y. Thus, if $x^2 + y^2 + z^2 - 1 = 0$, then

$$z = \sqrt{1 - x^2 - y^2} \qquad \text{or} \qquad z = -\sqrt{1 - x^2 - y^2},$$

both functions being defined for $x^2 + y^2 \leqslant 1$. Either function is said to be *implicitly defined* by the equation $x^2 + y^2 + z^2 - 1 = 0$.

Similarly, an equation
$$F(x, y, z, w) = 0$$

may define one or more implicit functions w of x, y, z. If two such equations are given:

$$F(x, y, z, w) = 0, \qquad G(x, y, z, w) = 0, \qquad (9\text{–}131)$$

it is usually possible (at least in theory) to reduce the equations by elimination to the form

$$w = f(x, y), \qquad z = g(x, y),$$

i.e., to obtain two functions of two variables. In general, if m equations in n unknowns are given ($m < n$), it is possible to solve for m of the variables in terms of the remaining $n - m$ variables; the *number of dependent variables equals the number of equations.*

The main question to be considered here is of the following type. Suppose a particular solution of the m simultaneous equations is known, e.g., a quadruple (x_1, y_1, z_1, w_1) of values satisfying (9–131); then we seek to determine the behavior of the m dependent variables as functions of the independent variables near the given point; e.g., for (9–131), we wish to study $f(x, y)$ and $g(x, y)$ near (x_1, y_1), given that $f(x_1, y_1) = w_1$ and $g(x_1, y_1) = z_1$. Determining behavior of a function near a point will consist here of finding the first partial derivatives at the point; when the first derivatives are known, the total differential can be found and hence a *linear approximation* to the function is known.

To analyze an equation of form (9–130), we assume that $z = f(x, y)$ is a differentiable function which satisfies the equation, so that

$$F(x, y, f(x, y)) = 0. \qquad (9\text{–}132)$$

We assume that this relation holds for (x, y) in a domain D and that the points (x, y, z), for (x, y) in D and $z = f(x, y)$, all lie in a domain where F is differentiable. Then from (9–130) or (9–132) we obtain:

$$F_x dx + F_y dy + F_z dz = 0. \tag{9–133}$$

Here z is the function $f(x, y)$, so that $dz = f_x dx + f_y dy$. From (9–133) we deduce:

$$dz = -\frac{F_x}{F_z} dx - \frac{F_y}{F_z} dy,$$

so that (provided $F_z \neq 0$ at the points considered)

$$f_x = \frac{\partial z}{\partial x} = -\frac{F_x}{F_z}, \qquad f_y = \frac{\partial z}{\partial y} = -\frac{F_y}{F_z}. \tag{9–134}$$

These are the desired expressions for the derivatives.

EXAMPLE 1 $x^2 + y^2 + z^2 = 1$ or $x^2 + y^2 + z^2 - 1 = 0$. We imitate the procedure of the preceding paragraph:

$$2x\,dx + 2y\,dy + 2z\,dz = 0, \qquad dz = -\frac{x}{z}dx - \frac{y}{z}dy,$$

$$\frac{\partial z}{\partial x} = -\frac{x}{z}, \qquad \frac{\partial z}{\partial y} = -\frac{y}{z} \quad (z \neq 0).$$

These equations apply to each differentiable function that satisfies the given equation, in particular to the function $z = (1 - x^2 - y^2)^{1/2}$. At $x = 1/2$, $y = 1/2$, we find $z = 1/\sqrt{2}$, and hence

$$\frac{\partial z}{\partial x} = -\frac{x}{z} = -\frac{\sqrt{2}}{2}, \qquad \frac{\partial z}{\partial y} = -\frac{y}{z} = -\frac{\sqrt{2}}{2}. \quad \blacktriangleleft$$

We remark that the same results can be obtained by taking *partial derivatives* instead of differentials in the given equation. From (9–130), with z considered as a function of x and y, we differentiate with respect to x and then with respect to y to obtain

$$F_x + F_z\frac{\partial z}{\partial x} = 0, \qquad F_y + F_z\frac{\partial z}{\partial y} = 0,$$

from which (9–134) again follows. In the case of Example 1, we obtain

$$2x + 2z\frac{\partial z}{\partial x} = 0, \qquad 2y + 2z\frac{\partial z}{\partial y} = 0.$$

For the pair of equations (9–131), we assume that differentiable functions $w = f(x, y)$, $z = g(x, y)$ satisfy the equations, and can then take differentials:

$$F_x dx + F_y dy + F_z dz + F_w dw = 0,$$
$$G_x dx + G_y dy + G_z dz + G_w dw = 0.$$

We consider these equations as simultaneous *linear* equations for dz and dw and solve by elimination or by determinants. Cramer's rule (Section 5–2) yields

$$dz = \frac{\begin{vmatrix} -F_x dx - F_y dy & F_w \\ -G_x dx - G_y dy & G_w \end{vmatrix}}{\begin{vmatrix} F_z & F_w \\ G_z & G_w \end{vmatrix}}, \qquad dw = \frac{\begin{vmatrix} F_z & -F_x dx - F_y dy \\ G_z & -G_x dx - G_y dy \end{vmatrix}}{\begin{vmatrix} F_z & F_w \\ G_z & G_w \end{vmatrix}},$$

and hence

$$dz = -\frac{\begin{vmatrix} F_x & F_w \\ G_x & G_w \end{vmatrix}}{\begin{vmatrix} F_z & F_w \\ G_z & G_w \end{vmatrix}} dx - \frac{\begin{vmatrix} F_y & F_w \\ G_y & G_w \end{vmatrix}}{\begin{vmatrix} F_z & F_w \\ G_z & G_w \end{vmatrix}} dy,$$

$$dw = -\frac{\begin{vmatrix} F_z & F_x \\ G_z & G_x \end{vmatrix}}{\begin{vmatrix} F_z & F_w \\ G_z & G_w \end{vmatrix}} dx - \frac{\begin{vmatrix} F_z & F_y \\ G_z & G_y \end{vmatrix}}{\begin{vmatrix} F_z & F_w \\ G_z & G_w \end{vmatrix}} dy.$$

Thus we can read off partial derivatives. Since the determinants appearing are Jacobian determinants, we can write the derivatives in terms of Jacobians:

$$\frac{\partial z}{\partial x} = -\frac{\partial(F, G)/\partial(x, w)}{\partial(F, G)/\partial(z, w)}, \qquad \frac{\partial z}{\partial y} = -\frac{\partial(F, G)/\partial(y, w)}{\partial(F, G)/\partial(z, w)},$$

$$\frac{\partial w}{\partial x} = -\frac{\partial(F, G)/\partial(z, x)}{\partial(F, G)/\partial(z, w)}, \qquad \frac{\partial w}{\partial y} = -\frac{\partial(F, G)/\partial(z, y)}{\partial(F, G)/\partial(z, w)}. \tag{9–135}$$

Here we must assume that the Jacobian $\partial(F, G)/\partial(z, w)$ in the denominator is different from 0 at the points considered.

EXAMPLE 2 $2x^2 + y^2 + z^2 - zw = 0,$
$x^2 + y^2 + 2z^2 + zw - 8 = 0.$

We observe that the equations are satisfied for $x = 1$, $y = 1$, $z = 1$, $w = 4$, and seek differentiable functions $z(x, y)$, $w(x, y)$ that satisfy the equations near this point. If such functions exist, then

$$4x\, dx + 2y\, dy + (2z - w)\, dz - z\, dw = 0,$$
$$2x\, dx + 2y\, dy + (4z + w)\, dz + z\, dw = 0.$$

By elimination we find

$$6x\, dx + 4y\, dy + 6z\, dz = 0,$$
$$6x(2z + w)\, dx + 4y(z + w)\, dy - 6z^2\, dw = 0$$

and hence

$$dz = -\frac{x}{z}\, dx - \frac{2y}{3z}\, dy,$$

$$dw = \frac{x(2z + w)}{z^2}\, dx + \frac{2y(z + w)}{3z^2}\, dy,$$

from which we can read off partial derivatives: $\partial z/\partial x = -x/z$ and so on. We could have applied (9–135) directly. For example,

$$\frac{\partial z}{\partial x} = -\frac{\begin{vmatrix} 4x & -z \\ 2x & z \end{vmatrix}}{\begin{vmatrix} 2z-w & -z \\ 4z+w & z \end{vmatrix}} = \frac{-6xz}{6z^2} = -\frac{x}{z}.$$

We observe that the determinant in the denominator equals $6z^2$, and this is different from 0 near $z = 1$. ◄

We could also have taken partial derivatives. By differentiating with respect to x, we obtain

$$4x + (2z - w)\frac{\partial z}{\partial x} - z\frac{\partial w}{\partial x} = 0,$$

$$2x + (4z + w)\frac{\partial z}{\partial x} + z\frac{\partial w}{\partial x} = 0.$$

Elimination gives $\partial z/\partial x = -x/z$, $\partial w/\partial x = x(2z + w)/z^2$. Taking differentials saves time, since all partial derivatives are obtained at once.

The reasoning can be generalized to an arbitrary set of m equations in $m + n$ unknowns. We give some illustrations:

One equation in two unknowns: $F(x, y) = 0$. Here

$$\frac{dy}{dx} = -\frac{F_x}{F_y}, \qquad F_y \neq 0. \tag{9–136}$$

One equation in three unknowns: $F(x, y, z) = 0$. Here, as in (9–134),

$$\frac{\partial z}{\partial x} = -\frac{F_x}{F_z}, \qquad \frac{\partial z}{\partial y} = -\frac{F_y}{F_z}, \qquad F_z \neq 0. \tag{9–137}$$

Two equations in three unknowns: $F(x, y, z) = 0$, $G(x, y, z) = 0$. Here

$$\frac{dz}{dx} = -\frac{\partial(F, G)/\partial(y, x)}{\partial(F, G)/\partial(y, z)}, \qquad \frac{dy}{dx} = -\frac{\partial(F, G)/\partial(x, z)}{\partial(F, G)/\partial(y, z)}, \tag{9–138}$$

where

$$\frac{\partial(F, G)}{\partial(y, z)} = \begin{vmatrix} F_y & F_z \\ G_y & G_z \end{vmatrix} \neq 0.$$

Three equations in five unknowns: $F(x, y, z, u, v) = 0$, $G(x, y, z, u, v) = 0$, $H(x, y, z, u, v) = 0$. Here

$$\frac{\partial x}{\partial u} = -\frac{\partial(F, G, H)/\partial(u, y, z)}{\partial(F, G, H)/\partial(x, y, z)}, \qquad \frac{\partial x}{\partial v} = -\frac{\partial(F, G, H)/\partial(v, y, z)}{\partial(F, G, H)/\partial(x, y, z)},$$

$$\frac{\partial y}{\partial u} = -\frac{\partial(F, G, H)/\partial(x, u, z)}{\partial(F, G, H)/\partial(x, y, z)}, \qquad \frac{\partial y}{\partial v} = -\frac{\partial(F, G, H)/\partial(x, v, z)}{\partial(F, G, H)/\partial(x, y, z)}, \tag{9–139}$$

$$\frac{\partial z}{\partial u} = -\frac{\partial(F, G, H)/\partial(x, y, u)}{\partial(F, G, H)/\partial(x, y, z)}, \qquad \frac{\partial z}{\partial v} = -\frac{\partial(F, G, H)/\partial(x, y, v)}{\partial(F, G, H)/\partial(x, y, z)}.$$

For greater precision the partial derivatives appearing should have subscripts attached. For example, in (9–139), $\partial x / \partial u$ should be $(\partial x / \partial u)_v$.

The entire preceding discussion has been based on the assumption that implicit functions are in fact defined by the equations. This assumption is not always fulfilled. For example, the equations

$$x^2 + y^2 + u^2 + v^2 + 1 = 0, \qquad x^2 - y^2 + 2uv = 0$$

define no functions at all. It can be shown that if the equations are satisfied at one point and the Jacobian determinant appearing in the denominator is not 0 at the point, then the implicit equations do define functions with partial derivatives, as indicated, in a neighborhood of the point. This is the *implicit-function theorem* (see CLA, Section 12–15, for more details).

9–14 LEVEL CURVES AND LEVEL SURFACES

The level curves of a function $F(x, y)$ are the curves $F(x, y) = c$, where c is an arbitrary constant. For example, the level curves of $x^2 + y^2$ are the circles $x^2 + y^2 = c$. The level curves of the function $y^2 - x^3 + 2x^2$ are shown in Fig. 9–31.

Let $F(x, y)$ have continuous first partial derivatives in the open region D. At each point (x_0, y_0) of D, $F(x_0, y_0) = c$ for some choice of c. Thus we expect a level curve $F(x, y) = c$ to pass through the point. To ensure that a well-defined level curve does indeed pass through the point we require that grad $F \neq \mathbf{0}$ at (x_0, y_0), since this ensures that either $F_x(x_0, y_0) \neq 0$ or $F_y(x_0, y_0) \neq 0$.

If, for example, $F_y(x_0, y_0) \neq 0$, then the implicit-function theorem of the preceding section ensures that near (x_0, y_0) the equation $F(x, y) = c$ is satisfied precisely at the points (x, y) of the graph of a function $y = f(x)$, with $y' = -F_x/F_y$.

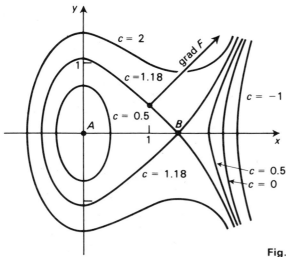

Fig. 9–31. Level curves of $y^2 - x^3 + 2x^2$.

This equation, or the equivalent equation

$$F_x dx + F_y dy = 0,$$

shows that grad F is *normal to the level curve at each point*, as illustrated in Fig. 9–31. If $F_y(x_0, y_0) = 0$ but $F_x(x_0, y_0) \ne 0$, then we come to a similar conclusion: near (x_0, y_0) the level curve is the graph of $x = g(y)$. If both partial derivatives are 0 at the point, then grad $F = \mathbf{0}$ at the point, and we cannot expect a well-defined level curve through the point. This is illustrated by points A and B of Fig. 9–31. We remark that points at which grad F is $\mathbf{0}$ are *critical points* of F and are potentially points at which F has a local maximum or minimum (Section 9–9). A similar discussion holds for a function $F(x, y, z)$ with *level surfaces* $F(x, y, z) = c$.

For each point where grad $F \ne \mathbf{0}$ there is such a surface through the point, representable as $z = f(x, y)$ if $F_z \ne 0$ at the point. Again grad F is normal to the level surface at each point. Points where grad $F = \mathbf{0}$ (the critical points of F) are exceptional.

Level curves and level surfaces find wide application. For example, the daily weather map shows level curves of temperature and barometric pressure. A topographical map shows level curves of altitude. Meteorology uses surfaces of constant barometric pressure called isobaric surfaces. The gravitational-force field about the earth has a potential $U(x, y, z)$ (see Section 9–8 above) and the level surfaces of U are called *equipotential surfaces*; they are roughly concentric spheres about the earth.

9 – 15 INVERSE FUNCTIONS

A pair of functions

$$y_1 = f_1(x_1, x_2), \qquad y_2 = f_2(x_1, x_2) \tag{9–150}$$

can be regarded as a mapping from the $x_1 x_2$-plane to the $y_1 y_2$-plane. Under appropriate conditions, this mapping is a one-to-one correspondence between an open region D in the $x_1 x_2$-plane and an open region D' in the $y_1 y_2$-plane (see Section 9–10 and Fig. 9–24). There is then an *inverse mapping* that takes each point (y_1, y_2) in D' to the unique point (x_1, x_2) in D such that Eqs. (9–150) hold. The inverse mapping is given by functions

$$x_1 = g_1(y_1, y_2), \qquad x_2 = g_2(y_1, y_2). \tag{9–151}$$

It may be difficult to solve Eqs. (9–150) to obtain these functions explicitly. Nevertheless, we can consider Eqs. (9–150) in the form

$$f(x_1, x_2) - y_1 = 0, \qquad f_2(x_1, x_2) - y_2 = 0 \tag{9–152}$$

as *implicit equations* for the functions (9–151). We can then seek partial derivatives of these functions as in Section 9–13. With

$$F_1(x_1, x_2, y_1, y_2) = f_1(x_1, y_1) - y_1 \quad \text{and} \quad F_2(x_1, x_2, y_1, y_2) = f_2(x_1, x_2) - y_2$$

we have, for example,

$$\frac{\partial x_1}{\partial y_1} = -\frac{\partial(F_1, F_2)/\partial(y_1, x_2)}{\partial(F_1, F_2)/\partial(x_1, x_2)} = -\frac{\begin{vmatrix} -1 & f_{1x_1} \\ 0 & f_{2x_2} \end{vmatrix}}{\partial(f_1, f_2)/\partial(x_1, x_2)} = \frac{\partial f_2/\partial x_2}{\partial(f_1, f_2)/\partial(x_1, x_2)}.$$

$$(9\text{-}153)$$

The Jacobian determinant in the denominator is simply the Jacobian of the mapping (9–150). As in Section 9–13, we assume it is not 0 at the points considered.

EXAMPLE 1 $y_1 = x_1^2 + 3x_1 x_2 + 2x_2^2 = f_1(x_1, x_2)$, $y_2 = 2x_1^2 - x_2^2 = f_2(x_1, x_2)$.
We write the equations as

$$x_1^2 + 3x_1 x_2 + 2x_2^2 - y_1 = 0,$$
$$2x_1^2 - x_2^2 - y_2 = 0,$$

which are implicit equations for x_1, x_2 as functions of y_1, y_2. Hence

$$\frac{\partial x_1}{\partial y_1} = -\frac{\begin{vmatrix} -1 & 4x_1 \\ 0 & -2x_2 \end{vmatrix}}{\begin{vmatrix} 2x_1 + 3x_2 & 3x_1 + 4x_2 \\ 4x_1 & -2x_2 \end{vmatrix}} = \frac{2x_2}{20x_1 x_2 + 6x_2^2 + 12x_1^2}$$

and the other partial derivatives can be obtained in the same way. At the point $x_1 = 1$, $x_2 = 1$, the Jacobian determinant in the denominator is not 0 and we can thus be sure that the inverse functions are well defined in some neighborhood of the corresponding point $y_1 = 6$, $y_2 = 1$. ◀

The fact that the functions (9–151) are solutions of Eqs. (9–150) means that

$$f_1[g_1(y_1, y_2), g_2(y_1, y_2)] \equiv y_1, \qquad f_2[g_1(y_1, y_2), g_2(y_1, y_2)] \equiv y_2. \,(9\text{-}154)$$

By the chain rule for mappings (Section 9–12), the Jacobian matrix of

$$f_1[g_1(y_1, y_2), g_2(y_1, y_2)], \quad f_2[g_1(y_1, y_2), g_2(y_1, y_2)]$$

is

$$\left(\frac{\partial f_i}{\partial x_j}\right)\left(\frac{\partial g_i}{\partial y_j}\right). \tag{9-155}$$

By Eqs. (9–154), these functions reduce to y_1, y_2 (as functions of y_1, y_2). Therefore, the product (9–155) must also equal the Jacobian matrix of the functions y_1, y_2 (of y_1 and y_2); that is, it must equal the identity matrix $I = \begin{bmatrix} 1 & 0 \\ 0 & 1 \end{bmatrix}$. Thus *the Jacobian matrix of a mapping is the inverse matrix of the Jacobian matrix of the inverse mapping.* (The two matrices must be evaluated at corresponding points (x_1, x_2), (y_1, y_2)). Taking determinants, we obtain the useful rule

$$\frac{\partial(y_1, y_2)}{\partial(x_1, x_2)} \cdot \frac{\partial(x_1, x_2)}{\partial(y_1, y_2)} = \det I = 1. \tag{9-156}$$

The preceding discussion extends to invertible mappings

$$y_i = f_i(x_1, \ldots, x_n), \quad i = 1, \ldots, n.$$

We obtain inverse functions

$$x_i = g_i(y_1, \ldots, y_n), \quad i = 1, \ldots, n.$$

The Jacobian matrices $\left(\dfrac{\partial y_i}{\partial x_j}\right)$ and $\left(\dfrac{\partial x_i}{\partial y_j}\right)$ are inverses of each other and

$$\frac{\partial(y_1, \ldots, y_n)}{\partial(x_1, \ldots, x_n)} \cdot \frac{\partial(x_1, \ldots, x_n)}{\partial(y_1, \ldots, y_n)} = 1. \tag{9–157}$$

══════════════════════════**Problems (Section 9-15)**══════════════════════════

1. Given that

$$2x + y - 3z - 2u = 0, \qquad x + 2y + z + u = 0,$$

find the following partial derivatives:

$$\left(\frac{\partial x}{\partial y}\right)_z, \quad \left(\frac{\partial y}{\partial x}\right)_u, \quad \left(\frac{\partial z}{\partial u}\right)_x, \quad \left(\frac{\partial y}{\partial z}\right)_x.$$

2. Given that

$$x^2 + y^2 + z^2 - u^2 + v^2 = 1, \qquad x^2 - y^2 + z^2 + u^2 + 2v^2 = 21,$$

a) Find du and dv in terms of dx, dy, dz at the point $x = 1$, $y = 1$, $z = 2$, $u = 3$, $v = 2$.
b) Find $(\partial u/\partial x)_{y,z}$ and $(\partial v/\partial y)_{x,z}$ at this point.
c) Find approximately the values of u and v for $x = 1.1$, $y = 1.2$, $z = 1.8$.

3. For the transformation: $x = r\cos\theta$, $y = r\sin\theta$ from rectangular to polar coordinates, verify the relations:
a) $dx = \cos\theta\,dr - r\sin\theta\,d\theta$, $dy = \sin\theta\,dr + r\cos\theta\,d\theta$
b) $dr = \cos\theta\,dx + \sin\theta\,dy$, $d\theta = -\dfrac{\sin\theta}{r}dx + \dfrac{\cos\theta}{r}dy$

c) $\left(\dfrac{\partial x}{\partial r}\right)_\theta = \cos\theta$, $\left(\dfrac{\partial x}{\partial r}\right)_y = \sec\theta$, $\left(\dfrac{\partial r}{\partial x}\right)_y = \cos\theta$, $\left(\dfrac{\partial r}{\partial x}\right)_\theta = \sec\theta$

d) $\dfrac{\partial(x, y)}{\partial(r, \theta)} = r$, $\dfrac{\partial(r, \theta)}{\partial(x, y)} = \dfrac{1}{r}$

4. Given that

$$x^2 - y\cos(uv) + z^2 = 0,$$
$$x^2 + y^2 - \sin(uv) + 2z^2 = 2,$$
$$xy - \sin u\cos v + z = 0,$$

find $(\partial x/\partial u)_v$ and $(\partial x/\partial v)_u$ at the point $x = 1$, $y = 1$, $u = \pi/2$, $v = 0$, $z = 0$.

5. For $z = F(x, y)$ as given, plot the level curves of F, showing critical points. Also show grad F at a few points and verify that it is orthogonal to the level curves at the points.
a) $z = x - 2y$ b) $z = 2x^2 + y^2$ c) $z = x^2 - 2y^2$
d) $z = y^2 - x^3 + 4x$ e) $z = x^2 e^{-y^2}$

6. Find $(\partial u/\partial x)_y$ if $x^2 - y^2 + u^2 + 2v^2 = 1$, $x^2 + y^2 - u^2 - v^2 = 2$.

7. Given the mapping $x = u - 2v, \quad y = 2u + v$.
   a) Write the equations of the inverse mapping.
   b) Evaluate the Jacobian of the mapping and that of the inverse mapping.

8. Given the mappting $x = u^2 - v^2, \quad y = 2uv$,

   a) Compute its Jacobian   b) Evaluate $\left(\dfrac{\partial u}{\partial x}\right)_y$ and $\left(\dfrac{\partial v}{\partial x}\right)_y$

9. Given the mapping $x = f(u, v), \quad y = g(u, v)$ with Jacobian $J = \partial(x, y)/\partial(u, v)$, show that for the inverse functions

$$\frac{\partial u}{\partial x} = \frac{1}{J}\frac{\partial y}{\partial v}, \qquad \frac{\partial u}{\partial y} = -\frac{1}{J}\frac{\partial x}{\partial v}, \qquad \frac{\partial v}{\partial x} = -\frac{1}{J}\frac{\partial y}{\partial u}, \qquad \frac{\partial v}{\partial y} = \frac{1}{J}\frac{\partial x}{\partial u}.$$

Use these results to check Problem 3(b) above.

10. Given the mapping $x = f(u, v, w), \quad y = g(u, v, w)$, and $z = h(u, v, w)$ with Jacobian $J = \partial(x, y, z)/\partial(u, v, w)$, show that for the inverse functions

$$\frac{\partial u}{\partial x} = \frac{1}{J}\frac{\partial(y, z)}{\partial(v, w)}, \qquad \frac{\partial u}{\partial y} = \frac{1}{J}\frac{\partial(z, x)}{\partial(v, w)}, \qquad \frac{\partial u}{\partial z} = \frac{1}{J}\frac{\partial(x, y)}{\partial(v, w)},$$

$$\frac{\partial v}{\partial x} = \frac{1}{J}\frac{\partial(y, z)}{\partial(w, u)}, \qquad \frac{\partial v}{\partial y} = \frac{1}{J}\frac{\partial(z, x)}{\partial(w, u)}, \qquad \frac{\partial v}{\partial z} = \frac{1}{J}\frac{\partial(x, y)}{\partial(w, u)},$$

$$\frac{\partial w}{\partial x} = \frac{1}{J}\frac{\partial(y, z)}{\partial(u, v)}, \qquad \frac{\partial w}{\partial y} = \frac{1}{J}\frac{\partial(z, x)}{\partial(u, v)}, \qquad \frac{\partial w}{\partial z} = \frac{1}{J}\frac{\partial(x, y)}{\partial(u, v)}.$$

11. For the transformation (9–109) from rectangular to spherical coordinates,
    a) Compute the Jacobian $\partial(x, y, z)/\partial(\rho, \varphi, \theta)$
    b) Evaluate $\partial\rho/\partial y, \partial\varphi/\partial z, \partial\theta/\partial x$ for the inverse transformation (see Problem 10).

12. Prove that if $F(x, y, z) = 0$, then $\left(\dfrac{\partial z}{\partial x}\right)_y \left(\dfrac{\partial x}{\partial y}\right)_z \left(\dfrac{\partial y}{\partial z}\right)_x = -1$.

13. Prove that, if $x = f(u, v), \ y = g(u, v)$, then

$$\left(\frac{\partial x}{\partial u}\right)_v \left(\frac{\partial u}{\partial x}\right)_y = \left(\frac{\partial y}{\partial v}\right)_u \left(\frac{\partial v}{\partial y}\right)_x, \qquad \left(\frac{\partial x}{\partial v}\right)_u \left(\frac{\partial v}{\partial x}\right)_y = \left(\frac{\partial u}{\partial y}\right)_x \left(\frac{\partial y}{\partial u}\right)_v, \qquad \left(\frac{\partial x}{\partial y}\right)_u \left(\frac{\partial y}{\partial x}\right)_u = 1.$$

14. In *thermodynamics* the variables p (pressure), T (temperature), U (internal energy), and V (volume) occur. For each substance these are related by two equations, so that any two of the four variables can be chosen as independent, the other two then being dependent. In addition, the second law of thermodynamics implies the relation

$$\frac{\partial U}{\partial V} - T\frac{\partial p}{\partial T} + p = 0, \tag{9–158}$$

when V and T are independent. Show that this relation can be written in each of the following forms:

a) $\dfrac{\partial T}{\partial V} + T\dfrac{\partial p}{\partial U} - p\dfrac{\partial T}{\partial U} = 0, \qquad U, V$ independent

b) $T - p\dfrac{\partial T}{\partial p} + \dfrac{\partial(T, U)}{\partial(V, p)} = 0, \qquad V, p$ independent

c) $\dfrac{\partial U}{\partial p} + T\dfrac{\partial V}{\partial T} + p\dfrac{\partial V}{\partial p} = 0,$ p, T independent

d) $\dfrac{\partial T}{\partial p} - T\dfrac{\partial V}{\partial U} + p\dfrac{\partial(V, T)}{\partial(U, p)} = 0,$ U, p independent

e) $T\dfrac{\partial(p, V)}{\partial(T, U)} - p\dfrac{\partial V}{\partial U} - 1 = 0,$ T, U independent

HINT: The relation (9–158) implies that, if
$$dU = a\,dV + b\,dT, \qquad dp = c\,dV + e\,dT$$
are the expressions for dU and dp in terms of dV and dT, then $a - Te + p = 0$. To prove (a), for example, we assume relations
$$dT = \alpha\,dV + \beta\,dU, \qquad dp = \gamma\,dV + \delta\,dU.$$
If these are solved for dU and dp in terms of dV and dT, then we obtain expressions for a and e in terms of α, β, γ, δ. If these expressions are substituted in the equation $a - Te + p = 0$, we have an equation in $\alpha, \beta, \gamma, \delta$. Since $\alpha = \partial T/\partial V$, etc., the relation of form (a) is obtained. The others are proved in the same way.

15. Let equations $F_1(x_1, x_2, x_3, x_4) = 0$, $F_2(x_1, x_2, x_3, x_4) = 0$ be given. At a certain point, where the equations are satisfied, it is known that
$$\frac{\partial F_i}{\partial x_j} = \begin{bmatrix} 3 & 1 & 0 & 2 \\ 5 & 1 & -1 & 4 \end{bmatrix}.$$

a) Evaluate $(\partial x_1/\partial x_3)_{x_4}$ and $(\partial x_1/\partial x_4)_{x_3}$ at the point.
b) Evaluate $(\partial x_1/\partial x_3)_{x_2}$ and $(\partial x_4/\partial x_3)_{x_2}$ at the point.
c) Evaluate $\partial(x_1, x_2)/\partial(x_3, x_4)$ and $\partial(x_3, x_4)/\partial(x_1, x_2)$ at the point.

9 – 16 HIGHER DERIVATIVES OF COMPOSITE FUNCTIONS

Let $z = f(x, y)$ and $x = g(t)$, $y = h(t)$, so that z can be expressed in terms of t alone. The derivative dz/dt can then be evaluated by the chain rule (9–45) of Section 9–4 above:
$$\frac{dz}{dt} = \frac{\partial z}{\partial x}\frac{dx}{dt} + \frac{\partial z}{\partial y}\frac{dy}{dt}. \tag{9–160}$$

By applying the product rule we obtain the following expression for the second derivative:
$$\frac{d^2z}{dt^2} = \frac{d}{dt}\left(\frac{dz}{dt}\right) = \frac{\partial z}{\partial x}\frac{d^2x}{dt^2} + \frac{dx}{dt}\frac{d}{dt}\left(\frac{\partial z}{\partial x}\right) + \frac{\partial z}{\partial y}\frac{d^2y}{dt^2} + \frac{dy}{dt}\frac{d}{dt}\left(\frac{\partial z}{\partial y}\right).$$

To evaluate the expressions $(d/dt)\,(\partial z/\partial x)$ and $(d/dt)\,(\partial z/\partial y)$, we use (9–160) again, this time applied to $\partial z/\partial x$ and $\partial z/\partial y$ rather than to z:

$$\frac{d}{dt}\left(\frac{\partial z}{\partial x}\right) = \frac{\partial^2 z}{\partial x^2}\frac{dx}{dt} + \frac{\partial^2 z}{\partial y\,\partial x}\frac{dy}{dt}, \qquad \frac{d}{dt}\left(\frac{\partial z}{\partial y}\right) = \frac{\partial^2 z}{\partial x\,\partial y}\frac{dx}{dt} + \frac{\partial^2 z}{\partial y^2}\frac{dy}{dt}.$$

Thus we find the rule:

$$\frac{d^2 z}{dt^2} = \frac{\partial z}{\partial x}\frac{d^2 x}{dt^2} + \frac{\partial^2 z}{\partial x^2}\left(\frac{dx}{dt}\right)^2 + 2\frac{\partial^2 z}{\partial x\,\partial y}\frac{dx}{dt}\frac{dy}{dt} + \frac{\partial^2 z}{\partial y^2}\left(\frac{dy}{dt}\right)^2 + \frac{\partial z}{\partial y}\frac{d^2 y}{dt^2}. \tag{9-161}$$

This is a new chain rule.

Similarly, if $z = f(x, y)$, $x = g(u, v)$, $y = h(u, v)$, we have

$$\frac{\partial z}{\partial u} = \frac{\partial z}{\partial x}\frac{\partial x}{\partial u} + \frac{\partial z}{\partial y}\frac{\partial y}{\partial u}, \qquad \frac{\partial^2 z}{\partial u^2} = \frac{\partial z}{\partial x}\frac{\partial^2 x}{\partial u^2} + \frac{\partial}{\partial u}\left(\frac{\partial z}{\partial x}\right)\frac{\partial x}{\partial u} + \frac{\partial z}{\partial y}\frac{\partial^2 y}{\partial u^2} + \frac{\partial}{\partial u}\left(\frac{\partial z}{\partial y}\right)\frac{\partial y}{\partial u}.$$

$$\tag{9-162}$$

Applying (9–162) again, we get

$$\frac{\partial}{\partial u}\left(\frac{\partial z}{\partial x}\right) = \frac{\partial^2 z}{\partial x^2}\frac{\partial x}{\partial u} + \frac{\partial^2 z}{\partial y\,\partial x}\frac{\partial y}{\partial u}, \qquad \frac{\partial}{\partial u}\left(\frac{\partial z}{\partial y}\right) = \frac{\partial^2 z}{\partial x\,\partial y}\frac{\partial x}{\partial u} + \frac{\partial^2 z}{\partial y^2}\frac{\partial y}{\partial u},$$

so that

$$\frac{\partial^2 z}{\partial u^2} = \frac{\partial z}{\partial x}\frac{\partial^2 x}{\partial u^2} + \left(\frac{\partial^2 z}{\partial x^2}\right)\left(\frac{\partial x}{\partial u}\right)^2 + 2\frac{\partial^2 z}{\partial x\,\partial y}\frac{\partial x}{\partial u}\frac{\partial y}{\partial u} + \frac{\partial^2 z}{\partial y^2}\left(\frac{\partial y}{\partial u}\right)^2 + \frac{\partial z}{\partial y}\frac{\partial^2 y}{\partial u^2}.$$

$$\tag{9-163}$$

It should be remarked that (9–163) is a special case of (9–161), since v is treated throughout as a constant.

Rules for $\partial^2 z/\partial u\,\partial v$, $\partial^2 z/\partial v^2$ and for higher derivatives can be formed, analogous to (9–161) and (9–163). These rules are important, but in most practical cases it is better to use only the chain rules (9–45) and (9–46), applying them repeatedly if necessary. One reason for this is that the formulas are simplified if the occurring derivatives are expressed in terms of the right variables, and a complete description of all possible cases would be too involved to be useful.

The possible variations can be illustrated by the following example, which involves only functions of one variable.

Let $y = f(x)$ and $x = e^t$. Then

$$\frac{dy}{dt} = \frac{dy}{dx}\frac{dx}{dt} = \frac{dy}{dx}e^t.$$

Hence

$$\frac{d^2 y}{dt^2} = \frac{d}{dt}\left(\frac{dy}{dx}\right)e^t + \frac{dy}{dx}e^t = \frac{d^2 y}{dx^2}\frac{dx}{dt}e^t + \frac{dy}{dx}e^t = \frac{d^2 y}{dx^2}e^{2t} + \frac{dy}{dx}e^t.$$

We could also write

$$\frac{dy}{dt} = \frac{dy}{dx}e^t = x\frac{dy}{dx}$$

and then

$$\frac{d^2y}{dt^2} = \frac{d}{dt}\left(x\frac{dy}{dx}\right) = \frac{d}{dx}\left(x\frac{dy}{dx}\right)\frac{dx}{dt} = x\frac{d}{dx}\left(x\frac{dy}{dx}\right) = \frac{d^2y}{dx^2}x^2 + \frac{dy}{dx}x.$$

The second method is clearly simpler than the first; the answers obtained are equivalent because of the equation $x = e^t$.

$9-17$ THE LAPLACIAN IN POLAR, CYLINDRICAL, AND SPHERICAL COORDINATES

The Laplacian operator was introduced in Section 8–1, its importance for physical laws was discussed, and corresponding partial differential equations were studied. We recall that the Laplacian of $w = f(x, y)$ is

$$\nabla^2 w = \frac{\partial^2 w}{\partial x^2} + \frac{\partial^2 w}{\partial y^2} \tag{9-170}$$

and that of $w = f(x, y, z)$ is

$$\nabla^2 w = \frac{\partial^2 w}{\partial x^2} + \frac{\partial^2 w}{\partial y^2} + \frac{\partial^2 w}{\partial z^2}. \tag{9-171}$$

A function w (with continuous second partial derivatives) in an open region D such that $\nabla^2 w \equiv 0$ in D is said to be *harmonic*. A function w such that $\nabla^4 w = \nabla^2(\nabla^2 w) \equiv 0$ in D is said to be *biharmonic*.

For the problems of Chapter 8 and other purposes it is important to express $\nabla^2 w$ in polar, cylindrical, and spherical coordinates. We do so here, as an application of the methods of the preceding section. First we consider the two-dimensional Laplacian (9–170) and its expression in terms of polar coordinates r, θ. Thus we are given $w = f(x, y)$ and $x = r\cos\theta$, $y = r\sin\theta$, and we wish to express $\nabla^2 w$ in terms of r, θ, and derivatives of w with respect to r and θ. The solution is as follows. By the chain rule we have

$$\frac{\partial w}{\partial x} = \frac{\partial w}{\partial r}\frac{\partial r}{\partial x} + \frac{\partial w}{\partial \theta}\frac{\partial \theta}{\partial x}, \qquad \frac{\partial w}{\partial y} = \frac{\partial w}{\partial r}\frac{\partial r}{\partial y} + \frac{\partial w}{\partial \theta}\frac{\partial \theta}{\partial y}. \tag{9-172}$$

To evaluate $\partial r/\partial x$, $\partial \theta/\partial x$, $\partial r/\partial y$, $\partial \theta/\partial y$ we use the equations

$$dx = \cos\theta\, dr - r\sin\theta\, d\theta, \qquad dy = \sin\theta\, dr + r\cos\theta\, d\theta.$$

These can be solved for dr and $d\theta$ by determinants or by elimination to give

$$dr = \cos\theta\, dx + \sin\theta\, dy, \qquad d\theta = -\frac{\sin\theta}{r}dx + \frac{\cos\theta}{r}dy.$$

Hence

$$\frac{\partial r}{\partial x} = \cos\theta, \qquad \frac{\partial r}{\partial y} = \sin\theta, \qquad \frac{\partial \theta}{\partial x} = -\frac{\sin\theta}{r}, \qquad \frac{\partial \theta}{\partial y} = \frac{\cos\theta}{r}.$$

Thus (9–172) can be written as follows:

$$\frac{\partial w}{\partial x} = \cos\theta\frac{\partial w}{\partial r} - \frac{\sin\theta}{r}\frac{\partial w}{\partial \theta}, \qquad \frac{\partial w}{\partial y} = \sin\theta\frac{\partial w}{\partial r} + \frac{\cos\theta}{r}\frac{\partial w}{\partial \theta}. \tag{9-173}$$

These equations provide general rules for expressing derivatives with respect to x or y in terms of derivatives with respect to r and θ. By applying the first equation to the function $\partial w/\partial x$, we find that

$$\frac{\partial^2 w}{\partial x^2} = \frac{\partial}{\partial x}\left(\frac{\partial w}{\partial x}\right) = \cos\theta\frac{\partial}{\partial r}\left(\frac{\partial w}{\partial x}\right) - \frac{\sin\theta}{r}\frac{\partial}{\partial \theta}\left(\frac{\partial w}{\partial x}\right);$$

by (9–173), this can be written as

$$\frac{\partial^2 w}{\partial x^2} = \cos\theta\frac{\partial}{\partial r}\left(\cos\theta\frac{\partial w}{\partial r} - \frac{\sin\theta}{r}\frac{\partial w}{\partial \theta}\right) - \frac{\sin\theta}{r}\frac{\partial}{\partial \theta}\left(\cos\theta\frac{\partial w}{\partial r} - \frac{\sin\theta}{r}\frac{\partial w}{\partial \theta}\right).$$

The rule for differentiation of a product gives finally

$$\frac{\partial^2 w}{\partial x^2} = \cos^2\theta\frac{\partial^2 w}{\partial r^2} - \frac{2\sin\theta\cos\theta}{r}\frac{\partial^2 w}{\partial r\partial \theta} + \frac{\sin^2\theta}{r^2}\frac{\partial^2 w}{\partial \theta^2} + \frac{\sin^2\theta}{r}\frac{\partial w}{\partial r}$$

$$+ \frac{2\sin\theta\cos\theta}{r^2}\frac{\partial w}{\partial \theta}. \tag{9–174}$$

In the same manner we find

$$\frac{\partial^2 w}{\partial y^2} = \frac{\partial}{\partial y}\left(\frac{\partial w}{\partial y}\right) = \sin\theta\frac{\partial}{\partial r}\left(\sin\theta\frac{\partial w}{\partial r} + \frac{\cos\theta}{r}\frac{\partial w}{\partial \theta}\right) + \frac{\cos\theta}{r}\frac{\partial}{\partial \theta}\left(\sin\theta\frac{\partial w}{\partial r} + \frac{\cos\theta}{r}\frac{\partial w}{\partial \theta}\right),$$

$$\frac{\partial^2 w}{\partial y^2} = \sin^2\theta\frac{\partial^2 w}{\partial r^2} + \frac{2\sin\theta\cos\theta}{r}\frac{\partial^2 w}{\partial r\partial \theta} + \frac{\cos^2\theta}{r^2}\frac{\partial^2 w}{\partial \theta^2} + \frac{\cos^2\theta}{r}\frac{\partial w}{\partial r} - \frac{2\sin\theta\cos\theta}{r^2}\frac{\partial w}{\partial \theta}. \tag{9–175}$$

Adding (9–174) and (9–175), we conclude:

$$\nabla^2 w = \frac{\partial^2 w}{\partial x^2} + \frac{\partial^2 w}{\partial y^2} = \frac{\partial^2 w}{\partial r^2} + \frac{1}{r^2}\frac{\partial^2 w}{\partial \theta^2} + \frac{1}{r}\frac{\partial w}{\partial r}; \tag{9–176}$$

this is the desired result.

Equation (9–176) at once allows us to write the expression for the three-dimensional Laplacian in cylindrical coordinates, since the transformation of coordinates

$$x = r\cos\theta, \qquad y = r\sin\theta, \qquad z = z$$

involves only x and y and in the same way as above. We find:

$$\nabla^2 w = \frac{\partial^2 w}{\partial x^2} + \frac{\partial^2 w}{\partial y^2} + \frac{\partial^2 w}{\partial z^2} = \frac{\partial^2 w}{\partial r^2} + \frac{1}{r^2}\frac{\partial^2 w}{\partial \theta^2} + \frac{1}{r}\frac{\partial w}{\partial r} + \frac{\partial^2 w}{\partial z^2}. \tag{9–177}$$

A procedure similar to the above gives the three-dimensional Laplacian in spherical coordinates:

$$\nabla^2 w = \frac{\partial^2 w}{\partial x^2} + \frac{\partial^2 w}{\partial y^2} + \frac{\partial^2 w}{\partial z^2} = \frac{\partial^2 w}{\partial \rho^2} + \frac{1}{\rho^2}\frac{\partial^2 w}{\partial \varphi^2} + \frac{1}{\rho^2\sin^2\varphi}\frac{\partial^2 w}{\partial \theta^2} + \frac{2}{\rho}\frac{\partial w}{\partial \rho} + \frac{\cot\varphi}{\rho^2}\frac{\partial w}{\partial \varphi}. \tag{9–178}$$

(see Problem 8 below).

9–18 HIGHER DERIVATIVES OF IMPLICIT FUNCTIONS

In Section 9–13 above procedures are given for obtaining differentials or first partial derivatives of functions implicitly defined by simultaneous equations. Since the results are in the form of *explicit* expressions for the first partial derivatives, these derivatives can be differentiated explicitly. An example will illustrate the situation. Let x and y be defined as functions of u and v by the implicit equations:

$$x^2 + y^2 + u^2 + v^2 = 1, \qquad x^2 + 2y^2 - u^2 + v^2 = 1.$$

Then with

$$F(x, y, u, v) = x^2 + y^2 + u^2 + v^2 - 1, \qquad G(x, y, u, v) = x^2 + 2y^2 - u^2 + v^2 - 1,$$

$$\frac{\partial x}{\partial u} = -\frac{\dfrac{\partial(F, G)}{\partial(u, y)}}{\dfrac{\partial(F, G)}{\partial(x, y)}} = -\frac{\begin{vmatrix} 2u & 2y \\ -2u & 4y \end{vmatrix}}{\begin{vmatrix} 2x & 2y \\ 2x & 4y \end{vmatrix}} = -\frac{3u}{x},$$

$$\frac{\partial y}{\partial u} = -\frac{\dfrac{\partial(F, G)}{\partial(x, u)}}{\dfrac{\partial(F, G)}{\partial(x, y)}} = -\frac{\begin{vmatrix} 2x & 2u \\ 2x & -2u \end{vmatrix}}{\begin{vmatrix} 2x & 2y \\ 2x & 4y \end{vmatrix}} = \frac{2u}{y}.$$

Hence

$$\frac{\partial^2 x}{\partial u^2} = \frac{3u}{x^2}\frac{\partial x}{\partial u} - \frac{3}{x} = \frac{3u}{x^2}\left(\frac{-3u}{x}\right) - \frac{3}{x} = -\frac{9u^2}{x^3} - \frac{3}{x},$$

$$\frac{\partial^2 y}{\partial u^2} = -\frac{2u}{y^2}\frac{\partial y}{\partial u} + \frac{2}{y} = -\frac{2u}{y^2}\left(\frac{2u}{y}\right) + \frac{2}{y} = -\frac{4u^2}{y^3} + \frac{2}{y}.$$

========================Problems (Section 9–18)========================

1. Find the indicated partial derivatives:

   a) $\dfrac{\partial^2 w}{\partial x^2}$ and $\dfrac{\partial^2 w}{\partial y^2}$ if $w = \dfrac{1}{\sqrt{x^2 + y^2}}$

   b) $\dfrac{\partial^2 w}{\partial x^2}$ and $\dfrac{\partial^2 w}{\partial y^2}$ if $w = \arctan\dfrac{y}{x}$

   c) $\dfrac{\partial^3 w}{\partial x \partial y^2}$ and $\dfrac{\partial^3 w}{\partial x^2 \partial y}$ if $w = e^{x^2 - y^2}$

   Formula (9–176) can be used as a check for (a) and (b).

2. Show that the following functions are harmonic in x and y:

   a) $e^x \cos y$ b) $x^3 - 3xy^2$ c) $\log \sqrt{x^2 + y^2}$

3. Show that every harmonic function is biharmonic.

4. Show that the following functions are biharmonic in x and y: $xe^x \cos y$, $x^4 - 3x^2 y^2$.

5. a) Prove the identity $\nabla^2(uv) = u\nabla^2 v + v\nabla^2 u + 2\nabla u \cdot \nabla v$ for functions u and v of x and y.

   b) Prove the identity of (a) for functions of x, y and z.

   c) Prove that if u and v are harmonic in two or three dimensions, then $w = xu + v$ is biharmonic.   HINT: Use the identity of (a) and (b).

d) Prove that if u and v are harmonic in two or three dimensions, then $w = r^2 u + v$ is biharmonic, where $r^2 = x^2 + y^2$ for two dimensions and $r^2 = x^2 + y^2 + z^2$ for three dimensions.

6. Establish a chain rule similar to (9–163) for $\partial^2 z/\partial u\, \partial v$.

7. Use the rule (9–163), applied to $\partial^2 w/\partial x^2$ and $\partial^2 w/\partial y^2$, to prove (9–176)

8. Prove (9–178).

HINT: Use (9–177) to express $\nabla^2 w$ in cylindrical coordinates; then note that the equations of transformation from (z, r) to (ρ, φ) are the same as those from (x, y) to (r, θ).

9. Prove that the biharmonic equation in x and y becomes

$$w_{rrrr} + \frac{2}{r^2} w_{rr\theta\theta} + \frac{1}{r^4} w_{\theta\theta\theta\theta} + \frac{2}{r} w_{rrr} - \frac{2}{r^3} w_{r\theta\theta} - \frac{1}{r^2} w_{rr} + \frac{4}{r^4} w_{\theta\theta} + \frac{1}{r^3} w_r = 0$$

in polar coordinates (r, θ).   HINT: Use (9–176).

10. Find $\partial^2 u/\partial x^2$ if u and v are functions of x and y defined by the equations

$$xy + uv = 1, \quad xu + yv = 1.$$

11. If u and v are inverse functions of the system $x = u^2 - v^2$, $y = 2uv$, show that u is harmonic.

12. A basic tool in the solution of differential equations is that of introducing new variables by appropriate substitution formulas. The introduction of polar coordinates in the Laplace equation in Section 9–17 illustrates this. The substitution can involve independent or dependent variables or both; in each case it must be indicated which of the new variables are to be treated as independent and which as dependent. Make the indicated substitutions in the following differential equations:

a) $\dfrac{dy}{dx} = \dfrac{2x}{y + x^2}$; new variables y (dependent) and $u = x^2$ (independent)

b) $\dfrac{dy}{dx} = \dfrac{2x - y + 1}{x + y - 4}$; new variables $v = y - 3$ (dependent) and $u = x - 1$ (independent)

c) $x^2 \dfrac{d^3 y}{dx^3} + 3x \dfrac{d^2 y}{dx^2} + \dfrac{dy}{dx} = 0$; new variables y (dependent) and $t = \ln x$ (independent)

d) $\dfrac{d^2 y}{dx^2} + \left(\dfrac{dy}{dx}\right)^3 = 0$; new variables x (dependent) and y (independent)

e) $\dfrac{d^2 y}{dx^2} - 4x \dfrac{dy}{dx} + y(3x^2 - 2) = 0$; new variables $v = e^{-x^2} y$ (dependent) and x (independent)

f) $a \dfrac{\partial u}{\partial x} + b \dfrac{\partial u}{\partial y} = 0$ $(a, b$ constants); new variables u (dependent), $z = bx - ay$ (independent), and $w = ax + by$ (independent)

g) $\dfrac{\partial^2 u}{\partial x^2} - \dfrac{\partial^2 u}{\partial y^2} = 0$; new variables u (dependent), $z = x + y$ (independent), and $w = x - y$ (independent)

Chapter 10 Vector Integral Calculus

10–1 INTRODUCTION

In this chapter we pursue our study of vector fields, with emphasis on integration processes. We study *multiple integrals* and some of their properties and applications; then we turn to *line integrals*. For example, the work done by a force with components $X(x, y, z), Y(x, y, z), Z(x, y, z)$ on a particle that moves from A to B along path C is given by a line integral

$$\int_{C \, A}^{B} X \, dx + Y \, dy + Z \, dz. \qquad (10\text{--}10)$$

Analysis of rate of flow of a fluid across a surface requires similar *surface integrals*

$$\int\!\!\int_{\mathscr{S}} L \, dy \, dz + M \, dz \, dx + N \, dx \, dy. \qquad (10\text{--}11)$$

If a vector field $\mathbf{u} = X\mathbf{i} + Y\mathbf{j} + Z\mathbf{k}$ is given in space, we can form the line integral (10–10). If the path C is a simple closed curve, then the *Stokes theorem* asserts that the line integral about C can be expressed as a surface integral (10–11) over a surface $\mathscr{S}$ bounded by C, where L, M, N are the components of the *curl* of the vector $\mathbf{u}$:

$$
\begin{aligned}
\text{curl } \mathbf{u} &= L\mathbf{i} + M\mathbf{j} + N\mathbf{k} \\
&= \left(\frac{\partial Z}{\partial y} - \frac{\partial Y}{\partial z} \right)\mathbf{i} + \left(\frac{\partial X}{\partial z} - \frac{\partial Z}{\partial x} \right)\mathbf{j} + \left(\frac{\partial Y}{\partial x} - \frac{\partial X}{\partial y} \right)\mathbf{k}.
\end{aligned}
$$

Thus curl $\mathbf{u}$ is a new vector field, important in physical problems.

Similarly, if $\mathscr{S}$ bounds a solid R in space, the *Gauss theorem* asserts that the surface integral (10–11) over $\mathscr{S}$ can be replaced by a triple integral over R of the scalar

$$\frac{\partial L}{\partial x} + \frac{\partial M}{\partial y} + \frac{\partial N}{\partial z},$$

which is called the *divergence* of the vector field $L\mathbf{i} + M\mathbf{j} + N\mathbf{k}$.

The Stokes and Gauss theorems have their counterparts in two dimensions, both of which are forms of *Green's theorem* for line integrals.

As we shall see, the Stokes and Gauss theorems can be used to deduce fundamental equations in such fields as fluid mechanics and heat conduction.

10–2 MULTIPLE INTEGRALS

We review briefly the concepts of multiple integrals, stressing double and triple integrals.

Let $z = f(x, y)$ be defined in a closed region R. We are here interested only in simple regions R, such as one bounded by a single closed curve, as in Fig. 10–1. We then consider subdivisions of R by lines parallel to the axes, as in the figure. Of the rectangles formed, we retain only those contained in R and number them by index i, where $i = 1, \ldots, n$. We let $\Delta_i A$ be the area of the ith rectangle and choose a point (x_i^*, y_i^*) in that rectangle. We let h be the largest diagonal of the n rectangles; h is called the *mesh* of the subdivision. Then we define

$$\int\int_R f(x, y)\, dA = \int\int_R f(x, y)\, dx\, dy = \lim_{h \to 0} \sum_{i=1}^{n} f(x_i^*, y_i^*)\Delta_i A$$

if the limit exists. that is, if for given $\varepsilon > 0$, we can choose $\delta > 0$ such that for all

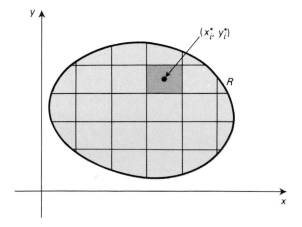

Fig. 10–1. Double integral over region R

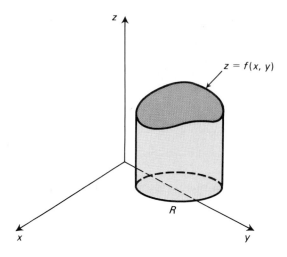

$z = f(x, y)$

R

subdivisions with mesh $h < \delta$, we have

$$\left| c - \sum_{i=1}^{n} f(x_i^*, y_i^*)\Delta_i A \right| < \varepsilon$$

for some number c; this number is then the value of the double integral. We can then show that the double integral exists if $f(x, y)$ is continuous in R and R satisfies reasonable conditions (see CLA, pp. 997–1008).

The two most common applications of the double integral are with regard to volume and mass. If $f(x, y) \geq 0$ in R, we interpret $\iint_R f(x, y)\,dx\,dy$ as the volume of the solid filling the region below the surface $z = f(x, y)$ and above the xy-plane for (x, y) in R, as in Fig. 10–2. Similarly, if $f(x, y) \geq 0$ in R, we interpret $f(x, y)$ as density (mass per unit area), $f(x, y)\,dA$ as mass of the area element, and the double integral as the total mass of the two-dimensional object filling R; the object can be thought of as a sheet of metal.

The double integral has the following properties:

$$\iint_R [c_1 f_1(x, y) + c_2 f_2(x, y)]\,dA = c_1 \iint_R f_1(x, y)\,dA + c_2 \iint_R f_2(x, y)\,dA,$$

$$\tag{10-20}$$

if c_1 and c_2 are constants;

$$\iint_{R_1} f(x, y)\,dA + \iint_{R_2} f(x, y)\,dA = \iint_R f(x, y)\,dA, \tag{10-21}$$

if R is decomposed into two parts R_1, R_2 overlapping only on a boundary curve;

$$mA \leqslant \iint\limits_R f(x, y)\,dA \leqslant MA, \qquad (10\text{-}22)$$

if $m \leqslant f(x, y) \leqslant M$ in R and R has total area A;

$$\left| \iint\limits_R f(x, y)\,dA \right| \leqslant \iint\limits_R |f(x, y)|\,dA; \qquad (10\text{-}23)$$

$$\iint\limits_R f(x, y)\,dA = A f(x^*, y^*) \text{ (mean-value property)}, \qquad (10\text{-}24)$$

if f is continuous in R and (x^*, y^*) is properly chosen in R; if $f(x, y)$ is continuous in R, $f(x, y) \geqslant 0$ in R and

$$\iint\limits_R f(x, y)\,dA = 0,$$

then $f(x, y) \equiv 0$ in R. (For proofs see CLA, pp. 997–1008.)

Iterated integral

If R has the special form: $a \leqslant x \leqslant b$, $p(x) \leqslant y \leqslant q(x)$ (Fig. 10-3), then

$$\iint\limits_R f(x, y)\,dA = \int_a^b \int_{p(x)}^{q(x)} f(x, y)\,dy\,dx. \qquad (10\text{-}25)$$

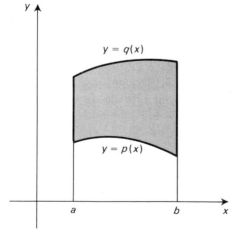

Fig. 10-3. Region for iterated integral.

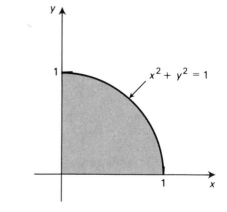

Fig. 10-4. Special region (quarter circle).

Thus if R is the quarter circle of Fig. 10-4, then

$$\iint_R y^3 \, dA = \int_0^1 \int_0^{\sqrt{1-x^2}} y^3 \, dy \, dx = \int_0^1 \frac{y^4}{4} \Big|_0^{\sqrt{1-x^2}} dx = \frac{1}{4} \int_0^1 (1-x^2)^2 \, dx = \frac{2}{15}.$$

Also, by interchanging the roles of x and y,

$$\iint_R y^3 \, dA = \int_0^1 \int_0^{\sqrt{1-y^2}} y^3 \, dx \, dy = \int_0^1 y^3 \sqrt{1-y^2} \, dy = \frac{1}{2} \int_0^1 (u^{1/2} - u^{3/2}) \, du = \frac{2}{15}.$$

Here we used the substitution $u = 1 - y^2$, $y = \sqrt{1-u}$.

Triple integrals

Here we have a function $w = f(x, y, z)$ defined in a closed region R in xyz-space. The region R is now bounded by one or more surfaces. Typical regions are the region bounded by a spherical surface, the region between two concentric spheres, a solid ellipsoid, a solid torus. By subdividing R with planes parallel to the coordinate planes, we can obtain n rectangular solids, indexed by i and contained in R; we denote the volume of the ith rectangular solid by $\Delta_i V$ and choose a point (x_i^*, y_i^*, z_i^*) in it. The mesh of the subdivision we denote by h; this is the largest diagonal of all the n rectangular solids. Then

$$\iiint_R f(x, y, z) \, dV = \iiint_R f(x, y, z) \, dx \, dy \, dz = \lim_{h \to 0} \sum_{i=1}^n f(x_i^*, y_i^*, z_i^*) \Delta_i V,$$

where the concept of limit is the same as for the double integral. Again the triple integral exists if $f(x, y, z)$ is continuous and R satisfies reasonable conditions.

The triple integral again gives the volume: $\iiint_R dV = $ Volume of R (just as $\iint_R dA = $ Area of R). It also gives the mass of the solid with density $f(x, y, z)$.

Both double and triple integrals have many other applications: for example, to moments, moments of inertia, total gravitational or electrostatic potential of an object.

The triple integral has properties similar to properties (10–20)–(10–24) for the double integral. For special regions it can also be expressed as an iterated integral. For the region R of Fig. 10–5, where $a \leqslant x \leqslant b$, $p(x) \leqslant y \leqslant q(x)$, $\varphi(x, y) \leqslant z \leqslant \psi(x, y)$, we have

$$\iiint_R f(x, y, z) \, dV = \int_a^b \int_{p(x)}^{q(x)} \int_{\varphi(x, y)}^{\psi(x, y)} f(x, y, z) \, dz \, dy \, dx. \tag{10–26}$$

Occasionally we may have a region in space of form

$$\varphi(x, y) \leqslant z \leqslant \psi(x, y), \qquad (x, y) \text{ in } R_{xy}, \tag{10–27}$$

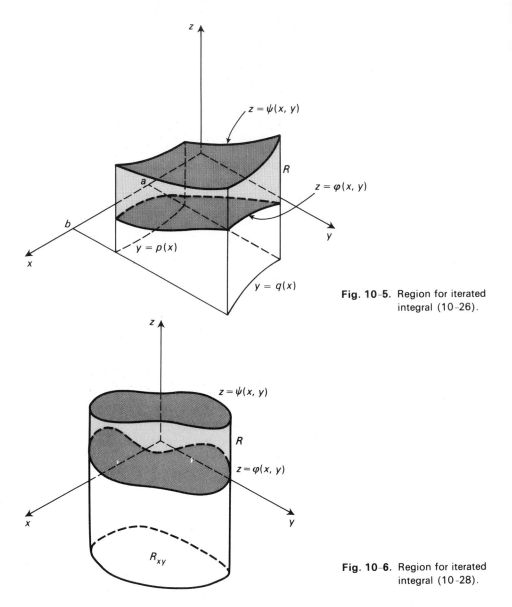

Fig. 10-5. Region for iterated integral (10-26).

Fig. 10-6. Region for iterated integral (10-28).

where R_{xy} is a region in the xy-plane. Thus φ and ψ are both defined in R_{xy} and the solid R is the region between the two surfaces $z = \varphi(x, y), z = \psi(x, y)$, as in Fig. 10-6. Corresponding to (10–27), we then have

$$\iiint\limits_R f(x, y, z)dV = \iint\limits_{R_{xy}} \left[\int_{\varphi(x, y)}^{\psi(x, y)} f(x, y, z)dz \right] dA. \qquad (10\text{–}28)$$

Thus the triple integral is reduced to a simple integral and a double integral.

Alternate subdivisions

Instead of subdividing R by lines or planes, we can use curves or surfaces to form the n elements of areas $\Delta_i A$ or volumes $\Delta_i V$. The integral is obtained as before, with h interpreted as the largest "diameter" of the n elements.

Vector integrals

As in Section 9–3 (see especially Eq. (9–36)), we can form double or triple integrals of vector functions of two or three variables: for example, if $\mathbf{F}(x, y) = f(x, y)\mathbf{i} + g(x, y)\mathbf{j}$, then

$$\iint_R \mathbf{F}(x, y)\, dA = \iint_R f(x, y)\, dA\, \mathbf{i} + \iint_R g(x, y)\, dA\, \mathbf{j}. \tag{10–29}$$

This is useful in representing total gravitational force (see Example 2 in the next section).

n-fold multiple integrals

Quadruple, quintuple, and, in general, n-fold multiple integrals have important applications, notably in computation of probabilities (Chapter 15). The definitions and properties parallel those for double and triple integrals, with area and volume replaced by higher-dimensional volume.

──────────────**Problems (Section 10–2)**──────────────

1. Evaluate the following integrals:

   a) $\displaystyle\iint_R (x^2 + y^2)\, dx\, dy$, where R is the triangle with vertices $(0, 0)$, $(1, 0)$, $(1, 1)$

   b) $\displaystyle\iiint_R u^2 v^2 w\, du\, dv\, dw$, where R is the region: $u^2 + v^2 \leqslant 1, 0 \leqslant w \leqslant 1$

   c) $\displaystyle\iint_R r^3 \cos\theta\, dr\, d\theta$, where R is the region: $1 \leqslant r \leqslant 2, \dfrac{\pi}{4} \leqslant \theta \leqslant \pi$

2. Express the following in terms of multiple integrals and reduce to iterated integrals, but do not evaluate:
   a) The mass of a sphere whose density is proportional to the distance from one diametral plane;
   b) The coordinates of the center of mass of the sphere of part (a);
   c) The moment of inertia about the x axis of the solid filling the region $0 \leqslant z \leqslant 1 - x^2 - y^2, 0 \leqslant x \leqslant 1, 0 \leqslant y \leqslant 1 - x$ and having density proportional to xy.

3. The moment of inertia of a solid about an arbitrary line L is defined as

$$I_L = \int\int\int_R d^2 f(x, y, z)\, dx\, dy\, dz,$$

where f is density and d is the distance from a general point (x, y, z) of the solid to the line L. Prove the *parallel-axis theorem*:

$$I_L = I_{\bar{L}} + Mh^2,$$

where $\bar{L}$ is a line parallel to L through the center of mass, M is the mass, and h is the distance between L and $\bar{L}$.

HINT: Take $\bar{L}$ to be the z axis.

4. Let L be a line through the origin O with direction cosines l, m, n. Prove that

$$I_L = I_x l^2 + I_y m^2 + I_z n^2 - 2I_{xy} lm - 2I_{yz} mn - 2I_{zx} ln,$$

where

$$I_{xy} = \int\int\int_R xy f(x, y, z)\, dx\, dy\, dz, \qquad I_{yz} = \int\int\int_R yzf, \ldots$$

The new integrals are called *products of inertia*. The locus

$$I_x x^2 + I_y y^2 + I_z z^2 - 2(I_{xy} xy + I_{yz} yz + I_{zx} zx) = 1$$

is an ellipsoid called the *ellipsoid of inertia*.

10-3 LEIBNITZ'S RULE FOR INTEGRALS

If in a definite integral or multiple integral the integrand depends on a variable, say t, in addition to the variable(s) of integration, then the integral also depends on t. Leibnitz's rule asserts that, with respect to the variable t, *the derivative of the integral equals the integral of the derivative*. Thus

$$\frac{d}{dt} \int_a^b f(x, t)\, dx = \int_a^b \frac{\partial f(x, t)}{\partial t}\, dx, \tag{10-30}$$

$$\frac{d}{dt} \int\int_R f(x, y, t)\, dA = \int\int_R \frac{\partial f(x, y, t)}{\partial t}\, dA, \tag{10-31}$$

$$\frac{d}{dt} \int\int\int_R f(x, y, z, t)\, dV = \int\int\int_R \frac{\partial f(x, y, z, t)}{\partial t}\, dV. \tag{10-32}$$

The equations are valid for an interval of t provided f and $\partial f/\partial t$ are continuous for t in this interval and for $(x, y, \ldots)$ in the interval or region of integration. (See AC, pp. 286–287 for proof.)

EXAMPLE 1 $\dfrac{d}{dt}\displaystyle\int_0^\pi \sin xt\,dx = \int_0^\pi x\cos xt\,dx.$ Here we can check the result, for the integral on the left is equal to $(1 - \cos \pi t)/t$; its derivative equals $(\pi t \sin \pi t + \cos \pi t - 1)/t^2$, which is the value of the integral on the right. (Here $t = 0$ is exceptional; it can be shown that the equality is still valid for $t = 0$.) ◄

EXAMPLE 2   As in Problem 9(b) following Section 9–8, the gravitational force **F** exerted on a particle of mass m at $P(x, y, z)$ by a particle of mass M at $P_0(x_0, y_0, z_0)$ is $\mathbf{F} = -kMm\mathbf{r}/|\mathbf{r}|^3$ where $\mathbf{r}$ is the vector $\overrightarrow{P_0 P}$ and $\mathbf{F} = -\text{grad } U$ (where $U = -kMm/|\mathbf{r}|$), so that U is the potential energy associated with F. To find the total gravitational force **F** exerted by a solid (filling region R) on a particle of unit mass at the point P outside the solid (see Fig. 10–7), we can subdivide the solid as for a triple integral, find the force due to each element, add and pass to the limit. If $\delta(x, y, z)$ is the (continuous) density function, then the ith piece has mass $\delta(x_i', y_i', z_i')\Delta_i V$ for appropriate choice of $P_i(x_i', y_i', z_i')$ in the ith piece. The force due to this piece is then approximately

$$-k\delta(x_i', y_i', z_i')\Delta_i V \frac{\mathbf{r} - \mathbf{r}_i'}{|\mathbf{r} - \mathbf{r}_i'|^3},$$

where $\mathbf{r} = \overrightarrow{OP}$ and $\mathbf{r}_i' = \overrightarrow{OP}_i$, as in Fig. 10-7. If we sum and pass to the limit in the familiar fashion, we obtain the total force as a vector integral:

$$\mathbf{F} = -k \iiint_R \delta(x', y', z') \frac{\mathbf{r} - \mathbf{r}'}{|\mathbf{r} - \mathbf{r}'|^3}\,dx'\,dy'\,dz',$$

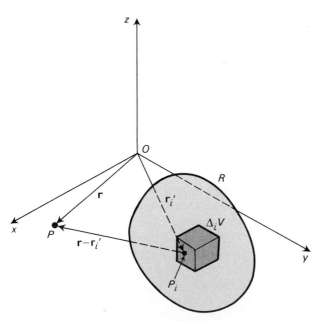

Fig. 10-7. Gravitational force due to a solid.

where $\mathbf{r}' = x'\mathbf{i} + y'\mathbf{j} + z'\mathbf{k}$ and $\mathbf{r} = x\mathbf{i} + y\mathbf{j} + z\mathbf{k}$. As above, the potential energy associated with the ith piece is

$$\Delta_i U = -k\,\delta(x_i', y_i', z_i')\frac{1}{|\mathbf{r} - \mathbf{r}_i'|}\,\Delta_i V.$$

If we sum and pass to the limit, we expect the total potential energy to be

$$U(x, y, z) = -k \iiint\limits_{R} \frac{\delta(x', y', z')}{|\mathbf{r} - \mathbf{r}'|}\,dx'\,dy'\,dz'. \tag{10-33}$$

Leibnitz's rule allows us to write:

$$-\operatorname{grad} U = k \iiint\limits_{R} \operatorname{grad} \frac{\delta(x', y', z')}{|\mathbf{r} - \mathbf{r}'|}\,dx'\,dy'\,dz';$$

here the gradient is taken with respect to x, y, z, with x', y', z' held fixed. But as in Problem 9(b) following Section 9–8,

$$
\begin{aligned}
\operatorname{grad} \frac{1}{|\mathbf{r} - \mathbf{r}'|} &= \operatorname{grad} \frac{1}{\sqrt{(x - x')^2 + (y - y')^2 + (z - z')^2}} \\
&= -\frac{(x - x')\mathbf{i} + (y - y')\mathbf{j} + (z - z')\mathbf{k}}{[(x - x')^2 + (y - y')^2 + (z - z')^2]^{3/2}} \\
&= -\frac{\mathbf{r} - \mathbf{r}'}{|\mathbf{r} - \mathbf{r}'|^3}.
\end{aligned}
$$

Hence

$$-\operatorname{grad} U = -k \iiint\limits_{R} \delta(x', y', z')\frac{\mathbf{r} - \mathbf{r}'}{|\mathbf{r} - \mathbf{r}'|^3}\,dx'\,dy'\,dz' = \mathbf{F}.$$

Thus the total potential energy is given by (10–33). (As always, U is determined only up to an additive constant.) ◄

Definite integral with limits depending on a parameter

Leibnitz's rule can be combined with the chain rule and the fundamental theorem of calculus:

$$\frac{d}{dt}\int_a^t f(x)\,dx = f(t) \tag{10-34}$$

to find the derivative of an integral

$$\int_{p(t)}^{q(t)} f(x, t)\,dx$$

which depends upon t both in the integrand and in the limits of integration. We write

$$u = q(t), \qquad v = p(t), \qquad w = t,$$

so that our integral equals $z = \int_v^u f(x, w)\,dx$, and u, v, w are all functions of t. Then by the chain rule

$$\frac{dz}{dt} = \frac{\partial z}{\partial u}\frac{du}{dt} + \frac{\partial z}{\partial v}\frac{dv}{dt} + \frac{\partial z}{\partial w}\frac{dw}{dt}. \qquad (10\text{-}35)$$

We evaluate the three partial derivatives in turn. By the fundamental theorem (10-34):

$$\frac{\partial z}{\partial u} = \frac{\partial}{\partial u}\int_v^u f(x, w)\,dx = f(u, w);$$

again by the fundamental theorem:

$$\frac{\partial z}{\partial v} = \frac{\partial}{\partial v}\int_v^u f(x, w)\,dx = -\frac{\partial}{\partial v}\int_u^v f(x, w)\,dx = -f(v, w);$$

by Leibnitz's rule:

$$\frac{\partial z}{\partial w} = \frac{\partial}{\partial w}\int_v^u f(x, w)\,dw = \int_v^u \frac{\partial f(x, w)}{\partial w}\,dw.$$

Accordingly, by Eq. (10–35):

$$\frac{dz}{dt} = \frac{d}{dt}\int_{p(t)}^{q(t)} f(x, t)\,dx$$

$$= f(q(t), t)q'(t) - f(p(t), t)p'(t) + \int_{p(t)}^{q(t)} \frac{\partial f(x, t)}{\partial t}\,dx. \qquad (10\text{-}36)$$

(Throughout we have assumed the continuity of the appearing functions.)

EXAMPLE 3 $\dfrac{d}{dt}\displaystyle\int_t^{e^t} (x^2 + t^2)^3\,dx = (e^{2t} + t^2)^3\,e^t - (2t^2)^3 + \int_t^{e^t} 3(x^2 + t^2)^2\,2t\,dt.$

◀

The rule (10-36) can be extended to multiple integrals over a variable region of integration; there are important applications to fluid motion (see Problems 5 and 6 following Section 10–19).

The fundamental theorem (10–34) has an important counterpart for multiple integrals. By the definition of the derivative, we can write the theorem as follows:

$$\lim_{\Delta t \to 0} \frac{1}{\Delta t}\int_t^{t+\Delta t} f(x)\,dx = f(t).$$

For double integrals we then have

$$\lim_{h \to 0} \frac{1}{\Delta A}\iint_R f(x, y)\,dA = f(x, y), \qquad (10\text{-}37)$$

provided f is continuous in an open region containing (x, y). Here R is a *variable region* of "diameter" less than h and area ΔA containing (x, y) (Fig. 10–8). This follows at once from the mean-value property (10–24), since the integral equals

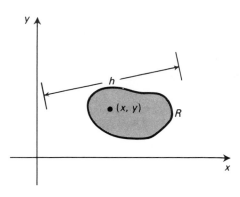

Fig. 10-8. Limit rule (10-37).

$f(x^*, y^*)\Delta A$ and, as $h \to 0$, (x^*, y^*) must approach (x, y), so that $f(x^*, y^*) \to f(x, y)$. There is an analogous rule for triple integrals and other multiple integrals.

═══════════════════════════════Problems (Section 10-3)═══════════════════════════════

1. Obtain the indicated derivatives in the form of integrals:

a) $\dfrac{d}{dt} \displaystyle\int_{\pi/2}^{\pi} \dfrac{\cos(xt)}{x} dx$

b) $\dfrac{d}{dt} \displaystyle\int_{1}^{2} \dfrac{x^2}{(1-tx)^2} dx$

c) $\dfrac{d}{du} \displaystyle\int_{1}^{2} \ln(xu) dx$

d) $\dfrac{d^n}{dy^n} \displaystyle\int_{1}^{2} \dfrac{\sin x}{x-y} dx$

2. Obtain the indicated derivatives:

a) $\dfrac{d}{dx} \displaystyle\int_{1}^{x} t^2\, dt$

b) $\dfrac{d}{dt} \displaystyle\int_{1}^{t^2} \sin(x^2) dx$

c) $\dfrac{d}{dt} \displaystyle\int_{t^3}^{2} \ln(1+x^2) dx$

d) $\dfrac{d}{dx} \displaystyle\int_{x}^{\tan x} e^{-t^2} dt$

3. Prove the following:

a) $\dfrac{d}{d\alpha} \displaystyle\int_{\sin\alpha}^{\cos\alpha} \ln(x+\alpha) dx = \ln\dfrac{\cos\alpha+\alpha}{\sin\alpha+\alpha} - \left[\sin\alpha\ln(\cos\alpha+\alpha) + \cos\alpha\ln(\sin\alpha+\alpha)\right]$

b) $\dfrac{d}{du} \displaystyle\int_{0}^{\pi/(2u)} u\sin ux\, dx = 0$

c) $\dfrac{d}{dy} \displaystyle\int_{y}^{y^2} e^{-x^2y^2} dx = 2ye^{-y^6} - e^{-y^4} - 2y\displaystyle\int_{y}^{y^2} x^2 e^{-x^2y^2} dx$

4. a) Evaluate $\displaystyle\int_{0}^{1} x^n \ln x\, dx$ by differentiating both sides of the equation

$\displaystyle\int_{0}^{1} x^n\, dx = \dfrac{1}{n+1}$ with respect to n $(n > -1)$.

b) Evaluate $\displaystyle\int_{0}^{\infty} x^n e^{-ax} dx$ by repeated differentiation of $\displaystyle\int_{0}^{\infty} e^{-ax} dx$ $(a > 0)$.

c) Evaluate $\displaystyle\int_0^\infty \frac{dy}{(x^2+y^2)^n}$ by repeated differentiation of $\displaystyle\int_0^\infty \frac{dy}{x^2+y^2}$.

HINT: In (b) and (c) the improper integrals are of a type to which Leibnitz's rule is applicable (see AC, p. 448). The result of (a) can be explicitly verified.

5. Consider a one-dimensional fluid motion, the flow taking place along the x axis. Let $v = v(x, t)$ be the velocity at position x at time t, so that, if x is the coordinate of a fluid particle at time t, we have $dx/dt = v$. If $f(x, t)$ is any scalar associated with the flow (velocity, acceleration, density, etc.), the variation of f along the flow can be studied with the aid of the Stokes derivative:

$$\frac{Df}{Dt} = \frac{\partial f}{\partial x}\frac{dx}{dt} + \frac{\partial f}{\partial t}$$

(see Problem 10 following Section 9–5). A piece of the fluid occupying an interval $a_0 \leqslant x \leqslant b_0$ when $t = 0$ will occupy an interval $a(t) \leqslant x \leqslant b(t)$ at time t, where $da/dt = v(a, t)$, $db/dt = v(b, t)$. The integral

$$F(t) = \int_{a(t)}^{b(t)} f(x, t)\,dx$$

is then an integral of f over a definite piece of the fluid, whose position varies with time; if f is density, this is the mass of the piece. Show that

$$\frac{dF}{dt} = \int_{a(t)}^{b(t)} \left[\frac{\partial f}{\partial t}(x, t) + \frac{\partial}{\partial x}(fv)\right]dx = \int_{a(t)}^{b(t)} \left(\frac{Df}{Dt} + f\frac{\partial v}{\partial x}\right)dx.$$

This is generalized to arbitrary three-dimensional flows in Section 10-19.

6. Let $f(\alpha)$ be continuous for $0 \leqslant \alpha \leqslant 2\pi$. Let

$$u(r, \theta) = \frac{1}{2\pi}\int_0^{2\pi} f(\alpha)\frac{1 - r^2}{1 + r^2 - 2r\cos(\theta - \alpha)}\,d\alpha$$

for $r < 1$, r and θ being polar coordinates. Show that u is harmonic for $r < 1$. This is the *Poisson integral formula*.

HINT: In polar coordinates, the Laplace equation becomes $u_{rr} + r^{-2}u_{\theta\theta} + r^{-1}u_r = 0$, as in Section 8–9 (see Section 9–17).

10–4 CHANGE OF VARIABLES IN MULTIPLE INTEGRALS

For a double integral $\iint_R F(x, y)\,dA$ it is often convenient to make a change of variables by setting

$$x = f(u, v), \qquad y = g(u, v). \tag{10-40}$$

This defines a mapping from the uv-plane to the xy-plane. We may be able to choose a closed region R_{uv} so that, as (u, v) varies over R_{uv}, (x, y) varies over all of R. Under suitable conditions, the double integral is then expressible as a double integral over R_{uv}. The crucial formula is as follows:

$$\iint_R F(x, y)\, dx\, dy = \iint_{R_{uv}} F[f(u, v),\, g(u, v)] \left| \frac{\partial(x, y)}{\partial(u, v)} \right| du\, dv. \qquad (10\text{-}41)$$

This is valid, in particular, if the functions f, g have continuous first partial derivatives in an open region containing R_{uv} and the mapping (10-40) is a one-to-one mapping of R_{uv} onto R (see Sections 9–10 and 9–15). (For proof under more general conditions, see AC, pp. 352–354.)

The main idea behind the formula is suggested in Eq. (9–104) of Section 9–10: the absolute value of the Jacobian $\partial(x, y)/\partial(u, v)$ equals the ratio of corresponding areas $\Delta A_{xy}/\Delta A_{uv}$ for the linear approximation of the mapping (10–40) at a point of R_{uv}. Thus for very small areas near the point, $|\partial(x, y)/\partial(u, v)|$ should closely approximate $\Delta A_{xy}/\Delta A_{uv}$ for the given mapping (10–40). If, for example, R_{uv} is a rectangle as in Fig. 10–9, then we can subdivide R_{uv} by parallels to the axes. The corresponding curves in the xy-plane determine a subdivision of R by *curvilinear rectangles*. A typical term in the approximating sum for the double integral $\int\int_R F(x, y)\, dA$ is then of form $F(x^*, y^*)\Delta A_{xy}$. Here $x^* = f(u^*, v^*)$, $y^* = g(u^*, v^*)$ for appropriate (u^*, v^*) in the corresponding piece of R_{uv} of area ΔA_{uv}. Also, as above,

$$\Delta A_{xy} \sim \left| \frac{\partial(x, y)}{\partial(u, v)} \right| \Delta A_{uv},$$

where the Jacobian is evaluated, say, at (u^*, v^*). Hence

$$F(x^*, y^*)\Delta A_{xy} \sim F[f(u^*, v^*),\, g(u^*, v^*)] \left| \frac{\partial(x, y)}{\partial(u, v)} \right| \Delta A_{uv}.$$

If we now sum and pass to the limit, it is plausible that the exact equality (10–41) is obtained.

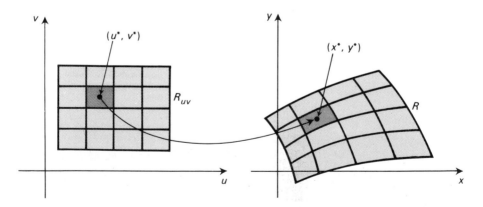

Fig. 10–9. Change of variables in a double integral.

As in Section 9–10, u, v can be interpreted as *curvilinear coordinates* in R. The most common case of (10–40) is the case of polar coordinates r, θ (replacing u, v). The *curved rectangles* of Fig. 10–9 are now bounded by line segments on rays passing through the origin and arcs of circles with center at the origin. The equations and Jacobian are

$$x = r \cos \theta, \qquad y = r \sin \theta, \qquad \frac{\partial(x, y)}{\partial(r, \theta)} = r.$$

Thus we obtain the much-used formula

$$\iint_R F(x, y)\, dx\, dy = \iint_{R_{r\theta}} F(r \cos \theta,\, r \sin \theta)\, r\, dr\, d\theta. \qquad (10\text{–}42)$$

The second integral can often be expressed as an iterated integral. The limits for r and θ can be obtained by constructing $R_{r\theta}$ or, more simply, directly from the figure R in the xy-plane.

EXAMPLE 1   The moment of inertia of a semicircular disc of radius a and of constant density 1 about its bounding diameter is $\iint_R y^2\, dx\, dy$, where the disc R lies in the first and second quadrants of the xy-plane, as in Fig. 10–10. In polar coordinates, R is described by $0 \leqslant r \leqslant a$, $0 \leqslant \theta \leqslant \pi$, so that $R_{r\theta}$ is a rectangle in the $r\theta$-plane. The integral becomes

$$\int_0^\pi \int_0^a r^2 \sin^2 \theta\, r\, dr\, d\theta = \frac{\pi}{2} \cdot \frac{a^4}{4} = \frac{\pi a^4}{8}. \quad \blacktriangleleft$$

Under analogous conditions, the basic formula (10–41) extends to triple integrals and multiple integrals in general. For example,

$$\iiint_R F(x, y, z)\, dx\, dy\, dz$$

$$= \iiint_{R_{uvw}} F[f(u, v, w),\, g(u, v, w),\, h(u, v, w)] \cdot \left| \frac{\partial(x, y, z)}{\partial(u, v, w)} \right| du\, dv\, dw, \qquad (10\text{–}43)$$

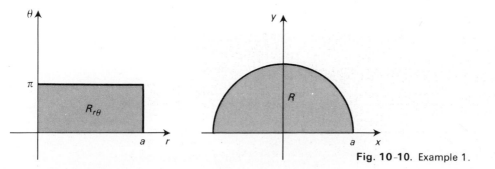

Fig. 10-10. Example 1.

where the mapping is given by

$$x = f(u, v, w), \ y = g(u, v, w), \ z = h(u, v, w) \quad \text{in } R_{uvw}. \tag{10-44}$$

The most common cases are those of triple integrals in cylindrical and spherical coordinates:

$$\iiint_R F(x, y, z)\,dx\,dy\,dz = \iiint_{R_{r\theta z}} F(r \cos\theta, r \sin\theta, z)\,r\,dr\,d\theta\,dz, \tag{10-45}$$

$$\iiint_R F(x, y, z)\,dx\,dy\,dz$$

$$= \iiint_{R_{\rho\varphi\theta}} F(\rho \sin\varphi \cos\theta, \rho \sin\varphi \sin\theta, \rho \cos\varphi) \cdot \rho^2 \sin\varphi\,d\rho\,d\varphi\,d\theta. \tag{10-46}$$

EXAMPLE 2 In Example 2 of Section 10–3 we found the gravitational potential of a solid at an external point $P(x, y, z)$ (see Eq. (10–33)). We now apply this result to a solid which is homogeneous (with constant density δ) and fills the region between two concentric spheres of radii a, b, where $0 < a < b$. We consider the case when P is inside the solid (see Fig. 10–11). The rectangular coordinates are chosen with origin at the center of the spheres and such that P has rectangular coordinates $(0, 0, c)$. We then use spherical coordinates for (x', y', z'). Thus in (10–33) we get:

$$|\mathbf{r} - \mathbf{r}'| = [(x - x')^2 + (y - y')^2 + (z - z')^2]^{1/2}$$
$$= (x'^2 + y'^2 + z'^2 - 2cz' + c^2)^{1/2} = (\rho^2 - 2c\rho \cos\phi + c^2)^{1/2}.$$

Hence by Eqs. (10–33) and (10–46),

$$U(0, 0, c) = -k\delta \int_0^{2\pi} \int_a^b \int_0^\pi \frac{\rho^2 \sin\varphi}{(\rho^2 - 2c\rho \cos\varphi + c^2)^{1/2}}\,d\varphi\,d\rho\,d\theta$$
$$= -2\pi k\delta(b^2 - a^2)$$

(the details are left to Problem 7 below). Thus $U(0, 0, c)$ is independent of c. Hence $U(0, 0, c) = U(0, 0, 0)$, and U must be identically constant inside the smaller sphere.

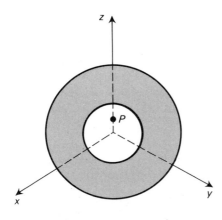

Fig. 10-11. Example 2.

Therefore, $\mathbf{F} = -\operatorname{grad} U = 0$ in this region, and there is no resultant gravitational force. Thus if a hollow spherical region could be somehow created at the center of the earth (which is approximately a homogeneous solid sphere), then this region would be approximately gravitation free. ◄

================Problems (Section 10-4)================

1. Evaluate the following integrals with the aid of the substitution suggested:

a) $\displaystyle\iint\limits_{R_{xy}} (1 - x^2 - y^2)\, dx\, dy$, where R_{xy} is the region $x^2 + y^2 \leqslant 1$, using $x = r\cos\theta$, $y = r\sin\theta$

b) $\displaystyle\iint\limits_{R_{xy}} (x - y^2)\sin^2(x + y)\, dx\, dy$, where R_{xy} is the parallelogram with successive vertices $(\pi, 0)$, $(2\pi, \pi)$, $(\pi, 2\pi)$, $(0, \pi)$, using $u = x - y$, $v = x + y$

2. Verify that the transformation $u = 2xy$, $v = x^2 - y^2$ defines a one-to-one mapping of the square $0 \leqslant x \leqslant 1$, $0 \leqslant y \leqslant 1$ onto a region of the uv-plane. Express the integral

$$\iint\limits_{R_{xy}} \sqrt[3]{x^4 - 6x^2y^2 + y^4}\, dx\, dy$$

over the square as an iterated integral in u and v.

3. Transform the integrals given, using the substitutions indicated:

a) $\displaystyle\int_0^1 \int_0^x \ln(1 + x^2 + y^2)\, dy\, dx$, $\quad x = u + v$, $\quad y = u - v$

b) $\displaystyle\int_0^1 \int_{1-x}^{1+x} \sqrt{1 + x^2 y^2}\, dy\, dx$, $\quad x = u$, $\quad y = u + v$

4. Verify the correctness of (10–46) as a special case of (10–43). Show the geometric meaning of the volume element $\rho^2 \sin\varphi\, \Delta\rho\, \Delta\varphi\, \Delta\theta$.

5. Transform to cylindrical coordinates but do not evaluate:

a) $\displaystyle\iiint\limits_{R_{xyz}} x^2 y\, dx\, dy\, dz$, where R_{xyz} is the region $x^2 + y^2 \leqslant 1$, $0 \leqslant z \leqslant 1$

b) $\displaystyle\int_0^1 \int_0^{\sqrt{1-x^2}} \int_0^{1+x+y} (x^2 + y^2)\, dz\, dy\, dx$

6. Transform to spherical coordinates but do not evaluate:

a) $\displaystyle\iiint\limits_{R_{xyz}} x^2 y\, dx\, dy\, dz$, where R_{xyz} is the sphere: $x^2 + y^2 + z^2 \leqslant a^2$

b) $\displaystyle\int_{-1}^1 \int_{-\sqrt{1-x^2}}^{\sqrt{1-x^2}} \int_{\sqrt{x^2+y^2}}^1 (x^2 + y^2 + z^2)\, dz\, dy\, dx$

7. Carry out the details in the evaluation of $U(0, 0, c)$ in Example 2 of Section 10–4.

10–5 LINE INTEGRALS IN THE PLANE

We introduce the concept of line integral by evaluating the work done by a force **F** acting on a particle moving from A to B on a smooth path C in the xy-plane (Fig. 10–12). The force **F** is allowed to vary continuously along the path. We write

$$\mathbf{F} = P(x, y)\mathbf{i} + Q(x, y)\mathbf{j}$$

for the force acting at position (x, y). The total work W done can be represented as a sum of work elements $\Delta_i W$, obtained by subdividing the path C by points $A(x_0, y_0)$, $(x_1, y_1), \ldots, B(x_n, y_n)$ and by considering the work $\Delta_1 W$ done in going from (x_0, y_0) to (x_1, y_1), $\Delta_2 W$ from (x_1, y_1) to (x_2, y_2), and so on. We write $\Delta_i x = x_i - x_{i-1}$, $\Delta_i y = y_i - y_{i-1}$, so that for $\Delta_i W$ the associated displacement vector is $\Delta_i x \mathbf{i} + \Delta_i y \mathbf{j}$. If the points (x_i, y_i) are sufficiently closely spaced, the force **F** is approximately constant during each displacement and the displacement itself approximately follows the line segment from the initial point to the terminal point. For a constant force **F** and displacement on a line segment $\overrightarrow{P_1 P_2}$, we know that the work done by **F** is $\mathbf{F} \cdot \overrightarrow{P_1 P_2}$ (Section 9–2). Therefore, approximately,

$$\Delta_i W = \mathbf{F}_i \cdot (\Delta_i x \mathbf{i} + \Delta_i y \mathbf{j}),$$

where $\mathbf{F}_i$ is the force applied, for example, at (x_i, y_i). Thus, approximately,

$$\Delta_i W = [P(x_i, y_i)\mathbf{i} + Q(x_i, y_i)\mathbf{j}] \cdot (\Delta_i x \mathbf{i} + \Delta_i y \mathbf{j})$$
$$= P(x_i, y_i)\Delta_i x + Q(x_i, y_i)\Delta_i y.$$

If we sum and pass to the limit, as the spacing of the (x_i, y_i) approaches 0, we expect the total work to be

$$W = \lim \sum_{i=1}^{n} [P(x_i, y_i)\Delta_i x + Q(x_i, y_i)\Delta_i y].$$

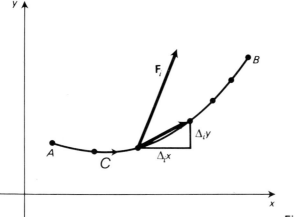

Fig. 10–12. Work as a line integral.

We denote this limit by the symbol for a line integral:

$$W = \int_{C}^{B}\! P(x, y)\,dx + Q(x, y)\,dy = \lim \sum_{i=1}^{n} [P(x_i, y_i)\Delta_i x + Q(x_i, y_i)\Delta_i y].$$

$$(10\text{--}50)$$

(The limits A, B can be omitted when they are clear from the context.) Normally the path C is assumed to be given by parametric equations

$$x = f(t), \qquad y = g(t), \qquad a \leqslant t \leqslant b, \qquad (10\text{--}51)$$

with $t = a$ corresponding to point A and $t = b$ to point B. The path C may cross itself; the parameter t serves to indicate which route to follow at a crossing, and in general gives a definite direction of traversal of C, namely, that of increasing t (Fig. 10–13). The subdivision of C by points (x_i, y_i) is achieved by subdividing the interval $[a, b]$ by points $a = t_0 < t_1 < \cdots < t_n = b$; then $\Delta_i x$ and $\Delta_i y$ are increments of x and y corresponding to the increment $\Delta_i t - t_i - t_{i-1}$. We can write:

$$P(x_i, y_i)\Delta_i x + Q(x_i, y_i)\Delta_i y = \left[P(x_i, y_i)\frac{\Delta_i x}{\Delta_i t} + Q(x_i, y_i)\frac{\Delta_i y}{\Delta_i t} \right]\Delta_i t.$$

If we make this replacement in the sum in Eq. (10–50) and pass to the limit, we obtain the rule

$$\int_{C}^{B}\! P(x, y)\,dx + Q(x, \dot y)\,dy = \int_{a}^{b} \left\{ P[f(t), g(t)]\frac{dx}{dt} + Q[f(t), g(t)]\frac{dy}{dt} \right\} dt.$$

$$(10\text{--}52)$$

Thus we have reduced the line integral to an ordinary integral.

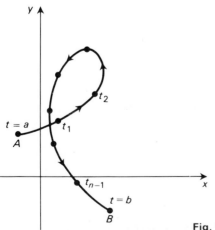

Fig. 10-13. Path C for line integral and its subdivision.

EXAMPLE 1   Evaluate $\int_{(0,0)}^{(1,1)} y^2\,dx + x^2\,dy$, where C is (a) the path $x = t$, $y = t$,
$0 \leqslant t \leqslant 1$, (b) the path $x = t$, $y = t^2$, $0 \leqslant t \leqslant 1$ (see Fig. 10-14). By Eq. (10-52) for path (a), the integral equals

$$\int_0^1 (t^2 \cdot 1 + t^2 \cdot 1)\,dt = \frac{2}{3},$$

and for path (b) it equals

$$\int_0^1 (t^4 \cdot 1 + t^2 \cdot 2t)\,dt = \frac{7}{10}. \quad \blacktriangleleft$$

In the case of a particle moving on a path, as in Fig. 10–12, with t as time,

$$P\frac{dx}{dt} + Q\frac{dy}{dt} = (P\mathbf{i} + Q\mathbf{j}) \cdot \left(\frac{dx}{dt}\mathbf{i} + \frac{dy}{dt}\mathbf{j}\right) = \mathbf{F} \cdot \mathbf{v},$$

where $\mathbf{v}$ is the velocity vector of the particle. Therefore, Eq. (10–52) can be written as

$$\text{Work} = W = \int_A^B P\,dx + Q\,dy = \int_a^b (\mathbf{F} \cdot \mathbf{v})\,dt. \qquad (10\text{–}53)$$

The line integral has the familiar properties concerning the integral of a sum and the integral of a constant times the integrand. Also if C is broken into two parts by an intermediate point E, so that C consists of a path C_1 from A to E and a path C_2 from E to B (Fig. 10–15), then

$$\int_A^E P\,dx + Q\,dy + \int_E^B P\,dx + Q\,dy = \int_A^B P\,dx + Q\,dy. \qquad (10\text{–}54)$$

The parameter along the path C can be changed freely as long as the path is followed in the chosen direction. Often x can serve as a parameter, that is, the equations $x = x$, $y = g(x)$ can be used; similarly, y can serve as a parameter, that is, $x = f(y)$, $y = y$ can be used. Or x may be used for one part of the path, y for another.

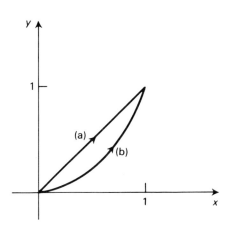

Fig. 10-14. Example 1.

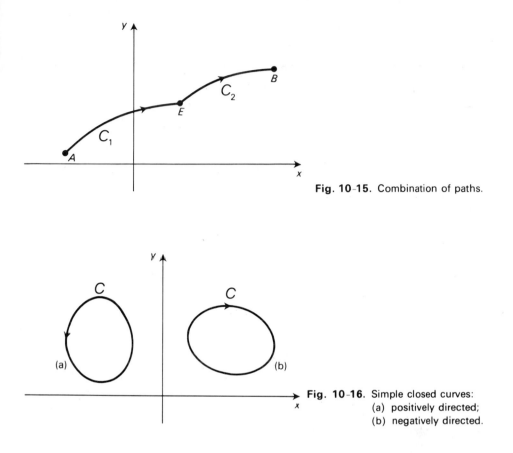

Fig. 10-15. Combination of paths.

Fig. 10-16. Simple closed curves:
(a) positively directed;
(b) negatively directed.

As above, we generally require that the parameter t increases along the path for the chosen direction. At times it is convenient to have t steadily *decreasing* along the path, so that $a > b$ and $t_0 > t_1 > \ldots > t_n$ in any subdivision. This does not affect the value of the line integral.

In many cases the path C is a *simple closed curve*, that is, in tracing C, (x, y) ends up by returning to its initial point but passes through no other point twice. Here there are two possibilities, suggested in Fig. 10–16; in (a) we say that C is traced in the *positive* direction, in (b) that C is traced in the *negative* direction. For the corresponding integrals we write

$$\oint P\,dx + Q\,dy, \qquad \oint P\,dx + Q\,dy,$$

respectively. There is clearly no need to give an initial point or final point. We often omit the arrow for positively oriented paths.

EXAMPLE 2 Evaluate $\oint_C (x^2 - y^2)dx + (x^2 + y^2)dy$ if C is the square with vertices $A(0, 0)$, $B(1, 0)$, $D(1, 1)$, $E(0, 1)$, as in Fig. 10–17.

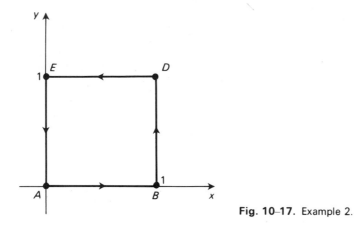

Fig. 10–17. Example 2.

We consider the integral as the sum of four integrals: from A to B, from B to D, from D to E, and from E to A. From A to B we use x as parameter and observe that $y = 0$ on this portion, so that $dy = (dy/dx)dx = 0$. Hence we obtain

$$\int_0^1 x^2 \, dx = \frac{1}{3}.$$

Similarly, we write

$$\int_B^D P \, dx + Q \, dy = \int_0^1 (1 + y^2) dy = \frac{4}{3}, \quad y \text{ parameter, } x = 1,$$

$$\int_D^E P \, dx + Q \, dy = \int_1^0 (x^2 - 1) dx = \frac{2}{3}, \quad x \text{ parameter, } y = 1,$$

$$\int_E^A P \, dx + Q \, dy = \int_1^0 y^2 \, dy = -\frac{1}{3}, \quad y \text{ parameter, } x = 0.$$

Summing, we obtain

$$\oint_C (x^2 - y^2) dx + (x^2 + y^2) dy = \frac{1}{3} + \frac{4}{3} + \frac{2}{3} - \frac{1}{3} = 2. \quad \blacktriangleleft$$

We observe that if C is a path from A to B and C' is the same path but in the opposite direction, so that C' goes from B to A, then

$$\int_{\substack{B \\ C'}}^A P \, dx + Q \, dy = -\int_{\substack{A \\ C}}^B P \, dx + Q \, dy. \tag{10–55}$$

This follows at once from the definition, since $\Delta_i x$ and $\Delta_i y$ are reversed in sign if the direction is reversed.

Arc length as parameter

For most paths it is permitted to use arc length s as parameter. Then by (10–52)

$$\int_C P(x, y)dx + Q(x, y)dy = \int_0^L \left\{ P[x(s), y(s)]\frac{dx}{ds} + Q[x(s), y(s)]\frac{dy}{ds} \right\}ds,$$

where L is the total length of the path. Here, as in Section 9–7 (Fig. 9–18), $(dx/ds)\mathbf{i} + (dy/ds)\mathbf{j} = \mathbf{T}$ is a unit vector tangent to the path in the direction of increasing s. If we write as above $\mathbf{F} = P\mathbf{i} + Q\mathbf{j}$, then

$$P[x(s), y(s)]\frac{dx}{ds} + Q[x(s), y(s)]\frac{dy}{ds} = (P\mathbf{i} + Q\mathbf{j}) \cdot \mathbf{T} = F_T,$$

where F_T is the component of $\mathbf{F}$ in the direction of $\mathbf{T}$ (Fig. 10–18). We conclude:

$$\int_C P(x, y)dx + Q(x, y)dy = \int_0^L F_T ds, \qquad \mathbf{F} = P\mathbf{i} + Q\mathbf{j}. \qquad (10\text{–}56)$$

The last integral is sometimes regarded as a line integral:

$$\int_0^L F_T ds = \int_C F_T ds. \qquad (10\text{–}57)$$

By Eq. (10–56) we conclude: *the work W done by force* $\mathbf{F}$ *equals the integral of the tangential component of* $\mathbf{F}$ *with respect to arc length on the path.* Thus the normal component of $\mathbf{F}$ does no work.

Let us examine the normal component further. Let $\mathbf{N} = \mathbf{T} \times \mathbf{k}$, so that $\mathbf{N}$ is obtained from $\mathbf{T}$ by rotating the vector by 90° in the clockwise direction, as in Fig. 10–18. If we similarly rotate $\mathbf{F}$ by 90°, we obtain a vector

$$\mathbf{G} = \mathbf{F} \times \mathbf{k} = (P\mathbf{i} + Q\mathbf{j}) \times \mathbf{k} = Q\mathbf{i} - P\mathbf{j}.$$

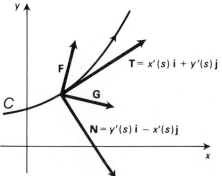

Fig. 10–18. Normal and tangential components.

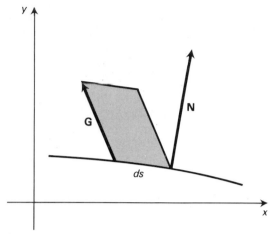

Fig. 10-19. Meaning of flux integral $\int G_N ds$ for fluid motion.

By geometry, $F_T = G_N$, the normal component of G. Therefore,

$$\int_C P\,dx + Q\,dy = \int_C F_T\,ds = \int_C G_N\,ds, \qquad G = Q\mathbf{i} - P\mathbf{j}. \qquad (10\text{-}58)$$

We can also relabel G as $P_1\mathbf{i} + Q_1\mathbf{j}$, so that $P_1 = Q$, $Q_1 = -P$. Then (10-58) becomes

$$\int_C -Q_1\,dx + P_1\,dy = \int_C G_N\,ds, \qquad G = P_1\mathbf{i} + Q_1\mathbf{j}. \qquad (10\text{-}58')$$

The integral $\int_C G_N\,ds$ is known as the *flux integral* of the vector field G across C.

An important application occurs in the study of two-dimensional fluid motion, with G the velocity field (at a given instant). For a short piece of C, of length ds, the fluid crossing this piece in one unit of time sweeps out a parallelogram whose area is $G_N\,ds$, as suggested in Fig. 10-19. (The area is considered negative if G_N is negative.) Therefore $\int_C G_N\,ds$ *gives the net rate at which fluid is crossing C, towards the side of the normal N, in units of area per unit time.*

=============================Problems (Section 10-5)=============================

1. Evaluate the following integrals along the straight-line paths joining the end points:

   a) $\displaystyle\int_{(0,0)}^{(2,2)} y^2\,dx$ **b)** $\displaystyle\int_{(2,1)}^{(1,2)} y\,dx$ **c)** $\displaystyle\int_{(1,1)}^{(2,1)} x\,dy$

2. Evaluate the following line integrals:

   a) $\displaystyle\int_{(0,-1)}^{(0,1)} y^2\,dx + x^2\,dy$, where C is the semicircle $x = \sqrt{1 - y^2}$

b) $\int_{(0,0)}^{(2,4)} y\,dx + x\,dy$, where C is the parabola $y = x^2$

c) $\int_{(1,0)}^{(0,1)} \dfrac{y\,dx - x\,dy}{x^2 + y^2}$, where C is the curve $x = \cos^3 t$, $y = \sin^3 t$, $0 \leqslant t \leqslant \dfrac{\pi}{2}$

HINT: Set $u = \tan^3 t$ in the integral for t.

3. Evaluate the following line integrals:

a) $\oint_C y^2\,dx + xy\,dy$, where C is the square with vertices $(1,\ 1),\ (-1,\ 1),\ (-1,\ -1),$ $(1,\ -1)$

b) $\oint_C y\,dx - x\,dy$, where C is the circle $x^2 + y^2 = 1$

c) $\oint_C x^2 y^2\,dx - xy^3\,dy$, where C is the triangle with vertices $(0,\ 0),\ (1,\ 0),\ (1,\ 1)$

4. Evaluate the following line integrals:

a) $\oint_C (x^2 - y^2)\,dx$, where C is the circle $x^2 + y^2 = 4$

b) $\int_{(0,0)}^{(1,1)} x\,ds$, where C is the line $y = x$

c) $\int_{(0,0)}^{(1,1)} ds$, where C is the parabola $y = x^2$

5. Show that if $f(x, y) \geqslant 0$ on C, then the integral $\int_C f(x, y)\,ds$ can be interpreted as the area of the cylindrical surface $0 \leqslant z \leqslant f(x, y)$, where (x, y) is on C.

6. If $\mathbf{v} = (x^2 + y^2)\mathbf{i} + (2xy)\mathbf{j}$, evaluate $\int_C v_T\,ds$ for the following paths:

a) From $(0,0)$ to $(1,1)$ on the line $y = x$,
b) From $(0,0)$ to $(1,1)$ on line $y = x^2$,
c) From $(0,0)$ to $(1,1)$ on the broken line with corner at $(1, 0)$.

7. Evaluate $\int_C v_N\,ds$ for the vector $\mathbf{v}$ given in Problem 6 on the paths (**a**), (**b**), (**c**) where N is chosen as the normal $\mathbf{T} \times \mathbf{k}$ as in Fig. 10–18.

8. Let a particle of mass m subject to force $\mathbf{F}$ move on a path C from A to B, as in Fig. 10–12. Show that *the work W done equals the gain in kinetic energy of the particle.*

HINT: As in Eq. (10–53), $W = \int_a^b \mathbf{F} \cdot \mathbf{v}\,dt$. Replace $\mathbf{F}$ by $m(d\mathbf{v}/dt)$, in accordance with Newton's second law, and observe that

$$m\frac{d\mathbf{v}}{dt} \cdot \mathbf{v} = \frac{m}{2}\frac{d}{dt}(\mathbf{v} \cdot \mathbf{v}).$$

9. The gravitational force near a point on the earth's surface is represented approximately by the vector $-mg\mathbf{j}$, where the y axis points upward. Show that the work done by this force on a body moving in a vertical plane from height h_1 to height h_2 along any path is equal to $mg(h_1 - h_2)$.

10. Show that the earth's gravitational potential $U = -kMm/r$ is equal to the negative of the work done by the gravitational force $\mathbf{F} = -(kMm/r^2)(\mathbf{r}/r)$ in bringing the particle to its present position from infinite distance along the ray passing through the earth's center.

10–6 GREEN'S THEOREM

The following theorem and its generalizations are fundamental in the theory of line integrals.

GREEN'S THEOREM   *Let D be a domain of the xy plane and let C be a piecewise smooth simple closed curve in D whose interior is also in D. Let P(x, y) and Q(x, y) be functions defined and continuous with continuous first partial derivatives in D. Then*

$$\oint_C P\,dx + Q\,dy = \iint_R \left(\frac{\partial Q}{\partial x} - \frac{\partial P}{\partial y}\right)dx\,dy, \qquad (10\text{-}60)$$

where R is the closed region bounded by C.

The theorem will be proved first for the case in which R is representable in both of the forms:

$$a \leqslant x \leqslant b, \qquad f_1(x) \leqslant y \leqslant f_2(x), \qquad (10\text{-}61)$$
$$c \leqslant y \leqslant d, \qquad g_1(y) \leqslant x \leqslant g_2(y), \qquad (10\text{-}62)$$

as in Fig. 10–20.

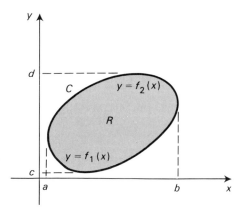

Fig. 10-20. Special region for Green's theorem.

The double integral $\iint\limits_{R}\dfrac{\partial P}{\partial y}\,dx\,dy$ can by (10-61) be written as an iterated integral:

$$\iint\limits_{R}\frac{\partial P}{\partial y}\,dx\,dy=\int_{a}^{b}\int_{f_{1}(x)}^{f_{2}(x)}\frac{\partial P}{\partial y}\,dy\,dx.$$

We can now integrate:

$$\iint\limits_{R}\frac{\partial P}{\partial y}\,dx\,dy=\int_{a}^{b}\{P[x,f_{2}(x)]-P[x,f_{1}(x)]\}\,dx$$

$$=-\int_{b}^{a}P[x,f_{2}(x)]\,dx-\int_{a}^{b}P[x,f_{1}(x)]\,dx$$

$$=-\oint_{C}P(x,\,y)\,dx.$$

In the same way $\iint\limits_{R}\dfrac{\partial Q}{\partial x}\,dx\,dy$ can be written as an iterated integral, with the aid of (10–62), and we conclude that

$$\iint\limits_{R}\frac{\partial Q}{\partial x}\,dx\,dy=\oint Q\,dy.$$

Adding the two double integrals, we find:

$$\iint\limits_{R}\left(\frac{\partial Q}{\partial x}-\frac{\partial P}{\partial y}\right)dx\,dy=\oint_{C}P\,dx+Q\,dy.$$

The theorem is thus proved for the special type of region R.

Suppose next that R is not itself of this form, but can be decomposed into a finite number of such regions: $R_{1},R_{2},\ldots,R_{n}$ by suitable lines or arcs, as in Fig. 10–21. Let $C_{1},C_{2},\ldots,C_{n}$ denote the corresponding boundaries. Then Eq. (10-60) can be applied to each region separately. Adding, we obtain the equation:

$$\oint_{C_{1}}(P\,dx+Q\,dy)+\oint_{C_{2}}(\quad)+\;\ldots\;+\oint_{C_{n}}(\quad)$$

$$=\iint\limits_{R_{1}}\left(\frac{\partial Q}{\partial x}-\frac{\partial P}{\partial y}\right)dx\,dy+\;\ldots\;+\iint\limits_{R_{n}}(\quad)\,dx\,dy.$$

But the sum of the integrals on the left is just $\oint_{C}P\,dx+Q\,dy$, since the integrals along the added arcs are taken once in each direction and hence cancel each other; the

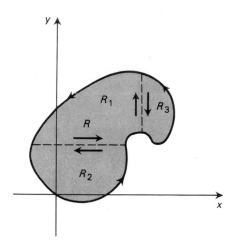

Fig. 10-21. Decomposition of region into special regions.

remaining integrals add up to precisely the integral around C in the positive direction. The integrals on the right add up to

$$\iint_R \left(\frac{\partial Q}{\partial x} - \frac{\partial P}{\partial y} \right) dx\, dy$$

and hence

$$\oint_C P\, dx + Q\, dy = \iint_R \left(\frac{\partial Q}{\partial x} - \frac{\partial P}{\partial y} \right) dx\, dy.$$

The proof thus far covers all regions R of interest in practical problems. To prove the theorem for the most general region R it is necessary to approximate this region by those of the special type just considered and then to use a limiting process. (For details, see O. D. KELLOGG, *Foundations of Potential Theory*, Berlin, Springer, 1929.)

EXAMPLE 1   Let C be the circle $x^2 + y^2 = 1$. Then

$$\oint_C 4xy^3\, dx + 6x^2 y^2\, dy = \iint_R (12xy^2 - 12xy^2)\, dx\, dy = 0. \quad \blacktriangleleft$$

EXAMPLE 2   Let C be the ellipse $x^2 + 4y^2 = 4$. Then

$$\oint_C (2x - y)\, dx + (x + 3y)\, dy = \iint_R (1 + 1)\, dx\, dy = 2\, A,$$

where A is the area of R. Since the ellipse has semi-axes $a = 2$, $b = 1$, the area is $\pi ab = 2\pi$ and the value of the line integral is 4π. $\quad \blacktriangleleft$

EXAMPLE 3 Let C be the ellipse $4x^2 + y^2 = 4$. Then Green's theorem is not applicable to the integral:

$$\oint_C \frac{y}{x^2 + y^2}\, dx - \frac{x}{x^2 + y^2}\, dy,$$

since P and Q fail to be continuous at the origin. ◄

EXAMPLE 4 Let C be the square with vertices $(1, 1), (1, -1), (-1, -1), (-1, 1)$. Then

$$\oint_C (x^2 + 2y^2)\, dx = -\iint_R 4y\, dx\, dy = -4A\bar{y},$$

where $(\bar{x}, \bar{y})$ is the centroid of the square. Hence the value of the line integral is 0.

As a special case of Green's theorem, we have

$$\oint_C x\, dy = -\oint_C y\, dx = \text{Area of } R, \qquad (10\text{-}63)$$

for by Green's theorem both line integrals equal

$$\iint_R 1\, dx\, dy = \text{Area of } R.$$

If the two equal line integrals are averaged, another expression for area is obtained:

$$\frac{1}{2}\oint_C -y\, dx + x\, dy = \text{Area of } R. \qquad (10\text{-}64)$$

This can also be checked by Green's theorem. ◄

10–7 DIVERGENCE OF A VECTOR FIELD.
THE DIVERGENCE THEOREM IN THE PLANE

In Section 10–5 we introduced the concept of flux of a vector field across a curve. We here denote our vector field by $\mathbf{v} = P\mathbf{i} + Q\mathbf{j}$ and consider a smooth closed curve C as in Green's theorem, with $\mathbf{n}$ chosen as the *outer* normal (Fig. 10–22). If we interpret $\mathbf{v}$ as velocity of a plane fluid motion, then (as in Section 10–5) $\oint_C v_n\, ds$ gives the net rate, in units of area per unit time, at which fluid is crossing C toward the side of $\mathbf{n}$, that is, it is the net rate at which fluid is leaving R. We observe that, for an incompressible fluid, this net rate is 0, since the area of fluid entering R must equal the area of fluid leaving R (by incompressibility). On the other hand, for an expanding fluid, such as air being heated, the rate is positive. We would like to measure somehow the rate of expansion, point by point.

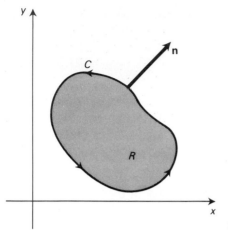

Fig. 10–22. Flux integral around a closed region.

To do so, we consider a very small region R enclosing a particular point $P_0(x_0, y_0)$. We then define the divergence of the vector field as

$$\text{div } \mathbf{v} = \lim \frac{1}{\text{Area of } R} \oint_C v_n \, ds, \tag{10–70}$$

where the limit is taken as R shrinks to (x_0, y_0), as in Eq. (10–37) above.

If, for example, $\mathbf{v} = b(x\mathbf{i} + y\mathbf{j})$ and b is a positive constant, then the fluid motion is a radial expansion away from the origin. If we take C to be a circle of radius ε with center at the origin (Fig. 10–23), then $v_n = |\mathbf{v}| = b\sqrt{x^2 + y^2} = b\varepsilon$ at each point of C.

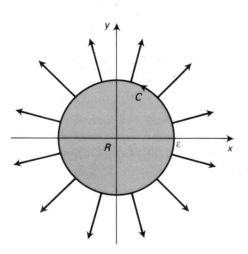

Fig. 10–23. Divergence for radial expansion.

Thus

$$\frac{1}{\text{Area of } R} \oint_C v_n \, ds = \frac{1}{\pi\varepsilon^2} \cdot \oint_C b\varepsilon \, ds = \frac{1}{\pi\varepsilon^2} b\varepsilon \cdot 2\pi\varepsilon = 2b.$$

If we take the limit as $\varepsilon \to 0$, we again obtain $2b$. Thus for such a pure expansion, the divergence of $\mathbf{v}$ at the origin is $2b$. If b is a negative constant, we have a contraction and the divergence of $\mathbf{v}$ at the origin is the negative number $2b$.

To calculate div $\mathbf{v}$, we now have the simple rule: for $\mathbf{v} = P\mathbf{i} + Q\mathbf{j}$,

$$\text{div } \mathbf{v} = \frac{\partial P}{\partial x} + \frac{\partial Q}{\partial y}. \tag{10-71}$$

To prove this, we first use Eq. (10-58') above to write

$$\oint_C v_n \, ds = \oint_C -Q \, dx + P \, dy$$

and then apply Green's theorem:

$$\oint_C -Q \, dx + P \, dy = \iint_R \left(\frac{\partial P}{\partial x} + \frac{\partial Q}{\partial y} \right) dx \, dy.$$

Now we divide by A, the area of R, and pass to the limit:

$$\lim \frac{1}{A} \oint_C v_n \, ds = \lim \frac{1}{A} \iint_R \left(\frac{\partial P}{\partial x} + \frac{\partial Q}{\partial y} \right) dx \, dy = \frac{\partial P}{\partial x} + \frac{\partial Q}{\partial y} \text{ at } (x_0, y_0).$$

Here we used Eq. (10-137) and assumed continuity of the functions involved.

EXAMPLE 1 Let $\mathbf{v} = (x^2 + y^2)(x\mathbf{i} + y\mathbf{j})$. Here again we have a radial expansion away from the origin, but the rate of expansion increases as we move away from the origin. We find:

$$\text{div } \mathbf{v} = \frac{\partial}{\partial x}(x^3 + xy^2) + \frac{\partial}{\partial y}(x^2 y + y^3) = 4(x^2 + y^2). \blacktriangleleft$$

EXAMPLE 2 $\mathbf{v} = -\omega y\mathbf{i} + \omega x\mathbf{j}$, $\omega = \text{const}$. Here $\mathbf{v}$ is perpendicular to the radius vector $x\mathbf{i} + y\mathbf{j}$ at each point so that we expect the divergence to be 0, at least at the origin. By (10-71) we find div $\mathbf{v} \equiv 0$. We are in fact dealing with an incompressible flow, a rotation about the origin with angular velocity ω (see Problem 5 below). $\blacktriangleleft$

The divergence appears in many other physical contexts. For example, the divergence of the electric force $\mathbf{E}$ satisfies the equation div $\mathbf{E} = 4\pi\rho$, where ρ is the charge density. In many problems there is a vector field that is the gradient of a scalar W; then the divergence of the vector field is (in the two-dimensional case):

$$\text{div grad } W = \frac{\partial}{\partial x}\left(\frac{\partial W}{\partial x} \right) + \frac{\partial}{\partial y}\left(\frac{\partial W}{\partial y} \right) = \frac{\partial^2 W}{\partial x^2} + \frac{\partial^2 W}{\partial y^2},$$

i.e., the *Laplacian* of W, commonly denoted by $\nabla^2 W$. In various equilibrium problems we have $\nabla^2 W \equiv 0$, that is, W is harmonic (this is illustrated for various cases in Chapter 8).

With the aid of divergence, Green's theorem can now be written in vector form:

$$\oint_C v_n \, ds = \iint_R \text{div } \mathbf{v} \, dx \, dy \tag{10-72}$$

for, as above, we can take $\mathbf{v} = P\mathbf{i} + Q\mathbf{j}$, so that the left side is $\oint_C -Q \, dx + P \, dy$, and then apply Green's theorem and (10-71). Equation (10-72) is the *divergence theorem* (or *Gauss theorem*) for the two-dimensional case (see Section 10-14 for the three-dimensional case.)

10–8 CURL OF A VECTOR FIELD. STOKES THEOREM IN THE PLANE

As in Eq. (10-57) we consider the integral of form $\oint_C v_T \, ds$, where $\mathbf{v} = P\mathbf{i} + Q\mathbf{j}$. When $\mathbf{v}$ is a force field $\mathbf{F}$, the integral represents the work done by $\mathbf{F}$. The integral also occurs in the study of planar fluid motion, with $\mathbf{v}$ the velocity and C a simple closed curve. Then $\oint_C v_T \, ds$ is called the *circulation* of the velocity field on C. If for example C is a circle and $\mathbf{v}$ is tangent to C in the direction of the unit tangent vector $\mathbf{T}$, then v_T is positive for all s and the circulation is positive. In general, a positive (negative) circulation on C indicates a tendency of the fluid to circulate around C in the positive (negative) direction.

As for the divergence, it is natural to ask for the circulation at a point (x_0, y_0) obtained as an appropriate limit when C shrinks to the point. We obtain the desired limit by considering circulation per unit area, that is, by considering the limit of $\frac{1}{A} \oint_C v_T \, ds$, where A is the area of region R enclosed by C.

With the aid of Green's theorem and Eq. (10-37) we find that

$$\text{Circulation at } (x_0, y_0) = \lim \frac{1}{A} \oint_C v_T \, ds$$

$$= \lim \frac{1}{A} \oint_C P \, dx + Q \, dy$$

$$= \lim \frac{1}{A} \iint_R \left(\frac{\partial Q}{\partial x} - \frac{\partial P}{\partial y} \right) dx \, dy$$

$$= \frac{\partial Q}{\partial x} - \frac{\partial P}{\partial y} \quad \text{at } (x_0, y_0).$$

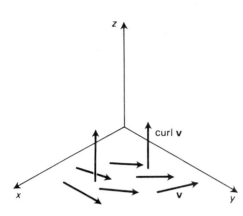

Fig. 10-24. Curl of a planar vector field v.

As for the divergence, the limit is taken as in Eq. (10-37), with the diameter of R approaching 0.

The circulation at a point is closely related to the *curl* of the planar vector field v. The curl of v is the vector field that assigns to each point (x_0, y_0) the vector perpendicular to the xy-plane whose z-component is the circulation of v at (x_0, y_0). Thus,

$$\text{curl } v = \left(\frac{\partial Q}{\partial x} - \frac{\partial P}{\partial y}\right) k. \tag{10-80}$$

In connection with the three-dimensional theory to follow (Section 10–15), it is simpler here to think of $v = P(x, y)i + Q(x, y)j$ as a vector field in xyz-space whose z-component happens to be 0 and whose x and y components are independent of z (indeed, that is how such planar fields normally arise in applications). Then curl v is also a vector field in space, whose x and y components are 0 and whose z-component depends only on x and y (see Fig. 10–24). The circulatory property measured by curl v is then the tendency to rotate in the positive direction about an axis parallel to the z-axis.

EXAMPLE 1 $v = -\omega y i + \omega x j$, as in Example 2 of the preceding section. Here, as noted above, we have a pure rotation about the z-axis with angular velocity ω. We find:

$$\text{curl } v = \left[\frac{\partial}{\partial x}(\omega x) - \frac{\partial}{\partial y}(-\omega y)\right] k = 2\omega k.$$

Thus the larger the angular velocity ω, the larger the curl. If ω is negative, we can think of the axis of rotation as reversed in direction. ◄

The fact that our fluid particles are moving around concentric circles in the positive direction does not by itself imply that the curl has a positive z-component, as shown by the following example.

EXAMPLE 2 $v = -\dfrac{cy}{x^2 + y^2}i + \dfrac{cx}{x^2 + y^2}j$, where $c = $ const. Again v is perpendicular to the radius vector at each point, and we verify that the particles follow

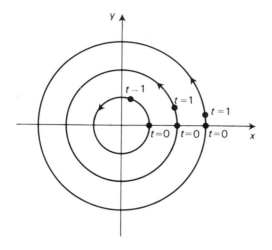

Fig. 10-25. Example 2.

circular paths with center at the origin (see Problem 6(a) below). However, the angular velocity on the circle of radius r is c/r^2 radians per unit time. Thus the rotation speeds up as the origin is approached, as suggested in Fig. 10-25 that shows the positions at $t = 0, 1$ of particles starting on the same radial line at $t = 0$. We find:

$$\operatorname{curl} \mathbf{v} = \left(\frac{\partial}{\partial x} \frac{cx}{x^2 + y^2} + \frac{\partial}{\partial y} \frac{cy}{x^2 + y^2} \right) \mathbf{k} = \mathbf{0}.$$

Thus the curl is $\mathbf{0}$ and there is no net rotational motion at any point. (We observe that $\mathbf{v}$ and hence curl $\mathbf{v}$ are undefined at the origin.) This can be understood if we consider $\oint v_T ds$ about a closed curve C formed of two circular arcs and two segments on rays (see Problem 6(b) below). ◄

We have again emphasized fluid motion, but the curl appears also in a great variety of other physical theories. For example, in electromagnetic theory we have the equation

$$\operatorname{curl} \mathbf{E} = -\frac{1}{c} \frac{\partial \mathbf{H}}{\partial t}$$

relating the electric force $\mathbf{E}$ and the magnetic-field strength $\mathbf{H}$.

It is natural to ask about the curl of a gradient field. Here we have a very simple result:

$$\operatorname{curl} \operatorname{grad} U \equiv \mathbf{0}, \tag{10-81}$$

since

$$\operatorname{curl} \operatorname{grad} U = \left(\frac{\partial}{\partial x} \frac{\partial U}{\partial y} - \frac{\partial}{\partial y} \frac{\partial U}{\partial x} \right) \mathbf{k} = \left(\frac{\partial^2 U}{\partial x \partial y} - \frac{\partial^2 U}{\partial y \partial x} \right) \mathbf{k} = \mathbf{0}$$

(continuity of the derivatives is assumed). Furthermore, curl $\mathbf{v} \equiv \mathbf{0}$ *only if* $\mathbf{v} = \operatorname{grad} U$ *for some* U (in a suitable region). This important rule is discussed in Section 10-9. Fields $\mathbf{v}$ for which curl $\mathbf{v} \equiv \mathbf{0}$ are called *irrotational*.

With the aid of the curl, we can now express Green's theorem in a second vector form:

$$\oint_C v_T \, ds = \iint_R \mathrm{curl}_z \, \mathbf{v} \, dx \, dy. \tag{10–82}$$

Here we have written $\mathrm{curl}_z \mathbf{v}$ for the z-component of curl $\mathbf{v}$, with $\mathbf{v} = P(x, y)\mathbf{i} + Q(x, y)\mathbf{j}$ as above. The left side of Eq. (10–82) is equal to the left side of Green's theorem, as in Eq. (10–60). By Eq. (10–80), the right side is the same as the right side of Green's theorem. Therefore, (10–82) is established. This equality is known as the *Stokes theorem* for two-dimensional fields (see Section 10–15 for the three-dimensional case).

===================================Problems (Section 10–8)====================

1. Evaluate by Green's theorem:

   a) $\oint_C ay \, dx + bx \, dy$ on any path;

   b) $\oint_C e^x \sin y \, dx + e^x \cos y \, dy$ around the rectangle with vertices $(0, 0)$, $(1, 0)$, $(1, \frac{1}{2}\pi)$, $(0, \frac{1}{2}\pi)$;

   c) $\oint (2x^3 - y^3)dx + (x^3 + y^3)dy$ around the circle $x^2 + y^2 = 1$;

   d) $\oint_C u_T \, ds$, where $\mathbf{u} = \mathrm{grad}\,(x^2 y)$ and C is the circle $x^2 + y^2 = 1$;

   e) $\oint_C v_n \, ds$, where $\mathbf{v} = (x^2 + y^2)\mathbf{i} - 2xy\mathbf{j}$, and C is the circle $x^2 + y^2 = 1$, $\mathbf{n}$ being the outer normal;

   f) $\oint_C f(x) \, dx + g(y) \, dy$ on any path.

2. If $\mathbf{r} = x\mathbf{i} + y\mathbf{j}$ is the position vector of an arbitrary point (x, y), show that

   $$\tfrac{1}{2}\oint_C r_n \, ds = \text{Area enclosed by } C,$$

   $\mathbf{n}$ being the outer normal to C.

3. Find the divergence of the following vector fields in the plane:

   a) $\mathbf{v} = e^x \sin 2y\mathbf{i} + e^x \cos 2y\mathbf{j}$

   b) $\mathbf{v} = (x^3 - 3xy^2)\mathbf{i} + (y^3 - 3x^2 y)\mathbf{j}$

   c) $\mathbf{v} = \mathrm{grad}\, \ln(x^2 + y^2)$

   d) $\mathbf{v} = \mathrm{grad}\, \dfrac{1}{x^2 + y^2}$

4. (a), (b), (c), (d). Find the curl of the vector fields of Problem 3.

5. In Example 2 of Section 10–7, $dx/dt = -\omega y$, $dy/dt = \omega x$ for each fluid particle. Show that this implies that $dr/dt = 0$ and $d\theta/dt = \omega$, where (r, θ) are polar coordinates of the particle, and that each particle moves with angular velocity ω along a circle with center at the origin.

6. a) For Example 2 of Section 10–8 show that each particle moves with angular velocity c/r^2 along a circle with center at the origin, where r is the radius of the circle.

   HINT: Proceed as in Problem 5.

b) For the same example, show that $\oint v_T \, ds = 0$ on a path C bounding a set $\alpha \leqslant \theta \leqslant \alpha + \varepsilon$, $b \leqslant r \leqslant b + \varepsilon$ and deduce from this that $\operatorname{curl} \mathbf{v} \equiv \mathbf{0}$.

7. Prove the rules:

a) div, grad, and curl are *linear operators*, that is, if c_1 and c_2 are scalars, then

$$\operatorname{div}(c_1 \mathbf{u}_1 + c_2 \mathbf{u}_2) = c_1 \operatorname{div} \mathbf{u}_1 + c_2 \operatorname{div} \mathbf{u}_2,$$

$$\operatorname{grad}(c_1 U_1 + c_2 U_2) = c_1 \operatorname{grad} U_1 + c_2 \operatorname{grad} U_2,$$

$$\operatorname{curl}(c_1 \mathbf{u}_1 + c_2 \mathbf{u}_2) = c_1 \operatorname{curl} \mathbf{u}_1 + c_2 \operatorname{curl} \mathbf{u}_2.$$

b) $\operatorname{grad}(fg) = f \operatorname{grad} g + g \operatorname{grad} f.$

c) $\operatorname{div}(f\mathbf{u}) = f \operatorname{div} \mathbf{u} + \operatorname{grad} f \cdot \mathbf{u}.$

8. Prove the relations (under appropriate continuity conditions):

a) $\displaystyle\oint_C \frac{\partial U}{\partial n} \, ds = \iint_R \nabla^2 U \, dx \, dy$, where $\mathbf{n}$ is an outer normal to C

b) $\displaystyle\oint_C \nabla U \cdot \mathbf{T} \, ds = 0$

c) If U is harmonic inside C, then $\displaystyle\oint_C \frac{\partial U}{\partial n} \, ds = 0$

d) $\displaystyle\oint_C U \frac{\partial V}{\partial n} \, ds = \iint_R U \nabla^2 V \, dx \, dy + \iint_R (\nabla U \cdot \nabla V) \, dx \, dy$

HINT: Use the result of Problem 7(c).

e) $\displaystyle\oint_C \left(U \frac{\partial V}{\partial n} - V \frac{\partial U}{\partial n} \right) ds = \iint_R (U \nabla^2 V - V \nabla^2 U) \, dx \, dy$

f) If U and V are harmonic inside C, then $\displaystyle\oint_C \left(U \frac{\partial V}{\partial n} - V \frac{\partial U}{\partial n} \right) ds = 0$

10–9 INDEPENDENCE OF PATH. SIMPLY CONNECTED OPEN REGIONS

Throughout the following discussion $P(x, y)$ and $Q(x, y)$ will be assumed to have continuous first partial derivatives in the open region D.

The line integral $\int P \, dx + Q \, dy$ is said to be *independent of path in D* if, for every pair of endpoints A and B in D, the value of line integral $\int_A^B P \, dx + Q \, dy$ is the same for all paths C from A to B in D. The value of the line integral will then in general depend on the choice of A and B but not on the choice of the path joining them. Thus, as in Fig. 10–26, the integrals on C_1, C_2, C_3 have the same value.

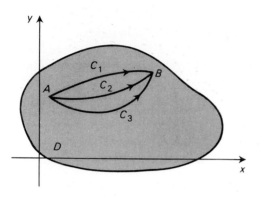

Fig. 10-26. Independence of path.

Independence of path is assured if $P\mathbf{i}+Q\mathbf{j} = \text{grad } U$ *for some* U, since then, in terms of a parameter t,

$$\int_A^B P(x,\,y)\,dx + Q(x,\,y)\,dy = \int_a^b \left[\frac{\partial U}{\partial x}\frac{dx}{dt} + \frac{\partial U}{\partial y}\frac{dy}{dt} \right] dt$$

$$= \int_a^b \frac{dU}{dt}\,dt = U(x(t),\,y(t)) \Big|_{t=a}^{t=b} = U(B) - U(A),$$

which is the difference of the values of U at A and B. We can also write, more concisely,

$$\int_A^B P\,dx + Q\,dy = \int_A^B \frac{\partial U}{\partial x}\,dx + \frac{\partial U}{\partial y}\,dy = \int_A^B dU = U(B) - U(A). \quad (10\text{-}90)$$

EXAMPLE 1 $\int_{(1,\,2)}^{(5,\,6)} 2xy\,dx + x^2\,dy = \int_{(1,\,2)}^{(5,\,6)} d(x^2 y) = x^2 y \big|_{(1,\,2)}^{(5,\,6)} = 150 - 2 = 148.$
No path was given; the value depends only on the values of $U = x\,y$ at the endpoints. ◄

We can show further that, conversely, *if* $\int P\,dx + Q\,dy$ *is independent of path in* D, *then* $P\mathbf{i}+Q\mathbf{j} = \text{grad } U$ *for some* U. We choose a fixed point $(x_0,\,y_0)$ in D and let

$$U(x,\,y) = \int_{(x_0,\,y_0)}^{(x,\,y)} P\,dx + Q\,dy. \quad (10\text{-}91)$$

Here we need not give the path, since the line integral is independent of path in D; to evaluate $U(x,\,y)$, we simply use any convenient path. If in particular we fix y and use a path that ends in a line segment from $(x_1,\,y)$ to $(x,\,y)$, as in Fig. 10-27, then

$$U(x,\,y) = \int_{(x_0,\,y_0)}^{(x_1,\,y)} P\,dx + Q\,dy + \int_{(x_1,\,y)}^{(x,\,y)} P\,dx + Q\,dy = \text{const} + \int_{x_1}^x P(t,\,y)\,dt,$$

where t is used instead of x as integration variable. Thus

$$\frac{\partial U}{\partial x} = \frac{d}{dx}\int_{x_1}^x P(t,\,y)\,dt = P(x,\,y), \qquad y = \text{const},$$

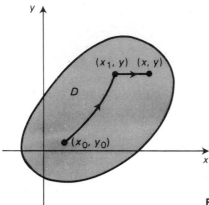

Fig. 10-27. Independence of path and gradient field.

by the fundamental theorem (10–34). Similarly $\partial U/\partial y = Q(x, y)$, and accordingly grad $U = P\mathbf{i} + Q\mathbf{j}$.

REMARK 1: If grad $U = 0\mathbf{i} + 0\mathbf{j} = \mathbf{0}$ in D, then by Eq. (10-90)

$$U(B) - U(A) = \int_A^B 0\,dx + 0\,dy = 0,$$

no matter how A and B are chosen in D. Thus $U(x, y) \equiv$ const in D.

It now follows that, if $\int P\,dx + Q\,dy$ is independent of path in D, then all the functions U such that grad $U = P\mathbf{i} + Q\mathbf{j}$ in D are given by

$$U = \int_{(x_0, y_0)}^{(x, y)} P\,dx + Q\,dy + \text{const}, \tag{10-92}$$

for we saw above that $U_1(x, y) = \int_{(x_0, y_0)}^{(x, y)} P\,dx + Q\,dy$ has gradient $P\mathbf{i} + Q\mathbf{j}$. If also $U(x, y)$ has gradient $P\mathbf{i} + Q\mathbf{j}$, then $U(x, y) - U_1(x, y)$ has gradient $\mathbf{0}$ by linearity (see Problem 7(a) following Section 10-8). Therefore $U(x, y) - U_1(x, y) = $ const or $U(x, y) = U_1(x, y) + $ const as in (10-92). Conversely, all functions U of form (10-92) have gradient $P\mathbf{i} + Q\mathbf{j}$, as proved above for (10-91).

If we are now given $\int P\,dx + Q\,dy$ and recognize $P\,dx + Q\,dy$ as dU for some U, then we know that we have independence of path. However, it is not always simple to recognize that $P = \partial U/\partial x, Q = \partial U/\partial y$ for some U. We need a test to that end and can provide one:

If $P\mathbf{i} + Q\mathbf{j} = $ grad U in D, *then*

$$\frac{\partial P}{\partial y} = \frac{\partial Q}{\partial x} \quad \text{in } D. \tag{10-93}$$

This is simply the statement that curl grad $U \equiv \mathbf{0}$, already verified in Section 10-8.

So far our test goes only one way: if $P\mathbf{i} + Q\mathbf{j}$ is a gradient, then Eq. (10-93) must hold. We need the converse: *if $\partial P/\partial y = \partial Q/\partial x$ in D, then $P\mathbf{i} + Q\mathbf{j}$ is a gradient.*

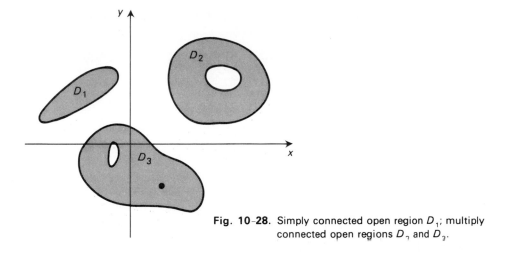

Fig. 10-28. Simply connected open region D_1; multiply connected open regions D_2 and D_3.

However, this new rule is valid only if we require that D be *simply connected*, that is, that D has no "holes." Figure 10-28 illustrates the concept: the open region D_1 is simply connected, has no holes; the open regions D_2 and D_3 have holes and are *multiply connected*. We would say: D_2, with one hole, is *doubly connected*; D_3, with two holes, is *triply connected*. In the case of D_3, one of the holes is a single point. For us the crucial property of a simply connected open region D is that every simple closed curve C in D has its interior in D. This clearly holds for D_1 and fails for D_2 and D_3.

We now formulate our rule: *if $\partial P/\partial y = \partial Q/\partial x$ in D and D is simply connected, then $P\mathbf{i} + Q\mathbf{j} = \text{grad } U$ for some U in D or, equivalently, $\int P\,dx + Q\,dy$ is independent of path in D.*

To justify the rule, we consider two paths C_1, C_2 from A to B in D, as in Fig. 10-29. The path C formed of C_1 and C_2' (obtained from C_2 by reversing

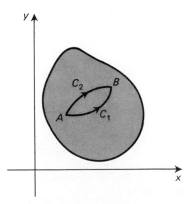

Fig. 10-29. Justification of test for gradient field.

direction) is then a simple closed path to which Green's theorem applies; the interior of C is in D since D is simply connected. Thus

$$\int_A^B P\,dx + Q\,dy + \int_B^A P\,dx + Q\,dy$$
$$= \oint_C P\,dx + Q\,dy = \iint_R \left(\frac{\partial Q}{\partial x} - \frac{\partial P}{\partial y}\right) dx\,dy = 0.$$

Accordingly,

$$\int_A^B P\,dx + Q\,dy = -\int_B^A P\,dx + Q\,dy = \int_A^B P\,dx + Q\,dy.$$

This proves independence for paths that do not cross each other. A repetition of the argument covers the case of paths that cross a finite number of times, and a limit process covers the general case.

EXAMPLE 2 The integral $\int 2xy^3\,dx + 3x^2y^2\,dy$ is independent of path in the whole plane, since

$$\frac{\partial P}{\partial y} - \frac{\partial Q}{\partial x} = 6xy^2 - 6xy^2 = 0.$$

To find a function F whose gradient is $P\mathbf{i} + Q\mathbf{j}$, set

$$F = \int_{(0,\,0)}^{(x,\,y)} 2xy^3\,dx + 3x^2y^2\,dy$$
$$= \int_{(0,\,0)}^{(x,\,0)} 2xy^3\,dx + 3x^2y^2\,dy + \int_{(x,\,0)}^{(x,\,y)} 2xy^3\,dx + 3x^2y^2\,dy,$$

using a broken line path, as in Fig. 10–30. Along the first part $dy = 0$ and $y = 0$, so that the integral reduces to 0; along the second part $dx = 0$ and x is constant, so that

$$F = \int_0^y 3x^2y^2\,dy = x^2y^3.$$

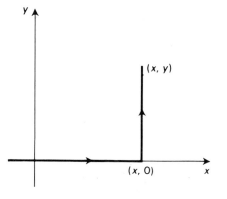

Fig. **10–30.** Example 2.

The general solution is then $x^2 y^3 + C$. To evaluate the integral on any path from $(1, 2)$ to $(3, -2)$, we then write

$$\int_{(1,\,2)}^{(3,\,-2)} 2xy^3\,dx + 3x^2 y^2\,dy = \int_{(1,\,2)}^{(3,\,-2)} d(x^2 y^3) = x^2 y^3 \Big|_{(1,\,2)}^{(3,\,-2)} = -80. \blacktriangleleft$$

EXAMPLE 3 The integral $\displaystyle\int \frac{-y\,dx + x\,dy}{x^2 + y^2}$ is independent of path in any simply connected domain D not containing the origin, for

$$\frac{\partial P}{\partial y} = \frac{\partial Q}{\partial x} = \frac{y^2 - x^2}{(x^2 + y^2)^2}$$

except at the origin. To find the function F whose differential is $P\,dx + Q\,dy$, we can proceed as in Example 2, using as path a broken line from $(1, 0)$ to (x, y). However, $P\,dx + Q\,dy$ is a familiar differential, namely that of the polar coordinate angle θ:

$$d\theta = d\left(\arctan\frac{y}{x}\right) = \frac{-y\,dx + x\,dy}{x^2 + y^2}.$$

The arc tangent is to be avoided if possible, because we cannot in general be restricted to principal values and also because of the awkwardness of arc tan ∞ for the y axis. We would therefore write

$$\int_A^B \frac{-y\,dx + x\,dy}{x^2 + y^2} = \int_A^B d\theta = \theta_B - \theta_A.$$

However, in order for θ to be a well-defined continuous function in a domain including the path (which in particular requires that θ be *single valued*), it is necessary to remain within a simply connected domain not containing the origin. Two such domains are shown in Figs. 10–31(a) and (b). In Fig. 10–31(c) the path is

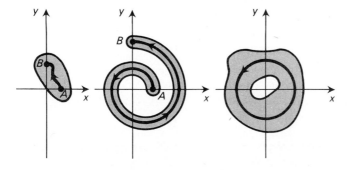

(a) $\displaystyle\int_A^B d\theta = \frac{1}{2}\pi$ (b) $\displaystyle\int_A^B d\theta = \frac{5}{2}\pi$ (c) $\displaystyle\oint \frac{-y\,dx + x\,dy}{x^2 + y^2} = 2\pi$ **Fig. 10-31.** Example 3.

the circle $x^2 + y^2 = 1$ and the domain is doubly connected. Since $x^2 + y^2 = 1$ on path,

$$\oint \frac{-y\,dx + x\,dy}{x^2 + y^2} = \oint -y\,dx + x\,dy = \iint_R 2\,dx\,dy = 2\pi;$$

here Green's theorem, which is not applicable to the given integral, could be applied after the line integral had been simplified by using the relation $x^2 + y^2 = 1$ on the path. The line integral can also be regarded as a sum of integrals from $\theta = 0$ to $\theta = \pi$ and from $\theta = \pi$ to $\theta = 2\pi$, giving $\pi + \pi = 2\pi$. A similar procedure can be followed in general, and we conclude that, for any path C not through $(0,0)$,

$$\int_{\substack{A \\ C}}^{B} \frac{-y\,dx + x\,dy}{x^2 + y^2} = \text{Total increase in } \theta \text{ from } A \text{ to } B,$$

as θ varies *continuously* on the path C. The integral is not independent of path but depends on the number of times C goes around the origin. ◄

EXAMPLE 4   The integral $\displaystyle\int \frac{x\,dx + y\,dy}{x^2 + y^2}$ is independent of path in the *doubly connected* domain consisting of the xy-plane minus the origin. First of all,

$$\frac{\partial P}{\partial y} = \frac{\partial Q}{\partial x} = \frac{-2xy}{(x^2 + y^2)^2},$$

but, as Example 3 shows, this would guarantee independence of path only in a *simply connected* domain. However, for this example we have additional information:

$$\frac{x\,dx + y\,dy}{x^2 + y^2} = d\ln \sqrt{x^2 + y^2},$$

and the function $\ln \sqrt{x^2 + y^2} = \ln r$ is well defined (i.e., *single valued*), with continuous derivatives except at the origin. Hence

$$\int_A^B \frac{x\,dx + y\,dy}{x^2 + y^2} = \int_A^B d\ln r = \ln r \Big|_A^B = \ln \frac{r_B}{r_A}$$

for any path from A to B not through the origin—in particular, then, for the paths of Fig. 10–31(a) and (b). On the circle of Fig. 10–31(c) the integral is 0, since for such a closed path we can choose $A = B$ and hence have

$$U(B) - U(A) = \ln r_B - \ln r_A = 0. \quad ◄$$

REMARK 2:   The argument above, based on Fig. 10–29, can be used to show that $\int P\,dx + Q\,dy$ is independent of path in the open region D (simply connected or not) if and only if $\oint P\,dx + Q\,dy = 0$ on every simple closed path C in D. This is illustrated by Example 4, in which D is doubly connected.

Case of multiply connected domains

Let D be doubly connected, as in Fig. 10–32, and let $\partial P/\partial y = \partial Q/\partial x$ in D. Then we do not expect independence of path or, equivalently, that $\oint P\,dx + Q\,dy = 0$ on every simple closed path in D. However, for the path C_1 of Fig. 10–32, necessarily $\oint_{C_1} P\,dx + Q\,dy = 0$, for C_1 does not enclose the hole and hence lies in a simply connected part of D, to which we can apply Green's theorem to conclude as above that the integral equals 0. This procedure is not valid for the paths C_2 or C_3 of Fig. 10–32. However, we can conclude that

$$\oint_{C_2} P\,dx + Q\,dy = \oint_{C_3} P\,dx + Q\,dy \qquad (10\text{–}94)$$

and, in general, that $\oint P\,dx + Q\,dy$ has the same value for all simple closed paths C that enclose the hole. To prove Eq. (10–94), we introduce auxiliary arcs AB and EF, as shown. Then the simple closed path $ABKEFLA$ lies in D and does not enclose the hole. Therefore, as for C_1, the integral $\oint P\,dx + Q\,dy$ around this path is 0. The same holds for the path $BAMFENB$. If we add the line integrals on these two paths, then we see that the integrals along the auxiliary arcs cancel out, since each appears once in each direction. We are left with integrals that together give

$$\oint_{C_2} P\,dx + Q\,dy + \oint_{C_3'} P\,dx + Q\,dy = 0,$$

where C_3' is C_3 with direction reversed. Therefore, Eq. (10–94) follows.

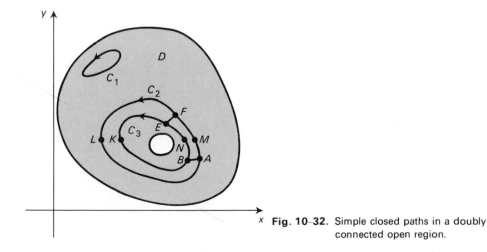

Fig. 10–32. Simple closed paths in a doubly connected open region.

Our conclusion is illustrated by Example 3 above, for which

$$\oint_C \frac{-y\,dx + x\,dy}{x^2 + y^2} = 2\pi$$

for every simple closed curve C enclosing the origin.

10–10 **APPLICATIONS TO THERMODYNAMICS**

Let a certain volume V of a gas be given, enclosed in a container and subject to a pressure p. It is known from experiment that for each kind of gas there is an *equation of state*

$$f(p, V, T) = 0, \tag{10–100}$$

connecting pressure, volume, and temperature T. For an *ideal gas* (low density and high temperature), Eq. (10–100) takes the special form:

$$pV = RT, \tag{10–100'}$$

where R is constant (the same for all gases, if one mole of gas is used).

   With each gas is also associated a scalar U, the total internal energy; this is analogous to the kinetic energy plus potential energy considered above. For each gas, U is given as a definite function of the *state*, hence of p and V:

$$U = U(p, V). \tag{10–101}$$

The particular equation (10–101) depends on the gas considered. For an ideal gas, Eq. (10–101) takes the form:

$$U = c_V \frac{pV}{R} = c_V T, \tag{10–101'}$$

where c_V is constant, the *specific heat* at constant volume.

   A particular *process* gone through by a gas is a succession of changes in the state, so that p and V, and hence T and U; become functions of time t. The state at time t can be represented by a point (p, V) on a pV diagram (Fig. 10–33) and the process by a curve C, with t as parameter. During such a process it is possible to measure the *amount of heat Q* received by the gas. The first law of thermodynamics is equivalent to the statement that

$$\frac{dQ}{dt} = \frac{dU}{dt} + p\frac{dV}{dt}, \tag{10–102}$$

where $Q(t)$ is the amount of heat received up to time t.

   Hence the amount of heat introduced in a particular process is given by an integral:

$$Q(h) - Q(0) = \int_0^h \left(\frac{dU}{dt} + p\frac{dV}{dt} \right) dt. \tag{10–103}$$

Now by (10–101) dU is expressible in terms of dp and dV:

$$dU = \left(\frac{\partial U}{\partial p} \right)_V dp + \left(\frac{\partial U}{\partial V} \right)_p dV. \tag{10–104}$$

Hence (10–103) can be written as a line integral:

$$Q(h) - Q(0) = \int_0^h \left\{ \left(\frac{\partial U}{\partial p} \right)_V \frac{dp}{dt} + \left[\left(\frac{\partial U}{\partial V} \right)_p + p \right] \frac{dV}{dt} \right\} dt$$

$$= \int_C \left(\frac{\partial U}{\partial p} \right)_V dp + \left[\left(\frac{\partial U}{\partial V} \right)_p + p \right] dV, \tag{10–105}$$

or, with (10–104) understood, simply thus:

$$Q(h) - Q(0) = \int_C dU + p \, dV. \tag{10–106}$$

For (10–105) or (10–106) to be independent of the path C, we must have

$$\frac{\partial}{\partial V} \left(\frac{\partial U}{\partial p} \right) = \frac{\partial}{\partial p} \left(\frac{\partial U}{\partial V} + p \right),$$

that is,

$$\frac{\partial^2 U}{\partial V \partial p} = \frac{\partial^2 U}{\partial p \partial V} + 1.$$

Since this is impossible (when U has continuous second derivatives), the heat introduced is dependent on the path. For a simple closed path C_1, as in Fig. 10–33, the heat introduced is $\oint_{C_1} dU + p \, dV$. Since U is a given function of p and V, $\oint dU = 0$; thus the heat introduced reduces to $\oint_{C_1} p \, dV$.

This integral is precisely the area integral $\int y \, dx$ considered in Section 10–6 with p replacing y and V replacing x. Hence for such a counterclockwise cycle the heat introduced is *negative*; there is a heat *loss*, equal to the area enclosed (a unit of area corresponding to a unit of *energy*). The integral $\int p \, dV$ can also be interpreted as the mechanical *work* done by the gas on the surrounding medium or as the negative of the work done on the gas by the surrounding medium. For the process of the curve

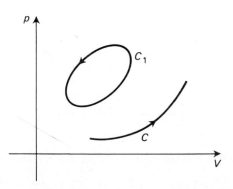

Fig. 10–33. Thermodynamic processes.

C_1 considered above, the heat loss equals the work done on the gas; the total energy remains unchanged, in agreement with the law of energy conservation expressed by the thermodynamics law (10–102).

While the integral $\int dU + p\,dV$ is not independent of path, it is an experimental law that the integral

$$\int \frac{1}{T}dU + \frac{p}{T}dV$$

is independent of path. We can accordingly introduce a scalar S whose differential is the expression being integrated:

$$dS = \frac{1}{T}dU + \frac{p}{T}dV = \frac{1}{T}\frac{\partial U}{\partial p}dp + \frac{1}{T}\left(\frac{\partial U}{\partial V}+p\right)dV; \qquad (10\text{–}107)$$

S is termed the *entropy*. In the first equation here, we can consider U and V as independent variables; in the second, p and V can be considered independent. Thus the first equation gives

$$\left(\frac{\partial S}{\partial U}\right)_V = \frac{1}{T}, \qquad \left(\frac{\partial S}{\partial V}\right)_U = \frac{p}{T},$$

and hence

$$\frac{\partial^2 S}{\partial V \partial U} = \frac{\partial}{\partial V}\left(\frac{1}{T}\right) = \frac{\partial}{\partial U}\left(\frac{p}{T}\right) = \frac{\partial^2 S}{\partial U \partial V}.$$

Accordingly, we find

$$T\frac{\partial p}{\partial U} - p\frac{\partial T}{\partial U} + \frac{\partial T}{\partial V} = 0, \qquad U, V \text{ independent.} \qquad (10\text{–}108)$$

A similar relation is obtainable from the second equation (10–107), and others are obtainable by choosing different independent variables. All these equations are simply different forms of the condition $\partial P/\partial y = \partial Q/\partial x$ for independence of path (see Problem 14 following Section 9–15).

The second law of thermodynamics states first the existence of the entropy S and second the fact that, for any closed system, $dS/dt \geq 0$, that is, the entropy can never decrease.

====================**Problems (Section 10-10)**====================

1. Determine by inspection a function $F(x, y)$ whose differential has the given value, and integrate the corresponding line integral:

   a) $dF = 2xy\,dx + x^2\,dy,$ $\displaystyle\int_{(0,0)}^{(1,1)} 2xy\,dx + x^2\,dy$, where C is the curve $y = x^{3/2}$

   b) $dF = ye^{xy}\,dx + xe^{xy}\,dy,$ $\displaystyle\int_{(0,0)}^{(\pi,0)} ye^{xy}\,dx + xe^{xy}\,dy$, where C is the curve $y = \sin^3 x$

   c) $dF = \dfrac{x\,dx + y\,dy}{(z^2 + y^2)^{3/2}},$ $\displaystyle\int_{(1,0)}^{(e^{2\pi},0)} \dfrac{x\,dx + y\,dy}{(x^2 + y^2)^{3/2}}$, where C is the curve $x = e^t \cos t$, $y = e^t \sin t$

2. Test for independence of path and evaluate the following integrals:

a) $\displaystyle\int_{(1,-2)}^{(3,4)} \frac{y\,dx - x\,dy}{x^2}$ on the line $y = 3x - 5$

b) $\displaystyle\int_{(0,2)}^{(1,3)} \frac{3x^2}{y}\,dx - \frac{x^3}{y^2}\,dy$ on the parabola $y = 2 + x^2$

3. Evaluate the following integrals:

a) $\displaystyle\oint [\sin(xy) + xy\cos(xy)]\,dx + x^2\cos(xy)\,dy$ on the circle $x^2 + y^2 = 1$

b) $\displaystyle\oint \frac{y\,dx - (x-1)\,dy}{(x-1)^2 + y^2}$ on the circle $x^2 + y^2 = 4$

c) $\displaystyle\oint y^3\,dx - x^3\,dy$ on the square $|x| + |y| = 1$

d) $\displaystyle\oint xy^6\,dx + (3x^2 y^5 + 6x)\,dy$ on the ellipse $x^2 + 4y^2 = 4$

4. Determine all values of the integral $\displaystyle\int_{(1,0)}^{(2,2)} \frac{-y\,dx + x\,dy}{x^2 + y^2}$ on a path not passing through the origin.

5. Show that the following functions are independent of path in the xy-plane and evaluate them:

a) $\displaystyle\int_{(1,1)}^{(x,y)} 2xy\,dx + (x^2 - y^2)\,dy$ b) $\displaystyle\int_{(0,0)}^{(x,y)} \sin y\,dx + x\cos y\,dy$

6. Evaluate $\displaystyle\oint \frac{x^2 y\,dx - x^3\,dy}{(x^2 + y^2)^2}$ around the square with vertices $(\pm 1, \pm 1)$. Note that $\partial P/\partial y = \partial Q/\partial x$.

7. Let D be as in Fig. 10-32, let $P(x, y)$ and $Q(x, y)$ have continuous first partial derivatives in D. Show that

$$\oint_{C_2} P\,dx + Q\,dy - \oint_{C_3} P\,dx + Q\,dy = \iint_R \left(\frac{\partial Q}{\partial x} - \frac{\partial P}{\partial y}\right)dx\,dy,$$

where R is the closed region bounded by C_2 and C_3.

HINT: Use auxiliary arcs as in the proof of Eq. (10–94) to reduce the problem to two problems to which Green's theorem is applicable.

8. Show that for an ideal gas

a) The law (10–105) becomes $\displaystyle Q(h) - Q(0) = \int_C p\left(1 + \frac{c_V}{R}\right)dV + V\,dp$

b) $S = \ln p^{c_V} V^{R + c_V} + \text{const}$

9. Show that, on the basis of the laws of thermodynamics, the line integral $\int S\,dT + p\,dV$ is independent of the path in the TV plane. The integrand is minus the differential of the *free energy* F.

10–11 LINE INTEGRALS IN SPACE

Here the discussion of Section 10-5 can be repeated with allowances for paths in xyz-space. If C is such a path from A to B, as in Fig. 10–34, then the line integral is

$$\int_{\substack{A \\ C}}^{B} X\,dx + Y\,dy + Z\,dz,$$ where $X(x, y, z)$, $Y(x, y, z)$, $Z(x, y, z)$ are defined on C. The

line integral can be interpreted as the *work* W done by the force $\mathbf{F} = X\mathbf{i} + Y\mathbf{j} + Z\mathbf{k}$ on a particle that moves from A to B. The value of the integral is then

$$W = \lim \sum_{i=1}^{n} [X(x_i, y_i, z_i)\Delta_i x + Y(\ldots)\Delta_i y + Z(\ldots)\Delta_i z], \qquad (10\text{-}110)$$

as in Eq. (10–50) and in subsequent discussion. The path C is usually given in parametric form

$$x = f(t), \qquad y = g(t), \qquad z = h(t), \qquad a \leqslant t \leqslant b \qquad (10\text{-}111)$$

and the subdivision of C by points (x_i, y_i, z_i) is obtained as before by subdividing the interval $a \leqslant t \leqslant b$.

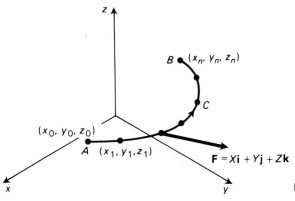

Fig. 10-34. Line integral in space.

With reference to the parameter, the line integral equals

$$\int_{a}^{b} \left\{ X[f(t), g(t), h(t)]\frac{dx}{dt} + Y[\ldots]\frac{dy}{dt} + Z[\ldots]\frac{dz}{dt} \right\} dt, \qquad (10\text{-}112)$$

under the usual continuity conditions. We can interpret (10–112) as

$$\int_{a}^{b} \mathbf{F} \cdot \mathbf{v}\,dt, \qquad (10\text{-}112')$$

where $\mathbf{v}$ is the velocity vector of the moving particle in terms of time t.

We can again change parameter freely on C, without affecting the value of the line integral, as in Section 10–5. If arc length s is chosen as parameter, the line integral becomes

$$\int_0^L \left(X\frac{dx}{ds} + Y\frac{dy}{ds} + Z\frac{dz}{ds} \right) ds = \int_0^L (X\cos\alpha + Y\cos\beta + Z\cos\gamma)\, ds$$

$$= \int_0^L \mathbf{F}\cdot\mathbf{T}\, ds = \int_0^L F_T\, ds, \qquad (10\text{--}113)$$

where $\mathbf{T} = \cos\alpha\,\mathbf{i} + \cos\beta\,\mathbf{j} + \cos\gamma\,\mathbf{k} = (dx/ds)\mathbf{i} + (dy/ds)\mathbf{j} + (dz/ds)\mathbf{k}$ is the unit tangent vector in the direction of motion (Section 9–3). We again write the integral with respect to arc length as a line integral

$$\int_A^B F_T\, ds. \qquad (10\text{--}113')$$

For a simple closed path C we have a line integral $\oint_C X\,dx + Y\,dy + Z\,dz$, but there is no way of indicating by an arrow a *positively directed path*.

The basic properties of line integrals carry over from Section 10–5.

10–12 SURFACES IN SPACE. SURFACE AREA. ORIENTABILITY.

A surface in space is often given in the form $z = z(x, y)$, as in Fig. 10–35. Here (x, y) varies over a closed region R_{xy} in the xy-plane. In calculus, the formula

$$S = \iint_{R_{xy}} \sqrt{1 + \left(\frac{\partial z}{\partial x}\right)^2 + \left(\frac{\partial z}{\partial y}\right)^2}\, dx\, dy \qquad (10\text{--}120)$$

is developed for the total area of the surface. This can also be written as

$$S = \iint_{R_{xy}} \sec\gamma\, dA, \qquad (10\text{--}120')$$

where γ is the angle between a unit upper normal $\mathbf{n}$ to the surface and the unit vector $\mathbf{k}$ (Fig. 10–35). As in Section 9–4, we have

$$\mathbf{n} = \left(-\frac{\partial z}{\partial x}\mathbf{i} - \frac{\partial z}{\partial y}\mathbf{j} + \mathbf{k} \right) \cdot \frac{1}{\sqrt{\left(\dfrac{\partial z}{\partial x}\right)^2 + \left(\dfrac{\partial z}{\partial y}\right)^2 + 1}}, \qquad (10\text{--}121)$$

so that

$$\cos\gamma = \mathbf{n}\cdot\mathbf{k} = \frac{1}{\sqrt{\left(\dfrac{\partial z}{\partial x}\right)^2 + \left(\dfrac{\partial z}{\partial y}\right)^2 + 1}}$$

and hence the integrand of Eq. (10–120) is indeed $\sec\gamma$.

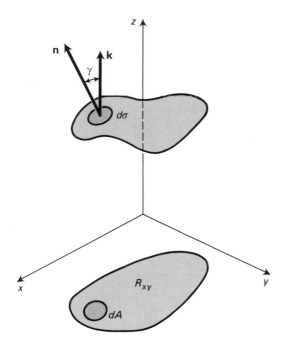

Fig. 10–35. Surface in space.

From Eq. (10–120′) we can write

$$d\sigma = \sec \gamma \, dA \tag{10-122}$$

for $d\sigma$, the surface area element. Thus

$$dA = \cos \gamma \, d\sigma. \tag{10-123}$$

This relation asserts that when we project our surface onto the xy-plane, the area of each small piece of the surface is multiplied by $\cos \gamma = \cos \angle (\mathbf{n}, \mathbf{k})$. Since $\mathbf{n}$ and $\mathbf{k}$ are normal to the surface and to the xy-plane, respectively, $\angle (\mathbf{n}, \mathbf{k})$ equals the angle between the tangent plane to the surface and the xy-plane. By geometry, for perpendicular projection of one plane onto another the areas are multiplied by $\cos \theta$, where θ is the angle between the planes. Thus Eq. (10–123) extends this result to projection of a small piece of a curved surface.

In Section 9–11 we studied representation of a surface in space in parametric form:

$$x = f(u, v), \qquad y = g(u, v), \qquad z = h(u, v), \tag{10-124}$$

and were led to an expression for a normal vector (not necessarily a unit vector):

$$\mathbf{n} = \mathbf{P}_1 \times \mathbf{P}_2, \tag{10-125}$$

where $\mathbf{P}_1 = x_u \mathbf{i} + y_u \mathbf{j} + z_u \mathbf{k}$, $\mathbf{P}_2 = x_v \mathbf{i} + y_v \mathbf{j} + z_v \mathbf{k}$, and for the surface area element:

$$d\sigma = |\mathbf{n}| \, du \, dv. \tag{10-126}$$

Thus, if (u, v) varies over a closed region R_{uv} in the uv-plane, the corresponding surface in xyz-space has total area

$$S = \iint\limits_{R_{uv}} |\mathbf{n}| \, du \, dv. \tag{10–127}$$

To calculate $|\mathbf{n}|$, it is common practice to introduce the quantities

$$\begin{cases} E = \left(\dfrac{\partial x}{\partial u}\right)^2 + \left(\dfrac{\partial y}{\partial u}\right)^2 + \left(\dfrac{\partial z}{\partial u}\right)^2 = |\mathbf{P}_1|^2, \\[2mm] F = \dfrac{\partial x}{\partial u}\dfrac{\partial x}{\partial v} + \dfrac{\partial y}{\partial u}\dfrac{\partial y}{\partial v} + \dfrac{\partial z}{\partial u}\dfrac{\partial z}{\partial v} = \mathbf{P}_1 \cdot \mathbf{P}_2, \\[2mm] G = \left(\dfrac{\partial x}{\partial v}\right)^2 + \left(\dfrac{\partial y}{\partial v}\right)^2 + \left(\dfrac{\partial z}{\partial v}\right)^2 = |\mathbf{P}_2|^2. \end{cases} \tag{10–128}$$

Then we can verify (see Problem 10 below) that

$$|\mathbf{n}| = \sqrt{EG - F^2}, \tag{10–129}$$

so that Eq. (10–127) can be replaced by

$$S = \iint\limits_{R_{uv}} \sqrt{EG - F^2} \, du \, dv. \tag{10–127'}$$

For the special case of parametric equations

$$x = u, \qquad y = v, \qquad z = h(u, v),$$

the parameter plane is the same as the xy-plane. For this case we find

$$E = 1 + \left(\frac{\partial z}{\partial u}\right)^2, \qquad F = \frac{\partial z}{\partial u}\frac{\partial z}{\partial v}, \qquad G = 1 + \left(\frac{\partial z}{\partial v}\right)^2,$$

$$EG - F^2 = 1 + \left(\frac{\partial z}{\partial u}\right)^2 + \left(\frac{\partial z}{\partial v}\right)^2,$$

so that Eq. (10–127') gives

$$S = \iint\limits_{R_{uv}} \sqrt{1 + \left(\frac{\partial z}{\partial u}\right)^2 + \left(\frac{\partial z}{\partial v}\right)^2} \, du \, dv.$$

Since u is the same as x, and v is the same as y, we have simply recovered Eq. (10–120).

Orientability of surfaces

We now want to consider surfaces in space that are smooth or, more generally, *piecewise smooth*. For example, the lateral faces of a triangular prism together form a piecewise smooth surface (Fig. 10–36). A spherical surface is smooth, but the total surface of a circular cylinder is only piecewise smooth, being formed of three smooth pieces.

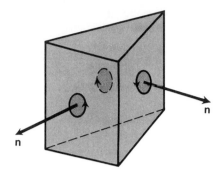

Fig. 10-36. Piecewise smooth surface.

To each such surface we would like to assign an *orientation*, the counterpart of direction for curves. One way to achieve this is to assign positive directions to simple closed curves, as suggested in Fig. 10-36; here it makes sense to consider only *small curves*, each contained in one smooth piece, and some care is needed to ensure that the procedure is consistent as we cross from one smooth piece to another. The procedure in fact leads to a definite positive direction of traversal of the boundary curve for each piece (rectangle in Fig. 10-36); adjacent pieces induce *opposite* directions along common portions of the boundary.

Another related procedure is to choose a unit normal vector varying continuously within each piece. The unit normal vector determines positive directions for simple closed curves, as suggested in Fig. 10-36. This matches the direction in which a right-handed screw must be turned to advance in the direction of the normal.

For a smooth surface, such as a sphere, the outer normal (or the inner normal) can be used to achieve an orientation. The same applies to the boundary of a cylinder, as it does to the boundary of any solid object in space.

In all these cases two orientations are possible, depending upon the choice of normal. However, for some surfaces no consistent choice of orientation can be made: the surfaces are *nonorientable*. The most famous example is the Möbius strip (Fig. 10-37). As we go around the surface once, we are forced to return with orientation reversed.

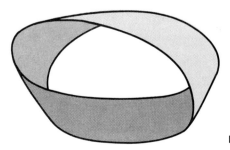

Fig. 10-37. Möbius strip, a nonorientable surface.

10–13 SURFACE INTEGRALS

Let $\mathscr{S}$ be a surface as above and let $H(x, y, z)$ be a function defined on $\mathscr{S}$. Then the surface integral of H over $\mathscr{S}$ is defined as follows:

$$\iint_{\mathscr{S}} H(x, y, z) \, d\sigma = \lim \sum_{i=1}^{n} H(x_i, y_i, z_i) \Delta_i \sigma. \qquad (10\text{–}130)$$

The surface $\mathscr{S}$ is here assumed subdivided into n pieces, $\Delta_i \sigma$ denotes the area of the ith piece and, in the limit process, the pieces all become arbitrarily small, in an appropriate sense. The subdivision and the limit process are best described in terms of a parametric representation (10–124); then the process is similar to that for a double integral over R_{uv}. (As pointed out above, representation $z = z(x, y)$ is a special case of parametric representation.)

If the representation $z = z(x, y)$ is used, the integral is reducible to a double integral in the xy-plane.

$$\iint_{\mathscr{S}} H \, d\sigma = \iint_{R} H\left[x, y, z(x, y)\right] \sec \gamma \, dx \, dy, \qquad (10\text{–}131)$$

where, as above, γ is the angle between the upper normal and the positive z-axis. This follows from the formula $d\sigma = \sec \gamma \, dx \, dy$.

If the parametric representation (10–124) is used, the integral has the expression

$$\iint_{\mathscr{S}} H \, d\sigma = \iint_{R_{uv}} H[f(u, v), g(u, v), h(u, v)] \sqrt{EG - F^2} \, du \, dv. \qquad (10\text{–}132)$$

This follows similarly from the expression $d\sigma = \sqrt{EG - F^2} \, du \, dv$ for the area element.

The integral $\iint H \, d\sigma$ is analogous to a line integral $\int f(s) \, ds$. In order to obtain a surface integral analogous to the line integral $\int X \, dx + Y \, dy + Z \, dz$, we consider an oriented smooth surface $\mathscr{S}$ with continuous *unit* normal vector $\mathbf{n}$:

$$\mathbf{n} = \cos \alpha \, \mathbf{i} + \cos \beta \, \mathbf{j} + \cos \gamma \, \mathbf{k}.$$

Let $L(x, y, z)$, $M(x, y, z)$, $N(x, y, z)$ be functions defined and continuous on $\mathscr{S}$. Then, by definition,

$$\iint_{\mathscr{S}} L \, dy \, dz = \iint_{\mathscr{S}} L \cos \alpha \, d\sigma,$$

$$\iint_{\mathscr{S}} M \, dz \, dx = \iint_{\mathscr{S}} M \cos \beta \, d\sigma, \qquad (10\text{–}133)$$

$$\iint_{\mathscr{S}} N \, dx \, dy = \iint_{\mathscr{S}} N \cos \gamma \, d\sigma.$$

If these are added, we obtain the general surface integral:

$$\iint_{\mathscr{S}} L\,dy\,dz + M\,dz\,dx + N\,dx\,dy = \iint_{\mathscr{S}} (L\cos\alpha + M\cos\beta + N\cos\gamma)\,d\sigma.$$

$$(10\text{--}134)$$

This can at once be written in terms of vectors:

$$\iint_{\mathscr{S}} L\,dy\,dz + M\,dz\,dx + N\,dx\,dy = \iint_{\mathscr{S}} (\mathbf{v}\cdot\mathbf{n})\,d\sigma,\qquad (10\text{--}135)$$

where $\mathbf{v} = L\mathbf{i} + M\mathbf{j} + N\mathbf{k}$.

Thus the surface integral $\iint_{\mathscr{S}} L\,dy\,dz + M\,dz\,dx + N\,dx\,dy$ is the integral over the surface of the *normal* component of the vector $\mathbf{v} = L\mathbf{i} + M\mathbf{j} + N\mathbf{k}$. As in Section 10–7, the surface integral can be interpreted as the *flux integral* of the vector field $\mathbf{v}$ across the surface $\mathscr{S}$.

We have emphasized surfaces consisting of one smooth piece, representable in parametric form or (as a special case) in a form such as $z = z(x, y)$. For surfaces that are pieced together, as in the prism surface of Fig. 10–35, the integrals over the pieces are simply added.

Equations (10–131) and (10–132) show how the evaluation of a surface integral can be reduced to a usual double integral. If this is followed through, the general useful rules are obtained:

I. *If $\mathscr{S}$ is given in the form $z = f(x, y)$, (x, y) in R_{xy}, with unit normal vector $\mathbf{n}$, then*

$$\iint_{\mathscr{S}} L\,dy\,dz + M\,dz\,dx + N\,dx\,dy = \pm \iint_{R_{xy}} \left(-L\frac{\partial z}{\partial x} - M\frac{\partial z}{\partial y} + N\right)dx\,dy,$$

$$(10\text{--}136)$$

with the $+$ sign when $\mathbf{n}$ is the upper normal and the $-$ sign when $\mathbf{n}$ is the lower normal.

II. *If $\mathscr{S}$ is given in parametric form $x = f(u, v)$, $y = g(u, v)$, $z = h(u, v)$, (u, v) in R_{uv}, with unit normal vector $\mathbf{n}$, then*

$$\iint_{\mathscr{S}} L\,dy\,dz + M\,dz\,dx + N\,dx\,dy$$

$$= \pm \iint_{R_{uv}} \left[L\frac{\partial(y, z)}{\partial(u, v)} + M\frac{\partial(z, x)}{\partial(u, v)} + N\frac{\partial(x, y)}{\partial(u, v)}\right]du\,dv,\qquad (10\text{--}137)$$

with the $+$ or $-$ sign corresponding to

$$\mathbf{n} = \pm \frac{\mathbf{P}_1 \times \mathbf{P}_2}{|\mathbf{P}_1 \times \mathbf{P}_2|},\qquad (10\text{--}138)$$

where $\mathbf{P}_1 = x_u\mathbf{i} + y_u\mathbf{j} + z_u\mathbf{k}$, $\mathbf{P}_2 = x_v\mathbf{i} + y_v\mathbf{j} + z_v\mathbf{k}$.

We shall verify the first rule, the procedure for the second rule being similar. Here $z = f(x, y)$ so that the unit normal vector is

$$\mathbf{n} = \pm\left(-\frac{\partial z}{\partial x}\mathbf{i} - \frac{\partial z}{\partial y}\mathbf{j} + \mathbf{k}\right)\sqrt{1 + \left(\frac{\partial z}{\partial x}\right)^2 + \left(\frac{\partial z}{\partial y}\right)^2}^{\,-1}$$

with the $+$ or $-$ sign depending on whether $\mathbf{n}$ is upper or lower normal. The denominator here is $\sec \gamma'$, where γ' is the angle between the *upper* normal and $\mathbf{k}$. Thus

$$\mathbf{n} = \pm\left(-\frac{\partial z}{\partial x}\mathbf{i} - \frac{\partial z}{\partial y}\mathbf{j} + \mathbf{k}\right)\frac{1}{\sec \gamma'}.$$

By (10–135),

$$\iint\limits_{\mathscr{S}} L\,dy\,dz + M\,dz\,dx + N\,dx\,dy = \iint\limits_{\mathscr{S}} [(L\mathbf{i} + M\mathbf{j} + N\mathbf{k})\cdot \mathbf{n}]\,d\sigma$$

$$= \pm\iint\limits_{\mathscr{S}}\frac{-L\dfrac{\partial z}{\partial x} - M\dfrac{\partial z}{\partial y} + N}{\sec \gamma'}\,d\sigma = \pm\iint\limits_{R_{xy}}\left(-L\frac{\partial z}{\partial x} - M\frac{\partial z}{\partial y} + N\right)dx\,dy,$$

since $d\sigma = \sec \gamma'\,dx\,dy$. Accordingly, (10–136) is proved.

EXAMPLE 1   Evaluate $\iint x\,dy\,dz + y\,dz\,dx + z\,dx\,dy$ over the surface $z = 1 - x^2 - y^2$ for $x^2 + y^2 \leqslant 1$, oriented by the upper normal. ◀

Solution   Here $L = x$, $M = y$, $N = z$. We apply (10–136) and obtain as value of the surface integral

$$\iint\limits_{R_{xy}} [(-x)(-2x) + (-y)(-2y) + z]\,dx\,dy.$$

On the surface, $z = 1 - x^2 - y^2$. Thus this can be reduced to

$$\iint\limits_{R_{xy}} (1 + x^2 + y^2)\,dx\,dy = \int_0^{2\pi}\int_0^1 (1 + r^2)r\,dr\,d\theta = \frac{3\pi}{2}.$$

The surface can be represented parametrically as follows:

$$x = v\cos u, \qquad y = v\sin u, \qquad z = 1 - v^2, \qquad 0 \leqslant u \leqslant 2\pi, \qquad 0 \leqslant v \leqslant 1.$$

Thus

$$\mathbf{P}_1 = -v\sin u\mathbf{i} + v\cos u\mathbf{j},$$
$$\mathbf{P}_2 = \cos u\mathbf{i} + \sin u\mathbf{j} - 2v\mathbf{k},$$
$$\mathbf{P}_1 \times \mathbf{P}_2 = -2v^2\cos u\mathbf{i} - 2v^2\sin u\mathbf{j} - v\mathbf{k},$$

so that $\mathbf{n}$ must be $-(\mathbf{P}_1 \times \mathbf{P}_2)/|\mathbf{P}_1 \times \mathbf{P}_2|$. The three components of $\mathbf{P}_1 \times \mathbf{P}_2$ are the three Jacobian determinants appearing in Eq. (10–137). Hence this equation gives the following expression for the surface integral:

$$-\iint\limits_{R_{uv}} [(v\cos u)(-2v^2\cos u) + (v\sin u)(-2v^2\sin u) + (1-v^2)(-v)]\,du\,dv$$

$$= -\iint\limits_{R_{uv}} (-2v^3 - v + v^3)\,du\,dv = \int_0^1 \int_0^{2\pi} (v + v^3)\,du\,dv = \frac{3\pi}{2}.$$

We observe that u, v are polar coordinates θ, r, as used in the evaluation above.

═══════════════════════════════**Problems (Section 10-13)**═══════════════════════════════

1. Evaluate the line integrals:

   a) $\displaystyle\int_{\substack{(1,0,0)\\C}}^{(1,0,2\pi)} z\,dx + x\,dy + y\,dz$, where C is the curve $x = \cos t$, $y = \sin t$, $z = t$, $0 \leqslant t \leqslant 2\pi$;

   b) $\displaystyle\int_{(1,0,1)}^{(2,3,2)} x^2\,dx - xz\,dy + y^2\,dz$ on the straight line joining the two points;

   c) $\displaystyle\int_{(1,1,0)}^{(0,0,\sqrt{2})} x^2 yz\,ds$ on the curve $x = \cos t$, $y = \cos t$, $z = \sqrt{2}\sin t$, $0 \leqslant t \leqslant \dfrac{\pi}{2}$;

   d) $\displaystyle\int_C u_T\,ds$, where $\mathbf{u} = 2xy^2 z\mathbf{i} + 2x^2 yz\mathbf{j} + x^2 y^2\mathbf{k}$ and C is the circle $x^2 + y^2 = 1$, $z = 2$.

2. If $\mathbf{u} = \operatorname{grad} F$ in a domain D, then show that

   a) $\displaystyle\int_{(x_1,y_1,z_1)}^{(x_2,y_2,z_2)} u_T\,ds = F(x_2, y_2, z_2) - F(x_1, y_1, z_1)$, where the integral is along any path in D joining the two points;

   b) $\displaystyle\int_C u_T\,ds = 0$ on any closed path in D.

3. Let a wire be given as a curve C in space. Let its density (mass per unit length) be $\delta = \delta(x, y, z)$, where (x, y, z) is a variable point in C. Justify the following formulas:

   a) Length of wire $= \displaystyle\int_C ds = L$;

   b) Mass of wire $= \displaystyle\int_C \delta\,ds = M$;

   c) Center of mass of the wire is $(\bar{x}, \bar{y}, \bar{z})$ where

   $$M\bar{x} = \int_C x\,\delta\,ds, \qquad M\bar{y} = \int_C y\,\delta\,ds, \qquad M\bar{z} = \int_C z\,\delta\,ds;$$

d) Moment of inertia of the wire about the z axis is

$$I_z = \int_C (x^2 + y^2)\delta \, ds.$$

4. Formulate and justify the formulas analogous to those of Problem 3 for the surface area, mass, center of mass, and moment of inertia of a thin curved sheet of metal forming a surface S in space.

5. Evaluate the following surface integrals:

a) $\iint_S x \, dy \, dz + y \, dz \, dx + z \, dx \, dy$, where S is the triangle with vertices $(1, 0, 0)$, $(0, 1, 0)$,

$(0, 0, 1)$ and the normal points away from $(0, 0, 0)$;

b) $\iint_S dy \, dz + dz \, dx + dx \, dy$, where S is the hemisphere $z = \sqrt{1 - x^2 - y^2}$, $x^2 + y^2 \leqslant 1$,

and the normal is the upper normal;

c) $\iint_S (x \cos \alpha + y \cos \beta + z \cos \gamma) \, d\sigma$ for the surface of part (b);

d) $\iint_S x^2 z \, d\sigma$, where S is the cylindrical surface $x^2 + y^2 = 1, 0 \leqslant z \leqslant 1$.

6. Evaluate the surface integrals of Problem 5 using the parametric representations:

a) $x = u + v$, $y = u - v$, $z = 1 - 2u$
b) $x = \sin u \cos v$, $y = \sin u \sin v$, $z = \cos u$
c) Same parameters as in (b)
d) $x = \cos u$, $y = \sin u$, $z = v$.

7. A steady fluid motion in space has velocity vector

$$\mathbf{v} = (x^2 + y^2)\mathbf{i} + (y^2 + z^2)\mathbf{j} + (1 - 2xz - 2yz)\mathbf{k}.$$

Evaluate the flux integral of $\mathbf{v}$ across each given surface in the direction of the given normal:

a) Circular region $x^2 + y^2 \leqslant 1$, $z = 0$, with normal $\mathbf{k}$;
b) Hemisphere $x^2 + y^2 + z^2 = 1$, $z \geqslant 0$, with normal $x\mathbf{i} + y\mathbf{j} + z\mathbf{k}$.

8. The flux integral for solar heating is $I = \iint_{\mathscr{S}} v_n \, d\sigma$, where $\mathbf{v}$ can be taken as independent of x, y, z. The integral gives the rate (calories per unit time) at which heat energy is provided to the collecting surface; the direction of $\mathbf{v}$ is that of the solar radiation. If $\mathbf{v}$ varies with time t, then so does I and $\int_a^b I(t) \, dt$ is the total heat provided for $a \leqslant t \leqslant b$. Let

$$\mathbf{v} = -c\left(\frac{3}{5}\cos\frac{\pi t}{12}\mathbf{i} + \frac{4}{5}\cos\frac{\pi t}{12}\mathbf{j} + \sin\frac{\pi t}{12}\mathbf{k}\right), \qquad 0 \leqslant t \leqslant 6, \quad c = \text{const} > 0,$$

and let the collector be a plane surface of area A. Compare the amount of heat received over the given interval if the collector is fixed and normal to $\mathbf{v}$ for $t = 3$, with the amount received if the collector turns so as to be normal to $\mathbf{v}$ for $0 \leqslant t \leqslant 6$.

9. a) Let a surface S with $z = f(x, y)$ be defined by an implicit equation $F(x, y, z) = 0$. Show that the surface integral $\iint H\, d\sigma$ over S becomes

$$\iint\limits_{R_{xy}} \sqrt{\left(\frac{\partial F}{\partial x}\right)^2 + \left(\frac{\partial F}{\partial y}\right)^2 + \left(\frac{\partial F}{\partial z}\right)^2}\; H \left|\frac{\partial F}{\partial z}\right|^{-1} dx\, dy,$$

provided $\partial F/\partial z \neq 0$.

b) Prove that, for the surface of part (a) with $\mathbf{n} = \nabla F/|\nabla F|$,

$$\iint\limits_{S} (\mathbf{v} \cdot \mathbf{n})\, d\sigma = \iint\limits_{R_{xy}} (\mathbf{v} \cdot \nabla F) \left|\frac{\partial F}{\partial z}\right|^{-1} dx\, dy.$$

10. a) Prove the vector identity $|\mathbf{u} \times \mathbf{v}|^2 = \begin{vmatrix} \mathbf{u} \cdot \mathbf{u} & \mathbf{u} \cdot \mathbf{v} \\ \mathbf{v} \cdot \mathbf{u} & \mathbf{v} \cdot \mathbf{v} \end{vmatrix}.$

b) Use the result of part (a) to deduce Eq. (10–129) from Eqs. (10–125) and (10–128).

11. a) Prove (10–137) under the conditions stated.

b) Derive (10–136) as a special case of (10–137).

10–14 GAUSS'S THEOREM. DIVERGENCE OF A VECTOR FIELD IN 3-DIMENSIONAL SPACE

For line integrals in the plane we proved Green's theorem, then defined the divergence of a vector field at a point and used Green's theorem to prove the rule:

$$\text{div } \mathbf{v} = \frac{\partial v_x}{\partial x} + \frac{\partial v_y}{\partial y}$$

for planar vector fields. We now parallel this procedure for surface integrals in space. The counterpart of Green's theorem is the following theorem:

GAUSS'S THEOREM   *Let $\mathbf{v} = L\mathbf{i} + M\mathbf{j} + N\mathbf{k}$ be a vector field in an open region D of space. Let L, M, N be continuous and have continuous first partial derivatives in D. Let $\mathcal{S}$ be a piecewise smooth surface in D that forms the complete boundary of a bounded closed region R in D. Let $\mathbf{n}$ be the unit outer normal vector of $\mathcal{S}$ with respect to R. Then*

$$\iint\limits_{\mathcal{S}} L\, dy\, dz + M\, dz\, dx + N\, dx\, dy = \iiint\limits_{R} \left(\frac{\partial L}{\partial x} + \frac{\partial M}{\partial y} + \frac{\partial N}{\partial z}\right) dx\, dy\, dz.$$

$$(10\text{–}140)$$

PROOF   It will be proved that

$$\iint\limits_{\mathcal{S}} N\, dx\, dy = \iiint\limits_{R} \frac{\partial N}{\partial z}\, dx\, dy\, dz, \qquad (10\text{–}141)$$

the proofs of the two equations

$$\iint\limits_{\mathscr{S}} L \, dy \, dz = \iint\limits_{R}\int \frac{\partial L}{\partial x} \, dx \, dy \, dz, \tag{10-142}$$

$$\iint\limits_{\mathscr{S}} M \, dz \, dx = \iint\limits_{R}\int \frac{\partial M}{\partial y} \, dx \, dy \, dz \tag{10-143}$$

being exactly the same.

It should first be remarked that the normal **n**, defined as the outer normal of $\mathscr{S}$ with respect to R, necessarily varies continuously on each smooth part of $\mathscr{S}$, so that the surface integral in (10–140) is well defined. The orientation defined by **n** on each part of $\mathscr{S}$ in fact determines an orientation of all of $\mathscr{S}$, so that $\mathscr{S}$ must be orientable.

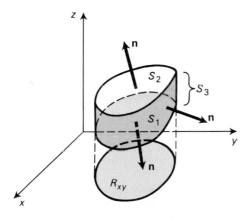

Fig. 10-38. Gauss's theorem.

We assume now that R is representable in the form:

$$f_1(x, y) \leqslant z \leqslant f_2(x, y), \qquad (x, y) \text{ in } R_{xy}, \tag{10-144}$$

where R_{xy} is a bounded closed region in the xy plane (cf. Fig. 10–38) bounded by a simple closed curve C. The surface $\mathscr{S}$ is then composed of three parts:

$$\mathscr{S}_1 : z = f_1(x, y), \quad (x, y) \text{ in } R_{xy}, \qquad \mathscr{S}_2 : z = f_2(x, y), \quad (x, y) \text{ in } R_{xy},$$
$$\mathscr{S}_3 : f_1(x, y) \leqslant z \leqslant f_2(x, y), \quad (x, y) \text{ on } C;$$

$\mathscr{S}_2$ forms the "top" of $\mathscr{S}$, $\mathscr{S}_1$ the "bottom," $\mathscr{S}_3$ the "sides" (the portion $\mathscr{S}_3$ may degenerate into a curve, as for a sphere). Now by definition (10–133) above

$$\iint\limits_{\mathscr{S}} N \, dx \, dy = \iint\limits_{\mathscr{S}} N \cos \gamma \, d\sigma,$$

where $\gamma = \measuredangle(\mathbf{n}, \mathbf{k})$. Along $\mathscr{S}_3$, $\gamma = \pi/2$, so that $\cos \gamma = 0$; along $\mathscr{S}_2$, $\gamma = \gamma'$, where γ'

is the angle between the *upper* normal and **k**; along $\mathscr{S}_1$, $\gamma = \pi - \gamma'$. Since $d\sigma = \sec\gamma'\, dx\, dy$ on $\mathscr{S}_1$ and $\mathscr{S}_2$, we have

$$\iint_{\mathscr{S}} N\, dx\, dy = \iint_{\mathscr{S}_1} N\, dx\, dy + \iint_{\mathscr{S}_2} N\, dx\, dy + \iint_{\mathscr{S}_3} N\, dx\, dy$$

$$= -\iint_{R_{xy}} N \cos\gamma' \sec\gamma'\, dx\, dy + \iint_{R_{xy}} N \cos\gamma' \sec\gamma'\, dx\, dy,$$

where $z = f_1(x, y)$ in the first integral and $z = f_2(x, y)$ in the second:

$$\iint_{\mathscr{S}} N\, dx\, dy = \iint_{R_{xy}} \{N[x, y, f_2(x, y)] - N[x, y, f_1(x, y)]\}\, dx\, dy. \quad (10\text{-}145)$$

Now the triple integral on the right of (10–141) can be evaluated as follows:

$$\iiint_{R} \frac{\partial N}{\partial z}\, dx\, dy\, dz = \iint_{R_{xy}} \left[\int_{f_1(x, y)}^{f_2(x, y)} \frac{\partial N}{\partial z}\, dz \right] dx\, dy$$

$$= \iint_{R_{xy}} \{N[x, y, f_2(x, y)] - N[x, y, f_1(x, y)]\}\, dx\, dy.$$

This is the same as (10–145); hence

$$\iint_{\mathscr{S}} N\, dx\, dy = \iiint_{R} \frac{\partial N}{\partial z}\, dx\, dy\, dz.$$

This is the result desired for the particular R assumed. For any region R that can be cut up into a finite number of pieces of this type by piecewise smooth auxiliary surfaces, the theorem can be proved by adding the result for each part separately (cf. the proof of Green's theorem in Section 10–6). The surface integrals over the auxiliary surfaces cancel in pairs and the sum of the surface integrals is precisely that over the complete boundary $\mathscr{S}$ of R; the volume integrals over the pieces add up to that over R. The result can finally be extended to the most general R envisaged in the theorem by a limit process (see O. D. KELLOGG, *Foundations of Potential Theory*, Berlin, Springer, 1929).

Once (10–141) has been established, (10–142) and (10–143) follow by merely relabeling the variables. By adding these three equations we obtain (10–140).

Definition of divergence

We can repeat the discussion of Section 10–7 for the three-dimensional case. The flux integral $\iint_{\mathscr{S}} v_n\, d\sigma$, where $\mathscr{S}$ is the complete boundary of a solid R, measures the net rate at which the fluid of velocity **v** is leaving R, in units of volume per unit of time.

This leads us to the definition of the divergence of **v** at a point:

$$\text{div } \mathbf{v} = \lim \frac{1}{V} \iint_{\mathscr{S}} v_n d\sigma, \qquad V = \text{Volume of } R, \qquad (10\text{–}146)$$

where, as usual, in the limit process R shrinks to a point (x_0, y_0, z_0).

If $\mathbf{v} = L\mathbf{i} + M\mathbf{j} + N\mathbf{k}$, then Gauss's theorem allows us to write

$$\frac{1}{V} \iint_{\mathscr{S}} v_n d\sigma = \frac{1}{V} \iint_{\mathscr{S}} L\, dy\, dz + M\, dz\, dx + N\, dx\, dy$$

$$= \frac{1}{V} \iiint_{R} \left(\frac{\partial L}{\partial x} + \frac{\partial M}{\partial y} + \frac{\partial N}{\partial z} \right) dx\, dy\, dz.$$

If we now pass to the limit and apply the analog of rule (10–37) for triple integrals, we conclude that at (x_0, y_0, z_0):

$$\text{div } \mathbf{v} = \frac{\partial L}{\partial x} + \frac{\partial M}{\partial y} + \frac{\partial N}{\partial z}.$$

Thus in general, for $\mathbf{v} = v_x \mathbf{i} + v_y \mathbf{j} + v_z \mathbf{k}$,

$$\text{div } \mathbf{v} = \frac{\partial v_x}{\partial x} + \frac{\partial v_y}{\partial y} + \frac{\partial v_z}{\partial z}, \qquad (10\text{–}147)$$

an obvious extension of (10–71) to three-dimensional space.

We can now restate Gauss's theorem in vector form: for $\mathscr{S}$, $\mathbf{n}$, and R as above we have

$$\iint_{\mathscr{S}} v_n d\sigma = \iiint_{R} \text{div } \mathbf{v}\, dV, \qquad (10\text{–}140')$$

for by Eq. (10–135) the left side of Eq. (10–140′) is the same as the left side of Eq. (10–140) and by Eq. (10–147) the right side of Eq. (10–140′) is the same as the right side of Eq. (10–140). Equation (10–140′) is called *the divergence theorem*.

As in two dimensions, the divergence measures the rate of expansion for fluid motions and $\text{div } \mathbf{v} \equiv 0$ for the velocity vector $\mathbf{v}$ of an incompressible flow. The divergence occurs in many other physical contexts. As in two dimensions, for gradient fields we have

$$\text{div grad } U = \frac{\partial}{\partial x}\left(\frac{\partial U}{\partial x}\right) + \frac{\partial}{\partial y}\left(\frac{\partial U}{\partial y}\right) + \frac{\partial}{\partial z}\left(\frac{\partial U}{\partial z}\right)$$

$$\qquad (10\text{–}148)$$

$$= \frac{\partial^2 U}{\partial x^2} + \frac{\partial^2 U}{\partial y^2} + \frac{\partial^2 U}{\partial z^2} = \nabla^2 U,$$

the Laplacian of U.

10–15 STOKES' THEOREM. CURL OF A VECTOR FIELD IN SPACE

For a planar field $\mathbf{v} = P(x, y)\mathbf{i} + Q(x, y)\mathbf{j}$ we defined the curl in Section 10–8 as the vector field

$$\lim \left(\frac{1}{A} \oint_C v_T \, ds \right) \mathbf{k},$$

where C is a simple closed path in the xy-plane, shrinking to a point (x, y) in the limit. This suggests that for an arbitrary vector field

$$\mathbf{v} = X(x, y, z)\mathbf{i} + Y(x, y, z)\mathbf{j} + Z(x, y, z)\mathbf{k}$$

in space we define the curl of $\mathbf{v}$ to be the vector field whose x, y and z components are obtained in similar fashion. Thus at $P_0(x_0, y_0, z_0)$ the z-component should be obtained as before as

$$\lim \frac{1}{A} \oint_C v_T \, ds,$$

where C is a simple closed path in the plane $z = z_0$, positively directed with respect to the normal $\mathbf{k}$, as in our definition of orientation in Section 10–12 (see Fig. 10–39(c)). For the x-component, C lies in the plane $x = x_0$ and is positively directed with respect to the normal $\mathbf{i}$ (Fig. 10–39(a)); for the y-component, C lies in the plane $y = y_0$ and is positively directed with respect to the normal $\mathbf{j}$ (Fig. 10–39(b)).

For the z-component, we can calculate as before:

$$\frac{1}{A} \oint_C v_T \, ds = \frac{1}{A} \oint_C X(x, y, z_0)dx + Y(x, y, z_0)dy$$

$$= \frac{1}{A} \iint_R \left(\frac{\partial Y}{\partial x} - \frac{\partial X}{\partial y} \right) dx \, dy.$$

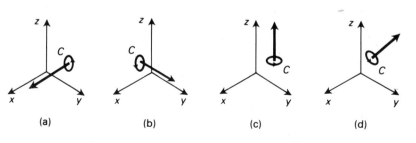

(a)   (b)   (c)   (d)

Fig. 10-39. Components of curl of $\mathbf{v}$.

Here we treat the plane $z = z_0$ as a plane with coordinates x, y and apply Green's theorem. A passage to the limit thus gives the z-component

$$\frac{\partial Y}{\partial x} - \frac{\partial X}{\partial y}$$

evaluated at (x_0, y_0, z_0). In the same way we obtain the x- and y-components

$$\frac{\partial Z}{\partial y} - \frac{\partial Y}{\partial z}, \qquad \frac{\partial X}{\partial z} - \frac{\partial Z}{\partial x}.$$

Thus in general we obtain the formula

$$\text{curl } \mathbf{v} = \left(\frac{\partial Z}{\partial y} - \frac{\partial Y}{\partial z}\right)\mathbf{i} + \left(\frac{\partial X}{\partial z} - \frac{\partial Z}{\partial x}\right)\mathbf{j} + \left(\frac{\partial Y}{\partial x} - \frac{\partial X}{\partial y}\right)\mathbf{k}. \qquad (10\text{–}150)$$

However, our definition has given special consideration to the directions $\mathbf{i}, \mathbf{j}, \mathbf{k}$. For it to have a physical meaning unrelated to choice of coordinate axes, curl $\mathbf{v}$ should have component

$$\lim \frac{1}{A} \oint_C v_T \, ds$$

in the direction of the normal $\mathbf{n}$, for a limit process at P_0 in a plane through P_0 with normal $\mathbf{n}$, as in Fig. 10–39(d). This is true, by our definition, if $\mathbf{n}$ is $\mathbf{i}, \mathbf{j}$, or $\mathbf{k}$. How do we know it is true for other choices of $\mathbf{n}$?

Our question can be answered affirmatively with the aid of Stokes' theorem, which we proceed to state:

STOKES THEOREM   *Let $\mathbf{v} = X\mathbf{i} + Y\mathbf{j} + Z\mathbf{k}$ be a vector field in the open region D of xyz-space. Let $\mathscr{S}$ be a smooth orientable surface in D bounded by a smooth simple closed curve C. Let the orientation on $\mathscr{S}$ be given by a normal $\mathbf{n}$ and let C be positively directed with relation to this normal. Let $X(x, y, z)$, $Y(x, y, z)$, $Z(x, y, z)$ have continuous first partial derivatives in D. Then*

$$\oint_C X \, dx + Y \, dy + Z \, dz = \iint_{\mathscr{S}} (\text{curl } \mathbf{v})_n \, d\sigma. \qquad (10\text{–}151).$$

(For a proof the reader is referred to AC, pp. 345–346.) The hypotheses are illustrated in Fig. 10-40. It should be remarked that the essential step in the proof is a reduction to Green's theorem with the aid of a suitable projection on a plane.

We can now show that our definition of curl has the desired property, for if we choose an arbitrary plane through P_0 with normal $\mathbf{n}$, then for each choice of C as above

$$\frac{1}{A} \oint_C v_T \, ds = \frac{1}{A} \oint_C X \, dx + Y \, dy + Z \, dz = \frac{1}{A} \iint_{\mathscr{S}} (\text{curl } \mathbf{v})_n \, d\sigma,$$

by Stokes' theorem. Here $\mathscr{S}$ is the planar surface enclosed by C and oriented by $\mathbf{n}$. If

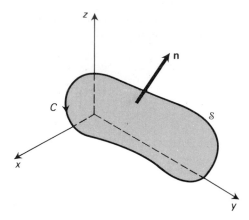

Fig. 10-40. Stokes' theorem.

we pass to the limit, as in (10–37), we obtain

$$\lim \frac{1}{A} \oint v_T \, ds = (\text{curl } \mathbf{v})_n \quad \text{at } P_0,$$

as claimed. Thus the curl is related to circulation around arbitrary axes in space, just as it is related to circulation around an axis with direction $\mathbf{k}$ for planar fields $P\mathbf{i} + Q\mathbf{j}$. For velocity fields the curl measures the extent to which the motion is a rotation about an axis.

In electromagnetic theory, in the absence of conductors, the curl and divergence both appear in *Maxwell's equations*:

$$\text{div } \mathbf{E} = 4\pi\rho, \qquad \text{div } \mathbf{H} = 0,$$

$$\text{curl } \mathbf{E} = -\frac{1}{c}\frac{\partial \mathbf{H}}{\partial t}, \qquad \text{curl } \mathbf{H} = \frac{1}{c}\frac{\partial \mathbf{E}}{\partial t}. \tag{10–152}$$

Here $\mathbf{E}$ is the electric-force vector, $\mathbf{H}$ is the magnetic-field strength; both in general vary with time t; ρ is the charge density, c is a universal constant. In magnetostatics there is the equation

$$\text{curl } \mathbf{H} = \frac{4\pi}{c}\mathbf{J}, \tag{10–153}$$

relating $\mathbf{H}$ to a current density vector $\mathbf{J}$. Integration of $\mathbf{J} \cdot \mathbf{n}$ across a surface $\mathscr{S}$ (like the flux integral previously considered for velocity fields) gives the total current I crossing $\mathscr{S}$. Thus from Stokes' theorem:

$$\oint_C H_T \, ds = \frac{4\pi}{c}\iint_{\mathscr{S}} \mathbf{J} \cdot \mathbf{n} \, d\sigma = \frac{4\pi}{c} I. \tag{10–154}$$

This is known as *Ampere's law*.

Remark on orientation of space The definition of curl is dependent on our way of relating the positive direction on C to the normal $\mathbf{n}$. This in turn is determined by the way xyz-space is oriented, that is, which ordered triples of vectors are considered positive, as in Section 9–2. Thus, if $\mathbf{T}$ is the unit tangent vector to C in the chosen direction and $\mathbf{N}$ is the outer normal to C relative to the surface $\mathscr{S}$ enclosed, then $\mathbf{N}$, $\mathbf{T}$, $\mathbf{n}$ must form a positive triple.

If we reverse the orientation of space, $\mathbf{n}$ would have to be replaced by the opposite vector $-\mathbf{n}$, while curl $\mathbf{v}$ would be replaced by its negative.

10–16 VECTOR IDENTITIES

We have already used the symbol ∇ (called *nabla*) for the gradient and its square ∇^2 for the Laplacian. These notations are suggested by an operator notation:

$$\nabla = \frac{\partial}{\partial x}\mathbf{i} + \frac{\partial}{\partial y}\mathbf{j} + \frac{\partial}{\partial z}\mathbf{k}. \tag{10-160}$$

From this, it is natural to write

$$\nabla^2 = \nabla \cdot \nabla = \frac{\partial^2}{\partial x^2} + \frac{\partial^2}{\partial y^2} + \frac{\partial^2}{\partial z^2}. \tag{10-161}$$

Also the divergence div $\mathbf{v}$ can be written as

$$\operatorname{div} \mathbf{v} = \frac{\partial}{\partial x}v_x + \frac{\partial}{\partial y}v_y + \frac{\partial}{\partial z}v_z = \nabla \cdot \mathbf{v}; \tag{10-162}$$

and the curl can be written as

$$\operatorname{curl} \mathbf{v} = \left(\frac{\partial}{\partial y}v_z - \frac{\partial}{\partial z}v_y\right)\mathbf{i} + \left(\frac{\partial}{\partial z}v_x - \frac{\partial}{\partial x}v_z\right)\mathbf{j} + \left(\frac{\partial}{\partial x}v_y - \frac{\partial}{\partial y}v_x\right)\mathbf{k}$$

$$= \begin{vmatrix} \mathbf{i} & \mathbf{j} & \mathbf{k} \\ \dfrac{\partial}{\partial x} & \dfrac{\partial}{\partial y} & \dfrac{\partial}{\partial z} \\ v_x & v_y & v_z \end{vmatrix} = \nabla \times \mathbf{v}. \tag{10-163}$$

These notations are very convenient and they are suggestive of various identities to follow. However, we must not forget that ∇ is a differential operator and, when applied to a product, must somehow be applied to each factor while the other one is untouched.

We now consider identities satisfied by these operations; proofs are left to the problems. First, all the operators are *linear*, that is, when applied to a sum, the result is the sum of the results obtained from each term separately and constants can be factored out. For example,

$$\operatorname{grad}(f + g) = \nabla(f + g) = \nabla f + \nabla g,$$
$$\operatorname{grad}(cf) = \nabla(cf) = c\nabla f.$$

There are similar rules for div, curl, and ∇^2.

In the rules to follow we assume continuity of all partial derivatives that arise.
Gradient, divergence, and curl of products. We have the rules:

$$\text{grad}(fg) = f \,\text{grad}\, g + g \,\text{grad}\, f,$$

$$\text{div}(f\mathbf{u}) = f \,\text{div}\, \mathbf{u} + \text{grad}\, f \cdot \mathbf{u}, \tag{10–164}$$

$$\text{curl}(f\mathbf{u}) = f \,\text{curl}\, \mathbf{u} + \text{grad}\, f \times \mathbf{u}.$$

Curl of a gradient.   As in two dimensions, we have

$$\text{curl grad}\, f = \nabla \times (\nabla f) = \mathbf{0}. \tag{10–165}$$

This is suggested by the vector rule $\mathbf{v} \times \mathbf{v} = \mathbf{0}$. There is an important converse: if curl
$\mathbf{v} = \mathbf{0}$, then $\mathbf{v} = \text{grad}\, f$ for some f (see Section 10–17). A vector field $\mathbf{v}$ such that curl
$\mathbf{v} \equiv \mathbf{0}$ is said to be *irrotational*.

Divergence of a curl.   Here we find:

$$\text{div curl}\, \mathbf{v} = \nabla \cdot (\nabla \times \mathbf{v}) = 0. \tag{10–166}$$

This is suggested by the vector rule: $\mathbf{u} \cdot \mathbf{u} \times \mathbf{v} = 0$ (Section 9–2). Again there is a
converse: if $\text{div}\, \mathbf{u} \equiv 0$, then $\mathbf{u} = \text{curl}\, \mathbf{v}$ for some $\mathbf{v}$ (see Section 10–17). A vector field $\mathbf{u}$
such that $\text{div}\, \mathbf{u} \equiv 0$ is called *solenoidal*.

Divergence of a vector product.   We have

$$\text{div}(\mathbf{u} \times \mathbf{v}) = \nabla \cdot (\mathbf{u} \times \mathbf{v}) = \mathbf{v} \cdot \text{curl}\, \mathbf{u} - \mathbf{u} \cdot \text{curl}\, \mathbf{v}. \tag{10–167}$$

Divergence of a gradient.   As in Section 10–14,

$$\text{div grad}\, f = \nabla \cdot (\nabla f) = \nabla^2 f.$$

A function f for which $\nabla^2 f \equiv 0$ in the open region D is called *harmonic* in D.
Curl of a curl.   Here an expansion into components yields the relation:

$$\text{curl curl}\, \mathbf{u} = \text{grad div}\, \mathbf{u} - (\nabla^2 u_x \mathbf{i} + \nabla^2 u_y \mathbf{j} + \nabla^2 u_z \mathbf{k}). \tag{10–168}$$

If the Laplacian of a vector $\mathbf{u}$ is defined to be the vector

$$\nabla^2 \mathbf{u} = \nabla^2 u_x \mathbf{i} + \nabla^2 u_y \mathbf{j} + \nabla^2 u_z \mathbf{k}, \tag{10–169}$$

then (10–168) becomes

$$\text{curl curl}\, \mathbf{u} = \text{grad div}\, \mathbf{u} - \nabla^2 \mathbf{u}. \tag{10–168'}$$

This identity can be written as an expression for the *gradient of a divergence:*

$$\text{grad div}\, \mathbf{u} = \text{curl curl}\, \mathbf{u} + \nabla^2 \mathbf{u}.$$

These identities cover the principal ones of interest in applications except those
for *gradient of a scalar product* and *curl of a vector product* (see Problems 11 and 12
below).

═══════════════════════**Problems (Section 10-16)**═══════════════

1. Evaluate by the divergence theorem:

   a) $\displaystyle\iint_{\mathscr{S}} x \, dy \, dz + y \, dz \, dx + z \, dx \, dy$, where $\mathscr{S}$ is the sphere $x^2 + y^2 + z^2 = 1$ and $\mathbf{n}$ is the
   outer normal;

b) $\iint\limits_{\mathscr{S}} v_n \, d\sigma$, where $\mathbf{v} = x^2\mathbf{i} + y^2\mathbf{j} + z^2\mathbf{k}$, $\mathbf{n}$ is the outer normal, and $\mathscr{S}$ is the surface of

the cube $0 \leqslant x \leqslant 1$, $0 \leqslant y \leqslant 1$, $0 \leqslant z \leqslant 1$.

2. Let $\mathscr{S}$ be the boundary surface of a region R in space and let $\mathbf{n}$ be its outer normal. Prove the formulas:

a) $V = \iint\limits_{\mathscr{S}} x \, dy \, dz = \iint\limits_{\mathscr{S}} y \, dz \, dx = \iint\limits_{\mathscr{S}} z \, dx \, dy = \dfrac{1}{3} \iint\limits_{\mathscr{S}} x \, dy \, dz + y \, dz \, dx + z \, dx \, dy,$

where V is the volume of R;

b) $\iint\limits_{\mathscr{S}} x^2 \, dy \, dz + 2xy \, dz \, dx + 2xz \, dx \, dy = 6V\bar{x}$, where $(\bar{x}, \bar{y}, \bar{z})$ is the centroid of R;

c) $\iint\limits_{\mathscr{S}} \operatorname{curl} \mathbf{v} \cdot \mathbf{n} \, d\sigma = 0$, where $\mathbf{v}$ is an arbitrary vector field.

3. Evaluate by Stokes' theorem:

a) $\displaystyle\int\limits_{C} u_T \, ds$, where C is the circle $x^2 + y^2 = 1$, $z = 2$, directed so that y increases for

positive x, and $\mathbf{u}$ is the vector $-3y\mathbf{i} + 3x\mathbf{j} + \mathbf{k}$;

b) $\displaystyle\int\limits_{C} 2xy^2z \, dx + 2x^2 yz \, dy + (x^2 y^2 - 2z) dz$ around   the   curve $x = \cos t$, $y = \sin t$,

$z = \sin t$, $0 \leqslant t \leqslant 2\pi$, directed with increasing t.

4. Let C be a simple closed *plane* curve in space. Let $\mathbf{n} = a\mathbf{i} + b\mathbf{j} + c\mathbf{k}$ be a unit vector normal to the plane of C and let the direction on C match that of $\mathbf{n}$. Prove that

$$\frac{1}{2} \int\limits_{C} (bz - cy) \, dx + (cx - az) \, dy + (ay - bx) \, dz$$

equals the plane area enclosed by C. What does the integral reduce to when C is in the xy plane?

5. Under the hypotheses of Problem 2, let $\mathbf{r} = x\mathbf{i} + y\mathbf{j} + z\mathbf{k}$, $r = |\mathbf{r}|$, $\mathbf{u} = r^{2p}\mathbf{r}$. Show that for $p \geqslant 1$, $\iint\limits_{\mathscr{S}} u_n \, d\sigma = (3 + 2p) \iiint\limits_{R} r^{2p} \, dV$.

6. Show that the result of Problem 5 is valid for $p < 1$ if $(0, 0, 0)$ is outside R and conclude that for $p = -3/2$ the surface integral is 0.

7. Prove that grad, div, and curl are linear operators.

8. Prove the rules:
   a) (10–164)   b) (10–165)   c) (10–166)   d) (10–167)   e) (10–168).

9. Prove the following identities (see HINT on next page):
   a) $\operatorname{div}[\mathbf{u} \times (\mathbf{v} \times \mathbf{w})] = (\mathbf{u} \cdot \mathbf{w}) \operatorname{div} \mathbf{v} - (\mathbf{u} \cdot \mathbf{v}) \operatorname{div} \mathbf{w} + \operatorname{grad}(\mathbf{u} \cdot \mathbf{w}) \cdot \mathbf{v} - \operatorname{grad}(\mathbf{u} \cdot \mathbf{v}) \cdot \mathbf{w}$
   b) $\operatorname{div}(\operatorname{grad} f \times f \operatorname{grad} g) = 0$
   c) $\operatorname{curl}(\operatorname{curl} \mathbf{v} + \operatorname{grad} f) = \operatorname{curl} \operatorname{curl} \mathbf{v}$ d) $\nabla^2 f = \operatorname{div}(\operatorname{curl} \mathbf{v} + \operatorname{grad} f)$

HINT: These should be established by means of the identities already found in this chapter and not by expanding into components.

10. The scalar product $\mathbf{u} \cdot \nabla$, with $\mathbf{u}$ on the *left* of the operator ∇, is defined as the operator

$$\mathbf{u} \cdot \nabla = u_x \frac{\partial}{\partial x} + u_y \frac{\partial}{\partial y} + u_z \frac{\partial}{\partial z}.$$

This is quite unrelated to $\nabla \cdot \mathbf{u} = \text{div } \mathbf{u}$. The operator $\mathbf{u} \cdot \nabla$ can be applied to a scalar f:

$$(\mathbf{u} \cdot \nabla) f = u_x \frac{\partial f}{\partial x} + u_y \frac{\partial f}{\partial y} + u_z \frac{\partial f}{\partial z} = \mathbf{u} \cdot (\nabla f).$$

Thus an associative law holds. The operator $\mathbf{u} \cdot \nabla$ can also be applied to a vector $\mathbf{v}$:

$$(\mathbf{u} \cdot \nabla) \mathbf{v} = u_x \frac{\partial \mathbf{v}}{\partial x} + u_y \frac{\partial \mathbf{v}}{\partial y} + u_z \frac{\partial \mathbf{v}}{\partial z},$$

where the partial derivatives $\partial \mathbf{v}/\partial x, \ldots$ are defined just as is $d\mathbf{v}/dt$ in Section 9-3; hence we have

$$\frac{\partial \mathbf{v}}{\partial x} = \frac{\partial v_x}{\partial x} \mathbf{i} + \frac{\partial v_y}{\partial x} \mathbf{j} + \frac{\partial v_z}{\partial x} \mathbf{k}.$$

a) Show that, if $\mathbf{u}$ is a unit vector, then $(\mathbf{u} \cdot \nabla) f = \nabla_u f$;
b) Evaluate $[(\mathbf{i} - \mathbf{j}) \cdot \nabla] f$;
c) Evaluate $[(x\mathbf{i} - y\mathbf{j}) \cdot \nabla] (x^2 \mathbf{i} - y^2 \mathbf{j} + z^2 \mathbf{k})$.

11. Prove the identity (cf. Problem 10):

$$\text{grad } (\mathbf{u} \cdot \mathbf{v}) = (\mathbf{u} \cdot \nabla) \mathbf{v} + (\mathbf{v} \cdot \nabla) \mathbf{u} + (\mathbf{u} \times \text{curl } \mathbf{v}) + (\mathbf{v} \times \text{curl } \mathbf{u}).$$

12. Prove the identity (cf. Problem 10):

$$\text{curl } (\mathbf{u} \times \mathbf{v}) = \mathbf{u} \text{ div } \mathbf{v} - \mathbf{v} \text{ div } \mathbf{u} + (\mathbf{v} \cdot \nabla) \mathbf{u} - (\mathbf{u} \cdot \nabla) \mathbf{v}.$$

13. Let $\mathbf{n}$ be the unit outer normal vector to the sphere $x^2 + y^2 + z^2 = 9$ and let $\mathbf{u}$ be the vector $(x^2 - z^2)(\mathbf{i} - \mathbf{j} + 3\mathbf{k})$. Evaluate $\partial/\partial n$ (div $\mathbf{u}$) at $(2, 2, 1)$.

14. A rigid body is rotating about the z-axis with angular velocity ω. Show that a typical particle of the body follows a path

$$\overrightarrow{OP} = r \cos (\omega t + \alpha)\mathbf{i} + r \sin (\omega t + \alpha)\mathbf{j} + z\mathbf{k},$$

where α, r, and z are constant, and that at each instant the velocity is

$$\mathbf{v} = \omega \times \overrightarrow{OP},$$

where $\omega = \omega \mathbf{k}$ (the *angular velocity vector* of the motion). Evaluate div $\mathbf{v}$ and curl $\mathbf{v}$.

15. A steady fluid motion has velocity $\mathbf{u} = y\mathbf{i}$. Show that all points that move do so on straight lines and that the flow is incompressible. Determine the volume occupied at time $t = 1$ by the points which at time $t = 0$ fill the cube bounded by the coordinate planes and the planes $x = 1$, $y = 1$, $z = 1$.

16. A steady fluid motion has velocity $\mathbf{u} = x\mathbf{i}$. Show that all points either do not move or else move on straight lines. Determine the volume occupied at time $t = 1$ by the points which at time $t = 0$ fill the cube of Problem 15.

HINT: Show that the paths of the individual points are given by $x = c_1 e^t$, $y = c_2$, $z = c_3$, where c_1, c_2, c_3 are constants. Is the flow incompressible?

17. Let $\mathcal{S}$ be the boundary surface of a region R, with outer normal $\mathbf{n}$, as in the divergence theorem above. Let $f(x, y, z)$ and $g(x, y, z)$ be functions defined and continuous, with

continuous first and second derivatives, in a domain D containing R. Prove the following relations:

a) $\displaystyle\iint_{\mathscr{S}} f\frac{\partial g}{\partial n}\,d\sigma = \iiint_R f\nabla^2 g\,dx\,dy\,dz + \iiint_R (\nabla f\cdot\nabla g)\,dx\,dy\,dz;$

HINT: Use the identity $\nabla\cdot(f\mathbf{u}) = \nabla f\cdot\mathbf{u} + f(\nabla\cdot\mathbf{u})$.

b) If g is harmonic in D, then $\displaystyle\iint_{\mathscr{S}}\frac{\partial g}{\partial n}\,d\sigma = 0.$

HINT: Put $f = 1$ in (a).

c) If f is harmonic in D, then

$$\iint_{\mathscr{S}} f\frac{\partial f}{\partial n}\,d\sigma = \iiint_R |\nabla f|^2\,dx\,dy\,dz.$$

d) If f is harmonic in D and $f\equiv 0$ on $\mathscr{S}$, then $f\equiv 0$ in R.

e) If f and g are harmonic in D and $f\equiv g$ on $\mathscr{S}$, then $f\equiv g$ in R.

HINT: Use (d).

f) If f is harmonic in D and $\partial f/\partial n = 0$ on $\mathscr{S}$, then f is constant in R.

g) If f and g are harmonic in D and $\partial f/\partial n = \partial g/\partial n$ on $\mathscr{S}$, then $f = g + \text{const}$ in R.

h) If f and g are harmonic in R, and

$$\frac{\partial f}{\partial n} = -f + h, \quad \frac{\partial g}{\partial n} = -g + h \text{ on } \mathscr{S}, \quad h = h(x,\,y,\,z),$$

then

$$f\equiv g \text{ in } R.$$

i) If f and g both satisfy the same *Poisson equation* in R,

$$\nabla^2 f = -4\pi h, \quad \nabla^2 g = -4\pi h, \quad h = h(x,\,y,\,z),$$

and $f = g$ on $\mathscr{S}$, then $f\equiv g$ in R.

j) $\displaystyle\iint_{\mathscr{S}}\left(f\frac{\partial g}{\partial n} - g\frac{\partial f}{\partial n}\right)d\sigma = \iiint_R (f\nabla^2 g - g\nabla^2 f)\,dx\,dy\,dz.$

HINT: Use (a).

k) If f and g are harmonic in R, then $\displaystyle\iint_{\mathscr{S}}\left(f\frac{\partial g}{\partial n} - g\frac{\partial f}{\partial n}\right)d\sigma = 0.$

l) If f and g satisfy the equations $\nabla^2 f = hf$, $\nabla^2 g = hg$, $h = h(x,\,y,\,z)$, in R, then

$$\iint_{\mathscr{S}}\left(f\frac{\partial g}{\partial n} - g\frac{\partial f}{\partial n}\right)d\sigma = 0.$$

REMARK: Parts (a) and (j) are known as *Green's first and second identitites*, respectively.

10–17 INTEGRALS INDEPENDENT OF PATH. IRROTATIONAL FIELDS AND SOLENOIDAL FIELDS

Since the generalization of Green's theorem to space takes two different forms (Gauss's theorem and Stokes' theorem), the discussion of Section 10-9 can be generalized in two ways: one for line integrals, one for surface integrals. In the case of line integrals, the results for two dimensions carry over with minor modifications. The surface integrals require a somewhat different treatment.

For the discussion of line integrals we let $\mathbf{u} = X\mathbf{i} + Y\mathbf{j} + Z\mathbf{k}$, where $X(x, y, z)$, $Y(x, y, z)$, and $Z(x, y, z)$ are continuous and have continuous first partial derivatives in an open region D in space. We then consider

$$\int_C u_T \, ds = \int_C X \, dx + Y \, dy + Z \, dz$$

on various paths in D.

Independence of path is defined as in Section 10-9 and the same proof shows that *the integral $\int u_T \, ds$ is independent of path in D precisely when* $\mathbf{u}$ *is a gradient field,* that is,

$$\mathbf{u} = \operatorname{grad} F \tag{10-170}$$

for some $F(x, y, z)$ *in* D. The function F (or $-F$) is then a *potential* for $\mathbf{u}$; F is determined only up to an additive constant. When (10–170) holds, then, as in Section 10–9,

$$\int_A^B u_T \, ds = \int_A^B \operatorname{grad} F \cdot \mathbf{T} \, ds = \int_A^B dF = F(B) - F(A). \tag{10-171}$$

As in Section 10–9, it can also be shown that *the integral $\int u_T \, ds$ is independent of path in D precisely when* $\oint_C u_T \, ds = 0$ *on every simple closed curve in* D.

If the integral $\int u_T \, ds$ is independent of path in D, then necessarily

$$\operatorname{curl} \mathbf{u} \equiv \mathbf{0} \quad \text{in} \quad D, \tag{10-172}$$

that is, $\mathbf{u}$ is irrotational. This follows at once from (10–170), since as in Section 10–16, always curl grad $F \equiv \mathbf{0}$.

Condition (10–172) is the extension to space of the condition $\partial P/\partial y = \partial Q/\partial x$ of Section 10-9. As in that section, this is the crucial test. However, as in that case, having curl $\mathbf{u} \equiv \mathbf{0}$ in D does not ensure that $\mathbf{u} = \operatorname{grad} F$ in all of D, but only in each *simply connected* part of D. For space we define an open region to be simply connected if each (piecewise smooth) simple closed path C in the region is the boundary of a surface $\mathscr{S}$ to which Stokes' theorem applies. If curl $\mathbf{u} \equiv \mathbf{0}$ in such a region, then

$$\oint_C u_T \, ds = \int\int_{\mathscr{S}} \operatorname{curl} \mathbf{u} \cdot \mathbf{n} \, d\sigma = 0,$$

for proper choice of the unit normal $\mathbf{n}$ on $\mathscr{S}$. Therefore $\int u_T \, ds = 0$ on all simple closed paths and hence, as above, the integral is independent of path and $\mathbf{u} = \operatorname{grad} F$ for some F.

Examples of simply connected regions in space are the following: the interior of a sphere, the interior of a cube, the region bounded by two concentric spheres. The interior of a solid torus is not simply connected.

For a simply connected open region D we can summarize the results by stating that the following conditions are all equivalent:

$$\int u_T \, ds \text{ is independent of path in } D,$$

$$\oint u_T \, ds = 0 \text{ on simple closed paths,}$$

$$\mathbf{u} = \operatorname{grad} F, \qquad \operatorname{curl} \mathbf{u} \equiv \mathbf{0} \text{ in } D.$$

A theory similar to the preceding one can be developed for surface integrals. However, the main interest is in the counterpart of the statement: every irrotational field (in a simply connected region) is a gradient field. The counterpart is the assertion: *every solenoidal field is the curl of another field*; that is, *if* div $\mathbf{u} \equiv 0$ *in D, then* $\mathbf{u} = \operatorname{curl} \mathbf{v}$ *in D for some* $\mathbf{v}$. As in the previous case, some restriction must be imposed on D. We shall assume that D is convex, that is, if points A, B are in D, then the line segment AB is in D. The interior of a sphere, ellipsoid, or cube is convex, as is all of xyz-space. We also assume that the coordinates are chosen so that the origin O is in D. Then the vector field $\mathbf{v}$ whose curl is $\mathbf{u}$ can be chosen as follows:

$$\mathbf{v}(x, y, z) = \int_0^1 t\mathbf{u}(xt, yt, zt) \times (x\mathbf{i} + y\mathbf{j} + z\mathbf{k}) \, dt. \qquad (10\text{-}173)$$

If the vector field $\mathbf{u}$ is *homogeneous of degree n*, that is, $\mathbf{u}(xt, yt, zt) = t^n\mathbf{u}(x, y, z)$ (see Problem 9 following Section 9–5), the formula can be simplified further:

$$\mathbf{v} = \int_0^1 t^{n+1}\mathbf{u}(x, y, z) \times (x\mathbf{i} + y\mathbf{j} + z\mathbf{k}) \, dt = \frac{1}{n+2}(\mathbf{u} \times \mathbf{r}),$$
$$\qquad (10\text{-}174)$$
$$\mathbf{r} = x\mathbf{i} + y\mathbf{j} + z\mathbf{k}.$$

The vector field $\mathbf{v}$ is not unique, since we can add to a particular $\mathbf{v}$ whose curl is $\mathbf{u}$ any vector field whose curl is $\mathbf{0}$, that is, we can add any gradient field, grad f (see Problem 3 below).

For a proof that (10–173) does solve the equation curl $\mathbf{v} = \mathbf{u}$, see AC, pp. 349–351.

EXAMPLE 1 Let $\mathbf{u} = y^2\mathbf{i} + z^2\mathbf{j} + x^2\mathbf{k}$. Then div $\mathbf{u} = 0$, so that $\mathbf{u} = \text{curl } \mathbf{v}$ for some $\mathbf{v}$. (Here D is all of xyz-space.) By (10–173), we can take

$$\mathbf{v} = \int_0^1 t(t^2 y^2\mathbf{i} + t^2 z^2\mathbf{j} + t^2 x^2\mathbf{k}) \times (x\mathbf{i} + y\mathbf{j} + z\mathbf{k})\, dt$$

$$= \int_0^1 t^3 [(z^3 - x^2 y)\mathbf{i} + (x^3 - y^2 z)\mathbf{j} + (y^3 - xz^2)\mathbf{k}]\, dt$$

$$= [(z^3 - x^2 y)\mathbf{i} + (x^3 - y^2 z)\mathbf{j} + (y^3 - xz^2)\mathbf{k}] \int_0^1 t^3\, dt$$

$$= \frac{1}{4}[(z^3 - x^2 y)\mathbf{i} + (x^3 - y^2 z)\mathbf{j} + (y^3 - xz^2)\mathbf{k}]. \quad \blacktriangleleft$$

Here $\mathbf{u}$ is homogeneous of degree 2, so that we could have used (10–174) with $n = 2$.

═══════════════════════════**Problems (Section 10-17)**═══════════════════════════

1. By showing that the integrand is an exact differential, evaluate

   a) $\displaystyle\int_{(1, 1, 2)}^{(3, 5, 0)} yz\, dx + xz\, dy + xy\, dz$ on any path;

   b) $\displaystyle\int_{(1, 0, 0)}^{(1, 0, 2\pi)} \sin yz\, dx + xz\cos yz\, dy + xy\cos yz\, dz$ on the helix $x = \cos t$, $y = \sin t$, $z = t$.

2. Let $\mathbf{u} = \dfrac{-y}{x^2 + y^2}\mathbf{i} + \dfrac{x}{x^2 + y^2}\mathbf{j} + z\mathbf{k}$ and let D be the interior of the torus obtained by rotating the circle $(x - 2)^2 + z^2 = 1$, $y = 0$ about the z axis. Show that curl $\mathbf{u} = \mathbf{0}$ in D but $\int_C u_T\, ds \neq 0$ when C is the circle $x^2 + y^2 = 4$, $z = 0$. Determine the possible values of the integral $\int_{(2, 0, 0)}^{(0, 2, 0)} u_T\, ds$ on a path in D.

3. a) Show that, if $\mathbf{v}$ is one solution of the equation curl $\mathbf{v} = \mathbf{u}$ for given $\mathbf{u}$ in a simply connected domain D, then all solutions are given by $\mathbf{v} + \text{grad } f$, where f is an arbitrary differentiable scalar in D.

   b) Find all vectors $\mathbf{v}$ such that curl $\mathbf{v} = \mathbf{u}$ if

   $$\mathbf{u} = (2xyz^2 + xy^3)\mathbf{i} + (x^2 y^2 - y^2 z^2)\mathbf{j} - (y^3 z + 2x^2 yz)\mathbf{k}.$$

4. Show that, if f and g are scalars with continuous second partial derivatives in a domain D, then $\mathbf{u} = \nabla f \times \nabla g$ is solenoidal in D. (It can be shown that every solenoidal vector has such a representation, at least in a suitably restricted domain.)

5. Show that, if $\iint_{\mathscr{S}} u_n\, d\sigma = 0$ for every oriented spherical surface S in a domain D and the components of $\mathbf{u}$ have continuous derivatives in D, then $\mathbf{u}$ is solenoidal in D. Does the converse hold?

═══

10–18 PHYSICAL APPLICATIONS: FLUIDS, HEAT CONDUCTION, THE VIBRATING MEMBRANE

In this and the following section we illustrate how line and surface integrals are used in several branches of physics. In particular, it will be seen how the Laplace equation, wave equation, and heat equation arise; the methods of solving these equations are studied in Chapter 8.

Fluid motion

We have seen the significance of the flux integral

$$\iint_{\mathscr{S}} u_n \, d\sigma$$

for a fluid motion of velocity **u** in space. One can also let $\mathbf{v} = \rho\mathbf{u}$, where $\rho = \rho(x, y, z, t)$ is the fluid density (mass per unit volume.). Then the corresponding flux integral

$$\iint_{\mathscr{S}} v_n \, d\sigma = \iint_{\mathscr{S}} \rho u_n \, d\sigma \tag{10–180}$$

measures the net rate at which *mass* is crossing $\mathscr{S}$, in the direction of the normal **n**. If we apply this to a solid region R bounded by $\mathscr{S}$, with **n** the outer normal (Fig. 10–41), then the integral (10–180) gives the net rate at which mass is leaving R; hence it is also equal to minus the rate of change of the total mass of R. Therefore,

$$\iint_{\mathscr{S}} v_n \, d\sigma = -\frac{\partial}{\partial t} \iiint_R \rho \, dV = -\iiint_R \frac{\partial \rho}{\partial t} \, dV.$$

For the last equality we used Leibnitz's rule. By the divergence theorem, this can be written as

$$\iiint_R \operatorname{div} \mathbf{v} \, dV = -\iiint_R \frac{\partial \rho}{\partial t} \, dV$$

or

$$\iiint_R \left(\operatorname{div} \mathbf{v} + \frac{\partial \rho}{\partial t} \right) dV = 0.$$

This equality must hold for *every* choice of R in the region of fluid motion, so that the integrand must be identically 0, as follows from the counterpart of Eq. (10–37) for triple integrals. Therefore,

$$\operatorname{div} \mathbf{v} + \frac{\partial \rho}{\partial t} = 0 \quad \text{or} \quad \operatorname{div}(\rho\mathbf{u}) + \frac{\partial \rho}{\partial t} = 0. \tag{10–181}$$

This is the *continuity equation* of fluid mechanics. By the identities (10–164) this can also be written as

$$\frac{\partial \rho}{\partial t} + \operatorname{grad} \rho \cdot \mathbf{u} + \rho \operatorname{div} \mathbf{u} = 0. \tag{10–181'}$$

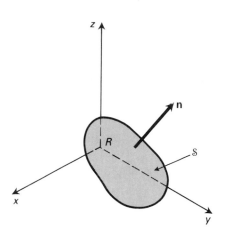

Fig. 10–41. Derivation of continuity equation.

Here the first two terms are equal to

$$\frac{\partial \rho}{\partial t} + \frac{\partial \rho}{\partial x}\frac{dx}{dt} + \frac{\partial \rho}{\partial y}\frac{dy}{dt} + \frac{\partial \rho}{\partial z}\frac{dz}{dt}, \tag{10-182}$$

where $x = x(t)$, $y = y(t)$, $z = z(t)$ is the path of a fluid particle passing through the given point at the given time t. Quantity (10–182) can be interpreted as the rate of change of ρ as we move with the particle; it is called the *Stokes' total derivative* of ρ and denoted by $D\rho/Dt$. Thus the continuity equation can be written as

$$\frac{D\rho}{Dt} + \rho \operatorname{div} \mathbf{u} = 0.$$

For an incompressible flow, $D\rho/Dt = 0$, and hence div $\mathbf{u} = 0$, as previously deduced. Thus for an incompressible flow the velocity field $\mathbf{u}$ is solenoidal.

If also curl $\mathbf{u} = \mathbf{0}$ in D, so that $\mathbf{u}$ is irrotational and D is simply connected, then, as in Section 10–17, $\mathbf{u} = \operatorname{grad} \varphi$ for some $\varphi(x, y, z, t)$. Thus.

$$\operatorname{div} \operatorname{grad} \varphi = 0 \quad \text{or} \quad \nabla^2 \varphi = 0$$

and φ is *harmonic* (for each fixed t).

Heat conduction

Let $T(x, y, z, t)$ be the temperature at the point (x, y, z) of a body at time t. If heat is being conducted in the body, the flow of heat can be represented by a vector $\mathbf{u}$ such that the flux integral $\iint_{\mathscr{S}} u_n \, d\sigma$ for each oriented surface $\mathscr{S}$ represents the number of calories crossing $\mathscr{S}$ in the direction of the given normal per unit of time. The simplest law of thermal conduction postulates that

$$\mathbf{u} = -k \operatorname{grad} \mathbf{T}, \tag{10-183}$$

with $k > 0$; k is usually treated as a constant. Equation (10–183) implies that heat flows in the direction of decreasing temperature and the rate of flow is proportional to the temperature gradient $|\operatorname{grad} T|$.

If $\mathscr{S}$ is a closed surface, forming the boundary of a region R in the body, then

$$\iint_{\mathscr{S}} u_n \, d\sigma = \iiint_R \operatorname{div} \mathbf{u} \, dx \, dy \, dz,$$

by the divergence theorem. Hence the total amount of heat *entering* R is

$$-\iint_{\mathscr{S}} u_n \, d\sigma = \iiint_R k \operatorname{div} \operatorname{grad} T \, dx \, dy \, dz.$$

On the other hand, the rate at which heat is being absorbed per unit mass can also be measured by $c(\partial T/\partial t)$, where c is the specific heat; the rate at which R is receiving heat is then

$$\iiint_R c\rho \frac{\partial T}{\partial t} \, dx \, dy \, dz,$$

where ρ is the density. Equating the two expressions, we find

$$\iiint_R \left(c\rho \frac{\partial T}{\partial t} - k \operatorname{div} \operatorname{grad} T \right) dx \, dy \, dz = 0.$$

Since this must hold for an *arbitrary* solid region R, the function integrated (if continuous) must be zero everywhere. Hence

$$c\rho \frac{\partial T}{\partial t} - k \operatorname{div} \operatorname{grad} T = 0 \quad \text{or} \quad c\rho \frac{\partial T}{\partial t} - k\nabla^2 T = 0. \qquad (10\text{–}184)$$

This is the *heat equation*, the fundamental equation for heat conduction. If the body is in temperature equilibrium, then $\partial T/\partial t = 0$, and we conclude that $\nabla^2 T = 0$, that is, T is *harmonic*.

The vibrating membrane

We consider a membrane idealized as a surface $u = u(x, y)$ in space (with u replacing z as the vertical coordinate). The membrane is allowed to move, hence $u = u(x, y, t)$ and (x, y) varies over a fixed region R_0 in the xy-plane. We consider only "small vibrations," so that u and its derivatives u_x, u_y are very small in absolute value. In the equilibrium state, $u \equiv 0$, so that the membrane coincides with the planar region R_0. In this state it is assumed that a constant tension c is exerted throughout the membrane, so that a constant force $c \, ds$ is exerted on an element of arc ds on the boundary of a portion R of the membrane; the force acts normal to the boundary (Fig. 10–42). In the state of motion, at time t, the portion R becomes a nearly planar portion of the surface $u = u(x, y, t)$ and it is assumed that the tension force on the arc

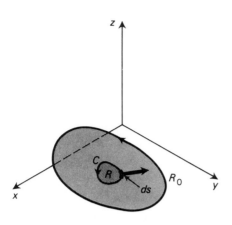

Fig. 10-42. Membrane in equilibrium state.

ds is again $c\,ds$, normal to the arc in the plane tangent to the surface. The tangent plane has a unit normal

$$\mathbf{n} = \frac{-u_x\mathbf{i} - u_y\mathbf{j} + \mathbf{k}}{\sqrt{u_x^2 + u_y^2 + 1}}$$

as usual. For small $|u_x|$, $|u_y|$, we make the approximation

$$\mathbf{n} = -u_x\mathbf{i} - u_y\mathbf{j} + \mathbf{k}.$$

The unit tangent vector $\mathbf{T}$ to the chosen arc equals

$$\frac{dx}{ds}\mathbf{i} + \frac{dy}{ds}\mathbf{j} + \frac{du}{ds}\mathbf{k},$$

and the direction of the force exerted is $\mathbf{T} \times \mathbf{n}$ (Fig. 10-43). Hence the force is $c\,ds\,(\mathbf{T} \times \mathbf{n})$, and the vertical component of the force is $c\,ds\,\mathbf{T} \times \mathbf{n} \cdot \mathbf{k}$, which we find to be $c(-u_y\,dx + u_x\,dy)$. If we "sum" around the boundary of the deflected region, we

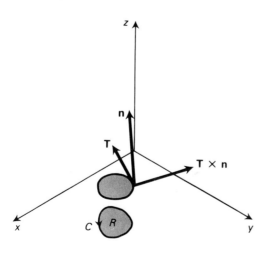

Fig. 10-43. Force on deflected portion of membrane.

obtain the total force as a line integral; since u_x and u_y can be expressed in terms of x and y, this line integral is $c \oint_C -u_y \, dx + u_x \, dy$, where C is the boundary of the region R in the xy-plane. By Green's theorem the total vertical component of force due to tension is

$$c \int\int_R (u_{xx} + u_{yy}) \, dx \, dy = c \int\int_R \nabla^2 u \, dx \, dy.$$

We ignore other forces (such as gravity). We now apply Newton's second law to each small portion of the deflected surface and "sum." If $\rho(x, y)$ is the density of the membrane, then $\rho(x, y) \, dx \, dy$ is the mass of the area element $dx \, dy$ of R of the portion of the deflected surface above R. The mass times the vertical component of acceleration for this portion of the deflected surface is $\rho(x, y) \, dx \, dy \, (\partial^2 u / \partial t^2)$. Hence the total mass times vertical component of acceleration for the deflected surface above R is

$$\int\int_R \rho(x, y) \frac{\partial^2 u}{\partial t^2} \, dx \, dy.$$

If we equate this to the total vertical component of force we obtain the equation

$$\int\int_R \left[c \nabla^2 u - \rho(x, y) \frac{\partial^2 u}{\partial t^2} \right] dx \, dy = 0.$$

This holds for an arbitrary portion R. Therefore

$$c \nabla^2 u - \rho(x, y) \frac{\partial^2 u}{\partial t^2} = 0.$$

This is the partial differential equation of the vibrating membrane. For $\rho(x, y)$ = const it is a wave equation in two dimensions (Section 8–13).

10–19   MORE PHYSICAL APPLICATIONS: ELECTROMAGNETISM

We now turn to one other branch of physics—electromagnetism. We have already (Section 10–15) made reference to Maxwell's equations. In the absence of conductors, these are

$$\operatorname{div} \mathbf{E} = 4\pi\rho, \qquad \operatorname{div} \mathbf{H} = 0,$$

$$\operatorname{curl} \mathbf{E} = -\frac{1}{c} \frac{\partial \mathbf{H}}{\partial t}, \qquad \operatorname{curl} \mathbf{H} = \frac{1}{c} \frac{\partial \mathbf{E}}{\partial t}, \qquad (10-190)$$

where ρ is the charge density and c is a universal constant.

In the electrostatic case, $\mathbf{H} \equiv 0$, so that $\mathbf{E}$ does not depend on time and

$$\operatorname{curl} \mathbf{E} = 0. \qquad (10-191)$$

Hence (in a simply connected domain), $\mathbf{E}$ is the gradient of a potential:

$$\mathbf{E} = - \operatorname{grad} \psi;$$

ψ is termed the *electrostatic potential*. The function ψ must then satisfy the *Poisson equation*:

$$\operatorname{div} \operatorname{grad} \psi = -4\pi\rho. \tag{10-192}$$

In any domain free of charge, ψ is therefore a *harmonic* function.

The function ψ can be computed by Coulomb's law for given charge distributions. Thus for a point charge e at the origin,

$$\psi = \frac{e}{r} + \text{const}, \qquad r = \sqrt{x^2 + y^2 + z^2}, \tag{10-193}$$

and ψ is obtained by simple addition for a sum of point charges. If the charge is distributed along a wire C and ρ_s is the density (charge per unit length), then

$$\psi(x_1, y_1, z_1) = \int_C \frac{\rho_s \, ds}{r_1} + \text{const}, \tag{10-194}$$

where $r_1 = \sqrt{(x-x_1)^2 + (y-y_1)^2 + (z-z_1)^2}$. If the charge is spread out over a surface S, then ψ is given by the surface integral

$$\psi(x_1, y_1, z_1) = \iint_S \frac{\rho_a}{r_1} \, d\sigma + \text{const}, \tag{10-195}$$

where ρ_a is the charge density (charge per unit area).

In empty space (vacuum), $\rho = 0$. We can write, from Eqs. (10-190),

$$\frac{1}{c} \frac{\partial^2 \mathbf{E}}{\partial t^2} = \frac{\partial}{\partial t} \operatorname{curl} \mathbf{H} = \operatorname{curl} \frac{\partial \mathbf{H}}{\partial t} = - \operatorname{curl} \operatorname{curl} \mathbf{E}$$

by interchanging the order of differentiation. Since now $\operatorname{div} \mathbf{E} = 0$, we can use the identity (10-168') to write

$$\frac{1}{c} \frac{\partial^2 \mathbf{E}}{\partial t^2} = -c(\operatorname{grad} \operatorname{div} \mathbf{E} - \nabla^2 \mathbf{E}) = c\nabla^2 \mathbf{E}.$$

Thus in three-dimensional space each component of $\mathbf{E}$ satisfies the wave equation

$$\frac{\partial^2 u}{\partial t^2} - c^2 \nabla^2 u = 0.$$

═══════════════════════**Problems (Section 10–19)**═══════════════════════

1. Let D be a simply connected domain in the xy plane and let $\mathbf{w} = u\mathbf{i} - v\mathbf{j}$ be the velocity vector of an irrotational incompressible flow in D. (This is the same as an irrotational incompressible flow in a three-dimensional domain whose projection is D and for which the z component of velocity is 0, while the x and y components of velocity are independent of z.) Show that the following properties hold:

a) u and v satisfy the Cauchy-Riemann equations $\dfrac{\partial u}{\partial x} = \dfrac{\partial v}{\partial y}$, $\dfrac{\partial u}{\partial y} = -\dfrac{\partial v}{\partial x}$ in D;

b) u and v are harmonic in D;
c) $\int u\,dx - v\,dy$ and $\int v\,dx + u\,dy$ are independent of the path in D;
d) There are scalars φ and ψ in D such that

$$\frac{\partial \varphi}{\partial x} = u, \qquad \frac{\partial \varphi}{\partial y} = -v \qquad \text{and} \qquad \frac{\partial \psi}{\partial x} = v, \qquad \frac{\partial \psi}{\partial y} = u;$$

e) With $F = \varphi \mathbf{i} - \psi \mathbf{j}$ and φ, ψ as in (d), $\text{div } F = 0$ and $\text{curl } F = 0$ in D;
f) φ and ψ are harmonic in D;
g) $\text{grad } \varphi = \mathbf{w}$, ψ is constant on each stream line.
The function φ is the *velocity potential*, ψ is the *stream function*.

2. Let a wire occupying the line segment from $(0, -c)$ to $(0, c)$ in the xy plane have a constant charge density equal to ρ. Show that the electrostatic potential due to this wire at a point (x_1, y_1) of the xy plane is given by

$$\psi = \rho \ln \frac{\sqrt{x_1^2 + (c - y_1)^2} + c - y_1}{\sqrt{x_1^2 + (c + y_1)^2} - c - y_1} + k,$$

where k is an arbitrary constant. Show that, if k is chosen so that $\psi(1, 0) = 0$, then, as c becomes infinite, ψ approaches the limiting value $-2\rho \ln |x_1|$. This is the potential of an infinite wire with uniform charge.

3. Find the temperature distribution in a solid whose boundaries are two parallel planes, d units apart, kept at temperatures T_1, T_2 respectively.

HINT: Take the boundaries to be the planes $x = 0$, $x = d$ and note that, by symmetry, T must be independent of y and z.

4. Consider a fluid motion in space. A particle occupying position (x_0, y_0, z_0) at time $t = 0$ occupies position (x, y, z) at time t. Thus x, y, z become functions of x_0, y_0, z_0, t:

$$x = \varphi(x_0, y_0, z_0, t), \qquad y = \psi(x_0, y_0, z_0, t), \qquad z = \chi(x_0, y_0, z_0, t).$$

Let the ∇ symbol be used as follows:

$$\nabla = \frac{\partial}{\partial x_0}\mathbf{i} + \frac{\partial}{\partial y_0}\mathbf{j} + \frac{\partial}{\partial z_0}\mathbf{k}$$

and let J denote the Jacobian

$$\frac{\partial(x, y, z)}{\partial(x_0, y_0, z_0)}.$$

Let $\mathbf{v}$ denote the velocity vector:

$$\mathbf{v} = \frac{\partial x}{\partial t}\mathbf{i} + \frac{\partial y}{\partial t}\mathbf{j} + \frac{\partial z}{\partial t}\mathbf{k} = \frac{\partial \varphi}{\partial t}\mathbf{i} + \frac{\partial \psi}{\partial t}\mathbf{j} + \frac{\partial \chi}{\partial t}\mathbf{k}.$$

a) Show that $J = \nabla x \cdot \nabla y \times \nabla z$.
b) Show that

$$\frac{\partial x_0}{\partial x} = \mathbf{i} \cdot \frac{\nabla y \times \nabla z}{J}, \qquad \frac{\partial y_0}{\partial x} = \mathbf{j} \cdot \frac{\nabla y \times \nabla z}{J}, \qquad \frac{\partial z_0}{\partial x} = \mathbf{k} \cdot \frac{\nabla y \times \nabla z}{J},$$

and obtain similar expressions for

$$\frac{\partial x_0}{\partial y}, \quad \frac{\partial y_0}{\partial y}, \quad \frac{\partial z_0}{\partial y}, \quad \frac{\partial x_0}{\partial z}, \quad \frac{\partial y_0}{\partial z}, \quad \frac{\partial z_0}{\partial z}.$$

HINT: See Problem 10 following Section 9–15.

c) Show that

$$\frac{\partial J}{\partial t} = \nabla v_z \cdot \nabla y \times \nabla z + \nabla v_y \cdot \nabla z \times \nabla x + \nabla v_z \cdot \nabla x \times \nabla y.$$

d) Show that $\operatorname{div} \mathbf{v} = \dfrac{1}{J} \dfrac{\partial J}{\partial t}$.

HINT: By the chain rule,

$$\frac{\partial v_x}{\partial x} = \frac{\partial v_x}{\partial x_0} \frac{\partial x_0}{\partial x} + \frac{\partial v_x}{\partial y_0} \frac{\partial y_0}{\partial x} + \frac{\partial v_z}{\partial z_0} \frac{\partial z_0}{\partial x}.$$

Use the result of (b) to show that

$$\frac{\partial v_x}{\partial x} = \frac{\nabla v_x \cdot \nabla y \times \nabla z}{J}$$

and obtain similar expressions for $\partial v_y/\partial y$, $\partial v_z/\partial z$. Add the results and use the result of (c).

REMARK: The Jacobian J can be interpreted as the ratio of the volume occupied by a small piece of the fluid at time t to the volume occupied by this piece when $t = 0$, as in Fig. 10–44. Hence by (d) the divergence of the velocity vector can be interpreted as measuring the percent of change in this ratio per unit time or simply as the *rate of change of volume per unit volume* of the moving piece of fluid.

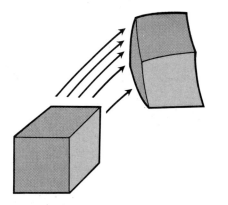

Figure 10–44.

5. Consider a piece of the fluid of Problem 4 (not necessarily a "small" piece) occupying a region $R = R(t)$ at time t and a region $R_0 = R(0)$ when $t = 0$. Let $F(x, y, z, t)$ be a function differentiable throughout the part of space concerned.

a) Show that

$$\iiint_{R(t)} F(x,\,y,\,z,\,t)\,dx\,dy\,dz = \iiint_{R_0} F[\varphi(x_0,\,y_0,\,z_0,\,t),\,\dots\,]\,J\,dx_0\,dy_0\,dz_0.$$

HINT: Use Eq. (10–43); the Jacobian J is necessarily positive here.

b) Show that

$$\frac{d}{dt}\iiint_{R(t)} F(x,\,y,\,z,\,t)\,dx\,dy\,dz = \iiint_{R(t)} \left[\frac{\partial F}{\partial t} + \operatorname{div}(F\mathbf{v})\right]dx\,dy\,dz.$$

HINT: Use (a) and apply Leibnitz's rule of Section 10–3 to differentiate the right-hand side. Use the result of part (d) of Problem 4 to simplify the result. Then return to the original variables by (a) again.

6. Let $\rho = \rho(x,\,y,\,z,\,t)$ be the density of the fluid motion of Problems 4 and 5. The integral

$$\iiint_{R(t)} \rho\,dx\,dy\,dz$$

represents the mass of the fluid filling $R(t)$. The conservation of mass implies that this integral is constant:

$$\frac{d}{dt}\iiint_{R(t)} \rho\,dx\,dy\,dz = 0.$$

Use this result and that of Problem 5(b) to establish the *continuity equation*

$$\frac{\partial \rho}{\partial t} + \operatorname{div}(\rho\mathbf{v}) = 0.$$

HINT: Cf. the derivation of the heat equation (10–184) above.

Chapter 11 Analytic Functions of a Complex Variable

Complex numbers and functions have appeared at various points in the earlier chapters, for example, in solving linear differential equations in Chapter 1 and in the Fourier transform of Chapter 4. In this chapter we develop the theory systematically and provide important applications to stability theory (the Nyquist criterion) and to solution of boundary-value problems for partial differential equations (application of conformal mapping). For convenience we number the theorems throughout this chapter.

11–1 THE COMPLEX-NUMBER SYSTEM

This is reviewed in Section 1–9. The main definitions are given again in Fig. 11–1. Complex numbers are added and multiplied as follows:

$$(a + bi) + (c + di) = (a + c) + (b + d)i, \tag{11–10}$$

$$(a + bi) \cdot (c + di) = ac - bd + i(ad + bc). \tag{11–11}$$

The addition is the same as vector addition, as in Fig. 11–1. The figure also shows the meaning of subtraction. From this we see that

$$|z_2 - z_1| = \text{Distance between } z_1 \text{ and } z_2. \tag{11–12}$$

Hence the set of z such that $|z - z_0| < \varepsilon$ is an ε–*neighborhood* of z_0, as in Fig. 11–1. From the figure we also see the *triangle inequality*:

$$|z_1 + z_2| \leqslant |z_1| + |z_2|. \tag{11–13}$$

We can represent z in terms of its polar coordinates r, θ:

$$z = r(\cos\theta + i\sin\theta) = re^{i\theta}, \qquad r = |z|, \qquad \theta = \arg z. \tag{11–14}$$

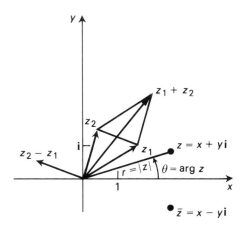

ε-neighborhood

Fig. 11–1. The complex-number system.

It follows that for $z_1 = r_1 e^{i\theta_1}$ and $z_2 = r_2 e^{i\theta_2}$

$$z_1 z_2 = r_1 r_2 e^{i(\theta_1 + \theta_2)},$$

so that

$$|z_1 z_2| = |z_1| |z_2|, \tag{11-15}$$

$$\arg z_1 z_2 = \arg z_1 + \arg z_2 \quad \text{(up to multiples of } 2\pi).$$

The conjugate $\bar{z} = x - iy$ obeys the rules:

$$\overline{z_1 + z_2} = \bar{z}_1 + \bar{z}_2, \qquad \overline{z_1 z_2} = \bar{z}_1 \bar{z}_2, \qquad z\bar{z} = |z|^2, \qquad |\bar{z}| = |z|. \tag{11-16}$$

11–2 COMPLEX FUNCTIONS OF A REAL VARIABLE

These are introduced in Section 1–10 and used extensively in connection with Fourier and Laplace transforms in Chapter 4. Here we need them mainly to represent paths in the complex plane. For $z = x + iy$, the equation

$$z = F(t) = f(t) + ig(t), \qquad a \leqslant t \leqslant b, \tag{11-20}$$

is equivalent to two equations

$$x = f(t), \qquad y = g(t), \qquad a \leqslant t \leqslant b,$$

which are parametric equations of a path in the xy-plane or z-plane. For example,

$$z = 5e^{it}, \qquad 0 \leqslant t \leqslant 2\pi,$$

is equivalent to

$$x = 5 \cos t, \qquad y = 5 \sin t, \qquad 0 \leqslant t \leqslant 2\pi,$$

and gives a parametrization of a circular path with center at $z = 0$ and radius 5.

Generally $F(t)$ will be assumed continuous. This is equivalent to requiring that $f(t)$ and $g(t)$ are continuous. Further, we normally assume that the path is piecewise smooth, that is, that $F'(t) = f'(t) + ig'(t)$ is continuous except for a finite number of jump discontinuities. (As in Section 9–3, $F'(t) = x'(t) + iy'(t)$ is a *tangent vector* to the path.)

Occasionally we need to take limits of a function $F(t)$ as $t \to t_0$ (or as $t \to t_0 \pm$ or as $t \to \pm \infty$). By definition,

$$\lim F(t) = \lim f(t) + i \lim g(t),$$

provided both limits exist. Also we need

$$\lim F(t) = \infty.$$

This is defined to mean:

$$\lim |F(t)| = \infty.$$

Thus approaching infinity in the complex plane means receding to infinite distance from the origin. There is *only one* complex number ∞; this number is discussed in Section 11 14.

The following are examples of limits:

$$\lim_{t \to 0} \frac{1 - e^{it}}{t} = \lim_{t \to 0} \left(\frac{1 - \cos t}{t} - i \frac{\sin t}{t} \right) = 0 - i = -i,$$

since $(1 - \cos t)/t$ has limit 0 and $(\sin t)/t$ has limit 1, as can be verified by L'Hospital's rule;

$$\lim_{t \to 0} \frac{e^{it}}{t} = \infty,$$

since

$$\left| \frac{e^{it}}{t} \right| = \frac{|e^{it}|}{|t|} = \frac{1}{|t|},$$

and this real function has limit ∞ as $t \to 0$.

As in Section 1–10, the calculus applies to complex functions $F(t)$ without essential change.

11–3 COMPLEX-VALUED FUNCTIONS OF A COMPLEX VARIABLE. LIMITS AND CONTINUITY

We turn to the general complex-valued function of a complex variable. These functions will be our principal concern for the remainder of this chapter. We write:

$$w = f(z),$$

where $z = x + iy$, $w = u + iv$, to indicate such a function. An example is the function

$$w = z^2, \qquad \text{for all } z.$$

Here we can also write:

$$u + iv = (x + iy)^2 = x^2 - y^2 + 2ixy,$$

so that (on taking real and imaginary parts)

$$u = x^2 - y^2, \qquad v = 2xy.$$

In a similar manner, *every* complex function $w = f(z)$ is equivalent to a pair of real functions:

$$u = u(x, y) = \operatorname{Re}[f(z)], \qquad v = v(x, y) = \operatorname{Im}[f(z)],$$

of the two real variables x, y. Also from such a pair of real functions, defined on the same set, we obtain a complex function of z. For example,

$$u = x^2 + xy + y^2, \qquad v = xy^3$$

is equivalent to the complex function

$$w = f(z) = x^2 + xy + y^2 + xy^3 i,$$

for which $f(1 + 2i) = 1 + 2 + 4 + 8i = 7 + 8i$.

The function $e^z = e^{x+iy}$ was defined in Section 1–10 for all z:

$$e^z = e^x (\cos y + i \sin y). \tag{11–30}$$

From this relation we deduce that

$$e^{iy} = \cos y + i \sin y, \qquad e^{-iy} = \cos y - i \sin y$$

and hence

$$\cos y = \frac{e^{iy} + e^{-iy}}{2}, \qquad \sin y = \frac{e^{iy} - e^{-iy}}{2i}.$$

These equations suggest the general definitions for all complex z:

$$\cos z = \frac{e^{iz} + e^{-iz}}{2}, \qquad \sin z = \frac{e^{iz} - e^{-iz}}{2i}. \tag{11–31}$$

We also define $\cosh z$ and $\sinh z$ as usual:

$$\cosh z = \frac{e^z + e^{-z}}{2}, \qquad \sinh z = \frac{e^z - e^{-z}}{2}. \tag{11–32}$$

These functions are now also defined for all z.

From Eq. (11–30) we deduce the corresponding pair of real functions defined for all (x, y):

$$u = \operatorname{Re} e^z = e^x \cos y, \qquad v = \operatorname{Im} e^z = e^x \sin y. \tag{11–33}$$

From the definitions (11–31) and (11–32) we find the analogous pairs for the other functions:

$$
\begin{array}{lll}
w = \sin z: & u = \sin x \cosh y, & v = \cos x \sinh y, \\
w = \cos z: & u = \cos x \cosh y, & v = -\sin x \sinh y, \\
w = \sinh z: & u = \sinh x \cos y, & v = \cosh x \sin y, \\
w = \cosh z: & u = \cosh x \cos y, & v = \sinh x \sin y.
\end{array}
\tag{11–34}
$$

The proofs are left as exercises (Problem 1 below). In (11–34) each function is defined for all z, that is, for all (x, y).

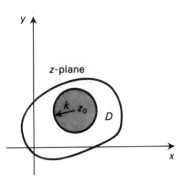

Fig. 11–2. Open region and neighborhood.

In general, we assume $w = f(z)$ to be defined in an open region* D in the z-plane, as suggested in Fig. 11–2. If z_0 is a point of D, we can then find a circular neighborhood $|z - z_0| < k$ about z_0 in D. If $f(z)$ is defined in such a neighborhood, except perhaps at z_0, then we write

$$\lim_{z \to z_0} f(z) = w_0 \tag{11–35}$$

if for every $\varepsilon > 0$ we can choose $\delta > 0$, so that

$$|f(z) - w_0| < \varepsilon \text{ for } 0 < |z - z_0| < \delta. \tag{11–36}$$

If $f(z_0)$ is defined and equals w_0, and (11–36) holds, then we call $f(z)$ *continuous at* z_0.

THEOREM 1   *The function $w = f(z)$ is continuous at $z_0 = x_0 + iy_0$ if and only if $u(x, y) = \mathrm{Re}\,[f(z)]$ and $v(x, y) = \mathrm{Im}\,[f(z)]$ are continuous at (x_0, y_0).*

Thus $w = z^2 = x^2 - y^2 + 2ixy$ is continuous for all z, since $u = x^2 - y^2$ and $v = 2xy$ are continuous for all (x, y). Theorem 1 follows from the fact that $w = u + iv$ is close to $w = u_0 + iv_0$ if and only if u is close to u_0 and v is close to v_0.

THEOREM 2   *The sum, product, and quotient of continuous functions of z are continuous, except for division by zero; a continuous function of a continuous function is continuous. Similarly, if the limits exist,*

$$\lim_{z \to z_0} [f(z) + g(z)] = \lim_{z \to z_0} f(z) + \lim_{z \to z_0} g(z). \tag{11–37}$$

These properties are proved as for real variables. (It is assumed in Theorem 2 that the functions are defined in appropriate regions.)

It follows from Theorem 2 that polynomials in z are continuous for all z, and each rational function is continuous except where the denominator is zero. From Theorem 1 it follows that

$$e^z = e^x \cos y + ie^x \sin y$$

---

* Since the region is usually open, we shall generally omit the word *open* in this chapter. An open region is also called a *domain*.

is continuous for all z. Hence, by Theorem 2, so also are the functions

$$\sin z = \frac{e^{iz} - e^{-iz}}{2i}, \qquad \cos z = \frac{e^{iz} + e^{-iz}}{2}.$$

We write

$$\lim_{z \to z_0} f(z) = \infty \quad \text{if} \quad \lim_{z \to z_0} |f(z)| = +\infty,$$

that is, if for each real number K there is a positive δ such that $|f(z)| > K$ for $0 < |z - z_0| < \delta$. Similarly, if $f(z)$ is defined for $|z| > R$, for some R, then $\lim_{z \to \infty} f(z) = c$ if for each $\varepsilon > 0$ we can choose a number R_0 such that $|f(z) - c| < \varepsilon$ for $|z| > R_0$. All these definitions emphasize that there is but *one* complex number ∞ and that "approaching ∞" is equivalent to receding from the origin.

11–4 DERIVATIVES AND DIFFERENTIALS

Let $w = f(z)$ be given in D and let z_0 be a point of D. Then w is said to have a derivative $f'(z_0)$ if

$$\lim_{\Delta z \to 0} \frac{f(z_0 + \Delta z) - f(z_0)}{\Delta z} = f'(z_0).$$

In appearance this definition is the same as that for functions of a real variable, and it will be seen that the derivative does have the usual properties. However, it will also be shown that if $w = f(z)$ has a continuous derivative in a domain D, then $f(z)$ has a number of additional properties; in particular, the second derivative $f''(z)$, third derivative $f'''(z)$, etc., must also exist in D.

The reason for the remarkable consequences of possession of a derivative lies in the fact that the increment Δz is allowed to approach zero in any manner. If we restricted Δz so that $z_0 + \Delta z$ approached z_0 along a particular line, then we would obtain a *directional derivative*. But here the limit obtained is required to be the *same for all directions*, so that the directional derivative has the same value in all directions. Moreover, $z_0 + \Delta z$ may approach z_0 in a quite arbitrary manner, for example along a spiral path. The limit of the ratio $\Delta w / \Delta z$ must be the same for all manners of approach.

We say that $f(z)$ has a *differential* $dw = c \, \Delta z$ at z_0 if

$$f(z_0 + \Delta z) - f(z_0) = c \, \Delta z + \varepsilon \, \Delta z,$$

where ε depends on Δz and is continuous at $\Delta z = 0$, with value zero when $\Delta z = 0$.

THEOREM 3   *If $w = f(z)$ has a differential $dw = c \, \Delta z$ at z_0, then w has a derivative $f'(z_0) = c$. Conversely, if w has a derivative at z_0, then w has a differential at z_0: $dw = f'(z_0) \, \Delta z$.*

This is proved just as for real functions. We also write $\Delta z = dz$, as for real variables, so that

$$dw = f'(z) \, dz, \qquad \frac{dw}{dz} = f'(z). \qquad (11\text{--}40)$$

From Theorem 3 we see that existence of the derivative $f'(z_0)$ implies continuity of f at z_0, for

$$f(z_0 + \Delta z) - f(z_0) = c\,\Delta z + \varepsilon\,\Delta z \to 0 \quad \text{as} \quad \Delta z \to 0.$$

THEOREM 4   *If w_1 and w_2 are functions of z which have differentials in D, then*

$$d(w_1 + w_2) = dw_1 + dw_2,$$
$$d(w_1 w_2) = w_1\,dw_2 + w_2\,dw_1, \qquad\qquad (11\text{--}41)$$
$$d\frac{w_1}{w_2} = \frac{w_2\,dw_1 - w_1\,dw_2}{w_2^2}, \qquad w_2 \neq 0.$$

If w_2 is a differentiable function of w_1, and w_1 is a differentiable function of z, then wherever $w_2[w_1(z)]$ is defined

$$\frac{dw_2}{dz} = \frac{dw_2}{dw_1}\cdot\frac{dw_1}{dz}. \qquad\qquad (11\text{--}42)$$

These rules are proved as in elementary calculus. We can now prove as usual the basic rule:

$$\frac{d}{dz}z^n = nz^{n-1}, \qquad n = 1, 2, \ldots \qquad\qquad (11\text{--}43)$$

Furthermore, the derivative of a constant is zero.

══════════════════════════**Problems (Section 11-4)**══════════════════════════

1. For each of the following write the given function as two real functions of x and y and determine where the given function is continuous:

    a) $w = (1 + i)z^2$ b) $w = \dfrac{z}{z+i}$

    c) $w = \tan z = \dfrac{\sin z}{\cos z}$ d) $w = \dfrac{e^{-z}}{z+1}$

    e) $w = e^z$ f) $w = \sin z$ g) $w = \cos z$

    h) $w = \sinh z$ i) $w = \cosh z$ j) $w = e^z \cos z$

2. Evaluate each of the following limits:

    a) $\lim\limits_{z \to \pi i} \dfrac{\sin z + z}{e^z + 2}$ b) $\lim\limits_{z \to 0} \dfrac{z^2 - z}{2z}$

    c) $\lim\limits_{z \to 0} \dfrac{\cos z}{z}$ d) $\lim\limits_{z \to \infty} \dfrac{z}{z^2 + 1}$

3. Differentiate each of the following complex functions:

    a) $w = z^3 + 5z + 1$ b) $w = \dfrac{1}{z-1}$

    c) $w = [1 + (z^2 + 1)^3]^7$ d) $w = \dfrac{z}{(z+1)^3}$

4. Prove the rule (11–43).

5. a) Prove that $\sin z$ and $\cos z$ are equal to zero only for z real, hence for $z = n\pi$ and for $z = (\pi/2) + n\pi$, respectively.

   b) Find the zeros of $\sinh z$ and $\cosh z$.

11–5 INTEGRALS

The complex integral $\int f(z)\,dz$ is defined as a line integral, and its properties are closely related to those of the integral $\int P\,dx + Q\,dy$ (see Chapter 10).

Let C be a path from A to B in the complex plane: $x = x(t)$, $y = y(t)$, $a \leqslant t \leqslant b$. We assume C to have a direction, usually that of increasing t. We subdivide the interval $a \leqslant t \leqslant b$ into n parts by $t_0 = a, t_1, \ldots, t_n = b$. We let $z_j = x(t_j) + iy(t_j)$ and $\Delta_j z = z_j - z_{j-1}$, $\Delta_j t = t_j - t_{j-1}$. We choose an arbitrary value t_j^* in the interval $t_{j-1} \leqslant t \leqslant t_j$ and set $z_j^* = x(t_j^*) + iy(t_j^*)$. These quantities are all shown in Fig. 11–3. We then write

$$\int_C f(z)\,dz = \int_A^B f(z)\,dz = \lim_{\substack{n \to \infty \\ \max \Delta_j t \to 0}} \sum_{j=1}^{n} f(z_j^*)\Delta_j z. \tag{11–50}$$

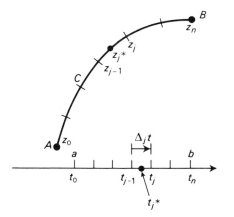

Fig. 11-3. Complex line integral.

If we take real and imaginary parts in (11–50), we find

$$\int_C f(z)\,dz = \lim \sum (u + iv)\,(\Delta x + i\,\Delta y)$$

$$= \lim \left[\sum (u\,\Delta x - v\,\Delta y) + i \sum (v\,\Delta x + u\,\Delta y) \right];$$

that is,

$$\int_C f(z)\,dz = \int_C (u+iv)\,(dx+i\,dy) \tag{11–51}$$

$$= \int_C (u\,dx - v\,dy) + i \int_C (v\,dx + u\,dy). \tag{11–52}$$

The complex line integral is thus simply a combination of two real line integrals. Hence we can apply all the theory of real line integrals. In the following, each path is assumed to be *piecewise smooth*, that is, $x(t)$ and $y(t)$ are to be continuous with piecewise continuous derivatives.

THEOREM 5 *If $f(z)$ is continuous in region D, then the integral (11–50) exists for C in D and*

$$\int_C f(z)\,dz = \int_a^b \left(u\frac{dx}{dt} - v\frac{dy}{dt} \right) dt + i \int_a^b \left(v\frac{dx}{dt} + u\frac{dy}{dt} \right) dt. \tag{11–53}$$

We now write our path as $z = z(t)$, as in Section 11–2. If we introduce the derivative

$$\frac{dz}{dt} = \frac{dx}{dt} + i\frac{dy}{dt}$$

of z with respect to the real variable t, and also use the theory of integrals of such functions (Section 1–10) we can write (11–53) more concisely:

$$\int_C f(z)\,dz = \int_a^b f[z(t)]\frac{dz}{dt}\,dt. \tag{11–54}$$

EXAMPLE 1 Let C be the path $x = 2t$, $y = 3t$, $1 \leqslant t \leqslant 2$. Let $f(z) = z^2$. Then

$$\int_C z^2\,dz = \int_1^2 (2t + 3it)^2\,(2 + 3i)\,dt = (2 + 3i)^3 \int_1^2 t^2\,dt$$

$$= -107\tfrac{1}{3} + 21i. \;\blacktriangleleft$$

EXAMPLE 2 Let C be the circular path $x = \cos t$, $y = \sin t$, $0 \leqslant t \leqslant 2\pi$. This can be written more concisely as $z = e^{it}$, $0 \leqslant t \leqslant 2\pi$. Since $dz/dt = ie^{it}$,

$$\int_C \frac{1}{z}\,dz = \int_0^{2\pi} e^{-it}\,ie^{it}\,dt = i \int_0^{2\pi} dt = 2\pi i. \;\blacktriangleleft$$

Further properties of complex integrals follow from those of real integrals:

THEOREM 6 *Let $f(z)$ and $g(z)$ be continuous in a region D. Let C be a piecewise smooth path in D. Then*

$$\int_C [f(z) + g(z)]\, dz = \int_C f(z)\, dz + \int_C g(z)\, dz,$$

$$\int_C k f(z)\, dz = k \int_C f(z)\, dz, \qquad k = \text{const},$$

$$\int_C f(z)\, dz = \int_{C_1} f(z)\, dz + \int_{C_2} f(z)\, dz,$$

where C is composed of a path C_1 from z_0 to z_1 and a path C_2 from z_1 to z_2, and

$$\int_C f(z)\, dz = -\int_{C'} f(z)\, dz,$$

where C' is obtained from C by reversing direction on C.

Upper estimates for the absolute value of a complex integral are obtained by the following theorem.

THEOREM 7 *Let $f(z)$ be continuous on C, let $|f(z)| \leq M$ on C, and let*

$$L = \int_C ds = \int_a^b \sqrt{\left(\frac{dx}{dt}\right)^2 + \left(\frac{dy}{dt}\right)^2}\, dt$$

be the length of C. Then

$$\left| \int_C f(z)\, dz \right| \leq \int_C |f(z)|\, ds \leq M \cdot L. \tag{11–55}$$

PROOF. The line integral $\int_C |f(z)|\, ds$ is defined as a limit:

$$\int_C |f(z)|\, ds = \lim \sum |f(z_j^*)|\, \Delta_j s,$$

where $\Delta_j s$ is the length of the jth arc of C. Now

$$|f(z_j^*) \Delta_j z| = |f(z_j^*)| \cdot |\Delta_j z| \leq |f(z_j^*)| \cdot \Delta_j s,$$

for $|\Delta_j z|$ represents the *chord* of the arc $\Delta_j s$. Hence

$$\left| \sum f(z_j^*) \Delta_j z \right| \leq \sum |f(z_j^*) \Delta_j z| \leq \sum |f(z_j^*)| \Delta_j s$$

by repeated application of the triangle inequality (11–13). Passing to the limit, we conclude that

$$\left| \int_C f(z)\, dz \right| \leq \int_C |f(z)|\, ds. \tag{11–56}$$

Also, if $|f| \leqslant M = \text{const}$,

$$\sum |f(z_j^*)| \, \Delta_j s \leqslant \sum M \Delta_j s = M \cdot L.$$

Hence

$$\int_C |f(z)| \, ds \leqslant M \cdot L. \tag{11–57}$$

Inequalities (11–55) follow from (11–56) and (11–57).

=========== **Problems (Section 11-5)** ===========

1. Evaluate the following integrals:

a) $\displaystyle\int_0^{1+i} (x^2 - iy^2) \, dz$ on the straight line from 0 to $1 + i$.

b) $\displaystyle\int_0^{\pi} z \, dz$ on the curve $y = \sin x$.   c) $\displaystyle\int_1^{1+i} \frac{dz}{z}$ on the line $x = 1$.

2. Write each of the following integrals in the form $\int u \, dx - v \, dy + i \int v \, dx + u \, dy$; then show that each of the two real integrals is independent of path in the xy-plane.
   a) $\int (z + 1) \, dz$ b) $\int e^z \, dz$ c) $\int z^4 \, dz$ d) $\int \sin z \, dz$

3. a) Evaluate $\oint dz/z$ on the circle $|z| = R$. The positive direction is understood.

   b) Show that $\oint dz/z = 0$ on every simple closed path not meeting or enclosing the origin.

   c) Show that $\oint dz/z^2 = 0$ on every simple closed path not passing through the origin.

11–6 ANALYTIC FUNCTIONS. CAUCHY-RIEMANN EQUATIONS

A function $w = f(z)$, defined in a domain D, is said to be an *analytic function* in D if w has a continuous derivative in D. Almost the entire theory of functions of a complex variable is confined to the study of such functions. Furthermore, almost all functions used in the applications of mathematics to physical problems are analytic functions or are derived from such.

It will be seen that possession of a continuous derivative implies possession of a continuous second derivative, third derivative, etc., and, in fact, convergence of the Taylor series

$$f(z_0) + f'(z_0) \frac{(z - z_0)}{1!} + f''(z_0) \frac{(z - z_0)^2}{2!} + \cdots$$

in a neighborhood of each z_0 of D. Thus we can define an analytic function as one so representable by Taylor series, and this definition is often used. The two definitions are equivalent, for convergence of the Taylor series in a neighborhood of each z_0 implies continuity of the derivatives of all orders.

While it is possible to construct continuous functions of z that are not analytic (examples will be given below), it is impossible to construct a function $f(z)$ possessing a derivative, but not a continuous one, in D. In other words, if $f(z)$ has a derivative in D, the derivative is necessarily continuous, so that $f(z)$ is analytic. Therefore we could define an analytic function as one merely possessing a derivative in domain D, and this definition is also often used. (For a proof that existence of the derivative implies its continuity, see Chapter III of "Conformal Mapping" by Z. NEHARI, McGraw-Hill, New York, 1952.)

THEOREM 8 *If $w = u + iv = f(z)$ is analytic in D, then u and v have continuous first partial derivatives in D and satisfy the Cauchy–Riemann equations*

$$\frac{\partial u}{\partial x} = \frac{\partial v}{\partial y}, \qquad \frac{\partial u}{\partial y} = -\frac{\partial v}{\partial x} \qquad (11\text{--}60)$$

in D. Furthermore,

$$\frac{dw}{dz} = \frac{\partial u}{\partial x} + i\frac{\partial v}{\partial x} = \frac{\partial v}{\partial y} + i\frac{\partial v}{\partial x} = \frac{\partial u}{\partial x} - i\frac{\partial u}{\partial y} = \frac{\partial v}{\partial y} - i\frac{\partial u}{\partial y}. \qquad (11\text{--}61)$$

Conversely, if $u(x, y)$ and $v(x, y)$ have continuous first partial derivatives in D and satisfy the Cauchy–Riemann equations (11–60), then $w = u + iv = f(z)$ is analytic in D.

PROOF: Suppose $f'(z_0)$ exists. Then $\Delta w/\Delta z$ has a limit $c = a + ib$ as $\Delta z \to 0$ or, equivalently, as in Theorem 3,

$$\Delta w = c\Delta z + \varepsilon\Delta z, \qquad (11\text{--}62)$$

where $\varepsilon \to 0$ as $\Delta z \to 0$. We write $\varepsilon = \varepsilon_1 + i\varepsilon_2$ and $\Delta z = \Delta x + i\Delta y$, as in Fig. 11–4. Then (11–62) becomes

$$\Delta w = (a + ib)(\Delta x + i\Delta y) + (\varepsilon_1 + i\varepsilon_2)(\Delta x + i\Delta y).$$

Hence if $\Delta w = \Delta u + i\Delta v$, by comparing real and imaginary parts, we obtain

$$\Delta u = a\Delta x - b\Delta y + \varepsilon_1\Delta x - \varepsilon_2\Delta y,$$
$$\Delta v = b\Delta x + a\Delta y + \varepsilon_2\Delta x + \varepsilon_1\Delta y. \qquad (11\text{--}63)$$

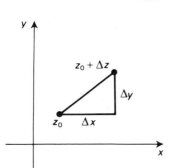

Fig. 11–4. Complex derivative.

Here $\varepsilon_1 \to 0$ and $\varepsilon_2 \to 0$ as $\Delta x \to 0$ and $\Delta y \to 0$. Thus, as in Section 9–4, u and v have differentials at (x_0, y_0):

$$du = a\Delta x - b\Delta y, \qquad dv = b\Delta x + a\Delta y,$$

and at this point

$$a = \frac{\partial u}{\partial x}, \qquad -b = \frac{\partial u}{\partial y}, \qquad b = \frac{\partial v}{\partial x}, \qquad a = \frac{\partial v}{\partial y}.$$

Therefore,

$$\frac{\partial u}{\partial x} = \frac{\partial v}{\partial y}, \qquad \frac{\partial u}{\partial y} = -\frac{\partial v}{\partial x},$$

and the Cauchy–Riemann equations are proved. Furthermore,

$$\frac{dw}{dz} = a + ib = \frac{\partial u}{\partial x} + i\frac{\partial v}{\partial x} = \frac{\partial v}{\partial y} - i\frac{\partial u}{\partial y} = \cdots$$

as in (11–61); this shows that u and v have continuous partial derivatives.

Conversely, if the Cauchy–Riemann equations hold, then we can effectively reverse the steps to show that (11–62) holds, so that $f'(z_0)$ exists. (See AC, pp. 594–595 for details).

REMARK: The Cauchy–Riemann equations (11–60) occur in a variety of physical problems. For example, $u\mathbf{i} - v\mathbf{j}$ can be interpreted as a vector field in the plane representing the velocity of a steady two-dimensional flow. If u, v satisfy the Cauchy–Riemann equations, then this flow is irrotational and incompressible (Sections 10–7 and 10–8), since

$$\text{curl}(u\mathbf{i} - v\mathbf{j}) = \left(-\frac{\partial v}{\partial x} - \frac{\partial u}{\partial y} \right)\mathbf{k}, \qquad \text{div}(u\mathbf{i} - v\mathbf{j}) = \frac{\partial u}{\partial x} - \frac{\partial v}{\partial y},$$

and both of these reduce to zero by (11–60).

Thus *every analytic function $f(z)$ gives rise to a possible irrotational, incompressible planar flow.* We return to this question in Section 11–29.

Theorem 8 provides a perfect test for analyticity: if $f(z)$ is analytic, then the Cauchy–Riemann equations hold; if the equations hold (and the derivatives concerned are continuous), then $f(z)$ is analytic.

EXAMPLE 1 $w = z^2 = x^2 - y^2 + i \cdot 2xy$. Here $u = x^2 - y^2$, $v = 2xy$. Thus

$$\frac{\partial u}{\partial x} = 2x = \frac{\partial v}{\partial y}, \qquad \frac{\partial u}{\partial y} = -2y = -\frac{\partial v}{\partial x}$$

and w is analytic for all z. ◀

EXAMPLE 2 $w = \dfrac{x}{x^2 + y^2} - \dfrac{iy}{x^2 + y^2}$. Here

$$\frac{\partial u}{\partial x} = \frac{y^2 - x^2}{(x^2 + y^2)^2} = \frac{\partial v}{\partial y}, \qquad \frac{\partial u}{\partial y} = \frac{-2xy}{(x^2 + y^2)^2} = -\frac{\partial v}{\partial x}.$$

Hence w is analytic except for $x^2 + y^2 = 0$, that is, for $z = 0$. ◀

EXAMPLE 3 $w = x - iy = \bar{z}$. Here $u = x$, $v = -y$ and

$$\frac{\partial u}{\partial x} = 1, \qquad \frac{\partial v}{\partial y} = -1, \qquad \frac{\partial u}{\partial y} = 0 = \frac{\partial v}{\partial x}.$$

Thus w is not analytic in any region. ◄

EXAMPLE 4 $w = x^2 y^2 + 2x^2 y^2 i$. Here

$$\frac{\partial u}{\partial x} = 2xy^2, \qquad \frac{\partial v}{\partial y} = 4x^2 y, \qquad \frac{\partial u}{\partial y} = 2x^2 y, \qquad \frac{\partial v}{\partial x} = 4xy^2. ◄$$

The Cauchy–Riemann equations give $2xy^2 = 4x^2 y$, $2x^2 y = -4xy^2$. These equations are satisfied only along the lines $x = 0$, $y = 0$. There is *no* region in which the Cauchy–Riemann equations hold, hence no region in which $f(z)$ is analytic. Functions analytic only at certain points are not considered unless these points form a region.

The terms *analytic at a point* or *analytic along a curve* are used, apparently in contradiction to the remark just made. However, we say that $f(z)$ is *analytic at the point* z_0 only if there is a region containing z_0 within which $f(z)$ is analytic. Similarly, $f(z)$ is *analytic along a curve* C only if $f(z)$ is analytic in a region containing C.

> THEOREM 9 *The sum, product, and quotient of analytic functions is analytic (provided in the last case the denominator is not equal to zero at any point of the region under consideration). All polynomials are analytic for all z. Every rational function is analytic in each region containing no root of the denominator. An analytic function of an analytic function is analytic.*

This follows from Theorem 4.

We readily verify (Problem 1 below) that the Cauchy–Riemann equations are satisfied for $u = \text{Re}(e^z)$, $v = \text{Im}(e^z)$. Hence e^z is analytic for all z. It then follows from Theorem 9 that $\sin z$, $\cos z$, $\sinh z$, and $\cosh z$ are analytic for all z, while $\tan z$, $\sec z$, and $\cot z$ are analytic except for certain points (Problem 6 below). Furthermore, the usual formulas for derivatives hold (Problem 3):

$$\frac{d}{dz} e^z = e^z, \qquad \frac{d}{dz} \sin z = \cos z, \qquad \dots \qquad (11\text{--}64)$$

Two basic theorems of more advanced theory are useful at this point (see Chapter IV of Vol. II, Part 1 of *A Course in Mathematical Analysis* by E. GOURSAT, Ginn and Co., N. Y. 1916.)

> THEOREM 10 *Given a function $f(x)$ of the real variable x, $a \leqslant x \leqslant b$, there is at most one analytic function $f(z)$ that reduces to $f(x)$ when z is real.*

> THEOREM 11 *If $f(z)$, $g(z)$, ... are functions which are all analytic in a region D which includes part of the real axis, and $f(z)$, $g(z)$, ... satisfy an algebraic identity when z is real, then these functions satisfy the same identity for all z in D.*

Theorem 10 implies that our definitions of e^z, $\sin z$, ... are the only ones that yield analytic functions and agree with the definitions for real variables.

Because of Theorem 11, we can be sure that all familiar identities of trigonometry, namely,

$$\sin^2 z + \cos^2 z = 1, \qquad \sin\left(\frac{\pi}{2} - z\right) = \cos z, \qquad \ldots \qquad (11-65)$$

continue to hold for complex z. A general algebraic identity is formed by replacing the variables $w_1, \ldots, w_n$ in an algebraic equation by functions $f_1(z), \ldots, f_n(z)$. Thus, in the two examples given, we have

$$w_1^2 + w_2^2 - 1 = 0, \qquad w_1 = \sin z, \qquad w_2 = \cos z,$$

$$w_1 - w_2 = 0, \qquad w_1 = \sin\left(\frac{\pi}{2} - z\right), \qquad w_2 = \cos z.$$

To prove identities such as

$$e^{z_1} \cdot e^{z_2} = e^{z_1 + z_2}, \qquad (11-66)$$

it may be necessary to apply Theorem 11 several times (see Problems 4 and 5 below).

It should be remarked that while e^z is written as a power of e, it is best not to think of it as such. Thus $e^{1/2}$ has only one value, not two, as would a usual complex root. To avoid confusion with the general power function to be defined below, we often write $e^z = \exp z$ and refer to e^z as the *exponential function of z*.

To obtain the real and imaginary parts of $\sin z$, we use the identity

$$\sin(z_1 + z_2) = \sin z_1 \cos z_2 + \cos z_1 \sin z_2,$$

which holds, by the reasoning described above, for all complex z_1 and z_2. Hence $\sin(x + iy) = \sin x \cos iy + \cos x \sin iy$. Now from the definitions (Section 11–3).

$$\sinh y = -i \sin iy, \qquad \cosh y = \cos iy. \qquad (11-67)$$

Hence

$$\sin z = \sin x \cosh y + i \cos x \sinh y. \qquad (11-68)$$

Similarly, we prove, as in (11–34), that

$$\cos z = \cos x \cosh y - i \sin x \sinh y,$$
$$\sinh z = \sinh x \cos y + i \cosh x \sin y, \qquad (11-69)$$
$$\cosh z = \cosh x \cos y + i \sinh x \sin y.$$

Conformal mapping.   A complex function $w = f(z)$ can be considered a *mapping* from the xy-plane to the uv-plane, as in Section 9–10. In the case of analytic function $f(z)$, this mapping has a special property: it is a *conformal mapping*. By this we mean that two curves in the xy-plane, meeting at (x_0, y_0) at angle α, correspond to two curves meeting at the corresponding point (u_0, v_0) at the *same angle* α (in value and in sense—positive or negative). This means that a small triangle in the xy-plane corresponds to a *similar* small (curvilinear) triangle in the uv-plane. (The properties described fail at the exceptional points where $f'(z) = 0$.) Furthermore, every conformal mapping from the xy-plane to the uv-plane is given by an analytic function. For a discussion of conformal mapping and its applications, see Sections 11–25 to 11–27.

=========================Problems (Section 11–6)=========================

1. Verify that the following are analytic functions of z:
   a) $2x^3 - 3x^2y - 6xy^2 + y^3 + i(x^3 + 6x^2y - 3xy^2 - 2y^3)$
   b) $w = e^z = e^x \cos y + ie^x \sin y$
   c) $w = \sin z = \sin x \cosh y + i \cos x \sinh y$

2. Test each of the following for analyticity:
   a) $x^3 + y^3 + i(3x^2y + 3xy^2)$ b) $\sin x \cos y + i \cos x \sin y$
   c) $3x + 5y + i(3y - 5x)$

3. Prove the following properties directly from the definitions of the functions:
   a) $\dfrac{d}{dz} e^z = e^z$
   b) $\dfrac{d}{dz} \sin z = \cos z, \dfrac{d}{dz} \cos z = -\sin z$
   c) $\sin (z + \pi) = -\sin z,$
   d) $\sin (-z) = -\sin z, \cos (-z) = \cos z$

4. Prove the identity $e^{z_1 + z_2} = e^{z_1} \cdot e^{z_2}$ by application of Theorem 11.
   HINT: Let $z_2 = b$, a fixed real number, and $z_1 = z$, a variable complex number. Then $e^{z+b} = e^z \cdot e^b$ is an identity connecting analytic functions which is known to be true for z real. Hence it is true for all complex z. Now proceed similarly with the identity $e^{z_1 + z} = e^{z_1} \cdot e^z$.

5. Prove the following identities by application of Theorem 11 (see Problem 4):
   a) $\cos(z_1 + z_2) = \cos z_1 \cos z_2 - \sin z_1 \sin z_2$
   b) $e^{iz} = \cos z + i \sin z$
   c) $(e^z)^n = e^{nz}, n = 0, 1, 2, \ldots.$

6. Determine where the following functions are analytic (see Problem 5 following Section 11–4):
   a) $\tan z = \dfrac{\sin z}{\cos z}$
   b) $\cot z = \dfrac{\cos z}{\sin z}$
   c) $\tanh z = \dfrac{\sinh z}{\cosh z}$
   d) $\dfrac{\sin z}{z}$
   e) $\dfrac{e^z}{z \cos z}$
   f) $\dfrac{e^z}{\sin z + \cos z}$

===

11–7 THE FUNCTIONS In z, a^z, z^a, $\sin^{-1} z$, $\cos^{-1} z$

The function $w = \ln z$ is defined as the inverse of the exponential function $z = e^w$. We write $z = re^{i\theta}$, in terms of polar coordinates r, θ, and $w = u + iv$, so that

$$re^{i\theta} = e^{u + iv} = e^u e^{iv},$$
$$e^u = r, \qquad v = \theta + 2k\pi, \qquad k = 0, \pm 1, \ldots$$

Accordingly,

$$w = \ln z = \ln r + i(\theta + 2k\pi) = \ln |z| + i \arg z, \qquad (11\text{–}70)$$

where ln r is the real logarithm of r. Thus ln z is a multiple-valued function of z, with infinitely many values except for $z = 0$. We can select one value of θ for each z and obtain a single-valued function ln $z = \ln r + i\theta$; however, θ cannot be chosen to depend continuously on z for all $z \neq 0$, since θ will increase by 2π each time the origin is encircled in the positive direction.

   If we concentrate on an appropriate portion of the z-plane, we can choose θ to vary continuously within the region. For example, the inequalities $-\pi < \theta < \pi$, $r > 0$ together describe a region (see Fig. 11–5) and also tell us how to assign the values of θ within the region. With θ so restricted, $\ln r + i\theta$ then defines a *branch* of ln z in the region chosen; this particular branch is called the *principal value* of ln z and is denoted by Ln z. The points on the negative real axis are excluded from the region, but we usually assign the values Ln $z = \ln |x| + i\pi$ on this line. Within the region of Fig. 11–5, Ln z is *an analytic function of z* (Problem 4 below). Other branches of ln z are obtained by varying the choice of θ or of the region. For example, in the region of Fig. 11–5, we might choose θ so that $\pi < \theta < 3\pi$, or so that $-3\pi < \theta < -\pi$. The inequalities $0 < \theta < 2\pi$, $\pi/2 < \theta < 5\pi/2, \ldots$ also suggest other regions and choices of θ. We can verify that so long as angle θ varies continuously in the region, ln $z = \ln r + i\theta$ is analytic there. The most general region possible here is an arbitrary simply-connected region not containing the origin.

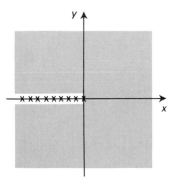

Fig. 11–5. Region for ln z.

   As a result of this discussion, it appears that ln z is formed of many branches, each analytic in some region not containing the origin. The branches fit together in a simple way; in general, we can get from one branch to another by moving around the origin a sufficient number of times, while varying the choice of ln z continuously. We say that the branches form *analytic continuations* of each other (see Section 11–23).

   We can further verify that for each branch of ln z, the rule

$$\frac{d}{dz} \ln z = \frac{1}{z}$$

remains valid. The familiar identities are also satisfied (Problems 4 and 5 below).

The *general exponential function* a^z is defined, for $a \neq 0$, by the equation

$$a^z = e^{z \ln a} = \exp(z \ln a). \tag{11-71}$$

Thus for $z = 0$, $a^0 = 1$. Otherwise, $\ln a = \ln |a| + i \arg a$, and we obtain many values: $a^z = \exp\{z[\ln|a| + i(\alpha + 2n\pi)]\}$, $n = 0, \pm 1, \pm 2, \ldots$, where α denotes one choice of $\arg a$. For example,

$$(1 + i)^i = \exp\left\{ i \left[\ln \sqrt{2} + i \left(\frac{\pi}{4} + 2n\pi \right) \right] \right\}$$

$$= e^{-(\pi/4) - 2n\pi} (\cos \ln \sqrt{2} + i \sin \ln \sqrt{2}).$$

If z is a positive integer m, a^z reduces to a^m and has only one value. The same holds for $z = -m$, and we have

$$a^{-m} = \frac{1}{a^m}. \tag{11-72}$$

If $z = 1/n$, where n is a positive integer, we find that a^z has n distinct values:

$$a^{1/n} = \sqrt[n]{a} = |a|^{1/n} \left[\cos \left(\frac{\theta}{n} + \frac{2k\pi}{n} \right) + i \sin \left(\frac{\theta}{n} + \frac{2k\pi}{n} \right) \right],$$

$$k = 0, 1, \ldots, n - 1. \tag{11-73}$$

Here $|a|^{1/n}$ is the real nth root of $|a|$. If $z = m/n$, a fraction in lowest terms, with n positive, then $a^z = a^{m/n}$ again has n distinct values, which are the nth roots of a^m. Equation (11–73) is valid also for $a = 0$, but then all n roots reduce to 0. Also $0^{m/n} = 0$ if m/n is a positive fraction.

If a fixed choice of $\ln a$ is made in (11–71), then a^z is simply e^{cz}, $c = \ln a$, and is hence an analytic function of z for all z. Each choice of $\ln a$ determines such a function.

If a and z are interchanged in (11–71), we obtain the *general power function*,

$$z^a = e^{a \ln z}. \tag{11-74}$$

If an analytic branch of $\ln z$ is chosen as above, then this function becomes an analytic function of an analytic function and is hence analytic in the region chosen. In particular, the *principal value* of z^a is defined as the analytic function $z^a = e^{a \operatorname{Ln} z}$, in terms of the principal value of $\ln z$.

For example, if $a = 1/2$, we have

$$z^{1/2} = e^{(1/2) \ln z} = e^{(1/2)(\ln r + i\theta)} = e^{(1/2)\ln r} e^{(1/2)i\theta}$$

$$= \sqrt{r} \left(\cos \frac{\theta}{2} + i \sin \frac{\theta}{2} \right),$$

as in Eq. (11–73). If $\operatorname{Ln} z$ is used, then $\sqrt{z} = f_1(z)$ becomes analytic in the region of Fig. 11–5. A second analytic branch $f_2(z)$ in the same region is obtained by requiring that $\pi < \theta < 3\pi$. These are the only two analytic branches that can be obtained in this region. It should be remarked that these two branches are related by the equation $f_2(z) = -f_1(z)$, for f_2 is obtained from f_1 by increasing θ by 2π, which replaces $e^{(1/2)i\theta}$ by

$$e^{(1/2)i(\theta + 2\pi)} = e^{\pi i} e^{(1/2)i\theta} = -e^{(1/2)i\theta}.$$

The functions $\sin^{-1} z$ and $\cos^{-1} z$ are defined as the inverses of $\sin z$ and $\cos z$. We then find

$$\sin^{-1} z = \frac{1}{i} \ln \left[iz \pm \sqrt{1 - z^2} \right],$$

$$\cos^{-1} z = \frac{1}{i} \ln \left[z \pm i \sqrt{1 - z^2} \right]. \tag{11–75}$$

The proofs are left to the exercises (Problem 2). It can be shown that analytic branches of both these functions can be defined in each simply connected region not containing the points ± 1. For each z other than ± 1, there are two choices of $\sqrt{1 - z^2}$ and for each of these choices there is an infinite sequence of choices of the logarithm, differing by multiples of $2\pi i$.

═══════════════════════════════Problems (Section 11–7)═══════════════════════════════

1. Obtain all values of each of the following:
   a) $(-1)^{1/2}$ b) $(-1)^{1/4}$ c) $(-1 + 3i)^{1/2}$ d) $(-1 - i)^{1/5}$
   e) $\ln 2$ f) $\ln i$ g) $\ln (1 - i)$ h) i^i
   i) $(1 + i)^{2/3}$ j) $i^{\sqrt{2}}$ k) $\sin^{-1} 1$ l) $\cos^{-1} 2$

2. Prove the formulas (11–75).

   HINT: If $w = \sin^{-1} z$, then $2iz = e^{iw} - e^{-iw}$; multiply by e^{iw} and solve the resulting equation as a quadratic for e^{iw}.

3. a) Evaluate $\sin^{-1} 0$, $\cos^{-1} 0$.
   b) Find all roots of $\sin z$ and $\cos z$.

   HINT: Compare part (a).

4. Show that each branch of $\ln z$ is analytic in each region where θ varies continuously and that $(d/dz) \ln z = 1/z$.

   HINT: Show from the equations $x = r \cos \theta$, $y = r \sin \theta$ that $\partial \theta / \partial x = -y/r^2$ and $\partial \theta / \partial y = x/r^2$. Show that the Cauchy–Riemann equations hold for $u = \ln r$, $v = \theta$.

5. Prove the following identities in the sense that, for proper selection of values of the multiple-valued functions concerned, the equation is correct for each allowed choice of the variables:
   a) $\ln (z_1 \cdot z_2) = \ln z_1 + \ln z_2$, $z_1 \neq 0$, $z_2 \neq 0$
   b) $e^{\ln z} = z$ $(z \neq 0)$ c) $\ln e^z = z$
   d) $\ln z_1^{z_2} = z_2 \ln z_1$, $z_1 \neq 0$

6. For each of the following determine all analytic branches of the multiple-valued function in the region given:
   a) $\ln z$, $x < 0$ b) $\sqrt[3]{z}$, $x > 0$

7. Prove that for the analytic function z^a (principal value),
   $$(d/dz) z^a = (az^a)/z = az^{a-1}.$$

8. Plot the functions $u = \operatorname{Re}(\sqrt{z})$ and $v = \operatorname{Im}(\sqrt{z})$ as functions of x and y and show the two branches described in the text.

11–8 INTEGRALS OF ANALYTIC FUNCTIONS. CAUCHY INTEGRAL THEOREM

All paths in the integrals concerned here, as elsewhere in the chapter, are assumed to be piecewise smooth.

The following theorem is fundamental for the theory of analytic functions:

THEOREM 12   *(Cauchy integral theorem) If $f(z)$ is analytic in a simply connected region D, then*

$$\oint_C f(z)\, dz = 0$$

on every simple closed path C in D (Fig. 11–6).

PROOF   We have, by (11–51) and with the positive direction understood,

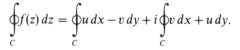

$$\oint_C f(z)\, dz = \oint_C u\, dx - v\, dy + i \oint_C v\, dx + u\, dy.$$

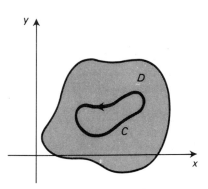

Fig. 11–6. Cauchy integral theorem.

The two real integrals are equal to zero (see Section 10–9) provided u and v have continuous derivatives in D and

$$\frac{\partial u}{\partial y} = -\frac{\partial v}{\partial x}, \qquad \frac{\partial v}{\partial y} = \frac{\partial u}{\partial x}.$$

These are just the Cauchy–Riemann equations. Hence

$$\oint_C f(z)\, dz = 0 + i \cdot 0 = 0.$$

This theorem can be stated in an equivalent form:

THEOREM 12′   *If $f(z)$ is analytic in the simply connected region D, then $\int f(z)\, dz$ is independent of the path in D.*

For independence of path and equaling zero on closed paths are equivalent properties of line integrals. If C is a path from z_1 to z_2, we can now write

$$\int_C f(z)\,dz = \int_{z_1}^{z_2} f(z)\,dz,$$

the integral being the same for all paths C from z_1 to z_2 in D.

THEOREM 13   *Let $f(z) = u + iv$ be defined in region D and let u and v have continuous partial derivatives in D. If*

$$\oint_C f(z)\,dz = 0 \qquad\qquad (11\text{–}80)$$

on every simple closed path C in D, then $f(z)$ is analytic in D.

PROOF   The condition (11–80) implies that

$$\oint_C u\,dx - v\,dy = 0, \qquad \oint_C v\,dx + u\,dy = 0$$

on all simple closed paths C, that is, the two real line integrals are independent of path in D. Therefore, as in Section 10–9,

$$\frac{\partial u}{\partial y} = -\frac{\partial v}{\partial x}, \qquad \frac{\partial v}{\partial y} = \frac{\partial u}{\partial x};$$

since the Cauchy–Riemann equations hold, f is analytic.

THEOREM 14   *If $f(z)$ is analytic in D, then*

$$\int_{z_1}^{z_2} f'(z)\,dz = f(z)\bigg|_{z_1}^{z_2} = f(z_2) - f(z_1) \qquad\qquad (11\text{–}81)$$

on every path in D from z_1 to z_2. In particular,

$$\oint f'(z)\,dz = 0$$

on every closed path in D.

PROOF   By (11–51) above,

$$\int_{z_1}^{z_2} f'(z)\,dz = \int_{z_1}^{z_2} \left(\frac{\partial u}{\partial x} + i\frac{\partial v}{\partial x} \right)(dx + i\,dy)$$

$$= \int_{z_1}^{z_2} \frac{\partial u}{\partial x}\,dx + \frac{\partial u}{\partial y}\,dy + i \int_{z_1}^{z_2} \frac{\partial v}{\partial x}\,dx + \frac{\partial v}{\partial y}\,dy$$

$$= \int_{z_1}^{z_2} du + i\,dv = (u + iv)\bigg|_{z_1}^{z_2} = f(z_2) - f(z_1).$$

This rule is the basis for evaluation of simple integrals, just as in elementary calculus. Thus we have

$$\int_{i}^{1+i} z^2\, dz = \frac{z^3}{3}\Big|_{i}^{1+i} = \frac{(1+i)^3 - i^3}{3} = -\tfrac{2}{3}+i,$$

$$\int_{i}^{-i} \frac{1}{z^2}\, dz = -\frac{1}{z}\Big|_{i}^{-i} = -i-i = -2i.$$

In the first of these any path can be used; in the second, any path not through the origin.

THEOREM 15   *If $f(z)$ is analytic in D and D is simply connected, then*

$$F(z) = \int_{z_1}^{z} f(z)\, dz, \qquad z_1 \text{ fixed in } D, \qquad (11\text{--}82)$$

is an indefinite integral of $f(z)$, that is, $F'(z) = f(z)$. Thus $F(z)$ is itself analytic.

PROOF   Since $f(z)$ is analytic in D and D is simply connected, $\int_{z_1}^{z} f(z)\, dz$ is independent of path and defines a function F that depends only on the upper limit z. We have, further, $F = U + iV$, where

$$U = \int_{z_1}^{z} u\, dx - v\, dy, \qquad V = \int_{z_1}^{z} v\, dx + u\, dy$$

and both integrals are independent of path. Hence $dU = u\, dx - v\, dy$, $dV = v\, dx + u\, dy$. Thus U and V satisfy the Cauchy–Riemann equations, so that $F = U + iV$ is analytic and

$$F'(z) = \frac{\partial U}{\partial x} + i\frac{\partial V}{\partial x} = u + iv = f(z).$$

Cauchy's theorem for multiply connected regions If $f(z)$ is analytic in a multiply connected region D, then we cannot conclude that

$$\oint f(z)\, dz = 0$$

on every simple closed path C in D. Thus, if D is the doubly connected region of Fig. 11–7 and C is the curve C_1 shown, then the integral around C need not be zero. However, by introducing cuts, as in Section 10–9, we can reason that

$$\oint_{C_1} f(z)\, dz = \oint_{C_2} f(z)\, dz, \qquad (11\text{--}83)$$

that is, the integral has the same value on all paths that go around the inner "hole" once in the positive direction. For a triply connected region, as in Fig. 11–8, we obtain the equation

$$\oint_{C_1} f(z)\, dz = \oint_{C_2} f(z)\, dz + \oint_{C_3} f(z)\, dz. \qquad (11\text{--}84)$$

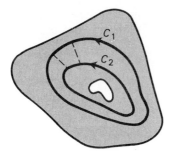

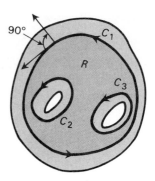

Fig. 11–7. Cauchy theorem for doubly
connected region.

Fig. 11–8. Cauchy theorem for triply
connected region.

This can be written in the form

$$\oint_{C_1} f(z)\,dz + \oint_{C_2} f(z)\,dz + \oint_{C_3} f(z)\,dz = 0. \qquad (11\text{–}85)$$

Equation (11–85) states that the integral of $f(z)$ around the complete boundary B of
the region R bounded by C_1, C_2, C_3 is zero, when we integrate in the *positive*
direction on the outer boundary curve C_1 and in the *negative* direction on the inner
boundary curves C_2, C_3. In each case the direction chosen is such that the outer
normal (pointing away from the region R) is $90°$ behind the tangent vector in the
direction of integration, as suggested on C_1 in Fig. 11–8.

The result clearly extends to the case of n boundary curves:

THEOREM 16   (*Cauchy's theorem for multiply connected regions*)   *Let $f(z)$
be analytic in a region D and let $C_1, \ldots, C_n$ be n simple closed curves in D
which together form the boundary B of a closed region R contained in D. Then*

$$\int_B f(z)\,dz = 0,$$

*where the direction of integration on B is such that the outer normal is $90°$
behind the tangent vector in the direction of integration.*

11–9 CAUCHY'S INTEGRAL FORMULA

Now let D be a simply connected region and let z_0 be a fixed point of D. If $f(z)$ is
analytic in D, the function $f(z)/(z - z_0)$ will fail to be analytic at z_0. Hence

$$\oint \frac{f(z)}{z - z_0}\,dz$$

will in general not be zero on a path C enclosing z_0. However, as above, this integral will have the same value on all paths C about z_0. To determine this value, we choose C to be a circle $z = z_0 + Re^{i\theta}$, $0 \leqslant \theta \leqslant 2\pi$. Then by (11–54) the integral becomes

$$\int_0^{2\pi} \frac{f(z_0 + Re^{i\theta})}{Re^{i\theta}} i Re^{i\theta}\, d\theta = i \int_0^{2\pi} f(z_0 + Re^{i\theta})\, d\theta.$$

Since f is analytic, hence continuous, the integral must thus be continuous in R. But we saw that the value of the integral is the same for all paths C enclosing z_0. Therefore, the value is independent of R. We now find the value by letting R approach 0:

$$\oint_C \frac{f(z)}{z - z_0}\, dz = \lim_{R \to 0} i \int_0^{2\pi} f(z_0 + Re^{i\theta})\, d\theta = i \int_0^{2\pi} f(z_0)\, d\theta = 2\pi i\, f(z_0).$$

We thus obtain the following fundamental result:

THEOREM 17 (*Cauchy integral formula*) *Let $f(z)$ be analytic in a region D. Let C be a simple closed curve in D, within which $f(z)$ is analytic, and let z_0 be inside C. Then*

$$f(z_0) = \frac{1}{2\pi i} \oint_C \frac{f(z)}{z - z_0}\, dz. \tag{11–90}$$

We remark that D need not itself be simply connected; the reasoning given above applies to a simply connected part of D, as suggested in Fig. 11–9.

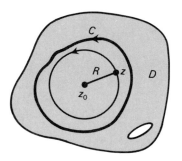

Fig. 11–9. Cauchy integral formula.

The integral formula (11–90) is remarkable in that it expresses the values of the function $f(z)$ at points z_0 inside the curve C in terms of the values along C alone. If C is taken as a circle $z = z_0 + Re^{i\theta}$, then (11–90) reduces to the following:

$$f(z_0) = \frac{1}{2\pi} \int_0^{2\pi} f(z_0 + Re^{i\theta})\, d\theta. \tag{11–91}$$

Thus the *value of an analytic function at the center of a circle equals the average (arithmetic mean) of the values on the circumference.*

Just as with the Cauchy integral theorem, the Cauchy integral formula can be extended to multiply connected regions. Under the hypotheses of Theorem 16,

$$f(z_0) = \frac{1}{2\pi i} \int_B \frac{f(z)}{z - z_0} \, dz = \frac{1}{2\pi i} \left[\oint_{C_1} \frac{f(z)}{z - z_0} \, dz + \oint_{C_2} \frac{f(z)}{z - z_0} \, dz + \cdots \right], \quad (11\text{--}92)$$

where z_0 is any point inside the closed region R bounded by C_1 (the outer boundary), $C_2, \ldots, C_n$. The proof is left as an exercise (see Problem 6 below).

════════════════════════Problems (Section 11-9)════════════════════

1. Evaluate the following integrals:

   a) $\oint z^2 \sin z \, dz$ on the ellipse $x^2 + 2y^2 = 1$;

   b) $\oint \dfrac{z^2}{z+1} \, dz$ on the circle $|z - 2| = 1$;

   c) $\displaystyle\int_1^{2i} z e^z \, dz$ on the line segment joining the endpoints;

   d) $\displaystyle\int_{1+i}^{1-i} \frac{1}{z^2} \, dz$ on the parabola $2y^2 = x + 1$.

2. a) Evaluate $\int_{-i}^{i} (dz/z)$ on the path $z = e^{it}$, $-\pi/2 \leqslant t \leqslant \pi/2$, with the aid of the relation $(\ln z)' = 1/z$, for an appropriate branch of $\ln z$.
   b) Evaluate $\int_i^{-i} (dz/z)$ on the path $z = e^{it}$, $\pi/2 \leqslant t \leqslant 3\pi/2$, as in part (a).
   c) Why does the relation $(\ln z)' = 1/z$ not imply that the sum of the two integrals of parts (a) and (b) is zero?

3. A certain function $f(z)$ is known to be analytic except for $z = 1$, $z = 2$, $z = 3$, and it is known that $\oint_{C_k} f(z) \, dz = a_k$, $k = 1, 2, 3$, where C_k is a circle of radius $1/2$ with center at $z = k$. Evaluate $\oint f(z) \, dz$ on each of the following paths:
   a) $|z| = 4$ b) $|z| = 2.5$ c) $|z - 2.5| = 1$

4. Evaluate each of the following with the aid of the Cauchy integral formula:

   a) $\oint \dfrac{z}{z - 3} \, dz$ on $|z| = 5$ b) $\oint \dfrac{e^z}{z^2 - 3z} \, dz$ on $|z| = 1$

   c) $\oint \dfrac{z + 2}{z^2 - 1} \, dz$ on $|z| = 2$ d) $\oint \dfrac{\sin z}{z^2 + 1} \, dz$ on $|z| = 2$

   HINT: for(c) and (d): Expand the rational function in partial fractions.

5. Prove that if $f(z)$ is analytic in region D and $f'(z) \equiv 0$, then $f(z) \equiv$ const.

   HINT: Apply Theorem 14.

6. Prove (11-92) under the hypotheses stated.

11–10 POWER SERIES AS ANALYTIC FUNCTIONS

We now proceed to enlarge the class of specific analytic functions still further by showing that every power series

$$\sum_{n=0}^{\infty} c_n(z - z_0)^n = c_0 + c_1(z - z_0) + \cdots + c_n(z - z_0)^n + \cdots$$

converging for some values of z other than $z = z_0$ represents an analytic function.

For the theory of series of complex numbers, see Section 2–20. The following fundamental theorem for complex power series is proved as for real series (see AC, Section 6–15).

> **THEOREM 18** *Every power series $\sum_{n=0}^{\infty} c_n(z - z_0)^n$ has a radius of convergence r^* such that the series converges absolutely when $|z - z_0| < r^*$, and diverges when $|z - z_0| > r^*$. The series converges uniformly for $|z - z_0| \leqslant r_1$, provided $r_1 < r^*$.*

The number r^* can be zero, in which case the series converges only for $z = z_0$, a positive number, or ∞, in which case the series converges for all z.

The number r^* can be evaluated as follows:

$$r^* = \lim_{n \to \infty} \left| \frac{c_n}{c_n + 1} \right|, \text{ if the limit exists,}$$

$$r^* = \lim_{n \to \infty} \frac{1}{\sqrt[n]{|c_n|}}, \text{ if the limit exists,} \tag{11–100}$$

and in any case by the formula

$$r^* = \frac{1}{\overline{\lim_{n \to \infty}} \sqrt[n]{|c_n|}}. \tag{11–101}$$

where $\overline{\lim}$ is the "upper limit" (see AC, p. 375).

As for real variables, no general statement can be made about convergence on the boundary of the region of convergence. This boundary (when $r^* \neq 0, r^* \neq \infty$) is a circle $|z - z_0| = r^*$, termed the *circle of convergence* (Fig. 11–10). The series may converge at some points, all points, or no points of this circle.

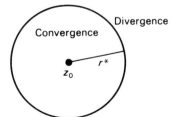

Fig. 11–10. Circle of convergence of a power series.

EXAMPLE 1 $\sum_{n=1}^{\infty} (z^n/n^2)$. The first formula (11–100) gives

$$r^* = \lim_{n \to \infty} \frac{(n+1)^2}{n^2} = 1.$$

The series converges absolutely on the circle of convergence, for when $|z| = 1$, the series of absolute value is the convergent series $\sum (1/n^2)$. ◄

EXAMPLE 2 $\sum_{n=0}^{\infty} z^n$. This complex geometric series converges for $|z| < 1$, as (11–100) shows. We have further

$$\sum_{n=0}^{\infty} z^n = \frac{1}{1-z}, \qquad |z| < 1,$$

as for real variables. On the circle of convergence, the series diverges everywhere, since the nth term fails to converge to zero. ◄

The following theorems are proved as for real variables.

THEOREM 19   *A power series with nonzero convergence radius represents a continuous function within the circle of convergence.*

THEOREM 20   *A power series can be integrated term by term within the circle of convergence, that is, if $r^* \neq 0$ and*

$$f(z) = \sum_{n=0}^{\infty} c_n(z - z_0)^n, \qquad |z - z_0| < r^*,$$

then, for every path C inside the circle of convergence

$$\int_{C}^{z_2}{}_{z_1} f(z)\, dz = \sum_{n=0}^{\infty} c_n \int_{z_1}^{z_2} (z - z_0)^n\, dz = \sum_{n=0}^{\infty} c_n \frac{(z - z_0)^{n+1}}{n+1} \Big|_{z_1}^{z_2},$$

or in terms of indefinite integrals,

$$\int f(z)\, dz = \sum_{n=0}^{\infty} c_n \frac{(z - z_0)^{n+1}}{n+1} + \text{const}, \qquad |z - z_0| < r^*.$$

THEOREM 21   *A power series can be differentiated term by term, that is, if $r^* \neq 0$ and*

$$f(z) = \sum_{n=0}^{\infty} c_n(z - z_0)^n, \qquad |z - z_0| < r^*,$$

then

$$f'(z) = \sum_{n=1}^{\infty} n c_n (z - z_0)^{n-1}, \qquad |z - z_0| < r^*,$$

$$f''(z) = \sum_{n=2}^{\infty} n(n-1) c_n (z - z_0)^{n-2}, \qquad |z - z_0| < r^*, \quad etc.$$

Hence every power series with nonzero convergence radius defines an analytic function $f(z)$ within the circle of convergence, and the power series is the Taylor series of $f(z)$:

$$c_n = \frac{f^{(n)}(z_0)}{n!}.$$

THEOREM 22   *If two power series $\sum_{n=0}^{\infty} c_n (z - z_0)^n$, $\sum_{n=0}^{\infty} C_n (z - z_0)^n$ have nonzero convergence radii and have equal sums wherever both series converge, then the series are identical, that is,*

$$c_n = C_n, \qquad n = 0, 1, 2, \ldots$$

11–11   POWER SERIES EXPANSION OF GENERAL ANALYTIC FUNCTION

In Section 11–10 it was shown that every power series with nonzero convergence radius represents an analytic function. We now proceed to show that all analytic functions are obtainable in this way. If a function $f(z)$ is analytic in a region D of general shape, we cannot expect to represent $f(z)$ by one power series, for the power series converges only in a circular region. However, we can show that for each circular region D_0 in D, there is a power series converging in D_0 whose sum is $f(z)$. Thus several (perhaps infinitely many) power series are needed to represent $f(z)$ throughout all of D.

THEOREM 23   *Let $f(z)$ be analytic in the region D. Let z_0 be in D and let R be the radius of the largest circle with center at z_0 and having its interior in D. Then there is a power series*

$$\sum_{n=0}^{\infty} c_n (z - z_0)^n$$

which converges to $f(z)$ for $|z - z_0| < R$. Furthermore,

$$c_n = \frac{f^{(n)}(z_0)}{n!} = \frac{1}{2\pi i} \oint_C \frac{f(z)}{(z - z_0)^{n+1}} \, dz, \qquad (11\text{--}110)$$

where C is a simple closed path in D enclosing z_0 and within which $f(z)$ is analytic.

PROOF   For simplicity we take $z_0 = 0$. The general case can then be obtained by the substitution $z' = z - z_0$. Let the circle $|z| = R$ be the largest circle with center at z_0 and its interior within D; the radius R is then positive or $+\infty$ (in which case D is the whole z-plane). Let z_1 be a point within this circle, so that $|z_1| < R$. Choose R_2 so that $|z_1| < R_2 < R$ (see Fig. 11–11). Then $f(z)$ is analytic in a region including the circle C_2: $|z| = R_2$ plus interior. Hence by the Cauchy integral formula,

$$f(z_1) = \frac{1}{2\pi i} \oint_{C_2} \frac{f(z)}{z - z_1}\, dz.$$

Now the factor $1/(z - z_1)$ can be expanded in a geometric series:

$$\frac{1}{z - z_1} = \frac{1}{z(1 - z_1/z)} = \frac{1}{z}\left(1 + \frac{z_1}{z} + \cdots + \frac{z_1^n}{z^n} + \cdots\right).$$

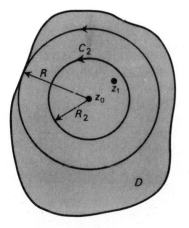

Fig. 11–11. Taylor series of an analytic function.

The series can be considered a power series in powers of $1/z$ for fixed z_1. It converges for $|z_1/z| < 1$ and converges uniformly for $|z_1/z| \leqslant |z_1|/R_2 < 1$.

If we multiply by $f(z)$, we find

$$\frac{f(z)}{z - z_1} = \frac{f(z)}{z} + z_1 \frac{f(z)}{z^2} + \cdots + z_1^n \frac{f(z)}{z^{n+1}} + \cdots;$$

since $f(z)$ is continuous for $|z| = R_2$, the series remains uniformly convergent on C_2. Hence we can integrate term by term on C_2:

$$\frac{1}{2\pi i} \oint_{C_2} \frac{f(z)}{z - z_1}\, dz = \frac{1}{2\pi i} \oint_{C_2} \frac{f(z)}{z}\, dz + \frac{z_1}{2\pi i} \oint_{C_2} \frac{f(z)}{z^2}\, dz + \cdots + \frac{z_1^n}{2\pi i} \oint_{C_2} \frac{f(z)}{z^{n+1}}\, dz + \cdots$$

The left-hand side is precisely $f(z_1)$, by the integral formula. Hence

$$f(z_1) = \sum_{n=0}^{\infty} c_n z_1^n, \qquad c_n = \frac{1}{2\pi i} \oint_{C_2} \frac{f(z)}{z^{n+1}} \, dz.$$

The path C_2 can be replaced by any path C as described in the theorem, since $f(z)/z^{n+1}$ is analytic in D except for $z = z_0 = 0$.

By Theorem 21, the series obtained is the Taylor series of f, so that

$$c_n = \frac{f^{(n)}(z_0)}{n!}, \qquad z_0 = 0.$$

The theorem is now completely proved.

The consequences of this theorem are far-reaching. First of all, not only does it guarantee that every analytic function is representable by power series, but it ensures that the Taylor series converges to the function within each circular region within the region in which the function is given. Thus, *without further analysis*, we at once conclude that

$$e^z = 1 + z + \frac{z^2}{2!} + \cdots + \frac{z^n}{n!} + \cdots,$$

$$\sin z = z - \frac{z^3}{3!} + \frac{z^5}{5!} + \cdots + (-1)^n \frac{z^{2n+1}}{(2n+1)!} + \cdots,$$

$$\cos z = 1 - \frac{z^2}{2!} + \cdots + (-1)^n \frac{z^{2n}}{(2n)!} + \cdots$$

for all z. A variety of other familiar expansions can be obtained in the same way.

It should be recalled that a function $f(z)$ is defined to be analytic in a region D if $f(z)$ has a continuous derivative $f'(z)$ in D (Section 11–6). By Theorem 23, $f(z)$ must have derivatives of all orders at every point of D. In particular, the derivative of an analytic function is itself analytic:

THEOREM 24   *If $f(z)$ is analytic in region D, then $f'(z), f''(z), \ldots, f^{(n)}(z), \ldots$ exist and are analytic in D. Furthermore, for each n*

$$f^{(n)}(z_0) = \frac{n!}{2\pi i} \oint_C \frac{f(z)}{(z - z_0)^{n+1}} \, dz, \qquad (11\text{--}111)$$

where C is any simple closed path in D enclosing z_0 and within which $f(z)$ is analytic.

Equation (11–111) is a restatement of (11–110).

Circle of convergence of the Taylor series   Theorem 23 guarantees convergence of the Taylor series of $f(z)$ about each z_0 in D in the largest circular region $|z - z_0| < R$ in D, as shown in Fig. 11–11. However, this does not mean that R is the radius of convergence r^* of the series, for r^* can be larger than R, as suggested

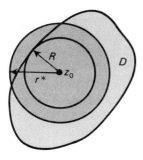

Fig. 11-12. Analytic continuation.

in Fig. 11–12. When this happens, the function $f(z)$ can be prolonged into a larger region, while retaining analyticity. For example, if $f(z) = \operatorname{Ln} z \ (0 < \theta < \pi)$ is explained in a Taylor series about the point $z = -1 + i$, the series has convergence radius $\sqrt{2}$, whereas $R - 1$ (see Problem 4(c) below).

The process of prolonging the function suggested here is called *analytic continuation.*

Harmonic functions   Let $w = f(z)$ be analytic in D and let
$$u(x, y) = \operatorname{Re}[f(z)], \qquad v(x, y) = \operatorname{Im}[f(z)].$$
By Theorem 8 (Section 11–6), $f'(z) = u_x + iv_x = v_y - iu_y = \ldots$ and, by Theorem 24, $f'(z)$ is analytic in D, so that
$$f''(z) = u_{xx} + iv_{xx} = v_{yx} - iu_{yx} = \cdots .$$
Thus $u_{xx}, u_{xy}, v_{xx}, \ldots$ are also continuous in D. In general, Theorem 24 shows that u and v have partial derivatives of all orders in D. Furthermore, by taking real and imaginary parts in the Taylor series of $f(z)$, as given in Theorem 23, we obtain convergent power series for $u(x, y)$ and $v(x, y)$. For example, if $f(z) = \sum c_n z^n$ for $|z| < R$, with $c_n = a_n + ib_n$, then
$$u(x, y) = a_0 + (a_1 x - b_1 y) + (a_2 x^2 - 2b_2 xy - a_2 y^2) + \cdots , \qquad x^2 + y^2 < R^2.$$
Also, by the Cauchy–Riemann equations, $u_x = v_y$, $u_y = -v_x$, so that
$$\frac{\partial^2 u}{\partial x^2} + \frac{\partial^2 u}{\partial y^2} = \frac{\partial^2 v}{\partial x \, \partial y} - \frac{\partial^2 v}{\partial y \, \partial x} = 0,$$
that is, $u(x, y)$ is *harmonic* in D. Similarly $v(x, y)$ is harmonic in D.

Harmonic functions u, v satisfying the Cauchy–Riemann equations are said to be *conjugate* harmonic functions; more precisely, v is said to be conjugate to u. If $u(x, y)$ is a harmonic function in a simply-connected region D, then we can construct a harmonic function $v(x, y)$ conjugate to u (see Problem 7 below).

We saw in Sections 8–9 and 10–18 that harmonic functions $u(x, y)$ represent electrostatic potentials and equilibrium temperature distributions. Thus all these harmonic functions can be obtained by taking the real part of analytic functions (see Section 11–28 below for further discussion).

═══════════════════════════Problems (Section 11–11)═══════════════════════

1. Determine the radius of convergence of each of the following series:

   a) $\displaystyle\sum_{n=0}^{\infty} \frac{z^n}{n^2+1}$ **b)** $\displaystyle\sum_{n=0}^{\infty} \frac{(z-1)^n}{3^n}$ **c)** $\displaystyle\sum_{n=0}^{\infty} n!z^n$ **d)** $\displaystyle\sum_{n=0}^{\infty} \frac{z^n}{n!}$

2. Given the series $\sum_{n=1}^{\infty} (z^n/n)$, show that
   a) The series has radius of convergence 1;
   b) The series diverges for $z = 1$;
   c) The series converges for $z = i$ and for $z = -1$. It can be shown that the series converges for $|z| = 1$, except for $z = 1$.

3. By means of (11–111), evaluate each of the following:

   a) $\displaystyle\oint \frac{ze^z}{(z-1)^4} dz$ on $|z| = 2$ **b)** $\displaystyle\oint \frac{\sin z}{z^4} dz$ on $|z| = 1$ **c)** $\displaystyle\oint \frac{dz}{z^3(z+4)}$ on $|z| = 2$

4. Expand in a Taylor series about the point indicated, and determine the radius of convergence r^* and the radius R of the largest circle within which the series converges to the function:
   a) $\sin z$ about $z = 0$ **b)** $1/(z-1)$ about $z = 2$
   c) $\operatorname{Ln} z$ $(0 < \theta < \pi)$ about $z = -1 + i$

5. *Cauchy's inequalities.* Let $f(z)$ be analytic in a domain including the circle $C: |z - z_0| = R$ and interior, and let $|f(z)| \leqslant M = \text{const}$ on C. Prove that

$$|f^{(n)}(z_0)| \leqslant \frac{Mn!}{R^n}, \qquad n = 0, 1, 2, \dots$$

   HINT: Apply (11–111) and Theorem 7.

6. A function $f(z)$ which is analytic in the whole z-plane is termed an *entire* function or an *integral* function. Examples are polynomials, e^z, $\sin z$, $\cos z$. Prove *Liouville's theorem:* If $f(z)$ is an entire function and $|f(z)| \leqslant M$ for all z, where M is constant, then $f(z)$ reduces to a constant.

   HINT: Take $n = 1$ in the Cauchy inequalities of Problem 5 to show that $f'(z_0) = 0$ for every z_0.

7. Let $u(x, y)$ be harmonic in the simply-connected region D and let (x_0, y_0) be a point of D.
   Show that $\displaystyle\int_{(x_0, y_0)}^{(x, y)} -\frac{\partial u}{\partial y} dx + \frac{\partial u}{\partial x} dy$ is independent of path in D and hence defines a function
   $v(x, y)$. Show that v is a harmonic conjugate of u.

═══

11–12 POWER SERIES IN POSITIVE AND NEGATIVE POWERS;LAURENT EXPANSION

We have shown that every power series $\sum a_n(z - z_0)^n$ with nonzero convergence radius represents an analytic function and that every analytic function can be built up out of such series. It thus appears unnecessary to seek other explicit expressions for analytic functions. However, the power series represent functions only in circular regions and are hence awkward for representing a function in a more complicated type of region. It is therefore worthwhile to consider other types of representations. A series of form

$$\sum_{n=1}^{\infty} \frac{b_n}{(z-z_0)^n} = \frac{b_1}{z-z_0} + \cdots + \frac{b_n}{(z-z_0)^n} + \cdots \qquad (11\text{--}120)$$

will also represent an analytic function in a region in which the series converges, for the substitution $z_1 = 1/(z-z_0)$ reduces the series to an ordinary power series $\sum_{n=1}^{\infty} b_n z_1^n$. If this series converges for $|z_1| < r_1^*$, then its sum is an analytic function $F(z_1)$; hence the series (11–120) converges for

$$|z-z_0| > \frac{1}{r_1^*} = r_0^* \qquad (11\text{--}121)$$

to the analytic function $g(z) = F(1/(z-z_0))$. The value $z_1 = 0$ corresponds to $z = \infty$, in a limiting sense; accordingly we can also say that $g(z)$ is analytic at ∞ and $g(\infty) = 0$. This will be justified more fully in Section 11–14.

The domain of convergence of the series (11–120) is the region (11–121), which is the *exterior* of a circle. It can happen that $r_1^* = \infty$, in which case the series converges for all z except z_0; if $r_1^* = 0$, the series diverges for all z (except $z = \infty$, as above).

If we add to a series (11–120) a usual power series

$$\sum_{n=0}^{\infty} a_n(z-z_0)^n = a_0 + a_1(z-z_0) + \cdots,$$

converging for $|z-z_0| < r_2^*$, we obtain a sum

$$\sum_{n=1}^{\infty} \frac{b_n}{(z-z_0)^n} + \sum_{n=0}^{\infty} a_n(z-z_0)^n. \qquad (11\text{--}122)$$

If $r_0^* < r_2^*$, the sum converges and represents an analytic function $f(z)$ in the *annular region* $r_0^* < |z-z_0| < r_2^*$, for each series has an analytic sum in this region so that the sum of the two series is analytic there. We can write this sum in the more compact form (after some relabeling)

$$f(z) = \sum_{n=-\infty}^{\infty} a_n(z-z_0)^n, \qquad (11\text{--}123)$$

though this should be interpreted as the sum of two series, as in (11–122).

In this way we build up a new class of analytic functions, each defined in a ring-shaped region. Every function analytic in such a region can be obtained in this way.

THEOREM 25 (*Laurent's theorem*) *Let* $f(z)$ *be analytic in the ring* $R_1 < |z - z_0| < R_2$. *Then*

$$f(z) = \sum_{n=-\infty}^{\infty} a_n(z - z_0)^n = [a_0 + a_1(z - z_0) + \cdots]$$

$$+ \left[\frac{a_{-1}}{z - z_0} + \frac{a_{-2}}{(z - z_0)^2} + \cdots\right],$$

where

$$a_n = \frac{1}{2\pi i} \oint_C \frac{f(z)}{(z - z_0)^{n+1}} \, dz \qquad (11-124)$$

and C is any simple closed curve separating $|z| = R_1$ *from* $|z| = R_2$. *The series converges uniformly for* $R_1 < k_1 \leqslant |z - z_0| \leqslant k_2 < R_2$.

PROOF   For simplicity we take $z_0 = 0$. Let z_1 be any point of the ring and choose r_1, r_2 so that $R_1 < r_1 < |z_1| < r_2 < R_2$ as in Fig. 11–13. We then apply the Cauchy integral formula in the general form of Eq. (11–92) to the region bounded by $C_1 : |z| = r_1$ and $C_2 : |z| = r_2$. Hence

$$f(z_1) = \frac{1}{2\pi i} \oint_{C_2} \frac{f(z)}{z - z_i} \, dz - \frac{1}{2\pi i} \oint_{C_1} \frac{f(z)}{z - z_1} \, dz.$$

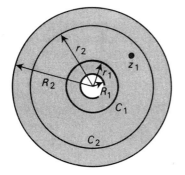

Fig. 11-13. Laurent's theorem.

The first term can be replaced by a power series

$$\sum_{n=0}^{\infty} a_n z_1^n, \qquad a_n = \frac{1}{2\pi i} \oint_{C_2} \frac{f(z)}{z^{n+1}} \, dz,$$

as in the proof of Theorem 23 (Section 11–11). For the second term, the series expansion

$$\frac{1}{z - z_1} = -\frac{1}{z_1}\left(\frac{1}{1 - z/z_1}\right) = -\frac{1}{z_1} - \frac{z}{z_1^2} - \frac{z^2}{z_1^3} - \cdots$$

valid for $|z_1| > |z| = r_1$, leads similarly to the series

$$\sum_{n=1}^{\infty} \frac{b_n}{z_1^n} = \sum_{n=-\infty}^{-1} a_n z_1^n, \qquad a_n = \frac{1}{2\pi i} \oint_{C_1} \frac{f(z)}{z^{n+1}} \, dz.$$

Hence

$$f(z_1) = \sum_{n=-\infty}^{\infty} a_n z_1^n, \qquad a_n = \frac{1}{2\pi i} \oint_{C_1} \frac{f(z)}{z^{n+1}} \, dz.$$

The path C_2 or C_1 can be replaced by any path C separating $|z| = R_1$ from $|z| = R_2$, since the function integrated is analytic throughout the annulus. The uniform convergence follows as for ordinary power series (Theorem 18). The theorem is now established.

Laurent's theorem continues to hold when $R_1 = 0$ or $R_2 = \infty$ or both. In the case $R_1 = 0$, the Laurent expansion represents a function $f(z)$ analytic in a *deleted neighborhood* of z_0, that is, in the circular region $|z - z_0| < R_2$ minus its center z_0. If $R_2 = \infty$, we can say similarly that the series represents $f(z)$ in a *deleted neighborhood* of $z = \infty$.

11–13 ISOLATED SINGULARITIES OF AN ANALYTIC FUNCTION. ZEROS AND POLES

Let $f(z)$ be defined and analytic in region D. We say that $f(z)$ has an *isolated singularity* at the point z_0 if $f(z)$ is analytic throughout a neighborhood of z_0 except at z_0 itself; that is, to use the term mentioned at the end of the preceding section, $f(z)$ is analytic in a deleted neighborhood of z_0, but not at z_0. The point z_0 is then a boundary point of D and would be called an *isolated boundary point* (see Fig. 11–14).

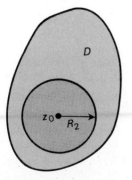

Fig. **11–14.** Isolated singularity.

A deleted neighborhood $0 < |z - z_0| < R_2$ forms a special case of the annular region for which Laurent's theorem is applicable. Hence in this deleted neighborhood $f(z)$ has a representation as a Laurent series:

$$f(z) = \sum_{n=-\infty}^{\infty} a_n (z - z_0)^n.$$

The form of this series leads to a classification of isolated singularities into three fundamental types:

Case I. No terms in negative powers of $z - z_0$ appear. In this case the series is a Taylor series and represents a function analytic in a neighborhood of z_0. Thus the singularity can be removed by setting $f(z_0) = a_0$. We call this a *removable singularity* of $f(z)$. It is illustrated by

$$\frac{\sin z}{z} = 1 - \frac{z^2}{3!} + \frac{z^4}{5!} - \cdots$$

at $z = 0$. In practice, we automatically remove the singularity by defining the function properly.

Case II. A finite number of negative powers of $z - z_0$ appear. Thus we have

$$f(z) = \frac{a_{-N}}{(z - z_0)^N} + \cdots + \frac{a_{-1}}{z - z_0} + a_0 + \cdots + a_n (z - z_0)^n + \cdots,$$

$$N \geqslant 1, \, a_{-N} \neq 0. \tag{11-130}$$

Here $f(z)$ is said to have a *pole of order N* at z_0. We can write

$$f(z) = \frac{1}{(z - z_0)^N} g(z), \qquad g(z) = a_{-N} + a_{-N+1}(z - z_0) + \cdots, \tag{11-131}$$

so that $g(z)$ is analytic for $|z - z_0| < R_2$ and $g(z_0) \neq 0$. Conversely, every function $f(z)$ representable in the form (11-131) has a pole of order N at z_0. Poles are illustrated by rational functions of z, such as

$$f(z) = \frac{z - 2}{(z^2 + 1)(z - 1)^3}, \tag{11-132}$$

which has poles of order 1 at $\pm i$ and of order 3 at $z = 1$.

The rational function

$$\frac{a_{-N}}{(z - z_0)^N} + \cdots + \frac{a_{-1}}{z - z_0} = p(z) \tag{11-133}$$

is called the *principal part* of $f(z)$ at the pole z_0. Thus $f(z) - p(z)$ is analytic at z_0.

EXAMPLE 1 $f(z) = \dfrac{e^z \cos z}{z^3}$ at $z = 0$. To obtain the Laurent series, we expand the numerator in a Taylor series:

$$e^z \cos z = \left(1 + z + \frac{z^2}{2!} + \cdots\right)\left(1 - \frac{z^2}{2!} + \cdots\right) = 1 + z - \frac{z^3}{3} + \cdots$$

Hence

$$\frac{e^z \cos z}{z^3} = \frac{1}{z^3} + \frac{1}{z^2} - \frac{1}{3} + \cdots$$

Here the first two terms form the principal part; the pole is of order 3. ◄

EXAMPLE 2 $f(z) = \dfrac{z}{(z+1)^2 (z^3 + 2)}$ at $z = -1$. We expand $\dfrac{z}{z^3 + 2}$ in a Taylor series about $z = -1$:

$$w = \frac{z}{z^3 + 2}, \qquad w' = \frac{2 - 2z^3}{(z^3 + 2)^2}, \qquad w'' = \frac{6(z^5 - 4z^2)}{(z^3 + 2)^3}, \ldots,$$

$$\frac{z}{z^3 + 2} = -1 + 4(z + 1) - 15(z + 1)^2 + \cdots,$$

$$\frac{z}{(z+1)^2 (z^3 + 2)} = \frac{-1}{(z+1)^2} + \frac{4}{z+1} - 15 + \cdots$$

The first two terms form the principal part; the pole is of order 2. ◄

Case III. Infinitely many negative powers of $z - z_0$ appear. In this case $f(z)$ is said to have an *essential singularity* at z_0. This is illustrated by the function

$$f(z) = e^{1/z} = 1 + \frac{1}{z} + \frac{1}{2!}\frac{1}{z^2} + \frac{1}{3!}\frac{1}{z^3} + \cdots,$$

which has an essential singularity at $z = 0$.

In Case I, $f(z)$ has a finite limit at z_0 and accordingly $|f(z)|$ is bounded near z_0, that is, there is a real constant M such that $|f(z)| < M$ for z sufficiently close to z_0.

In Case II, $\lim_{z \to z_0} f(z) = \infty$, and it is customary to assign the value ∞ (complex) to $f(z)$ at a pole. At an essential singularity, $f(z)$ has a very complicated discontinuity. In fact, for every complex number c, we can find a sequence z_n converging to z_0 such that $\lim_{n \to \infty} f(z_n) = c$ (see **Problem 8** below). It follows from this that if $|f(z)|$ is bounded near z_0, then z_0 must be a removable singularity, and if $\lim f(z) = \infty$ at z_0, then z_0 must be a pole.

If $f(z)$ is analytic at a point z_0 and $f(z_0) = 0$, then z_0 is termed a *root* or *zero* of $f(z)$. Thus the zeros of $\sin z$ are the numbers $n\pi$ $(n = 0, \pm 1, \pm 2, \ldots)$. The Taylor series about z_0 has the form

$$f(z) = a_N (z - z_0)^N + a_{N+1} (z - z_0)^{N+1} + \cdots,$$

where $N \geqslant 1$ and $a_N \neq 0$, or else $f(z) \equiv 0$ in a neighborhood of z_0. It will be seen that the latter case can occur only if $f(z) \equiv 0$ throughout the domain in which it is given. If now $f(z)$ is not identically zero, then

$$f(z) = (z - z_0)^N \varphi(z),$$

$$\varphi(z) = a_N + a_{N+1}(z - z_0) + \cdots,$$

$$\varphi(z_0) = \frac{f^{(N)}(z_0)}{N!} = a_N \neq 0.$$

We say that $f(z)$ has a zero of *order N* or *multiplicity N* at z_0. For example, $1 - \cos z$ has a zero of order 2 at $z = 0$, since

$$1 - \cos z = \frac{z^2}{2} - \frac{z^4}{24} + \cdots$$

If $f(z)$ has a zero of order N at z_0, then $F(z) = 1/f(z)$ has a pole of order N at z_0, and conversely. For if f has a zero of order N, then

$$f(z) = (z - z_0)^N \, \varphi(z)$$

as above, with $\varphi(z_0) \neq 0$. It follows from continuity that $\varphi(z) \neq 0$ in a sufficiently small neighborhood of z_0. Hence $g(z) = 1/\varphi(z)$ is analytic in the neighborhood and $g(z_0) \neq 0$. Now in this neighborhood, except for z_0,

$$F(z) = \frac{1}{f(z)} = \frac{1}{(z - z_0)^N \, \varphi(z)} = \frac{g(z)}{(z - z_0)^N},$$

so that F has a pole at z_0. The converse is proved in the same way.

It remains to consider the case when $f \equiv 0$ in a neighborhood of z_0. This is covered by the following theorem.

THEOREM 26. *The zeros of an analytic function are isolated, unless the function is identically zero, that is, if $f(z)$ is analytic in domain D and $f(z)$ is not identically zero, then for each zero z_0 of $f(z)$ there is a deleted neighborhood of z_0 in which $f(z) \neq 0$.*

(For a proof see Section 6–2 of IAF.)

11–14 THE COMPLEX NUMBER ∞

The complex number ∞ has been introduced several times in connection with limiting processes, for example, in the discussion of poles in the preceding section. In each case ∞ has appeared in a natural way as the limiting position of a point receding indefinitely from the origin. We can incorporate this number into the complex-number system with special algebraic rules:

$$\frac{z}{\infty} = 0 \quad (z \neq \infty), \qquad z \pm \infty = \infty \quad (z \neq \infty), \qquad \frac{z}{0} = \infty \quad (z \neq 0),$$

$$z \cdot \infty = \infty \quad (z \neq 0), \qquad \frac{\infty}{z} = \infty \quad (z \neq \infty). \tag{11–140}$$

Expressions such as $\infty + \infty$, $\infty - \infty$, and ∞/∞ are not defined.

A function $f(z)$ is said to be analytic in a deleted neighborhood of ∞ if $f(z)$ is analytic for $|z| > R_1$ for some R_1. In this case the Laurent expansion with $R_2 = \infty$ and $z_0 = 0$ is available, and we have

$$f(z) = \sum_{n = -\infty}^{\infty} a_n z^n, \qquad |z| > R_1.$$

If there are no *positive* powers of z here, $f(z)$ is said to have a *removable singularity* at ∞ and we make f *analytic at* ∞ by defining $f(\infty) = a_0$:

$$f(z) = a_0 + \frac{a_{-1}}{z} + \cdots + \frac{a_{-n}}{z^n} + \cdots, \qquad |z| > R_1. \qquad (11\text{–}141)$$

This is clearly equivalent to the statement that if we set $z_1 = 1/z$, then $f(z)$ becomes a function of z_1 with removable singularity at $z_1 = 0$.

If a finite number of positive powers occurs, we have, with $N \geqslant 1$,

$$f(z) = a_N z^N + \cdots + a_1 z + a_0 + \frac{a_{-1}}{z} + \cdots = z^N \varphi(z),$$

$$\varphi(z) = a_N + \frac{a_{N-1}}{z} + \cdots, \qquad (11\text{–}142)$$

where $\varphi(z)$ is analytic at ∞ and $\varphi(\infty) = a_N \neq 0$. In this case $f(z)$ is said to have a *pole of order N at* ∞. The same holds for $f(1/z_1)$ at $z_1 = 0$. Furthermore,

$$\lim_{z \to \infty} f(z) = \infty. \qquad (11\text{–}143)$$

If infinitely many positive powers appear, $f(z)$ is said to have an *essential singularity* at $z = \infty$.

If $f(z)$ is analytic at ∞ as in (11–141) and $f(\infty) = a_0 = 0$, then $f(z)$ is said to have a *zero* at $z = \infty$. If f is not identically zero, then necessarily some $a_{-N} \neq 0$ and

$$f(z) = \frac{a_{-N}}{z^N} + \frac{a_{-N-1}}{z^{N+1}} + \cdots = \frac{1}{z^N} g(z), \qquad |z| > R_1,$$

$$g(z) = a_{-N} + \frac{a_{-N-1}}{z} + \cdots \qquad (11\text{–}144)$$

Thus $g(z)$ is analytic at ∞ and $g(\infty) = a_{-N} \neq 0$. We say that $f(z)$ has a zero of order (or multiplicity) N at ∞. We can show that if $f(z)$ has a zero of order N at ∞, then $1/f(z)$ has a pole of order N at ∞, and conversely.

The significance of the complex number ∞ can be shown geometrically by the device of *stereographic projection*, i.e., a projection of the plane onto a sphere tangent to the plane at $z = 0$, as shown in Fig. 11–15. The sphere is given in

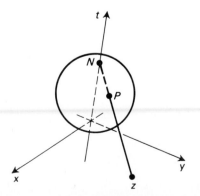

Fig. 11–15. Stereographic projection.

xyt-space by the equation

$$x^2 + y^2 + (t - \tfrac{1}{2})^2 = \frac{1}{4}, \tag{11-145}$$

so that the radius is $1/2$. The letter N denotes the *north pole* of the sphere, the point $(0, 0, 1)$. If N is joined to an arbitrary point z in the xy-plane, the line segment Nz will meet the sphere at one other point P, which is the projection of z on the sphere. For example, the points of the circle $|z| = 1$ project on the "equator" of the sphere, i.e., the great circle $t = 1/2$. As z recedes to infinite distance from the origin, P approaches N as limiting position. Thus N *corresponds to the complex number* ∞.

We refer to the z-plane plus the number ∞ as the *extended z-plane*. To emphasize that ∞ is *not* included, we refer to the *finite z-plane*.

═══════════════════════════**Problems (Section 11-14)**═══════════════════

1. For each of the following expand in a Laurent series at the isolated singularity given and state the type of singularity:

   a) $\dfrac{e^z - 1}{z}$ at $z = 0$ **b)** $\dfrac{1}{z^2(z - 3)}$ at $z = 0$

   c) $\dfrac{z - \cos z}{z}$ at $z = 0$ **d)** $\csc z$ at $z = 0$

   HINT: For (d): Write

   $$\csc z = \frac{1}{\sin z} = \frac{1}{z - (z^3/3!) + \cdots} = \frac{a_{-1}}{z} + a_0 + \cdots$$

   and determine the coefficients $a_{-1}, a_0, a_1, \ldots$, so that

   $$1 = (z - z^3/3! + \cdots) \cdot (a_{-1} z^{-1} + a_0 + a_1 z + \cdots).$$

2. For each of the following find the principal part at the pole given:

   a) $\dfrac{z^2 + 3z + 1}{z^4}, \quad z = 0$ **b)** $\dfrac{z^2 - 2}{z(z + 1)}, \quad z = 0$

   c) $\dfrac{e^z \sin z}{(z - 1)^2}, \quad z = 1$ **d)** $\dfrac{1}{z^2(z^3 + z + 1)}, \quad z = 0$

3. For each of the following expand in a Laurent series at $z = \infty$ and state the type of singularity:

   a) $\dfrac{1}{1 - z} = -\dfrac{1}{z} \dfrac{1}{1 - (1/z)}$ **b)** $\dfrac{z^2}{z + 2}$ **c)** $e^z + e^{1/z}$

4. Let $f(z)$ be a rational function in lowest terms:

   $$f(z) = \frac{a_0 z^n + a_1 z^{n-1} + \cdots + a_n}{b_0 z^m + b_1 z^{m-1} + \cdots + b_m}.$$

   The *degree d* of $f(z)$ is defined to be the larger of m and n. Assuming the fundamental theorem of algebra, show that $f(z)$ has precisely d zeros and d poles in the extended z-plane, a pole or zero of order N being counted as N poles or zeros.

   HINT: Take the cases $m = n$, $m < n$, $m > n$ in turn.

5. For each of the following locate all zeros and poles in the extended plane (compare with Problem 4):

a) $\dfrac{z}{z-1}$ b) $\dfrac{z-1}{z^2+3z+2}$ c) $\dfrac{z^3+3z^2+3z+1}{z}$

6. Let $A(z)$ and $B(z)$ be analytic at $z = z_0$; let $A(z_0) \neq 0$ and let $B(z)$ have a zero of order N at z_0, so that

$$f(z) = \frac{A(z)}{B(z)} = \frac{a_0 + a_1(z-z_0) + \cdots}{b_N(z-z_0)^N + b_{N+1}(z-z_0)^{N+1} + \cdots}$$

has a pole of order N at z_0. Show that the principal part of $f(z)$ at z_0 is

$$\frac{a_0}{b_N}\frac{1}{(z-z_0)^N} + \frac{a_1 b_N - a_0 b_{N+1}}{b_N^2}\frac{1}{(z-z_0)^{N-1}} + \cdots$$

and obtain the next term explicitly.

HINT: Set

$$\frac{a_0 + a_1(z-z_0) + \cdots}{b_N(z-z_0)^N + b_{N+1}(z-z_0)^{N+1} + \cdots} = \frac{C_{-N}}{(z-z_0)^N} + \frac{C_{-N+1}}{(z-z_0)^{N-1}} + \cdots$$

Multiply across and solve for $C_{-N}, C_{-N+1}, \ldots$

7. Prove *Riemann's theorem: If $|f(z)|$ is bounded in a deleted neighborhood of an isolated singularity z_0, then z_0 is a removable singularity of $f(z)$.*

HINT: Proceed as in Problem 6 following Section 11–11 with the aid of (11–124).

8. Prove the *Theorem of Weierstrass and Casorati: If z_0 is an essential singularity of $f(z)$, c is an arbitrary complex number, and $\varepsilon > 0$, then $|f(z) - c| < \varepsilon$ for some z in every neighborhood of z_0.*

HINT: If the property fails, then $1/[f(z) - c]$ is analytic and bounded in absolute value in a deleted neighborhood of z_0. Now apply Problem 7 and conclude that $f(z)$ has a pole or removable singularity at z_0.

11–15 RESIDUES

Let $f(z)$ be analytic throughout a region D except for an isolated singularity at a certain point z_0 of D. The integral $\oint f(z)\,dz$ will not in general be zero on a simple closed path in D. However, the integral will have the same value on all curves C that enclose z_0 and no other singularity of f. This value, divided by $2\pi i$, is known as the *residue* of $f(z)$ at z_0 and is denoted by $\operatorname{Res}[f(z), z_0]$. Thus

$$\operatorname{Res}[f(z), z_0] = \frac{1}{2\pi i}\oint_C f(z)\,dz, \qquad (11\text{–}150)$$

where the integral is taken over any path C within which $f(z)$ is analytic except at z_0 (Fig. 11–16).

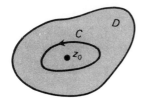

Fig. 11–16. Residue.

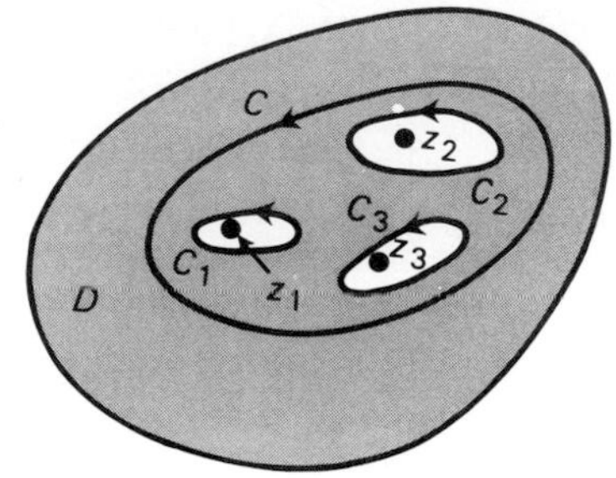

Fig. 11–17. Cauchy residue theorem.

THEOREM 27 *The residue of $f(z)$ at z_0 is given by the equation*
$$\operatorname{Res}[f(z), z_0] = a_{-1}, \qquad (11\text{–}151)$$
where
$$f(z) = \cdots + \frac{a_{-N}}{(z-z_0)^N} + \cdots + \frac{a_{-1}}{z-z_0} + a_0 + a_1(z-z_0) + \cdots$$
$$(11\text{–}152)$$

is the Laurent expansion of $f(z)$ at z_0.

PROOF By (11–124),
$$a_{-1} = \frac{1}{2\pi i} \oint_C f(z)\,dz,$$

where C is chosen as in the definition of residue. Hence (11–151) follows at once.

If C is a simple closed path in D, within which $f(z)$ is analytic except for isolated singularities at $z_1, \ldots, z_k$, then by Theorem 16

$$\oint_C f(z)\,dz = \oint_{C_1} f(z)\,dz + \cdots + \oint_{C_k} f(z)\,dz,$$

where C_1 encloses only the singularity at z_1, C_2 encloses only $z_2, \ldots$, as in Fig. 11–17. We thus obtain the following basic theorem:

THEOREM 28 *(Cauchy's residue theorem)* *If $f(z)$ is analytic in a region D and C is a simple closed curve in D within which $f(z)$ is analytic except for isolated singularities at $z_1, \ldots, z_k$, then*

$$\oint_C f(z)\,dz = 2\pi i\left\{\operatorname{Res}[f(z), z_1] + \cdots + \operatorname{Res}[f(z), z_k]\right\}. \quad (11\text{–}153)$$

This theorem permits rapid evaluation of integrals on closed paths, whenever it is possible to compute the coefficient a_{-1} of the Laurent expansion at each singularity inside the path. Various techniques for obtaining the Laurent expansion are illustrated in the problems preceding this section. However, if we wish only the term in $(z - z_0)^{-1}$ of the expansion, various simplifications are possible. We give several rules here:

Rule I.   At a simple pole z_0 (i.e., a pole of first order),

$$\text{Res}\left[f(z), z_0\right] = \lim_{z \to z_0} (z - z_0)f(z).$$

Rule II.   At a pole z_0 of order N ($N = 2, 3, \ldots$),

$$\text{Res}\left[f(z), z_0\right] = \lim_{z \to z_0} \frac{g^{(N-1)}(z)}{(N-1)!},$$

where $g(z) = (z - z_0)^N f(z)$.

Rule III.   If $A(z)$ and $B(z)$ are analytic in a neighborhood of z_0, $A(z_0) \neq 0$, and $B(z)$ has a zero at z_0 of order 1, then $f(z) = A(z)/B(z)$ has a pole of first order at z_0 and

$$\text{Res}\left[f(z), z_0\right] = \frac{A(z_0)}{B'(z_0)}.$$

Rule IV.   If $A(z)$ and $B(z)$ are as in Rule III, but $B(z)$ has a zero of second order at z_0, so that $f(z)$ has a pole of second order at z_0, then

$$\text{Res}\left[f(z), z_0\right] = \frac{6A'B'' - 2AB'''}{3B''^2}, \tag{11–154}$$

where A and the derivatives A', B'', B''' are evaluated at z_0.

PROOFS OF RULES: Let $f(z)$ have a pole of order N:

$$f(z) = \frac{1}{(z - z_0)^N}\left[a_{-N} + a_{-N+1}(z - z_0) + \cdots\right] = \frac{1}{(z - z_0)^N} g(z),$$

where

$$g(z) = (z - z_0)^N f(z), \qquad g(z_0) = a_{-N}$$

and g is analytic at z_0. The coefficient of $(z - z_0)^{-1}$ in the Laurent series for $f(z)$ is the coefficient of $(z - z_0)^{N-1}$ in the Taylor series for $g(z)$. This coefficient, which is the residue sought, is

$$\frac{g^{(N-1)}(z_0)}{(N-1)!} = \lim_{z \to z_0} \frac{g^{(N-1)}(z)}{(N-1)!}.$$

For $N = 1$ this gives Rule 1; for $N = 2$ or higher, we obtain Rule II. Rules III and IV follow from the identity of Problem 6 following Section 11–14:

$$\frac{A(z)}{B(z)} = \frac{a_0 + a_1(z - z_0) + \cdots}{b_N(z - z_0)^N + b_{N+1}(z - z_0)^{N+1} + \cdots}$$

$$= \frac{a_0}{b_N}\frac{1}{(z - z_0)^N} + \frac{a_1 b_N - a_0 b_{N+1}}{b_N^2}\frac{1}{(z - z_0)^{N-1}} + \cdots$$

For a first order pole, $N = 1$ and the residue is

$$\frac{a_0}{b_1} = \frac{A(z_0)}{B'(z_0)}.$$

For a second order pole, $N = 2$ and the residue is $[(a_1 b_2 - a_0 b_3)/b_2^2]$. Since

$$a_0 = A(z_0), \qquad a_1 = A'(z_0), \qquad b_2 = \frac{B''(z_0)}{2!}, \qquad b_3 = \frac{B'''(z_0)}{3!},$$

this reduces to the expression (11–154).

EXAMPLE 1 $\displaystyle\oint_{|z|=2} \frac{ze^z}{z^2-1} \, dz = 2\pi i \{\operatorname{Res}[f(z), 1] + \operatorname{Res}[f(z), -1]\}.$

Since $f(z)$ has first-order poles at ± 1, we find by Rule I that

$$\operatorname{Res}[f(z), 1] = \lim_{z \to 1} (z-1) \cdot \frac{ze^z}{z^2-1} = \lim_{z \to 1} \frac{ze^z}{z+1} = \frac{e}{2},$$

$$\operatorname{Res}[f(z), -1] = \lim_{z \to -1} (z+1) \cdot \frac{ze^z}{z^2-1} = \lim_{z \to -1} \frac{ze^z}{z-1} = \frac{-e^{-1}}{-2}.$$

Accordingly,

$$\oint_{|z|=2} \frac{ze^z}{z^2-1} \, dz = 2\pi i \left(\frac{e}{2} + \frac{e^{-1}}{2} \right) = 2\pi i \cosh 1.$$

Rule III could also have been used:

$$\operatorname{Res}[f(z), 1] = \frac{ze^z}{2z}\bigg|_{z=1} = \frac{e}{2}, \qquad \operatorname{Res}[f(z), -1] = \frac{ze^z}{2z}\bigg|_{z=-1} = \frac{-e^{-1}}{-2}.$$

This is simpler than Rule I, since the expression $A(z)/B'(z)$, once computed, serves for all poles of the prescribed type. ◄

EXAMPLE 2 $\displaystyle\oint_{|z|=2} \frac{z}{z^4-1} \, dz = 2\pi i \{\operatorname{Res}[f(z), 1] + \operatorname{Res}[f(z), -1]$

$$+ \operatorname{Res}[f(z), i] + \operatorname{Res}[f(z), -i]\}.$$

All poles are of first order. Rule III gives $A(z)/B'(z) = z/(4z^3) = 1/(4z^2)$ as the expression for the residue at any one of the four points. Moreover, $z^4 = 1$ at each pole, so that

$$\frac{1}{4z^2} = \frac{z^2}{4z^4} = \frac{z^2}{4}.$$

Hence

$$\oint_{|z|=2} \frac{z}{z^4-1} \, dz = \frac{2\pi i}{4}(1 + 1 - 1 - 1) = 0. \ ◄$$

EXAMPLE 3 $\displaystyle\oint_{|z|=2} \frac{e^z}{z(z-1)^2} \, dz = 2\pi i \{\operatorname{Res}[f(z), 0] + \operatorname{Res}[f(z), 1]\}.$

At the first-order pole $z = 0$, application of Rule 1 gives the residue 1.

At the second-order pole $z = 1$, Rule II gives

$$\text{Res}\left[f(z), 1\right] = \frac{d}{dz}\left(\frac{e^z}{z}\right)\bigg|_{z=1} = \frac{e^z(z-1)}{z^2}\bigg|_{z=1} = 0.$$

Rule IV could also be used, with $A = e^z$, $B = z^3 - 2z^2 + z$, to give

$$\text{Res}\left[f(z), 1\right] = \frac{6e^z(6z-4) - 2e^z \cdot 6}{3(6z-4)^2}\bigg|_{z=1} = 0.$$

Accordingly,

$$\oint_{|z|=2} \frac{e^z}{z(z-1)^2}\, dz = 2\pi i(1+0) = 2\pi i. \quad \blacktriangleleft$$

11–16 RESIDUE AT INFINITY

Let $f(z)$ be analytic for $|z| > R$. The *residue of $f(z)$* at ∞ is defined as follows:

$$\text{Res}\left[f(z), \infty\right] = \frac{1}{2\pi i}\oint_C f(z)\, dz,$$

where the integral is taken in the *negative* direction on a simple closed path C, in the region of analyticity of $f(z)$, *outside* of which $f(z)$ has no singularity other than ∞. This is suggested in Fig. 11–18. Theorem 27 has an immediate extension to this case:

THEOREM 29   *The residue of $f(z)$ at ∞ is given by the equation*

$$\text{Res}\left[f(z), \infty\right] = -a_{-1}, \tag{11–160}$$

where a_{-1} is the coefficient of z^{-1} in the Laurent expansion of $f(z)$ at ∞:

$$f(z) = \cdots + \frac{a_{-n}}{z^n} + \ldots + \frac{a_{-1}}{z} + a_0 + a_1 z + \cdots \tag{11–161}$$

The proof is the same as for Theorem 27. It should be stressed that the presence of a nonzero residue at ∞ is not related to presence of a pole or essential singularity at ∞. That is, $f(z)$ can have a nonzero residue whether or not there is a pole or essential singularity, for the pole or essential singularity at ∞ is due to the *positive powers* of z, not to negative powers (Section 11–14 above). Thus the function $e^{1/z} = 1 + z^{-1} + (2!z^2)^{-1} + \cdots$ is analytic at ∞, but has the residue -1 there.

Cauchy's residue theorem has also an extension to include ∞:

THEOREM 30   *Let $f(z)$ be analytic in a domain D which includes a deleted neighborhood of ∞. Let C be a simple closed path in D outside of which $f(z)$ is analytic except for isolated singularities at $z_1, \ldots, z_k$. Then*

$$\oint_C f(z)\, dz = 2\pi i\{\text{Res}\left[f(z), z_1\right] + \cdots + \text{Res}\left[f(z), z_k\right] + \text{Res}\left[f(z), \infty\right]\}.$$

$$\tag{11–162}$$

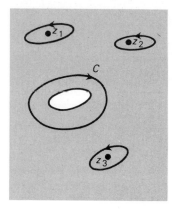

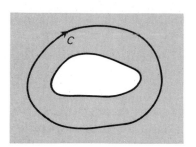

Fig. 11–18. Residue at infinity. **Fig. 11–19.** Residue theorem for exterior.

The proof, which is like that of Theorem 28, is left as an exercise (Problem 4, following Section 11–18). It is to be emphasized that the integral on C is taken in the *negative* direction (see Fig. 11–19) and that the *residue at ∞ must be included* on the right.

For a particular integral $\oint_C f(z)\,dz$ on a simple closed path C, we have now two modes of evaluation: the integral equals $2\pi i$ times the sum of the residues inside the path (provided there are only a finite number of singularities there), and it also equals *minus* $2\pi i$ times the sum of the residues outside the path plus that at ∞ (provided there are only a finite number of singularities in the exterior domain). We can evaluate the integral both ways to check results. The principle involved here is summarized in the following theorem:

THEOREM 31 *If $f(z)$ is analytic in the extended z-plane except for a finite number of singularities, then the sum of all residues of $f(z)$ (including ∞) is zero.*

To evaluate the residues at ∞, we can formulate a set of rules like the ones above. However, the following two rules are adequate for most purposes:

Rule V. If $f(z)$ has a zero of first order at ∞, then

$$\operatorname{Res}\left[f(z), \infty\right] = - \lim_{z \to \infty} z f(z).$$

If $f(z)$ has a zero of second or higher order at ∞, the residue at ∞ is zero.

Rule VI. $\operatorname{Res}\left[f(z), \infty\right] = -\operatorname{Res}\left[\dfrac{1}{z^2} f(1/z), 0\right].$

The proof of Rule V is left as an exercise (Problem 8 following Section 11–18). To prove Rule VI, we write

$$f(z) = \cdots + a_n z^n + \cdots + a_1 z + a_0 + \frac{a_{-1}}{z} + \frac{a_{-2}}{z^2} + \cdots, \qquad |z| > R.$$

Then for $0 < |z| < R^{-1}$,

$$f\left(\frac{1}{z}\right) = \cdots + \frac{a_n}{z^n} + \cdots + \frac{a_1}{z} + a_0 + a_{-1}z + a_{-2}z^2 + \cdots,$$

$$\frac{1}{z^2}f\left(\frac{1}{z}\right) = \cdots + \frac{a_0}{z^2} + \frac{a_{-1}}{z} + a_{-2} + \cdots$$

Hence $\mathrm{Res}\left[\frac{1}{z^2}f\left(\frac{1}{z}\right), 0\right] = a_{-1}$, and the rule follows. This result reduces the problem to evaluation of a residue at zero, to which **Rules I through IV** are applicable.

EXAMPLE 1 We consider the integral $\displaystyle\oint_{|z|=2} \frac{z}{z^4 - 1}\, dz$ of Example 2 in the preceding section. There is no singularity outside the path other than ∞, and at ∞ the function has a zero of order 3; hence the integral is zero. ◄

EXAMPLE 2 $\displaystyle\oint_{|z|=2} \frac{1}{(z+1)^4(z^2-9)(z-4)}\, dz.$

Here there is a fourth-order pole inside the path, at which evaluation of the residue is tedious. Outside the path there are first-order poles at ± 3 and 4 and a zero of order 7 at ∞. Hence by Rule I the integral equals

$$-2\pi i\left[\frac{1}{4^4 6(-1)} + \frac{1}{(-2)^4(-6)(-7)} + \frac{1}{5^4 \cdot 7}\right]. \quad ◄$$

11–17 LOGARITHMIC RESIDUES; ARGUMENT PRINCIPLE

Let $f(z)$ be analytic in a region D. Then $f'(z)/f(z)$ is analytic in D except at the zeros of $f(z)$. If an analytic branch of $\ln f(z)$ is chosen in part of D, necessarily excluding the zeros of $f(z)$, then

$$\frac{d}{dz}\ln f(z) = \frac{f'(z)}{f(z)}. \tag{11–170}$$

For this reason the expression f'/f is termed the *logarithmic derivative* of $f(z)$. Its value is demonstrated by the following theorem:

THEOREM 32 *Let $f(z)$ be analytic in region D. Let C be a simple closed path in D within which $f(z)$ is analytic except for a finite number of poles, and let $f(z) \neq 0$ on C. Then*

$$\frac{1}{2\pi i}\oint_C \frac{f'(z)}{f(z)}\, dz = N_0 - N_p,$$

where N_0 is the total number of zeros of f inside C and N_p is the total number of poles of f inside C, zeros and poles being counted according to multiplicities.

PROOF   The logarithmic derivative f'/f has isolated singularities precisely at the zeros and poles of f. At a zero z_0,

$$f(z) = (z - z_0)^N g(z), \qquad g(z_0) \neq 0,$$
$$f'(z) = (z - z_0)^N g'(z) + N(z - z_0)^{N-1} g(z),$$
$$\frac{f'(z)}{f(z)} = \frac{(z - z_0)^N g'(z) + N(z - z_0)^{N-1} g(z)}{(z - z_0)^N g(z)} = \frac{g'(z)}{g(z)} + \frac{N}{z - z_0}.$$

Hence the logarithmic derivative has a pole of first order, with residue N equal to the multiplicity of the zero. A similar analysis applies to each pole of f, with N replaced by $-N$. The theorem then follows from the Cauchy residue theorem (Theorem 28), provided we show that there are only a finite number of singularities. The poles of f are finite in number by assumption, and it is easily shown that there can be only a finite number of zeros (see Section 6–6 of IAF). Thus the theorem is proved.

REMARKS   Since

$$\frac{f'(z)}{f(z)}\, dz = d \ln f(z) = d \ln w = d\,(\ln |w| + i \arg w),$$

we have

$$\frac{1}{2\pi i} \int_C \frac{f'(z)}{f(z)}\, dz = \frac{1}{2\pi i} \int_C d \ln |w| + \frac{1}{2\pi} \int_C d \arg w.$$

Since $w \neq 0$ on C, $\ln |w|$ is continuous on C and the first integral on the right is zero. The second measures the total change in $\arg w$, divided by 2π, as w traces the path C_w, the image of C, in the w-plane. Hence it also measures the number of times C_w winds about the origin of the w-plane: in Fig. 11–20 the number is $+2$.

The statement

$$\frac{1}{2\pi} \cdot \left(\text{Increase in } \arg f(z) \text{ on path} \right) = N_0 - N_p \qquad (11\text{–}171)$$

is known as the *argument principle*. This is of great value in finding zeros and poles of analytic functions.

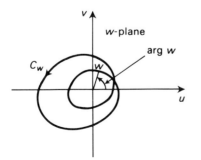

Fig. 11–20. Argument principle.

11–18 PARTIAL FRACTION EXPANSION OF RATIONAL FUNCTIONS

The theory of analytic functions provides a simple proof of the familiar rules for partial fraction expansions (Section 4–6).

Let

$$f(z) = \frac{a_0 z^n + \cdots + a_n}{b_0 z^m + \cdots + a_m}, \qquad a_0 \neq 0, \qquad b_0 \neq 0 \tag{11–180}$$

be given. We assume that $n < m$, so that f is a proper fraction. We also assume that the numerator and denominator have no common zeros. Let $z_1, z_2, \ldots, z_N$ be the *distinct zeros* of the *denominator* (no repetitions); these are the poles of $f(z)$. At $z_1, f(z)$ has a Laurent expansion:

$$f(z) = p_1(z) + g_1(z),$$

$$p_1(z) = \frac{A_{-k_1}}{(z - z_1)^{k_1}} + \frac{A_{-k_1+1}}{(z - z_1)^{k_1-1}} + \cdots + \frac{A_{-1}}{z - z_1}. \tag{11–181}$$

Here $p_1(z)$ is the principal part of $f(z)$ at z_1 and $g_1(z)$ is analytic at z_1; k_1 is the order of the pole at z_1. Similar expressions hold at the other poles.

The partial fraction expansion of $f(z)$ is now simply the identity

$$f(z) = p_1(z) + p_2(z) + \cdots + p_N(z). \tag{11–182}$$

To justify this, we let

$$F(z) = f(z) - [p_1(z) + p_2(z) + \cdots + p_N(z)].$$

Now $f(z) - p_1(z)$ has a removable singularity at z_1, while $p_2(z), \ldots, p_N(z)$ are analytic at z_1. Hence $F(z)$ has a removable singularity at z_1. In general, $F(z)$ has only removable singularities for finite z. At $\infty, f(z), p_1(z), \ldots, p_N(z)$ all have zeros; hence $F(z)$ has a zero at ∞. But $F(z)$ is a rational function with no poles. Hence $F(z)$ must be a polynomial. Thus $F(z)$ has a pole at ∞ unless F is constant; since we know $F = 0$ at ∞, F must be a constant, namely zero. This proves (11–182).

If $f(z)$ has only simple poles, the principal part at each pole z_j is only $A_j/(z - z_j)$, where A_j is the residue of f at z_j, and hence in this case

$$f(z) = \frac{A_1}{z - z_1} + \cdots + \frac{A_m}{z - z_m}, \qquad A_j = \text{Res}[f, z_j], \tag{11–183}$$

is the partial fraction expansion. If we write $f(z) = A(z)/B(z)$, then Rule III can be applied:

$$A_j = \frac{A(z_j)}{B'(z_j)}, \qquad f(z) = \sum_{j=1}^{m} \frac{A(z_j)}{B'(z_j)} \frac{1}{z - z_j}. \tag{11–184}$$

At a multiple pole z_j of order k, we can write

$$f(z) = \frac{1}{(z - z_j)^k} \varphi(z).$$

The principal part $p_j(z)$ is then

$$p_j(z) = \frac{\varphi(z_j)}{(z - z_j)^k} + \frac{\varphi'(z_j)}{1!(z - z_j)^{k-1}} + \cdots + \frac{\varphi^{(k-1)}(z_j)}{(k-1)!(z - z_j)}. \tag{11–185}$$

Hence $p_j(z)$ can be found without knowledge of the other poles.

EXAMPLE 1 $f(z) = \dfrac{z^2 + 1}{z^3 + 4z^2 + 3z} = \dfrac{z^2 + 1}{z(z+1)(z+3)}.$

There are simple poles at $0, -1, -3$. By (11–184),

$$f(z) = \frac{A(0)}{B'(0)}\frac{1}{z} + \frac{A(-1)}{B'(-1)}\frac{1}{z+1} + \frac{A(-3)}{B'(-3)}\frac{1}{z+3};$$

with $A = z^2 + 1$, $B = z^3 + 4z^2 + 3z$, $B' = 3z^2 + 8z + 3$, we find

$$f(z) = \frac{1}{3}\frac{1}{z} + \frac{2}{-2}\frac{1}{z+1} + \frac{10}{6}\frac{1}{z+3}. \quad \blacktriangleleft$$

EXAMPLE 2 $f(z) = \dfrac{z}{(z-1)^2(z^3 + z + 1)}.$

At the pole $z = 1$, we write

$$f = \frac{1}{(z-1)^2}\varphi(z), \qquad \varphi = \frac{z}{z^3 + z + 1}.$$

Since $\varphi(1) = 1/3$, $\varphi'(1) = -1/9$, the principal part at 1 is

$$\frac{1}{3}\frac{1}{(z-1)^2} - \frac{1}{9}\frac{1}{z-1}.$$

The cubic $z^3 + z + 1$ has one real root z_1 and two complex roots z_2, z_3. These are all simple. Hence we can write

$$f(z) = \frac{A(z_1)}{B'(z_1)}\frac{1}{z - z_1} + \frac{A(z_2)}{B'(z_2)}\frac{1}{z - z_2}$$

$$+ \frac{A(z_3)}{B'(z_3)}\frac{1}{z - z_3} + \frac{1}{3}\frac{1}{(z-1)^2} - \frac{1}{9}\frac{1}{z-1},$$

where $A(z) = z$, $B(z) = (z-1)^2(z^3 + z + 1)$. At the poles z_1, z_2, z_3, $B'(z)$ reduces to $z^2 + z + 7$. by the relation $z^3 + z + 1 = 0$. $\blacktriangleleft$

═══════════════════ **Problems (Section 11-18)** ═══════════════════

1. Evaluate the following integrals on the paths given:

a) $\displaystyle\oint \frac{z\,dz}{(z-1)(z-3)}$, $|z| = 2$ b) $\displaystyle\oint \frac{e^{3z}\,dz}{z^2 + 4}$, $|z| = 3$

c) $\displaystyle\oint \frac{\sin z}{z^4}\,dz$, $|z| = 1$ d) $\displaystyle\oint \frac{z\,dz}{z^3 + z + 1}$, $|z| = 10$

e) $\displaystyle\oint \frac{dz}{(z+1)^4(z+3)}$, $|z| = 2$ f) $\displaystyle\oint \frac{dz}{(z+1)^5(z+3)}$, $|z| = 4$

g) $\displaystyle\oint \frac{2z+2}{z^2 + 2z + 2}\,dz$, $|z| = 2$ h) $\displaystyle\oint \frac{3z^2 - 6z + 1}{z^3 - 3z^2 + z - 3}\,dz$, $|z| = 2$

2. Expand each of the following in partial fractions:

a) $\dfrac{1}{z^2 - 4}$ b) $\dfrac{z+1}{(z-1)(z-2)(z-3)}$ c) $\dfrac{z^2}{z^5 + 1}$ d) $\dfrac{1}{z^n - 1}$

e) $\dfrac{z}{(z-1)^2(z+1)^2}$ f) $\dfrac{\varphi(z)}{z^n(z+1)}, \ \varphi = c_0 + c_1 z + \cdots + c_n z^n$

g) $\dfrac{1}{(z-1)^3(z^4+z+1)}$ h) $\dfrac{1}{(z^2+1)(z^2+2z+2)}$

3. Prove the Fundamental Theorem of Algebra: *Every polynomial of degree at least 1 has a zero.*

   HINT: Show that $\text{Res}\,[f'(z)/f(z), \infty]$ is not 0, but is in fact minus the degree n of the polynomial $f(z)$. Then use Theorem 32 to show that f has n zeros.

4. Prove Theorem 30.

5. Formulate and prove Theorem 32 for integration around the boundary B of a region R in D, bounded by simple closed curves $C_1, \ldots, C_k$.

6. Prove that under the hypotheses of Theorem 32, if $g(z)$ is analytic in D and within C, then

$$\frac{1}{2\pi i} \oint_C \frac{g(z)f'(z)}{f(z)} \, dz = \sum_{k=1}^{n} g(z'_k) - \sum_{l=1}^{m} g(z''_l),$$

   where $z'_1, \ldots, z'_n$ are the zeros of f, and $z''_1, \ldots, z''_m$ are the poles of f inside C, repeated according to multiplicity.

7. Extend Rule IV of Section 11–15 to the case in which $A(z)$ has a first-order zero at z_0 and $B(z)$ has a second-order zero.

8. Prove Rule V of Section 11–16.

11–19 APPLICATION OF RESIDUES TO EVALUATION OF REAL INTEGRALS

A variety of real definite integrals between special limits can be evaluated with the aid of residues. For example, an integral $\int_0^{2\pi} R(\sin\theta, \cos\theta)\,d\theta$, where R is a rational function of $\sin\theta$ and $\cos\theta$, is converted to a complex line integral by the substitution:

$$z = e^{i\theta}, \qquad dz = ie^{i\theta}\,d\theta = iz\,d\theta,$$

$$\cos\theta = \frac{e^{i\theta} + e^{-i\theta}}{2} = \frac{1}{2}\left(z + \frac{1}{z}\right),$$

$$\sin z = \frac{e^{i\theta} - e^{-i\theta}}{2i} = \frac{1}{2i}\left(z - \frac{1}{z}\right);$$

the path of integration is the circle: $|z| = 1$.

EXAMPLE 1 $\displaystyle\int_0^{2\pi} \frac{1}{\cos\theta + 2}\,d\theta.$ The substitution reduces this to

$$\oint_{|z|=1} \frac{-2i}{z^2 + 4z + 1}\,dz = 4\pi\,\text{Res}\,[(z^2 + 4z + 1)^{-1}, -2 + \sqrt{3}].$$

since $-2+\sqrt{3}$ is the only root of the denominator inside the circle. Accordingly,

$$\int_0^{2\pi} \frac{1}{\cos\theta+2}\, d\theta = \frac{2\pi}{\sqrt{3}}. \quad \blacktriangleleft$$

The substitution can be summarized in the one rule:

$$\int_0^{2\pi} R\,(\sin\theta,\,\cos\theta)\, d\theta = \oint_{|z|=1} R\left(\frac{z^2-1}{2iz},\frac{z^2+1}{2z}\right)\frac{dz}{iz}. \qquad (11\text{--}190)$$

The complex integral can be evaluated by residues, provided R has no poles on the circle $|z|=1$.

A second example is provided by integrals of the type $\int_{-\infty}^{\infty} f(x)\,dx$. We recall that existence of such an improper integral is equivalent to existence of the integrals from $-\infty$ to 0 and from 0 to ∞ separately. More generally, it may be possible to use the *principal value*

$$(P)\int_{-\infty}^{\infty} f(x)\,dx = \lim_{R\to\infty}\int_{-R}^{R} f(x)\,dx,$$

if this limit exists. $\blacktriangleleft$

EXAMPLE 2 $\displaystyle\int_{-\infty}^{\infty}\frac{dx}{x^4+1}$. This integral can be regarded as a line integral of $f(z)=1/(z^4+1)$ along the real axis. The path is not closed (unless we adjoin ∞), but we show that it acts like a closed path "enclosing" the upper half-plane, so that the integral along the path equals the sum of the residues in the upper half-plane.

To establish this, we consider the integral of $f(z)$ on the semicircular path C_R shown in Fig. 11–21. When R is sufficiently large, the path encloses the two poles of $f(z)$:

$$z_1 = \exp\left(\frac{1}{4}i\pi\right), \qquad z_2 = \exp\left(\frac{3}{4}i\pi\right).$$

Hence

$$\oint_{C_R} f(z)\,dz = 2\pi i\,\{\,\mathrm{Res}[\,f(z),z_1\,]+\mathrm{Res}[\,f(z),z_2\,]\,\}.$$

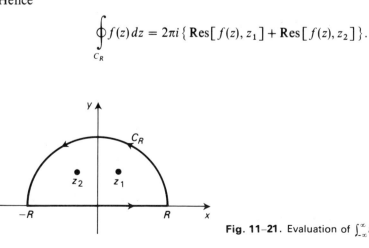

Fig. 11–21. Evaluation of $\int_{-\infty}^{\infty} f(x)\,dx$ by residues.

As R increases, the integral on C_R cannot change, since it always equals the sum of the residues times $2\pi i$. Hence

$$\oint_{C_R} f(z)\,dz = \lim_{R\to\infty} \oint_{C_R} f(z)\,dz = \lim_{R\to\infty} \int_{-R}^{R} \frac{dx}{x^4+1} + \lim_{R\to\infty} \int_{D_R} \frac{1}{z^4+1}\,dz,$$

where D_R is the semicircle $z = Re^{i\theta}$, $0 \le \theta \le \pi$. The limit of the first term is the integral desired (since the limits at $+\infty$ and $-\infty$ exist separately, as required). The limit of the second term is 0, since $|z|^4 = |z^4| = |z^4+1-1| \le |z^4+1|+1$, so that on D_R

$$\left| \frac{1}{z^4+1} \right| \le \frac{1}{|z|^4-1} = \frac{1}{R^4-1}$$

and by Theorem 7

$$\left| \int_{D_R} \frac{1}{z^4+1}\,dz \right| \le \frac{\pi R}{R^4-1}.$$

Accordingly,

$$\int_{C_R} f(z)\,dz = \int_{-\infty}^{\infty} \frac{dx}{x^4+1} = 2\pi i\,\{\, \mathrm{Res}[\,f(z), z_1\,] + \mathrm{Res}[\,f(z), z_2\,]\,\}.$$

By Rule III above, the sum of the residues is

$$\frac{1}{4z_1^3} + \frac{1}{4z_2^3} = -\frac{1}{4}(z_1+z_2) = -\frac{\sqrt{2}}{4}\,i$$

and hence

$$\int_{-\infty}^{\infty} \frac{dx}{x^4+1} = \frac{\pi\sqrt{2}}{2}. \quad \blacktriangleleft$$

At times it is more convenient to follow a similar procedure in the lower half-plane. We formulate a general principle covering both cases.

THEOREM 33   *Let $f(z)$ be analytic in a region D that includes the real axis and, except for a finite number of points, (a) all of the half-plane $y > 0$ or (b) all of the half-plane $y < 0$. Let*

$$(P) \int_{-\infty}^{\infty} f(x)\,dx \tag{11–191}$$

exist. In case (a),

$$(P)\int_{-\infty}^{\infty} f(x)\,dx = 2\pi i \cdot (\text{Sum of residues of } f \text{ in the upper half-plane}) + A,$$

$$\tag{11–192}$$

where

$$A = \lim_{R\to\infty} \int_{0}^{\pi} f(Re^{i\theta})\,iRe^{i\theta}\,d\theta. \tag{11–193}$$

In case (b),

$$(P) \int_{-\infty}^{\infty} f(x)\,dx = -2\pi i \cdot (\text{Sum of residues in the lower half-plane}) - B,$$

$$\tag{11-194}$$

where

$$B = \lim_{R \to \infty} \int_{\pi}^{2\pi} f(Re^{i\theta}) i\,Re^{i\theta}\,d\theta. \tag{11-195}$$

The theorem is proved as in Example 2 above, in which the principal value is not needed and in which (11–192) is used, with $A = 0$. As in the analysis of Example 2, we see that generally $A = 0$ *and* $B = 0$ *if* $f(z)$ *is analytic at* ∞ *and* $f(z)$ *has a zero of second order or higher at* ∞. If f has a zero of only first order at ∞, then

$$f(z) = \frac{a_{-1}}{z} + \frac{a_{-2}}{z^2} + \cdots = \frac{a_{-1}}{z} + f_1(z),$$

where $f_1(z)$ has a zero of second order or higher at ∞. Thus for $f_1(z)$ the corresponding limits A and B are 0, and for $f(z)$ we obtain

$$A = \lim_{R \to \infty} \int_0^{\pi} \frac{a_{-1}}{Re^{i\theta}} i\,Re^{i\theta}\,d\theta = \pi i a_{-1},$$

$$B = \lim_{R \to \infty} \int_{\pi}^{2\pi} \frac{a_{-1}}{Re^{i\theta}} i\,Re^{i\theta}\,d\theta = \pi i a_{-1}.$$

Hence, *if* $f(z)$ *has a zero of first order at* ∞,

$$A = B = -\pi i \cdot (\text{Residue of } f(z) \text{ at } \infty).$$

EXAMPLE 3 $(P) \displaystyle\int_{-\infty}^{\infty} (e^{1/(x-i)} - 1)\,dx.$ Here $f(z) = e^{1/(z-i)} - 1$ and f is analytic except at $z = i$. At ∞, $f(z)$ has a zero of first order since we have the Laurent series for $f(z)$:

$$e^{1/(z-i)} - 1 = \frac{1}{z-i} + \frac{1}{2!}\frac{1}{(z-i)^2} + \cdots, \qquad 0 < |z - i| < \infty;$$

thus as $z \to \infty$, $f(z) \to 0$ and $zf(z)$ has limit 1. Accordingly, f has residue -1 at ∞ and $A = B = \pi i$. From the Laurent series we see that f has residue 1 at i. Accordingly,

$$(P) \int_{-\infty}^{\infty} (e^{1/(x-i)} - 1)\,dx = 2\pi i + \pi i = 3\pi i.$$

If we take real and imaginary parts, we obtain

$$(P) \int_{-\infty}^{\infty} \left(e^{x/(x^2+1)} \cos \frac{1}{(x^2+1)} - 1 \right) dx = 0,$$

$$(P) \int_{-\infty}^{\infty} e^{x/(x^2+1)} \sin \frac{1}{(x^2+1)}\,dx = 3\pi.$$

REMARK. For this example, we tacitly assumed that the integral exists as a principal value. This must be the case and generally holds whenever $f(z)$ has a first-order zero at ∞, since in testing for existence of the principal value we can ignore the integral from $-a$ to a for a fixed a and ask whether

$$\lim_{R \to \infty} \left[\int_{-R}^{-a} f(x)\, dx + \int_{a}^{R} f(x)\, dx \right]$$

exists. For a sufficiently large, $f(x)$ can here be replaced by its Laurent series at ∞:

$$f(x) = \frac{a_{-1}}{x} + \frac{a_{-2}}{x^2} + \cdots = \frac{a_{-1}}{x} + \frac{g(x)}{x^2}.$$

We examine the two terms in turn. For the first term we have

$$\int_{-R}^{-a} \frac{a_{-1}}{x}\, dx + \int_{a}^{R} \frac{a_{-1}}{x}\, dx = a_{-1}\left(\ln \frac{a}{R} + \ln \frac{R}{a}\right) = 0.$$

For the second term we reason that $g(x)$ has a limit (namely, a_{-2}) as $x \to \infty$, and therefore $|g(x)/x^2| \leqslant \text{const}/x^2$, so that the integrals

$$\int_{a}^{\infty} \frac{g(x)}{x^2}\, dx = \lim_{R \to \infty} \int_{a}^{R} \frac{g(x)}{x^2}\, dx, \qquad \int_{-\infty}^{-a} \frac{g(x)}{x^2}\, dx = \lim_{R \to \infty} \int_{-R}^{-a} \frac{g(x)}{x^2}\, dx$$

exist separately. We conclude that the integral (11–191) exists. The same argument shows that when f has a second-order zero at ∞, the integral of f from $-\infty$ to ∞ exists, even without forming the principal value. ◀

Integrals of Fourier type   These are integrals of form $(P) \int_{-\infty}^{\infty} g(x)e^{i\alpha x}\, dx$, where α is a real parameter. We assume that $f(z) = g(z)e^{i\alpha z}$ satisfies the hypotheses of Theorem 33. It can be shown that, if $g(z)$ has a zero of first order or higher at ∞, then $A = 0$ for $\alpha > 0$ and $B = 0$ for $\alpha < 0$ (see OMLS, pp. 273–277). If further $g(z)$ has singularities only at z'_j ($j = 1, \ldots, k$) in the upper half-plane at z''_j ($j = 1, \ldots, l$) in the lower half-plane, then we can now write

$$(P) \int_{-\infty}^{\infty} g(x)e^{i\alpha x}\, dx = \begin{cases} 2\pi i \displaystyle\sum_{j=1}^{k} \text{Res}\,[g(z)e^{i\alpha z}, z'_j], & \alpha > 0 \\[2em] -2\pi i \displaystyle\sum_{j=1}^{l} \text{Res}\,[g(z)e^{i\alpha z}, z''_j], & \alpha < 0. \end{cases} \qquad (11\text{–}196)$$

For $\alpha = 0$, we are again in the case where $f(z)$ has a zero of first or higher order at ∞. We consider application of (11–196) to Fourier transforms in the next section.

EXAMPLE 3   The integrals $\displaystyle\int_{-\infty}^{\infty} \frac{x \cos x}{x^2 + 1}\, dx$ and $\displaystyle\int_{-\infty}^{\infty} \frac{x \sin x}{x^2 + 1}\, dx$ both exist (AC, page 444). Hence

$$\int_{-\infty}^{\infty} \frac{x e^{ix}}{x^2 + 1}\, dx = 2\pi i\, \text{Res}\left[\frac{z e^{iz}}{z^2 + 1}, i \right] = \frac{\pi i}{e}.$$

Taking real and imaginary parts, we find

$$\int_{-\infty}^{\infty} \frac{x \cos x}{x^2 + 1}\, dx = 0, \qquad \int_{-\infty}^{\infty} \frac{x \sin x}{x^2 + 1}\, dx = \frac{\pi}{e}.$$

Since the first integral is an integral of an *odd* function, the value of 0 could be predicted. ◀

================================Problems (Section 11–19)================================

1. Evaluate the following integrals:

a) $\displaystyle\int_{0}^{2\pi} \frac{1}{5 + 3 \sin \theta}\, d\theta$

b) $\displaystyle\int_{0}^{2\pi} \frac{1}{5 - 4 \cos \theta}\, d\theta$

c) $\displaystyle\int_{0}^{2\pi} \frac{1}{(\cos \theta + 2)^2}\, d\theta$

d) $\displaystyle\int_{0}^{2\pi} \frac{1}{(3 + \cos^2 \theta)^2}\, d\theta$

2. Evaluate the following integrals:

a) $\displaystyle\int_{-\infty}^{\infty} \frac{1}{x^2 + x + 1}\, dx$

b) $\displaystyle\int_{-\infty}^{\infty} \frac{1}{(x^2 + 1)\,(x^2 + 4)}\, dx$

c) $\displaystyle\int_{-\infty}^{\infty} \frac{1}{x^6 + 1}\, dx$

d) $\displaystyle\int_{0}^{\infty} \frac{1}{(x^2 + 1)^2}\, dx$

3. Evaluate the following integrals:

a) $\displaystyle\int_{-\infty}^{\infty} \frac{\cos x}{x^2 + 4}\, dx$

b) $\displaystyle\int_{-\infty}^{\infty} \frac{\sin 2x}{x^2 + x + 1}\, dx$

c) $\displaystyle\int_{0}^{\infty} \frac{x^3 \sin x}{x^4 + 1}\, dx$

d) $\displaystyle\int_{0}^{\infty} \frac{x^2 \cos 3x}{(x^2 + 1)^2}\, dx$

11–20   APPLICATION OF RESIDUES TO INVERSE FOURIER TRANSFORMS

It is clear from the rules (11–196) that residues can be used both for Fourier transforms and for inverse Fourier transforms. Here we emphasize the latter as the more common application.

Let $f(t)$ be as in Chapter 4, a piecewise smooth function defined for $-\infty < t < \infty$, having an absolutely convergent integral over this interval (Section 4–13). Thus function $\varphi(\omega) = \Phi[f]$ is well defined and continuous for $-\infty < \omega < \infty$, and (for proper choice of $f(t)$ at jumps)

$$f(t) = \frac{1}{2\pi}(P)\int_{-\infty}^{\infty} \varphi(\omega)e^{i\omega t}\, d\omega, \qquad -\infty < t < \infty. \qquad (11\text{–}200)$$

In the preceding section we saw a way of evaluating such integrals, with ω replaced by x, so that the integration is along the *real* axis of an $x + iy$ plane.

For notational consistency, we now prefer to interpret ω as the *imaginary part* of a complex variable $s = \sigma + i\omega$ and to regard the integral in (11–200) as one along the *imaginary* axis of the *s*-plane. Along that axis $s = i\omega$ and $ds = i\,d\omega$. If $\varphi(\omega) = F(i\omega)$ for an analytic function $F(s)$, then (11–200) can be written:

$$f(t) = \frac{1}{2\pi i}(P)\int_C F(s)e^{st}\,ds,\qquad\qquad(11\text{--}200')$$

where C is the imaginary axis of the *s*-plane, directed upward.

We have in effect rotated the procedures of Section 11–19 by 90° and hence can easily translate the results. We emphasize the counterpart of the formulas (11–196):

If $F(s)$ is analytic along the imaginary axis, $F(s)$ has a zero at infinity and $F(s)$ has singularities only at $s'_1, \ldots, s'_k$ in the left half-plane ($\sigma < 0$) and at $s''_1, \ldots, s''_l$ in the right half-plane, then

$$f(t) = \sum_{j=1}^{k} \operatorname{Res}\left[e^{st}F(s),\, s'_j\right]\quad\text{for } t > 0,\qquad\qquad(11\text{--}201)$$

$$f(t) = -\sum_{j=1}^{l} \operatorname{Res}\left[e^{st}F(s),\, s''_j\right]\quad\text{for } t < 0,\qquad\qquad(11\text{--}202)$$

$$f(0) = \sum_{j=1}^{k} \operatorname{Res}\left[F(s),\, s'_j\right] + \frac{1}{2}\operatorname{Res}\left[F(s),\, \infty\right]$$

$$= -\sum_{j=1}^{l} \operatorname{Res}\left[F(s),\, s''_j\right] - \frac{1}{2}\operatorname{Res}\left[F(s),\, \infty\right].\qquad\qquad(11\text{--}203)$$

EXAMPLE 1 $\Phi[f] = 1/(i\omega - 2)$; $F(s) = 1/(s-2)$. Then $F(s)$ has a pole of first order at $s = 2$ and a zero at ∞. Hence there is no point s_j for $\sigma < 0$, and $f(t) = 0$ for $t > 0$. For $t < 0$,

$$f(t) = -\operatorname{Res}\left[\frac{e^{st}}{s-2},\, 2\right] = -e^{2t}.$$

Hence $f(t)$ has a jump at $t = 0$. Equation (11–203) gives the value $f(0)$ (average of left and right limits) as $\frac{1}{2}\operatorname{Res}\left[1/(s-2),\, \infty\right]$. Now

$$\frac{1}{s-2} = \frac{1}{s(1-2/s)} = \frac{1}{s} + \frac{2}{s^2} + \cdots$$

Hence the residue is -1 and $f(0) = -1/2$. ◄

EXAMPLE 2 $\Phi[f] = \dfrac{1}{-\omega^2 - 4} = \dfrac{1}{(i\omega)^2 - 4}.$

We choose

$$F(s) = \frac{1}{s^2 - 4}.$$

Thus $F(s)$ has poles of first order at $s = \pm 2$ and a zero of *second* order at ∞. Hence there is zero residue at ∞. By Rule III of Section 11–15,

$$f(t) = \operatorname{Res}\left[\frac{e^{st}}{s^2 - 4}, -2\right] = \frac{e^{st}}{2s}\bigg|_{-2} = \frac{e^{-2t}}{-4} \quad \text{for } t \geqslant 0,$$

$$f(t) = -\operatorname{Res}\left[\frac{e^{st}}{s^2 - 4}, +2\right] = -\frac{e^{st}}{2s}\bigg|_{2} = -\frac{e^{2t}}{4} \quad \text{for } t \leqslant 0.$$

Thus $f(t)$ is continuous at $t = 0$, with value $-1/4$. The continuity at $t = 0$ will clearly hold whenever $F(s)$ is rational with a zero of second or higher order at ∞. ◄

EXAMPLE 3 $\Phi[f] = e^{1/(i\omega - 1)} - 1$; $F(s) = e^{1/(s-1)} - 1$. The function $1/(s - 1)$ is analytic at $s = \infty$, with value zero there; hence $e^{1/(s-1)}$ is also analytic at $s = \infty$, with value 1, and $F(s)$ has a zero at ∞. $F(s)$ has a singularity only at $s = 1$:

$$F(s) = \frac{1}{s - 1} + \frac{1}{2!}\frac{1}{(s-1)^2} + \cdots + \frac{1}{n!(s-1)^n} + \cdots$$

Hence $f(t) = 0$ for $t > 0$. For $t < 0$, $f(t) = -\operatorname{Res}[e^{st}F(s), 1]$. Now,

$$e^{st}F(s) = e^t\left[1 + \frac{t(s-1)}{1!} + \frac{t^2(s-1)^2}{2!} + \cdots\right]\left[\frac{1}{s-1} + \frac{1}{2!}\frac{1}{(s-1)^2} + \cdots\right].$$

Since both series converge absolutely for $s \neq 1$, we can multiply and arrange terms in any order. If we collect the terms in $(s-1)^{-1}$, we find

$$\frac{e^t}{s-1}\left[1 + \frac{t}{2!} + \frac{t^2}{2!3!} + \cdots \frac{t^n}{n!(n+1)!} + \cdots\right].$$

Hence for $t < 0$,

$$f(t) = -\operatorname{Res}[e^{st}F(s), 1] = -e^t\left[1 + \frac{t}{2!} + \frac{t^2}{2!3!} + \cdots + \frac{t^n}{n!(n+1)!} + \cdots\right].$$

$$(11\text{–}204)$$

This gives $f(t)$ as an infinite series. We can also write $f(t)$ as a contour integral:

$$f(t) = -\frac{1}{2\pi i}\oint_{C_0} e^{st}\left[e^{1/(s-1)} - 1\right]ds,$$

where C_0 is a simple closed path (e.g., a circle) enclosing the singularity $s = 1$. If C_0 is the circle $|s - 1| = 1/2$, then $\operatorname{Re}(s) \geqslant 1/2$ on C_0. Hence as $t \to -\infty$, $e^{st} \to 0$ uniformly on C_0, so that $f(t) \to 0$ as $t \to -\infty$. The series (11–204) gives $f(t)$ easily for small negative t, while for large negative t, $f(t) \to 0$. Accordingly we obtain the graph of Fig. 11–22.

For $t = 0$, we can compute the value by residues, but we can predict the result: $f(0) = -1/2$. ◄

This example suggests various new techniques that become available. In particular, it may be of value to replace the residue expressions given above by the

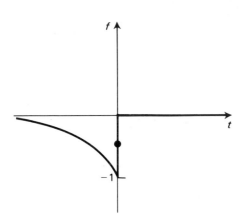

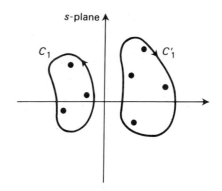

Fig. 11–22. Inverse Fourier transform
of $e^{1\,(i\omega-1)}-1$

Fig. 11–23. Paths for evaluation of inverse
Fourier transform.

corresponding contour integrals. Thus (11–201) can also be written as

$$f(t) = \frac{1}{2\pi i} \oint_{C_1} e^{st} F(s)\,ds, \qquad t > 0, \qquad (11\text{–}201')$$

where C_1 is a path enclosing all the points s_j (Fig. 11–23). Similarly, (11–202)
becomes

$$f(t) = \frac{1}{2\pi i} \oint_{C_1'} e^{st} F(s)\,ds, \quad t < 0. \qquad (11\text{–}202')$$

Thus the path of integration along the imaginary axis has been deformed into the
two types of closed paths.

═══════════════════════**Problems (Section 11–20)**═══════════════════════

1. Evaluate the inverse Fourier transforms by residues for the following functions of ω:

    a) $\dfrac{1}{(i\omega - 2)(i\omega + 1)}$
    b) $\dfrac{1}{\omega^2 + 1}$

    c) $\dfrac{1}{(i\omega - a)^n}$, $\operatorname{Re}(a) \neq 0$, $n = 1, 2, \ldots$
    d) $\dfrac{1}{\omega^4 + 1}$

2. Represent the inverse Fourier transforms of the following functions by means of contour
    integrals:

    a) $\exp\left[\dfrac{1}{1+\omega^2}\right] - 1$
    b) $\sin\dfrac{1}{i\omega - 2}$

11–21 THE LAPLACE TRANSFORM AS AN ANALYTIC FUNCTION

In Sections 4–1 to 4–5 the Laplace transform is defined as

$$F(s) = \mathscr{L}[f] = \int_0^\infty f(t)\, e^{-st}\, dt, \qquad s = \sigma + i\omega,$$

and its properties are developed. In particular, it is pointed out that there is an abscissa of absolute convergence σ^*, such that $\int_0^\infty |f(t)e^{-st}|\, dt$ converges for $\sigma > \sigma^*$. If $\sigma^* < \infty$, we thus have a half-plane D of absolute convergence, as in Fig. 4–6, and $F(s)$ is defined in the open region D. Furthermore, $F(s)$ *is analytic in* D. This follows at once from the rules

$$\mathscr{L}[tf(t)] = -F'(s), \qquad \mathscr{L}[t^2 f(t)] = F''(s), \qquad \cdots$$

(Eq. (4–36) in Section 4–3), which are valid for $\sigma > \sigma^*$. These rules show that $F'(s)$ is continuous in D and hence $F(s)$ is analytic in D. We observe that

$$\mathscr{L}[t] = \frac{1}{s}, \quad \sigma > 0; \qquad \mathscr{L}[t^n] = \frac{n!}{s^{n+1}}, \quad n = 1, 2, \ldots, \quad \sigma > 0;$$

$$\mathscr{L}[e^t] = \frac{1}{s-1}, \quad \sigma > 1; \qquad \mathscr{L}[\sin t] = \frac{1}{s^2+1}, \quad \sigma > 0. \tag{11–210}$$

These are all examples of functions $F(s)$ analytic in half-planes $\sigma > \sigma^*$. In each of these cases $F(s)$ is obtained from a function $F_1(s)$ analytic in a region D_1 containing much more than the half-plane D; we say that the function $F_1(s)$ is an *analytic continuation* of $F(s)$ (see Section 11–23). We often refer to the function $F_1(s)$ as the Laplace transform of the function $f(t)$, even though, strictly speaking, $\mathscr{L}[f]$ is defined only in the half-plane $\sigma > \sigma^*$. Thus, for example, we say that the Laplace transform of e^t is $1/(s-1)$, an analytic function defined except at $s = 1$, where it has a pole of first order.

It is natural to ask which analytic functions can appear as Laplace transforms. One class of such functions consists of the functions having a zero at ∞:

$$F(s) = \frac{a_{-1}}{s} + \frac{a_{-2}}{s^2} + \cdots, \qquad |s| > R_0. \tag{11–211}$$

If we take inverse transforms term by term, we obtain the function

$$f(t) = a_{-1} + a_{-2}t + \cdots + a_{-n}\frac{t^{n-1}}{(n-1)!} + \cdots \tag{11–212}$$

It can be verified that this series converges for *all* t, real or complex, and that $\mathscr{L}[f] = F(s)$ for $\sigma > R_0$. Furthermore, for all t, $f(t)$ satisfies an inequality of form

$$|f(t)| < ce^{R|t|}, \text{ where } c, R \text{ are positive constants.} \tag{11–213}$$

The function $f(t)$ defined by Eq. (11–212) for all t is analytic for all complex t; such a function, analytic for all t, is called an *entire function*. Because of the inequality

(11–213), $f(t)$ is said to be an entire function of *exponential type*. Thus we can say: if $F(s)$ is analytic and has a zero at $s = \infty$, then $\mathscr{L}[f] = F(s)$, where $f(t)$ is given by the series (11–212) and can thereby be defined for all complex t to become an entire function of exponential type. It can be shown further that every entire function of exponential type has a Laplace transform $F(s)$ as in (11–211), with a zero at ∞ (see OMLS, Section 6–9).

The transforms (11–210) all provide examples of such $f(t)$ and $F(s)$. Thus e^t is of exponential type and its Laplace transform is

$$\frac{1}{s-1} = \frac{1}{s}\frac{1}{1-(1/s)} = \frac{1}{s} + \frac{1}{s^2} + \cdots + \frac{1}{s^n} + \cdots, \qquad |s| > 1.$$

Inverting term by term, we obtain the usual power series for e^t.

If $F(s)$ has a zero at ∞ and $F(s)$ is analytic for all s except at a finite number of points $s_1, \ldots, s_k$, then the corresponding inverse Laplace transform $f(t)$ for $t \geq 0$ (with $f(0)$ equal to the average of left and right limits) can be obtained by residues:

$$f(t) = \sum_{j-1}^{k} \text{Res}[F(s)e^{st}, s_j], \qquad t > 0,$$

$$f(0) = \sum_{j=1}^{k} \text{Res}[F(s), s_j] + \frac{1}{2}\text{Res}[F(s), \infty]. \qquad (11\text{–}214)$$

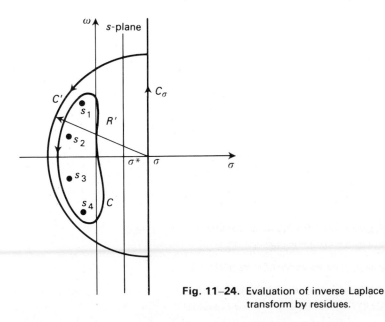

Fig. 11–24. Evaluation of inverse Laplace transform by residues.

This is simply an application of the reasoning leading to the formulas (11–201) and (11–202). As in Eq. (4–173), we have

$$f(t) = \frac{1}{2\pi} \int_{-\infty}^{\infty} F(s)e^{st}\, d\omega = \frac{1}{2\pi i} \int_{C_\sigma} F(s)e^{st}\, ds,$$

where $s = \sigma + i\omega$, σ is fixed, $\sigma > \sigma^*$, and C_σ is the line $\mathrm{Re}\, s = \sigma$ directed upward (Fig. 11–24). As in Section 11–19, for $t > 0$ the integral of $F(s)e^{st}$ on a semicircular arc C', as in the figure, approaches 0 as the radius R' approaches ∞, so that the first equation (11–214) follows. The result for $t = 0$ is also obtained as in Section 11–19.

For $t > 0$, we can also write the formula as follows:

$$f(t) = \frac{1}{2\pi i} \oint_C F(s)\, e^{st}\, ds, \tag{11–215}$$

where C is a simple closed curve enclosing all of $s_1, \ldots, s_k$ as in Fig. 11–24.

EXAMPLE 1   Let $F(s) = 1/s$. Then for $t > 0$

$$f(t) = \frac{1}{2\pi i} \oint_C \frac{e^{st}}{s}\, ds = \mathrm{Res}\left[\frac{e^{st}}{s}, 0\right] = 1,$$

while

$$f(0) = \mathrm{Res}\left[\frac{1}{s}, 0\right] + \frac{1}{2}\mathrm{Res}\left[\frac{1}{s}, \infty\right] = 1 - \frac{1}{2} = \frac{1}{2}.$$

Of course $f(0) = (1/2)\lim f(t)$ as $t \to 0^+$, since $f(t) = 0$ for $t < 0$. ◄

EXAMPLE 2   Let $F(s) = 1/(s-1)^3$. We find, for $t > 0$,

$$f(t) = \frac{1}{2\pi i} \oint_C \frac{e^{st}}{(s-1)^3}\, ds = \mathrm{Res}\left[\frac{e^{st}}{(s-1)^3}, 1\right] = t^2 \frac{e^t}{2!}.$$

For $t = 0$, the value must be 0. ◄

In both examples, $f(t)$ could be represented for $t > 0$ as the sum of a series (11–212).

======================Problems (Section 11–21)======================

1. Evaluate by residues the inverse Laplace transforms of the following functions:

   a) $\dfrac{1}{s^2 - 1}$ b) $\dfrac{1}{s^3 + 1}$ c) $\dfrac{1}{(s-1)^2}$ d) $\dfrac{s^2}{(s^2 - 1)^2}$

   e) $\dfrac{s}{(s+1)(s+2)}$ f) $\dfrac{s^5}{s^6 - 1}$

2. For each of the following find the inverse Laplace transforms as power series and as contour integrals:

   a) $se^{1/s} - 1 - s$ b) $\sin\dfrac{1}{s}$ c) $\displaystyle\sum_{n=1}^{\infty} \frac{n^2 + 1}{s^n}$

3. For each of the following prove that $f(t)$ is of exponential type, and represent $f(t)$ as an integral as in (11–215):

a) t^2 **b)** te^t **c)** $e^t - e^{2t}$ **d)** $\displaystyle\sum_{n=0}^{\infty} \frac{t^n}{n! + n^2}$

11–22 THE NYQUIST CRITERION

We consider a linear differential equation of order n with constant coefficients:

$$a_0 \frac{d^n x}{dt^n} + \cdots + a_n x = f(t). \tag{11–220}$$

We write $V(s) = a_0 s^n + \cdots + a_n$, so that the characteristic equation is the equation $V(s) = 0$. We know that a system described by (11–220) is *stable* precisely when all characteristic roots have *negative* real part. Below we shall also use the *transfer function* $Y(s) = 1/V(s)$ and *frequency response function* $Y(i\omega)$ (see Section 3–8).

Since $V(s)$ is an analytic function of the complex variable s, the number N_0 of zeros of $V(s)$ in a region bounded by a curve C can be found by the argument principle (Section 11–17):

$$N_0 = \frac{1}{2\pi i} \oint_C \frac{V'(s)}{V(s)} ds = \frac{1}{2\pi} \cdot \text{(Total increment in arg } V \text{ about } C\text{)}. \tag{11–221}$$

To study stability we wish to determine whether $V(s)$ has any zeros in the right half-plane, $\text{Re}(s) \geqslant 0$. To this end we choose C as a semicircle of radius R:

$$\text{Re } s = 0, \quad -R \leqslant \text{Im } s \leqslant R; \qquad |s| = R, \quad -\frac{\pi}{2} \leqslant \arg s \leqslant \frac{\pi}{2}$$

(see Fig. 11–25). If R is sufficiently large, C will enclose all roots of $V(s)$ in the right half-plane. If, in accordance with (11–221), the total increment of arg V about C is

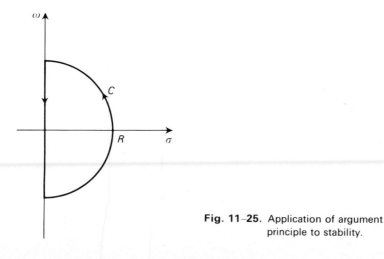

Fig. 11–25. Application of argument principle to stability.

zero, then there are no zeros of V in the right half-plane, and the corresponding differential equation (11–220) is stable. (It should be noted that if V has a zero on the imaginary axis, then arg V becomes undefined at this point; because of the root with $\text{Re}(s) = 0$, the system must be *unstable*.)

It is important to know how large R must be chosen to ensure that all roots of V in the right half-plane lie inside C. The following theorem provides an answer.

THEOREM 34   *The polynomial* $V(s) = a_0 s^n + \cdots + a_n$ *has no roots for which* $|s| \geqslant R$, *if* $R = 1 + M/|a_0|$ *and* M *is the largest of* $|a_1|, \ldots, |a_n|$.

PROOF   Let s be a root and let $|s| > 1$. Then

$$1 = -\frac{1}{a_0 s}\left(a_1 + \frac{a_2}{s} + \cdots + \frac{a_n}{s^{n-1}}\right),$$

so that

$$1 \leqslant \frac{1}{|a_0|\,|s|}\left(|a_1| + \frac{|a_2|}{|s|} + \cdots + \frac{|a_n|}{|s|^{n-1}}\right)$$

$$\leqslant \frac{M}{|a_0|\,|s|}\left(1 + \frac{1}{|s|} + \cdots + \frac{1}{|s|^{n-1}}\right)$$

$$< \frac{M}{|a_0|\,|s|}\sum_{k=0}^{\infty}\frac{1}{|s|^k} = \frac{M}{|a_0|}\frac{1}{|s|-1}.$$

In the last step we took advantage of the fact that the geometric series $\sum |s|^{-k}$ converges for $|s| > 1$ and can be summed by the familiar rule. We have finally

$$1 < \frac{M}{|a_0|}\frac{1}{|s|-1} \qquad \text{or} \qquad |s| < 1 + \frac{M}{a_0},$$

as asserted.

In computing the total change of arg V on C we can reason as in Section 11–17, that as s traces C, $w = V(s)$ traces a curve C_w in the w-plane. The number of zeros inside C is then equal to the number of times C_w encircles the origin of the w-plane in the positive direction. Hence it is sufficient to graph C_w and determine whether it encircles $w = 0$. In graphing C_w we can simplify the work by noting that along the circular part of C, $V(s)$ is approximately equal to $a_0 s^n = a_0 R^n e^{in\theta}$, for R large. Hence this part of C corresponds to an approximately circular arc in the w-plane, along which arg w varies from $-n\pi/2$ to $+n\pi/2$. This need not be graphed in detail, but can be joined to the graph of the part of C_w corresponding to one diameter of C to give a closed path in the w-plane.

EXAMPLE $V = s^2 + 2s - 3$. Here of course we immediately verify that the system is unstable. Along the diameter of C, $V = -3 - \omega^2 + 2i\omega$ and the path $w = V(s)$ is easily traced (Fig. 11–26). Along the circular part, $w = s^2$ approximately, and

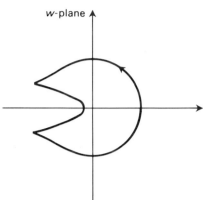

w-plane

Fig. 11-26. Nyquist diagram for $V(s) = s^2 + 2s - 3$.

accordingly we trace most of the circle $|w| = R^2$, $-\pi < \arg w < \pi$. In accordance with Theorem 34, R is chosen as 4. The resulting curve C_w encircles the origin $w = 0$ once; hence there is one zero inside C, and the system is unstable. ◄

A diagram such as that of Fig. 11-26 is known as a *Nyquist diagram*, and the determination of stability by determining how many times C_w encircles $w = 0$ is known as the *Nyquist criterion*. The essential part of the Nyquist diagram is a graph of $V(i\omega)$. Since $V(i\omega) = 1/Y(i\omega)$, this is essentially a graph of the *frequency response function*. Indeed we can graph $w = Y(s) = 1/V(s)$ for s on C, and can determine stability in the same way, since $\arg Y = -\arg V$, and if the total change of $\arg Y$ is zero on C, then the total change of $\arg V$ is zero. If $Y(s)$, rather than $V(s)$, is graphed, then the circular part of C corresponds to a *small* circular arc $w = 1/(a_0 R^n e^{in\theta})$, rather than a large one.

Stability determined from graph of $\arg V(i\omega)$

The analysis can be considerably simplified by noting that all the information is contained in the graph of $V(i\omega)$ or, even better, in the graph of $\arg V(i\omega)$. We formulate a precise form of the rule as follows:

THEOREM 35   *Let $V(s) = a_0 s^n + \cdots + a_0$ be a polynomial with real coefficients, with $a_0 > 0$. Let M be the largest of $|a_1|, |a_2|, \ldots, |a_n|$; let $R = 1 + 3M/|a_0|$. Let $\arg V(i\omega)$ be defined as a continuous function of ω for $\omega \geqslant 0$. Let α be the total change in $\arg V(i\omega)$ as ω goes from 0 to R:*

$$\alpha = \arg V(iR) - \arg V(0).$$

Let k be the integer closest to $\alpha \div (\pi/2)$. Then $-n \leqslant k \leqslant n$. If $k = n$, $V(s)$ has all its roots in the left half-plane. If $k < n$, then $V(s)$ has $(n-k)/2$ roots in the right half-plane.

REMARK.   As pointed out above, when V has a root on the imaginary axis, the procedure breaks down and the system is unstable.

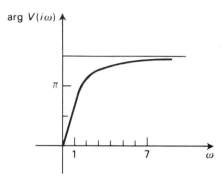

Fig. 11–27. Graph of arg $V(i\omega)$,
$$V(s) = s^3 + 2s^2 + s + 1$$

We illustrate Theorem 35 by an example: $V(s) = s^3 + 2s^2 + 2s + 1$. Here $V(i\omega) = 1 - 2\omega^2 + i(2\omega - \omega^3)$, and it is clear that $V(i\omega)$ is never zero, so that

$$\arg V(i\omega) = \arctan \frac{2\omega - \omega^3}{1 - 2\omega^2}$$

can be defined as a continuous function of ω. Here $M = 2$, $a_0 = 1$, $R = 7$, and we need to graph arg $V(i\omega)$ in the interval $0 \leqslant \omega \leqslant 7$. For $\omega = 0$, $V = 1$, and arg $V(i\omega)$ can be chosen as zero. Continuity then dictates the choice of arg $V(i\omega)$ for $0 \leqslant \omega \leqslant 7$. The graph is shown in Fig. 11–27. Hence α is very close to $3\pi/2$, so that $2\alpha/\pi$ is very close to 3; $k = 3 = n$. Hence $V(s)$ has no roots in the right half-plane. For a proof of Theorem 35, see OMLS, pp. 417–419.

Nyquist criterion for a rational function

In various problems the transfer function $Y(s)$ appears as a general rational function of s rather than as $1/V(s)$, where V is a polynomial. In these cases $Y = P(s)/Q(s)$, where P and Q are polynomials. The characteristic equation is $Q(s) = 0$, and stability is determined by whether its roots all lie in the left half-plane.

We can apply the methods given above to determine whether the roots of Q lie in the left half-plane. We can also apply the procedures to $Y(s)$ rather than to $Q(s)$. The roots of Q now correspond to *poles* of $Y(s)$, and we wish to know how many poles $Y(s)$ has in the right half-plane. If N_0, N_p are the number of zeros and poles of Y inside C, then by the argument principle,

$$N_0 - N_p = \frac{1}{2\pi i} \oint_C \frac{Y'(s)}{Y(s)} ds = \frac{1}{2\pi} \cdot (\text{Total increment of arg } Y(s) \text{ about } C).$$

$$(11\text{--}222)$$

If R is chosen large enough (see Theorem 34 above), all zeros and poles of $Y(s)$ in the right half-plane lie inside C. However, formula (11–222) will permit us to determine only the difference $N_0 - N_p$. If N_0 is known, then N_p can be found; if N_0 is known to be zero, then

$$N_p = -\frac{1}{2\pi} \cdot (\text{Total increment of arg } Y(s)).$$

In any case, if $N_0 - N_p$ is negative, $Y(s)$ must have poles in the right half-plane, so that the system is unstable.

In graphing $w = Y(s)$ for s on C, we can again obtain the image of the semicircular portion of C as an approximately circular arc. Thus if

$$Y(s) = \frac{a_0 s^n + \cdots + a_n}{b_0 s^m + \cdots + b_m},$$

then Y is approximately $(a_0/b_0)s^{n-m}$ for large R. In general, n will be less than m, so that the arc will be on a circle of small radius $|a_0/b_0|R^{n-m}$. Along the diameter of C, $Y(s) = Y(i\omega)$, the frequency response function. Accordingly, it is essentially the graph of the frequency response function that determines whether or not the system is stable.

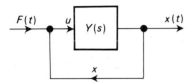

$$F(t) \quad u \quad \boxed{Y(s)} \quad x(t)$$

$$x$$

Fig. 11–28. System with feedback.

In studying physical systems with *feedback*, the point of view under consideration leads us to important applications. In the system illustrated in Fig. 11—28, there is an input $F(t)$ and an output $x(t)$. The output is compared to the input to yield the *error* $u(t) = F(t) - x(t)$. The error is then fed as input to a mechanism with transfer function $Y(s)$, and x is the corresponding output. Therefore, all initial values being zero,

$$\mathscr{L}[u] = \mathscr{L}[F] - \mathscr{L}[x], \qquad \mathscr{L}[x] = Y(s)\mathscr{L}[u],$$

so that

$$\mathscr{L}[x] = Y(s)(\mathscr{L}[F] - \mathscr{L}[x]), \qquad \mathscr{L}[x] = \frac{Y(s)}{1 + Y(s)}\mathscr{L}[F].$$

Hence the final transfer function is

$$Y_1(s) = \frac{Y(s)}{1 + Y(s)}.$$

This equation gives a relation between the *open-loop* transfer function $Y(s)$ and the *closed-loop* transfer function $Y_1(s)$. In many cases the open-loop transfer function is well known, and in particular the Nyquist diagram of $Y(s)$ for the curve C may be known. Finding the number of poles of $Y_1(s)$ inside C is now reduced to finding the number of zeros of $1 + Y(s)$ inside C. The argument principle can be applied to the graph of $Y(s)$ translated 1 unit to the right, and we can reason as before; or we can simply count the total number of times the graph of $Y(s)$ encloses the point -1. This gives the number of zeros of $1 + Y(s)$ inside C minus the number of poles of $Y(s)$ inside C. In many cases it is known that $Y(s)$ has no poles in the right half-plane, that is, that the open loop by itself is stable. In such a case, the closed loop is stable if and only if the point -1 is not enclosed by the graph of $Y(s)$, s on C.

═══════════════════**Problems (Section 11–22)**═══════════════════

1. Find the image of the curve C of Fig. 11–25 under the transformation $w = s^3 + 2$ for the cases $R = 1, 2, 3$. Use these graphs to determine the number of roots of $s^3 + 2 = 0$ in the right half-plane.

2. Graph $\arg V(i\omega)$ against ω for $\omega \geqslant 0$ and use the results to determine stability for the following choices of V:

   a) $s^3 + 2s^2 + 3s + 1$ **b)** $s^4 + 3s^3 + s^2 + s + 8$

3. Let the system of Fig. 11–28 have open-loop transfer function

$$Y(s) = \frac{s^2 + s + 7}{s^3 + 2s^2 + s + 1}.$$

Show that the open loop is stable, but the closed loop is unstable.

11–23 MULTIPLE-VALUED ANALYTIC FUNCTIONS. ANALYTIC CONTINUATION. RIEMANN SURFACES

We have encountered a number of multiple-valued analytic functions: $\ln z$, $z^{1/2}$, $\sin^{-1} z$, etc. (see especially Section 11–7). In each case analytic branches of the function in appropriate regions could be selected. Each such branch is a single-valued analytic function, to which the standard theory applies.

For a better understanding of the relation between various branches of a particular function we introduce the concept of analytic continuation. Let $f(z)$ and $g(z)$ be defined and analytic in open regions D_1, D_2 respectively and let the regions D_1, D_2 overlap. If then $f(z) = g(z)$ at every common point of the two regions, then we say: $g(z)$ is a *direct analytic continuation* of $f(z)$ (and $f(z)$ is a direct analytic continuation of $g(z)$). If $f_1(z), f_2(z), \ldots, f_n(z)$ are such that $f_2(z)$ is a direct analytic continuation of $f_1(z), f_3(z)$ of $f_2(z), \ldots, f_n(z)$ of $f_{n-1}(z)$, then we say that $f_n(z)$ is an *indirect analytic continuation* of $f_1(z)$, or simply, $f_n(z)$ is an analytic continuation of $f_1(z)$. The corresponding regions $D_1, D_2, \ldots, D_n$ must then form a chain as in Fig. 11–29. We observe that it may happen that D_3 overlaps D_1, for example, without $f_3(z) = f_1(z)$ at common points.

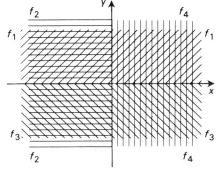

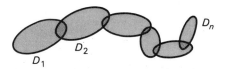

Fig. 11–29. Analytic continuation.

Fig. 11–30. Branches of $z^{1/2}$.

As an example we consider branches of the square-root function $w = z^{1/2}$ (see Fig. 11–30):

$$f_1(z) = r^{1/2}\left(\cos\frac{\theta}{2} + i\sin\frac{\theta}{2}\right), \qquad r > 0, \qquad 0 < \theta < \pi,$$

$$f_2(z) = r^{1/2}\left(\cos\frac{\theta}{2} + i\sin\frac{\theta}{2}\right), \qquad r > 0, \qquad \frac{\pi}{2} < \theta < \frac{3\pi}{2},$$

$$f_3(z) = r^{1/2}\left(\cos\frac{\theta}{2} + i\sin\frac{\theta}{2}\right), \qquad r > 0, \qquad \pi < \theta < 2\pi,$$

$$f_4(z) = r^{1/2}\left(\cos\frac{\theta}{2} + i\sin\frac{\theta}{2}\right), \qquad r > 0, \qquad \frac{3\pi}{2} < \theta < \frac{5\pi}{2},$$

$$f_5(z) = r^{1/2}\left(\cos\frac{\theta}{2} + i\sin\frac{\theta}{2}\right), \qquad r > 0, \qquad 2\pi < \theta < 3\pi.$$

Here $\theta = \arg z$, $r = |z|$. We see that $f_1(z) = f_2(z)$ in the common points, where $\pi/2 < \theta < \pi$ and hence $f_2(z)$ is a direct analytic continuation of $f_1(z)$. The same applies to the successive pairs: $f_2, f_3; f_3, f_4; f_4, f_5$. We observe that $f_5(z)$ and $f_1(z)$ are both defined in the upper half-plane and are unequal for every z, since for $f_5(z)$ angle 0 is 2π greater than it is for $f_1(z)$, so that $\cos 0/2$ and $\sin 0/2$ have opposite signs; thus $f_5(z) = -f_1(z)$ for every z. We can clearly continue the process to get $f_6(z) = -f_2(z)$, $f_7(z) = -f_3(z)$, $f_8(z) = -f_4(z)$ and finally $f_9(z) = f_1(z)$. From this point on the process repeats itself.

By a *complete analytic function* we now mean an analytic function $f_1(z)$ in a region D_1 plus all other analytic functions obtained from $f_1(z)$ by indirect analytic continuation. The function $w = z^{1/2}$ can be regarded as a complete analytic function formed of $f_1(z), \ldots, f_8(z)$ as above and all other functions obtained from $f_1(z)$ by indirect continuation (this includes branches of the square-root function in other sectors, circular regions, and so on). In fact, for every region D that contains no simple closed curve enclosing the origin there are exactly two branches of $z^{1/2}$ in D, opposite in sign, and all analytic continuations of $f_1(z)$ are obtainable in this way.

It is important to observe that analytic continuation, when possible, is *unique*. For example, let D_1 and D_2 overlap, let $f(z)$ be analytic in D_1, let $g(z)$ and $h(z)$ be analytic in D_2 and let $f(z) \equiv g(z), f(z) \equiv h(z)$ in the common part of D_1, D_2. Thus g and h would both be direct analytic continuations of f. Then $g(z) \equiv h(z)$ in all of D_2, for $g(z) - h(z)$ is analytic in D_2 and $g(z) - h(z) \equiv 0$ in the common part of D_2, D_1 (clearly an open set). Therefore, by Theorem 26, $g(z) - h(z) \equiv 0$ in D_2, that is, $g(z) \equiv h(z)$.

At times we refer to analytic continuation *on a path* C from z_0 to z_n. This is an indirect continuation from D_1 to D_n, as suggested in Fig. 11–31. Thus the points z_0, $z_1, \ldots, z_n$ are successive points on C; the part of C between z_0 and z_1 is in D_1; the part between z_1 and z_2 is in $D_2, \ldots$, that between z_{n-1} and z_n is in D_n. The uniqueness proof just given can be extended to show that if $f_1(z)$ in D_1 is given and continuation of $f_1(z)$ from z_0 to z_n on C is possible, then the values of the

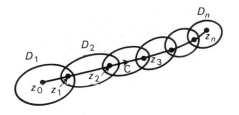

Fig. 11–31. Analytic continuation on a path.

continuation of f along C are uniquely determined. To make this precise, we let C be given in terms of a parameter t: $z = z(t)$, $t_0 \leqslant t \leqslant t_n$. Then the continuation uniquely determines functional values $f[z(t)]$. It also follows from this that the values of the continuation in a small neighborhood of z_n are uniquely determined.

Continuation along a path often arises by integration. For example, the values of $\ln z$ can be defined by an integral:

$$\ln z = \int_1^z \frac{d\zeta}{\zeta} \quad \text{on} \quad C{:}\zeta = \zeta(t), \quad a \leqslant t \leqslant b.$$

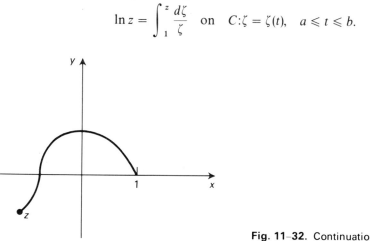

Fig. 11–32. Continuation of $\ln z$ on a path.

Here the path C must not pass through $z = 0$ (Fig. 11–32). As the endpoint z itself varies on C, we obtain the values of $\ln z$ as they would be obtained by analytic continuation along C of the branch of the logarithm defined near $z = 1$ and having the value 0 there.

Permanence of functional equations

If several analytic functions satisfy an algebraic identity in region D_1 and all can be continued analytically from D_1 to D_2, then their continuations satisfy the same identity in D_2. For example, if $f_1(z) = \operatorname{Ln} z^2$ and $g_1(z) = \operatorname{Ln} z$ in $D_1{:}\operatorname{Re} z > 0$, then f_1 and g_1 are analytic in D_1 and $f_1(z) = 2g_1(z)$ in D_1, as usual. It follows that the identity $f_2(z) = 2g_2(z)$ is satisfied by direct analytic continuations f_2, g_2 of f_1, g_1 from D_1 to D_2, for $f_2(z) - 2g_2(z)$ is then a continuation of $f_1 - 2g_1$. Since

$f_1(z) - 2g_1(z) \equiv 0$ in D_1, it follows as above that $f_2(z) - 2g_2(z) \equiv 0$ in D_2. By repeating the argument, we conclude that the identity is also preserved under indirect analytic continuation. By similar reasoning, the conclusion also applies to differential equations. For example, if $w = f(z)$ is a solution in D of the differential equation

$$p_0(z)w'' + p_1(z)w' + p_2(z)w = 0,$$

where $p_0(z)$, $p_1(z)$, $p_2(z)$ are, say, polynomials, then every analytic continuation of $f(z)$ must satisfy the same differential equation. All these statements are described as cases of *permanence of functional equations*.

Riemann surfaces

We illustrate the concept by the example of $w = z^{1/2}$ discussed above (Fig. 11–30). We observe that for each z other than $z = 0$ there are two values of the function, opposite in sign. This suggests inventing a double z-plane; each of two replicas is to be used for one assignment of values. However, the two z-planes must somehow fit together to provide a smooth transition between the two assignments of values. To achieve this, we cut each copy of the z-plane along the positive real axis, which serves as a *branch line*. We use the two z-planes, thus cut open, and call them sheet I and sheet II. On sheet I we set

$$w = z^{1/2} = r^{1/2}\left(\cos\frac{\theta}{2} + i\sin\frac{\theta}{2}\right), \qquad 0 < \theta < 2\pi, \qquad r > 0,$$

and on sheet II we set

$$w = z^{1/2} = r^{1/2}\left(\cos\frac{\theta}{2} + i\sin\frac{\theta}{2}\right), \qquad 2\pi < \theta < 4\pi, \qquad r > 0.$$

For each z the two values are negatives of each other.

We now imagine these two sheets brought close together and attached along the edges of the cut. However, the lower edge of the cut on sheet I is attached to the upper edge on sheet II; the upper edge of the cut on sheet I is attached to the lower edge of the cut on sheet II. The process is suggested in Fig. 11–33. It cannot be done physically in space without interpenetration of the two surfaces. However, we can imagine a gluing process that will yield smooth transitions. The final double z-plane is suggested in Fig. 11–33 (c). As we go around the origin in sheet I, with θ increasing from 0 to 2π, we approach the lower edge of the branch line. We may now cross the branch line and pass into sheet II, θ is passing smoothly through 2π. If we continue to go around the origin in sheet II, with θ going to 4π, we eventually reach the branch line again and can cross it, re-entering sheet I.

The double-sheeted surface is called the *Riemann surface* of $w = z^{1/2}$. It has the useful property that the function is *single-valued* on the surface. Furthermore, by assigning the values of w as above we have really described the complete analytic function, for every branch of the function can be pictured as a single-valued function in an open region on the Riemann surface.

We remark that at $z = 0$ we can assign the unique value $w = 0$; the point $z = 0$ is called a *branch point* of the function $w = z^{1/2}$.

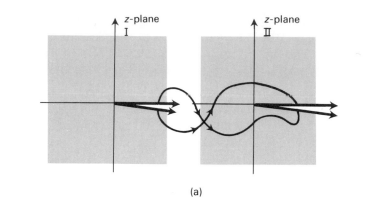

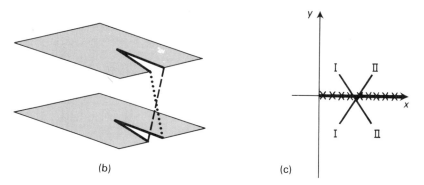

Fig. 11–33. (a) The two sheets of the Riemann surface of $w = z^{1/2}$; (b) process of attaching the sheets; (c) final diagram for the Riemann surface.

A similar procedure can be used for $w = z^{1/n}$ for each positive integer n. We need n sheets and can again use the positive real axis as a branch line, $z = 0$ as a branch point. The case $n = 3$ is suggested in Fig. 11–34.

For $w = \ln z$, we need *infinitely* many sheets, as suggested in Fig. 11–35.

We remark that there is a certain freedom in choosing the branch line. For the examples considered here it could be any ray from $z = 0$ to $z = \infty$ or even any curvilinear path proceeding from 0 to ∞ without crossing itself.

11–24 RESIDUES INVOLVING MULTIPLE-VALUED FUNCTIONS

We illustrate the ideas involved by examples.

EXAMPLE 1 $\int_0^\infty dx/(x^2 + 1)$. This can of course be integrated to yield $\pi/2$. However, to show our method, we proceed differently. We consider the function $\ln z = \ln r + i\theta, 0 < \theta < 2\pi$, on one sheet of its Riemann surface (Fig. 11–35). We can

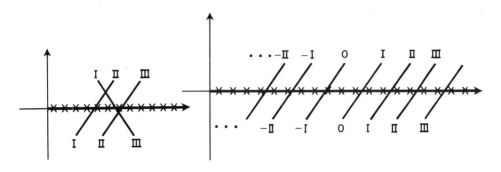

Fig. 11–34. Riemann surface of $w = z^{1/n}$.   **Fig. 11–35.** Riemann surface of $w = \ln z$.

even allow θ to approach 0 and 2π as limits as we approach the branch line from above and below. In the first case we obtain $\ln x + 0i$ and in the second we obtain $\ln x + 2\pi i$, at the point $x + 0i$.

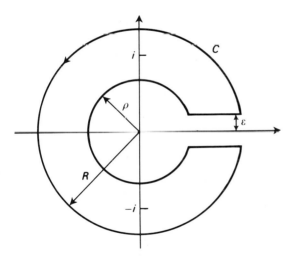

Fig. 11–36. Path for Example 1.

We now integrate the function $f(z) = (1 + z^2)^{-1} \ln z$ along the simple closed path C shown in Fig. 11–36. Here ε and ρ are small positive numbers and R is a large positive number. The function $f(z)$ is analytic inside and on C except for poles at $\pm i$, so that

$$\oint_C f(z)\,dz = 2\pi i \{\operatorname{Res}[f(z), i] + \operatorname{Res}[f(z), -i]\}$$

$$= 2\pi i \left[\frac{\ln i}{2i} + \frac{\ln(-i)}{2(-i)}\right]$$

$$= 2\pi i \left[\frac{i\pi/2}{2i} + \frac{i3\pi/2}{2(-i)}\right] = -\pi^2 i.$$

If we now let $\varepsilon \to 0^+$, then the integral around C (which always equals $-\pi^2 i$) approaches as limit the sum of four integrals: one around the circle $|z| = R$ in the positive direction, one around the circle $|z| = \rho$ in the negative direction, one along the x-axis of $(1 + x^2)^{-1} \ln x$ from ρ to R, and the last one along the x-axis of $(1 + x^2)^{-1} (\ln x + 2\pi i)$ from R to ρ. These last two integrals have sum

$$\int_\rho^R \left(\frac{\ln x}{1 + x^2} - \frac{\ln x + 2\pi i}{1 + x^2} \right) dx = -2\pi i \int_\rho^R \frac{dx}{1 + x^2}.$$

We now let $\rho \to 0$ and $R \to \infty$. We verify below that the integrals on the circles $|z| = R$ and $|z| = \rho$ have limit 0. Hence we conclude that

$$-2\pi i \int_0^\infty \frac{dx}{1 + x^2} = -\pi^2 i, \qquad \int_0^\infty \frac{dx}{1 + x^2} = \frac{\pi}{2}.$$

It remains to verify that the integrals on $|z| = R$ and $|z| = \rho$ approach 0. Since $(1 + z^2)^{-1}$ has a zero of second order at ∞, we can write

$$\left| \frac{1}{1 + z^2} \right| \leqslant \frac{\text{const}}{|z|^2}$$

for large $|z|$. Also $|\ln z| = |\ln r + i\theta| \leqslant |\ln r| + |\theta|$. Hence by Theorem 7 (Section 11–5) we have

$$\left| \oint_{|z| = R} f(z)\, dz \right| \leqslant \text{const} \oint_{|z| = R} \frac{|\ln z|}{R^2}\, ds \leqslant \text{const} \oint \frac{\ln R + 2\pi}{R^2}\, ds \leqslant \text{const} \frac{\ln R + 2\pi}{R^2} 2\pi R,$$

and the last expression has limit 0 as $R \to \infty$. Similarly, $(1 + z^2)^{-1}$ has limit 1 as $z \to 0$, so that $|(1 + z^2)^{-1}| \leqslant 2$ for $|z|$ small. Thus

$$\left| \oint_{|z| = \rho} f(z)\, dz \right| \leqslant 2 \oint \left| \ln z \right| ds \leqslant 2 \oint (\ln \rho + 2\pi)\, ds \leqslant 4\pi\rho\,(\ln\rho + 2\pi)$$

and the limit of the last expression is 0 as $\rho \to 0$.

This method is applicable generally to evaluate $\int_0^\infty R(x)\, dx$, where $R(z)$ is a rational function with no poles on the positive x-axis or at 0 and with a zero of second order or higher at ∞. ◀

EXAMPLE 2 $\displaystyle \int_0^\infty \frac{x^\beta}{1 + x^2}\, dx$, where $0 < \beta < 1$. Here we use

$$f(z) = \frac{z^\beta}{1 + z^2} = \frac{e^{\beta \ln z}}{1 + z^2},$$

where $\ln z$ is chosen as in Example 1. We can use the same path C as for Example 1, since $f(z)$ is analytic inside and on C except for poles at $\pm i$. We find

$$\oint_C f(z)\, dz = 2\pi i \left[\frac{e^{\beta \ln i}}{2i} + \frac{e^{\beta \ln(-i)}}{-2i} \right] = \pi \left[e^{\beta\,(\pi/2)i} - e^{\beta\,(3\pi/2)i} \right]$$

Again we let $\varepsilon \to 0$. The integrals on the two line segments give

$$\int_\rho^R \frac{e^{\beta \ln x}}{1+x^2}\, dx + \int_R^\rho \frac{e^{\beta(\ln x + 2\pi i)}}{1+x^2}\, dx = \int_\rho^R \frac{e^{\beta \ln x}}{1+x^2}(1 - e^{2\pi\, i\beta})\, dx$$

$$= (1 - e^{2\pi\, i\beta}) \int_\rho^R \frac{x^\beta}{1+x^2}\, dx.$$

Again we let $\rho \to 0$ and $R \to \infty$ and verify below that the integrals on the corresponding circles have limit 0. We conclude that

$$(1 - e^{2\pi\, i\beta}) \int_0^\infty \frac{x^\beta}{1+x^2}\, dx = \pi\left[e^{\beta(\pi/2)i} - e^{\beta(3\pi/2)i}\right]$$

After solving this equation for the integral sought and some trigonometry, we find

$$\int_0^\infty \frac{x^\beta}{1+x^2}\, dx = \pi\, \frac{\sin(\beta\pi/2)}{\sin \beta\pi}.$$

It remains to verify that the integrals on the circles approach 0 as stated. We have

$$|z|^\beta = \left|e^{\beta(\ln r + i\theta)}\right| = \left|e^{\beta \ln r} e^{i\,\beta\theta}\right| = e^{\beta \ln r} = r^\beta.$$

Hence on the circle $|z| = R$, as for Example 1,

$$\left|\oint f(z)\, dz\right| \leqslant \text{const} \oint \frac{R^\beta}{R^2}\, ds = \text{const} \cdot \frac{R^\beta}{R^2} \cdot 2\pi R,$$

and this has limit 0 as $R \to \infty$ since $0 < \beta < 1$. Similarly, on the circle $|z| = \rho$, as for Example 1,

$$\left|\oint f(z)\, dz\right| \leqslant 2\oint \rho^\beta ds = 2\rho^\beta \cdot 2\pi\rho,$$

and again the limit is 0.

The same method applies to $\int_0^\infty x^\beta R(x)\, dx$, $-1 < \beta < 1$, where $R(z)$ is a rational function with no poles on the positive real axis or at $x = 0$, with a zero of at least second order at ∞. ◄

EXAMPLE 3 $\displaystyle\int_0^1 \sqrt{\frac{x}{1-x}}\, \frac{1}{x^2+3x+2}\, dx.$ Here we need a Riemann surface for the function $[z(1-z)^{-1}]^{1/2}$. We can write $z^{1/2} = r^{1/2}e^{i\theta/2}$ as usual. We also introduce polar coordinates ρ, φ at $z = 1$ as in Fig. 11–37. Then we can write $z - 1 = \rho e^{i\varphi}$, so that $1 - z = \rho e^{i(\varphi + \pi)}$. We can now define two analytic branches of $[z(1-z)^{-1}]^{1/2}$ in the z-plane minus the line segment from 0 to 1:

I: $0 < \theta < 2\pi,$ $0 < \varphi < 2\pi,$ $w = \left(\dfrac{r}{\rho}\right)^{1/2} e^{i(\theta - \varphi - \pi)/2} = g_1(z),$

II: $0 < \theta < 2\pi,$ $2\pi < \varphi < 4\pi,$ $w = \left(\dfrac{r}{\rho}\right)^{1/2} e^{i(\theta - \varphi - \pi)/2} = g_2(z).$

These correspond to two sheets of the Riemann surface of the function, connected across a branch line along the line segment as shown in Fig. 11–37. The two

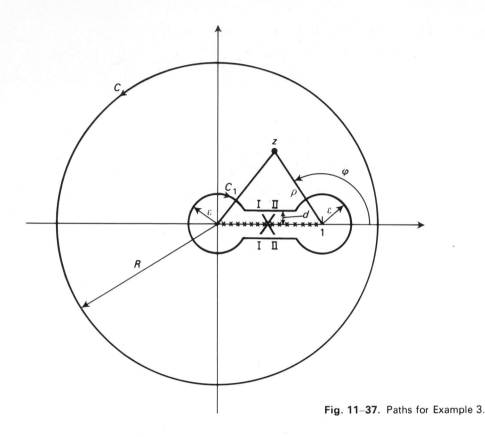

Fig. 11–37. Paths for Example 3.

functional values for each z are negatives of each other, since the φ values differ by 2π. On crossing the segment from 0 to 1, we pass from one sheet to the other. For example, when the segment is crossed from above to below starting in sheet I, θ has limit 0, φ has limit π, and the functional value has the same limit as that obtained from $\theta \to 2\pi$, $\varphi \to 3\pi$ as the segment is approached from below in sheet II. A similar analysis shows that there is *no* change of sheet in crossing the x-axis between 1 and ∞ and between $-\infty$ and 0 ($\theta - \varphi$ is continuous in both cases). Thus $g_1(z)$ and $g_2(z)$ can be treated as analytic for all z except for the segment from 0 to 1.

Using the branch $g_1(z)$, we form

$$f(z) = \frac{g_1(z)}{z^2 + 3z + 2} = \frac{g_1(z)}{(z + 1)(z + 2)}$$

and integrate around a circle $C: |z| = R$ as shown in Fig. 11–37. Along C, for large R,

$$|f(z)| = \left| \frac{z}{1 - z} \right|^{1/2} \left| \frac{1}{z^2 + 3z + 2} \right| \leqslant \frac{\text{const}}{R^2},$$

since $|z/(1 - z)| \to 1$ as $z \to \infty$. Hence, as above, we conclude that the integral around C has limit 0 as $R \to \infty$. But the integral around C is independent of R, so that the integral around C must be 0. By the Cauchy theorem, the integral around C plus that around the path C_1 shown equals

$$2\pi i \{\mathrm{Res}\,[f(z), -1] + \mathrm{Res}\,[f(z), -2]\} = 2\pi i \left[\frac{g_1(-1)}{1} + \frac{g_1(-2)}{-1} \right]$$

$$= 2\pi i \left[\left(\frac{1}{2}\right)^{1/2} e^{-\pi i/2} - \left(\frac{2}{3}\right)^{1/2} e^{-\pi i/2} \right] = 2\pi \left[\left(\frac{1}{2}\right)^{1/2} - \left(\frac{2}{3}\right)^{1/2} \right].$$

If we let the distance d (Fig. 11–37) approach 0, the integral around C_1 becomes the integral along the upper edge (in sheet I) of the branch line from ε to $1-\varepsilon$ plus that (in sheet I) along the lower edge from $1-\varepsilon$ to ε, plus integrals of $f(z)$ around circles of radius ε at 0 and 1. As in Example 1, we verify that the integral around each circle approaches 0 as $\varepsilon \to 0$. Hence

$$\left[\int_0^1 f(z)\,dz \right]_{\substack{\theta = 0. \\ \varphi = \pi}} + \left[\int_1^0 f(z)\,dz \right]_{\substack{\theta = 2\pi. \\ \varphi = \pi}} = 2\pi \left[\left(\frac{1}{2}\right)^{1/2} - \left(\frac{2}{3}\right)^{1/2} \right].$$

For $\theta = 0$ and $\varphi = \pi$,

$$g_1(z) = \left(\frac{x}{1-x}\right)^{1/2} e^{-\pi i} = -\left(\frac{x}{1-x}\right)^{1/2}.$$

For $\theta = 2\pi$ and $\varphi = \pi$,

$$g_1(z) = \left(\frac{x}{1-x}\right)^{1/2} e^{0i} = \left(\frac{x}{1-x}\right)^{1/2}.$$

Hence,

$$-2\int_0^1 \left(\frac{x}{1-x}\right)^{1/2} \frac{dx}{x^2 + 3x + 2} = 2\pi \left[\left(\frac{1}{2}\right)^{1/2} - \left(\frac{2}{3}\right)^{1/2} \right],$$

so that the integral sought equals $\pi[(2/3)^{1/2} - (1/2)^{1/2}]$. ◀

===========================Problems (Section 11–24)===========================

1. Show that $f(z)$ and $g(z)$ are direct analytic continuations of each other:

a) $f(z) = \displaystyle\sum_{n=0}^{\infty} z^n, \quad g(z) = \dfrac{1}{1-z}$

b) $f(z) = \displaystyle\sum_{n=0}^{\infty} z^n, \quad g(z) = \displaystyle\sum_{n=0}^{\infty} \dfrac{(z+1)^n}{2^{n+1}}$

c) $f(z) = \displaystyle\int_0^{\infty} t^2 e^{-zt}\,dt$ (Laplace transform), $g(z) = 2/z^3$.

d) $f(z) = \displaystyle\int_0^{\infty} e^{(1-z)t}\,dt, \quad g(z) = \dfrac{1}{z-1}$

2. Describe the Riemann surface of each function with the aid of suitable branch lines:

a) $f(z) = \sqrt{z-1}$ b) $f(z) = \sqrt{z^2 - 1}$

c) $f(z) = \sqrt{z(z^2 - 1)}$ d) $f(z) = \ln{(z^2 - 1)}$

3. Evaluate with the aid of residues:

a) $\displaystyle\int_0^\infty \frac{dx}{(x+1)(x+2)(x+3)}$ b) $\displaystyle\int_0^\infty \frac{dx}{(x^2+1)(x^2+4)}$

4. Evaluate with the aid of residues:

a) $\displaystyle\int_0^\infty \frac{x^\beta}{(x+1)(x+2)}\,dx, \; -1 < \beta < 1,$

b) $\displaystyle\int_0^\infty \frac{x^\beta}{(x+1)^2}\,dx, \; -1 < \beta < 1,$

c) $\displaystyle\int_0^1 \sqrt{\frac{x}{1-x}}\,\frac{1}{x^2+1}\,dx,$

d) $\displaystyle\int_0^1 \sqrt{x(1-x)}\,\frac{1}{(x+1)(x+2)(x+3)(x+4)}\,dx$

11–25 CONFORMAL MAPPING

As pointed out in Section 11–6, a function $w = f(z)$ can be interpreted as a *transformation* or *mapping* from the z plane to the w plane. The term *mapping* is really justified only when the correspondence between z values and w values is one-to-one, i.e., to each z of a region D_z corresponds just one $w = f(z)$ of a region D_w, and conversely. The region D_w is then a distorted picture or *image* of the region D_z; circles in D_z correspond to closed curves in D_w, triangles in D_z to curvilinear triangles in D_w, as illustrated in Fig. 11–38.

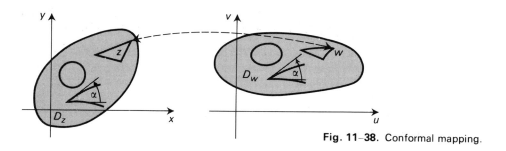

Fig. 11–38. Conformal mapping.

If $f(z)$ is analytic, such a mapping has an additional property: that of being conformal. A mapping from D_z to D_w is termed *conformal* if to each pair of curves in D_z intersecting at angle α there corresponds a pair of curves in D_w intersecting at angle α. The mapping is termed *conformal* and *sense-preserving* if the angles are equal and have the same sign, as illustrated in Fig. 11–38.

THEOREM 36   *Let* $w = f(z)$ *be analytic in the region* D_z, *and map* D_z *in a one-to-one fashion on a region* D_w. *If* $f'(z) \neq 0$ *in* D_z, *then* $f(z)$ *is conformal and sense-preserving.*

PROOF   Let $z(t) = x(t) + iy(t)$ be parametric equations of a smooth curve through z_0 in D_z. By proper choice of parameter (e.g., by using arc length) we can ensure that the tangent vector

$$\frac{dz}{dt} = \frac{dx}{dt} + i\frac{dy}{dt}$$

is not 0 at z_0. The given curve corresponds to a curve $w = w(t)$ in the w plane, with tangent vector

$$\frac{dw}{dt} = \frac{dw}{dz}\frac{dz}{dt}.$$

(This chain rule is established in the usual way). Hence

$$\arg\frac{dw}{dt} = \arg\frac{dw}{dz} + \arg\frac{dz}{dt}.$$

This equation asserts that, at $w_0 = f(z_0)$, the argument of the tangent vector differs from that of dz/dt by the angle $\arg f'(z_0)$, which is *independent of the particular curve chosen through* z_0. Accordingly, as the direction of the curve through z_0 is varied, the direction of the corresponding curve through w_0 must vary through the same angle (in magnitude and sign). The theorem is thus established.

Conversely, it can be shown that all conformal and sense-preserving maps $w = f(z)$ are given by analytic functions (see P. FRANKLIN, *A Treatise on Advanced Calculus*, pp. 425–428, New York: Wiley, 1940). From this geometric characterization it is also clear that the inverse of a one-to-one conformal, sense-preserving mapping has the same property and *is itself analytic.*

If $f'(z_0) = 0$ at a point z_0 of D_z, then $\arg f'(z_0)$ has no meaning and the argument above breaks down. It can be shown that conformality breaks down at z_0 and, in fact, the transformation is not one-to-one in any neighborhood of z_0 (see IAF, p. 112).

In practice, the term *conformal* is used loosely to mean *conformal and sense-preserving*; that will be done here. It should be noted that a reflection, such as the mapping $w = \bar{z}$, is conformal but sense-reversing.

Tests for one-to-one-ness   For the applications of conformal mapping, it is crucial that the mapping be one-to-one in the region chosen. In most examples, the mapping will also be defined and continuous on the boundary of the region; failure of one-to-one-ness on the boundary is less serious.

As a first step we should verify that $f'(z) \neq 0$ in the open region D_z considered, since, if $f'(z) = 0$ at a point, there is no chance that the mapping is one-to-one. However, even if $f'(z) \neq 0$ in D_z, the mapping may fail to be one-to-one and additional tests must be applied. Usually one of the following four tests suffices.

I. *Explicit formula for the inverse function.*   If an explicit formula for the inverse function $z = z(w)$ is available and it can be shown that, by this formula, there is at most one z in D_z for each w, then the mapping must be one-to-one. For example, $w = z^2$ is one-to-one in the *first quadrant* of the z plane, for to each w there is at most one square root $\sqrt{w}$ in the quadrant. As z varies over D_z, w varies over D_w: the upper half-plane $v > 0$, as will be seen below.

II. *Analysis of level curves of u and v.* We can verify one-to-one-ness and at the same time obtain a very clear picture of the mapping by plotting the level curves: $u(x, y) = c_1, v(x, y) = c_2$. If for given c_1, c_2 the loci $u = c_1, v = c_2$ intersect at most once in D_z, then the mapping from D_z to the uv plane is one-to-one.

III. *One-to-one-ness on the boundary.* If D_z is bounded by a simple closed curve C, if $w = f(z)$ is analytic in D_z plus C and is one-to-one on C, then $w = f(z)$ is one-to-one on D_z (see IAF, p. 112).

IV. *Direct method.* We verify that $f(z_1) = f(z_2)$ for z_1 and z_2 in D_z implies that $z_1 = z_2$. For example, $w = z + (1/z)$ is one-to-one for $|z| > 1$ since $z_1 + (1/z_1) = z_2 + (1/z_2)$ implies that

$$z_1 - z_2 + \frac{1}{z_1} - \frac{1}{z_2} = 0 \quad \text{or} \quad (z_1 - z_2)\left(1 - \frac{1}{z_1 z_2}\right) = 0.$$

Hence $z_1 = z_2$ or $z_1 z_2 = 1$. But $z_1 z_2 = 1$ is impossible for $|z_1| > 1$ and $|z_2| > 1$. Hence $z_1 = z_2$ and the mapping is one-to-one.

11–26 EXAMPLES OF CONFORMAL MAPPING

EXAMPLE 1 *Translations.* The general form is $w = z + a + bi$ (a, b real constants). Each point z is displaced through the vector $a + bi$, as in Fig. 11–39. ◄

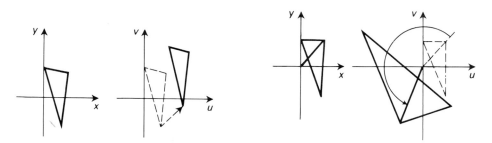

Fig. 11–39. Translation.   Fig. 11–40. Rotation–stretching.

EXAMPLE 2 *Rotation-stretching.* The general form is $w = Ae^{i\alpha}z$ (A, α real constants, $A > 0$). If we write $z = re^{i\theta}, w = \rho e^{i\varphi}$, so that ρ, φ are polar coordinates in the w plane, then we have $\rho = Ar, \varphi = \theta + \alpha$.

Thus distances from the origin are stretched in the ratio A to 1, while all figures are rotated about the origin through the angle α (Fig. 11–40). ◄

EXAMPLE 3   *The general linear integral transformation* $w = az + b$ (a, b complex constants). This is equivalent to a rotation-stretching, as in Example 2, with $a = Ae^{i\alpha}$, followed by a translation through the vector b.   ◄

EXAMPLE 4   *The reciprocal transformation* $w = 1/z$. In polar coordinates we have $\rho = 1/r$, $\varphi = -\theta$.

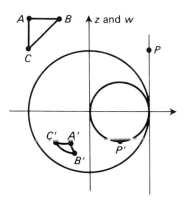

Fig.11–41. Transformation $w = 1/z$.

Thus this transformation involves both a reflection in the real axis and "inversion" in the circle of radius 1 about the origin (Fig. 11–41). Figures outside the circle correspond to smaller ones inside. It can be shown that circles (including straight lines, as "circles through ∞") correspond to circles (Problem 5 below). Thus the line $x = 1$ becomes the circle

$$\left(u - \frac{1}{2}\right)^2 + v^2 = \frac{1}{4}. \quad ◄$$

EXAMPLE 5   *The general linear fractional transformation*

$$w = \frac{az + b}{cz + d} \quad (a, b, c, d \text{ complex constants}); \qquad \begin{vmatrix} a & b \\ c & d \end{vmatrix} \neq 0. \quad (11\text{–}260)$$

If $ad - bc$ were equal to 0, w would reduce to a constant; hence this is ruled out. Examples 1, 2, 3, and 4 are special cases of (11–260). Moreover, the general transformation (11–260) is equivalent to a succession of transformations:

$$z_1 = cz + d, \qquad z_2 = \frac{1}{z_1}, \qquad w = \frac{a}{c} + \frac{bc - ad}{c} z_2 \qquad (11\text{–}261)$$

of the types of Examples 3 and 4; if $c = 0$, (11–260) is already of the type of Example 3.

The transformation (11–260) is analytic except for $z = -d/c$. It is one-to-one, for Eq. (11–260) can be solved for z, to give the inverse:

$$z = \frac{-dw + b}{cw - a}, \qquad (11\text{–}262)$$

which is single valued; w has a pole at $z = -d/c$ and z has a pole at $w = a/c$; in other words, $z = -d/c$ corresponds to $w = \infty$ and $z = \infty$ corresponds to $w = a/c$. If these values are included, then the transformation (11–260) is a *one-to-one transformation of the extended plane on itself.* When $c = 0$, $z = \infty$ corresponds to $w = \infty$. ◄

Since the transformations (11–261) all have the property of mapping circles (including straight lines) on circles, the general transformation (11–260) has also this property. By considering special regions bounded by circles and lines we obtain a variety of interesting one-to-one mappings. The following three are important cases (see IAF, pp. 134–135).

EXAMPLE 6 *Unit circle on unit circle.* The transformations

$$w = e^{i\alpha} \frac{z - z_0}{1 - \bar{z}_0 z} \quad (\alpha \text{ real, } |z_0| < 1)$$

all map $|z| \leqslant 1$ on $|w| \leqslant 1$, and every linear fractional (or even one-to-one conformal) transformation of $|z| \leqslant 1$ on $|w| \leqslant 1$ has this form. ◄

EXAMPLE 7 *Half-plane on half-plane.* The transformations

$$w = \frac{az + b}{cz + d} \quad (a, b, c, d, \text{ real and } ad - bc > 0)$$

all map $\text{Im}(z) \geqslant 0$ on $\text{Im}(w) \geqslant 0$, and every linear fractional transformation of $\text{Im}(z) \geqslant 0$ on $\text{Im}(w) \geqslant 0$ has this form. ◄

EXAMPLE 8 *Half-plane on unit circle.* The transformations

$$w = e^{i\alpha} \frac{z - z_0}{z - \bar{z}_0} \quad (\alpha \text{ real, } \text{Im}(z_0) > 0)$$

all map $\text{Im}(z) \geqslant 0$ on $|w| \leqslant 1$ and every linear fractional transformation of $\text{Im}(z) \geqslant 0$ on $|w| \leqslant 1$ has this form. ◄

EXAMPLE 9 *The transformation $w = z^2$.* Here the transformation is not one-to-one in the whole z plane, for the inverse is $z = \sqrt{w}$ and has two values for each w. In polar coordinates we have $\rho = r^2$, $\varphi = 2\theta$.

Thus each sector in the z plane with vertex at the origin and angle α is mapped onto a sector in the w plane with vertex at $w = 0$ and angle 2α. In particular the half-plane $D_z : \text{Im}(z) > 0$ is mapped on the w plane minus the positive real axis, as shown in Fig. 11–42. Since the two square roots of w are negatives of each other, each w has at most one square root in D_z, so that the mapping is one-to-one. The structure of the level curves

$$u = x^2 - y^2 = \text{const}, \qquad v = 2xy = \text{const},$$

shown in Fig. 11–42, also reveals the one-to-one-ness, in accordance with Test II of Section 11–25.

It should be noted that $dw/dz = 0$ for $z = 0$ and that the mapping is not conformal at this point; the angles between curves are *doubled*. ◄

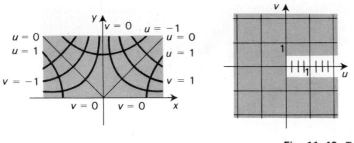

Fig. 11–42. Transformation $w = z^2$.

EXAMPLE 10 *The transformations* $w = z^n$ $(n = 2, 3, 4, \ldots)$. Here the inverse $z = w^{1/n}$ is *n*-valued. In polar coordinates $\rho = r^n$, $\varphi = n\theta$, so that each sector is expanded by a factor of *n*. We obtain a one-to-one mapping by restricting to a sector: $0 < \arg z < 2\pi/n$, as illustrated in Fig. 11–43. The one-to-one-ness follows from the explicit formula:

$$r = \sqrt[n]{\rho}, \qquad \theta = \frac{\varphi}{n}, \qquad 0 < \varphi < 2\pi$$

for the inverse function, or from one of the other tests. Actually *n* need not be an integer; for *n* any positive real number, exactly the same results hold. ◄

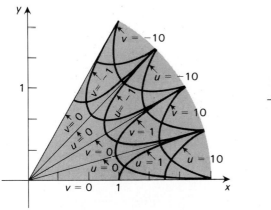

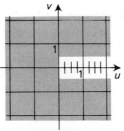

Fig. 11–43. Transformation $w = z^n$.

EXAMPLE 11 *The exponential transformation* $w = e^z$. Here the inverse $z = \ln w$ has infinitely many values. In polar coordinates:

$$\rho = e^x, \qquad \varphi = y + 2n\pi \quad (n = 0, 1, 2, \ldots).$$

This shows that lines $x = \text{const}$ in the *z* plane become circles $\rho = \text{const}$ in the

w plane, while lines $y =$ const become rays $\varphi =$ const. A one-to-one mapping is obtained by restricting z to a strip: $-\pi < y < \pi$, which is then mapped onto the w plane minus the negative real axis, as shown in Fig. 11–44. ◀

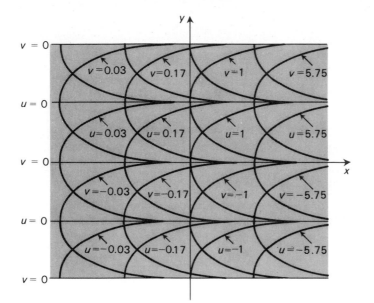

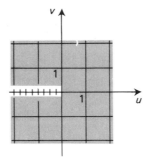

Fig. 11-44. Transformation $w = e^z$.

EXAMPLE 12 *The transformation* $w = \sin z$. Here the inverse

$$z = \sin^{-1} w = \frac{1}{i} \ln \left(iw + \sqrt{1 - w^2} \right)$$

is infinitely multiple valued. A one-to-one mapping is obtained by restricting z to the strip $-\pi/2 < x < \pi/2$, as the structure of the level curves of Fig. 11–45 reveals. ◀

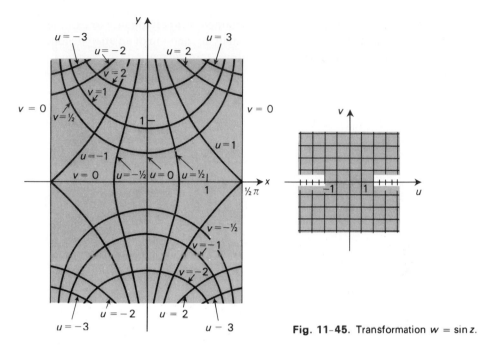

Fig. 11-45. Transformation $w = \sin z$.

EXAMPLE 13 *The transformation*

$$w = z + \frac{1}{z}.$$

The inverse is the two-valued solution of the quadratic equation

$$z^2 - zw + 1 = 0.$$

The mapping is one-to-one in the region $|z| > 1$, as the accompanying Fig. 11-46 reveals. Also Test IV applies, as shown in Section 11–25. ◄

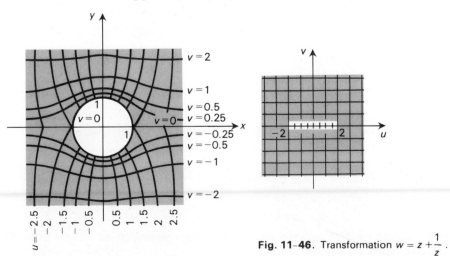

Fig. 11-46. Transformation $w = z + \dfrac{1}{z}$.

A variety of other mappings can be obtained by applying first one of the above mappings and then a second one, since an analytic function of an analytic function is analytic. Also each of the inverses of the above transformations is analytic and one-to-one in the appropriate region in the w plane.

═══════════════════════════**Problems (Section 11–26)**═══════════════════════

1. Determine the images of the circle $|z-1| = 1$ and the line $y = 1$ under the following transformations:

   a) $w = 2z$ **b)** $w = z + 3i - 1$ **c)** $w = 2iz$

   d) $w = \dfrac{1}{z}$ **e)** $w = \dfrac{z+i}{z-i}$ **f)** $w = z^2$

2. For each of the following transformations verify that the transformation is one-to-one in the given region, determine the corresponding region in the w plane, and plot the level curves of u and v:

   a) $w = \sqrt{z} = \sqrt{r}\, e^{i\theta/2}$ $\quad -\pi < \theta < \pi$ $\qquad$ **b)** $w = \dfrac{z-i}{z+i}$, $|z| < 1$

   c) $w = \dfrac{1}{z}$, $1 < x < 2$ $\qquad\qquad\qquad\qquad$ **d)** $w = \operatorname{Ln} z$, $\operatorname{Im}(z) > 0$

   e) $w = \operatorname{Ln} \dfrac{z-1}{z+1}$, $\operatorname{Im}(z) > 0$. $\qquad$ HINT: Use Test I.

   f) $w = z + \dfrac{1}{z}$, $\operatorname{Im}(z) > 0$. $\qquad$ HINT: Use Test IV.

   g) $w = z - \dfrac{1}{z}$, $|z| > 1$. $\qquad$ HINT: Use Test IV.

3. Verify the mapping onto the domain shown and the level curves of u and v for the transformations of

   a) Fig. 11–42 $\qquad$ **b)** Fig. 11–43 $\qquad$ **c)** Fig. 11–44

   d) Fig. 11–45 $\qquad$ **e)** Fig. 11–46

4. By combining particular transformations given above determine a one-to-one conformal mapping of
   a) The quadrant $x > 0$, $y > 0$ onto the region $|w| < 1$;
   b) The sector $0 < \theta < \pi/3$ on the quadrant $u > 0$, $v > 0$;
   c) The half-plane $y > 0$ on the strip $0 < v < \pi$;
   d) The half-strip $-\pi/2 < x < \pi/2$, $y > 0$ on the quadrant $u > 0$, $v > 0$;
   e) The region $r > 1$, $0 < \theta < \pi$ on the strip $0 < v < \pi$;
   f) The strip $1 < x + y < 2$ on the half-plane $v > 0$;
   g) The half-plane $x + y + 1 > 0$ on the quadrant $u > 0$, $v > 0$.

5. **a)** Show that the equation of an arbitrary circle or straight line can be written in the form
   $$a z \bar{z} + b z + \bar{b}\bar{z} + c = 0 \qquad (a, c \text{ real}).$$

   b) Using the result of (a), show that circles or lines become circles or lines under the transformation $w = 1/z$.

6. If $w = f(z)$ is a linear fractional function (11–260), show that $f'(z) \neq 0$.

7. A point where $f'(z) = 0$ is called a *critical point* of the analytic function $f(z)$. Locate the critical points of the following functions and show that none lies within the region of the corresponding mapping in Examples 11, 12, 13:

   a) $w = e^z$ **b)** $w = \sin z$ **c)** $w = z + \dfrac{1}{z}$

8. **a)** Show that the determinant of coefficients of the inverse (11–262) of the transformation (11–260) is not zero. Thus the *inverse of a linear fractional transformation is linear fractional*.

   b) Show that, if

   $$w_1 = \frac{az + b}{cz + d}, \qquad w_2 = \frac{a_1 w_1 + b_1}{c_1 w_1 + d_1}$$

   are linear fractional transformations $w_1 = f(z)$, $w_2 = g(w_1)$, then $w_2 = g[f(z)]$ is linear fractional. Thus a *linear fractional transformation followed by a linear fractional transformation gives rise to a linear fractional transformation*.

11–27 APPLICATIONS OF CONFORMAL MAPPING. THE DIRICHLET PROBLEM

The following problem, known as the *Dirichlet* problem, arises in a variety of situations in fluid dynamics, electric field theory, heat conduction, and elasticity: *given a region D, find a function $u(x, y)$ harmonic in D and having given values on the boundary of D* (see Sections 8–8 to 8–11).

The statement is somewhat loose as to the boundary values. It will be seen immediately how it can be made more precise.

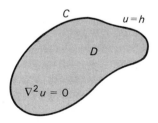

Fig. 11–47. Dirichlet problem.

Let D be a simply connected region bounded by a simple closed curve C, as in Fig. 11–47. Then assignment of boundary values is achieved by giving a function $h(z)$ for z on C and requiring that $u = h$ on C. If h is continuous, the natural formulation of the problem is to require that $u(x, y)$ be harmonic in D, continuous in D plus C, and equal to h on C. If h is piecewise continuous, it is natural to require that u be harmonic in D, continuous in D plus C, except where h is discontinuous, and equal to h except at the points of discontinuity.

If C is a circle $|z| = R$, the problem is solved by the Poisson integral formula of Section 8–10. For example, if $R = 1$, h can be written as $h(\theta)$ and

$$u(r, \theta) = \frac{1}{2\pi} \int_0^{2\pi} \frac{(1 - r^2)h(\theta')}{1 + r^2 - 2r \cos(\theta' - \theta)} d\theta' \tag{11–270}$$

defines a function harmonic in $|z| < 1$. If $h(\theta)$ is continuous and we define $u(1,\theta)$ to be $h(\theta)$, then u can be shown to be continuous for $|z| \leqslant 1$ and therefore satisfies all conditions. If $h(\theta)$ is discontinuous, the same procedure is successful and (11–270) again provides a solution.

While the formula (11–270) does provide a solution of the problem, we must also ask: is this the only solution? The answer is *yes*, when h is continuous for $|z| = 1$; the answer is *no* if h has jump discontinuities. In the latter case (11–270) provides a solution $u(x, y)$ which is bounded for $|z| < 1$, and it can be shown that this is the *only bounded solution*. With the supplementary requirement that the solution be bounded, (11–270) provides the one and only solution.

Now let D be a region bounded by C as in Fig. 11–47 and let us suppose that a one-to-one conformal mapping $z_1 = f(z)$ of D onto the circular region $|z_1| < 1$ has been found and that this mapping is also continuous and one-to-one on D plus C, taking C to $|z_1| = 1$. By assigning to each point z_1 on $|z_1| = 1$ the value of h at the corresponding point on C, we obtain boundary values $h_1(z_1)$ on $|z_1| = 1$. Let $u(z_1)$ solve the Dirichlet problem for $|z_1| < 1$ with these boundary values. *Then $u[f(z)]$ is harmonic in D and solves the given Dirichlet problem in D,* for $u(z_1)$ can be written as $\text{Re}[F(z_1)]$ where F is analytic and $u[f(z)] = \text{Re}\{F[f(z)]\}$; i.e., $u[f(z)]$ is the real part of an analytic function in D. Hence u is harmonic in D. Since continuous functions of continuous functions are continuous, $u[f(z)]$ has the proper behavior on the boundary C and hence does solve the problem.

Accordingly, conformal mapping appears as a powerful tool for solution of the Dirichlet problem. For any region that can be mapped conformally and one-to-one on the circle $|z| < 1$, the problem is explicitly solved by (11–270). It can be shown that *every simply connected region D can be mapped in a one-to-one conformal manner on the circular region $|z| < 1$, provided D does not consist of the entire z-plane. Furthermore, if D is bounded by a simple closed curve C, the mapping can always be defined on C so as to remain continuous and one-to-one.* For proofs of these theorems and of the asserted properties of the Poisson integral formula (11–270), see KELLOGG.

11–28 DIRICHLET PROBLEM FOR THE HALF-PLANE

Since the transformation

$$iz = \frac{z_1 - i}{z_1 + i} \tag{11–280}$$

maps the region $|z_1| < 1$ on the half-plane $\text{Im}(z) > 0$, the Dirichlet problem for the half-plane is reducible to that for the circle as above. However, it is simpler to treat the half-plane by itself. We shall develop the equivalent of (11–270) for the half-plane. Accordingly, if a region D can be mapped on the half-plane, the Dirichlet problem for D will be immediately solved.

We consider first several examples.

EXAMPLE 1 $u(x, y)$ harmonic for $y > 0$; $u = \pi$ for $y = 0$, $x < 0$; $u = 0$ for $y = 0$, $x > 0$, as shown in Fig. 11–48. The function

$$u = \arg z = \theta, \qquad 0 \leqslant \theta \leqslant \pi$$

clearly satisfies all conditions. It is harmonic in the upper half-plane, since

$$\arg z = \theta = \operatorname{Im}(\operatorname{Ln} z);$$

it is continuous on the boundary $y = 0$, except for $x = 0$, and has the correct boundary values. Furthermore $0 < u < \pi$ in the upper half-plane, so that the solution is bounded. ◀

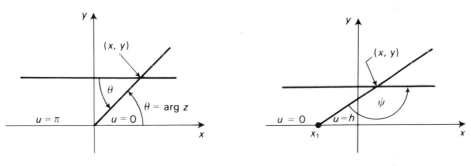

Figure 11-48.   Figure 11-49.

EXAMPLE 2 $u(x, y)$ harmonic for $y > 0$; $u = h = $ const for $y = 0$, $x > 0$; $u = 0$ for $y = 0$, $x < 0$. The solution is obtained as in Example 1:

$$u = h\left(1 - \frac{\arg z}{\pi}\right), \qquad 0 \leqslant \arg z \leqslant \pi.$$

This function is again harmonic and has the proper boundary values. ◀

EXAMPLE 3 $u(x, y)$ harmonic for $y > 0$; $u = h = $ const for $y = 0$, $x > x_1$; $u = 0$ for $y = 0$, $x < x_1$ (see Fig. 11-49). A translation reduces this to Example 2:

$$u = h\left[1 - \frac{\arg(z - z_1)}{\pi}\right],$$

$$0 \leqslant \arg(z - z_1) \leqslant \pi;$$

$$z_1 = x_1 + 0i. \quad ◀$$

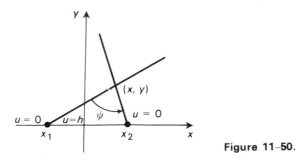

Figure 11–50.

EXAMPLE 4 $u(x, y)$ harmonic for $y > 0$; $u = h = $ const for $y = 0$, $x_1 < x < x_2$; $u = 0$ for $y = 0$, $x < x_1$ and $x > x_2$ (see Fig. 11–50). The solution is obtained by subtracting two solutions of the type of Example 3:

$$u = h \left[1 - \frac{\arg (z - z_1)}{\pi} \right] - h \left[1 - \frac{\arg (z - z_2)}{\pi} \right]$$

$$= \frac{h}{\pi} \left[\arg (z - z_2) - \arg (z - z_1) \right] = \frac{h}{\pi} \arg \frac{z - z_2}{z - z_1}. \qquad (11\text{–}281)$$

The result has an interesting geometric interpretation, for

$$\arg \frac{z - z_2}{z - z_1} = \psi, \qquad 0 < \psi < \pi,$$

where ψ is the angle shown in Fig. 11–50. Accordingly $u = h\psi / \pi$. The boundary values can be directly verified on the figure. ◄

REMARK   Examples 1, 2, and 3 can be regarded as limiting cases of Example 4. In Example 1, $x_1 = -\infty$, $x_2 = 0$, and $h = \pi$, so that ψ becomes the angle θ, as shown in Fig. 11–48, and $u = \psi = \theta$. In Example 3, $x_2 = +\infty$; the angle ψ is shown in Fig. 11–49.

EXAMPLE 5 $u(x, y)$ harmonic for $y > 0$; $u = h_0 = $ const for $y = 0$, $x < x_0$; $u = h_1$ for $y = 0$, $x_0 < x < x_1$; $u = h_2$ for $y = 0$, $x_1 < x < x_2$; $\dots$; $u = h_n$ for $y = 0$, $x_{n-1} < x < x_n$; $u = h_{n+1}$ for $y = 0$, $x > x_n$ as in Fig. 11–51. The solution is obtained by addition of solutions of problems like those in the previous examples:

$$u = \frac{1}{\pi} \left[h_0 \arg (z - z_0) + h_1 \arg \frac{z - z_1}{z - z_0} + \cdots + h_n \arg \frac{z - z_n}{z - z_{n-1}} \right.$$

$$\left. + h_{n+1} \{\pi - \arg (z - z_n)\} \right]; \qquad (11\text{–}282)$$

$$u = \frac{1}{\pi} \left[h_0 \psi_0 + h_1 \psi_1 + \cdots + h_{n+1} \psi_{n+1} \right],$$

$$0 \leqslant \psi_0 \leqslant \pi, \qquad 0 \leqslant \psi_1 \leqslant \pi, \cdots \qquad (11\text{–}283)$$

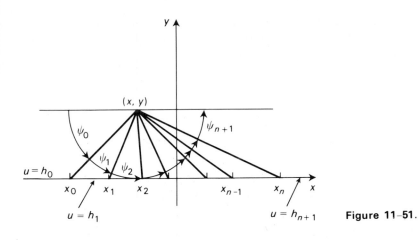

Figure 11-51.

The angles $\psi_0, \cdots, \psi_n$ are shown in Fig. 11–51. The sum of these angles is π; therefore u is the weighted mean of the numbers $h_0, h_1, \cdots, h_{n+1}$ and hence

$$h' \leqslant u \leqslant h'',\qquad(11\text{--}284)$$

where h' is the smallest of these numbers and h'' is the largest. ◄

Now let $h_0 = 0, h_{n+1} = 0$. Let z be fixed in the upper half-plane and let $t + 0i$ be a variable point on the x axis; write $g(t) = \arg[z - (t + 0i)]$, where the angle is always taken between 0 and π. Then (11–282) can be written thus:

$$u = \frac{1}{\pi}\{h_1[g(t_1) - g(t_0)] + h_2[g(t_2) - g(t_1)] + \cdots + h_n[g(t_n) - g(t_{n-1})]\}.$$

This formula suggests passage to the limit $n \to \infty$; the expression suggests an integral. In fact, the expression can be interpreted (with the aid of the law of the mean) as a sum

$$u = \frac{1}{\pi}[h(t_1)g'(t_1^*)\Delta_1 t + \cdots + h(t_n)g'(t_n^*)\Delta_n t]$$

which (under appropriate assumptions) converges to an integral as $n \to \infty$:

$$u = \frac{1}{\pi}\int_\alpha^\beta h(t)g'(t)\,dt.$$

Now

$$g(t) = \arg[z - (t + 0i)] = \arctan\frac{y}{x - t},$$

so that

$$g'(t) = \frac{y}{(x - t)^2 + y^2}.$$

We are thus led to the formula:

$$u(x, y) = \frac{1}{\pi}\int_\alpha^\beta \frac{h(t)y}{(x - t)^2 + y^2}\,dt\qquad(11\text{--}285)$$

as the expression for a harmonic function in the upper half-plane with boundary values $h(t)$ for $z = t + i0$, $\alpha < t < \beta$, and $u = 0$ otherwise on the boundary. We get complete generality by letting $\alpha = -\infty$, $\beta = +\infty$:

$$u(x, y) = \frac{1}{\pi} \int_{\infty}^{\infty} \frac{h(t)y}{(x - t)^2 + y^2} \, dt. \tag{11-286}$$

This integral is easily shown to converge, provided h is piecewise continuous and bounded. *The formula* (11–286) *is precisely the Poisson integral formula for the half-plane.* In fact a change of variables transforms (11–270) into (11–286) (see IAF, p. 145).

Just as important as the general formula (11–286) is the formula (11–282), which can be considered as the rectangular sum for evaluation of the integral (11–286).

EXAMPLE 6 $u(x, y)$ harmonic in the half-strip of Fig. 11–52, with the boundary values shown. We seek the mapping of this half-strip onto a half-plane and find it to be given by $z_1 = \sin z$, as in Example 12 of Section 11–26. Under this mapping, $z = -\pi/2$ maps on $z_1 = -1$, $z = \pi/2$ on $z_1 = 1$, and the whole boundary onto the real axis of the z_1 plane. The new problem in the z_1 plane requires a function u harmonic in the upper half-plane, with boundary values $u = 2$ for $-1 < x_1 < 1$, $u = 0$ for $x_1 > 1$ and for $x_1 < 0$. This is solved as in Example 4 by (11–281). Hence

$$u = \frac{2}{\pi} \arg \frac{z_1 - 1}{z_1 + 1}$$

and, in the z plane,

$$u = \frac{2}{\pi} \arg \frac{\sin z - 1}{\sin z + 1}.$$

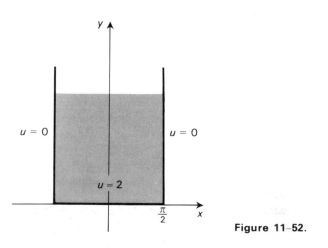

$u = 0$

$u = 0$

$u = 2$

$\frac{\pi}{2}$

Figure 11–52.

To apply this formula, the simplest procedure is to use the diagram of Fig. 11–45, showing the mapping from the z plane to the z_1 plane. For each z, we locate the corresponding $z_1 = \sin z$, measure the angle

$$\psi = \arg \frac{z_1 - 1}{z_1 + 1},$$

as in Fig. 11–50, and divide by $\pi/2$. The answer can also be written explicitly in real form:

$$u = \frac{2}{\pi} \arctan \left[\operatorname{Im} \left\{ \frac{\sin z - 1}{\sin z + 1} \right\} \div \operatorname{Re} \left\{ \frac{\sin z - 1}{\sin z + 1} \right\} \right]$$

$$= \frac{2}{\pi} \arctan \frac{-2 \cos x \sinh y}{\sin^2 x \cosh^2 y + \cos^2 x \sinh^2 y - 1}. \blacktriangleleft$$

EXAMPLE 7 $u(x, y)$ harmonic for $|z| < 1$, $u = 1$ for $r = 1$, $\alpha < \theta < \beta$, as in Fig. 11–53; $u = 0$ on the rest of the boundary. The function

$$z_1 = e^{(\alpha - \beta)i/2} \frac{z - e^{i\beta}}{z - e^{i\alpha}}$$

maps $|z| < 1$ on the half-plane $\operatorname{Im}(z_1) > 0$. The point $e^{i\alpha}$ has as image $z_1 = \infty$; the point $e^{i\beta}$ has as image $z_1 = 0$; the arc $\alpha < \theta < \beta$ corresponds to the negative real axis in the z_1 plane. Thus the solution is given by

$$u = \frac{1}{\pi} \arg z_1 = \frac{1}{\pi} \arg \left[e^{(\alpha - \beta)i/2} \frac{z - e^{i\beta}}{z - e^{i\alpha}} \right]. \qquad (11\text{–}287)$$

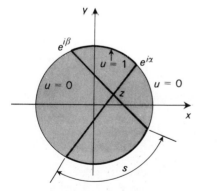

Figure 11–53.

This rather clumsy formula is the formula for the circle corresponding to (11–281) for the half-plane. By use of geometry, this can be written in a much simpler form:

$$u = \frac{s}{2\pi}, \qquad (11\text{–}288)$$

where s is the arc length of the arc on $|z| = 1$ determined by the chords through $e^{i\alpha}$ and z and through $e^{i\beta}$ and z (see Fig. 11–53). $\blacktriangleleft$

This problem could also have been solved directly by the Poisson integral formula (11–270)

$$u(r_0, \theta_0) = \frac{1}{2\pi} \int_\alpha^\beta \frac{1 - r^2}{1 + r^2 - 2r \cos(\theta_0 - \theta)} \, d\theta. \qquad (11\text{–}289)$$

The integration, though awkward, can be carried out.

========================Problems (Section 11–28)========================

1. Solve the following boundary-value problems:
   a) u harmonic and bounded in the first quadrant (see Fig. 11–54) and

$$\lim_{y \to 0+} u(x, y) = 1 \text{ for } 0 < x < 1, \qquad \lim_{y \to 0+} u(x, y) = 0 \text{ for } x > 1,$$

$$\lim_{x \to 0+} u(x, y) = 1 \text{ for } 0 < y < 1, \qquad \lim_{x \to 0+} u(x, y) = 0 \text{ for } y > 1.$$

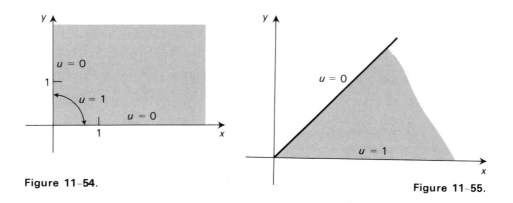

Figure 11–54.

Figure 11–55.

   b) u harmonic and bounded in the sector $0 < \theta < \pi/4$, with boundary values 1 on the x axis and 0 on the line $y = x$, as in Fig. 11–55.
   c) u harmonic and bounded for $0 < \theta < 2\pi$, with limiting value of 1 as z approaches the positive real axis from the upper half-plane and -1 as z approaches the positive real axis from the lower half-plane.
   d) u harmonic and bounded in the strip $0 < y < 1$ with boundary values 0 for $y = 0$, $x < 0$ and for $y = 1$, $x < 0$ and boundary values 2 for $y = 0$, $x > 0$ and for $y = 1$, $x > 0$.
   e) u harmonic and bounded in the region $0 < \theta < \pi, r > 1$ with boundary values of 1 on the real axis and -1 on the circle.

2. a) Verify that the transformation $w = 2 \operatorname{Ln} z - z^2$ maps the half-plane $\operatorname{Im}(z) > 0$ in a one-to-one conformal manner on the w plane minus the lines $v = 2\pi, u < -1$, and $v = 0$, $u < -1$.
   b) Find the electrostatic potential U between two condenser plates that are idealized as two half-planes perpendicular to the uv-plane along the lines $v = a, u < 0$, and $v = 0$, $u < 0$, if the potential difference between the plates is U_0; that is, solve the boundary-

value problem $U(u, a) = U_0$ for $u < 0$ and $U(u, 0) = 0$ for $u < 0$, $U(u, v)$ harmonic in the remaining portion of the uv-plane.

3. a) Show that the transformation $w = z^2$ maps the region $\text{Im}(z) > 1$ in a one-to-one conformal manner on the parabolic region

$$u < \frac{v^2}{4} - 1.$$

b) Let a solid be idealized as an infinite cylinder perpendicular to the uv-plane, whose cross section in the uv-plane is the region $u \leqslant v^2$. Let this solid be in temperature equilibrium, with temperatures T_1 maintained on the part of the boundary surface where $v > 0$ and T_2 where $v < 0$. Find the temperature distribution inside the solid, that is, solve the boundary-value problem: $T(u, v)$ harmonic for $u < v^2$ with boundary values $T(u, v) = T_1$ for $u = v^2$, $v > 0$, and $T(u, v) = T_2$ for $u = v^2$, $v < 0$.

11–29 CONFORMAL MAPPING IN HYDRODYNAMICS

Let $\mathbf{V} = u(x, y)\mathbf{i} - v(x, y)\mathbf{j}$ be the velocity field of a time-independent planar fluid motion in the region D. The conditions that the flow be irrotational and incompressible are then curl $\mathbf{V} = \mathbf{0}$ and div $\mathbf{V} - 0$. As in Section 11 6, these become

$$\frac{\partial u}{\partial y} + \frac{\partial v}{\partial x} = 0 \qquad \text{and} \qquad \frac{\partial u}{\partial x} - \frac{\partial v}{\partial y} = 0,$$

which are just the Cauchy–Riemann equations. Therefore, $f(z) = u + iv$ is analytic in D if and only if the corresponding flow $\mathbf{V} = u\mathbf{i} - v\mathbf{j}$ is irrotational and incompressible.

If we restrict attention to a simply connected region, then $f(z)$ has an indefinite integral $F(z)$ (determined up to a constant). If we write

$$F = \varphi(x, y) + i\psi(x, y),$$

then

$$F'(z) = \frac{\partial \varphi}{\partial x} - i\frac{\partial \varphi}{\partial y} = u + iv.$$

Hence

$$\frac{\partial \varphi}{\partial x} = u, \qquad \frac{\partial \varphi}{\partial y} = -v \qquad \text{or} \qquad \text{grad } \varphi = u\mathbf{i} - v\mathbf{j} = \mathbf{V}.$$

The function φ is termed the *velocity potential* and $F(z)$ is termed the *complex velocity potential*. We can write

$$\overline{F'(z)} = \frac{\partial \varphi}{\partial x} + i\frac{\partial \varphi}{\partial y} = u - iv.$$

hence *the conjugate of the derivative of the complex velocity potential is the velocity vector.*

The lines $\varphi(x, y) = $ const are called *equipotential lines;* they are orthogonal to the velocity vector grad φ at each point. The lines $\psi(x, y) = $ const are called *stream lines;*

the velocity vector is tangent to such a line at each point, so that these lines can be considered as paths of actual fluid particles.

Conformal mapping can be applied to hydrodynamics problems in several ways. First of all, particular problems can be formulated as boundary-value problems and solved with the aid of conformal mapping as in the preceding section. Second, starting with a known flow pattern, we can obtain a variety of others in an empirical way by simply applying different conformal transformations.

Here we treat briefly only one example of the first type. We consider the problem of flow past an obstacle, as suggested in Fig. 11–56. The region D of the flow is the exterior of a (piecewise smooth) simple closed curve C. Since the region is not simply connected, we cannot be sure that there is a (single-valued) complex potential $F(z)$. It will turn out that, if we make an appropriate assumption about the flow "at ∞," then $F(z)$ does exist. The natural assumption is that the flow approaches a uniform flow with constant velocity at ∞. Hence $f(z) = F'(z)$ is analytic at ∞:

$$f(z) = a_0 + \frac{a_{-1}}{z} + \frac{a_{-2}}{z^2} + \cdots, \qquad |z| > R,$$

$$F(z) = \text{const} + a_0 z + a_{-1} \ln z - \frac{a_{-2}}{z} + \cdots, \qquad |z| > R.$$

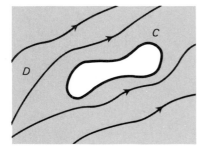

Fig. 11-56. Flow past an obstacle.

If $F(z)$ is to be single-valued, the term in $\ln z$ must not be present; hence we also assume the flow to be such that $a_{-1} = 0$, so that

$$f(z) = a_0 + \frac{a_{-2}}{z^2} + \cdots,$$

$$F(z) = \text{const} + a_0 z - a_{-2} z^{-1} + \cdots.$$

$$(11–290)$$

The constant a_0 is precisely the value of f at ∞, so that $\bar{a}_0$ is the velocity of the limiting uniform flow.

The stream function $\psi = \text{Im}[F(z)]$ is constant along C, since (in the absence of viscosity) the velocity vector must be tangent to C. It would be natural to formulate a boundary-value problem for ψ, but it is simpler to remark that the assumption

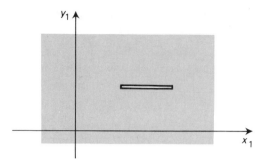

Fig. 11–57. Slit-region.

(11–290) about the behavior of $F(z)$ at ∞ and the condition $\text{Im}[F(z)] = \text{const}$ on C imply that $z_1 = F(z)$ *maps D in a one-to-one conformal manner onto a region D_1 of the z_1 plane*. This is established by the argument principle (Problem 3 below). Since $\text{Im}(z_1) = y_1$ is constant on C, the image of C must be a line segment $y_1 = \text{const}$ in the $x_1 y_1$ plane; thus D_1 *consists of the entire xy plane minus a slit*, as in Fig. 11–57. Conversely, if $F(z)$ is analytic in D and maps D in one-to-one fashion on such a slit-region D_1, then F must have a pole of first order at ∞ and necessarily $\psi = \text{Im}[F(z)]$ is constant on C; thus every mapping of D onto a slit-region D_1 provides an appropriate complex velocity potential $F(z)$.

EXAMPLE   Let C be the circle $|z| = 1$. Then

$$z_1 = a_0 \left(z + \frac{1}{z} \right) = F(z) \quad (a_0 \text{ real})$$

maps D onto a slit-region D_1, the slit lying on the real axis (cf. Example 13 of Section 11–26). Hence this is an appropriate potential for the flow past the circle. The stream function is

$$\psi = a_0 y - a_0 \frac{y}{x^2 + y^2} ;$$

the stream lines are shown in Fig. 11–46 (Section 11–26).   ◄

It can be shown that, given C and the velocity $\bar{a}_0$ at ∞, a mapping function $F(z)$ having the expansion (11–290) at ∞ exists and is uniquely determined up to an additive constant which has no effect on the velocity vector $\overline{F'(z)}$. A similar theorem holds for flow past several obstacles, bounded by curves $C_1, \ldots, C_n$; the complex velocity potential maps the region of the flow onto the z_1 plane minus n slits. (For proofs the reader is referred to Chapter VII of *Conformal Mapping* by Z, NEHARI, McGraw-Hill, New York, 1952.)

11–30 APPLICATIONS OF CONFORMAL MAPPING IN THE THEORY OF ELASTICITY

Two-dimensional problems in the theory of elasticity are reducible to solution of the biharmonic equation

$$\nabla^4 U = \nabla^2 (\nabla^2 U) = 0 \tag{11–300}$$

(Section 8–2). The function U is *Airy's stress function*; its second derivatives

$$\frac{\partial^2 U}{\partial x^2}, \qquad \frac{\partial^2 U}{\partial x\,\partial y}, \qquad \frac{\partial^2 U}{\partial y^2}$$

give components of the *stress tensor*, which describes the forces acting on an arbitrary plane cross section of the solid being studied.

Solution of (11–300) in a region D is equivalent to solution of two equations:

$$\nabla^2 U = P, \qquad \nabla^2 P = 0.$$

The solutions P of the second equation are harmonic functions. Furthermore, if U_1, U_2 satisfy the first equation for given P, then

$$\nabla^2(U_1 - U_2) = P - P = 0;$$

hence the solutions of the first equation are of form $U_1 + W$, where U_1 is a particular solution and W is harmonic. Now let harmonic functions u and v be chosen, if possible, so that

$$\frac{\partial u}{\partial x} = P = \frac{\partial v}{\partial y}.$$

Then

$$\nabla^2(xu + yv) = x\nabla^2 u + 2\frac{\partial u}{\partial x} + y\nabla^2 v + 2\frac{\partial v}{\partial y} = 4P.$$

Hence

$$U_1 = \frac{1}{4}(xu + yv)$$

is the desired particular solution, provided u and v can be found. If D is simply connected, we can choose $Q(x, y)$ to be a harmonic conjugate of $P(x, y)$, so that $P + iQ$ is analytic in D (see end of Section 11–11); then

$$u + iv = f(z) = \int F(z)\,dz$$

defines (up to additive constants) harmonic functions u and v such that

$$\frac{\partial u}{\partial x} = \frac{\partial v}{\partial y} = P$$

and U_1 can be written as follows:

$$U_1 = \frac{1}{4}(xu + yv) = \frac{1}{4}\operatorname{Re}[\bar{z}f(z)].$$

Finally, the solutions U of (11–300) can be written thus:

$$U = U_1 + W = \operatorname{Re}\left[\frac{\bar{z}f(z)}{4} + g(z)\right].$$

The factor of $1/4$ can be absorbed in $f(z)$ and we have the conclusion: *If $f(z)$ and $g(z)$ are analytic in region D, then*

$$U = \operatorname{Re}[\bar{z}f(z) + g(z)] \tag{11–301}$$

is biharmonic in D; if D is simply connected, then all biharmonic functions in D can be represented in this form.

The boundary-value problems for the stress function U can be formulated in terms of the analytic functions f and g. If D is mapped conformally on a second region D_1, the problem is transformed into a boundary-value problem for D_1. By mapping onto a simple region D_1, such as the half-plane or unit circle, we reduce the problem to a simpler one. Hence, just as for the Dirichlet problem, conformal mapping is a powerful aid. The importance of (11–301) is that it expresses U in terms of analytic functions, which *remain analytic* if a conformal change of variables is made. A biharmonic function does not in general remain biharmonic under such a transformation.

For further details on the applications to elasticity the reader is referred to Chapter V of *Mathematical Theory of Elasticity*, by I. S. SOKOLNIKOFF, Mimeographed Lectures at Brown University, 1941; also to a paper by V. MORKOVIN in *Quarterly of Applied Mathematics, vol. 2*, 350–352, 1944.

11–31 FURTHER APPLICATIONS OF CONFORMAL MAPPING

In general, conformal mapping can be helpful in solving all boundary-value problems associated with the Laplace equation or the more general *Poisson equation*

$$\frac{\partial^2 u}{\partial x^2} + \frac{\partial^2 u}{\partial y^2} = g(x, y) \tag{11–310}$$

in the plane. The crucial fact is that harmonic functions remain harmonic under conformal mapping. While the boundary values may be transformed in a complicated manner, this disadvantage is usually offset by the possibility of simplifying the region by an appropriate mapping.

We mention here one example of such possibilities. Let it be required to find a function $u(x, y)$ harmonic in a given region D bounded by a smooth simple closed curve C and satisfying the boundary conditions; u has given values $h(x, y)$ on an arc of C; $\partial u/\partial n = 0$ on the rest of C, where n is the normal vector to C. The condition $\partial u/\partial n = 0$ is equivalent to the condition that the conjugate function v is constant on C; hence this condition is invariant under conformal mapping.

To solve the problem, we map D, if possible, on the *quarter-plane* $x_1 > 0, y_1 > 0$, so that the arc on which $\partial u/\partial n = 0$ becomes the y_1 axis. The function $h(x, y)$ becomes a function $h_1(x_1)$, giving the values of u for $y_1 = 0$. We now solve the boundary value problem: $u(x_1, y_1)$ harmonic in the upper half-plane; $u(x_1, 0) = h(x_1)$ for $x_1 > 0$ and $u(x_1, 0) = h(-x_1)$ for $x_1 < 0$. In other words, we *reflect* the boundary values in the line $x_1 = 0$. The function u obtained is harmonic in the quarter-plane and has the correct boundary values for $x_1 > 0$. Furthermore, (11–286) shows that $u(x_1, y_1) = u(-x_1, y_1)$, that is, u shows the same symmetry as the boundary values. This implies that $\partial u/\partial x_1 = 0$ for $x_1 = 0$, that is, $\partial u/\partial n = 0$ on the boundary of the quarter-plane. Accordingly, u satisfies all conditions. If we return to the xy-plane, u

becomes a function of x and y that solves the given boundary-value problem. This can be shown to be the only bounded solution, provided h is piecewise continuous.

=========================Problems (Section 11–31)=========================

1. Prove that

$$F(z) = a_0 \left(z e^{i\alpha} + \frac{1}{z e^{i\alpha}} \right) + \text{const} \quad (\alpha \text{ real})$$

maps $D\colon |z| > 1$ in a one-to-one conformal manner on a slit region D_1 (this is the most general such map). Interpret F as a complex velocity potential.

2. Show that the vector

$$\mathbf{V} = \left(1 + \frac{y}{x^2 + y^2} - \frac{x^2 - y^2}{(x^2 + y^2)^2} \right) \mathbf{i} + \left(\frac{-x}{x^2 + y^2} - \frac{2xy}{(x^2 + y^2)^2} \right) \mathbf{j}$$

can be interpreted as the velocity of an irrotational, incompressible flow past the obstacle bounded by the circle $x^2 + y^2 = 1$. Find the complex velocity potential and the stream function, and plot several stream lines.

3. Let $w = F(z)$ be analytic in the region D outside the simple closed curve C and have a pole of first order at ∞. Let $F(z)$ be continuous in D plus C and let $\text{Im}[F(z)] = \text{const}$ on C. Show that $F(z)$ maps D in a one-to-one fashion on a slit-domain.

   HINT: Show that here the argument principle of Section 11–17 takes the form: the increase in arg $F(z)$ as z traces C in the *negative* direction is 2π times $(N_0 - N_p)$, where N_0 and N_p are the numbers of zeros and poles of $F(z)$ in D plus the point $z = \infty$. Show that the increase in arg $[F(z) - w_0]$ is 0 for w_0 not on the image of C. Hence $N_0 - N_p = 0$; but $N_p = 1$, so that $N_0 = 1$.

4. Let $U(x, y)$ be biharmonic for $x^2 + y^2 < 1$. Show that U can be expanded in a Taylor series in this domain, so that U is *analytic* in x and y.

5. Find the equilibrium temperature distribution T in the half-strip $0 < x < 1$, $y > 0$ if the edge $x = 0$ is maintained at a temperature T_0, the edge $x = 1$ is maintained at a temperature T_1, while the edge $y = 0$ is insulated ($\partial T / \partial n = 0$).

Chapter 12 Special Functions

12–1 GOAL OF THIS CHAPTER

At various points in previous chapters we have seen that particular problems arising in physical applications can be solved by means of Legendre polynomials, Bessel functions, and other special functions.

It is our purpose here to present some key properties of these functions. There are many books giving encyclopedic coverage of the properties. Among these we cite the work of ERDELYI as particularly useful.

As stressed in Chapter 3 (Sections 3–9 to 3–13), many special functions arise as *orthogonal systems*. In each such case we have a weight function $w(x)$ and a corresponding inner product and norm

$$(f, g) = \int_a^b f(x) \, g(x) \, w(x) \, dx, \qquad \|f\| = (f, f)^{1/2}$$

over a fixed interval, possibly infinite. An orthogonal system $\{f_n(x)\}$ satisfies the conditions

$$(f_m, f_n) = 0 \quad \text{for} \quad m \neq n, \qquad (f_n, f_n) = \|f_n\|^2 > 0.$$

An orthogonal system $\{f_n\}$ is *complete* if for each function $g(x)$ (say piecewise continuous) the corresponding Fourier series with respect to $\{f_n\}$ converges to $g(x)$ in the mean:

$$\|g(x) - (c_1 f_1 + \cdots + c_n f_n)\|^2 \to 0 \quad \text{as} \quad n \to \infty, \qquad c_k = \frac{(g, f_k)}{\|f_k\|^2}.$$

The orthogonal systems of special functions to be considered here are all complete. In many cases this assertion is a special case of the general theorems on Sturm–Liouville boundary-value problems (Section 3–14).

12–2 LINEAR DIFFERENTIAL EQUATIONS OF THE SECOND ORDER

The special functions to be studied (with the exception of the gamma function and beta function) all satisfy differential equations of the form

$$a_0(x)y'' + a_1(x)y' + a_2(x)y = 0, \tag{12–20}$$

where $a_0(x)$, $a_1(x)$, $a_2(x)$ are polynomials in x. Such equations are discussed in Sections 2–22 and 2–23. We recall the principal results and provide some additional information about such equations.

We assume that $a_0(x)$, $a_1(x)$, $a_2(x)$ have no common roots since we can always cancel common factors. Then a value x_0 such that $a_0(x_0) \neq 0$ is an *ordinary point*, a value x_0 such that $a_0(x_0) = 0$ is a *singular point*.

If x_0 is an ordinary point, then solutions $y_1(x)$, $y_2(x)$ of Eq. (12–20) can be obtained in the form of power series in powers of $x - x_0$, converging for $|x - x_0| < b$ for some b (possibly infinite), and forming two linearly independent solutions of Eq. (12–20) in this interval.

If x_0 is a singular point, then x_0 is said to be a *regular singular point* if either of the following sets of conditions holds:

$$a_0(x_0) = 0 \quad \text{and} \quad a_0'(x_0) \neq 0 \quad \text{or}$$
$$a_0(x_0) = 0, \quad a_0'(x_0) = 0, \quad a_0''(x_0) \neq 0, \quad a_1(x_0) = 0. \tag{12–21}$$

(These correspond to the forms (2–234a), (2–234b) of Section 2–23 for $x_0 = 0$). For a regular singular point, we can obtain at least one solution of the form

$$(x - x_0)^m \left[1 + \sum_{n=1}^{\infty} c_n(x - x_0)^n \right]. \tag{12–22}$$

Here m is chosen to be a root of the *indicial equation* $f(m) = 0$, a quadratic equation in m. For all the equations to be considered here, $f(m)$ is found by evaluating

$$[a_0(x)D^2 + a_1(x)D + a_2(x)] (x - x_0)^m.$$

We find that this expression can be written as

$$f(m) (x - x_0)^{m+h} + g(m) (x - x_0)^{m+h+k}, \quad \text{with } k > 0.$$

When the indicial equation has distinct roots m_1, m_2, we obtain two corresponding solutions (12–22), provided that $m_2 - m_1$ is not divisible by k. The two solutions $y_1(x)$, $y_2(x)$ obtained are valid and linearly independent for $0 < |x - x_0| < b$ for some b. One or both of m_1, m_2 may be negative or fractional, so that at x_0 itself the solutions may not be valid.

For an illustration of the procedure, see Example 1 in Section 2–23. We shall provide further illustrations in this chapter.

If $m_1 = m_2$ or, more generally, if $m_1 - m_2$ is divisible by k, the procedure described normally produces only one solution of form (12–22). If we call that one

$y_1(x)$ (with exponent m_1), then it can be shown that a second linearly independent solution is obtainable in the form

$$y(x) = cy_1(x) \ln(x - x_0) + (x - x_0)^m \sum_{n=0}^{\infty} b_n(x - x_0)^n. \qquad (12\text{–}23)$$

Here $m = m_1$ and $c \neq 0$ if the roots are equal, and $m = m_2$ (the smaller root) when $m_1 - m_2$ is divisible by k. The number c and the constants $b_1, b_2, \ldots$ can be obtained by substitution in the differential equation. (For details on this method, see ODE, pp. 366–386.)

Complex aspects The results described can be extended to complex values of x and y. We write the differential equation in the form

$$\frac{d^2w}{dz^2} + p(z)\frac{dw}{dz} + q(z)w = 0, \qquad (12\text{–}24)$$

where z and w are complex variables and $p(z)$ and $q(z)$ are analytic functions of z, which may have singularities.

A point z_0 at which $p(z)$ and $q(z)$ are both analytic is an *ordinary point* of Eq. (12–24). At an ordinary point we can obtain solutions $w_1(z)$, $w_2(z)$ that are analytic and linearly independent for $|z - z_0| < R$ for some R.

A point z_0 at which $p(z)$ and/or $q(z)$ has an isolated singularity is a *singular point* of Eq. (12–24). This singular point is *regular* if the only singularities occurring at z_0 are poles and the pole of $p(z)$ (if there is one) is of first order, that of $q(z)$ (if there is one) is of first or second order. (These correspond to the cases (12–21)). At a regular singular point z_0 we can obtain at least one solution of the form

$$w = (z - z_0)^m \left[1 + \sum_{n=1}^{\infty} c_n(z - z_0)^n \right], \qquad (12\text{–}25)$$

where m is a root of an *indicial equation* $f(m) = 0$. Here the series converges for $|z - z_0| < R$ for some R and the solution is valid for $0 < |z - z_0| < R$. However, if m is not an integer, $(z - z_0)^m$ is many-valued (Section 11–23) and we have to select a branch of this function; with each such branch (12–25) yields a solution.

As for the real case, we may be able to find two linearly independent solutions of form (12–25), or we may be forced to use the form

$$cw_1(z) \ln(z - z_0) + (z - z_0)^m \sum_{n=0}^{\infty} b_n(z - z_0)^n$$

for the second solution. For the logarithm we are of course forced to choose branches as in Section 11–23.

We can also study Eq. (12–24) for large $|z|$, that is, at $z = \infty$. To that end, we can make a substitution $z_1 = 1/z$ and obtain a new equation for w in terms of z_1, to be studied near $z_1 = 0$. If 0 is a regular singular point of the new equation, we can obtain solutions as above. If we then replace z_1 by $1/z$, we obtain solutions $w(z)$ valid for large $|z|$, for example, of form

$$z^{-m}\left(1 + \sum_{n=1}^{\infty} \frac{c_n}{z^n}\right). \tag{12–26}$$

========================Problems (Section 12–2)========================

1. Find linearly independent solutions of form $x^m\left(1 + \sum_{n=1}^{\infty} c_n x^n\right)$:

    a) $(1 - x^2)y'' - 2xy' + 6y = 0$ (Legendre equation of order 2),

    b) $x^2 y'' + xy' + \left(x^2 - \frac{1}{9}\right)y = 0$ $\left(\text{Bessel equation of order } \frac{1}{3}\right)$.

2. Find linearly independent solutions $w_1(z)$, $w_2(z)$ of form $z^m\left(1 + \sum_{n=1}^{\infty} c_n z^n\right)$:

    a) $2z(1 - z)w'' + (1 - 5z)w' - w = 0$
    b) $3z(1 - z)w'' + (1 - 5z)w' + w = 0$

12–3 THE GAMMA FUNCTION

The function $\Gamma(x)$ is an extension of the factorial operation to a continuous range of x. In particular,

$$\Gamma(n+1) = n!, \qquad n = 0, 1, 2, \ldots . \tag{12–30}$$

We define $\Gamma(x)$ for all real x except $0, -1, -2, -3, \ldots$, and $\Gamma(x)$ turns out to be continuous except at 0 and the negative integers, where $\Gamma(x)$ has $\pm \infty$ limits from the left and right, as indicated in Fig. 12–1. The function $\Gamma(x)$ turns out to satisfy the rule

$$\Gamma(x + 1) = x\Gamma(x) \tag{12–31}$$

called the *functional equation of the Γ-function*. For $x = n$, a positive integer, the rule (12–31) follows from Eq. (12–30).

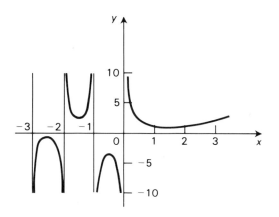

Fig. 12–1. Function $\Gamma(x)$.

In order to obtain such a function, we start with the observation that

$$\int_0^\infty t^k e^{-t}\,dt = k!$$

for k a positive integer, as can be verified by integration by parts. Hence we can define

$$\Gamma(x) = \int_0^\infty t^{x-1} e^{-t}\,dt \tag{12–32}$$

for arbitrary x, as long as the integral has meaning, and the rule (12–30) is satisfied. The integral in (12–32) is improper. We break it into two parts:

$$\int_0^1 t^{x-1} e^{-t}\,dt + \int_1^\infty t^{x-1} e^{-t}\,dt.$$

The second term is finite for all x, since the exponential decay factor e^{-t} dominates. The first term has a discontinuity at $t = 0$ if $x < 1$, but still exists as an improper integral if $x > 0$, since the integrand behaves like Const $\times\, t^{x-1}$ for t near 0 and

$$\int_0^1 t^{x-1}\,dt = \frac{t^x}{x}\bigg|_0^1 = \frac{1}{x}, \qquad x > 0.$$

Thus we can define

$$\Gamma(x) = \int_0^\infty t^{x-1} e^{-t}\,dt, \qquad x > 0. \tag{12–33}$$

We can verify that this function is continuous.

Furthermore, for $x > 0$, by integration by parts, we get

$$\Gamma(x+1) = \int_0^\infty t^x e^{-t}\,dt = -t^x e^{-t}\bigg|_0^\infty + x\int_0^\infty t^{x-1} e^{-t}\,dt$$

$$= x\int_0^\infty t^{x-1} e^{-t}\,dt = x\Gamma(x).$$

Accordingly, the rule (12–31) is satisfied.

We write the rule in the form

$$\Gamma(x) = \frac{\Gamma(x+1)}{x} \tag{12–31'}$$

and let $x \to 0^+$. On the right, $\Gamma(x+1) \to \Gamma(1) = 1$, and hence the right side has limit $+\infty$. Therefore,

$$\lim_{x\to 0^+} \Gamma(x) = \infty.$$

We can use the equation (12–31') to extend the definition of $\Gamma(x)$ to negative x. For if $-1 < x < 0$, then $0 < x+1 < 1$, so that the right side is defined, and we use this as value of $\Gamma(x)$. For example,

$$\Gamma\left(-\frac{1}{3}\right) = \frac{\Gamma(2/3)}{-1/3} = -3\Gamma\left(\frac{2}{3}\right) = -3\cdot 1.3544 = -4.063.$$

Clearly $\Gamma(x)$ is thus defined as a *negative* function between -1 and 0. As x

approaches 0^-, Eq. (12–31′) shows that $\Gamma(x) \to -\infty$; similarly, as x approaches -1^+, $\Gamma(x) \to -\infty$.

Having extended the definition to the interval $-1 < x < 0$, we can repeat the process and extend the definition to the interval $-2 < x < -1$, then to $-3 < x < -2$, and so on. From (12–31′) we see that the signs alternate in successive intervals and, from the limits already found, $\Gamma(x) \to \pm\infty$ as x approaches 0, $-1, -2, \ldots$, with opposite signs from the left and the right, as in Fig. 12–1.

If $\Gamma(x)$ has been tabulated for $1 \leqslant x \leqslant 2$, then the values for $0 < x < 1$ can be found in the same way from (12–31′); for example, $\Gamma(0.5) = \Gamma(1.5)/0.5 = 2\Gamma(1.5)$. Also from the same table, $\Gamma(x)$ can be found for $2 \leqslant x \leqslant 3$, then for $3 \leqslant x \leqslant 4$, and so on, using (12–31):

$$\Gamma(2.5) = 1.5\,\Gamma(1.5), \qquad \Gamma(3.5) = 2.5\,\Gamma(2.5).$$

For convenience we give some values in Table 12–1 for $1 \leqslant x \leqslant 2$. We have seen that such a table allows us to calculate Γ for all real x. The table shows that $\Gamma(x)$ has a minimum of about 0.8857 in this interval at $x = 1.45$. It can be shown that this is the only minimum for positive x and that $\Gamma'(x) < 0$ below this value, $\Gamma'(x) > 0$ above this value. Since $\Gamma(n+1) = n!$, $\Gamma(x)$ has limit ∞ as $x \to \infty$.

Table 12–1

x	$\Gamma(x)$	x	$\Gamma(x)$
1.00	1.0000	1.55	0.8889
1.05	0.9735	1.60	0.8935
1.10	0.9514	1.65	0.9001
1.15	0.9330	1.70	0.9086
1.20	0.9180	1.75	0.9191
1.25	0.9064	1.80	0.9314
1.30	0.8975	1.85	0.9456
1.35	0.8912	1.90	0.9618
1.40	0.8873	1.95	0.9799
1.45	0.8857	2.00	1.0000
1.50	0.8862		

We can also define $\Gamma(z)$ for complex z. First we use the analog of Eq. (12–32):

$$\Gamma(z) = \int_0^\infty t^{z-1} e^{-t}\, dt. \tag{12–34}$$

Here $t^{z-1} = e^{(z-1)\ln t}$, with $\ln t$ real, and this is well defined for $t > 0$ and all complex z; it is analytic in z for all complex z. The integral exists for $\mathrm{Re}\, z > 0$, as can be verified. Furthermore, the same argument shows that

$$\Gamma(z+1) = z\Gamma(z) \tag{12–31''}$$

for complex z, $\text{Re}\,z > 0$. If, as before, we write this as

$$\Gamma(z) = \frac{\Gamma(z+1)}{z}, \qquad (12\text{–}31''')$$

then we can again use the relation to study limits and to extend the definition. First we can verify from Eq. (12–34) that $\Gamma(z)$ is continuous for $\text{Re}\,z > 0$, in fact is analytic in z in this half-plane; this is suggested by the fact, noted above, that the integrand is analytic in z. If we now let z approach $z_0 = 0 + iy$, $y_0 \neq 0$, on the imaginary axis, then $1 + z$ approaches $1 + iy_0$, as suggested in Fig. 12–2. But $\Gamma(z)$ is analytic at $1 + iy_0$. Since $z_0 \neq 0$, the right side of Eq. (12–31''') is analytic for z in a neighborhood of z_0. Therefore Eq. (12–31''') can be used to define $\Gamma(z)$ as an analytic function in a neighborhood of z_0. Since for $\text{Re}\,z > 0$ we recover the values of $\Gamma(z)$ we already had, this process is an *analytic continuation* of $\Gamma(z)$ (see Section 11–23).

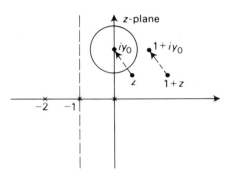

Fig. 12–2. Study of domain of $\Gamma(z)$.

As a result of it, we extend the definition of $\Gamma(z)$ to the domain $\text{Re}(z) > -1$, except for $z = 0$. Repeating the process, we extend the definition to the whole z-plane, except for the points 0, -1, -2, $\ldots$, and obtain an analytic function $\Gamma(z)$ with isolated singularities at $z = 0$, -1, -2, $\ldots$. The same reasoning as before shows that $\Gamma(z) \to \infty$ (complex) as z approaches each one of these points. Therefore (see Section 11–13) $\Gamma(z)$ has a pole at each point. From Eq. (12–31''') we see that $\Gamma(z)$ has a pole of first order at $z = 0$, since the numerator has nonzero value $\Gamma(1)$ at $z = 0$. A similar reasoning shows that *all the poles are of first order.*

The function $\Gamma(z)$ appears in certain Laplace transforms. For example, for $\text{Re}\,z > -1$, $\text{Re}\,s > 0$,

$$\mathscr{L}[t^z] = \int_0^\infty t^z e^{-st}\,dt = \frac{\Gamma(z+1)}{s^{z+1}} \qquad (12\text{–}35)$$

(Problem 4 below); for $z = n$, a positive integer, this is the familiar formula (Section 4-4) $\mathscr{L}[t^n] = n!/s^{n+1}$. Another related integral formula is (Problem 5 below)

$$\Gamma(z) = 2 \int_0^\infty e^{-t^2} t^{2z-1}\,dt. \qquad (12\text{–}36)$$

For large positive x, $\Gamma(x)$ can be evaluated with the aid of *Stirling's formula*:

$$\Gamma(x) = x^{x-1/2} e^{-x} \sqrt{2\pi} \, \exp\left[\frac{1}{12x} + \frac{g(x)}{x^3}\right],\tag{12-37}$$

where $g(x) \to 0$ as $x \to \infty$. This is commonly used to approximate $n!$ for large n. For example, we approximate $20! = 2.4329 \cdot 10^{18}$ by

$$\Gamma(21) = 21^{20.5} e^{-21} \sqrt{2\pi} \, e^{1/252} = 2.4323 \cdot 10^{18}.$$

12–4 THE BETA FUNCTION

The function $B(x, y)$ is defined first by the equation

$$B(x, y) = \int_0^1 t^{x-1} (1-t)^{y-1} \, dt.\tag{12-40}$$

The definition is valid if x and y are real and positive, since the improper integral exists in all these cases. The same reasoning applies if x and y are allowed to be complex with $\operatorname{Re} x > 0$ and $\operatorname{Re} y > 0$, and as usual $t^{x-1} = \exp[(x-1)\ln t]$, $(1-t)^{y-1} = \exp[(y-1)\ln(1-t)]$.

We have the important formula:

$$B(x, y) = \frac{\Gamma(x)\,\Gamma(y)}{\Gamma(x+y)}.\tag{12-41}$$

The right side is meaningful for all complex x, y, provided that none of x, y, $x+y$ falls at a pole of the gamma function (so x, y, $x+y$ must not be 0 or a negative integer). This formula in effect permits an analytic continuation of the beta function to the complex values x, y described (see Problem 7 below).

═══════════════════════ **Problems (Section 12–4)** ═══════════════════════

1. With the aid of Table 12–1 evaluate:

   a) $\Gamma(2.6)$ **b)** $\Gamma(5.8)$ **c)** $\Gamma(0.3)$ **d)** $\Gamma(-1.7)$

   e) $\Gamma(-3.5)$ **f)** $B(3,5)$ **g)** $B(1.2, 4.3)$

2. Prove the rules (state exceptions):

   a) $\Gamma(z+2) = (z+1) z \, \Gamma(z)$

   b) $\Gamma(z+3) = (z+2)(z+1) z \Gamma(z)$

   c) $\Gamma(z+n+1) = (z+n)(z+n-1) \cdots z\Gamma(z)$

3. a) Show that at $z = 0$, $\Gamma(z)$ has residue 1.

   HINT: As in Section 11–15, $\operatorname{Res}[\Gamma(z), z_0] = \lim (z - z_0)\Gamma(z)$ as $z \to z_0$. Write $z\Gamma(z) = \Gamma(z+1)$.

   b) Show that at $z = -1$, $\Gamma(z)$ has residue -1.

   HINT: By Problem 2(a) $(z+1)\Gamma(z) = \Gamma(z+2)/z$.

   c) Show that at $z = -n$, $\Gamma(z)$ has residue $(-1)^n/n!$

4. Prove the rule (12–35) for s real and positive.

    HINT: Set $u = st$ in the integral. (The validity for all s such that $\operatorname{Re} s > 0$ follows from the fact that the Laplace transform is analytic in s for $\operatorname{Re} s > 0$ and from theorems on uniqueness of analytic continuation (Section 11–23).)

5. Prove the rule (12–36).
6. Use Stirling's formula to approximate 10!
7. Prove the validity of (12–41) for $\operatorname{Re} x > 0$, $\operatorname{Re} y > 0$.

    HINT: Represent $\Gamma(x)$ and $\Gamma(y)$ by formula (12–36), then interpret $\Gamma(x)\,\Gamma(y)$ as a double integral over the first quadrant in a tu-plane. Change to polar coordinates r, θ and set $t = \sin^2 \theta$ to obtain finally

$$2 \int_0^\infty r^{2x + 2y - 1}\, e^{-r^2}\, dr \int_0^1 (1 - t)^{y-1}\, t^{x-1}\, dt,$$

which equals $\Gamma(x + y)\, B(x, y)$.

12–5 ORTHOGONAL POLYNOMIALS

We consider a fixed interval $a \leqslant x \leqslant b$ on the x-axis, which we first take to be finite, and a weight function $w(x)$ such that $w(x) > 0$ on the interval, except at a finite number of points, and such that the integrals $\int_a^b x^k w(x)\, dx$, $\quad k = 0, 1, \ldots$, exist and are finite. As we shall see, $w(x)$ may have discontinuities at a or b.

    It then follows that $\int_a^b f(x) w(x)\, dx$ exists for every continuous (or piecewise continuous) f on the interval and hence that the inner product

$$(f, g) = \int_a^b f(x)\, g(x)\, w(x)\, dx$$

exists for such f and g. This inner product has the usual properties (Section 3–10) and can be used to obtain a norm $\|f\| = (f, f)^{1/2}$.

    We recall the Gram–Schmidt process (Section 3–11). This assigns to a sequence $f_0, f_1, \ldots, f_n, \ldots$ of functions on the interval $a \leqslant x \leqslant b$ a new sequence

$$\begin{aligned} g_0 &= f_0, \qquad g_1 = k_{10} f_0 + f_1, \quad \ldots, \\ g_n &= k_{n0} f_0 + \cdots + k_{n, n-1} f_{n-1} + f_n, \quad \ldots \end{aligned} \tag{12–50}$$

It is assumed that, for each n, $f_0, \ldots, f_n$ are linearly independent on the interval. It then follows that, for a unique choice of the constants k_{ij}, $\{g_n\}$ is an orthogonal system.

    The process was illustrated in Section 3–12 to obtain the Legendre polynomials. In that case $w(x) \equiv 1$ and the interval was $-1 \leqslant x \leqslant 1$, the sequence $\{f_n\}$ was $\{x^n\}$.

    We now allow an arbitrary weight function $w(x)$ and interval $a \leqslant x \leqslant b$, but continue to use the sequence $x^0 = 1, x^1, x^2, \ldots, x^n, \ldots$ as sequence $\{f_n\}$. The Gram–Schmidt process (12–50) then yields functions

$$g_n(x) = k_{n0} + k_{n1} x + \cdots + k_{n, n-1} x^{n-1} + x^n, \qquad n = 0, 1, 2, \ldots,$$

which are clearly polynomials, g_n being of degree n. We refer to the g_n as a *system of orthogonal polynomials.* They are uniquely determined by the weight function $w(x)$ and interval $a \leqslant x \leqslant b$. In practice, we allow multiplication of each g_n by a nonzero scalar factor c_n to obtain a sequence

$$p_n(x) = c_n k_{n0} + \cdots + c_n x^n, \qquad n = 0,1,2,\ldots,$$

of polynomials, each of degree n, forming an orthogonal system on the given interval.

EXAMPLE 1   The interval is $-1 \leqslant x \leqslant 1$ and $w(x) = (1-x)^\alpha (1+x)^\beta$, where α and β are constants, $\alpha > -1, \beta > -1$.

The restrictions on α, β ensure that $x^k w(x)$ has a finite integral over the interval $-1 \leqslant x \leqslant 1$ for $k = 0,1,2,\ldots$ The resulting polynomials $p_n(x)$, for proper choice of the scalar factors c_n, are called the *Jacobi polynomials* and denoted by $P_n^{\alpha,\beta}(x)$. In particular, for $w(x) \equiv 1$ we obtain the Legendre polynomials $P_n(x)$:

$$P_n(x) = P_n^{0,0}(x).$$

The scalar multiplier c_n for $P_n^{\alpha,\beta}(x)$ is chosen to be 1 for $n = 0$; otherwise it is

$$c_n = \frac{1}{2^n n!} (\alpha + \beta + 2n)(\alpha + \beta + 2n - 1) \cdots (\alpha + \beta + n + 1). \tag{12–51}$$

The reason for the choice lies in the history of discovery of these polynomials. ◄

We can use the Gram–Schmidt process to obtain the Jacobi polynomials explicitly. However, it is easier to use a recursion formula (to be discussed below).

Besides the Legendre polynomials, there are other important special cases of the Jacobi polynomials modified by scalar factors:

The Chebyshev polynomials of the first kind

$$T_n(x) = \frac{2^{2n}(n!)^2}{(2n)!} P_n^{-1/2, -1/2}(x);$$

The *Gegenbauer polynomials*

$$C_n^\gamma(x) = \frac{(n+2\gamma-1)(n+2\gamma-2)\cdots(2\gamma)}{(n+\gamma-1/2)(n+\gamma-3/2)\cdots(\gamma+1/2)} P_n^{\gamma-1/2, \gamma-1/2},$$

where γ is a constant, $\gamma > -1/2$.

Case of infinite interval   The preceding discussion extends to the case of infinite intervals. The function $w(x)$ has to be such that $x^k w(x)$ has a finite integral over the interval. The inner products (f, g) are then defined for polynomials f, g and for appropriate continuous or piecewise continuous f, g, such that the integrals exist. The Gram–Schmidt process can again be applied and, from the functions $1, x, \ldots, x^n, \ldots$, produces a system of orthogonal polynomials $g_n(x) = x^n + \cdots$. Again we may multiply by scalar factors c_n.

EXAMPLE 2   The interval is $0 \leqslant x < \infty$ and $w(x) = x^\alpha e^{-x}$, where α is a constant greater than -1. The resulting polynomials multiplied by the scalar factors $(-1)^n/n!$ are the *Laguerre polynomials* $L_n^\alpha(x)$. In practice, other scalar factors are sometimes used. ◄

EXAMPLE 3 The interval is $-\infty < x < \infty$ and $w(x) = e^{-x^2/2}$. The resulting polynomials are the *Hermite polynomials* $H_n(x)$. We have not multiplied by scalar factors here, so that $H_n(x) = x^n + \cdots$. Some definitions use scalar factors. ◀

The polynomials described in this section are often referred to as the *classical orthogonal polynomials*. (See Section 3–13 for some of their special properties.)

12–6 RECURSION FORMULA FOR ORTHOGONAL POLYNOMIALS

We first observe that, by the Gram–Schmidt process (12–50), f_n can be expressed as a linear combination of $g_0, \ldots, g_n$. Thus, for each system of orthogonal polynomials $p_n(x)$, x^n can be expressed as a linear combination of $p_0(x), p_1(x), \ldots, p_n(x)$. Since $p_n(x)$ is orthogonal to $p_0(x), \ldots, p_{n-1}(x)$, it follows that $p_n(x)$ is orthogonal to x^k for $k < n$ and, more generally, that $p_n(x)$ is orthogonal to every polynomial of degree less than n.

From this property we can deduce that there is always a recursion formula of form

$$p_{n+1}(x) - (A_n x + B_n) p_n(x) + C_n p_{n-1}(x) = 0 \qquad (12\text{–}60)$$

for appropriate constants A_n, B_n, C_n.

For, if $p_n(x) = c_n x^n + \cdots$ as above, we let $A_n = c_{n+1}/c_n$. Then in $p_{n+1}(x) - A_n x p_n(x)$ the term of degree $n + 1$ cancels, so that this is a polynomial of degree n at most. By the remark above, it can be expressed as a linear combination of $p_0(x), \ldots, p_n(x)$, say

$$p_{n+1} - A_n x p_n = \sum_{k=0}^{n} h_k p_k. \qquad (12\text{–}61)$$

We take the inner product of both sides with p_k for $0 \leqslant k \leqslant n$. By orthogonality,

$$-A_n (x p_n, p_k) = h_k (p_k, p_k), \qquad k = 0, 1, \ldots, n. \qquad (12\text{–}62)$$

But $(x p_n, p_k) = (p_n, x p_k) = 0$ for $k \leqslant n - 2$, since $x p_k$ is then a polynomial of degree less than n. Therefore $h_k = 0$ for $k \leqslant n - 2$, and (12–61) is the desired recursion formula.

For the Legendre polynomials we find, as in Sections 3–12 and 3–13, the recursion formula

$$(n + 1) P_{n+1}(x) - (2n + 1) x P_n(x) + n P_{n-1}(x) = 0. \qquad (12\text{–}63)$$

Other recursion formulas are given in Section 3–13.

12–7 OTHER PROPERTIES OF ORTHOGONAL POLYNOMIALS

We let $\{p_n\}$ be a system of orthogonal polynomials, as above, $p_n(x) = c_n x^n + \cdots$. As a polynomial of degree n, $p_n(x)$ has n roots, some of which could conceivably be complex. However, the roots turn out to be real and simple and to lie between a and b

(excluding the endpoints). For suppose $p_n(x)$ has m roots $x_1, \ldots, x_m$ of odd multiplicity in the interval and that $m < n$. Let $q(x) = (x - x_1) \cdots (x - x_m)$, where $q(x) \equiv 1$ if $m = 0$. Then the degree of q is less than n, so that $(p_n, q) = 0$. But $p_n q$ has only roots of even multiplicity between a and b, hence cannot change sign; therefore

$$(p_n, q) = \int_a^b p_n(x) q(x) w(x) \, dx \neq 0.$$

There is a contradiction and, accordingly, $m = n$; thus all roots are simple and lie between a and b.

It can also be shown that, for each $n, p_n(x)$ and $p_{n+1}(x)$ have no common zeros and the zeros of p_n, p_{n+1} alternate on the interval from a to b.

If $m > n$, then between any two zeros of $p_n(x)$ there is at least one zero of $p_m(x)$. Furthermore, for each interval $\alpha \leqslant x \leqslant \beta$ contained in the interval $a \leqslant x \leqslant b$, there is an N such that $p_n(x)$ has a zero between α and β for $n > N$ (for proofs see HOCHSTADT, pp. 14–26).

Completeness of orthogonal polynomials

For simplicity we consider only the case of a continuous $g(x)$ and show that

$$E_n = \|g(x) - (c_1 p_1 + \cdots + c_n p_n)\|^2 \to 0 \qquad \text{as} \qquad n \to \infty, \qquad c_k = (g, p_k),$$

where $\{p_n\}$ is an orthonormal system of polynomials. We let

$$\int_a^b w(x) \, dx = K^2, \qquad K > 0.$$

By a famous theorem of Weierstrass (Section 13–8), we can choose a polynomial $q(x)$ such that $|g(x) - q(x)| < \varepsilon/K$, for given $\varepsilon > 0$ and $a \leqslant x \leqslant b$. Then

$$\|g - q\|^2 = \int_a^b (g(x) - q(x))^2 \, w(x) \, dx \leqslant \frac{\varepsilon^2}{K^2} \int_a^b w(x) \, dx \leqslant \varepsilon^2.$$

Next we use the fact that, for each n, x^n is a linear combination of $p_0(x), \ldots, p_n(x)$ to express the polynomial $q(x)$ as such a linear combination:

$$q(x) = a_0 p_0(x) + \cdots + a_N p_N(x),$$

where N is the degree of q. Hence

$$\|g - [a_0 p_0(x) + \cdots + a_N p_N(x)]\| \leqslant \varepsilon.$$

As in Section 3–10, the partial sum $c_0 p_0(x) + \cdots + c_N p_N(x)$ gives the best approximation of g, in sense of norm, among all linear combinations of $p_0, \ldots, p_N$. Therefore

$$E_N = \|g - [c_0 p_0(x) + \cdots + c_N p_N(x)]\| \leqslant \varepsilon. \qquad (12\text{–}70)$$

As in Eq. (3–107),

$$\|g - (c_0 p_0 + \cdots + c_n p_n)\|^2 = \|g\|^2 - (c_0^2 + \cdots + c_n^2)$$

and hence E_n decreases as n increases. Therefore, by (12–70), $E_n \leqslant \varepsilon$ for $n \geqslant N$. This shows that $E_n \to 0$ as $n \to \infty$.

Generating functions

If a function of t depending on a parameter x is expanded in a power series in t, the coefficients of the successive powers of t form a sequence of functions of x. The given function is said to be a *generating function* for the sequence of functions of x. For example,

$$(1+t)^x = 1 + xt + \frac{x(x-1)}{2!}t^2 + \cdots + \frac{x(x-1)\cdots(x-n+1)}{n!}t^n$$
$$+ \cdots, \qquad |t| < 1,$$

so that $(1+t)^x$ is a generating function for the sequence

$$f_n(x) = \frac{x(x-1)\cdots(x-n+1)}{n!}.$$

We can verify the following expansion that provides a generating function for the Legendre polynomials:

$$\frac{1}{(1-2tx+t^2)^{1/2}} = \sum_{n=0}^{\infty} P_n(x)t^n, \qquad |t| < 1. \tag{12–71}$$

To prove this, we let $g(x, t)$ denote the left-hand side, so that $g(x, t) = (1 - 2tx + t^2)^{-1/2}$. We proceed formally and comment on convergence below. We have

$$\frac{\partial g}{\partial t} = (x - t)(1 - 2tx + t^2)^{-3/2}$$

and hence

$$(1 - 2tx + t^2)\frac{\partial g}{\partial t} = (x - t)g.$$

We set $g(x, t) = \sum_{n=0}^{\infty} p_n(x)t^n$ and substitute in the previous equation, then equate coefficients of t^n on both sides:

$$(1 - 2tx + t^2) \sum_{n=1}^{\infty} np_n t^{n-1} = (x - t) \sum_{n=0}^{\infty} p_n t^n,$$

$$(n + 1)p_{n+1} - (2n + 1)xp_n + np_{n-1} = 0, \qquad n = 2, 3, \ldots$$

We see that $p_n(x)$ satisfies the same recursion formula as $P_n(x)$. Also $p_0(x) = g(x, 0) = 1$, $p_1(x) = g_t(x, 0) = x$. Hence $p_0 = P_0$, $p_1 = P_1$, and by the recursion formula $p_n = P_n$ for all n.

To analyze the convergence, we use complex-variable theory and consider the branch of the analytic function $g(x, z) = (1 - 2zx + z^2)^{-1/2}$ that reduces to 1 for $z = 0$ (Section 11–23). For fixed x, $g(x, z)$ is analytic at $z = 0$ and hence can be expanded in a power series as above. If, further, x is real, where $-1 \leqslant x \leqslant 1$, then g is seen to be analytic for $|z| < 1$, so that the series converges for $|t| < 1$.

Table of polynomials

For reference we list here the first five members of several systems of orthogonal polynomials:

Legendre: $P_0(x) = 1$, $P_1(x) = x$, $P_2(x) = \frac{1}{2}(3x^2 - 1)$, $P_3(x) = \frac{1}{2}(5x^3 - 3x)$,

$$P_4(x) = \frac{1}{8}(35x^4 - 30x^2 + 3).$$

Chebyshev: $T_0(x) = 1$, $T_1(x) = x$, $T_2(x) = 2x^2 - 1$, $T_3(x) = 4x^3 - 3x$,
$T_4(x) = 8x^4 - 8x^2 + 1$.

Laguerre: $L_0^\alpha(x) = 1$, $L_1^\alpha(x) = \alpha + 1 - x$, $L_2^\alpha(x) = \dfrac{(\alpha + 2)\,(\alpha + 1)}{2} - (\alpha + 2)x + \dfrac{x^2}{2}$,

$$L_3^\alpha(x) = \frac{(\alpha + 3)\,(\alpha + 2)\,(\alpha + 1)}{6} - \frac{(\alpha + 3)\,(\alpha + 2)\,x}{2} + \frac{(\alpha + 3)x^2}{2} - \frac{x^3}{6},$$

$$L_4^\alpha(x) = \frac{(\alpha + 4)\,(\alpha + 3)\,(\alpha + 2)\,(\alpha + 1)}{24} - \frac{(\alpha + 4)\,(\alpha + 3)\,(\alpha + 2)}{6}x$$

$$+ \frac{(\alpha + 4)\,(\alpha + 3)}{4}x^2 - \frac{\alpha + 4}{6}x^3 + \frac{x^4}{24}.$$

Hermite: $H_0(x) = 1$, $H_1(x) = x$, $H_2(x) = x^2 - 1$, $H_3(x) = x^3 - 3x$,
$H_4(x) = x^4 - 6x^2 + 3$.

12–8 DIFFERENTIAL EQUATIONS FOR ORTHOGONAL POLYNOMIALS

It can be shown that each of the classical orthogonal polynomials satisfies a differential equation $a_0(x)y'' + a_1(x)y' + a_2(x)y = 0$ with polynomial coefficients.

We show this in detail for the Legendre polynomials. Let $y(x) = P_n(x)$. Then

$$[(1 - x^2)y']' = (1 - x^2)y'' - 2xy'.$$

Here the right side is a polynomial of degree at most n, hence it can be expressed as a linear combination of $P_0(x), \ldots, P_n(x)$, so that

$$[(1 - x^2)y']' = b_0 P_0 + \cdots + b_n P_n.$$

We take the inner product of both sides with P_k for $k < n$ and obtain, by orthogonality,

$$\int_{-1}^{1} P_k(x)[(1 - x^2)y']' dx = b_k(P_k, P_k).$$

We integrate by parts twice, using the fact that $1 - x^2 = 0$ at $x = \pm 1$:

$$b_k(P_k, P_k) = \int_{-1}^{1} P_k(x)[(1 - x^2)y']' dx = -\int_{-1}^{1} P_k'(x)\,(1 - x^2)y'\,dx$$

$$= \int_{-1}^{1} y(x)[(1 - x^2)P_k'(x)]'\,dx.$$

The expression $[(1 - x^2)P_k']'$ is, as above, a polynomial of degree at most, k, hence is orthogonal to $y(x) = P_n(x)$. Thus the last integral is 0 and, accordingly, $b_k = 0$ for $k < n$. Therefore,

$$[(1 - x^2)P_n'(x)]' = b_n P_n.$$

This is the differential equation sought. To find b_n explicitly, we compare terms of degree n on both sides and find $b_n = -n(n + 1)$, so that

$$[(1 - x^2)P_n'(x)]' + n(n + 1)P_n = 0.$$

Thus $y = P_n(x)$ satisfies the differential equation

$$(1 - x^2)y'' - 2xy' + n(n + 1)y = 0. \tag{12–80}$$

For applications, it is important to consider the more general differential equation

$$(1 - x^2)y'' - 2xy' + \alpha(\alpha + 1)y = 0, \tag{12–81}$$

where α is arbitrary. Equation (12–81) is called the *Legendre differential equation*, its solutions are called *Legendre functions*.

Equation (12–81) has singular points at $x = \pm 1$, both of which are regular. To study the one at $x = 1$, we proceed as in Section 12–2 and form

$$[(1 - x^2)D^2 - 2xD + \alpha(\alpha + 1)] (x - 1)^m.$$

This reduces to

$$-2m^2(x - 1)^{m-1} + [\alpha(\alpha + 1) - m(m + 1)] (x - 1)^m.$$

Thus $f(m) = -2m^2$ and the indicial equation has a double root 0, 0. Thus there are solutions.

$$y_1(x) = 1 + \sum_{n=1}^{\infty} c_n(x - 1)^n,$$
$$\tag{12–82}$$
$$y_2(x) = cy_1(x) \ln(x - 1) + 1 + \sum_{n=1}^{\infty} b_n(x - 1)^n.$$

If this is done in detail, as in Section 2–23, we find that the series for $y_1(x)$ reduces to a polynomial of degree N precisely when $\alpha = N$ and is otherwise convergent for $|x - 1| < 2$, with infinite limit as $x \to -1$. In the function $y_2(x)$, c is not 0, so that $y_2(x)$ becomes infinite at $x = 1$. Thus the only solutions of Eq. (12–81) that are continuous for $-1 \leqslant x \leqslant 1$ are those obtained for $\alpha = N$, where $N = 0, 1, 2, \ldots$, and they are constants times the Legendre polynomials.

Another form of Eq. (12–81) is obtained by setting $x = \cos \varphi$. We find (Problem 7 below) that

$$\sin \varphi \frac{d^2 y}{d\varphi^2} + \cos \varphi \frac{dy}{d\varphi} + \alpha(\alpha + 1) \sin \varphi\, y = 0. \tag{12–81'}$$

Normally, we consider this equation for $0 \leqslant \varphi \leqslant \pi$ which corresponds to

$-1 \leqslant x \leqslant 1$. By our discussion above, the only solutions continuous for $0 \leqslant \varphi \leqslant \pi$ are those for $\alpha = N$, $N = 0, 1, 2, \ldots$, and they are constants times $P_N(\cos \varphi)$.

================================Problems (Section 12–8)================================

Use the table of polynomials at the end of Section 12–7 for the explicit expressions needed for the particular orthogonal polynomials occurring.

1. Verify the assertion that $p_n(x)$ has n zeros between a and b
   a) For $P_0(x), \ldots, P_4(x)$;   b) For $T_0(x), \ldots, T_4(x)$.

2. Verify the assertion that the zeros of $p_n(x)$ and $p_{n+1}(x)$ alternate:
   a) For $L_2^1(x), L_3^1(x)$,   b) For $H_3(x), H_4(x)$.

3. *Rodrigues formulas.* For the Legendre polynomials and Hermite polynomials, they are

$$P_n(x) = \frac{(-1)^n}{2^n n!} D^n[(1 - x^2)^n], \qquad H_n(x) = (-1)^n e^{x^2/2} D^n(e^{-x^2/2}).$$

   (See Section 3–13 for other such formulas.) Use these to obtain
   a) $P_0(x), \ldots, P_4(x)$ b) $H_0(x), \ldots, H_4(x)$

4. The Hermite polynomials have the recursion formula $H_{n+1}(x) - xH_n(x) + nH_{n-1}(x) = 0$. Use this and the Rodrigues formula of Problem 3 to prove the following:
   a) $H'_n(x) = n H_{n-1}(x)$
   b) $H_{n+1}(0) = -n H_{n-1}(0)$
   c) $H_{n+1}(x) = x H_n(x) - H'_n(x)$

5. Use the results of Problem 4 to calculate $H_2(x), H_3(x), H_4(x)$ from $H_0(x) = 1, H_1(x) = x$.

6. Obtain the solution $y_1(x)$, as in Eqs. (12–82), of Eq. (12–81) and show that $y_1(x)$ reduces to a polynomial when $\alpha = N$, a positive integer or 0.

7. Show that Eq. (12–81) becomes Eq. (12–81') if the substitution $x = \cos \varphi$ is made.

8. *Axially symmetric solutions of Laplace equation.* Consider the Laplace equation $\nabla^2 u = 0$ in spherical coordinates ρ, φ, θ (Section 9–17). We seek axially symmetric solutions, i.e., solutions independent of θ. Thus u satisfies the equation

$$\frac{\partial^2 u}{\partial \rho^2} + \frac{1}{\rho^2} \frac{\partial^2 u}{\partial \varphi^2} + \frac{2}{\rho} \frac{\partial u}{\partial \rho} + \frac{\cot \varphi}{\rho^2} \frac{\partial u}{\partial \varphi} = 0.$$

   Variables are separated as in Chapter 8: $u = R(\rho)\Phi(\varphi)$. Show that solutions of this form exist that are continuous in space, with $\Phi(\varphi) = P_n(\cos \varphi)$, $n = 0, 1, 2, \ldots$, and $R(\rho) = \rho^n$.

9. By taking $x = 0, 1, -1$ in Eq. (12–71), show that $P_{2n+1}(0) = 0, P_{2n}(0) = 2^{-n} \cdot 1 \cdot 3 \cdots (2n - 1)/n!, P_n(1) = 1, P_n(-1) = (-1)^n$.

10. Show that the gravitational potential at $P(x, y, z)$ of a mass m at $Q(x_0, y_0, z_0)$ can be expressed for $R > \rho$ in a series

$$-km \sum_{n=0}^{\infty} \frac{P_n(\cos \psi)}{R^{n+1}} \rho^n,$$

   where $\rho = |\overrightarrow{OQ}|, R = |\overrightarrow{OP}|, \psi = \angle (\overrightarrow{OP}, \overrightarrow{OQ})$.

11. The Schroedinger equation for stationary states of the hydrogen atom has the form

$$\nabla^2 u + \left(\lambda + \frac{c}{\rho} \right) u = 0,$$

where $u = u(\rho, \varphi, \theta)$ in spherical coordinates and λ, c are constants. Show that solutions depending on ρ alone exist and that, if $\lambda = -c^2/[4(n+1)^2]$, then $u = e^{-a\rho/2}L_n^1(a\rho)$ is a solution for $a = c/(n+1)$.

HINT: $y = L_n^\alpha(x)$ satisfies $xy'' + (\alpha + 1 - x)y' + ny = 0$.

12. The one-dimensional Schroedinger equation for stationary states of a linear oscillator has the form $u'' + (\lambda - c^2x^2)u = 0$. Show that if $\lambda = (2n+1)c$, then $u = e^{-a^2x^2/4}H_n(x)$ is a solution for $a = (2c)^{1/2}$.

HINT: $y = H_n(x)$ satisfies $y'' - xy' + ny = 0$

12–9 THE ASSOCIATED LEGENDRE FUNCTIONS

These are the functions

$$P_n^m(x) = (-1)^m(1-x^2)^{m/2}D^mP_n(x),$$ (12–90)
$$m = 0, 1, \ldots, n, \qquad n = 0, 1, 2, \ldots, \qquad |x| \leqslant 1.$$

Thus

$$P_2^0(x) = P_2(x) = \frac{3}{2}x^2 - \frac{1}{2},$$

$$P_2^1(x) = -(1-x^2)^{1/2} \cdot 3x,$$

$$P_2^2(x) = (1-x^2) \cdot 3.$$

For m even, they are polynomials; for all m and n, they are continuous when $-1 \leqslant x \leqslant 1$.

The functions $P_n^m(x)$ are used in the next section in constructing *spherical harmonics*. Here we develop a few important properties.

Differential equation   When $-1 < x < 1, y = P_n^m(x)$ satisfies the differential equation

$$(1-x^2)y'' - 2xy' + \left[n(n+1) - \frac{m^2}{1-x^2}\right]y = 0.$$ (12–91)

This can be derived from the Legendre differential equation (12–80) by differentiating it m times.

For the application to spherical harmonics, it is required that we make the substitution $x = \cos\varphi$, $-\pi \leqslant \varphi \leqslant \pi$. We verify from Eq. (12–91) that $u(\varphi) = P_n^m(\cos\varphi)$ satisfies the differential equation

$$u'' + \cot\varphi\, u' + [n(n+1) - m^2\csc^2\varphi]u = 0$$ (12–92)

(see Problem 2 following Section 12–10). Further, it can be shown that

$$\int_{-1}^1 (1-x^2)^mD^mP_n(x)D^mP_r(x)dx = 0, \qquad n \neq r.$$ (12–93)

where $m \leqslant n, m \leqslant r$, as in (12–90).

Equation (12–93) states that for fixed m, the functions $\{D^m P_n(x)\}$ for $n \geqslant m$ form an orthogonal system on the interval $-1 \leqslant x \leqslant 1$ with respect to the weight function

$$w(x) = (1 - x^2)^m = (1 - x)^m (1 + x)^m.$$

The functions $D^m P_n(x)$ are polynomials (of degree 0 for $n = m$, degree 1 for $n = m + 1, \ldots,$ degree k for $n = m + k, \ldots$). Hence we are dealing with a case of Jacobi polynomials, with $\alpha = m, \beta = m$; they can be regarded as Gegenbauer polynomials with $\gamma = m + (1/2)$. To get precise agreement with our definitions above, we need to multiply by appropriate scalar factors. By comparing the coefficients of x^{n-m}, we find that

$$D^m P_n = \frac{(n + m)!}{n! \, 2^m} P_{n-m}^{m,m}. \tag{12–94}$$

Further, we find that

$$(D^m P_n, D^m P_n) = \int_{-1}^{1} (1 - x^2)^m D^m P_n D^m P_n \, dx = \frac{2(n + m)!}{(n - m)! \, (2n + 1)}. \tag{12–95}$$

12–10 SPHERICAL HARMONICS

We consider the Laplace equation in space:

$$\nabla^2 U = U_{xx} + U_{yy} + U_{zz} = 0 \tag{12–100}$$

or, in spherical coordinates ρ, φ, θ (see Section 9–17),

$$U_{\rho\rho} + \frac{1}{\rho^2} U_{\varphi\varphi} + \frac{1}{\rho^2 \sin^2 \varphi} U_{\theta\theta} + \frac{2}{\rho} U_\rho + \frac{\cot \varphi}{\rho^2} U_\varphi = 0. \tag{12–100'}$$

We shall verify that Eq. (12–100') has solutions $\rho^n u_{mn}(\theta, \varphi)$, $\rho^n v_{mn}(\theta, \varphi)$, where $n = 0, 1, 2, \ldots,$ and

$$u_{mn}(\theta, \varphi) = \cos m\theta \, P_n^m (\cos \varphi), \qquad m = 0, 1, \ldots, n,$$

$$v_{mn}(\theta, \varphi) = \sin m\theta \, P_n^m (\cos \varphi), \qquad m = 1, \ldots, n. \tag{12–101}$$

The functions (12–101) or the functions $\rho^n u_{mn}(\theta, \varphi)$, $\rho^n v_{mn}(\theta, \varphi)$ are called *spherical harmonics*.

We are led to these functions in solving the Laplace equation (12–100') by separating variables:

$$U = \mathscr{R}(\rho) \Phi(\varphi) \Theta(\theta).$$

As in Chapter 8, we obtain the equations

$$(\rho^2 \, \mathscr{R}')' - \alpha \, \mathscr{R} = 0, \qquad \Theta'' + \beta \Theta = 0,$$

$$(\sin \varphi \, \Phi')' + (\alpha \sin \varphi - \beta \cos \varphi) \Phi = 0,$$

where α and β are constants to be chosen. The first equation has a solution $\mathscr{R} = \rho^n$ if $\alpha = n(n + 1)$, and this is well-behaved in space if $n = 0, 1, 2, \ldots$ There are other solutions (in particular, ρ^{-n-1} for the same α), but it can be verified that only the solutions ρ^n lead to solutions $U(x, y, z)$ that are expandable in a power series in x, y, z (in fact, as we shall see, $U(x, y, z)$ is a polynomial).

In the second equation we must take $\beta = m^2$, with $m = 0, 1, 2, \ldots$, in order to obtain periodicity of period 2π in θ. For such β, $\Theta(\theta)$ is a linear combination of $\cos m\theta$ and $\sin m\theta$. For $m = 0$, $\Theta = c_1 + c_2\theta$, and we must take $c_2 = 0$ to have periodicity.

The third equation now becomes

$$\Phi'' + \cot \varphi\, \Phi' + \left[n(n+1) - m^2 \cos^2\varphi \right] \Phi = 0.$$

This we recognize as the differential equation (12–92) satisfied by $u(\varphi) = P_n^m(\cos \varphi)$, provided $m \leqslant n$. For $m > n$, solutions can be found, but again they do not lead to the desired analyticity of the solution.

Thus finally we have obtained the solutions $\rho^n u_{mn}(\theta, \varphi)$, $\rho^n v_{mn}(\theta, \varphi)$ as in (12–101). Since

$$P_0^0(x) = 1, \qquad P_1^0(x) = x, \qquad P_1^1(x) = -(1-x^2)^{1/2}, \qquad P_2^0(x) = \frac{3}{2}x^2 - \frac{1}{2}, \ldots,$$

we find

$$\rho^0 u_{00} = 1, \qquad \rho^1 u_{01} = \rho \cos \varphi, \qquad \rho^1 u_{11} = -\rho \cos \theta \sin \varphi,$$

$$\rho^1 v_{11} = -\rho \sin \theta \sin \varphi, \qquad \rho^2 u_{02} = \rho^2 \left(\frac{3}{2}\cos^2 \varphi - \frac{1}{2} \right), \ldots$$

In rectangular coordinates, these are

$$1, \quad z, \quad -x, \quad -y, \quad z^2 - \frac{1}{2}(x^2 + y^2), \quad \ldots$$

In general, the functions $\rho^n u_{mn}$, $\rho^n v_{mn}$ become *homogeneous polynomials* in x, y, z of degree n, that is sums of terms $a\, x^i y^j z^k$, with $i + j + k = n$. As solutions of the Laplace equation, these polynomials are *harmonic*. It can be shown that every homogeneous harmonic polynomial in x, y, z of degree n is a linear combination of the $2n + 1$ polynomials $\rho^n u_{mn}$, $\rho^m v_{mn}$ in rectangular coordinates.

We now consider all functions f continuous on the sphere $\mathscr{S}: x^2 + y^2 + z^2 = 1$, and express them in spherical coordinates as $f(\theta, \varphi)$. We denote by Z the collection of all these functions. Then Z can be considered a vector space (Section 5–17), with the usual operations of addition and multiplication by real scalars. Furthermore, we can introduce an inner product in Z:

$$(f, g) = \int\!\!\int_{\mathscr{S}} fg\, d\sigma = \int_0^\pi \int_0^{2\pi} f(\theta, \varphi) g(\theta, \varphi) \sin \varphi\, d\theta\, d\varphi. \qquad (12\text{–}102)$$

For, as in Section 9–11, on the sphere of radius ρ, $d\sigma = \rho^2 \sin \varphi\, d\varphi\, d\theta$. We see at once that this inner product has the usual properties (Section 3–10) and can be used to define norms as usual: $\|f\| = (f, f)^{1/2}$.

The functions u_{mn}, v_{mn} together form an *orthogonal system* in Z:

$$u_{00}, \quad u_{01}, \quad u_{11}, \quad v_{11}, \quad u_{02}, \quad u_{12}, \quad v_{12}, \quad u_{22}, \quad v_{22}, \quad \ldots \qquad (12\text{–}103)$$

For example,
$$(u_{mn}, u_{rs}) = [\cos(m\theta)P_n^m(\cos\varphi), \cos(r\theta)P_s^r(\cos\varphi)].$$
If we substitute in Eq. (12–102) we see that, by the orthogonality of $\cos(m\theta)$, $\cos(r\theta)$ on $0 \leqslant \theta \leqslant 2\pi$, $(u_{mn}, u_{rs}) = 0$ for $m \neq r$. For $m = r$, the integral reduces to
$$\int_0^{2\pi} \cos^2(m\theta)d\theta \int_0^{\pi} P_n^m(\cos\varphi) P_s^m(\cos\varphi)\sin\varphi\,d\varphi.$$
The first factor equals π (or 2π for $m = 0$). If we use the definition (12–90) and set $x = \cos\varphi$ in the second integral, it becomes
$$\int_{-1}^1 (1 - x^2)^m D^m P_n(x)D^m P_s(x)\,dx.$$
By Eq. (12–93), this is 0 for $n \neq s$. Thus $(u_{mn}, u_{rs}) = 0$ except for $m = r$, $n = s$. Similarly, $(v_{mn}, v_{rs}) = 0$, except for $m = r$, $n = s$, and $(u_{mn}, v_{rs}) = 0$ for all cases (Problem 6 below). Thus the functions (12–103) form an orthogonal system, as asserted.

We can also find the norms of these functions. For example,
$$\|u_{mn}\|^2 = (u_{mn}, u_{mn}).$$
If we evaluate the inner product, as in the preceding paragraph, it becomes
$$\int_0^{2\pi} \cos^2(m\theta)\,d\theta \int_{-1}^1 (1 - x^2)^m [D^m P_n(x)]^2 dx.$$
The first factor is π (or 2π for $m = 0$). the second is $2(n+m)![(n-m)!\,(2n+1)]^{-1}$ by Eq. (12–95). Therefore
$$\|u_{mn}\|^2 = \frac{2\pi(n+m)!}{(n-m)!\,(2n+1)}, \quad 1 \leqslant m \leqslant n; \qquad \|u_{0n}\|^2 = \frac{4\pi}{2n+1}. \tag{12–104}$$
Similarly,
$$\|v_{mn}\|^2 = \frac{2\pi(n+m)!}{(n-m)!\,(2n+1)}, \quad 1 \leqslant m \leqslant n. \tag{12–104'}$$

It can be shown further that the system (12–103) is *complete* (see JACKSON, pp. 137–138). Hence an arbitrary function f in Z can be expanded in a Fourier series with respect to spherical harmonics:
$$\begin{aligned} f(\theta, \varphi) = a_{00}u_{00}(\theta, \varphi) &+ a_{01}u_{01}(\theta, \varphi) + a_{11}u_{11}(\theta, \varphi) \\ + b_{11}v_{11}(\theta, \varphi) &+ \cdots + a_{mn}u_{mn}(\theta, \varphi) + b_{mn}v_{mn}(\theta, \varphi) + \cdots \end{aligned} \tag{12–105}$$

Here
$$a_{mn} = \frac{(f, u_{mn})}{\|u_{mn}\|^2}, \qquad b_{mn} = \frac{(f, v_{mn})}{\|v_{mn}\|^2}, \tag{12–105'}$$

where the denominators are given by Eq. (12–104) and (12–104'). The series on the right of Eq. (12–105) is called the *Laplace series* of $f(\theta, \varphi)$. By completeness, this series converges to f at least in the mean: $\|f - S_k(\theta, \varphi)\| \to 0$ as $k \to \infty$ where $S_k(\theta, \varphi)$ is the sum of the first k terms of the series.

We can use this series to solve the Dirichlet problem for $\nabla^2 U = 0$ in the spherical region $x^2 + y^2 + z^2 < 1$. Let $U = f$ for $x^2 + y^2 + z^2 = 1$, where f is continuous, and express f as $f(\theta, \varphi)$, then expand f as in Eq. (12–105) as a Laplace series. Multiply the terms in u_{mn} and v_{mn} by ρ^n to obtain

$$U = \sum_{\substack{m \leqslant n \\ 0 \leqslant n < \infty}} \left[a_{mn} \rho^n u_{mn}(\theta, \phi) + b_{mn} \rho^n v_{mn}(\theta, \phi) \right]. \qquad (12\text{–}106)$$

Each term in the series is thus a harmonic function, expressible as a harmonic polynomial in x, y, and z. It can be verified that the series converges for $\rho < 1$ and may be repeatedly differentiated term by term with respect to ρ, φ, and θ. Hence

$$\nabla^2 U = \sum \nabla^2 \left[a_{mn} \rho^n u_{mn} + b_{mn} \rho^n v_{mn} \right] = 0,$$

so that U is harmonic for $\rho < 1$. For $\rho = 1$, U is given by the Laplace series for f, which converges to f in the mean. A more detailed study shows that U is continuous for $\rho \leqslant 1$, with $U = f$ for $\rho = 1$. In fact, it can be shown that the expansion (12–106) is equivalent to

$$U = \sum_{n=0}^{\infty} \rho^n S_n(\theta, \varphi),$$

$$S_n(\theta, \varphi) = \frac{2n+1}{4\pi} \int_0^\pi \int_0^{2\pi} f(\theta', \varphi') P_n(\cos \gamma) \sin \varphi' \, d\theta' \, d\varphi', \qquad (12\text{–}106')$$

where γ is the arc opposite N of the spherical triangle formed by the points (θ, φ), (θ', φ'), and $N(0, 0)$, the north pole (Fig. 12–3). By spherical trigonometry.

$$\cos \gamma = \cos \varphi \cos \varphi' + \sin \varphi \sin \varphi' \cos (\theta' - \theta).$$

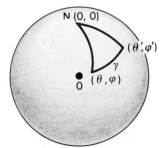

Fig. 12–3. Meaning of angle γ.

In Eq. (12–106'), the term $\rho^n S_n(\theta, \varphi)$ is the sum of all $2n+1$ terms with given n in (12–106), and is hence expressible as a homogeneous harmonic polynomial of degree

n in x, y, z. Further we show that (12–106′) is equivalent to a *Poisson integral formula* for U:

$$U(\rho, \varphi, \theta) = \frac{1}{4\pi} \int \int_{\mathscr{S}} f(\varphi', \theta') \frac{1 - \rho^2}{d^3} \, d\sigma, \qquad (12\text{–}107)$$

where d is the distance from the point (ρ, φ, θ) to the point $(1, \varphi', \theta')$ and $\mathscr{S}$ is the surface $\rho = 1$. (For details, see JACKSON, pp. 132–136.)

=========================Problems (Section 12–10)=========================

1. Find $P_3^0(x)$, $P_3^1(x)$, $P_3^2(x)$, $P_3^3(x)$.

2. Set $x = \cos \varphi$ and $y(\cos \varphi) = u(\varphi)$ in Eq. (12–91) to obtain Eq. (12–92).

3. Find all homogeneous harmonic polynomials in x, y, z
   a) of degree 2   b) of degree 3.

4. Find u_{00}, u_{01}, u_{11}, u_{02}, u_{12}, u_{22}, v_{11}, v_{12}, v_{22} as functions of φ and θ.

5. Express $\rho^n u_{mn}$ and $\rho^n v_{mn}$ in rectangular cordinates x, y, z for $n = 3$ and $m = 1$, $m = 2$.

6. a) Show that $(v_{mn}, v_{rs}) = 0$ except for $m = r$ and $n = s$.
   b) Show that $(u_{mn}, v_{rs}) = 0$.
   c) Prove (12–104′).

7. Find the equilibrium temperature $U(\rho, \varphi \, \theta)$ inside the sphere $\rho = 1$ if for $\rho = 1$
$$U = 2 + \cos \varphi - 3 \sin \theta \sin \varphi - 3 \cos \theta \sin \varphi \cos \varphi$$

12–11 BESSEL FUNCTIONS

We define

$$J_\nu(z) = \sum_{m=0}^{\infty} \frac{(-1)^m (z/2)^{\nu + 2m}}{m! \, \Gamma(\nu + m + 1)} \qquad (12\text{–}110)$$

to be the *Bessel function of the first kind and of order* ν. Here ν may be an arbitrary real or complex number. If ν is a negative integer, $\Gamma(\nu + m + 1) = \infty$ for certain values of m; we interpret $1/\Gamma(\nu + m + 1)$ to be 0 in these cases, and hence drop the corresponding terms from the series. For large m, the rule

$$\Gamma(\nu + m + 1) = (\nu + m)\Gamma(\nu + m)$$

shows that the factor $\Gamma(\nu + m + 1)$ in the denominator becomes infinite as $m \to \infty$. Thus the series converges for all complex z. It can be written as z^ν times the sum of an ordinary power series, converging for all z, and hence represents an analytic function in each domain in which z^ν is chosen to be analytic. If $\nu = n$, where $n = 0, 1, 2, \ldots$, we obtain $J_n(z)$, a function analytic for all z. In Fig. 12–4 we graph $J_0(x)$, $J_1(x)$, $J_4(x)$, $J_5(x)$ for x real.

Separation of variables for the wave equation and related equations (Section 8–13) leads to a differential equation that can be reduced to the

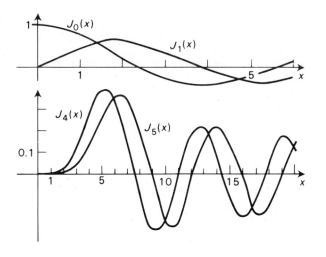

Fig. 12–4. Bessel functions.

form of the *Bessel equation*

$$z^2 w'' + zw' + (z^2 - v^2)w = 0. \tag{12–111}$$

This has singular points only at 0 and ∞; the first is regular, the second irregular. The indicial equation at $z = 0$ is $\mu^2 - v^2 = 0$. We verify that $J_v(z)$ and $J_{-v}(z)$ are solutions. If v is not an integer, these are linearly independent and provide the general solution in the form

$$w = c_1 J_v(z) + c_2 J_{-v}(z)$$

(see Problem 1 below).

To cover the case when v is an integer, it is convenient to introduce the *Neumann function or Bessel function of the second kind*

$$Y_v(z) = \frac{\cos v\pi\, J_v(z) - J_{-v}(z)}{\sin v\pi}. \tag{12–112}$$

This becomes indeterminate when v is an integer, but we can verify that the singularity is removable, so that $Y_v(z)$ is defined for all v and is analytic in z, except perhaps at $z = 0$, with branches as for $J_v(z)$. Furthermore, $J_v(z)$ and $Y_v(z)$ are linearly independent, even if v is an integer, so that they can be used to form the general solution of Eq. (12–111) (see WHITTAKER-WATSON, Chapter 17).

It should be remarked that, when v is a positive integer N or 0, the method of Section 12–2 leads to two solutions of the differential equation of form

$$w_1(z) = z^N \left(1 + \sum_{n=1}^{\infty} c_n z^n \right),$$

$$w_2(z) = c w_1(z) \ln z + z^{-N} \sum_{n=0}^{\infty} b_n z^n,$$

where $c \neq 0$. Only the first is continuous at $z = 0$; in fact, it is a constant times $J_N(z)$ and is analytic for all z. The second must be a linear combination of $J_N(z)$ and $Y_N(z)$. Accordingly, $Y_N(z)$ must be discontinuous at $z = 0$.

For the case of integral order, we have the *generating function* equation:

$$e^{z[t-(1/t)]/2} = \sum_{n=-\infty}^{\infty} J_n(z)\, t^n, \qquad 0 < |t| < \infty. \tag{12-113}$$

For fixed z, the function on the left is analytic for $0 < |t| < \infty$ and hence can be expanded in a Laurent series, which is the series on the right (see Problem 2 following Section 12). If we set $t = e^{i\theta}$ in Eq. (12–113), it takes the form of a Fourier series expansion

$$e^{iz \sin \theta} = \sum_{n=-\infty}^{\infty} J_n(z) e^{in\theta}. \tag{12-113'}$$

From the formulas for coefficients of a Laurent series (Section 11–12) and of a Fourier series (Section 3–7), we deduce the useful formulas:

$$J_n(z) = \frac{1}{2\pi i} \oint_{|t|=1} \frac{1}{t^{n+1}} e^{z(t-1/t)/2}\, dt, \tag{12-114}$$

$$J_n(z) = \frac{1}{2\pi} \int_0^{2\pi} e^{iz \sin \theta}\, e^{-in\theta}\, d\theta \tag{12-114'}$$

for $n = 0, \pm 1, \pm 2, \ldots$.

There are many other special identities satisfied by the functions $J_n(z)$ (see Problems 3 and 4 following Section 12–12).

For $J_v(z)$ for general v we also have identities (see Problem 4 below). We here stress properties related to series expansions in Bessel functions.

The differential equation (12–111) can be written in another form: if c is a nonzero constant, then

$$\frac{d^2}{dz^2} \left[\sqrt{z}\, J_v(cz) \right] + \left(c^2 + \frac{1-4v^2}{4z^2} \right) \sqrt{z}\, J_v(cz) = 0. \tag{12-115}$$

To verify this, we expand the first term and then multiply by $z^{3/2}$. The equation becomes

$$c^2 z^2 J_v''(cz) + cz\, J_v'(cz) + (c^2 z^2 - v^2) J_v(cz) = 0.$$

If we now set $\zeta = cz$, this reduces to Eq. (12–111) (with ζ as independent variable).

We now take $v \geq 0$ and consider the real-valued function $J_v(x)$ for $x \geq 0$. For $\alpha > 0$, $\beta > 0$, we have the relations

$$(\alpha^2 - \beta^2) \int_0^1 x J_v(\alpha x) J_v(\beta x) dx = \beta J_v(\alpha) J_v'(\beta) - \alpha J_v(\beta) J_v'(\alpha), \quad (12–116)$$

$$(\alpha^2 - \beta^2) \int_0^1 x J_v(\alpha x) J_v(\beta x) dx = \alpha J_v(\beta) J_{v+1}(\alpha) - \beta J_v(\alpha) J_{v+1}(\beta)$$

$$(12–116')$$

$$2\alpha \int_0^1 x J_v^2(\alpha x) dx = J_v(\alpha) J_{v+1}(\alpha) + \alpha [J_v(\alpha) \; J_{v+1}'(\alpha) - J_{v+1}(\alpha) J_v'(\alpha)].$$

$$(12–117)$$

To prove these we write $u(x) = \sqrt{x}\, J_v(\alpha x)$, $v(x) = \sqrt{x}\, J_v(\beta x)$, so that by (12–115), we have

$$u'' + \left(\alpha^2 + \frac{1 - 4v^2}{4x^2}\right) u = 0,$$

$$v'' + \left(\beta^2 + \frac{1 - 4v^2}{4x^2}\right) v = 0.$$

We multiply by $-v$, u respectively, add and integrate from b to 1, where $b > 0$. Since $uv'' - vu'' = (uv' - vu')'$, this leads to the equation

$$(uv' - vu')|_b^1 + (\beta^2 - \alpha^2) \int_b^1 uv\, dx = 0.$$

Here we use the definition of $u(x)$, $v(x)$ to find that

$$u(x)v'(x) - v(x)u'(x) = x[\beta J_v(\alpha x) J_v'(\beta x) - \alpha J_v(\beta x) J_v'(\alpha x)].$$

Since $v \geq 0$, this expression has a limit as $x \to 0$. We find from (12–110) that the limit is 0. Evaluating at b and 1 and letting $b \to 0$, we thus obtain Eq. (12–116). The rule (12–116') follows from this if we use the identity $J_v'(z) = (v/z)J_v(z) - J_{v+1}(z)$ (see Problem 4 (d) below).

To obtain (12–117), we fix α and let $\beta \to \alpha$ in the relation

$$(\alpha + \beta) \int_0^1 x J_v(\alpha x) J_v(\beta x) dx = \frac{\alpha J_v(\beta) J_{v+1}(\alpha) - \beta J_v(\alpha) J_{v+1}(\beta)}{\alpha - \beta},$$

which follows from (12–116'). If we apply L'Hospital's rule for the form 0/0, we obtain (12–117).

Zeros of $J_v(x)$ The differential equation satisfied by $u = \sqrt{x}\, J_v(x)$ is, as above,

$$u'' + \left(1 + \frac{1 - 4v^2}{4x^2}\right) u = 0.$$

If x is large, this is well approximated by $u'' + u = 0$, whose solutions are sinusoidal.

Therefore, we expect that $\sqrt{x} J_\nu(x)$ should be approximately sinusoidal, for large x, and have zeros about π apart. This can be justified. We consider the sinusoidal character in the next section. Here we stress the fact that $J_\nu(x)$ has zeros $\lambda_{\nu 1} < \lambda_{\nu 2} < \cdots < \lambda_{\nu n} < \cdots$ on the interval $0 < x < \infty$, where $\lambda_{\nu n} \to \infty$ as $n \to \infty$. (For proof see Chapter 15 of WATSON.) Each zero is simple, since $J_\nu(x_0) = 0$ and $J_\nu'(x_0) = 0$ imply that $J_\nu(x) \equiv 0$ by the existence theorem for ordinary differential equations. Also J_ν and $J_{\nu+1}$ have no common zeros, since by Problem 4 (d) below, $J_\nu'(x_0) = 0$ at such a common zero x_0. It can also be shown that between two successive zeros of $J_\nu(x)$ there is one and only one zero of $J_{\nu-1}(x)$ and one and only one zero of $J_{\nu+1}(x)$ (see Problem 5 below)

The resemblance to sines and cosines suggests that the functions $\sqrt{x} J_\nu(\lambda_{\nu n} x)$ form an orthogonal system, like the functions $\sin(n\pi x)$. That is, *the system* $\{J_\nu(\lambda_{\nu n} x)\}$ *is orthogonal with weight function* $w(x) = x$. This is true for the interval $0 \leqslant x \leqslant 1$. To prove it, we need only take $\alpha = \lambda_{\nu n}$, $\beta = \lambda_{\nu m}$ in Eq. (12–116).

We can also compute the norm squared $[J_\nu(\lambda_{\nu n} x), J_\nu(\lambda_{\nu n} x)]$. We apply Eq. (12–117) with $\alpha = \lambda_{\nu n}$ and obtain

$$[J_\nu(\lambda_n x), J_\nu(\lambda_n x)] = -\frac{1}{2} J_{\nu+1}(\lambda_{\nu n}) J_\nu'(\lambda_{\nu n}).$$

By Problem 4 (d) below, with $z = \lambda_{\nu n}$, $-J_\nu'(\lambda_{\nu n}) = J_{\nu+1}(\lambda_{\nu n})$. Therefore, finally,

$$\|J_\nu(\lambda_{\nu n} x)\|^2 = \frac{1}{2}[J_{\nu+1}(\lambda_{\nu n})]^2. \tag{12–118}$$

We can show further that the orthogonal system $\{J_\nu(\lambda_{\nu n} x)\}$ is *complete*, so that the *Fourier–Bessel series* can be formed as in Chapter 3 (for proofs see Chapter 18 of WATSON).

12–12 HANKEL FUNCTIONS. ASYMPTOTIC SERIES

The first and second Hankel functions are

$$H_\nu^{(1)}(z) = J_\nu(z) + i\, Y_\nu(z), \qquad H_\nu^{(2)}(z) = J_\nu(z) - i\, Y_\nu(z). \tag{12–120}$$

Since $J_\nu(z)$, $Y_\nu(z)$ are linearly independent, as above, so are $H_\nu^{(1)}(z)$ and $H_\nu^{(2)}(z)$, and hence they can be used to form the general solution of the Bessel equation as before. From Eqs. (12–120), we deduce the equations

$$J_\nu(z) = \frac{1}{2}[H_\nu^{(1)}(z) + H_\nu^{(2)}(z)], \tag{12–121}$$

$$Y_\nu(z) = \frac{1}{2i}[H_\nu^{(1)}(z) - H_\nu^2(z)]. \tag{12–122}$$

Asymptotic series   A power series, not necessarily convergent, may give information about a function in the following way: the nth partial sum of the series may give a very good approximation of the function, provided we stay close to the center of expansion of the series.

We here make this precise for expansions *at infinity*, and do it for complex z. We say that the series $\sum_{k=0}^{\infty} a_k z^{-k}$ is an *asymptotic series* for $f(z)$ at $z = \infty$, in symbols,

$$f(z) \approx \sum_{k=0}^{\infty} \frac{a_k}{z^k} \qquad \text{as} \qquad z \to \infty,$$

if

$$\lim_{|z| \to \infty} z^n \left\{ f(z) - \sum_{k=0}^{n} \frac{a_k}{z^k} \right\} = 0, \qquad n = 0, 1, 2, \ldots \tag{12–123}$$

In most applications we usually require that z lie in a certain sector $\alpha < \arg z < \beta$ and then refer to an asymptotic series for $f(z)$ at ∞ in the sector.

The condition (12–123) states that

$$f(z) = a_0 + \frac{a_1}{z} + \cdots + \frac{a_n}{z^n} + \frac{g_n(z)}{z^n},$$

where $g_n(z) \to 0$ as $|z| \to \infty$. Thus the first $n + 1$ terms of the series approximate $f(z)$, with an error that goes to zero faster than const/z^n as $|z| \to \infty$.

For example, it can be shown that

$$f(z) = \int_0^{\infty} \frac{e^{-t}}{1 - (t/z)} \, dt \approx \sum_{k=0}^{\infty} \frac{k!}{z^k} \qquad \text{as} \qquad z \to \infty.$$

The series diverges for all z, but the partial sums still approximate $f(z)$ for large z, as in (12–123). The idea is suggested in Fig. 12–5 for z real, $z = x$. To decrease the error, we must take x large (in contrast to making n large for a convergent series).

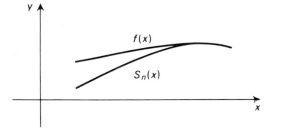

Fig. 12–5. Functions $f(x)$ versus nth partial sum $S_n(x)$ of asymptotic series.

It is convenient to extend the notation for asymptotic series as follows. We write

$$F(z) \approx p(z) \sum_{k=0}^{\infty} \frac{a_k}{z^k} + q(z) \sum_{k=0}^{\infty} \frac{b_k}{z^k}$$

to mean

$$F(z) = p(z) f(z) + q(z) g(z),$$

where

$$f(z) \approx \sum_{k=0}^{\infty} \frac{a_k}{z^k} \qquad \text{and} \qquad g(z) \approx \sum_{k=0}^{\infty} \frac{b_k}{z^k} \qquad \text{as} \qquad |z| \to \infty.$$

For $J_\nu(z)$ and $Y_\nu(z)$ we have then the following asymptotic relations for z approaching ∞ in a sector $-(\pi/2)+\varepsilon < \arg z < (\pi/2)-\varepsilon$, where $\varepsilon > 0$:

$$J_\nu(z) \approx \sqrt{\frac{2}{\pi z}}\left[\cos\left(z-\nu\frac{\pi}{2}-\frac{\pi}{4}\right)\left(1-\frac{c_2}{4z^2}+\frac{c_4}{16z^4}+\cdots\right)\right.$$

$$\left.+\sin\left(z-\nu\frac{\pi}{2}-\frac{\pi}{4}\right)\left(-\frac{c_1}{2z}+\frac{c_3}{8z^3}+\cdots\right)\right], \tag{12-124}$$

$$Y_\nu(z) \approx \sqrt{\frac{2}{\pi z}}\left[\sin\left(z-\nu\frac{\pi}{2}-\frac{\pi}{4}\right)\left(1-\frac{c_2}{4z^2}+\frac{c_4}{16z^4}+\cdots\right)\right.$$

$$\left.+\cos\left(z-\nu\frac{\pi}{2}-\frac{\pi}{4}\right)\left(\frac{c_1}{2z}-\frac{c_3}{8z^3}+\cdots\right)\right], \tag{12-125}$$

where

$$c_n = \frac{1}{n!}\frac{\Gamma\left(\nu+\frac{1}{2}+n\right)}{\Gamma\left(\nu+\frac{1}{2}-n\right)}.$$

From Eqs. (12–120) we can then obtain asymptotic series for the Hankel functions. The formula (12–124) gives the first approximation to $J_\nu(z)$ as

$$\sqrt{\frac{2}{\pi z}}\cos\left(z-\nu\frac{\pi}{2}-\frac{\pi}{4}\right), \tag{12-126}$$

with zeros at $z = (\pi/4)(3+4k+2\nu)$, $k = -1, 0, 1, 2, \ldots$ For $\nu = 2$, this gives $3\pi/4$, $7\pi/4$, $11\pi/4, \ldots$ or 2.3562, 5.4978, 8.6394, 11.7810, $\ldots$ From tables, the zeros are 5.135, 8.417, 11.620. The first zero estimated, 2.3562, is wrong; the asymptotic formula is good only for *large* $|z|$.

In Fig. 12–4 we graphed $J_0(x), \ldots, J_5(x)$; for further graphs and tables, see JAHNKE–EMDE and WATSON.

========================= **Problems (Section 12–12)**=========================

1. a) Verify that Eq. (12–110) defines a solution of the Bessel equation (12–111).
   b) Verify that for ν not an integer, $J_\nu(z)$ and $J_{-\nu}(z)$ are linearly independent in each open region in which they are defined as analytic functions.
   c) Show that, if n is an integer, $J_{-n}(z) = (-1)^n J_n(z)$, so that J_{-n} and J_n are not linearly independent.
   d) Show that, if n is an integer, $J_n(-z) = (-1)^n J_n(z)$, so that $J_n(z)$ is odd or even depending on whether n is odd or even.
2. Prove the generating-function relation (12–113).

   HINT: Multiply the equations

$$e^{zt/2} = 1 + \frac{zt}{2} + \left(\frac{zt}{2}\right)^2\frac{1}{2!} + \cdots,$$

$$e^{-z/(2t)} = 1 - \frac{z}{2t} + \left(\frac{z}{2t}\right)^2\frac{1}{2!} + \cdots.$$

3. Let n and m be integers. Prove the identities (see HINTS below):

a) $J_n(z) = \dfrac{1}{2\pi} \displaystyle\int_0^\pi \cos(z \sin\theta - n\theta)\, d\theta$

b) $\cos(z \sin\theta) = J_0(z) + 2 \displaystyle\sum_{m=1}^\infty J_{2m}(z) \cos(2m\theta)$

c) $\sin(z \sin\theta) = 2 \displaystyle\sum_{m=0}^\infty J_{2m+1}(z) \sin[(2m+1)\theta]$

d) $e^{iz \cos\theta} = \displaystyle\sum_{n=-\infty}^\infty i^n J_n(z) e^{in\theta}$

e) $\cos(z \cos\theta) = J_0(z) + 2 \displaystyle\sum_{m=1}^\infty (-1)^m J_{2m}(z) \cos(2m\theta)$

f) $\sin(z \cos\theta) = 2 \displaystyle\sum_{m=0}^\infty (-1)^m J_{2m+1}(z) \cos[(2m+1)\theta]$

HINTS: For (a) use (12–114′). For (b) write

$$\cos(z \sin\theta) = (e^{iz \sin\theta} + e^{-iz \sin\theta})/2,$$

use (12–113′) and the result of Problem 1(c). For (c) proceed as in (b). For (d) replace θ by $\theta + (\pi/2)$ in (12–113′). For (e) and (f) use the result of (d).

4. Prove the following identities:

a) $2J_v'(z) = J_{v-1}(z) - J_{v+1}(z)$

b) $\dfrac{2v}{z} J_v(z) = J_{v-1}(z) + J_{v+1}(z)$

c) $J_{v-1}(z) = J_v'(z) + \dfrac{v}{z} J_v(z)$ or $z^v J_{v-1}(z) = \dfrac{d}{dz}[z^v J_v(z)]$

d) $J_{v+1}(z) = \dfrac{v}{z} J_v(z) - J_v'(z)$ or $-z^{-v} J_{v+1}(z) = \dfrac{d}{dz}[z^{-v} J_v(z)]$.

HINTS: Use the definition (12–110) for (a) and (b). Use (a) and (b) to obtain (c) and (d).

5. Prove that between two successive zeros α and β of $J_v(x)$ there is one and only one zero of $J_{v-1}(x)$ and there is one and only one zero of $J_{v+1}(x)$.

HINT: $J_v'(\alpha)$ and $J_v'(\beta)$ must have opposite signs. Hence show by Problems 4(c) and (d) that both J_{v-1} and J_{v+1} have opposite signs α and β.

6. Compare the graph of $J_4(x)$ in Fig. 12–4 with that given by the asymptotic formula (12–124) with $v = 4$.

7. Prove the following:

a) $J_{1/2}(z) = \sqrt{\dfrac{2}{\pi z}} \sin z$

b) $J_{-1/2}(z) = \sqrt{\dfrac{2}{\pi z}} \cos z$

c) $J_{3/2}(z) = \sqrt{\dfrac{2}{\pi z}} \left(\dfrac{\sin z}{z} - \cos z \right)$

12–13 THE HYPERGEOMETRIC FUNCTION

The hypergeometric differential equation is the equation

$$z(1-z)w'' + [c - (1+a+b)]w' - abw = 0, \tag{12-130}$$

where a, b, c are constants. The differential equation has regular singular points at 0, 1, ∞.

The solutions of the equation are expressible in terms of the *hypergeometric function*, which is the function

$$F(a, b; c; z) = 1 + \frac{ab}{1 \cdot c} z + \frac{a(a+1)b(b+1)}{2!c(c+1)} z^2 + \cdots$$

$$+ \frac{a(a+1) \cdots (a+n-1)b(b+1) \cdots (b+n-1)}{n!c(c+1) \cdots (c+n-1)} z^n + \cdots \tag{12-131}$$

The series is well-defined, provided c is not a negative integer, and converges for $|z| < 1$. It provides a solution $w_1(z)$ of Eq. (12–130) in this region. If c is not an integer, a second, linearly independent solution is

$$w_2(z) = z^{1-c} F(a+1-c, \quad b+1-c; \quad 2-c; \quad z). \tag{12-132}$$

Solutions in other regions are obtainable by analytic continuation of these solutions.

The hypergeometric function is important as a unifying device. A host of other functions are expressible in terms of it. For example,

$$P_n(t) = F\left(-n, \quad n+1; \quad 1; \quad \frac{1-t}{2}\right), \tag{12-133}$$

and, in fact, all Jacobi polynomials are expressible in terms of the hypergeometric function. The Laguerre and Hermite polynomials and the Bessel functions are expressible in terms of the confluent hypergeometric function

$$_1F_1(a, c, z) = 1 + \frac{a}{1 \cdot c} z + \frac{a(a+1)}{2!c(c+1)} z^2 + \cdots$$

$$+ \frac{a(a+1) \cdots (a+n-1)}{n! \, c(c+1) \cdots (c+n-1)} z^n + \cdots \tag{12-134}$$

This is a solution of the *confluent hypergeometric equation*

$$zw'' + (c-z)w' - aw = 0. \tag{12-135}$$

We can consider Eq. (12–135) as a kind of limiting case of the hypergeometric equation (12–130). For further information on these functions see HOCHSTADT, Chapter 4, and ERDELYI, Chapters II, IV, V, VI.

═══════════════════**Problems (Section 12–13)**═══════════════════

1. Let c not be a negative integer:
   a) Show that the series (12–131) converges for $|z| < 1$.
   b) Show that $F(a, b; c; z)$ is a solution of Eq. (12–131).

2. Verify Eq. (12–133) for $n = 0, 1, 2, 3$.

3. Show that

a) $\dfrac{d}{dz} F(a, b; c; z) = \dfrac{ab}{c} F(a+1, b+1; c+1; z)$,

b) $\dfrac{d^2}{dz^2} F(a, b; c; z) = \dfrac{a(a+1)b(b+1)}{c(c+1)} F(a+2, b+2; c+2; z)$,

c) $(a-b)F(a, b; c; z) = aF(a+1, b; c; z) - bF(a, b+1; c; z)$.

4. Show that, if c is not a negative integer, the series (12–134) converges for all z and is a solution of Eq. (12–135).

Chapter 13
Numerical Analysis

13-1 INTRODUCTION

Many problems of applied mathematics can be solved only approximately, by numerical methods. In this chapter we consider some of the principal techniques needed. The following topics are covered:

Solution of equations and finding eigenvalues of matrices (Sections 13–2 to 13–5);

Interpolation and approximation by polynomials (Sections 13–6 to 13–8);

Interpolation and approximation by trigonometric polynomials, fast Fourier transform (13–9 to 13–11);

Numerical differentiation and integration (13–12 to 13–13);

Solution of initial-value and boundary-value problems for ordinary differential equations (Sections 13–14 to 13–15);

Solution of partial differential equations (Sections 13–16 to 13–18).

These six topics are treated essentially independently, so that those of interest can be selected and studied alone.

The subject of numerical analysis has undergone (and continues to undergo) a rapid development because of the phenomenal progress in digital computing. In this chapter we do not consider details of computer programming. However, computer questions are continually in the background and greatly influence the discussion.

The effect of *round-off error* is an example. In calculations (even with a hand-held calculator) we can carry only a limited number of decimal places and are forced to round off by some rule. In small-scale calculations this usually does not lead to serious error. However, in computations involving millions of steps the accumulated effect of rounding off can completely invalidate the answer.

Round off can be important even in a small-scale problem. The notorious example is the effect of subtracting two numbers that are very close together: $82.5374 \div (13.76201 - 13.76194)$ gives $1\,179\,105.8$ if we do not round off, $825\,374$ if we round off at the fourth decimal place, and is undefined if we round off at the third place.

A related problem is that of sensitivity to given data. It may happen that a minute change in the data leads to a very large change in the solution to a problem. When this arises, normal computational procedures usually become very unreliable. The difficulty is illustrated by ill-conditioned systems of linear equations (see Section 13–2).

The speed of the modern computer has rendered possible the solution of problems of great complexity, involving huge numbers of arithmetical operations. We still must be concerned about cost and are thus forced to compare alternative methods in terms of efficiency of computer use. In most cases the number of multiplications and divisions gives a reliable index of computer time required. Hence we seek methods that reduce the number of such operations.

13–2 SOLUTION OF SIMULTANEOUS LINEAR EQUATIONS

This topic is considered in Sections 5–12 and 5–13. We here present ways of improving the elimination process of Section 5–13, called the *Gauss–Jordan process*.

We consider the typical case of three equations in three unknowns:

$$\begin{aligned}
a_{11}x_1 + a_{12}x_2 + a_{13}x_3 &= k_1, \\
a_{21}x_1 + a_{22}x_2 + a_{23}x_3 &= k_2, \\
a_{31}x_1 + a_{32}x_2 + a_{33}x_3 &= k_3.
\end{aligned} \tag{13–20}$$

The Gauss–Jordan process operates on the matrix

$$\begin{bmatrix}
a_{11} & a_{12} & a_{13} & k_1 \\
a_{21} & a_{22} & a_{23} & k_2 \\
a_{31} & a_{32} & a_{33} & k_3
\end{bmatrix}.$$

By successive operations of dividing rows by numbers, interchanging rows, and subtracting a multiple of one row from another, we try to reduce this matrix to the form

$$\begin{bmatrix}
1 & 0 & 0 & c_1 \\
0 & 1 & 0 & c_2 \\
0 & 0 & 1 & c_3
\end{bmatrix}.$$

This process is equivalent to replacement of the given system (13–20) by the system

$$x_1 = c_1, \qquad x_2 = c_2, \qquad x_3 = c_3,$$

which give the unknowns x_1, x_2, x_3. If the matrix (a_{ij}) is nonsingular, the process described can be carried out in full.

An alternative to the Gauss–Jordan method is the simple *Gauss method*. Here we use the same steps, but seek only to produce zeros *below the diagonal*, that is, to achieve a form

$$\begin{bmatrix} 1 & b_{12} & b_{13} & c_1 \\ 0 & 1 & b_{23} & c_2 \\ 0 & 0 & 1 & c_3 \end{bmatrix}.$$

The corresponding set of equations is

$$x_1 + b_{12}x_2 + b_{13}x_3 = c_1, \qquad x_2 + b_{23}x_3 = c_2, \qquad x_3 = c_3.$$

We now obtain x_3 from the last equation, substitute this value in the next-to-last equation, and solve for x_2, substitute the known values of x_2 and x_3 in the first equation and obtain x_1.

It turns out that less multiplications and divisions are needed for the Gauss process than for the Gauss–Jordan process. Hence the simple Gauss process is preferred.

EXAMPLE 1 $2x_1 + 3x_2 + x_3 = 6,$
$\qquad\qquad\quad 5x_1 + 2x_2 - 3x_3 = 4,$
$\qquad\qquad\quad\ x_1 + x_2 + 3x_3 = 5.$

The matrix and its successive stages follow:

$$\begin{bmatrix} 2 & 3 & 1 & 6 \\ 5 & 2 & -3 & 4 \\ 1 & 1 & 3 & 5 \end{bmatrix}, \qquad \begin{bmatrix} 1 & 1.5 & 0.5 & 3 \\ 5 & 2 & -3 & 4 \\ 1 & 1 & 3 & 5 \end{bmatrix},$$

$$\begin{bmatrix} 1 & 1.5 & 0.5 & 3 \\ 0 & -5.5 & -5.5 & -11 \\ 0 & -0.5 & 2.5 & 2 \end{bmatrix}, \qquad \begin{bmatrix} 1 & 1.5 & 0.5 & 3 \\ 0 & 1 & 1 & 2 \\ 0 & -0.5 & 2.5 & 2 \end{bmatrix},$$

$$\begin{bmatrix} 1 & 1.5 & 0.5 & 3 \\ 0 & 1 & 1 & 2 \\ 0 & 0 & 3 & 3 \end{bmatrix}.$$

From the corresponding equations

$$x_1 + 1.5x_2 + 0.5x_3 = 3, \qquad x_2 + x_3 = 2, \qquad 3x_3 = 3$$

we obtain $x_3 = 1$, then $x_2 = 1$, and finally $x_1 = 1$. In this example there is no rounding off and the answer is exact. ◄

Tridiagonal case   When the coefficient matrix (a_{ij}) is such that all elements are zero except the diagonal elements a_{ii} and those immediately to the left and right of the diagonal elements, the coefficient matrix is said to be *tridiagonal*. For example,

$$\begin{bmatrix} 2 & 1 & 0 & 0 \\ 4 & 5 & 1 & 0 \\ 0 & 6 & 4 & 2 \\ 0 & 0 & 2 & 6 \end{bmatrix}$$

is a tridiagonal matrix. Such a matrix is reduced to a tridiagonal matrix with zeros everywhere below the diagonal by successively multiplying the rows by factors and subtracting from the next rows. In the above 4×4 matrix we multiply the first row by 2 and subtract from the second row, which becomes $(0, 3, 1, 0)$, then subtract twice the new second row from the third row to obtain a new third row $(0, 0, 2, 2)$, then subtract this third row from the fourth row to obtain a new fourth row $(0, 0, 0, 4)$.

For corresponding equations

$$2x_1 + x_2 = 3, \qquad 4x_1 + 5x_2 + x_3 = 5,$$
$$6x_2 + 4x_3 + 2x_4 = 9, \qquad 2x_3 + 6x_4 = 21$$

the same operations can be carried out on the augmented matrix, with fifth column col $(3, 5, 9, 21)$. The result is the matrix

$$\begin{bmatrix} 2 & 1 & 0 & 0 & 3 \\ 0 & 3 & 1 & 0 & -1 \\ 0 & 0 & 2 & 2 & 11 \\ 0 & 0 & 0 & 4 & 10 \end{bmatrix}$$

Solving the last equation $4x_4 = 10$, then substituting in the next to last $2x_3 + 2x_4 = 1$, and so on, we obtain $x_4 = 2.5$, $x_3 = 3$, $x_2 = -1.333$, $x_1 = 2.167$.

Tridiagonal matrices have important applications for ordinary and partial-differential equations (see Sections 13–15 and 13–16).

Pivoting In the elimination process the order of the equations has an effect on round-off errors, and it has been found that the best strategy is to use as first equation the one with the largest coefficient of x_1 (largest in absolute value). This coefficient a_{11} serves as a *pivot* for elimination of x_1. The second row of the new matrix is obtained from the second row of the given matrix by subtracting from it a_{21}/a_{11} times the first row. The larger a_{11} is, the less round-off error is introduced.

Similarly, in eliminating x_2, we may wish to interchange the second and third rows to achieve a large pivot. This idea extends to equations in more unknowns. In Example 1, it would thus have been normal to interchange the first and second equations before starting the first stage of the elimination.

Scaling In many cases accuracy can be improved by multiplying some or all the equations given by numbers, in order to reduce discrepancies in scale; for example, one equation might have coefficients between -10 and 10, another between 1000 and 2000, and it is better to have all coefficients reasonably close together. Also the coefficients of one unknown, say x_2, may be out of line; then replacement of x_2 by a new unknown $x_2' = cx_2$, for appropriate c, will help. These changes affect the choice of pivot. In fact, the main question is to determine which is the best pivot. (For a discussion, see *Theory and Applications of Numerical Analysis* by G. M. PHILLIPS and P. J. TAYLOR, Academic Press, New York, 1973.)

Ill-conditioned systems The following simple example illustrates another source of error:

EXAMPLE 2 $69x_1 + 68x_2 = 67$, $68x_1 + 67x_2 = 66$. We are led to the matrix process,

$$\begin{bmatrix} 69 & 68 & 67 \\ 68 & 67 & 66 \end{bmatrix}, \quad \begin{bmatrix} 1 & 0.986 & 0.971 \\ 68 & 67 & 66 \end{bmatrix}, \quad \begin{bmatrix} 1 & 0.986 & 0.971 \\ 0 & -0.048 & -0.028 \end{bmatrix}.$$

We now obtain $x_2 = 0.028 \div 0.048 = 0.583$, and hence $x_1 = 0.396$. These answers might be expected to be correct to three significant figures. However, the correct values are $x_1 = -1$, $x_2 = 2$, as can be verified. The accuracy can be improved by taking $68/69 = 0.98551$, $67/69 = 0.97101$. This leads to $x_2 = 0.02868 \div 0.01468 = 1.9537$ and hence $x_1 = -0.95438$. Thus we had to carry five significant figures in the crucial division in order to obtain an answer correct to one significant figure.

This is a case of extreme sensitivity to the data (the coefficients in the equations). The difficulty can be seen geometrically if the two equations are graphed as straight lines in the $x_1 x_2$-plane, as suggested in Fig. 13–1; the lines are almost coincident. The slightest change in the coefficients can shift the intersection point by a large distance. This example illustrates the so-called *ill-conditioned equations.* ◄

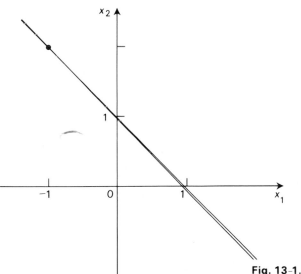

Fig. 13-1. Grazing lines (as in Example 2).

A similar problem occurs also for more unknowns. The only practical way to handle the difficulty is to carry extra decimal places as was done for Example 2. The *presence* of the difficulty is also revealed by large changes in the answer as the number of decimal places carried is increased.

Determinants   Linear equations can of course be solved by determinants, with the aid of Cramer's rule, as in Section 5–1. However, evaluation of determinants is time consuming and the elimination methods described above are hence preferable.

In fact, Gaussian elimination, slightly modified, can be used to evaluate a determinant. Only the steps of interchanging rows and subtracting a multiple of one row from another are allowed. The first of these steps multiplies the determinant of the matrix by -1; the second step does not affect the determinant.

EXAMPLE 3   We seek det A, where $A = \begin{bmatrix} 3 & 5 & 2 \\ 1 & 7 & 8 \\ 4 & 11 & 5 \end{bmatrix}$. The steps are as follows:

$$\begin{bmatrix} 3 & 5 & 2 \\ 1 & 7 & 8 \\ 4 & 11 & 5 \end{bmatrix}, \quad \begin{bmatrix} 1 & 7 & 8 \\ 3 & 5 & 2 \\ 4 & 11 & 5 \end{bmatrix}, \quad \begin{bmatrix} 1 & 7 & 8 \\ 0 & -16 & -22 \\ 0 & -17 & -27 \end{bmatrix}, \quad \begin{bmatrix} 1 & 7 & 6 \\ 0 & -16 & -22 \\ 0 & 0 & -3.625 \end{bmatrix}.$$

The last matrix is *upper triangular*, and its determinant (58) is the product of the diagonal elements. Since there was one row interchange, det $A = -58$. ◄

In general, row interchanges achieve the effect of pivoting and may be needed to reach desired accuracy. Also one may want to achieve the effect of scaling by dividing a row or column of A by a number. This number then must multiplied by the determinant of the new matrix.

Homogeneous systems   If $k_1 = k_2 = k_3 = 0$ in the system (13–20), then we are dealing with a homogeneous system. As in Section 5–12 such a system has always the *trivial solution* $x_i = 0$ for all i. Other solutions exist when the equations are dependent, that is, when the matrix $A = (a_{ij})$ has less than maximum rank or, equivalently, when det $A = 0$. In this case solutions are again found by elimination, as in Sections 5–12 and 5–13. The Gauss process can be followed, but one or more rows of zeros will be obtained, and the final solution will give some of the unknowns in terms of the remaining ones.

EXAMPLE 4 $5x_1 + x_2 - x_3 = 0$, $5x_1 + 14x_2 - 9x_3 = 0$, $3x_1 - 2x_2 + x_3 = 0$.

We apply the elimination process and thereby determine whether there are nontrivial solutions:

$$\begin{bmatrix} 5 & 1 & -1 \\ 5 & 14 & -9 \\ 3 & -2 & 1 \end{bmatrix}, \quad \begin{bmatrix} 1 & 0.2 & -0.2 \\ 5 & 14 & -9 \\ 3 & -2 & 1 \end{bmatrix}, \quad \begin{bmatrix} 1 & 0.2 & -0.2 \\ 0 & 13 & -8 \\ 0 & -2.6 & 1.6 \end{bmatrix},$$

$$\begin{bmatrix} 1 & 0.2 & -0.2 \\ 0 & 1 & -0.6154 \\ 0 & -2.6 & 1.6 \end{bmatrix}, \quad \begin{bmatrix} 1 & 0.2 & -0.2 \\ 0 & 1 & -0.6154 \\ 0 & 0 & 0 \end{bmatrix}.$$

Hence $x_2 = 0.6154x_3$, $x_1 = -0.2x_2 + 0.2x_3 = 0.07692x_3$. ◄

In theory, such a problem should be solved exactly, without rounding off, for the slightest error can change a zero in the lower right-hand corner to a number different from 0 and hence produce only the trivial solution. However, in most applications, it is known in advance that the precise entries in the matrix should yield a zero determinant and hence it is justified to ignore entries in the last row that are sufficiently close to 0. (For more discussion of linear equations see GERALD).

=============================**Problems (Section 13–2)**=============================

Throughout seek to obtain solutions precise to three significant figures.

1. Solve by the Gauss elimination procedure:
   a) $x_1 + 3x_2 = 1$, $2x_1 - x_2 = 5$
   b) $2x_1 + x_2 = 8$, $7x_1 + 2x_2 = 0$
   c) $3x_1 - x_2 + x_3 = 5$, $11x_1 + 3x_2 - 7x_3 = 2$, $x_1 + 5x_2 - 2x_3 = 1$
   d) $2x_1 + 4x_2 - 7x_3 = 2$, $x_1 + 7x_2 + 9x_3 = 13$, $12x_1 - 3x_2 - 5x_3 = 7$

2. For the system $x_1 + 793x_2 = 25$, $17x_1 + 25x_2 = 19$ solve by the Gauss procedure, without changing the order of the equations, and then solve with the equations interchanged. This illustrates how pivoting can help.

3. Comment on why each of the following systems is ill-conditioned. Solve by elimination and compare with the answer given:
   a) $3x_1 - x_2 = 5$, $85x_1 - 28x_2 = 11$
   b) $x_1 + 27x_2 = 1$, $11x_1 + 296x_2 = 5$
   c) $5x_1 + 4x_2 + 3x_3 = 12$, $50x_1 + 40x_2 + 31x_3 = 121$, $35x_1 + 27x_2 + 21x_3 = 83$
   d) $22x_1 + 55x_2 - 32x_3 = 13$, $4x_1 - x_2 + x_3 = 5$, $6x_1 + 4x_2 - 2x_3 = 6$

4. a) For a general system (13–20) in three unknowns show that the Gauss–Jordan process requires 18 multiplications and divisions, while the simple Gauss process requires 17 multiplications and divisions.
   b) Find the analogous numbers for four equations in four unknowns.

5. Evaluate the determinant of each matrix by elimination:

$$\text{a)} \begin{bmatrix} 5 & 3 & 1 \\ 3 & 7 & 11 \\ 8 & 4 & -1 \end{bmatrix} \qquad \text{b)} \begin{bmatrix} 3.21 & 7.43 & 1.08 \\ 4.05 & -2.16 & 3.12 \\ 2.58 & 1.42 & 7.05 \end{bmatrix}$$

6. Find all solutions:
   a) $7x_1 + 2x_2 - x_3 = 0$, $x_1 + 3x_2 + x_3 = 0$, $5x_1 - 23x_2 - 20x_3 = 0$
   b) $3x_1 - x_2 + 5x_3 = 0$, $x_1 + 4x_2 - x_3 = 0$, $18x_1 - 19x_2 + 38x_3 = 0$

13–3 **SOLUTION OF ALGEBRAIC EQUATIONS**

We here consider the problem of solving an equation of degree n:

$$p(x) = a_0 x^n + a_1 x^{n-1} + \cdots + a_n = 0, \qquad a_0 \neq 0. \qquad (13\text{–}30)$$

Thus $p(x)$ is a polynomial. We restrict attention to the case of real coefficients.

It is known from algebra (see also Problem 3 following Section 11–18) that such an equation has n roots, some of which may be complex and some of which may be repeated. If $x_1, \ldots, x_n$ are the roots, then $p(x) = a_0(x - x_1) \ldots (x - x_n)$.

In particular, if one or more roots have been found, then they can be factored out, so that the remaining roots are obtained from an equation $q(x) = 0$ of lower degree. When this is done numerically, errors arise in the coefficients of the lower-degree polynomial $q(x)$, and ultimately the roots of $q(x)$ have to be corrected in order to obtain the desired accuracy for the roots of Eq. (13–30).

The process of factoring can be achieved by *synthetic division*, discussed in many college algebra texts. The process finds $q(x) = b_0 x^{n-1} + \cdots + b_{n-1}$ such that $p(x) = q(x)(x-c) + R$, where $q(x)$ is quotient, R is the remainder, by the rules

$$b_0 = a_0, \qquad b_1 = a_1 + b_0 c, \qquad b_2 = a_2 + b_1 c, \ldots, \qquad b_n = a_n + b_{n-1} c,$$

and $b_n = R$. We verify from (13–31) that $R = p(c)$ and, by differentiation, that $q(c) = p'(c)$. The calculation can be written out as follows:

$$
\begin{array}{c|ccccccc}
c & a_0 & a_1 & a_2 & \cdots & a_{n-1} & a_n \\
  &   & b_0 c & b_1 c & & & \\
\hline
  & b_0 & b_1 & b_2 & \cdots & b_{n-1} & b_n
\end{array}
\tag{13–31}
$$

Repetition of the process with $b_0, b_1, \ldots$ instead of $a_0, a_1, \ldots$ gives a remainder $q(c) = p'(c)$ and, by similar reasoning, further repetition gives successive remainders $p''(c)/2!$, $p'''(c)/3!$, $\ldots$

EXAMPLE 1 Let $p(x) = x^3 - 5x^2 + 3x + 4$ and let $c = 4$. Then we obtain the calculation of Table 13–1.

Table 13–1

$$
\begin{array}{c|cccc}
4 & 1 & -5 & 3 & 4 \\
  &   & 4 & -4 & -4 \\
\hline
  & 1 & -1 & -1 & 0 = p(4) \\
  &   & 4 & 12 & \\
\hline
  & 1 & 3 & 11 = p'(4) & \\
  &   & 4 & & \\
\hline
p'''(4)/6 = 1 & & 7 = p''(4)/2 & &
\end{array}
$$

Since the first remainder is 0, $p(x)$ has 4 as a root and $p(x) = (x-4)(x^2 - x - 1)$. The successive derivatives are also shown. We can check the result by verifying the Taylor's series expression:

$$p(x) = x^3 - 5x^2 + 3x + 4 = 0 + 11(x-4) + 7(x-4)^2 + (x-4)^3. \quad \blacktriangleleft$$

In searching for roots of Eq. (13–30), the following rule is useful: *Equation* (13–30) *has no root x for which $|x| \geqslant R$, if $R = 1 + (M/|a_0|)$ and M is the largest of* $|a_1|, \ldots, |a_n|$. (For proof, see Section 11–22.) The rule applies to both real and complex roots.

In particular, to find real roots, we can restrict our attention to the interval $-R < x < R$. For example, let $p(x) = x^3 - 3x^2 - x + 2$, so that $a_0 = 1$, $M = 3$, and $R = 4$; we need to study only the interval $-4 < x < 4$. By synthetic division as above, we evaluate $p(x)$ for $-4, -3, \ldots, 4$; we obtain successive values $-106, -49, -16, -1, 2, -1, -4, -1, 4$. Each change of sign indicates passage through a root.

Hence there are roots between -1 and 0, between 0 and 1, between 3 and 4. By evaluating $p(x)$ at the midpoint of each of these intervals, we can corner the roots further. The process can be repeated indefinitely to get the desired accuracy. This is the method of *bisection*. Near a root of even multiplicity, no change of sign occurs, so that the method will not reveal the root; for a similar reason, no change of sign occurs between a and b if there is an even number of roots between a and b.

An alternative way to improve accuracy is *Newton's method*. Here, starting with an initial estimate x_1 for a root, we obtain successive estimates $x_2, x_3, \ldots, x_n, \ldots$ by the rule

$$x_{n+1} = x_n - \frac{p(x_n)}{p'(x_n)}. \tag{13–32}$$

(For a detailed discussion see CLA, pp. 444–459.) The method is widely used for solving both algebraic and transcendental equations (such as $e^x - 2 + x = 0$). For the case of an algebraic equation (13–30), the sequence $\{x_n\}$ will converge to a root, provided only that the initial estimate x_1 is close enough to the root. This is true even if the root is multiple.

For example, the polynomial $p(x) = x^3 - x^2 - x + 1$ has a double root at $x = 1$, since $p(x) = (x - 1)^2 (x + 1)$. If we start with $x_1 = 0.5$, then we obtain x_2 by Eq. (11–32), using synthetic division as in Table 13–2.

Table 13–2

0.5	1	-1	-1	1
		0.5	-0.25	-0.625
	1	-0.5	-1.25	0.375
		0.5	0	
	1	0	-1.25	

From the table we read off: $p(0.5) = 0.375$, $p'(0.5) = -1.25$, and therefore

$$x_2 = 0.5 - \frac{0.375}{-1.25} = 0.8.$$

Similarly, we find $x_3 = 0.91$, $x_4 = 0.95$. (The rate of convergence is slower for a double root than for a simple root; the error is roughly halved at each step. For a simple root, the value of $p'(x_n)$ changes very slowly for large n and it can even be treated as constant for large n.)

An entirely different method for finding the roots is that of Bernoulli. We form the sequence $x_0, x_1, \ldots$ by the rules

$$x_0 = 1, \qquad a_0 x_1 + a_1 x_0 = 0, \qquad a_0 x_2 + a_1 x_1 + a_2 x_0 = 0, \qquad \ldots,$$
$$a_0 x_{n-1} + a_1 x_{n-2} + \cdots + a_{n-1} x_0 = 0,$$
$$a_0 x_{n+k} + a_1 x_{n+k-1} + \cdots + a_n x_k = 0 \quad \text{for} \quad k = 0, 1, 2, \ldots$$

Here the first n equations successively determine $x_0, x_1, \ldots, x_{n-1}$ and the last equation is a difference equation or recursion formula that determines x_n, $x_{n+1}, \ldots$ successively. We now form the ratios x_{k+1}/x_k and study their behavior for large k. If Eq. (13–30) happens to have a simple real root of maximum absolute value (larger than the absolute values of all other real or complex roots), then it can be shown that x_{k+1}/x_k has that real root as limit.

EXAMPLE 2 $x^3 - 3x^2 - x + 2 = 0$. We have

$$x_0 = 1, \ x_1 - 3x_0 = 0, \qquad x_2 - 3x_1 - x_0 = 0, \qquad x_3 - 3x_2 - x_1 + 2x_0 = 0,$$
$$x_4 - 3x_3 - x_2 + 2x_1 = 0, \qquad \ldots, \qquad x_{3+k} - 3x_{2+k} - x_{1+k} + 2x_k = 0, \qquad \ldots$$

Therefore, $x_1 = 3, x_2 = 10, x_3 = 31, x_4 = 97, x_5 = 302, x_6 = 941, \ldots$ Accordingly, for $k = 0, 1, 2, \ldots$, the ratios x_{k+1}/x_k are 3, 3.33, 3.1, 3.129, 3.113, 3.116, $\ldots$ These appear to be approaching a limit. By Newton's method, we change the 3.116 to 3.115 and verify that this is a root correct to three decimal places. ◄

If there is a real root of maximum absolute value that is *multiple*, then the sequence x_{k+1}/x_k again converges to that root. The determination of the multiplicity requires an examination of derivatives at the root: if the multiplicity is l, then $p'(x) = 0, \ldots, p^{(l-1)}(x) = 0$ and $p^{(l)}(x) \neq 0$ at the root.

If the maximum absolute value is achieved by a pair of conjugate complex roots, then, for large k, these can be found approximately as the roots of the quadratic equation

$$\begin{vmatrix} z^2 & z & 1 \\ x_k & x_{k-1} & x_{k-2} \\ x_{k+1} & x_k & x_{k-1} \end{vmatrix} = 0. \tag{13–33}$$

EXAMPLE 3 $x^3 + 3x^2 + 7x + 5 = 0$. Here $x_0 = 1, x_1 + 3x_0 = 0, x_2 + 3x_1 + 7x_0 = 0$, $x_3 + 3x_2 + 7x_1 + 5x_0 = 0, \ x_{3+k} + 3x_{2+k} + 7x_{1+k} + 5x_k = 0$ for $k = 1, 2, \ldots$, so that we obtain the sequence $1, -3, 2, 10, -29, 7, 132, -300, -59, 1617, -2938, \ldots$ Apparently, the ratios x_{k+1}/x_k do not have a limit. We suspect a pair of complex roots and form the equation (13–33):

$$\begin{vmatrix} z^2 & z & 1 \\ 1617 & -59 & -300 \\ -2938 & 1617 & -59 \end{vmatrix} = 0$$

or $z^2 + 1.9993z + 4.997 = 0$. The original equation can be factored as $(x+1)$ $(x^2 + 2x + 5) = 0$. Thus we have approximately reproduced the quadratic factor and obtained the approximations $-0.9997 \pm 1.9997i$ to the roots $-1 \pm 2i$. ◄

Having found a quadratic factor, we can remove it by long division. To this end, a modified synthetic division can be used. We illustrate this for division of a sixth-degree polynomial $p(x) = a_0x^6 + \cdots + a_6$ by $x^2 - rx - q$ in Table 13–3. In each column we add, so that $b_0 = a_0, b_1 = a_1 + b_0r, \ldots$ The result shows that

$$a_0x^6 + \cdots + a_6 = (b_0x^4 + \cdots + b_4)(x^2 - rx - q) + Rx + S.$$

Table 13-3

	a_0	a_1	a_2	a_3	a_4	a_5	a_6
q			$b_0 q$	$b_1 q$	$b_2 q$	$b_3 q$	$b_4 q$
r		$b_0 r$	$b_1 r$	$b_2 r$	$b_3 r$	$b_4 r$	
	b_0	b_1	b_2	b_3	b_4	R	S

When R and S are 0, we have an exact quadratic factor. For example 3 we could divide by $x^2 + 2x + 5$ as in Table 13–4. The quotient is $x + 1$. In general,

$$p(x) = a_0 x^n + \cdots + a_n = (b_0 x^{n-2} + \cdots + b_{n-2})(x^2 - rx - q) + Rx + S.$$
$$(13\text{-}34)$$

Table 13-4

	1	3	7	5
-5			-5	-5
-2		-2	-2	
	1	1	0	0

The procedure is also of value in applying Newton's method to improve the accuracy of a complex root. If $a + bi$ is an estimate for such a root, then we form the corresponding quadratic polynomial

$$[x - (a + bi)][x - (a - bi)] = (x - a)^2 + b^2 = x^2 - 2ax + a^2 + b^2,$$

so that $r = 2a$ and $q = -(a^2 + b^2)$. We then use synthetic division as in Table 13–3 to obtain R and S and the polynomial $p_1(x) = b_0 x^{n-2} + \cdots + b_n$. Then we verify from (13–34) that

$$p(a + bi) = Ra + S + iRb,$$
$$p'(a + bi) = R + 2ibp_1(a + bi).$$
$$(13\text{-}35)$$

With these values, the next estimate $a + bi - [p(a + bi)/p'(a + bi)]$ can be found as in the real case.

EXAMPLE 4 We use the same equation as in Example 3, but stop at $k = 4$, expecting a poor approximation. The quadratic is

$$\begin{vmatrix} z^2 & z & 1 \\ 2 & -3 & 1 \\ 10 & 2 & -3 \end{vmatrix} = 7z^2 + 16z + 34 = 0.$$

The roots are $-1.1 \pm 1.9i$. We use Newton's method to correct these. With $a = -1.1, b = 1.9$, we obtain $r = -2.2, q = -4.82$. We then carry out synthetic division

Table 13–5

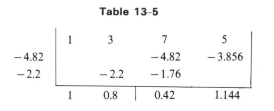

	1	3	7	5
-4.82			-4.82	-3.856
-2.2		-2.2	-1.76	
	1	0.8	0.42	1.144

as in Table 13–5, obtaining $R = 0.42$, $S = 1.144$, $p_1(x) = x + 0.8$. Equation (13–35) now gives

$$p(-1.1 + 1.9i) = 0.682 + 0.798i,$$
$$p'(-1.1 + 1.9i) = -6.8 - 1.14i.$$

Then by Newton's formula our next approximation is

$$-1.1 + 1.9i + \frac{0.682 + 0.798i}{6.8 + 1.14i} = -0.983 + 1.998i.$$

This is much closer to the true value $-1 + 2i$. ◄

Bernoulli's method has been extended (especially by Rutishauser) to yield approximations for *all* the roots of the equation. The procedure, known as the *quotient–difference scheme*, is described by HENRICI.

One other way of finding real or complex roots is Muller's method, described in the next section.

======= **Problems (Section 13–3)** =======

1. Use synthetic division to verify that $p(x) = x^4 - 18x^2 + 32x - 15$ has roots $1, 1, 3, -5$ and to obtain the corresponding factorization.

2. Evaluate with the aid of synthetic division:
   a) $p(3)$ for $p(x) = 7x^3 + 5x^2 + 2x + 1$
   b) $p(2)$ for $p(x) = x^4 - 11x^3 + 7x^2 - 5x + 2$
   c) $p(7)$ and $p'(7)$ for $p(x) = 2x^3 - x^2 + 1$
   d) $p(5)$ and $p'(5)$ for $p(x) = 7x^4 + 3x^3 - 2x + 3$

3. Obtain the real roots to within 0.25 by the method of bisection:
   a) $x^3 + 4x^2 - x - 3 = 0$ **b)** $5x^3 - 5x^2 + x - 2 = 0$
   c) $x^4 - x - 1 = 0$ **d)** $x^5 + 2x^3 - 1 = 0$

4. Use Newton's method with the given starting value x_1 to obtain approximations x_2, x_3 to a root of the given equation:
   a) $x^3 + x^2 - 3 = 0$, $x_1 = 2$
   b) $x^3 + 2x^2 + x - 1 = 0$, $x_1 = 1$
   c) $e^x - 2 + x = 0$, $x_1 = 0$
   d) $\sin x + x \sin x - 1 = 0$, $x_1 = 1$

5. For each of the following equations there is a single real root of largest absolute value. Use Bernoulli's method to find the root to three significant figures:
   a) $x^3 - 5x^2 - x + 5 = 0$ **b)** $x^3 - 7x^2 - x + 6 = 0$
   c) $x^4 - 8x^3 - 8x^2 - 8x - 8 = 0$ **d)** $45x^4 - 36x^3 - 8x^2 - 4x - 1 = 0$

6. For each of the following equations there is a pair of conjugate complex roots of maximum absolute value. Use Bernoulli's method to find these roots to three significant figures:
   a) $x^3 - 3x^2 + x + 5 = 0$
   b) $2x^5 + x^4 + 20x^3 + 10x^2 + 18x + 9 = 0$

7. Use synthetic division to carry out the indicated division:
   a) $(x^3 + 5x^2 + 3x + 2) \div (x^2 - 5x - 6)$
   b) $(x^5 + 7x^4 - x^3 + 2x^2 - 3x + 3) \div (x^2 - x - 2)$

8. Use Newton's method once to correct the given approximate value $a + bi$ for a complex root of the given polynomial:
   a) $x^3 - 3x^2 + x + 5 = 0$, $a + bi = 2.5 + 1.3i$ (root is $2 + i$),
   b) $2x^5 + x^4 + 20x^3 + 10x^2 + 2x + 1 = 0$, $a + bi = 1 + 4i$ (root is $3i$).

13–4 MULLER'S METHOD

This is one more procedure for finding the roots of an equation $f(x) = 0$. We start with three different estimates x_0, x_1, x_2 for a root. Then, with $y = f(x)$, we evaluate $y_0 = f(x_0)$, $y_1 = f(x_1)$, $y_2 = f(x_2)$, so that (x_0, y_0), (x_1, y_1), (x_2, y_2) are three points on the graph of f. Next we determine a quadratic function $ax^2 + bx + c$ whose graph passes through the same three points (we are thus interpolating by a quadratic). The quadratic equation $ax^2 + bx + c = 0$ has two roots. We choose one of these, x^*, according to a definite rule. Then we replace one of x_0, x_1, x_2 by x^* to obtain a new triple, that is relabeled as x_0, x_1, x_2, and the process is repeated; in the replacement, we reject the x_i that is furthest from x^*.

We consider this in detail for the case of real roots; the procedure can be modified to take care of complex roots (the only additional information needed is a rule for choosing the root x^*).

Let then x_i, y_i be given as above ($i = 0, 1, 2$). We shall assume the x_i numbered so that $x_2 < x_0 < x_1$. By a translation, our quadratic function can be written as

$$a(x - x_0)^2 + b(x - x_0) + c.$$

We want this to agree with $y = f(x)$ for $x = x_i$ ($i = 0, 1, 2$). Hence $c = y_0$ and

$$a(x_1 - x_0)^2 + b(x_1 - x_0) + c = y_1,$$
$$a(x_2 - x_0)^2 + b(x_2 - x_0) + c = y_2.$$

Since $c = y_0$, we have two linear equations for a, b, for which the corresponding matrix is

$$\begin{bmatrix} (x_1 - x_0)^2 & x_1 - x_0 & y_1 - y_0 \\ (x_2 - x_0)^2 & x_2 - x_0 & y_2 - y_0 \end{bmatrix}. \tag{13–40}$$

These linear equations can be solved by elimination for a, b. Then our quadratic equation is

$$a(x - x_0)^2 + b(x - x_0) + c = 0,$$

and its solutions are

$$x^* = x_0 + \frac{-b \pm \sqrt{b^2 - 4ac}}{2a}.$$

We multiply numerator and denominator of the fraction by $-b \mp \sqrt{b^2 - 4ac}$ to write this in another way (which gives more accurate results in computation):

$$x^* = x_0 - \frac{2c}{b \pm \sqrt{b^2 - 4ac}}. \tag{13-41}$$

This equation defines our next approximation x^*, whereby we choose

$$\text{plus sign if } b \geqslant 0, \qquad \text{minus sign if } b < 0. \tag{13-41'}$$

We now form a new set x_0, x_1, x_2, using x^* as one of the three and the old x_0, x_j as the other two, where the old x_i rejected is further from x^* than is the old x_j. We repeat the process, obtaining successive values $x_1^*, x_2^*, \ldots$ Under reasonable conditions, these converge to a root.

EXAMPLE 1 $x^3 - 3x^2 - x + 2 = 0$. This is the same as Example 1 of Section 13–3. We seek the root near 3 and select $x_0 = 2$, $x_1 = 3$, $x_2 = 1$. With $y = x^3 - 3x^2 - x + 2$, we find $y_0 = -4$, $y_1 = -1$, $y_2 = -1$. Hence $c = -4$ and the equations for a, b have matrix (13–40) as follows:

$$\begin{bmatrix} 1 & 1 & 3 \\ 1 & -1 & 3 \end{bmatrix}$$

We solve easily to obtain $b = 0$, $a = 3$. Now (13–41), (13–41') give

$$x^* = 2 - \frac{-8}{\sqrt{48}} = 3.15.$$

Thus we reject 1 from our first set of x_i and take new x_i thus: $x_0 = 3$, $x_1 = 3.15$, $x_2 = 2$. Accordingly, $c = -1$ and a, b are determined by the matrix

$$\begin{bmatrix} 0.0225 & 0.15 & 1.338 \\ 1 & -1 & -3 \end{bmatrix}$$

We find $a = 5.148$, $b = 8.148$. Then (13–41), (13–41') give

$$x^* = 3 - \frac{-2}{8.148 + \sqrt{86.65}} = 3.115.$$

As in Section 13–3, this is correct to three decimal places.

The geometric meaning of the process is suggested in Fig. 13–2 that gives the graph of $y = f(x) = x^3 - 3x^2 - x + 2$ and also shows the graph (a parabola) of the first quadratic obtained: $y = 3(x - 2)^2 - 4$. We observe that, since $b = 0$, the parabola is symmetric about the line $x = 2$. The case $b = 0$ is a borderline one for the rule (11–41'). In that case the minus sign could also be used. This would give us a different choice for x^* above, namely 0.85; this value is a good approximation for the root of $f(x) = 0$ between 0 and 1. Repetition of the process with this choice of x^* gives a sequence converging to this root.

From the geometric interpretation it is clear that the process will do well if the three initial points (x_i, y_i) are chosen reasonably close to a point where the graph of

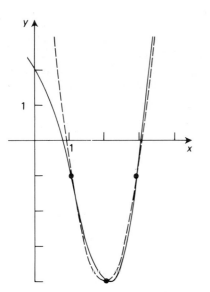

Fig. 13-2. Muller's method.

$y = f(x)$ crosses the x-axis. We can easily come up with examples and choices of (x_i, y_i) for which the parabola does not yield a better approximation; in fact, the parabola can fail to meet the x-axis ($b^2 - 4ac < 0$). Thus some caution is needed. However, experience with the method has been good. It has the advantage of simplicity. Also computations of functional values $f(x_i)$ are minimized, and the derivative is not needed, while in Newton's method it is needed. As remarked above, the process is easily adapted to finding complex roots. (For further discussion, see GERALD.) ◄

13-5 FINDING EIGENVALUES AND EIGENVECTORS

Let A be a square matrix. Then, as in Section 5-7, the eigenvalues of A are the solutions λ of the characteristic equation

$$\det (A - \lambda I) = 0. \tag{13-50}$$

If A is $m \times n$, this is an equation of degree n. Each root λ of this equation is an eigenvalue, and for such a λ there are corresponding eigenvectors $\mathbf{v}$, that is, vectors $\mathbf{v}$ such that $\mathbf{v} \neq \mathbf{0}$ and such that

$$A\mathbf{v} = \lambda\mathbf{v}. \tag{13-51}$$

It may happen that λ is complex; in that case, $\mathbf{v}$ will normally have complex components.

If an eigenvalue λ is found, then determination of the corresponding eigenvectors $\mathbf{v}$ is, by (13-51), equivalent to finding the nontrivial solutions of

$$(A - \lambda I)\mathbf{v} = \mathbf{0}, \tag{13-52}$$

that is, of n homogeneous equations in n unknowns. This can be carried out as in Section 13–2. Thus the principal difficulty is in finding the eigenvalues, especially if n is large. We here stress this aspect of the problem. Below we also describe an iteration method for finding eigenvalues and eigenvectors at the same time.

The natural way of solving (13–50) would be to expand the determinant and explicitly find the algebraic equation. For example, in case of a typical 3×3 matrix we would proceed as follows:

$$\begin{vmatrix} -3-\lambda & 1 & 2 \\ 3 & -\lambda & 1 \\ -2 & 1 & 4-\lambda \end{vmatrix} = 0, \quad (-3-\lambda)\begin{vmatrix} -\lambda & 1 \\ 1 & 4-\lambda \end{vmatrix} - 3\begin{vmatrix} 1 & 2 \\ 1 & 4-\lambda \end{vmatrix} - 2\begin{vmatrix} 1 & 2 \\ -\lambda & 1 \end{vmatrix} = 0.$$

Here we expanded by minors of the first column. Evaluating the 2×2 determinants and collecting terms, we finally obtain

$$-\lambda^3 + \lambda^2 + 12\lambda - 5 = 0. \tag{13–53}$$

This works well for $n = 3$ and would be manageable even with numbers with many digits. However, for large n, the expansion by minors requires a very large number of steps ($n!$, to be precise). More important, in the process of computing the coefficients there is inevitably a great loss of accuracy due to round off. Hence the expansion is not used, except for n small (say, up to 5).

An alternative method is to apply the Muller procedure of Section 13-4 directly to Eq. (13–50), which we write as $f(\lambda) = 0$. Thus we take three starting values λ_0, λ_1, λ_2 and proceed as in Section 13-4. We must now evaluate the $f(\lambda_i)$ and $f(\lambda_i) = \det(A - \lambda_i I)$. However, now the λ_i are given numbers and we can use elimination, as in Section 13–2. We illustrate the process by an example.

EXAMPLE 1 $A = \begin{bmatrix} -3 & 1 & 2 \\ 3 & 0 & 1 \\ -2 & 1 & 4 \end{bmatrix}$.

This is the same as the matrix of the example above, with characteristic equation (13–53). However, we proceed as described, without expanding, and let

$$f(\lambda) = \det(A - \lambda I) = \begin{vmatrix} -3-\lambda & 1 & 2 \\ 3 & -\lambda & 1 \\ -2 & 1 & 4-\lambda \end{vmatrix}.$$

We start with $\lambda_0 = 0$, $\lambda_1 = 1$, $\lambda_2 = -1$. (We comment below on how to find starting values.) Then

$$y_0 = f(\lambda_0) = f(0) = \begin{vmatrix} -3 & 1 & 2 \\ 3 & 0 & 1 \\ -2 & 1 & 4 \end{vmatrix} = \begin{vmatrix} -3 & 1 & 2 \\ 0 & 1 & 3 \\ 0 & 0.333 & 2.67 \end{vmatrix}$$

$$= \begin{vmatrix} -3 & 1 & 2 \\ 0 & 1 & 3 \\ 0 & 0 & 1.67 \end{vmatrix} = -5.01.$$

(Of course we could obtain the exact value, -5, but wish to illustrate the standard procedure that has round-off errors.) Similarly, we find $y_1 = 7, y_2 = -15$ (no round off). Thus the Muller procedure leads to the matrix

$$\begin{bmatrix} 1 & 1 & 12.01 \\ 1 & -1 & -9.99 \end{bmatrix}$$

for a, b, and we know $c = -5.01$. We find $a = 1.01, b = 11$. Thus

$$\lambda^* = -\frac{-10.02}{11 + \sqrt{141.24}} = 0.438.$$

We now repeat the process, using $\lambda_0 = 0.438, \lambda_1 = 1, \lambda_2 = 0$, and obtaining a new $\lambda^* = 0.4078$.

To find the remaining eigenvalues, we could proceed in the same way; we only need proper starting values. It is also tempting to try to factor out the known root 0.4078. We cannot use synthetic division, since $f(\lambda)$ is known only as a determinant.

We combine the two goals of the preceding paragraph by letting

$$f_1(\lambda) = \frac{f(\lambda)}{\lambda - 0.4078}.$$

Even though we cannot divide out on the right, we can use the equation to *evaluate* the reduced polynomial $f_1(\lambda)$ (of degree one less than that of f). We use Muller's procedure on $f_1(\lambda)$; since the degree is lower than that of f, the procedure should work even better. At least, it will be easier to work with $f_1(\lambda)$ until we have a close approximation to the root; then we can return to $f(\lambda)$ for further correction.

As starting values we try $\lambda_2 = 3, \lambda_0 = 4, \lambda_1 = 5$. Thus we compute

$$y_0 = f_1(\lambda_0) = f_1(4) = \frac{f(4)}{4 - 0.4078}$$

and, similarly, compute $y_1 = f_1(\lambda_1), y_2 = f(\lambda_2)$. The work is as before, and after two stages produces $\lambda^* = 3.8071$. We can now divide out both roots:

$$f_2(\lambda) = \frac{f(\lambda)}{(\lambda - 0.4078)(\lambda - 3.8071)}$$

and proceed similarly. If the roots were precise, $f_2(\lambda)$ would be a linear function, and we would obtain $a = 0$ for a "quadratic" interpolating polynomial. In general, when we know all but two of the roots, we can use two rules of algebra to obtain the missing ones: the sum of the roots of $a_0\lambda^n + \cdots + a_n = 0$ is $-a_1/a_0$; the product of the roots is $(-1)^n a_n/a_0$. These follow from writing the equation as

$$a_0(\lambda - \lambda_1) \cdots (\lambda - \lambda_n) = 0$$

and considering the coefficients of λ^{n-1}, λ^0. For a characteristic equation

$$\begin{vmatrix} a_{11} - \lambda & a_{12} & \cdots \\ a_{21} & a_{22} - \lambda & \cdots \\ \cdots & \cdots & \cdots \end{vmatrix} = 0$$

we see easily that $a_0 = (-1)^n$, $a_1 = (-1)^{n+1}(a_{11} + \cdots + a_{nn})$, $a_n = \det A$. Therefore,

$$\lambda_1 + \lambda_2 + \cdots + \lambda_n = a_{11} + \cdots + a_{nn},$$
$$\lambda_1 \lambda_2 \cdots \lambda_n = \det A. \tag{13-54}$$

(The expression $a_{11} + a_{22} + \cdots + a_{nn}$ is known as the *trace* of matrix A). If now all but two of $\lambda_1, \ldots, \lambda_n$ are known, then Eq. (13-54) provides two equations for the missing roots. If, say, λ_{n-1}, λ_n are not known, then

$$\lambda_{n-1} + \lambda_n = B, \qquad \lambda_{n-1}\lambda_n = C$$

are known, and λ_{n-1}, λ_n are the roots of the quadratic equation

$$\lambda^2 - B\lambda + C = 0.$$

By this method or direct solution, as above, we ultimately obtain the three eigenvalues of the given matrix: 0.4078, 3.8071, -3.2149. ◄

Starting values for eigenvalues Here a *theorem of Gershgorin* is useful (see pp. 636–637 of *Applied and Computational Complex Analysis* by P. Henrici, vol. 1, Wiley, New York, 1974). This asserts that if λ is an eigenvalue of the $n \times n$ matrix $A = (a_{ij})$, then at least one of the following relations must hold:

$$|a_{11} - \lambda| \leqslant |a_{12}| + |a_{13}| + \cdots + |a_{1n}|,$$
$$|a_{22} - \lambda| \leqslant |a_{21}| + |a_{23}| + \cdots + |a_{2n}|,$$
$$\vdots \tag{13-55}$$
$$|a_{nn} - \lambda| \leqslant |a_{n1}| + |a_{n2}| + \cdots + |a_{n,n-1}|.$$

This rule is known as the *circle theorem* since in the complex case it says that the eigenvalues are contained in the union of the n circular regions (13-55). When the eigenvalues are real, the regions become closed intervals on the real λ-axis.

For Example 1, Eq. (13-55) becomes

$$|-3 - \lambda| \leqslant 1 + 2 = 3, \qquad |-\lambda| \leqslant 3 + 1 = 4, \qquad |4 - \lambda| \leqslant 2 + 1 = 3.$$

Since the roots happen to be real, they must lie in the union of the intervals $-6 \leqslant \lambda \leqslant 0$, $-4 \leqslant \lambda \leqslant 4$, $1 \leqslant \lambda \leqslant 7$. Hence starting values should be chosen between -6 and 7.

We illustrate finding an eigenvector by taking $\lambda = 0.4078$ in Example 1. For this value of λ, $A - \lambda I$ becomes

$$\begin{bmatrix} -3.4078 & 1 & 2 \\ 3 & -0.4078 & 1 \\ -2 & 1 & 3.5922 \end{bmatrix}.$$

This is then the coefficient matrix for the equation $(A - \lambda I)\mathbf{v} = \mathbf{0}$. We reduce by Gauss elimination to

$$\begin{bmatrix} 1 & -0.2934 & -0.5868 \\ 0 & 1 & 5.8433 \\ 0 & 0 & 0.0041 \end{bmatrix}.$$

We know that, if λ were a precise eigenvalue, then det $(A - \lambda I)$ would be 0. Hence the last row should consist solely of zeros. We thus ignore this row and obtain $v = (v_1, v_2, v_3)$ from the first two rows. They give $v_2 = -5.8433v_3$, $v_1 = -1.1276v_3$, with v_3 arbitrary. Thus the eigenvectors are the nonzero scalar multiples of $(-1.1276, -5.8433, 1)$.

Iteration method   This is a method used normally for finding real eigenvalues and associated eigenvectors of matrix A. We select an arbitrary nonzero initial vector v_0 and form $v_1 = Av_0$, $v_2 = Av_1 = A^2v_0, \ldots, v_k = A^kv_0$. Then we divide each of these vectors by the component of largest absolute value to obtain vectors u_0, $u_1, \ldots, u_k, \ldots$ This sequence of vectors should approach an eigenvector as limit. Convergence is assured if A has a real eigenvalue λ greater in absolute value than all other eigenvalues (as in the Bernoulli method of Section 13–3). The idea can be made clear by the following simple example.

EXAMPLE 2 $A = \begin{bmatrix} 5 & 0 \\ 0 & 1 \end{bmatrix}$. So A has eigenvalues 5, 1. We take $v_0 = \text{col}\,(1, 1)$, so that

$$v_1 = \begin{bmatrix} 5 & 0 \\ 0 & 1 \end{bmatrix}\begin{bmatrix} 1 \\ 1 \end{bmatrix} = \begin{bmatrix} 5 \\ 1 \end{bmatrix}, \quad v_2 = \begin{bmatrix} 25 \\ 1 \end{bmatrix}, \quad v_3 = \begin{bmatrix} 125 \\ 1 \end{bmatrix}, \quad \ldots, \quad v_k = \begin{bmatrix} 5^k \\ 1 \end{bmatrix}$$

and hence (as column vectors)

$$u_0 = (1, 1), \quad u_1 = (1, 0.2), \quad u_2 = (1, 0.04), \quad u_3 = (1, 0.008), \quad \ldots$$

In general, $u_k = (1, 5^{-k})$ and u_k converges to $(1, 0)$, which is an eigenvector of A associated with the eigenvalue 5, largest in absolute value. As the calculation shows, at large k, A^kv_0 is influenced mainly by the largest (in absolute value) eigenvalue of A. ◀

EXAMPLE 3 $A = \begin{bmatrix} 2 & 6 \\ 1 & 1 \end{bmatrix}$. With $v_0 = \text{col}\,(1, 1)$, we find $v_1 = Av_0 = \text{col}\,(8, 2)$, $v_2 = Av_1 = \text{col}\,(28, 10)$, $v_3 = Av_2 = (116, 38)$, $v_4 = \text{col}\,(460, 154)$; thus, as column vectors, $u_0 = (1, 1)$, $u_1 = (1, 0.25)$, $u_2 = (1, 0.36)$, $u_3 = (1, 0.328)$, $u_4 = (1, 0.3348)$. This appears to be converging. It can be verified that $v = \text{col}\,(1, 0.333 \ldots)$ is an eigenvector and $Av = 4v$.

Although convergence is rapid in this example, in practice it is often found to be slow. The method can be extended to finding the eigenvalue of smallest absolute value, also to finding all eigenvalues, real and complex.

===============Problems (Section 13–5)===============

1. Use Muller's method in the following:
   a) Find a root of $5x^3 - x^2 - 6x + 1 = 0$ using starting values $x_0 = 2$, $x_1 = 3$, $x_2 = 1$.
   b) Find a root of $x^3 - 2x - 5 = 0$ using starting values $x_0 = 2$, $x_1 = 3$, $x_2 = 1$.
   c) Find all real roots of $x^5 + x^3 - 1 = 0$.
   d) Find all real roots of $xe^x - 1 = 0$.

2. Find all eigenvalues with the aid of Muller's method:

a) $\begin{bmatrix} 15 & 1 & 0 \\ 1 & 7 & -1 \\ 1 & 0 & 4 \end{bmatrix}$ b) $\begin{bmatrix} 0 & 2 & 1 \\ 2 & 3 & 4 \\ 1 & 4 & 5 \end{bmatrix}$

HINT for (a): Use $\lambda_0 = 4$, $\lambda_1 - 5$, $\lambda_2 = 3$ as starting values.

3. a) Apply Gershgorin's theorem to restrict the location of the eigenvalues for the matrix of Problem 2(a). *Note*: This matrix has exceptionally large diagonal values compared to the other entries; we say that it has *dominant diagonal*. In such a case, the theorem is especially helpful.
   b) Apply Gershgorin's theorem to the matrix

$$\begin{bmatrix} 7 & 0 & 1 \\ 2 & 5 & 0 \\ 1 & 3 & 8 \end{bmatrix}$$

   (eigenvalues are 9.2, $5.4 \pm 1.2i$).

4. Find eigenvectors corresponding to the eigenvalues for each matrix by using answers to Problem 2:
   a) As in Problem 2(a)   b) As in Problem 2(b)

5. Use iteration to find an eigenvalue and corresponding eigenvector for
   a) The matrix of Problem 2(a)   b) The matrix of Problem 2(b).

13–6 INTERPOLATION BY POLYNOMIALS

If a function $f(x)$ is known for several values of x, we can find a polynomial that agrees with f for these values and use the polynomial to approximate the values of f at other choices of x. The polynomial chosen is said to be an *interpolating polynomial* for f at the chosen values of x. (Technically we are *extrapolating* if we use the polynomial to approximate f outside of the smallest interval containing the given value of x.)

A very common case of interpolation is *linear interpolation*; here $f(x_0), f(x_1)$ are used to obtain the linear function

$$y = \frac{f(x_1) - f(x_0)}{x_1 - x_0}(x - x_0) + f(x_0).$$

This assigns to an x halfway between x_0 and x_1 the value y halfway between $f(x_0)$ and $f(x_1)$.

Quadratic interpolation is used in Muller's method (Section 13–4). There the values $f(x_0), f(x_1), f(x_2)$ were used to obtain a polynomial

$$y = a(x - x_0)^2 + b(x - x_0) + c.$$

These examples suggest that, *given* $f(x_0), f(x_1), \ldots, f(x_n)$, *where* $x_0, x_1, \ldots, x_n$ *are all different, there is a unique polynomial* $p_n(x)$ *of degree at most* n *such that*

$$p_n(x_i) = f(x_i) \qquad \text{for} \qquad i = 0, 1, \ldots, n. \tag{13–60}$$

If we write $p_n(x) = a_0 x^n + \cdots + a_n$, then Eq. (13–60) is equivalent to $n+1$ simultaneous linear equations for the $n+1$ coefficients $a_0, \ldots, a_n$. In the special case when all $f(x_i)$ are 0, these equations are *homogeneous* and the equations are satisfied only for $a_0 = 0, \ldots a_n = 0$, since a polynomial of degree at most n has at most n zeros, except when the polynomial is the zero polynomial. Thus the determinant of coefficients in (13–60) is not 0 (Section 5–1) and the equations have a unique solution for $a_0, \ldots, a_n$. If a_n happens to be 0, $p_n(x)$ has degree less than n.

The argument just given also shows that to find the interpolating polynomial we can simply solve $n+1$ linear equations. (That is what we did for Muller's method.) However, there are simpler ways, especially for large n. We here present one method, that of *Newton*: $p_n(x)$ is given as follows:

$$p_n(x) = f[x_0] + f[x_0, x_1](x - x_0) + f[x_0, x_1, x_2](x - x_0)(x - x_1)$$

$$+ \cdots + f[x_0, x_1, \ldots, x_n](x - x_0) \cdots (x - x_{n-1}). \tag{13-61}$$

Here

$$f[x_0] = f(x_0), \qquad f[x_0, x_1] = \frac{f(x_1) - f(x_0)}{x_1 - x_0}, \tag{13-62}$$

and in general

$$f[x_0, x_1, \ldots, x_k] = \frac{f[x_1, \ldots, x_k] - f[x_0, \ldots, x_{k-1}]}{x_k - x_0}, \qquad k = 1, 2, \ldots \tag{13-63}$$

Formula (13–63) allows us to compute successively $f[x_0, x_1]$, $f[x_0, x_1, x_2]$, $\ldots$, $f[x_0, x_1, \ldots, x_k]$ for any given k, where $f[x_0, \ldots, x_k]$ is called the *divided difference* of f at $x_0, \ldots, x_k$.

We can verify that the divided difference does not depend on the order of numbering $x_0, \ldots, x_k$. For example, $f[x_0, x_1, x_2] = f[x_1, x_2, x_0]$.

Equation (13–61) clearly defines $p_n(x)$ as a polynomial of degree at most n. Also we see at once that $p_n(x_0) = f(x_0)$. The condition $p_n(x_1) = f(x_1)$ is, by (13–62), the same as

$$f(x_1) = f(x_0) + \frac{f(x_0) - f(x_1)}{x_1 - x_0}(x_1 - x_0),$$

and this is true. Similarly, with the aid of (13–63), we verify that $p_n(x_i) = f(x_i)$ for $i = 0, 1, \ldots, n$.

The computation of the divided differences can be done systematically, as in Table 13–6, in which we omit the symbol f before the brackets, writing $[x_0, x_1, \ldots, x_k]$ for $f[x_0, x_1, \ldots, x_k]$. For each particular $[x_l, \ldots, x_k]$ we can pursue the two diagonals back to the left, ending at $[x_l]$, $[x_k]$; this tells us that $[x_l, \ldots, x_k]$ is obtained by dividing the difference of the immediate predecessors by $x_k - x_l$. In the table this is done for $[x_0, x_1, \ldots, x_4]$, showing that this is obtained by dividing $[x_1, x_2, x_3, x_4] - [x_0, x_1, x_2, x_3]$ by $x_4 - x_0$.

Table 13-6

x_0--$[x_0]$
$\quad\quad [x_0, x_1]$
$x_1\ [x_1]\quad\quad\quad [x_0, x_1, \bar{x}_2]$
$\quad\quad [x_1, x_2]\quad\quad\quad [x_0, x_1, \bar{x_2}, x_3]$
$x_2\ [x_2]\quad\quad\quad [x_1, x_2, x_3]\quad\quad\quad [x_0, x_1, \bar{x}_2, x_3, x_4]$
$\quad\quad [x_2, x_3]\quad\quad\quad [x_1, x_2, x_3, \bar{x}_4]\quad\quad\quad\quad\quad\quad [x_0, x_1, x_2, x_3, x_4, x_5]$
$x_3\ [x_3]\quad\quad\quad [x_2, x_3, x_4]\quad\quad\quad [x_1, x_2, x_3, x_4, x_5]$
$\quad\quad [x_3, x_4]\quad\quad\quad [x_2, x_3, x_4, x_5]$
x_4--$[x_4]$--$\quad [x_3, x_4, x_5]$
$\quad\quad [x_4, x_5]$
$x_5\ [x_5]$

EXAMPLE 1 Find the interpolating polynomial of degree at most four if $f(0) = 0$, $f(1) = 1$, $f(2) = 16$, $f(3) = 81$, $f(4) = 256$. The solution is illustrated in Table 13–7. By Eq. (13–61) and Table 13-6, the diagonal marked in Table 13-7 gives the successive coefficients in $p_4(x)$:

$$p_4(x) = 0 + 1(x - 0) + 7(x - 0)(x - 1) + 6(x - 0)(x - 1)(x - 2)$$
$$+ 1(x - 0)(x - 1)(x - 2)(x - 3).$$

This simplifies to $p_4(x) = x^4$, which agrees with the given values. ◄

From Table 13–7 we could of course get other information. For example, the $p_3(x)$ agreeing with f at 1, 2, 3, 4 is

$$p_3(x) = 1 + 15(x - 1) + 25(x - 1)(x - 2) + 10(x - 1)(x - 2)(x - 3).$$

Table 13-7

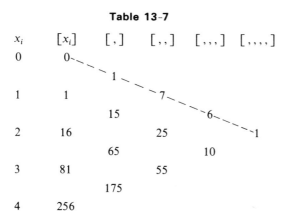

x_i	$[x_i]$	$[\,,]$	$[\,,,]$	$[\,,,,]$	$[\,,,,,]$
0	0				
		1			
1	1		7		
		15		6	
2	16		25		1
		65		10	
3	81		55		
		175			
4	256				

It should be observed that Table 13-6 can be built up from the first two given columns by calculating $[x_0, x_1]$, then $[x_1, x_2]$, then $[x_0, x_1, x_2]$; next $[x_2, x_3]$, then

$[x_1, x_2, x_3]$, then $[x_0, x_1, x_2, x_3]$, and so on. Thus we are successively adding the diagonals sloping up to the right. This procedure is of value if we are trying to interpolate to f at a particular new point $\bar{x}$. Then we can compute $p_k(\bar{x})$ for successive choices $k = 1, 2, \ldots$, using $x_0, x_1, \ldots, x_k$. Thus we continually add an x-value and raise the degree by one. We stop when we are satisfied that $p_k(\bar{x})$ is close enough to the "correct" value of f.

If this procedure is followed, then after the diagonal starting at x_k has been found, we calculate $\psi_k(\bar{x})$ and $p_k(\bar{x})$ by the recursion rules

$$\psi_k(\bar{x}) = \psi_{k-1}(\bar{x})\,(\bar{x} - x_{k-1}), \qquad \psi_0(\bar{x}) = 1,$$
$$p_k(\bar{x}) = p_{k-1}(\bar{x}) + f[x_0, \ldots, x_k]\psi_k(\bar{x}), \qquad p_0(\bar{x}) = f(x_0). \tag{13–64}$$

We carry this out for an extension of Example 1, using $\bar{x} = 2.5$, as shown in Table 13–8. Here we calculate $[x_0, x_1] = 1$, then

$$\psi_1(2.5) = \psi_0(2.5)\,(2.5 - 0) = 2.5$$

and

$$p_1(2.5) = p_0(2.5) + [x_0, x_1]\psi_1(2.5) = 2.5.$$

Next we calculate $[x_1, x_2] = 15$, $[x_0, x_1, x_2] = 7$; then

$$\psi_2(2.5) = \psi_1(2.5)\,(2.5 - 1) = 3.75$$

and

$$p_2(2.5) = p_1(2.5) + [x_0, x_1, x_2]\psi_2(2.5) = 28.75.$$

Thus up to this point we have used only the data above the dashed broken line in Table 13–8. At the next stage we move down one step.

Table 13–8

k	$p_k(2.5)$	$\psi_k(2.5)$	x_k	$[\ ]$	$[,]$	$[,,]$	$[,,,]$	$[,,,,]$	$[,,,,,]$
0	0	1	0						
					1				
1	2.5	2.5	1	1		7			
					15		6		
2	28.75	3.75	2	16		25		1	
					65		10		0
3	40	1.875	3	81		55		1	
					175		14		
4	39.0625	−0.9375	4	256		97			
					369				
5	39.0625		5	625					

We observe that the successive values of $p_k(2.5)$ seem to settle on 39.0625 (the exact value of 2.5^4). Thus by gradually increasing the degree and the number of points used we may be able to achieve high accuracy. However, success is not guaranteed! The process is most effective on a table of a standard function, such as e^x. For an empirical function, results may fluctuate widely.

Interpolation using derivatives

It can be shown that for a function f with continuous derivatives through the kth order near a point $\bar{x}$, $f[x_0, x_1, \ldots, x_k]$ has a limit as all $x_0, x_1, \ldots, x_k$ approach $\bar{x}$:

$$\lim f[x_0, x_1, \ldots, x_k] = \frac{f^{(k)}(\bar{x})}{k!} \quad \text{as} \quad x_0, x_1, \ldots, x_k \to \bar{x}. \quad (13\text{–}65)$$

By the mean-value theorem, this is clear for $k = 1$:

$$f[x_0, x_1] = \frac{f(x_1) - f(x_0)}{x_1 - x_0} = \frac{f'(\xi)(x_1 - x_0)}{x_1 - x_0} = f'(\xi),$$

where ξ is between x_0 and x_1. As x_0 and x_1 approach $\bar{x}$, ξ must approach $\bar{x}$ and the rule follows. A similar argument gives the rule for all k.

By this rule, it is now possible to define $f[x_0, \ldots, x_k]$ when several of the x_i coincide. For example, if $x_0 = x_1 < x_2$, then

$$f[x_0, x_1] = f'(x_0), \qquad f[x_1, x_2] = \frac{f(x_2) - f(x_1)}{x_2 - x_1},$$

$$f[x_0, x_1, x_2] = \frac{f[x_1, x_2] - f[x_0, x_1]}{x_2 - x_0} = \frac{f[x_1, x_2] - f'(x_0)}{x_2 - x_0}, \quad (13\text{–}66)$$

where $x_1 = x_0$. The Newton formula (13–61) *remains valid when some x_i coincide*. It gives a polynomial $p_n(x)$ of degree at most n which agrees with $f(x_i)$ at each x_i and is such that $p'_n(x_i) = f'(x_i), \ldots, p_n^{(k)}(x_i) = f^{(k)}(x_i)$ whenever $k + 1$ of the points have one value x_i.

If all $n + 1$ points coincide, the Newton formula reduces to the Taylor formula:

$$p_n(x) = f(x_0) + f'(x_0)(x - x_0) + \cdots + \frac{f^{(n)}(x_0)}{n!}(x - x_0)^n.$$

This defines $p_n(x)$ as the beginning of a Taylor series for f, and we see at once that $p_n^{(k)}(x_0) = f^{(k)}(x_0)$ for $k = 0, 1, \ldots, n$.

As another example, we obtain $p_2(x)$ for given $x_0 = x_1 < x_2$. Then formulas (13–66) apply and we obtain

$$p_2(x) = f(x_0) + f'(x_0)(x - x_0) + \frac{f[x_0, x_2] - f'(x_0)}{x_2 - x_0}(x - x_0)^2 \quad (13\text{–}67)$$

This provides a polynomial $p_2(x)$ with given values at x_0, x_2 and given derivative $f'(x_0)$ at x_0.

As a final example, we find a cubic $p_3(x)$ with given values and first derivatives at x_0, x_2. We take $x_0 = x_1, x_2 = x_3$, so that $f[x_0, x_1] = f'(x_0), f[x_2, x_3] = f'(x_2)$. Then

$$f[x_0, x_1, x_2] = \frac{f[x_1, x_2] - f'(x_0)}{x_2 - x_0}, \qquad f[x_1, x_2, x_3] = \frac{f'(x_2) - f[x_1, x_2]}{x_3 - x_1}.$$

Thus with $x_3 - x_0 = x_2 - x_0 = x_3 - x_1 = b$,

$$f[x_0, x_1, x_2, x_3] = \frac{f'(x_2) - 2f[x_1, x_2] + f'(x_0)}{b^2},$$

$$p_3(x) = f(x_0) + f'(x_0)(x - x_0) + \frac{f[x_1, x_2] - f'(x_0)}{b}(x - x_0)^2$$

$$+ \frac{f'(x_2) - 2f[x_1, x_2] + f'(x_0)}{b^2}(x - x_0)^2(x - x_2). \qquad (13\text{–}68)$$

Table 13-9

x_i	$f(x_i)$	$[\,,]$	$[\,,,]$	$[\,,,,]$
0	3			
		5		
0	4		-1	
		4		9
1	7		8	
		12		
1	7			

The divided differences can be calculated in table form, as before. Thus for $f(0) = 3$, $f'(0) = 5$, $f(1) = 7$, $f'(1) = 12$, we obtain the Table 13–9 and read off:

$$p_3(x) = 3 + 5x - x^2 + 9x^2(x - 1) = 3 + 5x - 10x^2 + 9x^3.$$

We verify that $p_3(0) = 3$, $p_3'(0) = 5$, $p_3(1) = 7$, $p_3'(1) = 12$.

=========================== **Problems (Section 13-6)** ===========================

1. Find the interpolating polynomial of the given degree (or less) for the functional values given:
   a) $f(0) = 1$, $f(1) = 2$, $f(2) = 13$, $f(3) = 46$, degree 3;
   b) $f(3) = 100$, $f(4) = 258$, $f(6) = 940$, $f(9) = 3328$, degree 3;
   c) $f(0) = 6$, $f(1) = 2$, $f(2) = 2$, $f(3) = 42$, $f(4) = 182$, degree 4;
   d) $f(0) = 1$, $f(1) = 3$, $f(3) = 2$, $f(4) = 1$, $f(5) = 1$, degree 4.

2. Use the procedure of Table 13–8 to obtain successive values $p_k(\bar{x})$ of interpolating polynomials $p_k(x)$ for $\bar{x}$ and k as given:
   a) $f(0) = 3$, $f(1) = 3$, $f(3) = 105$, $f(4) = 315$, $f(5) = 743$, $f(6) = 1503$, $\bar{x} = 2$, $k = 1, 2, 3, 4, 5$;
   b) $f(1.0) = 2.7183$, $f(1.1) = 3.0042$, $f(1.2) = 3.3201$, $f(1.3) = 3.6693$, $f(1.4) = 4.0552$, $f(1.5) = 4.4817$, $\bar{x} = 1.25$, $k = 1, 2, 3, 4, 5$. (The values are those of e^x.)

3. Find the polynomial $p(x)$ as specified:
   a) of degree $\leqslant 2$, $p(1) = 4$, $p'(1) = 5$, $p(2) = 15$;
   b) of degree $\leqslant 2$, $p(3) = 21$, $p(5) = 29$, $p'(5) = 4$;
   c) of degree $\leqslant 3$, $p(0) = 1$, $p'(0) = 0$, $p(1) = 0$, $p'(1) = 0$;
   d) of degree $\leqslant 3$, $p(4) = 1$, $p'(4) = 2$, $p''(4) = -1$, $p(5) = 0$.

4. Verify the rules:
   a) $f[x_0, x_1] = f[x_1, x_0]$ b) $f[x_0, x_1, x_2] = f[x_2, x_0, x_1]$
   c) $f[x_0, x_1, x_2, x_3] = f[x_3, x_2, x_1, x_0]$.

5. *Lagrange interpolation formula.* Given distinct values $x_0, \ldots, x_k$, we set

$$l_0(x) = \frac{(x - x_1)(x - x_2) \cdots (x - x_k)}{(x_0 - x_1)(x_0 - x_2) \cdots (x_0 - x_k)},$$

$$l_1(x) = \frac{(x - x_0)(x - x_2) \cdots (x - x_k)}{(x_1 - x_0)(x_1 - x_2) \cdots (x_1 - x_k)},$$

$$\cdots$$

$$l_k(x) = \frac{(x - x_0)(x - x_1) \cdots (x - x_{k-1})}{(x_k - x_0)(x_k - x_1) \cdots (x_k - x_{k-1})}.$$

   a) Verify that $l_i(x_j) = 0$ for $j \neq i$, $l_i(x_i) = 1$.
   b) Verify that $\sum_{i=1}^{k} f(x_i)l_i(x) = p_k(x)$ is a polynomial of degree at most k interpolating $f(x)$ at $x_0, \ldots, x_k$.
   c) Use these formulas to solve Problem 1(a) above.
   d) Use these formulas to solve Problem 1(b) above.
   e) Show that $l_i(x) = \dfrac{q(x)}{(x - x_i)q'(x_i)}$, where $q(x) = (x - x_0)(x - x_1) \cdots (x - x_k)$.

13–7 APPROXIMATION BY POLYNOMIALS

The interpolating polynomial $p_k(x)$ to $f(x)$ will give high accuracy at the x-values used (that is, $p_k(x_i) = f(x_i)$ at each point, within computational error). However, away from these values, there may be large errors, as suggested in Fig. 13–3. Thus the interpolating polynomial is not necessarily a good approximation to $f(x)$ over an interval $a \leqslant x \leqslant b$.

In this section we point out several other ways of approximating a function over such an interval by a polynomial (or some other useful simple function).

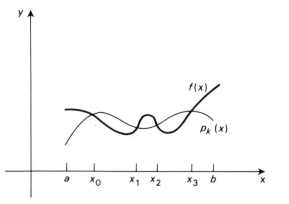

Fig. 13–3. Approximation by the interpolating polynomial.

Least squares approximation

In this method we choose a set of functions $\varphi_1(x), \ldots, \varphi_n(x)$ and seek coefficients $c_1, \ldots, c_n$ such that $c_1\varphi_1(x) + \cdots + c_n\varphi_n(x)$ approximates $f(x)$ in the sense that the total square error

$$E = \int_a^b \{f(x) - [c_1\varphi_1(x) + \cdots + c_n\varphi_n(x)]\}^2 dx \qquad (13\text{–}70)$$

is minimized; if $\varphi_1(x), \ldots, \varphi_n(x)$ are polynomials, we obtain a least-squares approximation of f by a polynomial. If $\varphi_1(x), \ldots, \varphi_n(x)$ are linearly independent over the interval $a \leqslant x \leqslant b$ and f is continuous (or piecewise continuous), then it can be shown that the minimum is attained at a unique set of values $c_1, \ldots, c_n$, which are the solutions of the simultaneous equations

$$\frac{\partial E}{\partial c_1} = 0, \quad \cdots, \quad \frac{\partial E}{\partial c_n} = 0. \qquad (13\text{–}71)$$

If we apply Leibnitz's rule for differentiating under the integral sign (Section 10–3), then equations (13–71) become simultaneous linear equations for $c_1, \ldots, c_n$:

$$\sum_{j=1}^n b_{ij} c_j = k_i, \qquad i = 1, \ldots, n,$$

$$\qquad (13\text{–}72)$$

$$b_{ij} = b_{ji} = \int_a^b \varphi_i(x)\varphi_j(x)\,dx, \qquad k_i = \int_a^b f(x)\varphi_i(x)\,dx.$$

We can express the b_{ij} and k_i as inner products, as in Section 3–10:

$$b_{ij} = (\varphi_i, \varphi_j), \qquad k_i = (f, \varphi_i). \qquad (13\text{–}73)$$

We observe from (13–72) or (13–73) that $B = (b_{ij})$ is a *symmetric* matrix (Section 5–15).

If the $\varphi_i(x)$ are pairwise *orthogonal* over the interval $a \leqslant x \leqslant b$, so that $b_{ij} = 0$ for $i \neq j$, then we are dealing with the methods of Sections 3–9 and 3–10. The equations (13–72) are then easy to solve and, as in Section 3–10, we obtain the c_i as *Fourier coefficients* of f with respect to the orthogonal system $\varphi_1(x), \ldots, \varphi_n(x)$. (The fact that we are dealing with a finite set of functions does not change the reasoning.) Thus

$$c_i = \frac{k_i}{b_{ii}} = \frac{(f, \varphi_i)}{(\varphi_i, \varphi_i)}, \qquad i = 1, \ldots, n. \qquad (13\text{–}74)$$

When $\varphi_1, \ldots, \varphi_n$ are not orthogonal, we can always "orthogonalize" them as in Section 3–11, that is, produce an orthogonal system $\psi_1, \ldots, \psi_n$ such that ψ_i is a linear combination of $\varphi_1, \ldots, \varphi_i$ for each i. Then the whole question of least-squares approximation can be restated in terms of $\psi_1, \ldots, \psi_n$. For this reason, in practice we generally use orthogonal functions.

If we have an infinite sequence of orthogonal functions $\{\varphi_n(x)\}$ that is *complete*, then, as in Section 3–10, the minimum square error E_n for approximation by $c_1\varphi_1 + \cdots + c_n\varphi_n$ approaches 0 as $n \to \infty$. Examples of complete systems are given in Chapters 3 and 12.

Weight function

As in Section 3–9, all the preceding discussion extends to the problem of minimizing the total weighted square error

$$E = \int_a^b \{f(x) - [c_1\varphi_1(x) + \cdots + c_n\varphi_n(x)]\}^2 w(x)\,dx, \qquad (13\text{–}70')$$

where the weight function $w(x)$ is an appropriate nonnegative function on the interval. The inner products are given by

$$(p, q) = \int_a^b p(x)\,q(x)\,w(x)\,dx.$$

EXAMPLE 1   Obtain least-squares approximations of form $c_0 + c_1 x$ and of form $c_0 + c_1 x + .c_2 x^2$ to $f(x) = e^x$ on the interval $-1 \leqslant x \leqslant 1$. Take $w(x) \equiv 1$.
Here $\varphi_0(x) = 1$, $\varphi_1(x) = x$, $\varphi_2(x) = x^2$ (we start numbering with 0, for convenience). Thus

$$b_{00} = \int_{-1}^1 1\,dx = 2, \qquad b_{01} = b_{10} = \int_{-1}^1 x\,dx = 0,$$

$$b_{02} = b_{20} = b_{11} = \int_{-1}^1 x^2\,dx = \frac{2}{3},$$

$$b_{12} = b_{21} = \int_{-1}^1 x^3\,dx = 0, \qquad b_{22} = \int_{-1}^1 x^4\,dx = \frac{2}{5},$$

$$k_0 = \int_{-1}^1 e^x\,dx = 2.3504, \qquad k_1 = \int_{-1}^1 x e^x\,dx = 0.7358,$$

$$k_2 = \int_{-1}^1 x^2 e^x\,dx = 0.8789,$$

Hence for the linear approximation, Eqs. (13–72) become

$$2c_0 = 2.3504, \qquad 0.6667c_1 = 0.7358,$$

and the linear approximation is $y = 1.1752 + 1.1036x$. For the quadratic approximation, the equations become

$$2c_0 + 0.6667c_2 = 2.3504, \qquad 0.6667c_1 = 0.7358, \qquad 0.6667c_0 + 0.4c_2 = 0.8789.$$

These are easily solved and give the quadratic approximation

$$y = 0.9961 + 1.1036x + 0.5372x^2.$$

Both the linear and quadratic approximations are shown in Fig. 13-4. ◀
The functions φ_0, φ_1, φ_2 of the example do not form an orthogonal system since $(\varphi_0, \varphi_1) = (\varphi_1, \varphi_2) = 0$ but $(\varphi_0, \varphi_2) \neq 0$. We can orthogonalize them as in Section 3–10 to obtain an orthogonal system formed of three polynomials of degrees 0, 1, 2; up to constant factors, these are Legendre polynomials. Hence we can also use the first three Legendre polynomials $P_0(x) = 1$, $P_1(x) = x$, $P_2(x) = (3x^2 - 1)/2$. We can thus seek our second-degree least-squares approxima-

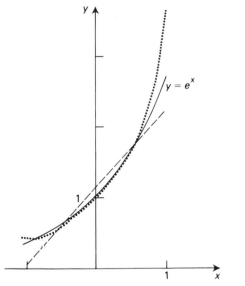

Fig. 13-4. Approximations to e^x.

tions in the form $c_0 P_0(x) + c_1 P_1(x) + c_2 P_2(x)$, where, in accordance with (11-74), $c_i = (f, P_i)/(P_i, P_i)$ (as in Section 3–12, where $(P_i, P_i) = 2/(2i+1)$). More simply, we can write

$$1 = P_0(x), \qquad x = P_1(x), \qquad x^2 = \frac{2}{3} P_2(x) + \frac{1}{3} = \frac{2}{3} P_2(x) + \frac{1}{3} P_0(x).$$

Then our previous polynomial $0.9961 + \cdots$ of degree 2 becomes

$$1.1752\, P_0(x) + 1.1036\, P_1(x) + 0.3579\, P_2(x).$$

For higher-degree approximations it is simpler to start with the expression $\sum c_i P_i(x)$ and use (13–74).

Discrete case

In many applications $f(x)$ is known, from computation or measurement, only for a finite set of values of x, say $x_0, \ldots, x_k$. The interpolating polynomial then provides values of f for other x. Without further knowledge, we cannot say that these values are in error. However, experience shows that the interpolating polynomial of high degree is not a reasonable way of completing the function from the limited initial knowledge; it tends to be highly sensitive to inaccuracies in $f(x_i)$ and to produce strange oscillations. Furthermore, if f is given for, say, 100 values of x, we would be working with a polynomial of degree 99; this is cumbersome in itself, but the function f might easily be quite simple (say, approximately linear) and an approximation by a polynomial of degree 99 would shed no light on this simplicity.

Accordingly, it is standard practice to approximate f by least squares by a polynomial $p_n(x) = c_0 + c_1 x + \cdots + c_n x^n$ of low degree. The approximation is chosen so that the total square error at all x_i

$$E = \sum_{m=0}^{k} [(p_n(x_m) - f(x_m)]^2 = \sum_{m=0}^{k} [c_0 + c_1 x_m + \cdots + c_n x_m^n - f(x_m)]^2$$

$$(13\text{--}70'')$$

is minimized. Thus the integral in (13–70) is now replaced by a sum and φ_i is now x^i. It is easily seen that the previous reasoning can be repeated with new inner products $(p, q) = \sum_{m=0}^{k} p(x_m) q(x_m)$. As before, $c_0, \ldots, c_n$ are determined by simultaneous linear equations

$$\sum_{j=0}^{n} b_{ij} c_j = k_i, \qquad i = 0, 1, \ldots, n,$$

$$b_{ij} = (x^i, x^j) = \sum_{m=0}^{k} x_m^i x_m^j = \sum_{m=0}^{k} x_m^{i+j}, \qquad (13\text{--}72')$$

$$k_i = (f, x^i) = \sum_{m=0}^{k} f(x_m) x_m^i.$$

Again $B = (b_{ij})$ is symmetric.

EXAMPLE 2 Let $f(0) = 11, f(1) = 14, f(2) = 16, f(3) = 18, f(4) = 21, f(5) = 25$. Find least-squares linear and quadratic approximations to f.

For the linear approximation $c_0 + c_1 x$ we have, with $x_0 = 0, x_1 = 1, x_2 = 2, x_3 = 3, x_4 = 4, x_5 = 5$,

$$b_{00} = \sum x_m^0 = 6, \qquad b_{01} = b_{10} = \sum x_m^1 = 15, \qquad b_{11} = \sum x_m^2 = 55,$$

$$k_0 = \sum f(x_m) = 105, \qquad k_1 = \sum f(x_m) x_m = 309.$$

Thus our equations are $6c_0 + 15c_1 = 105, 15c_0 + 55c_1 = 309$. From these we obtain $c_0 = 10.857, c_1 = 2.657$ and the linear approximation $y = 10.857 + 2.657x$. For $x = 0, 1, \ldots$ this gives $y = 10.857, 13.514, 16.171, 18.828, 21.485, 24.142$. These do not differ greatly from the given values $f(x_m)$.

For the quadratic approximation we have:

$$b_{02} = b_{11} = b_{20} = \sum x_m^2 = 55, \qquad b_{12} = b_{21} = \sum x_m^3 = 225,$$

$$b_{22} = \sum x_m^4 = 979, \qquad k_2 = \sum f(x_m) x_m^2 = 1201.$$

Thus our equations are

$$6c_0 + 15c_1 + 55c_2 = 105,$$
$$15c_0 + 55c_1 + 225c_2 = 309,$$
$$55c_0 + 225c_1 + 979c_2 = 1201.$$

We find $c_0 = 11.449, c_1 = 1.755, c_2 = 0.181$. Thus our quadratic approximation is

$$y = 11.449 + 1.755x + 0.181x^2.$$

For $x = 0, 1, \ldots$ this gives the values $y = 11.449, 13.3811, 15.683, 18.343, 21.365, 24.749$. The agreement is much better than in the linear case. ◄

Other applications

Attempt is often made to fit data by an equation such as $y = ae^{bx}$, $y = ax^b$, $y = a \ln bx$. Here a, b are to be adjusted to achieve the best fit. However, in each case the function does not depend linearly on a, b. Rather than deal with a nonlinear least-squares problem, we normally introduce new variables:

for $y = ae^{bx}$: $z = \ln y$, so that $z = \ln a + bx$;

for $y = ax^b$: $z = \ln y$, $u = \ln x$, so that $z = \ln a + bu$;

for $y = a \ln bx$: $u = \ln x$, so that $y = a \ln b + au$.

In all three cases we are now trying to approximate by a linear function in the new variables. The given data can be restated in terms of the new variables and we have a least-squares problem for linear approximation as above. A graphical aid is often used to reveal what is happening, by plotting the given data on log–log paper or semi-log paper. Special graph paper in effect makes the changes of variable.

13–8 UNIFORM APPROXIMATION BY POLYNOMIALS

In considering the approximation of a function $f(x)$ over an interval $a \leqslant x \leqslant b$ by a polynomial $p_n(x)$, we may want to consider the maximum absolute error:

$$E_0 = \max_{a \leqslant x \leqslant b} |p_n(x) - f(x)|. \qquad (13\text{–}80)$$

If f is continuous, this has meaning. If $p_n(x)$ is the nth partial sum of a Fourier series expansion of f in terms of orthogonal polynomials, and this series happens to be uniformly convergent, then E_0 can be made as small as we want by choosing n sufficiently large. In fact, Weierstrass proved that for every continuous f a polynomial $p_n(x)$ can be found such that E_0 is less than a given positive ε. This is the *Weierstrass approximation theorem*. It can be deduced from the Fourier series result mentioned. For, by change of coordinates, we can assume that the interval is $-1 \leqslant x \leqslant 1$. It is not hard to show that f can be approximated, with maximum absolute error less than $\varepsilon/2$, by a continuous piecewise linear function $g(x)$ (see Fig. 13–5). For such a g an expansion can be formed in Legendre polynomials (Section 3–12) and it can be verified that the series is uniformly convergent to $g(x)$. Then, for n sufficiently large, the nth partial sum $p_n(x)$ of this series will satisfy

$$\max_{-1 \leqslant x \leqslant 1} |g(x) - p_n(x)| < \varepsilon/2.$$

It then follows that

$$E_0 = \max |f(x) - p_n(x)| < \frac{\varepsilon}{2} + \frac{\varepsilon}{2} = \varepsilon.$$

In general, we say that $p_n(x)$ is a uniform approximation of f, within ε, if $E_0 < \varepsilon$, where E_0 is given by (13–80). The problem is to find such an approximation for given ε. Also we want the degree of $p_n(x)$ to be as small as possible.

There are many ways of obtaining such a uniform approximation to a continuous f, in particular, the way suggested by the proof outlined above for the

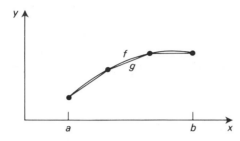

Fig. 13-5. Approximation of f by a piecewise linear function.

Weierstrass approximation theorem. If f is itself sufficiently smooth, its expansion in Legendre polynomials will be uniformly convergent, and partial sums of this series can be used. If f is analytic, a Taylor series for f about a suitable x_0 will be uniformly convergent on the interval $a \leqslant x \leqslant b$, if this interval lies inside the interval of convergence of the power series.

An alternative to expansion in Legendre polynomials is to expand in other systems of orthogonal polynomials. Especially useful is the expansion on $-1 \leqslant x \leqslant 1$ in terms of the Chebyshev polynomials $T_n(x)$ (these are discussed in Chapter 12). They have some special properties that make them especially desirable for such expansion and corresponding approximation. They form a complete orthogonal system on the interval $-1 \leqslant x \leqslant 1$ with weight function $(1 - x^2)^{-1/2}$. The polynomial $T_n(x)$ is of degree n, $T_n(x) = 2^{n-1}x^n + \cdots$ Also $T_n(x) = \cos(n \arccos x)$, $-1 \leqslant x \leqslant 1$, so that $-1 \leqslant T_n(x) \leqslant 1$ for $-1 \leqslant x \leqslant 1$, and $T_n(x) = \pm 1$ at $n+1$ points in the interval; in particular, $T_n(1) = 1$, $T_n(-1) = (-1)^n$. The function $T_n(x)$ has thus maximum absolute value 1 in the interval $-1 \leqslant x \leqslant 1$; it can be shown that every other polynomial of degree n with leading coefficient 2^{n-1} has maximum absolute value *greater* than 1. Thus $T_n(x)$ is in a sense best possible. We can say: for $|\delta| < \varepsilon$ and $p_n(x) = T_n(x)$, $\delta p_n(x)$ provides a uniform approximation, within ε, of the constant function $y = 0$, on the interval $-1 \leqslant x \leqslant 1$, and no other polynomial $p_n(x)$ of degree n with leading coefficient 2^{n-1} has this property.

It follows from this discussion that if

$$f(x) = c_0 T_0(x) + c_1 T_1(x) + \cdots + c_n T_n(x) + \cdots \quad \text{for } -1 \leqslant x \leqslant 1$$

and the series $\sum |c_n|$ converges, then the nth partial sum of the series provides a uniform approximation of $f(x)$, within ε_n, on the interval, where $\varepsilon_n \leqslant |c_{n+1}| + |c_{n+2}| + \cdots$, so that $\varepsilon_n \to 0$ as $n \to \infty$.

Because of these convergence properties, we may find it desirable to rearrange a given polynomial approximation of f to obtain an expansion in Chebyshev polynomials. To this end, we can use the rules

$$1 = T_0, \quad x = T_1, \quad x^2 = \frac{1}{2}(T_0 + T_2), \quad x^3 = \frac{1}{4}(3T_1 + T_3),$$

$$x^4 = \frac{1}{8}(3T_0 + 4T_2 + T_4),$$

and in general:

$$x^n = \frac{1}{2^{n-1}}\left[T_n + \binom{n}{1}T_{n-2} + \binom{n}{2}T_{n-4} + \cdots + \binom{n}{k-1}T_2\right] + \frac{1}{2^n}\binom{n}{k}T_0$$

for n even ($n = 2k$),

$$x^n = \frac{1}{2^{n-1}}\left[T_n + \binom{n}{1}T_{n-2} + \binom{n}{2}T_{n-4} + \cdots + \binom{n}{k}T_1\right] \text{ for } n \text{ odd } (n = 2k+1).$$

$$(13-81)$$

For example, from the series representation

$$\sin x = x - \frac{x^3}{6} + \frac{x^5}{120} + \cdots$$

we deduce that

$$\sin x = T_1(x) - \frac{1}{24}(3T_1 + T_3) + \frac{1}{1920}(T_5 + 5T_3 + 10T_1) + \cdots$$

If we drop the terms after the one with degree 5 in the power series, we obtain an approximation (uniform on $-1 \leqslant x \leqslant 1$, within $1/7!$) of $\sin x$ by a linear combination of $T_1(x)$, $T_3(x)$, $T_5(x)$:

$$\sin x \sim \frac{169}{192}T_1(x) - \frac{5}{128}T_3(x) + \frac{1}{1920}T_5(x).$$

On the interval $-1 \leqslant x \leqslant 1$, the last term is in absolute value at most $1/1920 = 0.00052$. We drop the term and now have in effect a uniform approximation by a third-degree polynomial, within $(1/7!) + (1/1920) = 0.0007279$. This third-degree approximation is almost as good as the initial fifth-degree approximation. Hence we have *economized* in degree, at small loss in accuracy. For example, $\sin 1 = 0.84147$; the three terms of the power series give the value 0.84167, while the cubic approximation

$$\sin x \sim \frac{169}{192}T_1(x) - \frac{5}{128}T_3(x)$$

gives 0.84115; the cubic approximation from the power series $x - (x^3/6)$ gives 0.83333. Thus the cubic approximation in terms of Chebyshev polynomials is much better. Typically, a partial sum of a power series $\sum c_n x^n$ gives high accuracy for x close to 0 and much less accuracy for x far from 0; the series of Chebyshev polynomials tends to be uniformly good across the interval $-1 \leqslant x \leqslant 1$.

If the given interval $a \leqslant x \leqslant b$ is other than $-1 \leqslant x \leqslant 1$, a substitution $x' = (2x - b - a)/(b - a)$ reduces our given $f(x)$ to a function $F(x')$ on the interval $-1 \leqslant x' \leqslant 1$.

The evaluation of a sum $c_n T_n(x) + \cdots + c_0 T_0(x)$ can be achieved by a modified synthetic division. The process to be described is based on a recursion formula

$$T_{n+1}(x) = 2xT_n(x) - T_{n-1}(x), \qquad n \geqslant 1, \tag{13-82}$$

for the Chebyshev polynomials (see Section 3–13). We indicate the procedure for $n = 6$. As in Table 13–10, we compute $d_6, d_5, \ldots$ by adding the indicated columns. It should be noted that we multiply d_1 by x (not by $2x$) in the last column.

Table 13–10

c_6	c_5	c_4	c_3	c_2	c_1	c_0
		$-d_6$	$-d_5$	$-d_4$	$-d_3$	$-d_2$
	$2xd_6$	$2xd_5$	$2xd_4$	$2xd_3$	$2xd_2$	xd_1
d_6	d_5	d_4	d_3	d_2	d_1	d_0 = Value sought

EXAMPLE 1 We evaluate $\sin x$ for $x = 1$ using $0.88021\,T_1(x) - 0.03906\,T_3(x) + 0.00052\,T_5(x)$ as above ($169/192 = 0.88021$, etc.). The work proceeds as in Table 13–11. The value found is 0.84167. ◀

Table 13–11

0.00052	0	−0.03906	0	0.88021	0
		−0.00052	−0.00104	0.03750	0.07604
	0.00104	0.00208	−0.07500	−0.15208	0.76563
0.00052	0.00104	−0.03750	−0.07604	0.76563	0.84167

Piecewise polynomial approximations

Above we referred to the piecewise linear approximation to a function (see Fig. 13–5). Here we are interpolating by a linear function between successive values of x. Instead of linear interpolation, we can use higher-degree interpolation, for example, the cubic obtained by interpolating at x_{i-2}, x_{i-1}, x_i, x_{i+1} to obtain an approximation for $x_{i-1} \leqslant x \leqslant x_i$. This would give a *piecewise cubic* approximation (some modification would have to be made at the ends of the given interval). Commonly the interpolation is done at x_{i-1}, x_{i-1}, x_i, x_i by using first derivatives (known or estimated) at x_{i-1} and x_i in addition to the function values. This provides a piecewise cubic approximation with a continuous first derivative. This is called a *piecewise Hermite approximation.* An example is suggested in Fig. 13–6. No new methods are needed; Eq. (13–68) is an explicit formula for the cubic polynomial. The process can be generalized, for example, to piecewise quintic approximation, using first and second derivatives, hence achieving an approximation with continuous first and second derivatives.

All these piecewise polynomial approximations have the property that they can be refined by subdividing the interval $a \leqslant x \leqslant b$ further, as in calculating a definite integral. If the function f is sufficiently smooth, they converge uniformly to f as the maximum length of a subdivision interval approaches 0. For piecewise Hermite approximation, the condition on f is that its fourth derivative be bounded for $a \leqslant x \leqslant b$.

There is one other kind of approximation by piecewise polynomial functions, the *spline* approximation. We illustrate the idea by the *cubic spline.* Here we choose a

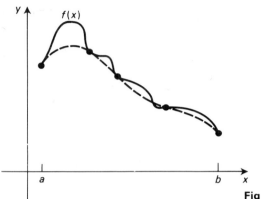

Fig. 13-6. Piecewise Hermite approximation.

piecewise cubic function $g(x)$ that agrees at successive subdivision points $a = x_0$, $x_1, \ldots, x_n = b$ and has continuous first and second derivatives for $a \leqslant x \leqslant b$. Conditions must also be imposed at a and b: as an example, we require that $g' = f'$ at these points.

EXAMPLE 2   Let $f(x) = x/(x+1)$ for $0 \leqslant x \leqslant 2$, let $x_0 = 0, x_1 = 1, x_2 = 2$. Thus $g(0) = 0, g(1) = 0.5, g(2) = 0.6667, g'(0) - 1, g'(2) - 0.1111$. We could obtain g as a piecewise Hermite function if we knew $g'(1)$. Let this missing value be denoted by s. Then we can apply Eq. (13–68) to obtain $g(x)$ for $0 \leqslant x \leqslant 1$:

$$g(x) = g(0) + g'(0)x + \frac{g[0,1] - g'(0)}{1}x^2 + \frac{g'(1) - 2g[0,1] + g'(0)}{1}x^2(x-1)$$

$$= 0 + x + (0.5 - 1)x^2 + (s - 1 + 1)x^2(x-1)$$

$$= x + (-s - 0.5)x^2 + sx^3.$$

Similarly, for $1 \leqslant x \leqslant 2$, we find

$$g(x) = 0.5 + s(x-1) + (0.1667 - s)(x-1)^2 + (s - 0.2222)(x-1)^2(x-2).$$

To achieve continuity of g'' at $x = 1$, we compute $g''(1)$ from both formulas and equate the results. This gives

$$4s - 1 = 0.7778 - 4s \qquad \text{or} \qquad s = 0.2222.$$

Accordingly,

$$g(x) = \begin{cases} x - 0.7222x^2 + 0.2222x^3, & 0 \leqslant x \leqslant 1, \\ 0.5 + 0.2222(x-1) - 0.0556(x-1)^2, & 1 \leqslant x \leqslant 2. \end{cases}$$

We verify that g is a very close approximation to f. For example, $f(0.5) = 0.3333$, $g(0.5) = 0.347; f(1.5) = 0.6, g(1.5) = 0.625$.

For more than two subintervals, the slopes $s_1, s_2, \ldots$ at the interior subdivision points must be supplied. The requirement that g'' be continuous at each such point gives as many linear equations as there are slopes to be found. It can be shown that these always have a unique solution.

===================== **Problems (Section 13–8)** =====================

1. Obtain least-squares approximations of $f(x)$, as requested, with weight function $w(x) \equiv 1$, and graph:
   a) $f(x) = 5x^3 - 4x^4$, $0 \leqslant x \leqslant 1$, by a linear function $c_0 + c_1 x$;
   b) $f(x) = x^4$, $-1 \leqslant x \leqslant 1$, by

$$c_0 + c_1 x + c_2 \left(\frac{3}{2} x^2 - \frac{1}{2} \right) = c_0 P_0(x) + c_1 P_1(x) + c_2 P_2(x);$$

   c) $f(x) = \sin x$, $0 \leqslant x \leqslant \pi$ by a cubic of form $c_1 x + c_2 x^2 + c_3 x^3$

   HINT: Use the fact that

$$\int_0^\pi x^k \sin x \, dx = \begin{cases} \pi \text{ for } k = 1, \\ \pi^2 - 4 \text{ for } k = 2, \\ \pi^3 - 6\pi \text{ for } k = 3, \end{cases}$$

   d) $f(x) = x^3 - 3x + 2$, $0 \leqslant x \leqslant 5$ by a quadratic $c_0 + c_1 x + c_2 x^2$.

2. a) Proceed as in Problem 1(a) with weight function $w(x) = x$.
   b) Proceed as in Problem 1(d) with weight function $w(x) = x(5 - x)$.

3. Find the least-squares approximation $p(x)$ to the given data as requested and graph:
   a) $f(0) = 5$, $f(1) = 8$, $f(2) = 10$, $f(3) = 14$, $p(x) = c_0 + c_1 x$
   b) $f(5.6) = 4.1$, $f(8.5) = 6.3$, $f(9.1) = 7.0$, $f(11.5) = 8.5$, $p(x) = c_0 + c_1 x$
   c) $f(1) = 6$, $f(2) = 10$, $f(3) = 20$, $f(4) = 31$, $f(5) = 55$, $p(x) = c_0 + c_1 x + c_2 x^2$
   d) $f(2.3) = 6.7$, $f(2.9) = 5.4$, $f(3.1) = 4.8$, $f(3.5) = 4.1$, $p(x) = c_0 + c_1 x + c_2 x^2$.

4. Follow the procedure used in Section 13–8 for $\sin x$ to obtain a uniform approximation to the given function for $-1 \leqslant x \leqslant 1$ by a linear combination of Chebyshev polynomials as suggested:
   a) $\cos x$ by $c_0 T_0 + c_1 T_1 + c_2 T_2$ (use the power series up to degree 4);
   b) e^x by $c_0 T_0 + \cdots + c_4 T_4$ (use the power series up to degree 4);
   c) $x + \dfrac{x^2}{10} + \dfrac{x^3}{50} + \dfrac{x^4}{150}$ by $c_0 T_0 + \cdots + c_3 T_3$;
   d) $\dfrac{1}{2 - x}$ by $c_0 + \cdots + c_3 T_3$ (use power series through degree 4).

5. Evaluate by modified synthetic division as requested (see Table 13–10):
   a) $7 T_0(x) + 11 T_1(x) - 13 T_2(x) + 4 T_3(x)$ for $x = 2$;
   b) $1.3 T_0(x) + 5.2 T_1(x) + 1.4 T_2(x) - 8.3 T_3(x) + T_4(x)$ for $x = 1.3$.

6. Find a cubic spline approximation to $f(x)$ as requested and comment on the accuracy:
   a) $f(x) = \dfrac{1}{x^2 + 1}$ for $0 \leqslant x \leqslant 2$ with $x_0 = 0$, $x_1 = 1$, $x_2 = 2$;
   b) $f(x) = x \sin x$ for $0 \leqslant x \leqslant 3\pi$ with $x_0 = 0$, $x_1 = \pi$, $x_2 = 2\pi$, $x_3 = 3\pi$.

13–9 FOURIER METHODS

Instead of ordinary polynomials, we can use *trigonometric polynomials* of form

$$p(x) = \frac{a_0}{2} + a_1 \cos x + b_1 \sin x + \cdots + a_N \cos Nx + b_N \sin Nx. \quad (13\text{--}90)$$

Questions of interpolation and approximation can then be studied.

Interpolation

We here consider only the case of equally spaced values x_k and, with (13–90) in mind, take them as

$$x_k = \frac{2\pi k}{2N+1}, \qquad k = 0, 1, 2, \ldots, 2N+1. \quad (13\text{--}91)$$

Further, we assume that the function $f(x)$ has period 2π, so that $f(0) = f(2\pi)$. Thus $x_{2N+1} = 2\pi$ and $x_0 = 0$ give the same value, and only $2N+1$ different values of f are required.

It can be shown that there is a unique interpolating trigonometric polynomial (13–90) for such data. Furthermore, the coefficients a_m, b_m are given by the formulas

$$a_m = \frac{2}{2N+1} \sum_{k=0}^{2N} f(x_k)\cos mx_k, \qquad m = 0, 1, \ldots, N,$$

$$\quad (13\text{--}92)$$

$$b_m = \frac{2}{2N+1} \sum_{k=0}^{2N} f(x_k)\sin mx_k, \qquad m = 1, 2, \ldots, N.$$

These rules parallel those for Fourier series. In fact, as in the case of Fourier series, orthogonality is in the background: the functions $1, \cos x, \ldots, \cos Nx, \sin x, \ldots,$ $\sin Nx$, when restricted to the finite set x_k defined by (13–91), form an orthogonal system with respect to summation. That is, if we defined (φ, ψ) for two such functions as

$$(\varphi, \psi) = \sum_{k=0}^{2N} \varphi(x_k)\,\psi(x_k),$$

then $(\cos mx, \cos nx) = 0$ and $(\sin mx, \sin nx) = 0$ for $m \neq n$ and $(\cos mx, \sin nx) = 0$, where m, n are integers between 0 and N inclusive. Furthermore,

$$(1, 1) = 2N+1, \qquad (\cos mx, \cos mx) = \sum_{k=0}^{2N} \cos^2(mx_k) = \frac{2N+1}{2},$$

$$\quad (13\text{--}93)$$

$$(\sin mx, \sin mx) = \frac{2N+1}{2}, \qquad m = 1, 2, \ldots, N$$

(see Problem 3 following Section 13–11). In view of these relations, the formulas (13–93) follow in the same way as the analogous ones for Fourier series (Section 3–2).

EXAMPLE 1 Let $f(x) = x(2\pi - x)$, $0 \le x \le 2\pi$, with $N = 2$, so that we have five interpolation points

$$x_0 = 0, \qquad x_1 = \frac{2\pi}{5}, \qquad x_2 = \frac{4\pi}{5}, \qquad x_3 = \frac{6\pi}{5}, \qquad x_4 = \frac{8\pi}{5}.$$

We find $f(x_0) = 0$, $f(x_1) = f(x_4) = 6.3165$, $f(x_2) = f(x_3) = 9.4748$. Hence

$$a_0 = \frac{2}{5}\left[f(x_0)\cos 0 x_0 + \cdots + f(x_4)\cos 0 x_4 \right] = 12.6330,$$

$$a_1 = \frac{2}{5}\left[f(x_0)\cos x_0 + \cdots + f(x_4)\cos x_4 \right] = -4.5707,$$

$$a_2 = \frac{2}{5}\left[f(x_0)\cos 2x_0 + \cdots + f(x_4)\cos 2x_4 \right] = -1.746,$$

and, similarly, $b_1 = b_2 = 0$. Accordingly, the interpolating polynomial is

$$p(x) = 6.3165 - 4.5707 \cos x - 1.746 \cos 2x.$$

This agrees very well with $f(x)$, as shown in Fig. 13–7. It can be shown that if f is continuous, periodic, and piecewise smooth, then the corresponding interpolating polynomials $p_N(x)$ approach $f(x)$ uniformly as $N \to \infty$. They are thus a very useful alternative to Fourier series. ◄

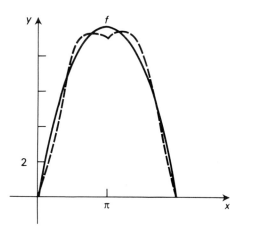

Fig. 13-7. Interpolation by trigonometric polynomial.

For certain purposes it is sometimes preferable to have an *even* number $2N$ of interpolating points x_k. To this end, the procedure can be modified as follows: we use

$$x_k = \frac{2\pi k}{2N} = \frac{\pi k}{N}, \qquad k = 0, 1, \ldots, 2N - 1. \tag{13–91'}$$

We can utilize only $2N$ of the $2N + 1$ functions previously used. We are in fact forced to use $\cos mx$ for $m = 0, 1, \ldots, N$ and $\sin mx$ for $m = 1, \ldots, N - 1$; the function $\sin Nx$ is of no help since $\sin Nx_k = 0$ for all x_k of (13–91'). Thus we have $f(x_k) = p(x_k)$ for all k, where

$$p(x) = \frac{a_0}{2} + a_1 \cos x + b_1 \sin x + \cdots + \frac{a_N}{2} \cos Nx.$$

The coefficients a_m, b_m are given by

$$a_m = \frac{1}{N} \sum_{k=0}^{2N-1} f(x_k) \cos mx_k, \qquad m = 0, 1, \ldots, N,$$

$$b_m = \frac{1}{N} \sum_{k=0}^{2N-1} f(x_k) \sin mx_k, \qquad m = 1, 2, \ldots, N-1.$$

(13–92')

The results and convergence properties are similar to those of the preceding case.

Least-squares approximation

We know from Section 3–6 that, for a periodic $f(x)$ with period 2π, each partial sum of the Fourier series of f provides the trigonometric polynomial $p(x)$ that minimizes the square error $\int_0^{2\pi} [f(x) - q(x)]^2 \, dx$ for all trigonometric polynomials $q(x)$ of the same form as $p(x)$.

As in Section 13–8, we now assume that $f(x)$ is known only at points (13–91), or (13–91'), and can seek the trigonometric polynomial

$$p(x) = \frac{a_0}{2} + a_1 \cos x + b_1 \sin x + \cdots + a_M \cos Mx + b_M \sin Mx \quad (13\text{–}93)$$

which minimizes the square error $\sum [f(x_k) - q(x_k)]^2$ among all trigonometric polynomials $q(x)$ of the same form as $p(x)$. Here $M \leqslant N$ and, for $M = N$, we get the interpolating trigonometric polynomial again, for which the total square error is 0.

As for Fourier series, we find that the minimum square error is attained by a unique $p(x)$ as in (13–93), where a_m and b_m are again given by (13–92) or (13–92'), except that $m \leqslant M$. Thus the least square error is achieved by a *partial sum* of the interpolating polynomial. For instance, in Example 1, $6.3165 - 4.5707 \cos x$ gives the best approximation to $f(x)$, in the sense of least-squares error at the points $x_0, \ldots, x_4$, among all functions of form $c_0 + c_1 \cos x + c_2 \sin x$.

13–10 FOURIER METHODS: COMPLEX FORMULATION

The preceding results have a complex counterpart: Say, we wish to interpolate $f(x)$, given for $0 \leqslant x \leqslant 2\pi$ with $f(0) = f(2\pi)$, at the points

$$x_k = \frac{2\pi k}{N}, \qquad k = 0, 1, \ldots, N-1, \qquad (13\text{–}100)$$

by a linear combination of the complex functions e^{inx} for $n = 0, \ldots, N-1$; that is, by

$$q(x) = \sum_{n=0}^{N-1} c_n e^{inx}. \tag{13–101}$$

Here f itself may be complex-valued. We wish to have

$$f(x_k) = q(x_k) = \sum_{n=0}^{N-1} c_n e^{inx_k}, \qquad k = 0, 1, \ldots, N-1. \tag{13–102}$$

If (13–102) holds, then we multiply by e^{-imx_k} and sum over k:

$$\sum_{k=0}^{N-1} f(x_k) e^{-imx_k} = \sum_{k=0}^{N-1} \sum_{n=0}^{N-1} c_n e^{i(n-m)x_k}.$$

On the right we can interchange the order of summation. Since

$$\sum_{k=0}^{N-1} e^{i(n-m)x_k} = \sum_{k=0}^{N-1} e^{i(n-m)2\pi k/N} = \sum_{k=0}^{N-1} r^k,$$

$$r = e^{i(m-n)2\pi/N},$$

we have, as usual for finite geometric series,

$$\sum_{k=0}^{N-1} r^k = \frac{1-r^N}{1-r}, \qquad r \neq 1$$

If $n = m$, $r = e^0 = 1$, and $\sum_{k=0}^{N-1} r^k = N$. Otherwise $r \neq 1$ but

$$1 - r^N = 1 - e^{i(n-m)2\pi} = 0,$$

so that $\sum_{k=0}^{N-1} r^k = 0$. We conclude that

$$\sum_{k=0}^{N-1} e^{i(n-m)x_k} = \begin{cases} N \text{ for } n = m, \\ 0 \text{ for } n \neq m. \end{cases} \tag{13–103}$$

It follows that

$$\sum_{k=0}^{N-1} f(x_k) e^{-imx_k} = \sum_{n=0}^{N-1} c_n \sum_{k=0}^{N-1} e^{i(n-m)x_k} = c_m N$$

or

$$c_m = \frac{1}{N} \sum_{k=0}^{N-1} f(x_k) e^{-imx_k}, \qquad m = 0, 1, \ldots, N-1. \tag{13–104}$$

This is the rule for coefficients in our interpolation relation (13–102).

If $f(x)$ happens to be real-valued, then the coefficients c_m are further restricted:

$$c_{N-m} = \bar{c}_m, \qquad m = 0, 1, \ldots, \qquad N-1 \qquad (f \text{ real}). \qquad (13\text{–}105)$$

For, by (13–104)

$$c_{N-m} = \frac{1}{N} \sum_{k=0}^{N-1} f(x_k) e^{-i(N-m)x_k} = \frac{1}{N} \sum_{k=0}^{N-1} f(x_k) e^{imx_k} = \bar{c}_m,$$

since $f(x_k)$ is real and the conjugate of e^{-ai} is e^{ai} for a real.

With $f(x)$ real-valued, we can proceed to recover our formulas in real form, as in Section 13–9. For (13–90), (13–92) we have $2N+1$ interpolation points (13–91). We therefore replace N by $2N+1$ in (13–102), (13–104), (13–105). Thus

$$f(x_k) = \sum_{n=0}^{2N} c_n e^{inx_k}, \text{ where } c_{2N+1-n} = \bar{c}_n; \qquad (13\text{–}106)$$

accordingly,

$$f(x_k) = \sum_{n=0}^{N} c_n e^{inx_k} + \sum_{n=N+1}^{2N} c_n e^{inx_k}.$$

We now set $m = 2N+1-n$ in the second sum and use (13–106) and the relation $e^{i(2n+1)x_k} = e^{2\pi ki} = 1$:

$$f(x_k) = \sum_{n=0}^{N} c_n e^{inx_k} + \sum_{m=1}^{N} c_{2N+1-m} e^{i(2N+1-m)x_k}$$

$$= \sum_{n=0}^{N} c_n e^{inx_k} + \sum_{m=1}^{N} \bar{c}_m e^{-imx_k}.$$

In the second sum we can replace the summation index by n and then combine terms:

$$f(x_k) = c_0 + \sum_{n=1}^{N} (c_n e^{inx_k} + \bar{c}_n e^{-inx_k}). \qquad (13\text{–}107)$$

By (13–103), c_0 is real (since $f(x)$ is real-valued) and can be denoted by $a_0/2$. From (13–104), with N replaced by $2N+1$, we can also write

$$c_n = \frac{a_n - ib_n}{2}, \qquad n = 1, \ldots, N,$$

$$a_n = \frac{2}{2N+1} \sum_{k=0}^{2N} f(x_k) \cos nx_k, \qquad b_n = \frac{2}{2N+1} \sum_{k=0}^{2N} f(x_k) \sin nx_k.$$

From (13–107) and these formulas, we recover (13–90), (13–92). In similar fashion we recover the rules (13–91'), (13–92').

REMARK. The general formulas (13–100)–(13–104) give us a complex exponential polynomial $q(x)$ as in (13–101) interpolating to $f(x)$ at the x_k as in Eq. (13–100). For many purposes we want mainly the representation (13–102): that is, an expansion of the function f, sampled at the points x_k, in terms of the functions e^{inx_k}. However, if we also use $q(x)$ at values of x other than the x_k, then (13–101) should be modified, as in recovering the real cases for N odd and even; the modification is needed to ensure convergence of the interpolation polynomials to $f(x)$ as $N \to \infty$.

If N is odd, we replace N by $2N + 1$, and write as above following Eq. (13–106):

$$\sum_{n=0}^{2N} c_n e^{inx_k} = \sum_{n=0}^{N} c_n e^{inx_k} + \sum_{m=1}^{N} c_{2N+1-m} e^{i(2N+1-m)x_k}$$

$$= \sum_{n=0}^{N} c_n e^{inx_k} + \sum_{n=1}^{N} c_{2N+1-n} e^{-inx_k}.$$

Thus we take

$$q(x) = \sum_{n=0}^{N} c_n e^{inx} + \sum_{n=1}^{N} c_{2N+1-n} e^{-inx}. \tag{13–101'}$$

This is a linear combination of $e^{0iNx} = 1, e^{\pm ix}, \ldots, e^{\pm iNx}$. If N is even, we replace N by $2N$ and obtain similarly

$$q(x) = \sum_{n=0}^{N} c_n e^{inx} + \sum_{n=1}^{N-1} c_{2N-n} e^{-inx}. \tag{13–101''}$$

This is a linear combination of $e^{0ix}, e^{\pm ix}, \ldots, e^{\pm i(N-1)x}, e^{iNx}$.

The importance of these modifications is shown by taking $f(x) = e^{-ix}$. Here (13–104) gives

$$c_n = \frac{1}{N} \sum_{k=0}^{N-1} e^{-ix} e^{-inx_k} = \frac{1}{N} \sum_{k=0}^{N} e^{-i(n+1)x_k}.$$

This sum is again one of the form $\sum r^k$, and we verify that it equals 0, except for $n + 1 = N$, in which case it equals N, so that $c_{N-1} = 1$. Thus $q(x)$ would be $e^{(N-1)ix} = \cos(N-1)x + i \sin(N-1)x$. For fixed x this has no limit as $N \to \infty$. However, $q(x_k) = e^{-ix_k} = f(x_k)$, since

$$e^{(N-1)ix_k} = e^{Nix_k} e^{-ix_k}.$$

Thus $q(x)$ does interpolate to $f(x)$, but is a poor approximation to f. If we modify the procedure as above, to include e^{inx} for negative n, we obtain a modified $q(x)$, namely e^{-ix} (for $N \geqslant 2$), so that $q(x) \equiv f(x)$.

The essential idea is that, in the complex procedure, for approximation purposes we should include e^{inx} for both positive and negative n; by including all such n with $|n| \leqslant N$ and N sufficiently large, we obtain a good approximation by the interpolating polynomial.

Error due to aliasing

For $f(x)$ of period 2π, the complete Fourier series representation is

$$f(x) = \sum_{n=-\infty}^{\infty} C_n e^{inx}, \qquad C_n = \frac{1}{2\pi} \int_0^{2\pi} f(x) e^{-inx} \, dx \qquad (13\text{–}108)$$

(see Section 3–7). We can regard the formula (13–104) as one giving approximate values c_m for C_m $(m = 0, 1, \ldots, N-1)$. We now ask how accurate the approximation is. From (13–108),

$$f(x_k) = \sum_{n=-\infty}^{\infty} C_n e^{inx_k},$$

so that by (13–104)

$$c_m = \frac{1}{N} \sum_{k=0}^{N-1} \sum_{n=-\infty}^{\infty} C_n e^{i(n-m)x_k}.$$

We interchange the order of summation and observe that, as in (13–103),

$$\sum_{k=0}^{N-1} e^{i(n-m)x_k} = \begin{cases} N \text{ for } n = m + lN, \\ 0, \text{ otherwise,} \end{cases}$$

where l may be $0, \pm 1, \pm 2, \ldots$ We deduce that

$$c_m = \frac{1}{N} \sum_{\substack{n=-\infty \\ n=m+lN}}^{\infty} NC_n = \sum_{l=-\infty}^{\infty} C_{m+lN}$$

$$= C_m + C_{m+N} + C_{m-N} + C_{m+2N} + C_{m-2N} + \cdots \qquad (13\text{–}109)$$

The terms in $e^{i(m+N)x}$, $e^{i(m-N)x}$, $e^{i(m+2N)x}$, $\ldots$ are also affecting c_m, since, for $x = x_k$, each one reduces to e^{imx_k}. This effect is termed *aliasing*.

A simple example illustrates its meaning. Let $f(x) = 3e^{ix} + 5e^{11ix}$, with $x_k = 2\pi k/10$. Then

$$f(x_k) = 3e^{ix_k} + 5e^{11ix_k} = 8e^{ix_k},$$

since $e^{11ix_k} = e^{10ix_k} e^{ix_k}$ and $e^{10ix_k} = 1$. Thus we find $c_1 = 8$, whereas $C_1 = 3$.

In general, Eq. (13–109) shows that the size of the error depends on how fast the C_n approach 0. If N is large, the effect should be small.

13–11 FAST FOURIER TRANSFORM

For given $f(x)$ and N the calculation of all the coefficients c_m in (13–104) requires N^2 multiplications (of pairs of complex numbers). For N large, a reduction in the number of multiplications is very much desired. The fast Fourier transform succeeds in achieving such a reduction for appropriate choice of N.

The principal idea is to choose N to be factorable: $N = AB$, where A and B are positive integers greater than 1. The number of multiplications can then be reduced to $AB(A + B + 1)$. If, for example, $N = 10^3$ and $A = 10^2$, $B = 10$, then $N^2 = 10^6$, but $AB(A + B + 1) = 1.1 \cdot 10^5$. Hence we have reduced the number almost by a factor of 10.

If A or B is factorable, the process can be repeated to achieve further economy. One common procedure is to choose N as a high power of 2, so that repeated factorizations are possible. In order to derive the procedure, we let

$$x_k = \frac{2\pi k}{N}, \qquad f(x_k) = h(k), \qquad k = 0, 1, \ldots, N-1, \qquad (13\text{–}110)$$

Thus $h(k)$ is a function of the integer variable k. We can thus write the sum in Eq. (13–104) as

$$H(m) = \sum_{k=0}^{N-1} h(k)e^{-i2\pi km/N}, \qquad m = 0, 1, \ldots, N-1. \qquad (13\text{–}111)$$

We call $H(m)$ the *discrete Fourier transform* of $h(k)$ and write $\Phi_N[h(k)] = H(m)$. It can be verified that Φ_N resembles the Fourier transform of Chapter 4 in many ways (see Problem 7 below). To simplify, we also write $e(u) = e^{-2\pi iu}$, so that (13–111) now reads

$$H(m) = \sum_{k=0}^{N-1} h(k)e\left(\frac{km}{N}\right), \qquad (13\text{–}111')$$

and $e(u)$ obeys the rules

$$e(u + v) = e(u)e(v), \qquad e(l) = 1 \text{ if } l \text{ is an integer.} \qquad (13\text{–}112)$$

We now wish to calculate $H(m)$ for $m = 0, 1, \ldots, N-1$. We let $N = AB$ as above. Then the numbers from 0 to $N-1$ inclusive can be represented as

$$0, \quad 1, \quad \ldots, \quad B-1, \quad B, \quad B+1, \quad \ldots, \quad B+(B-1), \quad 2B, \quad 2B+1, \quad \ldots,$$
$$(A-1)B, \quad (A-1)B+1, \quad \ldots, \quad (A-1)B+(B-1),$$

as in Fig. 13–8. Thus we obtain all $k = 0, 1, \ldots, N-1$ as

$$k = aB + b, \qquad 0 \leqslant a \leqslant A-1, \qquad 0 \leqslant b \leqslant B-1.$$

Each pair (a, b) determines a unique k, and k is obtained from only one pair (a, b). In the same way we can write the numbers $0, 1, \ldots, N-1$ as

$$m = \beta A + \alpha, \qquad 0 \leqslant \beta \leqslant B-1, \qquad 0 \leqslant \alpha \leqslant A-1.$$

Fig. 13-8. Grouping of integers from
0 to $N - 1$ ($N = 12$, $A = 4$, $B = 3$).

We now use these representations and the rule (13–112) to write

$$e\left(\frac{km}{N}\right) = e\left(\frac{(aB+b)(\beta A+\alpha)}{AB}\right) = e\left(a\beta + \frac{b\beta}{B} + \frac{a\alpha}{A} + \frac{b\alpha}{AB}\right)$$

$$= e(a\beta)e\left(\frac{b\beta}{B}\right)e\left(\frac{a\alpha}{A}\right)e\left(\frac{b\alpha}{AB}\right).$$

Here the first factor equals 1, since $a\beta$ is an integer. Now (13–111′) becomes, by summing over a and b,

$$H(\beta A + \alpha) = \sum_{b=0}^{B-1} \sum_{a=0}^{A-1} h(aB+b)e\left(\frac{b\beta}{B}\right)e\left(\frac{a\alpha}{A}\right)e\left(\frac{b\alpha}{AB}\right)$$

$$= \sum_{b=0}^{B-1} e\left(\frac{b\beta}{B}\right)\left[e\left(\frac{b\alpha}{AB}\right) \sum_{a=0}^{A-1} h(aB+b)e\left(\frac{a\alpha}{A}\right)\right]. \tag{13-113}$$

In the inner sum, α and b are held fixed. By analogy with (13–111′) we see that this can be interpreted as

$$\Phi_A[h(aB+b)] = H_b(\alpha), \tag{13-114}$$

where $h(aB + b)$ is regarded as a function of a ($a = 0, 1, \ldots, A - 1$) and $H_b(\alpha)$ as a function of α ($\alpha = 0, 1, \ldots, A - 1$). The quantity in brackets is now

$$e\left(\frac{b\alpha}{AB}\right)H_b(\alpha) = g_\alpha(b). \tag{13-115}$$

Thus

$$H(\beta A + \alpha) = \sum_{b=0}^{B-1} g_\alpha(b)e\left(\frac{b\beta}{B}\right) = \Phi_B[g_\alpha(b)]. \tag{13-116}$$

Accordingly, determination of $H(\beta A + \alpha)$ for $\alpha = 0, 1, \ldots, A - 1$, $\beta = 0, 1, \ldots, B - 1$ requires

I. Evaluating the discrete Fourier transforms $\Phi_A[h(aB+b)]$ as in (13–114) for $b = 0, 1, \ldots, B - 1$;

II. Multiplying the result of I for each b, α by $e(b\alpha/(AB))$ to obtain $g_\alpha(b)$;

III. Evaluating the discrete Fourier transforms $\Phi_B[g_\alpha(b)]$ for $\alpha = 0, 1, \ldots, A - 1$.

Evaluating $\Phi_A[h]$ in stage I requires A^2 multiplications; this must be done B times, hence in all A^2B multiplications are required. Stage II requires one multiplication for each choice of α, b, hence AB multiplications in all are required. Stage III involves calculation of $\Phi_B[g]$ for A choices of g, hence, as in stage I, BA^2 multiplications are required. In all, we need

$$A^2B + AB + BA^2 = AB(A + B + 1)$$

multiplications, as asserted.

It should be remarked that the first stage can be interpreted as a multiplication of the $B \times A$ matrix P, whose (b, a) entry is $h(aB + b)$, by the $A \times A$ matrix Q, whose (a, α) entry is $e(a\alpha/A)$. The product PQ is a $B \times A$ matrix indexed by b and α. At the second stage, the (b, α) entry of this matrix is multiplied by $e(b\alpha/(AB))$. The result is again a $B \times A$ matrix R indexed by b and α. At the third stage, this matrix R is multiplied on the left by the $B \times B$ matrix S, whose (β, b) entry is $e(b\beta/B)$.

In this matrix interpretation $h(k)$ is presented in order along the successive columns of matrix P and $H(m)$ is presented in order along the rows of the final $B \times A$ matrix SR. The following example will make this clear.

EXAMPLE 1   Let $f(x_k)$ have successive values 11, 15, 18, 21, 28, 22, 16, 14, 12, 15, 17, 15. Thus $h(0) = 11$, $h(1) = 15$, etc., and $N = 12 = 4 \times 3$. We thus take $A = 4$, $B = 3$. Then the first $B \times A$ matrix, with $h(aB + b)$ as the (b, a) entry, is

$$P = \begin{bmatrix} 11 & 21 & 16 & 15 \\ 15 & 28 & 14 & 17 \\ 18 & 22 & 12 & 15 \end{bmatrix}.$$

This is to be multiplied on the right by the $A \times A$ matrix

$$Q = \begin{bmatrix} e(0) & e(0) & e(0) & e(0) \\ e(0) & e(\tfrac{1}{4}) & e(\tfrac{1}{2}) & e(\tfrac{3}{4}) \\ e(0) & e(\tfrac{1}{2}) & e(1) & e(\tfrac{3}{2}) \\ e(0) & e(\tfrac{3}{4}) & e(\tfrac{3}{2}) & e(\tfrac{9}{4}) \end{bmatrix} = \begin{bmatrix} 1 & 1 & 1 & 1 \\ 1 & -i & -1 & i \\ 1 & -1 & 1 & -1 \\ 1 & i & -1 & -i \end{bmatrix}.$$

Here $e(u) = e^{-2\pi iu}$. We observe that Q must be symmetric (since $a\alpha/A = \alpha a/A$). We compute (48 multiplications)

$$PQ = \begin{bmatrix} 63 & -5 - 6i & -9 & -5 + 6i \\ 74 & 1 - 11i & -16 & 1 + 11i \\ 67 & 6 - 7i & -7 & 6 + 7i \end{bmatrix}.$$

Now we must multiply the (b, α) entry of this matrix by $e(b\alpha/12)$. The corresponding multipliers form the matrix

$$\begin{bmatrix} 1 & 1 & 1 & 1 \\ 1 & 0.866 - 0.5i & 0.5 - 0.866i & -i \\ 1 & 0.5 - 0.866i & -0.5 - 0.866i & -1 \end{bmatrix}.$$

We carry out the 12 multiplications (*this is not usual matrix multiplication!*) to obtain

$$R = \begin{bmatrix} 63 & -5-6i & -9 & -5+6i \\ 74 & -4.63-10.0i & -8+13.9i & 11-i \\ 67 & -3.06-8.7i & 3.5+6.06i & -6-7i \end{bmatrix}.$$

Finally we multiply on the left by S, whose (β, b) entry is $e(\beta b/3)$:

$$S = \begin{bmatrix} 1 & 1 & 1 \\ 1 & -0.5-0.866i & -0.5+0.866i \\ 1 & -0.5+0.866i & -0.5-0.866i \end{bmatrix}.$$

Matrix S is again symmetric. We obtain (by 36 multiplications):

$$SR = \begin{bmatrix} 204 & -12.7-24.7i & -13.5+19.9i & -2i \\ -7.5-6.06i & -2.3-4.72i & 0 & -2.3+4.72i \\ -7.5+6.06i & 2i & -13.5-19.9i & -12.7+24.7i \end{bmatrix}.$$

Accordingly, $H(0) = 204$, $H(1) = -22.7-24.7i$, $H(2) = -13.5+19.9i$, $H(3) = -10-2i$, $H(4) = -7.5-6.06i$, etc. We observe that $H(1) = \overline{H(11)}$, $H(2) = \overline{H(10)}$, $H(3) = \overline{H(9)}$, $H(4) = \overline{H(8)}$, $H(5) = \overline{H(7)}$. These follow from the rule (13–105), since $f(x_k) = h(k)$ is real-valued.

In all, we used $48 + 12 + 36 = 96$ multiplications, in agreement with our expression $AB(A + B + 1)$. ◀

============ **Problems (Section 13–11)** ============

1. Interpolate by a trigonometric polynomial:
   a) $f(x) = x^2(2\pi - x)/\pi^3$ for $0 \leqslant x \leqslant 2\pi$ at $x_k = 2\pi k/5$, $k = 0, 1, \ldots, 4$;
   b) $f(x) = (1/\pi)x \sin x$ for $0 \leqslant x \leqslant 2\pi$ at $x_k = 2\pi k/5$, $k = 0, 1, \ldots, 4$;
   c) $f(x) = x(2\pi - x)/\pi^2$ for $0 \leqslant x \leqslant 2\pi$ at $x_k = \pi k/3$, $k = 0, 1, \ldots, 5$;
   d) $f(x) = 1/(1 + \sin^2 x)$ for $0 \leqslant x \leqslant 2\pi$ at $x_k = \pi k/4$, $k = 0, \ldots, 7$.

2. For $f(x_k)$ as given, find the least-squares approximation $p(x)$ of the form requested:
   a) $x_k = 2\pi k/5$, $k = 0, 1, \ldots, 4$, $f(x_0) = 1$, $f(x_1) = 3$, $f(x_2) = 2$, $f(x_3) = 0$, $f(x_4) = 1$,
   $p(x) = (a_0/2) + a_1 \cos x + b_1 \sin x$;
   b) $x_k = 2\pi k/9$, $k = 0, 1, \ldots, 8$, $f(x_0) = 0$, $f(x_1) = 1$, $f(x_2) = 2$, $f(x_3) = 1$, $f(x_4) = 0$,
   $f(x_5) = 1$, $f(x_6) = 3$, $f(x_7) = 2$, $f(x_8) = 1$,
   $p(x) = (a_0/2) + a_1 \cos x + b_1 \sin x + a_2 \cos 2x + b_2 \sin 2x$.

3. Let $x_k = 2\pi k/(2N + 1)$ for $k = 0, \ldots, 2N$. Let m, n be integers, $0 \leqslant m \leqslant N$, $0 \leqslant n \leqslant N$. Prove the following:
   a) $\displaystyle\sum_{k=0}^{2N} \cos mx_k \cos nx_k = 0$ for $m \neq n$
   b) $\displaystyle\sum_{k=0}^{2N} \cos^2 mx_k = \frac{2N+1}{2}$
   c) $\displaystyle\sum_{k=0}^{2N} \sin mx_k \sin nx_k = 0$ for $m \neq n$

d) $\displaystyle\sum_{k=0}^{2N} \sin^2 mx_k = \frac{2N+1}{2}$ for $m \geqslant 1$

e) $\displaystyle\sum_{k=0}^{2N} \cos mx_k \sin nx_k = 0$

HINT: Use the identities $\cos u = (e^{iu} + e^{-iu})/2$, $\sin u = (e^{iu} - e^{-iu})/2i$ to reduce each sum to sums of exponentials and then use the geometric-series rule

$$1 + r + \cdots + r^h = (1 - r^{h+1})/(1 - r)$$

for $r \neq 1$ to evaluate the sums.

4. Let $x_k = 2\pi k/(2N) = \pi k/N$ for $k = 0, 1, \ldots, 2N - 1$. Show that

$$\sum_{k=0}^{2N-1} \cos^2 mx_k = \begin{cases} N & \text{for } m = 1, \ldots, N-1, \\ 2N & \text{for } m = 0 \text{ and } m = N. \end{cases}$$

5. Interpolate at the points $x_k = 2\pi k/5$ ($k = 0, 1, \ldots, 4$) to $f(x) = e^{ix^2/(2\pi)}$, $0 \leqslant x \leqslant 2\pi$, by a complex exponential polynomial $p(x) = \sum_{n=0}^{4} c_n e^{inx}$. Also obtain the corresponding $q(x)$ as in Eq. (13–101′) and compare the accuracy of approximation of f by $p(x)$ and $q(x)$ at $x = \pi/2$ and $x = \pi$.

6. Use the factorization $6 = 3 \cdot 2$ to compute the fast Fourier transform $H(m)$ corresponding to $h(k) = 3, 5, 8, 11, 7, 2$ for $k = 0, 1, \ldots, 5$, $N = 6$, $A = 3$, $B = 2$.

7. In studying the discrete Fourier transform $\Phi_N[h(k)]$ it is convenient to extend $h(k)$ to all integers k so as to have period N; thus $h(N) = h(0)$, $h(N+1) = h(1)$, $h(-1) = h(N-1), \ldots$ Also the corresponding $H(m)$ can be defined by Eq. (13–111) for all integers m.

a) Show that H also has period N and that, for each integer r,

$$H(m) = \sum_{k=r}^{k=r+N-1} h(k)e^{-2\pi ikm/N}.$$

b) Show that Φ_N is a linear operator: $\Phi_N[c_1 h_1 + c_2 h_2] = c_1 \Phi_N[h_1] + c_2 \Phi_N[h_2]$, where c_1, c_2 are complex constants.

c) Show that, if l is an integer and $\Phi_N[h(k)] = H(m)$, then

$$\Phi_N[h(k-l)] = e^{-2\pi ilm/N} H(m).$$

d) Define the convolution of $h_1(k)$, $h_2(k)$, both of period N, by

$$h_1 * h_2 = \sum_{l=0}^{N-1} h_1(l)h_2(k-l), \qquad k = 0, \pm 1, \pm 2, \ldots$$

Show that $h_1 * h_2$ also has period N and that $\Phi_N[h_1 * h_2] = \Phi_N[h_1]\Phi_N[h_2]$.

HINT: Substitute in the definition to obtain

$$\Phi_N[h_1 * h_2] = \sum_{k=0}^{N-1} \sum_{l=0}^{N-1} h_1(l)h_2(k-l)e^{-2\pi ikm/N},$$

interchange the order of summation, then set $s = k - l$ in the sum with respect to k and use the results of part (a).

8. a) Explain in detail why, in the matrix P of Section 13–11, the function $h(k)$ is presented in order along the successive columns; also why, in the matrix SR, $H(m)$ is presented in order along the successive rows.
   b) In Example 1 of Section 13–11, the matrix PQ has the property that the entries in the second column are the complex conjugates of the corresponding entries in the fourth column. Explain why this is so and state a general rule.

13–12 NUMERICAL DIFFERENTIATION

If $f(x)$ is known only for values $x_0, x_1, \ldots , x_k$ of x, then we can assign derivatives of f at these values (and other values) of x in a variety of ways. Unfortunately, slight errors in the values $f(x_i)$ can cause major errors in the derivatives; for the same reason, round-off errors can be disastrous.

Interpolation

We can interpolate to f at three successive values x_0, x_1, x_2 by a polynomial $p(x)$ of degree 2 and then choose as $f'(x)$ the derivative $p'(x)$ (at x_0, x_1, x_2 or throughout the smallest interval containing these three points). Most commonly, the values $x_0, x_1,$ x_2 are equally spaced: $x_0 = x_1 - h$, $x_2 = x_1 + h$. If we then interpolate as in Section 13–6 and evaluate $p'(x_1)$, we find

$$p'(x_1) = \frac{f(x_1 + h) - f(x_1 - h)}{2h} \sim f'(x_1). \tag{13–120}$$

This value is very commonly used as an estimate of $f'(x_1)$.

An alternative procedure is to interpolate by a linear function that agrees with f at, say x_1 and $x_2 = x_1 + h$. The corresponding estimate of the derivative at x_1 is then

$$\frac{f(x_1 + h) - f(x_1)}{h}; \tag{13–121}$$

this is the familiar $\Delta y / \Delta x$, or slope of the chord.

Even without the difficulties of round-off error, (13–120) is much preferable to (13–121) as an estimate of $f'(x_1)$. For example, if $f(x)$ is representable by a power series for small $|h|$:

$$f(x_1 + h) = f(x_1) + hf'(x_1) + \frac{h^2}{2!} f''(x_1) + \cdots , \tag{13–122}$$

then

$$\frac{f(x_1 + h) - f(x_1 - h)}{2h} = f'(x_1) + \frac{h^2}{3!} f'''(x_1) + \cdots ,$$

whereas

$$\frac{f(x_1 + h) - f(x_1)}{h} = f'(x_1) + \frac{h}{2!} f''(x_1) + \cdots$$

As $h \to 0$, the first series approaches its limit $f'(x_1)$ faster than the second.

However, if $f(x)$ is known only for $x_1, x_2, \ldots$ and $x_1 < x_2 < \cdots$ (or $x_1 > x_2 > \cdots$), then we are forced to use a *one-sided formula* such as (13–121) in preference to a *two-sided formula* such as (13–120).

For the *second derivative* of f at x_1, with f known at $x_1 \pm h$, we can again use the quadratic interpolating polynomial and obtain the estimate:

$$f''(x_1) \sim \frac{f(x_1 + h) - 2f(x_1) + f(x_1 - h)}{h^2}. \tag{13–123}$$

This can be written as

$$\frac{1}{h}\left[\frac{f(x_1 + h) - f(x_1)}{h} - \frac{f(x_1) - f(x_1 - h)}{h}\right].$$

The first quotient in brackets can be considered an estimate for $f'(x_1 + \frac{1}{2}h)$, the second for $f'(x_1 - \frac{1}{2}h)$; hence the whole expression becomes

$$\frac{1}{h}[f'(x_1 + \tfrac{1}{2}h) - f'(x_1 - \tfrac{1}{2}h)],$$

which estimates $f''(x_1)$ just as (13–120) estimates $f'(x_1)$ (f being replaced by f', h by $\frac{1}{2}h$).

If f is known only at x_1, $x_1 + h$, $x_1 + 2h$ and larger values of x, then for the second derivative at x_1 we are forced to use an estimate such as

$$f''(x_1) \sim \frac{f(x_1 + 2h) - 2f(x_1 + h) + f(x_1)}{h^2}, \tag{13–124}$$

obtained again by second-degree interpolation. A power-series analysis like the one above shows that this is not as good as (13–123).

The procedures can be extended to higher-order derivatives, in which case we are forced to interpolate by polynomials of increasing degrees and the results become increasingly sensitive to error.

EXAMPLE 1 Let $f(x) = \ln x$. From the values $\ln 1.9 = 0.64185$, $\ln 2 = 0.69315$, $\ln 2.1 = 0.74195$, we estimate by (13–120):

$$f'(2) \sim \frac{f(2.1) - f(1.9)}{0.2} = 0.5005.$$

On the other hand, (13–121) would give

$$f'(2) \sim \frac{f(2.1) - f(2.0)}{0.1} = 0.488,$$

which is not as good. The correct value is $1/2 = 0.5$. For the second derivative, (13–123) gives

$$f''(2) \sim \frac{f(2.1) - 2f(2) + f(1.9)}{0.01} = -0.25.$$

This happens to be exact. We now decrease h, using $\ln 1.99 = 0.68813$ and $\ln 2.01 = 0.69813$. By (13–120),

$$f'(2) \sim \frac{f(2.01) - f(1.99)}{0.02} = 0.500,$$

which is exact; Eq. (13–121) gives the poorer estimate

$$f'(2) \sim \frac{f(2.01) - f(2)}{0.1} = 0.498.$$

However, for $f''(2)$, Eq. (13–133) now gives

$$f''(2) \sim \frac{f(2.01) - f(2) + f(1.99)}{0.0001} = -0.4,$$

which is *not* as good as the value obtained with $h = 0.1$. To salvage the accuracy, we would have to know the values of f to several additional decimal places. ◄

For reference we give the commonly used estimates for third and fourth derivatives:

$$f'''(x_1) \sim \frac{f(x_1 + 3h) - 3f(x_1 + h) + 3f(x_1 - h) - f(x_1 - 3h)}{8h^3}, \qquad (13\text{–}125)$$

$$f^{iv}(x_1) \sim \frac{f(x_1 + 2h) - 4f(x_1 + h) + 6f(x_1) - 4f(x_1 - h) + f(x_1 - 2h)}{h^4}. \qquad (13\text{ }126)$$

In some cases differentiation by all the above formulas gives unacceptable results because of errors in the known values of f. We can try to salvage estimates for the derivatives by first *smoothing* the values of f. A simple way of doing this is to replace $f(x)$ by an average (or weighted average) of $f(x)$ with several nearby values, say $f(x \pm h)$ and $f(x \pm 2h)$. If this is done for a set of x-values, a new function $f_1(x)$ is obtained, whose derivatives estimate those of $f(x)$. Another procedure is to approximate f by least squares by a polynomial or trigonometric polynomial $f_1(x)$ (Sections 13–7 and 13–9) and again use the derivatives of $f_1(x)$.

13–13 NUMERICAL INTEGRATION

In contrast to differentiation, integration is generally easy to carry out with desired accuracy. For the definite integral of $f(x)$ (say, continuous) from a to b, where $a < b$, we have the basic double inequality

$$A(b - a) \leqslant \int_a^b f(x)\,dx \leqslant B(b - a) \qquad (13\text{–}130)$$

if it is known that $A \leqslant f(x) \leqslant B$ for $a \leqslant x \leqslant b$. From this we conclude that

$$\left| \int_a^b f(x)\,dx \right| \leqslant M(b - a), \qquad (13\text{–}131)$$

if it is known that $|f(x)| \leqslant M$ for $a \leqslant x \leqslant b$.

To evaluate the integral numerically, we can subdivide the interval from a to b (usually into equal subintervals) by successive points $x_0 = a, x_1, x_2, \ldots, x_n = b$. Then

$$\int_a^b f(x)\,dx = \int_a^{x_1} f(x)\,dx + \int_{x_1}^{x_2} f(x)\,dx + \cdots + \int_{x_{n-1}}^b f(x)\,dx. \qquad (13\text{–}132)$$

The rule (13–130) can then be applied to each of the intervals $[a, x_1], [x_1, x_2], \ldots$
Thus

$$A_1(x_1 - a) \leqslant \int_a^{x_1} f(x)\, dx \leqslant B_1(x_1 - a),$$

$$\vdots$$

$$A_n(b - x_{n-1}) \leqslant \int_{x_{n-1}}^b f(x)\, dx \leqslant B_n(b - x_{n-1}).$$

If the subintervals are short and f is known with reasonable accuracy in each interval
(so that A_1, B_1 are close together, A_2, B_2 are close together, and so on), then each
integral here is known quite accurately, and so is the sum, as in (13–132). In
particular, we can select a known value of f in each subinterval and use that to
estimate the corresponding integral, that is,

$$\int_a^{x_1} f(x)\, dx \sim f(x_1^*)(x_1 - a), \qquad a \leqslant x_1^* \leqslant x_1,$$

$$\int_{x_1}^{x_2} f(x)\, dx \sim f(x_2^*)(x_2 - x_1), \qquad x_1 \leqslant x_2^* \leqslant x_2, \quad \text{etc.}$$

Here the first value $f(x_1^*)(x_1 - a)$ lies between $A_1(x_1 - a)$ and $B_1(x_1 - a)$ and is hence
close to the integral over the subinterval; a similar statement applies to the other
expressions. Hence, upon adding, we obtain the estimate

$$\int_a^b f(x)\, dx \sim f(x_1^*)(x_1 - a) + f(x_2^*)(x_2 - x_1) + \cdots + f(x_n^*)(b - x_{n-1}).$$

$$(13\text{–}133)$$

This is known as the *rectangular rule*, since it represents the integral as the sum of the
areas of rectangles (shown in Fig. 13–9). We have considerable freedom in choosing
$x_1^*, \ldots, x_n^*$; they may be chosen as endpoints of the subintervals or at the midpoint
(usually a better choice). It is a basic theorem of calculus that, for continuous f, the

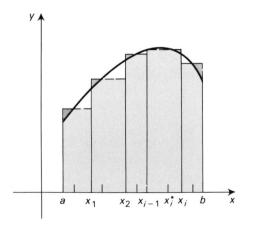

Fig. 13–9. Rectangular rule for integral.

sum on the right of (13–133) has the value on the left as limit $n \to \infty$ and the maximum length of a subinterval approaches zero. (This theorem shows that, although each term of (13–133) may differ from the integral over the subinterval, the cumulative error obtained by adding the terms can be made as small as desired by choosing all subintervals sufficiently small.)

Trapezoidal rule   If we join the successive points $(x_0, f(x))$, $(x_1, f(x_1))$, . . . , $(x_n, f(x_n))$ of the graph of f by line segments (Fig. 13–10), then we form trapezoids whose areas approximate the terms on the right of (13–132). Hence by geometry we are led to the estimates

$$\frac{f(x_0)+f(x_1)}{2}(x_1 - x_0), \quad \ldots, \quad \frac{f(x_{n-1})+f(x_n)}{2}(x_n - x_{n-1})$$

for these terms. This gives the *trapezoidal rule*

$$\int_a^b f(x)\,dx \sim \frac{1}{2}\{[f(x_0)+f(x_1)](x_1 - x_0) + \cdots + [f(x_{n-1})+f(x_n)](x_n - x_{n-1})\}.$$

$$(13\text{--}134)$$

For the case of equal subdivisions of length $h = (b - a)/n$, this becomes

$$\int_a^b f(x)\,dx \sim \frac{b-a}{2n}\{f(a)+2f(a+h)+ \cdots +2f[a + (n-1)h]+f(b)\}.$$

$$(13\text{--}134')$$

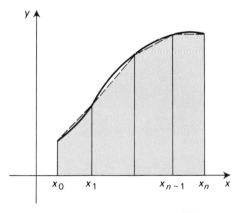

Fig. 13-10. Trapezoidal rule.

Simpson's rule   This rule is based on interpolation to $f(x)$ by a quadratic polynomial $p(x)$ at three successive values of x assumed equally spaced. Then the integral of $p(x)$ over the combined subinterval is taken as an approximation of the integral of f. If the successive values are $a, a+h, a+2h$, this gives Simpson's rule:

$$\int_a^{a+2h} f(x)\,dx \sim \frac{h}{3}[f(a)+4f(a+h)+f(a+2h)]. \qquad (13\text{--}135)$$

We can apply the same reasoning to the interval from a to b if we use an even number of subdivisions, chosen so that the first two, the next two, and so on, are of equal length. For the case of all equal subintervals of length h, we obtain

$$\int_a^b f(x)\,dx \sim \frac{h}{3}\{f(a) + 4f(a+h) + 2f(a+2h) + 4f(a+3h) + 2f(a+4h)$$

$$+ \cdots + 4f[a + (n-1)h] + f(b)\}. \tag{13–136}$$

EXAMPLE 1 We evaluate the interval of $f(x) = 1/x$ from 1 to 2, using the subdivision points 1, 1.2, 1.4, 1.6, 1.8, 2. The rectangular rule for midpoints gives

$$0.2f(1.1) + 0.2f(1.3) + \cdots + 0.2f(1.9) = 0.69191.$$

The trapezoidal rule with the same subdivision gives

$$\frac{1}{10}[f(1) + 2f(1.2) + \cdots + 2f(1.8) + f(2)] = 0.69564.$$

Simpson's rule (13–136) with subdivision at 1, 1.25, 1.5, 1.75, 2 gives

$$\frac{1}{12}[f(1) + 4f(1.25) + 2f(1.5) + 4f(1.75) + f(2)] = 0.69325.$$

The correct value is $\ln 2 = 0.69315$. We see that Simpson's rule clearly gives the best approximation. ◀

> REMARK: From its derivation, Simpson's rule (13–135) is correct if $f(x)$ coincides with a second-degree polynomial for $a \leqslant x \leqslant a + 2h$. It can be verified (see Problem 7 below) that the rule is also correct for third-degree polynomials.

Romberg integration   Let us *assume temporarily* that $b - a = 1$. Then Simpson's rule for the integral from a to $b = a + 2h$ gives the expression

$$S_1 = \frac{f(a) + 4f(a+h) + f(b)}{6}.$$

The trapezoidal rule for the whole interval (no subdivision) gives the expression

$$T_1 = \frac{f(a) + f(b)}{2}.$$

The trapezoidal rule for subdivision into two equal subintervals gives the expression

$$T_2 = \frac{f(a) + 2f(a+h) + f(b)}{4}.$$

It can now be verified by substitution that

$$S_1 = \frac{4}{3}T_2 - \frac{1}{3}T_1.$$

Thus Simpson's expression can be obtained as a *linear combination* of two trapezoidal expressions. We know that Simpson's rule gives the precise value for polynomials of degree at most three.

If we subdivide further into four equal intervals and apply the preceding reasoning to the first pair and second pair, we deduce that

$$S_2 = \frac{4}{3}T_3 - \frac{1}{3}T_2,$$

where S_2 is the Simpson expression and T_3 is the trapezoidal expression for the integral from a to b for this subdivision. We know that S_2 gives the precise value if f coincides with a polynomial of degree at most 3 in each of the two halves of the original interval.

Repeating the process by always doubling the number of subintervals we conclude that in general

$$S_n = \frac{4}{3}T_{n+1} - \frac{1}{3}T_n,$$

and S_n gives the correct value for the integral from a to b if $f(x)$ coincides with a polynomial of degree at most three in each of the 2^{n-1} equal subintervals.

In the same way we can derive a linear combination $C_n = pS_{n+1} + qS_n$ of two successive Simpson sums which is of higher-order accuracy; it is precise if $f(x)$ coincides with a polynomial of degree at most 5 in each of the 2^{n-1} equal subintervals. We find

$$p = \frac{16}{15} = \frac{4^2}{4^2 - 1}, \qquad q = -\frac{1}{15} = -\frac{1}{4^2 - 1}.$$

Thus

$$C_n = \frac{16}{15}S_{n+1} - \frac{1}{15}S_n = \frac{4^2}{4^2 - 1}S_{n+1} - \frac{1}{4^2 - 1}S_n.$$

The C_n are called *Cotes sums*.

Similarly, we obtain $D_n, E_n, \ldots$:

$$D_n = \frac{64}{63}C_{n+1} - \frac{1}{63}C_n = \frac{4^3}{4^3 - 1}C_{n+1} - \frac{1}{4^3 - 1}C_n,$$

$$E_n = \frac{256}{255}D_{n+1} - \frac{1}{255}D_n = \frac{4^4}{4^4 - 1}D_{n+1} - \frac{1}{4^4 - 1}D_n, \quad \text{etc.},$$

and the general pattern is clear. In each case we extend the accuracy by two: for D_n up to degree 7, for E_n up to degree 9, etc. In detail, for $b - a = 4h$, we get

$$C_1 = \frac{16}{15}S_2 - \frac{1}{15}S_1$$

$$= \frac{15f(a) + 64f(a+h) + 28f(a+2h) + 64f(a+3h) + 15f(b)}{90}, \quad (13\text{–}137)$$

and this is correct if f is of degree at most 5 over the whole interval from a to b.

We have been assuming $b - a = 1$, so that we could omit the factor $b - a$ in front of the previous expressions. Now we drop this assumption. All expressions have to be multiplied by $b - a$. Thus

$$(b-a)T_n, \quad (b-a)S_n, \quad (b-a)C_n, \quad (b-a)D_n, \quad (b-a)E_n, \quad \ldots$$

are all expressions for the integral of $f(x)$ from a to b, typically of increasing accuracy. In any case, for continuous f the basic theorem of calculus referred to above assures that as n becomes infinite, all approach the integral as limit. The Romberg method is a systematic way of calculating the expressions $T_n, S_n, \ldots$

For the T_n we can simplify matters by observing that

$$T_{n+1} = \frac{1}{2}(T_n + M_{n+1}),$$

where M_{n+1} is the average of the new values of f introduced in going from the nth stage to the $(n+1)$st stage. For example, if $a = 0$ and $b = 1$, then

$$T_1 = \frac{1}{2}\left[f(0) + f(1)\right],$$

$$T_2 = \frac{1}{4}\left[f(0) + 2f(\tfrac{1}{2}) + f(1)\right] = \frac{1}{2}\left\{\frac{1}{2}\left[f(0) + f(1)\right] + f(\tfrac{1}{2})\right\} = \frac{1}{2}(T_1 + M_2),$$

$$T_3 = \frac{1}{8}\left[f(0) + 2f(\tfrac{1}{4}) + 2f(\tfrac{1}{2}) + 2f(\tfrac{3}{4}) + f(1)\right]$$

$$= \frac{1}{2}\left\{\frac{1}{4}\left[f(0) + 2f(\tfrac{1}{2}) + f(1)\right] + \frac{1}{2}\left[f(\tfrac{1}{4}) + f(\tfrac{3}{4})\right]\right\} = \frac{1}{2}(T_2 + M_3),$$

and so on.

In going from the T_n to the S_n, from the S_n to the C_n, etc., it is useful to write

$$S_1 = T_2 + \frac{T_2 - T_1}{3}, \qquad S_2 = T_3 + \frac{T_3 - T_2}{3},$$

$$C_1 = S_2 + \frac{S_2 - S_1}{15}, \qquad C_2 = S_3 + \frac{S_3 - S_2}{15},$$

and so on, and to arrange these values as in Table 13–12. For the T_n the work can be arranged as suggested in Table 13–13 for the case $a = 0, b = 1$. Thus, on the first line of Table 13–13, M_1 is the average of the two values of f, and $M_1 = T_1$. On the second line M_2 is the value $f(\tfrac{1}{2})$, T_2 is the average of M_2 and T_1. On the third line M_3 is the average of the two f-values; T_3 is the average of M_3 and T_2.

Table 13–12

T_1				
T_2	S_1			
T_3	S_2	C_1		
T_4	S_3	C_2	D_1	
T_5	S_4	C_3	D_2	E_1

Table 13–13

$f(0)$		$f(1)$		M_1	T_1
$f(\tfrac{1}{2})$				M_2	T_2
$f(\tfrac{1}{4})$		$f(\tfrac{3}{4})$		M_3	T_3
$f(\tfrac{1}{8})$	$f(\tfrac{3}{8})$	$f(\tfrac{5}{8})$	$f(\tfrac{7}{8})$	M_4	T_4

EXAMPLE 2 $\int_0^{1.6} \sin x \, dx$. Corresponding to Table 13–13, we have Table 13–14. Here on the first line we have $f(0), f(1.6)$; on the second, $f(0.8)$; on the third, $f(0.4)$

Table 13–14

Stage					M	T
1	0	0.9957			0.49979	0.49979
2	0.71736				0.71736	0.60857
3	0.38942	0.93204			0.66073	0.63465
4	0.19867	0.56464	0.84147	0.98545	0.64756	0.64110
5	0.09983	0.29552	0.47943	0.64423	0.64432	0.64272
	0.78333	0.89121	0.96356	0.99749		

Table 13–15

Stage	T	S	C	D	E
1	0.49979				
2	0.60857	0.64483			
3	0.63465	0.64432	0.64324		
4	0.64110	0.64325	0.64324	0.64324	
5	0.64272	0.64325	0.64325	0.64325	0.64325

and $f(1.2)$, and so on. Using these T-values, we proceed as in Table 13–12; the results are given in Table 13–15. All these values have to be multiplied by $b - a = 1.6$ to give estimates for the integral. The best values should be obtained from the bottom line to the right. Hence we use the value $1.6 \cdot 0.64325 = 1.0292$ for the integral. The correct value is 1.029179. ◀

=================== **Problems (Section 13–13)** ===================

1. For $f(x) = e^x$ we have $f(1.98) = 7.2427, f(1.99) = 7.3155, f(2) = 7.3891, f(2.01) = 7.4633,$ $f(2.02) = 7.5383$. Estimate $f'(2)$ and compare with the exact value:
   a) Obtained by (13–120) with $h = 0.02$ and $h = 0.01$;
   b) Obtained by (13–121) with $h = 0.02$, $h = 0.01$, $h = -0.01$.

2. For $f(x) = e^x$, use the information given in Problem 1 to estimate $f''(2)$ and compare with the exact value:
   a) Obtained by (13–123) with $h = 0.02$ and $h = 0.01$;
   b) Obtained by (13–124) with $h = 0.01$ and $h = -0.01$.

3. a) Evaluate $\int_0^1 x^2\, dx$ by the trapezoidal rule (13–134′) and compare with the exact value, using $h = 1$, $h = 0.5$, $h = 0.25$.
   b) Evaluate the same integral by the rectangular rule (13–133), choosing the x_j^* as midpoints and using equal subdivisions as in part (a).

4. (a), (b). Proceed as in Problem 3 with $\displaystyle\int_0^1 \frac{dx}{1 + x}$.

5. Evaluate $\int_0^{\pi/2} \sin^{10} x \, dx$ by Simpson's rule (13–136), using $h = \pi/4$, then $h = \pi/8$. The

   exact value is $\dfrac{\pi}{2} \dfrac{1 \cdot 3 \cdot 5 \cdot 7 \cdot 9}{2 \cdot 4 \cdot 6 \cdot 8 \cdot 10}$.

6. Evaluate $\int_0^1 e^{-x^2} \, dx$ by Simpson's rule (13–136), using $h = 0.5$, then $h = 0.25$.

   HINT: Use $e^{-0.0625} = 0.93941$, $e^{-0.5625} = 0.56978$.

7. Show that Simpson's rule (13–135) is correct for $f(x) = 1, x, x^2,$ or x^3 and hence is correct for $f(x)$ a polynomial of degree at most 3.

8. The *improved trapezoidal rule* is

$$\int_a^{a+h} f(x)\,dx \sim \frac{h}{2}\left[f(a) + f(a+h)\right] - \frac{h^2}{12}\left[f'(a+h) - f'(a)\right].$$

   a) Use this rule to evaluate $\int_1^2 (1/x)\,dx$.

   b) Show, as in Problem 7, that the rule gives the precise value if $f(x)$ is a polynomial of degree at most 3.

9. Show that the Cotes formula (13–137), multiplied by $(b-a)$, gives the precise value for $\int_0^4 f(x)\,dx$ if $f(x)$ is a polynomial of degree at most 5.

10. Use Romberg integration to evaluate each of the following integrals to five decimal places.

   a) $\displaystyle\int_0^1 \frac{1}{1+x^2}\,dx,$ b) $\displaystyle\int_0^1 \frac{1}{1+x^3}\,dx$

   c) $\displaystyle\int_0^{1.6} \cos x \, dx,$ d) $\displaystyle\int_0^{1.6} e^{-x}\,dx$

13–14 INITIAL-VALUE PROBLEMS FOR ORDINARY DIFFERENTIAL EQUATIONS

We start with the first-order differential equation

$$y' = F(x, y), \tag{13–140}$$

where F satisfies continuity conditions such that the usual existence theorem (Section 1–1) holds. Thus there is a unique solution $y = f(x)$, $|x - x_0| \leq a$, of Eq. (13–140) such that $y = y_0$ for $x = x_0$, that is, $f(x_0) = y_0$. We describe methods for obtaining numerical estimates of $f(x)$. Typically we obtain numerical values for $f(x_0), f(x_1), f(x_2), \ldots$ with $x_0 < x_1 < x_2 < \cdots$ or $x_0 > x_1 > x_2 > \cdots$

The simplest numerical procedure is that of Section 1–3, usually called the *Euler method*. The differential equation is replaced by the difference equation

$$\Delta y = F(x, y)\,\Delta x \tag{13–141}$$

and $y_1 = f(x_1)$, $y_2 = f(x_2)$, $\ldots$ are calculated successively:

$$y_1 = y_0 + \Delta y = y_0 + F(x_0, y_0)\,\Delta x,$$
$$y_2 = y_1 + \Delta y = y_1 + F(x_1, y_1)\,\Delta x, \quad \text{etc.}$$

The size of Δx can be varied from step to step. If Δx is chosen small enough, then the values $y_1, y_2, \ldots$ are good estimates for $f(x_1), f(x_2), \ldots$ In particular, if $\Delta x = a/n$, then as $n \to \infty$ the estimates obtained for $f(x_0 + a) = f(x_0 + n\Delta x)$ approach the correct value of $f(x_0 + a)$ as limit.

In the following discussion we shall write h for Δx and generally assume that h remains the same from step to step. In practice we often compute a solution with chosen equal steps h, then try again with a smaller h, say half as large, to see whether the solution is changed significantly. We often find that a smaller h is needed for a certain x-interval because of large values of $F(x, y)$ or because of rapid changes in $F(x, y)$, whereas for other intervals a larger h can be used. The smaller the h, the more steps are needed to cover a given interval and hence the greater the danger of error accumulation due to round-off and other approximations. Therefore, a delicate balance is needed in selecting the best h.

Series solution   If $F(x, y)$ is expressible as a power series in x and y (Section 2–19) in a neighbourhood of (x_0, y_0), the solution $y = f(x)$ can be obtained by Taylor's series as in Section 2–22 (see Example 3). If this method is applied to the general case of Eq. (13–140), we obtain the following value for $y_1 = f(x_0 + h)$:

$$y_1 = f(x_0 + h) = y_0 + Fh + (F_x + F_y F)\frac{h^2}{2}$$

$$+ (F_{xx} + 2FF_{xy} + F^2 F_{yy} + F_x F_y + FF_y^2)\frac{h^3}{6}$$

$$+ [F_{xxx} + 3F_x F_{xy} + F_y F_{xx} + F_x F_y^2$$

$$+ F(3F_{xxy} + 5F_y F_{xy} + 3F_x F_{yy} + F_y^3)$$

$$+ F^2 (3F_{xyy} + 4F_y F_{yy}) + F^3 F_{yyy})]\frac{h^4}{24} + \cdots , \tag{13-142}$$

where $F, F_x = \partial F/\partial x, \ldots$ are all evaluated at (x_0, y_0). From this we see that the Euler method is equivalent to using only the first two terms of the series: $y_1 = y_0 + Fh$. The methods to follow in effect gain accuracy by adding more terms of the series, without actually computing derivatives of F.

Heun's method   For the special equation $y' = F(x)$, the value of y at $x_0 + a$ is $\int_{x_0}^{x_0 + a} F(x)dx$. It is easily seen that for this case the Euler method is a way of calculating the integral by a rectangular rule of Eq. (13–133). Heun's method is similarly related to the trapezoidal rule. For the value y_1 at $x = x_0 + h$ we use the estimate

$$y_1 = y_0 + \frac{m_0 + m_1}{2}h, \tag{13-143}$$

where $m_0 = y_0' = F(x_0, y_0)$ and m_1 is an estimate for $y_1' = F(x_1, y_1)$. Since we do not know y_1, we approximate it by $y_0 + m_0 h$ (the value given by the Euler method), and choose

$$m_1 = F(x_0 + h, y_0 + m_0 h). \tag{13-144}$$

The two formulas (13–143), (13–144) give y_1. Using (x_1, y_1) instead of (x_0, y_0), we can repeat the process to obtain (x_2, y_2), and so on.

EXAMPLE 1 $y' = xy^2 - y$, $y = 1$ for $x = 0$. We seek the solution for $0 \leqslant x \leqslant 1$. We try $h = 0.5$ and then $h = 0.25$. The results are shown in Tables 13–16 and 13–17. If, say, we want to have at least two significant figure accuracy, then we would not rely on the results of Table 13–17, but try a still smaller h, say 0.125. If this is done, we find that the results agree with those of Table 13–17, up to two significant figures. Hence we would conclude that $h = 0.25$ is small enough.

Table 13-16

x_0	y_0	$x_0 + h$	m_0	$y_0 + m_0 h$	m_1	$\Delta y = (m_0 + m_1)h/2$
0	1.000	0.5	−1.000	0.500	−0.375	−0.344
0.5	0.656	1.0	−0.441	0.435	−0.246	−0.172
1.0	0.484					

If we solve the problem by Taylor's series, we obtain

$$y = 1 - x + x^2 - x^3 + x^4 + \cdots$$

This suggests that the solution is given exactly by $y = 1/(1 + x)$ and substitution in the differential equation shows that this is correct. Accordingly, for $x = 1$ the value of y should be 0.5, in close agreement with Table 13–17.

Table 13-17

x_0	y_0	$x_0 + h$	m_0	$y_0 + m_0 h$	m_1	Δy
0	1.000	0.25	−1.000	0.75	−0.609	−0.201
0.25	0.799	0.50	−0.639	0.639	−0.435	−0.134
0.50	0.665	0.75	−0.444	0.554	−0.324	−0.096
0.75	0.569	1.00	−0.326	0.487	−0.250	−0.072 ◀
1.00	0.497					

Heun's method can be shown to be essentially equivalent to using the series solution (13–142) through the terms in h^2. It can also be considered a *predictor–corrector* method: first we predict y_1 to be $y_1^{(1)} = y_0 + m_0 h$, then we use the predicted value to obtain the corrected value

$$y_1^{(2)} = y_0 + \frac{h}{2}\left[m_0 + F(x_0 + h,\, y_1^{(1)}) \right].$$

Since $y_1^{(2)}$ should be better than $y_1^{(1)}$, we could use it to obtain an even better value

$$y_1^{(3)} = y_0 + \frac{h}{2}\left[m_0 + F(x_0 + h,\, y_1^{(2)}) \right],$$

and so on. In practice, we find that at most one such recorrection (using $y_1^{(3)}$) is worthwhile. If the results suggest further correction, it is better to use a smaller step.

In Example 1, with $h = 0.5$ and $x_0 = 0$, $y_0 = 1$, we found $y_1^{(1)} = 0.500$, $y_1^{(2)} = 0.656$. If we correct further, we obtain $y_1^{(3)} = 0.640$, so that we have actually made matters worse, and it is better to go to a smaller step.

Runge–Kutta method   We give one form of this method, a form that can be shown to agree with the series solution through the terms in h^4. We obtain y_1 as $y_0 + mh$, where m is a weighted average of four slopes $F(x, y)$. The formulas are as follows:

$$y_1 = y_0 + mh,$$
$$m = \frac{1}{6}(m_0 + 2m_1 + 2m_2 + m_3),$$
$$m_0 = F(x_0, y_0),$$
$$m_1 = F(x_0 + \frac{1}{2}h, y_0 + \frac{1}{2}m_0 h), \tag{13–145}$$
$$m_2 = F(x_0 + \frac{1}{2}h, y_0 + \frac{1}{2}m_1 h),$$
$$m_3 = F(x_0 + h, y_0 + m_2 h).$$

As an illustration we again take Example 1 with $h = 1$:

$$m_0 = F(0, 1) = -1, \qquad m_1 = F(0.5, 0.5) = -0.375,$$
$$m_2 = F(0.5, 0.8125) = -0.48242, \qquad m_3 = F(1, 0.51758) = -0.24969,$$
$$m = -0.49409, \qquad y_1 = 0.50591.$$

The result is almost as good as that obtained by Heun's method with four steps, as in Table 13–17.

Adams–Moulton method   This method assumes that we know four successive values y_0, y_1, y_2, y_3 of y corresponding to $x = x_0$, $x_0 + h = x_1$, $x_0 + 2h = x_2$, $x_0 + 3h = x_3$. The y-values can be found, for example, by the Runge–Kutta method. We then use the slopes $m_0 = F(x_0, y_0)$, $m_1 = F(x_1, y_1)$, $m_2 = F(x_2, y_2)$, $m_3 = F(x_3, y_3)$ to obtain an average slope

$$m^* = \frac{1}{24}(55m_3 - 59m_2 + 37m_1 - 9m_0).$$

With this slope we obtain a predicted y at $x_4 = x_0 + 4h$:

$$y_4^* = y_3 + m^* h.$$

This predicted value is then corrected by using $F(x_4, y_4^*) = m_4^*$ to produce a new average slope:

$$m = \frac{1}{24}(9m_4^* + 19m_3 - 5m_2 + m_1).$$

Finally we obtain the corrected value $y_4 = y_3 + mh$. If corrected and predicted values differ much, some adjustment is needed (the error in the corrected value should be about $1/14$ of the difference of corrected and predicted values)

EXAMPLE 2 $y' + y = 0$, $y = 1$ for $x = 0$. We take $h = 0.1$ and use the known solution $y = e^{-x}$ to obtain $y_1 = 0.90484$, $y_2 = 0.81873$, $y_3 = 0.74082$. Then

$m_0 = -1$, $m_1 = -0.90484$, $m_2 = -0.81873$, $m_3 = -0.74082$, so that $m^* = -0.70496$ and $y_4^* = 0.67359$. Thus $m_4^* = -0.67359$ and $m = -0.70621$, $y_4 = 0.67020$. The exact value is 0.67032. ◄

The Adams–Moulton formulas can be obtained by finding the value at y_4 of a polynomial of degree 4 of given value y_3 and slopes m_0, m_1, m_2, m_3 (predictor) and then finding the value at y_4 for given y_3, m_1, m_2, m_3, m_4 (corrector).

Systems of equations All the methods described above extend to systems of first-order equations and hence to nth-order equations. We illustrate the procedures for a pair of equations

$$\frac{dx}{dt} = F(x, y, t), \qquad \frac{dy}{dt} = G(x, y, t), \tag{13–146}$$

with given initial values x_0, y_0 for $t = t_0$. At $t = t_0 + h$, *Euler's method* gives

$$x = x_0 + F(x_0, y_0, t_0)h, \qquad y = y_0 + G(x_0, y_0, t_0)h.$$

Heun's method gives

$$x = x_0 + \frac{h}{2}[m_0 + F(x_0 + m_0h, y_0 + n_0h, t_0 + h)],$$

$$y = y_0 + \frac{h}{2}[n_0 + G(x_0 + m_0h, y_0 + n_0h, t_0 + h)],$$

where $m_0 = F(x_0, y_0, t_0)$, $n_0 = G(x_0, y_0, t_0)$. The *Runge–Kutta* method gives

$$x = x_0 + mh, \qquad y = y_0 + nh,$$

$$m = \frac{1}{6}(m_0 + 2m_1 + 2m_2 + m_3),$$

$$n = \frac{1}{6}(n_0 + 2n_1 + 2n_2 + n_3),$$

$$m_0 = F(x_0, y_0, t_0),$$

$$n_0 = G(x_0, y_0, t_0),$$

$$m_1 = F\left(x_0 + \frac{1}{2}m_0h, \, y_0 + \frac{1}{2}n_0h, \, t_0 + \frac{1}{2}h\right),$$

$$n_1 = G\left(x_0 + \frac{1}{2}m_0h, \, y_0 + \frac{1}{2}n_0h, \, t_0 + \frac{1}{2}h\right),$$

$$m_2 = F\left(x_0 + \frac{1}{2}m_1h, \, y_0 + \frac{1}{2}n_1h, \, t_0 + \frac{1}{2}h\right),$$

$$n_2 = G\left(x_0 + \frac{1}{2}m_1h, \, y_0 + \frac{1}{2}n_1h, \, t_0 + \frac{1}{2}h\right),$$

$$m_3 = F(x_0 + m_2h, \, y_0 + n_2h, \, t_0 + h),$$

$$n_3 = G(x_0 + m_2h, \, y_0 + n_2h, \, t_0 + h).$$

The *Adams–Moulton* method gives the predicted values x_4^*, y_4^* at $t_4 = t_0 + 4h$ in terms of the slopes at (x_0, y_0, t_0), (x_1, y_1, t_1), (x_2, y_2, t_2), (x_3, y_3, t_3), where $t_1 = t_0 + h, t_2 = t_0 + 2h, t_3 = t_0 + 3h$ and $x_1, y_1, \ldots$ are the values (assumed known) of x and y at t_1, t_2, t_3:

$$x_4^* = x_3 + m^*h, \qquad y_4^* = y_2 + n^*h,$$

$$m^* = \frac{1}{24}(55m_3 - 59m_2 + 37m_1 - 9m_0),$$

$$n^* = \frac{1}{24}(55n_3 - 59n_2 + 37n_1 - 9n_0),$$

$$m_i = F(x_i, y_i, t_i),$$

$$n_i = G(x_i, y_i, t_i) \text{ for } i = 0, 1, 2, 3.$$

Then the corrected values are

$$x_4 = x_3 + mh, \qquad y_4 = y_3 + nh,$$

$$m = \frac{1}{24}(9m_4^* + 19m_3 - 5m_2 + m_1),$$

$$n = \frac{1}{24}(9n_4^* + 19n_3 - 5n_2 + n_1),$$

where $m_4^* = F(x_4^*, y_4^*, t_4)$, $n_4^* = G(x_4^*, y_4^*, t_4)$. We illustrate the first two of these methods.

EXAMPLE 3 $\dfrac{d^2x}{dt^2} + \dfrac{dx}{dt} + x = e^{-t}$, $x = 1$ and $\dfrac{dx}{dt} = 2$ for $t = 0$. We seek x for $t = 1$. The corresponding system is

$$\frac{dx}{dt} = y, \qquad \frac{dy}{dt} = e^{-t} - x - y,$$

with $x = 1$ and $y = 2$ for $t = 0$. By the Euler method with $h = 1$,

$$x = 1 + 2 \cdot 1 = 3, \qquad y = 2 + (e^0 - 3) \cdot 1 = 0.$$

By Heun's method with $h = 1$,

$$x = 1 + \frac{1}{2}(2 + 0) = 2,$$

$$y = 2 + \frac{1}{2}(0 + e^{-1} - 3) = 0.6839.$$

Because of the large h, these values cannot be expected to be very accurate. For a more thorough discussion of numerical solution of ordinary differential equations, see GERALD.

════════════════════════════**Problems (Section 13-14)**════════════════════════════

1. Use Euler's method as indicated to find the value of y for $x = 1$:

   a) $\dfrac{dy}{dx} = 2x + y^2 - x^4$, $y = 0$ for $x = 0$, $h = 0.5$ (two steps, exact solution is $y = x^2$);

b) Same as (a) but $h = 0.25$ (four steps);

c) $\dfrac{dy}{dx} = \dfrac{x+1}{y^2}$, $y = 2$ for $x = 0$, $h = 0.5$ (exact solution is $y = [(3x^2/2) + 3x + 8]^{1/3}$).

d) Same as (c) but $h = 0.25$.

2. **(a)–(d)** Use Heun's method instead of Euler's method for Problem 1.

3. Use the Runge–Kutta method for Problem 1 (a) as follows:
   a) With $h = 1$ (one step);   **b)** With $h = 0.5$ (two steps).

4. The differential equation $y' = 2x + x^3 - xy$ has the solution $y = x^2$. Use the known values of y for $x = 1, 1.1, 1.2, 1.3$ to obtain y for $x = 1.4$ and 1.5 by the Adams–Moulton method.

5. Proceed as in Problem 4 with the equation $y' = 1 - y^2 - x^2 y^2$ and the known solution $y = x^{-1}$.

6. Use the Runge–Kutta method to find the values of x and $y = dx/dt$ for $t = 0.2$ for the solution of the differential equation

$$\frac{d^2 x}{dt^2} + x = 2e^t$$

such that $x = 0$ and $dx/dt = 1$ for $t = 0$ (exact solution is $x = e^t - \cos t$).

7. For the differential equation of Problem 6 use the known values of x and $y = dx/dt$ for $t = 0, 0.1, 0.2, 0.3$ to find the values of x and y for $t = 0.4$ by the Adams–Moulton method.

13–15 BOUNDARY-VALUE PROBLEMS FOR ORDINARY DIFFERENTIAL EQUATIONS

We have encountered such problems in Chapter 8, in cases in which explicit solution was possible. Here we consider more difficult problems, for which numerical methods are needed.

EXAMPLE 1 Find $y(x)$ for $0 \leqslant x \leqslant 1$ such that $y'' + x^4 y = 0$, $y(0) = 0$, $y(1) = 1$, (see Fig. 13–11).

Here we can reason as follows: If we knew $y'(0) = m_0$, we could find the desired solution by solving an initial-value problem, as in the preceding section. For $m_0 = 0$, we clearly get $y(x) \equiv 0$. For $m_0 = 1$, we get some function $y_0(x)$; let $y_0(1) = k_0$. If k_0

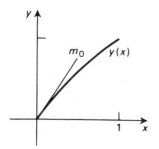

Fig. 13-11. Example 1.

happens to be 1, we are done. If not, we can simply choose h_0 so that $k_0 h_0 = 1$ and then use $m_0 = h_0$ for *multiplying m_0 by a constant leads to multiplication of the value of y at $x = 1$ by the same constant.* We justify this statement below. With initial values $y(0) = 0$ and $y'(0) = h_0$, we now obtain the solution $y(x)$ such that $y(1) = k_0 h_0 = 1$.

To justify our procedure, we observe that the given differential equation is homogeneous and linear. Its general solution can be written as

$$y = c_1 y_1(x) + c_2 y_2(x), \tag{13–150}$$

where $y_1(x)$, $y_2(x)$ are linearly independent solutions over the given interval. In particular, we can choose $y_1(x)$ such that $y_1(0) = 1$, $y_1'(0) = 0$, and $y_2(x)$ such that $y_2(0) = 0$, $y_2'(0) = 1$. With these choices, Eq. (13–150) gives the general solution, since we obtain the solution with arbitrary initial values y_0, y_0' for $x = 0$ as the function (13–150) with $c_1 = y_0$, $c_2 = y_0'$. For example,

$$y(0) = c_1 y_1(0) + c_2 y_2(0) = y_0 \cdot 1 + y_0' \cdot 0 = y_0.$$

It follows that those solutions with $y(0) = 0$ are given by

$$y = c_2 y_2(x), \qquad c_2 = y_0' = m_0.$$

Hence the value of y at $x = 1$ is proportional to m_0, as asserted.

We now solve the given equation numerically, with initial values $y(0) = 0$, $y'(0) = 1$. For simplicity, we use Euler's method with $h = 0.25$. Thus, following the procedure of Section 13–14, we consider the system

$$\frac{dy}{dx} = z, \qquad \frac{dz}{dx} = -x^4 y$$

and compute increments $\Delta y = z \, \Delta x$, $\Delta z = -x^4 y \Delta x$ with $\Delta x = h = 0.25$. For $x = 0$, we have $y = 0$, $z = y' = 1$, and $\Delta y = 0.25$, $\Delta z = 0$, so that at $x = 0.25$, $y = 0.25$, and $z = 1$. Similarly, we obtain $y = 0.50$, $z = 0.9998$ for $x = 0.5$; $y = 0.7499$, $z = 0.9920$ for $x = 0.75$, and $y = 0.9978$ for $x = 1$. We multiply this solution by $1/0.9978 = 1.0022$ to obtain the desired solution:

$$y(0) = 0, \quad y(0.25) = 0.2506, \quad y(0.5) = 0.5011, \quad y(0.75) = 0.7514, \quad y(1) = 1.$$

Of course, the accuracy can be greatly improved by using more steps or by using the Runge–Kutta method or Adams–Moulton method.

We could also have used power series. Substitution of $y = \sum c_n x^n$ in the differential equation and comparison of coefficients of each power of x leads to the solution:

$$y = c_1 \left(x - \frac{x^7}{7 \cdot 6} + \frac{x^{13}}{13 \cdot 12 \cdot 7 \cdot 6} - \frac{x^{17}}{17 \cdot 16 \cdot 13 \cdot 12 \cdot 7 \cdot 6} + \cdots \right).$$

The series clearly converges very rapidly for $0 \leqslant x \leqslant 1$. Here we took $c_0 = y(0) = 0$. The c_1 is our m_0. For $x = 1$, we find $y = 0.97634 c_1$; hence with $c_1 = 1.0242$ we obtain the desired solution such that $y(1) = 1$. In particular, $y(0.25) = 0.2560$, $y(0.75) = 0.7649$. These values are far more accurate than those obtained by the Euler method, but they show that the Euler method, simple as it is, can give good results. ◄

EXAMPLE 2 $y'' + x^4 y = 0$, $y(0) = 4$, $y(1) = 1$. Again we need to know $m_0 = y_0'$. However, the solutions with $y(0) = 4$ do not have value at $x = 1$ proportional to m_0. We can reason as above that the general solution is

$$y = y_0 y_1(x) + y_0' y_2(x),$$

where $y_1(x)$ is the solution with $y_1(0) = 1$, $y_1'(0) = 0$, and $y_2(x)$ the solution with $y_2(0) = 0$, $y_2'(0) = 1$. Here $y_0 = 4$ and $y_0' = m_0$. Hence

$$y(1) = 4y_1(1) + m_0 y_2(1),$$

and the value at $x = 1$ is a *linear function* of m_0. We must find both $y_1(1)$ and $y_2(1)$, then can choose m_0 so that $y(1) = 1$.

The function $y_2(x)$ was found in solving Example 1. We find $y_1(x)$ similarly by the Euler method: $y_1(0.25) = y_1(0.5) = 1$, $y_1(0.75) = 0.9976$, $y_1(1) = 0.9934$. Thus,

$$y(1) = 3.9904 + 0.9978 m_0.$$

Equating $y(1)$ to 1 gives $m_0 = -2.980$. Thus our desired solution is

$$y(x) = 4y_1(x) - 2.980 y_2(x),$$

so that $y(0) = 4$, $y(0.25) = 4 - 2.980 \cdot 0.2500 = 3.2550$, $y(0.5) = 2.5150$, $y(0.75) = 1.7560$, $y(1) = 1$. ◀

We remark that some problems of the type considered may have no solution or may have many solutions. For example, the problem $y'' + y = 0$, $y(0) = 0$, $y(\pi) = 1$ has no solution. Here the general solution is $y = c_1 \cos x + c_2 \sin x$, and the boundary conditions give $c_1 = 0$, $c_2 \sin \pi = 1$, so that c_2 cannot be found. The problem $y'' + y = 0$, $y(0) = 1$, $y(\pi) = 1$ is easily found to have infinitely many solutions $y = 1 + c \sin x$, where c is an arbitrary constant. Similar remarks apply to arbitrary linear differential equations with linear boundary conditions. The reason is that the general solution has the form

$$y = c_1 y_1(x) + \cdots + c_n y_n(x) + y^*(x).$$

By imposing linear boundary conditions we obtain linear equations for $c_1, \ldots, c_n$ as studied in Chapter 5. If these equations are homogeneous, then they have either the trivial solution $c_1 = 0, \ldots, c_n = 0$ or infinitely many solutions; if they are nonhomogeneous, then they may have a unique solution (and this holds when the corresponding homogeneous equations have only the trivial solution) or they may have infinitely many solutions or no solution (see Section 5–12).

When we solve the problems numerically, we change the coefficients in the linear equations slightly. Usually such slight changes will force the equations to have a solution; this corresponds to the effect of replacing parallel lines or planes by nearby intersecting lines or planes. Hence a solution found by a numerical method may not correspond to a solution of the given problem. However, if the given problem does have a unique solution, then small changes will usually preserve this property and will lead to a small change in the solution found.

From this discussion it is clear that some preliminary analysis is needed to ensure that a given boundary-value problem has a solution and that it can be found numerically. Very often physical reasoning or experience can provide the needed justification.

Difference equation method

An alternative method for solving problems such as Examples 1 and 2 is to subdivide the interval concerned by points x_i with $x_0 < x_1 < \ldots < x_n$ and approximate the derivatives at these points by difference expressions, as in Section 13–12. If the points are equally spaced at distance h and $y_i = y(x_i)$, then as in Eq. (13–123),

$$y''(x_i) \sim \frac{y_{i+1} - 2y_i + y_{i-1}}{h^2}.$$

By proceeding thus at each interior point x_i, we obtain $n-1$ equations for the $n-1$ unknown values $y_1, \ldots, y_{n-1}$. The equations for $i = 1$ and $i = n-1$ will use the known values y_0, y_n.

For Example 1, with $h = 0.25$, so that $h^2 = 1/16$, we obtain the three equations for y_1, y_2, y_3

$$16(y_2 - 2y_1 + y_0) + x_1^4 y_1 = 0, \qquad 16(y_3 - 2y_2 + y_1) + x_2^4 y_2 = 0,$$

$$16(y_4 - 2y_3 + y_2) + x_3^4 y_3 = 0,$$

with $y_0 = 0$, $y_4 = 1$, $x_1 = 0.25$, $x_2 = 0.5$, $x_3 = 0.75$. We solve easily to find $y_1 = 0.255$, $y_2 = 0.5096$, $y_3 = 0.762$, in fairly good agreement with the values found by the Euler method above.

It should be remarked that this procedure, when used for Example 1 with an arbitrary number of subdivision points leads to a *tridiagonal coefficient matrix*; that is, a matrix with all zeros except for the diagonal elements and along the two adjacent lines parallel to the diagonal. For such matrices, Gaussian elimination can be shortened as in Section 13–2 (see GERALD, pp. 129–133).

EXAMPLE 3 Choose λ so that the problem $y'' + (\lambda - x)y = 0$, $0 \leqslant x \leqslant 1$, $y(0) = 0$, $y(1) = 0$ has a solution, not identically 0, and find the solution.

This is a typical Sturm–Liouville problem, as discussed in Section 3–14. As pointed out there, under appropriate hypotheses on the coefficients, there is an infinite sequence $\lambda_0 \leqslant \lambda_1 \leqslant \lambda_2 \leqslant \ldots$ such that for $\lambda = \lambda_n$ the problem has a solution, and these give all solutions; each λ can appear at most twice in the sequence. The λ_n are the *eigenvalues* of the problem. Since the problem is *homogeneous*, for each λ_n there is a solution $y = \varphi_n(x)$, where $\varphi_n(x) \neq 0$, and $y = c\varphi_n(x)$ is also a solution for every choice of the nonzero constant c. If $\lambda_n = \lambda_{n+1}$, we obtain solutions $\varphi_n(x)$, $\varphi_{n+1}(x)$ which are linearly independent, and arbitrary linear combinations (excluding the trivial one reducing to 0) of these two solutions are solutions for this λ.

There are many methods for solving this problem numerically. Here we illustrate one, based on difference equations as above.

If we subdivide the interval $0 \leqslant x \leqslant 1$ by points x_i as above, we obtain the equations

$$\frac{y_{i+1} - 2y_i + y_{i-1}}{h^2} + (\lambda - x_i)y_i = 0, \qquad i = 1, \ldots, n-1,$$

where now $y_0 = 0$ and $y_n = 0$. These are simultaneous linear equations which can be written in the form

$$(A - \lambda I)y = 0,$$

where A is a symmetric square matrix of order $n-1$ and $\mathbf{y} = \mathrm{col}\,(y_1, \ldots, y_{n-1})$. This is the standard eigenvalue–eigenvector problem of Section 13–5 and can be treated by the methods of that section. For n small, we can use determinants; for large n, other methods are needed.

In any case, for each n we can obtain real eigenvalues $\lambda_0, \lambda_1, \ldots, \lambda_{n-2}$, arranged in increasing order. It can be shown that these should be treated as approximations to the true eigenvalues $\lambda_0, \ldots, \lambda_{n-2}$, and as $n \to \infty$ the approximate values converge to the true ones. Typically, the convergence is best for λ with small subscript (0, 1, 2, 3, for example), so that it is easiest to get a good approximation for the true eigenvalue λ_0, the lowest one.

For Example 3 we first take $n = 2$, so that $h = 0.5$ and there is just one unknown y_1, the value of y at $x = x_1 = 0.5$. The difference equation is

$$-\frac{2y_1}{h^2} + (\lambda - 1)y_1 = 0 \qquad \text{or} \qquad (\lambda - 9)y_1 = 0.$$

Thus we obtain $\lambda = 9$ only and y_1 can be chosen as 1, for example. Next we take $n = 4$, so that $h = 0.25$, and we have three unknowns y_1, y_2, y_3 at $x = 0.25, 0.5, 0.75$ respectively. The difference equations become

$$16\left(y_2 - 2y_1\right) + \left(\lambda - \frac{1}{4}\right)y_1 = 0,$$

$$16\left(y_3 - 2y_2 + y_1\right) + \left(\lambda - \frac{1}{2}\right)y_2 = 0,$$

$$16\left(-2y_3 + y_2\right) + \left(\lambda - \frac{3}{4}\right)y_3 = 0.$$

Equating the determinant of coefficients to 0, we obtain the cubic equation

$$\lambda^3 - 97.5\lambda^2 + 2649\lambda - 17690 = 0$$

with roots 10.0, 32.7, 54.3. Thus for the lowest eigenvalue we have the successive approximations 9 and 10.0. As a suggestion that we are approaching an eigenvalue, we observe that for $\lambda = 10.0$ the value of $\lambda - x$ varies between 9.0 and 10.0 over the interval $0 \leqslant x \leqslant 1$. Hence the solutions should be close to those of the equation

$$y'' + 9.5y = 0,$$

for which the coefficient of y is the average of 9.0 and 10.0. But this equation has the solution $y = \sin 3.082x$, which equals 0 at $x = 0$ and at $x = 1$ equals 0.016, a value very close to 0. For $\lambda = 10.0$ we find the eigenvector $(0.7191, 1, 0.6872)$ which closely approximates $y = \sin 3.082x$ at $x = 0.25$, $x = 0.75$. For $\lambda = 32.7$ we obtain the eigenvector $(-1.0, 0.028, 1)$; for $\lambda = 54.3$, we obtain $(-0.725, 1, -0.637)$.

In some problems the boundary conditions involve derivatives of the unknown function. In this case, when applying difference equations, we can imagine the function extended beyond the given interval. The boundary conditions and the difference equations at the endpoints then give equations by which the auxiliary values of the function can be eliminated. For example, for $y'' + xy = 0$ we might require $y'(0) = 1$ and $y(1) = 2$. We use the points $x_0 = 0$, $x_1 = h$, $x_2 = 2h$, ... as before but also use $x_{-1} = -h$ and, correspondingly, y_{-1}. Thus both y_0 and y_{-1} are unknown. At $x = 0$ we then have two equations

$$\frac{y_1 - 2y_0 + y_{-1}}{h^2} + 0 \cdot y_0 = 0, \qquad \frac{y_1 - y_{-1}}{2h} = 1.$$

The first expresses validity of the differential equation at the point. The second expresses the condition $y'(0) = 1$. We can eliminate y_{-1} from these two equations and then have n equations remaining for $y_0, y_1, \ldots, y_{n-1}$. ◄

The methods illustrated extend to a great variety of other linear boundary-value problems: we may use other forms of boundary conditions, the order of the differential equation may be higher (cases of order 4 occur in elasticity theory). In principle, nonlinear boundary-value problems can also be attacked in similar fashion; the difference-equation technique has been used effectively. However, in dealing with nonlinear problems we run into basic difficulties in establishing the existence and uniqueness of solutions and in computation of solutions when they exist.

═══════════════════ **Problems (Section 13-15)** ═══════════════════

1. Use Euler's method as indicated to solve the boundary-value problem given:
   a) $y'' + y = 0$, $y(0) = 0$, $y(\pi/2) = 1$, $0 \leqslant x \leqslant \pi/2$, use $h = \pi/8 = 0.3927$ (exact solution is $y = \sin x$);
   b) $y'' - 4y = 0$, $y(0) = 0$, $y(1) = 1$, $0 \leqslant x \leqslant 1$, use $h = 0.25$ (exact solution is $y = \sinh 2x/\sinh 2$);
   c) $y'' + xy = 0$, $y(0) = 2$, $y(1) = 3$, $0 \leqslant x \leqslant 1$, use $h = 0.2$;
   d) $y'' + xy = 0$, $y(0) = 1$, $y'(1) = -1$, $0 \leqslant x \leqslant 1$, use $h = 0.2$;
   e) $y'' + x^2 y = 0$, $y'(0) = 0$, $y'(1) = 1$, $0 \leqslant x \leqslant 1$, use $h = 0.2$;
   f) $y'' + x^2 y = 0$, $y(0) + y'(0) = 1$, $y(1) - y'(1) = 0$, $0 \leqslant x \leqslant 1$, use $h = 0.2$.

2. Use difference equations with h as above to solve the indicated part of Problem 1:
   a) Part (a)   b) Part (c)   c) Part (e)

3. Solve the eigenvalue problem by difference equations with h as indicated:
   a) $y'' + \lambda y = 0$, $0 \leqslant x \leqslant 1$, $y(0) = 0$, $y(1) = 0$, $h = 0.5$ (eigenvalues are $n^2 \pi^2$, eigenfunctions $\sin n\pi x$, $n = 1, 2, \ldots$);
   b) As in (a) with $h = 0.2$;
   c) $y'' + (\lambda - x^2)y = 0$, $0 \leqslant x \leqslant 1$, $y(0) = 0$, $y(1) = 0$, $h = 0.25$;
   d) $y'' + \lambda y = 0$, $0 \leqslant x \leqslant 1$, $y'(0) = 0$, $y(1) = 0$, $h = 1/3$, (eigenvalues are $(2n+1)^2 \pi^2/4$, eigenfunctions are $\cos (2n+1)\pi x/2$, $n = 0, 1, 2, \ldots$).
   e) $y'' + (\lambda - x^2)y = 0$, $0 \leqslant x \leqslant 1$, $y(0) = 0$, $y'(1) = 0$, $h = 0.25$.

13-16 NUMERICAL METHODS FOR PARTIAL DIFFERENTIAL EQUATIONS—ELLIPTIC EQUATIONS

As pointed out in Section 8-2, there are three basic types of second-order linear partial differential equations: elliptic, parabolic, hyperbolic. We shall discuss only such equations, beginning with the elliptic case and emphasizing equations with two independent variables. The principal tool is that of difference equations. (Another approach, the *finite-element method*, is discussed in Chapter 14.) For the difference-equation method we must approximate the partial derivatives by appropriate difference quotients. For a function $f(x, y)$, with $\Delta x = h$ and $\Delta y = k$, we typically use the following approximations:

$$f_x(x, y) \sim \frac{f(x+h, y) - f(x-h, y)}{2h},$$

$$f_y(x, y) \sim \frac{f(x, y+k) - f(x, y-k)}{2k},$$

$$f_{xx}(x, y) \sim \frac{f(x+h, y) - 2f(x, y) + f(x-h, y)}{h^2}, \tag{13-160}$$

$$f_{xy}(x, y) \sim \frac{f(x+h, y+k) - f(x-h, y+k) - f(x+h, y-k) + f(x-h, y-k)}{4hk},$$

$$f_{yy}(x, y) \sim \frac{f(x, y+k) - 2f(x, y) + f(x, y-k)}{k^2}.$$

There are similar expressions for higher derivatives and for functions of three or more variables (the formulas are derived from those of Section 13–12). In particular, the expression for $f_{xy}(x, y)$ is based on the difference expressions for first derivatives:

$$f_{xy} \sim \frac{f_x(x, y+k) - f_x(x, y-k)}{2k}$$

$$\sim \frac{1}{2k} \left[\frac{f(x+h, y+k) - f(x-h, y+k)}{2h} - \frac{f(x+h, y-k) - f(x-h, y-k)}{2h} \right].$$

For two independent variables, the prototype elliptic problem is that for the Laplace equation:

$$u_{xx} + u_{yy} = 0 \tag{13-161}$$

for the unknown function $u = f(x, y)$ in a region $\mathcal{D}$, with appropriate conditions on f on the boundary $\mathcal{C}$ of $\mathcal{D}$. Most commonly, $\mathcal{C}$ is a simple closed curve, $\mathcal{D}$ is the open region enclosed by $\mathcal{C}$, and f has given limiting values on $\mathcal{C}$. This is a *Dirichlet problem*. A similar problem arises for the general elliptic equation

$$Au_{xx} + Bu_{xy} + Cu_{yy} + Du_x + Eu_y + Fu = 0, \tag{13-162}$$

where $A = A(x, y)$, $B = B(x, y)$, ... and

$$B^2 - 4AC < 0 \quad \text{in } \mathcal{D}. \tag{13-163}$$

In the analytical methods of Chapter 8, only very special choices of the coefficients $A(x, y)$, $B(x, y)$, ..., $F(x, y)$ can be handled. For numerical methods, only some continuity conditions are needed. The condition (13–163) is also usually required to hold in a stronger form:

$$B^2 - 4AC \leqslant \delta < 0 \quad \text{in } \mathcal{D} \tag{13-163'}$$

for some constant δ. Instead of giving the values of f on $\mathcal{C}$, we may give the values of the normal derivative $\partial f/\partial n$ along $\mathcal{C}$ (Neumann problem), or some other condition involving f and $\partial f/\partial n$.

EXAMPLE 1 $u_{xx} + u_{yy} = 0$, $0 < x < 4$, $0 < y < 4$, $u = x^2 + 2x$ for $y = 0$,
$0 \leqslant x \leqslant 4$, $u = -2y - y^2$ for $x = 0$, $0 \leqslant y \leqslant 4$, $u = x^2 + 2x - 24$ for $y = 4$,
$0 \leqslant x \leqslant 4$, $u = 24 - y^2 - 2y$ for $x = 4$, $0 \leqslant y \leqslant 4$. The problem was constructed to
have the known solution $u = (x + 1)^2 - (y + 1)^2$. We observe that $B^2 - 4AC = -4$
in the region, so that (13–163′) is satisfied. We now subdivide the given region, a
square, by parallels to the axes spaced $\Delta x = h$ and $\Delta y = k$ apart, and consider the
values of u only at the points (x_i, y_j) where these lines meet; we call these points *mesh
points*. To be specific, we choose $h = 1$, $k = 1$ and consider u at the mesh points (i, j)
with $0 \leqslant i \leqslant 4, 0 \leqslant j \leqslant 4$. Here the values for $i = 0$ or 4 and for $j = 0$ or 4 are known,
as shown in Fig. 13–12. At the nine interior points we have the equations

$$u_{i+1,j} - 2u_{ij} + u_{i-1,j} + u_{i,j+1} - 2u_{ij} + u_{i,j-1} = 0, \qquad (13\text{–}164)$$

as follows from Eq. (13–160); here we have written u_{ij} for $u(i, j)$. There are nine linear
equations in nine unknowns. We write out a few of the equations:

for $i = 1, j = 1$: $u_{21} - 2u_{11} - 3 + u_{12} - 2u_{11} + 3 = 0$,
for $i = 2, j = 1$: $u_{31} - 2u_{21} + u_{11} + u_{22} - 2u_{21} + 8 = 0$,
for $i = 3, j = 2$: $16 - 2u_{32} + u_{22} + u_{33} - 2u_{32} + u_{31} = 0$.

Thus the equations are nonhomogeneous. It can be shown that they have a unique
solution. In fact, because the given problem has as known solution a *second-degree*
polynomial in x and y, the difference expressions *equal* the derivatives, and the
equations are satisfied exactly by

$$u(i, j) = (i + 1)^2 - (j + 1)^2,$$

as can be verified. The values of the solution are shown in Fig. 13–12.

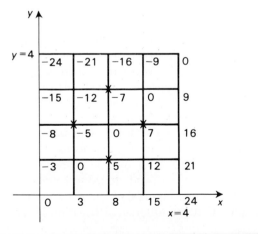

Fig. 13-12. Example 1.

If we had not known the solution, we could of course have found it by standard
elimination procedures, as in Section 13–2. For nine equations in nine unknowns,
this is not difficult, especially with the aid of a computer. Typically, of course, the

solution of the difference equations gives only an approximation to the values of the
solution $u(x, y)$ at the mesh points. However, it can be shown that, as $\Delta x = \Delta y \to 0$,
the maximum error approaches 0. ◄

All these remarks apply equally well to a general elliptic equation (13–162). For
the specific case of the Laplace equation, with $\Delta x = \Delta y$, the approximating
difference equation (13–164) has a special form which is helpful. We can write
(13–164) in the following way:

$$u(i,j) = \frac{1}{4}\left\{u(i+1, j) + u(i, j+1) + u(i-1, j) + u(i, j-1)\right\}. \quad (13\text{–}164')$$

Thus the value of u at each mesh point equals the average of the values of u at the
four adjacent mesh points; in Fig. 13–12, the adjacent points for $i = 2, j = 2$ are
marked by crosses. In Example 1, $h = k = 1$, but the statement in terms of averages
applies generally as long as $h = k$. This property is the basis of an iteration procedure
called *Liebmann's method* for solving the simultaneous equations for $u(x_i, y_j)$. Initial
guesses are made for all $u(x_i, y_j)$, then these are corrected by successively replacing
each $u(x_i, y_j)$ by the average of the adjacent values. The process is repeated by
"scanning" the set of mesh points (x_i, y_j). The corrected values then converge to the
solution sought. Such a procedure is helpful, especially if the number of mesh points
is large. (For more information on Liebmann's method and the related relaxation
method, see GERALD.)

Another iteration procedure is the *alternating-direction implicit method*. Here all
the values of u along each line $y = \text{const} = y_j$ are improved by regarding the
difference equation as that for the knowns $u(x_i, y_j)$, $i = 1, 2, \ldots$, in terms of the
values of u at other points, as found from the previous iteration. Thus Eq. (13–164)
can be written as

$$u^*_{i+1,j} - 4u^*_{ij} + u^*_{i-1,j} = -(u_{i, j+1} + u_{i, j-1}), \qquad i = 1, 2, \ldots,$$

by using the known values from the previous iteration on the right. These are
equations for $u^*_{1j}, u^*_{2j}, \ldots$ of tridiagonal form and are hence easily solved to give
corrected values u^*_{ij}. After sweeping through the lines $y = \text{const}$, the roles of x and y
are reversed and solutions are found for u along lines $x = \text{const}$ (alternating
directions). (For more details see GERALD.)

EXAMPLE 2 $u_{xx} + u_{yy} = xy$ in the square $0 \leqslant x \leqslant 3, 0 \leqslant y \leqslant 3$, with $\partial u/\partial n = 0$
on the boundary. The differential equation is now a nonhomogeneous elliptic
equation (of the form of a Poisson equation $\nabla^2 u = g(x, y)$) and the problem is a
Neumann problem. We again take $h = k = 1$, $u(i, j) = u_{ij}$ and are led to equations
for u_{ij} with $i = 1, 2, j = 1, 2$ (Fig. 13–13). However, the boundary values of u are not
known. In order to obtain them, we assume that the differential equation holds
along the boundary and apply it in difference-equation form at each boundary point.
This brings in the values of u at 16 points outside the square. The condition $\partial u/\partial n =$
0 in difference-equation form relates these 16 values to those at interior points:
$u(1, 1) = u(-1, 1), u(1, 2) = u(-1, 2), \ldots$; at the corners, $\partial u/n = 0$ is assumed to
mean $\partial u/\partial x = 0$ and $\partial u/\partial y = 0$. Thus the external values can be eliminated and we

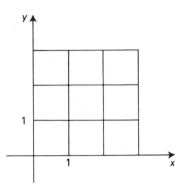

Fig. 13-13. Example 2.

are left with 16 linear equations for the values of u at the four interior points and 12 boundary points. The equations are nonhomogeneous and hence a unique solution might be expected. However, the corresponding homogeneous problem: $u_{xx} + u_{yy} = 0$ and $\partial u/\partial n = 0$ on the boundary, has a nontrivial solution: namely, $u \equiv \text{const} \neq 0$. It can be shown that there are no other nontrivial solutions. Therefore, the solutions of the given equation are determined only up to an additive constant. To obtain the solutions, we arbitrarily fix $u(0, 0)$ at 0 and then obtain a unique solution. To this solution we finally add an arbitrary constant.

The solution of the simultaneous equations is not difficult. We observe that the problem is unaffected by interchanging x and y. Hence we must have $u(x, y) = u(y, x)$, that is, $u(i, j) = u(j, i)$. This greatly reduces the number of unknowns. We find by very simple elimination the following values: $u(0, 0) = 0$, $u(1, 0) = 0$, $u(2, 0) = 0.1884$, $u(3, 0) = 0.5070$, $u(1, 1) = -0.1884$, $u(2, 1) = 0.1232$, $u(2, 1) = 0.1232$, $u(4, 1) = 0.8257$, $u(2, 2) = 0.3957$, $u(3, 2) = 5.042$, $u(3, 3) = 2.615$. The other values are found by interchanging x and y. ◄

If the region $\mathscr{D}$ is rectangular (with commensurable sides, parallel to the x and y axes), the procedures illustrated above can be applied. If $\mathscr{D}$ has a polygonal boundary with edges parallel to the axes, as in Fig. 13-14, they still apply, with obvious modifications.

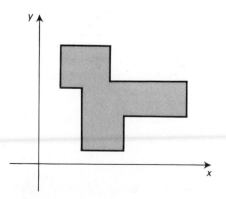

Fig. 13-14. Region bounded by polygon.

For a *circular region*, say $x^2 + y^2 < 1$, we can use polar coordinates r, θ and rewrite the differential equation (13–162) in terms of these coordinates by the equations (see Sections 9–16 and 9–17):

$$u_x = \cos\theta\, u_r - \frac{\sin\theta}{r}\, u_\theta, \qquad u_y = \sin\theta\, u_r + \frac{\cos\theta}{r}\, u_\theta,$$

$$u_{xx} = \cos^2\theta\, u_{rr} - \frac{2\sin\theta\cos\theta}{r}\, u_{r\theta} + \frac{\sin^2\theta}{r^2}\, u_{\theta\theta} + \frac{\sin^2\theta}{r}\, u_r + \frac{2\sin\theta\cos\theta}{r^2}\, u_\theta,$$

$$u_{xy} = \frac{2\sin\theta\cos\theta}{r}\, u_{rr} + \frac{\cos^2\theta - \sin^2\theta}{r}\, u_{r\theta} - \frac{\sin\theta\cos\theta}{r^2}\, u_{\theta\theta} - \frac{\sin\theta\cos\theta}{r}\, u_r$$
$$+ \frac{\sin^2\theta - \cos^2\theta}{r}\, u_\theta,$$

$$u_{yy} = \sin^2\theta\, u_{rr} + \frac{2\sin\theta\cos\theta}{r}\, u_{r\theta} + \frac{\cos^2\theta}{r^2}\, u_{\theta\theta} + \frac{\cos^2\theta}{r}\, u_r - \frac{2\sin\theta\cos\theta}{r^2}\, u_\theta.$$

Region $\mathscr{D}$ can then be subdivided by equally spaced rays from the origin and equally spaced circles centered at the origin, as illustrated in Fig. 13–15 (with the circles of radii 0, 1/3, 2/3, 1 and rays $\theta = k\pi/6$ for $k = 0, 1, \ldots, 11$). At a typical point (r, θ) of intersection of a ray and a circle we then replace $u_r, u_\theta, u_{rr}, u_{r\theta}, u_{\theta\theta}$ by difference expressions as in (13–160), with x replaced by r, y by θ, h by Δr, k by $\Delta\theta$, where Δr and $\Delta\theta$ are the spacings of circles and rays respectively. For example,

$$u_r \sim \frac{u(r + \Delta r, \theta) - u(r - \Delta r, \theta)}{2\Delta r}.$$

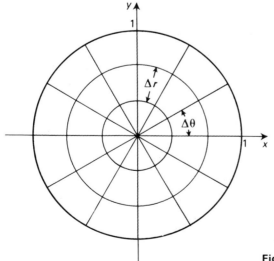

Fig. 13–15. Subdivision of circular region.

At $(0, 0)$, where $r = 0$, it is better to retain the expressions in x, y and use the rays $\theta = 0$, $\theta = \pm\pi/2$, $\theta = \pi$, so that

$$u_x \sim \frac{u(\Delta r, 0) - u(\Delta r, \pi)}{2\Delta r}, \qquad u_y \sim \frac{u(\Delta r, \pi/2) - u(\Delta r, 3\pi/2)}{2\Delta r},$$

$$u_{xx} \sim \frac{u(\Delta r, 0) - 2u(0, 0) + u(\Delta r, \pi)}{\Delta r^2},$$

$$u_{yy} \sim \frac{u(\Delta r, \pi/2) - 2u(0, 0) + u(\Delta r, 3\pi/2)}{\Delta r^2},$$

where on the right u is expressed in terms of polar coordinates. For $u_{xy}(0, 0)$ we can use the expression

$$\frac{u(\Delta x, \Delta y) - u(\Delta x, -\Delta y) - u(-\Delta x, \Delta y) + u(-\Delta x, -\Delta y)}{4\Delta x \Delta y},$$

where $\Delta x = \Delta r \cos\Delta\theta$, $\Delta y = \Delta r \sin\Delta\theta$; this assumes that the points $(\Delta r, +\Delta\theta)$, $(\Delta r, \pi \pm \Delta\theta)$ are mesh points.

By these formulas a given elliptic differential equation for $\mathscr{D}$ is replaced by a set of difference equations for the values of $u(r, \theta)$ at the mesh points. For a Dirichlet problem, $u(1, \theta)$ is known (for the circular region $x^2 + y^2 < 1$); for a Neumann problem, $\partial u/\partial r$ is known for $r = 1$.

For regions with other types of boundaries, the problem can often be reduced to the circular or rectangular case by introducing appropriate curvilinear coordinates. For example, in case of an elliptic region

$$\frac{x^2}{a^2} + \frac{y^2}{b^2} \leqslant 1,$$

the substitution $x_1 = x/a$, $y_1 = y/b$ reduces the problem to the circular case. For a region $\varphi(x) \leqslant y \leqslant \psi(x)$, $a \leqslant x \leqslant b$, as in Fig. 13–16, with $\varphi(x) < \psi(x)$ for $a \leqslant x \leqslant b$, the substitution

$$x_1 = x, \qquad y_1 = \frac{y - \varphi(x)}{\psi(x) - \varphi(x)}$$

reduces the problem to that for the rectangular region $a \leqslant x_1 \leqslant b$, $0 \leqslant y_1 \leqslant 1$. A variety of conformal mappings are also known (see Sections 11–25 and 11–26), and

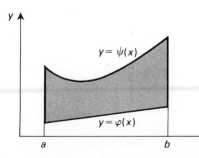

Fig. 13-16. Region reducible to a rectangular one.

they can also help in reducing the problem to a circular or rectangular case. (For further discussion of irregular boundaries see GERALD, pp. 361–370.)

The difference-equation procedures of this section also extend to three-dimensional problems, but the number of equations is usually much larger, for comparable accuracy. For higher dimensions, the problems easily become extremely complex and only the rapid pace of computer development renders them accessible.

13–17 PARABOLIC EQUATIONS

Here $B^2 - 4AC = 0$ for our basic second-order equation (13–162). The prototype equation is the heat equation. We consider a simple example to illustrate the special questions that arise.

EXAMPLE 1 $u_{xx} - u_t = 0$, $0 < x < \pi$, $0 < t < \infty$, $u(0, \ t) = 0$, $u(\pi, \ t) = 0$, $u(x, \ 0) = \sin x$. This problem is easily solved as in Section 8–5 to yield $u = e^{-t} \sin x$.

We want to try numerical methods and compare our results. To that end we subdivide the region $0 \leqslant x \leqslant \pi$, $0 \leqslant t < \infty$ by lines parallel to the axes at equal spacings Δx, Δt respectively, as suggested in Fig. 13–17. The values of u are known at all mesh points on the boundary. At an interior point (x, t), we have the difference equation

$$\frac{u(x + h, \ t) - 2u(x, \ t) + u(x - h, \ t)}{h^2} = \frac{u(x, \ t + k) - u(x, \ t)}{k}, \qquad (13\text{–}170)$$

where we have written h for Δx and k for Δt. On the right we have used a *forward difference quotient* to approximate $\partial u / \partial t$. This is natural for $t = 0$, since we have only values for $t = 0$ given; Eq. (13–170) then enables us to find $u(x, k)$ for each x in terms of the given values for $t = 0$. Having found these, we can again use the equation to

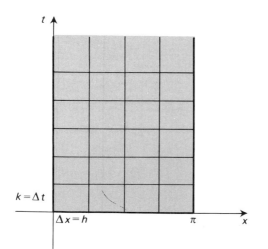

Fig. **13-17**. Subdivision for Example 1.

find $u(x, 2k)$, and so on. This *explicit method* is much like the Euler method for ordinary differential equations. It can be shown to give good results provided

$$r = \frac{\Delta t}{\Delta x^2} = \frac{k}{h^2} < \frac{1}{2}.$$

For greater values of the ratio r, the values found oscillate about the correct ones. This restriction on r requires many steps in time t to cover a large interval with high accuracy.

It has been found to be far more satisfactory to use an *implicit method*, the Crank–Nicholson method. Here we reason that the expression on the right of Eq. (13–170) should be treated as an approximation to $\partial u/\partial t$ at $(x, t + \frac{1}{2}k)$. The left side approximates $\partial^2 u/\partial x^2$ at (x, t). To obtain an approximation for $\partial^2 u/\partial x^2$ at $(x, t + \frac{1}{2}k)$, the difference expressions for t and for $t + k$ are averaged. This process leads to the equation

$$\frac{1}{2}\left[\frac{u(x + h, t) - 2u(x, t) + u(x - h, t)}{h^2} \right.$$
$$\left. + \frac{u(x + h, t + k) - 2u(x, t + k) + u(x - h, t + k)}{h^2} \right]$$
$$= \frac{u(x, t + k) - u(x, t)}{k}. \tag{13–171}$$

If we apply this equation for $t = 0$, we obtain a set of simultaneous equations (with tridiagonal coefficient matrix) for the unknown values $u(x, k)$, where $(x = h, 2h, \ldots, \pi - h)$. These equations are easily solved to give all $u(x, k)$. Then the process is repeated to give all $u(x, 2k)$, and so on.

The size of the ratio r is again important, but there is no critical value of r above which the method is unusable.

If, for example, we take $\Delta x = \Delta t = \pi/4$, then $r = 4/\pi = 1.27324$. We know $u(0, 0) = 0$, $u(h, 0) = 0.707$, $u(2h, 0) = 1$, $u(3h, 0) = 0.707$, $u(4h, 0) = u(\pi, 0) = 0$. Equation (13–171) gives us three equations for $u(h, k)$, $u(2h, k)$, $u(3h, k)$ which can be written thus:

$$-4.546\, u(h, k) + 1.273\, u(2h, k) = -0.887,$$
$$1.273\, u(h, k) - 4.546\, u(2h, k) + 1.273\, u(3h, k) = -1.254,$$
$$1.273\, u(2h, k) - 4.546\, u(3h, k) = -0.887.$$

Solving, we find

$$u(h, k) = u(3h, k) = 0.323, \qquad u(2h, k) = 0.457.$$

The exact values are 0.322 and 0.456. (The explicit method with the same r gives 0.180 and 0.254 respectively.) ◀

The method can be extended to more general parabolic equations in t and x. Extension to equations in t, x, and y, such as the heat equation

$$\frac{\partial u}{\partial t} = c^2 \left(\frac{\partial^2 u}{\partial x^2} + \frac{\partial^2 u}{\partial y^2} \right),$$

leads to more difficult simultaneous equations at each stage. A form of the alternating-direction implicit procedure can be used to recover a tridiagonal coefficient matrix (see GERALD, pp. 411–416, for details).

13–18 HYPERBOLIC EQUATIONS

Here the prototype equation is the wave equation

$$u_{tt} - c^2 u_{xx} = 0 \qquad (13\text{–}180)$$

and a typical problem is to solve Eq. (13–180) on the region $0 \leqslant x \leqslant 1, 0 \leqslant t < \infty$, with given values of $u(x, 0)$, $u_t(x, 0)$, $u(0, t)$, and $u(1, t)$. We subdivide the region by equally spaced parallels to the axes, as in Fig. 13–18. It turns out to be advantageous to choose $h = ck$ (with $h = \Delta x$ and $k = \Delta t$). Equation (13–180) is then replaced by the difference equation

$$\frac{u(x, t+k) - 2u(x, t) + u(x, t-k)}{k^2} = c^2 \frac{u(x+h, t) - 2u(x, t) + u(x-h, t)}{h^2},$$

which simplifies to

$$u(x, t+k) = u(x+h, t) + u(x-h, t) - u(x, t-k). \qquad (13\text{–}181)$$

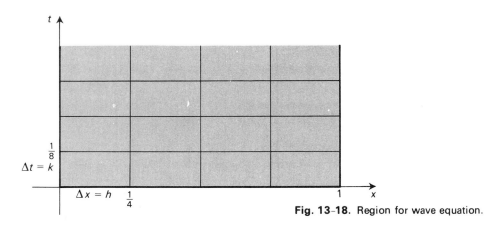

Fig. 13-18. Region for wave equation.

If now we know the values of u at the mesh points on two successive horizontal lines $t = \text{const}$, then Eq. (13–181) permits us to find u on the next higher line. To get started, we know u for $t = 0$. We can effectively give values for $t = -k$ by using the given initial values of u_t:

$$\frac{u(mh, k) - u(mh, -k)}{2k} = c_m = \text{Given value of } u_t(mh, 0).$$

This equation, along with Eq. (13–181) for $x = mh, t = 0$, gives two conditions from which $u(x, -k)$ can be eliminated to give

$$u(mh, k) = kc_m + \frac{1}{2}\left\{u[(m+1)h, 0] + u[(m-1)h, 0]\right\}. \qquad (13\text{–}182)$$

EXAMPLE 1 We consider Eq. (13–180) with $c = 2$ in the region of Fig. 13–18 and assume that $u(x, 0) = \sin \pi x$, $u_t(x, 0) = 0$, $u(0, t) = u(1, y) = 0$. We take $k = 1/8$ and hence $h = 1/4$. Thus the numbers c_m are 0 and Eq. (13–182) gives

$$u\left(\frac{m}{4}, \frac{1}{8}\right) = \frac{1}{2}\left[u\left(\frac{m+1}{4}, 0\right) + u\left(\frac{m-1}{4}, 0\right)\right]$$

for $m = 1, 2, 3$. Accordingly,

$$u\left(\frac{1}{4}, \frac{1}{8}\right) = \frac{1}{2}\left[\sin\frac{\pi}{2} + \sin 0\right] = \frac{1}{2},$$

$$u\left(\frac{1}{2}, \frac{1}{8}\right) = \frac{1}{2}\left[\sin\frac{3\pi}{4} + \sin\frac{\pi}{4}\right] = \frac{\sqrt{2}}{2} = 0.707,$$

$$u\left(\frac{3}{4}, \frac{1}{8}\right) = \frac{1}{2}\left[\sin\pi + \sin\frac{\pi}{2}\right] = \frac{1}{2}.$$

The values $u(0, 1/8) = u(1, 1/8) = 0$ are given. Next, by Eq. (13–181),

$$u\left(\frac{1}{4}, \frac{1}{4}\right) = u\left(\frac{1}{2}, \frac{1}{8}\right) + u\left(0, \frac{1}{8}\right) - u\left(\frac{1}{4}, 0\right) = \frac{\sqrt{2}}{2} + 0 - \frac{\sqrt{2}}{2} = 0,$$

$$u\left(\frac{1}{2}, \frac{1}{4}\right) = u\left(\frac{1}{4}, \frac{1}{8}\right) + u\left(\frac{3}{4}, \frac{1}{8}\right) - u\left(\frac{1}{2}, 0\right) = \frac{1}{2} + \frac{1}{2} - 1 = 0,$$

$$u\left(\frac{3}{4}, \frac{1}{4}\right) = u\left(1, \frac{1}{8}\right) + u\left(\frac{1}{2}, \frac{1}{8}\right) = u\left(\frac{3}{4}, 0\right) = 0 + \frac{\sqrt{2}}{2} - \frac{\sqrt{2}}{2} = 0.$$

The process can be repeated to obtain $u(x, t)$ for arbitrarily large t.

For this example, we find the exact solution as in Section 8–3:

$$u = \frac{1}{2}\left\{\sin[\pi(x - 2t)] + \sin[\pi(x + 2t)]\right\} = \sin(\pi x)\cos(2\pi t).$$

Thus $u(1/4, 1/8) = 1/2$, $u(1/2, 1/8) = \sqrt{2}/2$, $u(3/4, 1/8) = 1/2$, $u(x, 1/4) = 0, \ldots$, in precise agreement with the numerical results. The remarkable agreement is due to the special form of the solution (Section 8–4):

$$u = \frac{1}{2}\left[f(x + ct) + f(x - ct)\right].$$

The method extends to equations of form

$$A(x, y)u_{xx} + C(x, y)u_{yy} + D(x, y)u_x + E(x, y)u_y + F(x, y) = 0,$$

where $A(x, y) > 0$, $C(x, y) < 0$ in the region considered (so that $B^2 - 4AC = -4AC > 0$) and to problems in more variables, such as

$$u_{tt} - c^2(u_{xx} + u_{yy}) + p(x, y, t)u_t + q(x, y, t)u_x + r(x, y, t)u_y + s(x, y, t)u = 0,$$

in similar regions. ◄

======================Problems (Section 13–18):======================

1. Solve the boundary-value problem $u_{xx} + u_{yy} = 0$ for $0 \leqslant x \leqslant 3$, $0 \leqslant y \leqslant 4$, $u(x, 0) = x^3$, $u(x, 4) = x^3 - 48x$, $u(0, y) = 0$, $u(3, y) = 27 - 9y^2$ by considering u only at the mesh points (i, j) for $i = 0, 1, 2, 3, j = 0, 1, \ldots, 4$. (The exact solution is $u = x^3 - 3xy^2$.)

2. Solve the boundary-value problem $(x + 1)u_{xx} + u_{yy} = 0$ for $0 \leqslant x \leqslant 3$, $0 \leqslant y \leqslant 3$, $u(x, 0) = x$, $u(x, 3) = 3 + x$, $u(0, y) = y$, $u(3, y) = 3 + y$ by considering u only at the mesh points (i, j) for $i = 0, 1, 2, 3, j = 0, 1, 2, 3$.

3. Solve the boundary-value problem $u_{xx} + u_{yy} = 0$ for $x^2 + y^2 \leqslant 1$, with $\partial u/\partial n = \sin \theta$ at the point $(\cos \theta, \sin \theta)$, using polar coordinates and considering u only at the points (r, θ) with $r = 0, 1/2, 1, \theta = k\pi/4$ for $k = 0, 1, \ldots, 7$.

   HINT: Since the solution is determined only up to an additive constant, we can write it as $u_0(r, \theta) + \text{const}$, where $u_0(r, \theta)$ is a solution, which is chosen to have value 0 at the origin. By symmetry we see that

   $$u_0(r, 0) = u_0(r, \pi) = 0,$$

   $$u_0(r, \theta) = -u_0(r, -\theta) = u_0(r, \pi - \theta).$$

   REMARK: It can be shown that such a Neumann problem has solutions if and only if $\oint_C (\partial u/\partial n) \, ds = 0$, where C is the boundary curve. This condition is to be expected, since, as in Section 10–7,

   $$\oint_C \frac{\partial u}{\partial n} \, ds = \iint_R \nabla^2 u \, dx \, dy,$$

   where R is the region enclosed by C. Here $\nabla^2 u = 0$, so that the line integral must be 0. The condition is clearly satisfied for $\partial u/\partial n = \sin \theta$.

4. Solve the boundary-value problem $u_{xx} - 4u_t = 0$, $0 \leqslant x \leqslant 2$, $0 \leqslant t < \infty$, $u(0, t) = 0$, $u(2, t) = 0$, $u(x, 0) = 2x - x^2$, by considering u only at the mesh points $(0.5i, 0.5j)$ for $i = 0, 1, 2, 3, 4; j = 0, 1, 2$.

5. Solve the boundary-value problem $u_{tt} - u_{xx} = 0$, $0 \leqslant x \leqslant 1$, $0 \leqslant t < \infty$, $u(0, t) = u(1, t) = 0$, $u(x, 0) = x - x^2$, $u_t(x, 0) = 0$, by considering u only at the mesh points $(0.25i, 0.25j)$ for $i = 0, 1, 2, 3, 4; j = 0, 1, 2$.

6. Solve the boundary-value problem $u_{tt} - u_{xx} = 0$, $0 \leqslant x \leqslant 1$, $0 \leqslant t < \infty$, $u(0, t) = u(1, t) = 0$, $u(x, 0) = 0$, $u_t(x, 0) = \sin \pi x$, by considering u only at the mesh points $(0.25i, 0.25j)$ for $i = 0, 1, 2, 3, 4; j = 0, 1, 2$.

Chapter 14 The Finite-Element Method

14–1 OVERVIEW OF THE SUBJECT

The finite-element method is a numerical procedure for obtaining approximate solutions to boundary-value problems. It arose in engineering as an alternative to the difference-equation approach and was motivated by physical ideas. In this method, a physical object is thought of as composed of a finite number of pieces, each of which can be analyzed separately according to laws of physics. Thus each physical variable over the entire object can be expressed as a sum of finite elements, each element representing the contribution from one of the pieces. By combining the analyses of the separate pieces, a system of (usually linear) equations relating the finite elements has been deduced. Solution of the linear equations provides the desired approximate solution of the problem.

In setting up the system of equations, we can use several different procedures: least squares, weighted residuals, the Galerkin method, the Rayleigh–Ritz method. Of these the last two have long been known as powerful tools for solving boundary-value problems; they arise naturally when the problems are related to finding the maximum or minimum of certain integrals involving an unknown function. This last topic is the principal one in the *calculus of variations*; hence the procedures are called *variational methods*.

The variational interpretation has turned out to be extremely useful in many ways. It has led to alternative procedures, many giving additional physical information as fringe benefits. It has suggested important simplifications in the numerical work. It has served as a basis for estimating the error in the approximating solution and for proving that the error approached zero as smaller and smaller pieces are used.

Despite these advantages, there is a strong tendency in applying the finite-element method to ignore the variational aspects of a problem and to proceed directly to numerical techniques. A vast amount of practical experience has been gained, and this alone has provided confidence despite the absence of a deeper mathematical foundation.

In this chapter we give an introduction to the subject, emphasizing the practical side; references are given for more complete treatments and for the theory.

14–2 WEIGHTED RESIDUALS; THE GALERKIN METHOD

We illustrate the approach by the following example.

EXAMPLE 1 Consider the boundary-value problem

$$u'' - 4u = x, \qquad 0 \leqslant x \leqslant 1, \qquad u(0) = 0, \qquad u(1) = 0. \tag{14–20}$$

The exact solution is easily found to be

$$u = \frac{1}{4} \left(\frac{e^{2x} - e^{-2x}}{e^2 - e^{-2}} - x \right).$$

We seek to find this numerically by trying a linear combination of functions that satisfy the boundary conditions:

$$u = g(x) = \sum_{k=1}^{n} c_k \varphi_k(x), \qquad \varphi_k(0) = 0, \qquad \varphi_k(1) = 0, \tag{14–21}$$

where $c_1, \ldots, c_n$ are constants. Clearly, $g(x)$ will satisfy the boundary conditions in Eq. (14–20). We try to adjust the c_k so as to satisfy the differential equation with a small error. At each x there is an error (residual)

$$E(x) = g''(x) - 4g(x) - x = \sum_{k=1}^{n} c_k(\varphi_k'' - 4\varphi_k) - x. \tag{14–22}$$

We could require that $E(x_1) = 0, \ldots, E(x_n) = 0$ for n points $x_1, \ldots, x_n$ on the interval $0 \leqslant x \leqslant 1$; this would be the *collocation method*. This procedure leads to n equations for $c_1, \ldots, c_n$, that could be solved to give our approximate solution $g(x)$.

Here we emphasize another procedure, that of *weighted residuals*. We choose n functions (the *weights*) $w_1(x), \ldots, w_n(x)$ and require that $E(x)$ be *orthogonal* to all these functions:

$$\int_0^1 E(x) w_k(x) \, dx = 0, \qquad k = 1, \ldots, n. \tag{14–23}$$

If in fact the functions $w_1(x), \ldots, w_n(x)$ are the first n functions of an infinite sequence $\{w_n(x)\}$ forming a complete orthogonal system on the interval $0 \leqslant x \leqslant 1$, then the validity of (14–23) for *all* positive integers k would imply $E(x) \equiv 0$

(Section 3–10). There is a similar assertion if the sequence could be orthogonalized (Section 3–11) to yield a complete orthogonal system. Hence if (14–23) holds for large n, then $E(x)$ should be small. If we choose the $\varphi_k(x)$ and $w_k(x)$ skillfully, we can get very good results even for small n.

In any case, the conditions (14–23) lead to n equations for $c_1, \ldots, c_n$, which we can solve to get our approximate solution.

We illustrate the procedure, taking

$$\varphi_1(x) = x(x-1), \qquad \varphi_2(x) = x\left(x - \frac{1}{2}\right)(x-1), \qquad w_1(x) = 1, \qquad w_2(x) = x.$$

Then

$$g(x) = c_1(x^2 - x) + c_2\left(x^3 - \frac{3}{2}x^2 + \frac{1}{2}x\right),$$

$$E(x) = c_1(2 + 4x - 4x^2) + c_2(-3 + 4x + 6x^2 - 4x^3) - x,$$

$$\int_0^1 E(x)w_1(x)\,dx = \frac{8}{3}c_1 - \frac{1}{2}, \qquad \int_0^1 E(x)w_2(x)\,dx = \frac{4}{3}c_1 + \frac{8}{15}c_2 - \frac{1}{3}.$$

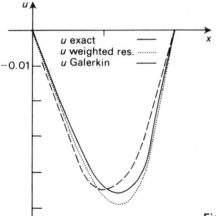

Fig. 14–1. Approximations to solution of Example 1.

Equating the last two quantities to 0, as in Eq. (14–23), and solving for c_1, c_2, gives $c_1 = 3/16$, $c_2 = 5/32$, so that $g(x) = (1/64)(10x^3 - 3x^2 - 7x)$. As illustrated in Fig. 14–1, $g(x)$ is a fairly good approximation to $u(x)$; for example,

$$g\left(\frac{1}{2}\right) = -0.0469, \qquad u\left(\frac{1}{2}\right) = -0.04399. \quad \blacktriangleleft$$

Galerkin method Here we choose the $\varphi_k(x)$ to be the same as the $w_k(x)$. Thus for Example 1, we would choose a sequence $\{\varphi_k(x)\}$ and form $g(x)$ and $E(x)$ as in

Eqs. (14–21) and (14–22), respectively. Then in Eq. (14–23) we use $w_k(x) = \varphi_k(x)$. Thus we obtain the equations

$$\int_0^1 \left[\sum_{k=1}^n c_k(\varphi_k'' - 4\varphi_k) - x \right] \varphi_l(x)\,dx = 0, \qquad l = 1, \ldots, n. \qquad (14\text{–}24)$$

The Galerkin method turns out to be related to the variational approach, hence it can be given a direct physical interpretation in many cases. This, in turn, leads to a better understanding of how the φ_k should be chosen and what accuracy can be expected.

If we apply the method to Example 1, using $\varphi_1(x)$, $\varphi_2(x)$ as above, we obtain the equations

$$-\frac{7}{15}c_1 + \frac{1}{12} = 0, \qquad -\frac{153}{420}c_2 + \frac{1}{120} = 0,$$

and hence $c_1 = 5/28$, $c_2 = 7/46$. Thus our new approximation is

$$u = 0.1786(x^2 - x) + 0.1522 \left(x^3 - \frac{3}{2}x^2 + \frac{1}{2}x \right).$$

This is again a good approximation: at $x = 1/2$ it gives the value -0.0446, which is closer to the exact value than that obtained above. The new approximation is also graphed in Fig. 14–1.

If the infinite sequence $\{\varphi_n(x)\}$ is a complete orthogonal system on the interval $0 \leqslant x \leqslant 1$ (or, when orthogonalized, forms a complete system), then the approximations $c_1\varphi_1(x) + \cdots + c_n\varphi_n(x)$, obtained as above for Example 1, can be shown to converge, at least in the mean, to the exact solution (see Chapter IV, Section 4, of the book by KANTOROVICH and KRYLOV cited in Section 14–10).

The method can be applied to a wide variety of boundary-value problems for ordinary and partial differential equations. The most common applications are to ordinary differential equations of form

$$[p(x)u']' + q(x)u = f(x), \qquad a \leqslant x \leqslant b,$$

where $p(x) > 0$ and $q(x) \geqslant 0$ for $a \leqslant x \leqslant b$, with boundary conditions

$$a_i u(a) + b_i u'(a) + c_i u(b) + d_i u'(b) = 0, \qquad i = 1, 2;$$

to partial differential equations of second order such as $\nabla^2 u = f$ in R, with Dirichlet or Neumann-type boundary conditions on the boundary of R; to more general boundary-value problems for elliptic second-order partial differential equations; and to fourth-order equations, such as $\nabla^4 u = g$ in R, with appropriate boundary conditions on the boundary of R. These problems include a major portion of those arising in continuum mechanics.

EXAMPLE 2   Consider the problem $\nabla^2 u(x, y) = 1$ in the square region $0 < x < 1, 0 < y < 1$, $u(x, y) = 0$ on the edges of the square.

Physically, this describes the displacement $u(x, y)$ of a square membrane, clamped at the edges, subject to a uniform pressure. The problem can be solved exactly by Fourier series, as in Chapter 8. We find

$$u = -\frac{16}{\pi^4} \sum_{n, m = 1}^{\infty} \frac{\sin(2n-1)\pi x \sin(2m-1)\pi y}{(2n-1)(2m-1)\left[(2n-1)^2 + (2m-1)^2\right]}.$$

We seek an approximate solution $g(x, y) = c_1\varphi_1(x, y) + c_2\varphi_2(x, y)$ with

$$\varphi_1(x, y) = xy(x-1)(y-1), \qquad \varphi_2 = xy(x-1)(y-1)(x+y).$$

Following the Galerkin procedure, we form the residual

$$E(x, y) = \nabla^2 g - 1 = c_1\nabla^2\varphi_1 + c_2\nabla^2\varphi_2 - 1$$

$$= c_1(2x^2 + 2y^2 - 2x - 2y) + c_2[2(x+y)^3 - 4x^2 - 4y^2 - 12xy + 2x + 2y] - 1$$

and require that

$$\iint_R E(x, y)\,\varphi_1(x, y)\,dx\,dy = 0, \qquad \iint_R E(x, y)\,\varphi_2(x, y)\,dx\,dy = 0.$$

After a fair amount of algebra, we find that $c_1 = -0.0142, c_2 = -0.0075$, and hence have the approximate solution

$$u(x, y) = -xy(x-1)(y-1)[0.0142 + 0.0075(x+y)].$$

At $x = 1/2, y = 1/2$, this has the value -0.00136, whereas from the exact solution we get

$$u\left(\frac{1}{2}, \frac{1}{2}\right) = -\frac{16}{\pi^4}\left(\frac{1}{2} - \frac{1}{30} - \frac{1}{30} + \cdots\right) \approx -0.072.$$

====================== **Problems (Section 14–2)**======================

1. Let the following boundary-value problem be given:

$$u'' + u = x, \qquad 0 \leqslant x \leqslant 1, \qquad u(0) = 0, \qquad u(1) = 0$$

and let

$$\varphi_1(x) = x(x-1), \qquad \varphi_2(x) = x\left(x - \frac{1}{2}\right)(x-1), \qquad \varphi_3(x) = x\left(x - \frac{1}{3}\right)\left(x - \frac{2}{3}\right)(x-1),$$

$$w_1(x) = 1, \qquad w_2(x) = x, \qquad w_3(x) = x^2.$$

Let $E(x)$ denote the residual $u'' + u - x$ for a trial function $u(x)$.

a) Verify that the exact solution is $u = x - \csc 1 \sin x = x - 1.1884 \sin x$;

b) Verify that $\varphi_1(x), \varphi_2(x), \varphi_3(x)$ satisfy the boundary conditions;

c) Choose c_1 and c_2 so that the residual $E(x)$ for $u = c_1\varphi_1 + c_2\varphi_2$ is orthogonal to w_1 and w_2 on the interval $0 \leqslant x \leqslant 1$ and compare the resulting function $u(x)$ with the exact solution.

d) Choose c_1, c_2, c_3 so that, for $u = c_1\varphi_1 + c_2\varphi_2 + c_3\varphi_3$, $E(x)$ is orthogonal to w_1, w_2, and w_3 on the interval $0 \leqslant x \leqslant 1$ and compare the resulting function $u(x)$ with the exact solution.

2. Let the following boundary-value problem be given:

$$u'' + xu = x^4 - x^3 + 6x - 2, \qquad 0 \leqslant x \leqslant 1, \qquad u(0) = 0, \qquad u(1) = 0.$$

Let $\varphi_1, \varphi_2, \varphi_3, w_1, w_2, w_3(x)$ be as in Problem 1 and let $E(x) = u'' + xu - x^4 + x^3 - 6x + 2$ for a trial function $u(x)$.

a) Verify that the exact solution is $u = x^3 - x^2$;

b) Proceed as in Problem 1(c) with the new residual $E(x)$;

c) Proceed as in Problem 1(d) with the new residual $E(x)$.

3. For the boundary-value problem and φ_1, φ_2 of Problem 1, use the Galerkin method to obtain an approximate solution of form $c_1 \varphi_1 + c_2 \varphi_2$ and compare with the exact solution.

4. For the boundary-value problem $u^{(iv)} + x^2 u = 1$, $-1 \leqslant x \leqslant 1$, $u(\pm 1) = 0$, $u'(\pm 1) = 0$, use the Galerkin method to find an approximate solution
$$u = c_1(x^4 - 2x^2 + 1) + c_2(x^6 - 2x^4 + x^2).$$

5. For the boundary-value problem $\nabla^2 u = 4(x^2 + y^2)^{1/2}$, $x^2 + y^2 \leqslant 1$, $u = 0$ for $x^2 + y^2 = 1$, use the Galerkin method to find an approximate solution
$$u = c_1(x^2 + y^2 - 1) + c_2(x^2 + y^2)(x^2 + y^2 - 1).$$

14–3 ROLE OF INTEGRATION BY PARTS

For the problem
$$u'' = f(x), \qquad a \leqslant x \leqslant b, \qquad u(a) = 0, \qquad u(b) = 0, \qquad (14\text{--}30)$$
the general method of weighted residuals leads to equations of form
$$\int_a^b [u'' - f(x)] w(x)\, dx = 0. \qquad (14\text{--}31)$$

Here we can integrate by parts:
$$\int_a^b u'' w(x)\, dx = u'(x)\, w(x) \Big|_a^b - \int_a^b u'(x)\, w'(x)\, dx. \qquad (14\text{--}32)$$

If $w(x)$ is chosen to satisfy the boundary conditions $w(a) = 0$, $w(b) = 0$, then (14–31) can be replaced by the equation
$$\int_a^b [u'(x)w'(x) + f(x)w(x)]\, dx = 0. \qquad (14\text{--}33)$$

We could also integrate by parts again:
$$\int_a^b u'(x) w'(x)\, dx = u(x) w'(x) \Big|_a^b - \int_a^b u(x) w''(x)\, dx.$$

Here if we assume that $u(x)$ is the exact solution of the given problem, the first term on the right is zero and hence (14–33) takes the form
$$\int_a^b [u(x) w''(x) - f(x) w(x)]\, dx = 0. \qquad (14\text{--}34)$$

The various forms obtainable in this way turn out to be very important in selecting weight functions $w(x)$. For example, in (14–33) we could use piecewise smooth functions $w(x)$. In the exact solution, $u'(x)$ would be continuous, so that the integrals in (14-33) could be evaluated. If an appproximate solution is used for $u(x)$, it also need be only piecewise smooth. These considerations are central to the finite-element method.

It should be stressed that, when we relax the continuity conditions on our functions after transforming the problem from (14-31) to (14-33), we are really extending the meaning of (14-31) in an essential way. The mathematical theory has shown that, for the cases of interest, the procedure is legitimate. In particular, a Galerkin method applied to (14-33), as above, will lead to approximations $g_n(x)$ which converge to the exact solution $u(x)$ of the problem (14-30).

There are other benefits in integrating by parts. If we consider the problem

$$u'' = f(x), \qquad a \leqslant x \leqslant b, \qquad u(a) = 0, \qquad u'(b) = 0 \qquad (14\text{-}35)$$

with a different boundary condition at b, then we can proceed as for the problem (14-30). We can now reason that if $u(x)$ in Eq. (14-32) were the exact solution to the new problem, then $u'(b)$ would be 0. Hence we can eliminate the first term on the right by requiring only $w(a) = 0$. This suggests that we use (14-33) and require it to hold for approximate solutions $u(x)$ and weight functions $w(x)$ that satisfy only the boundary condition at $x = a$. Our conclusion turns out to be correct. It can be described by saying that the condition $u(a) = 0$ is an *essential boundary condition* while the condition $u'(b) = 0$ is a *natural boundary condition*. If we follow a Galerkin procedure, with the φ_k obeying only the essential boundary condition, it can be shown that the approximate solutions approach as limit a function satisfying both the essential boundary condition and the natural boundary condition. A full justification of this result follows from the variational approach.

EXAMPLE 1 $u'' = x^{1/2}$, $0 \leqslant x \leqslant 1$, $u(0) = 0$, $u'(1) = 0$. Here we follow the Galerkin procedure, using $\varphi_k(x) = x^k$, $k = 1, 2, \ldots, n$. We also use Eq. (14-33), obtained through integration by parts as above. Here it is the equation $\int_0^1 (u'w' + x^{1/2}w)\,dx = 0$, where $u = g_n(x) = c_1 x + c_2 x^2 + \cdots + c_n x^n$ and $w = x$, $x^2, \ldots, x^n$. Thus we have the equations

$$\int_0^1 [(c_1 + 2c_2 x + \cdots + nc_n x^{n-1})kx^{k-1} + x^{(k+1/2)}]\,dx = 0,$$

for $k = 1, \ldots, n$. These, in turn, provide simultaneous linear equations for $c_1, \ldots, c_n$:

$$b_{k1}c_1 + \cdots + b_{kn}c_n = \beta_k, \qquad k = 1, \ldots, n,$$

$$b_{kl} = \frac{kl}{k+l-1}, \qquad \beta_k = -\frac{2}{2k+3}.$$

For $n = 1$, we have simply $b_{11}c_1 = \beta_1$ or $c_1 = -2/5$, and the approximate solution is $g_1(x) = -(2/5)x$. For $n = 2$, we have

$$c_1 + c_2 = -\frac{2}{5}, \qquad c_1 + \frac{4}{3}c_2 = -\frac{2}{7},$$

and hence $c_1 = -26/35$, $c_2 = 12/35$, $g_2(x) = -(26/35)x + (12/35)x^2$. Similarly for $n = 3$: we find $g_3(x) = (1/315)(-214x + 48x^2 + 40x^3)$. These all satisfy the boundary condition $u(0) = 0$. The natural boundary condition $u'(1) = 0$ should be approximately satisfied:

$$g_1'(1) = -\frac{2}{5}, \qquad g_2'(1) = -\frac{2}{35}, \qquad g_3'(1) = \frac{2}{315}, \qquad \text{etc.,}$$

and this suggests that $g_n'(1) \to 0$ as $n \to \infty$.

The exact solution is found to be

$$u = \frac{4}{15} x^{5/2} - \frac{2}{3} x.$$

At $x = 1$, this equals $-2/5$, and

$$g_1(1) = -\frac{2}{5}, \qquad g_2(1) = -\frac{2}{5}, \qquad g_3(1) = -\frac{116}{315} = -0.368.$$

The first two are exact, the third is close. At $x = 1/2$, the exact solution equals -0.28619, whereas

$$g_1\left(\frac{1}{2}\right) = -0.2, \qquad g_2\left(\frac{1}{2}\right) = -\frac{2}{7} = -0.28570, \qquad g_3\left(\frac{1}{2}\right) = -0.28571. \quad \blacktriangleleft$$

REMARK: The use of the functions $\varphi_k(x) = x^k$, as in Example 1, in the Galerkin method typically leads to ill-conditioned equations, not satisfactory for numerical solution (see Section 11–2). The functions can be orthogonalized, as in Section 3–11, and thereby it is usually possible to avoid the difficulty; however, this step lengthens the overall computation.

══════════════════**Problems (Section 14-3)**══════════════════

1. Follow the Galerkin procedure as in Example 1 above, using φ_1 and φ_2 as specified, to obtain an approximate solution and compare it with the exact solution:
   a) $u'' = e^x, 0 \leqslant x \leqslant 1, u(0) = 0, u'(1) = 0, \varphi_1(x) = x, \varphi_2(x) = x^2$;
   b) $u'' = \cos x, 0 \leqslant x \leqslant \pi, u(0) = 0, u'(\pi) = 0, \varphi_1(x) = \sin x, \varphi_2(x) = \sin 2x$;
   c) $u'' = \ln(1 + x), 0 \leqslant x \leqslant 1, u'(0) = 0, u(1) = 0, \varphi_1(x) = x - 1, \varphi_2(x) = (x - 1)^2$. Which is the essential boundary condition?
   d) $u'' = 2x - 1, 0 \leqslant x \leqslant 1, u'(0) = 0, u'(1) = 0, \varphi_1(x) = e^x, \varphi_2(x) = e^{2x}$ (see Problem 2 below).

2. Show that the boundary-value problem $u'' = f(x), a \leqslant x \leqslant b, u'(a) = 0, u'(b) = 0$ has a solution only if $\int_a^b f(x)\,dx = 0$ and that, when this condition holds, there is a solution, but it is not unique, since it is determined only up to an arbitrary additive constant.

3. The boundary-value problem $[p(x)u']' + q(x)u = f(x), a \leqslant x \leqslant b, u(a) = 0, u(b) = 0$, leads to a Galerkin equation $\int_a^b [(pu')' + qu - f]w\,dx = 0$. Show that, if w satisfies the given boundary conditions, this is equivalent to $\int_a^b (pu'w' - quw + fw)\,dx = 0$.

4. Proceed as in Problem 3 with the boundary-value problem $(pu'')'' + (qu')' + ru = f$, $a \leqslant x \leqslant b, u(a) = 0, u'(a) = 0, u(b) = 0, u'(b) = 0$, to obtain the equation

$$\int_a^b (pu''w'' - qu'w' + ruw - fw)\,dx = 0.$$

14–4 INTEGRATION BY PARTS IN TWO AND THREE DIMENSIONS

Integration by parts has its counterpart for higher dimensions. For example, in two dimensions

$$\iint_R f \frac{\partial g}{\partial x} dA = \oint_C fg \mathbf{n} \cdot \mathbf{i} \, ds - \iint_R g \frac{\partial f}{\partial x} dA, \tag{14–40}$$

where f and g are (continuously differentiable) functions in R plus C, curve C is a simple closed curve bounding region R, and $\mathbf{n} = (dy/ds)\mathbf{i} - (dx/ds)\mathbf{j}$ is the outer normal, as in Section 10–5 (see Fig. 14–2). The line integral can here be written as $\oint_C fg \, dy$, and Eq. (14–40) follows at once if we apply Green's theorem to $\oint (P \, dx + Q \, dy)$ with $P = 0$ and $Q = fg$. Similarly, we also have

$$\iint_R f \frac{\partial h}{\partial y} dA = \oint_C fh \, (\mathbf{n} \cdot \mathbf{j}) \, ds - \iint_R h \frac{\partial f}{\partial y} dA, \tag{14–41}$$

and addition of (14–40) and (14–41) gives

$$\iint_R f \left(\frac{\partial g}{\partial x} + \frac{\partial h}{\partial y} \right) dA = \oint_C f(g\mathbf{i} + h\mathbf{j}) \cdot \mathbf{n} \, ds - \iint_R \left(g \frac{\partial f}{\partial x} + h \frac{\partial f}{\partial y} \right) dA$$

or, with $\mathbf{v} = g\mathbf{i} + h\mathbf{j}$,

$$\iint_R f \operatorname{div} \mathbf{v} \, dA = \oint_C f\mathbf{v} \cdot \mathbf{n} \, ds - \iint_R \mathbf{v} \cdot \nabla f \, dA. \tag{14–42}$$

This rule also follows from the divergence theorem applied to $\oint f\mathbf{v} \cdot \mathbf{n} \, ds$, since $\operatorname{div} f\mathbf{v} = f \operatorname{div} \mathbf{v} + \mathbf{v} \cdot \nabla f$ (Section 10–7). We can also deduce (14–40) and (14–41) from (14–42) by taking $\mathbf{v} = g\mathbf{i} + 0\mathbf{j}$ and $\mathbf{v} = 0\mathbf{i} + h\mathbf{j}$, respectively.

If we take $\mathbf{v} = \nabla w$ in (14–42), we obtain the much-used Green's formula

$$\iint_R f \nabla^2 w \, dA = \oint_C f \frac{\partial w}{\partial n} ds - \iint_R \nabla f \cdot \nabla w \, dA. \tag{14–43}$$

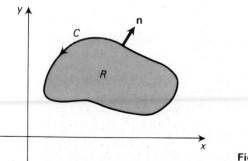

Fig. 14-2. Two-dimensional region.

Interchanging f and w and subtracting, we obtain another Green's formula:

$$\iint_R (f\nabla^2 w - w\nabla^2 f)\,dA = \oint_C \left(f\frac{\partial w}{\partial n} - w\frac{\partial f}{\partial n}\right) ds \qquad (14\text{-}44)$$

(see Problem 8 following Section 10–8). If we here take $f = \nabla^2 u$, we obtain another identity, useful in problems of elasticity:

$$\iint_R (\nabla^2 u \nabla^2 w - w\nabla^4 u)\,dA = \oint_C \left(\nabla^2 u\frac{\partial w}{\partial n} - w\frac{\partial \nabla^2 u}{\partial n}\right) ds. \qquad (14\text{-}45)$$

With the aid of these formulas, the Galerkin equations for two-dimensional problems can be transformed in important ways, as in the preceding section.

EXAMPLE 1 $\nabla^2 u = x^2 y^2$ for $0 < x < 1$, $0 < y < 1$; $u = 0$ for $x = 0$ or $y = 0$, $\partial u/\partial n = 0$ for $x = 1$ or $y = 1$.

Following Galerkin, we want $u(x, y)$ to be such that the boundary conditions hold and that $\iint_R (\nabla^2 u - x^2 y^2)w\,dA = 0$ for an appropriate sequence of functions w satisfying the boundary conditions. Here R is the given square region, C is its boundary. We transform this equation by the rule (14–43), with f replaced by w and w by u, obtaining the equation

$$\oint_C w\frac{\partial u}{\partial n}\,ds - \iint_R (\nabla u \cdot \nabla w + x^2 y^2 w)\,dA = 0.$$

If w and u satisfy the boundary conditions, then the first term is 0, since w is 0 on part of C and $\partial u/\partial n$ is 0 on the rest of C. By analogy with our experience for the example (14–35) above, we now expect that it is sufficient to make $w = 0$ for $x = 0$ or $y = 0$ (essential boundary conditions) and that we need not require $\partial w/\partial n = 0$ for $x = 1$ or $y = 1$ (natural boundary conditions). This can also be justified for the problem considered here.

Accordingly, we use the equation $\iint_R (\nabla u \cdot \nabla w + x^2 y^2 w)\,dA = 0$ and require that it hold for $u = c_1\varphi_1(x, y) + \cdots + c_n\varphi_n(x, y)$, where w is equal to any one of $\varphi_1(x, y), \ldots, \varphi_n(x, y)$. Here the functions $\varphi_k(x, y)$ should satisfy the essential boundary conditions, and the infinite sequence $\{\varphi_n(x, y)\}$ should be such that all functions $u(x, y)$ satisfying the essential boundary conditions can be arbitrarily well approximated, in terms of least-square error, by an appropriate linear combination of members of the sequence. We try only $\varphi_1(x, y) = xy$, $\varphi_2(x, y) = x^2 y + xy^2$, and are led to the two equations.

$$\iint_R [c_1(x^2 + y^2) + c_2(x^3 + 2x^2 y + 2xy^2 + y^3) + x^3 y^3]\,dA = 0,$$

$$\iint_R [c_1(x^3 + 2x^2 y + 2xy^2 + y^3) + c_2(x^4 + 4x^3 y + 8x^2 y^2 + 4xy^3 + y^4)$$
$$+ x^4 y^3 + x^3 y^4]\,dA = 0.$$

These give

$$\frac{2}{3}c_1 + \frac{7}{6}c_2 + \frac{1}{16} = 0, \qquad \frac{7}{6}c_1 + \frac{103}{45}c_2 + \frac{1}{10} = 0,$$

so that $c_1 = -114/712 = -0.1601$, $c_2 = 27/712 = 0.0379$, and our approximate solution is

$$u = -0.1601\,xy + 0.0379\,(x^2y + xy^2).$$

We do not attempt to find the exact solution. We observe that our solution does satisfy the essential boundary conditions, by construction. Also, when $x = 1$, for example,

$$\frac{\partial u}{\partial n} = \frac{\partial u}{\partial x} = -0.0843\,y + 0.0379\,y^2, \qquad 0 \leqslant y \leqslant 1.$$

This remains close to 0, having a maximum absolute value 0.046. Thus, as hoped, the natural boundary condition is well approximated. ◀

All integration-by-parts formulas (14–40) to (14–45) extend to three or more dimensions; the double integrals become triple integrals over a solid region R bounded by a surface $\mathscr{S}$ with outer normal $\mathbf{n}$; the line integrals become surface integrals over $\mathscr{S}$. Thus we have, under the appropriate continuity hypotheses,

$$\iiint_R f\frac{\partial g}{\partial x}\,dV = \iint_{\mathscr{S}} fg\,(\mathbf{n}\cdot\mathbf{i})\,d\sigma - \iiint_R g\frac{\partial f}{\partial x}\,dV, \qquad (14\text{--}40')$$

$$\iiint_R f\operatorname{div}\mathbf{v}\,dV = \iint_{\mathscr{S}} f\mathbf{v}\cdot\mathbf{n}\,d\sigma - \iiint_R \mathbf{v}\cdot\nabla f\,dV, \qquad (14\text{--}42')$$

$$\iiint_R f\nabla^2 w\,dV = \iint_{\mathscr{S}} f\frac{\partial w}{\partial n}\,d\sigma - \iiint_R \nabla f\cdot\nabla w\,dV, \qquad (14\text{--}43')$$

$$\iiint_R (f\nabla^2 w - w\nabla^2 f)\,dV = \iint_{\mathscr{S}}\left(f\frac{\partial w}{\partial n} - w\frac{\partial f}{\partial n}\right)d\sigma, \qquad (14\text{--}44')$$

$$\iiint_R (\nabla^2 u\nabla^2 w - w\nabla^4 u)\,dV = \iint_{\mathscr{S}}\left(\nabla^2 u\frac{\partial w}{\partial n} - w\frac{\partial\nabla^2 u}{\partial n}\right)d\sigma. \qquad (14\text{--}45')$$

The Galerkin method can be applied in the same way, and similar results are obtained about essential and natural boundary conditions.

14–5 SOME IMPORTANT PROBLEMS

We give here, for reference, the Galerkin formulations of several commonly occurring problems. The first two are one-dimensional, the others are two-dimensional. The formulations can be generalized to three dimensions, as suggested by the results of the preceding section.

A. $(pu')' + qu = f(x)$, $a \leqslant x \leqslant b$, $a_{11}u(a) + a_{12}u'(a) = 0$, $a_{21}u(b) + a_{22}u'(b) = 0$, where $p(x) > 0$ and $q(x) \geqslant 0$ for $a \leqslant x \leqslant b$, $a_{11}^2 + a_{12}^2 \neq 0$, $a_{21}^2 + a_{22}^2 \neq 0$. The Galerkin form is:

$$\int_a^b (pu'w' - quw + fw)\,dx = 0 \qquad (14\text{-}50)$$

for all w satisfying the essential boundary conditions. Here the essential boundary conditions are those for which $a_{i2} = 0$. The solution is unique except for the case $q(x) \equiv 0$, $a_{11} = 0$, $a_{21} = 0$. Then it is determined up to an additive constant.

B. $(pu'')'' + (qu')' + ru = f$, $a \leqslant x \leqslant b$, where two of the quantities u, u', u'' are 0 at a and two are 0 at b, $p(x) > 0$, $q(x) \leqslant 0$, and $r(x) \geqslant 0$ for $a \leqslant x \leqslant b$. The Galerkin form is:

$$\int_a^b (pu''w'' - qu'w' + ruw - fw)\,dx = 0 \qquad (14\text{-}51)$$

for all w satisfying the essential boundary conditions. Here only $u''(a) = 0$ or $u''(b) = 0$ is a natural boundary condition.

C. $\nabla^2 u + pu = f(x, y)$ in R, $u = 0$ on C, where $p(x, y) \geqslant 0$ in R. The Galerkin form is

$$\iint_R (\nabla u \cdot \nabla w - puw + fw)\,dA = 0 \qquad (14\text{-}52)$$

for all w such that $w = 0$ on C.

D. $\nabla^2 u + pu = f(x, y)$ in R, $u = g(x, y)$ on C, where $p(x, y) \geqslant 0$ in R. The Galerkin form is Eq. (14-52) for all w such that $w = 0$ on C. Here we try

$$u = u_0(x, y) + c_1\varphi_1(x, y) + \cdots + c_n\varphi_n(x, y),$$

where $u_0(x, y) = g(x, y)$ on the boundary and $\varphi_1(x, y), \ldots, \varphi_n(x, y)$ equal 0 on the boundary, and $w = \varphi_1, \varphi_2, \ldots, \varphi_n$ successively.

E. $\nabla^2 u + pu = f$ in R, $\partial u/\partial n = 0$ on C, where $p(x, y) \geqslant p_0 = \text{const} > 0$ in R. The Galerkin form is (14-52) for all w (no boundary condition).

F. $\nabla^2 u + pu = f$ in R, $\partial u/\partial n = g$ on C, where $p(x, y) \geqslant p_0 = \text{const} > 0$ in R. The Galerkin form is

$$\iint_R (\nabla u \cdot \nabla w - puw + fw)\,dA - \oint_C gw\,ds = 0 \qquad (14\text{-}53)$$

for all w.

G. $\nabla^2 u = f(x, y)$ in R, $\partial u/\partial n = 0$ on C. The Galerkin form is

$$\iint_R (\nabla u \cdot \nabla w + fw)\,dA = 0 \qquad (14\text{-}54)$$

for all w. It must be assumed that $\iint_R f\,dA = 0$; otherwise there is no solution. For if u is a solution, then by Eq. (14–43) with $f \equiv 1$ and w replaced by u, we get

$$\iint_R \nabla^2 u \, dA = \oint_C \frac{\partial u}{\partial n} \, ds = 0.$$

Furthermore, the solution is not unique, but is determined up to an additive constant.

H. $\nabla^2 u = f(x, y)$ in R, $\partial u / \partial n = g(x, y)$ on C. The Galerkin form is

$$\iint_R (\nabla u \cdot \nabla w + fw) \, dA - \oint_C gw \, ds = 0 \qquad (14\text{–}55)$$

for all w. As for problem G, it must be assumed that $\iint_R f \, dA = \oint_C g \, ds$, and again the solution is determined only up to an additive constant.

I. $\nabla^2 u = f(x, y)$ in R, $(\partial u / \partial n) + q(x, y)u = 0$ on C, where $q(x, y) > 0$. The Galerkin form is

$$\iint_R (\nabla u \cdot \nabla w + fw) \, dA - \oint_C quw \, ds = 0 \qquad (14\text{–}56)$$

for all w.

J. $\nabla^2 u + pu - f(x, y)$ in R, $u - 0$ on C_0, $\partial u / \partial n = 0$ on C_1, where $p(x, y) \geqslant 0$ in R, C_0 is formed of one or more arcs on C, C_1 is the rest of C. The Galerkin form is Eq. (14–52) for all w such that $w = 0$ on C_0.

K. $\nabla^4 u = f(x, y)$, $u = 0$ and $\partial u / \partial n = 0$ on C. The Galerkin form is

$$\iint_R (\nabla^2 u \nabla^2 w - fw) \, dA = 0 \qquad (14\text{–}57)$$

for all w such that $w = 0$ and $\partial w / \partial n = 0$ on C.

For a derivation of these results see pp. 1–20 of the book by CIARLET and Chapter 8 of the book by STAKGOLD cited in Section 14–10. In the following section we give a derivation for one typical case.

14–6 DIRICHLET PROBLEM (ILLUSTRATION OF THE VARIATIONAL APPROACH)

As an example of the way in which a variational approach leads to the Galerkin equations, we consider a Dirichlet problem for the Poisson equation:

$$\nabla^2 u = f(x, y) \text{ in } R, \qquad u = 0 \text{ on } C. \qquad (14\text{–}60)$$

This is a special case of problem C of the preceding section.

We assume that there is a solution $u(x, y)$. We assert that this solution provides the unique minimum of the integral

$$J = \iint_R (|\nabla v|^2 + 2fv) \, dA \qquad (14\text{–}61)$$

among all functions v in R having boundary values 0 on C. If v is such a function, we
let $U(x, y) = v(x, y) - u(x, y)$. Then

$$\iint_R (|\nabla v|^2 + 2fv)\, dA = \iint_R (\nabla v \cdot \nabla v + 2fv)\, dA$$

$$= \iint_R [(\nabla u + \nabla U) \cdot (\nabla u + \nabla U) + 2f(u + U)]\, dA$$

$$= \iint_R [|\nabla u|^2 + |\nabla U|^2 + 2\nabla u \cdot \nabla U + 2f(u + U)]\, dA.$$

If we apply identity (14–43) to the term in $\nabla u \cdot \nabla U$, we obtain

$$\iint_R \nabla u \cdot \nabla U\, dA = \oint_C U \frac{\partial u}{\partial n}\, ds - \iint_R U \nabla^2 u\, dA.$$

But $U = 0$ on C and $\nabla^2 u = f$ in R. Hence

$$\iint_R \nabla u \cdot \nabla U\, dA = -\iint_R fU\, dA.$$

Therefore,

$$\iint_R (|\nabla v|^2 + 2fv)\, dA = \iint_R (|\nabla u|^2 + 2fu)\, dA + \iint_R |\nabla U|^2\, dA \geq \iint_R (|\nabla u|^2 + 2fu)\, dA.$$

Thus u does minimize the integral J. Furthermore, by the last step, the minimum is
attained only for $\nabla U \equiv \mathbf{0}$ or $U \equiv$ const. Since $U = 0$ on C, U must be identically 0;
that is, the minimum is attained only for $v(x, y) \equiv u(x, y)$.

We now select $\varphi_1(x, y), \ldots, \varphi_n(x, y)$ in R, with 0 boundary values on C, and try
to choose $c_1, \ldots, c_n$ to minimize the integral J among all functions

$$v(x, y) = c_1\varphi_1 + \cdots + c_n\varphi_n. \tag{14–62}$$

If we replace v by this expression in (14–61), then J becomes a function of n real
variables $c_1, \ldots, c_n$, say $F(c_1, \ldots, c_n)$. As in Section 9–9, function F attains its
minimum (if it has one) at a point where $\partial F / \partial c_1 = 0, \ldots, \partial F / \partial c_n = 0$. Now

$$F(c_1, \ldots, c_n) = \iint_R (\nabla v \cdot \nabla v + 2fv)\, dA$$

$$= \iint_R [(c_1 \nabla \varphi_1 + \cdots + c_n \nabla \varphi_n) \cdot (c_1 \nabla \varphi_1 + \cdots + c_n \nabla \varphi_n)$$

$$+ 2f(c_1 \varphi_1 + \cdots + c_n \varphi_n)]\, dA$$

$$= \iint_R \left[\sum_{i,j=1}^{n} \nabla \varphi_i \cdot \nabla \varphi_j c_i c_j + 2f \sum_{i=1}^{n} (c_i \varphi_i) \right] dA. \tag{14–63}$$

Hence (see Problem 6 below)

$$\frac{\partial F}{\partial c_i} = 2 \iint\limits_{R} \left(\sum_{j=1}^{n} \nabla\varphi_i \cdot \nabla\varphi_j c_j + f\varphi_i \right) dA, \qquad i = 1, \ldots, n; \qquad (14\text{–}64)$$

accordingly, the condition for a minimum is

$$\iint\limits_{R} (\nabla v \cdot \nabla w + fw)\, dA = 0,$$

where v is as in Eq. (14–62) and $w = \varphi_1, \ldots, \varphi_n$. *Thus we obtain the Galerkin equations.* They are simply the conditions for a critical point of $F(c_1, \ldots, c_n)$.

When the Galerkin equations are derived in this way as conditions for minimizing an integral in a class of functions satisfying the boundary conditions, then the *Rayleigh–Ritz procedure* is followed. As indicated earlier, Galerkin equations can be obtained without reference to a minimization problem.

By an argument similar to that given above for J in general, we can show that the function $F(c_1, \ldots, c_n)$ has a unique minimum, provided that $\varphi_1, \ldots, \varphi_n$ are linearly independent over R. Further, we can show that, for proper choice of $\varphi_1, \ldots, \varphi_n, \ldots$ as described earlier in this chapter, when $n \to \infty$, the minimizing functions v_n approach u, which is the minimizing function for J and the unique solution of the Dirichlet problem (14–60).

Physical interpretation   The boundary-value problem (14–60) can be interpreted as that for the vertical displacement $u(x, y)$ of a membrane clamped along its boundary curve C and subject to an upward vertical pressure $f(x, y)$, as suggested in Fig. 14–3. The membrane has a unique equilibrium state for which the displacement function is the solution $u(x, y)$. For an arbitrary vertical-displacement function $v(x, y)$, it can be verified that J is the potential energy. Therefore our equilibrium solution gives the minimum potential energy, as is common in mechanics.

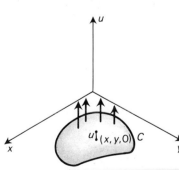

Fig. 14-3. Equilibrium of clamped membrane.

Other boundary-value problems as variational problems All the problems of Section 14–5 have variational counterparts. In fact, a large class of boundary-value problems can be treated uniformly as variational problems (see COURANT–HILBERT and the references given at the end of Section 14–5). This is not surprising in view of the minimum-potential-energy interpretation, that applies to a broad class of physical problems.

For some boundary-value problems, the variational form is less natural, even though such a form can always be found artificially. However, the Galerkin conditions, as we have seen, provide a reasonable way of seeking an approximate solution, even apart from any considerations of maximum or minimum.

═══════════════════════════**Problems (Section 14–6)**═══════════════════════════

1. a) Verify Eq. (14–40) for the case of the circular region R: $x^2 + y^2 \leqslant 1$, taking $f = x$, $g = x^2 + y^2$.
   b) Verify Eq. (14–43) for R, f as in part (a) and $w = x^2 - y^2$.

2. Follow the procedure of Example 1 in Section 14–4 to find approximate solutions of the given boundary-value problems, using φ_1, φ_2 as given:
   a) $\nabla^2 u = x + y$ for $0 < x < 1, 0 < y < 1$; $u = 0$ for $x = 1$ or $y = 1$, $\partial u/\partial n = 0$ for $x = 0$ or $y = 0$, $\varphi_1(x, y) = (x - 1)(y - 1)$, $\varphi_2(x, y) = (x + y)(x - 1)(y - 1)$;
   b) $\nabla^2 u = xy$ for $0 < x < 1$, $0 < y < 1$; $u = 0$ for $x = 0$ or $x = 1$, $\partial u/\partial n = 0$ for $y = 0$ or $y = 1$, $\varphi_1 = x^2 - x$, $\varphi_2 = (x + y)(x^2 - x)$.

3. Apply the Galerkin formulation of problem D in Section 14–5 to seek an approximate solution of the boundary-value problem $\nabla^2 u = 0$ in R: $x^2 + y^2 \leqslant 1$, $u = x^4 + y^4$ for $x^2 + y^2 = 1$, using
   $$u_0(x, y) = x^4 + y^4, \qquad \varphi_1(x, y) = x^2 + y^2 - 1, \qquad \varphi_2(x, y) = (x^2 + y^2)(x^2 + y^2 - 1).$$

4. Apply the Galerkin formulation of problem G in Section 14–5 to seek an approximate solution of the boundary-value problem $\nabla^2 u = x + y$ in R: $x^2 + y^2 \leqslant 1$, $\partial u/\partial n = 0$ on C, using $\varphi_1(x, y) = x^2$, $\varphi_2(x, y) = xy$, $\varphi_3(x, y) = y^2$. (Observe that the double integral of f over R is 0, as required.)

5. a) Imitate the proof given in Section 14–6 to show that the solution $u(x)$ of the boundary-value problem $u'' = f(x)$, $a \leqslant x \leqslant b$, $u(a) = 0$, $u(b) = 0$, provides the unique minimum of the integral $J = \int_a^b (v'^2 + 2fv) \, dx$ among all $v(x)$ having $v(a) = v(b) = 0$.
   b) Show further that minimizing J among all functions $v(x) = c_1 \varphi_1(x) + \cdots + c_n \varphi_n(x)$, where $\varphi_j(a) = 0$, $\varphi_j(b) = 0$ for $j = 1, \ldots, n$, leads to the Galerkin equation $\int_a^b (v'w' + fw) \, dx = 0$ for $w = \varphi_1, \ldots, \varphi_n$ (see Problem 6 below).

6. Show that for a quadratic form $(a_{ij} = a_{ji})$
   $$F(x_1, \ldots, x_n) = \sum_{i,j=1}^{n} a_{ij} x_i x_j = a_{11} x_1^2 + a_{12} x_1 x_2 + a_{22} x_2^2 + a_{21} x_2 x_1 + \cdots$$

   we have
   $$\frac{\partial F}{\partial x_i} = 2 \sum_{j=1}^{n} a_{ij} x_j, \qquad i = 1, \ldots, n,$$

   and hence obtain Eq. (14–64) from Eq. (14–63).

14–7 **FINITE ELEMENTS IN ONE DIMENSION**

We now introduce the finite-element method and for simplicity start with boundary-value problems in one dimension. Here finite-difference methods and other approaches (Section 13–15) are also successful, so that the finite-element methods do not help much (or even differ much, in many cases). Their real value is in two-dimensional and three-dimensional problems discussed in Section 14–8.

For a boundary-value problem for $u(x)$ on the interval $a \leqslant x \leqslant b$, the finite-element method is simply the application of the Galerkin procedure with the functions $\varphi_1(x), \ldots, \varphi_n(x)$ chosen to be *piecewise polynomial functions*. For a problem of the type of problem A in Section 14–5, we observe that the Galerkin conditions (14–50) can be applied if u and w are merely continuous and piecewise smooth; this means that u' and w' may have jump discontinuities. For such functions, the integrals remain meaningful, thus for such a problem continuous piecewise linear functions $\varphi_j(x)$, as in Fig. 14 4, are commonly used. We can also use smoother functions, such as a piecewise quadratic function (as in Fig. 14–5) or a piecewise cubic function with continuous first derivative. For problem B of Section 14–5, the Galerkin conditions (14–51) require the piecewise quadratic (or piecewise nth degree with $n > 3$), with continuous first derivative.

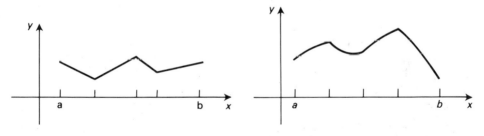

Fig. 14-4. Piecewise linear function.   **Fig. 14-5.** Piecewise quadratic function.

Here we begin with the piecewise linear form and illustrate the process.

EXAMPLE 1 $u'' = x^{1/2}, 0 \leqslant x \leqslant 1, u(0) = 0, u'(1) = 0$. This example is considered in Section 14–3; the exact solution is $u = (4/15)x^{5/2} - (2x/3)$. The Galerkin conditions have the form

$$\int_0^1 (u'w' + x^{1/2}w)\,dx = 0. \tag{14–70}$$

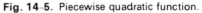

We subdivide the interval $0 \leqslant x \leqslant 1$ into four equal parts by the points $x_0 = 0$, $x_1 = 1/4, x_2 = 1/2, x_3 = 3/4, x_4 = 1$. It is also convenient to use the points $x_{-1} = -1/4$ and $x_5 = 5/4$. We let $\varphi_i(x)$ be the piecewise linear function equal to 1 for $x = x_i, i = 0, 1, 2, 3, 4$, and to 0 for all other x_j (Fig. 14–6). Functions φ_i are typical

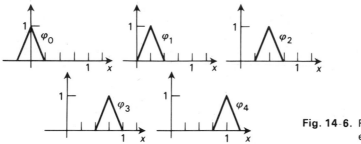

Fig. 14-6. Piecewise linear finite elements.

piecewise linear finite elements. Functions $\varphi_0(x)$ to $\varphi_4(x)$ are linearly independent for $0 \leqslant x \leqslant 1$ (Problem 2 below) and $u = c_0\varphi_0(x) + \cdots + c_4\varphi_4(x)$ is the piecewise linear function equal to c_j at x_j for $j = 0, 1, 2, 3, 4$; here the c_j are uniquely determined for a given piecewise linear function $u(x)$ which is linear between successive subdivision points.

We now impose the essential boundary condition: $u = 0$ for $x = 0$. This makes $c_0 = 0$ in our trial function $u(x)$, so that our final trial functions are

$$u = c_1\varphi_1(x) + \cdots + c_4\varphi_4(x). \tag{14-71}$$

We use these in the Galerkin procedure, taking $w = \varphi_1(x), \ldots, \varphi_4(x)$ in turn. Thus for $w = \varphi_i(x)$ we have the equation

$$\int_0^1 \left(\sum_{j=1}^4 c_j\varphi_j'\varphi_i' + x^{1/2}\varphi_i \right) dx = 0 \tag{14-72}$$

or

$$\sum_{j=1}^4 a_{ij}c_j = b_i, \qquad i = 1, 2, 3, 4, \tag{14-73}$$

where

$$a_{ij} = \int_0^1 \varphi_i'\varphi_j' \, dx, \qquad i, j = 1, 2, 3, 4,$$

$$b_i = -\int_0^1 x^{1/2}\varphi_i(x) \, dx, \qquad i = 1, 2, 3, 4.$$

We observe that in the four successive intervals into which the interval $0 \leqslant x \leqslant 1$ is subdivided, φ_1' has the values $4, -4, 0, 0$; φ_2' has the values $0, 4, -4, 0$; φ_3' has the values $0, 0, 4, -4$; φ_4' has the values $0, 0, 0, 4$. Accordingly, for example,

$$a_{11} = \int_0^{1/4} 16 \, dx + \int_{1/4}^{1/2} 16 \, dx = 8, \quad a_{12} = \int_{1/4}^{1/2} -16 \, dx = -4,$$

and similarly $a_{13} = 0, a_{14} = 0, a_{21} = 0, a_{22} = 8, a_{23} = -4, a_{24} = 0, \ldots, a_{44} = 4$.

Thus the a_{ij} form the matrix

$$A = \begin{bmatrix} 8 & -4 & 0 & 0 \\ -4 & 8 & -4 & 0 \\ 0 & -4 & 8 & -4 \\ 0 & 0 & -4 & 4 \end{bmatrix},$$

and, in vector notation, Eq. (14–73) is

$$Ac = b. \tag{14-73'}$$

The components b_i can be evaluated exactly. For simplicity, we calculate the integrals by the trapezoidal rule for each of the four subintervals. Thus

$$b_1 = -\int_0^{1/4} x^{1/2}\varphi_1\, dx - \int_{1/4}^{1/2} x^{1/2}\varphi_1\, dx$$

$$= -\frac{1}{8}\left[0 + \left(\frac{1}{4}\right)^{1/2}\right] - \frac{1}{8}\left[\left(\frac{1}{4}\right)^{1/2} + 0\right] = -\frac{1}{8}.$$

Similarly, we find $b_2 = -0.177$, $b_3 = -0.217$, $b_4 = -0.125$. Equation (14–73') is now easily solved to give

$$c_1 = -0.177, \qquad c_2 = -0.322, \qquad c_3 = -0.4075, \qquad c_4 = -0.439,$$

and our approximate solution $u(x)$ has these values at $x = 1/4, 1/2, 3/4, 1$, respectively. The exact solution has the values -0.158, -0.286, -0.370, -0.400, so that the agreement is fairly good.

We observe that the matrix A is *symmetric*. This is to be expected from the definition of the a_{ij} above. Also the matrix A is tridiagonal. For computational purposes we want many off-diagonal elements to be 0.

It can be verified that solution of the preceding problem by difference equations, as in Section 13–15, leads to the *same* linear system (14–73') (see Problem 3 below). Thus for this example we have gained nothing by introducing finite elements. ◄

We next introduce a piecewise cubic family of finite elements.

EXAMPLE 2 The problem is the same as in Example 1, but we use piecewise cubic functions $\varphi_i(x)$. By Hermite interpolation (Section 13–6), we can choose a piecewise cubic function that agrees in value and derivative with a given function at successive points $x_0 = 0$, $x_1, \ldots, x_n = 1$. Such functions should more closely approximate the given function than a piecewise linear one.

We use such piecewise cubics for the problem of Example 1, taking $x_0 = 0$, $x_1 = 1/2$, $x_2 = 1$. We also introduce $x_{-1} = -1/2$ and $x_3 = 3/2$. Consider the six successive numbers $\varphi_i(0)$, $\varphi_i'(0)$, $\varphi_i(1/2)$, $\varphi_i'(1/2)$, $\varphi_i(1)$, $\varphi_i'(1)$ for $i = 0, 1, \ldots, 5$ and let $\varphi_i(x)$ be the function equal to a cubic polynomial between successive points x_{-1}, $x_0, \ldots, x_3$ for which the $(i+1)$st member of this sequence is 1 and all others, as well as the value and derivative at x_{-1} and x_3, are 0. Thus $\varphi_i(x)$ has 0 value and 0 derivative at all five points except that $\varphi_0(0) = 1$, $\varphi_1'(0) = 1$, $\varphi_2(1/2) = 1$, $\varphi_3'(1/2) = 1$, $\varphi_4(1) = 1$, $\varphi_5'(1) = 1$. These functions are graphed in Fig. 14–7. They are examples of piecewise cubic finite elements.

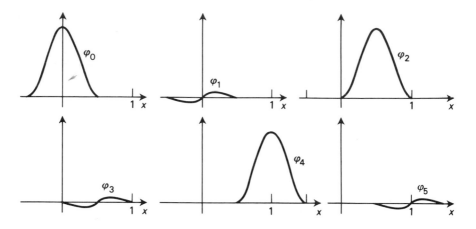

Fig. 14–7. Piecewise cubic finite elements.

It is clear that, if we consider each function to be 0 outside the interval from $-1/2$ to $3/2$, then

$$\varphi_2(x) = \varphi_0(x - \tfrac{1}{2}), \quad \varphi_4(x) = \varphi_0(x - 1), \quad \varphi_3(x) = \varphi_1(x - \tfrac{1}{2}), \quad \varphi_5(x) = \varphi_1(x - 1),$$

and that over the interval $-1/2 \leqslant x \leqslant 1/2$, $\varphi_0(x)$ is even and $\varphi_1(x)$ is odd while both these functions are 0 for $1/2 \leqslant x \leqslant 3/2$.

By Hermite interpolation (Section 13–6) we easily find that for $0 \leqslant x \leqslant 1/2$

$$\varphi_0(x) = 16x^3 - 12x^2 + 1 \quad \text{and} \quad \varphi_1(x) = 4x^3 - 4x^2 + x.$$

From the above rules it then follows that

$$\varphi_0(x) = -16x^3 - 12x^2 + 1, \quad -\tfrac{1}{2} \leqslant x \leqslant 0.$$
$$\varphi_1(x) = 4x^3 + 4x^2 + x, \quad -\tfrac{1}{2} \leqslant x \leqslant 0,$$
$$\varphi_2(x) = \begin{cases} -16x^3 + 12x^2, & 0 \leqslant x \leqslant \tfrac{1}{2} \\ 16x^3 - 36x^2 + 24x - 4, & \tfrac{1}{2} \leqslant x \leqslant 1 \end{cases}$$
$$\varphi_3(x) = \begin{cases} 4x^3 - 2x^2, & 0 \leqslant x \leqslant \tfrac{1}{2}, \\ 4x^3 - 10x^2 + 8x - 2, & \tfrac{1}{2} \leqslant x \leqslant 1, \end{cases}$$
$$\varphi_4(x) = -16x^3 + 36x^2 - 24x + 5, \quad \tfrac{1}{2} \leqslant x \leqslant 1.$$

The remaining values of $\varphi_4(x)$ and those of $\varphi_5(x)$ can also be found but will not be needed here (see Problem 4 below).

We now let $u(x) = c_0\varphi_0(x) + \ldots + c_5\varphi_5(x)$, $0 \leqslant x \leqslant 1$. This gives a collection of piecewise cubic functions that are continuous and have a continuous first derivative. Furthermore,

$$u(0) = c_0\varphi_0(0) = c_0, \quad u'(0) = c_1\varphi_1'(0) = c_1, \quad u\left(\frac{1}{2}\right) = c_2,$$

$$u'\left(\frac{1}{2}\right) = c_3, \quad u(1) = c_4, \quad u'(1) = c_5.$$

Hence there is a unique member of the collection with given value and derivative at 0, 1/2, 1.

For Example 2, we want the value at 0 to be 0, hence we require $c_0 = 0$. The boundary condition $u'(1) = 0$ is a natural one, but we may as well require it, so that $c_5 = 0$. Hence we seek approximate solutions

$$u = c_1\varphi_1(x) + c_2\varphi_2(x) + c_3\varphi_3(x) + c_4\varphi_4(x).$$

and apply Eq. (14–70) with such a $u(x)$ and with $w = \varphi_1(x), \ldots, \varphi_4(x)$.

As in Example 1, we are led to Eqs. (14–72), (14–73), (14–73'), where again

$$a_{ij} = \int_0^1 \varphi_i' \varphi_j' \, dx, \qquad b_i = -\int_0^1 x^{1/2} \varphi_i \, dx.$$

The calculation is more extensive than for Example 1, but is straightforward. For example,

$$a_{11} = \int_0^{1/2} (12x^2 - 8x + 1)^2 \, dx = \frac{1}{15} = 0.0667.$$

We find in all

$$A = \begin{bmatrix} 0.0667 & -0.1000 & -0.0167 & 0 \\ -0.1000 & 4.8000 & 0 & -2.4000 \\ -0.0167 & 0 & 0.1333 & -0.1000 \\ 0 & -2.4000 & -0.1000 & 2.4000 \end{bmatrix}, \qquad b = \begin{bmatrix} -0.0090 \\ -0.3468 \\ -0.0062 \\ -0.2300 \end{bmatrix}.$$

Solving $Ac = b$, we obtain that $c = \text{col}(-0.672, -0.286, -0.432, -0.400)$, and our approximate solution is

$$u = -0.672\varphi_1 - 0.286\varphi_2 - 0.432\varphi_3 - 0.400\varphi_4$$

$$= \begin{cases} -3.616x^3 + 0.120x^2 - 0.672x, & 0 \leqslant x \leqslant \frac{1}{2}, \\ 0.096x^3 + 0.216x^2 - 0.720x + 0.008, & \frac{1}{2} \leqslant x \leqslant 1. \end{cases}$$

This is a much better approximation of the exact solution than that obtained in Example 1. For instance, when $x = 0.25, 0.5, 0.75, 1$, we get $u = -0.158, -0.286, -0.370, -0.400$, respectively, in agreement with the exact values.

14–8 PROBLEMS IN TWO DIMENSIONS

For a boundary-value problem for $u(x, y)$ in a two-dimensional region R, we can seek a Galerkin formulation, as in Section 14–5 above. Then we can try $u = \sum c_i\varphi_i(x, y)$, $w = \varphi_1(x, y), \ldots,$ as before, using appropriate finite elements as the $\varphi_i(x, y)$. In order to obtain these elements we must subdivide R into small pieces R_k, $k = 1, \ldots, N$, and use functions φ_i which are 0 except in a few of these pieces. Typically, the R_k are chosen to be triangles, as in Fig. 14–8. Each polygonal region R can be *triangulated* in such a fashion: each two triangles either do not meet or have exactly one vertex in common or exactly one side in common. If R is not polygonal, it often suffices to approximate it very accurately by a polygonal region. There are

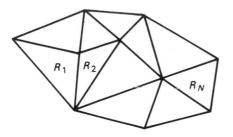

Fig. 14-8. Triangulated region R.

alternative ways of handling curvilinear boundaries (see pp. 34–79 of the book by
CIARLET cited in Section 14–10).

For the case of triangular pieces R_k we can use continuous piecewise linear
elements $\varphi_i(x, y)$ and thereby approximate our solution $u(x, y)$ by a continuous
piecewise linear function $\sum c_i \varphi_i(x, y)$. By analogy with the one-dimensional case, we
number the vertices of the subdivision by index i $(= 1, \ldots, m)$ and then choose φ_i to
be the piecewise linear function equal to 1 at the ith vertex and to 0 at all other
vertices (Fig. 14–9). Then the functions $\varphi_1, \ldots, \varphi_m$ are linearly independent over R
and form a basis for all continuous piecewise linear functions $u(x, y)$.

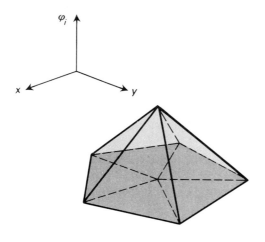

Fig. 14-9. Piecewise linear element.

Thus each such function $u(x, y)$ has a unique representation as

$$c_1 \varphi_1(x, y) + \cdots + c_m \varphi_m(x, y),$$

and we see that at the ith vertex u has value c_i. We observe that u has a piecewise
constant gradient vector ∇u and hence can be used in the Galerkin forms of
boundary-value problems for second-order partial differential equations, as in
Section 14–5.

Below we illustrate the method for a typical Dirichlet problem for the Laplace
equation. Before doing so, we evaluate some of the quantities that appear. We

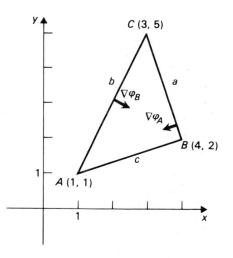

Fig. 14-10. Reference triangle.

consider a triangle ABC as in Fig. 14–10; we assume that the vertices A, B, C (in that order) follow the positive direction on the boundary of the triangle.

If $\varphi_A(x, y)$ is a linear function on this triangle such that $\varphi_A = 1$ at A, $\varphi_A = 0$ at B and C, then $\nabla\varphi_A$ is orthogonal to side BC (a level curve of φ_A), and since φ_A changes by 1 in going a distance h, the altitude on side BC, $\nabla\varphi_A$ has magnitude $1/h$. Thus

$$\nabla\varphi_A = \frac{1}{ah}\mathbf{k} \times \overrightarrow{BC}, \tag{14–80}$$

since the vector on the right has the correct direction and magnitude (see Problem 7(a) below). But $ah = 2\mathscr{A}$, where $\mathscr{A}$ is the area of the triangle, so that

$$\nabla\varphi_A = \frac{1}{2\mathscr{A}}\mathbf{k} \times \overrightarrow{BC}. \tag{14–80'}$$

If now φ_B is the analogous linear function such that $\varphi_B = 1$ at B and $\varphi_B = 0$ at A and C, then

$$\nabla\varphi_B = \frac{1}{2\mathscr{A}}\mathbf{k} \times \overrightarrow{CA},$$

so that

$$\nabla\varphi_A \cdot \nabla\varphi_B = \frac{1}{(2\mathscr{A})^2}(\mathbf{k} \times \overrightarrow{BC}) \cdot (\mathbf{k} \times \overrightarrow{CA}) = \frac{1}{4\mathscr{A}^2}\overrightarrow{BC} \cdot \overrightarrow{CA} \tag{14–81}$$

by vector identities (see Problem 7(b) below). But

$$\overrightarrow{BC} \cdot \overrightarrow{CA} = -ab\cos C, \qquad \mathscr{A} = \frac{1}{2}ab\sin C,$$

and thus by (14–81)

$$\nabla\varphi_A \cdot \nabla\varphi_B = -\frac{1}{2\mathscr{A}}\cot C. \tag{14–81'}$$

Accordingly, since the integrand is constant,

$$\iint\limits_{ABC} \nabla \varphi_A \cdot \nabla \varphi_B \, dx \, dy = \nabla \varphi_A \cdot \nabla \varphi_B \mathscr{A} = -\frac{1}{2} \cot C. \qquad (14\text{–}82)$$

For example, in Fig. 14–10 we verify that ABC is a $45°$ right triangle, so that $\cot C = 1$ and $\mathscr{A} = 5$. Also,

$$\varphi_A = \frac{3x + y - 14}{-10}, \qquad \varphi_B = \frac{4x - 2y - 2}{10}.$$

$$\nabla \varphi_A \cdot \nabla \varphi_B = \left(-\frac{3}{10}\mathbf{i} - \frac{1}{10}\mathbf{j}\right) \cdot \left(\frac{4}{10}\mathbf{i} - \frac{2}{10}\mathbf{j}\right) = -\frac{1}{10},$$

and Eq. (14–82) is satisfied.

We also deduce from Eq. (14–80') that $\nabla \varphi_A \cdot \nabla \varphi_A = a^2(4\mathscr{A}^2)$ and hence, as in (14–82),

$$\iint\limits_{ABC} \nabla \varphi_A \cdot \nabla \varphi_A \, dx \, dy = \frac{a^2}{4\mathscr{A}}. \qquad (14\text{–}83)$$

Now let R be a polygonal region as in Fig. 14–8, subdivided into triangles $R_1, \ldots, R_N$. We consider the Dirichlet problem $\nabla^2 u = 0$ in R, $u = g(x, y)$ on the boundary. This is a case of problem D in Section 14–5. Accordingly, we seek an approximate solution

$$u = u_0(x, y) + \sum_{j=1}^{n} c_j \varphi_j(x, y), \qquad (14\text{–}84)$$

where the φ_j correspond to the n inner vertices (and hence $\varphi_j = 0$ on the boundary), $u_0(x, y)$ is a function on R equal to $g(x, y)$ on the boundary. We may be able to find such a function $u_0(x, y)$ exactly. In most cases it is easier to take

$$u_0(x, y) = \sum_{j=n+1}^{m} g_j \varphi_j(x, y), \qquad (14\text{–}85)$$

where the boundary vertices are numbered $n+1, n+2, \ldots, m$ and g_j is the value of $g(x, y)$ at the jth vertex. This function $u_0(x, y)$ is a piecewise linear function agreeing with $g(x, y)$ at the boundary vertices. By using it instead of an exact $u_0(x, y)$ we are in effect using an approximate method to evaluate certain integrals below.

In accordance with Section 14–5, we use (14–84) and (14–85) in Eq. (14–52), with $w = \varphi_1(x, y), \ldots, \varphi_n(x, y)$, respectively. This leads to the equations

$$\iint\limits_{R} \left[\sum_{j=1}^{n} (c_j \nabla \varphi_i \cdot \nabla \varphi_j) + \sum_{j=n+1}^{m} (g_j \nabla \varphi_i \cdot \nabla \varphi_j) \right] dA = 0, \qquad i = 1, \ldots, n,$$

or

$$\sum_{j=1}^{n} a_{ij} c_j = b_i, \qquad i = 1, \ldots, n, \qquad (14\text{–}86)$$

where

$$a_{ij} = \iint_R \nabla \varphi_i \cdot \nabla \varphi_j \, dA,$$

$$(14\text{–}87)$$

$$b_i = - \sum_{j=n+1}^{m} \iint_R g_j \nabla \varphi_i \cdot \nabla \varphi_j \, dA.$$

We can write

$$a_{ij} = \sum_{k=1}^{N} e_{ijk}, \qquad e_{ijk} = \iint_{R_k} \nabla \varphi_i \cdot \nabla \varphi_j \, dA. \qquad (14\text{–}88)$$

For each k we get $e_{ijk} = 0$ except when i, j are numbers corresponding to vertices of R_k; when they do correspond to such vertices, e_{ijk} can be evaluated by Eqs. (14–82) and (14–83).

Similarly, we can write

$$-b_i = \sum_{j=n+1}^{m} d_{ij}, \qquad d_{ij} = \iint_R g_j \nabla \varphi_i \cdot \nabla \varphi_j \, dA,$$

$$(14\text{–}89)$$

$$d_{ij} = \sum_{k=1}^{N} f_{ijk}, \qquad f_{ijk} = \iint_{R_k} g_j \nabla \varphi_i \cdot \nabla \varphi_j \, dA = g_j e_{ijk}.$$

EXAMPLE 1   We choose R to be the region shown in Fig. 14–11, formed of 10 equilateral triangles with side 1. We seek the function $u(x, y)$ such that $\nabla^2 u = 0$ in R and $u = g(x, y)$ on the boundary of R, where g has the values shown and varies linearly between successive vertices. We number the vertices and triangles as shown, so that $i = 1, 2$ for the interior vertices, $i = 3, \ldots, 10$ for the boundary vertices.

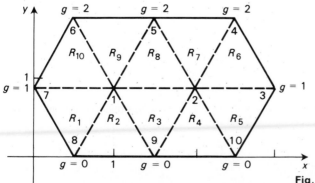

Fig. 14-11. Region for Example 1.

As above, our approximate solution is

$$u = u_0(x, y) + c_1\varphi_1(x, y) + c_2\varphi_2(x, y),$$

where

$$u_0(x, y) = \sum_{j=3}^{10} g_j\varphi_j(x, y)$$

and c_1, c_2 are the solutions of the equations

$$a_{11}c_1 + a_{12}c_2 = b_1, \qquad a_{21}c_1 + a_{22}c_2 = b_2,$$

and a_{ij} and b_i are given by (14–88) and (14–89).

We next compute the e_{ijk}. The area of each triangle is $\sqrt{3}/4$ and each angle is $\pi/3$. Hence, by (14–82) and (14–83),

$$e_{ijk} = -\frac{1}{2}\cot\frac{\pi}{3} = -\frac{1}{2\sqrt{3}}, \qquad i \neq j,$$

$$e_{iik} = \frac{1}{\sqrt{3}},$$

provided vertices i and j belong to the triangle R_k. Thus

$$e_{11k} = \frac{1}{\sqrt{3}} \quad \text{for } k = 1, 2, 3, 8, 9, 10, \qquad e_{11k} = 0 \text{ otherwise;}$$

$$e_{12k} = -\frac{1}{2\sqrt{3}} \quad \text{for } k = 3, 8, \qquad e_{12k} = 0 \text{ otherwise;}$$

$$e_{22k} = \frac{1}{\sqrt{3}} \quad \text{for } k = 3, \dots, 8, \qquad e_{22k} = 0 \text{ otherwise.}$$

Hence

$$a_{11} = a_{22} = \sum_{k=1}^{10} e_{11k} = \frac{6}{\sqrt{3}}, \qquad a_{12} = a_{21} = \sum_{k=1}^{10} e_{12k} = -\frac{1}{\sqrt{3}}.$$

Similarly $e_{ijk} = -1/(2\sqrt{3})$ for the following combinations of (i, j, k) with $i = 1$ or 2 and $j > 2$:

$$(1, 7, 1), (1, 8, 1), (1, 8, 2), (1, 9, 2), (1, 9, 3), (2, 9, 3), (2, 9, 4),$$
$$(2, 10, 4), (2, 10, 5), (2, 3, 5), (2, 3, 6), (2, 4, 6), (2, 4, 7),$$
$$(2, 5, 7), (1, 5, 8), (1, 5, 9), (1, 6, 9), (1, 6, 10), (1, 7, 10).$$

For all other such combinations $e_{ijk} = 0$. The f_{ijk} are obtained from the e_{ijk} by multiplying by g_j; thus

$$f_{ijk} = \begin{cases} 2e_{ijk}. & \text{for } j = 4, 5, 6, \\ e_{ijk} & \text{for } j = 3, 7 \\ 0 & \text{for } j = 8, 9, 10. \end{cases}$$

Accordingly,

$$-b_1 = \sum_{j=3}^{10} d_{1j} = \sum_{j=3}^{10}\sum_{k=1}^{10} f_{1jk} = 2\sum_{j=4}^{6}\sum_{k=1}^{10} e_{1jk} + \sum_{j=3}^{4}\sum_{k=1}^{10} e_{1jk}$$

$$= 2(e_{158} + e_{159} + e_{169} + e_{16,10}) + e_{171} + e_{17,10}$$

$$= 10 \cdot \frac{-1}{2\sqrt{3}} = -\frac{5}{3\sqrt{3}}.$$

Similarly, $b_2 = 5/\sqrt{3}$. The equations $a_{11}c_1 + a_{12}c_2 = b_1$, $a_{21}c_1 + a_{22}c_2 = b_2$ can now be solved to yield $c_1 = 1$, $c_2 = 1$. Thus our approximate solution is

$$u = \varphi_1(x, y) + \varphi_2(x, y) + \varphi_3(x, y) + 2[\varphi_4(x, y) + \varphi_5(x, y) + \varphi_6(x, y)] + \varphi_7(x, y)$$

and u has the value 1 at points 1 and 2.

In this problem, difference equations could have been used as in Section 13–15 to yield the same result. Also the problem has a simple exact solution $u = (2/\sqrt{3})y$ that coincides with our approximate solution (Problem 8 below). However, the finite-element method can be used for much more complicated regions, for which difference equations are not suited.

In this problem, the integration involved only constant integrands. In other problems (for example, one for $\nabla^2 u = f(x, y)$), there may be nonconstant integrands. In many cases the integrals can be calculated exactly. Approximate methods can also be used: for example, by interpolation replace the integrand by a piecewise linear function, which is easily integrated.

14–9 HIGHER-DEGREE APPROXIMATION IN TWO DIMENSIONS

By increasing the degree of the polynomials used, we can improve the accuracy of the approximation by means of our finite elements. Thus it is possible to use piecewise quadratic, piecewise cubic approximations, and so on. For problems in one dimension, the piecewise cubic approximations could be chosen to have continuous first derivative. The analogous condition for two dimensions requires polynomials of degree five or higher.

To illustrate the process, we consider a piecewise quadratic approximation. A quadratic polynomial in x, y has the form

$$p(x, y) = ax^2 + bxy + cy^2 + dx + ey + f. \tag{14–90}$$

There is a unique such polynomial $p(x, y)$ with given values at six points $(x_1, y_1), \ldots, (x_6, y_6)$, provided the points do not lie on a conic section (ellipse, parabola, hyperbola or, as degenerate cases, two parallel, coincident, or intersecting lines). To prove this we observe that the conditions $p(x_i, y_i) = k_i$ for $i = 1, \ldots, 6$ give six simultaneous linear equations for the six coefficients $a, \ldots, f$. The determinant of coefficients here cannot be 0, for if it were, then the homogeneous equations $p(x_i, y_i) = 0, i = 1, \ldots, 6$, would have a nontrivial solution for $a, \ldots, f$ and hence would determine a conic section $p(x, y) = 0$ (possibly degenerate) passing through all six points. Since the determinant is not 0, the equations $p(x_i, y_i) = k_i$ have a unique solution for $a, \ldots, f$, as asserted.

In particular, *we can choose $p(x, y)$ uniquely to have given values at the vertices and midpoints of the sides of a triangle*, as in Fig. 14–12, for a straight line meets a conic section in at most two points, except when the line is contained in the conic section (degenerate case). This follows from the fact that the intersection points are obtained by solving a quadratic equation in x or y. Hence if all six points $(x_1, y_1), \ldots, (x_6, y_6)$ are on a conic section, then the conic section would be degenerate and contain three lines, determined by the sides of the triangle. But a degenerate conic section is formed of at most two lines. Thus our assertion is proved.

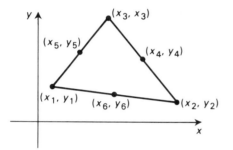

Fig. 14-12. Interpolation on a triangle.

If now R is a triangulated region, as in Fig. 14–13, and values are given at the vertices and midpoints of the sides of the triangles, then for each triangle R_k of the triangulation we can choose a quadratic polynomial $p_k(x, y)$ with the given values at the six selected points on the edges of R_k. The resulting piecewise quadratic function, equal to $p_k(x, y)$ in R_k, is necessarily continuous in R since if R_k and R_l have a common edge, then $p_k(x, y) = p_l(x, y)$ along the edge. This follows, as above, by considering the quadratic $p_k(x, y) - p_l(x, y)$; this is 0 at the three interpolation points on the edge and hence is 0 along the whole edge. Thus $p_k(x, y) = p_l(x, y)$ along the edge as asserted, and continuity follows at once for all points of the edge except,

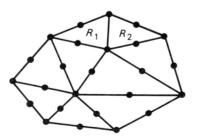
Fig. 14-13. Triangulated region.

perhaps, the endpoints. If a vertex is common to several triangles, the continuity of each $p_k(x, y)$ in its triangle R_k implies the continuity of the combined function at the vertex.

As in the piecewise linear case, we obtain a basis $\varphi_1(x, y), \ldots, \varphi_n(x, y)$ for all such piecewise quadratic functions on R by numbering all the interpolation points from 1 to n and choosing $\varphi_i(x, y)$ to be 1 at the ith point and 0 at all other points.

We give the form of a $\varphi_i(x, y)$ in a typical triangle R_k, which we take to be the triangle of Fig. 14-12. If the point i is not one of $(x_1, y_1), \ldots, (x_6, y_6)$, then $\varphi_i(x, y) \equiv 0$ in R_k, since it equals 0 at all six points. If, for example, the point i is the vertex (x_1, y_1), then we find (see Problem 9 below) that in R_k:

$$\varphi_i(x, y) = \frac{1}{4A_k^2} \begin{vmatrix} x - x_2 & y - y_2 \\ x_3 - x_2 & y_3 - y_2 \end{vmatrix} \begin{vmatrix} 2x - x_1 - x_2 & 2y - y_1 - y_2 \\ x_3 - x_2 & y_3 - y_2 \end{vmatrix}, \quad (14\text{-}91)$$

where A_k is the area of R_k. If the point i is the midpoint (x_4, y_4), then we find:

$$\varphi_i(x, y) = \frac{-1}{A_k^2} \begin{vmatrix} x - x_1 & y - y_1 \\ x_2 - x_1 & y_2 - y_1 \end{vmatrix} \begin{vmatrix} x - x_1 & y - y_1 \\ x_3 - x_1 & y_3 - y_1 \end{vmatrix}. \quad (14\text{-}92).$$

With the aid of these formulas we can calculate such expressions as $\nabla \varphi_i(x, y) \cdot \nabla \varphi_j(x, y)$ for various cases and hence proceed as in Section 14-8 to solve boundary-value problems of second order. Fourth-order problems require smoother elements, and hence we are forced to use polynomials of degree five.

In some cases we may be able to decompose R into squares or rectangles. Special elements have been devised for these cases. (For a detailed discussion of higher-order elements see pp. 40-79 of the book by CIARLET cited in Section 14-10. This book also discusses extension of the process to three-dimensional space.)

The application of the finite-element method to complicated regions can require lengthy calculations, for which a computer is essential. Many programs have been developed and valuable experience has been gained, showing the wide usefulness of the method.

═══════════════════**Problems (Section 14-9)**═══════════════════

1. Use the finite-element method as in Example 1 of Section 14-7 to find an approximate solution of the boundary-value problem given in the form $u = c_0\varphi_0(x) + \ldots + c_4\varphi_4(x)$:
   a) $u'' = 16e^x, 0 \leqslant x \leqslant 1, u(0) = u(1) = 0$;
   b) $u'' + xu = 1, 0 \leqslant x \leqslant 1, u'(0) = 0, u(1) = 0$.

   HINT for (b): As for problem A of Section 14-5, only the boundary condition $u(1) = 0$ is essential.

2. Let $x_1 < x_2 < \ldots < x_n$. For $i = 1, \ldots, n$ let $\varphi_i(x)$ be the continuous function on the interval $x_1 \leqslant x \leqslant x_n$ such that φ_i has the value 1 at x_i and the value 0 at all other x_j and such that φ_i is linear between successive points x_j, x_{j+1}. Show that $\varphi_1(x), \ldots, \varphi_n(x)$ are linearly independent on the interval.

3. Show that the approximate solution of Example 1 in Section 14–7 by difference equations, as in Section 13–15, that uses the same subdivision of the interval $0 \leqslant x \leqslant 1$, leads to the same linear system (14–73′) and hence to the same solution values at the subdivision points as that found by the finite element method.

HINT: As at the end of Section 13–15, interpret the condition $u'(1) = 0$ to mean $u(x_5) = u(x_3)$. Also apply the differential equation, in difference-equation approximation, at x_4.

4. With reference to Example 2 of Section 14–7, show that

$$\varphi_4(x) = 16x^3 - 60x^2 + 72x - 27, \quad 1 \leqslant x \leqslant \frac{3}{2},$$

$$\varphi_5(x) = \begin{cases} 4x^3 - 8x^2 + 5x - 1, & \frac{1}{2} \leqslant x \leqslant 1, \\ 4x^3 - 16x^2 + 21x - 9, & 1 \leqslant x \leqslant \frac{3}{2} \end{cases}$$

5. Solve Problem 1(a) approximately by piecewise-cubic finite elements, as done above in Example 2 of Section 14–7.

6. Follow the procedure of Example 1 in Section 14–8 to solve the Dirichlet problem $\nabla^2 u = 0$ in R, with $u = g(x, y)$ on the boundary of R, where R is the triangulated region shown in Fig. 14–14; the values of g at the vertices are given in the figure and g varies linearly between these vertices.

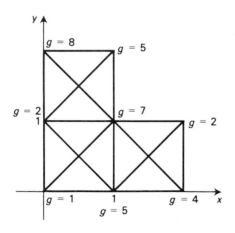

Fig. 14–14. Region for Problem 6.

7. With reference to the definition of φ_A in Section 14–8, prove
   a) Eq. (14–80)   b) Eq. (14–81)

8. Solve the boundary-value problem of Example 1 in Section 14–8 by difference equations, as in Section 13–15, and show that the same approximate solution is obtained; also show that $u = (2/\sqrt{3})y$ is the exact solution.

9. Verify that Eq. (14–91) defines a quadratic function of x, y that equals 1 at (x_1, y_1) and equals 0 at the other vertices and at the midpoints of the sides of the triangle of Fig. 14–12.

14–10 REFERENCES

Here we cite several books that shed more light on the ideas of this chapter.

PHILIPPE G. CIARLET, *Numerical Analysis of the Finite Element Method*, Les Presses de L'Université de Montreal, 1976.

JEROME J. CONNOR and C. A. BREBBIA, *Finite Element Techniques for Fluid Flow*, Newnes–Butterworths, London, 1976.

L. V. KANTOROVICH and V. I. KRYLOV, *Approximate Methods of Higher Analysis*, 4th ed., translated by Curtis D. Benster, Interscience Publishers, Inc., New York, 1958.

IVAR STAKGOLD, *Green's Functions and Boundary Value Problems*, John Wiley and Sons, New York, 1979.

Chapter 15
Probability
and Statistics

15–1 INTRODUCTION

When mathematical models are compared with physical experiments, one is confronted with uncertainties that require interpretation. Measurements give neither precise nor consistent results. The repeated measurements can simply be averaged, but even the averages are not consistent when further measurements are performed.

The problem can be analyzed by devising a theory that assigns probabilities to different measurement errors. Analysis of the simplest experiments, such as tossing coins and throwing dice, leads to such a probability theory. Years of experience have refined the theory into a powerful tool—statistics—permitting interpretation of very complex data.

An important application of the statistical methods is to allow conclusions to be drawn from data in a systematic way; this is the procedure of *statistical inference*. Another application is to permit discussion of physical systems if some of their components fluctuate in a random manner (*theory of noise and stochastic processes*).

In this chapter we present the principal concepts and methods and indicate some of their applications.

15–2 PROBABILITIES IN A FINITE SAMPLE SPACE

We consider a simple experiment with a finite number of possible outcomes. For example, we may toss a die; there are six outcomes, which we number 1, 2, . . . , 6. We may toss two dice; there are 36 outcomes, each of which can be described by an ordered pair of numbers: $(1, 1)$, $(1, 2)$, $(2, 1)$, $(1, 3)$, ..., $(6, 6)$. (We distinguish between $(1, 2)$ and $(2, 1)$, since in the first case the first die gives 1 and the second gives 2, while in the second case the first die gives 2 and the second gives 1.)

We call each outcome a *sample point* and the set of all possible outcomes a *sample space* $\mathscr{S}$ (see Fig. 15–1).

In probability theory, a number between 0 and 1 is assigned to each sample point. If there are n sample points $x_1, \ldots, x_n$, we write $\Pr(x_i)$ for the probability assigned to x_i and also write

$$P_i = \Pr(x_i).$$

We want the sum of all the probabilities to be 1. Thus our requirements are

$$0 \leqslant p_i \leqslant 1, \qquad i = 1, \ldots, n,$$
$$p_1 + p_2 + \cdots + p_n = 1. \tag{15–20}$$

For example, if we toss a die, we have the six sample points $x_1 = 1, x_2 = 2, \ldots,$ $x_6 = 6$. It is natural to let each $p_i = 1/6$. We justify this intuitively by saying that the chances that each one of the six values shows are 1 in 6. We can also say that if we toss the die many times, we expect a 1 or a 2, etc., to appear about 1/6 of the time. If we try this experimentally with an actual die, we find fairly good agreement. However, it is clear that the experimental results depend on how symmetrical the die is; a "loaded" die can give quite different results.

In the same way, if we toss two dice, there are 36 sample points and we assign each the probability 1/36.

The probabilities need not all be equal. For example, an urn may contain 11 white balls and 7 red balls. If we choose a ball at random, we have two possible outcomes: W (white) and R (red). It is natural here to take $\Pr(W) = 11/18$ and $\Pr(R) = 7/18$. If there were 18 balls, all white, then clearly we must take $\Pr(W) = 1$, $\Pr(R) = 0$. Thus a probability of 1 is associated with absolute certainty, a probability of 0 with impossibility. However these relationships require modification, as we shall see.

Having assigned probabilities $p_1, \ldots, p_n$ to sample points $x_1, \ldots, x_n$, as in (15–20), we can now also assign probabilities to *sets* of sample points: to each such

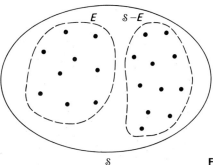

S

Fig. 15–1. Events as subsets of the sample space $\mathscr{S}$.

set E we assign the probability $\Pr(E)$ equal to the *sum of the probabilities of the sample points in E*. We think of E as an *event*. This interpretation is based on dividing the sample space $\mathscr{S}$ into two parts: one is E, the other is $\mathscr{S} - E$, the set of sample points not in E. If we then make our experiment, we may obtain a point of E; we say "The event E has occurred." It is clearly reasonable to assume that the probability of obtaining a point of E is the sum of the probabilities of the sample points in E.

As a specific example, we toss a die once and have $p_i = 1/6$ for $i = 1, \ldots, 6$ as above. Let E be the set consisting of 4, 5, 6. Then by our rule,

$$\Pr(E) = \frac{1}{6} + \frac{1}{6} + \frac{1}{6} = \frac{1}{2}.$$

We could say: if a die is tossed once, the chances of getting at least a 4 are 1 in 2. The "event" here is "Getting at least a 4." Similarly, if $E = 3$ and 6, $\Pr(E) = 1/3$; the event here could be described as "Getting a number divisible by 3."

By our assignment of probability to events, we can deduce some simple rules. Here we use the notation $A \cup B$ for the *union* of two sets (set of all points in A or B or both) and $A \cap B$ for the *intersection* (set of all points in both A and B); also we write ϕ for the empty set. Then our rules are:

$$\Pr(A \cup B) = \Pr(A) + \Pr(B) \quad \text{if} \quad A \cap B = \phi, \tag{15–21}$$

$$\Pr(A \cup B) = \Pr(A) + \Pr(B) - \Pr(A \cap B), \tag{15–22}$$

$$\Pr(E) + \Pr(\mathscr{S} - E) = 1, \tag{15–23}$$

$$\Pr(\phi) = 0, \qquad \Pr(\mathscr{S}) = 1 \tag{15–24}$$

The first rule is illustrated in Fig. 15–2(a); it follows at once from the way we assign probabilities to events. For the second rule, illustrated in Fig. 15–2(b), $\Pr(A) + \Pr(B)$ clearly counts each sample point in $A \cap B$ twice, so that we must subtract $\Pr(A \cap B)$ to obtain $\Pr(A \cup B)$. The other rules follow at once from the definition.

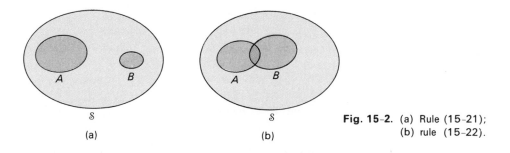

Fig. 15-2. (a) Rule (15–21); (b) rule (15–22).

We illustrate the rules by tossing a coin three times. If we write H for heads, T for tails, then there are eight outcomes:

HHH,   HHT,   HTH,   HTT,   THH,   THT,   TTH,   TTT.

Assuming that the coin is unbiased, we assign each sample point the probability 1/8.

Let A be the event "at least one head," B the event "at least one tail." Then

$$\Pr(A) = \frac{7}{8}, \quad \Pr(B) = \frac{7}{8}, \quad \Pr(A \cap B) = \frac{6}{8}, \quad \Pr(A \cup B) = 1.$$

Here $A \cup B = \mathscr{S}$, since at least one head or one tail must appear and

$$A \cap B = \{\text{HHT, HTH, HTT, THH, THT, TTH}\}.$$

We use the braces to enclose the sample points in the set $A \cap B$. Here $A \cap B$ must consist of the outcomes including at least one H and one T. The rule (15–22) is satisfied, since

$$\frac{7}{8} + \frac{7}{8} - \frac{6}{8} = 1.$$

15–3 CONDITIONAL PROBABILITIES

Let A, B be events. Then the conditional probability that A occurs, given that B occurs, is denoted by $\Pr(A|B)$. By saying "Given that B occurs," we are in effect limiting the outcomes to those sample points in B. Thus B has become the sample space. It is reasonable to assign to each sample point x_j in B the (conditional) probability $p_j/\Pr(B)$. Then if B consists of the points $x_1, \ldots, x_k$, the sum of these probabilities is

$$\frac{p_1 + \cdots + p_k}{\Pr(B)} = \frac{\Pr(B)}{\Pr(B)} = 1,$$

as we would want. If $A \cap B$ consists of only $x_1, \ldots, x_l$, then the conditional probability of A, given that B occurs, is

$$\Pr(A|B) = \frac{p_1 + \cdots + p_l}{\Pr(B)} = \frac{\Pr(A \cap B)}{\Pr(B)}. \tag{15–30}$$

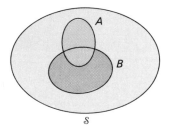

$\mathscr{S}$

Fig. 15-3. Meaning of Eq. (15–30).

These formulas all make sense only if $\Pr(B) > 0$; we shall assume that this inequality is valid. In the common case, when $p_1 = p_2 = \ldots$, Eq. (15–30) gives $\Pr(A|B)$ as the ratio of the number of sample points in $A \cap B$ to the number of sample points in B.

EXAMPLE 1   We choose a card from a bridge deck and note that it is a club. What is the probability that it is an ace? We assume that all cards have equal probabilities, so that there are 13 sample points in B and the probability of choosing the ace is $1/13$. ◄

EXAMPLE 2 We toss a die twice and are told that no 6 occurs. What is the probability that no odd number occurs? Since no 6 occurs, that is, $(1, 6), (6, 1), (2, 6),$ $(6, 2), \ldots, (5, 6), (6, 5), (6, 6)$ are excluded, the event B has $36 - 11 = 25$ points. There are 9 points in A, but only 4 of these are in $A \cap B$, namely $(2, 2), (2, 4), (4, 2)$ and $(4, 4)$. Hence $\Pr(A|B) = 4/25$. ◄

Very often we can find $\Pr(A|B)$ directly and then use Eq. (15–30) to find $\Pr(A \cap B)$:

$$\Pr(A \cap B) = \Pr(B)\Pr(A|B). \tag{15–31}$$

EXAMPLE 3 An urn contains three white balls, numbered 1, 2, 2 and four black balls, numbered 2, 4, 5, 5. If one ball is drawn at random, what is the probability that it is white and numbered 2? We take A to be the event "The number is 2," B the event "It is white." Then $\Pr(A|B) = 2/3$, $\Pr(B) = 3/7$ and

$$\Pr(A \cap B) = \frac{2}{3} \cdot \frac{3}{7} = \frac{2}{7}. \text{ ◄}$$

Independent events We say that A, B are independent if

$$\Pr(A|B) - \Pr(A). \tag{15–32}$$

Thus, whether B occurs or not is irrelevant to finding $\Pr(A)$. From Eq. (15–31) we conclude that, if A and B are independent, then

$$\Pr(A \cap B) = \Pr(A)\Pr(B). \tag{15–33}$$

Conversely, if Eq. (15–33) holds, then so does Eq. (15–32), as follows from (15–33) and (15–31). Thus Eq. (15–33) also characterizes the independence of A and B. By symmetry, we deduce from (15–33) that if A, B are independent, then

$$\Pr(B|A) = \Pr(B).$$

In Example 1 above, A and B are independent, since A (it is an ace) has probability $4/52 = 1/13 = \Pr(A|B)$. In Example 2 above, A and B are dependent (not independent), since A (no odd number occurs) has probability $9/36$; the dependence can be explained by saying that removing the 6's affects the probability of getting odd and even numbers.

At times, more than two events are referred to as independent. Events A, B, C can be defined as independent if

$$\Pr(A \cap B \cap C) = \Pr(A) \cap \Pr(B) \cap \Pr(C),$$

and a similar definition is given for four or more events. A typical application is to repeated experiments. If one tosses a die many times, the probability of getting a 1 on the 1st, 2nd, $\ldots$, nth toss should relate to independent events, so that the probability of getting a 1 every time should be $(1/6)^n$.

Total probability In many cases one finds it useful to analyze a probability problem by considering a number of alternatives which together exhaust all cases.

This process is equivalent to decomposing the sample space into nonoverlapping events $B_1, \ldots, B_k$, so that

$$\mathscr{S} = B_1 \cup B_2 \cup \ldots \cup B_k, \qquad B_j \cap B_l = \phi \quad \text{for} \quad j \neq l.$$

The process is suggested in Fig. 15-4. For an event A, we have

$$A = (A \cap B_1) \cup (A \cap B_2) \cup \ldots \cup (A \cap B_k),$$

as in Fig. 15-4. By repeated application of the rule (15–21). we conclude that

$$\Pr(A) = \sum_{j=1}^{k} \Pr(A \cap B_j). \tag{15–34}$$

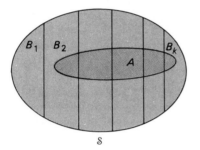

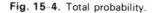

$\mathcal{S}$

Fig. 15-4. Total probability.

From the rule (15–31) we can also write:

$$\Pr(A) = \sum_{j=1}^{k} \Pr(B_j) \Pr(A \mid B_j). \tag{15–35}$$

This is the *rule of total probability*.

EXAMPLE 4 For the urn of Example 3, we could seek the probability of choosing a 2. We take B_1 and B_2 as the events of choosing a white or a black ball, respectively. Then by Eq. (15–35)

$$\Pr(A) = \Pr(B_1) \Pr(A \mid B_1) + \Pr(B_2) \Pr(A \mid B_2)$$
$$= \frac{3}{7} \cdot \frac{2}{3} + \frac{4}{7} \cdot \frac{1}{4} = \frac{3}{7}. \quad \blacktriangleleft$$

Bayes' theorem   We seek to deduce $\Pr(B \mid A)$ from $\Pr(A \mid B)$. As above,

$$\Pr(A \cap B) = \Pr(B) \Pr(A \mid B) = \Pr(A) \Pr(B \mid A).$$

Hence

$$\Pr(B \mid A) = \frac{\Pr(B) \Pr(A \mid B)}{\Pr(A)}. \tag{15–36}$$

Here $Pr(A)$ is often found by total probability, as above. Thus

$$Pr(B|A) = \frac{Pr(B)\,Pr(A|B)}{\displaystyle\sum_{j=1}^{k} Pr(B_j)\,Pr(A|B_j)}. \tag{15–36'}$$

Eqs. (15–36) and (15–36′) form Bayes' theorem.

EXAMPLE 5   A tool manufacturer ships equal quantities of hammers, pliers, and wrenches. From experience we assign probabilities of defective tools to be 0.005 for hammers, 0.003 for pliers, and 0.008 for wrenches. If a tool is found to be defective, what is the probability that it is a wrench?

Here we let B be the event "The tool is a wrench" and A be the event "The tool is defective." We know $Pr(A|B) = 0.008$. We seek $Pr(B|A)$. We let B_1 be the event "The tool is a hammer," B_2 the event "The tool is pliers," $B_3 = B$, i.e., "The tool is a wrench." Then

$$Pr(A|B_1) = 0.005, \qquad Pr(A|B_2) = 0.003, \qquad Pr(A|B_3) = 0.008.$$

Also $Pr(B_j) = \tfrac{1}{3}$, since equal quantities of each are shipped. Hence, by Eq. (15–36′),

$$Pr(B|A) = \frac{(1/3)(0.008)}{(1/3)(0.005 + 0.003 + 0.008)} = 1/2.$$

The result is reasonable, since in a shipment of 1000 of each tool, we would expect 5 defective hammers, 3 defective pliers, and 8 defective wrenches. ◄

15–4 PERMUTATIONS AND COMBINATIONS

We recall from algebra some useful rules about these quantities, which occur very often in probability questions. The number of *permutations* of n objects is $n!$ For example, the 3 objects a, b, c have the 6 permutations $abc, acb, bac, bca, cab, cba$. If we choose m objects out of n and consider all permutations of these, we obtain in all

$$\frac{n!}{(n-m)!} = n(n-1)\ldots(n-m+1). \tag{15–40}$$

For example, from the 4 objects a, b, c, d we obtain the $4!/2! = 12$ permutations $ab, ac, ad, ba, bc, bd, ca, cb, cd, da, db, dc$. If we disregard the order for each of these sets of m objects, we are counting *combinations* rather than permutations. Since each set of m objects has $m!$ permutations, there are

$$\frac{n!}{m!(n-m)!} = \frac{n(n-1)\ldots(n-m+1)}{m!} = \binom{n}{m} \tag{15–41}$$

combinations in all. Here $\binom{n}{m}$ is the general *binomial coefficient*. As an example, the number of combinations of two objects out of the four objects a, b, c, d is $\binom{4}{2} = 6$; the combinations (order disregarded) are ab, ac, ad, bc, bd, cd.

Next we consider the number of distinct permutations of n objects of which the first n_1 objects, the next n_2 objects, etc., and the final n_k objects are identical. For example, the permutations of *aaabb* are *aaabb, aabab, abaab, baaab, aabba, ababa, baaba, abbaa, babaa, bbaaa*, or 10 in all. In general, for $n = n_1 + n_2$, for each permutation of the n objects (regarded as different), we get $n_1!$ equivalent ones by permuting the n_1 objects and $n_2!$ equivalent ones by permuting the n_2 objects. Hence the number of different permutations is

$$\frac{n!}{n_1! n_2!}.$$

In our example, $n = 5$, $n_1 = 3$, $n_2 = 2$, and we get $5!/(3!2!) = 10$. Similarly, for $n = n_1 + n_2 + \cdots + n_k$, there are

$$\frac{n!}{n_1! n_2! \ldots n_k!} \qquad (15\text{-}42)$$

different permutations.

Finally, we determine how many combinations of m objects can be obtained from n objects if repetitions are allowed. For example, if an urn contains white balls and black balls and we choose three balls in turn, *replacing* the ball after each choice, then the possible combinations (order disregarded) are WWW, WWB, WBB, BBB. In general, if $n = 2$, as in the example, we see that there are $m + 1$ combinations (one each according to the number of W's). If $n = 3$, say W, B, and R (red), there are $m + 1$ choices for the number of reds; for 0 reds, there are $m + 1$ choices of the W, B combinations; for 1 red, there are m choices as the W, B combinations, and so on. Hence we get

$$(m + 1) + m + (m - 1) + \cdots + 1 = \frac{(m + 1)(m + 2)}{2}.$$

An inductive procedure leads to the general rule that there are

$$\frac{(m + 1)(m + 2) \cdots (m + n - 1)}{1 \cdot 2 \cdots (n - 1)} = \binom{m + n - 1}{n - 1} \qquad (15\text{-}43)$$

combinations.

═══════════════**Problems (Section 15-4)**═══════════════

1. A coin is tossed four times. Find the probability of
   a) Only two heads; b) At most two heads; c) More than two heads.

2. Two dice are tossed. Find the probability of
   a) Both the same; b) Being different; c) One even, one odd; d) At least one 4.

3. Two dice are tossed. Let A be the event "Both are even." Let B be the event "Both are greater than 3." Find
   a) $\Pr(A \cup B)$; b) $\Pr(A \cap B)$; c) $\Pr(A | B)$;
   d) $\Pr(B | A)$; e) Determine whether A and B are independent.

4. Two cards are chosen from a bridge deck. Let A be the event "Both are black (clubs or spades)," B the event "No red face card appears." Find
   a) $\Pr(A \cup B)$; b) $\Pr(A \cap B)$; c) $\Pr(A | B)$;
   d) $\Pr(B | A)$; e) Determine whether A and B are independent.

5. A manufacturer observes that the probability of producing a defective part on Monday is 0.002, on Tuesday is 0.003, on Wednesday is 0.002, on Thursday is 0.005, on Friday is 0.008. If equal quantities are produced each day, what is the probability that a part, chosen at random, is defective?

6. A person receives 2/3 of his phone calls from A, who calls three times as frequently in the afternoon as in the morning, and 1/3 of his calls from B, who calls four times as often in the morning as in the afternoon. If the person receives just one call on a given afternoon, what is the probability that the caller is A?

7. Solve Example 5 of Section 15–3 after modifying it to assume that the numbers of hammers, pliers, and wrenches shipped are in the proportion $10:2:3$.

8. What is the probability of being dealt a perfect hand (all of one suit) in bridge?

9. List all permutations of *abbccc* and verify that Eq. (15–42) is satisfied.

10. List all combinations of four choices of W, B, and R, allowing repetitions, and verify that Eq. (15–43) is satisfied.

15–5   PROBABILITY IN AN INFINITE SAMPLE SPACE

The results of physical measurements are real numbers which can, in principle, take on all real values. Hence in making a statistical analysis of measurements we assign probabilities in an infinite sample space. The same need arises in studying problems related to spatially distributed quantities. Take, for example, the problem of assigning probabilities of finding uranium in various geographical regions; in principle, there may be infinitely many regions, each containing infinitely many sample points.

In general, we are again dealing with a sample space $\mathscr{S}$ of infinitely many points—the sample points. However, we usually cannot develop our probabilities by simply assigning probabilities to the individual points. If we did, the total probability of $\mathscr{S}$ would have to be obtained by adding the probabilities of all points, and this would require an infinite sum. In some cases, this sum can be treated as an infinite series and a meaningful result is obtained. However, in many cases, the process fails. For example, if we try to assign probabilities to *all* numbers between 0 and 100 (which could be the heights, in inches, of individuals), we get an unmanageable sum.

The difficulty is resolved by requiring only that probabilities be assigned to certain subsets of $\mathscr{S}$. If, for example, $\mathscr{S}$ consists of all numbers between 0 and 100, we would assign probabilities to each interval $a \leqslant x \leqslant b$ and interpret $\Pr(a \leqslant x \leqslant b)$ as the probability that the sample point (height of an individual) lies between a and b, inclusive.

Having assigned probabilities to certain events $A, B, \ldots$, i.e., subsets of $\mathscr{S}$, we are then led to assign probabilities to other subsets of $\mathscr{S}$ obtained by applying set operations: $A \cup B$, $A \cap B$, $\mathscr{S} - A$. The choices are dictated by the basic rules (15–21)–(15–24). For example, having assigned probabilities to intervals, we must assign probabilities to unions of intervals. With some care, this can be done in a consistent fashion. For some purposes, we sometimes have to consider processes involving an infinite sequence of sets. A good example of this is the union $A_1 \cup A_2 \cup \cdots \cup A_n \cup \cdots$. If these sets are *pairwise disjoint*, we require that

$$\Pr(A_1 \cup A_2 \cup \cdots \cup A_n \cup \cdots) = \sum_{n=1}^{\infty} \Pr(A_n). \qquad (15\text{--}50)$$

This property is called *complete additivity*.

EXAMPLE 1   Let $\mathscr{S}$ be the interval $0 \leqslant x \leqslant 1$. We assign to each interval $a \leqslant x \leqslant b$ contained in interval $\mathscr{S}$ the probability $b - a$. Thus $\Pr(\mathscr{S}) = 1$, $\Pr(0.5 \leqslant x \leqslant 0.75) = 0.25$, $\Pr(0.1 \leqslant x \leqslant 0.7) = 0.6$. We can then assign other probabilities by the rules (15–21)–(15–24). For example,

$$\Pr\left(\left\{\frac{1}{8} \leqslant x \leqslant \frac{3}{8}\right\} \cup \left\{\frac{5}{8} \leqslant x \leqslant \frac{7}{8}\right\}\right) = \left(\frac{3}{8} - \frac{1}{8}\right) + \left(\frac{7}{8} - \frac{5}{8}\right) = \frac{1}{2}.$$

Also

$$\Pr\left(\left\{\frac{1}{2} \leqslant x \leqslant \frac{3}{4}\right\} \cup \left\{\frac{3}{4} \leqslant x \leqslant 1\right\}\right) = \Pr\left(\frac{1}{2} \leqslant x \leqslant 1\right) = \frac{1}{2}.$$

These two intervals have the point $x = 3/4$ in common. By the rule (15–22), we are forced to assign 0 probability to this point:

$$\Pr\left(x = \frac{3}{4}\right) = 0.$$

In general, for this example, $\Pr(x = a) = 0$; we can think of such a point as the interval $a \leqslant x \leqslant a$, and hence its probability is $a - a = 0$. (This does not mean that each individual point of $\mathscr{S}$ is *impossible* to attain, but rather that each one occurs so rarely that it is given probability 0.) ◄

EXAMPLE 2   Let $\mathscr{S}$ be the interval $0 \leqslant x \leqslant 1$ and let $\Pr(a \leqslant x \leqslant b) = \int_a^b 2x\,dx$. Thus

$$\Pr(\mathscr{S}) = \int_0^1 2x\,dx = 1,$$

$$\Pr\left(\frac{1}{2} \leqslant x \leqslant \frac{3}{4}\right) = \int_{1/2}^{3/4} 2x\,dx = \frac{9}{16} - \frac{1}{4} = \frac{5}{16}.$$

Again $\Pr(x = a) = 0$. This example suggests a generalization. To obtain probabilities on $\mathscr{S}$, the interval $0 \leqslant x \leqslant 1$, we choose $f(x)$ such that $f(x) \geqslant 0$ on the interval and $\int_0^1 f(x)\,dx = 1$. Then we assign $\Pr(a \leqslant x \leqslant b) = \int_a^b f(x)\,dx$.

Here $f(x)$ should be continuous or, say, piecewise continuous. We call $f(x)$ the *probability density* of the probability assignment. The same idea can be applied to other intervals $\mathscr{S}$, including infinite intervals. Another generalization is to choose $\mathscr{S}$ as a two-dimensional region, say, the square $-1 \leqslant x \leqslant 1$, $-1 \leqslant y \leqslant 1$ in the xy-plane, and select a probability density $f(x, y)$ such that $f(x, y) \geqslant 0$ and

$$\int_{-1}^{1} \int_{-1}^{1} f(x, y)\,dy\,dx = 1.$$

Then for each region R contained in $\mathscr{S}$ we set

$$\Pr(R) = \int\int_R f(x, y)\, dx\, dy.$$

Similar procedures can be followed in three or more dimensions. ◀

EXAMPLE 3 Let $\mathscr{S}$ be the set of all nonnegative integers: 0, 1, 2, ... Let

$$\Pr(x = n) = p_n = \frac{1}{2^{n+1}}.$$

Now we are naturally led to define $\Pr(E)$ as the sum of the probabilities of the sample points (integers) in E. In particular,

$$\Pr(x = 1, 2 \text{ or } 3) = p_1 + p_2 + p_3 = \frac{1}{4} + \frac{1}{8} + \frac{1}{16} = \frac{7}{16}.$$

Also

$$\Pr(\mathscr{S}) = \sum_{n=0}^{\infty} p_n = \sum_{n=0}^{\infty} \frac{1}{2^{n+1}} = 1,$$

by the rule for geometric series (Section 2–5). In this way $\Pr(A)$ is defined for *all* subsets of $\mathscr{S}$ and all rules are clearly satisfied, including complete additivity as in (15–50). ◀

It is helpful to think of all these probability assignments as comparable to distributions of mass. The probability assigned to a set is the total mass in the set. The probability density is then analogous to density in physics, i.e., mass per unit length, area, or volume. If a mass has continuous density, the mass at a point is 0. However, we may conceive of particles whose mass is concentrated at a point. In Example 3, we can think of particles located at $x = 0, 1, 2, \ldots$, with masses $1/2, 1/4, 1/8, \ldots$ For a probability assignment, the total mass is always taken as 1.

This analogy suggests going further in Example 3 and defining $\Pr(a \leqslant x \leqslant b)$ as the total "mass" between a and b, inclusive. For example,

$$\Pr(0.2 \leqslant x \leqslant 4.5) = 2^{-2} + 2^{-3} + 2^{-4} + 2^{-5} = \tfrac{15}{32}.$$

Similarly, we can define $\Pr(x \leqslant x_0)$ for any x_0.

In general, for a probability assignment on the x-axis we can define $\Pr(x \leqslant x_0)$ for each x_0. The value obtained, often denoted by $F(x_0)$, is called the *probability distribution function*. Thus $F(x)$ is a function of x that gives the probability assigned to the interval from $-\infty$ to x. The function $F(x)$ is necessarily monotone nondecreasing, with limit 0 as $x \to -\infty$ and limit 1 as $x \to +\infty$. The distribution function for Examples 1, 2, 3 is shown in Fig. 15–5.

If our probabilities for sets on the x-axis have a probability density $f(x)$, then clearly

$$F(x) = \int_{-\infty}^{x} f(t)\, dt, \qquad -\infty < x < \infty. \tag{15–51}$$

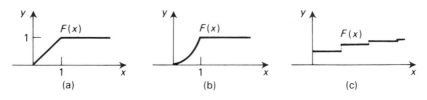

Fig. 15–5. Distribution functions: (a) Example 1; (b) Example 2; (c) Example 3.

Here we interpret $f(x)$ as 0 in each interval with 0 probability. Equation (15–51) is valid if $f(x)$ is continuous for all x or, more generally, if $f(x)$ is such that the integral in (15–51) always exists, perhaps, as an improper integral. From (15–51) it follows that, wherever $f(x)$ is continuous, $F'(x) = f(x)$.

15–6 RANDOM VARIABLES

For practical reasons, most probability questions are turned into questions about numbers. For example, instead of asking "What is the probability that the sum of two dice is even?" we denote the sum of the dice as X and seek $\Pr(\{2, 4, 6, 8, 10, 12\})$. Here X is called a *random variable*. It is clear that in this case the random variable is a function—a number X assigned to each sample point, and that we can determine the probability that X has a value in a set A of numbers by simply determining the probability of the set of all sample points for which X has a value in A. In our example, A is the set $\{2, 4, \ldots, 12\}$, and the corresponding set of sample points is $\{(1, 1), (1, 3), (2, 2), (3, 1), \ldots, (6, 6)\}$. We see that there are 18 sample points and hence the probability in question is $18/36 = 1/2$.

In general, a random variable X is a function defined on a sample space (on which probabilities have been assigned as above) with real-number values. Accordingly, for each random variable X we can assign probabilities to sets of real numbers. For example, $\Pr(a \leqslant X \leqslant b)$ is defined as the probability assigned to the set of sample points for which X lies between a and b, inclusive.

If we are tossing two dice and X is the sum, then X can equal only $2, 3, \ldots, 12$. Thus

$$\Pr(X = 2) = \frac{1}{36}, \qquad \Pr(X = 3) = \frac{2}{36}, \qquad \ldots, \qquad \Pr(X = 12) = \frac{1}{36}.$$

In the first case, we have only one sample point $(1, 1)$; in the second, two points $(1, 2)$ and $(2, 1)$. Thus on the X-axis we have "mass" only at the integers $2, 3, \ldots, 12$. However, as in Section 15–5, we can evaluate $\Pr(a \leqslant X \leqslant b)$ for all a, b as the total "mass" in this interval.

EXAMPLE 1 An archer is able to hit a circular target but finds that the probability of hitting a given region of the target is proportional to the area of the region. If the radius of the target is 1 meter, find the probability (a) of striking within 10 cm of the center, (b) of striking between 30 and 50 cm of the center.

Here the random variable X is the distance from the center and $\Pr(0 \leqslant X \leqslant r)$ $= c\pi r^2$, where c is a constant. For $r = 1$, this must be 1, so that $c = 1/\pi$. For (a) we want
$$\Pr(0 \leqslant X \leqslant 0.1) = 0.01;$$
for (b) we want
$$\Pr(0.3 \leqslant X \leqslant 0.5) = \Pr(0 \leqslant X \leqslant 0.5) - \Pr(0 \leqslant X \leqslant 0.3)$$
$$= 0.25 - 0.09 = 0.16 \quad \blacktriangleleft$$

Distribution function, density of a random variable   Since a random variable assigns probabilities to sets on an X-axis, we can, as in Section 15–5, define an associated distribution function $F(x) = \Pr(X \leqslant x)$. For a finite sample space, $F(x)$ is constant except for a finite number of jumps that occur at values $x_1, x_2, \ldots$ actually taken on by X. In an infinite sample space, $F(x)$ may be continuous and may have a density $f(x)$, so that Eq. (15–51) holds.

Discrete case and continuous case   In view of the applications, we shall henceforth consider only two cases for probabilities associated with a random variable X:

Discrete case.   Here X can take on only values of a sequence $x_1, x_2, \ldots$, finite or infinite. We have $\Pr(x_i) = p_i \geqslant 0$ and $p_1 + p_2 + \cdots = 1$ (an infinite series or finite sum). There is a distribution function
$$F(x) = \Pr(X \leqslant x) = \sum_{x_i \leqslant x} p_i,$$
that is, $F(x)$ is the sum of all p_i for $x_i \leqslant x$. This is the case of a set of particles at $x_1, x_2, \ldots$, with masses $p_1, p_2, \ldots$ and total mass 1.

Continuous case.   Here there is a continuous density $f(x) \geqslant 0$ and
$$\Pr(a \leqslant x \leqslant b) = \int_a^b f(x)\,dx.$$
There is also a distribution function $F(x)$ that satisfies (15–51) and $F(x) \to 0$ as $x \to -\infty$ and $F(x) \to 1$ as $x \to +\infty$, so that
$$F(\infty) = \int_{-\infty}^{\infty} f(x)\,dx = 1.$$

In the continuous case, $\Pr(X = x_0) = 0$ for every x_0; in the discrete case $\Pr(X = x_0) = 0$ when x_0 is not one of the values of the sequence $x_1, x_2, \ldots$

Mean, or expectation, of a random variable   Let the random variable X be given. Let X take on only a finite set of values $x_1, \ldots, x_n$ and let the probabilities $p_i = \Pr(X = x_i)$ be known for $i = 1, \ldots, n$. We define the *mean*, or *expectation*, of X, denoted by μ or $E(X)$, as the sum
$$\mu = E(X) = p_1 x_1 + \cdots + p_n x_n. \tag{15–60}$$
Since $p_1 + \cdots + p_n = 1$, μ is the weighted average of the values $x_1, \ldots, x_n$, where the weight p_i is assigned to x_i for $i = 1, \ldots, n$. We can think of p_i as the fraction of

times the value x_i would occur in many repetitions of an experiment. Thus, roughly, $p_i = k_i/N$, where k_i is the number of times x_i occurs and N is the number of repetitions. Hence the sum in (15–60) would be

$$\frac{k_1 x_1 + \cdots + k_n x_n}{N} = \frac{(x_1 + \cdots + x_1) + (x_2 + \cdots + x_2) + \cdots + (x_n + \cdots + x_n)}{N},$$

where there are k_1 terms in the first parenthesis, k_2 terms in the second, and so on; there are $k_1 + \cdots + k_n = N$ terms in all. Accordingly, the sum would be precisely the average value. Since we have only probabilities for the ratios k_i/N, we say that this is the *expected average value* or, more briefly, the *expected value*, or *expectation*.

If we interpret the p_i as masses of particles at the x_i, then the sum in Eq. (15–60) is the *first moment* about $x = 0$ of masses: the familiar equation for particles

$$\sum_{i=1}^{n} m_i x_i = M\bar{x}, \qquad M = m_1 + \cdots + m_n,$$

where $\bar{x}$ is the x-coordinate of the center of mass, shows that μ is the *center of mass* for our particles. (The total mass here is 1.)

The concept of expectation extends naturally to the discrete case with infinitely many values:

$$\mu = E(X) = \sum_{i=1}^{\infty} p_i x_i. \tag{15-61}$$

It also extends to the continuous case:

$$\mu = E(X) = \int_{-\infty}^{\infty} x f(x)\, dx, \tag{15-62}$$

where $f(x)$ is the associated probability density (so that $f(x)\, dx = dm$, the *mass element*). By a limit argument, this can be interpreted as the average value of X to be expected after many repetitions of the experiment. Also, μ can again be interpreted as a center of mass, and the sum or integral as a first moment about $x = 0$. (We assume here and throughout the chapter that improper integrals such as (15–62) are finite.)

EXAMPLE 2 Let X be the value that shows when one die is thrown. Hence there are six values $1, \ldots, 6$ with equal probabilities $1/6$. Therefore,

$$\mu = E(X) = \frac{1}{6} \cdot 1 + \frac{1}{6} \cdot 2 + \cdots + \frac{1}{6} \cdot 6 = 3.5. \quad \blacktriangleleft$$

EXAMPLE 3 For throwing two dice, let X be the sum. There are 11 values 2, 3, $\ldots$, 12 with probabilities 1/36, 2/36, 3/36, 4/36, 5/36, 6/36, 5/36, 4/36, 3/36, 2/36, 1/36 (obtained by counting how many pairs yield each sum). Thus

$$\mu = E(X) = \frac{1}{36} \cdot 2 + \frac{2}{36} \cdot 3 + \cdots + \frac{2}{36} \cdot 11 + \frac{1}{36} \cdot 12 = 7.$$

We have here 11 particles whose masses are symmetric with respect to $x = 7$, so that the result is to be expected. ◄

EXAMPLE 4 For the archer of Example 1, we found $\Pr(0 \leqslant X \leqslant x)$ to be x^2 for $x \leqslant 1$. Since negative values of X are excluded, we have $F(x) = 0$ for $x < 0$, $F(x) = x^2$ for $0 \leqslant x \leqslant 1$, $F(x) = 1$ for $x \geqslant 1$. Thus $f(x) = 2x$ for $0 \leqslant x \leqslant 1$, $f(x) = 0$ for $x < 0$ and for $x < 1$. Accordingly,

$$\mu = \int_0^1 x \cdot 2x \, dx = \frac{2}{3}.$$

Thus our archer who can hit the target but has no further control, so that he strikes "at random" in the circular target, should average 2/3 m from the center. ◄

Variance and standard deviation The standard deviation is the analog of *radius of gyration* about the center of mass. For the case of a finite set of values $x_1, \ldots, x_n$, the standard deviation is equal to σ, where $\sigma > 0$ and

$$\sigma^2 = p_1(x_1 - \mu)^2 + \cdots + p_n(x_n - \mu)^2 \qquad (15\text{--}63)$$

(the "total mass" is 1). For the case of an infinite sequence, σ is given by the equation

$$\sigma^2 = \sum_{i=1}^{\infty} p_i(x_i - \mu)^2, \qquad (15\ 64)$$

and for the continuous case it is given by

$$\sigma^2 = \int_{-\infty}^{\infty} (x - \mu)^2 f(x) \, dx. \qquad (15\text{--}65)$$

We call σ^2 the *variance* of the random variable X and denote the variance by $V(X)$. Thus

$$V(X) = \sigma^2 = \int_{-\infty}^{\infty} (x - \mu)^2 f(x) \, dx. \qquad (15\text{--}66)$$

As in mechanics, σ is a measure of how closely the mass is concentrated near the center of mass: a small value of σ indicates that values far from μ occur only rarely; a large value of σ indicates that such values occur often.

As illustrations we calculate σ for Examples 2 and 4. For Example 2,

$$\sigma^2 = \frac{1}{6} 2.5^2 + \frac{1}{6} 1.5^2 + \frac{1}{6} 0.5^2 + \frac{1}{6} 0.5^2 + \frac{1}{6} 1.5^2 + \frac{1}{6} 2.5^2,$$

so that $\sigma^2 = 2.92$ and $\sigma = 1.71$. For Example 4,

$$\sigma^2 = \int_0^1 \left(x - \frac{2}{3}\right)^2 \cdot 2x \, dx = \frac{1}{18}, \qquad \sigma = 0.236.$$

=======Problems (Section 15-6)=======

1. Let the sample space $\mathscr{S}$ be $\{1, 2, 3, \ldots\}$ and let n have probability $p_n = c/3^n$, where c is a constant.
    a) Show that $c = 2$.   b) Find $\Pr(\{3, 5, 7\})$.

c) Find the probability that n is even.   **d)** Find the probability that n is divisible by 5.

2. A coin is tossed repeatedly until heads appear. Let n be the number of tosses required to obtain heads (for the first time at the nth toss). Justify that $\Pr(\{n\}) = 2^{-n}$. What is the probability that heads never appear?

3. Probability assignments on the x-axis have probability density function $f(x)$ equal to 0 for $x < 0$ and to ce^{-x} for $x \geqslant 0$, where c is a constant.
   a) Show that $c = 1$.   **b)** Find $\Pr(\{0 \leqslant x \leqslant 1\})$.
   c) Find $\Pr(x \geqslant 2)$.   **d)** Find the distribution function $F(x)$ and graph.

4. Probability assignments on the x-axis have probability density function $f(x)$ equal to $c \sin^2 x$ for $0 \leqslant x \leqslant \pi$ and equal to 0 for all other x.
   a) Show that $c = 2/\pi$.   **b)** Find $\Pr(\{0 \leqslant x \leqslant \pi/2\})$.
   c) Find $\Pr(\{-\pi/2 \leqslant x \leqslant \pi/2\})$.   **d)** Find the distribution function $F(x)$ and graph.

5. Probability assignments in the xy-plane have probability density function $f(x, y)$ equal to $c|x - y|$ for $0 \leqslant x \leqslant 1$, $0 \leqslant y \leqslant 1$, and equal to 0 otherwise.
   a) Show that $c = 3$.
   b) Find $\Pr(A)$, where A is the triangle $0 \leqslant x \leqslant 1$, $0 \leqslant y \leqslant x$.
   c) Find $\Pr(B)$, where B is the rectangle $0 \leqslant x \leqslant \dfrac{1}{2}$, $0 \leqslant y \leqslant 1$.
   d) Find $\Pr(C)$, where C is the quarter-circle $x^2 + y^2 \leqslant 1$, $x \geqslant 0$, $y \geqslant 0$.

6. A coin is tossed three times. Let X be the number of heads that appear.
   a) Describe the probabilities associated with the random variable X.
   b) Find the distribution function of X.

7. Probabilities are assigned in the square $\mathscr{S}$: $0 \leqslant x \leqslant 1$, $0 \leqslant y \leqslant 1$ in the xy-plane in accordance with the probability density $f(x, y) = 4xy$.
   a) Find $\Pr(A)$ if A is the set of all (x, y) in $\mathscr{S}$ such that $y \leqslant x$.
   b) Find $\Pr(B)$ if B is the set of all (x, y) in $\mathscr{S}$ such that $y \leqslant 1 - x^2$.
   c) Let X be the random variable on $\mathscr{S}$ equal to $x + y$. Find $\Pr(0 \leqslant X \leqslant 1)$ and
   $$\Pr\left(X \geqslant \frac{1}{2}\right).$$
   d) Find the distribution function and probability density of the random variable of part (c).

8. Let the random variable X take on only the values $x_1, x_2, \ldots$ given, with probabilities $p_1, p_2, \ldots$ given. Find the mean and standard deviation of X.

   a) $x_1 = 0$, $x_2 = 1$, $p_1 = 1/3$, $p_2 = 2/3$.
   b) $x_1 = 1$, $x_2 = 2$, $x_3 = 3$, $p_1 = 1/5$, $p_2 = 3/5$, $p_3 = 1/5$.
   c) $x_k = k$ for $k = 1, \ldots, 5$, $p_k = (32/31) \cdot 2^{-k}$.

9. Let the random variable X have the given probability density $f(x)$. Find the mean and standard deviation of X.
   a) $f(x) = 0$ for $x < 0$, $f(x) = (3x/2) + (3x^2/4)$ for $0 \leqslant x \leqslant 1$, $f(x) = 0$ for $x > 1$;
   b) $f(x) = 0$ for $x < 0$, $f(x) = \sin x$ for $0 \leqslant x \leqslant \pi/2$, $f(x) = 0$ for $x > \pi/2$;
   c) $f(x) = 0$ for $x < 0$, $f(x) = 2e^{-2x}$ for $x \geqslant 0$;
   d) $f(x) = 0$ for $x < 1$, $f(x) = 3x^{-4}$ for $x \geqslant 1$.

15–7 **FUNCTIONS OF A RANDOM VARIABLE**

Let X be a random variable, so that X is a real-valued function defined on a sample space $\mathscr{S}$. Then X^2, X^3, e^X are all random variables, since each is also a real-valued function defined on $\mathscr{S}$. More generally, $\varphi(X)$ is again a random variable, if φ is a real-valued function of a real variable defined, say, for all real numbers (and continuous, where appropriate).

We now have the basic rule: *the expectation $E(\varphi(X))$ of the random variable $\varphi(X)$ can be calculated from the probabilities (discrete case) or probability density (continuous case) associated with X:* in the discrete case,

$$E(\varphi(X)) = \sum \varphi(x_j) p_j; \tag{15–70}$$

in the continuous case,

$$E(\varphi(X)) = \int_{-\infty}^{\infty} \varphi(x) f(x)\, dx. \tag{15–71}$$

We justify these rules under the assumption that φ is monotone strictly increasing, with inverse function ψ (with continuity as required below). In the discrete case, $\varphi(X)$ takes on the values $u_j = \varphi(x_j)$ with probabilities $p_j = \Pr(X = x_j)$ since the set of sample points for which $X = x_j$ is the *same* set as that for which $\varphi(X) = u_j$. Thus

$$E(\varphi(X)) = \sum_j (u_j p_j) = \sum [p_j \varphi(x_j)].$$

as asserted. In the continuous case, the distribution function of $\varphi(X)$ is the function $F_\varphi(u)$ such that

$$F_\varphi(u) = \Pr(\varphi(X) \leqslant u).$$

The set in $\mathscr{S}$ where $\varphi(X) \leqslant u$ is the *same* as the set for which $X \leqslant x$, provided $u = \varphi(x)$ or $x = \psi(u)$. Therefore,

$$F_\varphi(u) = \Pr(X \leqslant x) = \int_{-\infty}^{x} f(s)\, ds, \qquad x = \varphi(u),$$

where f is the density associated with X. Therefore, φ has density

$$f_\varphi(u) = \frac{d}{du} F_\varphi(u) = \frac{d}{dx} F_\varphi(u) \cdot \frac{dx}{du} = f(x) \cdot \psi'(u),$$

where $x = \psi(u)$, that is,

$$f_\varphi(u) = f[\psi(u)] \psi'(u).$$

Accordingly, $\varphi(X)$ has expectation

$$E(\psi(X)) = \int_{-\infty}^{\infty} u f_\varphi(u)\, du = \int_{-\infty}^{\infty} u f[\psi(u)] \psi'(u)\, du.$$

If we now change the variable

$$u = \varphi(x), \qquad x = \psi(u), \qquad dx = \psi'(u)\, du$$

in the last integral, we obtain $E(\varphi(X)) = \int_{-\infty}^{\infty} \varphi(x) f(x) dx$, as asserted. (For a proof under general conditions, see CRAMÉR, Chapters 15, 16.)

From formulas (15–70) and (15–71) it follows at once that finding expectations is a *linear* operation:

$$E(c_1\varphi_1(X) + c_2\varphi_2(X)) = c_1 E(\varphi_1(X)) + c_2 E(\varphi_2(X)), \qquad (15\text{–}72)$$

where c_1, c_2 are constants.

As a special case we can take $\varphi(X) = X$ and obtain (say, in the continuous case)

$$\mu = E(X) = \int_{-\infty}^{\infty} x f(x)\,dx,$$

as expected. If we take $\varphi(X) = k$, where k is a constant, then

$$E(\varphi(X)) = E(k) = \int_{-\infty}^{\infty} k f(x)\,dx = k \int_{-\infty}^{\infty} f(x)\,dx = k. \qquad (15\text{–}73)$$

This result is also to be expected, since $\varphi(X)$ now has a discrete distribution: all the mass is concentrated at the one point k.

Next we observe that $\sigma^2 = V(X)$, the variance of X, equals the expectation of $(X - \mu)^2$. For example, in the continuous case,

$$V(X) = \sigma^2 = \int_{-\infty}^{\infty} (x - \mu)^2 f(x)\,dx = E((X - \mu)^2). \qquad (15\text{–}74)$$

Now

$$(X - \mu)^2 = X^2 - 2\mu X + \mu^2.$$

Therefore, by linearity as in (15–72) and the rule for constants (15–73), we have

$$\sigma^2 = E((X - \mu)^2) = E(X^2) - 2\mu E(X) + \mu^2$$
$$= E(X^2) - 2\mu^2 + \mu^2 = E(X^2) - \mu^2.$$

Accordingly,

$$V(X) = \sigma^2 = E(X^2) - \mu^2 = E(X^2) - [E(X)]^2. \qquad (15\text{–}75)$$

Here

$$E(X^2) = \int_{-\infty}^{\infty} x^2 f(x)\,dx. \qquad (15\text{–}76)$$

All the above discussion applies in the discrete case also, with

$$E(X^2) = \sum_j x_j^2 p_j. \qquad (15\text{–}76')$$

Linearly related random variables   If X is a random variable and a, b are constants, with $a \neq 0$, then $Y = aX + b$ is also a random variable, linearly related to X. By linearity and the rule for a constant random variable, as in (15–73),

$$E(Y) = aE(X) + b \qquad \text{or} \qquad \mu_y = a\mu_x + b. \qquad (15\text{–}77)$$

Accordingly, $Y = \mu_y = a(X - \mu_x)$, so that

$$V(Y) = E((Y - \mu_y)^2) = E(a^2(X - \mu_x)^2)$$
$$= a^2 E((X - \mu_x)^2) = a^2 V(X).$$

Thus

$$\sigma_y^2 = V(Y) = a^2 \sigma_x^2 = a^2 V(X). \qquad (15\text{–}78)$$

In particular, we can take $Y = a(X - \mu_x)$; then (15–77) and (15–78) give

$$E(Y) = 0, \qquad V(Y) = a^2 V(X).$$

We also observe that if X is a discrete random variable, with probabilities $p_i = \Pr(X = x_i)$, $i = 1, 2, \ldots$, then $Y = aX + b$ is also a discrete random variable with probabilities $p_i = \Pr(Y = y_i)$, $y_i = ax_i + b$. If X is a continuous random variable with density $f(x)$, then Y is also continuous with density $g(y)$ such that $|a| g(y) = f(x)$ at corresponding values x, y: $y = ax + b$. Thus

$$|a| g(ax + b) = f(x) \qquad \text{or} \qquad g(y) = \frac{1}{|a|} f\left(\frac{y - b}{a}\right), \tag{15–79}$$

since if $a > 0$, then

$$\Pr(y_1 \leqslant Y \leqslant y_2) = \Pr(y_1 \leqslant aX + b \leqslant y_2)$$

$$= \Pr\left(\frac{y_1 - b}{a} \leqslant X \leqslant \frac{y_2 - b}{a}\right)$$

$$= \int_{x_1}^{x_2} f(x)\,dx = \int_{y_1}^{y_2} f\left(\frac{y - b}{a}\right)\frac{1}{a}\,dy = \int_{y_1}^{y_2} g(y)\,dy,$$

where $x_1 = (y_1 - b)/a$, $x_2 = (y_2 - b)/a$.

If $a < 0$, then x_1, x_2 are interchanged as limits of integration and a must be replaced by $|a|$ to obtain the final equality.

15–8 MOMENT-GENERATING FUNCTION

It is convenient here to define expectations of complex-valued functions. We are interested mainly in $E(e^{itX})$, where t is a real parameter, $-\infty < t < \infty$. We have

$$E(e^{itX}) = \int_{-\infty}^{\infty} e^{itx} f(x)\,dx = \chi(t),$$

$$\tag{15–80}$$

$$E(e^{itX}) = \sum_j e^{itx_j} p_j = \chi(t),$$

in the continuous and discrete cases, respectively. The first of these is a Fourier integral (Section 4–13); the second suggests a Fourier series, and can be so interpreted if the x_j are commensurable. We call $\chi(t)$ the *characteristic function* of X. This function is also a *moment-generating function*, that is, it can be used to derive the expectations

$$\mu_k = E(X^k) = \begin{cases} \displaystyle\int_{-\infty}^{\infty} x^k f(x)\,dx, \\[2mm] \displaystyle\sum_j x_j^k p_j, \end{cases} \qquad k = 0, 1, 2, \ldots, \tag{15–81}$$

called the *moments* of the random variable X. From (15–80) we obtain, by repeated differentiation,

$$\chi^{(k)}(t) = \int_{-\infty}^{\infty} (ix)^k e^{itx} f(x)\, dx,$$

$$\chi^{(k)}(t) = \sum_{j} (ix_j)^k e^{itx_j} p_j.$$

If we set $t = 0$ and divide both sides by i^k, we obtain the desired moments:

$$\mu_k = \frac{1}{i^k} \chi^{(k)}(0). \tag{15–82}$$

This formula is extremely useful and saves time, even for very simple distributions. We observe that $\mu_1 = \mu = E(X)$, $\mu_2 = E(X^2)$, so that, by Eq. (15–75),

$$\sigma^2 = \mu_2 - \mu_1^2 \tag{15–83}$$

EXAMPLE 1 For throwing one die, let X be the value that shows (as in Example 2, Section 15–6). Then

$$\chi(t) = \frac{1}{6}(e^{it} + e^{2it} + e^{3it} + e^{4it} + e^{5it} + e^{6it}).$$

Hence

$$\chi'(t) = \frac{i}{6}(e^{it} + 2e^{2it} + 3e^{3it} + 4e^{4it} + 5e^{5it} + 6e^{6it}),$$

$$\chi''(t) = \frac{-1}{6}(e^{it} + 4e^{2it} + 9e^{3it} + 16e^{4it} + 25e^{5it} + 36e^{6it}).$$

Now Eqs. (15–82) and (15–83) give

$$\mu_1 = \frac{1}{6}(1 + 2 + \cdots + 6) = \frac{7}{2},$$

$$\mu_2 = \frac{1}{6}(1 + 4 + \cdots + 36) = \frac{91}{6},$$

$$\sigma^2 = \frac{91}{6} - \frac{49}{4} = \frac{35}{12} = 2.92,$$

as at the end of Section 15–6. ◄

EXAMPLE 2 Let X have density $f(x) = e^{-x}$ for $x \geqslant 0$, $f(x) = 0$ for $x < 0$. Then

$$\chi(t) = \int_{0}^{\infty} e^{itx} e^{-x}\, dx = \frac{1}{1 - it},$$

$$\chi'(t) = \frac{i}{(1 - it)^2}, \qquad \chi''(t) = \frac{-2}{(1 + it)^3}.$$

Thus $\chi'(0) = i$, $\chi''(0) = -2$, so that $\mu_1 = 1$, $\mu_2 = 2$, $\sigma^2 = 1$. ◄

REMARK: In the continuous case, $f(x)$ is uniquely determined by its characteristic function, for, since χ is a Fourier integral,

$$f(x) = \frac{1}{2\pi} \int_{-\infty}^{\infty} \chi(t)\, e^{-ixt}\, dt,$$

as follows from the theory of Sections 4–13 and 4–16; (we assume that the integrals are absolutely convergent). A similar statement applies in the discrete case. In particular, if the x_j are all multiples (positive, negative, or 0) of one positive number c, then

$$p_j = \frac{c}{2\pi} \int_0^{2\pi/c} \chi(t)\, e^{-itx_j}\, dt.$$

by the theory of Fourier series Eq. (3–76).

15–9 BINOMIAL AND POISSON DISTRIBUTIONS

We here describe two basic distributions that cover a substantial part of the applications of probability.

Binomial distribution   An experiment has two outcomes, which (for convenience) we call *success* and *failure*. We assume that the probability of success is p, that of failure is $q = 1 - p$. We now repeat the experiment n times and ask for the probability of no success, 1 success, 2 successes, ..., k successes, where $k = 0$, 1, ..., n. We assume that the various repetitions do not affect each other, so that, for example, the events "First trial succeeds" and "Second trial fails" are independent. Thus the probability that both occur is the product of the two probabilities, or pq. A similar reasoning applies to other cases. Thus the probability of k successes followed by $n - k$ failures is $p^k q^{n-k}$. The same probability applies to each case of k successes and $n - k$ failures, no matter in what order they occur. The number of such cases is the number of different *combinations* of k objects and of n objects, or $\binom{n}{k}$, by Eq. (15–41). Thus we conclude:

$$\Pr(k \text{ successes, } n \text{ failures}) = \binom{n}{k} p^k q^{n-k}. \tag{15–90}$$

This equation defines the *binomial distribution* (with parameters p and n). A random variable X is said to have a binomial distribution if

$$\Pr(X = k) = \binom{n}{k} p^k q^{n-k}, \qquad k = 0, 1, \ldots, n, \tag{15–91}$$

and $\Pr(X = x) = 0$ for all other x. Thus X has a discrete distribution on the integers 0, 1, ..., n. The sum of the probabilities in (15–91) is 1, since, by the binomial formula,

$$\sum_{k=0}^{n} \binom{n}{k} p^k q^{n-k} = (p+q)^n = 1. \tag{15–92}$$

EXAMPLE 1 We toss a coin n times and count heads as successes, tails as failures. The number X of successes thus has a binomial distribution with $p = 1/2$, as in (15–91):

$$p_k = \Pr(X = k) = \binom{n}{k}\left(\frac{1}{2}\right)^k\left(\frac{1}{2}\right)^{n-k} = \binom{n}{k}\left(\frac{1}{2}\right)^n. \tag{15–93}$$

In Fig. 15–6 the probabilities are graphed for $n = 5$ and $n = 10$. We observe that (especially for $n = 10$) they suggest the famous *bell-shaped curve* of the normal distribution to be discussed below; the reason for the resemblance is explained in Section 15–13. ◄

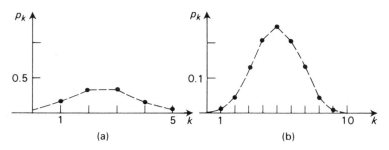

Fig. 15-6. Binomial distributions for $p = 1/2$ and (a) $n = 5$, (b) $n = 10$.

The *characteristic function* of the binomial distribution (15–90) is

$$\chi(t) = \sum_{k=0}^{n}\binom{n}{k}p^k q^{n-k}e^{ikt} = (pe^{it} + q)^n. \tag{15–94}$$

Accordingly,

$$\chi'(t) = n(pe^{it} + q)^{n-1}pie^{it}$$

$$\chi''(t) = n(n-1)(pe^{it} + q)^{n-2}(pie^{it})^2 + n(pe^{it} + q)^{n-1}(-pe^{it}).$$

Therefore, by Eq. (15–82),

$$\mu_1 = n(p+q)^{n-1}p = np, \qquad \mu_2 = np(np + q).$$

Here we used the relation $p + q = 1$. Accordingly, by Eq. (12–83), $\sigma^2 = \mu_2 - \mu_1^2 = npq$.

Therefore, *the binomial distribution* (15–91) *has mean and standard deviation*

$$\mu = np, \qquad \sigma = \sqrt{npq}. \tag{15–95}$$

The mean is as we would expect, for in n trials, on the average, the fraction of successes should be close to p, that is, X/n should be close to p.

Poisson distribution This is a discrete distribution on the integers $k = 0, 1, 2, \ldots$ such that

$$p_k = \Pr\{X = k\} = e^{-\lambda}\lambda^k/k!, \qquad k = 0, 1, 2, \ldots \tag{15–96}$$

Here λ is a fixed positive number called the *parameter* of the Poisson distribution. It can be shown that this distribution is a good model for such experiments as counting the number of telephone calls received in a given time interval or the number of α-particles emitted by a radioactive substance in a given time interval. In both these cases, λ is taken as at, where t is the length of the time interval and a is a positive constant. In Fig. 15–7 we graph a typical Poisson distribution with $\lambda = 5$. This again resembles both the binomial and the normal distribution.

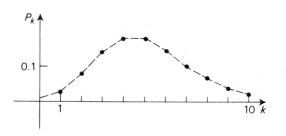

Fig. 15–7. Poisson distribution with parameter 5.

We observe that the sum of the probabilities p_k in (15–96) is 1 since

$$\sum_{k=0}^{\infty} p_k = e^{-\lambda} \sum_{k=0}^{\infty} \frac{\lambda^k}{k!} = e^{-\lambda} e^{\lambda} = 1.$$

The characteristic function for the Poisson distribution (15–96) is

$$\chi(t) = \sum_{k=0}^{\infty} e^{-\lambda} \frac{\lambda^k}{k!} e^{ikt} = e^{-\lambda} \sum_{k=0}^{\infty} \frac{(\lambda e^{it})^k}{k!} = e^{-\lambda} e^{\lambda e^{it}}.$$

From this we deduce easily (see Problem 7 below) that $\mu_1 = \lambda$, $\mu_2 = \mu^2 + \lambda$, and, accordingly, the mean and variance for the Poisson distribution are both equal to λ:

$$\mu = \sigma^2 = \lambda. \tag{15–97}$$

15–10 NORMAL DISTRIBUTION

A random variable X is said to have a normal distribution with parameters (μ, σ), where $\sigma > 0$, if X has probability density

$$f(x) = \frac{1}{\sqrt{2\pi}\,\sigma} e^{-(x-\mu)^2/(2\sigma^2)}. \tag{15–100}$$

The density function $f(x)$ is graphed for a typical case in Fig. 15–8. This distribution is the most important one for applications. One reason for its broad applicability is given in Section 15–13 below. Typically, repeated measurements of a physical quantity yield a normal distribution.

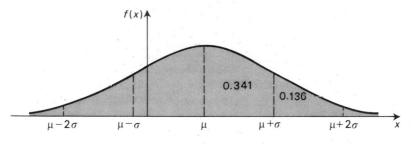

Fig. 15-8. Probability density for the normal distribution.

We observe that $f(x)$ is a true probability density, since

$$\int_{-\infty}^{\infty} f(x)\,dx = \frac{1}{\sqrt{2\pi}\,\sigma}\int_{-\infty}^{\infty} e^{-(x-\mu)^2/(2\sigma^2)}\,dx = \sqrt{\frac{1}{\pi}}\int_{-\infty}^{\infty} e^{-u^2}\,du = 1,$$

by the substitution $u = (x-\mu)/(\sqrt{2}\sigma)$.

The corresponding characteristic function is

$$\chi(t) = \frac{1}{\sqrt{2\pi}\,\sigma}\int_{-\infty}^{\infty} e^{-(x-\mu)^2/(2\sigma)^2}\, e^{ixt}\,dx.$$

If we set $v = x - \mu$ and $\omega = -t$, the integral becomes a Fourier transform and, with the aid of Entry 19 in Table 4–2, we find (see Problem 9 below) that

$$\chi(t) = \exp\left(i\mu t - \frac{\sigma^2 t^2}{2}\right). \tag{15–101}$$

Accordingly, $\chi'(0) = i\mu$, $\chi''(0) = -\mu^2 - \sigma^2$, and, as above, we deduce that the normal distribution has mean μ and variance σ^2, so that the notation μ, σ in (15–100) is justified.

We observe that if X has the normal distribution (μ, σ), then the random variable

$$Y = \frac{X - \mu}{\sigma} \tag{15–102}$$

has the density function

$$g(y) = \sigma f(\sigma y + \mu) = \frac{1}{\sqrt{2\pi}} e^{-y^2/2} \tag{15–103}$$

(as in Section 15–7 above), so that Y has the normal distribution $(0, 1)$, that is, mean 0 and variance 1. We say that Y has a *standardized normal distribution*.

As usual, the areas under the graph of $y = f(x)$ give probabilities for the corresponding intervals on the x-axis. In Fig. 15–8 we have indicated some of these probabilities. From them and the symmetry of the graph about $x = \mu$ we conclude that the probability of X differing from μ by at most σ is 0.682; the probability of X differing from μ by at most 2σ is 0.954. We also note that the probability of X differing from μ by at most 2.6σ is 99 %.

Tables of the distribution function $F(x)$ for the standardized normal distribution are given in most statistics books and sets of numerical tables. The probabilities and distribution function for the Poisson distribution and binomial distribution are also widely available.

For convenience we give here a very short table (Table 15–1) of $F(x)$ values for the standardized normal distribution.

Table 15–1

$$F(x) = (2\pi)^{-1/2} \int_{-\infty}^{x} e^{-t^2/2} \, dt$$

x	$F(x)$	x	$F(x)$
0.0	0.5000	0.75	0.7734
0.1	0.5398	1	0.8413
0.2	0.5793	1 5	0.9332
0.3	0.6179	2	0.9772
0.4	0.6554	2.5	0.9938
0.5	0.6915	3	0.9987

====================Problems (Section 15–10)====================

1. Let the discrete random variable X have values $-1, 0, 1$ with probabilities $1/4, 1/2, 1/4$. Describe the probabilities associated with each of the following random variables:
   a) $X + 1$ b) $2X$ c) e^X d) X^2
   e) $-X$ f) $X^2 - X$ g) $X^3 - X$

2. Let the continuous random variable X have density function $f(x)$ equal to $2x$ for $0 \leqslant x \leqslant 1$ and to 0 otherwise. Find the density function for each of the following random variables:
   a) $X + 1$ b) $2X$ c) e^X
   d) $-X$ e) X^2 f) $X^2 - X$

   HINT for (e):   For $U = X^2$, consider $\Pr(U \leqslant a)$ for $0 \leqslant a \leqslant 1$.

3. Find the moment-generating function for the random variable X of Problem 1 and use it to calculate $\mu_1, \mu_2, E(X), V(X)$.

4. Proceed as in Problem 3 with the random variable X of Problem 2.

5. Graph the binomial distribution, as in Fig. 15–6, for the cases indicated:

   a) $p = \dfrac{1}{2}, n = 8$ b) $p = \dfrac{1}{3}, n = 6$

   c) $p = \dfrac{1}{3}, n = 12$ d) $p = \dfrac{1}{3}, n = 16$

6. Try to reproduce the graph of Fig. 15–6(a) experimentally by repeating 10 times the experiment of tossing a coin five times and counting the number of heads for each experiment.

7. Prove that for the Poisson distribution (15–96), $E(X)$ and $V(X)$ are both equal to λ.

8. Let X have a Poisson distribution (15–96) with $\lambda = 1$. Determine the following probabilities:

   a) $\Pr(X = 1)$ **b)** $\Pr(X = 2)$ **c)** $\Pr(X \leqslant 5)$
   d) $\Pr(X \geqslant 3)$ **e)** $\Pr(4 \leqslant X \leqslant 7)$

9. For the normal distribution with density (15–100) verify that
   a) $\chi(t)$ is given by Eq. (15–101);
   b) $Y = (X - \mu)/\sigma$ has normal distribution (15–103).

10. Let X have a standardized normal distribution. With the aid of Table 15–1 find the following probabilities:

   a) $\Pr(X \leqslant 0)$ **b)** $\Pr(X \geqslant 1)$
   c) $\Pr(-1 \leqslant X \leqslant 1)$ **d)** $\Pr(-3 \leqslant X \leqslant 2)$

11. Let X have normal distribution $(3, 6)$. With the aid of Table 15–1 find the following probabilities:

   a) $\Pr(X \geqslant 9)$ **b)** $\Pr(X \geqslant 6)$
   c) $\Pr(9 \leqslant X \leqslant 12)$ **d)** $\Pr(1 \leqslant X \leqslant 2)$

15–11 JOINT DISTRIBUTION OF SEVERAL RANDOM VARIABLES

Let both X and Y be random variables defined on the same sample space $\mathscr{S}$ (Fig. 15–9). Therefore, for each point s of $\mathscr{S}$ we have an ordered pair of numbers (x, y), where $x = X(s)$, $y = Y(s)$, and we can ask for probabilities such as

$$\Pr(X = 1, \; Y = 2), \; \Pr(0 \leqslant X \leqslant 1, \; 1 \leqslant Y \leqslant 2).$$

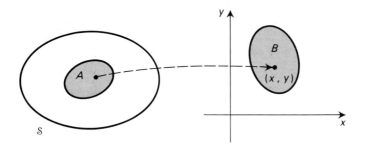

Fig. 15-9. Probabilities associated with a pair of random variables X, Y.

In each case, the conditions imposed restrict s to lie in a subset A of $\mathscr{S}$ and the probability of the event A is sought. We observe that the set where $X = 1$ and $Y = 2$

is the intersection of the set where $X = 1$ and the set where $Y = 2$, so that
$$\Pr(X = 1, \ Y = 2) = \Pr(\{X = 1\} \cap \{Y = 2\}).$$
If the events $X = 1$, $Y = 2$ are *independent*, we can write further
$$\Pr(X = 1, \ Y = 2) = \Pr(X = 1)\Pr(Y = 2). \tag{15–110}$$
Similarly, under the assumption of independence,
$$\Pr(0 \leqslant X \leqslant 1, 1 \leqslant Y \leqslant 2) = \Pr(0 \leqslant X \leqslant 1)\Pr(1 \leqslant Y \leqslant 2). \tag{15–111}$$
We can represent the pairs (x, y) of values of X and Y as points in the xy-plane. The above examples suggest that we are assigning probabilities to sets in the xy-plane. For the case of (15–110), the set consists of just one point $(1, 2)$; for the case of (15–111), the set is the square $0 \leqslant x \leqslant 1, 1 \leqslant y \leqslant 2$.

More generally, we can select an arbitrary set B in the xy-plane and consider the set A of all sample points s for which $(X(s), \ Y(s)) = (x, y)$ is in B. Then the probability that $(X, \ Y)$ is in B is taken as the probability of the set A:
$$\Pr((X, \ Y) \text{ in } B) = \Pr(A),$$
$$A = \text{Set of all } s \text{ for which } (X(s), \ Y(s)) \text{ is in } B.$$
In this way we can assign probabilities to a great variety of sets in the xy-plane. These probabilities are referred to as the *joint probability distribution* of X, Y.

As before, we concentrate on two cases: the *discrete case*, in which X takes on values $x_1, x_2, \ldots, x_k, \ldots$ with probabilities p'_k and Y takes on values y_1, $y_2, \ldots y_l, \ldots$ with probabilities p''_l, and $\sum_k p'_k = 1, \sum_l p''_l = 1$; and the *continuous case*, in which X and Y have probability densities $f(x)$ and $g(y)$.

In the discrete case, $(X, \ Y)$ can take on only the values $(x_k, \ y_l)$ for all allowable indices k, l, and each has probability p_{kl}, where necessarily
$$\sum_{k, l} p_{kl} = 1. \tag{15–112}$$
We observe that the set of sample point s, for which $X = x_k$ and $Y = y_l$, is the *intersection* of the sets A, B, where
$$A = \text{All } s \text{ for which } X = x_k,$$
$$B = \text{All } s \text{ for which } Y = y_l.$$
If the events A, B are independent, then
$$\Pr(A \cap B) = \Pr(A)\Pr(B).$$
We call X, Y independent (discrete) *random variables* if this holds for all k and l. *Thus when X, Y are independent,*
$$\Pr(X = x_k \text{ and } Y = y_l) = \Pr(X = x_k) \cdot \Pr(Y = y_l),$$
so that
$$P_{k,l} = p'_k p''_l \text{ for all } k, l. \tag{15–113}$$
A simple illustration of independence is provided by an urn with three balls numbered 1, four balls numbered 2, five balls numbered 3. Let X be the number

chosen by lot the first time, Y that chosen the second time. We assume the ball is *replaced* after the choice, so that the two choices do not influence each other. Thus, with $x_k = y_k = k$ for $k = 1, 2, 3$, Eq. (15–113) is satisfied:

$$p'_1 = p''_1 = \frac{1}{4}, \qquad p'_2 = p''_2 = \frac{1}{3}, \qquad p'_3 = p''_3 = \frac{5}{12},$$

$$p_{11} = \frac{1}{4}\cdot\frac{1}{4} = \frac{1}{16}, \qquad p_{12} = p_{21} = \frac{1}{12}, \qquad p_{13} = p_{31} = \frac{5}{48},$$

$$p_{22} = \frac{1}{9}, \qquad p_{23} = p_{32} = \frac{5}{36}, \qquad p_{33} = \frac{25}{144}.$$

If after the first choice the ball is not replaced, the random variables X, Y are *not* independent. For example, $p'_1 = 1/4$, $p'_2 = 1/3$, $p'_3 = 5/12$, but

$$\begin{aligned}
p''_1 = \Pr(Y = 1) &= \Pr(X = 1)\Pr(Y = 1 \mid X = 1)\\
&\quad + \Pr(X = 2)\Pr(Y = 1 \mid X = 2)\\
&\quad + \Pr(X = 3)\Pr(Y = 1 \mid X = 3),
\end{aligned}$$

by the rule (15–35). Now if $X = 1$, then after the first choice only two balls numbered 1 remain, so that

$$\Pr(Y = 1 \mid X = 1) = \frac{2}{11}.$$

Reasoning similarly about the other cases, we conclude that

$$p''_1 = \frac{1}{4}\cdot\frac{2}{11} + \frac{1}{3}\cdot\frac{3}{11} + \frac{5}{12}\cdot\frac{3}{11} = \frac{1}{4}.$$

However, by (15–31),

$$\begin{aligned}
p_{11} = \Pr(X = 1,\, Y = 1) &= \Pr(X = 1)\Pr(Y = 1 \mid X = 1)\\
&= \frac{1}{4}\cdot\frac{2}{11} = \frac{1}{22}.
\end{aligned}$$

Thus

$$p_{11} \neq p'_1 p''_1 = \frac{1}{4}\cdot\frac{1}{4} = \frac{1}{16}.$$

Accordingly, as asserted, X and Y are now not independent. This is to be expected, since the value of X influences that of Y.

In general, for such discrete random variables X, Y, the joint probabilities p_{kl} form a matrix. This is suggested in Fig. 15–10 for the case when X has values $x_1, \ldots, x_m$ and Y has values $y_1, \ldots, y_n$. By Eq. (15–112), the sum of all entries is 1.

$$
\begin{array}{c}
\text{Sums}\\
\begin{bmatrix}
p_{11} & p_{12} & \cdots & p_{1n}\\
p_{21} & p_{22} & \cdots & p_{2n}\\
\vdots & \vdots & & \vdots\\
p_{m1} & p_{m2} & \cdots & p_{mn}
\end{bmatrix}
\begin{matrix}
p'_1\\
p'_2\\
\vdots\\
p'_m
\end{matrix}
\end{array}
$$

Sums $p''_1 \quad p''_2 \quad \cdots \quad p''_n$

Fig. 15–10. Marginal probabilities.

The row sums are called *marginal probabilities*; they are the probabilities $p'_1, \ldots, p'_m$ associated with X. For example,

$$
\begin{aligned}
p_{11} + p_{12} + \cdots + p_{1n} &= \Pr(\{X = x_1\} \cap \{Y = y_1\}) + \cdots \\
&\quad + \Pr(\{X = x_1\} \cap \{Y = y_n\}) \\
&= \Pr(\{X = x_1\} \cap [\{Y = y_1\} \cup \{Y = y_2\} \cup \cdots \cup \{Y = y_n\}]) \\
&= \Pr(X = x_1) = p'_1.
\end{aligned}
$$

Here we used the rule (15–34) and the fact that the set in square brackets is the whole sample space $\mathscr{S}$. Similarly, the column sums are also *marginal* probabilities, and they are the probabilities associated with Y. As a check, we observe that both sets of marginal probabilities must add up to 1, by Eq. (15–112).

These last definitions can be extended to the case when X or Y or both have infinitely many values.

For the continuous case we assume that we have a continuous joint probability density $f(x, y)$ such that for each set A in the xy-plane

$$
\Pr(A) = \Pr((X, \ Y) \text{ lies in } A) = \iint_A f(x, y) \, dx \, dy. \tag{15–114}
$$

(Here A should be some set appropriate for double integrals.) The function $f(x, y)$ must be nonnegative and

$$
\int_{-\infty}^{\infty} \int_{-\infty}^{\infty} f(x, y) \, dx \, dy = 1, \tag{15–115}
$$

by analogy with Eq. (15–112). The random variables X, Y are said to be *independent* if, whenever A is a rectangle $a \leqslant x \leqslant b$, $c \leqslant y \leqslant d$,

$$
\Pr(A) = \Pr(a \leqslant X \leqslant b) \Pr(c \leqslant Y \leqslant d).
$$

This is equivalent to the equation

$$
\int_c^d \int_a^b f(x, y) \, dx \, dy = \int_a^b f(x) \, dx \int_c^d g(y) \, dy.
$$

Thus $f(x, y)$ and $f(x)g(y)$ must have the same integral over every rectangle, and we conclude that X and Y are independent if and only if

$$
f(x, y) = f(x)g(y) \quad \text{for all } (x, y). \tag{15–116}
$$

This relation is analogous to Eq. (15–113).

For a general $f(x, y)$, the *marginal probability densities* are

$$
f_1(x) = \int_{-\infty}^{\infty} f(x, y) \, dy, \qquad g_1(y) = \int_{-\infty}^{\infty} f(x, y) \, dx.
$$

However, the region $a \leqslant x \leqslant b$ is a strip in the xy-plane, and hence

$$
\Pr(a \leqslant X \leqslant b) = \int_a^b \int_{-\infty}^{\infty} f(x, y) \, dy \, dx = \int_a^b f_1(x) \, dx.
$$

Since this holds for all a, b, we conclude that $f_1(x) = f(x)$. Similarly, $g_1(y) = g(y)$. Thus the marginal densities are simply the densities associated with X and Y.

To test whether X, Y are independent, we can calculate $f(x)$ and $g(y)$ as marginal probabilities and then determine whether Eq. (15–116) is satisfied. Alternatively, we can determine whether $f(x, y)$ is factorable as $p(x) q(y)$; in that case, it is easily seen that, except for constant factors, $p(x)$ and $q(y)$ must be $f(x)$, $g(y)$, respectively. Furthermore, if such a factorization is possible, then for all x_1, x_2, y_1, y_2,

$$f(x_1, y_1) f(x_2, y_2) = f(x_1, y_2) f(x_2, y_1), \qquad (15–117)$$

since both sides are equal to $p(x_1) p(x_2) q(y_1) q(y_2)$. Conversely, if (15–117) holds, then $f(x, y)$ is factorable as $p(x) q(y)$; for example, if $f(x_2, y_2) \neq 0$ for a particular (x_2, y_2), then for all (x_1, y_1)

$$f(x_1, y_1) = \frac{f(x_1, y_2)}{f(x_2, y_2)} f(x_2, y_1) = p(x_1) q(y_1),$$

since x_2, y_2 are held fixed.

EXAMPLE 1 For shooting at a target, represented as the circular region $x^2 + y^2 \leqslant 1$ in the xy-plane, we assume the probability density

$$f(x, y) = \begin{cases} \dfrac{3}{\pi}(1 - \sqrt{x^2 + y^2}), & x^2 + y^2 \leqslant 1, \\ 0, & x^2 + y^2 > 1. \end{cases}$$

Thus

$$\int_{-\infty}^{\infty} \int_{-\infty}^{\infty} f(x, y) \, dx \, dy = \int_{0}^{2\pi} \int_{0}^{1} \frac{3}{\pi}(1 - r) r \, dr \, d\theta = 1,$$

as required. The probability of hitting within distance b of the center $(0 < b < 1)$ is

$$\int_{0}^{2\pi} \int_{0}^{b} \frac{3}{\pi}(1 - r) r \, dr \, d\theta = 6 \left(\frac{b^2}{2} - \frac{b^3}{3} \right) = b^2 (3 - 2b).$$

For $b = 1/3$, this is $7/27$; for $b = 2/3$, this is $20/27$.

In this example, X, Y are not independent. For instance, the test (15–117) fails, say for $x_1 = 0$, $y_1 = 0$, $x_2 = 1/2$, $y_2 = 1/2$; the equation reduces to $1 - \sqrt{1/2} = (1/2) \cdot (1/2)$, which is false. We can also compute marginal probabilities; for $0 < |x| \leqslant 1$:

$$f(x) = \frac{6}{\pi} \int_{0}^{\sqrt{1 - x^2}} \left(1 - \sqrt{x^2 + y^2} \right) dy = \frac{3}{\pi} \left(2\sqrt{1 - x^2} - x^2 \ln \frac{1 + \sqrt{1 - x^2}}{|x|} \right);$$

also, $f(0) = 3/\pi$ and $f(x) = 0$ for $|x| > 1$. By symmetry, $g(y)$ is obtained from $f(x)$ when x is replaced with y. ◀

EXAMPLE 2 Let $f(x, y) = \dfrac{1}{2\pi} e^{-(x^2 + y^2)/2}$.

This is the bivariate standard normal distribution for *uncorrelated* random variables X, Y. Clearly,

$$f(x, y) = \frac{1}{\sqrt{2\pi}} e^{-x^2/2} \cdot \frac{1}{\sqrt{2\pi}} e^{-y^2/2} = f(x) g(y);$$

X and Y are independent and each has a standard normal distribution (see also Problem 5 below). ◄

Case of three or more random variables The preceding results extend in a natural way to the study of more than two random variables, say, to $X_1, \ldots, X_n$. When these are all discrete, we have probabilities

$$p_{i_1 i_2 \ldots i_n} = \Pr(X_1 = x_{1,i_1}, \ldots, X_n = x_{n,i_n}),$$

whose sum is 1; when the random variables are independent,

$$p_{i_1 i_2 \ldots i_n} = p_{1,i_1} \, p_{2,i_2} \cdots p_{n,i_n}$$

where $p_{j,i_j} = \Pr(X_j = x_{j,i_j})$.

When the random variables are continuous, we have a joint probability density $f(x_1, \ldots, x_n)$ and

$$\Pr(A) = \underset{A}{\int \int} \cdots \int f(x_1, \ldots, x_n) \, dx_1 \cdots dx_n.$$

The variables are independent when $f(x_1, \ldots, x_n) = f_1(x_1) \ldots f_n(x_n)$, where X_j has probability density $f_j(x_j)$.

We are led naturally to such a sequence $X_1, \ldots, X_n$ of random variables in repeating an experiment whose outcome is a number X; X_j is the outcome of the jth trial.

15–12 FUNCTIONS OF SEVERAL RANDOM VARIABLES

If X and Y are random variables and $\varphi(x, y)$ is a function of x and y defined for all x and y, then we can form a new random variable $Z = \varphi(X, Y)$.

If, for example, X and Y are discrete, with values x_k, y_l, then Z is discrete, with values $z_{kl} = \varphi(x_k, y_l)$; some z_{kl}, of different indices, may coincide. To each z that is a value of φ for some x_k, y_l we assign the probability

$$\Pr(Z = z) = \sum_{k, l} p_{kl} \, \varepsilon_{kl}(z),$$

where $\varepsilon_{kl}(z) = 1$ if $\varphi(x_k, y_l) = z$ and $\varepsilon_{kl}(z) = 0$ if $\varphi(x_k, y_l) \neq z$. To all other z we assign the probability 0. Then the values of Z can be numbered as $z_1, z_2, \ldots$, and

$$\sum_k \Pr(Z = z_k) = 1.$$

If X, Y are continuous, with joint-probability density $f(x, y)$, then we can assign a probability $\Pr(a \leqslant Z \leqslant b)$ to each interval $[a, b]$ by the equation

$$\Pr(a \leqslant Z \leqslant b) = \Pr(A) = \underset{A}{\int \int} f(x, y) \, dx \, dy,$$

where A is the set of all (x, y) such that $a \leqslant \varphi(x, y) \leqslant b$. Under reasonable hypotheses, we can then verify that Z has a probability density $h(z)$.

EXAMPLE 1 Let X, Y have the joint density

$$f(x, y) = \frac{1}{2\pi} e^{-(x^2 + y^2)/2},$$

as in Example 2 of Section 15–11. Let $Z = (X^2 + Y^2)^{1/2}$; then Z takes on only nonnegative values. For $0 \leqslant a \leqslant b$,

$$\Pr(a \leqslant Z \leqslant b) = \iint\limits_{A} f(x, y) \, dx \, dy,$$

where A is the set $a \leqslant (x^2 + y^2)^{1/2} \leqslant b$. Hence this probability equals

$$\frac{1}{2\pi} \int_0^{2\pi} \int_a^b e^{-r^2/2} r \, dr \, d\theta = \int_a^b e^{-r^2/2} r \, dr = \frac{e^{-a^2/2} - e^{-b^2/2}}{2}.$$

We see from the middle expression that Z has density

$$h(z) = \begin{cases} 0, & z \leqslant 0, \\ ze^{-z^2/2}, & z > 0. \end{cases} \quad \blacktriangleleft$$

EXAMPLE 2 For X, Y as in Example 1, let $Z = X + Y$. Then

$$\Pr(a \leqslant Z \leqslant b) = \iint\limits_{A} f(x, y) \, dx \, dy = \Pr(A),$$

where A is the set $a \leqslant x + y \leqslant b$. The set A is shown in Fig. 15–11. From the symmetry of the function $f(x, y)$, it follows that $\Pr(A) = \Pr(A_1)$, where A_1 is

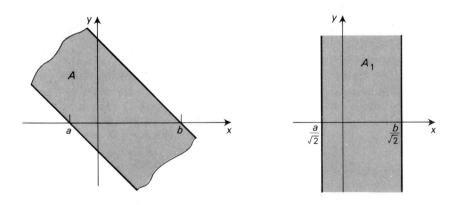

Fig. 15–11. Example 2.

obtained from A by a $45°$ rotation, as in Fig. 15–11. Therefore,

$$\Pr(a \leqslant Z \leqslant b) = \frac{1}{2\pi} \int_{a/\sqrt{2}}^{b/\sqrt{2}} \int_{-\infty}^{\infty} e^{-(x^2+y^2)/2} \, dy \, dx = \frac{1}{\sqrt{2\pi}} \int_{a/\sqrt{2}}^{b/\sqrt{2}} e^{-x^2/2} \, dx$$

and

$$\Pr\left(a \leqslant \frac{Z}{\sqrt{2}} \leqslant b\right) = \Pr(a\sqrt{2} \leqslant Z \leqslant b\sqrt{2}) = \frac{1}{\sqrt{2\pi}} \int_{a}^{b} e^{-x^2/2} \, dx.$$

Accordingly, $Z/\sqrt{2} = (X+Y)/\sqrt{2}$ has a standard normal distribution. ◀
   For functions $Z = \varphi(X, Y)$, we have the fundamental relations for expectations:

$$E(Z) = E[\varphi(X, Y)] = \int_{-\infty}^{\infty} \int_{-\infty}^{\infty} \varphi(x, y) \, f(x, y) \, dx \, dy \qquad (15\text{–}120)$$

in the continuous case and

$$E(Z) = \sum_{k,\,l} \varphi(x_k, \, v_l) \, p_{kl} \qquad (15\text{–}121)$$

in the discrete case. (The proof for the discrete case is like the proof of Eq. (15–70) in Section 15–7. For the continuous case, see CRAMÉR, Chapters 15, 16.)
   From these we deduce the rules:

$$E(\alpha X + \beta Y) = \alpha E(X) + \beta E(Y) \qquad (15\text{–}122)$$

$$E(\varphi(X, Y) + \psi(X, Y)) = E(\varphi(X, Y)) + E(\psi(X, Y)) \qquad (15\text{–}123)$$

and, if X, Y are *independent*,

$$E(XY) = E(X) \, E(Y), \qquad (15\text{–}124)$$

$$V(X + Y) = V(X) + V(Y). \qquad (15\text{–}125)$$

The rules (15–122) and (15–123) follow at once from (15–120) and (15–121). For (15–124) and (15–125) in the continuous case, for example, we have $f(x, y) = f(x) \, g(y)$ by independence, so that

$$E(XY) = \int_{-\infty}^{\infty} \int_{-\infty}^{\infty} xy f(x) g(y) \, dx \, dy = \int_{-\infty}^{\infty} y g(y) \left(\int_{-\infty}^{\infty} x f(x) \, dx \right) dy$$

$$= \int_{-\infty}^{\infty} x f(x) \, dx \int_{-\infty}^{\infty} y g(y) \, dy = E(X) \, E(Y),$$

$$V(X + Y) = E((X + Y)^2) - [E(X + Y)]^2$$

$$= E(X^2 + 2XY + Y^2) - [E(X) + E(Y)]^2$$

$$= E(X^2) + 2E(XY) + E(Y^2) - [E(X)]^2 - 2E(X) \, E(Y) - [E(Y)]^2.$$

Here the terms in $E(XY)$ and $E(X) \, E(Y)$ cancel by (15–124) and

$$V(X) = E(X^2) - [E(X)]^2, \qquad V(Y) = E(Y^2) - [E(Y)]^2,$$

so that the above expression reduces to $V(X) + V(Y)$, as asserted.

The rules (15–122) and (15–123) extend to any number of summands in any number of random variables $X_1, X_2, \ldots, X_n$. If $X_1, \ldots, X_n$ are independent, then (15–124) and (15–125) also have analogs:

$$E(X_1 X_2 \cdots X_n) = E(X_1)E(X_2) \cdots E(X_n), \qquad (15\text{–}124')$$

$$V(X_1 + X_2 + \cdots + X_n) = V(X_1) + V(X_2) + \cdots + V(X_n). \quad (15\text{–}125')$$

As in Example 2 above, we also verify that if $X_1, \ldots, X_n$ are independent, with standard normal distributions, then $Z = (X_1 + \cdots + X_n)/\sqrt{n}$ also has a standard normal distribution. More generally, if $X_1, \ldots, X_n$ are independent, with normal distributions $(\mu_1, \sigma_1), (\mu_2, \sigma_2), \ldots, (\mu_n, \sigma_n)$, then $Y = X_1 + \cdots + X_n$ has a normal distribution; by (15–122) and (15–125), Y has mean $\mu_1 + \cdots + \mu_n$ and variance $\sigma_1^2 + \cdots + \sigma_n^2$.

15–13 CENTRAL LIMIT THEOREM

In many physical problems an experiment is repeated under presumably unchanged conditions. The result of the experiment is a number X, a random variable, with mean μ and variance σ^2. For successive repetitions, a sequence of independent random variables $X_1, \ldots X_n, \ldots$, all having the *same* probability distribution is obtained. This process is often described as one of repeated *sampling* from a given population, and

$$\overline{X}_n = \frac{X_1 + \cdots + X_n}{n} \qquad (15\text{–}130)$$

is called the *sample mean*. It is itself a random variable and, as in the previous section, has mean and variance

$$E(\overline{X}_n) = \frac{1}{n}[E(X_1) + \cdots + E(X_n)] = \frac{1}{n}(\mu + \cdots + \mu) = \mu, \quad (15\text{–}131)$$

$$V(\overline{X}_n) = \frac{1}{n^2}[V(X_1) + \cdots + V(X_n)] = \frac{1}{n^2}(\sigma^2 + \cdots + \sigma^2) = \frac{\sigma^2}{n}. \quad (15\text{–}132)$$

This shows that $\overline{X}_n$ has the same mean as X, but its variance approaches 0 as $n \to \infty$. Thus the values of $\overline{X}_n$ can be expected to cluster more and more closely to μ as more experiments are carried out.

However, a much stronger statement can be made. As n increases, the distribution of $\overline{X}_n$ approaches a *normal distribution* with mean μ and variance σ^2/n. More precisely,

$$Z = \frac{\overline{X}_n - \mu}{\sigma/\sqrt{n}}$$

is a random variable of mean 0 and variance 1, and the distribution of Z approaches a *standard normal distribution* in the sense that

$$\lim_{n \to \infty} \Pr\{a \leqslant Z \leqslant b\} = \frac{1}{\sqrt{2\pi}} \int_a^b e^{-x^2} dx. \tag{15-133}$$

This statement is the *central limit theorem*. It can be restated in several ways:

$$\lim_{n \to \infty} \Pr\left\{a \leqslant \frac{\overline{X}_n - \mu}{\sigma/\sqrt{n}} \leqslant b\right\} = \frac{1}{\sqrt{2\pi}} \int_a^b e^{-x^2/2} dx, \tag{15-133'}$$

$$\lim_{n \to \infty} \Pr\left\{\mu + \frac{a\sigma}{\sqrt{n}} \leqslant \overline{X}_n \leqslant \mu + \frac{b\sigma}{\sqrt{n}}\right\} = \frac{1}{\sqrt{2\pi}} \int_a^b e^{-x^2/2} dx, \tag{15-133''}$$

$$\lim_{n \to \infty} \Pr\{n\mu + a\sigma \sqrt{n} \leqslant X_1 + \cdots + X_n \leqslant n\mu + b\sigma \sqrt{n}\} = \frac{1}{\sqrt{2\pi}} \int_a^b e^{-x^2/2} dx. \tag{15-133'''}$$

(For a proof see CRAMÉR, Chapter 17.)

Application to the binomial distribution   As in Section 15–9, we repeat n times an experiment with probability p of success, q of failure. We let $X = 1$ for success, $X = 0$ for failure. Thus X is a discrete random variable with values 0, 1, $p = \Pr(X = 1)$, $q = \Pr(X = 0)$. If we repeat the experiment n times, we have n discrete independent random variables $X_1, \ldots, X_n$ with the same distribution as X. Now $Y_n = X_1 + \cdots + X_n$ is also a random variable; its value is simply the total number of successes. Hence Y_n has precisely the binomial distribution:

$$\Pr(Y_n = k) = \binom{n}{k} p^k q^{n-k}, \qquad k = 0, 1, \ldots, n. \tag{15-134}$$

Also

$$E(Y_n) = E(X_1) + \cdots + E(X_n) = np,$$
$$V(Y_n) = V(X_1) + \cdots + V(X_n) = npq, \tag{15-135}$$

since X has mean p and variance pq, as can be easily calculated. These results were also obtained in Section 15–9.

Now we apply the central limit theorem and consider the distribution of $Y_n = X_1 + \cdots + X_n$. This is the total number of successes in n trials. We conclude that (15–133''') holds with $\mu = p$ and $\sigma = \sqrt{pq}$. Thus,

$$\lim_{n \to \infty} \Pr\{np + a \sqrt{npq} < Y_n < np + b \sqrt{npq}\} = \frac{1}{\sqrt{2\pi}} \int_a^b e^{-x^2/2} dx. \tag{15-136}$$

This result is known as *DeMoivre's theorem*.

Poisson distribution as a limit of binomial distribution   In the binomial distribution we take $p = p_n = \lambda_n/n$, where $\lambda_n \to \lambda$ as $n \to \infty$. Then for fixed k we get

$$\lim_{n \to \infty} \binom{n}{k} p_n^k (1 - p_n)^{n-k} = \lambda^k e^{-\lambda}/k!, \tag{15-137}$$

since

$$\binom{n}{k} p_n^k (1 - p_n)^{n-k} = \frac{n!}{k!\,(n-k)!}\,\frac{\lambda_n^k}{n^k}\left(1 - \frac{\lambda_n}{n}\right)^{n-k}$$

$$= \frac{1}{k!}\,\frac{n(n-1)\cdots(n-k+1)}{n \cdot n \cdots n}\,\frac{\lambda_n^k}{[1-(\lambda_n/n)]^k}\left[\left(1 - \frac{\lambda_n}{n}\right)^{n/\lambda_n}\right]^{\lambda_n}.$$

The last quantity in brackets approaches e^{-1} since $\lambda_n/n \to 0$, and $(1-x)^{1/x} \to e^{-1}$ as $x \to 0$. We thus obtain as limit

$$\frac{1}{k!}\,1 \cdot 1 \cdots 1 \cdot \frac{\lambda^k}{1^k}e^{-\lambda},$$

as asserted. Thus for small p and large n (but np not too large) we can approximate

$$\binom{n}{k} p^k (1-p)^{n-k} \text{ by } (np)^k e^{-np}/k!.$$

To illustrate, we take $p = 0.1$, $n = 20$, so that $np = 2$. For $k = 0, 1, 2, \ldots$, the binomial probabilities are 0.122, 0.270, 0.285, 0.190, 0.090, 0.032, ..., whereas the Poisson probabilities are 0.135, 0.271, 0.271, 0.180, 0.090, 0.036, ...

> REMARKS: The principal conclusion from the results of this section is that the normal distribution has an extremely wide range of applications. For the sample mean of an arbitrary population, the distribution is well approximated by a normal distribution: the larger the sample, the better the approximation. The binomial distribution really comes under this statement, and hence, for large n, it is also well approximated by a normal distribution. The Poisson distribution is in a sense a limiting case of a binomial distribution and therefore can also be approximated, with some care, by a normal distribution.

To illustrate, we consider the binomial distribution for $n = 20$, $p = 0.1$, $q = 0.9$ and ask for $\Pr(0.5 \leqslant Y_n \leqslant 4.5)$. From the values given above for $k = 1, \ldots, 4$, this is $0.270 + \cdots + 0.090 = 0.835$. If we approximate as above by the Poisson distribution, we obtain $0.271 + \cdots + 0.090 = 0.812$. If we use the normal approximation (15–136), we must choose a and b so that

$$2 + 1.34a = 0.5, \qquad 2 + 1.34b = 4.5.$$

Thus $a = -1.12$, $b = 1.86$ and

$$\Pr\{0.5 \leqslant Y_n \leqslant 4.5\} = \frac{1}{\sqrt{2\pi}} \int_{-1.12}^{1.86} e^{-x^2/2}\,dx.$$

By tables, the probability is 0.838. Thus the normal approximation is very good, while the Poisson approximation is not so good.

======================**Problems (Section 15-13)**======================

1. An urn contains three white balls, two of which are numbered 1 and one of which is numbered 2; it also contains five red balls, three of which are numbered 1 and two of

which are numbered 2. A ball is drawn; let $X = 1$ if the ball is white, let $X = 2$ if it is red. Let Y be the number on the ball.

a) Find $\Pr(X = 1)$, $\Pr(X = 2)$, $\Pr(Y = 1)$, $\Pr(Y = 2)$.

b) Find $p_{ij} = \Pr(X = i, Y = j)$ for $i, j = 1, 2$.

c) Determine whether X, Y are independent.

2. Let X, Y have the given joint probability matrix (p_{ij}), where $p_{ij} = \Pr(X = x_i, Y = y_j)$:

$$
\begin{bmatrix}
\dfrac{1}{16} & \dfrac{1}{32} & \dfrac{1}{32} \\[6pt]
\dfrac{1}{4} & \dfrac{1}{16} & \dfrac{1}{8} \\[6pt]
\dfrac{1}{8} & \dfrac{1}{16} & \dfrac{1}{16} \\[6pt]
\dfrac{1}{8} & \dfrac{1}{32} & \dfrac{1}{32}
\end{bmatrix}.
$$

a) Find $\Pr(X = x_1, Y = y_1)$, $\Pr(X = x_2, Y = y_3)$, $\Pr(X = x_3, Y = y_2)$.

b) Find $\Pr(X = x_2)$ and $\Pr(Y = y_2)$.

c) Find $\Pr(X = x_4$ and/or $Y = y_2)$.

d) Determine whether X, Y are independent.

3. Let X, Y have the joint-probability density $f(x, y) = x + y$ for $0 \leqslant x \leqslant 1$, $0 \leqslant y \leqslant 1$; $f(x, y) = 0$ otherwise. Find

a) $\Pr(1/3 < X < 2/3)$ b) $\Pr(X^2 + Y^2 < 1/4)$

c) $\Pr(1/4 \leqslant X \leqslant 1/2$ and $1/4 \leqslant Y \leqslant 1/2)$

d) The marginal probability densities for X and Y.

4. Let X be a random variable having probability density $f(x)$ equal to $2x$ for $0 \leqslant x \leqslant 1$ and to 0 otherwise; let $Y = X^2$.

a) Find $\Pr\left(0 < X \leqslant \dfrac{1}{2}, 0 < Y \leqslant \dfrac{1}{2}\right)$.   b) Find $\Pr\left(0 \leqslant X \leqslant \dfrac{1}{2}, \dfrac{1}{2} \leqslant Y \leqslant 1\right)$.

c) Find $\Pr(0 \leqslant X \leqslant 1, 0 \leqslant Y \leqslant 2 - X)$.

d) Show that the joint probability distribution of X, Y is concentrated on the curve C: $y = x^2$, $0 \leqslant x \leqslant 1$, in the xy-plane; to each arc $y = x^2$, $a \leqslant x \leqslant b$, on this curve we assign the probability $\int_a^b 2x\,dx = b^2 - a^2$ and, for every set A in the xy-plane, $\Pr(A) = \Pr(A \cap C)$.

e) Are X, Y independent?

5. *Nonsingular bivariate normal distribution.* This is the joint distribution of X, Y having density

$$
f(x, y) = \frac{1}{2\pi \sigma_x \sigma_y \sqrt{1 - \rho^2}} \exp\left[\frac{-Q(x, y)}{2(1 - \rho^2)}\right],
$$

where σ_x, σ_y, ρ, μ_x, μ_y are constants, $\sigma_x > 0$, $\sigma_y > 0$, $0 \leqslant \rho < 1$, and

$$
Q(x, y) = \frac{(x - \mu_x)^2}{\sigma_x^2} - 2\rho\,\frac{x - \mu_x}{\sigma_x}\,\frac{y - \mu_y}{\sigma_y} + \frac{(y - \mu_y)^2}{\sigma_y^2}.
$$

a) Verify that $\int_{-\infty}^{\infty} \int_{-\infty}^{\infty} f(x, y)\,dx\,dy = 1$.

HINT: Use polar coordinates $x - \mu_x = r \cos \theta$, $y - \mu_y = r \sin \theta$.

b) Find the marginal probability densities and show that X and Y are normally distributed, with means μ_x, μ_y and standard deviations σ_x, σ_y.

c) Show that X, Y are independent if and only if $\rho = 0$.

6. Let X and Y be independent random variables, each having a binomial distribution with $n = 5$, $p = 1/3$.

   a) Find $\Pr(X + Y = 0)$, $\Pr(X + Y = 3)$, $\Pr(4 \leqslant X + Y \leqslant 8)$.

   b) Find $\Pr(XY = 4)$, $\Pr(XY \geqslant 10)$.

   c) Find $\Pr(X = Y)$.

   d) Find $E(X + Y)$ and $V(X + Y)$.

7. Let X, Y be random variables having joint probability density $f(x, y)$ equal to $(3x/2) + (y/2)$ for $0 \leqslant x \leqslant 1$, $0 \leqslant y \leqslant 1$ and to 0 otherwise.

   a) Obtain the probability density for $Z = X + Y$.

   b) Obtain the probability density for $Z = XY$.

   c) Find $E(X + Y)$, $E(XY)$

   d) Find $V(X + Y)$ and $V(XY)$.

8. Let $X_1, \ldots, X_n$ be independent continuously distributed random variables. Let $Y_n = X_1 + \cdots + X_n$. Let each X_k have the characteristic function $\chi_k(t)$.

   a) Show that Y_n has the characteristic function $\chi(t) = \chi_1(t) \cdots \chi_n(t)$.

   b) In particular, let X_k have the normal distribution (μ_k, σ_k) for $k = 1, \ldots, n$. Show that

   $$\chi(t) = \exp\left(i\mu t - \frac{\sigma^2 t^2}{2}\right),$$

   where $\mu = \mu_1 + \cdots + \mu_m$, $\sigma^2 = \sigma_1^2 + \cdots + \sigma_n^2$. Conclude that Y_n has a normal distribution (μ, σ), also that $\overline{X}_n = Y_n/n$ is normally distributed with mean $(\mu_1 + \cdots + \mu_n)/n$ and variance $(\sigma_1^2 + \cdots + \sigma_n^2)/n^2$.

9. Repeated sampling is carried out n times for a given population, yielding a sample mean $\overline{X}_n$. Use the central limit theorem to evaluate (approximately) the requested probability for each of the following cases:

   a) $n = 5$, discrete population at 0 and 1 with $\Pr(0) = 3/5$ and $\Pr(1) = 2/5$; find $\Pr(0.5 \leqslant \overline{X}_n \leqslant 0.75)$.

   b) $n = 10$, discrete population at 0 and 1 with $\Pr(0) = 0.1$ and $\Pr(1) = 0.9$; find $\Pr(0.4 \leqslant \overline{X}_n \leqslant 0.6)$.

   c) $n = 20$, population has continuous density $2e^{-2x}$ for $x \geqslant 0$ and density 0 otherwise; find $\Pr(0.5 \leqslant \overline{X}_n \leqslant 0.6)$.

   d) $n = 50$, population has normal distribution $(3, 1)$; find $\Pr(3.1 \leqslant \overline{X}_n \leqslant 3.2)$.

15–14 ESTIMATION OF PARAMETERS

We consider a random variable X and the corresponding independent random variables $X_1, \ldots, X_n, \ldots$ obtained by repeated sampling. For example, X might be the weight of a tool produced in large numbers by a given manufacturer; $X_1, \ldots, X_n, \ldots$ are the weights of successive tools chosen at random.

We denote by μ the mean of X, and by σ the standard deviation of X. The problem of estimation is to obtain good approximations to these or other parameters of the probability distribution of X. Sometimes one of μ, σ can be considered known and the other one is sought. Sometimes the distribution of X is wholly unknown; in other cases, there are good reasons to assume it has a special form, usually normal (μ, σ).

To compute our estimates, we use actual measured values of $X_1, \ldots, X_n, \ldots$; we denote these by $x_1, \ldots, x_n, \ldots$ From them we find the *sample mean*

$$\bar{x}_n = \frac{x_1 + \cdots + x_n}{n} \tag{15-140}$$

and *sample variance*

$$s_n^2 = \frac{1}{n} \sum_{i=1}^{n} (x_i - \bar{x}_n)^2. \tag{15-141}$$

It can be verified that we can also write

$$s_n^2 = \left(\frac{1}{n} \sum_{i=1}^{n} x_i^2 \right) - \bar{x}_n^2 \tag{15-141'}$$

by analogy with Eq. (15–75).

a) Unbiased estimates of μ, σ   We can use $\bar{x}_n$ as an estimate of μ. In order to determine how good this estimate is, we observe that the possible values $\bar{x}_n$ themselves yield the random variable

$$\bar{X}_n = \frac{X_1 + \cdots + X_n}{n} \tag{15-142}$$

whose distribution is determined by that of X. In particular, by Eq. (15–131), $E(\bar{X}_n) = \mu = E(X)$.

In general, we call an estimate unbiased if it is the value of a random variable whose expectation equals the parameter being estimated. Thus, Eq. (15–140) provides an unbiased estimate of μ.

In the same way, we can use s_n^2 to estimate the variance σ^2 of X. We are now using a value of the random variable

$$S_n^2 = \frac{1}{n} \sum_{i=1}^{n} (X_i - \bar{X}_n)^2 = \frac{1}{n} \sum_{i=1}^{n} X_i^2 - \bar{X}_n^2.$$

However, S_n^2 is not unbiased; in fact

$$E(S_n^2) = \frac{n-1}{n} \sigma^2, \tag{15-143}$$

for by the rules (15–122), (15–75), (15–132),

$$E(S_n^2) = \frac{1}{n} \sum_{i=1}^{n} E(X_i^2) - E(\bar{X}_n^2)$$

$$= E(X^2) - [V(\bar{X}_n) + \{E(\bar{X}_n)\}^2]$$

$$= \sigma^2 + \mu^2 - \left(\frac{\sigma^2}{n} + \mu^2 \right) = \frac{n-1}{n} \sigma^2.$$

Thus to obtain an unbiased estimate we should multiply S_n^2, and hence s_n^2, by $n/(n-1)$.

By the central limit theorem (Section 15–13), we of course know much more about the validity of the estimate $\bar{x}_n$ for μ. For large n, $\bar{X}_n$ has approximately a normal distribution $(\mu, \sigma/\sqrt{n})$, as in Eq. (15–133′).

b) Maximum-likelihood estimate If the form of the distribution of X is known (in terms of certain parameters), then the values of these parameters can be estimated by selecting a set of values of the parameters that maximize the probability or probability density for the observed values $x_1, \ldots, x_n$. The resulting parameter values are called *maximum-likelihood* estimates of the true values.

For example, let X have a normal distribution (μ, σ), where σ is known, and we want to estimate μ. The probability density for $(X_1, \ldots, X_n)$ is $f(x_1)f(x_2) \cdots f(x_n)$, where

$$f(x) = \frac{1}{\sqrt{2\pi}\,\sigma} \exp\left[-\frac{(x-\mu)^2}{2\sigma^2}\right].$$

We here regard $x_1, \ldots, x_n$ as known values (the results of experiments) and seek to maximize the above probability as a function of μ, that is, we consider the *likelihood function*

$$z = L(x_1, \ldots, x_n, \mu) = \frac{1}{(\sqrt{2\pi}\,\sigma)^n} \exp\left[-\frac{1}{2\sigma^2}\sum_{i=1}^{n}(x_i - \mu)^2\right] \quad (15\text{–}144)$$

and try to maximize z as a function of μ for fixed $x_1, \ldots, x_n$; σ is also fixed (a given number). We find

$$\frac{dz}{d\mu} = \frac{z}{\sigma^2}\sum_{i=1}^{n}(x_i - \mu) = \frac{zn}{\sigma^2}(\bar{x}_n - \mu).$$

Thus $dz/d\mu = 0$ only for $\mu = \bar{x}_n$, the sample mean. Since clearly $z \to 0$ as $\mu \to \pm\infty$ and z is always positive, we have indeed found the value of μ yielding the maximum density. Accordingly, the sample mean is the maximum-likelihood estimate of μ.

In similar fashion, we find that if X has a normal distribution (μ, σ) with known μ, then the maximum-likelihood estimate of σ^2 is

$$\tilde{s}_n^2 = \frac{1}{n}\sum_{i=1}^{n}(x_i - \mu)^2 \quad (15\text{–}145)$$

(see Problem 4(a) below); $\tilde{s}_n^2$ differs from the population variance in that $\bar{x}_n$ is replaced by μ.

We can also obtain maximum-likelihood estimates for both μ and σ^2, when X has the normal distribution (μ, σ) with μ and σ unknown; the estimates found are $\bar{x}_n$ and s_n^2 (see Problem 4(b) below).

c) Confidence intervals   We can apply the central limit theorem to obtain (approximate) intervals in which the mean μ lies, with "high probability," or *confidence intervals* for μ. As an example, to obtain a 90% confidence interval for μ, *when the population standard deviation σ is known*, we first select a and b such that $\Pr(a \leqslant Z \leqslant b) = 0.9$ for a random variable Z having the normal distribution $(0, 1)$. Then, by the central limit theorem (15–133″), we get approximately (for large n)

$$\Pr\left\{\mu + \frac{a\sigma}{\sqrt{n}} \leqslant \overline{X}_n \leqslant \mu + \frac{b\sigma}{\sqrt{n}}\right\} = 0.9,$$

or

$$\Pr\left\{\overline{X}_n - \frac{b\sigma}{\sqrt{n}} \leqslant \mu \leqslant \overline{X}_n - \frac{a\sigma}{\sqrt{n}}\right\} = 0.9. \tag{15–146}$$

We can thus assert: if μ lies in the interval

$$I: \left[\overline{x}_n - \frac{b\sigma}{\sqrt{n}}, \ \overline{x}_n - \frac{a\sigma}{\sqrt{n}}\right],$$

then the measured value $\overline{x}_n$ of $\overline{X}_n$ lies in an interval of probability 90%; similarly, if μ lies outside the above interval I, then the measured value $\overline{x}_n$ of $\overline{X}_n$ lies in an interval of probability 10%. We would therefore say: μ lies in interval I at a confidence level of 90%.

The conclusion can also be phrased in terms of testing hypotheses. Here the natural hypothesis (called *null hypothesis*) would be that $\mu = \mu_0$, a given number. If μ_0 falls outside the interval I, we *reject the hypothesis* $\mu = \mu_0$ at the 10% significance level. If μ_0 lies in the interval I, then we *do not reject the hypothesis* $\mu = \mu_0$ at the 10% significance level.

The percentages used can of course be varied; often 95% and 99% are used instead of 90%.

d) Confidence intervals for σ^2   If X has a normal distribution (μ, σ), then it can be shown that nS_n^2/σ^2 has a χ^2 distribution with $n - 1$ degrees of freedom. Here the χ^2 distribution with n degrees of freedom is the continuous distribution with probability density

$$f(x) = \begin{cases} \dfrac{x^{(n-2)/2} e^{-x/2}}{2^{n/2}\, \Gamma(n/2)}, & x \geqslant 0, \\ 0, & x < 0 \end{cases} \tag{15–147}$$

(see Brunk, pp. 230–231). Thus if we choose $a > 0$ and $b > 0$ so that (with n replaced by $n - 1$) $\int_a^b f(x)dx = 0.95$, then we deduce a 95% confidence interval for σ^2 from the observed value s_n^2 of S_n^2:

$$\Pr\left(a \leqslant \frac{nS_n^2}{\sigma^2} \leqslant b\right) = 0.95$$

or

$$\Pr\left(\frac{nS_n^2}{b} \leqslant \sigma^2 \leqslant \frac{nS_n^2}{a}\right) = 0.95.$$

Accordingly, from the measured values we find that $[ns_n^2/b, \, ns_n^2/a]$ is the desired 95 % confidence interval for σ^2. The χ^2 distribution is widely available in tabulated form.

We observe that in thus estimating σ^2 we do not need to known the mean μ of X.

e) Application of student's t-distribution Under the same hypotheses as for (d), it can be shown that the random variable

$$\frac{(\overline{X}_n - \mu)\sqrt{n-1}}{S_n}$$

has *Student's t-distribution with $n-1$ degrees of freedom*. Here such a distribution with n degrees of freedom is the distribution with probability density

$$f(x) = \frac{\Gamma\left(\dfrac{n+1}{2}\right)\left(1 + \dfrac{x^2}{n}\right)^{-(n+1)/2}}{\Gamma\left(\dfrac{n}{2}\right)\sqrt{\pi n}}, \qquad -\infty < x < \infty.$$

Accordingly, by choosing a and b so that (with n replaced by $n-1$) $\int_a^b f(x)dx = 0.95$, we obtain a 95 % confidence interval for μ from the measured values $\bar{x}_n, \, s_n$:

$$\Pr\left[a \leqslant \frac{(\overline{X}_n - \mu)\sqrt{n-1}}{S_n} \leqslant b\right] = 0.95,$$

so that

$$\left[\bar{x}_n - \frac{bs_n}{\sqrt{n-1}}, \quad \bar{x}_n - \frac{as_n}{\sqrt{n-1}}\right]$$

is the desired 95 % confidence interval for μ. The t-distribution is also widely available in tables.

We observe that, in obtaining our confidence interval for μ, we do not need to know the standard deviation σ of X, in contrast to the procedure of (c) above.

f) Tests involving two random variables For two random variables X, Y, defined on the same sample space, we define the *correlation coefficient* ρ by the equation

$$\rho = E\left(\frac{X - \mu_x}{\sigma_x} \frac{Y - \mu_y}{\sigma_y}\right)$$

and verify that $-1 \leqslant \rho \leqslant 1$ always. For $\rho = 0$, X and Y are said to be *uncorrelated*; this is surely the case when X, Y are independent, but can also hold when they are dependent. For $\rho = \pm 1$, Y is a linear function of X, so that the joint probability distribution is concentrated on a line in the xy-plane. In general, we interpret values of ρ close to 0 as indicating little relationship between X and Y, and the values close to ± 1 as indicating a strong relationship. There are methods for obtaining confidence intervals for ρ from sample values $(x_1, y_1), \ldots, (x_n, y_n)$ of (X, Y) (see BRUNK, Chapter 12). There are also methods for testing the equality of μ_x, μ_y and the equality of σ_x, σ_y (see BRUNK, Chapter 14).

EXAMPLE Samples of a tool are weighed to give the following values (in grams): 100.49, 107.27, 103.66, 101.39, 115.93, 112.22, 101.50, 103.65, 111.85, 100.08, 116.23, 117.71, 100.41, 107.46. It is assumed that the weights are normally distributed. The sample mean and the sample variance are found to be $\bar{x}_{14} = 107.13$ and $s_{14}^2 = 39.625$, respectively. These are estimates for μ and σ^2. However, as in (a) above, we can improve the estimate for σ^2 by using

$$\frac{14}{13} s_{14}^2 = 42.673.$$

We now assume from previous experience that σ^2 should be 42.500, and seek a 95% confidence interval for μ as in (c) above, with 0.9 replaced by 0.95 and a, b correspondingly adjusted, so that $\Pr(a \leqslant Z \leqslant b) = 0.95$ for Z having the normal distribution $(0, 1)$. We can take $a = -b$, so that $\Pr(-\infty \leqslant Z \leqslant b) = 0.975$ and hence, from tables, $b = 1.96$, $a = -1.96$. Thus

$$\frac{b\sigma}{\sqrt{n}} = \frac{1.96 \cdot 6.519}{\sqrt{14}} = 3.41 = -\frac{a\sigma}{\sqrt{n}},$$

and our 95% confidence interval for μ is $[103.72, 110.54]$.

Next we seek a 95% confidence interval for σ^2. We use the χ^2-distribution as above, with 13 degrees of freedom, and find from tables that

$$\int_0^{5.01} f(x)\,dx = 0.025, \qquad \int_0^{24.7} f(x)\,dx = 0.975,$$

so that we can choose $a = 5.01$, $b = 24.7$. Thus

$$\frac{ns_n^2}{a} = \frac{14 \cdot 39.625}{5.01} = 110.72, \qquad \frac{ns_n^2}{b} = 22.46,$$

and our 95% confidence interval for σ^2 is $[22.46, 110.72]$.

Finally we use Student's t-distribution to obtain a 95% confidence interval for μ. We observe that the t-distribution is symmetric about $x = 0$, that is, $f(x)$ is even. Hence, as for the normal distribution, we can take $a = -b$ and choose b so that

$$\int_{-b}^{b} f(x)\,dx = 0.975.$$

We find from tables (13 degrees of freedom) that $b = 2.1620$. Hence, as in (e) above, we calculate

$$\frac{bs_n}{\sqrt{n-1}} = \frac{2.1620 \cdot 6.2918}{\sqrt{13}} = 3.7746 = -\frac{as_n}{\sqrt{n-1}}.$$

Thus our 95% confidence interval for μ is $[103.35, 110.79]$. This is very close to that found above under the assumption that $\sigma^2 = 42.500$.

══════════════════════**Problems (Section 15–14)**══════════════════════

1. Toss a coin 10 times and record 1 for heads, 0 for tails. Compute the sample mean and variance and compare with the theoretical values $\mu = 1/2$, $\sigma^2 = 1/4$ for a perfect coin.

2. Place nine balls (or slips of paper) numbered 1 and four numbered 0 in an urn and repeatedly draw and replace, recording the number drawn. Compute the sample mean and variance and compare with the theoretical values $\mu = 9/13$, $\sigma^2 = 36/169$.

3. Prove that the sample variance (15–141) can be written in the form (15–141').

4. Let X have a normal distribution (μ, σ).
   a) Show that the maximum-likelihood estimate of σ^2, for given μ, is equal to $\hat{s}_n^2$, as in Eq. (15–145).
   b) Show that the maximum-likelihood estimates for μ, σ, for both unknown, are $\bar{x}_n$ and s_n^2.

5. Let X have a binomial distribution, so that,

$$\Pr(X = k) = \binom{N}{k} p^k (1 - p)^{N-k}, \qquad k = 0, 1, \ldots, N.$$

Suppose that N is known and sampling yields the values $x_1, \ldots, x_n$ for X. Show that the maximum-likelihood estimate of p is $\bar{x}_n / N$.

6. Sampling yields the following values of a random variable X: 12.5, 11.9, 12.4, 12.1, 12.5, 11.8, 12.3, 12.1, 12.4.
   a) Find the sample mean and variance, and use them to estimate σ and μ for X.
   b) Use the central limit theorem to obtain a 95% confidence interval for μ if σ^2 is known to be 0.06.
   c) Assume that X has a normal distribution and use the χ^2-distribution to obtain a 95% confidence interval for σ^2.
   d) Assume that X has a normal distribution and use Student's t-distribution to obtain a 95% confidence interval for μ.

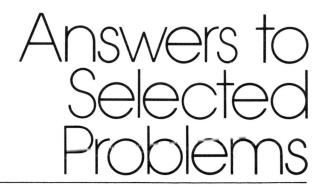

Answers to Selected Problems

Answers to Selected Problems

SECTION 1–2, PAGE 4

1. a) order 1, linear   c) order 2, nonlinear
e) order 4, linear   g) order 1, nonlinear

3. a) $y = e^{-x} \sin x$ c) $y = e^{-x} \sin x$

4. a) $y = 2e^{-2x} - e^{-x}$ c) $y = 2e^{1-x} - e^{2-2x}$

5. a) $y' = \cos x + [2(y - \sin x)/x]$ c) $y' = -x/y$ e) $y'' - y = 0$

SECTION 1–4, PAGE 10

1. a) $y = -\cos x + c$ c) $x^2 + y^2 = c$ e) $y = ce^{x^2}$
g) $y = cx$ i) $y = (1 + cx^2)/(1 - cx^2)$

3. a) $y = e^{x^2/2} - 1$ c) $y = \ln \dfrac{2}{\cos 2x + 1}$ e) $p = p_0 \exp \dfrac{-g\rho_0(y - y_0)}{p_0}$

g) $v = \dfrac{av_0 e^{at/m}}{bv_0 e^{at/m} + a - bv_0}$ i) $x = \dfrac{2(1 - e^t)}{1 - 2e^t}$

4. a) $v = \pm 8\sqrt{x},\ x = 16t^2$ c) $x = (3/\sqrt{2}\,t + 2\sqrt{2})^{2/3}$

5. a) $0, 0.01, 0.0561, 0.0905$ c) $-0.4, -0.745, -1.0135, -1.1998$

6. a) $y = 1, 0.99, 0.97, 0.9401$; $y' = -0.1, -0.2, -0.299, -0.396$

7. a) $y = -x \ln \ln cx$ **c)** $\ln (x^2 + y^2) + 2 \tan^{-1} (y/x) = c$

8. a) $r = ce^{-\theta}$

SECTION 1–6, PAGE 17

1. a) $x^2 + 4xy + 3y^2 = c$ **c)** $x \sin xy - \cos y = c$

     e) $1/(x^2 + y^2) = c$ **g)** $x^5 + 10x^3 y^2 + 5x^2 y^3 + y^5 = c$

2. a) $x^2 + xy + y^2 = 3$ **c)** no solution

3. a) $U = (k_1 x^2 + k_2 y^2)/2$, equipotential lines are ellipses

4. a) $(x^2 + y^2)^2 + 4xy = c$ **c)** $x^3 y + x^2 y^2 = c$ **e)** $(2x^2)^{-1} + e^{-xy} = c$

5. a) $y = cx$ **c)** $\cos x \sinh y = c$

SECTION 1–8, PAGE 25

1. a) $y = \dfrac{e^x}{2} + ce^{-x}$ **c)** $y = 1 + c \exp\left(-\dfrac{x^2}{2}\right)$

     e) $y = \dfrac{x}{2 \sin x} + \dfrac{\cos x}{4} + \dfrac{c}{\sin x}$

2. a) $y = \dfrac{e^x}{4} + ce^{-3x}$ **c)** $y = \dfrac{x}{3} - \dfrac{1}{2} + \dfrac{c}{x^2}$

     e) $y = \dfrac{\sec x + \tan x + c - x}{\sec x + \tan x}$

4. a) $x = e^{-3t}$ **c)** $x = e^{2t}$ **e)** $x = \dfrac{3}{2}(1 - e^{-2t})$ **g)** $x = t + \dfrac{1}{2}(e^{-2t} - 1)$

     i) $x = \dfrac{3}{5}(\sin 2t - 2 \cos 2t + 2e^{-t})$ **k)** $x = 0$ for $t \le 0$; $x = 1 - e^{-t}$ for $t \ge 0$

     m) $x = 0$ for $t \le 0$; $x = 1 - e^{-t}$ for $0 \le t \le 2$;

         $x = 3 - e^{-t} - 2e^{2-t}$ for $2 \le t \le 3$; $x = (3e^3 - 1 - 2e^2)e^{-t}$ for $t \ge 3$

5. $x = A(a^2 + \omega^2)^{-1}(a \cos \omega t + \omega \sin \omega t) + ce^{-at}$

7. $a = 1.54 \cdot 10^{-10}$ years^{-1} **8. a)** $14.72°$ **9.** $k^{-1} \ln m_1/(m_1 - m_0)$

SECTION 1–10, PAGE 30

1. a) $36 - 8i$ **c)** 5 **e)** $2^{14}(\sqrt{2} - i\sqrt{2})$ **g)** i **i)** $(-1)^n$

3. a) $6x - 5 + i(3x^2 - 14x)$ **c)** $3ie^{3ix}$ **e)** $e^{2ix}(2ix^2 + 4ix + 4 + 3i)$

4. a) $[12\pi^3 - 30\pi^2 + i(3\pi^4 - 28\pi^3)]/12$ **c)** $-(2 + 5i)(e^{2\pi} + 1)/29$

5. a) $2^{1/3}\left(\cos \dfrac{\pi}{4} + i \sin \dfrac{\pi}{4}\right)$, $2^{1/3}\left(\cos \dfrac{11\pi}{12} + i \sin \dfrac{11\pi}{12}\right)$,

     $2^{1/3}\left(\cos \dfrac{19\pi}{12} + i \sin \dfrac{19\pi}{12}\right)$

SECTION 1–11, PAGE 34

1. a) $y = c_1 e^{3x} + c_2 e^{-3x}$ c) $y = c_1 e^x + c_2 e^{-x/2}$ e) $y = e^{4x}(c_1 + c_2 x)$
   g) $y = c_1 + c_2 x$ i) $y = c_1 \cos x + c_2 \sin x$
   k) $y = c_1 \cos (x/2) + c_2 \sin (x/2)$
   m) $y = e^{-x}(c_1 \cos 3x + c_2 \sin 3x)$
   o) $y = e^{-x/2}[c_1 \cos (\sqrt{3}x/2) + c_2 \sin (\sqrt{3}x/2)]$

2. a) $y = 2e^{3x} + 2e^{-3x}$ c) $x = 3 \cos 5t$ e) $y = 0$

3. a) $x = 5 \sin (t + \gamma)$, $A = 5$, $\cos \gamma = 4/5$, $\sin \gamma = 3/5$
   c) $x = -3 \cos 3t$, $A = 3$, $\gamma = 3\pi/2$

4. a) $1/5$ rad/sec, 10π sec   c) $1/5$ rad/sec

7. a) $(e^4 - 1)^{-1}[(3e^2 - 2)e^{2x} + (2e^4 - 3e^2)e^{-2x}]$
   c) $y = c_2 \sin 3x$ (infinitely many solutions)

SECTION 1–12, PAGE 40

1. $y = c_1 e^{3x} + c_2 e^{-3x} + y^*(x)$, where $y^*(x)$ can be chosen as follows:
   a) $-2e^{2x}/5$ c) $(2/3)xe^{3x}$ e) $(2/15)(-3e^{2x} + 5xe^{3x})$
   g) $e^x(2 - x)$ i) $(e^{3x}/36)(3x^2 - x)$.

2. $x = c_1 \cos 2t + c_2 \sin 2t + x^*(t)$, where $x^*(t)$ can be chosen as follows:
   a) $-\dfrac{1}{7} \sin 5t$ c) $-\dfrac{1}{5}(2 \cos 3t + 3 \sin 3t)$ e) $-\dfrac{5}{4}t \cos 2t$

   g) $\dfrac{1}{16}(2t^2 \sin 2t + t \cos 2t)$ i) $(e^{-t}/17)(\cos 2t - 4 \sin 2t)$

3. $x = e^{-t}(c_1 \cos t + c_2 \sin t) + x^*(t)$, where $x^*(t)$ can be chosen as follows:
   a) $\cos t + 2 \sin t$ c) $7e^{-t}$ e) $(e^{-t}/2)t \sin t$

4. a) $y = (1/30)(25e^{3x} + 17e^{-3x} - 12e^{2x})$ c) $x = (1/7)(6 \sin 2t - \sin 5t)$
   e) $x = e^{-t}(-\cos t - 3 \sin t) + \cos t + 2 \sin t$

7. a) $y = 2x^2$ c) $y = (5/4)x \sin 2x + 3x^2 + 2xe^{3x}$

SECTION 1–13, PAGE 42

1. a) $y = xe^{2x}/4$ c) $y = \cos x \ln \cos x + x \sin x$
   e) $y = \dfrac{1}{2}(e^{-x} - e^x) \ln (1 + e^x) - \dfrac{1}{2} + \dfrac{1}{2}xe^x$

SECTION 1–14, PAGE 50

1. a) $x = \sqrt{2}e^{-t/2} \sin (2t + \pi)/4$ c) $x = e^{-2t} - e^{-3t}$
   e) $x = 0.600e^{-t/2} \sin (0.5t + 0.02) + 0.06 \sin (5t - 2.94)$
   g) $x = -0.517e^{-2t} - 0.441e^{-3t} + 0.096 \sin (5t + 0.797)$
   i) $x = \sin t - \dfrac{1}{2} \sin 2t$ k) $x = (5 \sin 2t - 2 \sin 5t)/21$

SECTION 1–15, PAGE 56

1. a) $y = c_1 e^x + c_2 e^{-2x} + c_3 e^{-5x}$ **c)** $y = e^x(c_1 + c_2 x) + c_3 e^{-x}$
   e) $y = c_1 e^{-x/2} + c_2 e^{-x} + e^{-x}(c_3 \cos x + c_4 \sin x)$
   g) $y = (c_1 + c_2 x + c_3 x^2) \cos 4x + (c_4 + c_5 x + c_6 x^2) \sin 4x$

2. a) $y = e^{-x}(1 + 3x + 2x^2)$ **c)** $y = (1/12)(-51 + 100x - 24x^2 + 12x^3 - x^4)$

3. a) $y = (5e^{2x}/24) + y_c(x)$, with $y_c(x) = c_1 e^x + c_2 e^{-2x} + c_3 e^{-4x}$
   c) $y = e^{2x}(12x - 17)/228 + y_c(x)$, with $y_c(x)$ as in (a)
   e) $y = e^x[(x^2/30) - (8x/225)] + y_c(x)$, with $y_c(x)$ as in (a)
   g) $y = (1/192)e^x(4x^4 - 4x^3 + 3x^2) + e^x(c_1 + c_2 x) + c_3 e^{-3x}$
   i) $y = (1/10)(3 \cos 3x + \sin 3x) + c_1 e^x + c_2 \cos 2x + c_3 \sin 2x$

4. $w = \dfrac{1}{\Delta}[\beta\gamma - \alpha\delta - \sqrt{2}\beta\delta) \sin \sqrt{2}x \sinh \sqrt{2}x$

$\qquad + (\sqrt{2}\alpha\gamma - \alpha\delta - \beta\gamma) \cos \sqrt{2}x \cosh \sqrt{2}x] + x^2$, where $\alpha = \sin\sqrt{2}$,
$\beta = \cos \sqrt{2}$, $\gamma = \sinh \sqrt{2}$, $\delta = \cosh \sqrt{2}$,
$\Delta = \alpha^2\gamma\delta + \alpha\beta\delta^2 + \beta^2\gamma\delta - \alpha\beta\gamma^2$

5. $x_1 = A_1 \sin(t + \gamma_1) + A_2 \sin(6^{-1/2}t + \gamma_2)$,
$\quad x_2 = -2A_1 \sin(t + \gamma_1) + (4A_2/3) \sin(6^{-1/2}t + \gamma_2)$

SECTION 1–18, PAGE 62

1. a) $y = c_1(x + 1) + c_2(x + 1)^3$ **2. b)** $3 - 2x$

3. (Solution can be modified by adding a solution of the related homogeneous
equation.)   **a)** $y = e^{5x}/12$ **c)** $y = (e^x/4)(3x + x^2)$

  e) $y = \dfrac{1}{10}[e^{3x}\int f(x)e^{-3x}dx + \cos x \int f(x)(3 \sin x - \cos x)dx$

$\qquad - \sin x \int f(x)(\sin x + 3 \cos x)dx]$
  g) $y = -x^2 - 3x - 3$
4. a) commute

============================ Chapter 2 ============================

SECTION 2–3, PAGE 70

1. a) converges to 0   **c)** converges to 0
   e) converges to 0   **g)** converges to 1

2. a) 2, 1.75, 1.73214, 1.73208, 1.732054   **c)** 0, 0.5, 0.44, 0.443

3. a) $n \geqslant 13$ **4. a)** both   **c)** both   **e)** both   **g)** both

5. a) decreasing   **c)** decreasing   **e)** increasing
   g) increasing   **i)** increasing   **k)** increasing

SECTION 2–6, PAGE 74

1. a) 2, 2.5, 2.667, 2.708, 2.717 (sum is e)
   c) $-1, 1, -2, 2, -3$ e) $-1, 7, -20, 44, -81$
2. a) 3/2   c) 5/2   e) 1   g) 35/6   i) 1/2   k) -2
3. a) diverges   c) test fails   e) diverges   g) test fails

SECTION 2–9, PAGE 80

1. a) converges   c) converges   e) converges   g) converges
2. a) converges   c) converges   e) converges
   g) diverges   i) converges   k) converges
3. a) converges   c) converges   e) converges   g) converges

SECTION 2–10, PAGE 84

1. a) absolute convergence   c) not absolute (conditional) convergence
   e) absolute convergence   g) absolute convergence
2. a) converges   c) diverges   e) diverges
   g) converges   i) converges
3. a) 0.18   c) 0.07

SECTION 2–12, PAGE 87

1. a) absolute convergence   c) absolute convergence
   e) diverges   g) diverges   i) test fails
2. a) absolute convergence   c) absolute convergence
   e) test fails (diverges by nth-term test)   g) test fails

SECTION 2–13, PAGE 90

1. a) 1.18, $1/125 < R_4 < 1/32$ c) 0.178, $-1/730 < R_5 < 0$
   e) 769/1944, $1/29160 < R_4 < (1/29160)(18/17)$
   g) 0.841, $0.072 < R_4 < 0.72$
2. a) 4   c) 100

SECTION 2–14, PAGE 94

1. a) limit is 1 for $x = 0$ and 0 for $x > 0$
   c) limit is 0 for $x > 1$ and for $x < -1$, limit is 1 for $x = 1$
   e) limit is $2/(2 - x)$ for $-2 < x < 2$

2. a) $-3/2 < x < 3/2$, $r^* = 3/2$ **c)** $-1 \leqslant x \leqslant 1$, $r^* = 1$
   e) all x, $r^* = \infty$ **g)** $-1 \leqslant x < 1$, $r^* = 1$ **i)** $x = 0$, $r^* = 0$
   k) $2 \leqslant x \leqslant 4$, $r^* = 1$ **m)** $1/2 < x < 7/2$
3. a) $x < -1$ and $x > 1$ **c)** $x > 0$ **e)** $x > -3/2$

SECTION 2–15, PAGE 98

2. a) $n = 10$ **c)** $n = 1000$

SECTION 2–17, PAGE 102

1. $-x - (x^2/2) - \cdots - (x^n/n) - \cdots$, $|x| < 1$
2. a) $1 - x + x^2 - x^3 + \cdots$
   c) $x - (x^3/3) + \cdots + (-1)^{n+1}[x^{2n-1}/(2n-1)] + \cdots$
   e) $2x + 2(x^3/3) + \cdots + 2[x^{2n-1}/(2n-1)] + \cdots$
   g) $2(x^2/2!) - 2^3(x^4/4!) + \cdots + (-1)^{n+1}2^{2n-1}[x^{2n}/(2n)!] + \cdots$
   i) $1 - 3^2(x^2/2!) + 3^4(x^4/4!) + \cdots + [(-1)^n 3^{2n}x^{2n}/(2n)!] + \cdots$
3. a) $x + x^2 + (x^3/3)$ **c)** $(x/2) - (x^2/4) - (x^3/12)$ **e)** $x - (x^3/3)$
   g) $1 - x^2$ **i)** $1 + x + (3x^2/2) + (13x^3/6)$ **k)** $x + (x^3/6)$
4. a) $1 + [(x-1)/2] - [(x-1)^2/8] + [(x-1)^3/16]$
   c) $1 - (2x/3) + (2x^2/9) - (22x^3/81)$

SECTION 2–20, PAGE 110

1. a) between 3.1 and 3.25   **c)** between 29.66 and 34
   e) between 4.29 and 4.37

4. a) converges to 0   **c)** converges to 0
5. a) absolute convergence   **c)** converges   **e)** absolute convergence
7. a) 1   **c)** ∞ **e)** 5

SECTION 2–22, PAGE 118

1. a) 1.853   **2. a)** 0.935

3. a) $y = c_1 \left[1 - \dfrac{x^2}{2} + \cdots + (-1)^n \dfrac{x^{2n}}{(2n)!} + \cdots \right]$

$\qquad + c_2 \left[x - \dfrac{x^3}{6} + \cdots + (-1)^{n+1} \dfrac{x^{2n-1}}{(2n-1)!} + \cdots \right]$

$\qquad = c_1 \cos x + c_2 \sin x$

c) $y = x + x^2 + \cdots + \dfrac{x^n}{(n-1)!} + \cdots + c\left(1 + x + \dfrac{x^2}{2} + \cdots\right) = xe^x + ce^x$

e) $y = c_1\left[1 - \dfrac{x^2}{2} + \dfrac{x^4}{2\cdot4} + \cdots + (-1)^n\dfrac{x^{2n}}{2\cdot4\cdots(2n)} + \cdots\right]$

$\qquad + c_2\left[x - \dfrac{x^3}{3} + \dfrac{x^5}{3\cdot5} + \cdots + (-1)^n\dfrac{x^{2n+1}}{3\cdot5\cdots(2n+1)} + \cdots\right]$

4. a) $y = 3(1+x^2) + 7\left[x + \dfrac{x^3}{1\cdot3} - \dfrac{x^5}{3\cdot5} + \cdots + (-1)^{n+1}\dfrac{x^{2n+1}}{(2n-1)(2n+1)} + \cdots\right]$

6. a) $y = 1 - x^2\dfrac{N(N+1)}{2!} + x^4\dfrac{N(N-2)(N+1)(N+3)}{4!} + \cdots$

$\qquad + (-1)^{N/2}x^N\dfrac{N(N-2)\cdots2(N+1)(N+3)\cdots(2N-1)}{N!}$

b) $y = x - x^3\dfrac{(N+2)(N-1)}{3!} + x^5\dfrac{(N+2)(N+4)(N-1)(N-3)}{5!} + \cdots$

$\qquad + (-1)^{(N-1)/2}x^N\dfrac{(N+2)(N+4)\cdots(2N-1)(N-1)(N-3)\cdots2}{N!}$

8. a) $y = x - \dfrac{x^3}{3} + \dfrac{2x^5}{15}$ 　c) $y = 1 - \dfrac{x^2}{2} + \dfrac{x^4}{8}$

7. $x = 1 + t + \dfrac{t^2}{2} - \dfrac{t^4}{8} + \cdots$

SECTION 2–23, PAGE 124

All summations go from 1 to ∞.
1. a) irregular singular point 　　c) ordinary point 　　e) ordinary point
2. a) regular singular point 　　c) regular singular point
　　e) regular singular point

3. a) $y_1(x) = x^{-1}\left[1 + \sum\dfrac{(-1)^nx^{3n}(-1)5\cdots(6n-7)}{1\cdot4\cdots(3n-2)\cdot3\cdot6\cdots(3n)}\right]$,

$\qquad y_2(x) = x\left[1 + \sum\dfrac{(-1)^nx^{3n}3\cdot9\cdots(6n-3)}{3\cdot6\cdots(3n)\cdot5\cdot8\cdots(3n+2)}\right]$

c) $y_1(x) = x^{-1}\left[1 + \sum\dfrac{(-1)^nx^{3n}1\cdot7\cdots(6n-5)}{2\cdot5\cdots(3n-1)\cdot3\cdot6\cdots(3n)}\right]$,

$\qquad y_2(x) = 1 + \sum\dfrac{(-1)^nx^{3n}3\cdot9\cdots(6n-3)}{3\cdot6\cdots(3n)\cdot4\cdot7\cdots(3n+1)}$

6. a) $y = c_1x^{-1} + c_2x^{-2}$ 　　c) $y = c_1x^2 + c_2x^2\ln x$

7. a) $y = c_1 x \left[1 + \sum \dfrac{(-1)^n x^{-3n}}{1 \cdot 3 \cdot 4 \cdot 6 \cdots (3n-2)(3n)} \right]$

$+ c_2 x^{-1} \left[1 + \sum \dfrac{(-1)^n x^{-3n}}{3 \cdot 5 \cdot 6 \cdot 8 \cdots (3n)(3n+2)} \right]$

===================Chapter 3==================

SECTION 3–2, PAGE 134

1. a) $\pi + 2 \displaystyle\sum_{n=1}^{\infty} \dfrac{(-1)^n}{n} \sin nx$ c) $\dfrac{\pi^2}{3} + 4 \displaystyle\sum_{n=1}^{\infty} \dfrac{(-1)^n \cos nx}{n^2}$

e) $\dfrac{e^\pi - e^{-\pi}}{\pi} \left[\dfrac{1}{2} + \displaystyle\sum_{n=1}^{\infty} \dfrac{(-1)^n \cos nx}{1+n^2} - \displaystyle\sum_{n=1}^{\infty} \dfrac{(-1)^n n \sin nx}{1+n^2} \right]$

SECTION 3–3, PAGE 141

1. a) Jumps at $-\pi, \pi$; no corners   c) corners at $-\pi, 0, \pi$; no jumps

4. a) 1   c) e

SECTION 3–5, PAGE 148

1. a) $\dfrac{1}{2} - \dfrac{2}{\pi} \displaystyle\sum_{n=1}^{\infty} (-1)^{n+1} \dfrac{\cos(2n-1)x}{2n-1}$ c) $\dfrac{2}{\pi} - \dfrac{4}{\pi} \displaystyle\sum_{n=1}^{\infty} \dfrac{\cos 2nx}{4n^2 - 1}$

2. a) $\dfrac{2}{\pi} \displaystyle\sum_{n=1}^{\infty} \dfrac{1 - \cos(n\pi/2)}{n} \sin nx$ c) $\dfrac{8}{\pi} \displaystyle\sum_{n=1}^{\infty} \dfrac{n}{4n^2 - 1} \sin 2nx$

3. a) $\dfrac{2}{\pi} \displaystyle\sum_{n=1}^{\infty} (-1)^{n+1} \dfrac{\sin n\pi x}{n}$

c) $\sinh 1 + 2 \sinh 1 \displaystyle\sum_{n=1}^{\infty} \dfrac{(-1)^n}{n^2\pi^2 + 1} (\cos n\pi x + n\pi \sin n\pi x)$

4. a) $\dfrac{3}{2} - \dfrac{4}{\pi^2} \displaystyle\sum_{n=1}^{\infty} \dfrac{\cos(2n-1)\pi x}{(2n-1)^2}$

5. a) $3 \sin x \, e^{-ct} + 5 \sin 2x \, e^{-4ct}$

c) $\dfrac{4}{\pi} \displaystyle\sum_{n=1}^{\infty} \dfrac{n}{n^4+4} [(-1)^{n-1} e^{\pi} - 1] \sin nx \, e^{-n^2 ct}$

SECTION 3–8, PAGE 158

1. a) $E_n = (2\pi^3/3) - 4\pi(1^{-2} + \cdots + n^{-2}); \quad 8.10, \ 4.96, \ 3.57$

2. c) $E_n = (\pi^3/6) - (16/\pi)[1^{-4} + 3^{-4} + \cdots + (2n-1)^{-4}]; \quad 0.075, \ 0.012, \ 0.004$

3. a) $\dfrac{2\pi^2}{3} - 2 \displaystyle\sum_{n=-\infty}^{\infty}{}' \dfrac{e^{inx}}{n^2}$ **c)** $\dfrac{2}{\pi} \displaystyle\sum_{n=-\infty}^{\infty} \dfrac{e^{inx}}{1-4n^2}$

4. c) $x^*(t) = \dfrac{2\pi^2}{3} - 2 \displaystyle\sum_{n=-\infty}^{\infty} \dfrac{e^{inx}}{n^2[(1-6n^2) + in(7-n^2)]}$

5. c) $x^*(t) = \dfrac{2}{\pi} \displaystyle\sum_{n=-\infty}^{\infty} \dfrac{e^{inx}}{(1-4n^2)[0.01 n^4 - 0.24n^2 + 1 + i(0.4n - 0.04n^3)]}$

SECTION 3–12, PAGE 171

3. a) $1/4$ **c)** $1/3$ **e)** $19/20$ **4. a)** $3f_1 - 4f_2$

5. $g_0 = 1, \ g_1 = x - \dfrac{1}{2}, \ g_2 = x^2 - x + \dfrac{1}{6}, \ g_3 = x^3 - \dfrac{3}{2}x^2 + \dfrac{3}{5}x - \dfrac{1}{20}$

7. a) $\dfrac{3}{2} P_1(x) - \dfrac{7}{8} P_3(x) + \dfrac{11}{16} P_5(x)$

c) $1.1752 P_0(x) + 1.1037 P_1(x) + 0.3579 P_2(x) + 0.0702 P_3(x) + 0.0110 P_4(x)$
$+ 0.0014 P_5(x)$

9. a) $x^2 = (1/3)P_0 + (2/3)P_2$

SECTION 3–15, PAGE 178

3. a) $0.86, \ 3.42, \ 6.44$

6. a) eigenvalues $0, 1, 4, \ldots, n^2, \ldots$;
eigenfunctions $1, \cos x, \cos 2x, \ldots, \cos nx, \ldots$

c) eigenvalues $\omega_0^2, \omega_1^2, \ldots, \omega_n^2, \ldots$, where $\omega_0, \omega_1, \omega_2, \ldots$ are successive positive solutions of $\tan \omega = -\omega$;
eigenfunctions $\sin \omega_0 x, \sin \omega_1 x, \ldots, \sin \omega_n x, \ldots$

8. a) $\pi - (4/\pi)\sum_{n=1}^{\infty} (2n-1)^{-2}[\cos (2n-1)x + \cos (2n-1)y]$

===============================Chapter 4===============================

SECTION 4-4, PAGE 189

1. a) $s^{-2}(1-e^{-s})$ **c)** $s^{-3}[2-e^{-s}(s^2+2s+2)] + (s+1)^{-1}e^{-s}$
e) $[(s+1)^2+4]^{-1}(2-e^{-2\pi(s+1)})$
2. a) $(s^2+3s+2)^{-1}(5s+8)$ **c)** $(s^2-4s+13)^{-1}(12s-39)$
e) $s^{-1}(s^2-4s+13)^{-2}(s^2-4s-5)$ **g)** $2e^{-s}(s+1)^{-3}$
i) $(1+e^{-s\pi})(s^2+1)^{-1}(1-e^{-s\pi})^{-1}$

SECTION 4-7, PAGE 197

2. a) $5+7t$ **c)** $3e^t+2e^{-t}$ **e)** $e^{-t}(t+3)$ **g)** $2\sin t + 3\cos t$
i) 0 for $0 \leqslant t < 1$, 2 for $1 \leqslant t < 2$, $2+3e^{t-2}$ for $t \geqslant 2$
3. a) $e^{2t}-e^t$ **c)** $3e^{2t}-e^t-2e^{-2t}$ **e)** $2e^t-2-t$
g) $(1/4)[e^{-2t}(4t^2+2t+1)-1]$ **i)** $1-\cos t$
k) $(1/10)[4+e^{-t}(26\cos 2t - 17\sin 2t)]$
4. a) $(1/24)(22e^{3t}-7e^{-3t}-15e^t)$
5. a) $(1/6)(4\cos t - 2\sin t - 4\cos 2t + \sin 2t)$
c) $(1/12)(8\sin t + 4\cos t - 4\sin 2t - \cos 2t - 3)$

SECTION 4-10, PAGE 205

1. a) $t^2/2$ for $t \geqslant 0$ **c)** $(e^{5t}-e^{2t})/3$ for $t \geqslant 0$
e) e^t-1 for $0 \leqslant t \leqslant 1$, e^t-e^{t-1} for $t \geqslant 1$
2. a) $e^t * e^{2t}$ **c)** $f(t) * e^{5t}$, where $f(t) = 0$ for $0 \leqslant t \leqslant 1$, $f(t) = e^{3(t-1)}$ for $t \geqslant 1$
e) $f(t) * e^t$, where $f(t) = 1$ for $0 \leqslant t \leqslant 2$, $f(t) = 0$ for $t \geqslant 2$
g) $f(t) * e^{2t}$, where $f(t) = t$ for $0 \leqslant t \leqslant 3$, $f(t) = 6-t$ for $3 \leqslant t \leqslant 6$, f has period 6 (triangular wave).
6. a) $f' = 2e^{2t}U(t) + \delta(t)$, $f'' = 4e^{2t}U(t) + 2\delta(t) + \delta'(t)$
c) $f' = \delta(t) - \delta(t-1)$, $f'' = \delta'(t) - \delta'(t-1)$
e) $f' = \cos t \{U(t) - U[t-(\pi/2)]\} - \delta[t-(\pi/2)]$,
$f'' = -\sin t \{U(t) - U[t-(\pi/2)]\} + \delta(t) - \delta'[t-(\pi/2)]$
7. a) $\delta(t)$ **c)** $e^6\delta(t-2)$ **e)** $\delta'(t) - 3\delta(t)$
g) $e^4[\delta''(t-2) - 4\delta'(t-2) + 4\delta(t-2)]$
8. a) 1 **c)** 1 **e)** -2 **g)** -2
9. Values of $\mathscr{L}[f']$: **a)** $(3s+2)/s$ **c)** $1-e^{-s}$ **e)** $2-3s$

10. a) $e^{3t}U(t)$ **c)** $2e^{2t}U(t)+\delta(t)$ **e)** $2U(t-1)$
  g) $[4+t^3+(t^4/4)+(t^6/60)]\,U(t)+2\delta'(t)+2\delta''(t)$
11. a) $\delta(t)-\delta'(t)$ **c)** $\delta(t)-2\sin t$ **e)** $2\delta(t-2)+5\delta'(t-2)$

SECTION 4–12, PAGE 214

1. a) $x=3t+e^{-2t}$ **c)** $x=-\sin t+\cosh t$
  e) $x=(1-2e^{-t}\sin t-e^{-t}\cos t)/2$ **g)** $x=-11e^{-2t}+8e^{-t}+4e^{-3t}$
  i) $x=f(t)-f(t-2)-h(t-2)$, where $f(t)=(1/16)\,(e^{2t}-e^{-2t}-4t)U(t)$,
    $h(t)=(1/4)\,(e^{2t}+e^{-2t}-2)U(t)$
  k) $x=e^{-t}\sin t * e^{-t^2}=\int_0^t e^{-u}\sin u\,e^{-(t-u)^2}\,du$
2. a) $1/(s+2),\ e^{-2t}U(t)$ **c)** $(s^2+2s+2)^{-1},\ e^{-t}\sin t\,U(t)$
  e) $(s^3+6s^2+11s+6)^{-1},\ (1/2)(e^{-t}-2e^{-2t}+e^{-3t})U(t)$
3. a) $u=u_0e^{-kt}+\int_0^t e^{-kv}g(t-v)dv$
4. a) $\theta=\theta_0\cos\omega t+(\theta_0'/\omega)\sin\omega t+(I\omega)^{-1}\int_0^t g(u)\sin\omega(t-u)du$,
    where $\omega=(k/I)^{1/2}$
5. b) $y=(5/2)-(e^{-2t}/2)-2\cos t-\sin t$
6. a) $x=5e^{-3t}U(t)$ **c)** $x=(e^{2t}+e^{-2t})U(t)$,
  e) $x=(-1/2)(e^t+3e^{-t}+6e^{-2t})U(t)$ **g)** $x=2\cos 3t-14\sin 3t+5\delta(t)$
7. Impulse response: **a)** $x=e^{-3t}U(t)$ **c)** $x=(1/4)(e^{2t}-e^{-2t})U(t)$
  e) $x=(1/3)\sin 3t\,U(t)$ Step response: **a)** $x=(1/3)(1-e^{-3t})U(t)$
  c) $(1/8)(e^{2t}+e^{-2t}-2)U(t)$ **e)** $(1/9)(1-\cos 3t)U(t)$

SECTION 4–17, PAGE 230

1. a) $f(t)=1/2$ for $-3<t<3$, $f(t)=0$ otherwise
  c) $f(t)=-e^{3t}$ for $t<0$, $f(t)=0$ for $t>0$
  e) $f(t)=-3ie^{(1-i)t}$ for $t<0$, $f(t)=-7ie^{(-1-i)t}$ for $t>0$
  g) $f(t)=-t^2e^{3t}$ for $t<0$, $f(t)=-5t^2e^{-3t}/2$ for $t>0$
  i) $f(t)=(1/3)(4e^{-4t}-e^{-t})$ for $t>0$, $f(t)=0$ for $t<0$
  k) $f(t)=(1/5)(e^{-3t/2}-e^{-2t/3})$ for $t>0$, $f(t)=0$ for $t<0$
  m) $f(t)=2e^{-t}(\cos t-1)$ for $t>0$, $f(t)=0$ for $t<0$.
2. a) $\dfrac{1}{2\pi}\displaystyle\int_{-\infty}^{\infty}\dfrac{e^{-\omega^2}}{\omega^2+1}e^{i\omega t}d\omega$ **c)** $\dfrac{1}{2\pi}\displaystyle\int_{-\infty}^{\infty}\dfrac{\cos\omega\,e^{i\omega t}}{\omega^2+3\omega+11}d\omega$
3. a) $1/2$ for $-1<t<1$, $1/4$ for $t=\pm 1/2$, 0 otherwise   **c)** $1/(t^2+1)$

SECTION 4–19, PAGE 234

1. a) $e^{-t}-e^{-2t}$ for $t\geqslant 0$, 0 for $t<0$ **c)** $(1/2)e^{-t}$
2. a) $\dfrac{1}{2\pi}\displaystyle\int_{-\infty}^{\infty}\pi e^{-|u|}\dfrac{2}{(u-\omega)^2+1}\,du$

SECTION 4–21, PAGE 242

1. a) $\dfrac{e^{3t}}{30}$ **c)** $\dfrac{6t-5}{36}$ **e)** $\dfrac{e^{2t}}{16}$ for $t<0$, $\dfrac{4-3e^{-2t}-4te^{-2t}}{16}$ for $t\geqslant 0$

2. a) $\dfrac{1}{2\pi}\displaystyle\int_{-\infty}^{\infty}\dfrac{2\sin\omega\,e^{i\omega t}}{\omega(-\omega^2+2i\omega+3)}\,d\omega$ **c)** $\dfrac{1}{2\pi}\displaystyle\int_{-\infty}^{\infty}\dfrac{-\pi i\omega e^{-2|\omega|}\,e^{i\omega t}}{4(-\omega^2+i\omega+1)}\,d\omega$

e) $\dfrac{1}{2\pi}\displaystyle\int_{-\infty}^{\infty}\dfrac{[e^{2(5-i\omega)}-e^{-2(5-i\omega)}]e^{i\omega t}}{(5-i\omega)(-i\omega^3-3\omega^2+4i\omega+2)}\,d\omega$

g) $\displaystyle\int_{-\infty}^{\infty}\dfrac{e^{i\omega t}}{(-\omega^2+2i\omega+1)(\omega^4+1)}\,d\omega$

3. a) $(-10\cos 5t-11\sin 5t)/221$ **c)** $1/3$ **e)** $3(e^{-t}-e^{-3t})U(t)/2$

4. b) $Y(s)=(LCs^2+R_2Cs+1)[LCR_1s^2+(L+R_1R_2C)s+R_1+R_2]^{-1}$

SECTION 4–22, PAGE 245

2. a) π

=Chapter 5=

SECTION 5–4, PAGE 254

1. a) A: 2 and 1, F: 2 and 3, H: 1 and 3, L: 3 and 3, P: 3 and 2

b) $a_{11}=1$, $a_{21}=3$, $c_{21}=4$, $c_{22}=1$, $d_{12}=-1$, $e_{21}=2$, $f_{11}=1$, $g_{23}=-1$, $g_{21}=-1$, $h_{12}=0$, $m_{23}=1$

c) C: $(2,3)$ and $(4,1)$, G: $(3,1,4)$ and $(-1,0,-1)$, L: $(3,1,0)$, $(2,5,6)$ and $(1,4,3)$, P: $(2,2)$, $(-1,-1)$, $(3,3)$

d) D: col $(1,2)$ and col $(-1,0)$, F: col $(1,2)$, col $(4,0)$ and col $(5,7)$, L: col $(3,2,1)$, col $(1,5,4)$ and col $(0,6,3)$, N: col $(1,0,7)$ and col $(4,3,1)$

2. a) $\begin{bmatrix}3\\3\\3\end{bmatrix}$ **c)** meaningless **e)** $\begin{bmatrix}-1&2\\1&4\\4&-2\end{bmatrix}$

g) $\begin{bmatrix}10&15\\20&5\end{bmatrix}$ **i)** $\begin{bmatrix}7&2\\14&12\end{bmatrix}$ **k)** meaningless

3. a) $D-C=\begin{bmatrix}-1&-4\\-2&-1\end{bmatrix}$

4. a) $X=\tfrac{1}{2}(N+P)=\begin{bmatrix}\frac{3}{2}&3\\-\frac{1}{2}&1\\5&2\end{bmatrix}$, $\qquad Y=\tfrac{1}{2}(N-P)=\begin{bmatrix}-\frac{1}{2}&1\\\frac{1}{2}&2\\2&-1\end{bmatrix}$

6. $W = 20$ lbs, $N = 10\sqrt{3}$ lbs

8. $I_1 = -30/31$ amp, $I_2 = 20/31$ amp, $I_3 = -50/31$ amp

9. a) 4   c) -76 e) 30

10. a) $x = 2$, $y = 1$ c) $x = 1$, $y = 1$, $z = -1$ e) $x = 0$, $y = 0$
   g) $x = 2z/3$, $y = -5z/3$ i) $x = 0$, $y = 0$, $z = 0$

SECTION 5–5, PAGE 260

1. a) meaningless   c) meaningless   e) $\begin{bmatrix} 8 & 16 \\ 6 & 12 \end{bmatrix}$ and $\begin{bmatrix} 10 & 5 \\ 20 & 10 \end{bmatrix}$

   g) L i) O k) $\begin{bmatrix} 3 & 15 \\ 44 & 29 \\ 22 & 19 \end{bmatrix}$ m) (7, 10, 5)

   o) N (O is 3×2)   q) $25E$ s) meaningless

2. a) $\begin{bmatrix} 3 & 1 & 9 & 1 & 9 \\ 5 & 2 & 16 & 2 & 16 \end{bmatrix}$ **3.** a) $y_1 = 17x_1 + x_2$, $y_2 = 31x_1 + 7x_2$

4. b) 1.919, 2.599, 2.839   **7.** It is true precisely when $AB = BA$

SECTION 5–6, PAGE 266

1. a) $\dfrac{1}{2}\begin{bmatrix} 4 & -5 \\ -2 & 3 \end{bmatrix}$ b) $\dfrac{1}{17}\begin{bmatrix} 6 & -7 \\ -1 & 4 \end{bmatrix}$ c) $\begin{bmatrix} 3 & -1 & 2 \\ -2 & 1 & -1 \\ -2 & 1 & -2 \end{bmatrix}$

2. a) $\dfrac{1}{2}\begin{bmatrix} -5 & 4 \\ 3 & -2 \end{bmatrix}$ c) $\dfrac{1}{34}\begin{bmatrix} 214 & -279 \\ -64 & 89 \end{bmatrix}$ e) $\frac{1}{2}\,\mathrm{col}\,(1, 1)$

   g) $\dfrac{1}{17}\begin{bmatrix} -29 & 36 & 33 \\ 19 & -6 & -14 \end{bmatrix}$ i) $(-7, 4, -6)$

3. a) $A^2 B$ c) B^4

8. All matrices whose inverses appear in the answers are assumed to be nonsingular.

   a) $X = \frac{1}{2}(A + B)$, $Y = \frac{1}{2}(A - B)$

   c) $X = B - A(A - C)^{-1}(B - D)$, $Y = (A - C)^{-1}(B - D)$

   e) $X = (B^{-1}A - E^{-1}D)^{-1}(B^{-1}C - E^{-1}F)$, $Y = (A^{-1}B - D^{-1}E)^{-1}$
   $(A^{-1}C - D^{-1}F)$

9. a) $u_1 = 10/24$, $u_2 = 11/24$, $u_3 = 58/24$, $u_4 = 29/24$

SECTION 5–7, PAGE 272

Throughout, k is an arbitrary nonzero real scalar; c is an arbitrary nonzero complex scalar.

1. a) $\lambda = 1$, $k(1, -2)$ and $\lambda = 5$, $k(1, 2)$
   c) $\lambda = 1$, $k(0, 2, 1)$; $\lambda = 2$, $k(1, 2, 0)$; $\lambda = 3$, $k(1, 1, -1)$

2. a) $C = \dfrac{1}{4}\begin{bmatrix} 2 & -1 \\ 2 & 1 \end{bmatrix}$, $\quad B = \begin{bmatrix} 1 & 0 \\ 0 & 5 \end{bmatrix}$

   c) $C = \begin{bmatrix} -2 & 1 & -1 \\ 3 & -1 & 2 \\ -2 & 1 & -2 \end{bmatrix}$, $\quad B = \begin{bmatrix} 1 & 0 & 0 \\ 0 & 2 & 0 \\ 0 & 0 & 3 \end{bmatrix}$

3. a) $\lambda = 1 + 2i$, $c(1, -2i)$ and $\lambda = 1 - 2i$, $c(1, 2i)$,

   $$C = \frac{1}{4i}\begin{bmatrix} 2i & -1 \\ 2i & 1 \end{bmatrix}, \quad B = \begin{bmatrix} 1 + 2i & 0 \\ 0 & 1 - 2i \end{bmatrix}$$

   c) $\lambda = 0$, $c(1, 0, 0)$; $\lambda = i$, $c(0, 2, 1 - i)$; $\lambda = -i$, $c(0, 2, 1 + i)$;

   $$C = \frac{1}{4i}\begin{bmatrix} 1 & 0 & 0 \\ 0 & 1 + i & -2 \\ 0 & -1 + i & 2 \end{bmatrix}, \quad B = \begin{bmatrix} 0 & 0 & 0 \\ 0 & i & 0 \\ 0 & 0 & -i \end{bmatrix}$$

4. a) $\lambda = 1$, all nonzero vectors   c) $\lambda = -1$, $k(1, 1)$

SECTION 5–9, PAGE 280

1. a) $(4, 2, 2, 2)$, $(8, 6, 2, -2)$, $(6, 4, 2, 0)$, $(-3, 0, -3, -6)$, **0**
   c) 4, 24, $\sqrt{14}$, $\sqrt{6}$
2. c) $(6, 2, -1, 3, 6)$
5. a) dependent   c) dependent
6. a) $a = 3$, $b = 0$, $c = -2$
8. a) a basis   c) not a basis   e) not a basis

SECTION 5–10, PAGE 287

2. a) $(2, 3)$, $(1, 5)$, $(3, 1)$, $(-1, 2)$ c) Range is R^2; T maps R^2 onto R^2.
3. a) $(2, 4)$, $(3, 6)$, $(-1, -2)$, $(-5, -10)$
   c) Range is all $t(1, 2)$; T does not map R^2 onto R^2.
4. a) $n - 3$, $m - 2$ c) Range is R^3; T maps R^3 onto R^2.

5. a) $n = 3$, $m = 2$ c) Range is all $t(1, 2)$; T does not map R^3 onto R^2.

6. a) $n = 2$, $m = 3$ b) All $t(1, -2)$, not one-to-one.

7. a) $n = 2$, $m = 3$
   c) Range is all $t_1(2, 1, 1) + t_2(1, 2, 2)$; T does not map R^2 onto R^3.

8. a) $n = 3$, $m = 3$
   c) Range is all $t_1(3, 1, 5) + t_2(1, 0, 2)$; T does not map R^3 onto R^3.

9. a) $n = 3$, $m = 3$ c) Range is R^3; T maps R^3 onto R^3.

10. Angle is $\pi/4$, $\|x\| = \|T(x)\|$.   **12.** a) Reflection in origin.

SECTION 5–11, PAGE 293

1. a) $r = 2$, $k = 0$ c) $r = 3$, $k = 0$ e) $r = 1$, $k = 1$ g) $r = 3$, $k = 1$

2. Ranges: a) R^2 c) t col $(1, 2, 3)$

3. Kernels: a) $(0, 0)$ c) t col $(5, -1)$

4. a) yes   c) yes

SECTION 5–14, PAGE 300

The answers can be given in different forms. Problem 4 indicates how to compare different forms.

1. a) $x_1 = 3$, $x_2 = 2$ c) $x_1 = 1$, $x_2 = 5$, $x_3 = 1$
   e) none   f) $x_1 = 3$, $x_2 = 3$, $x_3 = 1$
   h) $x_1 = 2 - 3x_2$, $x_3 = 1$, $x_4 = 0$
   j) $x_1 = (1/10)(7 + 2x_3 - 2x_4)$, $x_2 = (1/10)(-1 + 4x_3 - 4x_4)$
   l) none   n) $x_1 = -1 - 2x_2$, $x_3 = 2 + 3x_4$, $x_5 = 3$
   p) $x_2 = (1/2)(-1 + x_1 + x_4 - 2x_5)$, $x_3 = (1/2)(-3 + 5x_1 + 3x_4)$

2. a) $\dfrac{1}{9}\begin{bmatrix} -2 & 3 \\ 5 & -3 \end{bmatrix}$, rank 2   c) no inverse, rank 2

   e) $\dfrac{1}{2}\begin{bmatrix} -1 & 2 & 1 \\ 2 & -2 & -2 \\ 1 & -2 & 1 \end{bmatrix}$, rank 3   g) $\dfrac{1}{7}\begin{bmatrix} 1 & -3 & 4 & 1 \\ 2 & 1 & 1 & -5 \\ -4 & 5 & -2 & 3 \\ 3 & -2 & -2 & 3 \end{bmatrix}$

   i) no inverse, rank 3   k) $\begin{bmatrix} 1 & -1 & 1 & 0 & -2 \\ 0 & 1 & -2 & -\frac{1}{2} & \frac{11}{2} \\ 0 & 0 & 1 & -\frac{1}{2} & -\frac{1}{2} \\ 0 & 0 & 0 & \frac{1}{2} & -\frac{3}{2} \\ 0 & 0 & 0 & 0 & 1 \end{bmatrix}$

3. a) col $\left(-\frac{3}{2}, 2, \frac{1}{2}\right)$, col $\left(-\frac{9}{2}, 5, \frac{3}{2}\right)$, col $\left(\frac{1}{2}, 0, \frac{1}{2}\right)$

SECTION 5–16, PAGE 308

1. a) $\begin{bmatrix} 1 & 3 \\ 2 & 0 \\ 3 & 5 \end{bmatrix}$ c) col (1, 5, 0, 4) 2. a) $a = 1$

3. a) $\begin{bmatrix} 5 & 2 \\ 2 & 3 \end{bmatrix}$ c) $\begin{bmatrix} 1 & 2 & 3 \\ 2 & 3 & 1 \\ 3 & 1 & -1 \end{bmatrix}$

SECTION 5–18, PAGE 316

1. a) dependent   c) independent   e) independent
2. a) infinite   c) 2, basis $\cos 3x$, $\sin 3x$ e) infinite   g) infinite
3. c) diag $(1, 0, 0)$, diag $(0, 1, 0)$, diag $(0, 0, 1)$
5. a) range is U, kernel is function $u \equiv 0$, T is one-to-one;
   c) range is all $v(x) = b_0 x^3 + b_1 x^2 + b_2 x + b_3$ such that $b_0 - b_1 + b_2 - b_3 = 0$; kernel is all multiples of $x^3 - x^2 + x - 1$, mapping T is not one-to-one;
   e) range is all functions $v(x)$ with continuous derivative and such that $v(0) = 0$; kernel is function $u(x) \equiv 0$; mapping T is one-to-one.

=========================Chapter 6=========================

SECTION 6–2, PAGE 327

2. a) $\dfrac{dx_1}{dt} = x_2$, $\dfrac{dx_2}{dt} = 3x_2 - 5x_1 + 5 \sin 2t$

   c) $\dfrac{dx_1}{dt} = 5x_1 - 2x_2$, $\dfrac{dx_2}{dt} = x_3$, $\dfrac{dx_3}{dt} = -3x_3 - 2x_1 - 4x_2$

   e) $\dfrac{dx_1}{dt} = -\dfrac{D}{m} + g \sin x_2$, $\dfrac{dx_2}{dt} = -\dfrac{L}{mx_1} + \dfrac{g}{x_1} \cos x_2$, $\dfrac{dx_3}{dt} = x_4$, $\dfrac{dx_4}{dt} = -\dfrac{M}{I}$

3. a) independent   c) dependent
5. a) $\mathbf{x} = 2\mathbf{x}_1(t) - 3\mathbf{x}_2(t)$ c) $\mathbf{x} = \mathbf{x}_1(t) - 2\mathbf{x}_2(t) + \mathbf{x}_3(t)$

6. a) $\mathbf{x} = \begin{bmatrix} 5e^t \\ 2e^t \end{bmatrix} + c_1 \begin{bmatrix} 2e^{3t} \\ 5e^{3t} \end{bmatrix} + c_2 \begin{bmatrix} e^{4t} \\ 3e^{4t} \end{bmatrix}$

SECTION 6–4, PAGE 334

REMARK: The answers can be given in different forms. It can be shown that both expressions

$$\mathbf{x} = \mathbf{x}^*(t) + c_1 \mathbf{x}_1(t) + \cdots + c_n \mathbf{x}_n(t),$$
$$\mathbf{x} = \mathbf{y}^*(t) + c_1 \mathbf{y}_1(t) + \cdots + c_n \mathbf{y}_n(t)$$

represent the same general solution of an equation (6–40) if $\mathbf{x}^*(t)$ and $\mathbf{y}^*(t)$ are both particular solutions and if $X(t) = Q Y(t)$, where $X(t)$ is the (fundamental) matrix whose column vectors are $\mathbf{x}_1(t), \ldots, \mathbf{x}_n(t)$, $Y(t)$ is the matrix whose column vectors are $\mathbf{y}_1(t), \ldots, \mathbf{y}_n(t)$, and Q is a nonsingular constant $n \times n$ matrix (see CLA, vol. 2, p. 1160, Problem 2).

1. a) $c_1 e^{5t}\mathrm{col}(1, 2) + c_2 e^{-5t}\mathrm{col}(1, -3)$
   c) $c_1 e^t\mathrm{col}(1, -2, -1) + c_2 e^{-t}\mathrm{col}(1, 0, -1) + c_3 e^{2t}\mathrm{col}(4, -3, -1)$
   e) $c_1\mathrm{col}(2\cos 3t, \cos 3t + 3\sin 3t) + c_2\mathrm{col}(2\sin 3t, \sin 3t - 3\cos 3t)$

2. a) $-(1/4)e^t\mathrm{col}(1, 2) + c_1 e^{5t}\mathrm{col}(1, 2) + c_2 e^{-5t}\mathrm{col}(1, -3)$
   c) $(1/2)\mathrm{col}(4, -5 - 2t, -5) + c_1 e^t\mathrm{col}(1, -2, -1) + c_2 e^{-t}\mathrm{col}(1, 0, -1)$
   $+ c_3 e^{2t}\mathrm{col}(4, -3, -1)$
   e) $(1/8)\mathrm{col}(\sin t - \cos t, 4\sin t - \cos t) + c_1\,\mathrm{col}(2\cos 3t, \cos 3t + 3\sin 3t) +$
   $c_2\,\mathrm{col}(2\sin 3t, \sin 3t - 3\cos 3t)$

3. a) $(1/4)(e^{5t} - e^t)\mathrm{col}(1, 2)$
   c) $(1/6)\mathrm{col}(12 - 9e^t - 7e^{-t} + 4e^{-2t}, -15 - 6t + 18e^t - 3e^{2t},$
   $-15 + 9e^t + 7e^{-t} - e^{2t})$

4. a) $x = c_1 \sin(\omega_1 t + \alpha_1) + c_2 \sin(\omega_2 t + \alpha_2),$
   $y = (\omega_1^2 - 1)c_1 \sin(\omega_1 t + \alpha_1) + (\omega_2^2 - 1)c_2 \sin(\omega_2 t + \alpha_2),$
   where $\omega_1 = (3 + 5^{1/2})^{1/2}$, $\omega_2 = (3 - 5^{1/2})^{1/2}$

SECTION 6–5, PAGE 337

4. $\dfrac{d}{dt}\begin{bmatrix} I \\ q_1 \\ q_2 \end{bmatrix} = A\begin{bmatrix} I \\ q_1 \\ q_2 \end{bmatrix} + \begin{bmatrix} \mathscr{E}/L \\ 0 \\ 0 \end{bmatrix}$, $A = \dfrac{1}{RC_1 C_2}\begin{bmatrix} 0 & -RC_2/L & 0 \\ RC_1 C_2 & -C_2 & C_1 \\ 0 & C_2 & -C_1 \end{bmatrix}$

SECTION 6–6, PAGE 343

1. a) $\mathbf{x} = \mathrm{col}(2e^{3t}, e^{3t})$ c) $\mathbf{x} = e^{-t}\mathrm{col}(1 - 6t, 1 + 3t)$
   e) $\mathbf{x} = (e^t/2)\mathrm{col}(2t + 5t^2, -5t^2, 4t)$
   g) with $a = \sqrt{2}$, $x_1 = (1/8)[4t - 4 + (2 + a)e^{at} + (2 - a)e^{-at}]$,
   $x_2 = (1/8)[4t - 4 - (2 + a)e^{at} - (2 - a)e^{-at}] + e^t$

2. a) $\mathbf{x} = e^t\mathrm{col}[c_1(\cos t - 2\sin t) + c_2\sin t, -5c_1\sin t + c_2(2\sin t + \cos t)]$
   c) $\mathbf{x} = c_1\mathrm{col}(6e^{3t} - 5e^{5t}, 3e^{3t} - 3e^{5t}) + c_2\mathrm{col}(10e^{5t} - 10e^{3t}, 6e^{5t} - 5e^{3t})$
   or $\mathbf{x} = c_1 e^{5t}\mathrm{col}(5, 3) + c_2 e^{3t}\mathrm{col}(2, 1)$

SECTION 6-9, PAGE 350

2. a) $x = c_1 e^t \text{col}(1, 1) + c_2 e^{-t} \text{col}(3, 5)$
   c) $x = c_1 e^t \text{col}(1, 2, 3) + c_2 e^{2t} \text{col}(1, 3, 5) + c_3 e^{3t} \text{col}(1, -1, -2)$
   e) $x = c_1 e^{3t} \text{col}(t+1, -t) + c_2 e^{3t} \text{col}(1, -1)$

3. a) $x = c_1 e^{5t} \text{col}(2, 1) + c_2 \text{col}(1, -2)$
   c) $x = c_1 \text{col}(1, -2, 2) + c_2 e^{9t} \text{col}(2, 1, 0) + c_3 e^{9t} \text{col}(0, 1, 1)$

4. a) $x_1 = A_1 \sin(1.47t + \alpha_1)$, $x_2 = -6.6A_1 \sin(1.47t + \alpha_1)$ and
   $x_1 = A_2 \sin(0.59t + \alpha_2)$, $x_2 = 0.6A_2 \sin(0.59t + \alpha_2)$

5. b) $x = c_1 \sin(\alpha t + \varepsilon_1) \text{col}(1, \sqrt{2}, 1) + c_2 \sin(\beta t + \varepsilon_2) \text{col}(1, 0, -1)$
   $+ c_3 \sin(\gamma t + \varepsilon_3) \text{col}(1, -\sqrt{2}, 1)$,
   where $\alpha = (2 - \sqrt{2})^{1/2}$, $\beta = \sqrt{2}$, $\gamma = (2 + \sqrt{2})^{1/2}$

SECTION 6-10, PAGE 356

1. a) $J_1 = I_1$, $J_2 = I_2 = I_6$, $J_3 = I_3$, $I_4 = J_1 - J_2$, $I_5 = J_2 - J_3$
   b) $[L_1 D^2 + R_4 D + (1/C_1)]Q_1 - R_4 D Q_2 = \mathscr{E}_1$,
   $-R_4 D Q_1 + [L_2 D^2 + (R_4 + R_5)D + (1/C_2)]Q_2 - R_5 D Q_3 = 0$,
   $-R_5 D Q_2 + [L_3 D^2 + (R_3 + R_5)D + (1/C_3)]Q_3 = \mathscr{E}_3$

3. b) $2T = L_1 Q_1'^2 + L_2 Q_2'^2 + L_7 Q_3'^2 + (L_4 + L_7)Q_4'^2 - 2L_7 Q_3' Q_4'$,
   $2V = (Q_1^2/C_1) + Q_2^2[(1/C_2) + (1/C_8)] + (Q_4^2/C_8) - 2(Q_2 Q_4/C_8)$,
   $2F = (R_1 + R_5 + R_6)Q_1'^2 + (R_2 + R_5)Q_2'^2 + R_6 Q_3'^2 - 2R_5 Q_1' Q_2' - 2R_6 Q_1' Q_3'$

SECTION 6-12, PAGE 364

1. a) stable   b) $(s^2 + 3s + 2)^{-1} \begin{bmatrix} s-2 & 4 \\ -3 & s+5 \end{bmatrix}$

   c) $(2 - \omega^2 + 3i\omega)^{-1} \begin{bmatrix} i\omega - 2 & 4 \\ -3 & i\omega + 5 \end{bmatrix}$

2. a) $\dfrac{1}{10} \text{col}(3e^{3t}, e^{3t})$ c) $\dfrac{1}{5} \text{col}(14 \sin 2t - 8 \cos 2t, 9 \sin 2t - 3 \cos 2t)$

   e) $\displaystyle\sum_{n=-\infty}^{\infty} e^{int}(n^2+1)^{-1}(n^4 + 5n^2 + 4)^{-1} \begin{bmatrix} 20 - 7n^2 + i(-n^3 - 28n) \\ 24 - 3n^2 + i(-3n^3 - 30n) \end{bmatrix}$

3. a) $e^{-t} \begin{bmatrix} -3 & 4 \\ -3 & 4 \end{bmatrix} + e^{-2t} \begin{bmatrix} 4 & -4 \\ 3 & -3 \end{bmatrix}$

   b) $\begin{bmatrix} e^{-t} * (-3f_1 + 4f_2) + e^{-2t} * (4f_1 - 4f_2) \\ e^{-t} * (-3f_1 + 4f_2) + e^{-2t} * (3f_1 - 3f_2) \end{bmatrix}$

4. a) stable   b) $(s^3 + 3s^2 + 4s + 2)^{-1}$ $\begin{bmatrix} s^2 + 9s + 8 & -2 - 4s & 5 + 5s \\ 0 & s^2 + 2s + 2 & 0 \\ -10 - 10s & 4 + 6s & s^2 - 5s - 6 \end{bmatrix}$

5. a) $(1/10)e^t \, \mathrm{col}\,(4,\ 5,\ 0)$

   c) $(16354)^{-1}\,[\cos 5t \, \mathrm{col}\,(3268, 4403, -3450) + \sin 5t \, \mathrm{col}\,(220, 22015, 14080)]$

   e) $\sum e^{int}(n^4 + 1)^{-1}[2 - 3n^2 + i(4n - n^3)]^{-1}\,\mathrm{col}\,(45 - 3n^2 + 44in,\ 2 - 4n^2 + 4in,$
    $-52 - 3n^2 - 43in)$

6. a) $W = e^{-t} \begin{bmatrix} 7\sin t + \cos t & -2\cos t - 4\sin t + 2 & 5\sin t \\ 0 & 1 & 0 \\ -10\sin t & -2 + 2\cos t + 6\sin t & -7\sin t + \cos t \end{bmatrix}$

7. b) $X_1(s) = \sqrt{3}(as^2 + 2a + b)[Q(s)]^{-1}$, $X_2(s) = \sqrt{3}(bs^2 + 2b + a)[Q(s)]^{-1}$,
   where $Q(s) = (s^2 + 1)(s^2 + 3)^2$; no resonance for $b = a$

=====Chapter 7=====

SECTION 7–3, PAGE 384

2. a) unstable focus   c) unstable saddle point   e) neutrally stable center

3. a) center   c) stable focus

5. a) saddle point, unstable   c) stable focus

6. a) $(2, 1)$ and $(-2, -1)$, both unstable

7. a) stable focus   c) stable focus

SECTION 7–6, PAGE 394

1. a) $x = 3c\cos\omega t$, $y = 2c\cos\omega t + c\omega\sin\omega t$, $\omega = \sqrt{5}$, $c > 0$

   c) $y^2 = \cos x + c$, $|x - 2n\pi| \leqslant \cos^{-1}(-c)$, $n = 0, \pm 1, \ldots, -1 < c < 1$,
   (t obtainable by integration)

3. a) $\lambda = -8$, stable

SECTION 7–7, PAGE 397

2. b) saddle at $(0, 0)$, center at $(a_2/k_2, a_1/k_1)$

3. saddle at $(0, 0)$, $(2, 0)$, $(0, 3)$, stable focus at $(1, 1)$

SECTION 7–9, PAGE 404

1. $128a_1 + 3(a_0^2 - b_0^2)(3a_0^2 b_0 - b_0^3) - 6a_0 b_0(a_0^3 - 3a_0 b_0^2)$
$+ 96(a_0^2 b_1 + 3b_0^2 b_1 + 2a_0 a_1 b_0) = 0$
$128b_1 - 3(a_0^2 - b_0^2)(a_0^3 - 3a_0 b_0^2) - 6a_0 b_0(3a_0^2 b_0 - b_0^3)$
$- 96(3a_0^2 a_1 + 2a_1 b_0^2 + 2a_0 b_0 b_1) = 0.$

2. a) $a_0 \cos t + b_0 \sin t$, where $4a_0 - a_0^3 - a_0 b_0^2 = 0$, $4b_0 - b_0^3 - a_0^2 b_0 = 0$.

=Chapter 8=

SECTION 8–3, PAGE 416

2. a) elliptic   c) parabolic   e) elliptic   g) elliptic

3. a) $\dfrac{3}{4}\sin x \cos t - \dfrac{1}{4}\sin 3x \cos 3t$ c) $y = \dfrac{8}{\pi}\sum\limits_{n=1}^{\infty}(4n^2 - 1)^{-2}\sin 2nx \sin 2nt$

4. a) $(1/2)(1 + \cos 2x \cos 2t)$

SECTION 8–5, PAGE 424

All summations go from 1 to ∞.

6. a) $y = \dfrac{1}{2}(p_0 + q_0 t) + \sum[\cos nx(p_n \cos nct + q_n \sin nct)$
$+ \sin nx\,(P_n \cos nct + Q_n \sin nct)],$
where
$$P_n = \pi^{-1}\int_0^{2\pi} f(x)\cos nx\,dx, \quad n = 0, 1, \ldots,$$
$$P_n = \pi^{-1}\int_0^{2\pi} f(x)\sin nx\,dx, \quad n = 1, 2, \ldots,$$
$$q_0 = \pi^{-1}\int_0^{2\pi} g(x)dx, \quad q_n = (nc\,\pi)^{-1}\int_0^{2\pi} g(x)\cos nx\,dx \text{ for } n = 1, 2, \ldots,$$
$$Q_n = (nc\,\pi)^{-1}\int_0^{2\pi} g(x)\sin nx\,dx, \quad n = 1, 2, \ldots$$

b) $y = \sum\left[p_n \sin\dfrac{(2n-1)x}{2}\cos\dfrac{(2n-1)ct}{2} + q_n \sin\dfrac{(2n-1)x}{2}\sin\dfrac{(2n-1)ct}{2}\right],$
where for $n = 1, 2, \ldots$
$$p_n = \dfrac{2}{\pi}\int_0^{\pi} f(x)\sin\dfrac{(2n-1)x}{2}dx, \quad q_n = \dfrac{4}{\pi c(2n-1)}\int_0^{\pi} g(x)\sin\dfrac{(2n-1)x}{2}dx$$

c) $y = \sum (p_n \sin \omega_n x \cos c\omega_n t + q_n \sin \omega_n x \sin c\omega_n t)$, where $\omega_1, \omega_2, \ldots$ are the successive positive roots of the equation $\tan \omega + \omega = 0$,

$$p_n = k_n^{-1} \int_0^1 f(x) \sin \omega_n x \, dx,$$

$$q_n = (k_n \omega_n c)^{-1} \int_0^1 g(x) \sin \omega_n x \, dx,$$

$$k_n = \int_0^1 \sin^2 \omega_n x \, dx$$

7. $y(x, t) = \sum n^{-3} \sin nx \cos nct + \sum \dfrac{e^{-n} \sin nx}{n^4 + n^2 c^2} \left(e^{-n^2 t} - \cos nct + \dfrac{n \sin nct}{c} \right)$

8. $y(x, t) = \left(1 + \dfrac{x}{\pi}\right) \sin t + \dfrac{2}{\pi} \sum \dfrac{[2(-1)^n - 1][cn \sin t - \sin nct]}{n^2 c(1 - n^2 c^2)} \sin nx$

SECTION 8–7, PAGE 431

1. a) $e^{-9t} \sin 3x$ c) $(4/\pi) \sum_{n=1}^{\infty} (-1)^{(n-1)} (2n-1)^{-2} e^{-(2n-1)^2 c^2 t} \sin(2n-1)x$

2. a) $(8/\pi) \sum_{n=1}^{\infty} e^{-(2n-1)^2 t/4} (3 + 4n - 4n^2)^{-1} \cos(2n-1)x/2$

   c) $(e^{-t} - e^{-2t}) \sin x + \dfrac{1}{24}(e^{-2t} + 23 e^{-8t}) \sin 2x$

   $+ \sum_{n=3}^{\infty} (e^{-nt} - e^{-2n^2 t}) n^{-3} (2n-1)^{-1} \sin nx$

   e) $(x/\pi) \sin t - (2/\pi) \sum_{n=1}^{\infty} \{[n^4 + 2 - (-1)^n (2n^2 + 2)] e^{-n^2 t}$
   $+ (n^4 \cos t + n^2 \sin t)\} n^{-3} (1 + n^4)^{-1} (-1)^n \sin nx$

   g) $e^{-t}(1-x)\pi^{-1} + [e^{-t}(1 - 4\pi^{-1}) - te^{-t} + 2\pi^{-1} te^{-t}] \sin x$
   $+ 2 \sum_{n=2}^{\infty} [\pi n(n^2 - 1)]^{-1} \{e^{-t} + e^{-n^2 t}[2(-1)^n (n^2 - 1) - 2n^2]\} \sin nx$

3. a) $u_1(x, t) + u_3(x, t)$ c) $u_1(x, t) + \cdots + u_4(x, t)$

SECTION 8–11, PAGE 439

1. a) $\sin y \sinh(\pi - x)(\sinh \pi)^{-1} + 2 \sin 3y \sinh 3(\pi - x)(\sinh 3\pi)^{-1}$

   c) $2\pi^{-1} \sum_{n=1}^{\infty} \{[1 - (-1)^n] \sinh n\pi y$
   $+ (-1)^{n+1} \sinh n\pi(1 - y)\} n^{-1} (\sinh n\pi)^{-1} \sin n\pi x$
   $+ 2\pi^{-1} \sum_{n=1}^{\infty} \{[1 - (-1)^n] \sinh n\pi x$
   $+ (-1)^{n+1} \sinh n\pi(1 - x)\} n^{-1} (\sinh n\pi)^{-1} \sin n\pi y$

   e) $5r \sin \theta + r^2 \cos 2\theta$ g) $(1/2) + (2/\pi) \sum_{n=1}^{\infty} (2n-1)^{-1} r^{2n-1} \sin(2n-1)\theta$

2. a) $(2\pi)^{-1} \int_0^\pi \{(1 - r^2)/[1 + r^2 - 2r \cos(\varphi - \theta)]\} d\varphi$

3. $u = \sum_{n=1}^{\infty} b_n e^{-ny} \sin nx$, $b_n = (2/\pi) \int_0^\pi f(x) \sin nx \, dx$

SECTION 8–12, PAGE 443

1. $\sin x \sin t + (2/3) \sin 3x \sin 3t$

3. $f\left(t - \dfrac{x}{c}\right)$, where f is defined to be 0 for $t < 0$

5. $e^{-4t} \sin 2x - 7e^{-25t} \sin 5x$

7. $e^{t-x} - \dfrac{x}{2}\pi^{-1/2} \displaystyle\int_0^t e^{t-\tau} \tau^{-3/2} e^{-x^2/(4\tau)} d\tau$

9. $\pi^{-1} \displaystyle\int_{-\infty}^{\infty} y f(v)/[\,(x-v)^2 + y^2\,] dv$

SECTION 8–13, PAGE 450

1. $5e^{-10t} \sin x \sin 3y$

3. $\cos y\{(\pi^3/2)\cos t - (24/\pi)\sum_{n=1}^{\infty} n^{-4}[1-(-1)^n]\cos nx \cos (n^2+1)^{1/2}t\}$

5. $\sum_{n=1}^{\infty} p_n J_1(s_{\ln} r)\cos s_{\ln} t \sin \theta$, where $p_n = 2[J_2(s_{\ln})]^{-2}\int_0^1 (r-r^2)J_1^2(s_{\ln} r)dr$

==============================Chapter 9==============================

SECTION 9–3, PAGE 460

1. a) $\sqrt{11}$, $\sqrt{21}$, $\sqrt{2}$ b) $\cos^{-1} 6/\sqrt{42}$, $\cos^{-1} -4/\sqrt{22}$, $\cos^{-1} 15/\sqrt{231}$
   c) $c(\mathbf{i} + 2\mathbf{j} - \mathbf{k})$ d) $\sqrt{6}/2$

3. b) 3/7, 2/7, 6/7 and 2/3, 2/3, 1   **5.** b) $x - 2y + z + 2 = 0$

6. b) 147 000 erg   **8.** $5/\sqrt{266}$ **9.** c) $(1/3)\mathbf{i} + (5/4)\mathbf{j} + (17/6)\mathbf{k}$

SECTION 9–5, PAGE 470

1. a) $z_x = 4xy^2(x^2+y^2)^{-2}$, $z_y = -4x^2y(x^2+y^2)^{-2}$
   c) $z_{xx} = 12x^2 - 24xy$, $z_{yy} = 24xy - 12y^2$

2. a) 36.24   **3.** a) $z = 10x + 2y - 9$

5. a) $4xy(x^2+y^2)^{-2}[y(2t+3) - x(8t-5)]$

7. a) open region, boundary: points $(x, 0)$ for $x \geqslant 0$ and $(0, y)$ for $y \geqslant 0$
   c) open region, boundary: point $(0, 0)$ and points on $x^2 + y^2 = 1$
   e) not open region, open but not connected, boundary: points $(\pm 1, y)$

8. a) $w_x = 12x^2y^2z - 5y^3z^2$, $w_y = 8x^3yz - 15xy^2z^2$

SECTION 9–8, PAGE 477

1. a) 6, 10, 0   c) 0, $\sqrt{2}$, $-1/5$ **2.** a) $5/\sqrt{2}$ **5.** a) $-48/\sqrt{13}$

7. f has maximum value 3 at $(\pm 1, 0)$, minimum value -4 at $(0, \pm 1)$

8. a) $(3 - 5\sqrt{3})/2$ c) $10/3$

SECTION 9–9, PAGE 482

1. a) maximum at $(0, 0)$ c) saddle point at $(1, 1)$
   e) relative minimum at each point of the line $y = x$

2. a) max $= 1$ at $(0, 0)$ c) max $= \sqrt{3}$, min $= -\sqrt{3}$

3. a) positive definite   c) positive definite

4. 6   **5.** $(1/4, 1/4, 1/2)$ **6.** a) $(0, 0)$, stable   c) unstable

7. $z = 4$ at $(1/2, 1/2)$

SECTION 9–12, PAGE 494

1. a) $\begin{bmatrix} 2 & -3 \\ 1 & 2 \end{bmatrix}$ c) $\begin{bmatrix} x_2 x_3 & x_1 x_3 & x_1 x_2 \\ 2x_1 x_3 & 0 & x_1^2 \end{bmatrix}$

2. a) $9(x^2 + y^2)^2$ c) $2u^3 v^2 w$

3. a) $e^2 = 7.39$ b) 7.44 c) $du = edx, dv = edy, e^2 = 7.39$

5. Volume elements: a) $r\, dr\, d\theta\, dz$ b) $\rho^2 \sin\varphi\, d\rho\, d\varphi\, d\theta$

7. a) $\mathbf{n} = a(\cos u\, \mathbf{i} + \sin u\, \mathbf{j})$, $d\sigma = a\, du\, dv$
   c) $\mathbf{n} = a(b + a\cos v)(\cos u \cos v\, \mathbf{i} + \sin u \cos v\, \mathbf{j} + \sin v\, \mathbf{k})$,
   $d\sigma = a(b + a\cos v)\, du\, dv$

8. a) $\begin{bmatrix} x_2 - 3 & x_1 \\ 2x_2 + 2 & 2x_2 + 2x_1 - 1 \end{bmatrix} \begin{bmatrix} \cos 3u_2 & -3u_1 \sin 3u_2 \\ \sin 3u_2 & 3u_1 \cos 3u_2 \end{bmatrix}, \begin{bmatrix} -3 & 0 \\ 2 & 0 \end{bmatrix}$

   c) $\begin{bmatrix} e^{x_2} & x_1 e^{x_2} \\ e^{-x_2} & -x_1 e^{-x_2} \\ 2x_1 & 0 \end{bmatrix} \begin{bmatrix} 2u_1 & 1 \\ 4u_1 & -1 \end{bmatrix}, \begin{bmatrix} 6e^2 & 0 \\ -2e^{-2} & 2e^{-2} \\ 4 & 2 \end{bmatrix}$

SECTION 9–15, PAGE 503

1. $\left(\dfrac{\partial x}{\partial y}\right)_z = -\dfrac{5}{4}$, $\left(\dfrac{\partial y}{\partial x}\right)_u = -\dfrac{5}{7}$, $\left(\dfrac{\partial z}{\partial u}\right)_x = -\dfrac{5}{7}$, $\left(\dfrac{\partial y}{\partial z}\right)_x = \dfrac{1}{5}$

2. a) $du = \frac{1}{9}(dx + 3dy + 2dz)$, $dv = -\frac{1}{3}(dx + 2dz)$

   b) $\left(\dfrac{\partial u}{\partial x}\right)_{y,z} = \frac{1}{9}$, $\left(\dfrac{\partial v}{\partial y}\right)_{x,z} = 0$;   c) $u = 3.033$, $v = 2.1$

4. $\left(\dfrac{\partial x}{\partial u}\right)_v = 0,$ $\left(\dfrac{\partial x}{\partial v}\right)_u = \dfrac{\pi}{12}$ **6.** $\left(\dfrac{\partial u}{\partial x}\right)_y = \dfrac{3x}{u}$

7. a) $u = \frac{1}{5}(x+2y),$ $v = \frac{1}{5}(y-2x)$ b) $J = 5, J$ for inverse $= \frac{1}{5}$

8. a) $J = 4(u^2 + v^2)$ b) $\left(\dfrac{\partial u}{\partial x}\right)_y = \dfrac{u}{2(u^2+v^2)},$ $\left(\dfrac{\partial v}{\partial x}\right)_y = -\dfrac{v}{2(u^2+v^2)}$

11. a) $J = \rho^2 \sin \varphi$ b) $\dfrac{\partial \rho}{\partial y} = \sin\varphi \sin \theta,$ $\dfrac{\partial \varphi}{\partial z} = -\dfrac{\sin \varphi}{\rho},$ $\dfrac{\partial \theta}{\partial x} = -\dfrac{\sin \theta}{\rho \sin \varphi}$

15. a) $\frac{1}{2}, -1$ b) $-1, \frac{3}{2}$ c) $-1, -1$

SECTION 9–18, PAGE 509

1. a) $\dfrac{2x^2 - y^2}{(x^2+y^2)^{5/2}}, \dfrac{2y^2 - x^2}{(x^2+y^2)^{5/2}}$ b) $\dfrac{2xy}{(x^2+y^2)^2}, \dfrac{-2xy}{(x^2+y^2)^2}$

c) $4xe^{x^2-y^2}(2y^2 - 1),$ $-4ye^{x^2-y^2}(2x^2 + 1)$

6. $\dfrac{\partial z}{\partial x}\dfrac{\partial^2 x}{\partial u\, \partial v} + \dfrac{\partial z}{\partial y}\dfrac{\partial^2 y}{\partial u\, \partial v} + \dfrac{\partial^2 z}{\partial x^2}\dfrac{\partial x}{\partial u}\dfrac{\partial x}{\partial v} + \dfrac{\partial^2 z}{\partial x\, \partial y}\left(\dfrac{\partial x}{\partial u}\dfrac{\partial y}{\partial v} + \dfrac{\partial x}{\partial v}\dfrac{\partial y}{\partial u}\right) + \dfrac{\partial^2 z}{\partial y^2}\dfrac{\partial y}{\partial u}\dfrac{\partial y}{\partial v}$

10. $\dfrac{2(u^2 - y^2)}{(1 - 2ux)^3}(2u - 3u^2 x - xy^2)$

12. a) $\dfrac{dy}{du} = \dfrac{1}{y+u}$ b) $\dfrac{dv}{du} = \dfrac{2u-v}{u+v}$ c) $\dfrac{d^3 y}{dt^3} = 0$ d) $\dfrac{d^2 x}{dy^2} - 1 = 0$

e) $\dfrac{d^2 v}{dx^2} - x^2 v = 0$ f) $\dfrac{\partial u}{\partial w} = 0$ g) $\dfrac{\partial^2 u}{\partial z\, \partial w} = 0$

=Chapter 10=

SECTION 10–2, PAGE 517

1. a) $1/3$ b) $\pi/48$ c) $(-15\sqrt{2})/8$

SECTION 10–3, PAGE 522

1. a) $-\displaystyle\int_{\pi/2}^{\pi} \sin(xt)\, dx$ c) $\displaystyle\int_{1}^{2} \dfrac{1}{u}\, dx$

2. a) x^2 c) $-3t^2 \log(1+t^6)$

4. a) $\dfrac{-1}{(n+1)^2}$ c) $\dfrac{\pi}{2}\dfrac{1\cdot 3 \cdots (2n-3)}{2\cdot 4 \cdots (2n-2)}\dfrac{1}{x^{2n-1}}, x > 0$

SECTION 10–4, PAGE 527

1. a) $\dfrac{\pi}{2}$ b) $\dfrac{\pi^4}{3}$

3. a) $2\displaystyle\int_0^{1/2}\int_v^{1-v}\ln(1+2u^2+2v^2)\,du\,dv$ b) $\displaystyle\int_0^1\int_{1-2u}^1\sqrt{1+u^2(u+v)^2}\,dv\,du$

5. a) $\displaystyle\int_0^{2\pi}\int_0^1\int_0^1 r^4\cos^2\theta\sin\theta\,dz\,dr\,d\theta$

   b) $\displaystyle\int_0^{\pi/2}\int_0^1\int_0^{1+r(\cos\theta+\sin\theta)} r^3\cos 2\theta\,dz\,dr\,d\theta$

6. a) $\displaystyle\int_0^{2\pi}\int_0^\pi\int_0^a \rho^5\sin^4\varphi\cos^2\theta\sin\theta\,d\rho\,d\varphi\,d\theta$

   b) $\displaystyle\int_0^{2\pi}\int_0^{\pi/4}\int_0^{\sec\varphi} \rho^4\sin\varphi\,d\rho\,d\psi\,d\theta$

SECTION 10–5, PAGE 534

1. a) $8/3$ c) 0 **2.** a) $4/3$ c) $-\pi/2$ **3.** a) 0 c) $-1/4$

4. a) 0 c) $\dfrac{1}{4}[2\sqrt{5}+\ln(2+\sqrt{5})]$

6. a) $\dfrac{4}{3}$ c) $\dfrac{4}{3}$ **7.** a) 0 c) $\dfrac{4}{3}$

SECTION 10–8, PAGE 545

1. a) $(b-a)\times$ Area enclosed by C c) $3\pi/2$ e) 0

3. a) $-e^x\sin 2y$ c) 0 **4.** a) $(-e^x\cos 2y)\mathbf{k}$ c) $\mathbf{0}$

SECTION 10–10, PAGE 556

1. a) $F=x^2y$, integral $=1$ c) $F=-1/\sqrt{x^2+y^2}$, integral $=1-e^{-2\pi}$

2. a) $-\dfrac{10}{3}$ b) $\dfrac{1}{3}$ **3.** a) 0 c) -2

4. $\dfrac{\pi}{4}+2n\pi\,(n=0,\pm 1,\pm 2,\dots)$ **5.** a) $x^2y-\dfrac{1}{3}(y^3+2)$ **6.** $-\pi$

SECTION 10–13, PAGE 566

1. a) 3π b) $-5/3$ c) $1/2$ d) 0
5. a) $1/2$ b) π c) 2π d) $\pi/2$
7. a) π **8.** $6\,cA$ if **v** moves; $5.4\,cA$ if **v** is fixed

SECTION 10–16, PAGE 576

1. a) 4π b) 3   **3.** a) 6π b) 0
10. b) $\dfrac{\partial f}{\partial x} - \dfrac{\partial f}{\partial y}$ c) $2x^2\mathbf{i} + 2y^2\mathbf{j}$ **13.** $-2/3$
14. div **v** $= 0$, curl **v** $= 2\,\boldsymbol{\omega}$ **15.** Volume $= 1$ **16.** Volume $= e$

SECTION 10–17, PAGE 582

1. a) -2 b) 0   **2.** $\dfrac{\pi}{2} \pm 2n\pi$
3. b) $\mathbf{v} + \operatorname{grad} f$, where $\mathbf{v} = x^2y^2z\mathbf{i} - xy^3z\mathbf{j} + xy^2z^2\mathbf{k}$, or
$$\mathbf{v} = \frac{1}{6}[(3x^2y^2z - y^2z^3 + y^4z)\mathbf{i} - (2x^3yz + 2xy^3z - 2xyz^3)\mathbf{j}$$
$$+ (3xy^2z^2 + xy^4 - x^3y^2)\mathbf{k}]$$

SECTION 10–19, PAGE 588

3. $T = T_1 + \dfrac{T_2 - T_1}{d}\,x$

===Chapter 11===

SECTION 11–4, PAGE 599

1. a) $u = x^2 - y^2 - 2xy$, $v = x^2 - y^2 + 2xy$, all z
   c) $u = \tan x \operatorname{sech}^2 y\,[1 + \tan^2 x \tanh^2 y]^{-1}$
   $v = \tanh y \sec^2 x\,[1 + \tan^2 x \tanh^2 y]^{-1}$
   $z \ne (\pi/2) + n\pi,\ n = 0,\ \pm 1,\ \pm 2,\ \ldots$
   e) through (i) see (11–33) and (11–34), continuous for all z
2. a) $i(\pi + \sinh \pi)$ c) ∞
3. a) $3z^2 + 5$ c) $42[1 + (z^2 + 1)^3]^6 (z^2 + 1)^2 z$
5. b) $n\pi i$ and $[(\pi/2) + n\pi]i,\ n = 0,\ \pm 1,\ \ldots$

SECTION 11–5, PAGE 603

1. a) $\dfrac{2}{3}$ **c)** $\dfrac{1}{2}\ln 2 - i\dfrac{\pi}{4}$ **3. a)** $2\pi i$

SECTION 11–6, PAGE 608

2. a) Analytic nowhere   **c)** Analytic for all z
6. The functions are analytic except at the following points:
   a) $\frac{1}{2}\pi + n\pi$ **c)** $\frac{1}{2}\pi i + n\pi i$ **e)** $\frac{1}{2}\pi + n\pi$

SECTION 11–7, PAGE 611

1. a) $\exp(2k+1)\dfrac{i\pi}{3}$, $k = 0, 1, 2$ **c)** $\pm 2^{1/2}\exp(i\pi/3)$

   e) $0.693 + 2n\pi i$ **g)** $0.347 + i(\frac{7}{4}\pi + 2n\pi)$

   i) $\sqrt[3]{2}\exp\left(\dfrac{1}{6}\pi i + \dfrac{4n\pi}{3}i\right)$ **k)** $\frac{1}{2}\pi + 2n\pi$

   The range of n is $0, \pm 1, \pm 2, \ldots$, except in (i), where it is 0, 1, 2.
3. a) $n\pi$ and $(\pi/2) + n\pi$, $n = 0, \pm 1, \pm 2, \ldots$
6. a) $\ln r + i\theta$, $(\pi/2) + 2n\pi < \theta < (3\pi/2) + 2n\pi$, $n = 0, \pm 1, \pm 2, \ldots$
   b) $\sqrt[3]{r}\exp(i\theta/3)$, $-(\pi/2) + 2n\pi < \theta < (\pi/2) + 2n\pi$, $n = 0, 1, 2$

SECTION 11–9, PAGE 617

1. a) 0   **c)** $(2i-1)e^{2i}$ **2. a)** πi
3. a) $a_1 + a_2 + a_3$ **c)** $a_2 + a_3$ **4. a)** $6\pi i$ **c)** $2\pi i$

SECTION 11–11, PAGE 624

1. a) 1   **c)** 0   **3. a)** $4\pi ei/3$ **c)** $\pi i/32$

4. a) $\displaystyle\sum_{n=0}^{\infty} \dfrac{(-1)^n z^{2n+1}}{(2n+1)!}$, $r^* = R = \infty$ **b)** $\displaystyle\sum_{n=0}^{\infty} (-1)^n (z-2)^n$, $r^* = R = 1$

   c) $\dfrac{1}{2}\ln 2 + \dfrac{3}{4}\pi i - \displaystyle\sum_{n=1}^{\infty} \left(\dfrac{1+i}{2}\right)^n \dfrac{(z+1-i)^n}{n}$, $r^* = \sqrt{2}$, $R = 1$

SECTION 11–14, PAGE 632

1. a) $\sum_{n=1}^{\infty} z^{n-1}/n!$, removable   **c)** $-\dfrac{1}{z} + 1 + \sum_{n=0}^{\infty} \dfrac{(-1)^n z^{2n+1}}{(2n+2)!}$, pole of order 1

2. a) $\dfrac{1}{z^4} + \dfrac{3}{z^3} + \dfrac{1}{z^2}$ **c)** $\dfrac{e \sin 1}{(z-1)^2} + \dfrac{e(\cos 1 + \sin 1)}{z-1}$

3. a) $\sum_{n=1}^{\infty} \dfrac{-1}{z^n}$, removable, zero of first order

b) $z - 2 + \sum_{n=1}^{\infty} \dfrac{(-2)^{n+1}}{z^n}$, pole of order 1

c) $2 + \sum_{n=1}^{\infty} \dfrac{z^n}{n!} + \sum_{n=1}^{\infty} \dfrac{z^{-n}}{n!}$, essential

5. a) zero: 0, pole: 1   **c)** zeros: $-1, -1, -1$, poles: $0, \infty, \infty$

6. $\dfrac{a_2 b_N^2 - a_1 b_N b_{N+1} - a_0 b_{N+2} b_N + a_0 b_{N+1}^2}{b_N^3 (z - z_0)^{N-2}}$

SECTION 11–18, PAGE 642

1. a) $-\pi i$ **c)** $-\pi i/3$ **e)** $-\pi i/8$ **g)** $4\pi i$

2. a) $\dfrac{1}{4} \dfrac{1}{z-2} - \dfrac{1}{4} \dfrac{1}{z+2}$

c) $-\dfrac{1}{5} \left[\dfrac{z_2}{z - z_1} + \dfrac{z_5}{z - z_2} + \dfrac{z_3}{z - z_3} + \dfrac{z_1}{z - z_4} + \dfrac{z_4}{z - z_5} \right]$,
where $z_k = \exp[(2k-1)\pi i/5]$, $k = 1, \ldots, 5$

e) $\dfrac{1}{4} \dfrac{1}{(z-1)^2} - \dfrac{1}{4(z+1)^2}$

g) $\dfrac{1}{3} \dfrac{1}{(z-1)^3} - \dfrac{5}{9} \dfrac{1}{(z-1)^2} + \dfrac{7}{27} \dfrac{1}{z-1} + \sum_{j=1}^{4} \dfrac{g(z_j)}{z - z_j}$,
where $z_1, \ldots, z_4$ are the roots of $z^4 + z + 1 = 0$ and
$g = (-7z^3 + 5z^2 + 3z - 13)^{-1}$

h) $\dfrac{1}{10} \left(\dfrac{-2-i}{z-i} + \dfrac{-2+i}{z+i} + \dfrac{2-i}{z+1-i} + \dfrac{2+i}{z+1+i} \right)$

SECTION 11–19, PAGE 648

1. a) $\dfrac{\pi}{2}$ **c)** $\dfrac{4\pi}{3\sqrt{3}}$ **2. a)** $\dfrac{2\pi\sqrt{3}}{3}$ **c)** $\dfrac{2\pi}{3}$

3. a) $\tfrac{1}{2}\pi e^{-2}$ **b)** $-2\dfrac{\sqrt{3}}{3}\pi e^{-\sqrt{3}}\sin 1$ **c)** $\tfrac{1}{2}\pi e^{-(1/2)\sqrt{2}}\cos\left(\tfrac{1}{2}\sqrt{2}\right)$

   d) $-\tfrac{1}{2}\pi e^{-3}$

SECTION 11–20, PAGE 651

1. a) $-\tfrac{1}{3}e^{2t}[1-U(t)]-\tfrac{1}{3}e^{-t}U(t)$ with $U(t)$ as in Section 4–9.

   c) for Re $(a) > 0$, $-\dfrac{t^{n-1}e^{at}U(-t)}{(n-1)!}$; for Re $(a) < 0$, $\dfrac{t^{n-1}e^{at}U(t)}{(n-1)!}$

2. a) $\dfrac{1}{2\pi i}\oint_{C_1}\exp\dfrac{1}{1-s^2}\,e^{st}\,ds$ for $t > 0$,

$$-\dfrac{1}{2\pi i}\oint_{C_1'}\exp\dfrac{1}{1-s^2}\,e^{st}\,ds \text{ for } t < 0,$$

   where C_1 encloses -1 but not 1, and C_1' encloses 1 but not -1

   b) $f = 0$ for $t > 0$, $f = -\dfrac{1}{2\pi i}\oint_C \sin\dfrac{1}{s-2}\,e^{st}\,ds$ for $t < 0$,

   where C encloses 2

SECTION 11–21, PAGE 654

1. a) $\tfrac{1}{2}(e^t - e^{-t})$ **c)** te^t **e)** $2e^{-2t} - e^{-t}$

2. a) $\displaystyle\sum_{n=0}^{\infty}\dfrac{t^n}{n!(n+2)!}$ $\dfrac{1}{2\pi i}\oint_{|s|=R} e^{st}se^{1/s}\,ds,\ R > 0$

   b) $\displaystyle\sum_{n=0}^{\infty}\dfrac{(-1)^n t^{2n}}{(2n)!\cdot(2n+1)!}$, $\dfrac{1}{2\pi i}\oint_{|s|=R} e^{st}\sin\dfrac{1}{s}\,ds,\ R > 0$

   c) $\displaystyle\sum_{n=0}^{\infty}\dfrac{n^2+2n+2}{n!}t^n$, $\dfrac{1}{2\pi i}\oint_{|s|=R} e^{st}\displaystyle\sum_{1}^{\infty}\dfrac{n^2+1}{s^n}\,ds,\ R > 1$

3. a) $\dfrac{1}{2\pi i}\displaystyle\oint_{|s|=R}\dfrac{2e^{st}}{s^3}\,ds,\quad R>0$ b) $\dfrac{1}{2\pi i}\displaystyle\oint_{|s|=R}\dfrac{e^{st}}{(s-1)^2}\,ds,\ R>1$

c) $-\dfrac{1}{2\pi i}\displaystyle\oint_{|s|=R}\dfrac{e^{st}\,ds}{s^2-3s+2},\quad R>2$

d) $\dfrac{1}{2\pi i}\displaystyle\oint_{|s|=R}e^{st}\sum_{n=0}^{\infty}\dfrac{n!}{(n!+n^2)s^{n+1}}\,ds,\quad R>1$

SECTION 11–22, PAGE 660

1. Two roots in the right half-plane   2. a) stable   b) unstable

SECTION 11–24, PAGE 669

2. a) two sheets, connected across branch line $y=0$, $1<x<\infty$
   c) two sheets, connected across branch lines from -1 to 0 and from 1 to ∞ on the real axis

3. a) $\dfrac{1}{2}\ln\dfrac{4}{3}$ 4. a) $\dfrac{(2^\beta-1)\pi}{\sin\beta\pi}$ c) $\pi 2^{-1/4}\cos\dfrac{3\pi}{8}$

SECTION 11–26, PAGE 678

1. a) circle $|w-2|=2$, line $v=2$
   c) circle $|w-2i|=2$, line $u=-2$; circle $(u+1)^2+(v-2)^2=4$, line $u=1$

7. a) none   b) $\dfrac{1}{2}+n,\quad n=0,\pm1,\ldots$ c) ±1

SECTION 11–28, PAGE 686

1. a) $\dfrac{1}{\pi}\arg\dfrac{z^2-1}{z^2+1}$ b) $\dfrac{1}{\pi}(\pi-\arg z^4)$ c) $1-\dfrac{\theta}{\pi},\ 0<\theta<2\pi$

   d) $2-\dfrac{2}{\pi}\arg\dfrac{e^{\pi z}-1}{e^{\pi z}+1}$ e) $1-\dfrac{2}{\pi}\arg\left(\dfrac{z-1}{z+1}\right)^2$

   The argument functions are all taken between 0 and π.

2. b) $U = \dfrac{U_0}{\pi} \arg z$, where z is the inverse of the function

$$w = \frac{a}{2\pi}(2 \operatorname{Ln} z - z^2 + 1), \quad \operatorname{Im}(z) > 0$$

3. b) $T_1 + \dfrac{T_2 - T_1}{\pi} \arg(\sqrt{4w - 1} - i)$; the square root is chosen to have imaginary part greater than 1 and the arg is chosen between 0 and π.

SECTION 11–31, PAGE 692

2. $F(z) = z + \dfrac{1}{z} + i \ln z$ (not single-valued)

5. $T_0 + x(T_1 - T_0)$

===========Chapter 12===========

SECTION 12–2, PAGE 696

1. a) $1 - 3x^2$, $x \left[1 - \dfrac{2}{3}x^2 - \displaystyle\sum_{s=2}^{\infty} \dfrac{(s+1)x^{2s}}{(2s-1)(2s+1)} \right]$

2. a) $\displaystyle\sum_{s=0}^{\infty} z^s = \dfrac{1}{1-z}$, $z^{1/2}\left(1 + \displaystyle\sum_{s=1}^{\infty} z^s\right) = \dfrac{z^{1/2}}{1-z}$

SECTION 12–4, PAGE 700

1. a) 1.4296   **c)** 2.9917   **e)** 0.2701   **g)** 0.1321

6. 3.60×10^6

SECTION 12–8, PAGE 708

6. $1 + \displaystyle\sum_{s=1}^{\infty} (x-1)^s \dfrac{\alpha(\alpha-1)\cdots(\alpha-s+1)(\alpha+1)\cdots(\alpha+s)}{2^s 1^2 2^2 \cdots s^2}$

SECTION 12–10, PAGE 714

1. $\frac{1}{2}(5x^3 - 3x)$, $-\frac{1}{2}(1 - x^2)^{1/2}(15x^2 - 3)$, $15x(1 - x^2)$, $-15(1 - x^2)^{3/2}$

3. a) $ax^2 + by^2 - (a + b)z^2 + dxy + eyz + fxz$

4. $u_{00} = 1$, $u_{01} = \cos\varphi$, $u_{11} = -\cos\theta\sin\varphi$, $u_{02} = (1/2)(3\cos^2\varphi - 1)$,
   $u_{12} = -3\cos\theta\sin\varphi\cos\varphi$, $u_{22} = 3\sin^2\varphi\cos2\theta$, $v_{11} = -\sin\theta\sin\varphi$,
   $v_{12} = -3\sin\theta\sin\varphi\cos\varphi$, $v_{22} = 3\sin2\theta\sin^2\varphi$

5. $\rho^3 u_{13} = (1/2)(3x^3 + 3xy^2 - 12xz^2)$, $\rho^3 u_{23} = 15(x^2z - y^2z)$,
   $\rho^2 v_{13} = (1/2)(3y^3 + 3x^2y - 12yz^2)$, $\rho^2 v_{23} = 30xyz$

7. $U = 2 + \rho\cos\varphi - 3\rho\sin\theta\sin\varphi - 3\rho^2\cos\theta\sin\varphi\cos\varphi$

===========================Chapter 13============================

SECTION 13–2, PAGE 731

1. a) $x_1 = 2.29$, $x_2 = -0.429$ b) $x_1 = -5.33$, $x_2 = 18.7$
   c) $x_1 = 1.25$, $x_2 = 0.750$, $x_3 = -1.63$
   d) $x_1 = 1.064$, $x_2 = 0.9694$, $x_3 = 0.5722$

2. $x_1 = 1.073276$, $x_2 = 0.03017241$

3. a) $x_1 = -129$, $x_2 = -392$ b) $x_1 = -161$, $x_2 = 6$
   c) $x_1 = 1$, $x_2 = 1$, $x_3 = 1$ d) $x_1 = 1$, $x_2 = 1$, $x_3 = 2$

4. b) 40 and 36   5. a) -26 b) -203.21

6. a) $x_1 = 0.368x_3$, $x_2 = -0.789x_3$ b) $x_1 = -1.462x_3$, $x_2 = 0.616x_3$

SECTION 13–3, PAGE 736

2. a) 241   b) -52 c) 638, 280   d) 4743, 3723

3. a) $-4, -1, 1$ b) 1   c) $1, -1$ d) 0.5

4. a) 1.44, 1.4415   b) 0.625, 0.4858   c) 0.5, 0.4439   d) 0.64, 0.6401

5. a) 5.000   b) 7.02   c) 9.00   d) 1.06

6. a) $2 \pm i$ (exact)   b) $\pm 3i$ (exact)

7. a) $x + 10$, remainder $59x + 62$ b) $x^3 + 8x^2 + 9x + 27$, remainder $42x + 57$

8. a) $2.15 + 1.03i$ b) $0.665 + 3.45i$

SECTION 13–5, PAGE 743

1. a) 1.1153   b) 2.0945   c) 0.8375   d) 0.567

2. a) 3.9691, 6.9187, 15.1122   b) 0.6946, 8.6002, -1.2526

3. a) $|15 - \lambda| \leqslant 1, |7 - \lambda| \leqslant 2$ or $|4 - \lambda| \leqslant 1$
  b) $|\lambda - 7| \leqslant 1, |\lambda - 5| \leqslant 2, |\lambda - 8| \leqslant 4$ in the complex plane.

4. a) for 3.9691, c col $(1, -11.0309, -32.4335)$;
     for 6.9187, c col $(1, -8.0813, 0.3430)$;
     for 15.1122, c col $(1, 0.1122, 0.0898)$
  b) for 0.6946, c col $(1, 0.8139, -0.9783)$; for 8.6002, c col $(1, 2.6765, 3.2472)$;
     for -1.2526, c col $(1, -0.8033, 0.3541)$.

5. Obtain an approximation to an eigenvector for 15.1122 in (a), for 8.6002 in (b).

SECTION 13–6, PAGE 749

1. a) $2x^3 - x^2 + 1$ b) $5x^3 - 4x^2 + x - 2$ c) $x^4 - 5x^2 + 6$
  d) $1 + 2x - 0.8333x(x - 1) + 0.1667x(x - 1)(x - 3)$

2. a) 3, 37, 19, 23, 23   b) 3.433, 3.489, 3.490, 3.490, 3.490

3. a) $6x^2 - 7x + 5$ b) $4x + 9$ c) $1 - 3x^2 + 2x^3$

  d) $1 + (2x - 4) - \dfrac{1}{2}(x - 4)^2 - \dfrac{5}{2}(x - 4)^3$

SECTION 13–8, PAGE 760

1. a) $-0.2 + 1.3x$ b) $(1/5) + (4/7)P_2(x)$ c) $1.19x - 0.341x^2 - 0.0142x^3$
  d) $8.25 - 18x + 7.5x^2$

2. a) $-(2/7) + (10/7)x$ b) $-10.4 + 2.30x + 3.23x^2$

3. a) $4.9 + 2.9x$ b) $-0.0518 + 0.752x$
  c) $9.2002 - 5.6716x + 2.9286x^2$ d) $13.7 - 3.57x + 0.238x^2$

4. a) $(49/64)T_0 - (11/48)T_2$
  b) $(1/384)(486T_0 + 432T_1 + 104T_2 + 16T_3 + 2T_4)$
  c) $(21/400)T_0 + (203/200)T_1 + (4/75)T_2 + (1/200)T_3$
  d) $(1/256)(147T_0 + 76T_1 + 20T_2 + 4T_3)$

5. a) 42   b) -18.9896

6. a) $1 - 0.94x^2 + 0.44x^3$ for $0 \leqslant x \leqslant 1$,
     $0.5 - 0.56(x - 1) + 0.38(x - 1)^2 - 0.12(x - 1)^3$ for $1 \leqslant x \leqslant 2$
  b) $x^2(\pi - x)/(5\pi)$ for $0 \leqslant x \leqslant \pi$,
     $(\pi/5)(\pi - x) - (2/5)(x - \pi^2) + 3[(x - \pi)^3/5\pi]$ for $\pi \leqslant x \leqslant 2\pi$,
     $(4\pi/5)(x - 2\pi) + (7/5)(x - 2\pi)^2 - (11/5\pi)(x - 2\pi)^3$ for $2\pi \leqslant x \leqslant 3\pi$

SECTION 13–11, PAGE 771

1. a) $0.6400 - 0.4630\cos x - 0.3828\sin x - 0.1772\cos 2x - 0.0347\sin 2x$
  b) $-0.2753 - 0.0650\cos x + 0.3403\cos 2x + 1.000\sin x$
  c) $0.6482 - 0.4444\cos x - 0.1481\cos 2x - 0.1111\cos 3x$
  d) $0.7083 + 0.25\cos 2x + 0.0417\cos 4x$

2. a) $1.4 + 0.2472 \cos x + 1.2311 \sin x$
   b) $1.2222 - 0.1585 \cos x - 1.1640 \cos 2x - 0.4609 \sin x + 0.6037 \sin 2x$
5. $c_0 + c_1 e^{ix} + c_2 e^{2ix} + c_3 e^{3ix} + c_4 e^{4ix}$, where $c_0 = 0.2459 + 0.2186i$,
   $c_1 = 0.4394 - 0.7369i$, $c_2 = c_0$, $c_3 = 0.0344 + 0.1499i = c_4$
   or $c_0 + c_1 e^{ix} + c_2 e^{2ix} + c_4 e^{-ix} + c_3 e^{-2ix}$.
6. $36, \; -12 - 3.464i, \; 3 - 1.732i, \; 0, \; 3 + 1.732i, \; -12 + 3.464i$

SECTION 13–13, PAGE 781

1. a) 7.39, 7.39   b) 7.46, 7.42, 7.36   2. a) 7, 6   b) 8, 8
3. a) 0.5, 0.375, 0.34375 (exact value $0.3333 \ldots$)
   b) 0.25, 0.3125, 0.3281
4. a) 0.75, 0.70833, 0.69702 (exact value 0.69315)   b) 0.66667, 0.68571, 0.69122
5. 0.29452, 0.31193 (exact value 0.38656)
6. 0.74718, 0.74685   8. a) 0.6875 (exact value 0.69315)
10. a) $\pi/4 = 0.78540$ b) 0.83566   c) 0.99957   d) $1 - e^{-1.6} = 0.79810$

SECTION 13–14, PAGE 787

1. a) 0.4688   b) 0.6909   c) 2.291   d) 2.3070
2. a) 0.8659   b) 0.96534   c) 2.32089   d) 2.32073
3. a) 1.088104   b) 1.007743   4. 1.960548, 2.250410
5. 0.714254, 0.666094   6. 0.26472, 1.40072
7. 0.5708064, 1.881285

SECTION 13–15, PAGE 793

1. a) $y(\pi/8) = 0.2955$, $y(\pi/4) = 0.5914$, $y(3\pi/8) = 0.8412$
   c) $y(0.2) = 2.242$, $y(0.4) = 2.484$, $y(0.6) = 2.708$, $y(0.8) = 3.092$
   e) $y(0) = -4.19$, $y(0.2) = -4.19$, $y(0.4) = -4.19$, $y(0.6) = -4.18$,
      $y(0.8) = -4.15$, $y(1) = -4.06$
2. a) $y(\pi/8) = 0.3850$, $y(\pi/4) = 0.7107$, $y(3\pi/8) = 0.9267$
   c) $y(0) = y(0.2) = -3.055$, $y(0.4) = -3.050$, $y(0.6) = -3.026$,
      $y(0.8) = -2.958$, $y(1.0) = -2.814$
3. a) $\lambda = 8$, $y(0.5) = 1$
   c) $\lambda = 9.64$ and $(0.715, 1.00, 0.698)$, $\lambda = 32.3$ and $(-1.00, 0.0148, 1.00)$,
      $\lambda = 54.9$ and $(0.700, -1.00, 0.716)$
   e) $\lambda = 2.98$ and $(0.346, 0.581, 1.00)$, $\lambda = 18.4$ and $(1.00, 0.0319, -1.00)$,
      $\lambda = 34.2$ and $(0.472, -0.842, 1.00)$

SECTION 13-18, PAGE 804

1. $u(1, 1) = -2$, $u(2, 1) = 2$, $u(1, 2) = -11$, $u(2, 2) = -16$, $u(1, 3) = -26$, $u(2, 3) = -46$. (The values are exact—why?)

2. $u(1, 1) = 2$, $u(2, 1) = 3 = u(1, 2)$, $u(2, 2) = 4$

3. $u = u_0(r, \theta) + \text{const}$, where $u_0(0, 0) = 0$, $u_0(1/2, \pi/4) = 0.36$, $u_0(1/2, \pi/2) = 0.5$, $u_0(1, \pi/4) = 0.72$, $u_0(1, \pi/2) = 1.02$, other values determined by symmetry. Exact solution is $u_0(r, \theta) = y = r \sin \theta$.

4. $u(0.5, 0.5) = u(1.5, 0.5) = 0.179 = 5/28$, $u(1, 0.5) = 0.286 = 2/7$, $u(0.5, 1) = u(1.5, 1) = 0.078 = 23/196$, $u(1, 1) = 0.102 = 5/49$

5. $u(0.25, 0.25) = u(0.75, 0.25) = 1/8$, $u(0.5, 0.25) = 3/16$, $u(0.25, 0.5) = u(0.75, 0.5) = u(0.5, 0.5) = 0$

6. $u(0.25, 0.25) = u(0.75, 0.25) = \sqrt{2}/8$, $u(0.5, 0.25) = 1/4$, $u(0.25, 0.5) = u(0.75, 0.5) = 1/4$, $u(0.5, 0.25) = \sqrt{2}/4$

=================================Chapter 14=================================

SECTION 14-2, PAGE 809

1. c) $c_1 = 3/11$, $c_2 = 10/59$ 2. b) $c_1 = 1/2$, $c_2 = 1$ (gives exact solutions)

3. $c_1 = 1/2$, $c_2 = 1$ 5. $c_1 = 16/35$, $c_2 = 8/35$

SECTION 14-3, PAGE 812

1. a) $c_1 = 3e - 10$, $c_2 = 9 - 3e$ c) $c_1 = 7/6$, $c_2 = (19/12) - \ln 4$

SECTION 14-6, PAGE 820

2. a) $c_1 = -0.152$, $c_2 = -0.393$ 3. $u = x^4 + y^4 + (3/8)(4\varphi_1 - 5\varphi_2)$

SECTION 14-9, PAGE 833

1. a) $c_1 = -2.317$, $c_2 = -2.733$, $c_3 = -3.349$
   b) $c_0 = -0.449$, $c_1 = -0.416$, $c_2 = -0.398$, $c_3 = -0.264$

5. $c_1 = -11.51$, $c_2 = 3.352$, $c_3 = -4.013$, $c_5 = 16.20$

6. $u = \frac{11}{2}\varphi_1(x, y) + \frac{15}{4}\varphi_2(x, y) + \frac{9}{2}\varphi_3(x, y) + \sum_{j=4}^{11} g_j\varphi_j(x, y)$, where the vertices 1, 2, 3 are $(1/2, 3/2)$, $(1/2, 1/2)$, $(3/2, 1/2)$ respectively and the vertices 4 to 11 are boundary vertices.

===========================Chapter 15===========================

SECTION 15–4, PAGE 844

1. a) 3/8   c) 5/16   2. a) 1/6   c) 1/2
3. a) 7/18   c) 1/2   e) dependent
4. a) 0.78   c) 0.31   e) dependent
5. 0.0004   6. $\frac{15}{17}$ 7. $\frac{3}{10}$ 8. $4 \div \binom{52}{13}$

SECTION 15–6, PAGE 851

1. b) 182/2187   d) 1/121

3. b) $1 - e^{-1}$ d) 0 for $x \leqslant 0$, $1 - e^{-x}$ for $x \geqslant 0$

4. b) $\frac{1}{2}$ d) 0 for $x < 0$, $\dfrac{1}{\pi}(x - \sin x \cos x)$ for $0 \leqslant x \leqslant \pi$, 1 for $x \geqslant \pi$

5. b) $\frac{1}{2}$ d) $2\sqrt{2} - 2$
6. a) $\Pr(X = 0) = \Pr(X = 3) = 1/8$, $\Pr(X = 1) = \Pr(X - 2) = 3/8$
7. a) 1/2   c) 1/6, 11/12

8. a) $\mu = 2/3$, $\sigma = \sqrt{2}/3$ c) $\mu = 1.84$, $\sigma = 1.08$
9. a) $\mu = 11/16$, $\sigma = 0.229$ c) $\mu = 1/2$, $\sigma = 1/2$

SECTION 15–10, PAGE 861

1. a) values 1, 0, 2 with probabilities 1/4, 1/2, 1/4;
   c) values e^{-1}, 1, e with probabilities 1/4, 1/2, 1/4;
   e) values 1, 0, −1 with probabilities 1/4, 1/2, 1/4;
   g) value 0 with probability 1

2. a) $2u - 2$ for $1 \leqslant u \leqslant 2$, 0 otherwise;
   c) $(2/u) \ln u$ for $1 \leqslant u \leqslant e$, 0 otherwise;
   e) 1 for $0 \leqslant u \leqslant 1$, 0 otherwise

3. $\chi(t) = (\cos t + 1)/2$, $\mu_1 = 0$, $\mu_2 = 1/2 = V(X)$
4. $\chi(t) = 2t^{-2}[e^{it}(1 - it) - 1] = 1 + (2it/3) - (t^2/4) + \cdots$, $\mu_1 = 2/3$,
   $\mu_2 = 1/2$, $V(X) = 1/18$
8. a) e^{-1} c) 0.9994   e) 0.0189
10. a) 1/2   c) 0.6826   11. a) 0.1587   c) 0.0919

SECTION 15–13, **PAGE** 872

1. a) 3/8, 5/8, 5/8, 3/8   c) not independent

2. a) 1/16, 1/8, 1/16   c) 11/32   **3.** a) 1/3   c) 3/64

4. a) 1/4   c) 1   e) not independent

6. a) 0.0173, 0.260, 0.440   c) 0.263

7. a) x^2 for $0 \leqslant x \leqslant 1$, $x(2-x)$ for $1 \leqslant x \leqslant 2$, 0 otherwise;
c) 7/6, 1/3.

9. a) 0.259   c) 0.314

SECTION 15–14, **PAGE** 879

6. a) $\bar{x}_n = 12.22$, $s_n^2 = 0.0595$ c) $[0.027, 0.174]$

Index

Index